AIM/FAR 1995

AIRMAN'S INFORMATION MANUAL/FEDERAL AVIATION REGULATIONS

TAB/AERO STAFF

TAB **AERO**
Division of McGraw-Hill, Inc.
Blue Ridge Summit, PA 17294-0850

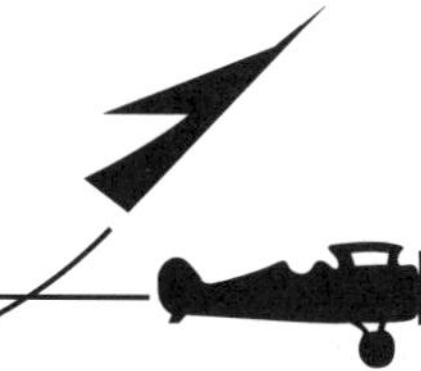

© 1995 by **TAB/AERO Books**, an imprint of TAB Books.
TAB Books is a division of McGraw-Hill, Inc.

AIM revised to August 18, 1994.
FAR revised to September 2, 1994.

Printed in the United States of America. All rights reserved. The publisher takes no responsibility for the use of any of the materials or methods described in this book, nor for the products thereof.

No copyright claim is being made for any material that has been taken from United States government sources.

pbk 1 2 3 4 5 6 7 8 9 FGR/FGR 9 9 8 7 6 5 4
hc 1 2 3 4 5 6 7 8 9 FGR/FGR 9 9 8 7 6 5 4

Library of Congress Card Number
70-186849

AIM/FAR 1995
ISSN 0886-9200
ISBN 0-07-063083-6 Hard
ISBN 0-07-063084-4 Paper

Aviation acquisitions editor: Jeff Worsinger
Editorial team: Norval G. Kennedy, Editor
Robert E. Ostrander, Executive Editor
Jodi L. Tyler, Indexer
Production team: Katherine G. Brown, Director
Wanda S. Ditch, Desktop operator
Rhonda Baker, Coding
Susan Hansford, Coding
Ollie A. Harmon, Coding
Lisa M. Mellot, Coding
Brenda Wilhide, Desktop Illustrator
Linda M. Cramer, Proofreading
Quality control team: N. Nadine McFarland
Donna M. Gladhill
Designer: Jaclyn J. Boone

AIM
0630844

From the editor

TAB/AERO's *AIM/FAR 1995* is a milestone edition for McGraw-Hill, Inc., because the entire contents of the book have been converted to desktop publishing to improve readability. Through the years, readers have commented about how hard it is to read small text, especially in the FARs. Desktop publishing at TAB Books, a division of McGraw-Hill, Inc., made this easier-to-read edition possible.

Every effort has been made to prepare the material in a fashion that serves your needs. Reader comments regarding readability were taken into account during the conversion to desktop publishing. Please continue to let TAB know how to improve the product. Your guidance is very worthwhile and appreciated; comments and suggestions are reviewed, and whenever possible, extra consideration is given to making a change for the better.

This edition continues the effort by TAB/McGraw-Hill, Inc., to annually publish the most up-to-date and most informative single manual for the general aviation community, from student pilots, to corporate pilots, to managers and staff of fixed-base operations. The AIM changes become effective every 112 days throughout the year; FAR changes become effective anytime throughout the year. AIM and FAR changes that have occurred since the last edition of TAB/AERO AIM/FAR have been shaded for easy reference.

Changes that occur within approximately 6 months after this edition is published will be included in the *AIM/FAR 1995 Midyear Update*. Be sure to request your free Update by filling in your name and address on the postage-paid card in the back of this book. The Update is your free insurance against the loss of important aviation safety information and regulations.

The *Midyear Update* request card is addressed to TAB/AERO for your convenience. Other correspondence should be mailed to:

Norval Kennedy
Senior Editor
McGraw-Hill, Inc.
13311 Monterey Lane
Blue Ridge Summit, PA 17294–0850

Acknowledgments

Credits on the copyright page note the personnel involved with the annual edition of each AIM/FAR. Thanks go to them for helping make possible this first edition "on desktop."

Thanks must also go to personnel not usually noted on the copyright page. Kelly Koons and Bonnie Paulson, with assistance from Donna Gladhill and Susan Hansford, efficiently and patiently contributed their expertise preparing the U.S. government's raw AIM and FAR text for desktop publishing. They met or exceeded crucial deadlines from February 1994 through September 1994 to scan, format, correct, and submit the text for edit; they simultaneously coped with scheduled AIM revisions and unpredictable FAR amendments.

Everyone helped bridge a wide gap from a legacy of TAB's annual AIM/FAR pasteup, which was a paper jigsaw puzzle, to desktop publishing, which resembles a fine portrait via digital technology. Thanks again to everyone for your extraordinary efforts to make *AIM/FAR 1995* an attractive and accurate aviation reference.

Key to TAB/AERO AIM/FAR features

Shading of text indicates material that has been updated, revised, or added since the last edition of TAB/AERO's AIM/FAR.

[Boldface brackets within text shading indicate specific words, phrases, sentences, or paragraphs that have been amended.**]**

[Boldface brackets without text shading indicate previous amendments.**]**

Contents at a glance

AIRMAN'S INFORMATION MANUAL

AIM contents 1

1 Navigation aids **1**
2 Aeronautical lighting and other airport visual aids **38**
3 Airspace **66**
4 Air traffic control **91**
5 Air traffic procedures **137**
6 Emergency procedures **180**
7 Safety of flight **196**
8 Medical facts for pilots **246**
9 Aeronautical charts and related publications **253**

Pilot/controller glossary 264
FAA's AIM paragraph index 324

FEDERAL AVIATION REGULATIONS

1 Definitions and abbreviations **335**
43 Maintenance, preventive maintenance, rebuilding, and alteration (with selected appendices) **343**
61 Certification: pilots and flight instructors **356**
67 Medical standards and certification **411**
71 Designation of federal airways, area low routes, controlled airspace, and reporting points (basic part) **417**
73 Special use airspace **421**
91 General operating and flight rules **424**
97 Standard instrument approach procedures **501**
99 Security control of air traffic **504**
103 Ultralight vehicles **508**
105 Parachute jumping **511**
121 Appendix I: Drug testing program; Appendix J: Alcohol misuse prevention program **515**
135 Air taxi operators and commercial operators: certification and operating requirements **532**
141 Pilot schools **621**
143 Ground instructors **639**

NATIONAL TRANSPORTATION SAFETY BOARD

830 Notification and reporting of aircraft accidents or incidents and overdue aircraft, and preservation of aircraft wreckage, mail, cargo, and records **641**

Flight forum 644
METAR: New meteorological report coding 647
Index 651
Completely cost-free *Midyear Update* request card Inside back cover

Airman's Information Manual

Chapter

Navigation radio aids 1

Lighting and visual aids 2

Airspace 3

Air traffic control 4

Air traffic procedures 5

Emergency procedures 6

Safety of flight 7

Medical facts 8

Charts and publications 9

Pilot/controller glossary

FEDERAL AVIATION ADMINISTRATION (FAA)

The Federal Aviation Administration is responsible for insuring the safe and efficient use of the Nation's airspace, by military as well as civil aviation, for fostering civil aeronautics and air commerce in the United States and abroad, and for supporting the requirements of national defense.

The activities required to carry out these responsibilities include: safety regulation; airspace management and the establishment, operation, and maintenance of a civil-military common system of air traffic control and navigation facilities; research and development in support of the fostering of a national system of airports, promulgation of standards and specifications for civil airports, and administration of Federal grants-in-aid for developing public airports; various joint and cooperative activities with the Department of Defense; and technical assistance (under State Department auspices) to other countries.

FLIGHT INFORMATION PUBLICATION POLICY

The following is in essence, the statement issued by the FAA Administrator and published in the December 10, 1964, issue of the Federal Register, concerning the FAA policy as pertaining to the type of information that will be published as NOTAMs and in the Airman's Information Manual.

1. It is a pilot's inherent responsibility that he be alert at all times for and in anticipation of all circumstances, situations and conditions which affect the safe operation of his aircraft. For example, a pilot should expect to find air traffic at any time or place. At or near both civil and military airports and in the vicinity of known training areas, a pilot should expect concentrated air traffic although he should realize concentrations of air traffic are not limited to these places.

2. It is the general practice of the agency to advertise by NOTAM or other flight information publications such information it may deem appropriate; information which the agency may from time to time make available to pilots is solely for the purpose of assisting them in executing their regulatory responsibilities. Such information serves the aviation community as a whole and not pilots individually.

3. The fact that the agency under one particular situation or another may or may not furnish information does not serve as a precedent of the agency's responsibility to the aviation community; neither does it give assurance that other information of the same or similar nature will be advertised nor does it guarantee that any and all information known to the agency will be advertised.

Consistent with the foregoing, it shall be the policy of the Federal Aviation Administration to furnish information only when, in the opinion of the agency, a unique situation should be advertised and not to furnish routine information such as concentrations of air traffic, either civil or military. The Airman's Information Manual will not contain informative items concerning everyday circumstances that pilots should, either by good practice or regulation, expect to encounter or avoid.

AIRMAN'S INFORMATION MANUAL (AIM) BASIC FLIGHT INFORMATION AND ATC PROCEDURES

This manual is designed to provide airmen with basic flight information and ATC procedures for use in the National Airspace System (NAS) of the United States. The information contained parallels the U.S. Aeronautical Information Publication (AIP) distributed internationally.

This manual contains the fundamentals required in order to fly in the U.S. NAS. It also contains items of interest to pilots concerning health and medical facts, factors affecting flight safety, a pilot/controller glossary of terms used in the Air Traffic Control System, and information on safety, accident, and hazard reporting.

This manual is complemented by other operational publications which are available upon separate subscription. These publications are:

Notices to Airmen publication—A publication containing current Notices to Airmen (NOTAM's) which are considered essential to the safety of flight as well as supplemental data affecting the other operational publications listed here. It also includes current FDC NOTAM's, which are regulatory in nature, issued to establish restrictions to flight or to amend charts or published Instrument Approach Procedures. This publication is issued every 14 days and is available through subscription from the Superintendent of Documents.

The Airport/Facility Directory, the Alaska Supplement, and the Pacific Chart Supplement—These publications contain information on airports, communications, navigational aids, instrument landing systems, VOR receiver check points, preferred routes, FSS/Weather Service telephone numbers, Air Route Traffic Control Center (ARTCC) frequencies, part-time surface areas, and various other pertinent special notices essential to air navigation. These publications are available upon subscription from the NOAA Distribution Branch (N/CG33), National Ocean Service, Riverdale, Maryland 20737.

AIRMAN'S INFORMATION MANUAL (AIM), FEDERAL AVIATION REGULATIONS, AND ADVISORY CIRCULARS

Federal Aviation Regulations—The FAA publishes the Federal Aviation Regulations (FAR's) to make readily available to the aviation community the regulatory requirements placed upon them. These *Regulations* are sold as individual *Parts* by the Superintendent of Documents.

The more frequently amended Parts are sold on subscription service with subscribers receiving Changes automatically as issued. Less active Parts are sold on a single-sale basis. Changes to single-sale Parts will be sold separately as issued. Information concerning these Changes will be furnished by the FAA through its *Status of Federal Aviation Regulations, AC 00-44.*

Advisory Circulars—The FAA issues advisory circulars (AC's) to inform the aviation public of nonregulatory material of interest. Unless incorporated into a regulation by reference, the contents of an advisory circular are not binding on the public. Advisory circulars are issued in a numbered subject system corresponding to the subject areas of the Federal Aviation Regulations (14 CFR Ch. 1).

The AC 00-2 Checklist contains advisory circulars that are for sale as well as those distributed free of charge by the Federal Aviation Administration.

NOTE—The above information relating to FAR's and AC's is extracted from AC 00-2, *Advisory Circular Checklist (and Status of Other FAA Publications).* Many of the FAR's and AC's listed in AC 00-2 are cross-referenced in the AIM. These regulatory and nonregulatory references cover a wide range of subjects and are a source of detailed information of value to airmen. AC 00-2 is issued annually and can be obtained free-of-charge from:

U.S. Department of Transportation
Distribution Requirements Section M-494.1
Washington, DC 20590

External References—All reference to Advisory Circulars and other FAA publications in the Airman's Information Manual include the FAA Advisory Circular or Manual identification numbers (when available); however, due to varied publication dates, the basic publication letter is not included (Example: The Order 7110.65, Air Traffic Control is referenced as 7110.65). To insure that you are consulting the latest information when referencing one of these other documents, use the FAA Advisory Circular AC: 00-2, Advisory Circular Checklist (and Status of Other FAA Publications).

AIRMAN'S INFORMATION MANUAL
EXPLANATION OF MAJOR CHANGES

1-13. INTERIM STANDARD MICROWAVE LANDING SYSTEM (ISMLS). Change was a recommendation from a user.

1-15. NAVAIDS WITH VOICE. Change was a recommendation from a user and clarifies the paragraph.

1-17. LORAN. Updates the Loran Section.

2-2. VISUAL GLIDESLOPE INDICATORS. This change clarifies the explanation of signals provided by the Pulsating visual approach slope indicators, which are also described in Figure 2-2[6]. Therefore, the change will provide a complete explanation of the system.

2-5. IN-RUNWAY LIGHTING. Change deletes paragraph 2-5c and revises the description of "Runway Centerline Lighting."

2-32. RUNWAY MARKINGS. Text explaining the change.

3-10. GENERAL. Addition clarifies that it is the pilot's responsibility to obtain all necessary authorizations prior to entering Class B, Class C, or Class D airspace even when the pilot is receiving ATC radar advisories.

3-13. CLASS C AIRSPACE. Change made to clarify airspace that is part of the Class C airspace area, from non-regulatory airspace surrounding the associated Class C airspace. Additionally, a note was added to point out communications and transponder requirements for Class C and D airspace.

3-14. CLASS D AIRSPACE. (1) Change to paragraph 3-14b made due to a suggestion received from a reader. Additionally, a note was added to point out communications and transponder requirements for Class C and D airspace. (2) This editorial change clarifies what happens to Class D airspace when the control tower is no longer operational. The airspace may convert to Class E airspace or to Class E and Class G airspace depending if the approach control has communications and weather services available.

4-9. TRAFFIC ADVISORY PRACTICES AT AIRPORTS WITHOUT OPERATING CONTROL TOWERS. Change coincides with recommended transmissions contained in Advisory Circular (AC) 90-66A.

4-21. AIRPORT RESERVATIONS OPERATIONS AND PROCEDURES. (1) Change reflects the new toll-free telephone numbers for the Airport Reservation Office (ARO) telephone and computer communications. (2) Change is due to the Air Traffic Control System Command Center's Airport Reservation Office transition to the Computerized Voice Reservation System (CVRS). The best method to disseminate information on system operation and procedures is through the AIM. (3) This is editorial change to correct Table 4-21 [1] CODES FOR TAIL NUMBER INPUT ONLY.

4-51. TOWER CONTROLLED AIRPORTS. Change updates paragraph and better defines operations at an airport where traffic control is being exercised.

4-53. TRAFFIC PATTERNS. Editorial change to Figure 4-53 [2].

4-57. BRAKING ACTION REPORTS AND ADVISORIES. Change deletes the reference to braking action reported as "good." Good braking action is the normal condition and should not be stated in a NOTAM.

4-84. SPECIAL VFR CLEARANCES. Change made to the reference to the FAR Part.

5-1. PREFLIGHT PREPARATION. Change adds reference to AIM, Section 4-21.

5-4. FLIGHT PLAN - VFR FLIGHTS. 5-7, FLIGHT PLAN - IFR FLIGHTS. This change will correct procedures for indicating aircraft type on flight plans and brings the FAA into line with other user services. It will require any person filing a flight plan to provide the proper aircraft type and, if unknown, to contact an FSS specialist for the proper computer compatible aircraft type.

5-23. DEPARTURE RESTRICTIONS, CLEARANCE VOID TIMES, HOLD FOR RELEASE, AND RELEASE TIMES. (1) Change clarifies the definition of the ATC instruction "hold for release." (2) Change recommends to pilots departing uncontrolled airports instrument flight rules (IFR) to obtain their IFR clearance on the ground when two way radio communications with the controlling facility are available and clarifies the definition of the ATC instruction "Hold for Release."

5-36. HOLDING. Change describes entry procedures into a standard holding pattern and recommends to pilots that they should triple the inbound drift correction on the outbound leg of a holding pattern to avoid making major turning adjustments.

5-45. APPROACH CLEARANCE. This change adds an explanatory paragraph which clarifies the name of an instrument approach stated in an air traffic control approach clearance when a component of the approach aid is inoperative or unusable.

5-57. VISUAL APPROACH. Change clarifies pilot responsibility for cloud clearance and visibility requirements. Editorial changes have also been made to better organize the information in this paragraph.

5-78. SPEED ADJUSTMENTS. Order 7110.65H, Air Traffic Control, Paragraph 5–100, Speed Adjustment, Application, has been revised to raise the altitude at which controllers must get pilot consent before issuing speed adjustments. That altitude has been raised from FL290 to FL390. This change revises the AIM to include that requirement. In addition, it clarifies that it is the pilot's responsibility and prerogative to refuse speed adjustments that he considers excessive or contrary to the aircraft's operating specifications.

5-80. VISUAL APPROACH. Change emphasizes the word "vector" by printing it in bold print in the sentence that refers to vectoring for a visual approach in the controller responsibilities. Editorial changes have also been made to better organize the information in this paragraph.

5-84. MINIMUM FUEL ADVISORY. This change adds "minimum fuel" to pertinent remarks for terminal and en route flight data entries and adds a suggested procedure for pilots in a minimum fuel status to use when changing frequencies.

7-1. NATIONAL WEATHER SERVICE AVIATION PRODUCTS. Change removes reference to Yukon Time Zone from Figure 7-1[1] and changes most time zones in the State of Alaska.

7-72. REPORTING BIRD STRIKES. Editorial change and FAA Form 5200-7 added to manual.

7-85. SEAPLANE SAFETY. Change clarifies the CG requirements for compliance with Inland Navigational Rules and the FAA approved life vests.

7-86. FLIGHT OPERATIONS IN VOLCANIC ASH. A-2 New Appendix (2). Volcanic ash clouds represent a significant hazard to aviation. Past encounters with volcanic ash clouds have resulted in flameout of all engines on at least two B747 aircraft. Volcanic ash is not normally detected by airborne or air traffic radar systems. If instrument flight conditions exist, or if after sunset, volcanic ash clouds may not be visually detected. The only way to safely operate when airborne volcanic ash is present is to avoid the area of the ash cloud. Many similarities exist between volcanic ash procedures and procedures used for other weather phenomenon; however, volcanic ash presents unique problems that must be explicitly addressed. These proposed changes identify volcanic ash as a hazard to aviation and as such, information pertaining to its presence must be solicited, gathered, and distributed in a timely manner. Pilots are requested to report volcanic activity in PIREPS. Specific air traffic control procedures are also identified in these changes.

7-87. EMERGENCY AIRBORNE INSPECTION OF OTHER AIRCRAFT. Change made to update paragraph in providing assistance to another aircraft in an emergency situation.

9-2 through 9-5. Changes made to update paragraph and to replace or delete certain charts.

Pilot/Controller Glossary

AREA NAVIGATION. Changes provide a phraseology example for clearing an aircraft for a GPS approach and provide an updated explanation of the GPS system.

CHARTED VISUAL FLIGHT PROCEDURE APPROACH. This change to the definition of CVFP clarifies pilot responsibility for cloud clearance and visibility requirements.

CONTROLLED AIRSPACE. This change corrects a problem with having two definitions of "outer area" in the Pilot/Controller Glossary.

GLOBAL POSITIONING SYSTEM. Same as **AREA NAVIGATION**.

SEE AND AVOID. This change to the definition of see and avoid in the PCG clarifies the actual intent of what is meant by see and avoid by removing any reference to visual meteorological conditions.

VISUAL APPROACH. This change to the definition of visual approach clarifies pilot responsibility for cloud clearance and visibility requirements.

VOICE SWITCHING AND CONTROL SYSTEM. The development and implementation of a new communications system switch requires an update to the glossary. There will be no operational impact from the addition of these terms but it will ensure the understanding of the terms being used in new system.

TABLE OF CONTENTS

(➤ indicates paragraphs updated for this edition)

Chapter 1. NAVIGATION AIDS

Section 1. AIR NAVIGATION RADIO AIDS

1-1. GENERAL 1
1-2. NONDIRECTIONAL RADIO BEACON (NDB) 1
1-3. VHF OMNI-DIRECTIONAL RANGE (VOR) 1
1-4. VOR RECEIVER CHECK 1
1-5. TACTICAL AIR NAVIGATION (TACAN) 2
1-6. VHF OMNI-DIRECTIONAL RANGE/TACTICAL AIR NAVIGATION (VORTAC) 2
1-7. DISTANCE MEASURING EQUIPMENT (DME) 3
1-8. NAVAID SERVICE VOLUMES 3
1-9. MARKER BEACON 4
1-10. INSTRUMENT LANDING SYSTEM (ILS) 5
1-11. SIMPLIFIED DIRECTIONAL FACILITY (SDF) 10
1-12. MICROWAVE LANDING SYSTEM (MLS) 10
➤1-13. INTERIM STANDARD MICROWAVE LANDING SYSTEM (ISMLS) 14
1-14. NAVAID IDENTIFIER REMOVAL DURING MAINTENANCE 14
➤1-15. NAVAIDS WITH VOICE 15
1-16. USER REPORTS ON NAVAID PERFORMANCE 15
➤1-17. LORAN 15
1-18. OMEGA AND OMEGA/VLF NAVIGATION SYSTEMS 31
1-19. VHF DIRECTION FINDER 32
1-20. INERTIAL NAVIGATION SYSTEM (INS) 32
1-21. DOPPLER RADAR 32
1-22. FLIGHT MANAGEMENT SYSTEM (FMS) 32
1-23. GLOBAL POSITIONING SYSTEM (GPS) 32
1-24. thru 1-29 RESERVED 33

Section 2. RADAR SERVICES AND PROCEDURES

1-30. RADAR 33
1-31. AIR TRAFFIC CONTROL RADAR BEACON SYSTEM (ATCRBS) 34
1-32. SURVEILLANCE RADAR 34
1-33. PRECISION APPROACH RADAR (PAR) 37

Chapter 2. AERONAUTICAL LIGHTING AND OTHER AIRPORT VISUAL AIDS

Section 1. AIRPORT LIGHTING AIDS

2-1. APPROACH LIGHT SYSTEMS (ALS) 38
➤2-2. VISUAL GLIDESLOPE INDICATORS 38
2-3. RUNWAY END IDENTIFIER LIGHTS (REIL) 40

2-4. RUNWAY EDGE LIGHT SYSTEMS 41
➤2-5. IN-RUNWAY LIGHTING 41
2-6. CONTROL OF LIGHTING SYSTEMS 42
2-7. PILOT CONTROL OF AIRPORT LIGHTING 42
2-8. AIRPORT (ROTATING) BEACONS 43
2-9. TAXIWAY LIGHTS 43
2-10. thru 2-19 RESERVED 43

Section 2. AIR NAVIGATION AND OBSTRUCTION LIGHTING

2-20. AERONAUTICAL LIGHT BEACONS 44
2-21. CODE BEACONS AND COURSE LIGHTS 44
2-22. OBSTRUCTION LIGHTS 44
2-23. thru 2-29. RESERVED 44

Section 3. AIRPORT MARKING AIDS AND SIGNS

2-30. GENERAL 44
2-31. AIRPORT PAVEMENT MARKINGS 45
➤2-32. RUNWAY MARKINGS 45
2-33. TAXIWAY MARKINGS 47
2-34. HOLDING POSITION MARKINGS 52
2-35. OTHER MARKINGS 57
2-36. AIRPORT SIGNS 60
2-37. MANDATORY INSTRUCTION SIGNS 60
2-38. LOCATION SIGNS 61
2-39. DIRECTION SIGNS 63
2-40. DESTINATION SIGNS 64
2-41. INFORMATION SIGNS 64
2-42. RUNWAY DISTANCE REMAINING SIGNS 64
2-43. AIRCRAFT ARRESTING DEVICES 64

Chapter 3. AIRSPACE

Section 1. GENERAL

3-1. GENERAL 66
3-2. GENERAL DIMENSIONS OF AIRSPACE SEGMENTS 66
3-3. BASIC VFR WEATHER MINIMUMS 67
3-4. VFR CRUISING ALTITUDES AND FLIGHT LEVELS 67
3-5. thru 3-9. RESERVED 67

Section 2. CONTROLLED AIRSPACE

➤3-10. GENERAL 67
3-11. CLASS A AIRSPACE 68
3-12. CLASS B AIRSPACE 68
➤3-13. CLASS C AIRSPACE 70
➤3-14. CLASS D AIRSPACE 71

➤3-15. CLASS E AIRSPACE 71
3-16. thru 3-19. RESERVED 74

Section 3. CLASS G AIRSPACE

3-20. GENERAL 74
3-21. VFR REQUIREMENTS 74
3-22. IFR REQUIREMENTS 74
3-23. thru 3-29. RESERVED 74

Section 4. SPECIAL USE AIRSPACE

3-30. GENERAL 74
3-31. PROHIBITED AREA 74
3-32. RESTRICTED AREA 75
3-33. WARNING AREA 75
3-34. MILITARY OPERATIONS AREAS (MOA) 75
3-35. ALERT AREA 75
3-36. CONTROLLED FIRING AREAS 75
3-37. thru 3-39. RESERVED 75

Section 5. OTHER AIRSPACE AREAS

3-40. AIRPORT ADVISORY AREA 76
3-41. MILITARY TRAINING ROUTES (MTR) 76
3-42. TEMPORARY FLIGHT RESTRICTIONS 76
3-43. FLIGHT LIMITATIONS/PROHIBITIONS 78
3-44. PARACHUTE JUMP AIRCRAFT OPERATIONS 78
3-45. PUBLISHED VFR ROUTES 78
3-46. TERMINAL RADAR SERVICE AREA (TRSA) 80
3-47. SUSPENSION OF THE TRANSPONDER AND AUTOMATIC ALTITUDE REPORTING EQUIPMENT REQUIREMENTS FOR OPERATIONS IN THE VICINITY OF DESIGNATED MODE C VEIL AIRPORTS 83

Chapter 4. AIR TRAFFIC CONTROL

Section 1. SERVICES AVAILABLE TO PILOTS

4-1. AIR ROUTE TRAFFIC CONTROL CENTERS 91
4-2. CONTROL TOWERS 91
4-3. FLIGHT SERVICE STATIONS 91
4-4. RECORDING AND MONITORING 91
4-5. COMMUNICATIONS RELEASE OF IFR AIRCRAFT LANDING AT AN AIRPORT NOT BEING SERVED BY AN OPERATING TOWER 91
4-6. PILOT VISITS TO AIR TRAFFIC FACILITIES 91
4-7. OPERATION TAKE-OFF AND OPERATION RAINCHECK 91
4-8. APPROACH CONTROL SERVICE FOR VFR ARRIVING AIRCRAFT 91
➤4-9. TRAFFIC ADVISORY PRACTICES AT AIRPORTS WITHOUT OPERATING CONTROL TOWERS 91

4-10. IFR APPROACHES/GROUND VEHICLE OPERATIONS 94
4-11. DESIGNATED UNICOM/MULTICOM FREQUENCIES 95
4-12. USE OF UNICOM FOR ATC PURPOSES 95
4-13. AUTOMATIC TERMINAL INFORMATION SERVICE (ATIS) 95
4-14. RADAR TRAFFIC INFORMATION SERVICE 96
4-15. SAFETY ALERT 97
4-16. RADAR ASSISTANCE TO VFR AIRCRAFT 98
4-17. TERMINAL RADAR SERVICES FOR VFR AIRCRAFT 98
4-18. TOWER EN ROUTE CONTROL (TEC) 100
4-19. TRANSPONDER OPERATION 101
4-20. HAZARDOUS AREA REPORTING SERVICE 102
➤4-21. SPECIAL TRAFFIC MANAGEMENT PROGRAMS 104
4-22. thru 4-29. RESERVED 106

Section 2. RADIO COMMUNICATIONS PHRASEOLOGY AND TECHNIQUES

4-30. GENERAL 107
4-31. RADIO TECHNIQUE 107
4-32. CONTACT PROCEDURES 107
4-33. AIRCRAFT CALL SIGNS 108
4-34. DESCRIPTION OF INTERCHANGE OR LEASED AIRCRAFT 109
4-35. GROUND STATION CALL SIGNS 110
4-36. PHONETIC ALPHABET 110
4-37. FIGURES 111
4-38. ALTITUDES AND FLIGHT LEVELS 111
4-39. DIRECTIONS 111
4-40. SPEEDS 112
4-41. TIME 112
4-42. COMMUNICATIONS WITH TOWER WHEN AIRCRAFT TRANSMITTER OR RECEIVER OR BOTH ARE INOPERATIVE 112
4-43. COMMUNICATIONS FOR VFR FLIGHTS 113
4-44. thru 4-49. RESERVED 113

Section 3. AIRPORT OPERATIONS

4-50. GENERAL 113
➤4-51. TOWER CONTROLLED AIRPORTS 113
4-52. VISUAL INDICATORS AT UNCONTROLLED AIRPORTS 114
➤4-53. TRAFFIC PATTERNS 115
4-54. UNEXPECTED MANEUVERS IN THE AIRPORT TRAFFIC PATTERN 116
4-55. USE OF RUNWAYS/DECLARED DISTANCES 117
4-56. LOW LEVEL WIND SHEAR ALERT SYSTEM (LLWAS) 117
➤4-57. BRAKING ACTION REPORTS AND ADVISORIES 117
4-58. RUNWAY FRICTION REPORTS AND ADVISORIES 118
4-59. INTERSECTION TAKEOFFS 118
4-60. SIMULTANEOUS OPERATIONS ON INTERSECTING RUNWAYS 118

4-61. LOW APPROACH 119
4-62. TRAFFIC CONTROL LIGHT SIGNALS 119
4-63. COMMUNICATIONS 119
4-64. GATE HOLDING DUE TO DEPARTURE DELAYS 120
4-65. VFR FLIGHTS IN TERMINAL AREAS 120
4-66. VFR HELICOPTER OPERATIONS AT CONTROLLED AIRPORTS 121
4-67. TAXIING 122
4-68. TAXI DURING LOW VISIBILITY 123
4-69. EXITING THE RUNWAY AFTER LANDING 123
4-70. PRACTICE INSTRUMENT APPROACHES 123
4-71. OPTION APPROACH 125
4-72. USE OF AIRCRAFT LIGHTS 125
4-73. FLIGHT INSPECTION/"FLIGHT CHECK" AIRCRAFT IN TERMINAL AREAS 125
4-74. HAND SIGNALS 126
4-75. thru 4-79. RESERVED 129

Section 4. ATC CLEARANCES/SEPARATIONS

4-80. CLEARANCE 130
4-81. CLEARANCE PREFIX 130
4-82. CLEARANCE ITEMS 130
4-83. AMENDED CLEARANCES 131
➤4-84. SPECIAL VFR CLEARANCES 131
4-85. PILOT RESPONSIBILITY UPON CLEARANCE ISSUANCE 132
4-86. IFR CLEARANCE VFR-ON-TOP 132
4-87. VFR/IFR FLIGHTS 133
4-88. ADHERENCE TO CLEARANCE 133
4-89. IFR SEPARATION STANDARDS 134
4-90. SPEED ADJUSTMENTS 134
4-91. RUNWAY SEPARATION 135
4-92. VISUAL SEPARATION 135
4-93. USE OF VISUAL CLEARING PROCEDURES 135
4-94. TRAFFIC ALERT AND COLLISION AVOIDANCE SYSTEM (TCAS I & II) 136

Chapter 5. AIR TRAFFIC PROCEDURES

Section 1. PREFLIGHT

➤5-1. PREFLIGHT PREPARATION 137
5-2. FOLLOW IFR PROCEDURES EVEN WHEN OPERATING VFR 138
5-3. NOTICE TO AIRMEN (NOTAM) SYSTEM 138
➤5-4. FLIGHT PLAN—VFR FLIGHTS 139
5-5. FLIGHT PLAN—DEFENSE VFR (DVFR) FLIGHTS 140
5-6. COMPOSITE FLIGHT PLAN (VFR/IFR) FLIGHTS) 140
➤5-7. FLIGHT PLAN—IFR FLIGHTS 141
5-8. IFR OPERATIONS TO HIGH ALTITUDE DESTINATIONS 145
5-9. FLIGHTS OUTSIDE THE UNITED STATES AND U.S. TERRITORIES 145

5-10. CHANGE IN FLIGHT PLAN 146
5-11. CHANGE IN PROPOSED DEPARTURE TIME 146
5-12. CLOSING VFR/DVFR FLIGHT PLANS 146
➤5-13. CANCELLING IFR FLIGHT PLAN 146
5-14. thru 5-19. RESERVED 146

Section 2. DEPARTURE PROCEDURES

5-20. PRE-TAXI CLEARANCE PROCEDURES 147
5-21. TAXI CLEARANCE 147
5-22. ABBREVIATED IFR DEPARTURE CLEARANCE (CLEARED ... AS FILED) PROCEDURES 147
➤5-23. DEPARTURE RESTRICTIONS, CLEARANCE VOID TIMES, HOLD FOR RELEASE, AND RELEASE TIMES 148
5-24. DEPARTURE CONTROL 148
5-25. INSTRUMENT DEPARTURES 149
5-26. thru 5-29. RESERVED 150

Section 3. EN ROUTE PROCEDURES

5-30. ARTCC COMMUNICATIONS 150
5-31. POSITION REPORTING 151
5-32. ADDITIONAL REPORTS 152
5-33. AIRWAYS AND ROUTE SYSTEMS 152
5-34. AIRWAY OR ROUTE COURSE CHANGES 153
5-35. CHANGEOVER POINTS (COP's) 154
➤5-36. HOLDING 154
5-37. thru 5-39. RESERVED 157

Section 4. ARRIVAL PROCEDURES

5-40. STANDARD TERMINAL ARRIVAL (STAR) 157
5-41. LOCAL FLOW TRAFFIC MANAGEMENT PROGRAM 157
5-42. APPROACH CONTROL 158
5-43. ADVANCE INFORMATION ON INSTRUMENT APPROACH 158
5-44. INSTRUMENT APPROACH PROCEDURE CHARTS 159
➤5-45. APPROACH CLEARANCE 160
5-46. INSTRUMENT APPROACH PROCEDURES 160
5-47. PROCEDURE TURN 161
5-48. TIMED APPROACHES FROM A HOLDING FIX 162
5-49. RADAR APPROACHES 163
5-50. RADAR MONITORING OF INSTRUMENT APPROACHES 164
5-51. SIMULTANEOUS ILS/MLS APPROACHES 164
5-52. PARALLEL ILS/MLS APPROACHES 165
5-53. SIMULTANEOUS CONVERGING INSTRUMENT APPROACHES 165
5-54. SIDE-STEP MANEUVER 165
5-55. APPROACH AND LANDING MINIMUMS 166
5-56. MISSED APPROACH 167

➤5-57. VISUAL APPROACH 167
5-58. CHARTED VISUAL FLIGHT PROCEDURES (CVFP) 168
5-59. CONTACT APPROACH 168
5-60. LANDING PRIORITY 169
5-61. OVERHEAD APPROACH MANEUVER 169
5-62. thru 5-69. RESERVED 169

Section 5. PILOT/CONTROLLER ROLES AND RESPONSIBILITIES

5-70. GENERAL 169
5-71. AIR TRAFFIC CLEARANCE 170
5-72. CONTACT APPROACH 170
5-73. INSTRUMENT APPROACH 170
5-74. MISSED APPROACH 171
5-75. RADAR VECTORS 171
5-76. SAFETY ALERT 171
5-77. SEE AND AVOID 172
➤5-78. SPEED ADJUSTMENTS 172
5-79. TRAFFIC ADVISORIES (Traffic Information) 172
➤5-80. VISUAL APPROACH 172
5-81. VISUAL SEPARATION 173
5-82. VFR-ON-TOP 173
5-83. INSTRUMENT DEPARTURES 173
➤5-84. MINIMUM FUEL ADVISORY 174
5-85. thru 5-89. RESERVED 174

Section 6. NATIONAL SECURITY AND INTERCEPTION PROCEDURES

5-90. NATIONAL SECURITY 174
5-91. INTERCEPTION PROCEDURES 175
5-92. LAW ENFORCEMENT OPERATIONS BY CIVIL AND MILITARY ORGANIZATIONS 176
5-93. INTERCEPTION SIGNALS 176
5-94. ADIZ BOUNDARIES AND DESIGNATED MOUNTAINOUS AREAS 176

Chapter 6. EMERGENCY PROCEDURES

Section 1. GENERAL

6-1. PILOT RESPONSIBILITIES AND AUTHORITY 180
6-2. EMERGENCY CONDITION-REQUEST ASSISTANCE IMMEDIATELY 180
6-3. thru 6-9. RESERVED 180

Section 2. EMERGENCY SERVICES AVAILABLE TO PILOTS

6-10. RADAR SERVICE FOR VFR AIRCRAFT IN DIFFICULTY 180
6-11. TRANSPONDER EMERGENCY OPERATION 180

6-12. DIRECTION FINDING INSTRUMENT APPROACH PROCEDURE 181
6-13. INTERCEPT AND ESCORT 181
6-14. EMERGENCY LOCATOR TRANSMITTERS 181
6-15. FAA SPONSORED EXPLOSIVES DETECTION (DOG/HANDLER TEAM) LOCATIONS 182
6-16. SEARCH AND RESCUE 182
6-17. thru 6-19. RESERVED 188

Section 3. DISTRESS AND URGENCY PROCEDURES

6-20. DISTRESS AND URGENCY COMMUNICATIONS 189
6-21. OBTAINING EMERGENCY ASSISTANCE 189
6-22. DITCHING PROCEDURES 190
6-23. SPECIAL EMERGENCY (AIR PIRACY) 193
6-24. FUEL DUMPING 194
6-25. thru 6-29. RESERVED 194

Section 4. TWO-WAY RADIO COMMUNICATIONS FAILURE

6-30. TWO-WAY RADIO COMMUNICATIONS FAILURE 194
6-31. TRANSPONDER OPERATION DURING TWO-WAY COMMUNICATIONS FAILURE 195
6-32. REESTABLISHING RADIO CONTACT 195

Chapter 7. SAFETY OF FLIGHT

Section 1. METEOROLOGY

➤7-1. NATIONAL WEATHER SERVICE AVIATION PRODUCTS 196
7-2. FAA WEATHER SERVICES 197
7-3. PREFLIGHT BRIEFING 197
7-4. EN ROUTE FLIGHT ADVISORY SERVICE (EFAS) 199
7-5. IN-FLIGHT WEATHER ADVISORIES 199
7-6. CATEGORICAL OUTLOOK 201
7-7. PILOTS AUTOMATIC TELEPHONE WEATHER ANSWERING SERVICE (PATWAS) AND TELEPHONE INFORMATION BRIEFING SERVICE (TIBS) 201
7-8. TRANSCRIBED WEATHER BROADCAST (TWEB) 201
7-9. IN-FLIGHT WEATHER BROADCASTS 202
7-10. WEATHER OBSERVING PROGRAMS 203
7-11. WEATHER RADAR SERVICES 211
7-12. ATC IN-FLIGHT WEATHER AVOIDANCE ASSISTANCE 212
7-13. RUNWAY VISUAL RANGE (RVR) 213
7-14. REPORTING OF CLOUD HEIGHTS 214
7-15. REPORTING PREVAILING VISIBILITY 214
7-16. ESTIMATING INTENSITY OF PRECIPITATION 214
7-17. ESTIMATING INTENSITY OF DRIZZLE 215
7-18. ESTIMATING INTENSITY OF SNOW 215
7-19. PILOT WEATHER REPORTS (PIREP'S) 215
7-20. PIREP'S RELATING TO AIRFRAME ICING 216
7-21. PIREP'S RELATING TO TURBULENCE 216

7-22. WIND SHEAR PIREP'S 216
7-23. CLEAR AIR TURBULENCE (CAT) PIREP'S 218
7-24. MICROBURSTS 218
7-25. THUNDERSTORMS 219
7-26. THUNDERSTORM FLYING 220
7-27. KEY TO AVIATION WEATHER OBSERVATIONS AND FORECASTS 221
7-28. AUTOB DECODING KEY 222
7-29. INTERNATIONAL CIVIL AVIATION ORGANIZATION (ICAO) TERMINAL FORECAST (TAF) 222
7-30. thru 7-39 RESERVED 228

Section 2. ALTIMETER SETTING PROCEDURES

7-40. GENERAL 228
7-41. PROCEDURES 229
7-42. ALTIMETER ERRORS 230
7-43. HIGH BAROMETRIC PRESSURE 230
7-44. LOW BAROMETRIC PRESSURE 230
7-45. thru 7-49. RESERVED 230

Section 3. WAKE TURBULENCE

7-50. GENERAL 231
7-51. VORTEX GENERATION 231
7-52. VORTEX STRENGTH 231
7-53. VORTEX BEHAVIOR 232
7-54. OPERATIONS PROBLEM AREAS 232
7-55. VORTEX AVOIDANCE PROCEDURES 233
7-56. HELICOPTERS 233
7-57. PILOT RESPONSIBILITY 234
7-58. AIR TRAFFIC WAKE TURBULENCE SEPARATIONS 234
7-59. thru 7-69. RESERVED 234

Section 4. BIRD HAZARDS, AND FLIGHT OVER NATIONAL REFUGES, PARKS, AND FORESTS

7-70. MIGRATORY BIRD ACTIVITY 235
7-71. REDUCING BIRD STRIKE RISKS 235
➤7-72. REPORTING BIRD STRIKES 235
7-73. REPORTING BIRD AND OTHER WILDLIFE ACTIVITIES 235
7-74. PILOT ADVISORIES ON BIRD AND OTHER WILDLIFE HAZARDS 235
7-75. FLIGHTS OVER CHARTED U.S. WILDLIFE REFUGES, PARKS, AND FOREST SERVICE AREAS 235
7-76. thru 7-79. RESERVED 235

Section 5. POTENTIAL FLIGHT HAZARDS

7-80. ACCIDENT CAUSE FACTORS 236
7-81. VFR IN CONGESTED AREAS 236

7-82. OBSTRUCTIONS TO FLIGHT 236
7-83. AVOID FLIGHT BENEATH UNMANNED BALLOONS 237
7-84. MOUNTAIN FLYING 237
➤7-85. SEAPLANE SAFETY 238
➤7-86. FLIGHT OPERATIONS IN VOLCANIC ASH 240
➤7-87. EMERGENCY AIRBORNE INSPECTION OF OTHER AIRCRAFT 240
7-88. thru 7-89. RESERVED 240

Section 6. SAFETY, ACCIDENT AND HAZARD REPORTS

7-90. AVIATION SAFETY REPORTING PROGRAM 241
7-91. AIRCRAFT ACCIDENT AND INCIDENT REPORTING 241
7-92. NEAR MIDAIR COLLISION REPORTING 242

Chapter 8. MEDICAL FACTS FOR PILOTS

Section 1. FITNESS FOR FLIGHT

8-1. FITNESS FOR FLIGHT 246
8-2. EFFECTS OF ALTITUDE 247
8-3. HYPERVENTILATION IN FLIGHT 248
8-4. CARBON MONOXIDE POISONING IN FLIGHT 249
8-5. ILLUSIONS IN FLIGHT 249
8-6. VISION IN FLIGHT 250
8-7. AEROBATIC FLIGHT 251
8-8. JUDGMENT ASPECTS OF COLLISION AVOIDANCE 251

Chapter 9. AERONAUTICAL CHARTS AND RELATED PUBLICATIONS

Section 1. TYPES OF CHARTS AVAILABLE

9-1. GENERAL 253
➤9-2. OBTAINING CIVIL AERONAUTICAL CHARTS 253
➤9-3. A FEW OF THE CHARTS AND PRODUCTS THAT ARE AVAILABLE 253
➤9-4. GENERAL DESCRIPTION OF EACH CHART SERIES 253
➤9-5. RELATED PUBLICATIONS 259
9-6. AUXILIARY CHARTS 263

PILOT/CONTROLLER GLOSSARY 264

FAA's AIM INDEX 324

Chapter 1. Navigation aids

Section 1. AIR NAVIGATION RADIO AIDS

1-1. GENERAL

Various types of air navigation aids are in use today, each serving a special purpose. These aids have varied owners and operators, namely: the Federal Aviation Administration (FAA), the military services, private organizations, individual states and foreign governments. The FAA has the statutory authority to establish, operate, maintain air navigation facilities and to prescribe standards for the operation of any of these aids which are used for instrument flight in federally controlled airspace. These aids are tabulated in the Airport/Facility Directory.

1-2. NONDIRECTIONAL RADIO BEACON (NDB)

a. A low or medium frequency radio beacon transmits nondirectional signals whereby the pilot of an aircraft properly equipped can determine his bearing and "home" on the station. These facilities normally operate in the frequency band of 190 to 535 kHz and transmit a continuous carrier with either 400 or 1020 Hz modulation. All radio beacons except the compass locators transmit a continuous three-letter identification in code except during voice transmissions.

b. When a radio beacon is used in conjunction with the Instrument Landing System markers, it is called a Compass Locator.

c. Voice transmissions are made on radio beacons unless the letter "W" (without voice) is included in the class designator (HW).

d. Radio beacons are subject to disturbances that may result in erroneous bearing information. Such disturbances result from such factors as lightning, precipitation static, etc. At night radio beacons are vulnerable to interference from distant stations. Nearly all disturbances which affect the ADF bearing also affect the facility's identification. Noisy identification usually occurs when the ADF needle is erratic. Voice, music or erroneous identification may be heard when a steady false bearing is being displayed. Since ADF receivers do not have a "flag" to warn the pilot when erroneous bearing information is being displayed, the pilot should continuously monitor the NDB's identification.

1-3. VHF OMNI-DIRECTIONAL RANGE (VOR)

a. VORs operate within the 108.0 to 117.95 MHz frequency band and have a power output necessary to provide coverage within their assigned operational service volume. They are subject to line-of-sight restrictions, and the range varies proportionally to the altitude of the receiving equipment. The normal service ranges for the various classes of VORs are given in paragraph 1-8d *NAVAID SERVICE VOLUMES.*

b. Most VORs are equipped for voice transmission on the VOR frequency. VORs without voice capability are indicated by the letter "W" (without voice) included in the class designator (VORW).

c. The only positive method of identifying a VOR is by its Morse Code identification or by the recorded automatic voice identification which is always indicated by use of the word "VOR" following the range's name. Reliance on determining the identification of an omnirange should never be placed on listening to voice transmissions by the Flight Service Station (FSS) (or approach control facility) involved. Many FSSs remotely operate several omniranges with different names. In some cases, none of the VORs have the name of the "parent" FSS. During periods of maintenance, the facility may radiate a T-E-S-T code (**- . ... -**) or the code may be removed.

d. Voice identification has been added to numerous VORs. The transmission consists of a voice announcement, "AIRVILLE VOR" alternating with the usual Morse Code identification.

e. The effectiveness of the VOR depends upon proper use and adjustment of both ground and airborne equipment.

1. Accuracy: The accuracy of course alignment of the VOR is excellent, being generally plus or minus 1 degree.

2. Roughness: On some VORs, minor course roughness may be observed, evidenced by course needle or brief flag alarm activity (some receivers are more susceptible to these irregularities than others). At a few stations, usually in mountainous terrain, the pilot may occasionally observe a brief course needle oscillation, similar to the indication of "approaching station." Pilots flying over unfamiliar routes are cautioned to be on the alert for these vagaries, and in particular, to use the "to/from" indicator to determine positive station passage.

(a) Certain propeller RPM settings or helicopter rotor speeds can cause the VOR Course Deviation Indicator to fluctuate as much as plus or minus six degrees. Slight changes to the RPM setting will normally smooth out this roughness. Pilots are urged to check for this modulation phenomenon prior to reporting a VOR station or aircraft equipment for unsatisfactory operation.

1-4. VOR RECEIVER CHECK

a. The FAA VOR test facility (VOT) transmits a test signal which provides users a convenient means to determine the operational status and accuracy of a VOR receiver while on the ground where a VOT is located. The airborne use of VOT is permitted; however, its use is strictly limited to those areas/altitudes specifically authorized in the Airport/Facility Directory or appropriate supplement.

b. To use the VOT service, tune in the VOT frequency on your VOR receiver. With the Course Deviation Indicator (CDI) centered, the omni-bearing selector should read 0 degrees with the to/from indication showing "from" or the omni-bearing selector should read 180 degrees with the to/from indication showing "to." Should the VOR receiver operate an RMI (Radio Magnetic Indicator), it will indicate 180 degrees on any OBS setting. Two means of identification are used. One is a series of dots and the other is a continuous tone. Information concerning an individual test signal can be obtained from the local FSS.

c. Periodic VOR receiver calibration is most important. If a receiver's Automatic Gain Control or modulation circuit deteriorates, it is possible for it to display acceptable accuracy and sensitivity close into the VOR or VOT and display out-of-tolerance readings when located at greater distances where weaker signal areas exist. The likelihood of this deterioration varies between receivers, and is generally considered a function of time. The best assurance of having an accurate receiver is periodic calibration. Yearly intervals are recommended at which time an authorized repair facility should recalibrate the receiver to the manufacturer's specifications.

d. Federal Aviation Regulations (Part 91.171) provides for certain VOR equipment accuracy checks prior to flight under instrument flight rules. To comply with this requirement and to ensure satisfactory operation of the airborne system, the FAA has provided pilots with the following means of checking VOR receiver accuracy:

1. VOT or a radiated test signal from an appropriately rated radio repair station.

2. Certified airborne check points.

3. Certified check points on the airport surface.

e. A radiated VOR test signal (VOT) from an appropriately rated radio repair station serves the same purpose as an FAA VOR signal and the check is made in much the same manner as a VOT with the following differences:

1. The frequency normally approved by the FCC is 108.0 MHz.

2. Repair stations are not permitted to radiate the VOR test signal continuously; consequently, the owner or operator must make arrangements with the repair station to have the test signal transmitted. This service is not provided by all radio repair stations. The aircraft owner or operator must determine which repair station in his local area provides this service. A representative of the repair station must make an entry into the aircraft logbook or other permanent record certifying to the radial accuracy and the date of transmission. The owner, operator or representative of the repair station may accomplish the necessary checks in the aircraft and make a logbook entry stating the results. It is necessary to verify which test radial is being transmitted and whether you should get a "to" or "from" indication.

f. Airborne and ground check points consist of certified radials that should be received at specific points on the airport surface or over specific landmarks while airborne in the immediate vicinity of the airport.

1. Should an error in excess of plus or minus 4 degrees be indicated through use of a ground check, or plus or minus 6 degrees using the airborne check, IFR flight shall not be attempted without first correcting the source of the error.

CAUTION: No correction other than the correction card figures supplied by the manufacturer should be applied in making these VOR receiver checks.

2. Locations of airborne check points, ground check points and VOTs are published in the Airport/ Facility Directory.

3. If a dual system VOR (units independent of each other except for the antenna) is installed in the aircraft one system may be checked against the other. Turn both systems to the same VOR ground facility and note the indicated bearing to that station. The maximum permissible variations between the two indicated bearings is 4 degrees.

1-5. TACTICAL AIR NAVIGATION (TACAN)

a. For reasons peculiar to military or naval operations (unusual siting conditions, the pitching and rolling of a naval vessel, etc.) the civil VOR/DME system of air navigation was considered unsuitable for military or naval use. A new navigational system, TACAN, was therefore developed by the military and naval forces to more readily lend itself to military and naval requirements. As a result, the FAA has been in the process of integrating TACAN facilities with the civil VOR/DME program. Although the theoretical, or technical principles of operation of TACAN equipment are quite different from those of VOR/DME facilities, the end result, as far as the navigating pilot is concerned, is the same. These integrated facilities are called VORTAC's.

b. TACAN ground equipment consists of either a fixed or mobile transmitting unit. The airborne unit in conjunction with the ground unit reduces the transmitted signal to a visual presentation of both azimuth and distance information. TACAN is a pulse system and operates in the UHF band of frequencies. Its use requires TACAN airborne equipment and does not operate through conventional VOR equipment.

1-6. VHF OMNI-DIRECTIONAL RANGE/TACTICAL AIR NAVIGATION (VORTAC)

a. A VORTAC is a facility consisting of two components, VOR and TACAN, which provides three individual services: VOR azimuth, TACAN azimuth and TACAN distance (DME) at one site. Although consisting of more than one component, incorporating more than one operating frequency, and using more than one antenna system, a VORTAC is considered to be a unified navigational aid. Both components of a VORTAC are envisioned as operating simultaneously and providing the three services at all times.

b. Transmitted signals of VOR and TACAN are each identified by three-letter code transmission and are interlocked so that pilots using VOR azimuth with TACAN distance can be assured that both signals being received are definitely from the same ground station. The frequency channels of the VOR and the TACAN at each VORTAC facility are "paired" in accordance with a national plan to simplify airborne operation.

1-7. DISTANCE MEASURING EQUIPMENT (DME)

a. In the operation of DME, paired pulses at a specific spacing are sent out from the aircraft (this is the interrogation) and are received at the ground station. The ground station (transponder) then transmits paired pulses back to the aircraft at the same pulse spacing but on a different frequency. The time required for the round trip of this signal exchange is measured in the airborne DME unit and is translated into distance (Nautical Miles) from the aircraft to the ground station.

b. Operating on the line-of-sight principle, DME furnishes distance information with a very high degree of accuracy. Reliable signals may be received at distances up to 199 NM at line-of-sight altitude with an accuracy of better than ½ mile or 3 percent of the distance, whichever is greater. Distance information received from DME equipment is SLANT RANGE distance and not actual horizontal distance.

c. DME operates on frequencies in the UHF spectrum between 962 MHz and 1213 MHz. Aircraft equipped with TACAN equipment will receive distance information from a VORTAC automatically, while aircraft equipped with VOR must have a separate DME airborne unit.

d. VOR/DME, VORTAC, ILS/DME, and LOC/DME navigation facilities established by the FAA provide course and distance information from collocated components under a frequency pairing plan. Aircraft receiving equipment which provides for automatic DME selection assures reception of azimuth and distance information from a common source when designated VOR/DME, VORTAC, ILS/DME, and LOC/DME are selected.

e. Due to the limited number of available frequencies, assignment of paired frequencies is required for certain military noncollocated VOR and TACAN facilities which serve the same area but which may be separated by distances up to a few miles. The military is presently undergoing a program to collocate VOR and TACAN facilities or to assign nonpaired frequencies to those that cannot be collocated.

f. VOR/DME, VORTAC, ILS/DME, and LOC/DME facilities are identified by synchronized identifications which are transmitted on a time share basis. The VOR or localizer portion of the facility is identified by a coded tone modulated at 1020 Hz or a combination of code and voice. The TACAN or DME is identified by a coded tone modulated at 1350 Hz. The DME or TACAN coded identification is transmitted one time for each three or four times that the VOR or localizer coded identification is transmitted. When either the VOR or the DME is inoperative, it is important to recognize which identifier is retained for the operative facility. A single coded identification with a repetition interval of approximately 30 seconds indicates that the DME is operative.

g. Aircraft equipment which provides for automatic DME selection assures reception of azimuth and distance information from a common source when designated VOR/DME, VORTAC and ILS/DME navigation facilities are selected. Pilots are cautioned to disregard any distance displays from automatically selected DME equipment when VOR or ILS facilities, which do not have the DME feature installed, are being used for position determination.

1-8. NAVAID SERVICE VOLUMES

a. Most air navigation radio aids which provide positive course guidance have a designated standard service volume (SSV). The SSV defines the reception limits of unrestricted NAVAIDS which are usable for random/unpublished route navigation.

b. A NAVAID will be classified as restricted if it does not conform to flight inspection signal strength and course quality standards throughout the published SSV. However, the NAVAID should not be considered usable at altitudes below that which could be flown while operating under random route IFR conditions (Part 91.177), even though these altitudes may lie within the designated SSV. Service volume restrictions are first published in the Notices to Airman (NOTAM) and then with the alphabetical listing of the NAVAID's in the Airport/Facility Directory.

c. Standard Service Volume limitations do not apply to published IFR routes or procedures.

d. VOR/DME/TACAN STANDARD SERVICE VOLUMES (SSV)

Standard service volumes (SSVs) are graphically shown in Figure 1-8[1], Figure 1-8[2], Figure 1-8[3], Figure 1-8[4], and Figure 1-8[5]. The SSV of a station is indicated by using the class designator as a prefix to the station type designation.

EXAMPLE: TVOR, LDME, and HVORTAC.

Within 25 NM, the bottom of the T service volume is defined by the curve in Figure 1-8[4]. Within 40 NM, the

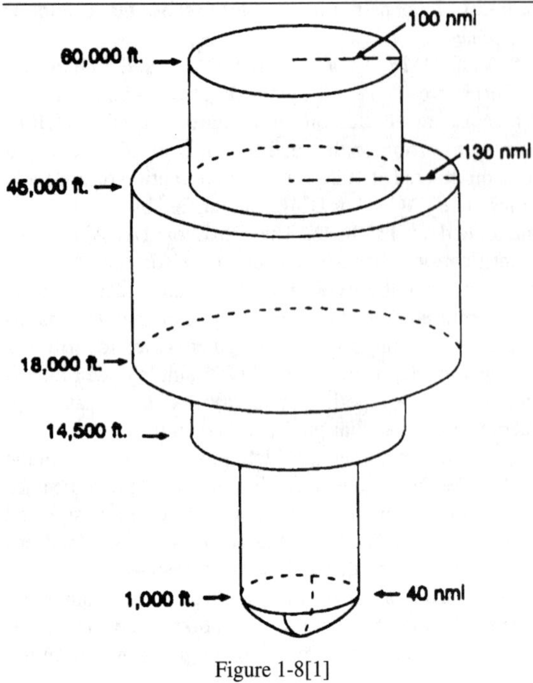

Figure 1-8[1]

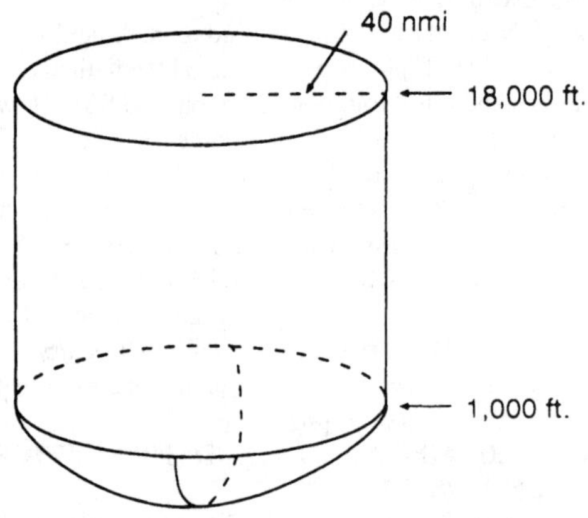

NOTE: All elevations shown are with respect to the station's site elevation (AGL). Coverage is not available in a cone of airspace directly above the facility.

Figure 1-8[2]

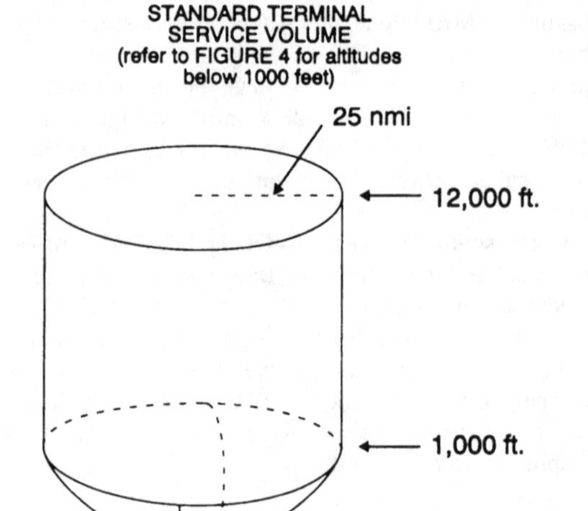

Figure 1-8[3]

Table 1-8[1]

SSV CLASS DESIGNATOR	ALTITUDE AND RANGE BOUNDARIES
T (Terminal)	From 1000 feet above ground level (AGL) up to and including 12,000 feet AGL at radial distances out to 25 NM.
L (Low Altitude)	From 1000 feet AGL up to and including 18,000 feet AGL at radial distances out to 40 NM.
H (High Altitude)	From 1000 feet AGL up to and including 14,500 feet AGL at radial distances out to 40 NM. From 14,500 feet AGL up to and including 60,000 feet at radial distances out to 100 NM. From 18,000 feet AGL up to and including 45,000 feet AGL at radial distances out to 130 NM.

Table 1-8[2]

CLASS	DISTANCE (RADIUS)
Compass Locator	15 NM
MH	25 NM
H	50 NM*
HH	75 NM

* Service ranges of individual facilities may be less than 50 nautical miles (NM). Restrictions to service volumes are first published as a Notice to Airmen and then with the alphabetical listing of the NAVAID in the Airport/Facility Directory.

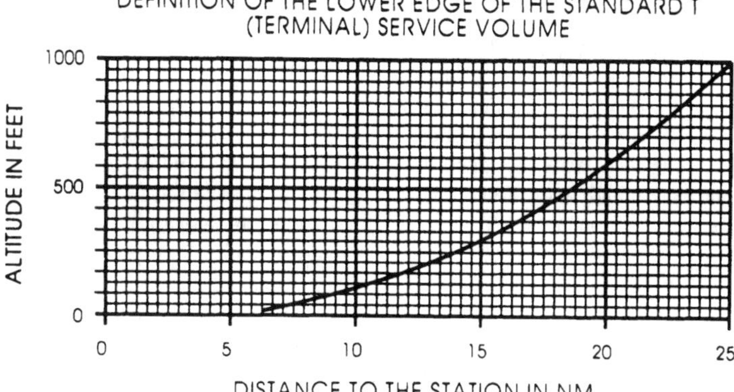

Figure 1-8[4]

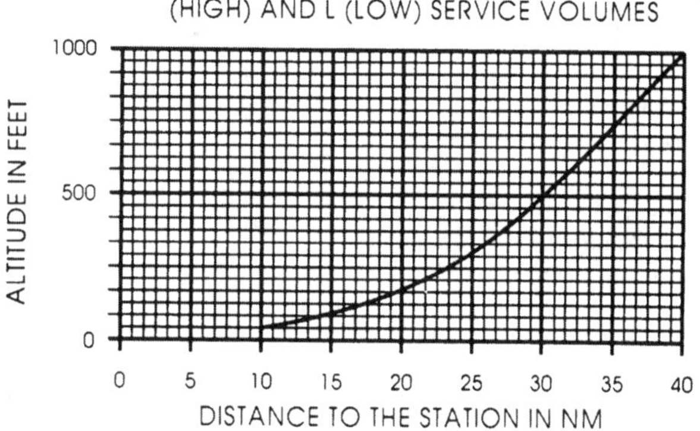

Figure 1-8[5]

bottoms of the L and H service volumes are defined by the curve in Figure 1-8[5]. (See Table 1-8[1]).

e. NONDIRECTIONAL RADIO BEACON (NDB) NDBs are classified according to their intended use. The ranges of NDB service volumes are shown in Table 1-8[2]. The distances (radius) are the same at all altitudes.

1-9. MARKER BEACON

a. Marker beacons serve to identify a particular location in space along an airway or on the approach to an instrument runway. This is done by means of a 75 MHz transmitter which transmits a directional signal to be received by aircraft flying overhead. These markers are generally used in conjunction with en route NAVAIDs and ILS as point designators.

b. There are three classes of en route marker beacons: Fan Marker (FM), Low Powered Fan Markers (LFM) and Z Markers. They transmit the letter "R" (dot dash dot) identification, or (if additional markers are in the same area) the letter "K," "P," "X," or "Z."

1. Class FMs are used to provide a positive identification of positions at definite points along the airways. The transmitters have a power output of approximately 100 watts. Two types of antenna array are used with class FMs.

(a) The first type used, and generally referred to as the standard type, produces an elliptical shaped pattern, which, at an elevation of 1,000 feet above the station, is about 4 NM wide and 12 NM long. At 10,000 feet the pattern widens to about 12 NM wide and 35 NM long.

(b) The second array produces a dumbbell or boneshaped pattern, which, at the "handle," is about three miles wide at 1,000 feet. The boneshaped marker is preferred at approach control locations where "timed" approaches are used.

2. The class LFM or low powered FMs have a rated power output of 5 watts. The antenna array produces a circular pattern which appears elongated at right angles to the airway due to the directional characteristics of the aircraft receiving antenna.

3. The Station Location, or Z-Marker, was developed to meet the need for a positive position indicator for aircraft operating under instrument flight conditions to show the pilot when he was passing directly over a low frequency navigational aid. The marker consists of a 5 watt transmitter and a directional antenna array which is located on the range plot between the towers or the loop antennas.

1-9b3 NOTE—ILS marker beacon information is included in paragraph 1-10 *INSTRUMENT LANDING SYSTEMS (ILS).*

1-10. INSTRUMENT LANDING SYSTEM (ILS)

a. GENERAL

1. The ILS is designed to provide an approach path for exact alignment and descent of an aircraft on final approach to a runway.

2. The ground equipment consists of two highly directional transmitting systems and, along the approach, three (or fewer) marker beacons. The directional transmitters are known as the localizer and glide slope transmitters.

3. The system may be divided functionally into three parts:

(a) Guidance information—localizer, glide slope

(b) Range information—marker beacon, DME

(c) Visual information—approach lights, touchdown and centerline lights, runway lights

4. Compass locators located at the Outer Marker (OM) or Middle Marker (MM) may be substituted for marker beacons. DME, when specified in the procedure, may be substituted for the OM.

5. Where a complete ILS system is installed on each end of a runway; (i.e. the approach end of Runway 4 and the approach end of Runway 22) the ILS systems are not in service simultaneously.

b. LOCALIZER

1. The localizer transmitter operates on one of 40 ILS channels within the frequency range of 108.10 to 111.95 MHz. Signals provide the pilot with course guidance to the runway centerline.

2. The approach course of the localizer is called the front course and is used with other functional parts, e.g., glide slope, marker beacons, etc. The localizer signal is transmitted at the far end of the runway. It is adjusted for a course width (full scale fly-left to a full scale fly-right) of 700 feet at the runway threshold.

3. The course line along the extended centerline of a runway, in the opposite direction to the front course is called the back course.

CAUTION: Unless the aircraft's ILS equipment includes reverse sensing capability, when flying inbound on the back course it is necessary to steer the aircraft in the direction opposite the needle deflection when making corrections from off-course to on-course. This "flying away from the needle" is also required when flying outbound on the front course of the localizer. DO NOT USE BACK COURSE SIGNALS for approach unless a BACK COURSE APPROACH PROCEDURE is published for that particular runway and the approach is authorized by ATC.

4. Identification is in International Morse Code and consists of a three-letter identifier preceded by the letter I (..) transmitted on the localizer frequency.

EXAMPLE: I-DIA

5. The localizer provides course guidance throughout the descent path to the runway threshold from a distance of 18 NM from the antenna between an altitude of 1,000 feet above the highest terrain along the course line and 4,500 feet above the elevation of the antenna site. Proper off-course indications are provided throughout the following angular areas of the operational service volume:

(a) To 10 degrees either side of the course along a radius of 18 NM from the antenna, and

(b) From 10 to 35 degrees either side of the course along a radius of 10 NM. (See Figure 1-10[1])

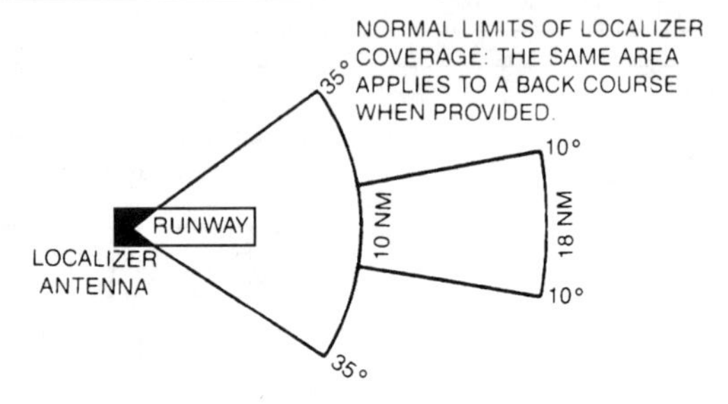

Figure 1-10[1]

6. Unreliable signals may be received outside these areas.

c. LOCALIZER-TYPE DIRECTIONAL AID

1. The Localizer-type Directional Aid (LDA) is of comparable use and accuracy to a localizer but is not part of a complete ILS. The LDA course usually provides a more precise approach course than the similar Simplified Directional Facility (SDF) installation, which may have a course width of 6 or 12 degrees. The LDA is not aligned with the runway. Straight-in minimums may be published where alignment does not exceed 30 degrees between the course and runway. Circling minimums only are published where this alignment exceeds 30 degrees.

d. GLIDE SLOPE/GLIDE PATH

1. The UHF glide slope transmitter, operating on one of the 40 ILS channels within the frequency range 329.15 MHz, to 335.00 MHz radiates its signals in the direction of the localizer front course. The term "glide path" means that portion of the glide slope that intersects the localizer.

CAUTION: False glide slope signals may exist in the area of the localizer back course approach which can cause the glide slope flag alarm to disappear and present unreliable glide slope information. Disregard all glide slope signal indications when making a localizer back course approach unless a glide slope is specified on the approach and landing chart.

2. The glide slope transmitter is located between 750 feet and 1,250 feet from the approach end of the runway (down the runway) and offset 250 to 650 feet from the runway centerline. It transmits a glide path beam 1.4 degrees wide. The signal provides descent information for navigation down to the lowest authorized decision height (DH) specified in the approved ILS approach procedure. The glidepath may not be suitable for navigation below the lowest authorized DH and any reference to glidepath indications below that height must be supplemented by visual reference to the runway environment. Glidepaths with no published DH are usable to runway threshold.

3. The glide path projection angle is normally adjusted to 3 degrees above horizontal so that it intersects the MM at about 200 feet and the OM at about 1,400 feet above the runway elevation. The glide slope is normally usable to the distance of 10 NM. However, at some locations, the glide slope has been certified for an extended service volume which exceeds 10 NM.

4. Pilots must be alert when approaching the glidepath interception. False courses and reverse sensing will occur at angles considerably greater than the published path.

5. Make every effort to remain on the indicated glide path (reference: Part 91.129(d)(2)). Exercise caution: avoid flying below the glide path to assure obstacle/terrain clearance is maintained.

6. The published glide slope threshold crossing height (TCH) *DOES NOT* represent the height of the actual glide path on-course indication above the runway threshold. It is used as a reference for planning purposes which represents the height above the runway threshold that an aircraft's glide slope antenna should be, if that aircraft remains on a trajectory formed by the four-mile-to-middle marker glidepath segment.

7. Pilots must be aware of the vertical height between the aircraft's glide slope antenna and the main gear in the landing configuration and, at the DH, plan to adjust the descent angle accordingly if the published TCH indicates the wheel crossing height over the runway threshold may not be satisfactory. Tests indicate a comfortable wheel crossing height is approximately 20 to 30 feet, depending on the type of aircraft.

e. DISTANCE MEASURING EQUIPMENT (DME)

1. When installed with the ILS and specified in the approach procedure, DME may be used:

(a) In lieu of the OM.

(b) As a back course (BC) final approach fix (FAF).

(c) To establish other fixes on the localizer course.

2. In some cases, DME from a separate facility may be used within Terminal Instrument Procedures (TERPS) limitations:

(a) To provide ARC initial approach segments.

(b) As a FAF for BC approaches.

(c) As a substitute for the OM.

f. MARKER BEACON

1. ILS marker beacons have a rated power output of 3 watts or less and an antenna array designed to produce an elliptical pattern with dimensions, at 1,000 feet above the antenna, of approximately 2,400 feet in width and 4,200 feet in length. Airborne marker beacon receivers with a selective sensitivity feature should always be operated in the "low" sensitivity position for proper reception of ILS marker beacons.

2. Ordinarily, there are two marker beacons associated with an ILS, the OM and MM. Locations with a Category II and III ILS also have an Inner Marker (IM). When an aircraft passes over a marker, the pilot will receive the following indications: (See Table 1-10[1]).

Table 1-10[1]

MARKER	CODE	LIGHT
OM	– – –	BLUE
MM	.–.–	AMBER
IM	. . . .	WHITE
BC	. . . .	WHITE

(a) The OM normally indicates a position at which an aircraft at the appropriate altitude on the localizer course will intercept the ILS glide path.

(b) The MM indicates a position approximately 3,500 feet from the landing threshold. This is also the position where an aircraft on the glide path will be at an altitude of approximately 200 feet above the elevation of the touchdown zone.

(c) The inner marker (IM) will indicate a point at which an aircraft is at a designated decision height (DH) on the glide path between the MM and landing threshold.

3. A back course marker normally indicates the ILS back course final approach fix where approach descent is commenced.

g. COMPASS LOCATOR

1. Compass locator transmitters are often situated at the MM and OM sites. The transmitters have a power of less than 25 watts, a range of at least 15 miles and operate between 190 and 535 kHz. At some locations, higher powered radio beacons, up to 400 watts, are used as OM compass locators. These generally carry Transcribed Weather Broadcast (TWEB) information.

2. Compass locators transmit two letter identification groups. The outer locator transmits the first two letters of the localizer identification group, and the middle locator transmits the last two letters of the localizer identification group.

h. ILS FREQUENCY

1. See Table 1-10[2] for frequency pairs allocated for ILS.

i. ILS MINIMUMS

1. The lowest authorized ILS minimums, with all required ground and airborne systems components operative, are

(a) Category I—Decision Height (DH) 200 feet and Runway Visual Range (RVR) 2,400 feet (with touchdown

Table 1-10[2]

Localizer mHz	Glide Slope
108.10	334.70
108.15	334.55
108.3	334.10
108.35	333.95
108.5	329.90
108.55	329.75
108.7	330.50
108.75	330.35
108.9	329.30
108.95	329.15
109.1	331.40
109.15	331.25
109.3	332.00
109.35	331.85
109.50	332.60
109.55	332.45
109.70	333.20
109.75	333.05
109.90	333.80
109.95	333.65
110.1	334.40
110.15	334.25
110.3	335.00
110.35	334.85
110.5	329.60
110.55	329.45
110.70	330.20
110.75	330.05
110.90	330.80
110.95	330.65
111.10	331.70
111.15	331.55
111.30	332.30
111.35	332.15
111.50	332.9
111.55	332.75
111.70	333.5
111.75	333.35
111.90	331.1
111.95	330.95

zone and centerline lighting, RVR 1800 Category A, B, C; RVR 2000 Category D).

(b) Category II—DH 100 feet and RVR 1,200 feet.

(c) Category IIIA—RVR 700 feet.

1-10i1c NOTE—Special authorization and equipment are required for Category II and IIIA.

j. INOPERATIVE ILS COMPONENTS

1. Inoperative localizer: When the localizer fails, an ILS approach is not authorized.

2. Inoperative glide slope: When the glide slope fails, the ILS reverts to a nonprecision localizer approach.

1-10j2 NOTE—Refer to the Inoperative Component Table in the U.S. Government Terminal Procedures Publication (TPP), for adjustments to minimums due to inoperative airborne or ground system equipment.

k. ILS COURSE DISTORTION

1. All pilots should be aware that disturbances to ILS localizer and glide slope courses may occur when surface vehicles or aircraft are operated near the localizer or glide slope antennas. Most ILS installations are subject to signal interference by either surface vehicles, aircraft or both. ILS CRITICAL AREAS are established near each localizer and glide slope antenna.

2. ATC issues control instructions to avoid interfering operations within ILS critical areas at controlled airports during the hours the Airport Traffic Control Tower (ATCT) is in operation as follows:

(a) Weather Conditions—Less than ceiling 800 feet and/or visibility 2 miles.

(1) LOCALIZER CRITICAL AREA—Except for aircraft that land, exit a runway, depart or miss approach, vehicles and aircraft are not authorized in or over the critical area when an arriving aircraft is between the ILS final approach fix and the airport. Additionally, when the ceiling is less than 200 feet and/or the visibility is RVR 2,000 or less, vehicle and aircraft operations in or over the area are not authorized when an arriving aircraft is inside the ILS MM.

(2) GLIDE SLOPE CRITICAL AREA—Vehicles and aircraft are not authorized in the area when an arriving aircraft is between the ILS final approach fix and the airport unless the aircraft has reported the airport in sight and is circling or side stepping to land on a runway other than the ILS runway.

(b) Weather Conditions—At or above ceiling 800 feet and/or visibility 2 miles.

(1) No critical area protective action is provided under these conditions.

(2) If an aircraft advises the tower that an AUTOLAND or COUPLED approach will be conducted, an advisory will be promptly issued if a vehicle or aircraft will be in or over a critical area when the arriving aircraft is inside the ILS MM.

EXAMPLE: GLIDE SLOPE SIGNAL NOT PROTECTED.

(3) Aircraft holding below 5,000 feet between the outer marker and the airport may cause localizer signal variations for aircraft conducting the ILS Approach. Accordingly, such holding is not authorized when weather or visibility conditions are less than ceiling 800 feet and/or visibility 2 miles.

(4) Pilots are cautioned that vehicular traffic not subject to ATC may cause momentary deviation to ILS course or glide slope signals. Also, critical areas are not protected at uncontrolled airports or at airports with an operating control tower when weather or visibility conditions are above those requiring protective measures. Aircraft conducting coupled or autoland operations should be especially alert in monitoring automatic flight control systems. (See Figure 1-10[2].)

ILS

[FAA INSTRUMENT LANDING SYSTEM]

STANDARD CHARACTERISTICS AND TERMINOLOGY

ILS approach charts should be consulted to obtain variations of individuals systems.

VHF LOCALIZER

Provides Horizontal Guidance.

108.10 to 111.95 MHz. Radiates about 100 watts. Horizontal polarization. Modulation frequencies 90 and 150 Hz. Modulation depth on course 20% for each frequency. Code identification (1020 Hz, 5%) and voice communication (modulated 50%) provided on same channel.

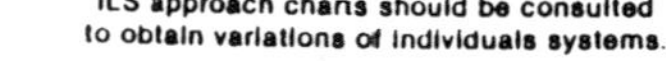

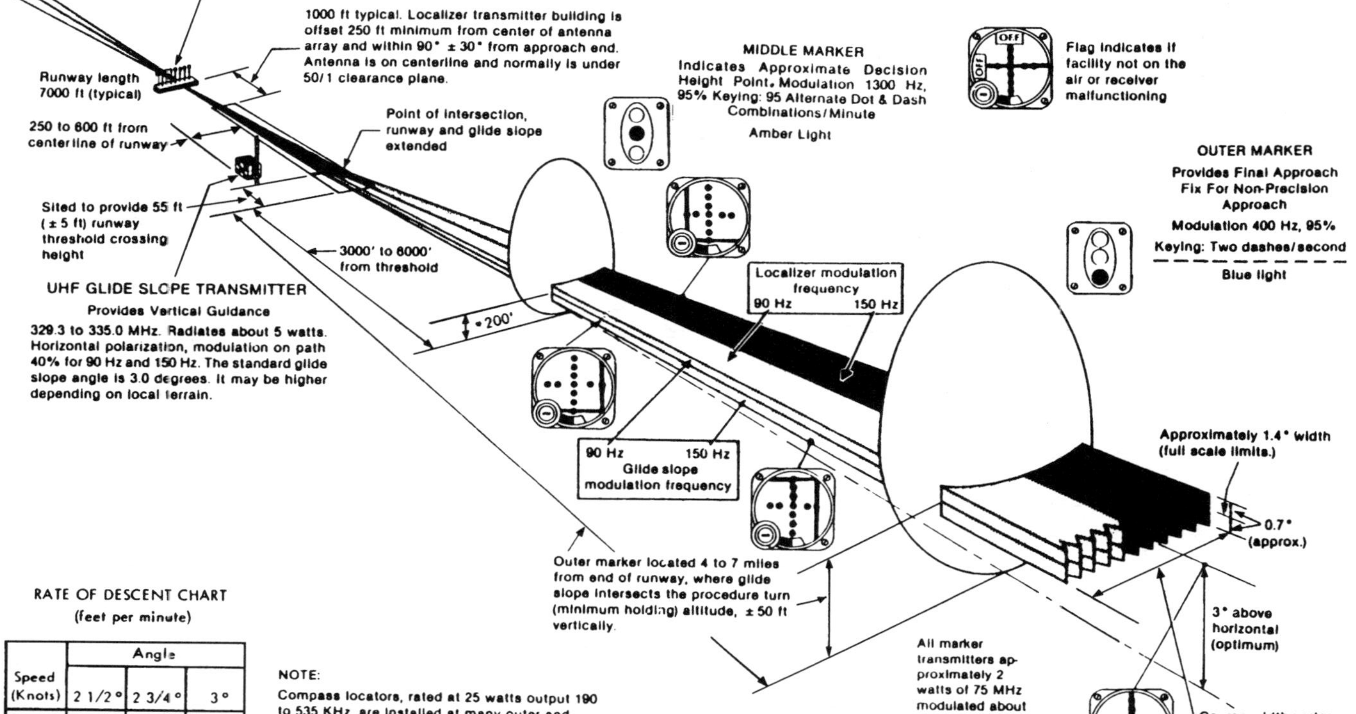

RATE OF DESCENT CHART

(feet per minute)

Speed (Knots)	Angle 2 1/2°	2 3/4°	3°
90	400	440	475
110	485	535	585
130	575	630	690
150	665	730	795
160	707	778	849

NOTE:

Compass locators, rated at 25 watts output 190 to 535 KHz, are installed at many outer and some middle markers. A 400 Hz or a 1020 Hz tone, modulating the carrier about 95%, is keyed with the first two letters of the ILS identification on the outer locator and the last two letters on the middle locator. At some locators, simultaneous voice transmissions from the control tower are provided, with appropriate reduction in identification percentage.

Figures marked with asterisk are typical. Actual figures vary with deviations in distances to markers, glide angles and localizer widths.

Figure 1-10[2]

1-11. SIMPLIFIED DIRECTIONAL FACILITY (SDF)

a. The SDF provides a final approach course similar to that of the ILS localizer. It does not provide glide slope information. A clear understanding of the ILS localizer and the additional factors listed below completely describe the operational characteristics and use of the SDF.

b. The SDF transmits signals within the range of 108.10 to 111.95 MHz.

c. The approach techniques and procedures used in an SDF instrument approach are essentially the same as those employed in executing a standard localizer approach except the SDF course may not be aligned with the runway and the course may be wider, resulting in less precision.

d. Usable off-course indications are limited to 35 degrees either side of the course centerline. Instrument indications received beyond 35 degrees should be disregarded.

e. The SDF antenna may be offset from the runway centerline. Because of this, the angle of convergence between the final approach course and the runway bearing should be determined by reference to the instrument approach procedure chart. This angle is generally not more than 3 degrees. However, it should be noted that inasmuch as the approach course originates at the antenna site, an approach which is continued beyond the runway threshold will lead the aircraft to the SDF offset position rather than along the runway centerline.

f. The SDF signal is fixed at either 6 degrees or 12 degrees as necessary to provide maximum flyability and optimum course quality.

g. Identification consists of a three-letter identifier transmitted in Morse Code on the SDF frequency. The appropriate instrument approach chart will indicate the identifier used at a particular airport.

1-12. MICROWAVE LANDING SYSTEM (MLS)

a. GENERAL

1. The MLS provides precision navigation guidance for exact alignment and descent of aircraft on approach to a runway. It provides azimuth, elevation, and distance.

2. Both lateral and vertical guidance may be displayed on conventional course deviation indicators or incorporated into multipurpose cockpit displays. Range information can be displayed by conventional DME indicators and also incorporated into multipurpose displays.

3. The MLS initially supplements and will eventually replace ILS as the standard landing system in the United States for civil, military and international civil aviation. The transition plan assures duplicate ILS and MLS facilities where needed to protect current users of ILS. At international airports ILS service is protected to the year 1995.

4. The system may be divided into five functions:

(a) Approach azimuth.

(b) Back azimuth.

(c) Approach elevation.

(d) Range.

(e) Data communications.

5. The *standard configuration* of MLS ground equipment includes:

(a) An azimuth station to perform functions (a) and (e) above. In addition to providing azimuth navigation guidance, the station transmits basic data which consists of information associated directly with the operation of the landing system, as well as advisory data on the performance of the ground equipment.

(b) An elevation station to perform function (c).

(c) Precision Distance Measuring Equipment (DME/P) to perform function (d). The DME/P provides continuous range information that is compatible with standard navigation DME but has improved accuracy and additional channel capabilities.

6. MLS Expansion Capabilities—The standard configuration can be expanded by adding one or more of the following functions or characteristics.

(a) Back azimuth—Provides lateral guidance for missed approach and departure navigation.

(b) Auxiliary data transmissions—Provides *additional* data, including refined airborne positioning, meteorological information, runway status, and other supplementary information.

(c) Larger proportional guidance.

7. MLS identification is a four-letter designation starting with the letter M. It is transmitted in International Morse Code at least six times per minute by the approach azimuth (and back azimuth) ground equipment.

b. APPROACH AZIMUTH GUIDANCE

1. The azimuth station transmits MLS angle and data on one of 200 channels within the frequency range of 5031 to 5091 MHz. (See Table 1-12[1]) for MLS angle and data channeling, and (See Table 1-12[2]) for the DME.

2. The equipment is normally located about 1,000 feet beyond the stop end of the runway, but there is considerable flexibility in selecting sites. For example, for heliport operations the azimuth transmitter can be collocated with the elevation transmitter.

3. The azimuth coverage (See Figure 1-12[1].) extends:

(a) Laterally, at least 40 degrees on either side of the runway.

(b) In elevation, up to an angle of 15 degrees—and to at least 20,000 feet.

(c) In range, to at least 20 NM.

c. BACK AZIMUTH GUIDANCE.

1. The back azimuth transmitter is essentially the same as the approach azimuth transmitter. However, the equipment transmits at a somewhat lower data rate because the guidance accuracy requirements are not as stringent as for the landing approach. The equipment operates on the same frequency as the approach azimuth but at a different time in the transmission sequence.

Table 1-12[1]

CHANNEL NUMBER	FREQUENCY (MHz)
500	5031.0
501	5031.3
502	5031.6
503	5031.9
504	5032.2
505	5032.5
—	—
—	—
—	—
—	—
510	5034.0
—	—
—	—
—	—
—	—
515	5035.5
—	—
—	—
—	—
—	—
520	5037.0
—	—
—	—
—	—
—	—
525	5038.5
—	—
—	—
—	—
—	—
530	5040.0
—	—
—	—
—	—
—	—
535	5041.5
—	—
—	—
—	—
—	—
540	5043.0
—	—
—	—
—	—
—	—
545	5044.5
—	—
—	—
—	—
—	—
550	5046.0
—	—
—	—
—	—
—	—
555	5047.5
—	—
—	—
—	—
—	—
560	5049.0
—	—
—	—
—	—
—	—
565	5050.5
—	—
—	—
—	—
—	—
570	5052.0
—	—
—	—
—	—
—	—
575	5053.5
—	—
—	—
—	—
—	—
580	5055.0
—	—
—	—
—	—
—	—
585	5056.5
—	—
—	—
—	—
—	—
590	5058.0
—	—
—	—
—	—
—	—
595	5059.5
—	—
—	—
—	—
—	—
600	5061.0
—	—
—	—
—	—
—	—
605	5062.5
—	—
—	—
—	—
—	—
610	5064.0
—	—
—	—
—	—
—	—
615	5065.5
—	—
—	—
—	—
—	—
620	5067.0
—	—
—	—
—	—
—	—
625	5068.5
—	—
—	—
—	—
—	—
630	5070.0
—	—
—	—
—	—
—	—
635	5071.5
—	—
—	—
—	—
—	—
640	5073.0
—	—
—	—
—	—
—	—
645	5074.5
—	—
—	—
—	—
—	—
650	5076.0
—	—
—	—
—	—
—	—
655	5077.5
—	—
—	—
—	—
—	—
660	5079.0
—	—
—	—
—	—
—	—
665	5080.5
—	—
—	—
—	—
—	—
670	5982.0
—	—
—	—
—	—
—	—
675	5083.5
—	—
—	—
—	—
—	—
680	5085.0
—	—
—	—
—	—
—	—
685	5086.5
—	—
—	—
—	—
—	—
690	5088.0
—	—
—	—
—	—
—	—
695	5089.5
696	5089.8
697	5090.1
698	5090.4
699	5090.7

Table 1-12[2]

DME CHANNEL (NUMBER)	VHF CHANNEL (MHz)	C-BAND CHANNEL (MHz)	ANGLE CHANNEL (NUMBER)	INTERROGATOR FREQUENCY (MHz)	NON-PRECISION INTERROGATOR PULSE CODE (USEC)	PRECISION INTERROGATOR PULSE CODE (USEC)	TRANSPONDER FREQUENCY (MHz)	TRANSPONDER PULSE CODE (USEC)
1X				1025	12		962	12
1Y				1025	36		1088	30
2X				1026	12		963	12
2Y				1026	36		1089	30
3X				1027	12		964	12
3Y				1027	36		1090	30
4X				1028	12		965	12
4Y				1028	36		1091	30
5X				1029	12		966	12
5Y				1029	36		1092	30
6X				1030	12		967	12
6Y				1030	36		1093	30
7X				1031	12		968	12
7Y				1031	36		1094	30
8X				1032	12		969	12
8Y				1032	36		1095	30
9X				1033	12		970	12
9Y				1033	36		1096	30
10X				1034	12		971	12
10Y				1034	36		1097	30
11X				1035	12		972	12
11Y				1035	36		1098	30
12X				1036	12		973	12
12Y				1036	36		1099	30
13X				1037	12		974	12
13Y				1037	36		1100	30
14X				1038	12		975	12
14Y				1038	36		1101	30
15X				1039	12		976	12
15Y				1039	36		1102	30
16X				1040	12		977	12
16Y				1040	36		1103	30
			—	1041	12	—	978	12
			540	1041	36	42	1104	30
			541	1041	21	27	1104	15
			500	1042	12	18	979	12
			501	1042	24	33	979	24
			542	1042	36	42	1105	30
			543	1042	21	27	1105	15
		—	—	1043	12	—	980	12

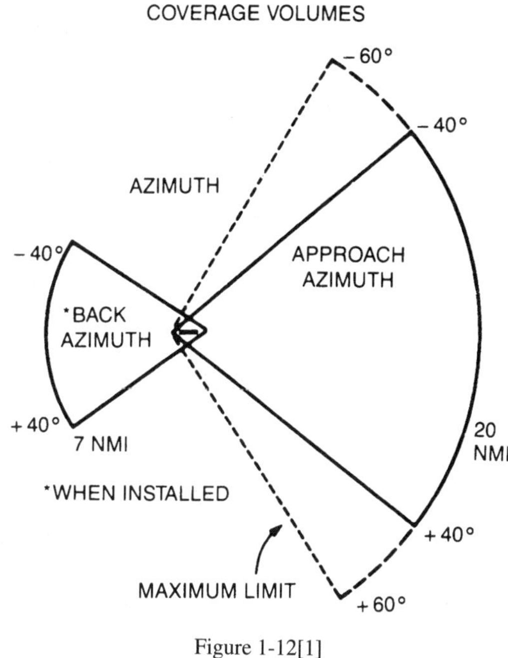

Figure 1-12[1]

2. The equipment is normally located about 1,000 feet in front of the approach end of the runway. On runways that have MLS service on both ends (e.g., Runway 9 and 27), the azimuth equipment can be switched in their operation from the approach azimuth to the back azimuth and vice versa, and thereby reduce the amount of equipment required.

3. The back azimuth provides coverage as follows:

(a) Laterally, at least 40 degrees on either side of the runway centerline.

(b) In elevation, up to an angle of 15 degrees.

(c) In range, to at least 7 NM from the runway stop end. (See Figure 1-12[1].)

1-12c3c NOTE—The actual coverage is normally the same as for the approach azimuth.

d. ELEVATION GUIDANCE

1. The elevation station transmits signals on the same frequency as the azimuth station. A single frequency is time-shared between angle and data functions.

2. The elevation transmitter is normally located about 400 feet from the side of the runway between runway threshold and the touchdown zone.

3. Elevation coverage is provided in the same airspace as the azimuth guidance signals:

(a) In elevation, to at least +15 degrees.

(b) Laterally, 40 degrees on either side of the runway centerline.

(c) In range, to at least 20 NM. (See Figure 1-12[2].)

e. RANGE GUIDANCE

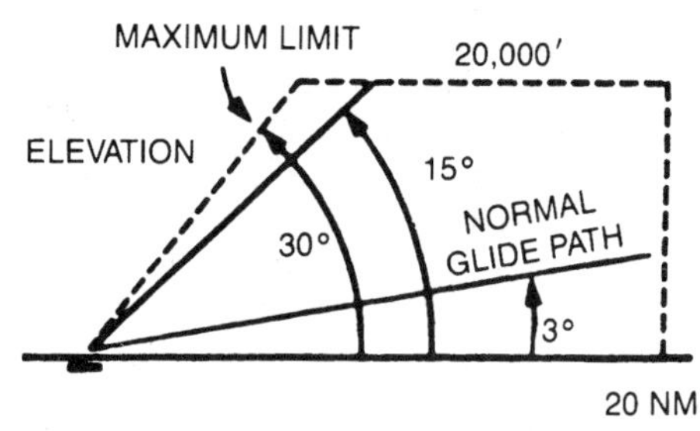

Figure 1-12[2]

1. The MLS Precision Distance Measuring Equipment (DME/P) functions the same as the navigation DME described in paragraph 1-7, but there are some technical differences. The beacon transponder operates in the frequency band 962 to 1105 MHz and responds to an aircraft interrogator. The MLS DME/P accuracy is improved to be consistent with the accuracy provided by the MLS azimuth and elevation stations.

2. A DME/P channel is paired with the azimuth and elevation channel. A complete listing of the 200 paired channels of the DME/P with the angle functions is contained in FAA Standard 022 (MLS Interoperability and Performance Requirements).

3. The DME/P is an integral part of the MLS and is installed at all MLS facilities unless a waiver is obtained. This occurs infrequently—and only at outlying, low density airports where marker beacons or compass locators are already in place.

f. DATA COMMUNICATIONS

1. The data transmission can include both basic and auxiliary data words. All MLS facilities transmit basic data. In the future facilities at some airports, including most high density airports, will also transmit auxiliary data.

2. Coverage limits—MLS data are transmitted throughout the azimuth (and back azimuth when provided) coverage sectors.

3. Basic data content—Representative data include:

(a) Station identification.

(b) Exact locations of azimuth, elevation and DME/P stations (for MLS receiver processing functions).

(c) Ground equipment performance level.

(d) DME/P channel and status.

4. Auxiliary data content—Representative data include:

(a) 3-D locations of MLS equipment.

(b) Waypoint coordinates.

(c) Runway conditions.

(d) Weather (e.g., RVR, ceiling, altimeter setting, wind, wake vortex, wind shear).

g. OPERATIONAL FLEXIBILITY. The MLS has the

capability to fulfill a variety of needs in the transition, approach, landing, missed approach and departure phases of flight. For example: Curved and segmented approaches; selectable glide path angles; accurate 3-D positioning of the aircraft in space; and the establishment of boundaries to ensure clearance from obstructions in the terminal area. While many of these capabilities are available to any MLS-equipped aircraft, the more sophisticated capabilities (such as curved and segmented approaches) are dependent upon the particular capabilities of the airborne equipment.

h. SUMMARY

1. Accuracy. The MLS provides precision three-dimensional navigation guidance accurate enough for all approach and landing maneuvers.

2. Coverage. Accuracy is consistent throughout the coverage volumes. (See Figure 1-12[3].)

COVERAGE VOLUMES

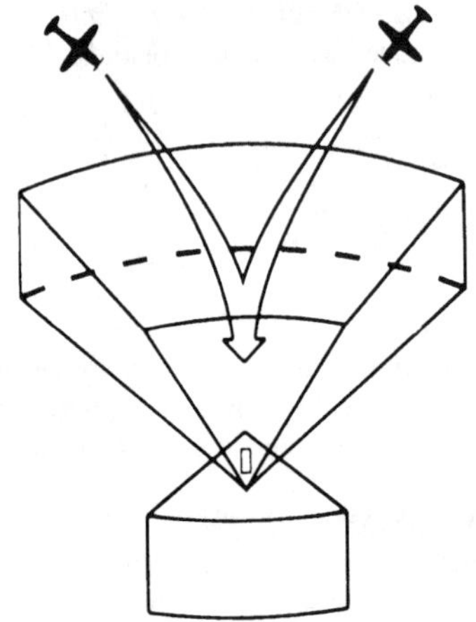

Figure 1-12[3]

3. Environment. The system has low susceptibility to interference from weather conditions and airport ground traffic.

4. Channels. MLS has 200 channels—enough for any foreseeable need.

5. Data. The MLS transmits ground-air data messages associated with the systems operation.

6. Range information. Continuous range information is provided with an accuracy of about 100 feet.

1-13. INTERIM STANDARD MICROWAVE LANDING SYSTEM (ISMLS)

a. The ISMLS is designed to provide approach information similar to the ILS for an aircraft on final approach to a runway. The system provides both lateral and vertical guidance which is displayed on a conventional course deviation indicator or approach horizon. Operational performance and coverage areas are similar to the ILS system.

b. ISMLS operates in the C-band microwave frequency range (about 5,000 MHz). ISMLS signals will not be received by unmodified VHF/UHF ILS receivers. Aircraft with ISMLS must be equipped with a C-band receiving antenna in addition to other special equipment mentioned below. The C-band antenna limits reception of the signal to an angle of about 50 degrees from the inbound course. An aircraft so equipped will not receive the ISMLS signal until flying a magnetic heading within 50 degrees either side of the inbound course. Because of this, ISMLS procedures are designed to restrict the use of the ISMLS signal until the aircraft is in position for the final approach. Transition to the ISMLS, holding and procedure turns at the ISMLS facility must be predicated on other navigation aids such as NDB, VOR, etc. Once established on the approach course inbound, the system can be flown the same as an ILS. No back course is provided.

c. The ISMLS consists of the following basic components:

1. C-Band (5000 MHz–5030 MHz) localizer.

2. C-Band (5220 MHz–5250 MHz) glide path.

3. VHF marker beacons (75 MHz).

4. A VHF/UHF ILS receiver modified to receive ISMLS signals.

5. C-Band antenna.

6. Converter unit.

7. A Microwave/ILS Mode Control.

d. Identification consists of a three letter Morse Code identifier preceded by the Morse Code for "M" (— —) (e.g., M—STP). The "M" distinguishes this system from ILS which is preceded by the Morse Code for "I" (..) (e.g., I—STP).

e. Approaches published in conjunction with the ISMLS are identified as "MLS Rwy-(Interim)." The frequency displayed on the ISMLS approach chart is a VHF frequency. ISMLS frequencies are tuned by setting the receiver to the listed VHF frequencies. When the ISMLS mode is selected, receivers modified to accept ISMLS signals receive a paired C-band frequency that is processed by the receiver.

CAUTION: Pilots should not attempt to fly ISMLS procedures unless the aircraft is so equipped.

1-14. NAVAID IDENTIFIER REMOVAL DURING MAINTENANCE

During periods of routine or emergency maintenance, coded identification (or code and voice, where applicable) is removed from certain FAA NAVAIDs. Removal of identification serves as a warning to pilots that the facility is officially off the air for tune-up or repair and may be unreliable even though intermittent or constant signals are received.

1-14 NOTE—During periods of maintenance VHF ranges may radiate a T—E—S—T code (- • ••• -).

1-15. NAVAIDS WITH VOICE

a. Voice equipped enroute radio navigational aids are under the operational control of either an FAA AFSS or an approach control facility. The voice communication is available on some facilities. The HIWAS broadcast capability on selected VOR sites is in the process of being implemented throughout the conterminous United States and does not provide voice communication. The availability of two-way voice communication and HIWAS is indicated in the Airport Facility Directory and aeronautical charts.

b. Unless otherwise noted on the chart, all radio navigation aids operate continuously except during shutdowns for maintenance. Hours of operation of facilities not operating continuously are annotated on charts and in the Airport/Facility Directory.

1-16. USER REPORTS ON NAVAID PERFORMANCE

a. Users of the National Airspace System (NAS) can render valuable assistance in the early correction of NAVAID malfunctions by reporting their observations of undesirable NAVAID performance. Although NAVAID's are monitored by electronic detectors, adverse effects of electronic interference, new obstructions or changes in terrain near the NAVAID can exist without detection by the ground monitors. Some of the characteristics of malfunction or deteriorating performance which should be reported are: erratic course or bearing indications; intermittent, or full, flag alarm; garbled, missing or obviously improper coded identification; poor quality communications reception; or, in the case of frequency interference, an audible hum or tone accompanying radio communications or NAVAID identification.

b. Reporters should identify the NAVAID, location of the aircraft, time of the observation, type of aircraft and describe the condition observed; the type of receivers in use is also useful information. Reports can be made in any of the following ways:

1. Immediate report by direct radio communication to the controlling Air Route Traffic Control Center (ARTCC), Control Tower, or FSS. This method provides the quickest result.

2. By telephone to the nearest FAA facility.

3. By FAA Form 8000-7, Safety Improvement Report, a postage-paid card designed for this purpose. These cards may be obtained at FAA FSSs, General Aviation District Offices, Flight Standards District Offices, and General Aviation Fixed Base Operations.

c. In aircraft that have more than one receiver, there are many combinations of possible interference between units. This can cause either erroneous navigation indications or, complete or partial blanking out of the communications. Pilots should be familiar enough with the radio installation of the particular airplanes they fly to recognize this type of interference.

1-17. LORAN

a. INTRODUCTION

1. Loran, which uses a network of land-based radio transmitters, was developed to provide an accurate system for LOng RAnge Navigation. The system was configured to provide reliable, all weather navigation for marine users along the U.S. coasts and in the Great Lakes. The current system, known as Loran-C, was the third version of four developed since World War II.

2. With an expanding user group in the general aviation community, the Loran coastal facilities were augmented in 1991 to provide signal coverage over the entire continental U.S. The Federal Aviation Administration (FAA) and the United States Coast Guard (USCG) are incorporating Loran into the National Airspace System (NAS) for supplemental en route and nonprecision approach operations. Loran-C is also supported in the Canadian airspace system. This guide is intended to provide an introduction to the Loran system, Loran avionics, the use of Loran for aircraft navigation, and to examine the possible future of Loran in aviation.

b. LORAN CHAIN

1. The 27 U.S. Loran transmitters that provide signal coverage for the continental U.S. and the southern half of Alaska are distributed from Caribou, Maine to Attu Island in the Aleutians. Station operations are organized into subgroups of four to six stations called "chains." One station in the chain is designated the "Master" and the others are "secondary" stations.

2. The Loran navigation signal is a carefully structured sequence of brief radio frequency pulses centered at 100 kilohertz. The sequence of signal transmissions consists of a pulse group from the Master (M) station followed at precise time intervals by groups from the secondary stations which are designated by the U.S. Coast Guard with the letters V, W, X, Y and Z. All secondary stations radiate pulses in groups of eight, but the Master signal for identification has an additional ninth pulse.

3. The time interval between the reoccurrence of the Master pulse group is the Group Repetition Interval (GRI). The GRI is the same for all stations in a chain and each Loran chain has a unique GRI. Since all stations in a particular chain operate on the same radio frequency, the GRI is the key by which a Loran receiver can identify and isolate signal groups from a specific chain.

EXAMPLE: Transmitters in the Northeast U.S. chain operate with a GRI of 99,600 microseconds which is shortened to 9,960 for convenience. The Master station (M) at Seneca NY controls: Secondary stations (W) at Caribou, ME; (X) at Nantucket, MA; (Y) at Carolina Beach, NC; and (Z) at Dana, IN. In order to keep chain operations precise, the system uses monitor receivers at Cape Elizabeth, ME, Sandy Hook, NJ and Plumbrook, OH. Monitor receivers continuously measure various aspects of the quality and accuracy of Loran signals and report system status to a control station where chain timing is maintained.

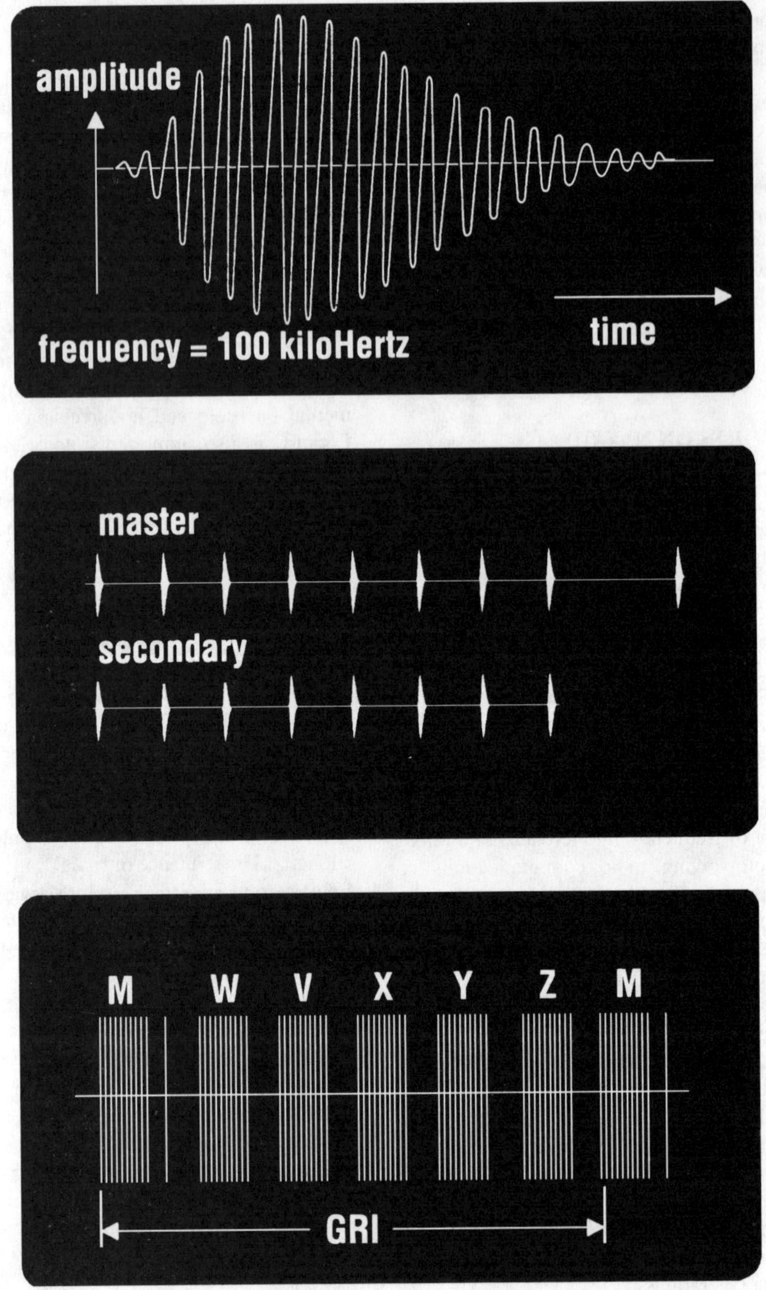

Figure 1-17[1]. Loran-C Pulse

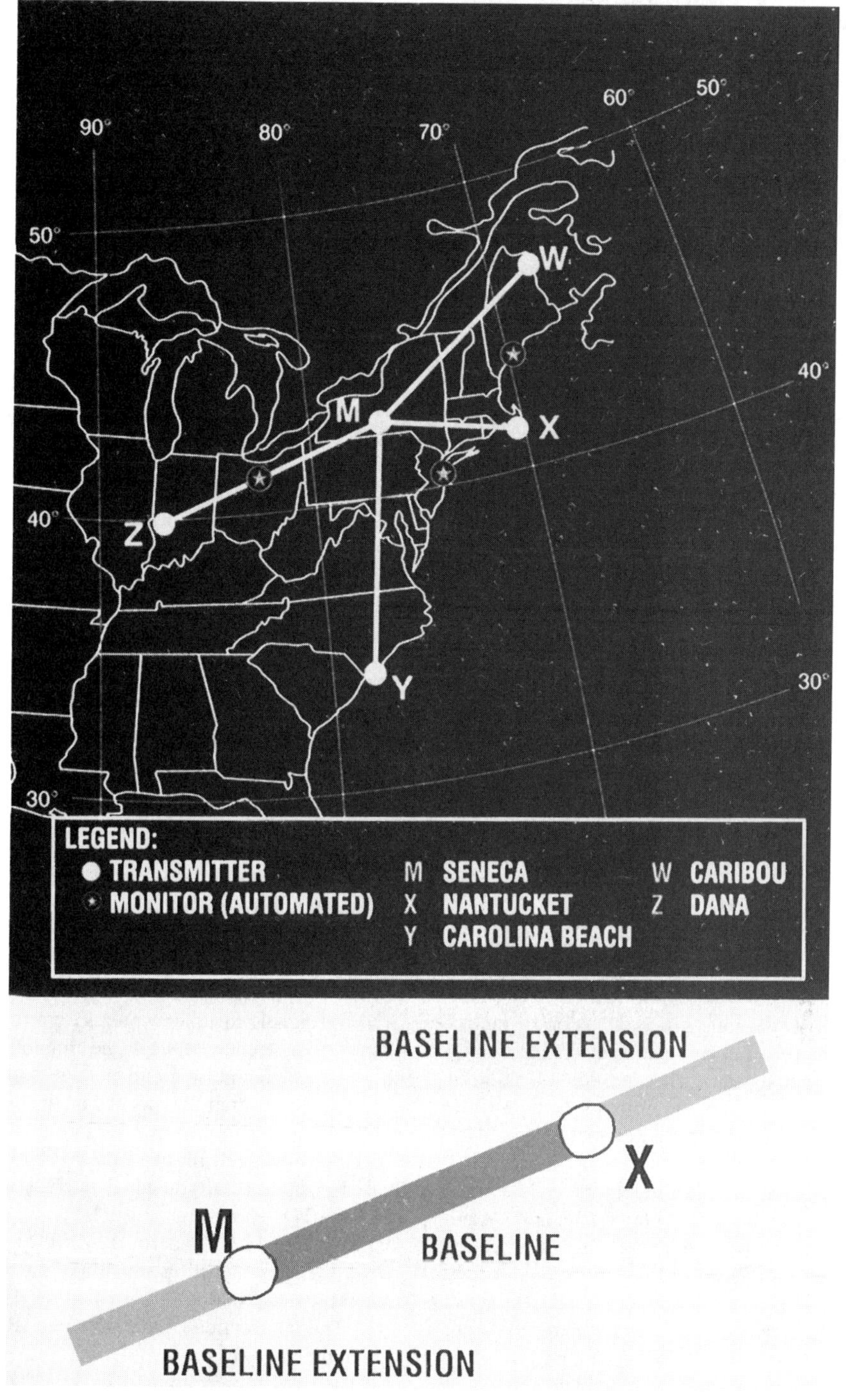

Figure 1-17[2]. Loran-C Northeast U. S. Chain

4. The line between the Master and each secondary station is the "baseline" for a pair of stations. Typical baselines are from 600 to 1,000 nautical miles in length. The continuation of the baseline in either direction is a "baseline extension."

5. Loran transmitter stations have time and control equipment, a transmitter, auxiliary power equipment, a building about 100 by 30 feet in size and an antenna that is about 700 feet tall. A station generally requires approximately 100 or more acres of land to accommodate guy lines that keep the antenna in position. Each Loran stations transmits from 400 to 1,600 kilowatts of signal power.

6. The USCG operates 27 stations, comprising eight chains, in the U.S. NAS. Four control stations, which monitor chain performance, have personnel on duty full time. The Canadian east and west coast chains also provide signal coverage over small areas of the NAS.

7. When a control station detects a signal problem that could affect navigation accuracy, an alert signal called "Blink" is activated. Blink is a distinctive change in the group of eight pulses that can be recognized automatically by a receiver so the user is notified instantly that the Loran system should not be used for navigation. In addition, other problems can cause signal transmissions from a station to be halted.

8. Each individual Loran chain provides navigation-quality signal coverage over an identified area as shown for the West Coast chain, GRI 9940. The chain Master station is at Fallon, NV and secondary stations are at George, WA; Middletown, CA; and Searchlight, NV. In a signal coverage area the signal strength relative to the normal ambient radio noise must be adequate to assure successful reception.

c. THE LORAN RECEIVER

1. Before a Loran receiver can provide navigation information for a pilot, it must successfully receive, or "acquire," signals from three or more stations in a chain. Acquisition involves the time synchronization of the receiver with the chain GRI, identification of the Master station signals from among those checked, identification of secondary station signals, and the proper selection of the point in each signal at which measurements should be made.

2. Signal reception at any site will require a pilot to provide location information such as approximate latitude and longitude, or the GRI to be used, to the receiver. Once activated, most receivers will store present location information for later use.

3. The basic measurements made by Loran receivers are the differences in time-of-arrival between the Master signal and the signals from each of the secondary stations of a chain. Each "time difference" (TD) value is measured to a precision of about 0.1 microseconds. As a rule of thumb, 0.1 microsecond is equal to about 100 feet.

4. An aircraft's Loran receiver must recognize three signal conditions: (1) usable signals, (2) absence of signals, and (3) signal Blink. The most critical phase of flight is during the approach to landing at an airport. During the approach phase the receiver must detect a lost signal, or a signal Blink, within 10 seconds of the occurrence and warn the pilot of the event.

5. Most receivers have various internal tests for estimating the probable accuracy of the current TD values and consequent navigation solutions. Tests may include verification of the timing alignment of the receiver clock with the Loran pulse, or a continuous measurement of the signal-to-noise ratio (SNR). SNR is the relative strength of the Loran signals compared to the local ambient noise level. If any of the tests fail, or if the quantities measured are out of the limits set for reliable navigation, then an alarm will be activated to alert the pilot.

6. Loran signals operate in the low frequency band around (100 kHz) that has been reserved for Loran use. Adjacent to the band, however, are numerous low frequency communications transmitters. Nearby signals can distort the Loran signals and must be eliminated by the receiver to assure proper operation. To eliminate interfering signals, Loran receivers have selective internal filters. These filters, commonly known as "notch filters" reduce the effect of interfering signals.

7. Careful installation of antennas, good metal-to-metal electrical bonding, and provisions for precipitation noise discharge on the aircraft are essential for the successful operation of Loran receivers. A Loran antenna should be installed on an aircraft in accordance with the manufacturer's instructions. Corroded bonding straps should be replaced, and static discharge devices installed at points indicated by the aircraft manufacturer.

d. LORAN NAVIGATION

1. An airborne Loran receiver has four major parts: (1) signal processor, (2) navigation computer, (3) control/display, and (4) antenna.

2. The signal processor acquires Loran signals and measures the difference between the time-of-arrival of each secondary station pulse group and the Master station pulse group. The measured TDs depend on the location of the receiver in relation to the three or more transmitters.

(a) The first TD will locate an aircraft somewhere on a line-of-position (LOP) on which the receiver will measure the same TD value.

(b) A second LOP is defined by a TD measurement between the Master station signal and the signal from another secondary station.

(c) The intersection of the measured LOPs is the position of the aircraft.

3. The navigation computer converts TD values to corresponding latitude and longitude. Once the time and position of the aircraft is established at two points, distance to destination, cross track error, ground speed, estimated time of arrival, etc., can be determined. Cross track error can be

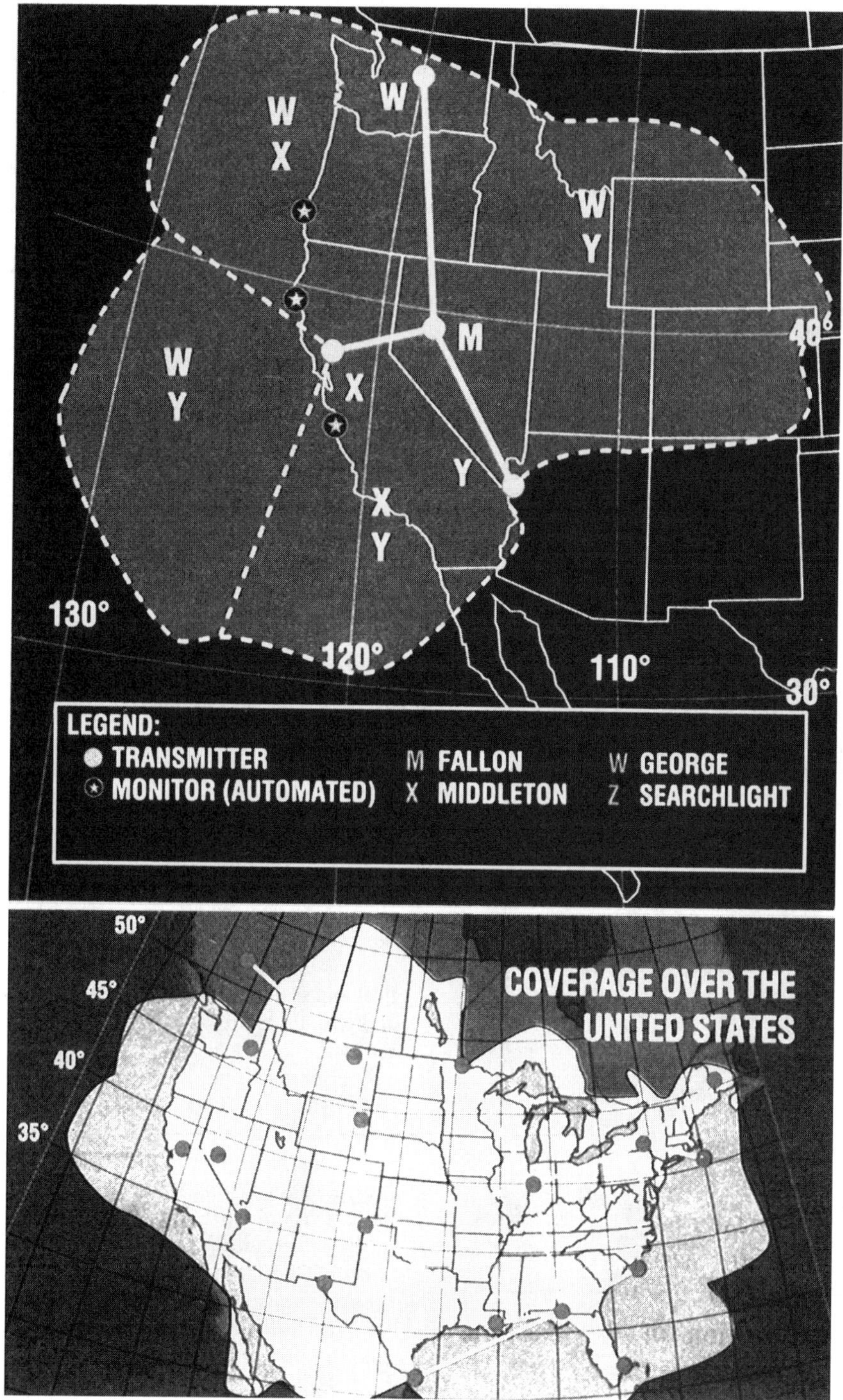

Figure 1-17[3]. Loran-C West Coast Chain

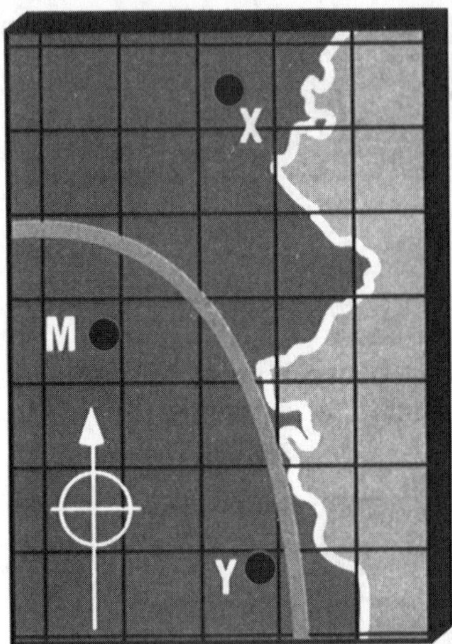

Figure 1-17[4]

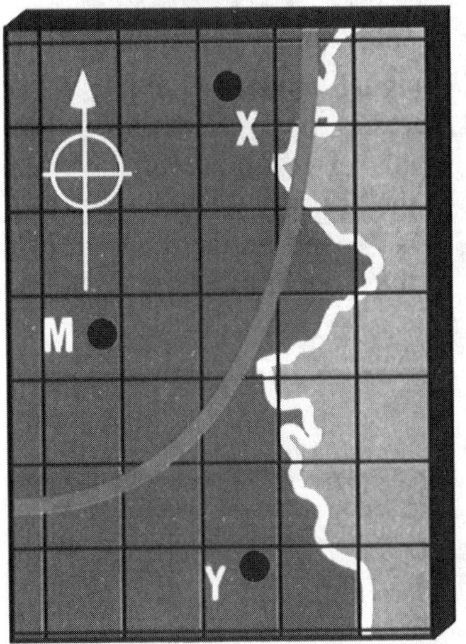

Figure 1-17[5]

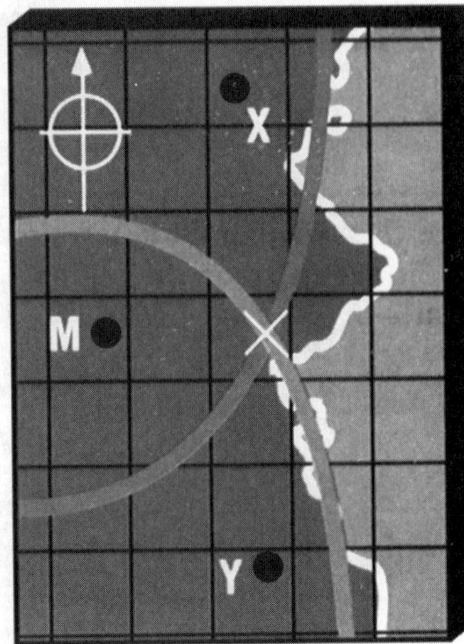

Figure 1-17[6]

displayed as the vertical needle of a course deviation indicator, or digitally, as decimal parts of a mile left or right of course. During a nonprecision approach, course guidance must be displayed to the pilot with a full scale deviation of ±0.30 nautical miles or greater.

4. Loran navigation for nonprecision approaches requires accurate and reliable information. During an approach the occurrence of signal Blink or loss of signal must be detected within 10 seconds and the pilot must be notified. Loran signal accuracy for approaches is 0.25 nautical miles, well within the required accuracy of 0.30 nautical miles. Loran signal accuracy can be improved by applying the correction values published with approach procedures.

5. Flying a Loran nonprecision approach is different from flying a VOR approach. A VOR approach is on a radial of the VOR station, with guidance sensitivity increasing as the aircraft nears the airport. The Loran system provides a linear grid, so there is constant guidance sensitivity everywhere in the approach procedure. Consequently, inaccuracies and ambiguities that occur during operations in close proximity to VORs (station passage, for example) do not occur in Loran approaches.

6. The navigation computer also provides storage for data entered by pilot or provided by the receiver manufacturer. The receiver's data base is updated at local maintenance facilities every 60 days to include all changes made by the FAA.

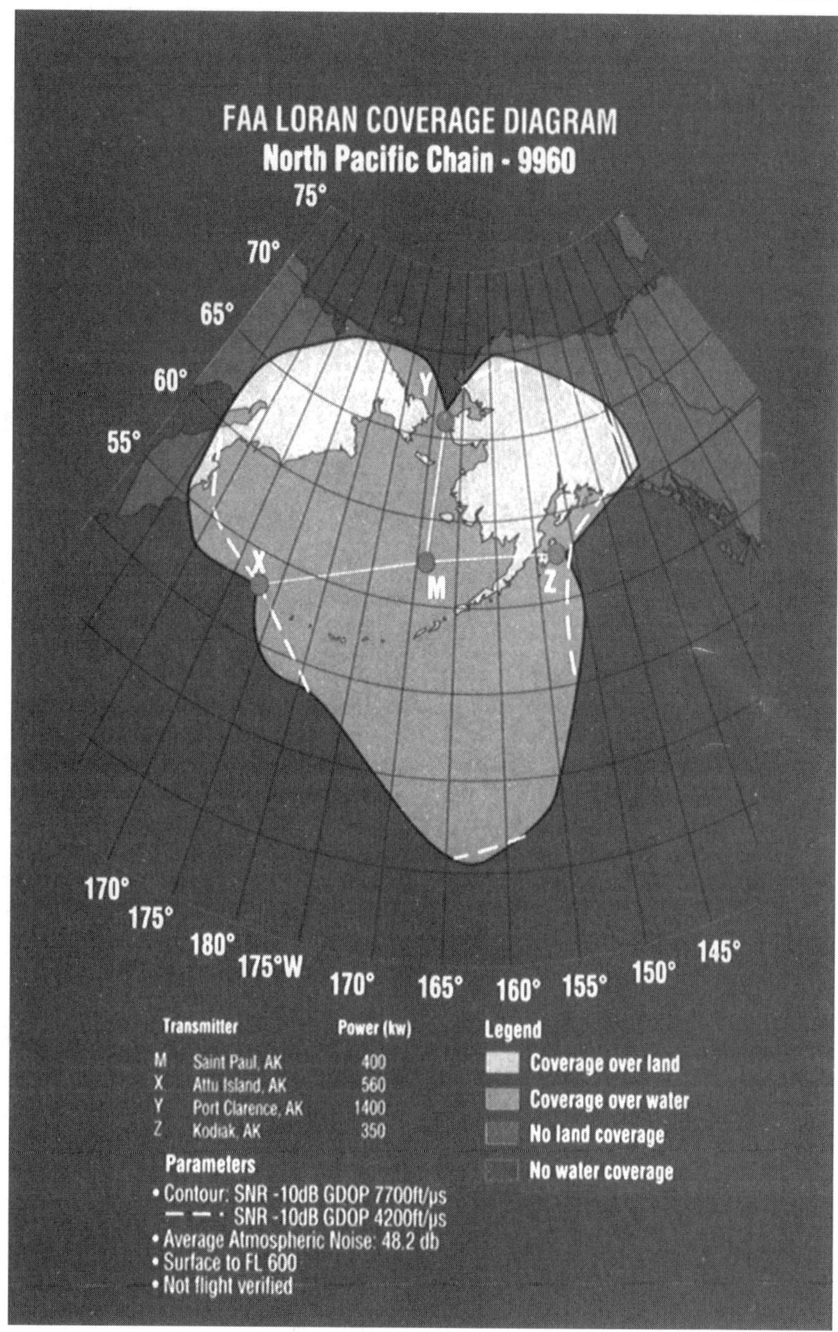

Figure 1-17[7]. North Pacific Chain

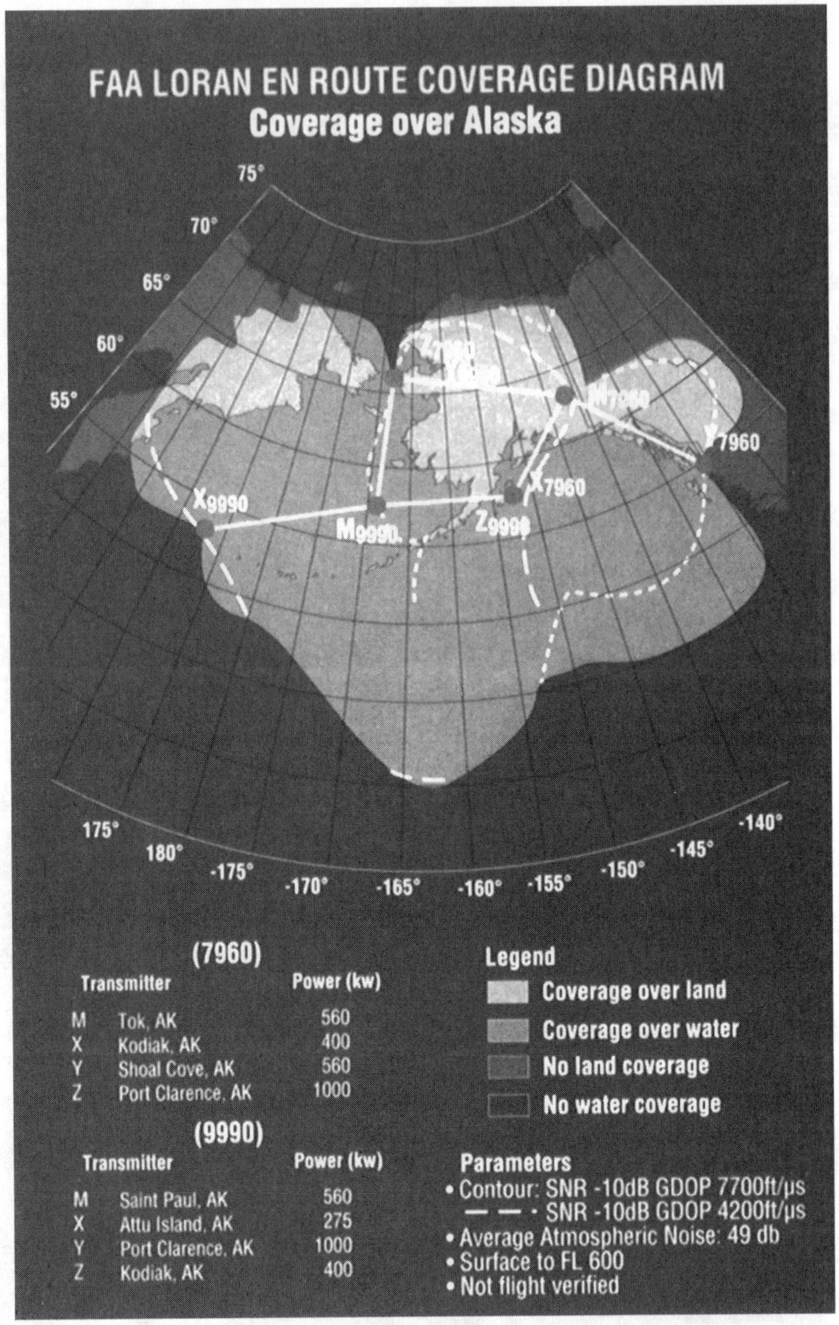

Figure 1-17[8]. Coverage Over Alaska

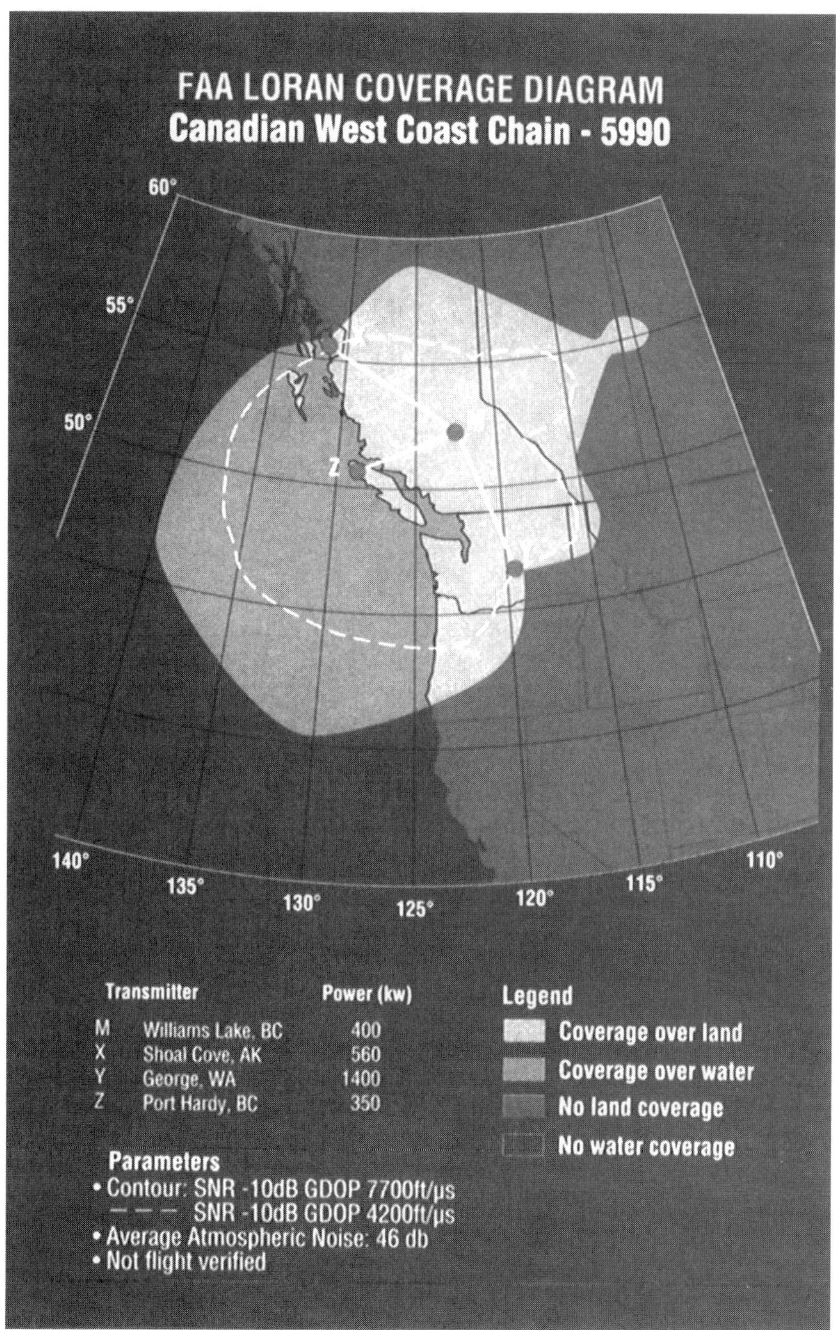

Figure 1-17[9]. Canadian West Coast Chain

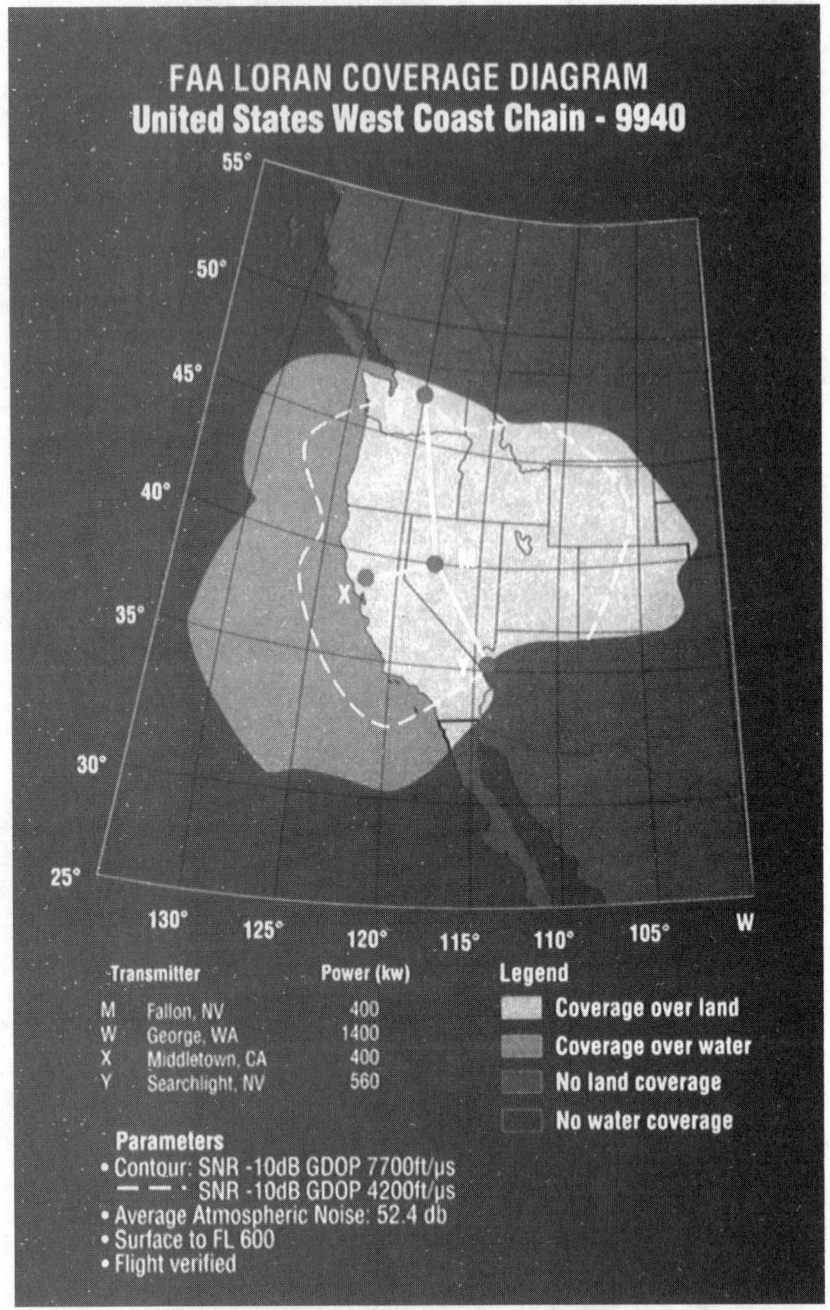

Figure 1-17[10]. United States West Coast Chain

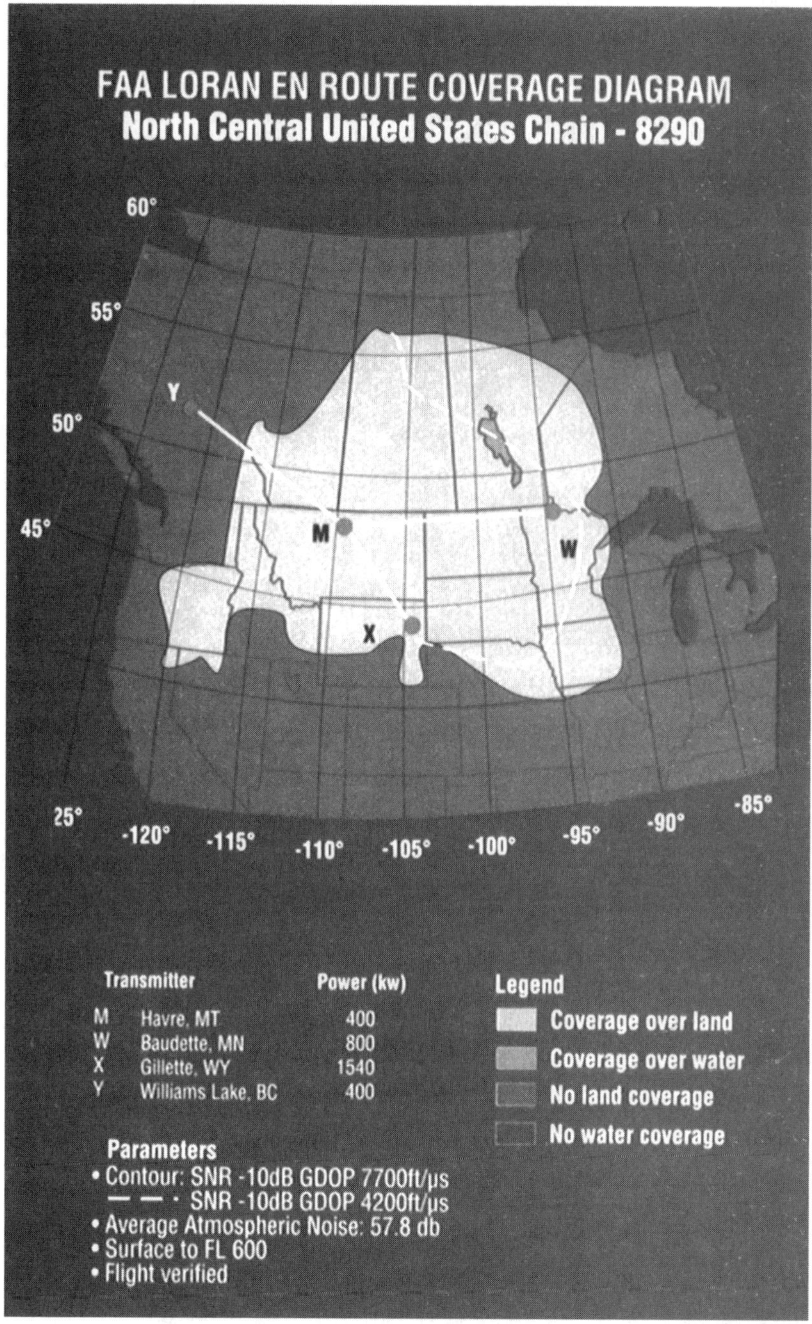

Figure 1-17[11]. North Central United States Chain

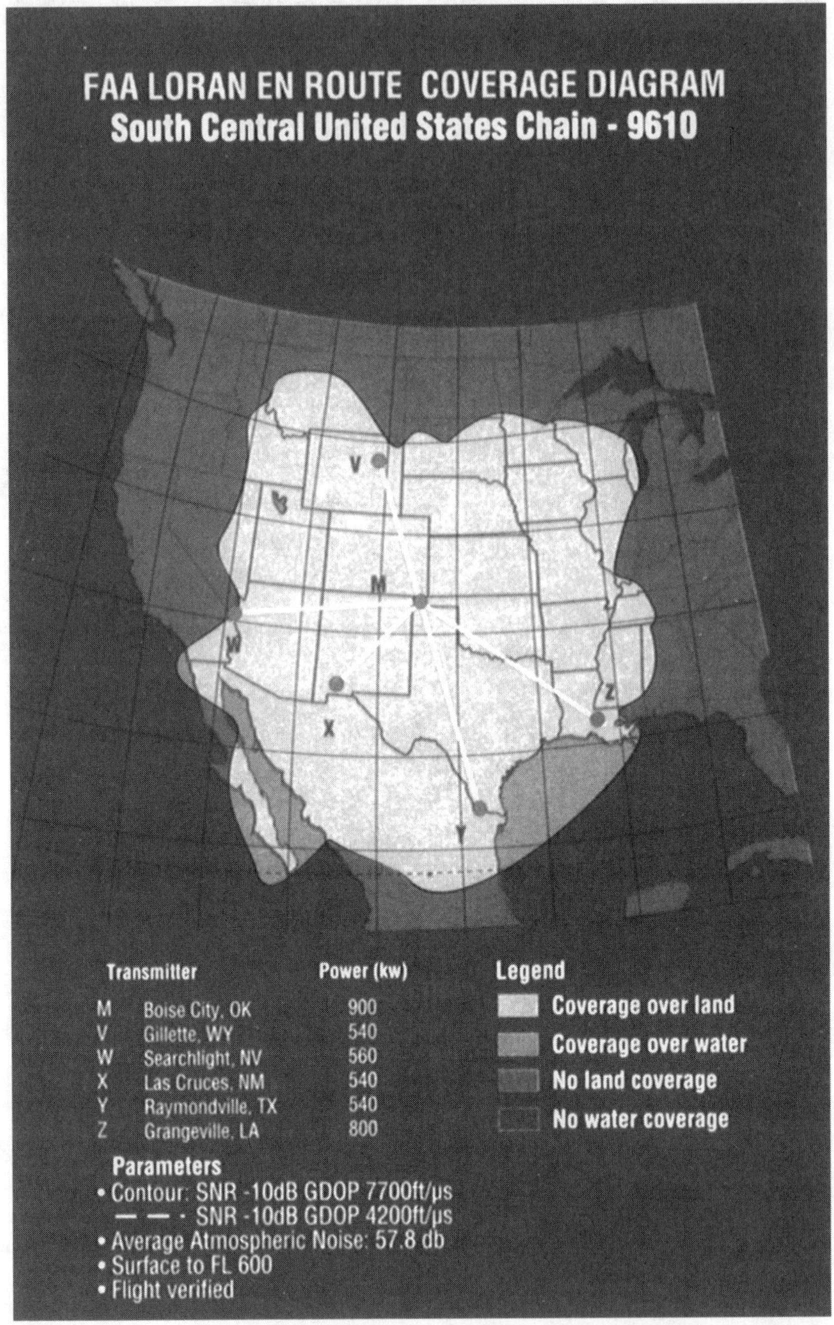

Figure 1-17[12]. South Central United States Chain

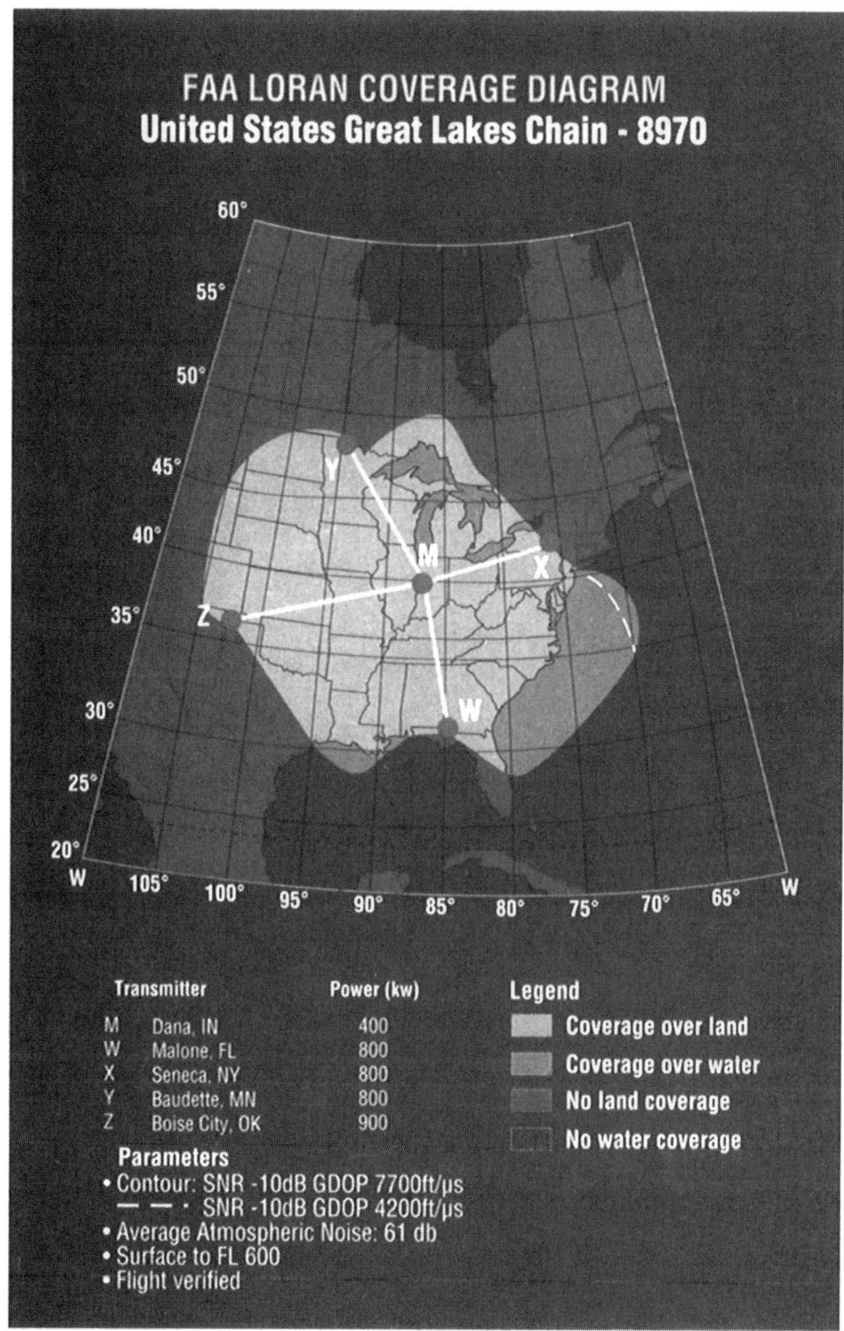

Figure 1-17[13]. United States Great Lakes Chain

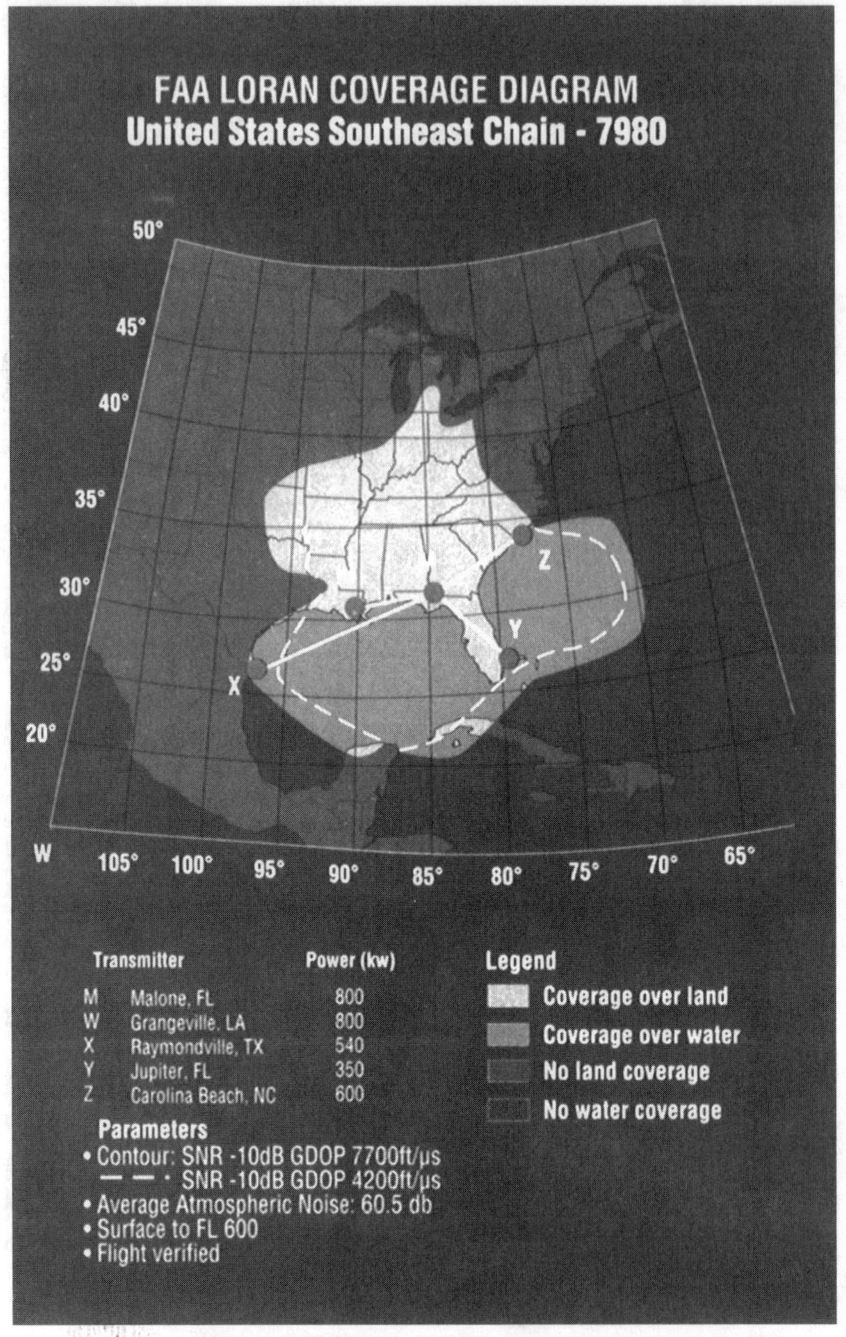

Figure 1-17[14. United States Southeast Chain

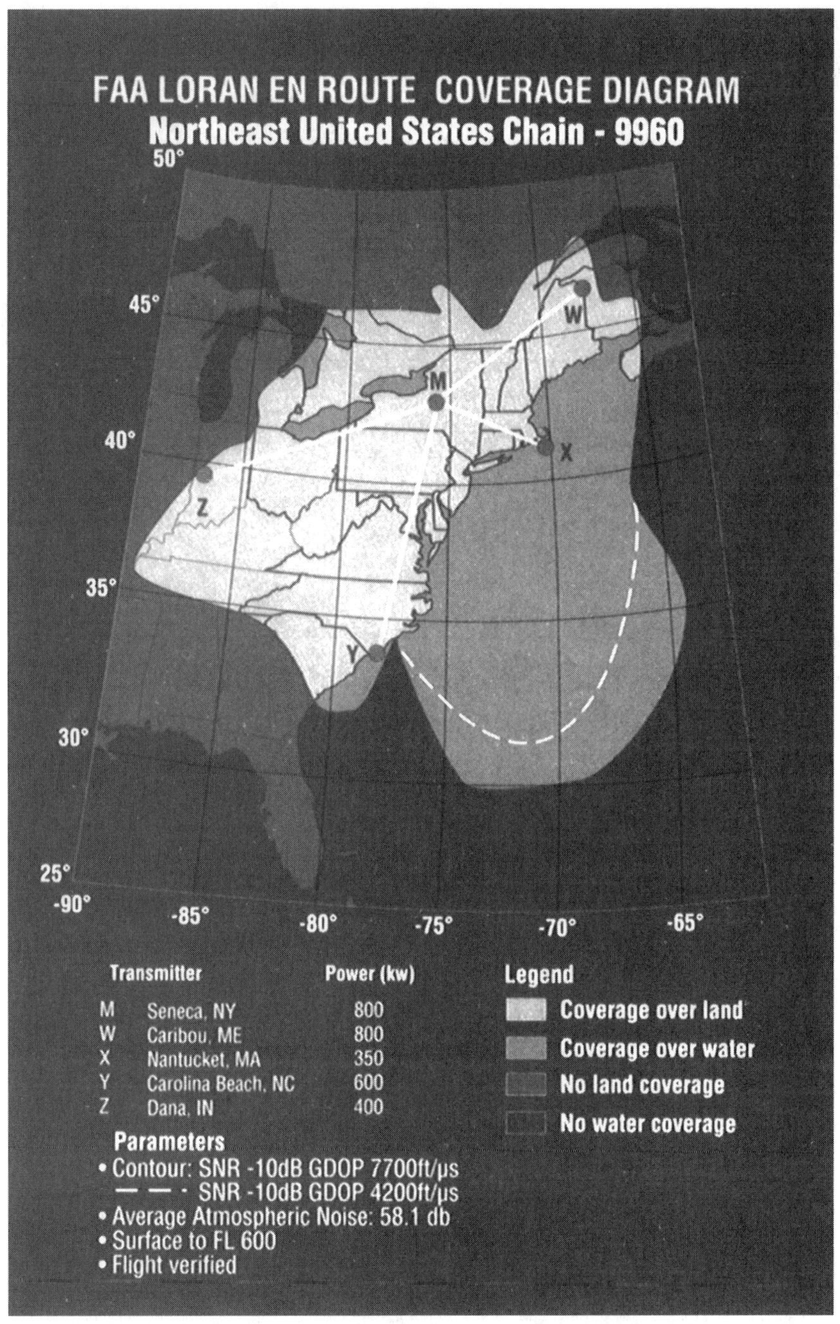

Figure 1-17[15]. Northeast United States Chain

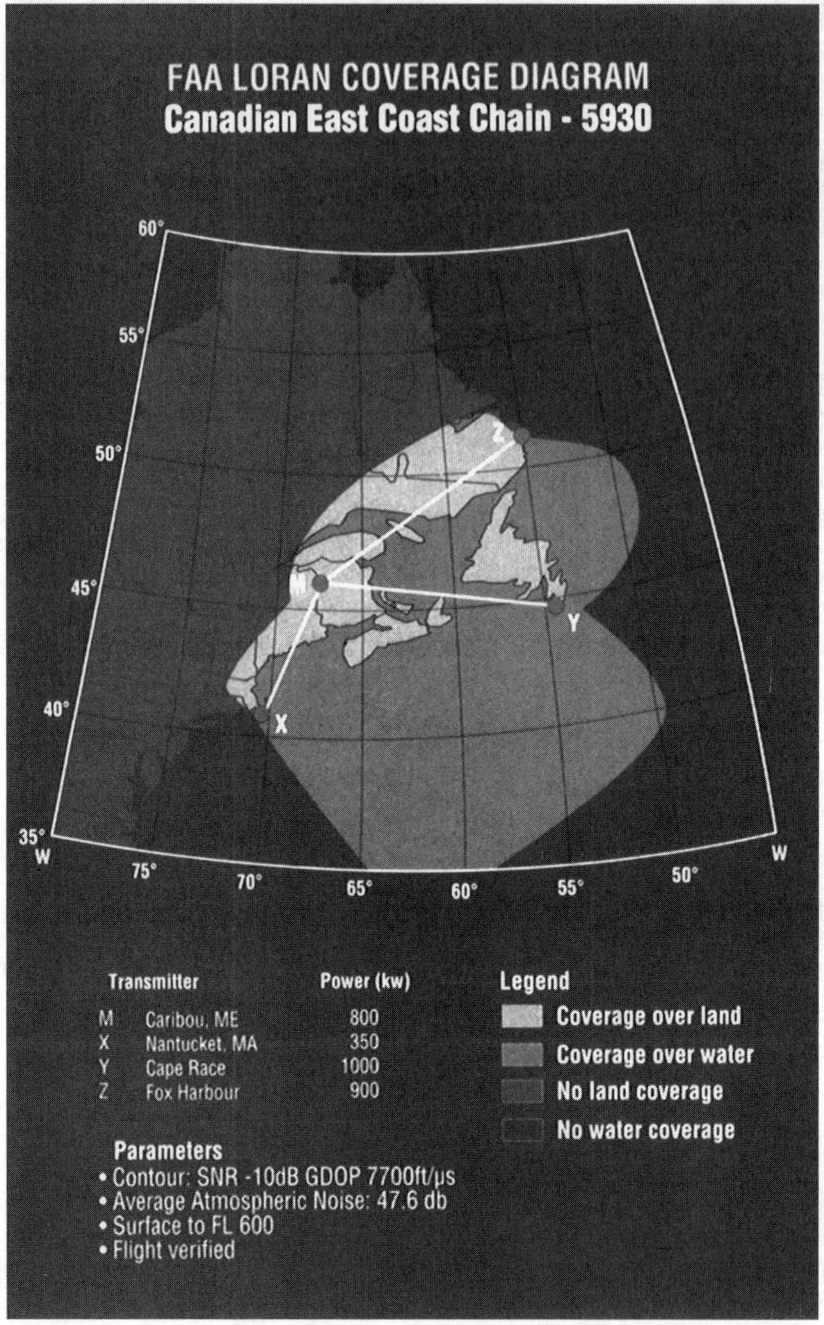

Figure 1-17[16]. Canadian East Coast Chain

e. Notices to Airmen (NOTAMs) are issued for LORAN—C chain or station outages. Domestic NOTAM (D)'s are issued under the identifier "LRN." International NOTAMs are issued under the KNMH series. Pilots may obtain these NOTAMs from FSS briefers upon request.

f. LORAN—C status information

Prerecorded telephone answering service messages pertaining to LORAN—C are available in Table 1-17[1] and Table 1-17[2].

Table 1-17[1]

RATE	CHAIN	TELEPHONE
5930	Canadian East Coast	709-454-3261*
7980	Southeast U.S.	904-569-5241
8970	Great Lakes	607-869-5395
9960	Northeast U.S.	607-869-5395

*St. Anthony, Newfoundland, Canada

Information can also be obtained directly from the office of the Coordinator of Chain Operations (COCO) for each chain. The following telephone numbers are for each COCO office:

1-18. OMEGA AND OMEGA/VLF NAVIGATION SYSTEMS

a. OMEGA

1. OMEGA is a network of eight transmitting stations located throughout the world to provide worldwide signal coverage. These stations transmit in the Very Low Frequency (VLF) band. Because of the low frequency, the signals are receivable to ranges of thousands of miles. The stations are located in Norway, Liberia, Hawaii (USA), North Dakota (USA), La Reunion, Argentina, Australia, and Japan.

2. Presently each station transmits on four basic navigational frequencies: 10.2 kHz, 11.05 kHz, 11.3 kHz, and 13.6 kHz, in sequenced format. This time sequenced format prevents interstation signal interference. With eight stations and a silent .2-second interval between each transmission, the entire cycle repeats every 10 seconds.

3. In addition to the four basic navigational frequencies, each station transmits a unique navigation frequency. An OMEGA station is said to be operating in full format when the station transmits on the basic frequencies plus the unique frequency. Unique frequencies are presently assigned as follows: (See Table 1-18[1].)

b. VLF

1. The U.S. Navy operates a communications system in the VLF band. The stations are located worldwide and transmit at powers of 500-1000 kW. Some airborne OMEGA receivers have the capability to receive and process these VLF signals for navigation in addition to OMEGA signals. The VLF stations generally used for navigation are located in Australia, Japan, England, Hawaii and on the U.S. mainland in Maine, Washington state, and Maryland.

2. Although the Navy does not object to the use of VLF communications signals for navigation, the system is not dedicated to navigation. Signal format, transmission, and other parameters of the VLF system are subject to change at the Navy's discretion. The VLF communications stations are individually shut down for scheduled maintenance for a few hours each week. Regular NOTAM service regarding the VLF system or station status is not available. However, the Naval Observatory provides a taped message concerning phase differences, phase values, and shutdown information for both the VLF communications network and the OMEGA system (phone 202-653-1757).

c. Operational Use of OMEGA and OMEGA/VLF

1. The OMEGA navigation network is capable of providing consistent fixing information to an accuracy of plus or minus 2 NM depending upon the level of sophistication of the receiver/processing system. OMEGA signals are affected by propagation variables which may degrade fix accuracy. These variables include daily variation of phase velocity, polar cap absorption, and sudden solar activity. Daily compen-

Table 1-17[2]

RATE	CHAIN	TELEPHONE	LOCATION
4990	Central Pacific	808-247-5591	Kaneohe, HI
5930	Canadian East Coast	709-454-2392	St. Anthony, NF
5990	Canadian West Coast	604-666-0472	Vancouver, BC
7930	North Atlantic	011-44-1-409-4758	London, UK
7960	Gulf of Alaska	907-487-5583	Kodiak, AK
7970	Norwegian Sea	011-44-1-409-4758	London, UK
7980	Southeast U.S.	205-899-5225	Malone, FL
7990	Mediterranean Sea	011-44-1-409-4758	London, UK
8290	North Central U.S.	707-987-2911	Middletown, CA
8970	Great Lakes	607-869-5393	Seneca, NY
9610	South Central U.S.	205-899-5225	Malone, FL
9940	West Coast U.S.	707-987-2911	Middletown, CA
9960	Northeast U.S.	607-869-5393	Seneca, NY
9970	Northwest Pacific	415-437-3224	San Francisco, CA
9990	North Pacific	907-487-5583	Kodiak, AK

Table 1-18[1]

STATION	LOCATION	FREQUENCY
Station A	Norway	12.1 kHz
Station B	Liberia	12.0 kHz
Station C	Hawaii	11.8 kHz
Station D	North Dakota	13.1 kHz
Station E	La Reunion	12.3 kHz
Station F	Argentina	12.9 kHz
Station G	Australia	13.0 kHz
Station H	Japan	12.8 kHz

sation for variation within the receiver/ processor, or occasional excessive solar activity and its effect on OMEGA, cannot be accurately forecast or anticipated. If an unusual amount of solar activity disturbs the OMEGA signal enlargement paths to any extent, the U.S. Coast Guard advises the FAA and an appropriate NOTAM is sent.

2. At 16 minutes past each hour, WWV (Fort Collins, Colorado) broadcasts a message concerning the status of each OMEGA station, signal irregularities, and other information concerning OMEGA. At 47 minutes past each hour, WWVH (Hawaii) broadcasts similar information. The U.S. Coast Guard provides a taped OMEGA status report (703-866-3801). NOTAMs concerning OMEGA are available through any FSS. OMEGA NOTAMs should be requested by OMEGA station name.

3. The FAA has recognized OMEGA and OMEGA/VLF systems as an additional means of enroute IFR navigation in the conterminous United States and Alaska when approved in accordance with FAA guidance information. Use of OMEGA or OMEGA/VLF requires that all navigation equipment otherwise required by the Federal Aviation Regulations be installed and operating. When flying RNAV routes, VOR and DME equipment is required.

4. The FAA recognizes the use of the Naval VLF communications system as a supplement to OMEGA, but not the sole means of navigation.

1-19. VHF DIRECTION FINDER

a. The VHF Direction Finder (VHF/DF) is one of the common systems that helps pilots without their being aware of its operation. It is a ground based radio receiver used by the operator of the ground station. FAA facilities that provide VHF/DF service are identified in the Airport/Facility Directory.

b. The equipment consists of a directional antenna system and a VHF radio receiver.

c. The VHF/DF receiver display indicates the magnetic direction of the aircraft from the ground station each time the aircraft transmits.

d. DF equipment is of particular value in locating lost aircraft and in helping to identify aircraft on radar. (Reference.—Direction Finding Instrument Approach Procedure, paragraph 6-12).

1-20. INERTIAL NAVIGATION SYSTEM (INS)

The Inertial Navigation System is a totally self-contained navigation system, comprised of gyros, accelerometers, and a navigation computer, which provides aircraft position and navigation information in response to signals resulting from inertial effects on system components, and does not require information from external references. INS is aligned with accurate position information prior to departure, and thereafter calculates its position as it progresses to the destination. By programming a series of waypoints, the system will navigate along a predetermined track. New waypoints can be inserted at any time if a revised routing is desired. INS accuracy is very high initially following alignment, and decays with time at the rate of about 1-2 nautical miles per hour. Position update alignment can be accomplished inflight using ground based references, and many INS systems now have sophisticated automatic update using dual DME and or VOR inputs. INS may be approved as the sole means of navigation or may be used in combination with other systems.

1-21. DOPPLER RADAR

Doppler Radar is a semiautomatic self-contained dead reckoning navigation system (radar sensor plus computer) which is not continuously dependent on information derived from ground based or external aids. The system employs radar signals to detect and measure ground speed and drift angle, using the aircraft compass system as its directional reference. Doppler is less accurate than INS or OMEGA however, and the use of an external reference is required for periodic updates if acceptable position accuracy is to be achieved on long range flights.

1-22. FLIGHT MANAGEMENT SYSTEM (FMS)

The Flight Management System is a computer system that uses a large data base to allow routes to be preprogrammed and fed into the system by means of a data loader. The system is constantly updated with respect to position accuracy by reference to conventional navigation aids. The sophisticated program and its associated data base insures that the most appropriate aids are automatically selected during the information update cycle.

1-23. GLOBAL POSITIONING SYSTEM (GPS)

The Global Positioning System is a space-base radio positioning, navigation, and time-transfer system being developed by Department of Defense. When fully deployed, the system is intended to provide highly accurate position and velocity information, and precise time, on a continuous global basis, to an unlimited number of properly equipped users. The system will be unaffected by weather, and will provide a worldwide common grid reference system. The GPS concept is predicated upon accurate and continuous knowledge of the spatial position of each satellite in the system with respect to time and distance from a transmitting satellite to the user. The GPS receiver automatically selects appropriate signals from the satellites in view and translates these into a three-dimensional position, velocity, and time. Predictable system accuracy for civil users is projected to be 100 meters horizontally. Performance standards and certification criteria have not yet been established.

1-24 thru 1-29. RESERVED

Section 2. RADAR SERVICES AND PROCEDURES

1-30. RADAR

a. Capabilities

1. Radar is a method whereby radio waves are transmitted into the air and are then received when they have been reflected by an object in the path of the beam. *Range* is determined by measuring the time it takes (at the speed of light) for the radio wave to go out to the object and then return to the receiving antenna. The *direction* of a detected object from a radar site is determined by the position of the rotating antenna when the reflected portion of the radio wave is received.

2. More reliable maintenance and improved equipment have reduced radar system failures to a negligible factor. Most facilities actually have some components duplicated—one operating and another which immediately takes over when a malfunction occurs to the primary component.

b. Limitations

1. It is very important for the aviation community to recognize the fact that there are limitations to radar service and that ATC controllers may not always be able to issue traffic advisories concerning aircraft which are not under ATC control and cannot be seen on radar. (See Figure 1-30[1].)

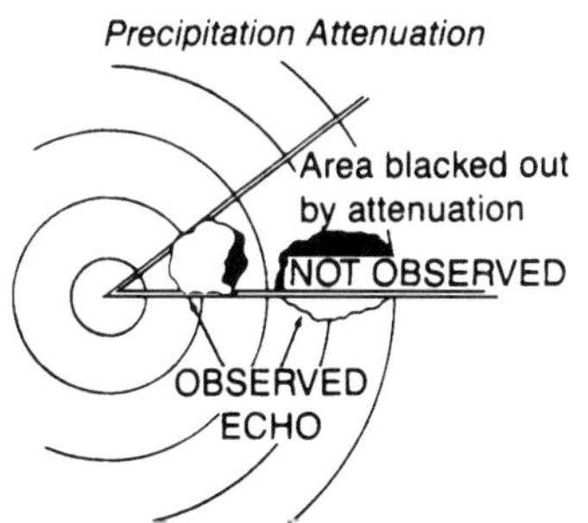
Precipitation Attenuation
Area blacked out by attenuation
NOT OBSERVED
OBSERVED ECHO
The nearby target absorbs and scatters so much of the out-going and returning energy that the radar does not detect the distant target.

Figure 1-30[1]

(a) The characteristics of radio waves are such that they normally travel in a continuous straight line unless they are:

(1) "Bent" by abnormal atmospheric phenomena such as temperature inversions;

(2) Reflected or attenuated by dense objects such as heavy clouds, precipitation, ground obstacles, mountains, etc.; or

(3) Screened by high terrain features.

(b) The bending of radar pulses, often called anomalous propagation or ducting, may cause many extraneous blips to appear on the radar operator's display if the beam has been bent toward the ground or may decrease the detection range if the wave is bent upward. It is difficult to solve the effects of anomalous propagation, but using beacon radar and electronically eliminating stationary and slow moving targets by a method called moving target indicator (MTI) usually negate the problem.

(c) Radar energy that strikes dense objects will be reflected and displayed on the operator's scope thereby blocking out aircraft at the same range and greatly weakening or completely eliminating the display of targets at a greater range. Again, radar beacon and MTI are very effectively used to combat ground clutter and weather phenomena, and a method of circularly polarizing the radar beam will eliminate some weather returns. A negative characteristic of MTI is that an aircraft flying a speed that coincides with the canceling signal of the MTI (tangential or "blind" speed) may not be displayed to the radar controller.

(d) Relatively low altitude aircraft will not be seen if they are screened by mountains or are below the radar beam due to earth curvature. The only solution to screening is the installation of strategically placed multiple radars which has been done in some areas.

(e) There are several other factors which affect radar control. The amount of reflective surface of an aircraft will determine the size of the radar return. Therefore, a small light airplane or a sleek jet fighter will be more difficult to see on radar than a large commercial jet or military bomber. Here again, the use of radar beacon is invaluable if the aircraft is equipped with an airborne transponder. All ARTCC radars in the conterminous U.S. and many airport surveillance radars have the capability to interrogate

MODE C and display altitude information to the controller from appropriately equipped aircraft. However, there are a number of airport surveillance radars that are still two dimensional (range and azimuth) only and altitude information must be obtained from the pilot.

(f) At some locations within the ATC en route environment, secondary-radar-only (no primary radar) gap filler radar systems are used to give lower altitude radar coverage between two larger radar systems, each of which provides both primary and secondary radar coverage. In those geographical areas served by secondary-radar only, aircraft without transponders cannot be provided with radar service. Additionally, transponder equipped aircraft cannot be provided with radar advisories concerning primary targets and weather. (See Pilot/Controller Glossary, RADAR/Radio Detection and Ranging).

(g) The controllers' ability to advise a pilot flying on instruments or in visual conditions of his proximity to another aircraft will be limited if the unknown aircraft is not observed on radar, if no flight plan information is available, or if the volume of traffic and workload prevent his issuing traffic information. The controller's first priority is given to establishing vertical, lateral, or longitudinal separation between aircraft flying IFR under the control of ATC.

c. FAA radar units operate continuously at the locations shown in the Airport/Facility Directory, and their services are available to all pilots, both civil and military. Contact the associated FAA control tower or ARTCC on any frequency guarded for initial instructions, or in an emergency, any FAA facility for information on the nearest radar service.

1-31. AIR TRAFFIC CONTROL RADAR BEACON SYSTEM (ATCRBS)

a. The ATCRBS, sometimes referred to as secondary surveillance radar, consists of three main components:

1. Interrogator. Primary radar relies on a signal being transmitted from the radar antenna site and for this signal to be reflected or "bounced back" from an object (such as an aircraft). This reflected signal is then displayed as a "target" on the controller's radarscope. In the ATCRBS, the *Interrogator*, a ground based radar beacon transmitter-receiver, scans in synchronism with the primary radar and transmits discrete radio signals which repetitiously requests all transponders, on the mode being used, to reply. The replies received are then mixed with the primary returns and both are displayed on the same radarscope.

2. Transponder. This airborne radar beacon transmitter-receiver automatically receives the signals from the interrogator and selectively replies with a specific pulse group (code) only to those interrogations being received on the mode to which it is set. These replies are independent of, and much stronger than a primary radar return.

3. Radarscope. The radarscope used by the controller displays returns from both the primary radar system and the ATCRBS. These returns, called targets, are what the controller refers to in the control and separation of traffic.

b. The job of identifying and maintaining identification of primary radar targets is a long and tedious task for the controller. Some of the advantages of ATCRBS over primary radar are:

1. Reinforcement of radar targets.

2. Rapid target identification.

3. Unique display of selected codes.

c. A part of the ATCRBS ground equipment is the decoder. This equipment enables the controller to assign discrete transponder codes to each aircraft under his control. Normally only one code will be assigned for the entire flight. Assignments are made by the ARTCC computer on the basis of the National Beacon Code Allocation Plan. The equipment is also designed to receive MODE C altitude information from the aircraft.

1-31c NOTE—Refer to figures with explanatory legends for an illustration of the target symbology depicted on radar scopes in the NAS Stage A (en route), the ARTS III (terminal) systems, and other nonautomated (broadband) radar systems. (See Figure 1-31[1] and Figure 1-31[2].)

d. It should be emphasized that aircraft transponders greatly improve the effectiveness of radar systems. (Reference—Transponder Operation, paragraph 4-19.)

1-32. SURVEILLANCE RADAR

a. Surveillance radars are divided into two general categories: Airport Surveillance Radar (ASR) and Air Route Surveillance Radar (ARSR).

1. ASR is designed to provide relatively short-range coverage in the general vicinity of an airport and to serve as an expeditious means of handling terminal area traffic through observation of precise aircraft locations on a radarscope. The ASR can also be used as an instrument approach aid.

2. ARSR is a long-range radar system designed primarily to provide a display of aircraft locations over large areas.

3. Center Radar Automated Radar Terminal Systems (ARTS) Processing (CENRAP) was developed to provide an alternative to a non-radar environment at terminal facilities should an Airport Surveillance Radar (ASR) fail or malfunction. CENRAP sends aircraft radar beacon target information to the ASR terminal facility equipped with ARTS. Procedures used for the separation of aircraft may increase under certain conditions when a facility is utilizing CENRAP because radar target information updates at a slower rate than the normal ASR radar. Radar services for VFR aircraft are also limited during CENRAP operations because of the additional workload required to provide services to IFR aircraft.

b. Surveillance radars scan through 360 degrees of az-

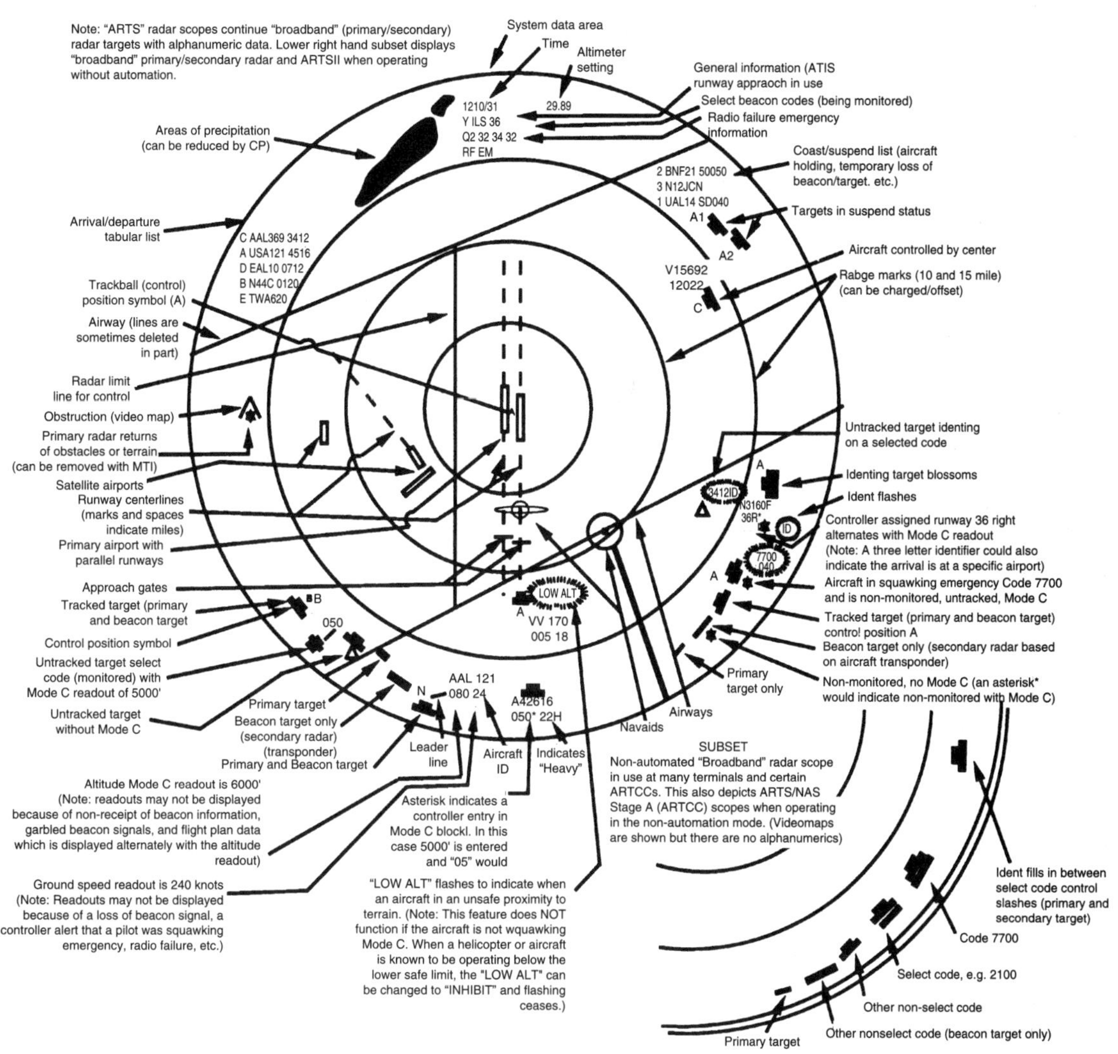
Note: "ARTS" radar scopes continue "broadband" (primary/secondary) radar targets with alphanumeric data. Lower right hand subset displays "broadband" primary/secondary radar and ARTSII when operating without automation.
System data area
Time
Altimeter setting
General information (ATIS runway appraoch in use
Select beacon codes (being monitored)
Radio failure emergency information
Coast/suspend list (aircraft holding, temporary loss of beacon/target. etc.)
Targets in suspend status
Aircraft controlled by center
Rabge marks (10 and 15 mile) (can be charged/offset)
Areas of precipitation (can be reduced by CP)
Arrival/departure tabular list
Trackball (control) position symbol (A)
Airway (lines are sometimes deleted in part)
Radar limit line for control
Obstruction (video map)
Primary radar returns of obstacles or terrain (can be removed with MTI)
Satellite airports
Runway centerlines (marks and spaces indicate miles)
Primary airport with parallel runways
Approach gates
Tracked target (primary and beacon target)
Control position symbol
Untracked target select code (monitored) with Mode C readout of 5000'
Untracked target without Mode C
Primary target
Beacon target only (secondary radar) (transponder)
Primary and Beacon target
Leader line
Aircraft ID
Indicates "Heavy"
Untracked target identing on a selected code
Identing target blossoms
Ident flashes
Controller assigned runway 36 right alternates with Mode C readout (Note: A three letter identifier could also indicate the arrival is at a specific airport)
Aircraft in squawking emergency Code 7700 and is non-monitored, untracked, Mode C
Tracked target (primary and beacon target) control position A
Beacon target only (secondary radar based on aircraft transponder)
Non-monitored, no Mode C (an asterisk* would indicate non-monitored with Mode C)
Primary target only
Airways
Navaids
1210/31
29.89
Y ILS 36
Q2 32 34 32
RF EM
2 BNF21 50050
3 N12JCN
1 UAL14 SD040
A1
A2
V15692
12022
C
C AAL369 3412
A USA121 4516
D EAL10 0712
B N44C 0120
E TWA620
3412ID
N3160F
36R*
ID
7700
040
LOW ALT
A
VV 170
005 18
B
050
AAL 121
N
080 24
A42616
050* 22H
SUBSET
Non-automated "Broadband" radar scope in use at many terminals and certain ARTCCs. This also depicts ARTS/NAS Stage A (ARTCC) scopes when operating in the non-automation mode. (Videomaps are shown but there are no alphanumerics)
Altitude Mode C readout is 6000' (Note: readouts may not be displayed because of non-receipt of beacon information, garbled beacon signals, and flight plan data which is displayed alternately with the altitude readout)
Ground speed readout is 240 knots (Note: Readouts may not be displayed because of a loss of beacon signal, a controller alert that a pilot was squawking emergency, radio failure, etc.)
Asterisk indicates a controller entry in Mode C blockl. In this case 5000' is entered and "05" would
"LOW ALT" flashes to indicate when an aircraft in an unsafe proximity to terrain. (Note: This feature does NOT function if the aircraft is not wquawking Mode C. When a helicopter or aircraft is known to be operating below the lower safe limit, the "LOW ALT" can be changed to "INHIBIT" and flashing ceases.)
Ident fills in between select code control slashes (primary and secondary target)
Code 7700
Select code, e.g. 2100
Other non-select code
Primary target
Other nonselect code (beacon target only)
ARTSIII radar scope with alphanumeric data. Note: A number of radar terminals do not have ART equiment. Those facilities and certain ARTCC outside the contiguous US would also have similar displays and certain services based on automation may not be available.

Figure 1-31[1]

RADAR SERVICES AND PROCEDURES

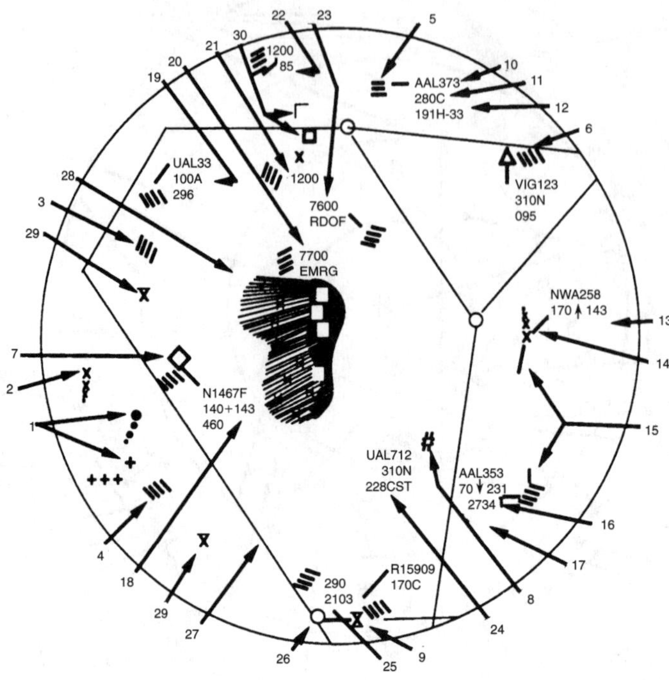

Target symbols

1 Uncorreleated primary radar target **+ ●**
2 *Correlated primary radar target **X**
3 Uncorrelated beacon target **/**
4 Correlated beacon target ****
5 Identing beacon target **≡**
(*Correlated means the association of radar data with the computer projected track of an identified aricraft)

Position symbols

6 Free track(no flight plan tracking) **Δ**
7 First track(flight plan tracking) **◊**
8 Coast(beacon target lost) **#**
9 Present position hold **⧗**

Data block information

10 *Aircraft identification
11 *Assigned altitude FL280, mode C altitude same or within ±200' of asgnd altitude
12 *Computer ID #191, Handoff is to Sector 33 (0-33 would mean handoff accepted) (*Nr's 10, 11, 12 constitute a "full data block")
13 Assigned altitude 17,000', aircraft is climbing, mode C readout was 14,300 when last beacon interrogation was received
14 Leader line connecting target symbol and data block
15 Track velocity and direction vector line (projected ahead of target)
16 Assigned altitude 7000, aircraft is descending, last mode C readout (or last reported altitude was 100' above FL 230
17 Transponder code shows in full data block only when different than assigned code
18 Aircraft is 300' above assigned altitude
19 Reported altitude (no mode C readout same as assigned. An "N" would indicate no reported altitude)
20 Transponder set on emergency code 7700 (EMRG flashes to attract attention)
21 Transponder code 1200 (VFR) with no mode C
22 Code 1200(VFR) with mode C and last altitude readout
23 Transponder set on Radio Failure code 7600, (RDOF flashes)
24 Computer ID #288, CST indicates target is in Coast status
25 Assigned altitude FL 290, transponder

Other symbols

26 Navigational Aid
27 Airway or jet route
28 Outline of weather returns based on primary radar (See Chapter 4, ARTCC Radar Weather Display. H's represent areas of high density precipitation which might be thunderstorms. Radial lines indicate lower density precipitation)
29 Obstruction
30 Airports Major: **□** Small: **Γ**

NAS Stage A Controllers View Plan Display. This figure illustrates the controller's radar scope (PVD) when operating in the full automatic (RDP) mode, which is normally 20 hours per day. (Note: When in automation mode, the display is similar to the broadband mode shown in the ARTS III Radar Scope figure. Certain ARTCC's outside the contiguous U.S. also operate in "broadband" mode.)

Figure 1-31[2]

imuth and present target information on a radar display located in a tower or center. This information is used independently or in conjunction with other navigational aids in the control of air traffic.

1-33. PRECISION APPROACH RADAR (PAR)

a. PAR is designed to be used as a *landing aid*, rather than an aid for sequencing and spacing aircraft. PAR equipment may be used as a primary landing aid, or it may be used to monitor other types of approaches. It is designed to display *range*, *azimuth* and *elevation* information.

b. Two antennas are used in the PAR array, one scanning a vertical plane, and the other scanning horizontally. Since the range is limited to 10 miles, azimuth to 20 degrees, and elevation to 7 degrees, only the final approach area is covered. Each scope is divided into two parts. The upper half presents altitude and distance information, and the lower half presents azimuth and distance.

Chapter 2. Aeronautical lighting and other airport visual aids

Section 1. AIRPORT LIGHTING AIDS

2-1. APPROACH LIGHT SYSTEMS (ALS)

a. Approach light systems provide the basic means to transition from instrument flight to visual flight for landing. Operational requirements dictate the sophistication and configuration of the approach light system for a particular runway.

b. Approach light systems are a configuration of signal light starting at the landing threshold and extending into the approach area a distance of 2400-3000 feet for precision instrument runways and 1400-1500 feet for nonprecision instrument runways. Some systems include sequenced flashing lights which appear to the pilot as a ball of light traveling towards the runway at high speed (twice a second). (See Figure 2-1[1] [Precision & Nonprecision Configurations].)

2-2. VISUAL GLIDESLOPE INDICATORS

a. Visual Approach Slope Indicator (VASI)

1. The VASI is a system of lights so arranged to provide visual descent guidance information during the approach to a runway. These lights are visible from 3–5 miles during the day and up to 20 miles or more at night. The visual glide path of the VASI provides safe obstruction clearance within plus or minus 10 degrees of the extended runway centerline and to 4 NM from the runway threshold. Descent, using the VASI, should not be initiated until the aircraft is visually aligned with the runway. Lateral course guidance is provided by the runway or runway lights.

2. VASI installations may consist of either 2, 4, 6, 12, or 16 light units arranged in bars referred to as near, middle, and far bars. Most VASI installations consist of 2 bars, near

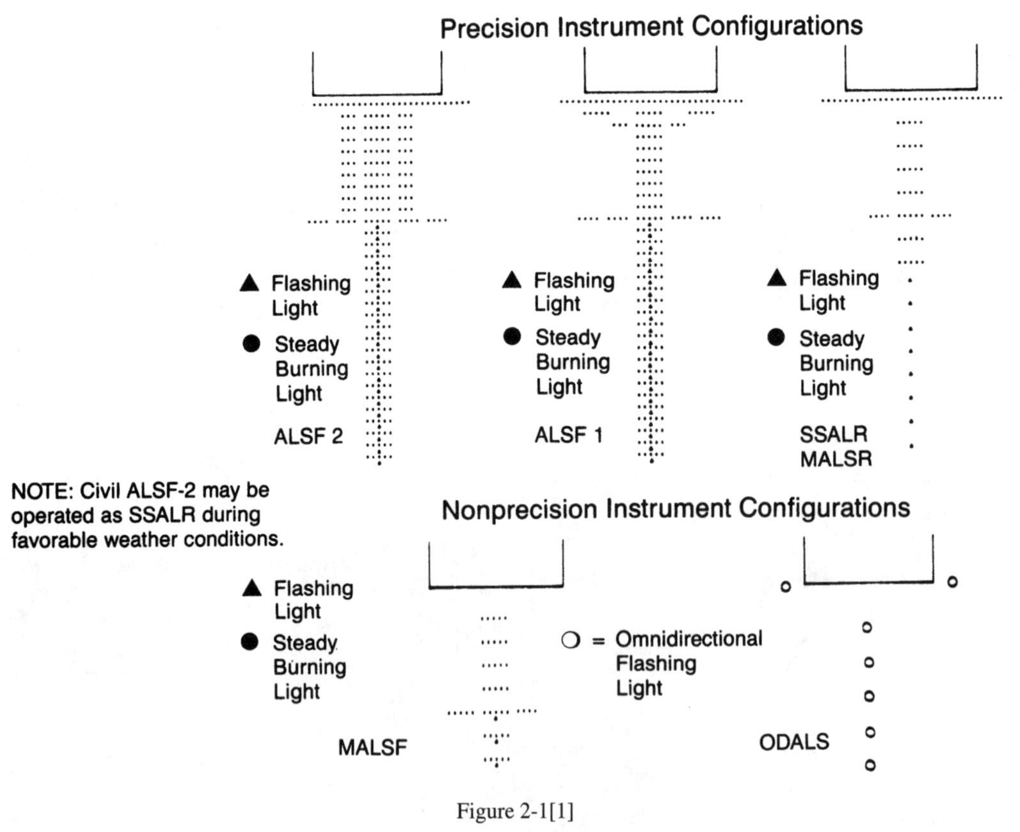

Figure 2-1[1]

and far, and may consist of 2, 4, or 12 light units. Some VASIs consist of three bars, near, middle, and far, which provide an additional visual glide path to accommodate high cockpit aircraft. This installation may consist of either 6 or 16 light units. VASI installations consisting of 2, 4, or 6 light units are located on one side of the runway, usually the left. Where the installation consists of 12 or 16 light units, the units are located on both sides of the runway.

3. Two-bar VASI installations provide one visual glide path which is normally set at 3 degrees. Three-bar VASI installations provide two visual glide paths. The lower glide path is provided by the near and middle bars and is normally set at 3 degrees while the upper glide path, provided by the middle and far bars, is normally ¼ degree higher. This higher glide path is intended for use only by high cockpit aircraft to provide a sufficient threshold crossing height. Although normal glide path angles are three degrees, angles at some locations may be as high as 4.5 degrees to give proper obstacle clearance. Pilots of high performance aircraft are cautioned that use of VASI angles in excess of 3.5 degrees may cause an increase in runway length required for landing and rollout.

4. The basic principle of the VASI is that of color differentiation between red and white. Each light unit projects a beam of light having a white segment in the upper part of the beam and red segment in the lower part of the beam. The light units are arranged so that the pilot using the VASIs during an approach will see the combination of lights shown below.

5. For 2-bar VASI (4 light units) See Figure 2-2[1].

6. For 3-bar VASI (6 light units) See Figure 2-2[2].

7. For other VASI configurations See Figure 2-2[3].

b. Precision Approach Path Indicator (PAPI)—The precision approach path indicator (PAPI) uses light units similar to the VASI but are installed in a single row of either two or four light units. These systems have an effective visual range of about 5 miles during the day and up to 20 miles at night. The row of light units is normally installed on the left side of the runway and the glide path indications are as depicted. (See Figure 2-2[4].)

c. Tri-color Systems—Tri-color visual approach slope indicators normally consist of a single light unit projecting a three-color visual approach path into the final approach area of the runway upon which the indicator is installed. The below glide path indication is red, the above glide path indication is amber, and the on glide path indication is green. These types of indicators have a useful range of approximately one-half to one mile during the day and up to five miles at night depending upon the visibility conditions. (See Figure 2-2[5].)

d. Pulsating Systems—Pulsating visual approach slope indicators normally consist of a single light unit projecting a two-color visual approach path into the final approach area of the runway upon which the indicator is installed. The on glide path indication is a steady white light. The slightly below glide path indication is a steady red light. If the aircraft descends further below the glide path the red light starts to pulsate. The above glide path indication is a

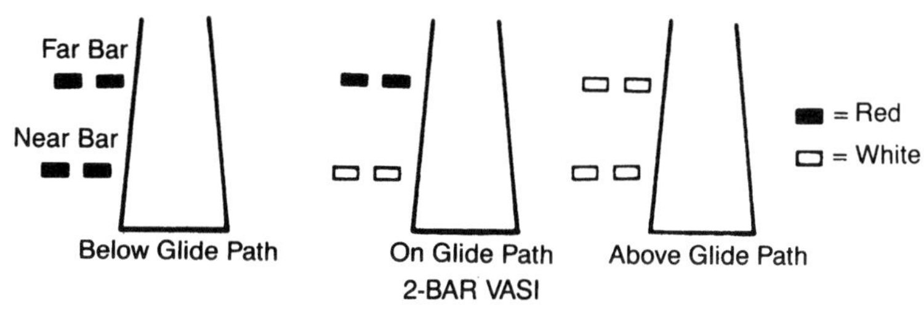

Figure 2-2[1]

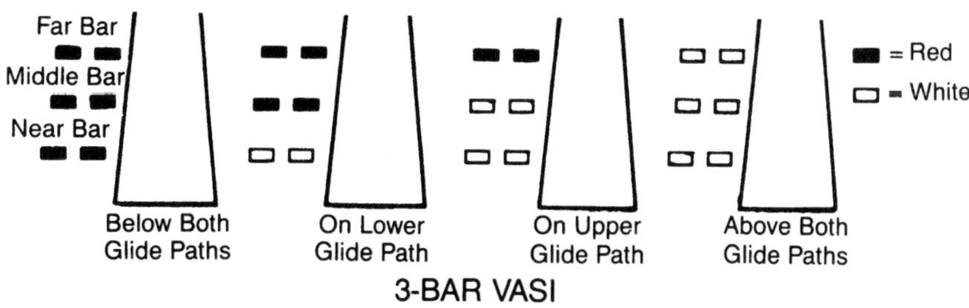

Figure 2-2[2]

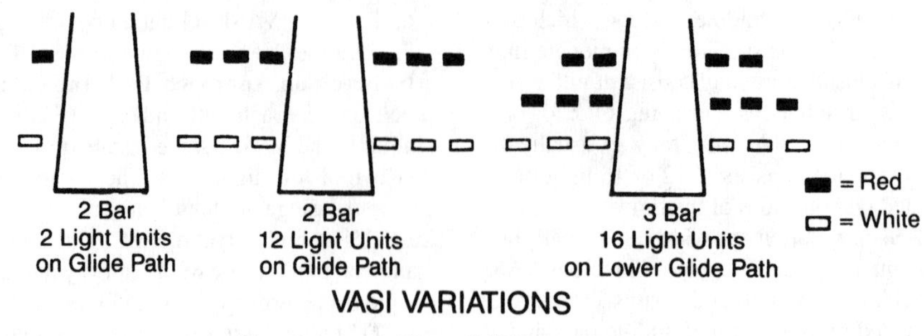

Figure 2-2[3]

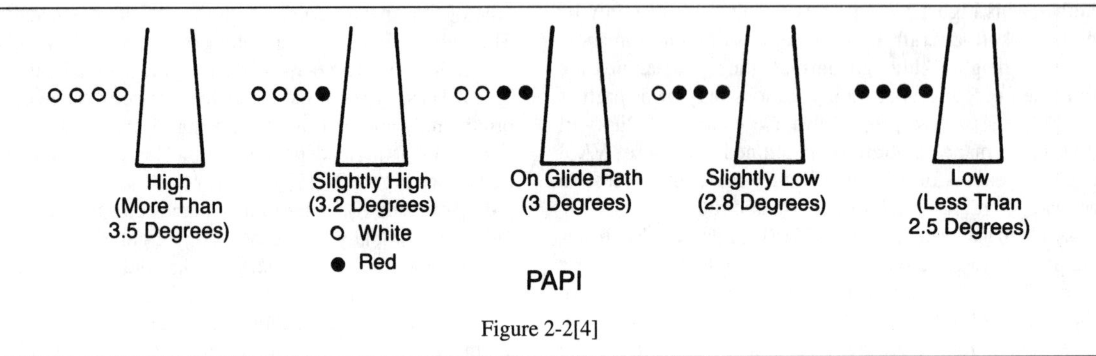

Figure 2-2[4]

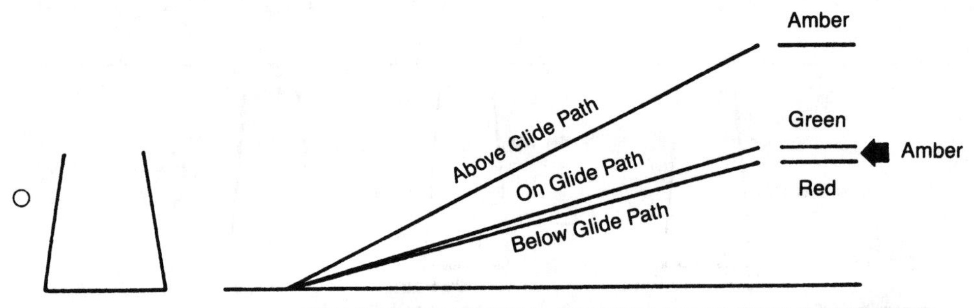

TRI-COLOR VISUAL APPROACH SLOPE INDICATOR

Caution: When the aircraft descends from green to red, the pilot may seek a dark amber color during the transition from green to red.

Figure 2-2[5]

pulsating white light. The pulsating rate increases as the aircraft gets further above or below the desired glide slope. The useful range of the system is about four miles during the day and up to ten miles night. (See Figure 2-2[6].)

e. Alignment of Elements Systems—Alignment of elements systems are installed on some small general aviation airports and are a low-cost system consisting of painted plywood panels, normally black and white or fluorescent orange. Some of these systems are lighted for night use. The useful range of these systems is approximately three-quarter miles. To use the system the pilot positions his aircraft so the elements are in alignment. The glide path indications are shown in Figure 2-2[7].

2-3. RUNWAY END IDENTIFIER LIGHTS (REIL)

REILs are installed at many airfields to provide rapid and positive identification of the approach end of a particular runway. The system consists of a pair of synchronized

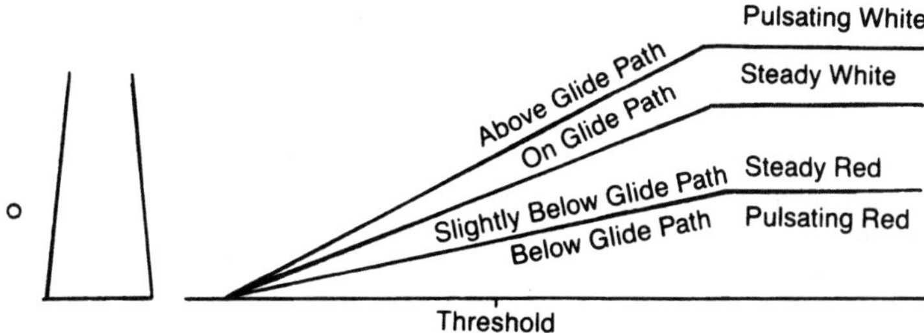

PULSATING VISUAL APPROACH SLOPE INDICATOR

Caution: When viewing the pulsating visual approach slope indicators in the pulsating white or pulsating red sectors, it is possible to mistake this lighting aid for another aircraft or a ground vehicle. Pilots should exercise caution when using this type of system.

Figure 2-2[6]

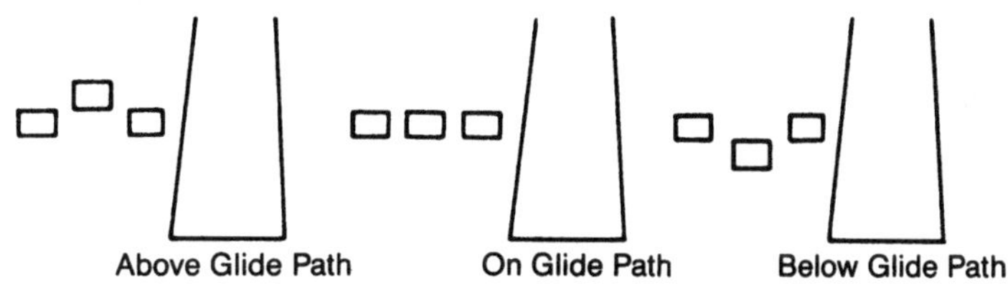

ALIGNMENT OF ELEMENTS

Figure 2-2[7]

flashing lights located laterally on each side of the runway threshold. REILs may be either omnidirectional or unidirectional facing the approach area. They are effective for:

1. Identification of a runway surrounded by a preponderance of other lighting.

2. Identification of a runway which lacks contrast with surrounding terrain.

3. Identification of a runway during reduced visibility.

2-4. RUNWAY EDGE LIGHT SYSTEMS

a. Runway edge lights are used to outline the edges of runways during periods of darkness or restricted visibility conditions. These light systems are classified according to the intensity or brightness they are capable of producing: they are the High Intensity Runway Lights (HIRL), Medium Intensity Runway Lights (MIRL), and the Low Intensity Runway Lights (LIRL). The HIRL and MIRL systems have variable intensity controls, whereas the LIRLs normally have one intensity setting.

b. The runway edge lights are white, except on instrument runways amber replaces white on the last 2,000 feet or half the runway length, whichever is less, to form a caution zone for landings.

c. The lights marking the ends of the runway emit red light toward the runway to indicate the end of runway to a departing aircraft and emit green outward from the runway end to indicate the threshold to landing aircraft.

2-5. IN-RUNWAY LIGHTING

Touchdown zone lights and runway centerline lights are installed on some precision approach runways to facilitate landing under adverse visibility conditions. Taxiway turnoff lights may be added to expedite movement of aircraft from the runway.

a. Touchdown Zone Lighting (TDZL)—two rows of transverse light bars disposed symmetrically about the runway centerline in the runway touchdown zone. The system starts 100 feet from the landing threshold and extends to 3000 feet from the threshold or the midpoint of the runway, whichever is the lesser.

b. Runway Centerline Lighting (RCLS)—flush centerline lights spaced at 50-foot intervals beginning 75 feet from the landing threshold and extending to within 75 feet of opposite end. Viewed from the landing threshold, the runway centerline lights are white until the last 3,000 feet of the runway. The white lights begin to alternate with the red for the next 2,000 feet, and for the last 1,000 feet of the runway, all lights are red.

c. Taxiway turnoff lights—flush lights spaced at 50 foot intervals defining the curved path of aircraft travel from the runway centerline to a point on the taxiway. These lights are steady burning and emit green light.

2-6. CONTROL OF LIGHTING SYSTEMS

a. Operation of approach light systems and runway lighting is controlled by the control tower (ATCT). At some locations the FSS may control the lights where there is no control tower in operation.

b. Pilots may request that lights be turned on or off. Runway edge lights, in-pavement lights and approach lights also have intensity controls which may be varied to meet the pilots request. Sequenced flashing lights (SFL) may be turned on and off. Some sequenced flashing light systems also have intensity control.

2-7. PILOT CONTROL OF AIRPORT LIGHTING

a. Radio control of lighting is available at selected airports to provide airborne control of lights by keying the aircraft's microphone. Control of lighting systems is often available at locations without specified hours for lighting and where there is no control tower or FSS or when the tower or FSS is closed (locations with a part-time tower or FSS) or specified hours. All lighting systems which are radio controlled at an airport, whether on a single runway or multiple runways, operate on the same radio frequency. (See Table 2-7[1] and Table 2-7[2].)

b. With FAA approved systems, various combinations of medium intensity approach lights, runway lights, taxiway lights, VASI and/or REIL may be activated by radio control. On runways with both approach lighting and runway lighting (runway edge lights, taxiway lights, etc.) systems, the approach lighting system takes precedence for air-to-ground radio control over the runway lighting system which is set at a predetermined intensity step, based on expected visibility conditions. Runways without approach lighting may provide radio controlled intensity adjustments of runway edge lights. Other lighting systems, including VASI, REIL, and taxiway lights may be either controlled with the runway edge lights or controlled independently of the runway edge lights.

c. The control system consists of a 3-step control responsive to 7, 5, and/or 3 microphone clicks. This 3-step control will turn on lighting facilities capable of either 3-step, 2-step or 1-step operation. The 3-step and 2-step lighting facilities can be altered in intensity, while the 1-step cannot. All lighting is illuminated for a period of 15

Table 2-7[1]—RUNWAYS WITH APPROACH LIGHTS

Lighting System	No. of Int. Steps	Status During Nonuse Period	Intensity Step Selected Per No. of Mike Clicks		
			3 Clicks	5 Clicks	7 Clicks
Approach Lights (Med. Int.)	2	Off	Low	Low	High
Approach Lights (Med. Int.)	3	Off	Low	Med	High
MIRL	3	Off or Low	⌑	⌑	⌑
HIRL	5	Off or Low	⌑	⌑	⌑
VASI	2	Off	◊	◊	◊

NOTES:
⌑ Predetermined intensity step.
◊ Low intensity for night use. High intensity for day use as determined by photocell control.

Table 2-7[2]—RUNWAYS WITHOUT APPROACH LIGHTS

Lighting System	No. of Int. Steps	Status During Nonuse Period	Intensity Step Selected Per No. of Mike Clicks		
			3 Clicks	5 Clicks	7 Clicks
MIRL	3	Off or Low	Low	Med.	High
HIRL	5	Off or Low	Step 1 or 2	Step 3	Step 5
LIRL	1	Off	On	On	On
VASI☆	2	Off	◊	◊	◊
REIL☆	1	Off	Off	On/Off	On
REIL☆	3	Off	Low	Med.	High

NOTES:
◊ Low intensity for night use. High intensity for day use as determined by photocell control.
☆ The control of VASI and/or REIL may be independent of other lighting systems.

minutes from the most recent time of activation and may not be extinguished prior to end of the 15 minute period (except for 1-step and 2-step REILs which may be turned off when desired by keying the mike 5 or 3 times respectively).

d. Suggested use is to always initially key the mike 7 times; this assures that all controlled lights are turned on to the maximum available intensity. If desired, adjustment can then be made, where the capability is provided, to a lower intensity (or the REIL turned off) by keying 5 and/or 3 times. Due to the close proximity of airports using the same frequency, radio controlled lighting receivers may be set at a low sensitivity requiring the aircraft to be relatively close to activate the system. Consequently, even when lights are on, always key mike as directed when overflying an airport of intended landing or just prior to entering the final segment of an approach. This will assure the aircraft is close enough to activate the system and a full 15 minutes lighting duration is available. Approved lighting systems may be activated by keying the mike (within 5 seconds) as indicated below: (See Table 2-7[3].)

Table 2-7[3]—RADIO CONTROL SYSTEM

Key Mike	Function
7 times within 5 seconds	Highest intensity available
5 times within 5 seconds	Medium or lower intensity (Lower REIL or REIL-off)
3 times within 5 seconds	Lowest intensity available (Lower REIL or REIL-off)

e. For all public use airports with FAA standard systems the Airport/Facility Directory contains the types of lighting, runway and the frequency that is used to activate the system. Airports with IAPs include data on the approach chart identifying the light system, the runway on which they are installed, and the frequency that is used to activate the system.

2-7e NOTE—Although the CTAF is used to activate the lights at many airports, other frequencies may also be used. The appropriate frequency for activating the lights on the airport is provided in the Airport/Facility Directory and the Standard Instrument Approach Procedures publications. It is not identified on the sectional charts.

f. Where the airport is not served by an IAP, it may have either the standard FAA approved control system or an independent type system of different specification installed by the airport sponsor. The Airport/Facility Directory contains descriptions of pilot controlled lighting systems for each airport having other than FAA approved systems, and explains the type lights, method of control, and operating frequency in clear text.

2-8. AIRPORT (ROTATING) BEACONS

a. The airport beacon has a vertical light distribution to make it most effective from one to 10 degrees above the horizon; however, it can be seen well above and below this peak spread. The beacon may be an omnidirectional capacitor-discharge device, or it may rotate at a constant speed which produces the visual effect of flashes at regular intervals. Flashes may be one or two colors alternately. The total number of flashes are:

1. 12 to 30 per minute for beacons marking airports, landmarks, and points on Federal airways.

2. 30 to 60 per minute for beacons marking heliports.

b. The colors and color combinations of beacons are:

1. White and Green—Lighted land airport

2. *Green alone—Lighted land airport

3. White and Yellow—Lighted water airport

4. *Yellow alone—Lighted water airport

5. Green, Yellow, and White—Lighted heliport

2-8b5 NOTE—*Green alone or yellow alone is used only in connection with a white-and-green or white-and-yellow beacon display, respectively.

c. Military airport beacons flash alternately white and green, but are differentiated from civil beacons by dual-peaked (two quick) white flashes between the green flashes.

d. In Class B, Class C, Class D and Class E surface areas, operation of the airport beacon during the hours of daylight often indicates that the ground visibility is less than 3 miles and/or the ceiling is less than 1,000 feet. ATC clearance in accordance with Part 91 is required for landing, takeoff and flight in the traffic pattern. Pilots should not rely solely on the operation of the airport beacon to indicate if weather conditions are IFR or VFR. At some locations with operating control towers, ATC personnel turn the beacon on or off when controls are in the tower. At many airports the airport beacon is turned on by a photoelectric cell or time clocks and ATC personnel cannot control them. There is no regulatory requirement for daylight operation and it is the pilot's responsibility to comply with proper preflight planning as required by Part 91.103.

2-9. TAXIWAY LIGHTS

a. Taxiway Edge Lights.—Taxiway edge lights are used to outline the edges of taxiways during periods of darkness or restricted visibility conditions. These fixtures emit blue light.

2-9a NOTE—At most major airports these lights have variable intensity settings and may be adjusted at pilot request or when deemed necessary by the controller.

b. Taxiway Centerline Lights.—Taxiway centerline lights are used to facilitate ground traffic under low visibility conditions. They are located along the taxiway centerline in a straight line on straight portions, on the centerline of curved portions, and along designated taxiing paths in portions of runways, ramp, and apron areas. Taxiway centerline lights are steady burning and emit green light.

2-10 thru 2-19 RESERVED

Section 2. AIR NAVIGATION AND OBSTRUCTION LIGHTING

2-20. AERONAUTICAL LIGHT BEACONS

a. An aeronautical light beacon is a visual NAVAID displaying flashes of white and/or colored light to indicate the location of an airport, a heliport, a landmark, a certain point of a Federal airway in mountainous terrain, or an obstruction. The light used may be a rotating beacon or one or more flashing lights. The flashing lights may be supplemented by steady burning lights of lesser intensity.

b. The color or color combination displayed by a particular beacon and/or its auxiliary lights tell whether the beacon is indicating a landing place, landmark, point of the Federal airways, or an obstruction. Coded flashes of the auxiliary lights, if employed, further identify the beacon site.

2-21. CODE BEACONS AND COURSE LIGHTS

a. CODE BEACONS

1. The code beacon, which can be seen from all directions, is used to identify airports and landmarks and to mark obstructions. The number of code beacon flashes are:

(a) Green coded flashes not exceeding 40 flashes or character elements per minute, or constant flashes 12 to 15 per minute, for identifying land airports.

(b) Yellow coded flashes not exceeding 40 flashes or character elements per minute, or constant flashes 12 to 15 per minute, for identifying water airports.

(c) Red flashes, constant rate, 12 to 40 flashes per minute for marking hazards.

b. COURSE LIGHTS

1. The course light, which can be seen clearly from only one direction, is used only with rotating beacons of the Federal Airway System: two course lights, back to back, direct coded flashing beams of light in either direction along the course of airway.

2-21b1 NOTE—Airway beacons are remnants of the "lighted" airways which antedated the present electronically equipped Federal Airways System. Only a few of these beacons exist today to mark airway segments in remote mountain areas. Flashes in Morse Code identify the beacon site.

2-22. OBSTRUCTION LIGHTS

a. Obstructions are marked/lighted to warn airmen of their presence during daytime and nighttime conditions. They may be marked/lighted in any of the following combinations:

1. Aviation Red Obstruction Lights. Flashing aviation red beacons and steady burning aviation red lights during nighttime operation. Aviation orange and white paint is used for daytime marking.

2. High Intensity White Obstruction Lights. Flashing high intensity white lights during daytime with reduced intensity for twilight and nighttime operation. When this type system is used, the marking of structures with red obstruction lights and aviation orange and white paint may be omitted.

3. Dual Lighting. A combination of flashing aviation red beacons and steady burning aviation red lights for nighttime operation and flashing high intensity white lights for daytime operation. Aviation orange and white paint may be omitted.

b. High intensity flashing white lights are being used to identify some supporting structures of overhead transmission lines located across rivers, chasms, gorges, etc. These lights flash in a middle, top, lower light sequence at approximately 60 flashes per minute. The top light is normally installed near the top of the supporting structure, while the lower light indicates the approximate lower portion of the wire span. The lights are beamed towards the companion structure and identify the area of the wire span.

c. High intensity, flashing white lights are also employed to identify tall structures, such as chimneys and towers, as obstructions to air navigation. The lights provide a 360 degree coverage about the structure at 40 flashes per minute and consist of from one to seven levels of lights depending upon the height of the structure. Where more than one level is used the vertical banks flash simultaneously.

2-23 thru 2-29. RESERVED

Section 3. AIRPORT MARKING AIDS AND SIGNS

2-30. GENERAL

a. Airport pavement markings and signs provide information that is useful to a pilot during takeoff, landing, and taxiing.

b. Uniformity in airport markings and signs from one airport to another enhances safety and improves efficiency. Pilots are encouraged to work with the operators of the airports they use to achieve the marking and sign standards described in this section.

c. Pilots who encounter ineffective, incorrect, or confusing markings or signs on an airport should make the operator of the airport aware of the problem. These situations may also be reported under the Aviation Safety Reporting Program as described in paragraph 7-90. Pilots may also report these situations to the FAA regional airports division.

d. The markings and signs described in this section of the AIM reflect the current FAA recommended standards.

2-30d NOTE—Refer to AC 150/5340-1 Standards for Airport Markings and to AC 150/5340-18 Airport Sign Standards.

2-31. AIRPORT PAVEMENT MARKINGS

a. *General.*—For the purpose of this presentation the Airport Pavement Marking have been grouped into the four areas:

1. Runway Markings.

2. Taxiway Markings.

3. Holding Position Markings.

4. Other Markings.

b. *Marking Colors.*—Markings for runways are white. Markings defining the landing area on a heliport are also white except for hospital heliports which use a red "H" on a white cross. Markings for taxiways, areas not intended for use by aircraft (closed and hazardous areas), and holding positions (even if they are on a runway) are yellow.

2-32. RUNWAY MARKINGS

a. *General.*—There are three types of markings for runways: visual, nonprecision instrument and precision instrument. Table 2-32[1] identifies the marking elements for each type of runway and Fig. 2-32[2] identifies runway threshold markings.

b. *Runway Designators.*—Runway numbers and letters are determined from the approach direction. The runway number is the whole number nearest one-tenth the magnetic azimuth of the centerline of the runway, measured clockwise from the magnetic north. The letters, differentiate between left (L), right (R), or center (C), parallel runways, as applicable:

1. For two parallel runways "L" "R"

2. For three parallel runways "L" "C" "R"

c. *Runway Centerline Marking.*—The runway centerline identifies the center of the runway and provides alignment guidance during takeoff and landings. The centerline consists of a line of uniformly spaced stripes and gaps.

d. *Runway Aiming Point Marking.*—The Aiming Point marking serves as a visual aiming point for a landing aircraft. These two rectangular markings consists of a broad white stripe located on each side of the runway centerline and approximately 1,000 feet from the landing threshold, as shown in Figure 2-32[1] Precision Instrument Runway Markings.

e. *Runway Touchdown Zone Markers.*—The touchdown zone markings identify the touchdown zone for landing operations and are coded to provide distance information in 500 feet (150m) increments. These markings consist of groups of one, two, and three rectangular bars symmetrically arranged in pairs about the runway centerline, as shown in Figure 2-32[1]. Precision Instrument Runway Markings. For runways having touchzone markings on both ends, those pairs of markings which extend to within 900 feet (270m) of the midpoint between the thresholds are eliminated.

f. *Runway Side Strip Marking.*—Runway side stripes delineate the edges of the runway. They provide a visual contrast between runway and the abutting terrain or shoulders. Side stripes consist of continuous white stripes located, on each side of the runway as shown in Figure 2-32[5].

g. *Runway Shoulder Markings.*—Runway shoulder stripes may be used to supplement runway side stripes to identify pavement areas contiguous to the runway sides that are not intended for use by aircraft. Runway Shoulder stripes are Yellow. (See Figure 2-32[3].)

h. *Runway Threshold Markings.*—Runway threshold markings come in two configurations. They either consist of eight longitudinal stripes of uniform dimensions disposed symmetrically about the runway centerline, as shown in Figure 2-32[1], or the number of stripes is related to the runway width as indicated in Table 2-32[2]. A threshold marking landing threshold may be relocated or displaced.

1. Relocated Threshold.—Sometimes construction, maintenance, or other activities require the threshold to be relocated towards the departure end of the runway. In these cases, where the relocation is temporary, a notam should be issued by the airport operator identifying the portion of the runway that is closed, e.g., First 2,000 feet of Runway 24 closed. Because the duration of the relocation can vary from a few hours to several months, methods for identifying the relocated threshold vary. One common practice is to use a ten feet wide white threshold bar across the width of

Table 2-32[1]—Runway Marking Elements

Marking Element Runway	Visual Runway	Nonprecision Instrument Runway	Precision Instrumc
Designation (par. 6)	X	X	X
Centerline (par. 7)	X	X	X
Threshold (par. 8)	X[1]	X	X
Aiming Point (par. 9)	X[2]	X	X
Touchdown Zone (par. 10)			X
Side Stripes (par. 11)			

[1] On runways used, or intended to be used, by international commercial transport.
[2] On runways 4,000 feet (1200 m) or longer used by jet aircraft.

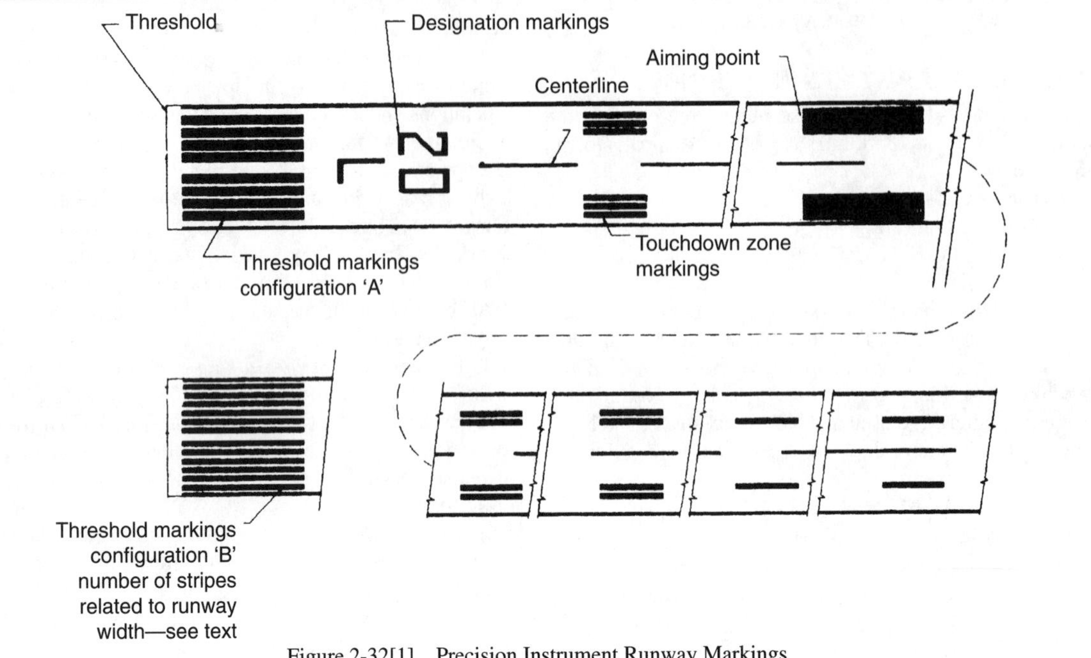

Figure 2-32[1]. Precision Instrument Runway Markings

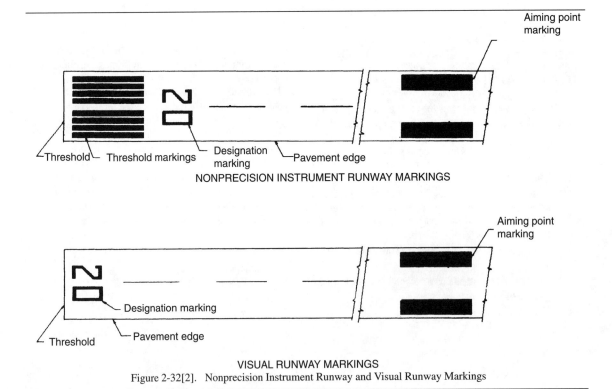

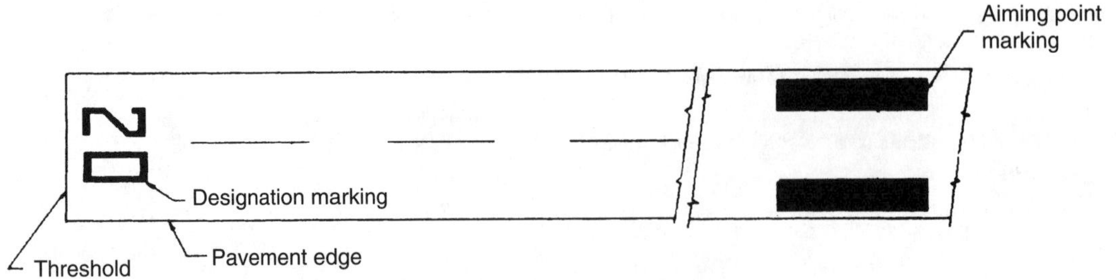

VISUAL RUNWAY MARKINGS

Figure 2-32[2]. Nonprecision Instrument Runway and Visual Runway Markings

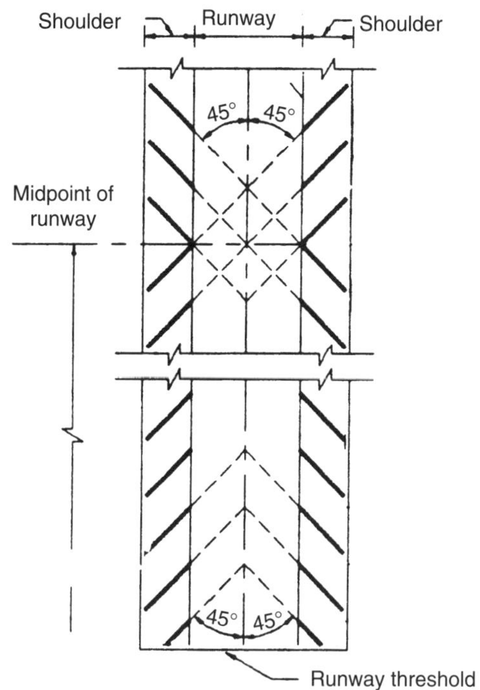

Figure 2-32[3]. Runway Shoulder Markings

the runway. Although the runway lights in the area between the old threshold and relocated threshold will not be illuminated, the runway markings in this area may or may not be obliterated, removed, or covered. (See Figure 2-32[4].)

2. Displaced Threshold.—A displaced threshold is a threshold located at a point on the runway other than the designated beginning of the runway. A ten feet wide white threshold bar is located across the width of the runway at the displaced threshold. White arrows are located along the centerline in the area between the beginning of the runway and displaced threshold. White arrow heads are located across the width of the runway just prior to the threshold bar, as shown in Figure 2-32[5].

i. *Runway Threshold BAR.*—A threshold bar delineates the beginning of the runway that is available for landing when the threshold has been relocated or displaced. A threshold bar is 10 feet (3m) in width and extends across the width of the runway, as shown in Figure 2-32[5].

Table 2-32[2]

Runway Width	Numbero Stripes
60 feet (18 m)	4
75 feet (23 m)	6
100 feet (30 m)	8
150 feet (45 m)	12
200 feet (60 m)	16

j. *Demarcation Bar.*—A demarcation bar delineates a runway with a displaced threshold from a blast pad, stopway or taxiway that precedes the runway. A demarcation bar is 3 feet (1m) wide and yellow, since it is not located on the runway as shown in Figure 2-32[6].

k. *Chevrons.*—These markings are used to show pavement areas aligned with the runway that are unusable for landing, takeoff, and taxiing. Chevrons are yellow. (See Figure 2-32[7].)

2-33. TAXIWAY MARKINGS

a. *General.*—All taxiways should have centerline markings, (See Paragraph 2-34), and runway holding position markings whenever they intersect a runway. Taxiway edge markings are present whenever there is a need to separate the taxiway from a pavement that is not intended for aircraft use or to delineate the edge of the taxiway. Taxiways may also have shoulder markings and holding position markings for Instrument Landing System/Microwave Landing System (ILS/MLS) critical areas, and taxiway/taxiway intersection markings.

b. *Taxiway Centerline.*—The taxiway centerline is a single continuous yellow line, 6 inches (15 cm) to 12 inches (30 cm) in width. This provides a visual cue to permit taxiing along a designated path. Ideally the aircraft should be kept centered over this line during taxi to ensure wing-tip clearance. (See Figure 2-33[1].)

c. *Taxiway Edge Markings.*—Taxiway edge markings are used to define the edge of the taxiway. They are primarily used when the taxiway edge does not correspond with the edge of the pavement.

There are two types of markings depending upon whether the aircraft is suppose to cross the taxiway edge:

1. Continuous Markings.—These consist of a continuous double yellow line, with each line being at least 6 inches (15 cm) in width spaced 6 inches (15 cm) apart. They are used to define the taxiway edge from the shoulder or some other abutting paved surface not intended for use by aircraft.

2. Dashed Markings.—These markings are used when there is an operational need to define the edge of a taxiway or taxilane on a paved surface where the adjoining pavement to the taxiway edge is intended for use by aircraft. e.g., an apron. Dashed taxiway edge markings consist of a broken double yellow line, with each line being at least 6 inches (15 cm) in width, spaced 6 inches (15 cm) apart (edge to edge). These lines are 15 feet (4.5 m) in length with 25 foot (7.5 m) gaps. (See Figure 2-33[2].)

d. *Taxi Shoulder Markings.*—Taxiways, holding bays, and aprons are sometimes provided with paved shoulders to prevent blast and water erosion. Although shoulders may have the appearance of full strength pavement they are not intended for use by aircraft, and may be unable to support an aircraft. Usually the taxiway edge marking will

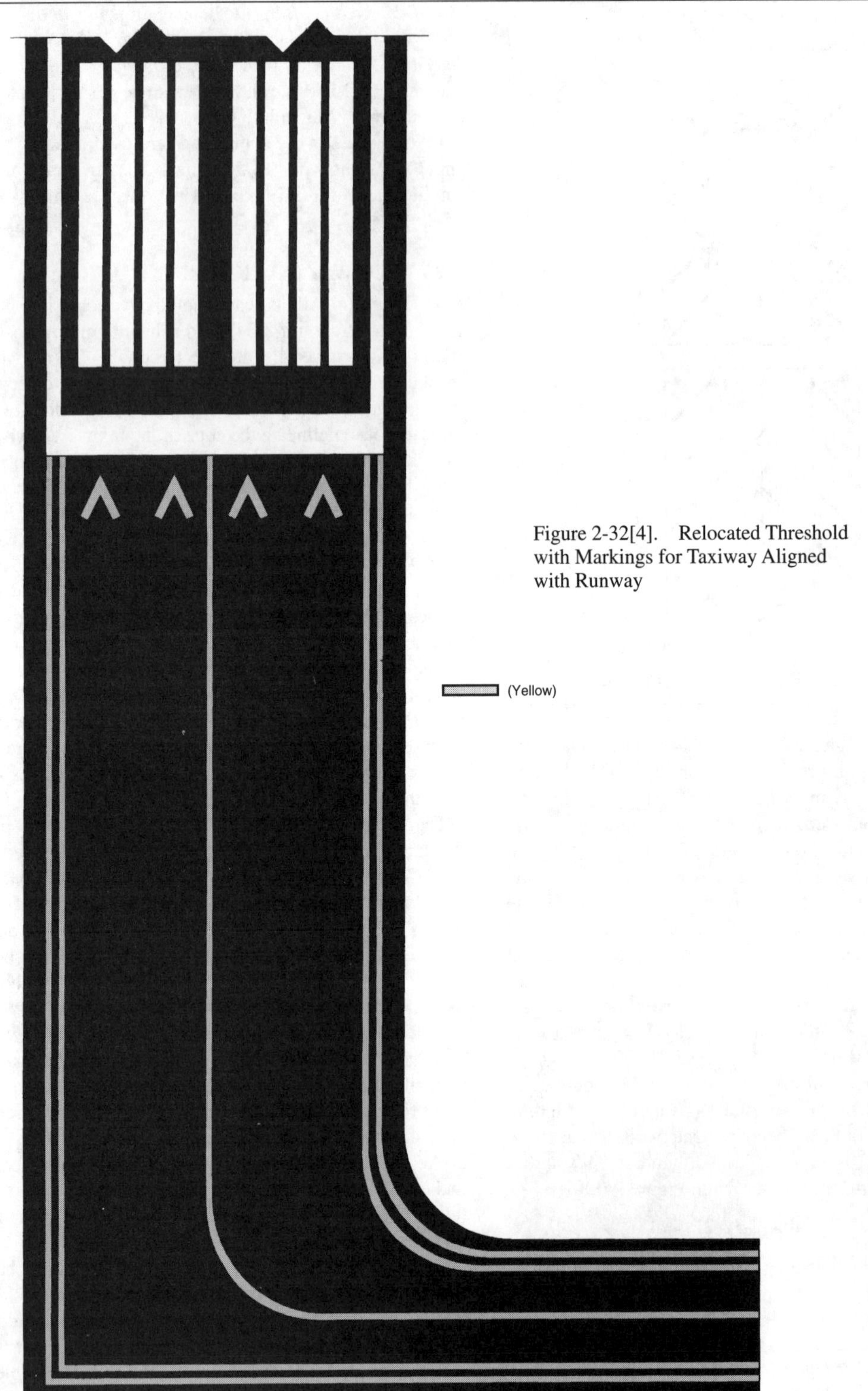

Figure 2-32[4]. Relocated Threshold with Markings for Taxiway Aligned with Runway

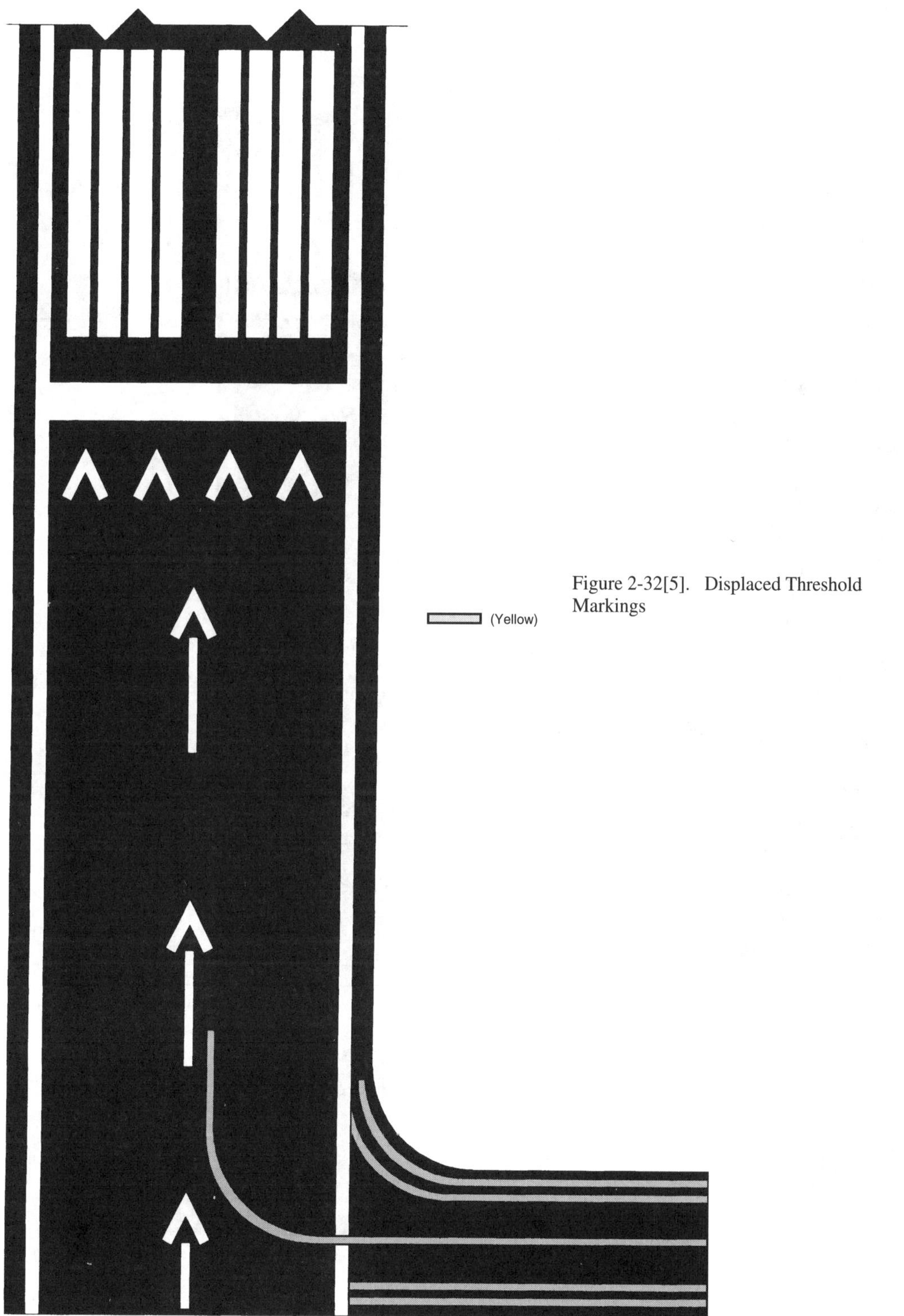

Figure 2-32[5]. Displaced Threshold Markings

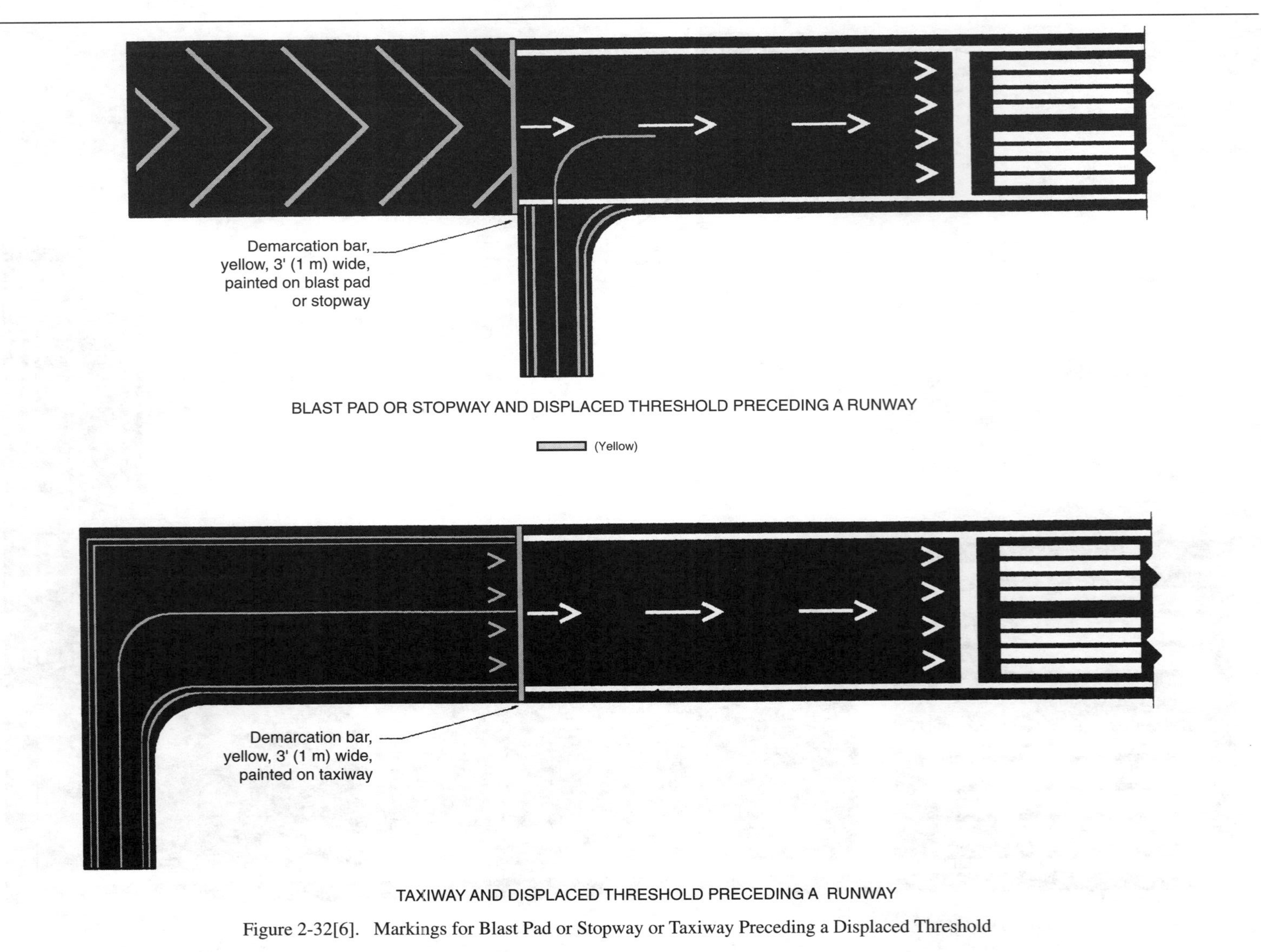

Figure 2-32[6]. Markings for Blast Pad or Stopway or Taxiway Preceding a Displaced Threshold

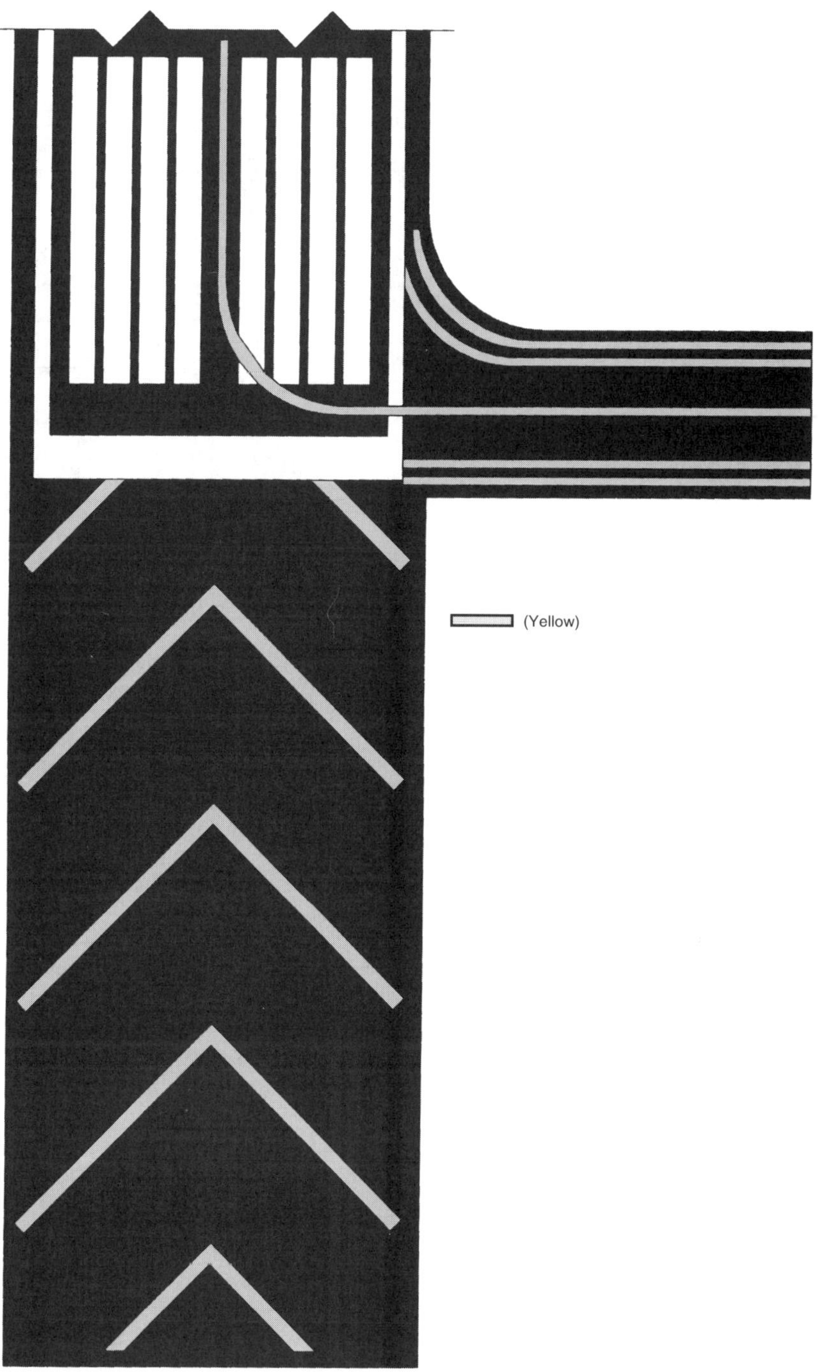

Figure 2-32[7]. Markings for Blast Pads and Stopways

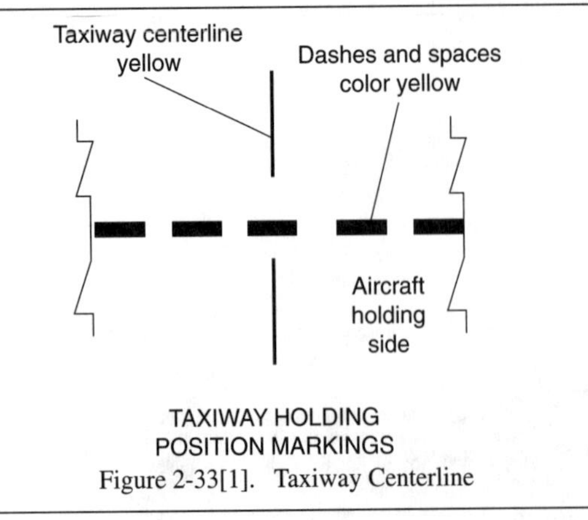

Figure 2-33[1]. Taxiway Centerline

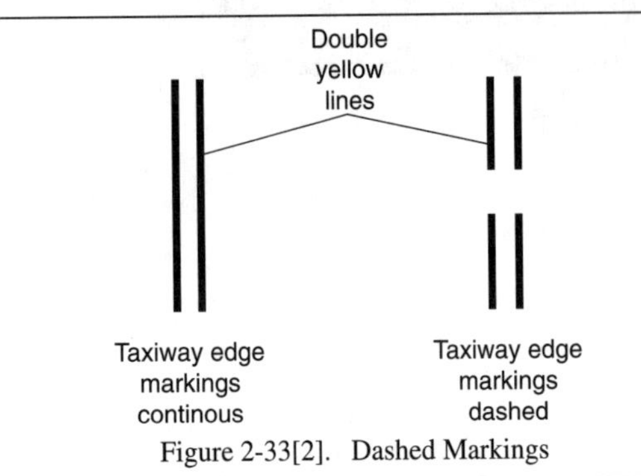

Figure 2-33[2]. Dashed Markings

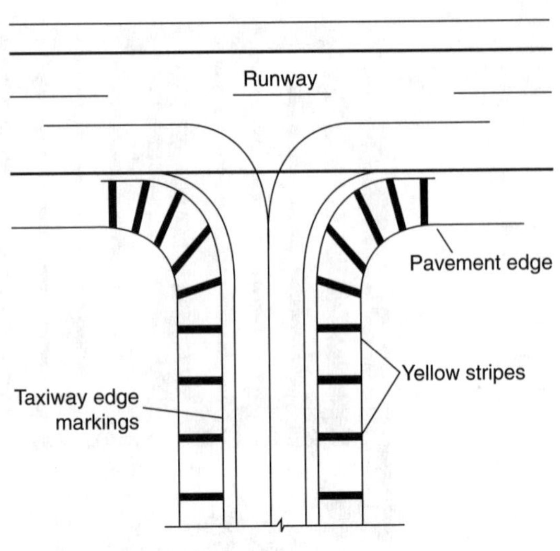

Figure 2-33[3]. Taxi Shoulder Markings

define this area. Where conditions exist such as islands or taxiway curves that may cause confusion as to which side of the edge stripe is for use by aircraft, taxiway shoulder markings may be used to indicate the pavement is unusable. Taxiway shoulder markings are yellow. (See Figure 2-33[3].)

e. *Surface Painted Taxiway Direction Signs.*—Surface painted taxiway direction signs have a yellow background with a black inscription, and are provided when it is not possible to provide taxiway direction signs at intersections, or when necessary to supplement such signs. These markings are located adjacent to the centerline with signs indicating turns to the left being on the left side of the taxiway centerline and sips indicating turns to the right being on the right side of the centerline. (See Figure 2-33[4].)

f. *Surface Painted Location Signs.*—Surface painted location signs have a black background with a yellow inscription. When necessary, these markings are used to supplement location signs located along side the taxiway and assist the pilot in confirming the designation of the taxiway on which the aircraft is located. These markings are located on the right side of the centerline. (See Figure 2-33[5].)

g. *Geographic Position Markings.*—These markings are located at points along low visibility taxi routes designated in the airport's Surface Movement Guidance Control System (SMGCS) plan. They are used to identify the location of taxiing aircraft during low visibility operations. Low visibility operations are those that occur when the runway visible range (RVR) is below 1200 feet (360m). They are positioned to the left of the taxiway centerline in the direction of taxiing. (See Figure 2-33[6].) The Geographic position marking is a circle comprised of an outer black ring contiguous to a white ring with a pink circle in the middle. When installed on asphalt or other dark-colored pavements, the white ring and the black ring are reversed, i.e., the white ring becomes the outer ring and the black ring becomes the inner ring. It is designated with either a number or a number and letter. The number corresponds to the consecutive position of the marking on the route.

2-34. HOLDING POSITION MARKINGS

a. *Runway Holding Position Markings.*—For runways these markings indicate where an aircraft is supposed to stop. They consist of four yellow lines two solid and two dashed, spaced six inches apart and extending across the width of the taxiway or runway. The solid lines are always on the side where the aircraft is to hold. There are three locations where runway holding position markings are encountered.

1. Runway Holding Position Markings on Taxiways.—These markings identify the locations on a taxiway where an aircraft is supposed to stop when it does not have clearance to proceed onto the runway. The runway holding position markings are shown in Figure 2-34[1].

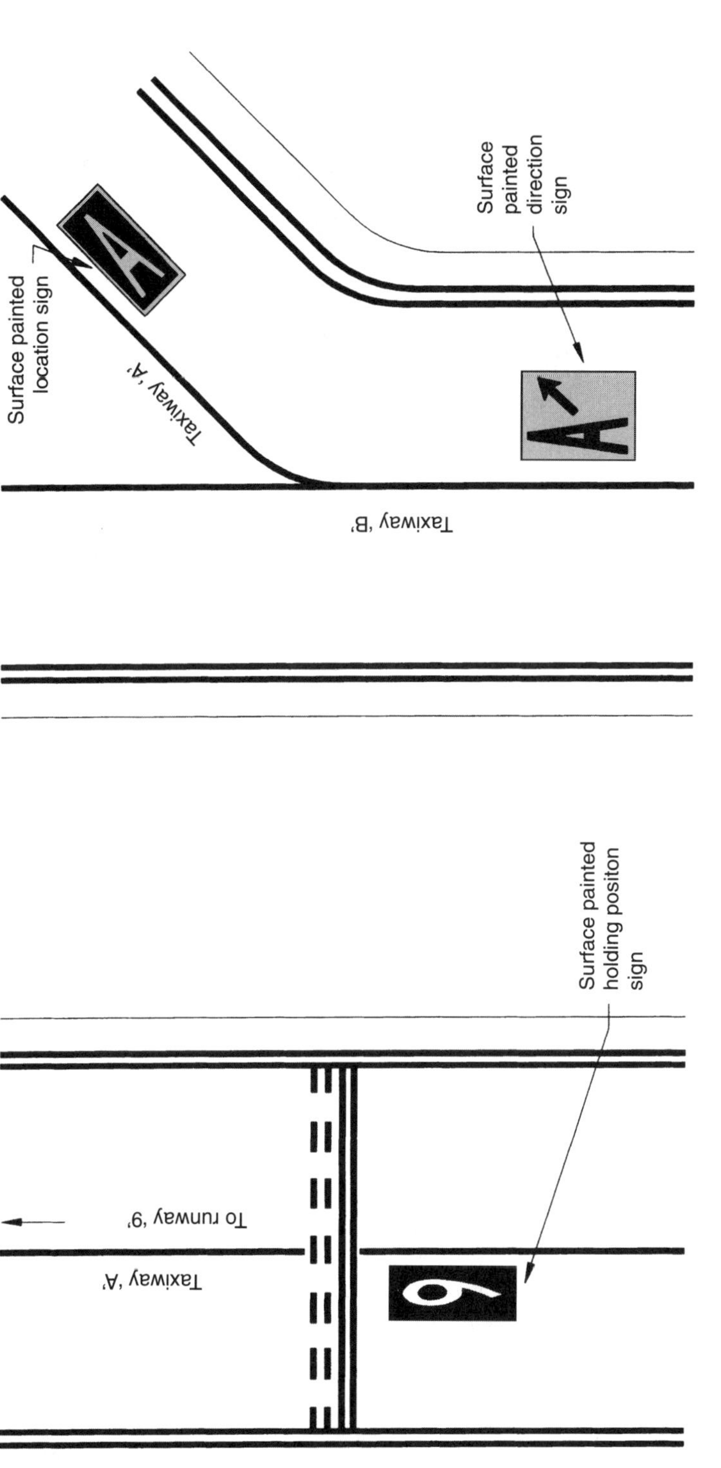

Figure 2-33[4]. Surface Painted Signs

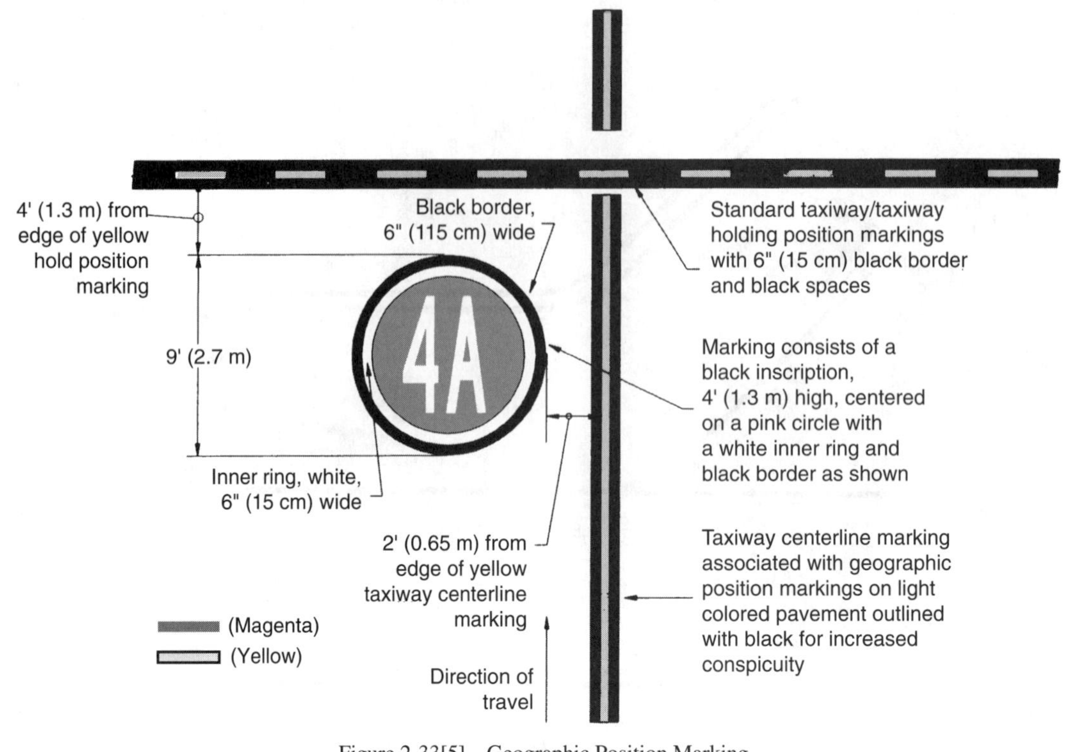

Figure 2-33[5]. Geographic Position Marking

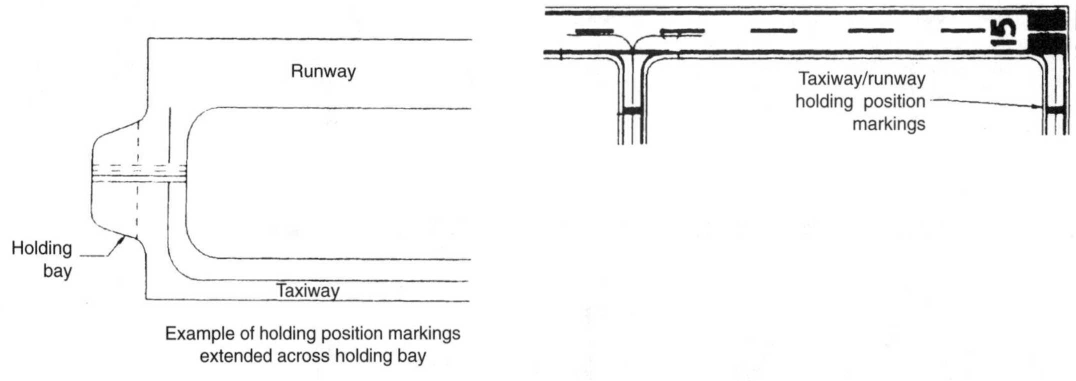

Figure 2-34[1]. Runway Holding Postion Markings on Taxiways

(a) When instructed by ATC "HOLD SHORT OF (runway "XX")" the pilot should stop so no part of his aircraft extends beyond the holding position marking. When approaching the holding position marking, a pilot should not cross the marking without ATC clearance at a controlled airport or without making sure of adequate separation from other aircraft at uncontrolled airports. An aircraft exiting a runway is not clear of the runway until all parts of the aircraft have crossed the applicable holding position marking.

2. Runway Holding Position Markings on Runways.— These markings are installed on runways only if the runway is normally used by air traffic control for "land, hold short" operations or taxiing operations and have operational significance only for those two types of operations. A sign with a white inscription on a red background is installed adjacent to these holding position markings. (See Figure 2-34[2].)

(a) The hold position markings are placed on runways prior to the intersection with another runway, or some des-

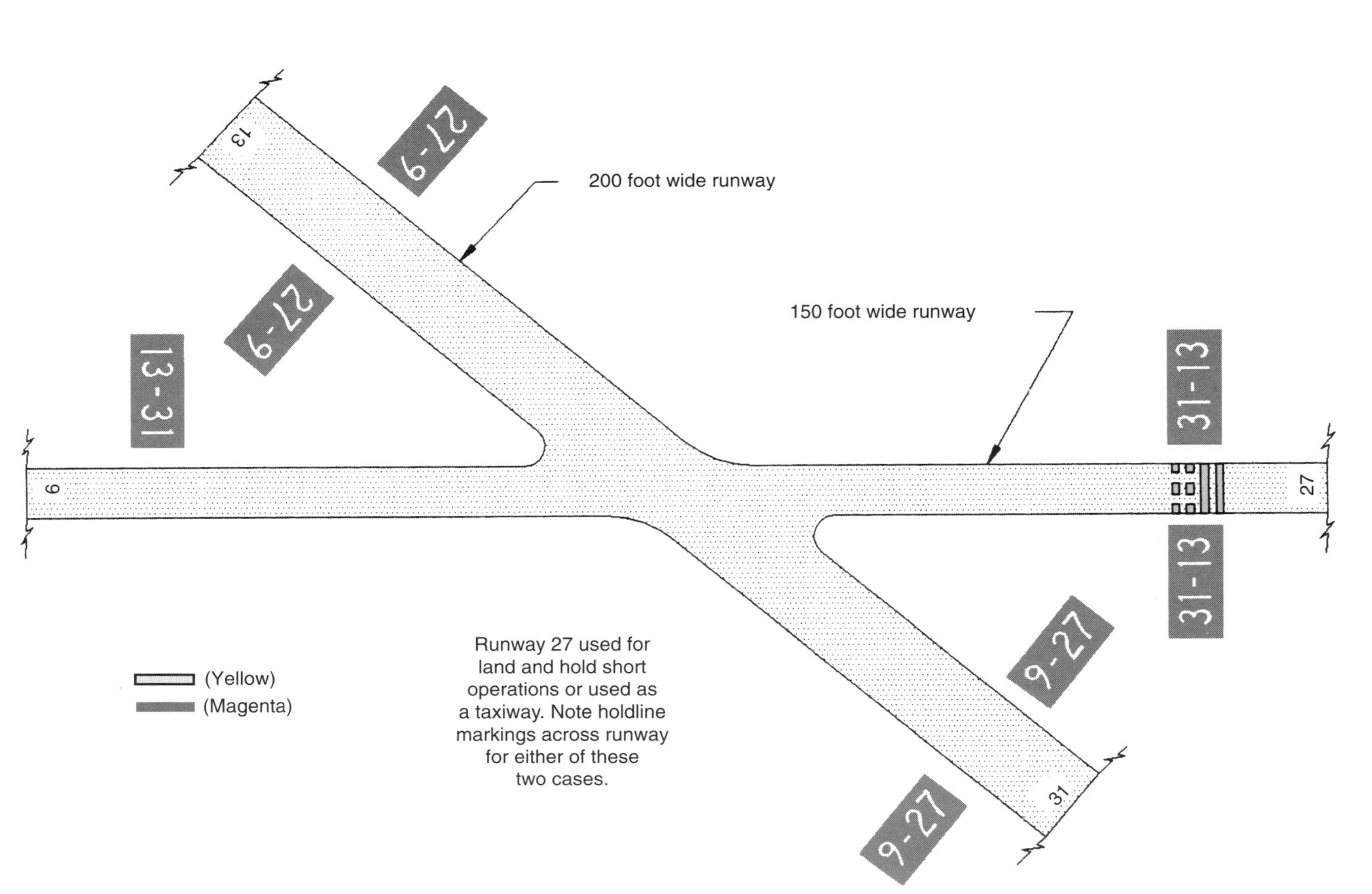

Figure 2-34[2]. Runway Holding Position Markings on Runways

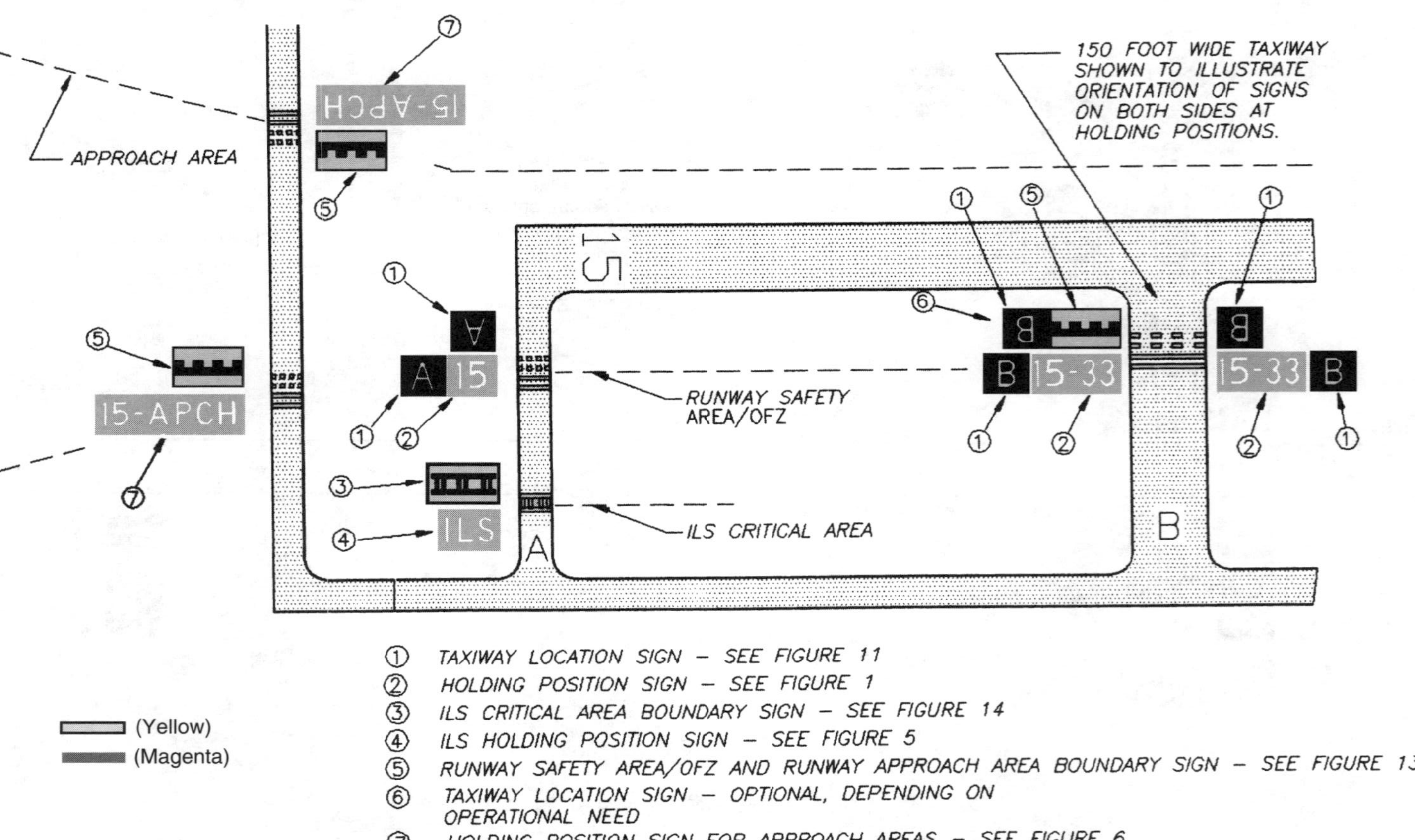

Figure 2-34[3]. Taxiways Located in Runway Approach Area

ignated point. Pilots receiving instructions "Clear to Land, Runway "XX" from Air Traffic Control are authorized to use the entire landing length of the runway and should disregard any holding position markings located on the runway. Pilots receiving and accepting instructions "Clear to Land Runway "XX," Hold Short of Runway "yy" from Air Traffic Control must either exit Runway "XX," or stop at the holding position prior to Runway "yy."

3. Taxiways Located in Runway Approach Areas.—These markings are used at some airports where it is necessary to hold an aircraft on a taxiway located in the approach or departure area of a runway so that the aircraft does not interfere with the operations on that runway. This marking is co-located with the Runway approach area holding position sign. (See Paragraph 2-34.c and Figure 2-34[2].) (See Figure 2-34[3].)

b. *Holding Position Markings for Instrument Landing System (ILS)*—Holding position markings for ILS/MLS critical areas consist of two yellow solid lines spaced two feet apart connected by pairs of solid lines spaced ten feet apart extending across the width of the taxiway as shown. (See Figure 2-34[4] [Holding Position Markings: ILS Critical Areas].) A sign with an inscription in white on a red background is installed adjacent to these hold position markings.

1. When the ILS critical area is being protected (ref para 1-10.k) the pilot should stop so no part of his aircraft extends beyond the holding position marking. When approaching the holding position marking, a pilot should not cross the marking without ATC clearance. ILS critical area is not clear until all parts of the aircraft have crossed the applicable holding position marking.

c. *Holding Position Markings for Taxiway/Taxiway Intersections*—Holding position markings for taxiway/taxiway intersections consist of a single dashed line extending across the width of the taxiway as shown. (See Figure 2-34[5].) They are installed on taxiways where air traffic control normally holds aircraft short of a taxiway intersection.

1. When instructed by ATC "HOLD SHORT OF (taxiway)" the pilot should stop so no part of his aircraft extends beyond the holding position marking. When the marking is not present the pilot should stop the aircraft at a point which provides adequate clearance from an aircraft on the intersecting taxiway.

d. *Surface Painted Holding Position Signs*—Surface painted holding position signs have a red background with a white inscription and supplement the signs located at the holding position. This type of marking is normally used where the width of the holding position on the taxiway is greater than 200 feet (60m). It is located to the left side of the taxiway centerline on the holding side and prior to the holding position marking. (See Figure 2-33[4].)

2-35. OTHER MARKINGS

a. *Vehicle Roadway Markings.*—The vehicle roadway markings are used when necessary to define a pathway for vehicle operations on or crossing areas that are also intended for aircraft. These markings consist of a white solid line to delineate each edge of the roadway and a dashed line to separate lanes within the edges of the roadway. In lieu of the solid lines, zipper markings may be used to delineate the edges of the vehicle roadway.

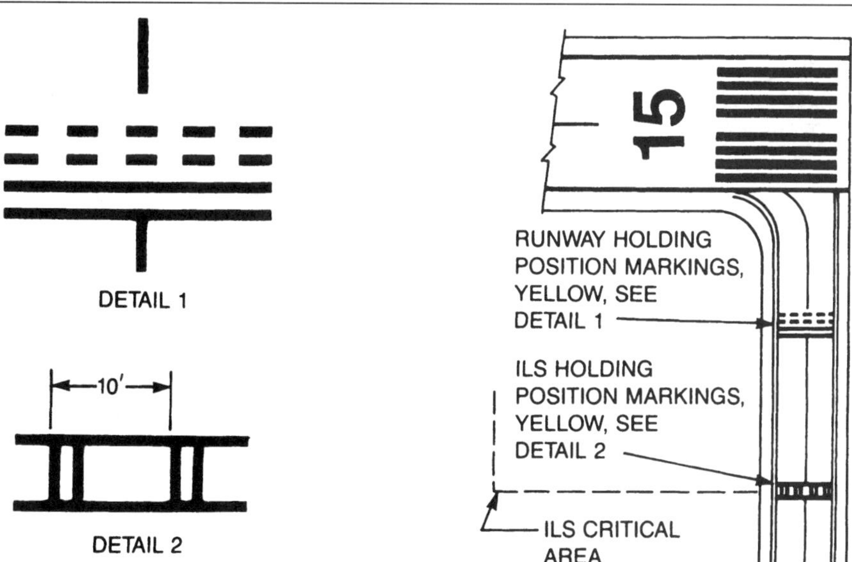

Figure 2-34[4]. Holding Position Markings: ILS Critical Area

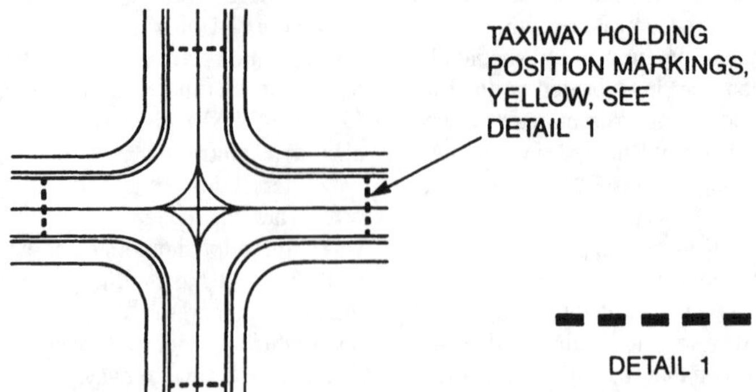

Figure 2-34[5]. Holding Position Markings: Taxiway/Taxiway Intersections

(See Figure 2-35[1] Vehicle Roadway Markings.) Details of the zipper markings are shown in Figure 2-35[2].

b. *VOR Receiver Checkpoint Markings.*—The VOR receiver checkpoint marking allows the pilot to check aircraft instruments with navigational aid signals. It consists of a painted circle with an arrow in the middle; the arrow is aligned in the direction of the checkpoint azimuth. This marking, and an associated sign, is located on the airport apron or taxiway at a point selected for easy access by aircraft but where other airport traffic is not be unduly obstructed. (See Figure 2-35[3].)

1. The associated sign contains the VOR station identification letter and course selected (published) for the check, the words "VOR CHECK COURSE," and DME

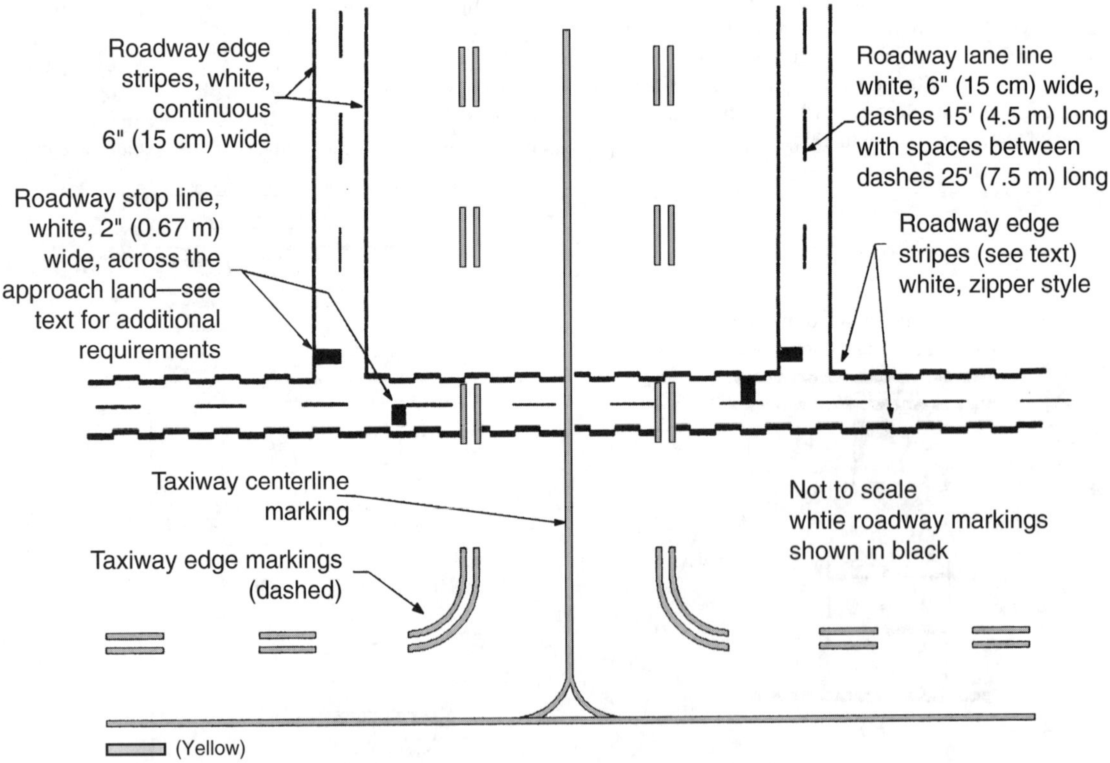

Figure 2-35[1]. Vehicle Roadway Markings

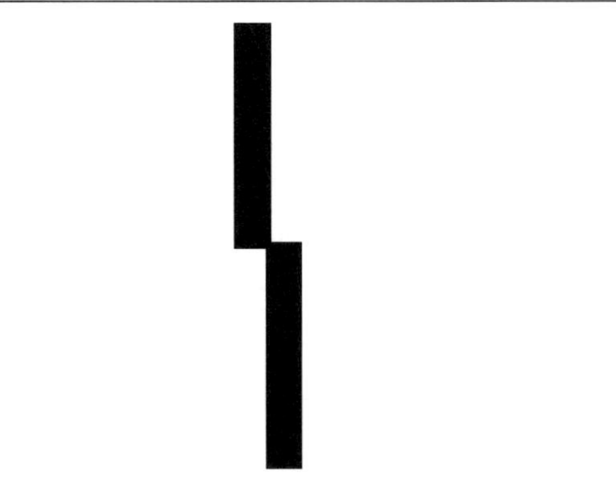

Figure 2-35[2]. Roadway Edge Stripes, White, Zipper Style

data (when applicable). The color of the letters and numerals are black on a yellow background.

EXAMPLE:

DCA 176-356

VOR CHECK COURSE

DME XXX

c. *Non-Movement Area Boundary Markings.*—These markings delineate the movement area, i.e., area under air traffic control. These markings are yellow and located on the boundary between the movement and non-movement area. The non-movement area boundary markings consist of two yellow lines (one solid and one dashed) 6 inches (15cm) in width. The solid line is located on the non-movement area side while the dashed yellow line is located on the movement area side. The non-movement boundary marking area is shown in Figure 2-35[4].

d. *Marking and Lighting of Permanently Closed Runways and Taxiways.*—For runways and taxiways which are permanently closed, the lighting circuits will be disconnected. The runway threshold, runway designation, and touchdown markings are obliterated and yellow crosses are placed at each end of the runway and at 1,000 foot intervals. (See Figure 2-35[5].)

e. *Temporarily Closed Runways and Taxiways.*—To provide a visual indication to pilots that a runway is temporarily closed, crosses are placed on the runway only at each end of the runway. The crosses are yellow in color. (See Figure 2-35[5].)

1. A raised lighted yellow cross may be placed on each runway end in lieu of the markings described in the paragraph "e" to indicate the runway is closed.

2. A visual indication may not be present depending on the reason for the closure, duration of the closure, airfield configuration and the existence and the hours of operation of an airport traffic control Tower. Pilots should check Notams and the Automated Terminal In-

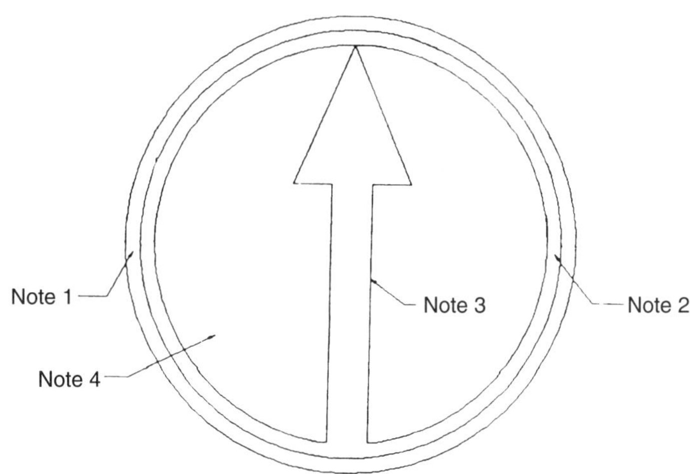

Notes
1 Paint white
2 Paint yellow
3 Paint yellow. Arrow to be aligned toward the facility
4 Paint interior of circle black (concrete surfaces only)
5 Circle may be bordered on inside and outside with 6" black band if necessary for contrast

Figure 2-35[3]. Ground Receiver Checkpoint Markings

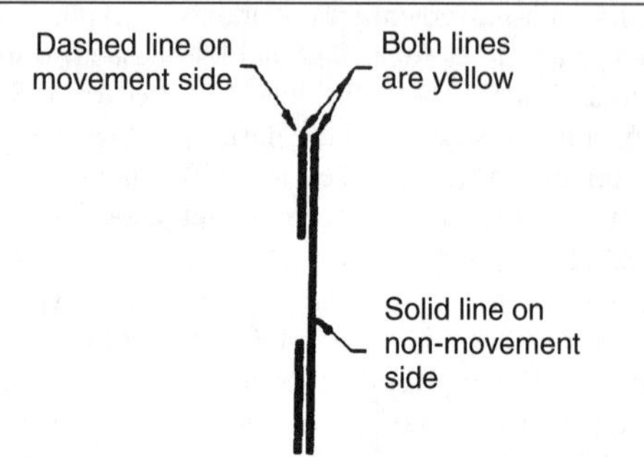

Figure 2-35[4]. Non-Movement Area Boundary Markings

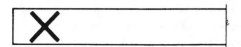

Figure 2-35[5]. Closed or Temporarily Closed Runway and Taxiway Markings

formation System (ATIS) for local Runway and taxiway closure information.

3. Temporarily closed taxiways are usually treated as hazardous areas, in which no part of an aircraft may enter, and are blocked with barricades. However, as an alternative a yellow cross may be installed at each entrance to the taxiway.

f. *Helicopter Landing Areas.*—The markings illustrated in Figure 2-35[6] (Heliport landing Areas) are used to identify the landing and takeoff area at a public use heliport and hospital heliport. The letter "H" in the markings is oriented to align with the intended direction of approach. Figure 2-35[6] also depicts the markings for a closed airport.

2-36. AIRPORT SIGNS

There are six types of signs installed on airfields: mandatory instruction signs, location signs, direction signs, destination signs, information signs, and runway distance remaining signs. The characteristics and use of these signs are discussed in paragraph 2-37 through paragraph 2-42.

2-36 NOTE—Refer to AC 150/5340-18C, Standards for Airport Sign Systems for detailed information on airport signs.

2-37. MANDATORY INSTRUCTION SIGNS

a. These signs have a red background with a white inscription and are used to denote:

1. An entrance to a runway or critical area and;

2. Areas where an aircraft is prohibited from entering. Typical mandatory signs and applications are:

b. Runway Holding Position Sign. This sign is located at the holding position on taxiways that intersect a runway or on runways that intersect other runways. The inscription on the sign contains the designation of the intersecting runway as shown in Figure 2-37[1]. The runway numbers on the sign are arranged to correspond to the respective runway threshold. For example, "15-33" indicates that the threshold for Runway 15 is to the left and the threshold for Runway 33 is to the right.

1. On taxiways that intersect the beginning of the takeoff runway, only the designation of the takeoff runway may appear on the sign as shown in Figure 2-37[2], while all other signs will have the designation of both runway directions.

2. If the sign is located on a taxiway that intersects the intersection of two runways, the designations for both runways will be shown on the sign along with arrows showing the approximate alignment of each runway as shown in Figure 2-37[3]. In addition to showing the approximate runway alignment, the arrow indicates the direction **to the threshold of the runway** whose designation is immediately next to the arrow.

3. A runway holding position sign on a taxiway will be installed adjacent to holding position markings on the taxiway pavement. On runways, holding position markings will be located only on the runway pavement adjacent to the sign, if the runway is normally used by air traffic control for

HELICOPTER LANDING AREA

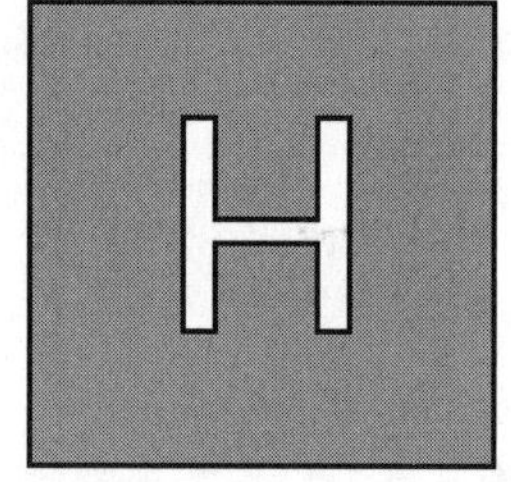

Recommended Marking for Civil Heliports

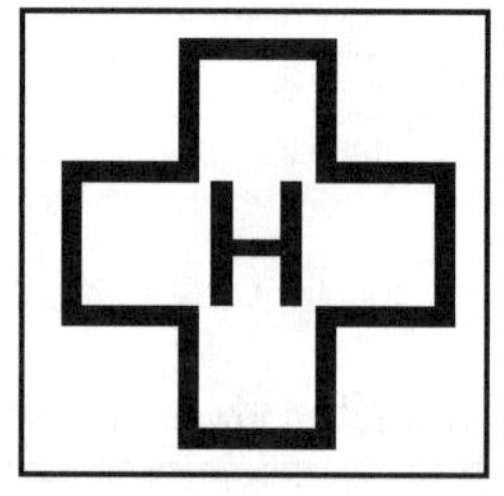

Recommended Marking for Hospital Heliports

Recommended Marking for Closed Heliports

(Yellow)

Figure 2-35[6]. Helicopter Landing Areas

Figure 2-37[1]. Runway Holding Position Sign

Figure 2-37[2]. Holding Position Sign at Beginning of Takeoff Runway

"Land, Hold Short" operations or as a taxiway. The holding position markings are described in paragraph 2-34a2.

c. Runway Approach Area Holding Position Sign. At some airports, it is necessary to hold an aircraft on a taxiway located in the approach or departure area for a runway so that the aircraft does not interfere with operations on that runway. In these situations a sign with the designation of the approach end of the runway followed by a "dash" (—) and letters "APCH" will be located at the holding position on the taxiway. Holding position markings in accordance with paragraph 2-34c will be located on the taxiway pavement. An example of this sign is shown in Figure 2-37[4]. In this example, the sign may protect the approach to Runway 15 and/or the departure for Runway 33.

d. ILS Critical Area Holding Position Sign. At some airports, when the instrument landing system is being used, it is necessary to hold an aircraft on a taxiway at a location other than the holding position described in paragraph 2-34b. In these situations the holding position sign for these operations will have the inscription "ILS" and be located adjacent to the holding position marking on the taxiway described in paragraph 2-34b. An example of this sign is shown in Figure 2-37[5].

e. No Entry Sign. This sign, shown in Figure 2-37[6], prohibits an aircraft from entering an area. Typically, this sign would be located on a taxiway intended to be used in only one direction or at the intersection of vehicle roadways with runways, taxiways or aprons where the roadway may be mistaken as a taxiway or other aircraft movement surface.

2-37 NOTE—The holding position sign provides the pilot with a visual cue as to the location of the holding position marking. The operational significance of holding position markings are described in the notes for paragraph 2-34a.

2-38. LOCATION SIGNS

a. Location signs are used to identify either a taxiway or runway on which the aircraft is located. Other location signs provide a visual cue to pilots to assist them in determining when they have exited an area. The various location signs are described below.

Figure 2-37[3]. Holding Position Sign for a Taxiway that Intersects the Intersection of Two Runways

Figure 2-37[4]. Holding Position Sign for a Runway Approach Area

Figure 2-37[5]. Holding Position Sign for ILS Critical Area

Figure 2-37[6]. Sign Prohibiting Aircraft Entry into an Area

b. Taxiway Location Sign. This sign has a black background with a yellow inscription and yellow border as shown in Figure 2-38[1]. The inscription is the designation of the taxiway on which the aircraft is located. These signs are installed along taxiways either by themselves or in conjunction with direction signs (See Figure 2-39[1]) or runway holding position signs (See Figure 2-38[2]).

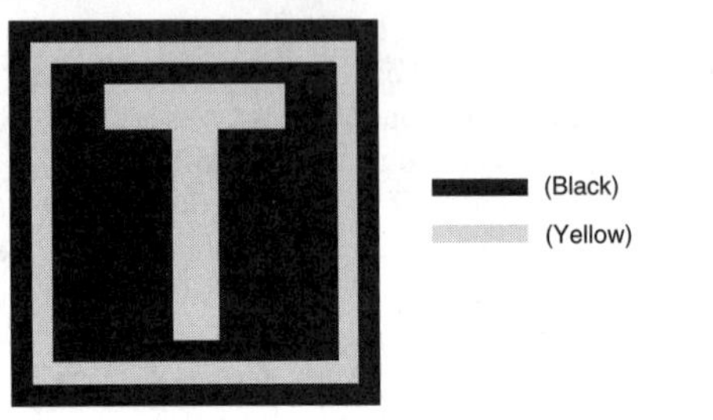

Figure 2-38[1]. Taxiway Location Sign

c. Runway Location Sign. This sign has a black background with a yellow inscription and yellow border as shown in Figure 2-38[3]. The inscription is the designation of the runway on which the aircraft is located. These signs are intended to complement the information available to pilots through their magnetic compass and typically are installed where the proximity of two or more runways to one another could cause pilots to be confused as to which runway they are on.

Figure 2-38[2]. Taxiway Location Sign Collocated with Runway Holding Position Sign

Figure 2-38[3]. Runway Location Sign

d. Runway Boundary Sign. This sign has a yellow background with a black inscription with a graphic depicting the pavement holding position marking as shown in Figure 2-38[4]. This sign, which faces the runway and is visible to the pilot exiting the runway, is located adjacent to the holding position marking on the pavement. The sign is intended to provide pilots with another visual cue which they can use as a guide in deciding when they are "clear of the runway."

e. ILS Critical Area Boundary Sign. This sign has a yellow background with a black inscription with a graphic

(Black)
(Yellow)

Figure 2-38[4]. Runway Boundary Sign

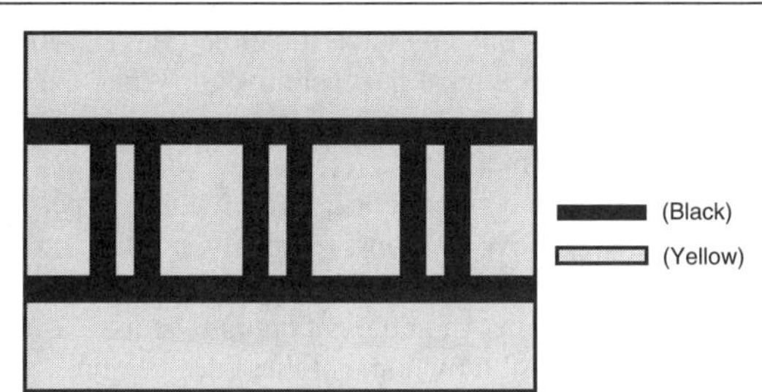
(Black)
(Yellow)

Figure 2-38[5]. ILS Critical Area Boundary Sign

depicting the ILS pavement holding position marking as shown in Figure 2-38[5]. This sign is located adjacent to the ILS holding position marking on the pavement and can be seen by pilots leaving the critical area. The sign is intended to provide pilots with another visual cue which they can use as a guide in deciding when they are "clear of the ILS critical area."

2-39. DIRECTION SIGNS

a. Direction signs have a yellow background with a black inscription. The inscription identifies the designation(s) of the intersecting taxiway(s) leading out of an intersection that a pilot would normally be expected to turn

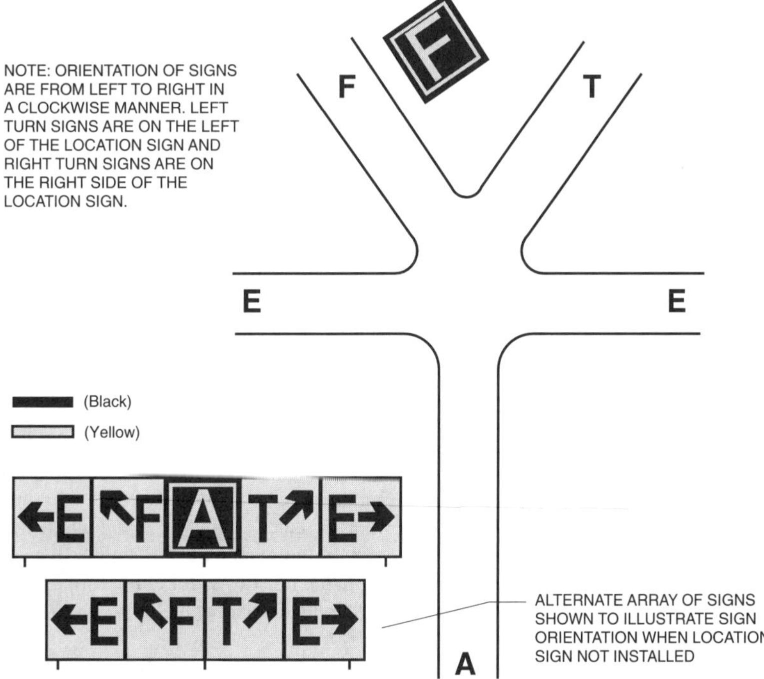
NOTE: ORIENTATION OF SIGNS ARE FROM LEFT TO RIGHT IN A CLOCKWISE MANNER. LEFT TURN SIGNS ARE ON THE LEFT OF THE LOCATION SIGN AND RIGHT TURN SIGNS ARE ON THE RIGHT SIDE OF THE LOCATION SIGN.
F
F
T
E
E
A
(Black)
(Yellow)
←E ↖F A T↗ E→
←E ↖F T↗ E→
ALTERNATE ARRAY OF SIGNS SHOWN TO ILLUSTRATE SIGN ORIENTATION WHEN LOCATION SIGN NOT INSTALLED

Figure 2-39[1]. Direction Sign Array with Location Sign on Far Side of Intersection

Figure 2-39[2]. Direction Sign for Runway Exit

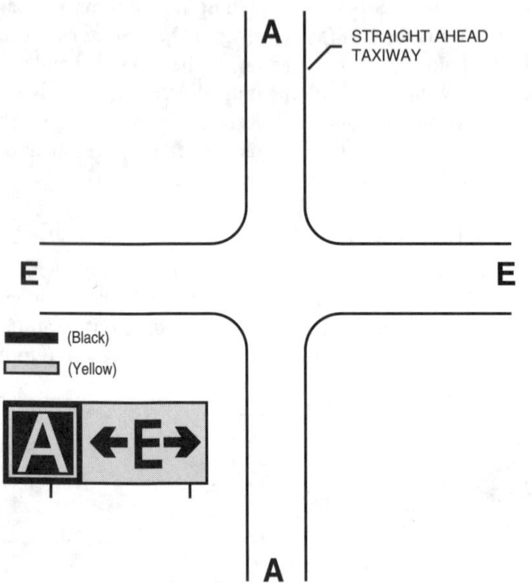

Figure 2-39[3]. Direction Sign Array for Simple

onto or hold short of. Each designation is accompanied by an arrow indicating the direction of the turn.

b. Except as noted in subparagraph e, each taxiway designation shown on the sign is accompanied by only one arrow. When more than one taxiway designation is shown on the sign each designation and its associated arrow is separated from the other taxiway designations by either a vertical message divider or a taxiway location sign as shown in Figure 2-39[1].

c. Direction signs are normally located on the left prior to the intersection. When used on a runway to indicate an exit, the sign is located on the same side of the runway as the exit. Figure 2-39[2] shows a direction sign used to indicate a runway exit.

d. The taxiway designations and their associated arrows on the sign are arranged clockwise starting from the first taxiway on the pilot's left, see Figure 2-39[1].

e. If a location sign is located with the direction signs, it is placed so that the designations for all turns to the left will be to the left of the location sign; the designations for continuing straight ahead or for all turns to the right would be located to the right of the location sign. See Figure 2-39[1].

f. When the intersection is comprised of only one crossing taxiway, it is permissible to have two arrows associated with the crossing taxiway as shown in Figure 2-39[3]. In this case, the location sign is located to the left of the direction sign.

2-40. DESTINATION SIGNS

a. Destination signs also have a yellow background with a black inscription indicating a destination on the airport. These signs always have an arrow showing the direction of the taxiing route to that destination. Figure 2-40[1] is an example of a typical destination sign. When the arrow on the destination sign indicates a turn, the sign is located prior to the intersection.

b. Destinations commonly shown on these types of signs include runways, aprons, terminals, military areas, civil aviation areas, cargo areas, international areas, and fixed base operators. An abbreviation may be used as the inscription on the sign for some of these destinations.

c. When the inscription for two or more destinations having a common taxiing route are placed on a sign, the destinations are separated by a "dot" (•) and one arrow would be used as shown in Figure 2-40[2]. When the inscription on a sign contains two or more destinations having different taxiing routes, each destination will be accompanied by an arrow and will be separated from the other destinations on the sign with a vertical black message divider as shown in Figure 2-40[3].

2-41. INFORMATION SIGNS

Information signs have a yellow background with a black inscription. They are used to provide the pilot with information on such things as areas that cannot be seen from the control tower, applicable radio frequencies, and noise abatement procedures. The airport operator determines the need, size, and location for these signs.

2-42. RUNWAY DISTANCE REMAINING SIGNS

Runway distance remaining signs have a black background with a white numeral inscription and may be installed along one or both side(s) of the runway. The number on the signs indicates the distance (in thousands of feet) of landing runway remaining. The last sign, i.e., the sign with the numeral "1" will be located at least 950 feet from the runway end. Figure 2-42[1] shows an example of a runway distance remaining sign.

2-43. AIRCRAFT ARRESTING DEVICES

a. Certain airports are equipped with a means of rapidly stopping military aircraft on a runway. This equipment,

Figure 2-40[1]. Destination Sign for Military Area

Figure 2-40[2]. Destination Sign for Common Taxiing Route to Two Runways

Figure 2-40[3]. Destination Sign for Different Taxiing Routes to Two Runways

normally referred to as EMERGENCY ARRESTING GEAR, generally consists of pendant cables supported over the runway surface by rubber "donuts." Although most devices are located in the overrun areas, a few of these arresting systems have cables stretched over the operational areas near the ends of a runway.

b. Arresting cables which cross over a runway require special markings on the runway to identify the cable location. These markings consist of 10 feet diameter solid circles painted "identification yellow," 30 feet on center, perpendicular to the runway centerline across the entire runway width. Additional details are contained in AC 150/5220-9, Aircraft Arresting Systems for Joint/Civil Military Airports.

2-43b NOTE—Aircraft operations on the runway are NOT restricted by the installation of aircraft arresting devices.

Figure 2-42[1]. Runway Distance Remaining Sign Indicating 3000 feet of Runway Remaining

Chapter 3. Airspace
Section 1. GENERAL

3-1. GENERAL

There are two categories of airspace or airspace areas; regulatory and nonregulatory. Within these two categories, there are controlled, uncontrolled, special use, and other airspace area types. The categories and types of airspace are dictated by: (1) the complexity or density of aircraft movements; (2) the nature of the operations conducted within the airspace; and (3) the level of safety required; and (4) the national and public interest. It is important that pilots be familiar with the operational requirements for each of the various types or classes of airspace. Subsequent sections will cover each category and class in sufficient detail to facilitate understanding.

3-2. GENERAL DIMENSIONS OF AIRSPACE SEGMENTS

Refer to Federal Aviation Regulations (FAR) for specific dimensions, exceptions, geographical areas covered, ex-

Table 3-3
BASIC VFR WEATHER MINIMUMS

Airspace	Flight Visibility	Distance from Clouds
Class A	Not Applicable	Not Applicable
Class B	3 statute miles	Clear of Clouds
Class C	3 statute miles	500 feet below 1,000 feet above 2,000 feet horizontal
Class D	3 statute miles	500 feet below 1,000 feet above 2,000 feet horizontal
Class E. Less than 10,000 feet MSL	3 statute miles	500 feet below 1,000 feet above 2,000 feet horizontal
At or above 10,000 feet MSL	5 statute miles	1,000 feet below 1,000 feet above 1 statute mile horizontal
Class G. 1,200 feet or less above the surface (regardless of MSL altitude). Day, except as provided in section 91.155(b).	1 statute mile	Clear of clouds
Night, except as provided in section 91.155(b).	3 statute miles	500 feet below 1,000 feet above 2,000 feet horizontal
More than 1,200 feet above the surface but less than 10,000 feet MSL. Day	1 statute mile	500 feet below 1,000 feet above 2,000 feet horizontal
Night	3 statute miles	500 feet below 1,000 feet above 2,000 feet horizontal
More than 1,200 feet above the surface and at or above 10,000 feet MSL.	5 statute miles	1,000 feet below 1,000 feet above 1 statute mile horizontal

Table 3-4
VFR CRUISING ALTITUDES AND FLIGHT LEVELS

If your magnetic course (ground track) is:	*And you are more than 3,000 feet above the surface but below 18,000 feet MSL, fly:*	*And you are above 18,000 feet MSL to FL 290 (except within CLASS A AIRSPACE (FAR Part 71.33)), fly:*
0° to 179°	Odd thousands MSL, plus 500 feet (3,500, 5,500, 7,500, etc.).	Odd Flight Levels plus 500 feet (FL 195, FL 215, FL 235, etc.).
180° to 359°	Even thousands MSL, plus 500 feet (4,500, 6,500, 8,500, etc.).	Even Flight Levels plus 500 feet (FL 185, FL 205, FL 225, etc.).

clusions, specific transponder or equipment requirements, and flight operations.

3-3. BASIC VFR WEATHER MINIMUMS

No person may operate an aircraft under basic VFR when the flight visibility is less, or at a distance from clouds that is less, than that prescribed for the corresponding altitude and class of airspace. (See Table 3-3.)

3-3 NOTE—Student pilots must comply with Part 61.89(a) (6) and (7).

3-4. VFR CRUISING ALTITUDES AND FLIGHT LEVELS

(See Table 3-4).

3-5 thru 3-9. RESERVED

Section 2. CONTROLLED AIRSPACE

3-10. GENERAL

a. Controlled Airspace: A generic term that covers the different classification of airspace (Class A, Class B, Class C, Class D, and Class E airspace) and defined dimensions within which air traffic control service is provided to IFR flights and to VFR flights in accordance with the airspace classification. (See Figure 3-10[1].)

b. IFR Requirements: IFR operations in any class of

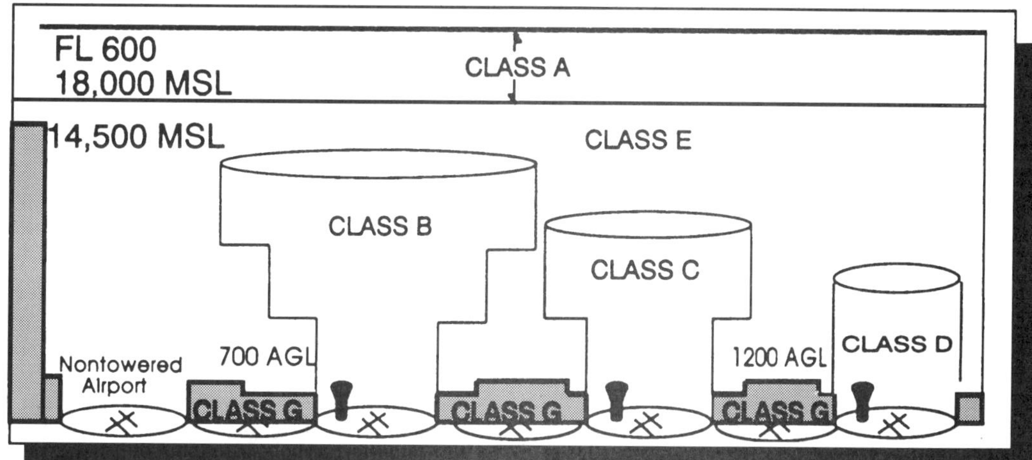

Figure 3-10[1]

controlled airspace requires that a pilot must file an IFR flight plan and receive an appropriate ATC clearance.

c. IFR Separation: Standard IFR separation is provided to all aircraft operating under IFR in controlled airspace.

d. VFR Requirements: It is the responsibility of the pilot to insure that ATC clearance or radio communication requirements are met prior to entry into Class B, Class C, or Class D airspace. The pilot retains this responsibility when receiving ATC radar advisories. See FAR Part 91.

e. Traffic Advisories: Traffic advisories will be provided to all aircraft as the controller's work situation permits.

f. Safety Alerts: Safety Alerts are mandatory services and are provided to ALL aircraft. There are two types of Safety Alerts, Terrain/Obstruction Alert and Aircraft Conflict/Mode C Intruder Alert.

1. Terrain/Obstruction Alert. A Terrain/Obstruction Alert is issued when, in the controller's judgment, an aircraft's altitude places it in unsafe proximity to terrain and/or obstructions.

2. Aircraft Conflict/Mode C Intruder Alert. An Aircraft Conflict/Mode C Intruder Alert is issued if the controller observes another aircraft which places it in an unsafe proximity. When feasible, the controller will offer the pilot an alternative course of action.

g. Ultralight Vehicles: No person may operate an ultralight vehicle within Class A, Class B, Class C, or Class D airspace or within the lateral boundaries of the surface area of Class E airspace designated for an airport unless that person has prior authorization from the ATC facility having jurisdiction over that airspace. See FAR Part 103.

h. Unmanned Free Balloons: Unless otherwise authorized by ATC, no person may operate an unmanned free balloon below 2,000 feet above the surface within the lateral boundaries of Class B, Class C, Class D, or Class E airspace designated for an airport. See FAR Part 101.

i. Parachute Jumps: No person may make a parachute jump, and no pilot in command may allow a parachute jump to be made from that aircraft, in or into Class A, Class B, Class C, or Class D airspace without, or in violation of, the terms of an ATC authorization issued by the ATC facility having jurisdiction over the airspace. See FAR Part 105.

3-11. CLASS A AIRSPACE

a. Definition: Generally, that airspace from 18,000 feet MSL up to and including FL600, including the airspace overlying the waters within 12 nautical miles of the coast of the 48 contiguous States and Alaska; and designated international airspace beyond 12 nautical miles of the coast of the 48 contiguous States and Alaska within areas of domestic radio navigational signal or ATC radar coverage, and within which domestic procedures are applied.

b. Operating Rules and Pilot/Equipment Requirements: Unless otherwise authorized, all persons must operate their aircraft under IFR. See FAR Part 71.33 and Part 91.167 through Part 91.193.

c. Charts: Class A airspace is not specifically charted.

3-12. CLASS B AIRSPACE

a. Definition: Generally, that airspace from the surface to 10,000 feet MSL surrounding the nation's busiest airports in terms of IFR operations or passenger enplanements. The configuration of each Class B airspace area is individually tailored and consists of a surface area and two or more layers (some Class B airspace areas resemble upside-down wedding cakes), and is designed to contain all published instrument procedures once an aircraft enters the airspace. An ATC clearance is required for all aircraft to operate in the area, and all aircraft that are so cleared receive separation services within the airspace. The cloud clearance requirement for VFR operations is "clear of clouds."

b. Operating Rules and Pilot/Equipment Requirements for VFR Operations: Regardless of weather conditions, an ATC clearance is required prior to operating within Class B airspace. Pilots should not request a clearance to operate within Class B airspace unless the requirements of FAR Part 91.215 and Part 91.131 are met. Included among these requirements are:

1. Unless otherwise authorized by ATC, aircraft must be equipped with an operable two-way radio capable of communicating with ATC on appropriate frequencies for that Class B airspace.

2. No person may take off or land a civil aircraft at the following primary airports within Class B airspace unless the pilot in command holds at least a private pilot certificate:

Andrews Air Force Base, MD
Atlanta Hartsfield Airport, GA
Boston Logan Airport, MA
Chicago O'Hare Intl Arpt, IL
Los Angeles Intl Airport, CA
Miami Intl Airport, FL
Newark Intl Airport, NJ
New York Kennedy Airport, NY
New York La Guardia Arpt, NY
San Francisco Intl Airport, CA
Washington National Arpt, DC
Dallas/Fort Worth Intl Arpt, TX

3. No person may take off or land a civil aircraft at an airport within Class B airspace or operate a civil aircraft within Class B airspace unless:

(a) The pilot in command holds at least a private pilot certificate; or,

(b) The aircraft is operated by a student pilot or recreational pilot who seeks private pilot certification and has met the requirements of FAR Part 61.95.

4. Unless otherwise authorized by ATC, each person operating a large turbine engine-powered airplane to or from a primary airport shall operate at or above the designated floors while within the lateral limits of Class B airspace.

5. Unless otherwise authorized by ATC, each aircraft must be equipped as follows:

(a) For IFR operations, an operable VOR or TACAN receiver; and

(b) For all operations, a two-way radio capable of communications with ATC on appropriate frequencies for that area; and

(c) Unless otherwise authorized by ATC, an operable radar beacon transponder with automatic altitude reporting equipment.

3-12b5(c) NOTE—ATC may, upon notification, immediately authorize a deviation from the altitude reporting equipment requirement; however, a request for a deviation from the 4096 transponder equipment requirement must be submitted to the controlling ATC facility at least one hour before the proposed operation. (Reference—Transponder Operation, paragraph 4-19.)

c. Charts: Class B airspace is charted on Sectional Charts, IFR En Route Low Altitude, and Terminal Area Charts.

d. Flight Procedures:

1. Flights: Aircraft within Class B airspace are required to operate in accordance with current IFR procedures. A clearance for a visual approach to a primary airport is not authorization for turbine powered airplanes to operate below the designated floors of the Class B airspace.

2. VFR Flights:

(a) Arriving aircraft must obtain an ATC clearance prior to entering Class B airspace and must contact ATC on the appropriate frequency, and in relation to geographical fixes shown on local charts. Although a pilot may be operating beneath the floor of the Class B airspace on initial contact, communications with ATC should be established in relation to the points indicated for spacing and sequencing purposes.

(b) Departing aircraft require a clearance to depart Class B airspace and should advise the clearance delivery position of their intended altitude and route of flight. ATC will normally advise VFR aircraft when leaving the geographical limits of the Class B airspace. Radar service is not automatically terminated with this advisory unless specifically stated by the controller.

(c) Aircraft not landing or departing the primary airport may obtain an ATC clearance to transit the Class B airspace when traffic conditions permit and provided the requirements of FAR Part 91.131 are met. Such VFR aircraft are encouraged, to the extent possible, to operate at altitudes above or below the Class B airspace or transit through established VFR corridors. Pilots operating in VFR corridors are urged to use frequency 122.750 MHz for the exchange of aircraft position information.

e. ATC Clearances and Separation: An ATC clearance is required to enter and operate within Class B airspace. VFR pilot's are provided sequencing and separation from other aircraft while operating within Class B airspace. (Reference Terminal VFR Radar Service, paragraph 4-17.)

3-12e NOTE 1—Separation and sequencing of VFR aircraft will be suspended in the event of a radar outage as this service is dependent on radar. The pilot will be advised that the service is not available and issued wind, runway information and the time or place to contact the tower.

3-12e NOTE 2—Separation of VFR aircraft will be suspended during CENRAP operations. Traffic advisories and sequencing to the primary airport will be provided on a workload permitting basis. The pilot will be advised when Center Radar Presentation (CENRAP) is in use.

1. VFR aircraft are separated from all VFR/IFR aircraft which weigh 19,000 pounds or less by a minimum of:

(a) Target resolution, or

(b) 500 feet vertical separation, or

(c) Visual separation

2. VFR aircraft are separated from all VFR/IFR aircraft which weigh more than 19,000 and turbojets by no less than:

(a) 1½ miles lateral separation, or

(b) 500 feet vertical separation, or

(c) Visual separation

3. This program is not to be interpreted as relieving pilots of their responsibilities to see and avoid other traffic operating in basic VFR weather conditions, to adjust their operations and flight path as necessary to preclude serious wake encounters, to maintain appropriate terrain and obstruction clearance or to remain in weather conditions equal to or better than the minimums required by FAR Part 91.155. Approach control should be advised and a revised clearance or instruction obtained when compliance with an assigned route, heading and/or altitude is likely to compromise pilot responsibility with respect to terrain and obstruction clearance, vortex exposure, and weather minimums.

4. ATC may assign altitudes to VFR aircraft that do not conform to FAR Part 91.159. **"RESUME APPROPRIATE VFR ALTITUDES"** will be broadcast when the altitude assignment is no longer needed for separation or when leaving Class B airspace. Pilots must return to an altitude that conforms to FAR Part 91.159.

f. Proximity operations: VFR aircraft operating in proximity to Class B airspace are cautioned against operating too closely to the boundaries, especially where the floor of the Class B airspace is 3,000 feet or less or where VFR cruise altitudes are at or near the floor of higher levels. Observance of this precaution will reduce the potential for encountering an aircraft operating at the altitudes of Class B floors. Additionally, VFR aircraft are encouraged to utilize the VFR Planning Chart as a tool for planning flight in proximity to Class B airspace. Charted VFR Flyway Planning charts are published on the back of the existing VFR Terminal Area Charts.

3-13. CLASS C AIRSPACE

a. Definition: Generally, that airspace from the surface to 4,000 feet above the airport elevation (charted in MSL) surrounding those airports that have an operational control tower, are serviced by a radar approach control, and that have a certain number of IFR operations or passenger enplanements. Although the configuration of each Class C airspace area is individually tailored, the airspace usually consists of a 5 NM radius core surface area that extends from the surface up to 4,000 feet above the airport elevation, and a 10 NM radius shelf area that extends from 1,200 feet to 4,000 feet above the airport elevation.

b. Outer Area: The normal radius will be 20NM, with some variations based on site specific requirements. The outer area extends outward from the primary airport and extends from the lower limits of radar/radio coverage up to the ceiling of the approach control's delegated airspace, excluding the Class C airspace and other airspace as appropriate.

c. Charts: Class C airspace is charted on Sectional Charts, IFR En Route Low Altitude, and Terminal Area Charts where appropriate.

d. Operating Rules and Pilot/Equipment Requirements:

1. Pilot Certification: No specific certification required.

2. Equipment:

(a) Two-way radio, and

(b) Unless otherwise authorized by ATC, an operable radar beacon transponder with automatic altitude reporting equipment.

3. Arrival or Through Flight Entry Requirements: Two-way radio communication must be established with the ATC facility providing ATC services prior to entry and thereafter maintain those communications while in Class C airspace. Pilots of arriving aircraft should contact the Class C airspace ATC facility on the publicized frequency and give their position, altitude, radar beacon code, destination, and request Class C service. Radio contact should be initiated far enough from the Class C airspace boundary to preclude entering Class C airspace before two-way radio communications are established.

3-13d3 NOTE 1—If the controller responds to a radio call with, "(aircraft callsign) standby," radio communications have been established and the pilot can enter the Class C airspace.

3-13d3 NOTE 2—If workload or traffic conditions prevent immediate provision of Class C services, the controller will inform the pilot to remain outside the Class C airspace until conditions permit the services to be provided.

EXAMPLE: *[aircraft callsign]* "REMAIN OUTSIDE THE CLASS C AIRSPACE AND STANDBY."

3-13d3 NOTE 3—It is important to understand that if the controller responds to the initial radio call WITHOUT using the aircraft identification, radio communications have not been established and the pilot may not enter the Class C airspace.

EXAMPLE: "AIRCRAFT CALLING DULLES APPROACH CONTROL, STANDBY."

4. Departures from:

(a) A primary or satellite airport with an operating control tower: Two-way radio communications must be established and maintained with the control tower, and thereafter as instructed by ATC while operating in Class C airspace.

(b) A satellite airport without an operating control tower: Two-way radio communications must be established as soon as practicable after departing with the ATC facility having jurisdiction over the Class C airspace.

5. Aircraft Speed: Unless otherwise authorized or required by ATC, no person may operate an aircraft at or below 2,500 feet above the surface within 4 nautical miles of the primary airport of a Class C airspace area at an indicated airspeed of more than 200 knots (230 mph).

e. Air Traffic Services: When two-way radio communications and radar contact are established, all participating VFR aircraft are:

1. Sequenced to the primary airport

2. Provided Class C services within the Class C airspace and the Outer Area.

3. Provided basic radar services beyond the outer area on a workload permitting basis. This can be terminated by the controller if workload dictates.

f. Aircraft Separation: Separation is provided within the Class C airspace and the Outer Area after two-way radio communications and radar contact are established. VFR aircraft are separated from IFR aircraft within the Class C airspace by any of the following:

1. Visual separation.

2. 500 feet vertical; except when operating beneath a heavy jet.

3. Target resolution.

3-13f NOTE 1—Separation and sequencing of VFR aircraft will be suspended in the event of a radar outage as this service is dependent on radar. The pilot will be advised that the service is not available and issued wind, runway information and the time or place to contact the tower.

3-13f NOTE 2—Separation of VFR aircraft will be suspended during CENRAP operations. Traffic advisories and sequencing to the primary airport will be provided on a workload permitting basis. The pilot will be advised when CENRAP is in use.

3-13f NOTE 3—Pilot participation is voluntary within the outer area and can be discontinued, within the outer area, at the pilot's request. Class C services will be provided in the outer area unless the pilot requests termination of the service.

3-13f NOTE 4—Some facilities provide Class C services only during published hours. At other times, terminal IFR radar service will be provided. It is important to note that the communications requirements for entry into the airspace and transponder Mode C requirements are in effect at all times.

g. Secondary Airports:

1. In some locations Class C airspace may overlie the Class D surface area of a secondary airport. In order to allow that control tower to provide service to aircraft, portions of the overlapping Class C airspace may be procedurally excluded when the secondary airport tower is in operation. Aircraft operating in these procedurally excluded areas will only be provided airport traffic control services when in communication with the secondary airport tower.

2. Aircraft proceeding inbound to a satellite airport will be terminated at a sufficient distance to allow time to change to the appropriate tower or advisory frequency. Class C services to these aircraft will be discontinued when the aircraft is instructed to contact the tower or change to advisory frequency.

3. Aircraft departing secondary controlled airports will not receive Class C services until they have been radar identified and two-way communications have been established with the Class C airspace facility.

h. This program is not to be interpreted as relieving pilots of their responsibilities to see and avoid other traffic operating in basic VFR weather conditions, to adjust their operations and flight path as necessary to preclude serious wake encounters, to maintain appropriate terrain and obstruction clearance or to remain in weather conditions equal to or better than the minimums required by FAR Part 91.155. Approach control should be advised and a revised clearance or instruction obtained when compliance with an assigned route, heading and/or altitude is likely to compromise pilot responsibility with respect to terrain and obstruction clearance, vortex exposure, and weather minimums. (See Table 3-13, CLASS C AIRSPACE by State.)

These areas are currently designated as CLASS C AIRSPACE and are depicted on Sectional Charts. The table will be updated as additional CLASS C AIRSPACE areas are implemented. Pilots should consult current sectional charts and NOTAM's for the latest information on services available. Pilots should be aware that some CLASS C AIRSPACE underlies or is adjacent to CLASS B AIRSPACE.

3-14. CLASS D AIRSPACE

a. Definition: Generally, that airspace from the surface to 2,500 feet above the airport elevation (charted in MSL) surrounding those airports that have an operational control tower. The configuration of each Class D airspace area is individually tailored and when instrument procedures are published, the airspace will normally be designed to contain the procedures.

b. Operating Rules and Pilot/Equipment Requirements:

1. Pilot Certification: No specific certification required.

2. Equipment: Unless otherwise authorized by ATC, an operable two-way radio is required.

3. Arrival or Through Flight Entry Requirements: Two-way radio communication must be established with the ATC facility providing ATC services prior to entry and thereafter maintain those communications while in the Class D airspace. Pilots of arriving aircraft should contact the control tower on the publicized frequency and give their position, altitude, destination, and any request(s). Radio contact should be initiated far enough from the Class D airspace boundary to preclude entering the Class D airspace before two-way radio communications are established.

3-14b3 NOTE 1—If the controller responds to a radio call with, [aircraft callsign] "STANDBY," radio communications have been established and the pilot can enter the Class D airspace.

3-14b3 NOTE 2—If workload or traffic conditions prevent immediate entry into Class D airspace, the controller will inform the pilot to remain outside the Class D airspace until conditions permit entry.

EXAMPLE: *[aircraft callsign]* "REMAIN OUTSIDE THE CLASS D AIRSPACE AND STANDBY."

3-14b3 NOTE 3—It is important to understand that if the controller responds to the initial radio call without using the aircraft callsign, radio communications have not been established and the pilot may not enter the Class D airspace.

EXAMPLE:"AIRCRAFT CALLING MANASSAS TOWER STANDBY."

3-14b3 NOTE 4—At those airports where the control tower does not operate 24 hours a day, the operating hours of the tower will be listed on the appropriate charts and in the airport facility directory (AFD). During the hours the tower is not in operation the Class E surface area rules or a combination of Class E rules to 700 feet above ground level and Class G rules to the surface will become applicable. Check the AFD for specifics.

4. Departures from:

(a) A primary or satellite airport with an operating control tower: Two-way radio communications must be established and maintained with the control tower, and thereafter as instructed by ATC while operating in the Class D airspace.

(b) A satellite airport without an operating control tower: Two-way radio communications must be established as soon as practicable after departing with the ATC facility having jurisdiction over the Class D airspace as soon as practicable after departing.

5. Aircraft Speed: Unless otherwise authorized or required by ATC, no person may operate an aircraft at or below 2,500 feet above the surface within 4 nautical miles of the primary airport of a Class D airspace area at an indicated airspeed of more than 200 knots (230 mph).

c. Class D airspace areas are depicted on Sectional and Terminal charts with blue segmented lines, and on IFR En Route Lows with a boxed [D].

d. Arrival extensions for instrument approach procedures may be Class D or Class E airspace. As a general rule, if all extensions are 2 miles or less, they remain part of the Class D surface area. However, if any one extension is greater than 2 miles, then all extensions become Class E.

e. Separation for VFR Aircraft: No separation services are provided to VFR aircraft.

3-15. CLASS E AIRSPACE

a. Definition: Generally, if the airspace is not Class A, Class B, Class C, or Class D, and it is controlled airspace, it is Class E airspace.

Table 3-13. Class C Airspace by State

STATE/CITY	AIRPORT
ALABAMA.	
Birmingham	Municipal
Huntsville	Madison Co-Carl T Jones Field.
Mobile	Bates Field
ALASKA.	
Anchorage	International
ARIZONA.	
Davis-Monthan	AFB
Tucson	International
ARKANSAS.	
Little Rock	Adams Field
CALIFORNIA.	
Beale	AFB
Burbank	Burbank - Glendale - Pasadena
Castle	AFB
Fresno	Air Terminal
McClellan	AFB
March	AFB
Mather	AFB
Monterey	Peninsula
Norton	AFB
Oakland	Metropolitan Int.
Ontario	International
Sacramento	Metropolitan
San Jose	International
Santa Ana	El Toro MCAS
Santa Ana	John Wayne/Orange County
Santa Barbara	Municipal
COLORADO.	
Colorado Springs	Municipal Airport.
CONNECTICUT.	
Windsor Locks	Bradley Int.
DELAWARE.	
DISTRICT OF COLUMBIA.	
FLORIDA.	
Daytona Beach	Regional
Fort Lauderdale	Hollywood Int.
Fort Myers	SW Florida Regional
Jacksonville	International
Palm Beach	International
Pensacola	NAS
Pensacola	Regional
Sarasota	Bradenton
Tallahassee	Municipal
Whiting	NAS
GEORGIA.	
Columbus	Metropolitan
Savannah	International
HAWAII.	
Kahului	International
IDAHO.	
Boise	Air Terminal
ILLINOIS.	
Champaign	U of Illinois Willard
Chicago	Midway
Moline	Quad City
Peoria	Greater Peoria
Springfield	Capital
INDIANA.	
Evansville	Evansville Dress Regional
Fort Wayne	Municipal
Indianapolis	International
South Bend	Michiana Regional
IOWA.	
Cedar Rapids	Municipal
Des Moines	International
KANSAS.	
Wichita	Mid-Continent
KENTUCKY.	
Covington	Greater Cincinnati International
Lexington	Blue Grass Field
Louisville	Standiford Field
LOUISIANA.	
Barksdale	Air Force Base
Baton Rouge	BTR Metro, Ryan Field
Lafayette	Regional
Shreveport	Regional
MAINE.	
Bangor	International
Portland	International Jetport
MARYLAND.	
MASSACHUSETTS.	
MICHIGAN.	
Flint	Bishop
Grand Rapids	Kent County
Lansing	Capital City
MINNESOTA.	
MISSISSIPPI.	
Columbus	Air Force Base
Jackson	Allen C Thompson Field
MISSOURI.	
MONTANA.	
NEBRASKA.	
Lincoln	Municipal
Omaha	Eppley Airfield
Offutt	Air Force Base
NEVADA.	
Reno	Cannon Int.
NEW HAMPSHIRE.	
Manchester	Manchester
NEW JERSEY.	
Atlantic City	City

STATE/CITY	AIRPORT
NEW MEXICO.	
Albuquerque	International
NEW YORK.	
Albany	County
Buffalo	Greater Buff. Int.
Islip	Long Island MacArthur
Rochester	Rochester-Monroe Co.
Syracuse	Hancock Int.
NORTH CAROLINA.	
Fayetteville	Municipal/Grannis Field
Greensboro	Highpoint-Winston-Salem Regional
Pope	Air Force Base
Raleigh	Raleigh-Durham
NORTH DAKOTA.	
OHIO.	
Akron	Akron-Canton Reg.
Columbus	Port Columbus Int.
Dayton	James M. Cox Int.
Toledo	Express
OKLAHOMA.	
Oklahoma City	Will Rogers World
Tinker	Air Force Base
Tulsa	International
OREGON.	
Portland	International
PENNSYLVANIA.	
Allentown	Bethlehem-Easton
PUERTO RICO.	
San Juan	International
RHODE ISLAND.	
Providence	Theodore Francis Green St.
SOUTH CAROLINA.	
Charleston	Air Force Base/International
Columbia	Metropolitan
Greer	Greenville-Spartanburg
Shaw	Air Force Base
SOUTH DAKOTA.	
TENNESSEE.	
Chattanooga	Lovell Field
Knoxville	McGhee Tyson
Nashville	Metropolitan
TEXAS.	
Abilene	Municipal
Amarillo	International
Austin	Robert Mueller Mun.
Corpus Christi	International
Dyess	Air Force Base
El Paso	International
Harlingen	Rio Grande Valley
Laughlin	Air Force Base
Lubbock	International
Midland	International
San Antonio	International
UTAH.	
VERMONT.	
Burlington	International
VIRGINIA.	
Richmond	Richard Evelyn Byrd Inter.
Norfolk	International
Roanoke	Regional/Woodrum Field
WASHINGTON.	
Fairchild	Air Force Base
Spokane	International
Whidbey Island	Naval Air Station
WEST VIRGINIA.	
Charleston	Yeager
WISCONSIN.	
Green Bay	Austin Straubel Field
Madison	Dane County Reg.
Milwaukee	General Mitchell Int.
WYOMING.	

b. Operating Rules and Pilot/Equipment Requirements:

1. Pilot Certification: No specific certification required.

2. Equipment: No specific equipment required by the airspace.

3. Arrival or Through Flight Entry Requirements: No specific requirements.

c. Charts: Class E airspace below 14,500 feet MSL is charted on Sectional, Terminal, World, and IFR En Route Low Altitude charts.

d. Vertical limits: Except for 18,000 feet MSL, Class E airspace has no defined vertical limit but rather it extends upward from either the surface or a designated altitude to the overlying or adjacent controlled airspace.

e. Types of Class E Airspace:

1. Surface area designated for an airport: When designated as a surface area for an airport, the airspace will be configured to contain all instrument procedures.

2. Extension to a surface area: There are Class E airspace areas that serve as extensions to Class B, Class C, and Class D surface areas designated for an airport. Such airspace provides controlled airspace to contain standard instrument approach procedures without imposing a communications requirement on pilots operating under VFR.

3. Airspace used for transition: There are Class E airspace areas beginning at either 700 or 1,200 feet AGL used to transition to/from the terminal or en route environment.

4. En Route Domestic Areas: There are Class E airspace areas that extend upward from a specified altitude and are en route domestic airspace areas that provide controlled airspace in those areas where there is a requirement to provide IFR en route ATC services but the Federal airway system is inadequate.

5. Federal Airways: The Federal airways are Class E airspace areas and, unless otherwise specified, extend upward from 1,200 feet to, but not including, 18,000 feet MSL. The colored airways are Green, Red, Amber, and Blue. The VOR airways are classified as Domestic, Alaskan, and Hawaiian.

6. Offshore Airspace Areas: There are Class E airspace areas that extend upward from a specified altitude to, but not including, 18,000 feet MSL and are designated as offshore airspace areas. These areas provide controlled airspace beyond 12 miles from the coast of the United States in those areas where there is a requirement to provide IFR en route ATC services and within which the United States is applying domestic procedures.

7. Unless designated at a lower altitude, Class E airspace begins at 14,500 MSL over the United States, including that airspace overlying the waters within 12 nautical miles of the coast of the 48 contiguous States and Alaska excluding the Alaska peninsula west of long. 160°00'00"W; and the airspace less than 1,500 feet above the surface of the earth.

e. Separation for VFR Aircraft: No separation services are provided to VFR aircraft.

3-16 thru 3-19. RESERVED

Section 3. CLASS G AIRSPACE

3-20. GENERAL

Class G airspace is that portion of the airspace that has not been designated as Class A, Class B, Class C, Class D and Class E airspace (uncontrolled).

3-21. VFR REQUIREMENTS

Rules governing VFR flight have been adopted to assist the pilot in meeting his responsibility to see and avoid other aircraft. Minimum flight visibility and distance from clouds required for VFR flight are contained in FAR Part 91.155. (See Table 3-3.)

3-22. IFR REQUIREMENTS

a. The FARs specify the pilot and aircraft equipment requirements for IFR flight. Pilots are reminded that in addition to altitude or flight level requirements, FAR Part 91.177 includes a requirement to remain at least 1,000 feet (2,000 feet in designated mountainous terrain) above the highest obstacle within a horizontal distance of 4 nautical miles from the course to be flown.

b. IFR Altitudes and Flight Levels. (See Table 3-22.)

3-23 thru 3-29. RESERVED

Table 3-22
IFR ALTITUDES AND FLIGHT LEVELS—CLASS G AIRSPACE

If your magnetic course (ground track) is:	*And you are below 18,000 feet MSL, fly:*	*And you are at or above 18,000 feet MSL but below FL 290, fly:*	*And you are at or above FL 290, fly 4,000 foot intervals:*
0° to 179°	Odd thousands MSL, (3,000, 5,000, 7,000, etc.).	Odd Flight Levels, (FL 190, 210, 230, etc.).	Beginning at FL 290, (FL 290, 330, 370, etc.).
180° to 359°	Even thousands MSL, (2,000, 4,000, 6,000, etc.).	Even Flight Levels, (FL 180, 200, 220, etc.).	Beginning at FL 310, (FL 310, 350, 390, etc.).

Section 4. SPECIAL USE AIRSPACE

3-30. GENERAL

a. Special use airspace consists of that airspace wherein activities must be confined because of their nature, or wherein limitations are imposed upon aircraft operations that are not a part of those activities, or both. Except for Controlled Firing Areas, special use airspace areas are depicted on aeronautical charts.

b. Prohibited and Restricted Areas are regulatory special use airspace and are established in FAR Part 73 through the rulemaking process.

c. Warning Areas, Military Operations Areas, Alert Areas, and Controlled Firing Areas are nonregulatory special use airspace.

d. All special use airspace descriptions are contained in FAA Order 7400.8.

e. Special use airspace (except CFA's) are charted on IFR and Visual charts and includes the hours of operation, altitudes, and the controlling agency.

3-31. PROHIBITED AREA

Prohibited Areas contain airspace of defined dimensions identified by an area on the surface of the earth within which the flight of aircraft is prohibited. Such areas are established for security or other reasons associated with the national welfare. These areas are published in the Federal Register and are depicted on aeronautical charts.

3-32. RESTRICTED AREA

a. Restricted Areas contain airspace identified by an area on the surface of the earth within which the flight of aircraft, while not wholly prohibited, is subject to restrictions. Activities within these areas must be confined because of their nature or limitations imposed upon aircraft operations that are not a part of those activities or both. Restricted areas denote the existence of unusual, often invisible, hazards to aircraft such as artillery firing, aerial gunnery, or guided missiles. Penetration of Restricted Areas without authorization from the using or controlling agency may be extremely hazardous to the aircraft and its occupants. Restricted Areas are published in the Federal Register and constitute FAR Part 73.

b. ATC facilities apply the following procedures when aircraft are operating on an IFR clearance (including those cleared by ATC to maintain VFR-ON-TOP) via a route which lies within joint-use restricted airspace.

1. If the restricted area is not active and has been released to the controlling agency (FAA), the ATC facility will allow the aircraft to operate in the restricted airspace without issuing specific clearance for it to do so.

2. If the restricted area is active and has not been released to the controlling agency (FAA), the ATC facility will issue a clearance which will ensure the aircraft avoids the restricted airspace unless it is on an approved altitude reservation mission or has obtained its own permission to operate in the airspace and so informs the controlling facility.

3-32b2 NOTE—The above apply only to joint-use restricted airspace and not to prohibited and nonjoint-use airspace. For the latter categories, the ATC facility will issue a clearance so the aircraft will avoid the restricted airspace unless it is on an approved altitude reservation mission or has obtained its own permission to operate in the airspace and so informs the controlling facility.

c. Restricted airspace is depicted on the En Route Chart appropriate for use at the altitude or flight level being flown. For joint-use restricted areas, the name of the controlling agency is shown on these charts. For all prohibited areas and nonjoint-use restricted areas, unless otherwise requested by the using agency, the phrase "NO A/G" is shown.

3-33. WARNING AREA

Warning Areas are airspace which may contain hazards to nonparticipating aircraft in international airspace. Warning Areas are established beyond the 3 mile limit. Though the activities conducted within Warning Areas may be as hazardous as those in Restricted Areas, Warning Areas cannot be legally designated as Restricted Areas because they are over international waters. Penetration of Warning Areas during periods of activity may be hazardous to the aircraft and its occupants. Official descriptions of Warning Areas may be obtained on request to the FAA, Washington, D.C.

3-34. MILITARY OPERATIONS AREAS (MOA)

a. MOAs consist of airspace of defined vertical and lateral limits established for the purpose of separating certain military training activities from IFR traffic. Whenever a MOA is being used, nonparticipating IFR traffic may be cleared through a MOA if IFR separation can be provided by ATC. Otherwise, ATC will reroute or restrict nonparticipating IFR traffic.

b. Most training activities necessitate acrobatic or abrupt flight maneuvers. Military pilots conducting flight in Department of Defense aircraft within a designated and active military operations area (MOA) are exempted from the provisions of FAR Part 91.303(c) and (d) which prohibit acrobatic flight within Federal airways and Class B, Class C, Class D, and Class E surface areas.

c. Pilots operating under VFR should exercise extreme caution while flying within a MOA when military activity is being conducted. The activity status (active/inactive) of MOA's may change frequently. Therefore, pilots should contact any FSS within 100 miles of the area to obtain accurate real-time information concerning the MOA hours of operation. Prior to entering an active MOA, pilots should contact the controlling agency for traffic advisories.

d. MOA's are depicted on Sectional, VFR Terminal Area, and En Route Low Altitude Charts.

3-35. ALERT AREA

Alert Areas are depicted on aeronautical charts to inform nonparticipating pilots of areas that may contain a high volume of pilot training or an unusual type of aerial activity. Pilots should be particularly alert when flying in these areas. All activity within an Alert Area shall be conducted in accordance with FARs, without waiver, and pilots of participating aircraft as well as pilots transiting the area shall be equally responsible for collision avoidance. Information concerning these areas may be obtained upon request to the FAA, Washington, D.C.

3-36. CONTROLLED FIRING AREAS

Controlled Firing Areas contain activities which, if not conducted in a controlled environment, could be hazardous to nonparticipating aircraft. The distinguishing feature of the Controlled Firing Area, as compared to other special use airspace, is that its activities are suspended immediately when spotter aircraft, radar, or ground lookout positions indicate an aircraft might be approaching the area. There is no need to chart Controlled Firing Areas since they do not cause a nonparticipating aircraft to change its flight path.

3-37 thru 3-39. RESERVED

Section 5. OTHER AIRSPACE AREAS

3-40. AIRPORT ADVISORY AREA

a. The airport advisory area is the area within 10 statute miles of an airport where a control tower is not operating but where an FSS is located. At such locations, the FSS provides advisory service to arriving and departing aircraft. (Reference—Traffic Advisory Practices at Airports Where a Tower is Not in Operation, paragraph 4-9.)

b. It is not mandatory that pilots participate in the Local Airport Advisory program, but it is strongly recommended that they do.

3-41. MILITARY TRAINING ROUTES (MTR)

a. National security depends largely on the deterrent effect of our airborne military forces. To be proficient, the military services must train in a wide range of airborne tactics. One phase of this training involves "low level" combat tactics. The required maneuvers and high speeds are such that they may occasionally make the see-and-avoid aspect of VFR flight more difficult without increased vigilance in areas containing such operations. In an effort to ensure the greatest practical level of safety for all flight operations, the MTR program was conceived.

b. The MTRs program is a joint venture by the FAA and the Department of Defense (DOD). MTR routes are mutually developed for use by the military for the purpose of conducting low-altitude, high-speed training. The routes above 1,500 feet above ground level (AGL) are developed to be flown, to the maximum extent possible, under IFR. The routes at 1,500 feet AGL and below are generally developed to be flown under Visual Flight Rules (VFR).

c. Generally, MTRs are established below 10,000 feet MSL for operations at speeds in excess of 250 knots. However, route segments may be defined at higher altitudes for purposes of route continuity. For example, route segments may be defined for descent, climbout, and mountainous terrain. There are IFR and VFR routes as follows:

1. IFR Military Training Routes—IR: Operations on these routes are conducted in accordance with IFRs regardless of weather conditions.

2. VFR Military Training Routes—VR: Operations on these routes are conducted in accordance with VFRs except, flight visibility shall be 5 miles or more; and flights shall not be conducted below a ceiling of less than 3,000 feet AGL.

d. Military training routes will be identified and charted as follows:

1. Route identification.

(a) MTRs with no segment above 1,500 feet AGL shall be identified by four number characters; e.g., IR1206, VR1207.

(b) MTRs that include one or more segments above 1,500 feet AGL shall be identified by three number characters; e.g., IR206, VR207.

(c) Alternate IR/VR routes or route segments are identified by using the basic/principal route designation followed by a letter suffix, e.g., IR008A, VR1007B, etc.

2. Route charting.

(a) IFR Low Altitude En Route Chart—This chart will depict all IR routes and all VR routes that accommodate operations above 1,500 feet AGL.

(b) VFR Planning Chart—This chart will depict routes (military training activities such as IR and VR regardless of altitude), MOAs, and restricted, warning and alert areas.

(c) Area Planning (AP/1B) Chart (DOD Flight Information Publication—FLIP). This chart is published by the DOD primarily for military users and contains detailed information on both IR and VR routes. (Reference—Auxiliary Charts, paragraph 9-6.)

e. The FLIP contains charts and narrative descriptions of these routes. This publication is available to the general public by single copy or annual subscription from the DIRECTOR, DMACSC, Attention: DOCP, Washington, DC 20315-0020. This DOD FLIP is available for pilot briefings at FSS and many airports.

f. Nonparticipating aircraft are not prohibited from flying within an MTR; however, extreme vigilance should be exercised when conducting flight through or near these routes. Pilots should contact FSSs within 100 NM of a particular MTR to obtain current information or route usage in their vicinity. Information available includes times of scheduled activity, altitudes in use on each route segment, and actual route width. Route width varies for each MTR and can extend several miles on either side of the charted MTR centerline. Route width information for IR and VR MTR's is also available in the FLIP AP/1B along with additional MTR (SR/AR) information. When requesting MTR information, pilots should give the FSS their position, route of flight, and destination in order to reduce frequency congestion and permit the FSS specialist to identify the MTR routes which could be a factor.

3-42. TEMPORARY FLIGHT RESTRICTIONS

a. General—This paragraph describes the types of conditions under which the FAA may impose temporary flight restrictions. It also explains which FAA elements have been delegated authority to issue a temporary flight restrictions NOTAM and lists the types of responsible agencies/offices from which the FAA will accept requests to establish temporary flight restrictions. The FAR is explicit as to what operations are prohibited, restricted, or allowed in a temporary flight restrictions area. Pilots are responsible to comply with FAR Part 91.137 when conducting flight in an area where a temporary flight restrictions area is in effect, and should check appropriate NOTAMs during flight planning.

b. The purpose for establishing a temporary flight restrictions area is to:

(1) Protect persons and property in the air or on the surface from an existing or imminent hazard associated with an incident on the surface when the presence of low flying aircraft would magnify, alter, spread, or compound that hazard (FAR Part 91.137(a)(1));

(2) Provide a safe environment for the operation of disaster relief aircraft (FAR Part 91.137(a)(2));

(3) Or prevent an unsafe congestion of sightseeing aircraft above an incident or event which may generate a high degree of public interest (FAR Part 91.137(a)(3)).

c. Except for hijacking situations, when the provisions of FAR Part 91.137(a)(1) or (a)(2) are necessary, a temporary flight restrictions area will only be established by or through the area manager at the Air Route Traffic Control Center (ARTCC) having jurisdiction over the area concerned. A temporary flight restrictions NOTAM involving the conditions of FAR Part 91.137(a)(3) will be issued at the direction of the regional air traffic division manager having oversight of the airspace concerned. When hijacking situations are involved, a temporary flight restrictions area will be implemented through the FAA Washington Headquarters Office of Civil Aviation Security. The appropriate FAA air traffic element, upon receipt of such a request, will establish a temporary flight restrictions area under FAR Part 91.137(a)(1).

d. The FAA accepts recommendations for the establishment of a temporary flight restrictions area under FAR Part 91.137(a)(1) from military major command headquarters, regional directors of the Office of Emergency Planning, Civil Defense State Directors, State Governors, or other similar authority. For the situations involving FAR Part 91.137(a)(2), the FAA accepts recommendations from military commanders serving as regional, subregional, or Search and Rescue (SAR) coordinators; by military commanders directing or coordinating air operations associated with disaster relief; or by civil authorities directing or coordinating organized relief air operations (includes representatives of the Office of Emergency Planning, U.S. Forest Service, and State aeronautical agencies). Appropriate authorities for a temporary flight restrictions establishment under FAR Part 91.137(a)(3) are any of those listed above or by State, county, or city government entities.

e. The type of restrictions issued will be kept to a minimum by the FAA consistent with achievement of the necessary objective. Situations which warrant the extreme restrictions of FAR Part 91.137(a)(1) include, but are not limited to: toxic gas leaks or spills, flammable agents, or fumes which if fanned by rotor or propeller wash could endanger persons or property on the surface, or if entered by an aircraft could endanger persons or property in the air; imminent volcano eruptions which could endanger airborne aircraft and occupants; nuclear accident or incident; and hijackings. Situations which warrant the restrictions associated with FAR Part 91.137(a)(2) include: forest fires which are being fought by releasing fire retardants from aircraft; and aircraft relief activities following a disaster (earthquake, tidal wave, flood, etc.). FAR Part 91.137(a)(3) restrictions are established for events and incidents that would attract an unsafe congestion of sightseeing aircraft.

f. The amount of airspace needed to protect persons and property or provide a safe environment for rescue/relief aircraft operations is normally limited to within 2,000 feet above the surface and within a two nautical mile radius. Incidents occurring within Class B, Class C, or Class D airspace will normally be handled through existing procedures and should not require the issuance of temporary flight restrictions NOTAM.

g. The FSS nearest the incident site is normally the "coordination facility." When FAA communications assistance is required, the designated FSS will function as the primary communications facility for coordination between emergency control authorities and affected aircraft. The ARTCC may act as liaison for the emergency control authorities if adequate communications cannot be established between the designated FSS and the relief organization. For example, the coordination facility may relay authorizations from the on-scene emergency response official in cases where news media aircraft operations are approved at the altitudes used by relief aircraft.

h. ATC may authorize operations in a temporary flight restrictions area under its own authority only when flight restrictions are established under FAR Part 91.137(a)(2) and (a)(3) and only when such operations are conducted under instrument flight rules (IFR). The appropriate ARTCC/air traffic control tower manager will, however, ensure that such authorized flights do not hamper activities or interfere with the event for which restrictions were implemented. However, ATC will not authorize local IFR flights into the temporary flight restrictions area.

i. To preclude misunderstanding, the implementing NOTAM will contain specific and formatted information. The facility establishing a temporary flight restrictions area will format a NOTAM beginning with the phrase "FLIGHT RESTRICTIONS" followed by: the location of the temporary flight restrictions area; the effective period; the area defined in statute miles; the altitudes affected; the FAA coordination facility and commercial telephone number; the reason for the temporary flight restrictions; the agency directing any relief activities and its commercial telephone number; and other information considered appropriate by the issuing authority.

EXAMPLE: FAR Part 91.137(a)(1):

The following NOTAM prohibits all aircraft operations except those specified in the NOTAM.

FLIGHT RESTRICTIONS MATTHEWS, VIRGINIA, EFFECTIVE IMMEDIATELY UNTIL 1200 GMT JANUARY 20, 1987. PURSUANT TO FAR 91.137(a)(1) TEMPORARY FLIGHT

RESTRICTIONS ARE IN EFFECT. RESCUE OPERATIONS IN PROGRESS. ONLY RELIEF AIRCRAFT OPERATIONS UNDER THE DIRECTION OF THE DEPARTMENT OF DEFENSE ARE AUTHORIZED IN THE AIRSPACE AT AND BELOW 5,000 FEET MSL WITHIN A TWO MILE RADIUS OF LASER AFB, MATTHEWS, VIRGINIA. COMMANDER, LASER AFB, IN CHARGE (897) 946-5543. STEENSON FSS IS THE FAA COORDINATION FACILITY (792) 555-6141.

EXAMPLE: FAR Part 91.137(a)(2):

The following NOTAM permits the on-site emergency response official to authorize media aircraft operations below the altitudes used by the relief aircraft.

FLIGHT RESTRICTIONS 25 MILES EAST OF BRANSOME, IDAHO, EFFECTIVE IMMEDIATELY UNTIL 2359 JANUARY 20, 1987. PURSUANT TO FAR 91.137(a)(2) TEMPORARY FLIGHT RESTRICTIONS ARE IN EFFECT WITHIN A FOUR MILE RADIUS OF THE INTERSECTION OF COUNTY ROADS 564 and 315 AT AND BELOW 3,500 FEET MSL TO PROVIDE A SAFE ENVIRONMENT FOR FIRE FIGHTING AIRCRAFT OPERATIONS. DAVIS COUNTY SHERIFF'S DEPARTMENT (792) 555-8122 IS IN CHARGE OF ON-SCENE EMERGENCY RESPONSE ACTIVITIES. GLIVINGS FSS (792) 555-1618 IS THE FAA COORDINATION FACILITY.

EXAMPLE: FAR Part 91.137(a)(3):

The following NOTAM prohibits sightseeing aircraft operations.

FLIGHT RESTRICTIONS BROWN, TENNESSEE, DUE TO OLYMPIC ACTIVITY. EFFECTIVE 1100 GMT JUNE 18, 1987, UNTIL 0200 GMT JULY 19, 1987. PURSUANT TO FAR 91.137(a)(3) TEMPORARY FLIGHT RESTRICTIONS ARE IN EFFECT WITHIN A THREE MILE RADIUS OF THE BASSETT SPORTS COMPLEX AT AND BELOW 2,500 FEET MSL. NORTON FSS (423) 555-6742 IS THE FAA COORDINATION FACILITY.

3-43. FLIGHT LIMITATIONS/PROHIBITIONS

a. Flight Limitations in the proximity of Space Flight Operations are designated in a Notice to Airman (NOTAM). FAR Part 91.143 provides protection from potentially hazardous situations for pilots and space flight crews and costly delays of shuttle operations.

b. Flight Restrictions in the proximity of Presidential and Other Parties are put into effect because numerous aircraft and large assemblies of persons may be attracted to areas to be visited or traveled by the President or Vice President, heads of foreign states, and other public figures. Such conditions may create a hazard to aircraft engaged in air commerce and to persons and property on the ground. In addition, responsible agencies of the United States Government may determine that certain regulatory actions should be taken in the interest of providing protection for these public figures. FAR Part 91.141 provides for the issuance of a regulatory NOTAM to establish flight restrictions where required in such cases.

3-44. PARACHUTE JUMP AIRCRAFT OPERATIONS

a. Procedures relating to parachute jump areas are contained in FAR Part 105. Tabulations of parachute jump areas in the U.S. are contained in the Airport/Facility Directory.

b. Pilots of aircraft engaged in parachute jump operations are reminded that all reported altitudes must be with reference to mean sea level, or flight level, as appropriate, to enable ATC to provide meaningful traffic information.

c. Parachute operations in the vicinity of an airport without an operating control tower—There is no substitute for alertness while in the vicinity of an airport. It is essential that pilots conducting parachute operations be alert, look for other traffic, and exchange traffic information as recommended in paragraph 4-9 *TRAFFIC ADVISORY PRACTICES AT AIRPORTS WITHOUT OPERATING CONTROL TOWERS*. In addition, pilots should avoid releasing parachutes while in an airport traffic pattern when there are other aircraft in that pattern. Pilots should make appropriate broadcasts on the designated Common Traffic Advisory Frequency (CTAF), and monitor that CTAF until all parachute activity has terminated or the aircraft has left the area. Prior to commencing a jump operation, the pilot should broadcast the aircraft's altitude and position in relation to the airport, the approximate relative time when the jump will commence and terminate, and listen to the position reports of other aircraft in the area.

3-45. PUBLISHED VFR ROUTES

Published VFR routes for transitioning around, under and through complex airspace such as Class B airspace were developed through a number of FAA and industry initiatives. All of the following terms, i.e., "VFR Flyway," "VFR Corridor," "Class B Airspace VFR Transition Route" and "Terminal Area VFR Route" have been used when referring to the same or different types of routes or airspace. The following paragraphs identify and clarify the functionality of each type of route, and specify where and when an ATC clearance is required.

a. VFR Flyways

1. VFR Flyways and their associated Flyway Planning charts were developed from the recommendations of a National Airspace Review Task Group. A VFR Flyway is defined as a general flight path not defined as a specific course, for use by pilots in planning flights into, out of, through or near complex terminal airspace to avoid Class B airspace. An ATC clearance is NOT required to fly these routes.

2. VFR Flyways are depicted on the reverse side of some of the VFR Terminal Area Charts (TAC), commonly referred to as Class B airspace charts. (See Figure 3-45[1].) Eventually all TAC's will include a VFR Flyway Planning Chart. These charts identify VFR flyways designed to help VFR pilots avoid major controlled traffic flows. They may

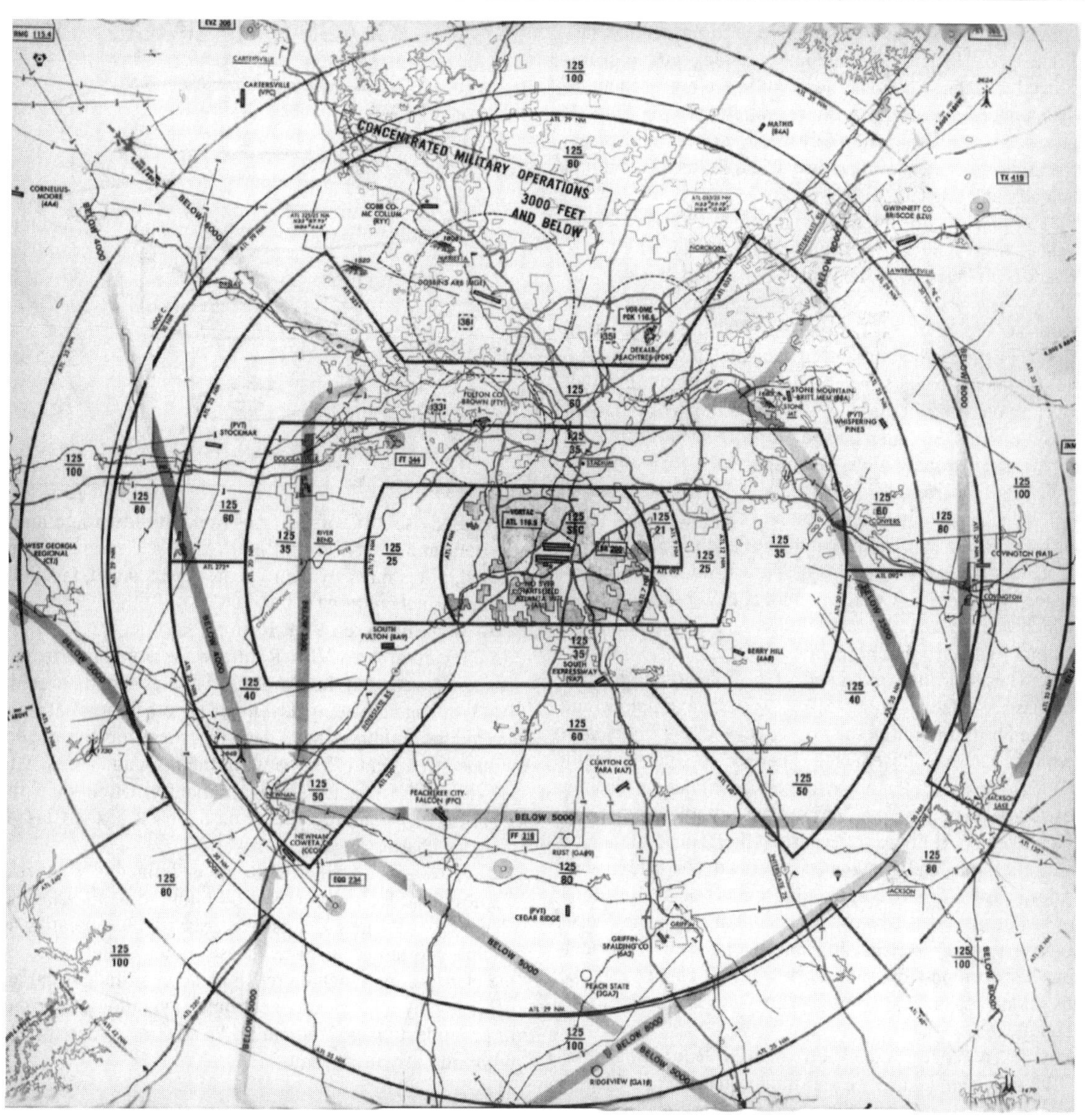
CONCENTRATED MILITARY OPERATIONS 3000 FEET AND BELOW
BELOW 5000
BELOW 4000
BELOW 6000
BELOW 8000

Figure 3-45[1]

further depict multiple VFR routings throughout the area which may be used as an alternative to flight within Class B airspace. The ground references provide a guide for improved visual navigation. These routes are not intended to discourage requests for VFR operations within Class B airspace but are designed solely to assist pilots in planning for flights under and around busy Class B airspace without actually entering Class B airspace.

3. It is very important to remember that these suggested routes are not sterile of other traffic. The entire Class B airspace, and the airspace underneath it, may be heavily congested with many different types of aircraft. Pilot adherence to VFR rules must be exercised at all times. Further, when operating beneath Class B airspace, communications must be established and maintained between your aircraft and any control tower while transiting the Class B, Class C, and Class D surface areas of those airports under Class B Airspace.

b. VFR Corridors

1. The design of a few of the first Class B airspace areas provided a corridor for the passage of uncontrolled traffic. A VFR corridor is defined as Airspace through Class B airspace, with defined vertical and lateral boundaries, in which aircraft may operate without an ATC clearance or communication with air traffic control.

2. These corridors are, in effect, a "hole" through Class B airspace. (See Figure 3-45[2].) A classic example would be the corridor through the Los Angeles Class B airspace, which has been subsequently changed to Special Flight Rules airspace (SFR). A corridor is surrounded on all sides by Class B airspace and does not extend down to the surface like a VFR Flyway. Because of their finite lateral and vertical limits, and the volume of VFR traffic using a corridor, extreme caution and vigilance must be exercised.

3. Because of the heavy traffic volume and the procedures necessary to efficiently manage the flow of traffic, it has not been possible to incorporate VFR corridors in the development or modifications of Class B airspace in recent years.

c. Class B Airspace VFR Transition Routes

1. To accommodate VFR traffic through certain Class B airspace, such as Seattle, Phoenix and Los Angeles, Class B Airspace VFR Transition Routes were developed. A Class B Airspace VFR Transition Route is defined as a specific flight course depicted on a Terminal Area Chart (TAC) for transiting a specific Class B airspace. These routes include specific ATC-assigned altitudes, and pilots must obtain an ATC clearance prior to entering Class B airspace on the route.

2. These routes, as depicted in Figure 3-45[3], are designed to show the pilot where to position his/her aircraft outside of, or clear of, the Class B airspace where an ATC clearance can normally be expected with minimal or no delay. Until ATC authorization is received, pilots must remain clear of Class B airspace. On initial contact, pilots should advise ATC of their position, altitude, route name desired, and direction of flight. After a clearance is received, pilots must fly the route as depicted and, most importantly, adhere to ATC instructions.

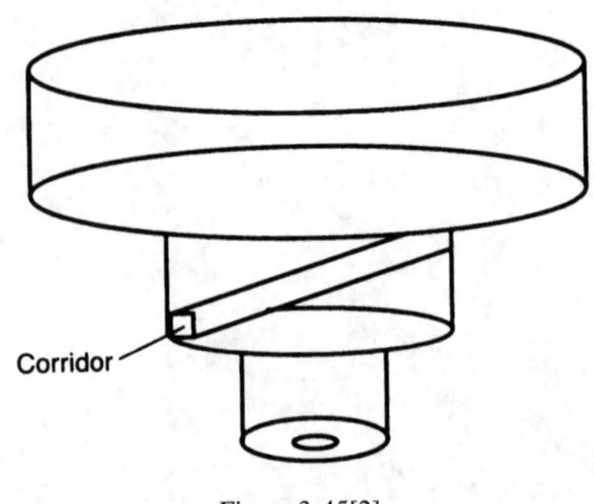

Figure 3-45[2]

d. Terminal Area VFR Routes

1. Terminal Area VFR Routes were developed from a concept evaluated in the Los Angeles Basin area in 1988–89, and are being developed for other terminal areas around the country. Charts depicting these routes were developed in a joint effort between the FAA and industry to provide more specific navigation information than the VFR Flyway Planning Charts on the back of the Class B airspace charts. (See Figure 3-45[4].)

2. A Terminal Area VFR Route is defined as a specific flight course for optional use by pilots to avoid Class B, Class C, and Class D airspace areas while operating in complex terminal airspace. These routes are depicted on the chart(s), may include recommended altitudes, and are described by reference to electronic navigational aids and/or prominent visual landmarks. An ATC clearance is NOT required to fly these routes.

3-46. TERMINAL RADAR SERVICE AREA (TRSA)

a. Background—The terminal radar service areas (TRSA's) were originally established as part of the Terminal Radar Program at selected airports. TRSA's were never controlled airspace from a regulatory standpoint because the establishment of TRSA's were never subject to the rulemaking process; consequently, TRSA's are not contained in FAR Part 71 nor are there any TRSA operating rules in Part 91. Part of the Airport Radar Service Area (ARSA) program was to eventually replace all TRSA'S. However, the ARSA requirements became relatively stringent and it was subsequently decided that TRSA's would have to meet ARSA criteria before they would be converted. TRSA's do

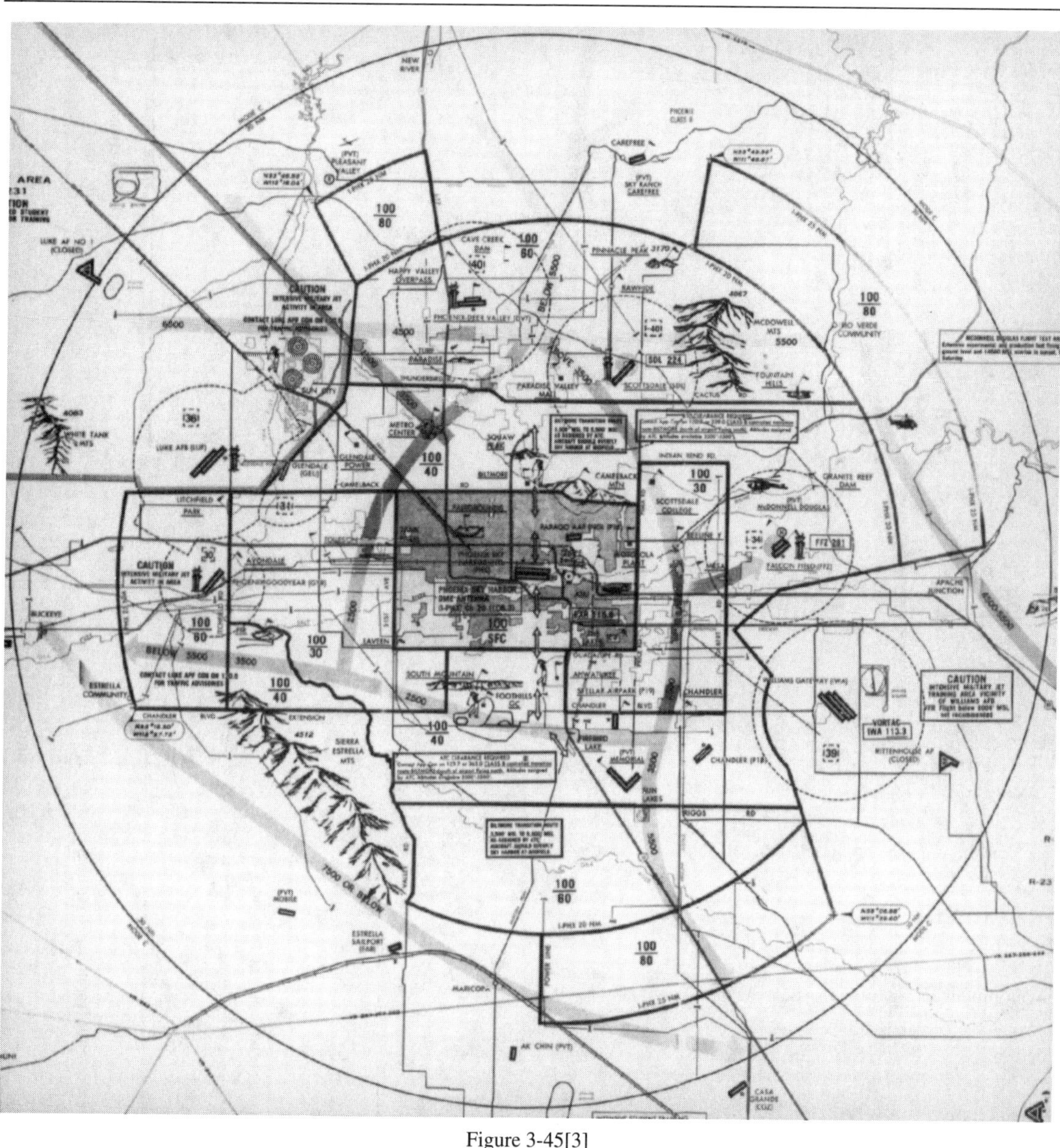
NEW RIVER
PLEASANT VALLEY
100
80
CAVE CREEK DAM
100
60
PINNACLE PEAK
HAPPY VALLEY OVERPASS
CAUTION
4500
6500
RAWHIDE
MCDOWELL MTS
5500
RIO VERDE COMMUNITY
FOUNTAIN HILLS
SCOTTSDALE
SDL 224
PARADISE
PARADISE VALLEY MALL
CACTUS
METRO CENTER
SQUAW PEAK
100
40
GLENDALE POWER
BILTMORE
CAMELBACK MTN
100
30
SCOTTSDALE COLLEGE
GRANITE REEF DAM
MCDONNELL DOUGLAS
LITCHFIELD PARK
AVONDALE
FALCON FIELD (FFZ)
FFZ 281
APACHE JUNCTION
BUCKEYE
100
80
BELOW 5500
3500
100
30
2500
100
SFC
SAVEN
SOUTH MOUNTAIN
FOOTHILLS
STELLAR AIRPARK
CHANDLER
ESTRELLA COMMUNITY
100
40
SIERRA ESTRELLA MTS
4512
100
40
CHANDLER
WILLIAMS GATEWAY
VORTAC
IWA 113.3
RITTENHOUSE AF (CLOSED)
CAUTION
SUN LAKES
RIGGS
100
80
7500 OR BELOW
ESTRELLA SAILPORT
100
80
MARICOPA
AK CHIN
CASA GRANDE

Figure 3-45[3]

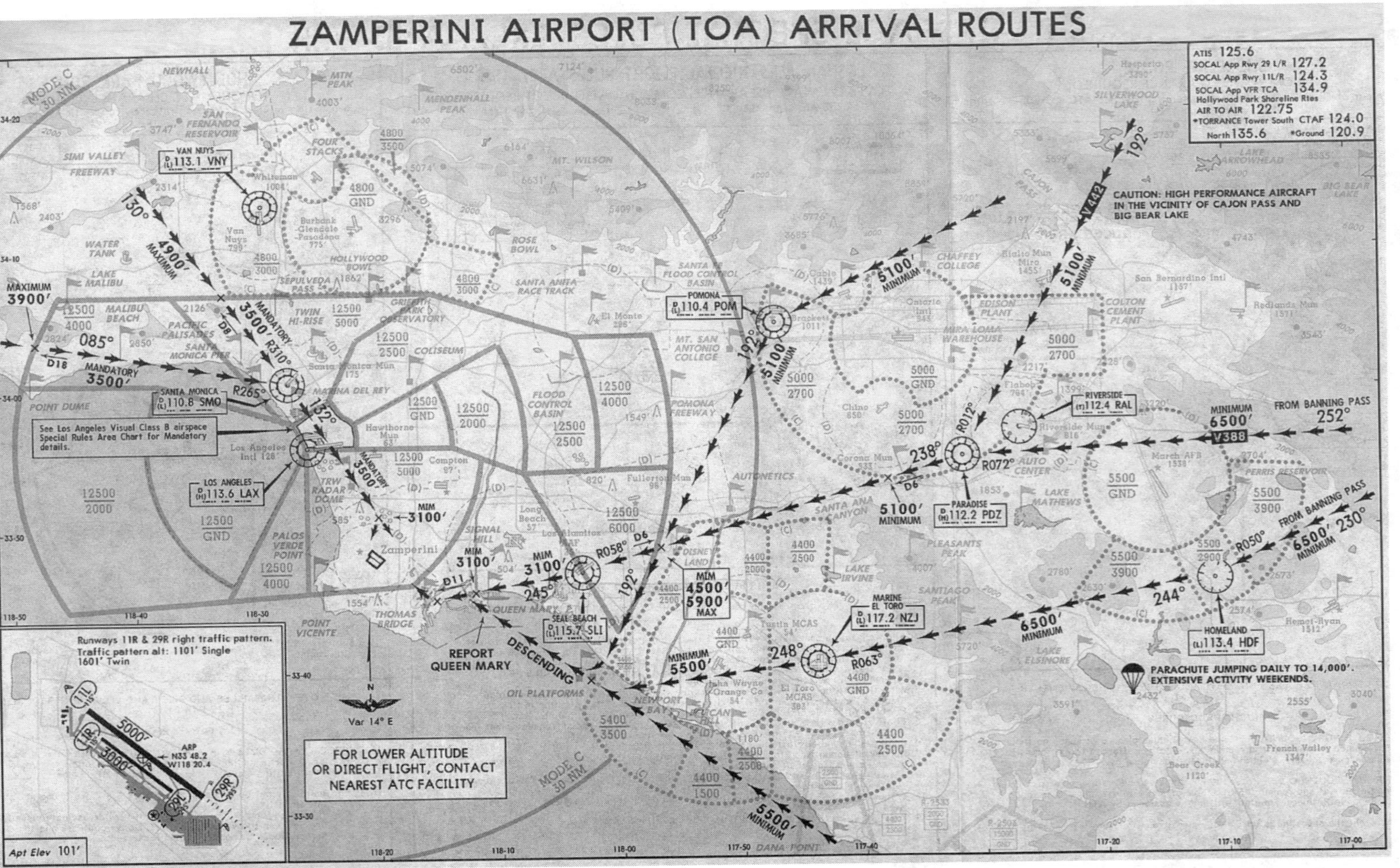

Copyright 1994, Jeppesen Sanderson, Inc.
Excerpted from VFR NavService, Los Angeles Basin VFR Routes,
by permission of Jeppesen Sanderson, Inc.

Reduced for illustrative purposes.
Not for navigation.

Figure 3-45[4]. Terminal Area VFR Routes

not fit into any of the U.S. Airspace Classes; therefore, they will continue to be non-Part 71 airspace areas where participating pilots can receive additional radar services which have been redefined as TRSA Service.

b. TRSA Areas—The primary airport(s) within the TRSA become Class D airspace. The remaining portion of the TRSA overlies other controlled airspace which is normally Class E airspace beginning at 700 or 1,200 feet and established to transition to/from the enroute/terminal environment.

c. Participation—Pilots operating under VFR are encouraged to contact the radar approach control and avail themselves of the TRSA Services. However, participation is voluntary on the part of the pilot. See Chapter 4 for details and procedures.

d. Charts—TRSA's are depicted on visual charts with a solid black line and altitudes for each segment. The Class D portion is charted with a blue segmented line.

3-47. SUSPENSION OF THE TRANSPONDER AND AUTOMATIC ALTITUDE REPORTING EQUIPMENT REQUIREMENTS FOR OPERATIONS IN THE VICINITY OF DESIGNATED MODE C VEIL AIRPORTS

a. Pursuant to Special Federal Aviation Regulation (SFAR) No. 62, aircraft operating in the vicinity of the airports listed below are excluded from the Mode C transponder equipment requirements of Federal Aviation Regulation FAR Part 91.215(b)(2). The exclusion from the Mode C transponder equipment requirement only applies to those operations at or below the altitude specified for each airport that are: (1) within a 2-nautical mile radius of a listed airport: and (2) along a direct route between that airport and the outer boundary of the Mode C veil. The routing must be consistent with established traffic patterns, noise abatement procedures, and safety. The designation of altitudes for each airport is not intended to supersede the provisions of FAR Part 91.119, Minimum Safe Altitudes. Routings to and from each airport are intentionally unspecified to permit the pilot, complying with FAR Part 91.119, to avoid operating over obstructions, noise-sensitive areas, etc. Further, should the pilot of an aircraft intending to operate into or out of an airport listed in the SFAR determine that the operation at or below the specified altitude is unsafe due to meteorological conditions, aircraft operating characteristics, or other factors, the pilot should seek relief from the Mode C transponder requirement via the ATC authorization process.

b. Airports at which the Mode C transponder equipment requirements of FAR Part 91.215(b)(2) do not apply are listed in Tables 3-47[1] thru [24].

1. Atlanta Class B Airspace Mode C veil. (See Table 3-47[1] Atlanta Class B Airspace Airports excluded from the Mode C Transponder Requirement.)

2. Boston Class B Airspace Mode C veil. (See Table 3-47[2] Boston Class B Airspace Airports excluded from the Mode C Transponder Requirement.)

3. Charlotte Class B Airspace Mode C veil. (See Table 3-47[3] Charlotte Class B Airspace Airports excluded from the Mode C Transponder Requirement.)

4. Chicago Class B Airspace Mode C veil. (See Table 3-47[4] Chicago Class B Airspace Airports excluded from the Mode C Transponder Requirement.)

5. Cleveland Class B Airspace Mode C veil. (See Table 3-47[5] Cleveland Class B Airspace Airports excluded from the Mode C Transponder Requirement.)

6. Dallas/Fort Worth Class B Airspace Mode C veil. (See Table 3-47[6] Dallas/Fort Worth Class B Airspace Airports excluded from the Mode C Transponder Requirement.)

7. Denver Class B Airspace Mode C veil. (See Table 3-47[7] Denver Class B Airspace Airports excluded from the Mode C Transponder Requirement.)

8. Detroit Class B Airspace Mode C veil. (See Table 3-47[8] Detroit Class B Airspace Airports excluded from the Mode C Transponder Requirement.)

9. Honolulu Class B Airspace Mode C veil. (See Table 3-47[9] Honolulu Class B Airspace Airports excluded from the Mode C Transponder Requirement.)

10. Houston Class B Airspace Mode C veil. (See Table 3-47[10] Houston Class B Airspace Airports excluded from the Mode C Transponder Requirement.)

11. Kansas City Class B Airspace Mode C veil. (See Table 3-47[11] Kansas City Class B Airspace Airports excluded from the Mode C Transponder Requirement.)

12. Las Vegas Class B Airspace Mode C veil. (See Table 3-47[12] Las Vegas Class B Airspace Airports excluded from the Mode C Transponder Requirement.)

13. Memphis Class B Airspace Mode C veil. (See Table 3-47[13] Memphis Class B Airspace Airports excluded from the Mode C Transponder Requirement.)

14. Minneapolis Class B Airspace Mode C veil. (See Table 3-47[14] Minneapolis Class B Airspace Airports excluded from the Mode C Transponder Requirement.)

15. New Orleans Class B Airspace Mode C veil. (See Table 3-47[15] New Orleans Class B Airspace Airports excluded from the Mode C Transponder Requirement.)

16. New York Class B Airspace Mode C veil. (See Table 3-47[16] New York Class B Airspace Airports excluded from the Mode C Transponder Requirement.)

17. Orlando Class B Airspace Mode C veil. (See Table 3-47[17] Orlando Class B Airspace Airports excluded from the Mode C Transponder Requirement.)

18. Philadelphia Class B Airspacc Mode C veil. (See Table 3-47[18] Philadelphia Class B Airspace Airports excluded from the Mode C Transponder Requirement.)

19. Phoenix Class B Airspace Mode C veil. (See Table 3-47[19] Phoenix Class B Airspace Airports excluded from the Mode C Transponder Requirement.)

20. St. Louis Class B Airspace Mode C veil. (See Table

3-47[20] St. Louis Class B Airspace Airports excluded from the Mode C Transponder Requirement.)

21. Salt Lake City Class B Airspace Mode C veil. (See Table 3-47[21] Salt Lake City Class B Airspace Airports excluded from the Mode C Transponder Requirement.)

22. Seattle Class B Airspace Mode C veil. (See Table 3-47[22] Seattle Class B Airspace Airports excluded from the Mode C Transponder Requirement.)

23. Tampa Class B Airspace Mode C veil. (See Table 3-47[23] Tampa Class B Airspace Airports excluded from the Mode C Transponder Requirement.)

24. Washington Tri-Area Mode C veil. (See Table 3-47[24] Washington Tri-Area Class B Airspace Airports excluded from the Mode C Transponder Requirement.)

(1) Airports within a 30-nautical-mile radius of The William B. Hartsfield Atlanta International Airport.

Table 3-47[1]

Airport Name	Arpt ID	Alt. (AGL)
Air Acres Airport, Woodstock, GA	5GA4	1,500
B & L Strip Airport, Hollonville, GA	GA29	1,500
Camfield Airport, McDonough, GA	GA36	1,500
Cobb County-McCollum Field Airport, Marietta, GA	RYY	1,500
Covington Municipal Airport, Covington, GA	9A1	1,500
Diamond R Ranch Airport, Villa Rica, GA	3GA5	1,500
Dresden Airport, Newnan, GA	GA79	1,500
Eagles Landing Airport Williamson, GA	5GA3	1,500
Fagundes Field Airport, Haralson, GA	6GA1	1,500
Gable Branch Airport, Haralson, GA	5GA0	1,500
Georgia Lite Flite Ultralight Airport, Acworth, GA	31GA	1,500
Griffin-Spalding County Airport, Griffin, GA	6A2	1,500
Howard Private Airport, Jackson, GA	GA02	1,500
Newnan Coweta County Airport, Newnan, GA	CCO	1,500
Peach State Airport, Williamson, GA	3GA7	1,500
Poole Farm Airport, Oxford, GA	2GA1	1,500
Powers Airport, Hollonville, GA	GA31	1,500
S & S Landing Strip Airport, Griffin, GA	8GA6	1,500
Shade Tree Airport, Hollonville, GA	GA73	1,500

(2) Airports within a 30-nautical-mile radius of the General Edward Lawrence Logan International Airport.

Table 3-47[2]

Airport Name	Arpt ID	Alt. (AGL)
Berlin Landing Area Airport, Berlin, MA	MA19	2,500
Hopedale Industrial Park Airport, Hopedale, MA	IB6	2,500
Larson's SPB, Tyngsboro, MA	MA74	2,500
Moore AAF, Ayer/Fort Devens, MA	AYE	2,500
New England Gliderport, Salem, NH	NH29	2,500
Plum Island Airport, Newburyport, MA	2B2	2,500
Plymouth Municipal Airport, Plymouth, MA	PYM	2,500
Taunton Municipal Airport, Taunton, MA	TAN	2,500
Unknown Field Airport, Southborough, MA	1MA5	2,500

(3) Airports within a 30-nautical-mile radius of the Charlotte/Douglas International Airport.

Table 3-47[3]

Airport Name	Arpt ID	Alt. (AGL)
Arant Airport, Wingate, NC	1NC6	2,500
Bradley Outernational Airport, China Grove, NC	NC29	2,500
Chester Municipal Airport, Chester, SC	9A6	2,500
China Grove Airport, China Grove, NC	76A	2,500
Goodnight's Airport, Kannapolis, NC	2NC8	2,500
Knapp Airport, Marshville, NC	3NC4	2,500
Lake Norman Airport, Mooresville, NC	14A	2,500
Lancaster County Airport, Lancaster, SC	LKR	2,500
Little Mountain Airport, Denver, NC	66A	2,500
Long Island Airport, Long Island, NC	NC26	2,500
Miller Airport, Mooresville, NC	8A2	2,500
U S Heliport, Wingate, NC	NC56	2,500
Unity Aerodrome Airport, Lancaster, SC	SC76	2,500
Wilhelm Airport, Kannapolis, NC	6NC2	2,500

(4) Airports within a 30-nautical-mile radius of the Chicago O'Hare International Airport.

Table 3-47[4]

Airport Name	Arpt ID	Alt. (AGL)
Aurora Municipal Airport, Chicago/Aurora, IL	ARR	1,200
Donald Alfred Gade Airport, Antioch, IL	IL11	1,200
Dr. Joseph W. Esser Airport, Hampshire, IL	7IL6	1,200
Flying M. Farm Airport, Aurora, IL	IL20	1,200
Fox Lake SPB, Fox Lake, IL	IS03	1,200
Graham SPB, Crystal Lake, IL	IS79	1,200
Herbert C. Mass Airport, Zion, IL	IL02	1,200
Landings Condominium Airport, Romeoville, IL	C49	1,200
Lewis University Airport, Romeoville, IL	LOT	1,200
McHenry Farms Airport, McHenry, IL	44IL	1,200
Olson Airport, Plato Center, IL	LL53	1,200
Redeker Airport, Milford, IL	IL85	1,200
Reid RLA Airport, Gilberts, IL	6IL6	1,200
Shamrock Beef Cattle Farm Airport, McHenry, IL	49LL	1,200
Sky Soaring Airport, Union, IL	55LL	1,200
Waukegan Regional Airport, Waukegan, IL	UGN	1,200
Wormley Airport, Oswego, IL	85LL	1,200

(5) Airports within a 30-nautical-mile radius of the Cleveland-Hopkins International Airport.

Table 3-47[5]

Airport Name	Arpt ID	Alt. (AGL)
Akron Fulton International Airport, Akron, OH	AKR	1,300
Bucks Airport, Newbury, OH	400H	1,300
Derecsky Airport, Auburn Center, OH	60I0	1,300
Hannum Airport, Streetsboro, OH	690H	1,300
Kent State University Airport, Kent, OH	1G3	1,300
Lost Nation Airport, Willoughby, OH	LNN	1,300
Mills Airport, Mantua, OH	OH06	1,300
Portage County Airport, Ravenna, OH	29G	1,300
Stoney's Airport, Ravenna, OH	OI32	1,300
Wadsworth Municipal Airport, Wadsworth, OH	3G3	1,300

(6) Airports within a 30-nautical-mile radius of the Dallas/Fort Worth International Airport.

Table 3-47[6]

Airport Name	Arpt ID	Alt. (AGL)
Beggs Ranch/Aledo Airport, Aledo, TX	TX15	1,800
Belcher Airport, Sanger, TX	TA25	1,800
Bird Dog Field Airport, Krum, TX	TA48	1,800
Boe-Wrinkle Airport, Azle, TX	28TS	1,800
Flying V Airport, Sanger, TX	71XS	1,800
Graham Ranch Airport, Celina, TX	TX44	1,800
Haire Airport, Bolivar, TX	TX33	1,800
Hartlee Field Airport, Denton, TX	1F3	1,800
Hawkin's Ranch Strip Airport, Rhome, TX	TA02	1,800
Horseshoe Lake Airport, Sanger, TX	TE24	1,800
Ironhead Airport, Sanger, TX	T58	1,800
Kezer Air Ranch Airport, Springtown, TX	61F	1,800
Lane Field Airport, Sanger, TX	58F	1,800
Log Cabin Airport, Aledo, TX	TX16	1,800
Lone Star Airpark Airport, Denton, TX	T32	1,800
Rhome Meadows Airport, Rhome, TX	TS72	1,800
Richards Airport, Krum, TX	TA47	1,800
Tallows Field Airport, Celina, TX	79TS	1,800
Triple S Airport, Aledo, TX	42XS	1,800
Warshun Ranch Airport, Denton, TX	4TA1	1,800
Windy Hill Airport, Denton, TX	46XS	1,800
Aero Country Airport, McKinney, TX	TX05	1,400

Table 3-47[6] ***Continued***

Bailey Airport, Midlothian, TX	7TX8	1,400
Bransom Farnt Airport, Burleson, TX	TX42	1,400
Carroll Air Park Airport, De Soto, TX	F66	1,400
Carroll Lake-View Airport, Venus, TX	70TS	1,400
Eagle's Nest Estates Airport, Ovilla, TX	2T36	1,400
Flying B Ranch Airport, Ovilla, TX	TS71	1,400
Lancaster Airport, Lancaster, TX	LNC	1,400
Lewis Farm Airport, Lucas, TX	6TX1	1,400
Markum Ranch Airport, Fort Worth, TX	TX79	1,400
McKinney Municipal Airport, McKinney, TX	TK1	1,400
O'Brien Airpark Airport, Waxahachie, TX	F25	1,400
Phil L. Hudson Municipal Airport, Mesquite, TX	HQZ	1,400
Plover Heliport, Crowley, TX	82Q	1,400
Venus Airport, Venus, TX	75TS	1,400

(7) Airports within a 30-nautical-mile radius of the Stapleton International Airport.

Table 3-47[7]

Airport Name	Arpt ID	Alt. (AGL)
Athanasiou Valley Airport, Blackhawk, CO	C007	1,200
Boulder Municipal Airport, Boulder, CO	1V5	1,200
Bowen Farms No. 2 Airport, Strasburg, CO	3C05	1,200
Carrera Airpark Airport, Mead, CO	93CO	1,200
Cartwheel Airport, Mead, CO	OC08	1,200
Colorado Antique Field Airport, Niwot, CO	8C07	1,200
Comanche Airfield Airport, Strasburg, CO	3C06	1,200
Comanche Livestock Airport, Strasburg, CO	59CO	1,200
Flying J Ranch Airport, Evergreen, CO	27CO	1,200
Frederick-Firestone Air Strip Airport, Frederick, CO	C058	1,200]
Frontier Airstrip Airport, Mead, CO	84CO	1,200
Hoy Airstrip Airport, Bennett, CO	76CO	1,200

Table 3-47[7] ***Continued***

J & S Airport, Bennett, CO	CD14	1,200
Kugel-Strong Airport, Platteville, CO	27V	1,200
Land Airport, Keenesburg, CO	C082	1,200
Lindys Airpark Airport, Hudson, CO	7C03	1,200
Marshdale STOL, Evergreen, CO	C052	1,200
Meyer Ranch Airport, Conifer, CO	5C06	1,200
Parkland Airport, Erie, CO	7C00	1,200
Pine View Airport, Elizabeth, CO	02V	1,200
Platte Valley Airport, Hudson, CO	18V	1,200
Rancho De Aereo Airport, Mead, CO	05CO	1,200
Spickard Farm Airport, Byers, CO	5CO4	1,200
Vance Brand Airport, Longmont, CO	2V2	1,200
Yoder Airstrip Airport, Bennett, CO	CD09	1,200

(8) Airports within a 30-nautical-mile radius of the Detroit Metropolitan Wayne County Airport.

Table 3-47[8]

Airport Name	Arpt ID	Alt. (AGL)
Al Meyers Airport, Tecumseh, MI	3TE	1,400
Brighton Airport, Brighton, MI	45G	1,400
Cackleberry Airport, Dexter, MI	2MI9	1,400
Erie Aerodome Airport, Erie, MI	05MI	1,400
Ham-A-Lot Field Airport, Petersburg, MI	MI48	1,400
Merillat Airport, Tecumseh, MI	34G	1,400
Rossettie Airport, Manchester, MI	75G	1,400
Tecumseh Products Airport, Tecumseh, MI	0D2	1,400

(9) Airports within a 30-nautical-mile radius of the Honolulu International Airport.

Table 3-47[9]

Airport Name	Arpt ID	Alt. (AGL)
Dillingham Airfield Airport, Mokuleia, HI	HDH	2,500

(10) Airports within a 30-nautical-mile radius of the Houston Intercontinental Airport.

Table 3-47[10]

Airport Name	**Arpt ID**	**Alt. (AGL)**
Ainsworth Airport, Cleveland, TX	0T6	1,200
Biggin Hill Airport, Hockley, TX	0TA3	1,200
Cleveland Municipal Airport, Cleveland, TX	6R3	1,200
Fay Ranch Airport, Cedar Lane, TX	0T2	1,200
Freeman Property Airport, Katy, TX	61T	1,200
Gum Island Airport, Dayton, TX	3T6	1,200
Harbican Airpark Airport, Katy, TX	9XS9	1,200
Harold Freeman Farm Airport, Katy, TX	8XS1	1,200
Hoffpauir Airport, Katy, TX	59T	1,200
Horn-Katy Hawk International Airport, Katy, TX	57T	1,200
Houston-Hull Airport, Houston, TX	SGR	1,200
Houston-Southwest Airport, Houston, TX	AXH	1,200
King Air Airport, Katy, TX	55T	1,200
Lake Bay Gall Airport, Cleveland, TX	0T5	1,200
Lake Bonanza Airport, Montgomery, TX	33TA	1,200
R W J Airpark Airport, Baytown, TX	54TX	1,200
Westheimer Air Park Airport, Houston, TX	5TA4	1,200

(11) Airports within a 30-nautical-mile radius of the Kansas City International Airport.

Table 3-47[11]

Airport Name	**Arpt ID**	**Alt. (AGL)**
Amelia Earhart Airport, Atchison, KS	K59	1,000
Booze Island Airport, St. Joseph, MO	64MO	1,000
Cedar Air Park Airport, Olathe, KS	51K	1,000
D'Field Airport, McLouth, KS	KS90	1,000
Dorei Airport, McLouth, KS	K69	1,000
East Kansas City Airport, Grain Valley, MO	3GV	1,000
Excelsior Springs Memorial Airport, Excelsior Springs, MO	3EX	1,000
Flying T Airport, Oskaloosa, KS	7KS0	1,000
Hermon Farm Airport, Gardner, KS	KS59	1,000
Hillside Airport, Stilwell, KS	63K	1,000
Independence Memorial Airport, Independence, MO	3IP	1,000
Johnson County Executive Airport, Olathe, KS	OJC	1,000
Johnson County Industrial Airport, Olathe, KS	IXD	1,000
Kimray Airport, Plattsburg, MO	7M07	1,000
Lawrence Municipal Airport, Lawrence, KS	LWC	1,000
Martins Airport, Lawson, MO	21MO	1,000
Mayes Homestead Airport, Polo, MO	37MO	1,000
McComas-Lee's Summit Municipal Airport, Lee's Summit, MO	K84	1,000
Mission Road Airport, Stilwell, KS	64K	1,000
Northwood Airport, Holt, MO	2M02	1,000
Plattsburg Airpark Airport, Plattsburg, MO	M028	1,000
Richards-Gebaur Airport, Kansas City, MO	GVW	1,000
Rosecrans Memorial Airport, St. Joseph, MO	STJ	1,000
Runway Ranch Airport, Kansas City, MO	2MO9	1,000
Sheller's Airport, Tonganoxie, KS	11KS	1,000
Shomin Airport, Oskaloosa, KS	0KS1	1,000
Stonehenge Airport, Williamstown, KS	71KS	1,000
Threshing Bee Airport, McLouth, KS	41K	1,000

(12) Airports within a 30-nautical-mile radius of the McCarran International Airport.

Table 3-47[12]

Airport Name	**Arpt ID**	**Alt. (AGL)**
Sky Ranch Estates Airport, Sandy Valley, NV	3L2	2,500

(13) Airports within a 30-nautical-mile radius of the Memphis International Airport.

Table 3-47[13]

Airport Name	Arpt ID	Alt. (AGL)
Bernard Manor Airport, Earle, AR	65M	2,500
Holly Springs-Marshall County Airport, Holly Springs, MS	M41	2,500
McNeely Airport, Earle, AR	M63	2,500
Price Field Airport, Joiner, AR	8OM	2,500
Tucker Field Airport, Hughes, AR	78M	2,500
Tunica Airport, Tunica, MS	30M	2,500
Tunica Municipal Airport, Tunica, MS	M97	2,500

(14) Airports within a 30-nautical-mile radius of the Minneapolis-St. Paul International Wold-Chamberlain Airport.

Table 3-47[14]

Airport Name	Arpt ID	Alt. (AGL)
Belle Plaine Airport, Belle Plaine, MN	7Y7	1,200
Carleton Airport, Stanton, MN	SYN	1,200
Empire Farm Strip Airport, Bongards, MN	MN15	1,200
Flying M Ranch Airport, Roberts, WI	78WI	1,200
Johnson Airport, Rockford, MN	MY86	1,200
River Falls Airport, River Falls, WI	Y53	1,200
Rusmar Farms Airport, Roberts, WI	WS41	1,200
Waldref SPB, Forest Lake, MN	9Y6	1,200
Ziermann Airport, Mayer, MN	MN71	1,200

(15) Airports within 30-nautical-mile radius of the New Orleans International/Moisant Field Airport.

Table 3-47[15]

Airport Name	Arpt ID	Alt. (AGL)
Bollinger SPB, Larose, LA	L38	1,500
Clovelly Airport, Cut Off, LA	LA09	1,500

(16) Airports within a 30-nautical-mile radius of the John F. Kennedy International Airport, the La Guardia Airport, and the Newark International Airport.

Table 3-47[16]

Airport Name	Arpt ID	Alt. (AGL)
Allaire Airport, Belmar/Farmingdale, NJ	BLM	2,000
Cuddihy Landing Strip Airport, Freehold, NJ	NJ60	2,000
Ekdahl Airport, Freehold, NJ	NJ59	2,000
Fla-Net Airport, Netcong, NJ	0NJ5	2,000
Forrestal Airport, Princeton, NJ	N21	2,000
Greenwood Lake Airport, West Milford, NJ	4N1	2,000
Greenwood Lake SPB, West Milford, NJ	6NJ7	2,000
Lance Airport, Whitehouse Station, NJ	6NJ8	2,000
Mar Bar L Farms, Englishtown, NJ	NJ46	2,000
Peekskill SPB, Peekskill, NY	7N2	2,000
Peters Airport, Somerville, NJ	4NJ8	2,000
Princeton Airport, Princeton/Rocky Hill, NJ	39N	2,000
Solberg-Hunterdon Airport, Readington, NJ	N51	2,000

(17) Airports within a 30-nautical-mile radius of the Orlando International Airport.

Table 3-47[17]

Airport Name	Arpt ID	Alt. (AGL)
Arthur Dunn Air Park Airport, Titusville, FL	X21	1,400
Space Center Executive Airport, Titusville, FL	TIX	1,400

(18) Airports within a 30-nautical-mile radius of the Philadelphia International Airport.

Table 3-47[18]

Airport Name	Arpt ID	Alt. (AGL)
Ginns Airport, West Grove, PA	78N	1,000
Hammonton Municipal Airport, Hammonton, NJ	N81	1,000
Li Calzi Airport, Bridgeton, NJ	N50	1,000
New London Airport, New London, PA	N01	1,000
Wide Sky Airpark Airport, Bridgeton, NJ	N39	1,000

(19) Airports within a 30-nautical-mile radius of the Phoenix Sky Harbor International Airport.

Table 3-47[19]

Airport Name	Arpt ID	Alt. (AGL)
Ak Chin Community Airfield Airport, Maricopa, AZ	E31	2,500
Boulais Ranch Airport, Maricopa, AZ	9E7	2,500
Estrella Sailport, Maricopa, AZ	E68	2,500
Hidden Valley Ranch Airport, Maricopa, AZ	AZ17	2,500
Millar Airport, Maricopa, AZ	2AZ4	2,500
Pleasant Valley Airport, New River, AZ	AZ05	2,500
Serene Field Airport, Maricopa, AZ	AZ31	2,500
Sky Ranch Carefree Airport, Carefree, AZ	E18	2,500
Sycamore Creek Airport, Fountain Hills, AZ	0AS0	2,500
University of Arizona, Maricopa Agricultural Center Airport, Maricopa, AZ	3AZ2	2,500

(20) Airports within a 30-nautical-mile radius of the Lambert/St. Louis International Airport.

Table 3-47[20]

Airport Name	Arpt ID	Alt. (AGL)
Blackhawk Airport, Old Monroe, MO	6M00	1,000
Lebert Flying L Airport, Lebanon, IL	3H5	1,000
Shafer Metro East Airport, St. Jacob, IL	3K6	1,000
Sloan's Airport, Elsberry, MO	OM08	1,000
Wentzville Airport, Wentzville, MO	M050	1,000
Woodliff Airpark Airport, Foristell, MO	98MO	1,000

(21) Airports within a 30-nautical-mile radius of the Salt Lake City International Airport.

Table 3-47[21]

Airport Name	Arpt ID	Alt. (AGL)
Bolinder Field-Tooele Valley Airport, Tooele, UT	TVY	2,500
Cedar Valley Airport, Cedar Fort, UT	UT10	2,500
Morgan County Airport, Morgan, UT	42U	2,500
Tooele Municipal Airport, Tooele, UT	U26	2,500

(22) Airports within a 30-nautical-mile radius of the Seattle Tacoma International Airport.

Table 3-47[22]

Airport Name	Arpt ID	Alt. (AGL)
Firstair Field Airport, Monroe, WA	WA38	1,500
Gower Field Airport, Olympia, WA	6WAZ	1,500
Harvey Field Airport, Snohomish, WA	S43	1,500

(23) Airports within a 30-nautical-mile radius of the Tampa In-ternational Airport.

Table 3-47[23]

Airport Name	Arpt ID	Alt. (AGL)
Hernando County Airport, Brooksville, FL	BKV	1,500
Lakeland Municipal Airport, Lakeland, FL	LAL	1,500
Zephyrhills Municipal Airport, Zephyrhills, FL	ZPH	1,500

(24) Airports within a 30-nautical-mile radius of the Washington National Airport, Andrews Air Force Base Airport, Baltimore Washington International Airport, and Dulles International Airport.

Table 3-47[24]

Airport Name	Arpt ID	Alt. (AGL)
Albrecht Airstrip Airport, Long Green, MD	MD48	2,000
Armacost Farms Airport, Hampstead, MD	MD38	2,000
Barnes Airport, Lisbon, MD	MD47	2,000
Bay Bridge Airport, Stevensville, MD	W29	2,000
Carroll County Airport, Westminster, MD	W54	2,000
Castle Marina Airport, Chester, MD	OW6	2,000
Clearview Airpark Airport, Westminster, MD	2W2	2,000
Davis Airport, Laytonsville, MD	W50	2,000
Fallston Airport, Fallston, MD	W42	2,000
Faux-Burhans Airport, Frederick, MD	3MD0	2,000
Forest Hill Airport, Forest Hill, MD	MD31	2,000

Table 3-47[24] *Continued*

Airport Name	Arpt ID	Alt. (AGL)
Fort Detrick Helipad Heliport, Fort Detrick (Frederick), MD	MD32	2,000
Frederick Municipal Airport, Frederick, MD	FDK	2,000
Fremont Airport, Kemptown, MD	MD41	2,000
Good Neighbor Farm Airport, Unionville, MD	MD74	2,000
Happy Landings Farm Airport, Unionville, MD	MD73	2,000
Harris Airport, Still Pond, MD	MD69	2,000
Hybarc Farm Airport, Chestertown, MD	MD19	2,000
Kennersley Airport, Church Hill, MD	MD23	2,000
Kentmorr Airpark Airport, Stevensville, MD	3W3	2,000
Montgomery County Airpark Airport, Gaithersburg, MD	GAI	2,000
Phillips AAF, Aberdeen, MD	APG	2,000
Pond View Private Airport, Chestertown, MD	0MD4	2,000
Reservoir Airport, Finksburg, MD	1W8	2,000
Scheeler Field Airport, Chestertown, MD	0W7	2,000
Stolcrest STOL, Urbana, MD	MD75	2,000
Tinsley Airstrip Airport, Butler, MD	MD17	2,000
Walters Airport, Mount Airy, MD	0MD6	2,000
Waredaca Farm Airport, Brookeville, MD	MD16	2,000
Weide AAF, Edgewood Arsenal, MD	EDG	2,000
Woodbine Gliderport, Woodbine, MD	MD78	2,000
Wright Field Airport, Chestertown, MD	MD11	2,000
Aviacres Airport, Warrenton, VA	3VA2	1,500
Birch Hollow Airport, Hillsboro, VA	W60	1,500
Flying Circus Aerodrome Airport, Warrenton, VA	3VA3	1,500
Fox Acres Airport, Warrenton, VA	15VA	1,500
Hartwood Airport, Somerville, VA	8W8	1,500
Horse Feathers Airport, Midland, VA	53VA	1,500
Krens Farm Airport, Hillsboro, VA	14VA	1,500
Scott Airpark Airport, Lovettsville, VA	VA61	1,500
The Grass Patch Airport, Lovettsville, VA	VA62	1,500
Walnut Hill Airport, Calverton, VA	58VA	1,500
Warrenton Air Park Airport, Warrenton, VA	9W0	1,500
Warrenton-Fauquier Airport, Warrenton, VA	W66	1,500
Whitman Strip Airport, Manassas, VA	0V5	1,500
Aqua-Land/Cliffton Skypark Airport, Newburg, MD	2W8	1,000
Buds Ferry Airport, Indian Head, MD	MD39	1,000
Burgess Field Airport, Riverside, MD	3W1	1,000
Chimney View Airport, Fredericksburg, VA	5VA5	1,000
Holly Springs Farni Airport, Nanjemoy, MD	MD55	1,000
Lanseair Farms Airport, La Plata, MD	MD97	1,000
Nyce Airport, Mount Victoria, MD	MD84	1,000
Parks Airpark Airport, Nanjemoy, MD	MD54	1,000
Pilots Cue Airport, Tompkinsville, MD	MD06	1,000
Quantico MCAF, Quantico, VA	NYG	1,000
Stewart Airport, St. Michaels, MD	MD64	1,000
U.S. Naval Weapons Center, Dahlgren Lab Airport, Dahlgren, VA	NDY	1,000

Chapter 4. Air traffic control

Section 1. SERVICES AVAILABLE TO PILOTS

4-1. AIR ROUTE TRAFFIC CONTROL CENTERS

Centers are established primarily to provide Air Traffic Service to aircraft operating on IFR flight plans within controlled airspace, and principally during the en route phase of flight.

4-2. CONTROL TOWERS

Towers have been established to provide for a safe, orderly and expeditious flow of traffic on and in the vicinity of an airport. When the responsibility has been so delegated, towers also provide for the separation of IFR aircraft in the terminal areas (Reference—Approach Control, paragraph 5-42).

4-3. FLIGHT SERVICE STATIONS

a. Flight Service Stations (FSS) are air traffic facilities which provide pilot briefings, en route communications and VFR search and rescue services, assist lost aircraft and aircraft in emergency situations, relay ATC clearances, originate Notices to Airmen, broadcast aviation weather and National Airspace System (NAS) information, receive and process IFR flight plans, and monitor NAVAIDS. In addition, at selected locations FSSs provide En Route Flight Advisory Service (Flight Watch), take weather observations, issue airport advisories, and advise Customs and Immigration of transborder flights.

b. Supplemental Weather Service Locations (SWSLs) are airport facilities staffed with contract personnel who take weather observations and provide current local weather to pilots via telephone or radio. All other services are provided by the parent FSS.

4-4. RECORDING AND MONITORING

a. Calls to air traffic control (ATC) facilities (ARTCCs, Towers, FSSs, Central Flow, and Operations Centers) over radio and ATC operational telephone lines (lines used for operational purposes such as controller instructions, briefings, opening and closing flight plans, issuance of IFR clearances and amendments, counter hijacking activities, etc.) may be monitored and recorded for operational uses such as accident investigations, accident prevention, search and rescue purposes, specialist training and evaluation, and technical evaluation and repair of control and communications systems.

b. Where the public access telephone is recorded, a beeper tone is not required. In place of the "beep" tone the FCC has substituted a mandatory requirement that persons to be recorded be given notice they are to be recorded and give consent. Notice is given by this entry, consent to record is assumed by the individual placing a call to the operational facility.

4-5. COMMUNICATIONS RELEASE OF IFR AIRCRAFT LANDING AT AN AIRPORT NOT BEING SERVED BY AN OPERATING TOWER

Aircraft operating on an IFR flight plan, landing at an airport not being served by a tower will be advised to change to the airport advisory frequency when direct communications with ATC is no longer required. Towers and centers do not have nontower airport traffic and runway in use information. The instrument approach may not be aligned with the runway in use; therefore, if the information has not already been obtained, pilots should make an expeditious change to the airport advisory frequency when authorized. (Reference—Advance Information on Instrument Approach, paragraph 5-43.)

4-6. PILOT VISITS TO AIR TRAFFIC FACILITIES

Pilots are encouraged to visit air traffic facilities (Towers, Centers and FSSs) and familiarize themselves with the ATC system. On rare occasions, facilities may not be able to approve a visit because of ATC workload or other reasons. It is therefore requested that pilots contact the facility prior to the visit and advise of the number of persons in the group, the time and date of the proposed visit and the primary interest of the group. With this information available, the facility can prepare an itinerary and have someone available to guide the group through the facility.

4-7. OPERATION TAKE-OFF AND OPERATION RAINCHECK

Operation Take-off is a program that educates pilots in how best to utilize the FSS modernization efforts and services available in Automated Flight Service Stations (AFSS), as stated in FAA Order 7230.17. Operation Raincheck is a program designed to familiarize pilots with the ATC system, its functions, responsibilities and benefits.

4-8. APPROACH CONTROL SERVICE FOR VFR ARRIVING AIRCRAFT

a. Numerous approach control facilities have established programs for arriving VFR aircraft to contact approach control for landing information. This information includes: wind, runway, and altimeter setting at the airport of intended landing. This information may be omitted if contained in the ATIS broadcast and the pilot states the appropriate ATIS code.

4-8a NOTE—Pilot use of "Have Numbers" does not indicate receipt of the ATIS broadcast. In addition, the controller will provide traffic advisories on a workload permitting basis.

b. Such information will be furnished upon initial contact with concerned approach control facility. The pilot will be requested to change to the *tower* frequency at a predetermined time or point, to receive further landing information.

c. Where available, use of this procedure will not hinder the operation of VFR flights by requiring excessive spacing between aircraft or devious routing.

d. Compliance with this procedure is not mandatory but pilot participation is encouraged. (Reference—Terminal Radar Services for VFR Aircraft, paragraph 4-17.)

4-8d NOTE—Approach control services for VFR aircraft are normally dependent on air traffic control radar. These services are not available during periods of a radar outage. Approach control services for VFR aircraft are limited when CENRAP is in use.

4-9. TRAFFIC ADVISORY PRACTICES AT AIRPORTS WITHOUT OPERATING CONTROL TOWERS

(See Table 4-9[1].)

a. Airport Operations Without Operating Control Tower

1. There is no substitute for alertness while in the vicinity of an airport. It is essential that pilots be alert and look for other traffic and exchange traffic information when approaching or departing an airport without an operating control tower. This is of particular importance since other aircraft may not have communication capability or, in some cases, pilots may not communicate their presence or intentions when operating into or out of such airports. To achieve the greatest degree of safety, it is essential that all radio-equipped aircraft transmit/receive on a common frequency identified for the purpose of airport advisories.

2. An airport may have a full- or part-time tower or Flight Service Station (FSS) located on the airport, a full-

Table 4-9[1]
SUMMARY OF RECOMMENDED COMMUNICATION PROCEDURES

COMMUNICATION/BROADCAST PROCEDURES

	FACILITY AT AIRPORT	FREQUENCY USE	OUTBOUND	INBOUND	PRACTICE INSTRUMENT APPROACH
1.	UNICOM (No Tower or FSS)	Communicate with UNICOM station on published CTAF frequency (122.7, 122.8, 122.725, 122.975, or 123.0). If unable to contact UNICOM station, use self-announce procedures on CTAF.	Before taxiing and before taxiing on the runway for departure.	10 miles out. Entering downwind, base, and final. Leaving the runway.	
2.	No Tower, FSS, or UNICOM	Self-announce on MULTICOM frequency 122.9.	Before taxiing and before taxiing on the runway for departure.	10 miles out. Entering downwind, base, and final. Leaving the runway.	Departing final approach fix (name) or on final approach segment inbound.
3.	No Tower in operation, FSS open	Communicate with FSS on CTAF frequency.	Before taxiing and before taxiing on the runway for departure.	10 miles out. Entering downwind, base, and final. Leaving the runway.	Approach completed/ terminated.
4.	FSS closed (No Tower)	Self-announce on CTAF.	Before taxiing and before taxiing on the runway for departure.	10 miles out. Entering downwind, base, and final. Leaving the runway.	
5.	Tower or FSS not in operation	Self-announce on CTAF.	Before taxiing and before taxiing on the runway for departure.	10 miles out. Entering downwind, base, and final Leaving the runway.	

or part-time UNICOM station or no aeronautical station at all. There are three ways for pilots to communicate their intention and obtain airport/traffic information when operating at an airport that does not have an operating tower: by communicating with an FSS, a UNICOM operator, or by making a self-announce broadcast.

b. Communicating on a Common Frequency.

1. The key to communicating at an airport without an operating control tower is selection of the correct common frequency. The acronym *CTAF* which stands for Common Traffic Advisory Frequency, is synonymous with this program. A CTAF is a frequency designated for the purpose of carrying out airport advisory practices while operating to or from an airport without an operating control tower. The CTAF may be a UNICOM, MULTICOM, FSS, or tower frequency and is identified in appropriate aeronautical publications.

2. The CTAF frequency for a particular airport is contained in the Airport/Facility Directory (A/FD), Alaska Supplement, Alaska Terminal Publication, Instrument Approach Procedure Charts, and Standard Instrument Departure (SID) charts. Also, the CTAF frequency can be obtained by contacting any FSS. Use of the appropriate CTAF, combined with a visual alertness and application of the following recommended good operating practices, will enhance safety of flight into and out of all uncontrolled airports.

c. Recommended Traffic Advisory Practices.

1. Pilots of inbound traffic should monitor and communicate as appropriate on the designated CTAF from 10 miles to landing. Pilots of departing aircraft should monitor/communicate on the appropriate frequency from start-up, during taxi, and until 10 miles from the airport unless the FARs or local procedures require otherwise.

2. Pilots of aircraft conducting other than arriving or departing operations at altitudes normally used by arriving and departing aircraft should monitor/communicate on the appropriate frequency while within 10 miles of the airport unless required to do otherwise by the FAR's or local procedures. Such operations include parachute jumping/dropping (Reference—Parachute Jump Aircraft Operations, paragraph 3-44, enroute, practicing maneuvers, etc.).

d. Local Airport Advisory provided by an FSS.

1. Local Airport Advisory (LAA) is a service provided by an FSS physically located on an airport which does not have a control tower or where the tower is operated on a part-time basis. The CTAF for FSSs which provide this service will be disseminated in appropriate aeronautical publications.

2. In communicating with a CTAF FSS, establish two-way communications before transmitting outbound/inbound intentions or information. An inbound aircraft should report approximately 10 miles from the airport, reporting altitude and aircraft type, location relative to the airport, state whether landing or overflight, and request airport advisory. Departing aircraft should state the aircraft type, full identification number, type of flight planned, i.e., VFR or IFR and the planned destination or direction of flight. Report before taxiing and before taxiing on the runway for departure. If communications with a UNICOM are necessary after initial report to FSS, return to FSS frequency for traffic update.

(a) Inbound

EXAMPLE:

VERO BEACH RADIO, CENTURION SIX NINER DELTA DELTA IS TEN MILES SOUTH, TWO THOUSAND, LANDING VERO BEACH. REQUEST AIRPORT ADVISORY.

(b) Outbound

EXAMPLE:

VERO BEACH RADIO, CENTURION SIX NINER DELTA DELTA, READY TO TAXI, VFR, DEPARTING TO THE SOUTHWEST. REQUEST AIRPORT ADVISORY.

3. A CTAF FSS provides wind direction and velocity, favored or designated runway, altimeter setting, known traffic, notices to airmen, airport taxi routes, airport traffic pattern information, and instrument approach procedures. These elements are varied so as to best serve the current traffic situation. Some airport managers have specified that under certain wind or other conditions designated runways be used. Pilots should advise the FSS of the runway they intend to use.

CAUTION: All aircraft in the vicinity of an airport may not be in communication with the FSS.

e. Information provided by Aeronautical Advisory Stations (UNICOM).

1. UNICOM is a nongovernment air/ground radio communication station which may provide airport information at public use airports where there is no tower or FSS.

2. On pilot request, UNICOM stations may provide pilots with weather information, wind direction, the recommended runway, or other necessary information. If the UNICOM frequency is designated as the CTAF, it will be identified in appropriate aeronautical publications.

3. Should LAA by an FSS or, Aeronautical Advisory Station (UNICOM) be unavailable, wind and weather information may be obtainable from nearby controlled airports via Automatic Terminal Information Service (ATIS) or Automated Weather Observing System (AWOS) frequency.

f. Self-Announce Position and/or Intentions.

1. General. *Self-announce* is a procedure whereby pilots broadcast their position or intended flight activity or ground operation on the designated CTAF. This procedure is used primarily at airports which do not have an FSS on the airport. The self-announce procedure should also be used if a pilot is unable to communicate with the FSS on the designated CTAF.

2. If an airport has a tower and it is temporarily closed, or operated on a part-time basis and there is no FSS on the airport or the FSS is closed, use the CTAF to self-announce your position or intentions.

3. Where there is no tower, FSS, or UNICOM station on the airport, use MULTICOM frequency 122.9 for self-announce procedures. Such airports will be identified in appropriate aeronautical information publications.

4. Practice Approaches. Pilots conducting practice instrument approaches should be particularly alert for other aircraft that may be departing in the opposite direction. When conducting any practice approach, regardless of its direction relative to other airport operations, pilots should make announcements on the CTAF as follows:

(a) departing the final approach fix, inbound (non precision approach) or departing the Outer Marker or fix used in lieu of the outer marker, inbound (precision approach);

(b) established on the final approach segment or immediately upon being released by ATC;

(c) upon completion or termination of the approach; and

(d) upon executing the missed approach procedure.

5. Departing aircraft should always be alert for arrival aircraft coming from the opposite direction.

6. Recommended Self-Announce Phraseologies: It should be noted that aircraft operating to or from another nearby airport may be making self-announce broadcasts on the same UNICOM or MULTICOM frequency. To help identify one: airport from another, the airport name should be spoken at the beginning and end of each self-announce transmission.

(a) Inbound

EXAMPLE:

STRAWN TRAFFIC, APACHE TWO TWO FIVE ZULU, (POSITION), (ALTITUDE), (DESCENDING) OR ENTERING DOWNWIND/BASE/FINAL (AS APPROPRIATE) RUNWAY ONE SEVEN FULL STOP, TOUCH-AND-GO, STRAWN.

STRAWN TRAFFIC APACHE TWO TWO FIVE ZULU CLEAR OF RUNWAY ONE SEVEN STRAWN.

(b) Outbound

EXAMPLE:

STRAWN TRAFFIC, QUEEN AIR SEVEN ONE FIVE FIVE BRAVO (LOCATION ON AIRPORT) TAXIING TO RUNWAY TWO SIX STRAWN.

STRAWN TRAFFIC, QUEEN AIR SEVEN ONE FIVE FIVE BRAVO DEPARTING RUNWAY TWO SIX. "DEPARTING THE PATTERN TO THE (DIRECTION), CLIMBING TO (ALTITUDE) STRAWN."

(c) Practice Instrument Approach

EXAMPLE:

STRAWN TRAFFIC, CESSNA TWO ONE FOUR THREE QUEBEC (POSITION FROM AIRPORT) INBOUND DESCENDING THROUGH (ALTITUDE) PRACTICE (NAME OF APPROACH) APPROACH RUNWAY THREE FIVE STRAWN.

STRAWN TRAFFIC, CESSNA TWO ONE FOUR THREE QUEBEC PRACTICE (TYPE) APPROACH COMPLETED OR TERMINATED RUNWAY THREE FIVE STRAWN.

g. UNICOM Communications Procedures.

1. In communicating with a UNICOM station, the following practices will help reduce frequency congestion, facilitate a better understanding of pilot intentions, help identify the location of aircraft in the traffic pattern, and enhance safety of flight:

(a) Select the correct UNICOM frequency.

(b) State the identification of the UNICOM station you are calling in each transmission.

(c) Speak slowly and distinctly.

(d) Report approximately 10 miles from the airport, reporting altitude, and state your aircraft type, aircraft identification, location relative to the airport, state whether landing or overflight, and request wind information and runway in use.

(e) Report on downwind, base, and final approach.

(f) Report leaving the runway.

2. Recommended UNICOM Phraseologies:

(a) Inbound

EXAMPLE:

FREDERICK UNICOM CESSNA EIGHT ZERO ONE TANGO FOXTROT 10 MILES SOUTHEAST DESCENDING THROUGH (ALTITUDE) LANDING FREDERICK, REQUEST WIND AND RUNWAY INFORMATION FREDERICK

FREDERICK TRAFFIC CESSNA EIGHT ZERO ONE TANGO FOXTROT ENTERING DOWNWIND/BASE/FINAL (AS APPROPRIATE) FOR RUNWAY ONE NINER (FULL STOP/TOUCH-AND-GO) FREDERICK.

FREDERICK TRAFFIC CESSNA EIGHT ZERO ONE TANGO FOXTROT CLEAR OF RUNWAY ONE NINER FREDERICK.

(b) Outbound

EXAMPLE:

FREDERICK UNICOM CESSNA EIGHT ZERO ONE TANGO FOXTROT (LOCATION ON AIRPORT) TAXING TO RUNWAY ONE NINER, REQUEST WIND AND TRAFFIC INFORMATION FREDERICK.

FREDERICK TRAFFIC CESSNA EIGHT ZERO ONE TANGO FOXTROT DEPARTING RUNWAY ONE NINER. "REMAINING IN THE PATTERN" OR "DEPARTING THE PATTERN TO THE (DIRECTION) (AS APPROPRIATE)" FREDERICK.

4-10. IFR APPROACHES/GROUND VEHICLE OPERATIONS

a. IFR Approaches.

When operating in accordance with an IFR clearance and ATC approves a change to the advisory frequency, make an expeditious change to the CTAF and employ the recommended traffic advisory procedures.

b. Ground Vehicle Operation.

Airport ground vehicles equipped with radios should monitor the CTAF frequency when operating on the airport movement area and remain clear of runways/taxiways being used by aircraft. Radio transmissions from ground vehicles should be confined to safety-related matters.

c. Radio Control of Airport Lighting Systems.

Whenever possible, the CTAF will be used to control airport lighting systems at airports without operating control towers. This eliminates the need for pilots to change frequencies to turn the lights on and allows a continuous listening watch on a single frequency. The CTAF is published on the instrument approach chart and in other appropriate aeronautical information publications. For further details concerning radio controlled lights, see AC 150/5340-27.

4-11. DESIGNATED UNICOM/MULTICOM FREQUENCIES

a. Communications Between Aircraft.

CAUTION—The Federal Communications Commission (FCC) requires an aircraft station license to operate on UNICOM/MULTICOM frequencies and usage must be in accordance with Part 87 of the FCC Rules (see Section 87.29 regarding license applications). Misuse of these frequencies may result in either the imposition of fines and/or revocation/suspension of FCC aircraft station license.

b. Frequency Use.

1. The following listing depicts UNICOM and MULTICOM frequency uses as designated by the Federal Communications Commission (FCC). (See Table 4-11[1].)

Table 4-11[1]

USE	FREQUENCY
Airports without an operating control tower.	122.700 122.725 122.800 122.975 123.000 123.050 123.075
(MULTICOM FREQUENCY) Activities of a temporary seasonal, emergency nature or search and rescue, as well as, airports with no tower, FSS, or UNICOM.	122.900
(MULTICOM FREQUENCY) Forestry management and fire suppression, fish and game management and protection, and environmental monitoring and protection.	122.925
Airports with a control tower or FSS on airport.	122.950

4-11b1 NOTE—In some areas of the country, frequency interference may be encountered from nearby airports using the same UNICOM frequency. Where there is a problem, UNICOM operators are encouraged to develop a "least interference" frequency assignment plan for airports concerned using the frequencies designated for airports without operating control towers. UNICOM licensees are encouraged to apply for UNICOM 25 kHz spaced channel frequencies. Due to the extremely limited number of frequencies with 50 kHz channel spacing, 25 kHz channel spacing should be implemented. UNICOM licensees may then request FCC to assign frequencies in accordance with the plan, which FCC will review and consider for approval.

4-11b1 NOTE—Wind direction and runway information may not be available on UNICOM frequency 122.950.

2. The following listing depicts other frequency uses as designated by the Federal Communications Commission (FCC). (See Table 4-11[2].)

Table 4-11[2]

USE	FREQUENCY
Air-to-air communications and private airports (not open to the public).	122.750 122.850
Air-to-air communications (general aviation helicopters). Nature or search and rescue, as well as, airports with no tower, FSS, or UNICOM.	123.025
Aviation instruction, Glider, Hot Air Balloon (not to be used for advisory service).	123.300 123.500

4-12. USE OF UNICOM FOR ATC PURPOSES

UNICOM service may be used for air traffic control purposes, only under the following circumstances:

1. Revision to proposed departure time.

2. Takeoff, arrival, or flight plan cancellation time.

3. ATC clearance, provided arrangements are made between the ATC facility and the UNICOM licensee to handle such messages.

4-13. AUTOMATIC TERMINAL INFORMATION SERVICE (ATIS)

a. ATIS is the continuous broadcast of recorded noncontrol information in selected high activity terminal areas. Its purpose is to improve controller effectiveness and to relieve frequency congestion by automating the repetitive transmission of essential but routine information. Pilots are urged to cooperate in the ATIS program as it relieves frequency congestion on approach control, ground control, and local control frequencies. The Airport Facility/Directory indicates airports for which ATIS is provided.

b. ATIS information includes the time of the latest weather sequence, ceiling, visibility, obstructions to visibility, temperature, dew point (if available), wind direction (magnetic), and velocity, altimeter, other pertinent remarks, instrument approach and runway in use. The ceiling/sky condition, visibility, and obstructions to vision may be omitted from the ATIS broadcast if the ceiling is above 5,000 feet and the visibility is more than 5 miles. ATIS is continuously broadcast on the voice feature of a TVOR/VOR/VORTAC located on or near the airport, or on a discrete VHF/UHF frequency. The departure runway will only be given if different from the landing runway except at locations having a separate ATIS for departure. The broadcast may include the appropriate frequency and instructions for VFR arrivals to make initial contact with approach control. Pilots of aircraft arriving or departing the terminal area can receive the continuous ATIS broadcast at times when cockpit duties are least pressing and listen to as many repeats as desired. ATIS broadcast shall be updated upon the receipt of any official hourly and special weather. A new recording will also be made when there is a change in other pertinent data such as runway change, instrument approach in use, etc.

EXAMPLE:

DULLES INTERNATIONAL INFORMATION SIERRA. 1300 ZULU WEATHER. MEASURED CEILING THREE THOUSAND OVERCAST. VISIBILITY THREE, SMOKE. TEMPERATURE SIX EIGHT. WIND THREE FIVE ZERO AT EIGHT. ALTIMETER TWO NINER NINER TWO. ILS RUNWAY ONE RIGHT APPROACH IN USE. LANDING RUNWAY ONE RIGHT AND LEFT. DEPARTURE RUNWAY THREE ZERO. ARMEL VORTAC OUT OF SERVICE. ADVISE YOU HAVE SIERRA.

c. Pilots should listen to ATIS broadcasts whenever ATIS is in operation.

d. Pilots should notify controllers on initial contact that they have received the ATIS broadcast by repeating the alphabetical code word appended to the broadcast.

EXAMPLE:

"INFORMATION SIERRA RECEIVED."

e. When the pilot acknowledges that he has received the ATIS broadcast, controllers may omit those items contained in the broadcast if they are current. Rapidly changing conditions will be issued by ATC and the ATIS will contain words as follows:

EXAMPLE:

"LATEST CEILING/VISIBILITY/ALTIMETER/WIND/(other conditions) WILL BE ISSUED BY APPROACH CONTROL/TOWER."

4-13e NOTE—The absence of a sky condition or ceiling and/or visibility on ATIS indicates a sky condition or ceiling of 5,000 feet or above and visibility of 5 miles or more. A remark may be made on the broadcast, "The weather is better than 5000 and 5," or the existing weather may be broadcast.

f. Controllers will issue pertinent information to pilots who do not acknowledge receipt of a broadcast or who acknowledge receipt of a broadcast which is not current.

g. To serve frequency limited aircraft, FSSs are equipped to transmit on the omnirange frequency at most en route VORs used as ATIS voice outlets. Such communication interrupts the ATIS broadcast. Pilots of aircraft equipped to receive on other FSS frequencies are encouraged to do so in order that these override transmissions may be kept to an absolute minimum.

h. While it is a good operating practice for pilots to make use of the ATIS broadcast where it is available, some pilots use the phrase "Have Numbers" in communications with the control tower. Use of this phrase means that the pilot has received wind, runway, and altimeter information ONLY and the tower does not have to repeat this information. It does not indicate receipt of the ATIS broadcast and should never be used for this purpose.

4-14. RADAR TRAFFIC INFORMATION SERVICE

This is a service provided by radar ATC facilities. Pilots receiving this service are advised of any radar target observed on the radar display which may be in such proximity to the position of their aircraft or its intended route of flight that it warrants their attention. This service is not intended to relieve the pilot of his responsibility for continual vigilance to see and avoid other aircraft.

a. Purpose of the Service:

1. The issuance of traffic information as observed on a radar display is based on the principle of assisting and advising a pilot that a particular radar target's position and track indicates it may intersect or pass in such proximity to his intended flight path that it warrants his attention. This is to alert the pilot to the traffic so that he can be on the lookout for it and thereby be in a better position to take appropriate action should the need arise.

2. Pilots are reminded that the surveillance radar used by ATC does not provide altitude information unless the aircraft is equipped with MODE C and the Radar Facility is capable of displaying altitude information.

b. Provisions of the Service:

1. Many factors, such as limitations of the radar, volume of traffic, controller workload and communications frequency congestion, could prevent the controller from providing this service. The controller possesses complete discretion for determining whether he is able to provide or continue to provide this service in a specific case. His reason against providing or continuing to provide the service in a particular case is not subject to question nor need it be communicated to the pilot. In other words, the provision of this service is entirely dependent upon whether the controller believes he is in a position to provide it. Traffic information is routinely provided to all aircraft operating on IFR Flight Plans except when the pilot advises he does not desire the service, or the pilot is operating within Class A airspace. Traffic information may be provided to flights not operating on IFR Flight Plans when requested by pilots of such flights.

4-14b1 NOTE—Radar ATC facilities normally display and monitor both primary and secondary radar when it is available, except that secondary radar may be used as the sole display source in Class A Airspace, and under some circumstances outside of Class A Airspace (beyond primary coverage and in en route areas where only secondary is available). Secondary radar may also be used outside Class A Airspace as the sole display source when the primary radar is temporarily unusable or out of service. Pilots in contact with the affected ATC facility are normally advised when a temporary outage occurs; i.e., "primary radar out of service; traffic advisories available on transponder aircraft only." This means simply that only the aircraft which have transponders installed and in use will be depicted on ATC radar indicators when the primary radar is temporarily out of service.

2. When receiving VFR radar advisory service, pilots should monitor the assigned frequency at all times. This is to preclude controllers' concern for radio failure or emergency assistance to aircraft under his jurisdiction. VFR radar advisory service does not include vectors away from conflicting traffic unless requested by the pilot. When advisory service is no longer desired, advise the controller before changing frequencies and then change your transponder code to 1200, if applicable. Pilots should also inform the controller when changing VFR cruising altitude. Except in programs where radar service is automatically terminated, the controller will advise the aircraft when radar is terminated.

4-14b2 NOTE—Participation by VFR pilots in formal programs implemented at certain terminal locations constitutes pilot request. This also applies to participating pilots at those locations where arriving VFR flights are encouraged to make their first contact with the tower on the approach control frequency.

c. Issuance of Traffic Information—Traffic information will include the following concerning a target which may constitute traffic for an aircraft that is:

1. Radar identified:

(a) Azimuth from the aircraft in terms of the 12 hour clock, or

(b) When rapidly maneuvering civil test or military aircraft prevent accurate issuance of traffic as in (a) above, specify the direction from an aircraft's position in terms of the eight cardinal compass points (N, NE, E, SE, S, SW, W, NW). This method shall be terminated at the pilot's request.

(c) Distance from the aircraft in nautical miles;

(d) Direction in which the target is proceeding; and

(e) Type of aircraft and altitude if known.

EXAMPLE:

Traffic 10 o'clock, 3 miles, west-bound (type aircraft and altitude, if known, of the observed traffic). The altitude may be known, by means of MODE C, but not verified with the pilot for accuracy. (To be valid for separation purposes by ATC, the accuracy of MODE C readouts must be verified. This is usually accomplished upon initial entry into the radar system by a comparison of the readout to pilot stated altitude, or the field elevation in the case of continuous readout being received from an aircraft on the airport.) When necessary to issue traffic advisories containing unverified altitude information, the controller will issue the advisory in the same manner as if it were verified due to the accuracy of these readouts. The pilot may upon receipt of traffic information, request a vector (heading) to avoid such traffic. The vector will be provided to the extent possible as determined by the controller provided the aircraft to be vectored is within the airspace under the jurisdiction of the controller.

2. Not radar identified:

(a) Distance and direction with respect to a fix;

(b) Direction in which the target is proceeding; and

(c) Type of aircraft and altitude if known.

EXAMPLE:

Traffic 8 miles south of the airport northeastbound, (type aircraft and altitude if known).

d. The examples depicted in the following figures point out the possible error in the position of this traffic when it is necessary for a pilot to apply drift correction to maintain this track. This error could also occur in the event a change in course is made at the time radar traffic information is issued.

1. In Figure 4-14[1] traffic information would be issued to the pilot of aircraft "A" as 12 o'clock. The actual position of the traffic as seen by the pilot of aircraft "A" would be 2 o'clock. Traffic information issued to aircraft "B" would also be given as 12 o'clock, but in this case, the pilot of "B" would see his traffic at 10 o'clock.

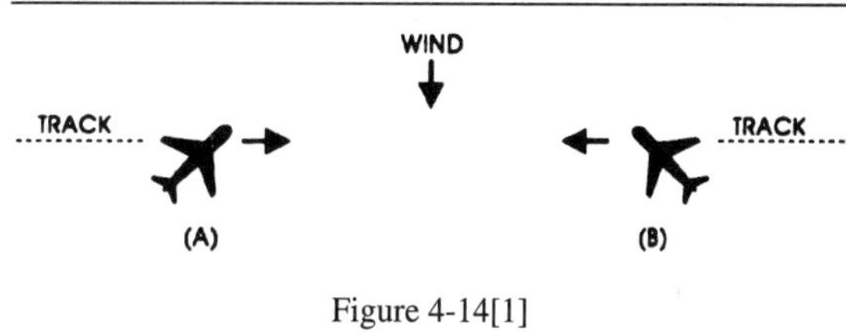

Figure 4-14[1]

2. In Figure 4-14[2] traffic information would be issued to the pilot of aircraft "C" as 2 o'clock. The actual position of the traffic as seen by the pilot of aircraft "C" would be 3 o'clock. Traffic information issued to aircraft "D" would be at an 11 o'clock position. Since it is not necessary for the pilot of aircraft "D" to apply wind correction (crab) to make good his track, the actual position of the traffic issued would be correct. Since the radar controller can only observe aircraft track (course) on his radar display, he must issue traffic advisories accordingly, and pilots should give due consideration to this fact when looking for reported traffic.

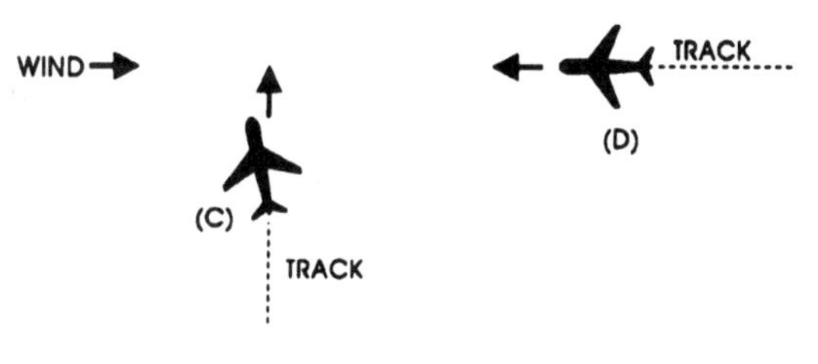

Figure 4-14[2]

4-15. SAFETY ALERT

A safety alert will be issued to pilots of aircraft being controlled by ATC if the controller is aware the aircraft is at an altitude which, in the controller's judgment, places the aircraft in unsafe proximity to terrain, obstructions or other aircraft. The provision of this service is contingent upon the capability of the controller to have an awareness of a situation involving unsafe proximity to terrain, obstructions and uncontrolled aircraft. The issuance of a safety alert cannot be mandated, but it can be expected on a reasonable, though intermittent basis. Once the alert is issued, it is solely the pilot's prerogative to determine what course of action, if any, he will take. This procedure is intended for use in time critical situations where aircraft safety is in question. Noncritical situations should be handled via the normal traffic alert procedures.

a. Terrain or Obstruction Alert:

1. The controller will immediately issue an alert to the pilot of an aircraft under his control when he recognizes that the aircraft is at an altitude which, in his judgment, may be in unsafe proximity to terrain/obstructions. The primary method of detecting unsafe proximity is through MODE C automatic altitude reports.

EXAMPLE:
LOW ALTITUDE ALERT, CHECK YOUR ALTITUDE IMMEDIATELY. THE, as appropriate, MEA/MVA/MOCA IN YOUR AREA IS (altitude) or, if past the final approach fix (non precision approach) or the Outer Marker or fix used in lieu of the outer marker (precision approach), THE, as appropriate, MDA/DH (if known) is (altitude).

2. Terminal ARTS IIA, III, and IIIA facilities have an automated function which, if operating, alerts the controller when a tracked MODE C equipped aircraft under his control is below or is predicted to be below a predetermined Minimum Safe Altitude. This function, called Minimum Safe Altitude Warning (MSAW), is designed solely as a controller aid in detecting potentially unsafe aircraft proximity to terrain/obstructions. The ARTS IIA, III, and IIIA facility will, when MSAW is operating, provide MSAW monitoring for all aircraft with an operating MODE C altitude encoding transponder that are tracked by the system and are:

(a) Operating on an IFR flight plan, or

(b) Operating VFR and have requested MSAW monitoring.

(3) Terminal AN/TPX-42A (number beacon decoder system) facilities have an automated function called Low Altitude Alert System (LAAS). Although not as sophisticated as MSAW, LAAS alerts the controller when a MODE C transponder equipped aircraft operating on an IFR flight plan is below a predetermined Minimum Safe Altitude.

4-15a3 NOTE—Pilots operating VFR may request MSAW or LAAS monitoring if their aircraft are equipped with MODE C transponders.

EXAMPLE:
APACHE THREE THREE PAPA REQUEST MSAW/LAAS.

b. Aircraft Conflict Alert:

1. The controller will immediately issue an alert to the pilot of an aircraft under his control if he is aware of another aircraft which is not under his control is at an altitude which, in the controller's judgment, places both aircraft in unsafe proximity to each other. With the alert, when feasible, the controller will offer the pilot the position of the traffic if time permits and an alternate course(s) of action. Any alternate course(s) of action the controller may recommend to the pilot will be predicated only on other traffic under his control.

EXAMPLE:
AMERICAN THREE, TRAFFIC ALERT, (position of traffic, if time permits), ADVISE YOU TURN RIGHT/LEFT HEADING (degrees) AND/OR CLIMB/DESCEND TO (altitude) IMMEDIATELY.

4-16. RADAR ASSISTANCE TO VFR AIRCRAFT

a. Radar equipped FAA ATC facilities provide radar assistance and navigation service (vectors) to VFR aircraft provided the aircraft can communicate with the facility, are within radar coverage, and can be radar identified.

b. Pilots should clearly understand that authorization to proceed in accordance with such radar navigational assistance does not constitute authorization for the pilot to violate FARs. In effect, assistance provided is on the basis that navigational guidance information issued is advisory in nature and the job of flying the aircraft safely, remains with the pilot.

c. In many cases, the controller will be unable to determine if flight into instrument conditions will result from his instructions. To avoid possible hazards resulting from being vectored into IFR conditions, pilots should keep the controller advised of the weather conditions in which he is operating and along the course ahead.

d. Radar navigation assistance (vectors) may be initiated by the controller when one of the following conditions exist:

1. The controller suggests the vector and the pilot concurs.

2. A special program has been established and vectoring service has been advertised.

3. In the controller's judgment the vector is necessary for air safety.

e. Radar navigation assistance (vectors) and other radar derived information may be provided in response to pilot requests. Many factors, such as limitations of radar, volume of traffic, communications frequency, congestion, and controller workload could prevent the controller from providing it. The controller has complete discretion for determining if he is able to provide the service in a particular case. His decision not to provide the service in a particular case is not subject to question.

4-17. TERMINAL RADAR SERVICES FOR VFR AIRCRAFT

a. Basic Radar Service

1. In addition to the use of radar for the control of IFR aircraft, all commissioned radar facilities provide the following basic radar services for VFR aircraft:

(a) Safety alerts.

(b) Traffic advisories.

(c) Limited radar vectoring (on a workload permitting basis).

(d) Sequencing at locations where procedures have been established for this purpose and/or when covered by a letter of agreement.

4-17a1 NOTE—When the Stage services were developed, two basic radar services (traffic advisories and limited vectoring) were identified as "Stage I." This definition became unnecessary and the term "Stage I" was eliminated from use. The term "Stage II" has been eliminated in conjunction with the airspace reclassification, and sequencing services to locations with local procedures and/or letters of agreement to provide this service have been included in basic services to VFR aircraft. These basic services will still be provided by all terminal radar facilities whether they include Class B, Class C,

Class D or Class E airspace. "Stage III" services have been replaced with "Class B" and "TRSA" service where applicable.

2. Vectoring service may be provided when requested by the pilot or with pilot concurrence when suggested by ATC.

3. Pilots of arriving aircraft should contact approach control on the publicized frequency and give their position, altitude, aircraft callsign, type aircraft, radar beacon code (if transponder equipped), destination, and request traffic information.

4. Approach control will issue wind and runway, except when the pilot states "have numbers" or this information is contained in the ATIS broadcast and the pilot states that the current ATIS information has been received. Traffic information is provided on a workload permitting basis. Approach control will specify the time or place at which the pilot is to contact the tower on local control frequency for further landing information. Radar service is automatically terminated upon being advised to contact the tower.

5. Sequencing for VFR aircraft is available at certain terminal locations (see locations listed in the Airport/Facility Directory). The purpose of the service is to adjust the flow of arriving VFR and IFR aircraft into the traffic pattern in a safe and orderly manner and to provide radar traffic information to departing VFR aircraft. Pilot participation is urged but is not mandatory. Traffic information is provided on a workload permitting basis. Standard radar separation between VFR or between VFR and IFR aircraft is not provided.

(a) Pilots of arriving VFR aircraft should initiate radio contact on the publicized frequency with approach control when approximately 25 miles from the airport at which sequencing services are being provided. On initial contact by VFR aircraft, approach control will assume that sequencing service is requested. After radar contact is established, the pilot may use pilot navigation to enter the traffic pattern or, depending on traffic conditions, approach control may provide the pilot with routings or vectors necessary for proper sequencing with other participating VFR and IFR traffic en route to the airport. When a flight is positioned behind a preceding aircraft and the pilot reports having that aircraft in sight, the pilot will be instructed to follow the preceding aircraft. THE ATC INSTRUCTION TO FOLLOW THE PRECEDING AIRCRAFT DOES NOT AUTHORIZE THE PILOT TO COMPLY WITH ANY ATC CLEARANCE OR INSTRUCTION ISSUED TO THE PRECEDING AIRCRAFT. If other "nonparticipating" or "local" aircraft are in the traffic pattern, the tower will issue a landing sequence. Radar service will be continued to the runway. If an arriving aircraft does not want the service, the pilot should state "NEGATIVE RADAR SERVICE" or make a similar comment, on initial contact with approach control.

(b) Pilots of departing VFR aircraft are encouraged to request radar traffic information by notifying ground control on initial contact with their request and proposed direction of flight.

EXAMPLE:
XRAY GROUND CONTROL, NOVEMBER ONE EIGHT SIX, CESSNA ONE SEVENTY TWO, READY TO TAXI, VFR SOUTHBOUND AT 2,500, HAVE INFORMATION BRAVO AND REQUEST RADAR TRAFFIC INFORMATION.

4-17b5 NOTE—Following takeoff, the tower will advise when to contact departure control.

(c) Pilots of aircraft transiting the area and in radar contact/communication with approach control will receive traffic information on a controller workload permitting basis. Pilots of such aircraft should give their position, altitude, aircraft callsign, aircraft type, radar beacon code (if transponder equipped), destination, and/or route of flight.

b. TRSA Service (Radar Sequencing and Separation Service for VFR Aircraft in a TRSA).

1. This service has been implemented at certain terminal locations. The service is advertised in the Airport/Facility Directory. The purpose of this service is to provide separation between all participating VFR aircraft and all IFR aircraft operating within the airspace defined as the Terminal Radar Service Area (TRSA). Pilot participation is urged but is not mandatory.

2. If any aircraft does not want the service, the pilot should state "NEGATIVE TRSA SERVICE" or make a similar comment, on initial contact with approach control or ground control, as appropriate.

3. TRSAs are depicted on sectional aeronautical charts and listed in the Airport/Facility Directory.

4. While operating within a TRSA, pilots are provided TRSA service and separation as prescribed in this paragraph. In the event of a radar outage, separation and sequencing of VFR aircraft will be suspended as this service is dependent on radar. The pilot will be advised that the service is not available and issued wind, runway information, and the time or place to contact the tower. Traffic information will be provided on a workload permitting basis.

5. Visual separation is used when prevailing conditions permit and it will be applied as follows:

(a) When a VFR flight is positioned behind a preceding aircraft and the pilot reports having that aircraft in sight, the pilot will be instructed by ATC to follow the preceding aircraft. THE ATC INSTRUCTION TO FOLLOW THE PRECEDING AIRCRAFT DOES NOT AUTHORIZE THE PILOT TO COMPLY WITH ANY ATC CLEARANCE OR INSTRUCTION ISSUED TO THE PRECEDING AIRCRAFT. Radar service will be continued to the runway.

(b) If other "nonparticipating" or "local" aircraft are in the traffic pattern, the tower will issue a landing sequence.

(c) Departing VFR aircraft may be asked if they can visually follow a preceding departure out of the TRSA. The pilot will be instructed to follow the other aircraft provided that the pilot can maintain visual contact with that aircraft.

6. Until visual separation is obtained, standard vertical or radar separation will be provided.

(a) 1000 feet vertical separation may be used between IFR aircraft.

(b) 500 feet vertical separation may be used between VFR aircraft, or between a VFR and an IFR aircraft.

(c) Radar separation varies depending on size of aircraft and aircraft distance from the radar antenna. The minimum separation used will be 1½ miles for most VFR aircraft under 12,500 pounds GWT. If being separated from larger aircraft, the minimum is increased appropriately.

7. Participating pilots operating VFR in a TRSA

(a) Must maintain an altitude when assigned by ATC unless the altitude assignment is to maintain at or below a specified altitude. ATC may assign altitudes for separation that do not conform to FAR Part 91.159. When the altitude assignment is no longer needed for separation or when leaving the TRSA, the instruction will be broadcast, "RESUME APPROPRIATE VFR ALTITUDES." Pilots must then return to an altitude that conforms to FAR Part 91.159 as soon as practicable.

(b) When not assigned an altitude, the pilot should coordinate with ATC prior to any altitude change.

8. Within the TRSA, traffic information on observed but unidentified targets will, to the extent possible, be provided all IFR and participating VFR aircraft. The pilot will be vectored to avoid the observed traffic, provided the aircraft to be vectored is within the airspace under the jurisdiction of the controller.

9. Departing aircraft should inform ATC of their intended destination and/or route of flight and proposed cruising altitude.

10. ATC will normally advise participating VFR aircraft when leaving the geographical limits of the TRSA. Radar service is not automatically terminated with this advisory unless specifically stated by the controller.

c. Class C Service. This service provides, in addition to basic radar service, approved separation between IFR and VFR aircraft, and sequencing of VFR arrivals to the primary airport.

d. Class B Service. This service provides, in addition to basic radar service, approved separation of aircraft based on IFR, VFR, and/or weight, and sequencing of VFR arrivals to the primary airport(s).

e. PILOT RESPONSIBILITY: THESE SERVICES ARE NOT TO BE INTERPRETED AS RELIEVING PILOTS OF THEIR RESPONSIBILITIES TO SEE AND AVOID OTHER TRAFFIC OPERATING IN BASIC VFR WEATHER CONDITIONS, TO ADJUST THEIR OPERATIONS AND FLIGHT PATH AS NECESSARY TO PRECLUDE SERIOUS WAKE ENCOUNTERS, TO MAINTAIN APPROPRIATE TERRAIN AND OBSTRUCTION CLEARANCE, OR TO REMAIN IN WEATHER CONDITIONS EQUAL TO OR BETTER THAN THE MINIMUMS REQUIRED BY FAR PART 91.155. WHENEVER COMPLIANCE WITH AN ASSIGNED ROUTE, HEADING AND/OR ALTITUDE IS LIKELY TO COMPROMISE PILOT RESPONSIBILITY RESPECTING TERRAIN AND OBSTRUCTION CLEARANCE, VORTEX EXPOSURE, AND WEATHER MINIMUMS, APPROACH CONTROL SHOULD BE SO ADVISED AND A REVISED CLEARANCE OR INSTRUCTION OBTAINED.

f. ATC services for VFR aircraft participating in terminal radar services are dependent on air traffic control radar. Services for VFR aircraft are not available during periods of a radar outage and are limited during CENRAP operations. The pilot will be advised when VFR services are limited or not available.

4-17f NOTE—Class B and Class C airspace are areas of regulated airspace. The absence of ATC radar does not negate the requirement of an ATC clearance to enter Class B airspace or two way radio contact with ATC to enter Class C airspace.

4-18. TOWER EN ROUTE CONTROL (TEC)

a. TEC is an ATC program to provide a service to aircraft proceeding to and from metropolitan areas. It links designated Approach Control Areas by a network of identified routes made up of the existing airway structure of the National Airspace System. The FAA initiated an expanded TEC program to include as many facilities as possible. The program's intent is to provide an overflow resource in the low altitude system which would enhance ATC services. A few facilities have historically allowed turbojets to proceed between certain city pairs, such as Milwaukee and Chicago, via tower en route and these locations may continue this service. However, the expanded TEC program will be applied, generally, for nonturbojet aircraft operating at and below 10,000 feet. The program is entirely within the approach control airspace of multiple terminal facilities. Essentially, it is for relatively short flights. Participating pilots are encouraged to use TEC for flights of two hours duration or less. If longer flights are planned, extensive coordination may be required within the multiple complex which could result in unanticipated delays.

b. Pilots requesting TEC are subject to the same delay factor at the destination airport as other aircraft in the ATC system. In addition, departure and en route delays may occur depending upon individual facility workload. When a major metropolitan airport is incurring significant delays, pilots in the TEC program may want to consider an alternative airport experiencing no delay.

c. There are no unique requirements upon pilots to use the TEC program. Normal flight plan filing procedures will ensure proper flight plan processing. Pilots should include the acronym "TEC" in the remarks section of the flight plan when requesting tower en route.

d. All approach controls in the system may not operate up to the maximum TEC altitude of 10,000 feet. IFR flight may be planned to any satellite airport in proximity to the major primary airport via the same routing.

4-19. TRANSPONDER OPERATION

a. GENERAL

1. Pilots should be aware that proper application of transponder operating procedures will provide both VFR and IFR aircraft with a higher degree of safety in the environment where high-speed closure rates are possible. Transponders substantially increase the capability of radar to see an aircraft and the MODE C feature enables the controller to quickly determine where potential traffic conflicts may exist. Even VFR pilots who are not in contact with ATC will be afforded greater protection from IFR aircraft and VFR aircraft which are receiving traffic advisories. Nevertheless, pilots should never relax their visual scanning vigilance for other aircraft.

2. Air Traffic Control Radar Beacon System (ATCRBS) is similar to and compatible with military coded radar beacon equipment. Civil MODE A is identical to military MODE 3.

3. Civil and military transponders should be adjusted to the "on" or normal operating position as late as practicable prior to takeoff and to "off" or "standby" as soon as practicable after completing landing roll, unless the change to "standby" has been accomplished previously at the request of ATC. IN ALL CASES, WHILE IN CONTROLLED AIRSPACE EACH PILOT OPERATING AN AIRCRAFT EQUIPPED WITH AN OPERABLE ATC TRANSPONDER MAINTAINED IN ACCORDANCE WITH FAR PART 91.413 **SHALL** OPERATE THE TRANSPONDER, INCLUDING MODE C IF INSTALLED, ON THE APPROPRIATE CODE OR AS ASSIGNED BY ATC. IN CLASS G AIRSPACE, THE TRANSPONDER SHOULD BE OPERATING WHILE AIRBORNE UNLESS OTHERWISE REQUESTED BY ATC.

4. If a pilot on an IFR flight cancels his IFR flight plan prior to reaching his destination, he should adjust his transponder according to VFR operations.

5. If entering a U.S. OFFSHORE AIRSPACE AREA from outside the U.S., the pilot should advise on first radio contact with a U.S. radar ATC facility that such equipment is available by adding "transponder" to the aircraft identification.

6. It should be noted by all users of the ATC Transponders that the coverage they can expect is limited to "line of sight." Low altitude or aircraft antenna shielding by the aircraft itself may result in reduced range. Range can be improved by climbing to a higher altitude. It may be possible to minimize antenna shielding by locating the antenna where dead spots are only noticed during abnormal flight attitudes.

b. TRANSPONDER CODE DESIGNATION

1. For ATC to utilize one or a combination of the 4096 discrete codes FOUR DIGIT CODE DESIGNATION will be used, e.g., code 2100 will be expressed as TWO ONE ZERO ZERO. Due to the operational characteristics of the rapidly expanding automated air traffic control system, THE LAST TWO DIGITS OF THE SELECTED TRANSPONDER CODE SHOULD ALWAYS READ "00" UNLESS SPECIFICALLY REQUESTED BY ATC TO BE OTHERWISE.

c. AUTOMATIC ALTITUDE REPORTING (MODE C)

1. Some transponders are equipped with a MODE C automatic altitude reporting capability. This system converts aircraft altitude in 100 foot increments to coded digital information which is transmitted together with MODE C framing pulses to the interrogating radar facility. The manner in which transponder panels are designed differs, therefore, a pilot should be thoroughly familiar with the operation of his transponder so that ATC may realize its full capabilities.

2. Adjust transponder to reply on the MODE A/3 code specified by ATC and, if equipped, to reply on MODE C with altitude reporting *capability activated* unless deactivation is directed by ATC or unless the installed aircraft equipment has not been tested and calibrated as required by FAR Part 91.217. If deactivation is required by ATC, turn off the altitude reporting feature of your transponder. An instruction by ATC to "STOP ALTITUDE SQUAWK, ALTITUDE DIFFERS (number of feet) FEET," may be an indication that your transponder is transmitting incorrect altitude information or that you have an incorrect altimeter setting. While an incorrect altimeter setting has no effect on the MODE C altitude information transmitted by your transponder (transponders are preset at 29.92), it would cause you to fly at an actual altitude different from your assigned altitude. When a controller indicates that an altitude readout is invalid, the pilot should initiate a check to verify that the aircraft altimeter is set correctly.

3. Pilots of aircraft with operating MODE C altitude reporting transponders should report exact altitude or Flight Level to the nearest hundred foot increment when establishing initial contact with an ATC facility. Exact altitude or flight level reports on initial contact provide ATC with information that is required prior to using MODE C altitude information for separation purposes. This will significantly reduce altitude verification requests.

d. TRANSPONDER IDENT FEATURE

1. The transponder shall be operated only as specified by ATC. Activate the "IDENT" feature only upon request of the ATC controller.

e. CODE CHANGES

1. When making routine code changes, pilots should avoid inadvertent selection of codes 7500, 7600 or 7700 thereby causing momentary false alarms at automated ground facilities. For example when switching from code 2700 to code 7200, switch first to 2200 then to 7200, NOT to 7700 and then 7200. This procedure applies to nondiscrete code 7500 and all discrete codes in the 7600 and 7700 series (i.e., 7600–7677, 7700–7777) which will trigger special indicators in automated facilities. Only nondiscrete code 7500 will be decoded as the hijack code.

2. Under no circumstances should a pilot of a civil aircraft operate the transponder on Code 7777. This code is reserved for military interceptor operations.

3. Military pilots operating VFR or IFR within restricted/warning areas should adjust their transponders to code 4000 unless another code has been assigned by ATC.

f. MODE C TRANSPONDER REQUIREMENTS

1. Specific details concerning requirements to carry and operate Mode C transponders, as well as exceptions and ATC authorized deviations from the requirements are found in FAR Part 91.215 and FAR Part 99.12.

2. In general, the FAR requires aircraft to be equipped with Mode C transponders when operating:

(a) at or above 10,000 feet MSL over the 48 contiguous states or the District of Columbia, excluding that airspace below 2,500 feet AGL;

(b) within 30 miles of a Class B airspace primary airport, below 10,000 feet MSL. Balloons, gliders, and aircraft not equipped with an engine driven electrical system are excepted from the above requirements when operating below the floor of Class A airspace and/or; outside of a Class B airspace and below the ceiling of the Class B Airspace (or 10,000 feet MSL, whichever is lower);

(c) within and above all Class C airspace, up to 10,000 feet MSL;

(d) within 10 miles of certain designated airports, excluding that airspace which is both outside the Class D surface area and below 1,200 feet AGL. Balloons, gliders and aircraft not equipped with an engine driven electrical system are excepted from this requirement.

3. FAR Part 99.12 requires all aircraft flying into, within, or across the contiguous U.S. ADIZ be equipped with a Mode C or Mode S transponder. Balloons, gliders and aircraft not equipped with an engine driven electrical system are excepted from this requirement.

4. Pilots shall ensure that their aircraft transponder is operating on an appropriate ATC assigned VFR/IFR code and MODE C when operating in such airspace. If in doubt about the operational status of either feature of your transponder while airborne, contact the nearest ATC facility or FSS and they will advise you what facility you should contact for determining the status of your equipment.

5. In-flight requests for "immediate" deviation from the transponder requirement may be approved by controllers only when the flight will continue IFR or when weather conditions prevent VFR descent and continued VFR flight in airspace not affected by the FAR. All other requests for deviation should be made by contacting the nearest Flight Service or Air Traffic facility in person or by telephone. The nearest ARTCC will normally be the controlling agency and is responsible for coordinating requests involving deviations in other ARTCC areas.

g. TRANSPONDER OPERATION UNDER VISUAL FLIGHT RULES (VFR)

1. Unless otherwise instructed by an Air Traffic Control Facility, adjust Transponder to reply on MODE 3/A code 1200 regardless of altitude.

2. Adjust transponder to reply on MODE C, with altitude reporting *capability activated* if the aircraft is so equipped, unless deactivation is directed by ATC or unless the installed equipment has not been tested and calibrated as required by FAR Part 91.217. If deactivation is required and your transponder is so designed, turn off the altitude reporting switch and continue to transmit MODE C framing pulses. If this capability does not exist, turn off MODE C.

h. RADAR BEACON PHRASEOLOGY

Air traffic controllers, both civil and military, will use the following phraseology when referring to operation of the Air Traffic Control Radar Beacon System (ATCRBS). Instructions by ATC refer only to MODE A/3 or MODE C operation and do not affect the operation of the transponder on other MODEs.

1. SQUAWK (number)—Operate radar beacon transponder on designated code in MODE A/3.

2. IDENT—Engage the "IDENT" feature (military I/P) of the transponder.

3. SQUAWK (number) and IDENT—Operate transponder on specified code in MODE A/3 and engage the "IDENT" (military I/P) feature.

4. SQUAWK STANDBY—Switch transponder to standby position.

5. SQUAWK LOW/NORMAL—Operate transponder on low or normal sensitivity as specified. Transponder is operated in "NORMAL" position unless ATC specifies "LOW." ("ON" is used instead of "NORMAL" as a master control label on some types of transponders.)

6. SQUAWK ALTITUDE—Activate MODE C with automatic altitude reporting.

7. STOP ALTITUDE SQUAWK—Turn off altitude reporting switch and continue transmitting MODE C framing pulses. If your equipment does not have this capability, turn off MODE C.

8. STOP SQUAWK (mode in use)—Switch off specified mode. (Used for military aircraft when the controller is unaware of military service requirements for the aircraft to continue operation on another MODE.)

9. STOP SQUAWK—Switch off transponder.

10. SQUAWK MAYDAY—Operate transponder in the emergency position (MODE A Code 7700 for civil transponder. MODE 3 Code 7700 and emergency feature for military transponder.)

11. SQUAWK VFR—Operate radar beacon transponder on code 1200 in the MODE A/3, or other appropriate VFR code.

4-20. HAZARDOUS AREA REPORTING SERVICE

a. Selected FSSs provide flight monitoring where regularly traveled VFR routes cross large bodies of water,

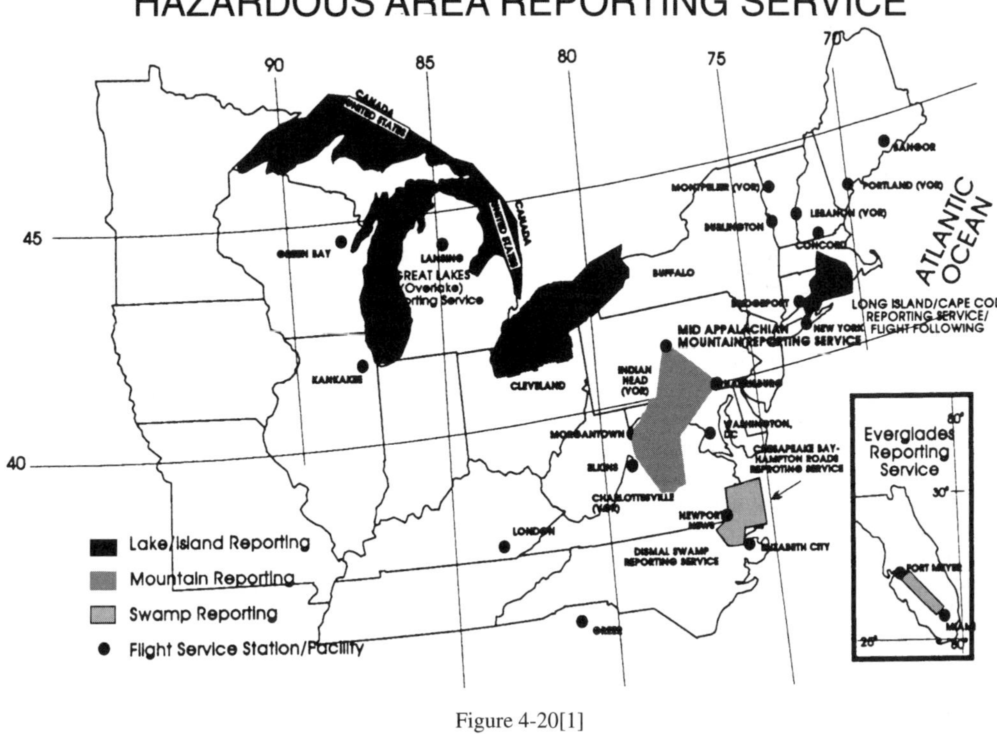

Figure 4-20[1]

swamps, and mountains. This service is provided for the purpose of expeditiously alerting Search and Rescue facilities when required. (See Figure 4-20[1].)

1. When requesting the service either in person, by telephone or by radio, pilots should be prepared to give the following information—type of aircraft, altitude, indicated airspeed, present position, route of flight, heading.

2. Radio contacts are desired at least every 10 minutes. If contact is lost for more than 15 minutes, Search and Rescue will be alerted. Pilots are responsible for canceling their request for service when they are outside the service area boundary. Pilots experiencing two-way radio failure are expected to land as soon as practicable and cancel their request for the service. The accompanying illustration titled "LAKE, ISLAND, MOUNTAIN AND SWAMP REPORTING SERVICE" depicts the areas and the FSS facilities involved in this program.

b. LONG ISLAND SOUND REPORTING SERVICE

The New York and Bridgeport AFSSs provide Long Island Sound Reporting service on request for aircraft traversing Long Island Sound.

1. When requesting the service pilots should ask for SOUND REPORTING SERVICE and should be prepared to provide the following appropriate information:

(a) Type and color of aircraft,

(b) The specific route and altitude across the sound including the shore crossing point,

(c) The overwater crossing time,

(d) Number of persons on board,

(e) True air speed.

2. Radio contacts are desired at least every 10 minutes, however, for flights of shorter duration a midsound report is requested. If contact is lost for more than 15 minutes Search and Rescue will be alerted. Pilots are responsible for canceling their request for the Long Island Sound Reporting Service when outside the service area boundary. Aircraft experiencing radio failure will be expected to land as soon as practicable and cancel their request for the service.

3. COMMUNICATIONS: Primary communications—pilots are to transmit on 122.1 MHz and listen on one of the following VOR frequencies:

(a) NEW YORK AFSS CONTROLS:

(1) Hampton RCO (FSS transmits and receives on 122.6 MHz).

(2) Calverton VORTAC (FSS transmits on 117.2 and receives on standard FSS frequencies).

(3) Kennedy VORTAC (FSS transmits on 115.9 and receives on 122.1 MHz).

(b) BRIDGEPORT AFSS CONTROLS:

(1) Madison VORTAC (FSS transmits on 110.4 and receives on 122.1 MHz).

Automated Terminal Information Service. This information can be obtained from any FSS or by referring to the HDTA teletype weather report. The code "VNA" at the end of the weather report indicates VFR arrival reservations are not authorized. The indication will not be made when IFR weather conditions exist.

5. The requirements for obtaining reservations pursuant to FAR Part 93, Subpart K, are mandatory. Failure to operate in accordance with the FAR may be grounds for enforcement action.

f. SPECIAL TRAFFIC MANAGEMENT PROGRAMS (STMP):

1. Special procedures may be established when a location requires special traffic handling to accommodate above normal traffic demand (e.g., the Indianapolis 500, Super Bowl, etc.) or reduced airport capacity (e.g., airport runway/taxiway closures for airport construction). The special procedures may remain in effect until the problem has been resolved or until local traffic management procedures can handle the situation and a need for special handling no longer exists.

2. CVRS may be used to allocate the reservations during an STMP. If CVRS is being used, the toll-free telephone numbers will be advertised by NOTAM. Be sure to check current NOTAM's to determine what airports are included in an STMP, days and times reservations are required, time limits for reservations requests, and who to contact for reservations.

g. MAKING HDTA/STMP RESERVATIONS USING THE CVRS:

1. *Computer Modem Users:* A Personal Computer (PC) may be used to make reservations on the CVRS. Equipment required is a computer with a modem capable of a 300 to 9600 baud rate and a communications software program. There are several communications software programs available from many computer stores. The type program required is one which is used to connect with a Bulletin Board System (BBS). The CVRS modem data is transmitted using No Parity, 8 data bits, and 1 stop bit (N,8,1). Be sure your computer software is set to these parameters.

2. When your computer connects with CVRS, you will be presented with a screen that will ask you to log on. If this is the first time you have logged onto the CVRS, you will be asked for your name, the city you are calling from, and a password. (Be sure to record your password for future use). CVRS uses your name and password to save your computer "set-up" so that the next time you call you will have the same display. After you have logged on, every thing you need to do involving a reservation is menu driven. There are also several files you can download which explain CVRS operations in greater detail.

3. *Telephone users:* When using the telephone to make a reservation, you are prompted for input of information about what you wish to do. All input is accomplished using the keypad or rotary dial on the telephone. The only problem with a telephone is that most keys have a letter and number associated with them. When the system asks for a date or time, it is expecting an input of numbers. A problem arises when entering a tail number. The system does not detect if you are entering a letter (alpha character) or a number. Therefore, when entering a tail number two keys are used to represent each letter or number. When entering a number, precede the number you wish by the number 0 (zero) i.e., 01, 02, 03, 04, . . . If you wish to enter a letter, first press the key on which the letter appears and then press 1, 2, or 3, depending upon whether the letter you desire is the first, second, or third letter on that key. For example to enter the letter "N" first press the "6" key because "N" is on that key, then press the "2" key because the letter "N" is the second letter on the "6" key. Since there are no keys for the letters "Q" and "Z" CVRS pretends they are on the number "1" key. Therefore, to enter the letter "Q", press 11, and to enter the letter "Z" press 12.

4-2lg3 Note—Users are reminded to enter the "N" character with their tail numbers. (See Table 4-21[1].)

4. Additional helpful key entries: (see Table 4-21[2].)

Table 4-21[1]

CODES FOR TAIL NUMBER INPUT ONLY

A-21	J-51	S-73	1-01
B-22	K-52	T-81	2-02
C-23	L-53	U-82	3-03
D-31	M-61	V-83	4-04
E-32	N-62	W-91	5-05
F-33	O-63	X-92	6-06
G-41	P-71	Y-93	7-07
H-42	Q-11	Z-12	8-08
I-43	R-72	0-00	9-09

Table 4-21[2]

#	After entering a tail number, depressing the "pound key" (#) twice will indicate the end of the tail number.
*2	Will take the user back to the start of the process.
*3	Will repeat the tail number used in a previous reservation.
*5	Will repeat the previous question.
*8	Tutorial mode: In the tutorial mode each prompt for input includes a more detailed description of what is expected as input. *8 is a toggle on/off switch. If you are in tutorial mode and enter *8, you will return to the normal mode.
*0	Expert mode: In the expert mode each prompt for input is brief with little or no explanation. Expert mode is also an on/off toggle.

4-22 thru 4-29. RESERVED

Section 2. RADIO COMMUNICATIONS PHRASEOLOGY AND TECHNIQUES

4-30. GENERAL

a. Radio communications are a critical link in the ATC system. The link can be a strong bond between pilot and controller or it can be broken with surprising speed and disastrous results. Discussion herein provides basic procedures for new pilots and also highlights safe operating concepts for all pilots.

b. The single, most important thought in pilot-controller communications is understanding. It is essential, therefore, that pilots acknowledge each radio communication with ATC by using the appropriate aircraft call sign. Brevity is important, and contacts should be kept as brief as possible, but the controller must know what you want to do before he can properly carry out his control duties. And you, the pilot, must know exactly what he wants you to do. Since concise phraseology may not always be adequate, use whatever words are necessary to get your message across. Pilots are to maintain vigilance in monitoring air traffic control radio communications frequencies for potential traffic conflicts with their aircraft especially when operating on an active runway and/or when conducting a final approach to landing.

c. All pilots will find the Pilot/Controller Glossary very helpful in learning what certain words or phrases mean. Good phraseology enhances safety and is the mark of a professional pilot. Jargon, chatter, and "CB" slang have no place in ATC communications. The Pilot/Controller Glossary is the same glossary used in the Air Traffic Control Order. We recommend that it be studied and reviewed from time to time to sharpen your communication skills.

4-31. RADIO TECHNIQUE

a. *Listen* before you transmit. Many times you can get the information you want through ATIS or by monitoring the frequency. Except for a few situations where some frequency overlap occurs, if you hear someone else talking, the keying of your transmitter will be futile and you will probably jam their receivers causing them to repeat their call. If you have just changed frequencies, pause, listen, and make sure the frequency is clear.

b. *Think* before keying your transmitter. Know what you want to say and if it is lengthy; e.g., a flight plan or IFR position report, jot it down.

c. The microphone should be very close to your lips and after pressing the mike button, a slight pause may be necessary to be sure the first word is transmitted. Speak in a normal, conversational tone.

d. When you release the button, wait a few seconds before calling again. The controller or FSS specialist may be jotting down your number, looking for your flight plan, transmitting on a different frequency, or selecting his transmitter to your frequency.

e. Be alert to the sounds *or the lack of sounds* in your receiver. Check your volume, recheck your frequency, and *make sure that your microphone is not stuck* in the transmit position. Frequency blockage can, and has, occurred for extended periods of time due to unintentional transmitter operation. This type of interference is commonly referred to as a "stuck mike," and controllers may refer to it in this manner when attempting to assign an alternate frequency. If the assigned frequency is completely blocked by this type of interference, use the procedures described for en route IFR radio frequency outage to establish or reestablish communications with ATC.

f. Be sure that you are within the performance range of your radio equipment and the ground station equipment. Remote radio sites do not always transmit and receive on all of a facility's available frequencies, particularly with regard to VOR sites where you can hear but not reach a ground station's receiver. Remember that higher altitudes increase the range of VHF "line of sight" communications.

4-32. CONTACT PROCEDURES

a. Initial Contact-

1. The terms *initial contact* or *initial callup* means the first radio call you make to a given facility or the first call to a different controller or FSS specialist within a facility. Use the following format:

(a) Name of the facility being called;

(b) Your *full* aircraft identification as filed in the flight plan or as discussed under Aircraft Call Signs below;

(c) The type of message to follow or your request if it is short, and

(d) The word "Over" if required.

EXAMPLE:
"NEW YORK RADIO, MOONEY THREE ONE ONE ECHO."

EXAMPLE:
"COLUMBIA GROUND, CESSNA THREE ONE SIX ZERO FOXTROT, I-F-R MEMPHIS."

EXAMPLE:
"MIAMI CENTER, BARON FIVE SIX THREE HOTEL, REQUEST V-F-R TRAFFIC ADVISORIES."

2. Many FSSs are equipped with RCO's and can transmit on the same frequency at more than one location. The frequencies available at specific locations are indicated on charts above FSS communications boxes. To enable the specialist to utilize the correct transmitter, advise the location and the frequency on which you expect a reply.

EXAMPLE:
St. Louis FSS can transmit on frequency 122.3 at either Farmington, MO, or Decatur, IL. If you are in the vicinity of Decatur, your callup should be "SAINT LOUIS RADIO, PIPER SIX

NINER SIX YANKEE, RECEIVING DECATUR ONE TWO TWO POINT THREE."

3. If radio reception is reasonably assured, inclusion of your request, your position or altitude, and the phrase "Have Numbers" or "Information Charlie received" (for ATIS) in the initial contact helps decrease radio frequency congestion. Use discretion, and do not overload the controller with information he does not need. If you do not get a response from the ground station, recheck your radios or use another transmitter, but keep the next contact short.

EXAMPLE:
"ATLANTA CENTER, DUKE FOUR ONE ROMEO, REQUEST V-F-R TRAFFIC ADVISORIES, TWENTY NORTHWEST ROME, SEVEN THOUSAND FIVE HUNDRED, OVER."

b. Initial Contact When your Transmitting and Receiving Frequencies are Different—

1. If you are attempting to establish contact with a ground station and you are receiving on a different frequency than that transmitted, indicate the VOR name or the frequency on which you expect a reply. Most FSSs and control facilities can transmit on several VOR stations in the area. Use the appropriate FSS call-sign as indicated on charts.

EXAMPLE:
New York FSS transmits on the Kennedy, the Hampton, and the Calverton VORTAC's. If you are in the Calverton area, your callup should be "NEW YORK RADIO, CESSNA THREE ONE SIX ZERO FOXTROT, RECEIVING CALVERTON V-O-R, OVER."

2. If the chart indicates FSS frequencies above the VORTAC or in the FSS communications boxes, transmit or receive on those frequencies nearest your location.

3. When unable to establish contact and you wish to call *any* ground station, use the phrase "ANY RADIO (tower) (station), GIVE CESSNA THREE ONE SIX ZERO FOXTROT A CALL ON (frequency) OR (V-O-R)." If an emergency exists or you need assistance, so state.

c. Subsequent Contacts and Responses to Callup from a Ground Facility—

Use the same format as used for the initial contact except you should state your message or request with the callup in one transmission. The ground station name and the word "Over" may be omitted if the message requires an obvious reply and there is no possibility for misunderstandings. *You should acknowledge all callups or clearances* unless the controller or FSS specialist advises otherwise. There are some occasions when the controller must issue time-critical instructions to other aircraft, and he may be in a position to observe your response, either visually or on radar. If the situation demands your response, take appropriate action or immediately advise the facility of any problem. Acknowledge with your aircraft identification and one of the words "Wilco," "Roger," "Affirmative," "Negative," or other appropriate remarks; e.g., "PIPER TWO ONE FOUR LIMA, ROGER." If you have been receiving services; e.g., VFR traffic advisories and you are leaving the area or changing frequencies, advise the ATC facility and terminate contact.

d. Acknowledgement of Frequency Changes—

1. When advised by ATC to change frequencies, acknowledge the instruction. If you select the new frequency without an acknowledgement, the controller's workload is increased because he has no way of knowing whether you received the instruction or have had radio communications failure.

2. At times, a controller/specialist may be working a sector with multiple frequency assignments. In order to eliminate unnecessary verbiage and to free the controller/specialist for higher priority transmissions, the controller/specialist may request the pilot "(Identification), change to my frequency 123.4." This phrase should alert the pilot that he is only changing frequencies, not controller/specialist, and that initial callup phraseology may be abbreviated.

EXAMPLE:
"UNITED TWO TWENTY-TWO ON ONE TWO THREE POINT FOUR."

e. Compliance with Frequency Changes—

When instructed by ATC to change frequencies, select the new frequency as soon as possible unless instructed to make the change at a specific time, fix, or altitude. A delay in making the change could result in an untimely receipt of important information. If you are instructed to make the frequency change at a specific time, fix, or altitude, monitor the frequency you are on until reaching the specified time, fix, or altitudes unless instructed otherwise by ATC. (Reference—ARTCC Communications, paragraph 5-30.)

4-33. AIRCRAFT CALL SIGNS

a. Precautions in the Use of Call Signs—

1. Improper use of call signs can result in pilots executing a clearance intended for another aircraft. Call signs should *never be abbreviated on an initial contact or at any time when other aircraft call signs have similar numbers/sounds or identical letters/number*; e.g., Cessna 6132F, Cessna 1622F, Baron 123F, Cherokee 7732F, etc.

EXAMPLE: Assume that a controller issues an approach clearance to an aircraft at the bottom of a holding stack and an aircraft with a similar call sign (at the top of the stack) acknowledges the clearance with the last two or three numbers of his call sign. If the aircraft at the bottom of the stack did not hear the clearance and intervene, flight safety would be affected, and there would be no reason for either the controller or pilot to suspect that anything is wrong. This kind of "human factors" error can strike swiftly and is extremely difficult to rectify.

2. Pilots, therefore, must be certain that aircraft identification is complete and clearly identified before taking action on an ATC clearance. ATC specialists will not abbreviate call signs of air carrier or other civil aircraft having authorized call signs. ATC specialists may initiate abbreviated call signs of other aircraft by using the *prefix and the last three digits/letters* of the aircraft identification after communications are established. The pilot may use the abbreviated call sign in subsequent contacts with the ATC specialist. When aware of similar/identical call signs, ATC specialists will take action to minimize errors by emphasizing certain numbers/letters, by repeating the entire call sign, by repeating the prefix, or by asking pilots to use a different call sign temporarily. Pilots should use the phrase "VERIFY CLEARANCE FOR (your complete call sign)" if doubt exists concerning proper identity.

3. Civil aircraft pilots should state the aircraft type, model or manufacturer's name, followed by the digits/letters of the registration number. When the aircraft manufacturer's name or model is stated, the prefix "N" is dropped; e.g., Aztec Two Four Six Four Alpha.

EXAMPLE:
BONANZA SIX FIVE FIVE GOLF.

EXAMPLE:
BREEZY SIX ONE THREE ROMEO EXPERIMENTAL (omit "Experimental" after initial contact).

4. Air Taxi or other commercial operators *not* having FAA authorized call signs should prefix their normal identification with the phonetic word "Tango. "

EXAMPLE:
TANGO AZTEC TWO FOUR SIX FOUR ALPHA.

5. Air carriers and commuter air carriers having FAA authorized call signs should identify themselves by stating the complete call sign (using group form for the numbers) and the word "heavy" if appropriate.

EXAMPLE:
UNITED TWENTY-FIVE HEAVY.

EXAMPLE:
MIDWEST COMMUTER SEVEN ELEVEN.

6. Military aircraft use a variety of systems including serial numbers, word call signs, and combinations of letters/numbers. Examples include Army Copter 48931, Air Force 61782, REACH 31792, Pat 157, Air Evac 17652, Navy Golf Alfa Kilo 21, Marine 4 Charlie 36, etc.

b. Air Ambulance Flights—

Because of the priority afforded air ambulance flights in the ATC system, extreme discretion is necessary when using the term "LIFEGUARD." It is only intended for those missions of an urgent medical nature and to be utilized only for that portion of the flight requiring expeditious handling. When requested by the pilot, necessary notification to expedite ground handling of patients, etc., is provided by ATC; however, when possible, this information should be passed in advance through non-ATC communications systems.

1. Civilian air ambulance flights responding to medical emergencies (first call to an accident scene, carrying patients, organ donors, organs, or other urgently needed lifesaving medical material) will be expedited by ATC when necessary. When expeditious handling is necessary, add the word "LIFEGUARD" in the remarks section of the flight plan. In radio communications, use the call sign "LIFEGUARD" followed by the aircraft registration letters/numbers.

2. Similar provisions have been made for the use of "AIR EVAC" and "MED EVAC" by military air ambulance flights, except that these military flights will receive priority handling only when specifically requested.

EXAMPLE:
LIFEGUARD TWO SIX FOUR SIX.

c. Student Pilots Radio Identification—

3. Air carrier and Air Taxi flights responding to medical emergencies will also be expedited by ATC when necessary. The nature of these medical emergency flights usually concerns the transportation of urgently needed lifesaving medical materials or vital organs. IT IS IMPERATIVE THAT THE COMPANY/PILOT DETERMINE, BY THE NATURE/URGENCY OF THE SPECIFIC MEDICAL CARGO, IF PRIORITY ATC ASSISTANCE IS REQUIRED. Pilots shall ensure that the word "LIFEGUARD" is included in the remarks section of the flight plan and use the call sign "LIFEGUARD" followed by the company name and flight number for all transmissions when expeditious handling is required. It is important for ATC to be aware of "LIFEGUARD" status, and it is the pilot's responsibility to ensure that this information is provided to ATC.

EXAMPLE:
LIFEGUARD DELTA THIRTY-SEVEN.

1. The FAA desires to help the student pilot in acquiring sufficient practical experience in the environment in which he will be required to operate. To receive additional assistance while operating in areas of concentrated air traffic, a student pilot need only identify himself as a student pilot during his initial call to an FAA radio facility.

EXAMPLE:
DAYTON TOWER, THIS IS FLEETWING ONE TWO THREE FOUR, STUDENT PILOT.

2. This special identification will alert FAA ATC personnel and enable them to provide the student pilot with such extra assistance and consideration as he may need. This procedure is not mandatory.

4-34. DESCRIPTION OF INTERCHANGE OR LEASED AIRCRAFT

a. Controllers issue traffic information based on familiarity with airline equipment and color/markings. When an air carrier dispatches a flight using another company's equipment and the pilot does not advise the terminal ATC facility, the possible confusion in aircraft identification can compromise safety.

b. Pilots flying an "interchange" or "leased" aircraft not bearing the colors/markings of the company operating the aircraft should inform the terminal ATC facility on first contact the name of the operating company and trip number, followed by the company name as displayed on the aircraft, and aircraft type.

EXAMPLE:
AIR CAL THREE ELEVEN, UNITED (INTERCHANGE/LEASE), BOEING SEVEN TWO SEVEN.

Table 4-35[1]

Facility	Call Sign
Airport UNICOM	"Shannon UNICOM"
FAA Flight Service Station	"Chicago Radio"
FAA Flight Service Station (En Route Flight Advisory Service (Weather))	"Seattle Flight Watch"
Airport Traffic Control Tower	"Augusta Tower"
Clearance Delivery Position (IFR)	"Dallas Clearance Delivery"
Ground Control Position in Tower	"Miami Ground"
Radar or Nonradar Approach Control Position	"Oklahoma City Approach"
Radar Departure Control Position	"St. Louis Departure"
FAA Air Route Traffic Control Center	"Washington Center"

4-35. GROUND STATION CALL SIGNS

Pilots, when calling a ground station, should begin with the name of the facility being called followed by the type of the facility being called as indicated in the Table 4-35[1].

4-36. PHONETIC ALPHABET

The International Civil Aviation Organization (ICAO) phonetic alphabet is used by FAA personnel when communications conditions are such that the information cannot be readily received without their use. ATC facilities may also request pilots to use phonetic letter equivalents when aircraft with similar sounding identifications are receiving communications on the same frequency. Pilots should use the phonetic alphabet when identifying their aircraft during initial contact with air traffic control facilities. Additionally, use the phonetic equivalents for single letters and to spell out groups of letters or difficult words during adverse communications conditions. (See Table 4-36[1].)

Table 4-36[1]

CHARACTER	MORSE CODE	TELEPHONY	PHONIC (PRONUNCIATION)
A	•—	Alfa	(AL-FAH)
B	—•••	Bravo	(BRAH-VOH)
C	—•—•	Charlie	(CHAR-LEE) or (SHAR-LEE)
D	—••	Delta	(DELL-TAH)
E	•	Echo	(ECK-OH)
F	••—•	Foxtrot	(FOKS-TROT)
G	——•	Golf	(GOLF)
H	••••	Hotel	(HOH-TEL)
I	••	India	(IN-DEE-AH)
J	•———	Juliett	(JEW-LEE-ETT)
K	—•—	Kilo	(KEY-LOH)
L	•—••	Lima	(LEE-MAH)
M	——	Mike	(MIKE)
N	—•	November	(NO-VEM-BER)
O	———	Oscar	(OSS-CAH)
P	•——•	Papa	(PAH-PAH)
Q	——•—	Quebec	(KEH-BECK)
R	•—•	Romeo	(ROW-ME-OH)
S	•••	Sierra	(SEE-AIR-RAH)
T	—	Tango	(TANG-GO)
U	••—	Uniform	(YOU-NEE-FORM) or (OO-NEE-FORM)
V	•••—	Victor	(VIK-TAH)
W	•——	Whiskey	(WISS-KEY)
X	—••—	Xray	(ECKS-RAY)
Y	—•——	Yankee	(YANG-KEY)
Z	——••	Zulu	(ZOO-LOO)
1	•————	One	(WUN)
2	••———	Two	(TOO)
3	•••——	Three	(TREE)
4	••••—	Four	(FOW-ER)
5	•••••	Five	(FIFE)
6	—••••	Six	(SIX)
7	——•••	Seven	(SEV-EN)
8	———••	Eight	(AIT)
9	————•	Nine	(NIN-ER)
0	—————	Zero	(ZEE-RO)

4-37. FIGURES

a. Figures indicating hundreds and thousands in round number, as for ceiling heights, and upper wind levels up to 9,900 shall be spoken in accordance with the following:

EXAMPLE:

Table 4-37[1]

500	FIVE HUNDRED

EXAMPLE:

Table 4-37[2]

4,500	FOUR THOUSAND FIVE HUNDRED

b. Numbers above 9,900 shall be spoken by separating the digits preceding the word "thousand."

EXAMPLE:

Table 4-37[3]

10,000	ONE ZERO THOUSAND

EXAMPLE:

Table 4-37[4]

13,500	ONE THREE THOUSAND FIVE HUNDRED

c. Transmit airway or jet route numbers as follows:

EXAMPLE:

Table 4-37[5]

V12	VICTOR TWELVE

EXAMPLE:

Table 4-37[6]

J533	J FIVE THIRTY-THREE

d. All other numbers shall be transmitted by pronouncing each digit.

EXAMPLE:

Table 4-37[7]

10	ONE ZERO

e. When a radio frequency contains a decimal point, the decimal point is spoken as "POINT."

EXAMPLE:

Table 4-37[8]

122.1	ONE TWO TWO POINT ONE

4-37e NOTE—ICAO Procedures require the decimal point be spoken as "DECIMAL," and FAA will honor such usage by military aircraft and all other aircraft required to use ICAO Procedures.

4-38. ALTITUDES AND FLIGHT LEVELS

a. Up to but not including 18,000 feet MSL, state the separate digits of the thousands plus the hundreds if appropriate.

EXAMPLE:

Table 4-38[1]

12,000	ONE TWO THOUSAND

EXAMPLE:

Table 4-38[2]

12,500	ONE TWO THOUSAND FIVE HUNDRED

b. At and above 18,000 feet MSL (FL 180), state the words "flight level" followed by the separate digits of the flight level.

EXAMPLE:

Table 4-38[3]

190	FLIGHT LEVEL ONE NINER ZERO

4-39. DIRECTIONS

The three digits of bearing, course, heading, or wind direction should always be magnetic. The word "true" must be added when it applies.

EXAMPLE:

Table 4-39[1]

(Magnetic course) 005	ZERO ZERO FIVE

EXAMPLE:

Table 4-39[2]

(True course) 050	ZERO FIVE ZERO TRUE

EXAMPLE:

Table 4-39[3]

(Magnetic bearing) 360	THREE SIX ZERO

EXAMPLE:

Table 4-39[4]

(Magnetic heading) 100	ONE ZERO ZERO

EXAMPLE:

Table 4-39[5]

(Wind direction) 220	TWO TWO ZERO

4-40. SPEEDS

The separate digits of the speed followed by the word "KNOTS." Except, controllers may omit the word "KNOTS" when using speed adjustment procedures; e.g., "REDUCE/INCREASE SPEED TO TWO FIVE ZERO."

EXAMPLES:

Table 4-40[1]

(Speed) 250	TWO FIVE ZERO KNOTS
(Speed) 190	ONE NINER ZERO KNOTS

The separate digits of the MACH Number preceded by "MACH."

EXAMPLES:

Table 4-40[2]

(Mach number) 1.5	MACH ONE POINT FIVE
(Mach number) 0.64	MACH POINT SIX FOUR
(Mach number) 0.7	MACH POINT SEVEN

4-41. TIME

a. FAA uses Coordinated Universal Time (UTC) for all operations. The term "Zulu" is used when ATC procedures require a reference to UTC.

EXAMPLE:

Table 4-41[1]

0920	ZERO NINER TWO ZERO

b. To Convert from Standard Time to Coordinated Universal Time:

Table 4-41[2]

Eastern Standard Time	Add 5 hours
Central Standard Time	Add 6 hours
Mountain Standard Time	Add 7 hours
Pacific Standard Time	Add 8 hours
Alaska Standard Time	Add 9 hours
Hawaii Standard Time	Add 10 hours

4-41b NOTE—For Daylight Time, subtract 1 hour.

c. The 24-hour clock system is used in radiotelephone transmissions. The hour is indicated by the first two figures and the minutes by the last two figures.

EXAMPLE:

Table 4-41[3]

0000	ZERO ZERO ZERO ZERO

EXAMPLE:

Table 4-41[4]

0920	ZERO NINER TWO ZERO

d. Time may be stated in minutes only (two figures) in radio telephone communications when no misunderstanding is likely to occur.

e. Current time in use at a station is stated in the nearest quarter minute in order that pilots may use this information for time checks. Fractions of a quarter minute less than 8 seconds are stated as the preceding quarter minute; fractions of a quarter minute of 8 seconds or more are stated as the succeeding quarter minute.

EXAMPLE:

Table 4-41[5]

0929:05	TIME, ZERO NINER TWO NINER

EXAMPLE:

Table 4-41[6]

0929:10	TIME, ZERO NINER TWO NINER AND ONE-QUARTER

4-42. COMMUNICATIONS WITH TOWER WHEN AIRCRAFT TRANSMITTER OR RECEIVER OR BOTH ARE INOPERATIVE

a. Arriving Aircraft—

1. Receiver inoperative—If you have reason to believe your receiver is inoperative, remain outside or above the Class D surface area until the direction and flow of traffic has been determined; then, advise the tower of your type aircraft, position, altitude, intention to land, and request that you be controlled with light signals. (Reference—Traffic Control Light Signals, paragraph 4-62.) When you are approximately 3 to 5 miles from the airport, advise the tower of your position and join the airport traffic pattern. From this point on, watch the tower for light signals. Thereafter, if a complete pattern is made, transmit your position downwind and/or turning base leg.

2. Transmitter inoperative—Remain outside or above the Class D surface area until the direction and flow of traffic has been determined; then, join the airport traffic pattern. Monitor the primary local control frequency as depicted on Sectional Charts for landing or traffic information, and look for a light signal which may be addressed to your aircraft. During hours of daylight, acknowledge tower transmissions or light signals by rocking your wings. At night, acknowledge by blinking the landing or navigation lights. To acknowledge tower transmissions during daylight hours, hovering helicopters will turn in the direction of the controlling facility and flash the landing light. While in flight, helicopters should show their acknowledgement of receiving a transmission by making shallow banks in opposite directions. At night, helicopters will acknowledge receipt of transmissions by flashing either the landing or the search light.

3. Transmitter and receiver inoperative—Remain outside or above the Class D surface area until the direction 4-

and flow of traffic has been determined; then, join the airport traffic pattern and maintain visual contact with the tower to receive light signals. Acknowledge light signals as noted above.

b. Departing Aircraft—If you experience radio failure prior to leaving the parking area, make every effort to have the equipment repaired. If you are unable to have the malfunction repaired, call the tower by telephone and request authorization to depart without two-way radio communications. If tower authorization is granted, you will be given departure information and requested to monitor the tower frequency or watch for light signals as appropriate. During daylight hours, acknowledge tower transmissions or light signals by moving the ailerons or rudder. At night, acknowledge by blinking the landing or navigation lights. If radio malfunction occurs after departing the parking area, watch the tower for light signals or monitor tower frequency.

4-42b NOTE—Refer to FAR Part 91.129 and Part 91.125.

4-43. COMMUNICATIONS FOR VFR FLIGHTS

a. FSSs and Supplemental Weather Service Locations (SWSLs) are allocated frequencies for different functions; for example, 122.0 MHz is assigned as the En Route Flight Advisory Service frequency at selected FSSs. In addition, certain FSSs provide Local Airport Advisory on 123.6 MHz. Frequencies are listed in the Airport/Facility Directory. If you are in doubt as to what frequency to use, 122.2 MHz is assigned to the majority of FSSs as a common en route simplex frequency.

4-43a NOTE—In order to expedite communications, state the frequency being used and the aircraft location during initial callup.

EXAMPLE:

DAYTON RADIO, THIS IS NOVEMBER ONE TWO THREE FOUR FIVE ON ONE TWO TWO POINT TWO, OVER SPRINGFIELD V-O-R, OVER.

b. Certain VOR voice channels are being utilized for recorded broadcasts; i.e., ATIS, HIWAS, etc. These services and appropriate frequencies are listed in the Airport/Facility Directory. On VFR flights, pilots are urged to monitor these frequencies. When in contact with a control facility, notify the controller if you plan to leave the frequency to monitor these broadcasts.

4-44 thru 4-49. RESERVED

Section 3. AIRPORT OPERATIONS

4-50. GENERAL

Increased traffic congestion, aircraft in climb and descent attitudes, and pilots' preoccupation with cockpit duties are some factors that increase the hazardous accident potential near the airport. The situation is further compounded when the weather is marginal—that is, just meeting VFR requirements. Pilots must be particularly alert when operating in the vicinity of an airport. This section defines some rules, practices, and procedures that pilots should be familiar with and adhere to for safe airport operations.

4-51. TOWER CONTROLLED AIRPORTS

a. When operating at an airport where traffic control is being exercised by a control tower, pilots are required to maintain two-way radio contact with the tower while operating within the Class B, Class C, and Class D surface area unless the tower authorizes otherwise. Initial callup should be made about 15 miles from the airport. Unless there is a good reason to leave the tower frequency before exiting the Class B, Class C, and Class D surface area, it is a good operating practice to remain on the tower frequency for the purpose of receiving traffic information. In the interest of reducing tower frequency congestion, pilots are reminded that it is not necessary to request permission to leave the tower frequency once outside of Class B, Class C, and Class D surface area. Not all airports with an operating control tower will have Class D airspace. These airports do not have weather reporting which is a requirement for surface based controlled airspace, previously known as a control zone. The controlled airspace over these airports will normally begin at 700 feet or 1200 feet above ground level and can be determined from the visual aeronautical charts. Pilots are expected to use good operating practices and communicate with the control tower as described in this section.

b. When necessary, the tower controller will issue clearances or other information for aircraft to generally follow the desired flight path (traffic patterns) when flying in Class B, Class C, and Class D surface areas and the proper taxi routes when opera ting on the ground. If not otherwise authorized or directed by the tower, pilots of fixed-wing aircraft approaching to land must circle the airport to the left. Pilots approaching to land in a helicopter must avoid the flow of fixed-wing traffic. However, in all instances, an appropriate clearance must be received from the tower before landing.

c. The following terminology for the various components of a traffic pattern has been adopted as standard for use by control towers and pilots. (See Figure 4-51[1].) (**NOTE**—This diagram is intended only to illustrate terminology used in identifying various components of a traffic pattern. *It should not be used as a reference or guide on how to enter a traffic pattern.*)

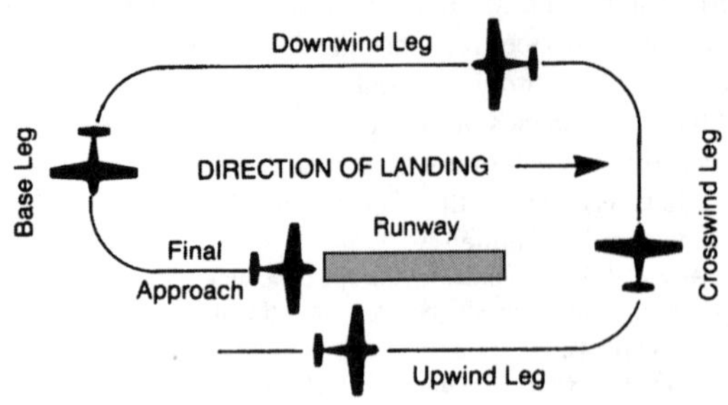

NOTE—**This diagram is intended only to illustrate terminology used in identifying various components of a traffic pattern. It should not be used as a reference or guide on how to enter a traffic pattern.**

Figure 4-51[1]

1. Upwind leg—A flight path parallel to the landing runway in the direction of landing.

2. Crosswind leg—A flight path at right angles to the landing runway off its takeoff end.

3. Downwind leg—A flight path parallel to the landing runway in the opposite direction of landing.

4. Base leg—A flight path at right angles to the landing runway off its approach end and extending from the downwind leg to the intersection of the extended runway centerline.

5. Final approach—A flight path in the direction of landing along the extended runway centerline from the base leg to the runway.

d. Many towers are equipped with a tower radar display. The radar uses are intended to enhance the effectiveness and efficiency of the local control, or tower, position. They are not intended to provide radar services or benefits to pilots except as they may accrue through a more efficient tower operation. The four basic uses are:

1. To determine an aircraft's exact location—This is accomplished by radar identifying the VFR aircraft through any of the techniques available to a radar position, such as having the aircraft *squawk ident.* Once identified, the aircraft's position and spatial relationship to other aircraft can be quickly determined, and standard instructions regarding VFR operation in Class B, Class C, and Class D surface areas will be issued. Once initial radar identification of a VFR aircraft has been established and the appropriate instructions have been issued, radar monitoring may be discontinued; the reason being that the local controller's primary means of surveillance in VFR conditions is visually scanning the airport and local area.

2. To provide radar traffic advisories—Radar traffic advisories may be provided to the extent that the local controller is able to monitor the radar display. Local control has primary control responsibilities to the aircraft operating on the runways, which will normally supersede radar monitoring duties.

3. To provide a direction or suggested heading—The local controller may provide pilots flying VFR with generalized instructions which will facilitate operations; e.g.,"PROCEED SOUTHWEST BOUND, ENTER A RIGHT DOWNWIND RUNWAY THREE ZERO," or provide a suggested heading to establish radar identification or as an advisory aid to navigation; e.g., "SUGGESTED HEADING TWO TWO ZERO, FOR RADAR IDENTIFICATION." In both cases, the instructions are advisory aids to the pilot flying VFR and are not radar vectors. PILOTS HAVE COMPLETE DISCRETION REGARDING ACCEPTANCE OF THE SUGGESTED HEADINGS OR DIRECTIONS AND HAVE SOLE RESPONSIBILITY FOR SEEING AND AVOIDING OTHER AIRCRAFT.

4. To provide information and instructions to aircraft operating within Class B, Class C, and Class D surface areas—In an example of this situation, the local controller would use the radar to advise a pilot on an extended downwind when to turn base leg.

4-51d4 NOTE—The above tower radar applications are intended to augment the standard functions of the local control position. There is no controller requirement to maintain constant radar identification. In fact, such a requirement could compromise the local controller's ability to visually scan the airport and local area to meet FAA responsibilities to the aircraft operating on the runways and within the Class B, Class C, and Class D surface areas. Normally, pilots will not be advised of being in radar contact since that continued status cannot be guaranteed and since the purpose of the radar identification is not to establish a link for the provision of radar services.

e. A few of the radar equipped towers are authorized to use the radar to ensure separation between aircraft in specific situations, while still others may function as limited radar approach controls. The various radar uses are strictly a function of FAA operational need. The facilities may be indistinguishable to pilots since they are all referred to as tower and no publication lists the degree of radar use. Therefore, WHEN IN COMMUNICATION WITH A TOWER CONTROLLER WHO MAY HAVE RADAR AVAILABLE, DO NOT ASSUME THAT CONSTANT RADAR MONITORING AND COMPLETE ATC RADAR SERVICES ARE BEING PROVIDED.

4-52. VISUAL INDICATORS AT UNCONTROLLED AIRPORTS

a. At those airports *without an operating control tower*, a segmented circle visual indicator system, if installed, is designed to provide traffic pattern information. (Reference—Traffic Advisory Practices at Airports without Operating Control Towers, paragraph 4-9.) The segmented circle system consists of the following components:

1. The segmented circle—Located in a position affording maximum visibility to pilots in the air and on the ground and providing a centralized location for other elements of the system.

2. The wind direction indicator—A wind cone, wind sock, or wind tee installed near the operational runway to indicate wind direction. The large end of the wind cone/wind sock points into the wind as does the large end (cross bar) of the wind tee. In lieu of a tetrahedron and where a wind sock or wind cone is collocated with a wind tee, the wind tee may be manually aligned with the runway in use to indicate landing direction. These signaling devices may be located in the center of the segmented circle and may be lighted for night use. Pilots are cautioned against using a tetrahedron to indicate wind direction.

3. The landing direction indicator—A tetrahedron is installed when conditions at the airport warrant its use. It may be used to indicate the direction of landings and takeoffs. A tetrahedron may be located at the center of a segmented circle and may be lighted for night operations. The small end of the tetrahedron points in the direction of landing. Pilots are cautioned against using a tetrahedron for any purpose other than as an indicator of landing direction. Further, pilots should use extreme caution when making runway selection by use of a tetrahedron in very light or calm wind conditions as the tetrahedron may not be aligned with the designated calm-wind runway. At airports with control towers, the tetrahedron should only be referenced when the control tower is not in operation. Tower instructions supersede tetrahedron indications.

4. Landing strip indicators—Installed in pairs as shown in the segmented circle diagram and used to show the alignment of landing strips.

5. Traffic pattern indicators—Arranged in pairs in conjunction with landing strip indicators and used to indicate the direction of turns when there is a variation from the normal left traffic pattern. (If there is no segmented circle installed at the airport, traffic pattern indicators may be installed on or near the end of the runway.)

b. Preparatory to landing at an airport without a control tower, or when the control tower is not in operation, the pilot should concern himself with the indicator for the approach end of the runway to be used. When approaching for landing, all turns must be made to the left unless a traffic pattern indicator indicates that turns should be made to the right. If the pilot will mentally enlarge the indicator for the runway to be used, the base and final approach legs of the traffic pattern to be flown immediately become apparent. Similar treatment of the indicator at the departure end of the runway will clearly indicate the direction of turn after takeoff.

c. When two or more aircraft are approaching an airport for the purpose of landing, the aircraft at the lower altitude has the right of way, but it shall not take advantage of this rule to cut in front of another which is on final approach to land, or to overtake that aircraft (FAR Part 91.113(f)).

4-53. TRAFFIC PATTERNS

At most airports and military air bases, traffic pattern altitudes for propeller-driven aircraft generally extend from 600 feet to as high as 1,500 feet above the ground. Also, traffic pattern altitudes for military turbojet aircraft sometimes extend up to 2,500 feet above the ground. Therefore, pilots of en route aircraft should be constantly on the alert for other aircraft in traffic patterns and avoid these areas whenever possible. Traffic pattern altitudes should be maintained unless otherwise required by the applicable distance from cloud criteria (FAR Part 91.155). (See Figure 4-53[1] and Figure 4-53[2].)

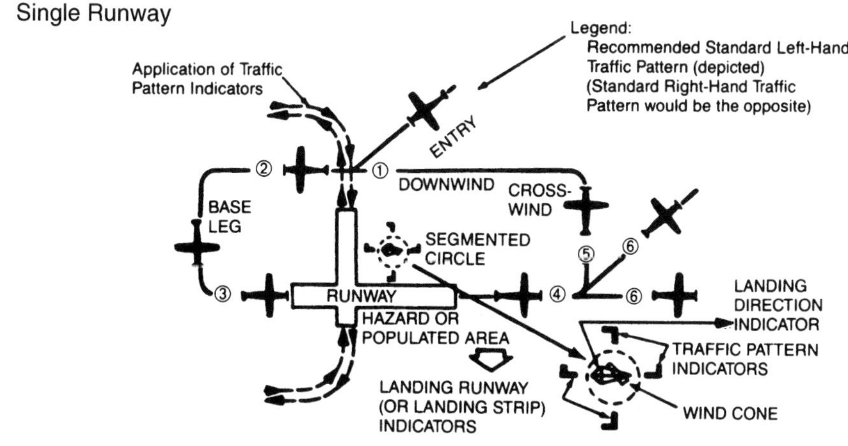

Figure 4-53[1]

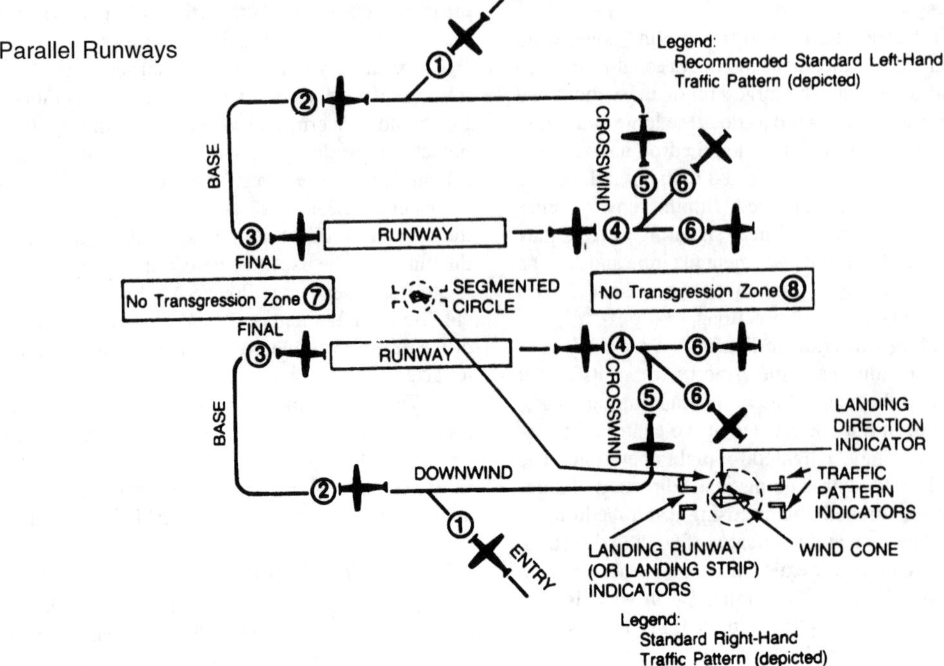

Key:

① Enter pattern in level flight, abeam the midpoint of the runway, at pattern altitude. (1000' AGL is recommended pattern altitude unless established otherwise.)

② Maintain pattern altitude until abeam approach end of the landing runway, on downwind leg.

③ Complete turn to final at least 1/4 mile from the runway.

④ Continue straight ahead until beyond departure end of runway.

⑤ If remaining in the traffic pattern, commence turn to crosswind leg beyond the departure end of the runway, within 300 feet of pattern altitude.

⑥ If departing the traffic pattern, continue straight out, or exit with a 45° left turn beyond the departure end of the runway, after reaching pattern altitude.

⑦ Do not overshoot final or continue on a track which will penetrate the final approach of the parallel runway.

⑧ Do not continue on a track which will penetrate the departure path of the parallel runway.

Figure 4-53[2]

4-54. UNEXPECTED MANEUVERS IN THE AIRPORT TRAFFIC PATTERN

There have been several incidents in the vicinity of controlled airports that were caused primarily by aircraft executing unexpected maneuvers. ATC service is based upon observed or known traffic and airport conditions. Controllers establish the sequence of arriving and departing aircraft by requiring them to adjust flight as necessary to achieve proper spacing. These adjustments can only be based on observed traffic, accurate pilot reports, and anticipated aircraft maneuvers. Pilots are expected to cooperate so as to preclude disrupting traffic flows or creating conflicting patterns. The pilot-in-command of an aircraft is directly responsible for and is the final authority as to the operation of his aircraft. On occasion it may be necessary for a pilot to maneuver his aircraft to maintain spacing with the traffic he has been sequenced to follow. The controller can anticipate minor maneuvering such as shallow "S" turns. The controller cannot, however, anticipate a major maneuver such as a 360 degree turn. If a pilot makes a 360 degree turn after he has obtained a landing sequence, the result is usually a gap in the landing interval and, more importantly, it causes a chain reaction which may result in a conflict with following traffic and an interruption of the sequence established by the tower or approach controller. Should a pilot decide he needs to make maneuvering turns to maintain spacing behind a preceding aircraft, he should always advise the controller if at all possible. Except when requested by the controller or in emergency situations, a 360 degree turn should never be executed in the traffic pattern or when receiving radar service without first advising the controller.

4-55. USE OF RUNWAYS/DECLARED DISTANCES

a. Runways are identified by numbers which indicate the nearest 10-degree increment of the azimuth of the runway centerline. For example, where the magnetic azimuth is 183 degrees, the runway designation would be 18; for a magnetic azimuth of 87 degrees, the runway designation would be 9. For a magnetic azimuth ending in the number 5, such as 185, the runway designation could be either 18 or 19. Wind direction issued by the tower is also magnetic and wind velocity is in knots.

b. Airport proprietors are responsible for taking the lead in local aviation noise control. Accordingly, they may propose specific noise abatement plans to the FAA. If approved, these plans are applied in the form of Formal or Informal Runway Use Programs for noise abatement purposes. (Reference—Pilot/Controller Glossary, Runway Use Program.)

1. At airports where no runway use program is established, ATC clearances may specify:

(a) The runway most nearly aligned with the wind when it is 5 knots or more;

(b) The "calm wind" runway when wind is less than 5 knots, or;

(c) Another runway if operationally advantageous.

4-55b1c NOTE—It is not necessary for a controller to specifically inquire if the pilot will use a specific runway or to offer him a choice of runways. If a pilot prefers to use a different runway from that specified or the one most nearly aligned with the wind, he is expected to inform ATC accordingly.

2. At airports where a runway use program is established, ATC will assign runways deemed to have the least noise impact. If in the interest of safety a runway different from that specified is preferred, the pilot is expected to advise ATC accordingly. ATC will honor such requests and advise pilots when the requested runway is noise sensitive. When use of a runway other than the one assigned is requested, pilot cooperation is encouraged to preclude disruption of traffic flows or the creation of conflicting patterns.

c. At some airports, the airport proprietor may declare that sections of a runway at one or both ends are not available for landing or takeoff. For these airports, the declared distance of runway length available for a particular operation is published in the Airport/Facility Directory. Declared distances (TORA, TODA, ASDA, and LDA) are defined in the Pilot/Controller Glossary. These distances are calculated by adding to the full length of paved runway any applicable clearway or stopway and subtracting from that sum the sections of the runway unsuitable for satisfying the required takeoff run, takeoff, accelerate/stop, or landing distance.

4-56. LOW LEVEL WIND SHEAR ALERT SYSTEM (LLWAS)

a. This computerized system detects the presence of a possible hazardous low-level wind shear by continuously comparing the winds measured by sensors installed around the periphery of an airport with the wind measured at the center field location. If the difference between the center field wind sensor and a peripheral wind sensor becomes excessive, a thunderstorm or thunderstorm gust front wind shear is probable. When this condition exists, the tower controller will provide arrival and departure aircraft with an advisory of the situation, which includes the center field wind plus the remote site location and wind.

b. Since the sensors are not all associated with specific runways, descriptions of the remote sites will be based on an eight-point compass system.

EXAMPLE:

DELTA ONE TWENTY FOUR CENTER FIELD WIND TWO SEVEN ZERO AT ONE ZERO. SOUTH BOUNDARY WIND ONE FOUR ZERO AT THREE ZERO.

c. An airport equipped with the Low Level Wind Shear Alert System is so indicated in the Airport/Facility Directory under *Weather Data Sources* for that particular airport.

4-57. BRAKING ACTION REPORTS AND ADVISORIES

a. When available, ATC furnishes pilots the quality of braking action received from pilots or airport management. The quality of braking action is described by the terms "good," "fair," "poor," and "nil," or a combination of these terms. When pilots report the quality of braking action by using the terms noted above, they should use descriptive terms that are easily understood, such as, "braking action poor the first/last half of the runway," together with the particular type of aircraft.

b. For NOTAM purposes, braking action reports are classified according to the most critical term used. Reports containing the term "fair" are classified as NOTAM(L). Reports containing the terms "poor" or "nil" are classified as NOTAM(D).

c. When tower controllers have received runway braking action reports which include the terms *poor* or *nil*, or whenever weather conditions are conducive to deteriorating or rapidly changing runway braking conditions, the tower will include on the ATIS broadcast the statement, "*BRAKING ACTION ADVISORIES ARE IN EFFECT.*"

d. During the time that braking action advisories are in effect, ATC will issue the latest braking action report for the runway in use to each arriving and departing aircraft. Pilots should be prepared for deteriorating braking conditions and should request current runway condition information if not volunteered by controllers. Pilots should also be prepared to provide a descriptive runway condition report to controllers after landing.

4-58. RUNWAY FRICTION REPORTS AND ADVISORIES

a. Friction is defined as the ratio of the tangential force needed to maintain uniform relative motion between two contacting surfaces (aircraft tires to the pavement surface) to the perpendicular force holding them in contact (distributed aircraft weight to the aircraft tire area). Simply stated, friction quantifies slipperiness of pavement surfaces.

b. The greek letter MU (pronounced "myew"), is used to designate a friction value representing runway surface conditions.

c. MU (friction) values range from 0 to 100 where zero is the lowest friction value and 100 is the maximum friction value obtainable. For frozen contaminants on runway surfaces, a MU value of 40 or less is the level when the aircraft braking performance starts to deteriorate and directional control begins to be less responsive. The lower the MU value, the less effective braking performance becomes and the more difficult directional control becomes.

d. At airports with friction measuring devices, airport management should conduct friction measurements on runways covered with compacted snow and/or ice.

1. Numerical readings may be obtained by using any FAA approved friction measuring device. It is not necessary to designate the type of friction measuring device since they provide essentially the same numerical reading when the values are 40 or less.

2. When the MU value for any one-third zone of an active runway is 40 or less, a report should be given to ATC by airport management for dissemination to pilots. The report will identify the runway, the time of measurement, MU values for each zone, and the contaminant conditions, e.g., wet snow, dry snow, slush, deicing chemicals, etc. Measurements for each one third zone will be given in the direction of takeoff and landing on the runway. A report should also be given when MU values rise above 40 in all zones of a runway previously reporting a MU below 40.

3. Airport management should initiate a NOTAM(D) when the friction measuring device is out of service.

e. When MU reports are provided by airport management, the ATC facility providing approach control or local airport advisory will provide the report to any pilot upon request.

f. Pilots should use MU information with other knowledge including aircraft performance characteristics, type, and weight, previous experience, wind conditions, and aircraft tire type (i.e., bias ply vs. radial constructed) to determine runway suitability.

g. No correlation has been established between MU values and the descriptive terms "good," "fair," "poor," and "nil" used in braking action reports.

4-59. INTERSECTION TAKEOFFS

a. In order to enhance airport capacities, reduce taxiing distances, minimize departure delays, and provide for more efficient movement of air traffic, controllers may initiate intersection takeoffs as well as approve them when the pilot requests. If for ANY reason a pilot prefers to use a different intersection or the full length of the runway or desires to obtain the distance between the intersection and the runway end, HE IS EXPECTED TO INFORM ATC ACCORDINGLY.

b. An aircraft is expected to taxi to (but not onto) the end of the assigned runway unless prior approval for an intersection departure is received from ground control.

c. Pilots should state their position on the airport when calling the tower for takeoff from a runway intersection.

EXAMPLE:

CLEVELAND TOWER, APACHE 3722P, AT THE INTERSECTION OF TAXIWAY OSCAR AND RUNWAY TWO THREE RIGHT, READY FOR DEPARTURE.

d. Controllers are required to separate small aircraft (12,500 pounds or less, maximum certificated takeoff weight) departing (same or opposite direction) from an intersection behind a large nonheavy aircraft on the same runway, by ensuring that at least a 3-minute interval exists between the time the preceding large aircraft has taken off and the succeeding small aircraft begins takeoff roll. To inform the pilot of the required 3-minute hold, the controller will state, "Hold for wake turbulence." If after considering wake turbulence hazards, the pilot feels that a lesser time interval is appropriate, he may request a waiver to the 3-minute interval. Pilots must initiate such a request by stating, "Request waiver to 3-minute interval," or by making a similar statement. Controllers may then issue a takeoff clearance if other traffic permits, since the pilot has accepted responsibility for his own wake turbulence separation.

e. The 3-minute interval is not required when the intersection is 500 feet or less from the departure point of the preceding aircraft and both aircraft are taking off in the same direction. Controllers may permit the small aircraft to alter course after takeoff to avoid the flight path of the preceding departure.

f. The 3-minute interval is mandatory behind a heavy aircraft in all cases.

4-60. SIMULTANEOUS OPERATIONS ON INTERSECTING RUNWAYS

a. Despite the many new and lengthened runways which have been added to the nation's airports in recent years, limited runway availability remains a major contributing factor to operational delays. Many high-density airports have gained operational experience with intersecting runways which clearly indicates that simultaneous operations are safe and feasible. Tower controllers may authorize simultaneous landings or a simultaneous landing and takeoff on intersecting runways when the following conditions are met:

1. The runways are dry and the controller has received no reports that braking action is less than good.

2. A simultaneous takeoff and landing operation may be conducted only in VFR conditions.

3. Instructions are issued to restrict one aircraft from entering the intersecting runway being used by another aircraft.

4. Traffic information issued is acknowledged by the pilots of both aircraft.

5. The measured distance from runway threshold to intersection is issued if the pilot requests it.

6. The conditions specified in 3, 4 and 5 are met at or before issuance of the landing clearance.

7. The distance from landing threshold to the intersection is adequate for the category of aircraft being held short. Controllers are provided a general table of aircraft category/minimum runway length requirements as a guide. Operators of STOL aircraft should identify their aircraft as such on initial contact with the tower, unless a Letter of Agreement concerning this fact, is in effect. WHENEVER A HOLD SHORT CLEARANCE IS RECEIVED, IT IS INCUMBENT ON THE PILOT TO DETERMINE HIS/HER ABILITY TO HOLD SHORT OF AN INTERSECTION AFTER LANDING WHEN INSTRUCTED TO DO SO. ADDITIONALLY, PILOTS SHOULD INCLUDE THE WORDS "HOLD SHORT OF (POINT)" IN THE ACKNOWLEDGEMENT OF SUCH CLEARANCES.

8. There is no tailwind for the landing aircraft restricted to hold short of the intersection.

b. THE SAFETY AND OPERATION OF AN AIRCRAFT REMAIN THE RESPONSIBILITY OF THE PILOT. IF FOR ANY REASON; e.g., DIFFICULTY IN DISCERNING LOCATION OF AN INTERSECTION AT NIGHT, INABILITY TO HOLD SHORT OF AN INTERSECTION, WIND FACTORS, ETC., A PILOT ELECTS TO USE THE FULL LENGTH OF THE RUNWAY, A DIFFERENT RUNWAY OR DESIRES TO OBTAIN THE DISTANCE FROM THE LANDING THRESHOLD TO THE INTERSECTION, HE IS EXPECTED TO PROMPTLY INFORM ATC ACCORDINGLY.

4-61. LOW APPROACH

a. A low approach (sometimes referred to as a low pass) is the go-around maneuver following an approach. Instead of landing or making a touch-and-go, a pilot may wish to go around (low approach) in order to expedite a particular operation (a series of practice instrument approaches is an example of such an operation). Unless otherwise authorized by ATC, the low approach should be made straight ahead, with no turns or climb made until the pilot has made a thorough visual check for other aircraft in the area.

b. When operating within a Class B, Class C, and Class D surface area, a pilot intending to make a low approach should contact the tower for approval. This request should be made prior to starting the final approach.

c. When operating to an airport, not within a Class B, Class C, and Class D surface area, a pilot intending to make a low approach should, prior to leaving the final approach fix inbound (nonprecision approach) or the outer marker or fix used in lieu of the outer marker inbound (precision approach), so advise the FSS, UNICOM, or make a broadcast as appropriate. (Reference—Traffic Advisory Practices at Airports Without Operating Control Towers, paragraph 4-9.)

4-62. TRAFFIC CONTROL LIGHT SIGNALS

a. The following procedures are used by ATCTs in the control of aircraft, ground vehicles, equipment, and personnel not equipped with radio. These same procedures will be used to control aircraft, ground vehicles, equipment, and personnel equipped with radio if radio contact cannot be established. ATC personnel use a directive traffic control signal which emits an intense narrow light beam of a selected color (either red, white, or green) when controlling traffic by light signals.

b. Although the traffic signal light offers the advantage that some control may be exercised over nonradio equipped aircraft, pilots should be cognizant of the disadvantages which are:

1. The pilot may not be looking at the control tower at the time a signal is directed toward him.

2. The directions transmitted by a light signal are very limited since only approval or disapproval of a pilot's anticipated actions may be transmitted. No supplement or explanatory information may be transmitted except by the use of the "General Warning Signal" which advises the pilot to be on the alert.

c. Between sunset and sunrise, a pilot wishing to attract the attention of the control tower should turn on a landing light and taxi the aircraft into a position, clear of the active runway, so that light is visible to the tower. The landing light should remain on until appropriate signals are received from the tower.

d. Air Traffic Control Tower Light Gun Signals: (See Table 4-62[1].)

e. During daylight hours, acknowledge tower transmissions or light signals by moving the ailerons or rudder. At night, acknowledge by blinking the landing or navigation lights. If radio malfunction occurs after departing the parking area, watch the tower for light signals or monitor tower frequency.

4-63. COMMUNICATIONS

a. Pilots of departing aircraft should communicate with the control tower on the appropriate ground control/clearance delivery frequency prior to starting engines to receive engine start time, taxi and/or clearance information. Unless otherwise advised by the tower, remain on that frequency during taxiing and runup, then change to local

Table 4-62[1]
ATCT Light Gun Signals

COLOR AND TYPE OF SIGNAL	*MEANING* MOVEMENT OF VEHICLES EQUIPMENT AND PERSONNEL	AIRCRAFT ON THE GROUND	AIRCRAFT IN FLIGHT
Steady green	Cleared to cross, proceed or go	Cleared for takeoff	Cleared to land
Flashing green	Not applicable	Cleared for taxi	Return for landing (to be followed by steady green at the proper time)
Steady red	STOP	STOP	Give way to other aircraft and continue circling
Flashing red	Clear the taxiway/runway	Taxi clear of the runway in use	Airport unsafe, do not land
Flashing white	Return to starting point on airport	Return to starting point on airport	Not applicable
Alternating red and green	Exercise extreme caution	Exercise extreme caution	Exercise extreme caution

control frequency when ready to request takeoff clearance. (Reference—Automatic Terminal Information Service (ATIS) for continuous broadcast of terminal information paragraph 4-13.)

b. The majority of ground control frequencies are in the 121.6-121.9 MHz bandwidth. Ground control frequencies are provided to eliminate frequency congestion on the tower (local control) frequency and are limited to communications between the tower and aircraft on the ground and between the tower and utility vehicles on the airport, providing a clear VHF channel for arriving and departing aircraft. They are used for issuance of taxi information, clearances, and other necessary contacts between the tower and aircraft or other vehicles operated on the airport. A pilot who has just landed should not change from the tower frequency to the ground control frequency until he is directed to do so by the controller. Normally, only one ground control frequency is assigned at an airport; however, at locations where the amount of traffic so warrants, a second ground control frequency and/or another frequency designated as a clearance delivery frequency, may be assigned.

c. A controller may omit the ground or local control frequency if the controller believes the pilot knows which frequency is in use. If the ground control frequency is in the 121 MHz bandwidth the controller may omit the numbers preceding the decimal point; e.g., 121.7, "CONTACT GROUND POINT SEVEN." However, if any doubt exists as to what frequency is in use, the pilot should promptly request the controller to provide that information.

d. Controllers will normally avoid issuing a radio frequency change to helicopters, known to be single-piloted, which are hovering, air taxiing, or flying near the ground. At times, it may be necessary for pilots to alert ATC regarding single pilot operations to minimize delay of essential ATC communications. Whenever possible, ATC instructions will be relayed through the frequency being monitored until a frequency change can be accomplished. You must promptly advise ATC if you are unable to comply with a frequency change. Also, you should advise ATC if you must land to accomplish the frequency change unless it is clear the landing, e.g., on a taxiway or in a helicopter operating area, will have no impact on other air traffic.

4-64. GATE HOLDING DUE TO DEPARTURE DELAYS

a. Pilots should contact ground control or clearance delivery prior to starting engines as gate hold procedures will be in effect whenever departure delays exceed or are anticipated to exceed 15 minutes. The sequence for departure will be maintained in accordance with initial call up unless modified by flow control restrictions. Pilots should monitor the ground control or clearance delivery frequency for engine startup advisories or new proposed start time if the delay changes.

b. The tower controller will consider that pilots of turbine powered aircraft are ready for takeoff when they reach the runway or warm-up block unless advised otherwise.

4-65. VFR FLIGHTS IN TERMINAL AREAS

Use reasonable restraint in exercising the prerogative of VFR flight, especially in terminal areas. The weather minimums and distances from clouds are minimums. Giving yourself a greater margin in specific instances is just good judgment.

a. Approach Area—Conducting a VFR operation in a Class B, Class C, Class D and Class E surface area when the official visibility is 3 or 4 miles is not prohibited, but good judgment would dictate that you keep out of the approach area.

b. Reduced Visibility—It has always been recognized that precipitation reduces forward visibility. Consequently, although again it may be perfectly legal to cancel your IFR flight plan at any time you can proceed VFR, it is good practice, when precipitation is occurring, to continue IFR operation into a terminal area until you are reasonably close to your destination.

c. Simulated Instrument Flights—In conducting simulated instrument flights, be sure that the weather is good enough to compensate for the restricted visibility of the safety pilot and your greater concentration on your flight instruments. Give yourself a little greater margin when your flight plan lies in or near a busy airway or close to an airport.

4-66. VFR HELICOPTER OPERATIONS AT CONTROLLED AIRPORTS

a. General—

1. The following ATC procedures and phraseologies recognize the unique capabilities of helicopters and were developed to improve service to all users. Helicopter design characteristics and user needs often require operations from movement areas and nonmovement areas within the airport boundary. In order for ATC to properly apply these procedures, it is essential that pilots familiarize themselves with the local operations and make it known to controllers when additional instructions are necessary.

2. Insofar as possible, helicopter operations will be instructed to avoid the flow of fixed-wing aircraft to minimize overall delays; however, there will be many situations where faster/larger helicopters may be integrated with fixed-wing aircraft for the benefit of all concerned. Examples would include IFR flights, avoidance of noise sensitive areas, or use of runways/taxiways to minimize the hazardous effects of rotor downwash in congested areas.

3. Because helicopter pilots are intimately familiar with the effects of rotor downwash, they are best qualified to determine if a given operation can be conducted safely. Accordingly, the pilot has the final authority with respect to the specific airspeed/altitude combinations. ATC clearances are in no way intended to place the helicopter in a hazardous position. It is expected that pilots will advise ATC if a specific clearance will cause undue hazards to persons or property.

b. Controllers normally limit ATC ground service and instruction to *movement* areas; therefore, operations from *nonmovement* areas are conducted at pilot discretion and should be based on local policies, procedures, or letters of agreement. In order to maximize the flexibility of helicopter operations, it is necessary to rely heavily on sound pilot judgment. For example, hazards such as debris, obstructions, vehicles, or personnel must be recognized by the pilot, and action should be taken as necessary to avoid such hazards. Taxi, hover taxi, and air taxi operations are considered to be ground movements. Helicopters conducting such operations are expected to adhere to the same conditions, requirements, and practices as apply to other ground taxiing and ATC procedures in the AIM.

1. The phraseology *taxi* is used when it is intended or expected that the helicopter will taxi on the airport surface, either via taxiways or other prescribed routes. *Taxi* is used primarily for helicopters equipped with wheels or in response to a pilot request. Preference should be given to this procedure whenever it is necessary to minimize effects of rotor downwash.

2. Pilots may request a *hover taxi* when slow forward movement is desired or when it may be appropriate to move very short distances. Pilots should avoid this procedure if rotor downwash is likely to cause damage to parked aircraft or if blowing dust/snow could obscure visibility. If it is necessary to operate above 25 feet AGL when hover taxiing, the pilot should initiate a request to ATC.

3. *Air taxi* is the preferred method for helicopter ground movements on airports provided ground operations and conditions permit. Unless otherwise requested or instructed, pilots are expected to remain below 100 feet AGL. However, if a higher than normal airspeed or altitude is desired, the request should be made prior to lift-off. The pilot is solely responsible for selecting a safe airspeed for the altitude/operation being conducted. Use of *air taxi* enables the pilot to proceed at an optimum airspeed/altitude, minimize downwash effect, conserve fuel, and expedite movement from one point to another. Helicopters should avoid overflight of other aircraft, vehicles, and personnel during air-taxi operations. Caution must be exercised concerning active runways and pilots must be certain that air taxi instructions are understood. Special precautions may be necessary at unfamiliar airports or airports with multiple/intersecting active runways. The taxi procedures given in: Taxiing, paragraph 4-67, Taxi During Low Visibility, paragraph 4-68, and Exiting the Runway After Landing, paragraph 4-69 also apply. (Reference—Pilot/Controller Glossary, Taxi, Hover Taxi, and Air Taxi.)

c. Takeoff and Landing Procedures—

1. Helicopter operations may be conducted from a runway, taxiway, portion of a landing strip, or any clear area which could be used as a landing site such as the scene of an accident, a construction site, or the roof of a building. The terms used to describe designated areas from which helicopters operate are: movement area, landing/takeoff area, apron/ramp, heliport and helipad (See Pilot/Controller Glossary). These areas may be improved or unimproved and may be separate from or located on an airport/heliport. ATC will issue takeoff clearances from *movement* areas other than active runways, or in diverse directions from active runways, with additional instructions as necessary. Whenever possible, takeoff clearance will be issued in lieu of extended hover/air taxi operations. Phraseology will be "CLEARED FOR TAKEOFF FROM (taxiway, helipad, runway number, etc.), MAKE RIGHT/LEFT TURN FOR (direction, heading, NAVAID radial) DEPARTURE/DEPARTURE ROUTE (number, name, etc.)." Unless requested by the pilot, downwind takeoffs will not be issued if the tailwind exceeds 5 knots.

2. Pilots should be alert to wind information as well as to wind indications in the vicinity of the helicopter. ATC should be advised of the intended method of departing. A pilot request to take off in a given direction indicates that the pilot is willing to accept the wind condition and controllers will honor the request if traffic permits. Departure points could be a significant distance from the control tower and it may be difficult or impossible for the controller to determine the helicopter's relative position to the wind.

3. If takeoff is requested from *nonmovement* areas, the phraseology "PROCEED AS REQUESTED" will be used. Additional instructions will be issued as necessary. The pilot is responsible for operating in a safe manner and should exercise due caution. When other known traffic is not a factor and takeoff is requested from an area not visible from the tower, an area not authorized for helicopter use, an unlighted area at night, or an area not on the airport, the phraseology "DEPARTURE FROM (location) WILL BE AT YOUR OWN RISK (with reason, and additional instructions as necessary)."

4. Similar phraseology is used for helicopter landing operations. Every effort will be made to permit helicopters to proceed direct and land as near as possible to their final destination on the airport. Traffic density, the need for detailed taxiing instructions, frequency congestion, or other factors may affect the extent to which service can be expedited. As with ground movement operations, a high degree of pilot/controller cooperation and communication is necessary to achieve safe and efficient operations.

4-67. TAXIING

a. General: Approval must be obtained prior to moving an aircraft or vehicle onto the movement area during the hours an Airport Traffic Control Tower is in operation.

1. Always state your position on the airport when calling the tower for taxi instructions.

2. The movement area is normally described in local bulletins issued by the airport manager or control tower. These bulletins may be found in FSSs, fixed base operators offices, air carrier offices, and operations offices.

3. The control tower also issues bulletins describing areas where they cannot provide ATC service due to nonvisibility or other reasons.

4. A clearance must be obtained prior to taxiing on a runway, taking off, or landing during the hours an Airport Traffic Control Tower is in operation.

5. When ATC clears an aircraft to "taxi to" an assigned takeoff runway, the absence of holding instructions authorizes the aircraft to "cross" all runways which the taxi route intersects except the assigned takeoff runway. It does not include authorization to "taxi onto" or "cross" the assigned takeoff runway at any point. In order to preclude misunderstandings in radio communications, ATC will not use the word "cleared" in conjunction with authorization for aircraft to taxi.

6. In the absence of holding instructions, a clearance to "taxi to" any point other than an assigned takeoff runway is a clearance to cross all runways that intersect the taxi route to that point.

7. Air traffic control will first specify the runway, issue taxi instructions, and then state any required hold short instructions, when authorizing an aircraft to taxi for departure. This does not authorize the aircraft to "enter" or "cross" the assigned departure runway at any point. AIR TRAFFIC CONTROLLERS ARE REQUIRED TO OBTAIN FROM THE PILOT A READBACK OF ALL RUNWAY HOLD SHORT INSTRUCTIONS.

b. ATC clearances or instructions pertaining to taxiing are predicated on known traffic and known physical airport conditions. Therefore, it is important that pilots clearly understand the clearance or instruction. Although an ATC clearance is issued for taxiing purposes, when operating in accordance with the FARs, it is the responsibility of the pilot to avoid collision with other aircraft. Since "the pilot-in-command of an aircraft is directly responsible for, and is the final authority as to, the operation of that aircraft" the pilot should obtain clarification of any clearance or instruction which is not understood. (Reference—General, paragraph 7-50.)

1. Good operating practice dictates that pilots acknowledge all runway crossing, hold short, or takeoff clearances unless there is some misunderstanding, at which time the pilot should query the controller until the clearance is understood. AIR TRAFFIC CONTROLLERS ARE REQUIRED TO OBTAIN FROM THE PILOT A READBACK OF ALL RUNWAY HOLD SHORT INSTRUCTIONS. Pilots operating a single pilot aircraft should monitor only assigned ATC communications after being cleared onto the active runway for departure. Single pilot aircraft should not monitor other than ATC communications until flight from Class B, Class C, or Class D surface area is completed. This same procedure should be practiced from after receipt of the clearance for landing until the landing and taxi activities are complete. Proper effective scanning for other aircraft, surface vehicles, or other objects should be continuously exercised in all cases.

2. If the pilot is unfamiliar with the airport or for any reason confusion exists as to the correct taxi routing, a request may be made for progressive taxi instructions which include step-by-step routing directions. Progressive instructions may also be issued if the controller deems it necessary due to traffic or field conditions; i.e., construction or closed taxiways.

c. At those airports where the U.S. Government operates the control tower and ATC has authorized noncompliance with the requirement for two-way radio communications while operating within the Class B, Class C, or Class D surface area, or at those airports where the U.S. Government does not operate the control tower and radio communications cannot be established, pilots shall obtain a clearance by visual light signal prior to taxiing on a runway and prior to takeoff and landing.

d. The following phraseologies and procedures are used in radio-telephone communications with aeronautical ground stations.

1. *Request for taxi instructions prior to departure:* State your aircraft identification, location, type of operation planned (VFR or IFR), and the point of first intended landing.

EXAMPLE:

Aircraft: "WASHINGTON GROUND, BEECHCRAFT ONE THREE ONE FIVE NINER AT HANGAR EIGHT, READY TO TAXI, I-F-R TO CHICAGO."

Tower: "BEECHCRAFT ONE THREE ONE FIVE NINER, WASHINGTON GROUND, TAXI TO RUNWAY THREE SIX, WIND ZERO THREE ZERO AT TWO FIVE, ALTIMETER THREE ZERO ZERO FOUR,"

or

Tower: "BEECHCRAFT ONE THREE ONE FIVE NINER, WASHINGTON GROUND, RUNWAY TWO SEVEN, TAXI VIA TAXIWAYS CHARLIE AND DELTA, HOLD SHORT OF RUNWAY THREE THREE LEFT."

Aircraft: "BEECHCRAFT ONE THREE ONE FIVE NINER, HOLD SHORT OF RUNWAY THREE THREE LEFT."

2. *Receipt of ATC clearance:* ARTCC clearances are relayed to pilots by airport traffic controllers in the following manner.

EXAMPLE:

Tower: BEECHCRAFT ONE THREE ONE FIVE NINER, CLEARED TO THE CHICAGO MIDWAY AIRPORT VIA VICTOR EIGHT, MAINTAIN EIGHT THOUSAND.

Aircraft: "BEECHCRAFT ONE THREE ONE FIVE NINER, CLEARED TO THE CHICAGO MIDWAY AIRPORT VIA VICTOR EIGHT, MAINTAIN EIGHT THOUSAND."

4-67d1 NOTE—Normally, an ATC IFR clearance is relayed to a pilot by the ground controller. At busy locations, however, pilots may be instructed by the ground controller to "CONTACT CLEARANCE DELIVERY" on a frequency designated for this purpose. No surveillance or control over the movement of traffic is exercised by this position of operation.

3. *Request for taxi instructions after landing:* State your aircraft identification, location, and that you request taxi instructions.

EXAMPLE:

Aircraft: "DULLES GROUND, BEECHCRAFT ONE FOUR TWO SIX ONE CLEARING RUNWAY ONE RIGHT ON TAXIWAY ECHO THREE, REQUEST CLEARANCE TO PAGE."

Tower: "BEECHCRAFT ONE FOUR TWO SIX ONE, DULLES GROUND, TAXI TO PAGE VIA TAXIWAYS ECHO THREE, ECHO ONE, AND ECHO NINER."

or

Aircraft: "ORLANDO GROUND, BEECHCRAFT ONE FOUR TWO SIX ONE CLEARING RUNWAY ONE EIGHT LEFT AT TAXIWAY BRAVO THREE, REQUEST CLEARANCE TO PAGE."

Tower: "BEECHCRAFT ONE FOUR TWO SIX ONE, ORLANDO GROUND, HOLD SHORT OF RUNWAY ONE EIGHT RIGHT."

Aircraft: "BEECHCRAFT ONE FOUR TWO SIX ONE, HOLD SHORT OF RUNWAY ONE EIGHT RIGHT."

4-68. TAXI DURING LOW VISIBILITY

a. Pilots and aircraft operators should be constantly aware that during certain low visibility conditions the movement of aircraft and vehicles on airports may not be visible to the tower controller. This may prevent visual confirmation of an aircraft's adherence to taxi instructions. Pilots should, therefore, exercise extreme vigilance and proceed cautiously under such conditions.

b. Of vital importance is the need for pilots to notify the controller when difficulties are encountered or at the first indication of becoming disoriented. Pilots should proceed with extreme caution when taxiing toward the sun. When vision difficulties are encountered pilots should immediately inform the controller.

4-69. EXITING THE RUNWAY AFTER LANDING

The following procedures should be followed after landing and reaching taxi speed.

a. Exit the runway without delay at the first available taxiway or on a taxiway as instructed by air traffic control (ATC).

b. Taxi clear of the runway unless otherwise directed by ATC. In the absence of ATC instructions the pilot is expected to taxi clear of the landing runway even if that requires the aircraft to protrude into or cross another taxiway, runway, or ramp area. This does not authorize an aircraft to cross a subsequent taxiway/runway/ramp after clearing the landing runway.

4-69b NOTE—The tower will issue the pilot with instructions which will normally permit the aircraft to enter another taxiway, runway, or ramp area when required to taxi clear of the runway.

c. Stop the aircraft after clearing the runway if instructions have not been received from ATC.

d. Immediately change to ground control frequency when advised by the tower and obtain a taxi clearance.

4-69d NOTE 1—The tower will issue instructions required to resolve any potential conflictions with other ground traffic prior to advising the pilot to contact ground control.

4-69d NOTE 2—A clearance from ATC to taxi to the ramp authorizes the aircraft to cross all runways and taxiway intersections. Pilots not familiar with the taxi route should request specific taxi instructions from ATC.

4-70. PRACTICE INSTRUMENT APPROACHES

a. Various air traffic incidents have indicated the necessity for adoption of measures to achieve more organized and controlled operations where practice instrument approaches are conducted. Practice instrument approaches are considered to be instrument approaches made by either a VFR aircraft not on an IFR flight plan or an aircraft on an IFR flight plan. To achieve this and thereby enhance air safety, it is Air Traffic Operations Service policy to provide for separation of such operations at locations where approach control facilities are located and, as resources permit, at certain other locations served by ARTCCs or parent approach control facilities. Pilot requests to practice instrument approaches may be approved by ATC subject to traffic and workload conditions. Pilots should anticipate that in some instances the controller may find it necessary to deny approval or withdraw previous approval when traffic conditions warrant. It must be clearly understood, however, that even though the controller may be providing separation, pilots on VFR flight plans are required to comply with basic visual flight rules (FAR Part 91.155). Application of ATC procedures or any action taken by the controller to avoid traffic conflictions does not relieve IFR and VFR pilots of their responsibility to see-and-avoid other traffic while operating in VFR conditions. (FAR Part 91.113) In addition to the normal IFR separation minimums (which includes visual separation) during VFR conditions, 500 feet vertical separation may be applied between VFR aircraft and between a VFR aircraft and the IFR aircraft. Pilots not on IFR flight plans desiring practice instrument approaches should always state 'Practice' when making requests to ATC. Controllers will instruct VFR aircraft requesting an instrument approach to maintain VFR. This is to preclude misunderstandings between the pilot and controller as to the status of the aircraft. If the pilot wishes to proceed in accordance with instrument flight rules, he must specifically request and obtain, an IFR clearance.

b. Before practicing an instrument approach, pilots should inform the approach control facility or the tower of the type of practice approach they desire to make and how they intend to terminate it, i.e., full-stop landing, touch-and-go, or missed or low approach maneuver. This information may be furnished progressively when conducting a series of approaches. Pilots on an IFR flight plan, who have made a series of instrument approaches to full stop landings should inform ATC when they make their final landing. The controller will control flights practicing instrument approaches so as to ensure that they do not disrupt the flow of arriving and departing itinerant IFR or VFR aircraft. The priority afforded itinerant aircraft over practice instrument approaches is not intended to be so rigidly applied that it causes grossly inefficient application of services. A minimum delay to itinerant traffic may be appropriate to allow an aircraft practicing an approach to complete that approach.

4-70b NOTE—A clearance to land means that appropriate separation on the landing runway will be ensured. A landing clearance does not relieve the pilot from compliance with any previously issued restriction.

c. At airports without a tower, pilots wishing to make practice instrument approaches should notify the facility having control jurisdiction of the desired approach as indicated on the approach chart. All approach control facilities and ARTCCs are required to publish a Letter to Airmen depicting those airports where they provide standard separation to both VFR and IFR aircraft conducting practice instrument approaches.

d. The Controller will provide approved separation between both VFR and IFR aircraft when authorization is granted to make practice approaches to airports where an approach control facility is located and to certain other airports served by approach control or an ARTCC. Controller responsibility for separation of VFR aircraft begins at the point where the approach clearance becomes effective, or when the aircraft enters Class B or Class C airspace, or a TRSA, whichever comes first.

e. VFR aircraft practicing instrument approaches are not automatically authorized to execute the missed approach procedure. This authorization must be specifically requested by the pilot and approved by the controller. Separation will not be provided unless the missed approach has been approved by ATC.

f. Except in an emergency, aircraft cleared to practice instrument approaches must not deviate from the approved procedure until cleared to do so by the controller.

g. At radar approach control locations when a full approach procedure (Procedure Turn, etc.,) cannot be approved, pilots should expect to be vectored to a final approach course for a practice instrument approach which is compatible with the general direction of traffic at that airport.

h. When granting approval for a practice instrument approach, the controller will usually ask the pilot to report to the tower prior to or over the final approach fix inbound (nonprecision approaches) or over the outer marker or fix used in lieu of the outer marker inbound (precision approaches).

i. When authorization is granted to conduct practice instrument approaches to an airport with a tower, but where approved standard separation is not provided to aircraft conducting practice instrument approaches, the tower will approve the practice approach, instruct the aircraft to maintain VFR and issue traffic information, as required.

j. When an aircraft notifies an FSS providing Local Airport Advisory to the airport concerned of the intent to conduct a practice instrument approach and whether or not separation is to be provided, the pilot will be instructed to contact the appropriate facility on a specified frequency prior to initiating the approach. At airports where separation is not provided, the FSS will acknowledge the message and issue known traffic information but will neither approve nor disapprove the approach.

k. Pilots conducting practice instrument approaches should be particularly alert for other aircraft operating in the local traffic pattern or in proximity to the airport.

4-71. OPTION APPROACH

The "Cleared for the Option" procedure will permit an instructor, flight examiner or pilot the option to make a touch-and-go, low approach, missed approach, stop-and-go, or full stop landing. This procedure can be very beneficial in a training situation in that neither the student pilot nor examinee would know what maneuver would be accomplished. The pilot should make his request for this procedure passing the final approach fix inbound on an instrument approach or entering downwind for a VFR traffic pattern. The advantages of this procedure as a training aid are that it enables an instructor or examiner to obtain the reaction of a trainee or examinee under changing conditions, the pilot would not have to discontinue an approach in the middle of the procedure due to student error or pilot proficiency requirements, and finally it allows more flexibility and economy in training programs. This procedure will only be used at those locations with an operational control tower and will be subject to ATC approval.

4-72. USE OF AIRCRAFT LIGHTS

a. Aircraft position and anticollision lights are required to be lighted on aircraft operated from sunset to sunrise. Anticollision lights, however, need not be lighted when the pilot-in-command determines that, because of operating conditions, it would be in the interest of safety to turn off the lights (FAR Part 91.209). For example, strobe lights should be turned off on the ground when they adversely affect ground personnel or other pilots, and in flight when there are adverse reflections from clouds.

b. An aircraft anticollision light system can use one or more rotating beacons and/or strobe lights, be colored either red or white, and have different (higher than minimum) intensities when compared to other aircraft. Many aircraft have both a rotating beacon and a strobe light system.

c. The FAA has a voluntary pilot safety program, *Operation Lights On*, to enhance the *see-and-avoid* concept. Pilots are encouraged to turn on their anticollision lights any time the engine(s) are running, day or night. Use of these lights is especially encouraged when operating on airport surfaces during periods of reduced visibility and when snow or ice control vehicles are or may be operating. Pilots are also encouraged to turn on their landing lights during takeoff; i.e., either after takeoff clearance has been received or when beginning takeoff roll. Pilots are further encouraged to turn on their landing lights when operating below 10,000 feet, day or night, especially when operating within 10 miles of any airport, or in conditions of reduced visibility and in areas where flocks of birds may be expected, i.e., coastal areas, lake areas, around refuse dumps, etc. Although turning on aircraft lights does enhance the *see-and-avoid* concept, pilots should not become complacent about keeping a sharp lookout for other aircraft. Not all aircraft are equipped with lights and some pilots may not have their lights turned on. Aircraft manufacturer's recommendations for operation of landing lights and electrical systems should be observed.

d. Prop and jet blast forces generated by large aircraft have overturned or damaged several smaller aircraft taxiing behind them. To avoid similar results, and in the interest of preventing upsets and injuries to ground personnel from such forces, the FAA recommends that air carriers and commercial operators turn on their rotating beacons anytime their aircraft engines are in operation. General Aviation pilots using rotating beacon equipped aircraft are also encouraged to participate in this program which is designed to alert others to the potential hazard. Since this is a voluntary program, exercise caution and do not rely solely on the rotating beacon as an indication that aircraft engines are in operation.

4-73. FLIGHT INSPECTION/"FLIGHT CHECK" AIRCRAFT IN TERMINAL AREAS

a. *Flight check* is a call sign used to alert pilots and air traffic controllers when an FAA aircraft is engaged in flight inspection/certification of NAVAIDs and flight procedures. Flight Check aircraft fly preplanned high/low altitude flight patterns such as grids, orbits, DME arcs, and tracks, including low passes along the full length of the runway to verify NAVAID performance. In most instances, these flight checks are being automatically recorded and/or flown in an automated mode.

b. Pilots should be especially watchful and avoid the flight paths of any aircraft using the call sign, "Flight Check" or "Flight Check Recorded." The latter call sign; e.g., "Flight Check 47 Recorded" indicates that automated flight inspections are in progress in terminal areas. These flights will normally receive special handling from ATC. Pilot patience and cooperation in allowing uninterrupted recordings can significantly help expedite flight inspections, minimize costly, repetitive runs, and reduce the burden on the U.S. taxpayer.

4-74. HAND SIGNALS

(See Figure 4-74[1][Signalman Directs Towing].)

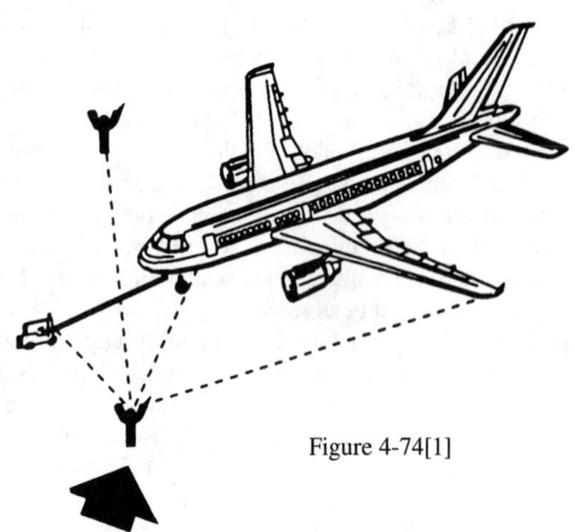

Figure 4-74[1]

(See Figure 4-74[3][Flagman Directs Pilot].)

Figure 4-74[3]

(See Figure 4-74[2][Signalman's Position].)

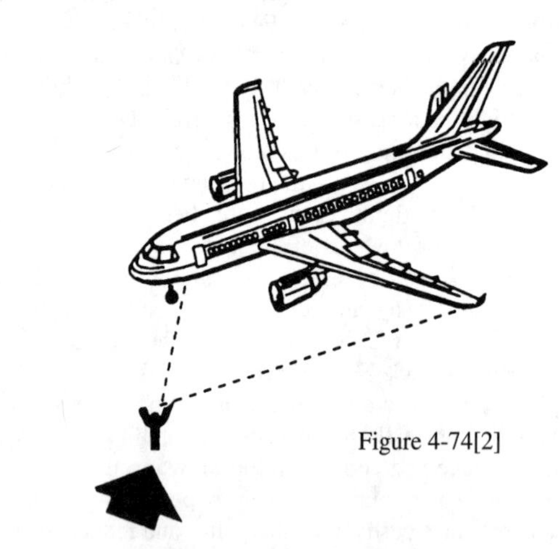

Figure 4-74[2]

(See Figure 4-74[4][All Clear].)

Figure 4-74[4]

(See Figure 4-74[5][Start Engine].)

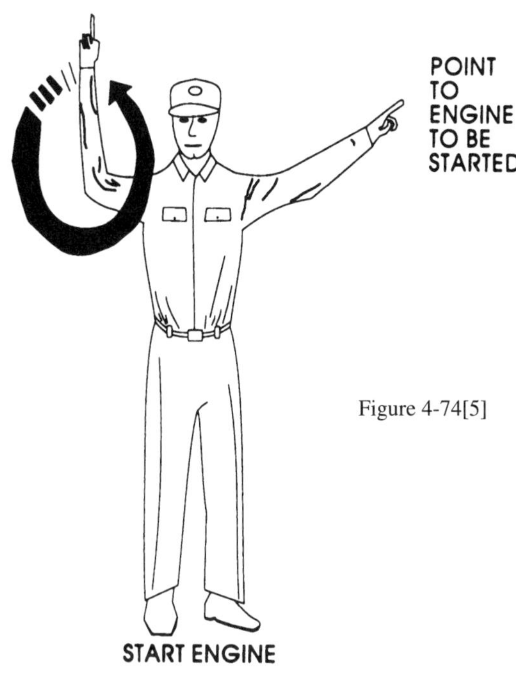

Figure 4-74[5]

(See Figure 4-74[6][Pull Chocks].)

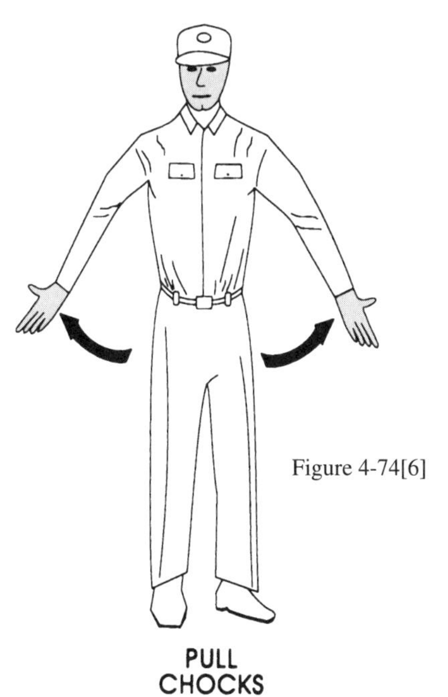

Figure 4-74[6]

(See Figure 4-74[7][Come Ahead].)

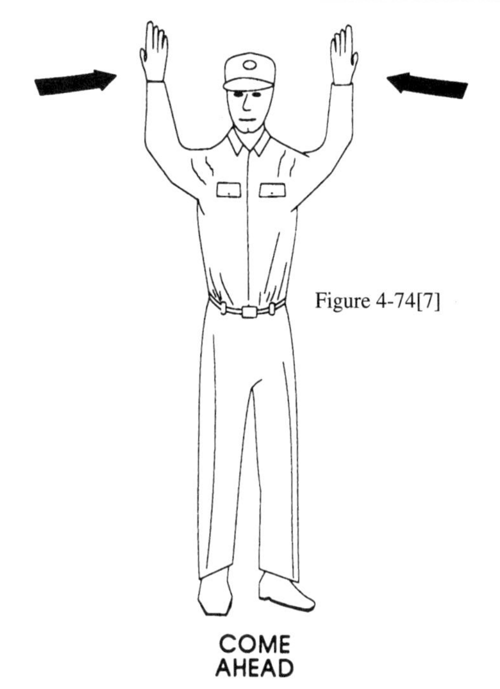

Figure 4-74[7]

(See Figure 4-74[8][Left Turn].)

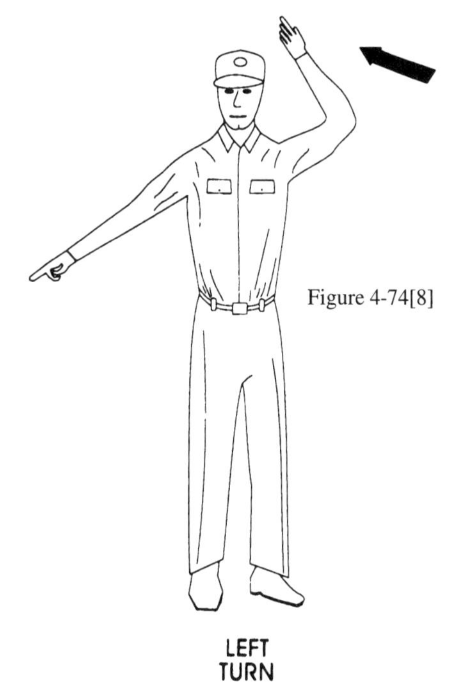

Figure 4-74[8]

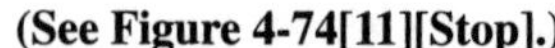

(See Figure 4-74[9][Right Turn].)

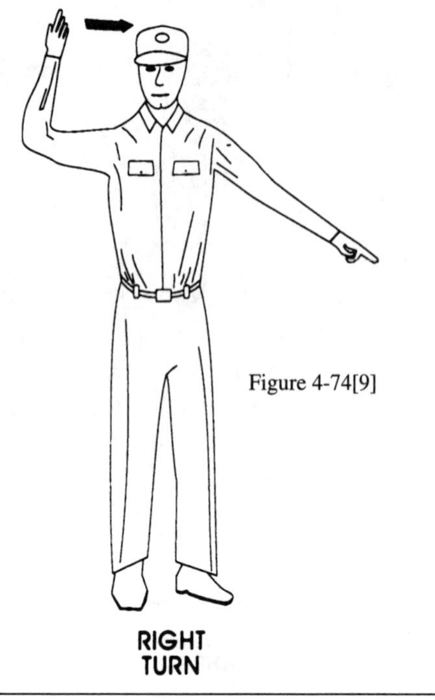

Figure 4-74[9]

(See Figure 4-74[10][Slow Down].)

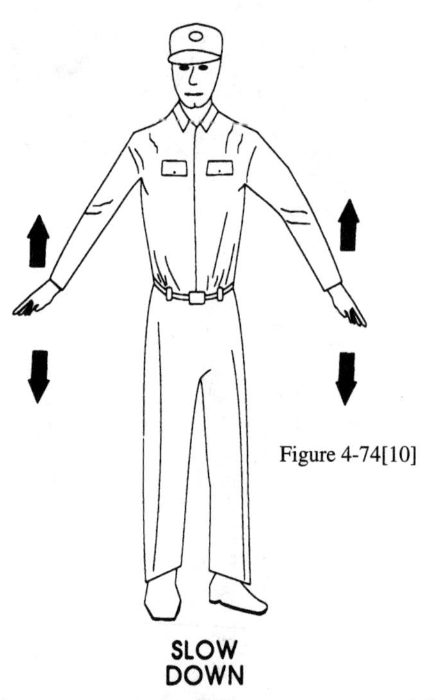

Figure 4-74[10]

(See Figure 4-74[11][Stop].)

Figure 4-74[11]

STOP

(See Figure 4-74[12][Insert Chocks].)

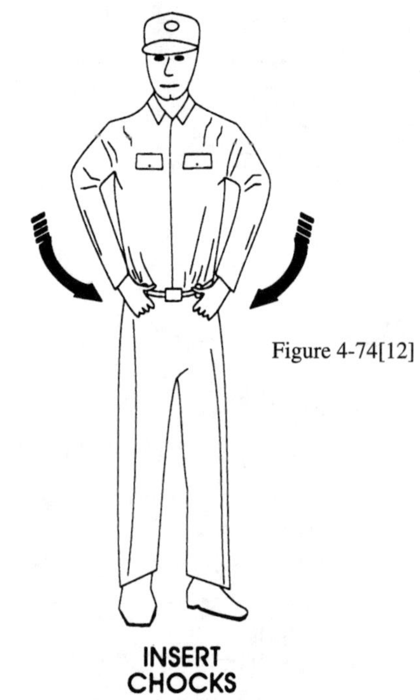

Figure 4-74[12]

(See Figure 4-74[13][Cut Engines].)

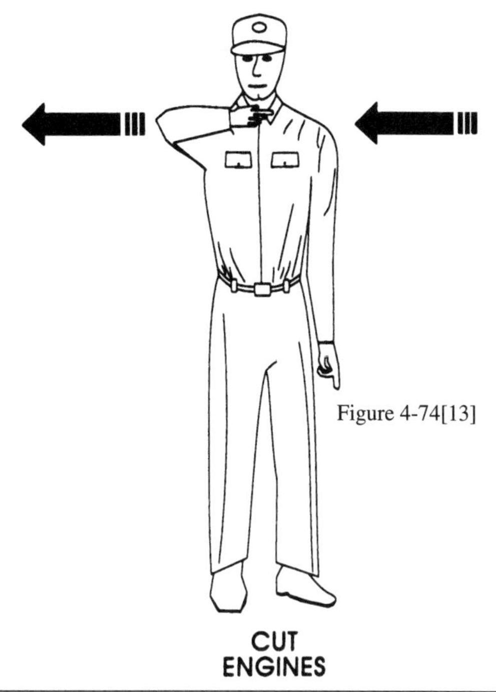

Figure 4-74[13]

(See Figure 4-74[14][Night Operation].)

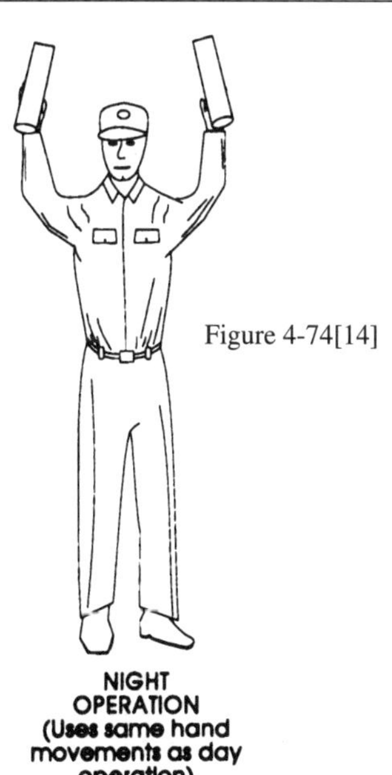

NIGHT
OPERATION
(Uses same hand
movements as day
operation)

Figure 4-74[14]

(See Figure 4-74[15][Emergency Stop].)

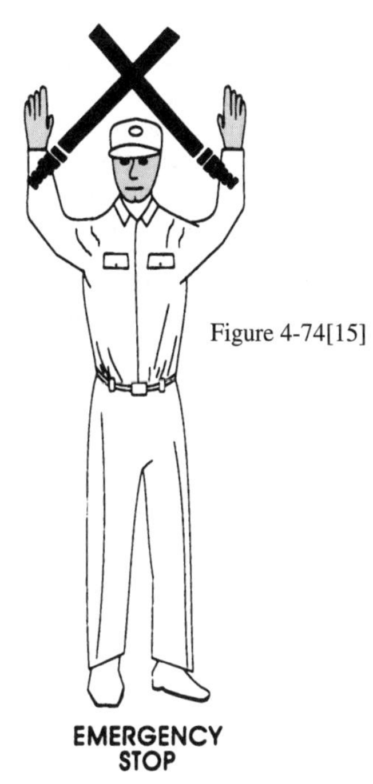

Figure 4-74[15]

4-75 thru 4-79 RESERVED.

Section 4. ATC CLEARANCES/SEPARATIONS

4-80. CLEARANCE

a. A clearance issued by ATC is predicated on known traffic and known physical airport conditions. An ATC clearance means an authorization by ATC, for the purpose of preventing collision between known aircraft, for an aircraft to proceed under specified conditions within controlled airspace. IT IS NOT AUTHORIZATION FOR A PILOT TO DEVIATE FROM ANY RULE REGULATION OR MINIMUM ALTITUDE NOR TO CONDUCT UNSAFE OPERATION OF HIS AIRCRAFT.

b. FAR Part 91.3(a) states: "The pilot-in-command of an aircraft is directly responsible for, and is the final authority as to, the operation of that aircraft." If ATC issues a clearance that would cause a pilot to deviate from a rule or regulation, or in the pilot's opinion, would place the aircraft in jeopardy, IT IS THE PILOT'S RESPONSIBILITY TO REQUEST AN AMENDED CLEARANCE. Similarly, if a pilot prefers to follow a different course of action, such as make a 360 degree turn for spacing to follow traffic when established in a landing or approach sequence, land on a different runway, take off from a different intersection, take off from the threshold instead of an intersection, or delay his operation, HE IS EXPECTED TO INFORM ATC ACCORDINGLY. When he requests a different course of action, however, the pilot is expected to cooperate so as to preclude disruption of traffic flow or creation of conflicting patterns. The pilot is also expected to use the appropriate aircraft call sign to acknowledge all ATC clearances, frequency changes, or advisory information.

c. Each pilot who deviates from an ATC clearance in response to a Traffic Alert and Collision Avoidance System resolution advisory shall notify ATC of that deviation as soon as possible. (Reference—Pilot/Controller Glossary, Traffic Alert and Collision Avoidance System.)

d. When weather conditions permit, during the time an IFR flight is operating, it is the direct responsibility of the pilot to avoid other aircraft since VFR flights may be operating in the same area without the knowledge of ATC. Traffic clearances provide standard separation only between IFR flights.

4-81. CLEARANCE PREFIX

A clearance, control information, or a response to a request for information originated by an ATC facility and relayed to the pilot through an air-to-ground communication station will be prefixed by "ATC clears," "ATC advises," or "ATC requests."

4-82. CLEARANCE ITEMS

ATC clearances normally contain the following:

a. Clearance Limit.—The traffic clearance issued prior to departure will normally authorize flight to the airport of intended landing. Under certain conditions, at some locations a short-range clearance procedure is utilized whereby a clearance is issued to a fix within or just outside of the terminal area and the pilot is advised of the frequency on which he will receive the long-range clearance direct from the center controller.

b. Departure Procedure.—Headings to fly and altitude restrictions may be issued to separate a departure from other air traffic in the terminal area. (Reference—Abbreviated IFR Departure Clearance Procedures, paragraph 5-22 and Instrument Departures, paragraph 5-25.) Where the volume of traffic warrants, SIDs have been developed.

c. Route of Flight—

1. Clearances are normally issued for the altitude or flight level and route filed by the pilot. However, due to traffic conditions, it is frequently necessary for ATC to specify an altitude or flight level or route different from that requested by the pilot. In addition, flow patterns have been established in certain congested areas or between congested areas whereby traffic capacity is increased by routing all traffic on preferred routes. Information on these flow patterns is available in offices where preflight briefing is furnished or where flight plans are accepted.

2. When required, air traffic clearances include data to assist pilots in identifying radio reporting points. It is the responsibility of the pilot to notify ATC immediately if his radio equipment cannot receive the type of signals he must utilize to comply with his clearance.

d. Altitude Data—

1. The altitude or flight level instructions in an ATC clearance normally require that a pilot "MAINTAIN" the altitude or flight level at which the flight will operate when in controlled airspace. Altitude or flight level changes while en route should be requested prior to the time the change is desired.

2. When possible, if the altitude assigned is different from the altitude requested by the pilot, ATC will inform the pilot when to expect climb or descent clearance or to request altitude change from another facility. If this has not been received prior to crossing the boundary of the ATC facility's area and assignment at a different altitude is still desired, the pilot should reinitiate his request with the next facility.

3. The term "cruise" may be used instead of "MAINTAIN" to assign a block of airspace to a pilot from the minimum IFR altitude up to and including the altitude specified in the cruise clearance. The pilot may level off at any intermediate altitude within this block of airspace. Climb/descent within the block is to be made at the discretion of the pilot. However, once the pilot starts descent and verbally reports leaving an altitude in the block, he may not return to that altitude without additional ATC clearance.

4-82d3 NOTE—Reference—Pilot/Controller Glossary, Cruise.

e. Holding Instructions—

1. Whenever an aircraft has been cleared to a fix other than the destination airport and delay is expected, it is the responsibility of the ATC controller to issue complete holding instructions (unless the pattern is charted), an EFC time, and his best estimate of any additional en route/terminal delay.

2. If the holding pattern is charted and the controller doesn't issue complete holding instructions, the pilot is expected to hold as depicted on the appropriate chart. When the pattern is charted, the controller may omit all holding instructions except the charted holding direction and the statement *AS PUBLISHED*, e.g., *"HOLD EAST AS PUBLISHED."* Controllers shall always issue complete holding instructions when pilots request them.

4-82e2 NOTE—Only those holding patterns depicted on U.S. Government or commercially produced (meeting FAA requirements) *Low/High Altitude Enroute*, and *Area* or *STAR* charts should be used.

3. If no holding pattern is charted and holding instructions have not been issued, the pilot should ask ATC for holding instructions prior to reaching the fix. This procedure will eliminate the possibility of an aircraft entering a holding pattern other than that desired by ATC. If the pilot is unable to obtain holding instructions prior to reaching the fix (due to frequency congestion, stuck microphone, etc.), he should hold in a standard pattern on the course on which he approached the fix and request further clearance as soon as possible. In this event, the altitude/flight level of the aircraft at the clearance limit will be protected so that separation will be provided as required.

4. When an aircraft is 3 minutes or less from a clearance limit and a clearance beyond the fix has not been received, the pilot is expected to start a speed reduction so that he will cross the fix, initially, at or below the maximum holding airspeed.

5. When no delay is expected, the controller should issue a clearance beyond the fix as soon as possible and, whenever possible, at least 5 minutes before the aircraft reaches the clearance limit.

6. Pilots should report to ATC the time and altitude/flight level at which the aircraft reaches the clearance limit and report leaving the clearance limit.

4-82e6 NOTE—In the event of two-way communications failure, pilots are required to comply with FAR Part 91.185.

4-83. AMENDED CLEARANCES

a. Amendments to the initial clearance will be issued at any time an air traffic controller deems such action necessary to avoid possible confliction between aircraft. Clearances will require that a flight "hold" or change altitude prior to reaching the point where standard separation from other IFR traffic would no longer exist.

4-83a NOTE—Some pilots have questioned this action and requested "traffic information" and were at a loss when the reply indicated "no traffic report." In such cases the controller has taken action to prevent a traffic confliction which would have occurred at a distant point.

b. A pilot may wish an explanation of the handling of his flight at the time of occurrence; however, controllers are not able to take time from their immediate control duties nor can they afford to overload the ATC communications channels to furnish explanations. Pilots may obtain an explanation by directing a letter or telephone call to the chief controller of the facility involved.

c. The pilot has the privilege of requesting a different clearance from that which has been issued by ATC if he feels that he has information which would make another course of action more practicable or if aircraft equipment limitations or company procedures forbid compliance with the clearance issued.

4-84. SPECIAL VFR CLEARANCES

a. An ATC clearance must be obtained *prior* to operating within a Class B, Class C, Class D or Class E surface area when the weather is less than that required for VFR flight. A VFR pilot may request and be given a clearance to enter, leave, or operate within most Class D and Class E surface areas and some Class B and Class C surface areas in Special VFR conditions, traffic permitting, and providing such flight will not delay IFR operations. All Special VFR flights must remain clear of clouds. The visibility requirements for Special VFR aircraft (other than helicopters) are:

1. At least 1 statute mile flight visibility for operations within Class B, Class C, Class D and Class E surface areas.

2. At least 1 statute mile ground visibility if taking off or landing. If ground visibility is not reported at that airport, the flight visibility must be at least 1 statute mile.

3. The restrictions in (1) and (2) do not apply to helicopters. Helicopters must remain clear of clouds and may operate in Class B, Class C, Class D and Class E surface areas with less than 1 statute mile visibility.

b. When a control tower is located within the Class B, Class C, or Class D surface area, requests for clearances should be to the tower. In a Class E surface area, a clearance may be obtained from the nearest tower, FSS, or center.

c. It is not necessary to file a complete flight plan with the request for clearance, but the pilot should state his intentions in sufficient detail to permit ATC to fit his flight into the traffic flow. The clearance will not contain a specific altitude as the pilot must remain clear of clouds. The controller may require the pilot to fly at or below a certain altitude due to other traffic, but the altitude specified will permit flight at or above the Minimum Safe Altitude. In addition, at radar locations, flights may be vectored if necessary for control purposes or on pilot request.

4-84c NOTE—The pilot is responsible for obstacle or terrain clearance (reference FAR Part 91.119).

d. Special VFR clearances are effective within Class B, Class C, Class D and Class E surface areas only. ATC does not provide separation after an aircraft leaves the Class B, Class C, Class D or Class E surface area on a Special VFR clearance.

e. Special VFR operations by fixed-wing aircraft are prohibited in some Class B and Class C surface areas due to the volume of IFR traffic. A list of these Class B and Class C surface areas is contained in FAR Part 91, Appendix D, Section 3. They are also depicted on Sectional Aeronautical Charts.

f. ATC provides separation between Special VFR flights and between these flights and other IFR flights.

g. Special VFR operations by fixed-wing aircraft are prohibited between sunset and sunrise unless the pilot is instrument rated and the aircraft is equipped for IFR flight.

4-85. PILOT RESPONSIBILITY UPON CLEARANCE ISSUANCE

a. Record ATC clearance—When conducting an IFR operation, make a written record of your clearance. The specified conditions which are a part of your air traffic clearance may be somewhat different from those included in your flight plan. Additionally, ATC may find it necessary to ADD conditions, such as particular departure route. The very fact that ATC specifies different or additional conditions means that other aircraft are involved in the traffic situation.

b. ATC Clearance/Instruction Readback—Pilots of airborne aircraft should read back *those parts* of ATC clearances and instructions containing altitude assignments or vectors as a means of mutual verification. The readback of the "numbers" serves as a double check between pilots and controllers and reduces the kinds of communications errors that occur when a number is either "misheard" or is incorrect.

1. Precede all readbacks and acknowledgements with the aircraft identification. This aids controllers in determining that the correct aircraft received the clearance or instruction. The requirement to include aircraft identification in all readbacks and acknowledgements becomes more important as frequency congestion increases and when aircraft with similar call signs are on the same frequency.

2. Read back altitudes, altitude restrictions, and vectors in the same sequence as they are given in the clearance or instruction.

3. Altitudes contained in charted procedures, such as SID's, instrument approaches, etc., should not be read back unless they are specifically stated by the controller.

c. It is the responsibility of the pilot to accept or refuse the clearance issued.

4-86. IFR CLEARANCE VFR-ON-TOP

a. A pilot on an IFR flight plan operating in VFR weather conditions, may request VFR-ON-TOP in lieu of an assigned altitude. This would permit the pilot to select an altitude or flight level of his choice (subject to any ATC restrictions).

b. Pilots desiring to climb through a cloud, haze, smoke, or other meteorological formation and then either cancel their IFR flight plan or operate VFR-ON-TOP may request a climb to VFR-ON-TOP. The ATC authorization shall contain either a top report or a statement that no top report is available, and a request to report reaching VFR-ON-TOP. Additionally, the ATC authorization may contain a clearance limit, routing and an alternative clearance if VFR-ON-TOP is not reached by a specified altitude.

c. A pilot on an IFR flight plan, operating in VFR conditions, may request to climb/descend in VFR conditions.

d. ATC may not authorize VFR-ON-TOP/VFR CONDITIONS operations unless the pilot requests the VFR operation or a clearance to operate in VFR CONDITIONS will result in noise abatement benefits where part of the IFR departure route does not conform to an FAA approved noise abatement route or altitude.

e. When operating in VFR conditions with an ATC authorization to "MAINTAIN VFR-ON-TOP/MAINTAIN VFR CONDITIONS" pilots on IFR flight plans must:

1. Fly at the appropriate VFR altitude as prescribed in FAR Part 91.159.

2. Comply with the VFR visibility and distance from cloud criteria in FAR Part 91.155 (BASIC VFR WEATHER MINIMUMS).

3. Comply with instrument flight rules that are applicable to this flight; i.e., minimum IFR altitudes, position reporting, radio communications, course to be flown, adherence to ATC clearance, etc.

4-86e3 NOTE—Pilots should advise ATC prior to any altitude change to insure the exchange of accurate traffic information.

f. ATC authorization to "MAINTAIN VFR-ON-TOP" is not intended to restrict pilots so that they must operate only *above* an obscuring meteorological formation (layer). Instead, it permits operation above, below, between layers, or in areas where there is no meteorological obscuration. It is imperative, however, that pilots understand that clearance to operate "VFR-ON-TOP/VFR CONDITIONS" does not imply cancellation of the IFR flight plan.

g. Pilots operating VFR-ON-TOP/VFR CONDITIONS may receive traffic information from ATC on other pertinent IFR or VFR aircraft. However, aircraft operating in Class B airspace/TRSA's shall be separated as required by FAA Order 7110.65.

4-86g NOTE—When operating in VFR weather conditions, it is the pilot's responsibility to be vigilant so as to see-and-avoid other aircraft.

h. ATC will not authorize VFR or VFR-ON-TOP operations in Class A airspace. (Reference—Class A Airspace, paragraph 3-11.)

4-87. VFR/IFR FLIGHTS

A pilot departing VFR, either intending to or needing to obtain an IFR clearance en route, must be aware of the position of the aircraft and the relative terrain/obstructions. When accepting a clearance below the MEA/MIA/MVA, pilots are responsible for their own terrain/obstruction clearance until reaching the MEA/MIA/MVA. If the pilot is unable to maintain terrain/obstruction clearance the controller will advise the pilot to state intentions.

4-88 ADHERENCE TO CLEARANCE

a. When air traffic clearance has been obtained under either Visual or Instrument Flight Rules, the pilot-in-command of the aircraft shall not deviate from the provisions thereof unless an amended clearance is obtained. When ATC issues a clearance or instruction, pilots are expected to execute its provisions upon receipt. ATC, in certain situations, will include the word "IMMEDIATELY" in a clearance or instruction to impress urgency of an imminent situation and expeditious compliance by the pilot is expected and necessary for safety. The addition of a VFR or other restriction; i.e., climb or descent point or time, crossing altitude, etc., does not authorize a pilot to deviate from the route of flight or any other provision of the ATC clearance.

b. When a heading is assigned or a turn is requested by ATC, pilots are expected to promptly initiate the turn, to complete the turn, and maintain the new heading unless issued additional instructions.

c. The term "AT PILOT'S DISCRETION" included in the altitude information of an ATC clearance means that ATC has offered the pilot the option to start climb or descent when he wishes. He is authorized to conduct the climb or descent at any rate he wishes and to temporarily level off at any intermediate altitude he may desire. However, once he has vacated an altitude, he may not return to that altitude.

d. When ATC has not used the term "AT PILOT'S DISCRETION" nor imposed any climb or descent restrictions, pilots should initiate climb or descent promptly on acknowledgement of the clearance. Descend or climb at an optimum rate consistent with the operating characteristics of the aircraft to 1,000 feet above or below the assigned altitude, and then attempt to descend or climb at a rate of between 500 and 1,500 fpm until the assigned altitude is reached. If at anytime the pilot is unable to climb or descend at a rate of at least 500 feet a minute, advise ATC. If it is necessary to level off at an intermediate altitude during climb or descent, advise ATC, except when leveling off at 10,000 feet MSL on descent, or 2,500 feet above airport elevation (prior to entering a Class B, Class C, or Class D surface area), when required for speed reduction (FAR Part 91.117).

4-88d NOTE—Leveling off at 10,000 feet MSL on descent or 2,500 feet above airport elevation (prior to entering a Class B, Class C, or Class D surface area) to comply with FAR Part 91.117 airspeed restrictions is commonplace. Controllers anticipate this action and plan accordingly. Leveling off at any other time on climb or descent may seriously affect air traffic handling by ATC. Consequently, it is imperative that pilots make every effort to fulfill the above expected actions to aid ATC in safely handling and expediting traffic.

e. If the altitude information of an ATC DESCENT clearance includes a provision to "CROSS (fix) AT" or "AT OR ABOVE/BELOW (altitude)," the manner in which the descent is executed to comply with the crossing altitude is at the pilot's discretion. This authorization to descend at pilot's discretion is only applicable to that portion of the flight to which the crossing altitude restriction applies, and the pilot is expected to comply with the crossing altitude as a provision of the clearance. Any other clearance in which pilot execution is optional will so state "AT PILOT'S DISCRETION."

EXAMPLE:

"UNITED FOUR SEVENTEEN, DESCEND AND MAINTAIN SIX THOUSAND."

4-88e NOTE—The pilot is expected to commence descent upon receipt of the clearance and to descend at the suggested rates until reaching the assigned altitude of 6,000 feet.

EXAMPLE:

"UNITED FOUR SEVENTEEN, DESCEND AT PILOT'S DISCRETION, MAINTAIN SIX THOUSAND."

4-88e NOTE—The pilot is authorized to conduct descent within the context of the term AT PILOT'S DISCRETION as described above.

EXAMPLE:

"UNITED FOUR SEVENTEEN, CROSS LAKEVIEW V-O-R AT OR ABOVE FLIGHT LEVEL TWO ZERO ZERO, DESCEND AND MAINTAIN SIX THOUSAND."

4-88e NOTE—The pilot is authorized to conduct descent AT PILOT'S DISCRETION until reaching Lakeview VOR. He must comply with the clearance provision to cross the Lakeview VOR at or above FL 200. After passing Lakeview VOR, he is expected to descend at the suggested rates until reaching the assigned altitude of 6,000 feet.

EXAMPLE:

"UNITED FOUR SEVENTEEN, CROSS LAKEVIEW V-O-R AT SIX THOUSAND, MAINTAIN SIX THOUSAND."

4-88e NOTE—The pilot is authorized to conduct descent AT PILOT'S DISCRETION, however, he must comply with the clearance provision to cross the Lakeview VOR at 6,000 feet.

EXAMPLE:

"UNITED FOUR SEVENTEEN, DESCEND NOW TO FLIGHT LEVEL TWO SEVEN ZERO, CROSS LAKEVIEW V-O-R AT OR BELOW ONE ZERO THOUSAND, DESCEND AND MAINTAIN SIX THOUSAND."

4-88e NOTE—The pilot is expected to promptly execute and complete descent to FL 270 upon receipt of the clearance. After reaching FL 270 he is authorized to descend "at pilot's discretion" until reaching Lakeview VOR. He must comply with the clearance provision to cross Lakeview VOR at or below 10,000 feet. After Lakeview VOR he is expected to descend at the suggested rates until reaching 6,000 feet.

EXAMPLE:
"UNITED THREE TEN, DESCEND NOW AND MAINTAIN FLIGHT LEVEL TWO FOUR ZERO, PILOT'S DISCRETION AFTER REACHING FLIGHT LEVEL TWO EIGHT ZERO."

4-88e NOTE—The pilot is expected to commence descent upon receipt of the clearance and to descend at the suggested rates until reaching flight level 280. At that point, the pilot is authorized to continue descent to flight level 240 within the context of the term "AT PILOT'S DISCRETION" as described above.

f. In case emergency authority is used to deviate from provisions of an ATC clearance, the pilot-in-command shall notify ATC as soon as possible and obtain an amended clearance. In an emergency situation which does not result in a deviation from the rules prescribed in FAR Part 91 but which requires ATC to give priority to an aircraft, the pilot of such aircraft shall, when requested by ATC, make a report within 48 hours of such emergency situation to the manager of that ATC facility.

g. The guiding principle is that the last ATC clearance has precedence over the previous ATC clearance. When the route or altitude in a previously issued clearance is amended, the controller will restate applicable altitude restrictions. If altitude to maintain is changed or restated, whether prior to departure or while airborne, and previously issued altitude restrictions are omitted, those altitude restrictions are canceled, including SID altitude restrictions.

EXAMPLE:
A departure flight receives a clearance to destination airport to maintain FL 290. The clearance incorporates a SID which has certain altitude crossing restrictions. Shortly after takeoff, the flight receives a new clearance changing the maintaining FL from 290 to 250. If the altitude restrictions are still applicable, the controller restates them.

EXAMPLE:
A departing aircraft is cleared to cross Fluky intersection at or above 3,000 feet, Gordonville VOR at or above 12,000 feet, maintain FL 200. Shortly after departure, the altitude to be maintained is changed to FL 240. If the altitude restrictions are still applicable, the controller issues an amended clearance as follows: "CROSS FLUKY INTERSECTION AT OR ABOVE THREE THOUSAND, CROSS GORDONVILLE V-O-R AT OR ABOVE ONE TWO THOUSAND, MAINTAIN FLIGHT LEVEL TWO FOUR ZERO."

EXAMPLE:
An arriving aircraft is cleared to his destination airport via V45 Delta VOR direct; he is cleared to cross Delta VOR at 10,000 feet, and then to maintain 6,000 feet. Prior to Delta VOR, the controller issues an amended clearance as follows: "TURN RIGHT HEADING ONE EIGHT ZERO FOR VECTOR TO RUNWAY THREE SIX I-L-S APPROACH, MAINTAIN SIX THOUSAND."

4-88g NOTE—Because the altitude restriction "cross Delta V-O-R at 10,000 feet" was omitted from the amended clearance, it is no longer in effect.

h. Pilots of turbojet aircraft equipped with afterburner engines should advise ATC prior to takeoff if they intend to use afterburning during their climb to the en route altitude. Often, the controller may be able to plan his traffic to accommodate a high performance climb and allow the pilot to climb to his planned altitude without restriction.

4-89. IFR SEPARATION STANDARDS

a. ATC effects separation of aircraft vertically by assigning different altitudes; longitudinally by providing an interval expressed in time or distance between aircraft on the same, converging, or crossing courses, and laterally by assigning different flight paths.

b. Separation will be provided between all aircraft operating on IFR flight plans except during that part of the flight (outside Class B airspace or a TRSA) being conducted on a VFR-ON-TOP/VFR CONDITIONS clearance. Under these conditions, ATC may issue traffic advisories, but it is the sole responsibility of the pilot to be vigilant so as to see and avoid other aircraft.

c. When radar is employed in the separation of aircraft at the same altitude, a minimum of 3 miles separation is provided between aircraft operating within 40 miles of the radar antenna site, and 5 miles between aircraft operating beyond 40 miles from the antenna site. These minima may be increased or decreased in certain specific situations.

4-89 NOTE—Certain separation standards are increased in the terminal environment when CENRAP is being utilized.

4-90. SPEED ADJUSTMENTS

a. ATC will issue speed adjustments to pilots of radar-controlled aircraft to achieve or maintain required or desired spacing.

b. ATC will express all speed adjustments in terms of knots based on indicated airspeed (IAS) in 10 knot increments except that at or above FL 240 speeds may be expressed in terms of Mach numbers in 0.01 increments. The use of Mach numbers is restricted to turbojet aircraft with Mach meters.

c. Pilots complying with speed adjustments are expected to maintain a speed within plus or minus 10 knots or 0.02 Mach number of the specified speed.

d. Unless pilot concurrence is obtained, ATC requests for speed adjustments will be in accordance with the following minimums:

1. To aircraft operating between FL 280 and 10,000 feet, a speed not less than 250 knots or the equivalent Mach number.

2. To turbine powered aircraft operating below 10,000 feet:

(a) A speed not less than 210 knots, except;

(b) Within 20 flying miles of the airport of intended landing, a speed not less than 170 knots.

3. Reciprocating engine or turboprop aircraft within 20 flying miles of the runway threshold of the airport of intended landing, a speed not less than 150 knots.

4. To departing aircraft:

(a) Turbine powered aircraft, a speed not less than 230 knots.

(b) Reciprocating engine aircraft, a speed not less than 150 knots.

e. When ATC combines a speed adjustment with a descent clearance, the sequence of delivery, with the word "then" between, indicates the expected order of execution;

EXAMPLE:

DESCEND AND MAINTAIN (altitude); THEN, REDUCE SPEED TO (speed).

EXAMPLE:

REDUCE SPEED TO (speed); THEN, DESCEND AND MAINTAIN (altitude)...................

4-90e NOTE—The maximum speeds below 10,000 feet as established in FAR Part 91.117 still apply. If there is any doubt concerning the manner in which such a clearance is to be executed, request clarification from ATC.

f. If ATC determines (before an approach clearance is issued) that it is no longer necessary to apply speed adjustment procedures, they will inform the pilot to resume normal speed. Approach clearances supersede any prior speed adjustment assignments, and pilots are expected to make their own speed adjustments, as necessary, to complete the approach. Under certain circumstances however, it may be necessary for ATC to issue further speed adjustments after approach clearance is issued to maintain separation between successive arrivals. Under such circumstances, previously issued speed adjustments will be restated if that speed is to be maintained or additional speed adjustments are requested. ATC must obtain pilot concurrence for speed adjustments after approach clearances are issued. Speed adjustments should not be assigned inside the final approach fix on final or a point 5 miles from the runway, whichever is closer to the runway.

g. The pilots retain the prerogative of rejecting the application of speed adjustment by ATC if the minimum safe airspeed for any particular operation is greater than the speed adjustment. IN SUCH CASES, PILOTS ARE EXPECTED TO ADVISE ATC OF THE SPEED THAT WILL BE USED.

h. Pilots are reminded that they are responsible for rejecting the application of speed adjustment by ATC if, in their opinion, it will cause them to exceed the maximum indicated airspeed prescribed by FAR Part 91.117(a). IN SUCH CASES, THE PILOT IS EXPECTED TO SO INFORM ATC. Pilots operating at or above 10,000 feet MSL who are issued speed adjustments which exceed 250 knots IAS and are subsequently cleared below 10,000 feet MSL are expected to comply with FAR Part 91.117(a).

i. For operations conducted below 10,000 feet MSL when outside the United States and beneath Class B airspace, airspeed restrictions apply to all U.S. registered aircraft. For operations conducted below 10,000 feet MSL when outside the United States within a Class B airspace, there are no speed restrictions.

j. For operations in a Class B, Class C, and Class D surface area, ATC is authorized to request or approve a speed greater than the maximum indicated airspeeds prescribed for operation within that airspace (FAR Part 91.117(b)).

k. When in communication with the ARTCC, pilots should, as a good operating practice, state any ATC assigned speed restriction on initial radio contact associated with an ATC communications frequency change.

4-91. RUNWAY SEPARATION

Tower controllers establish the sequence of arriving and departing aircraft by requiring them to adjust flight or ground operation as necessary to achieve proper spacing. They may "HOLD" an aircraft short of the runway to achieve spacing between it and an arriving aircraft; the controller may instruct a pilot to "EXTEND DOWNWIND" in order to establish spacing from an arriving or departing aircraft. At times a clearance may include the word "IMMEDIATE." For example: "CLEARED FOR IMMEDIATE TAKEOFF." In such cases "IMMEDIATE" is used for purposes of *air traffic separation*. It is up to the pilot to refuse the clearance if, in his opinion, compliance would adversely affect his operation. (Reference—Gate Holding Due to Departure Delays, paragraph 4-64.)

4-92. VISUAL SEPARATION

a. Visual separation is a means employed by ATC to separate aircraft only in terminal areas. There are two methods employed to effect this separation:

1. The tower controller sees the aircraft involved and issues instructions, as necessary, to ensure that the aircraft avoid each other.

2. A pilot sees the other aircraft involved and upon instructions from the controller provides his own separation by maneuvering his aircraft to avoid it. This may involve following in-trail behind another aircraft or keeping it in sight until it is no longer a factor.

b. A pilot's acceptance of instructions to follow another aircraft or provide visual separation from it is an acknowledgment that the pilot will maneuver his/her aircraft as necessary to avoid the other aircraft or to maintain in-trail separation. In operations conducted behind heavy jet aircraft, it is also an acknowledgment that the pilot accepts the responsibility for wake turbulence separation.

c. WHEN A PILOT HAS BEEN TOLD TO FOLLOW ANOTHER AIRCRAFT OR TO PROVIDE VISUAL SEPARATION FROM IT HE/SHE SHOULD PROMPTLY NOTIFY THE CONTROLLER IF VISUAL CONTACT WITH THE OTHER AIRCRAFT IS LOST OR CANNOT BE MAINTAINED OR IF THE PILOT CANNOT ACCEPT THE RESPONSIBILITY FOR THE SEPARATION FOR ANY REASON.

d. Pilots should remember, however, that they have a regulatory responsibility (FAR Part 91.113(a)) to see and avoid other aircraft when weather conditions permit.

4-93. USE OF VISUAL CLEARING PROCEDURES

a. Before Takeoff—Prior to taxiing onto a runway or landing area in preparation for takeoff, pilots should scan the approach areas for possible landing traffic, executing appropriate clearing maneuvers to provide him a clear view of the approach areas.

b. Climbs and Descents—During climbs and descents in flight conditions which permit visual detection of other traffic, pilots should execute gentle banks, left and right at a frequency which permits continuous visual scanning of the airspace about them.

c. Straight and Level—Sustained periods of straight and level flight in conditions which permit visual detection of other traffic should be broken at intervals with appropriate clearing procedures to provide effective visual scanning.

d. Traffic Pattern—Entries into traffic patterns while descending create specific collision hazards and should be avoided.

e. Traffic at VOR Sites—All operators should emphasize the need for sustained vigilance in the vicinity of VORs and airway intersections due to the convergence of traffic.

f. Training Operations—Operators of pilot training programs are urged to adopt the following practices:

1. Pilots undergoing flight instruction at all levels should be requested to verbalize clearing procedures (call out "clear" left, right, above, or below) to instill and sustain the habit of vigilance during maneuvering.

2. High-wing airplane: momentarily raise the wing in the direction of the intended turn and look.

3. Low-wing airplane: momentarily lower the wing in the direction of the intended turn and look.

4. Appropriate clearing procedures should precede the execution of all turns including chandelles, lazy eights, stalls, slow flight, climbs, straight and level, spins, and other combination maneuvers.

4-94. TRAFFIC ALERT AND COLLISION AVOIDANCE SYSTEM (TCAS I & II)

a. TCAS I provides proximity warning only, to assist the pilot in the visual acquisition of intruder aircraft. No recommended avoidance maneuvers are provided nor authorized as a direct result of a TCAS I warning. It is intended for use by smaller commuter aircraft holding 10 to 30 passenger seats, and general aviation aircraft.

b. TCAS II provides traffic advisories (TA's) and resolution advisories (RA's). Resolution advisories provide recommended maneuvers in a vertical direction (climb or descend only) to avoid conflicting traffic. Airline aircraft, and larger commuter and business aircraft holding 31 passenger seats or more, use TCAS II equipment.

1. Each pilot who deviates from an ATC clearance in response to a TCAS II RA shall notify ATC of that deviation as soon as practicable and expeditiously return to the current ATC clearance when the traffic conflict is resolved.

2. Deviations from rules, policies, or clearances should be kept to the minimum necessary to satisfy a TCAS II RA.

3. The serving IFR air traffic facility is not responsible to provide approved standard IFR separation to an aircraft after a TCAS II RA maneuver until one of the following conditions exists:

(a) The aircraft has returned to its assigned altitude and course.

(b) Alternate ATC instructions have been issued.

c. TCAS does not alter or diminish the pilot's basic authority and responsibility to ensure safe flight. Since TCAS does not respond to aircraft which are not transponder equipped or aircraft with a transponder failure, TCAS alone does not ensure safe separation in every case.

d. At this time, no air traffic service nor handling is predicated on the availability of TCAS equipment in the aircraft.

Chapter 5. Air traffic procedures

Section 1. PREFLIGHT

5-1. PREFLIGHT PREPARATION

a. Every pilot is urged to receive a preflight briefing and to file a flight plan. This briefing should consist of the latest or most current weather, airport, and en route NAVAID information. Briefing service may be obtained from an FSS either by telephone or interphone, by radio when airborne, or by a personal visit to the station. Pilots with a current medical certificate in the 48 contiguous States may access toll-free the Direct User Access Terminal System (DUATS) through a personal computer. DUATS will provide alpha-numeric preflight weather data and allow pilots to file domestic VFR or IFR flight plans. (Reference—FAA Weather Services, paragraph 7-2c(5) lists DUATS vendors.)

5-1a NOTE—Pilots filing flight plans via "fast file" who desire to have their briefing recorded, should include a statement at the end of the recording as to the source of their weather briefing.

b. The information required by the FAA to process flight plans is contained on FAA Form 7233-1, Flight Plan. (Reference—Flight Plan—VFR Flights, paragraph 5-4 and Flight Plan—IFR Flights, paragraph 5-7.) The forms are available at all flight service stations. Additional copies will be provided on request.

c. Consult an FSS or a Weather Service Office (WSO) for preflight weather briefing. Supplemental Weather Service Locations (SWSLs) do not provide weather briefing.

d. FSSs are required to advise of pertinent NOTAMs if a *standard* briefing is requested, but if they are overlooked, don't hesitate to remind the specialist that you have not received NOTAM information.

5-1d NOTE—NOTAMs which are known in sufficient time for publication and are of 7 days duration or longer are normally incorporated into the Notices to Airmen publication and carried there until cancellation time. FDC NOTAMs, which apply to instrument flight procedures, are also included in the Notices to Airmen publication up to and including the number indicated in the FDC NOTAM legend. Printed NOTAMs are not provided during a briefing unless specifically requested by the pilot since the FSS specialist has no way of knowing whether the pilot has already checked the Notices to Airmen publication prior to calling. Remember to *ask* for NOTAMs in the Notices to Airmen publication. This information is not normally furnished during your briefing. (Reference—Notice to Airmen (NOTAM) System, paragraph 5-3.)

e. Pilots are urged to use only the latest issue of aeronautical charts in planning and conducting flight operations. Aeronautical charts are revised and reissued on a regular scheduled basis to ensure that depicted data are current and reliable. In the conterminous U.S., Sectional Charts are updated each 6 months, IFR En route Charts each 56 days, and amendments to civil IFR Approach Charts are accomplished on a 56 day cycle with a change notice volume issued on the 28 day midcycle. Charts that have been superseded by those of a more recent date may contain obsolete or incomplete flight information. (Reference—General Description of each Chart Series, paragraph 9-4.)

f. When requesting a preflight briefing, identify yourself as a pilot and provide the following:

1. Type of flight planned; e.g., VFR or IFR.
2. Aircraft's number or pilot's name.
3. Aircraft type.
4. Departure Airport.
5. Route of flight.
6. Destination.
7. Flight altitude(s).
8. ETD and ETE.

g. Prior to conducting a briefing, briefers are required to have the background information listed above so that they may tailor the briefing to the needs of the proposed flight. The objective is to communicate a "picture" of meteorological and aeronautical information necessary for the conduct of a safe and efficient flight. Briefers use all available weather and aeronautical information to summarize data applicable to the proposed flight. They do not read weather reports and forecasts verbatim unless specifically requested by the pilot. Refer to paragraph 7-3 *PREFLIGHT BRIEFINGS* for those items of a weather briefing that should be expected or requested.

h. The Federal Aviation Administration (FAA) by Federal Aviation Regulation, Part 93, Subpart K, has designated High Density Traffic Airports (HDTA's) and has prescribed air traffic rules and requirements for operating aircraft (excluding helicopter operations) to and from these airports. (Reference—Airport/Facility Directory, Special Notices Section, and AIM, Section 4-21, for further details.)

i. In addition to the filing of a flight plan, if the flight will traverse or land in one or more foreign countries, it is particularly important that pilots leave a complete itinerary with someone directly concerned, keep that person advised of the flight's progress, and inform him that, if serious doubt arises as to the safety of the flight, he should first contact the FSS. (Reference—Flights Outside the United States and U.S. Territories, paragraph 5-9.)

j. Pilots operating under provisions of FAR Part 135 and not having an FAA assigned 3-letter designator, are urged to prefix the normal registration (N) number with the letter "T" on flight plan filing; e.g., TN1234B. (Reference—Aircraft Call Signs, paragraph 4-33.)

5-2. FOLLOW IFR PROCEDURES EVEN WHEN OPERATING VFR

a. To maintain IFR proficiency, pilots are urged to practice IFR procedures whenever possible, even when operating VFR. Some suggested practices include:

1. Obtain a complete preflight and weather briefing. Check the NOTAMs.

2. File a flight plan. This is an excellent low cost insurance policy. The cost is the time it takes to fill it out. The insurance includes the knowledge that someone will be looking for you if you become overdue at your destination.

3. Use current charts.

4. Use the navigation aids. Practice maintaining a good course—keep the needle centered.

5. Maintain a constant altitude which is appropriate for the direction of flight.

6. Estimate en route position times.

7. Make accurate and frequent position reports to the FSSs along your route of flight.

b. Simulated IFR flight is recommended (under the hood); however, pilots are cautioned to review and adhere to the requirements specified in FAR Part 91.109 before and during such flight.

c. When flying VFR at night, in addition to the altitude appropriate for the direction of flight, pilots should maintain an altitude which is at or above the minimum en route altitude as shown on charts. This is especially true in mountainous terrain, where there is usually very little ground reference. Do not depend on your eyes alone to avoid rising unlighted terrain, or even lighted obstructions such as TV towers.

5-3. NOTICE TO AIRMEN (NOTAM) SYSTEM

a. Time-critical aeronautical information which is of either a temporary nature or not sufficiently known in advance to permit publication on aeronautical charts or in other operational publications receives immediate dissemination via the National Notice to Airmen (NOTAM) System.

5-3a NOTE—NOTAM information is that aeronautical information that could affect a pilot's decision to make a flight. It includes such information as airport or primary runway closures, changes in the status of navigational aids, ILS's, radar service availability, and other information essential to planned enroute, terminal, or landing operations.

b. NOTAM information is classified into three categories. These are NOTAM (D) or distant, NOTAM (L) or local, and Flight Data Center (FDC) NOTAMs.

1. NOTAM (D) information is disseminated for all navigational facilities that are part of the National Airspace System (NAS), all public use airports, seaplane bases, and heliports listed in the Airport/Facility Directory (A/FD). The complete file of all NOTAM (D) information is maintained in a computer data base at the National Communications Center (NATCOM), located in Kansas City. This category of information is distributed automatically, appended to the hourly weather reports, via the Service A telecommunications system. Air traffic facilities, primarily FSSs, with Service A capability have access to the entire NATCOM data base of NOTAMs. These NOTAMs remain available via Service A for the duration of their validity or until published.

2. NOTAM (L)

(a) NOTAM (L) information includes such data as taxiway closures, personnel and equipment near or crossing runways, airport rotating beacon outages and airport lighting aids that do not affect instrument approach criteria, such as VASI.

(b) NOTAM (L) information is distributed locally only and is not attached to the hourly weather reports. A separate file of local NOTAMs is maintained at each FSS for facilities in their area only. NOTAM (L) information for other FSS areas must be specifically requested directly from the FSS that has responsibility for the airport concerned.

5-3a2b NOTE—DUATS vendors are not required to provide NOTAM L information.

3. FDC NOTAMs

(a) On those occasions when it becomes necessary to disseminate information which is regulatory in nature, the National Flight Data Center (NFDC), in Washington, DC, will issue an FDC NOTAM. FDC NOTAMs contain such things as amendments to published IAP's and other current aeronautical charts. They are also used to advertise temporary flight restrictions caused by such things as natural disasters or large-scale public events that may generate a congestion of air traffic over a site.

(b) FDC NOTAMs are transmitted via Service A only once and are kept on file at the FSS until published or canceled. FSSs are responsible for maintaining a file of current, unpublished FDC NOTAMs concerning conditions within 400 miles of their facilities. FDC information concerning conditions that are more than 400 miles from the FSS, or that is already published, is given to a pilot only on request.

5-3a3b NOTE 1—DUATS vendors will provide FDC NOTAMs only upon site-specific requests using a location identifier.

54a3b NOTE 2—NOTAM data may not always be current due to the changeable nature of National Airspace System components, delays inherent in processing information, and occasional temporary outages of the United States NOTAM System. While en route, pilots should contact FSSs and obtain updated information for their route of flight and destination.

c. An integral part of the NOTAM System is the biweekly Notices to Airmen publication (NTAP). Data is included in this publication to reduce congestion on the telecommunications circuits and, therefore, is not available via Service A. Once published, the information is not provided during pilot weather briefings unless specifically requested by the pilot. This publication contains two sections.

1. The first section consists of notices that meet the criteria for NOTAM (D) and are expected to remain in effect for an extended period and FDC NOTAMs that are current at the time of publication. Occasionally, some NOTAM (L) and other unique information is included in this section when it will contribute to flight safety.

2. The second section contains special notices that are either too long or concern a wide or unspecified geographic area and are not suitable for inclusion in the first section. The content of these notices vary widely and there are no specific criteria for their inclusion, other than their enhancement of flight safety.

3. The number of the last FDC NOTAM included in the publication is noted on the first page to aid the user in updating the listing with any FDC NOTAMs which may have been issued between the cut-off date and the date the publication is received. All information contained will be carried until the information expires, is canceled, or in the case of permanent conditions, is published in other publications, such as the A/FD.

4. All new notices entered, excluding FDC NOTAMs, will be published only if the information is expected to remain in effect for at least 7 days after the effective date of the publication.

d. NOTAM information is not available from a Supplemental Weather Service Location (SWSL).

5-4. FLIGHT PLAN—VFR FLIGHTS

a. Except for operations in or penetrating a Coastal or Domestic ADIZ or DEWIZ a flight plan is not required for VFR flight. (Reference—National Security, paragraph 5-90). However, it is strongly recommended that one be filed with an FAA FSS. This will ensure that you receive VFR Search and Rescue Protection. (Reference—Search and Rescue, paragraph 6-16g for the proper method of filing).

b. To obtain maximum benefits from the flight plan program, flight plans should be filed directly with the nearest FSS. For your convenience, FSSs provide aeronautical and meteorological briefings while accepting flight plans. Radio may be used to file if no other means are available.

5-4b NOTE—Some states operate aeronautical communications facilities which will accept and forward flight plans to the FSS for further handling.

c. When a "stopover" flight is anticipated, it is recommended that a separate flight plan be filed for each "leg" when the stop is expected to be more than 1 hour duration.

d. Pilots are encouraged to give their departure times directly to the FSS serving the departure airport or as otherwise indicated by the FSS when the flight plan is filed. This will ensure more efficient flight plan service and permit the FSS to advise you of significant changes in aeronautical facilities or meteorological conditions. When a VFR flight plan is filed, it will be held by the FSS until 1 hour after the proposed departure time unless:

1. The actual departure time is received.

2. A revised proposed departure time is received.

3. At a time of filing, the FSS is informed that the proposed departure time will be met, but actual time cannot be given because of inadequate communications (assumed departures).

e. On pilot's request, at a location having an active tower, the aircraft identification will be forwarded by the tower to the FSS for reporting the actual departure time. This procedure should be avoided at busy airports.

f. Although position reports are not required for VFR flight plans, periodic reports to FAA FSSs along the route are good practice. Such contacts permit significant information to be passed to the transiting aircraft and also serve to check the progress of the flight should it be necessary for any reason to locate the aircraft.

EXAMPLE:
BONANZA 314K, OVER KINGFISHER AT (time), VFR FLIGHT PLAN, TULSA TO AMARILLO.

EXAMPLE:
CHEROKEE 5133J, OVER OKLAHOMA CITY AT (time), SHREVEPORT TO DENVER, NO FLIGHT PLAN.

g. Pilots not operating on an IFR flight plan and when in level cruising flight, are cautioned to conform with VFR cruising altitudes appropriate to the direction of flight.

h. When filing VFR flight plans, indicate aircraft equipment capabilities by appending the appropriate suffix to aircraft type in the same manner as that prescribed for IFR flight. (Reference—Flight Plan—IFR Flights, paragraph 5-7). Under some circumstances, ATC computer tapes can be useful in constructing the radar history of a downed or crashed aircraft. In each case, knowledge of the aircraft's transponder equipment is necessary in determining whether or not such computer tapes might prove effective.

i. Flight Plan Form—(See Figure 5-4[1]).

j. Explanation of VFR Flight Plan Items—

Block 1. Check the type flight plan. Check both the VFR and IFR blocks if composite VFR/IFR.

Block 2. Enter your complete aircraft identification including the prefix "N" if applicable.

Block 3. Enter the designator for the aircraft, or if unknown, consult an FSS briefer.

Block 4. Enter your true airspeed (TAS).

Block 5. Enter the departure airport identifier code, or if unknown, the name of the airport.

Block 6. Enter the proposed departure time in Coordinated Universal Time (UTC). If airborne, specify the actual or proposed departure time as appropriate.

Block 7. Enter the appropriate VFR altitude (to assist the briefer in providing weather and wind information).

Block 8. Define the route of flight by using NAVAID identifier codes and airways.

Block 9. Enter the destination airport identifier code, or if unknown, the airport name.

Block 9 NOTE—Include the city name (or even the state name) if needed for clarity.

U.S. DEPARTMENT OF TRANSPORTATION
FEDERAAL AVIATION ADMINISTRATION
FLIGHT PLAN

(FAA USE ONLY) ☐ PILOT BRIEFING ☐ VNR | TIME STARTED | SPECIALIST INITIALS
☐ STOPOVER

1. TYPE: VFR / IFR / DVFR | 2. AIRCRAFT IDENTIFICATION | 3. AIRCRAFT TYPE/ SPECIAL EQUIPMENT | 4. TRUE AIRSPEED KTS | 5. DEPARTURE POINT | 6. DEPARTURE TIME: PROPOSED (Z) ACTUAL (Z) | 7. CRUISING ALTITUDE

8. ROUTE OF FLIGHT

9. DESTINATION (Name of airport and city) | 10. EST. TIME ENROUTE: HOURS MINUTES | 11. REMARKS

12. FUEL ON BOARD | 13. ALTERNATE AIRPORT(S) | 14. PILOT'S NAME, ADDRESS & TELEPHONE NUMBER & AIRCRAFT HOME BASE | 15. NUMBER ABOARD

17. DESTINATION CONTACT/TELEPHONE (OPTIONAL)

16. COLOR OF AIRCRAFT | CIVIL AIRCRAFT PILOTS. FAR 91 requires you file an IFR flight plan to operate under instrument flight rules in controlled airspace. Failure to file could result in a civil penalty not to exceed $1,000 for each violation (Section 901 of the Federal Aviation Act of 1958, as amended). Filing of a VFR flight plan is recommended as a good operating practice. See also Part 99 for requirements concerning DVFR flight plans

FAA Form 7233-1 (8-82) CLOSE VFR FLIGHT PLAN WITH ______________ FSS ON ARRIVAL

Figure 5-4[1]

Block 10. Enter your Estimated Time en Route in hours and minutes.

Block 11. Enter only those remarks pertinent to ATC or to the clarification of other flight plan information, such as the appropriate radiotelephony (call sign) associated with the designator filed in Block 2. Items of a personal nature are not accepted.

Block 12. Specify the fuel on board in hours and minutes.

Block 13. Specify an alternate airport if desired.

Block 14. Enter your complete name, address, and telephone number. Enter sufficient information to identify home base, airport, or operator.

5-4j Block 14 NOTE—This information is essential in the event of search and rescue operations.

Block 15. Enter total number of persons on board (POB) including crew.

Block 16. Enter the predominant colors.

Block 17. Record the FSS name for closing the flight plan. If the flight plan is closed with a different FSS or facility, state the recorded FSS name that would normally have closed your flight plan. **(Optional)**—Record a destination telephone number to assist search and rescue contact should you fail to report or cancel your flight plan within ½ hour after your Estimated Time of Arrival (ETA).

5-4j Block 17 NOTE—The information transmitted to the destination FSS will consist only of flight plans Blocks 2, 3, 9, and 10. Estimated time en route (ETE) will be converted to the correct estimated time of arrival (ETA).

5-5. FLIGHT PLAN—DEFENSE VFR (DVFR) FLIGHTS

VFR flights into a Coastal or Domestic ADIZ/DEWIZ are required to file DVFR flight plans for security purposes. Detailed ADIZ procedures are found in the National Security section of this chapter. (See FAR Part 99.)

5-6. COMPOSITE FLIGHT PLAN (VFR/IFR FLIGHTS)

a. Flight plans which specify VFR operation for one portion of a flight, and IFR for another portion, will be accepted by the FSS at the point of departure. If VFR flight is conducted for the first portion of the flight, the pilot should report his departure time to the FSS with which he filed his VFR/IFR flight plan; and, subsequently, close the VFR portion and request ATC clearance from the FSS nearest the point at which change from VFR to IFR is proposed. Regardless of the type facility you are communicating with (FSS, center, or tower), it is the pilot's responsibility to request that facility to "CLOSE VFR

FLIGHT PLAN." The pilot must remain in VFR weather conditions until operating in accordance with the IFR clearance.

b. When a flight plan indicates IFR for the first portion of flight and VFR for the latter portion, the pilot will normally be cleared to the point at which the change is proposed. Once the pilot has reported over the clearance limit and does not desire further IFR clearance, he should advise ATC to cancel the IFR portion of his flight plan. Then, he should contact the nearest FSS to activate the VFR portion of his flight plan. If the pilot desires to continue his IFR flight plan beyond the clearance limit, he should contact ATC at least 5 minutes prior to the clearance limit and request further IFR clearance. If the requested clearance is not received prior to reaching the clearance limit fix, the pilot will be expected to establish himself in a standard holding pattern on the radial or course to the fix unless a holding pattern for the clearance limit fix is depicted on a U.S. Government or commercially produced (meeting FAA requirements) Low or High Altitude En Route, Area or STAR Chart. In this case the pilot will hold according to the depicted pattern.

5-7. FLIGHT PLAN—IFR FLIGHTS

a. General—

1. Prior to departure from within, or prior to entering controlled airspace, a pilot must submit a complete flight plan and receive an air traffic clearance, if weather conditions are below VFR minimums. Instrument flight plans may be submitted to the nearest FSS or ATCT either in person or by telephone (or by radio if no other means are available). Pilots should file IFR flight plans at least 30 minutes prior to estimated time of departure to preclude possible delay in receiving a departure clearance from ATC. To minimize your delay in entering Class B, Class C, Class D and Class E surface area at destination when IFR weather conditions exist or are forecast at that airport, an IFR flight plan should be filed before departure. Otherwise, a 30 minute delay is not unusual in receiving an ATC clearance because of time spent in processing flight plan data. Traffic saturation frequently prevents control personnel from accepting flight plans by radio. In such cases, the pilot is advised to contact the nearest FSS for the purpose of filing the flight plan.

5-7a1 NOTE—There are several methods of obtaining IFR clearances at nontower, non-FSS, and outlying airports. The procedure may vary due to geographical features, weather conditions, and the complexity of the ATC system. To determine the most effective means of receiving an IFR clearance, pilots should ask the nearest FSS the most appropriate means of obtaining the IFR clearance.

2. When filing an IFR flight plan for a TCAS/heavy equipped aircraft, add the prefix T for TCAS, H for Heavy, or B for both TCAS and heavy to the aircraft type.

EXAMPLE:

H/DC10/U T/B727/A B/B747/R

3. When filing an IFR flight plan for flight in an aircraft equipped with a radar beacon transponder, DME equipment, TACAN-only equipment or a combination of both, identify equipment capability by adding a suffix to the AIRCRAFT TYPE preceded by a slant, as follows:

/X—no transponder.

/T—transponder with no altitude encoding capability.

/U—transponder with altitude encoding capability.

/D—DME, but no transponder.

/B—DME and transponder, but no altitude encoding capability.

/A—DME and transponder with altitude encoding capability.

/M—TACAN only, but no transponder.

/N—TACAN only and transponder, but with no altitude encoding capability.

/P—TACAN only and transponder with altitude encoding capability.

/C—RNAV and transponder, but with no altitude encoding capability.

/R—RNAV and transponder with altitude encoding capability.

/W—RNAV but no transponder.

/G—Flight Management System (FMS) and Electronic Flight Instrument System (EFIS) equipped aircraft with /R capability having a "Special Aircraft and Aircrew Authorization" issued by the FAA.

5-7a3 NOTE 1—Criteria for use of the /G designation is presently identified only for certain /R equipped air carrier aircraft and specially qualified crews. Authorization for use of the "/G" designation is obtained through the All-Weather Operations Branch of the FAA Flight Standards Service and an air carrier's certificate holding district office.

5-7a3 NOTE 2—The use of "/G" is limited to aircraft which operate totally within airspace controlled by U.S. air traffic control facilities.

4. It is recommended that pilots file the maximum transponder or navigation capability of their aircraft in the equipment suffix. This will provide ATC with the necessary information to utilize all facets of navigational equipment and transponder capabilities available. In the case of area navigation equipped aircraft, pilots should file the /C, /R, or /W capability of the aircraft even though an RNAV route or random RNAV route has not been requested. This will ensure ATC awareness of the pilot's ability to navigate point-to-point and may be utilized to expedite the flight.

5-7a4 NOTE—The suffix is not to be added to the aircraft identification or be transmitted by radio as part of the aircraft identification.

b. Airways and Jet Routes Depiction on Flight Plan—

1. It is vitally important that the route of flight be accurately and completely described in the flight plan. To simplify definition of the proposed route, and to facilitate ATC, pilots are requested to file via airways or jet routes established for use at the altitude or Flight Level planned.

2. If flight is to be conducted via designated airways or jet routes, describe the route by indicating the type and number designators of the airway(s) or jet route(s) requested. If more than one airway or jet route is to be used, clearly indicate points of transition. If the transition is made at an unnamed intersection, show the next succeeding NAVAID or named intersection on the intended route and the complete route from that point. Reporting points may be identified by using authorized name/code as depicted on appropriate aeronautical charts. The following two examples illustrate the need to specify the transition point when two routes share more than one transition fix.

EXAMPLE:
ALB J37 BUMPY J14 BHM

SPELLED OUT: From Albany, New York, via Jet Route 37 transitioning to Jet Route 14 at BUMPY intersection, thence via Jet Route 14 to Birmingham, Alabama.

EXAMPLE:
ALB J37 ENO J14 BHM

SPELLED OUT: From Albany, New York, via Jet Route 37 transitioning to Jet Route 14 at Kenton VORTAC (ENO) thence via Jet Route 14 to Birmingham, Alabama.

(a) The route of flight may also be described by naming the reporting points or NAVAID's over which the flight will pass, provided the points named are established for use at the altitude or flight level planned.

EXAMPLE:
BWI V44 SWANN V433 DQO

SPELLED OUT: From Baltimore-Washington International, via Victor 44 to Swann Intersection, transitioning to Victor 433 at Swann, thence via V433 to Dupont.

(b) When the route of flight is defined by named reporting points, whether alone or in combination with airways or jet routes, and the navigational aids (VOR, VORTAC, TACAN, NDB) to be used for the flight are a combination of different types of aids, enough information should be included to clearly indicate the route requested.

EXAMPLE:
LAX J5 LKV J3 GEG YXC FL 330 J500 VLR J515 YWG

SPELLED OUT: From Los Angeles International via Jet Route 5 Lakeview, Jet Route 3 Spokane, direct Cranbrook, British Columbia VOR/DME, Flight Level 330 Jet Route 500 to Langruth, Manitoba VORTAC, Jet Route 515 to Winnepeg, Manitoba.

(c) When filing IFR, it is to the pilot's advantage to file a preferred route.

5-72c NOTE—Preferred IFR routes are described and tabulated in the Airport/Facility Directory.

(d) ATC may issue a SID or a STAR, as appropriate. (Reference—Instrument Departure, paragraph 5-25 and Standard Terminal Arrival (STAR), paragraph 5-40.)

5-72d NOTE—Pilots not desiring a SID or STAR should so indicate in the remarks section of the flight plan as "NO SID" or "NO STAR."

c. Direct Flights—

1. All or any portions of the route which will not be flown on the radials or courses of established airways or routes, such as direct route flights, must be defined by indicating the radio fixes over which the flight will pass. Fixes selected to define the route shall be those over which the position of the aircraft can be accurately determined. Such fixes automatically become compulsory reporting points for the flight, unless advised otherwise by ATC. Only those navigational aids established for use in a particular structure; i.e., in the low or high structures, may be used to define the en route phase of a direct flight within that altitude structure.

2. The azimuth feature of VOR aids and that azimuth and distance (DME) features of VORTAC and TACAN aids are assigned certain frequency protected areas of airspace which are intended for application to established airway and route use, and to provide guidance for planning flights outside of established airways or routes. These areas of airspace are expressed in terms of cylindrical service volumes of specified dimensions called "class limits" or "categories." (Reference—Navaid Service Volumes, paragraph 1-8.) An operational service volume has been established for each class in which adequate signal coverage and frequency protection can be assured. To facilitate use of VOR, VORTAC, or TACAN aids, consistent with their operational service volume limits, pilot use of such aids for defining a direct route of flight in controlled airspace should not exceed the following:

(a) Operations above FL 450—Use aids not more than 200 NM apart. These aids are depicted on En Route High Altitude Charts.

(b) Operation off established routes from 18,000 feet MSL to FL 450—Use aids not more than 260 NM apart. These aids are depicted on En Route High Altitude Charts.

(c) Operation off established airways below 18,000 feet MSL—Use aids not more than 80 NM apart. These aids are depicted on Enroute Low Altitude Charts.

(d) Operation off established airways between 14,500 feet MSL and 17,999 feet MSL in the conterminous U.S.-(H) facilities not more than 200 NM apart may be used.

3. Increasing use of self-contained airborne navigational systems which do not rely on the VOR/VORTAC/TACAN system has resulted in pilot requests for direct routes which exceed NAVAID service volume limits. These direct route requests will be approved only in a radar environment, with approval based on pilot responsibility for navigation on the authorized direct route. Radar flight following will be provided by ATC for ATC purposes.

4. At times, ATC will initiate a direct route in a radar environment which exceeds NAVAID service volume limits. In such cases ATC will provide radar monitoring and navigational assistance as necessary.

5. Airway or jet route numbers, appropriate to the stratum in which operation will be conducted, may also be included to describe portions of the route to be flown.

EXAMPLE:
MDW V262 BDF V10 BRL STJ SLN GCK

SPELLED OUT: From Chicago Midway Airport via Victor 262 to Bradford, Victor 10 to Burlington, Iowa, direct St. Joseph, Missouri, direct Salina, Kansas, direct Garden City, Kansas.

5-7c5 NOTE—When route of flight is described by radio fixes, the pilot will be expected to fly a direct course between the points named.

6. Pilots are reminded that they are responsible for adhering to obstruction clearance requirements on those segments of direct routes that are outside of controlled airspace. The MEA's and other altitudes shown on Low Altitude IFR Enroute Charts pertain to those route segments within controlled airspace, and those altitudes may not meet obstruction clearance criteria when operating off those routes.

d. Area Navigation (RNAV)—

1. Random RNAV routes can only be approved in a radar environment. Factors that will be considered by ATC in approving random RNAV routes include the capability to provide radar monitoring and compatibility with traffic volume and flow. ATC will radar monitor each flight, however, navigation on the random RNAV route is the responsibility of the pilot.

2. To be certified for use in the National Airspace System, RNAV equipment must meet the specifications outlined in AC 90-45. The pilot is responsible for variations in equipment capability and must advise ATC if a RNAV clearance cannot be accepted as specified. The controller need only be concerned that the aircraft is RNAV equipped; if the flight plan equipment suffix denotes RNAV capability, the RNAV routing can be applied.

3. Pilots of aircraft equipped with operational area navigation equipment may file for random RNAV routes throughout the National Airspace System, where radar monitoring by ATC is available, in accordance with the following procedures.

(a) File airport-to-airport flight plans prior to departure.

(b) File the appropriate RNAV capability certification suffix in the flight plan.

(c) Plan the random route portion of the flight plan to begin and end over appropriate arrival and departure transition fixes or appropriate navigation aids for the altitude stratum within which the flight will be conducted. The use of normal preferred departure and arrival routes (SID/STAR), where established, is recommended.

(d) File route structure transitions to and from the random route portion of the flight.

(e) Define the random route by waypoints. File route description waypoints by using degree-distance fixes based on navigational aids which are appropriate for the altitude stratum.

(f) File a minimum of one route description waypoint for each ARTCC through whose area the random route will be flown. These waypoints must be located within 200 NM of the preceding center's boundary.

(g) File an additional route description waypoint for each turnpoint in the route.

(h) Plan additional route description waypoints as required to ensure accurate navigation via the filed route of flight. Navigation is the pilot's responsibility unless ATC assistance is requested.

(i) Plan the route of flight so as to avoid Prohibited and Restricted Airspace by 3 NM unless permission has been obtained to operate in that airspace and the appropriate ATC facilities are advised.

4. Pilots of aircraft equipped with latitude/longitude coordinate navigation capability, independent of VOR/TACAN references, may file for random RNAV routes at and above FL 390 within the conterminous United States using the following procedures.

(a) File airport-to-airport flight plans prior to departure.

(b) File the appropriate RNAV capability certification suffix in the flight plan.

(c) Plan the random route portion of the flight to begin and end over published departure/arrival transition fixes or appropriate navigation aids for airports without published transition procedures. The use of preferred departure and arrival routes, such as SID and STAR where established, is recommended.

(d) Plan the route of flight so as to avoid prohibited and restricted airspace by 3 NM unless permission has been obtained to operate in that airspace and the appropriate ATC facility is advised.

(e) Define the route of flight after the departure fix, including each intermediate fix (turnpoint) and the arrival fix for the destination airport in terms of latitude/longitude coordinates plotted to the nearest minute. The arrival fix must be identified by both the latitude/longitude coordinates and a fix identifier.

EXAMPLE:

MIA[1] SRQ[2] 3407/10615[3] 3407/11546 TNP[4] LAX[5]

[1] Departure airport.

[2] Departure fix.

[3] Intermediate fix (turning point).

[4] Arrival fix.

[5] Destination airport.

(f) Record latitude/longitude coordinates by four figures describing latitude in degrees and minutes followed by a solidus and five figures describing longitude in degrees and minutes.

(g) File at FL 390 or above for the random RNAV portion of the flight.

(h) Fly all routes/route segments on Great Circle tracks.

(i) Make any inflight requests for random RNAV clearances or route amendments to an en route ATC facility.

e. Flight Plan Form—See Figure 5-7[1].

f. Explanation of IFR Flight Plan Items—

Block 1. Check the type flight plan. Check both the VFR and IFR blocks if composite VFR/IFR.

Block 2. Enter your complete aircraft identification including the prefix "N" if applicable.

U.S. DEPARTMENT OF TRANSPORTATION
FEDERAAL AVIATION ADMINISTRATION
FLIGHT PLAN

(FAA USE ONLY) ☐ PILOT BRIEFING ☐ VNR ☐ STOPOVER | TIME STARTED | SPECIALIST INITIALS

1. TYPE: VFR, IFR, DVFR | 2. AIRCRAFT IDENTIFICATION | 3. AIRCRAFT TYPE/SPECIAL EQUIPMENT | 4. TRUE AIRSPEED KTS | 5. DEPARTURE POINT | 6. DEPARTURE TIME: PROPOSED (Z) ACTUAL (Z) | 7. CRUISING ALTITUDE

8. ROUTE OF FLIGHT

9. DESTINATION (Name of airport and city) | 10. EST. TIME ENROUTE: HOURS MINUTES | 11. REMARKS

12. FUEL ON BOARD | 13. ALTERNATE AIRPORT(S) | 14. PILOT'S NAME, ADDRESS & TELEPHONE NUMBER & AIRCRAFT HOME BASE | 15. NUMBER ABOARD

17. DESTINATION CONTACT/TELEPHONE (OPTIONAL)

16. COLOR OF AIRCRAFT | CIVIL AIRCRAFT PILOTS. FAR 91 requires you file an IFR flight plan to operate under instrument flight rules in controlled airspace. Failure to file could result in a civil penalty not to exceed $1,000 for each violation (Section 901 of the Federal Aviation Act of 1958, as amended). Filing of a VFR flight plan is recommended as a good operating practice. See also Part 99 for requirements concerning DVFR flight plans

FAA Form 7233-1 (8-82) CLOSE VFR FLIGHT PLAN WITH ______________ FSS ON ARRIVAL

Figure 5-7[1]

Block 3. Enter the designator for the aircraft, followed by a slant (/), and the transponder or DME equipment code letter; e.g., C-182/U. Heavy aircraft, add prefix "H" to aircraft type; example: H/DC10/U. Consult an FSS briefer for any unknown elements.

Block 4. Enter your computed true airspeed (TAS).

5-7f Block 4 NOTE—If the average TAS changes plus or minus 5 percent or 10 knots, whichever is greater, advise ATC.

Block 5. Enter the departure airport identifier code (or the name if the identifier is unknown).

5-7 Block 5 NOTE—Use of identifier codes will expedite the processing of your flight plan.

Block 6. Enter the proposed departure time in Coordinated Universal Time (UTC) (Z). If airborne, specify the actual or proposed departure time as appropriate.

Block 7. Enter the requested en route altitude or flight level.

5-7 Block 7 NOTE—Enter only the initial requested altitude in this block. When more than one IFR altitude or flight level is desired along the route of flight, it is best to make a subsequent request direct to the controller.

Block 8. Define the route of flight by using NAVAID identifier codes (or names if the code is unknown), airways, jet routes, and waypoints (for RNAV).

5-7 Block 8 NOTE—Use NAVAID's or WAYPOINT's to define direct routes and radials/bearings to define other unpublished routes.

Block 9. Enter the destination airport identifier code (or name if the identifier is unknown).

Block 10. Enter your Estimated Time en Route based on latest forecast winds.

Block 11. Enter only those remarks pertinent to ATC or to the clarification of other flight plan information, such as the appropriate radiotelephony (call sign) associated with the designator filed in Block 2. Items of a personal nature are not accepted. Do not assume that remarks will be automatically transmitted to every controller. Specific ATC or en route requests should be made directly to the appropriate controller.

Block 12. Specify the fuel on board, computed from the departure point.

Block 13. Specify an alternate airport if desired or required, but do not include routing to the alternate airport.

Block 14. Enter the complete name, address, and telephone number of pilot-in-command, or in the case of a formation flight, the formation commander. Enter sufficient information to identify home base, airport, or operator.

5-7 Block 14 NOTE—This information would be essential in the event of search and rescue operation.

Block 15. Enter the total number of persons on board including crew.

Block 16. Enter the predominant colors.

5-7 Block 16 NOTE—Close IFR flight plans with tower, approach control, or ARTCC, or if unable, with FSS. When landing at an airport with a functioning control tower, IFR flight plans are automatically canceled.

g. The information transmitted to the ARTCC for IFR flight plans will consist of only flight plan blocks 2, 3, 4, 5, 6, 7, 8, 9, 10, and 11.

h. A description of the International Flight Plan Form is contained in the International Flight Information Manual (IFIM).

5-8. IFR OPERATIONS TO HIGH ALTITUDE DESTINATIONS

Pilots planning IFR flights to airports located in mountainous terrain are cautioned to consider the necessity for an alternate airport even when the forecast weather conditions would technically relieve them from the requirement to file one (Reference: FAR Part 91.167 and paragraph 4-18(b)). The FAA has identified three possible situations where the failure to plan for an alternate airport when flying IFR to such a destination airport could result in a critical situation if the weather is less than forecast and sufficient fuel is not available to proceed to a suitable airport.

a. An IFR flight to an airport where the MDA's or landing visibility minimums for *ALL INSTRUMENT APPROACHES* are higher than the forecast weather minimums specified in FAR Part 91.167(b). For example, there are 11 high altitude airports in the United States with approved instrument approach procedures where all of the Minimum Descent Altitudes (MDA's) are greater than 2,000 feet and/or the landing visibility minimums are greater than 3 miles (Bishop, California; South Lake Tahoe, California; Ukiah, California; Aspen-Pitkin Co./Sardy Field, Colorado; Butte, Montana; Helena, Montana; Missoula, Montana; Chadron, Nebraska; Ely, Nevada; Klamath Falls, Oregon; and Omak, Washington). In the case of these 11 airports, it is possible for a pilot to elect, on the basis of forecasts, not to carry sufficient fuel to get to an alternate when the ceiling and/or visibility is actually lower than that necessary to complete the approach.

b. A small number of other airports in mountainous terrain have MDA's which are slightly (100 to 300 feet) below 2,000 feet AGL. In situations where there is an option as to whether to plan for an alternate, pilots should bear in mind that just a slight worsening of the weather conditions from those forecast could place the airport below the published IFR landing minimums.

c. An IFR flight to an airport which requires special equipment; i.e., DME, glide slope, etc., in order to make the available approaches to the lowest minimums. Pilots should be aware that all other minimums on the approach charts may require weather conditions better than those specified in FAR Part 91.167(b). An inflight equipment malfunction could result in the inability to comply with the published approach procedures or, again, in the position of having the airport below the published IFR landing minimums for all remaining instrument approach alternatives.

5-9. FLIGHTS OUTSIDE THE UNITED STATES AND U.S. TERRITORIES

a. When conducting flights, particularly extended flights, outside the U.S. and its territories, full account should be taken of the amount and quality of air navigation services available in the airspace to be traversed. Every effort should be made to secure information on the location and range of navigational aids, availability of communications and meteorological services, the provision of air traffic services, including alerting service, and the existence of search and rescue services.

b. Pilots should remember that there is a need to continuously guard the VHF emergency frequency 121.5 MHz when on long over-water flights, except when communications on other VHF channels, equipment limitations, or cockpit duties prevent simultaneous guarding of two channels. Guarding of 121.5 MHz is particularly critical when operating in proximity to Flight Information Region (FIR) boundaries, for example, operations on Route R220 between Anchorage and Tokyo, since it serves to facilitate communications with regard to aircraft which may experience in-flight emergencies, communications, or navigational difficulties. (Reference ICAO Annex 10, Vol II Paras 5.2.2.1.1.1 and 5.2.2.1.1.2.)

c. The filing of a flight plan, always good practice, takes on added significance for extended flights outside U.S. airspace and is, in fact, usually required by the laws of the countries being visited or overflown. It is also particularly important in the case of such flights that pilots leave a complete itinerary and schedule of the flight with someone directly concerned, keep that person advised of the flight's progress and inform him that if serious doubt arises as to the safety of the flight he should first contact the appropriate FSS. Round Robin Flight Plans to Mexico are not accepted.

d. All pilots should review the foreign airspace and entry restrictions published in the IFIM during the flight planning process. Foreign airspace penetration without official authorization can involve both danger to the aircraft and the imposition of severe penalties and inconvenience to both passengers and crew. A flight plan on file with ATC authorities does not necessarily constitute the prior permission required by certain other authorities. The possibility of fatal consequences cannot be ignored in some areas of the world.

e. Current NOTAMs for foreign locations must also be

reviewed. The publication International Notices to Airmen, published biweekly, contains considerable information pertinent to foreign flight. Current foreign NOTAMs are also available from the U.S. International NOTAM Office in Washington, D.C., through any local FSS.

f. When customs notification is required, it is the responsibility of the pilot to arrange for customs notification in a timely manner. The following guidelines are applicable:

1. When customs notification is required on flights to Canada and Mexico and a predeparture flight plan cannot be filed or an advise customs message (ADCUS) cannot be included in a predeparture flight plan, call the nearest en route domestic or International FSS as soon as radio communication can be established and file a VFR or DVFR flight plan, as required, and include as the last item the advise customs information. The station with which such a flight plan is filed will forward it to the appropriate FSS who will notify the customs office responsible for the destination airport.

2. If the pilot fails to include ADCUS in the radioed flight plan, it will be assumed that other arrangements have been made and FAA will not advise customs.

3. The FAA assumes no responsibility for any delays in advising customs if the flight plan is given too late for delivery to customs before arrival of the aircraft. It is still the pilot's responsibility to give timely notice even though a flight plan is given to FAA.

5-10. CHANGE IN FLIGHT PLAN

In addition to altitude or flight level, destination and/or route changes, increasing or decreasing the speed of an aircraft constitutes a change in a flight plan. Therefore, at any time the average true airspeed at cruising altitude between reporting points varies or is expected to vary from that given in the flight plan by *plus or minus 5 percent, or 10 knots, whichever is greater*, ATC should be advised.

5-11. CHANGE IN PROPOSED DEPARTURE TIME

a. To prevent computer saturation in the en route environment, parameters have been established to delete proposed departure flight plans which have not been activated. Most centers have this parameter set so as to delete these flight plans a minimum of 1 hour after the proposed departure time. To ensure that a flight plan remains active, pilots whose actual departure time will be delayed 1 hour or more beyond their filed departure time, are requested to notify ATC of their departure time.

b. Due to traffic saturation, control personnel frequently will be unable to accept these revisions via radio. It is recommended that you forward these revisions to the nearest FSS.

5-12. CLOSING VFR/DVFR FLIGHT PLANS

A pilot is responsible for ensuring that his VFR or DVFR flight plan is canceled (FAR Part 91.153 and FAR Part 91.169). You should close your flight plan with the nearest FSS, or if one is not available, you may request any ATC facility to relay your cancellation to the FSS. *Control towers do not automatically close VFR or DVFR flight plans* since they do not know if a particular VFR aircraft is on a flight plan. If you fail to report or cancel your flight plan within ½ hour after your ETA, search and rescue procedures are started.

5-13. CANCELING IFR FLIGHT PLAN

a. FAR Part 91.153 and FAR Part 91.169 includes the statement "When a flight plan has been filed, the pilot-in-command, upon canceling or completing the flight under the flight plan, shall notify the nearest FSS or ATC facility."

b. An IFR flight plan may be canceled at any time the flight is operating in VFR conditions outside Class A airspace by the pilot stating "CANCEL MY IFR FLIGHT PLAN" to the controller or air/ground station with which he is communicating. Immediately after canceling an IFR flight plan, a pilot should take necessary action to change to the appropriate air/ground frequency, VFR radar beacon code and VFR altitude or flight level.

c. ATC separation and information services will be discontinued, including radar services (where applicable). Consequently, if the canceling flight desires VFR radar advisory service, the pilot must specifically request it.

5-13c NOTE—Pilots must be aware that other procedures may be applicable to a flight that cancels an IFR flight plan within an area where a special program, such as a designated TRSA, Class C airspace, or Class B airspace, has been established.

d. If a DVFR flight plan requirement exists, the pilot is responsible for filing this flight plan to replace the canceled IFR flight plan. If a subsequent IFR operation becomes necessary, a new IFR flight plan must be filed and an ATC clearance obtained before operating in IFR conditions.

e. If operating on an IFR flight plan to an airport with a functioning control tower, the flight plan is automatically closed upon landing.

f. If operating on an IFR flight plan to an airport where there is no functioning control tower, the pilot must initiate cancellation of the IFR flight plan. This can be done after landing if there is a functioning FSS or other means of direct communications with ATC. In the event there is no FSS and air/ground communications with ATC is not possible below a certain altitude, the pilot should, weather conditions permitting, cancel his IFR flight plan while still airborne and able to communicate with ATC by radio. This will not only save the time and expense of canceling the flight plan by telephone but will quickly release the airspace for use by other aircraft.

5-14 thru 5-19. RESERVED

Section 2. DEPARTURE PROCEDURES

5-20. PRE-TAXI CLEARANCE PROCEDURES

a. Certain airports have established Pre-taxi Clearance programs whereby pilots of departing IFR aircraft may elect to receive their IFR clearances before they start taxiing for takeoff. The following provisions are included in such procedures:

1. Pilot participation is not mandatory.

2. Participating pilots call clearance delivery or ground control not more than 10 minutes before proposed taxi time.

3. IFR clearance (or delay information, if clearance cannot be obtained) is issued at the time of this initial call-up.

4. When the IFR clearance is received on clearance delivery frequency, pilots call ground control when ready to taxi.

5. Normally, pilots need not inform ground control that they have received IFR clearance on clearance delivery frequency. Certain locations may, however, require that the pilot inform ground control of a portion of his routing or that he has received his IFR clearance.

6. If a pilot cannot establish contact on clearance delivery frequency or has not received his IFR clearance before he is ready to taxi, he contacts ground control and informs the controller accordingly.

b. Locations where these procedures are in effect are indicated in the Airport/Facility Directory.

5-21. TAXI CLEARANCE

Pilots on IFR flight plans should communicate with the control tower on the appropriate ground control or clearance delivery frequency, prior to starting engines, to receive engine start time, taxi and/or clearance information.

5-22. ABBREVIATED IFR DEPARTURE CLEARANCE (CLEARED...AS FILED) PROCEDURES

a. ATC facilities will issue an abbreviated IFR departure clearance based on the ROUTE of flight filed in the IFR flight plan, provided the filed route can be approved with little or no revision. These abbreviated clearance procedures are based on the following conditions:

1. The aircraft is on the ground or it has departed VFR and the pilot is requesting IFR clearance while airborne.

2. That a pilot will not accept an abbreviated clearance if the route or destination of a flight plan filed with ATC has been changed by him or the company or the operations officer before departure.

3. That it is the responsibility of the company or operations office to inform the pilot when they make a change to the filed flight plan.

4. That it is the responsibility of the pilot to inform ATC in his initial call-up (for clearance) when the filed flight plan has been either:

(a) amended, or

(b) canceled and replaced with a new filed flight plan.

5-22a3b NOTE—The facility issuing a clearance may not have received the revised route or the revised flight plan by the time a pilot requests clearance.

b. The controller will issue a detailed clearance when he knows that the original filed flight plan has been changed or when the pilot requests a full route clearance.

c. The clearance as issued will include the destination airport filed in the flight plan.

d. ATC procedures now require the controller to state the SID name, the current number and the SID Transition name after the phrase "Cleared to (destination) airport" and prior to the phrase, "then as filed," for ALL departure clearances when the SID or SID Transition is to be flown. The procedures apply whether or not the SID is filed in the flight plan.

e. STARs, when filed in a flight plan, are considered a part of the filed route of flight and will not normally be stated in an initial departure clearance. If the ARTCC's jurisdictional airspace includes both the departure airport and the fix where a STAR or STAR Transition begins, the STAR name, the current number and the STAR Transition name MAY be stated in the initial clearance.

f. "Cleared to (destination) airport as filed" does NOT include the en route altitude filed in a flight plan. An en route altitude will be stated in the clearance or the pilot will be advised to expect an assigned or filed altitude within a given time frame or at a certain point after departure. This may be done verbally in the departure instructions or stated in the SID.

g. In both radar and nonradar environments, the controller will state "Cleared to (destination) airport as filed" or:

1. If a SID or SID Transition is to be flown, specify the SID name, the current SID number, the SID Transition name, the assigned altitude/Flight Level, and any additional instructions (departure control frequency, beacon code assignment, etc.) necessary to clear a departing aircraft via the SID or SID Transition and the route filed.

EXAMPLE:

NATIONAL SEVEN TWENTY CLEARED TO MIAMI AIRPORT INTERCONTINENTAL ONE DEPARTURE, LAKE CHARLES TRANSITION THEN AS FILED, MAINTAIN FLIGHT LEVEL TWO SEVEN ZERO.

2. When there is no SID or when the pilot cannot accept a SID, the controller will specify the assigned altitude or Flight Level, and any additional instructions necessary to clear a departing aircraft via an appropriate departure routing and the route filed.

5-22g2 NOTE—A detailed departure route description or a radar vector may be used to achieve the desired departure routing.

3. If it is necessary to make a minor revision to the filed route, the controller will specify the assigned SID or SID Transition (or departure routing), the revision to the filed route, the assigned altitude or flight level and any additional instructions necessary to clear a departing aircraft.

EXAMPLE:
JET STAR ONE FOUR TWO FOUR CLEARED TO ATLANTA AIRPORT, SOUTH BOSTON TWO DEPARTURE THEN AS FILED EXCEPT CHANGE ROUTE TO READ SOUTH BOSTON VICTOR 20 GREENSBORO, MAINTAIN ONE SEVEN THOUSAND.

4. Additionally, in a nonradar environment, the controller will specify one or more fixes, as necessary, to identify the initial route of flight.

EXAMPLE:
CESSNA THREE ONE SIX ZERO FOXTROT CLEARED TO CHARLOTTE AIRPORT AS FILED VIA BROOKE, MAINTAIN SEVEN THOUSAND.

h. To ensure success of the program, pilots should:

1. Avoid making changes to a filed flight plan just prior to departure.

2. State the following information in the initial call-up to the facility when no change has been made to the filed flight plan: Aircraft call sign, location, type operation (IFR) and the name of the airport (or fix) to which you expect clearance.

EXAMPLE:
"WASHINGTON CLEARANCE DELIVERY (or ground control if appropriate) AMERICAN SEVENTY SIX AT GATE ONE, IFR LOS ANGELES."

3. If the flight plan has been changed, state the change and request a full route clearance.

EXAMPLE:
"WASHINGTON CLEARANCE DELIVERY, AMERICAN SEVENTY SIX AT GATE ONE. IFR SAN FRANCISCO. MY FLIGHT PLAN ROUTE HAS BEEN AMENDED (or destination changed). REQUEST FULL ROUTE CLEARANCE."

4. Request verification or clarification from ATC if ANY portion of the clearance is not clearly understood.

5. When requesting clearance for the IFR portion of a VFR/IFR flight, request such clearance prior to the fix where IFR operation is proposed to commence in sufficient time to avoid delay. Use the following phraseology:

EXAMPLE:
"LOS ANGELES CENTER, APACHE SIX ONE PAPA, VFR ESTIMATING PASO ROBLES VOR AT THREE TWO, ONE THOUSAND FIVE HUNDRED, REQUEST IFR TO BAKERSFIELD."

5-23. DEPARTURE RESTRICTIONS, CLEARANCE VOID TIMES, HOLD FOR RELEASE, AND RELEASE TIMES

a. ATC may assign departure restrictions, clearance void times, hold for release, and release times, when necessary, to separate departures from other traffic or to restrict or regulate the departure flow.

1. CLEARANCE VOID TIMES—A pilot may receive a clearance, when operating from an airport without a control tower, which contains a provision for the clearance to be void if not airborne by a specific time. A pilot who does not depart prior to the clearance void time must advise ATC as soon as possible of his or her intentions. ATC will normally advise the pilot of the time allotted to notify ATC that the aircraft did not depart prior to the clearance void time. This time cannot exceed 30 minutes. Failure of an aircraft to contact ATC within 30 minutes after the clearance void time will result in the aircraft being considered overdue and search and rescue procedures initiated.

5-23a1 NOTE 1—Other IFR traffic for the airport where the clearance is issued is suspended until the aircraft has contacted ATC or until 30 minutes after the clearance void time or 30 minutes after the clearance release time if no clearance void time is issued.

5-23a1 NOTE 2—Pilots who depart at or after their clearance void time are not afforded IFR separation and may be in violation of FAR Part 91.173 which requires that pilots receive an appropriate ATC clearance before operating IFR in controlled airspace.

EXAMPLE:
CLEARANCE VOID IF NOT OFF BY (clearance void time) and, if required, IF NOT OFF BY (clearance void time) ADVISE (facility) NOT LATER THAN (time) OF INTENTIONS.

2. HOLD FOR RELEASE—ATC may issue "hold for release" instructions in a clearance to delay an aircraft's departure for traffic management reasons (i.e., weather, traffic volume, etc.). When ATC states in the clearance, "hold for release," the pilot may not depart utilizing that instrument flight rules (IFR) clearance until a release time or additional instructions are issued by ATC. This does not preclude the pilot from cancelling the IFR clearance with ATC and departing under visual flight rules (VFR); but an IFR clearance may not be available after departure. In addition, ATC will include departure delay information in conjunction with "hold for release" instructions.

EXAMPLE:
(aircraft identification) CLEARED TO (destination) AIRPORT AS FILED, MAINTAIN (altitude), and, if required (additional instructions or information), HOLD FOR RELEASE, EXPECT (time in hours and/or minutes) DEPARTURE DELAY.

3. RELEASE TIMES—A "release time" is a departure restriction issued to a pilot by ATC, specifying the earliest time an aircraft may depart. ATC will use "release times" in conjunction with traffic management procedures and/or to separate a departing aircraft from other traffic.

EXAMPLE:
(aircraft identification) RELEASED FOR DEPARTURE AT (time in hours and/or minutes).

b. If practical, pilots departing uncontrolled airports should obtain IFR clearances prior to becoming airborne when two-way communications with the controlling ATC facility is available.

5-24. DEPARTURE CONTROL

a. Departure Control is an approach control function responsible for ensuring separation between departures. So as to expedite the handling of departures, Departure Control may suggest a takeoff direction other than that which may normally have been used under VFR handling. Many times it is preferred to offer the pilot a runway that will require the fewest turns after takeoff to place the pilot on his filed course or selected departure route as quickly as possible. At many locations particular attention is paid to the use of preferential runways for local noise abatement programs, and route departures away from congested areas.

b. Departure Control utilizing radar will normally clear aircraft out of the terminal area using SIDs via radio navigation aids. When a departure is to be vectored immediately following takeoff, the pilot will be advised prior to takeoff of the initial heading to be flown but may not be advised of the purpose of the heading. Pilots operating in a radar environment are expected to associate departure headings with vectors to their planned route or flight. When given a vector taking his aircraft off a previously assigned nonradar route, the pilot will be advised briefly what the vector is to achieve. Thereafter, radar service will be provided until the aircraft has been reestablished "on-course" using an appropriate navigation aid and the pilot has been advised of his position or a handoff is made to another radar controller with further surveillance capabilities.

c. Controllers will inform pilots of the departure control frequencies and, if appropriate, the transponder code before takeoff. Pilots should not operate their transponder until ready to start the takeoff roll or change to the departure control frequency until requested. Controllers may omit the departure control frequency if a SID has or will be assigned and the departure control frequency is published on the SID.

5-25. INSTRUMENT DEPARTURES

a. STANDARD INSTRUMENT DEPARTURES (SID)

1. A SID is an ATC coded departure procedure which has been established at certain airports to simplify clearance delivery procedures.

2. Pilots of civil aircraft operating from locations where SID procedures are effective may expect ATC clearance containing a SID. Use of a SID requires pilot possession of at least the textual description of the SID procedures. Controllers may omit the departure control frequency if a SID clearance is issued and the departure control frequency is published on the SID. If the pilot does not possess a charted SID or a preprinted SID description or, for any other reason, does not wish to use a SID, he is expected to advise ATC. Notification may be accomplished by filing "NO SID" in the remarks section of the filed flight plan or by the less desirable method of verbally advising ATC.

3. All effective SIDs are published in textual and graphic form by the National Ocean Service in Terminal Procedures Publication (TPP).

4. SID procedures will be depicted in one of two basic forms.

(a) Pilot Navigation (Pilot NAV) SIDs: are established where the pilot is primarily responsible for navigation on the SID route. They are established for airports when terrain and safety related factors indicate the necessity for a pilot NAV SID. Some pilot NAV SIDs may contain vector instructions which pilots are expected to comply with until instructions are received to resume normal navigation on the filed/assigned route or SID procedure.

(b) Vector SIDs: are established where ATC will provide radar navigational guidance to a filed/assigned route or to a fix depicted on the SID.

b. OBSTRUCTION CLEARANCE DURING DEPARTURE

1. Published instrument departure procedures and SIDs assist pilots conducting IFR flight in avoiding obstacles during climbout to Minimum en Route Altitude (MEA). These procedures are established only at locations where instrument approach procedures are published. Standard instrument takeoff minimums and departure procedures are prescribed in FAR Part 91.175. Airports with takeoff minimums other than standard (one statute mile for aircraft having two engines or less and one-half statute mile for aircraft having more than two engines) are described in airport listings on separate pages titled IFR TAKE-OFF MINIMUMS AND DEPARTURE PROCEDURES, at the front of each U.S. Government Terminal Procedures Publication (TPP). The approach chart and SID chart for each airport where takeoff minimums are not standard and/or departure procedures are published is annotated with a special symbol *. The use of this symbol indicates that the separate listing should be consulted. These minimums also apply to SIDs unless the SIDs specify different minimums.

2. Obstacle clearance is based on the aircraft climbing at least 200 feet per nautical mile, crossing the end of the runway at least 35 feet AGL, and climbing to 400 feet above airport elevation before turning, unless otherwise specified in the procedure. A slope of 152 feet per nautical mile, starting no higher than 35 feet above the departure end of the runway, is assessed for obstacles. A minimum obstacle clearance of 48 feet per nautical mile is provided in the assumed climb gradient.

(a) If no obstacles penetrate the 152 feet per nautical mile slope, IFR departure procedures are not published.

(b) If obstacles do penetrate the slope, avoidance procedures are specified. These procedures may be: a ceiling and visibility to allow the obstacles to be seen and avoided; a climb gradient greater than 200 feet per nautical mile; detailed flight maneuvers; or a combination of the above. In extreme cases, IFR takeoff may not be authorized for some runways.

EXAMPLE:

Rwy 17, 300-1 or standard with minimum climb of 220 feet per NM to 1100.

3. Climb gradients are specified when required for obstacle clearance. Crossing restrictions in the SIDs may be established for traffic separation or obstacle clearance. When no gradient is specified, the pilot is expected to climb at least 200 feet per nautical mile to MEA unless required to level off by a crossing restriction.

EXAMPLE:

"CROSS ALPHA INTERSECTION AT OR BELOW 4000; MAINTAIN 6000." The pilot climbs at least 200 feet per nautical mile to 6000. If 4000 is reached before ALPHA, the pilot levels off at 4000 until passing ALPHA; then immediately resumes at least 200 feet per nautical mile climb.

4. Climb gradients may be specified to an altitude/fix, above which the normal gradient applies.

EXAMPLE:

"MINIMUM CLIMB 340 FEET PER NM TO 2700." The pilot climbs at least 340 feet per nautical mile to 2700, then at least 200 feet per NM to MEA.

5. Some IFR departure procedures require a climb in visual conditions to cross the airport (or an on-airport NAVAID) in a specified direction, at or above a specified altitude.

EXAMPLE:

"CLIMB IN VISUAL CONDITIONS SO AS TO CROSS THE McELORY AIRPORT SOUTHBOUND, AT OR ABOVE 6000, THEN CLIMB VIA KEEMMLING R-033 TO KEEMMLING VORTAC."

(a) When climbing in visual conditions it is the pilot's responsibility to see and avoid obstacles. Specified ceiling and visibility minimums will allow visual avoidance of obstacles until the pilot enters the standard obstacle protection area. Obstacle avoidance is not guaranteed if the pilot maneuvers farther from the airport than the visibility minimum.

(b) That segment of the procedure which requires the pilot to see and avoid obstacles ends when the aircraft crosses the specified point at the required altitude. Thereafter, standard obstacle protection is provided.

6. Each pilot, prior to departing an airport on an IFR flight should consider the type of terrain and other obstacles on or in the vicinity of the departure airport and:

(a) Determine whether a departure procedure and/or SID is available for obstacle avoidance.

(b) Determine if obstacle avoidance can be maintained visually or that the departure procedure or SID should be followed.

(c) Determine what action will be necessary and take such action that will assure a safe departure.

5-25b6c NOTE—The term **Radar Contact**, when used by the controller during departure, should not be interpreted as relieving pilots of their responsibility to maintain appropriate terrain and obstruction clearance. (Reference—Pilot/Controller Glossary, Radar Contact.)

Terrain/obstruction clearance is not provided by ATC until the controller begins to provide navigational guidance, i.e., Radar Vectors.

5-26 thru 5-29. RESERVED

Section 3. EN ROUTE PROCEDURES

5-30. ARTCC COMMUNICATIONS

a. Direct Communications, Controllers and Pilots—

1. ARTCC's are capable of direct communications with IFR air traffic on certain frequencies. Maximum communications coverage is possible through the use of Remote Center Air/Ground (RCAG) sites comprised of both VHF and UHF transmitters and receivers. These sites are located throughout the U.S. Although they may be several hundred miles away from the ARTCC, they are remoted to the various ARTCC's by land lines or microwave links. Since IFR operations are expedited through the use of direct communications, pilots are requested to use these frequencies strictly for communications pertinent to the control of IFR aircraft. Flight plan filing, en route weather, weather forecasts, and similar data should be requested through FSS's, company radio, or appropriate military facilities capable of performing these services.

2. An ARTCC is divided into sectors. Each sector is handled by one or a team of controllers and has its own sector discrete frequency. As a flight progresses from one sector to another, the pilot is requested to change to the appropriate sector discrete frequency.

b. ATC Frequency Change Procedures—

1. The following phraseology will be used by controllers to effect a frequency change:

EXAMPLE:

(Aircraft identification) CONTACT (facility name or location name and terminal function) (frequency) AT (time, fix, or altitude).

5-30b1 NOTE—Pilots are expected to maintain a listening watch on the transferring controller's frequency until the time, fix, or altitude specified. ATC will omit frequency change restrictions whenever pilot compliance is expected upon receipt.

2. The following phraseology should be utilized by pilots for establishing contact with the designated facility:

(a) When a position report will be made:

EXAMPLE:

(Name) CENTER, (aircraft identification), (position).

(b) When no position report will be made:

EXAMPLE:

(Name) CENTER, (aircraft identification), (position) ESTIMATING (reporting point and time) AT (altitude or Flight Level) (CLIMBING or DESCENDING) TO MAINTAIN (altitude or flight level).

(c) When operating in a radar environment and no position report is required: On initial contact, the pilot should inform the controller of the pilot's assigned altitude preceded by the words "level," or " climbing to," or "descending to," as appropriate; and the pilot's present vacating altitude, if applicable. Also, when on other than published routes, the pilot should include the presently assigned routing on initial contact with each air traffic controller.

EXAMPLES:

1. (Name) CENTER, (aircraft identification), LEVEL (altitude or flight level), HEADING (exact heading).

2. (Name) CENTER, (aircraft identification), LEAVING (exact altitude or flight level), CLIMBING TO (altitude or flight level), PROCEEDING TO (name) V-O-R VIA THE (VOR name) (number) RADIAL.

3. (Name) CENTER, (aircraft identification), LEAVING (exact altitude or flight level), DESCENDING TO (altitude or flight level), DIRECT (name) V-O-R.

5-30b2c NOTE—Exact altitude or flight level means to the nearest 100 foot increment. Exact altitude or flight level reports on initial contact provide ATC with information required prior to using MODE C altitude information for separation purposes.

3. At times controllers will ask pilots to verify that they are at a particular altitude. The phraseology used will be: "VERIFY AT (altitude)." In climbing or descending situations, controllers may ask pilots to "VERIFY ASSIGNED ALTITUDE AS (altitude)." Pilots should confirm that they are at the altitude stated by the controller or that the assigned altitude is correct as stated. If this is not the case, they should inform the controller of the actual altitude being maintained or the different assigned altitude.

CAUTION: Pilots should not take action to change their actual altitude or different assigned altitude to the altitude stated in the controller's verification request unless the controller specifically authorizes a change.

c. ARTCC Radio Frequency Outage. ARTCC's normally have at least one back-up radio receiver and transmitter system for each frequency, which can usually be placed into service quickly with little or no disruption of ATC service. Occasionally, technical problems may cause a delay but switchover seldom takes more than 60 seconds. When it appears that the outage will not be quickly remedied, the ARTCC will usually request a nearby aircraft, if there is one, to switch to the affected frequency to broadcast communications instructions. It is important, therefore, that the pilot wait at least 1 minute before deciding that the ARTCC has actually experienced a radio frequency failure. When such an outage does occur, the pilot should, if workload and equipment capability permit, maintain a listening watch on the affected frequency while attempting to comply with the following recommended communications procedures:

1. If two-way communications cannot be established with the ARTCC after changing frequencies, a pilot should attempt to recontact the transferring controller for the assignment of an alternative frequency or other instructions.

2. When an ARTCC radio frequency failure occurs after two-way communications have been established, the pilot should attempt to reestablish contact with the center on any other known ARTCC frequency, preferably that of the next responsible sector when practicable, and ask for instructions. However, when the next normal frequency change along the route is known to involve another ATC facility, the pilot should contact that facility, if feasible, for instructions. If communications cannot be reestablished by either method, the pilot is expected to request communications instructions from the FSS appropriate to the route of flight.

5-30c2 NOTE—The exchange of information between an aircraft and an ARTCC through an FSS is quicker than relay via company radio because the FSS has direct interphone lines to the responsible ARTCC sector. Accordingly, when circumstances dictate a choice between the two, during an ARTCC frequency outage, relay via FSS radio is recommended.

5-31. POSITION REPORTING

The safety and effectiveness of traffic control depends to a large extent on accurate position reporting. In order to provide the proper separation and expedite aircraft movements, ATC must be able to make accurate estimates of the progress of every aircraft operating on an IFR flight plan.

a. Position Identification—

1. When a position report is to be made passing a VOR radio facility, the time reported should be the time at which the first complete reversal of the "to/from" indicator is accomplished.

2. When a position report is made passing a facility by means of an airborne ADF, the time reported should be the time at which the indicator makes a complete reversal.

3. When an aural or a light panel indication is used to determine the time passing a reporting point, such as a fan marker, Z marker, cone of silence or intersection of range courses, the time should be noted when the signal is first received and again when it ceases. The mean of these two times should then be taken as the actual time over the fix.

4. If a position is given with respect to distance and direction from a reporting point, the distance and direction should be computed as accurately as possible.

5. Except for terminal area transition purposes, position reports or navigation with reference to aids not established for use in the structure in which flight is being conducted will not normally be required by ATC.

b. Position Reporting Points—FAR's require pilots to maintain a listening watch on the appropriate frequency and, unless operating under the provisions of (c), to furnish position reports passing certain reporting points. Reporting points are indicated by symbols on en route charts. The designated compulsory reporting point symbol is a solid triangle and the "on request" reporting point symbol is the open triangle. Reports passing an "on request" reporting point are only necessary when requested by ATC.

c. Position Reporting Requirements—

1. Flights along airways or routes—A position report is required by all flights regardless of altitude, including those operating in accordance with an ATC clearance specifying "VFR ON TOP," over each designated compulsory reporting point along the route being flown.

2. Flight along a Direct Route—Regardless of the altitude or flight level being flown, including flights operating in accordance with an ATC clearance specifying "VFR ON TOP," pilots shall report over each reporting point used in the flight plan to define the route of flight.

3. Flights in a Radar Environment—When informed by ATC that their aircraft are in "Radar Contact," pilots should discontinue position reports over designated reporting points. They should resume normal position reporting when ATC advises "RADAR CONTACT LOST" or "RADAR SERVICE TERMINATED."

5-31c3 NOTE—ATC will inform a pilot that he is in "RADAR CONTACT" (a) when his aircraft is initially identified in the ATC system; and (b) when radar identification is reestablished after radar service has been terminated or radar contact lost. Subsequent to being advised that the controller has established radar contact, this fact will not be repeated to the pilot when handed off to another controller. At times, the aircraft identity will be *confirmed* by the receiving controller; however, this should not be construed to mean that radar contact has been lost. The identity of transponder equipped aircraft will be confirmed by asking the pilot to "IDENT," "SQUAWK STANDBY," or to change codes. Aircraft without transponders will be advised of their position to confirm identity. In this case, *the pilot is expected to advise the controller if he disagrees with the position given.* If the pilot cannot confirm the accuracy of the position given because he is not tuned to the NAVAID referenced by the controller, the pilot should ask for another radar position relative to the NAVAID to which he is tuned.

d. Position Report Items—

1. Position reports should include the following items:

(a) Identification.

(b) Position.

(c) Time.

(d) Altitude or flight level (include actual altitude or flight level when operating on a clearance specifying VFR-ON-TOP.)

(e) Type of flight plan (not required in IFR position reports made directly to ARTCC's or approach control),

(f) ETA and name of next reporting point.

(g) The name only of the next succeeding reporting point along the route of flight, and

(h) Pertinent remarks.

5-32. ADDITIONAL REPORTS

a. The following reports should be made to ATC or FSS facilities without a specific ATC request:

1. At all times:

(a) When vacating any previously assigned altitude or flight level for a newly assigned altitude or flight level.

(b) When an altitude change win be made if operating on a clearance specifying VFR ON TOP.

(c) When *unable* to climb/descend at a rate of a least 500 feet per minute.

(d) When approach has been missed. (Request clearance for specific action; i.e., to alternative airport, another approach, etc.)

(e) Change in the average true airspeed (at cruising altitude) when it varies by 5 percent or 10 knots (whichever is greater) from that filed in the flight plan.

(f) The time and altitude or flight level upon reaching a holding fix or point to which cleared.

(g) When leaving any assigned holding fix or point.

5-32a1g NOTE—The reports in subparagraphs (f) and (g) may be omitted by pilots of aircraft involved in instrument training at military terminal area facilities when radar service is being provided.

(h) Any loss, in controlled airspace, of VOR, TACAN, ADF, low frequency navigation receiver capability, complete or partial loss of ILS receiver capability or impairment of air/ground communications capability. Reports should include aircraft identification, equipment affected, degree to which the capability to operate under IFR in the ATC system is impaired, and the nature and extent of assistance desired from ATC.

5-32a1h NOTE—Other equipment installed in an aircraft may effectively impair safety and/or the ability to operate under IFR. If such equipment (e.g., airborne weather radar) malfunctions and in the pilot's judgment either safety or IFR capabilities are affected, reports should be made as above.

(i) Any information relating to the safety of flight.

2. When not in radar contact:

(a) When leaving final approach fix inbound on final approach (nonprecision approach) or when leaving the outer marker or fix used in lieu of the outer marker inbound on final approach (precision approach).

(b) A corrected estimate at anytime it becomes apparent that an estimate as previously submitted is in error in excess of 3 minutes.

b. Pilots encountering weather conditions which have not been forecast, or hazardous conditions which have been forecast, are expected to forward a report of such weather to ATC. (Reference—Pilot Weather Reports (PIREPs), paragraph 7-19 and FAR Part 91.183(b) and (c).)

5-33. AIRWAYS AND ROUTE SYSTEMS

a. Two fixed route systems are established for air navigation purposes. They are the VOR and L/MF system, and the jet route system. To the extent possible, these route systems are aligned in an overlying manner to facilitate transition between each.

1. The VOR and L/MF Airway System consists of airways designated from 1,200 feet above the surface (or in some instances higher) up to but not including 18,000 feet MSL. These airways are depicted on En Route Low Altitude Charts.

5-33a1 NOTE—The altitude limits of a Victor airway should not be exceeded except to effect transition within or between route structures.

(a) Except in Alaska and coastal North Carolina, the VOR airways are predicated solely on VOR or VORTAC navigation aids; are depicted in blue on aeronautical charts; and are identified by a "V" (Victor) followed by the airway number (e.g., V12).

5-33a1a NOTE—Segments of VOR airways in Alaska and North Carolina (V56, V290) are based on L/MF navigation aids and charted in brown instead of blue on en route charts.

(1) A segment of an airway which is common to two or more routes carries the numbers of all the airways which coincide for that segment. When such is the case, a pilot filing a flight plan needs to indicate only that airway number of the route which he is using.

5-33a1a1 NOTE—A pilot who intends to make an airway flight, using VOR facilities, will simply specify the appropriate "Victor" airway(s) in his flight plan. For example, if a flight is to be made from Chicago to New Orleans at 8,000 feet, using omniranges only, the route may be indicated as "Departing from Chicago-Midway, cruising 8,000 feet via Victor 9 to Moisant International." If flight is to be conducted in part by means of L/MF navigation aids and in part on omniranges, specifications of the appropriate airways in the flight plan will indicate which types of facilities will be used along the described routes, and, for IFR flight, permit ATC to issue a traffic clearance accordingly. A route may also be described by specifying the station over which the flight will pass, but in this case since many VOR's and L/MF aids have the same name, the pilot must be careful to indicate which aid will be used at a particular location. This will be indicated in the route of flight portion of the flight plan by specifying the type of facility to be used after the location name in the following manner: Newark L/MF, Allentown VOR.

(2) With respect to position reporting, reporting points are designated for VOR Airway Systems. Flights using Victor Airways will report over these points unless advised otherwise by ATC.

(b) The L/MF airways (colored airways) are predicated solely on L/MF navigation aids and are depicted in brown on aeronautical charts and are identified by color name and number (e.g., Amber One). Green and Red airways are plotted east and west. Amber and Blue airways are plotted north and south.

5-33a1b NOTE—Except for G13 in North Carolina, the colored airway system exists only in the State of Alaska. All other such airways formerly so designated in the conterminous U.S. have been rescinded.

2. The Jet Route system consists of jet routes established from 18,000 feet MSL to FL 450 inclusive.

(a) These routes are depicted on En Route High Altitude Charts. Jet routes are depicted in black on aeronautical charts and are identified by a "J" (Jet) followed by the airway number (e.g., J 12). Jet routes, as VOR airways, are predicated solely on VOR or VORTAC navigation facilities (except in Alaska).

5-33a2a NOTE—Segments of jet routes in Alaska are based on L/MF navigation aids and are charted in brown color instead of black on en route charts.

(b) With respect to position reporting, reporting points are designated for Jet Route systems. Flights using Jet Routes will report over these points unless otherwise advised by ATC.

3. Area Navigation (RNAV) Routes—

(a) RNAV is a method of navigation that permits aircraft operations on any desired course within the coverage of station referenced navigation signals or within the limits of a self-contained system capability or combination of these.

(b) Fixed RNAV routes are permanent, published routes which can be flight planned for use by aircraft with RNAV capability. A previously established fixed RNAV route system has been terminated except for a few high altitude routes in Alaska.

(c) Random RNAV routes are direct routes, based on area navigation capability, between waypoints defined in terms of latitude/longitude coordinates, degree-distance fixes, or offsets from established routes/airways at a specified distance and direction. Radar monitoring by ATC is required on all random RNAV routes.

b. Operation above FL 450 may be conducted on a point-to-point basis. Navigational guidance is provided on an area basis utilizing those facilities depicted on the En Route High Altitude Charts.

c. Radar Vectors. Controllers may vector aircraft within controlled airspace for separation purposes, noise abatement considerations, when an operational advantage will be realized by the pilot or the controller, or when requested by the pilot. Vectors outside of controlled airspace will be provided only on pilot request. Pilots will be advised as to what the vector is to achieve when the vector is controller initiated and will take the aircraft off a previously assigned nonradar route. To the extent possible, aircraft operating on RNAV routes will be allowed to remain on their own navigation.

d. When flying in Canadian airspace, pilots are cautioned to review Canadian Air Regulations.

1. Special attention should be given to the parts which differ from U.S. FAR's.

(a) The Canadian Airways Class B airspace restriction is an example. Class B airspace is all controlled low level airspace above 12,500 feet MSL or the MEA, whichever is higher, within which only IFR and controlled VFR flights are permitted. (Low level airspace means an airspace designated and defined as such in the *Designated Airspace Handbook*.)

(b) Regardless of the weather conditions or the height of the terrain, no person shall operate an aircraft under VFR conditions within Class B airspace except in accordance with a clearance for VFR flight issued by ATC.

(c) The requirement for entry into Class B airspace is a student pilot permit (under the guidance or control of a flight instructor).

(d) VFR flight requires visual contact with the ground or water at all times.

2. Segments of VOR airways and high level routes in Canada are based on L/MF navigation aids and are charted in brown color instead of blue on en route charts.

5-34. AIRWAY OR ROUTE COURSE CHANGES

a. Pilots of aircraft are required to adhere to airways or routes being flown. Special attention must be given to this requirement during course changes. Each course change consists of variables that make the technique applicable in each case a matter only the pilot can resolve. Some variables which must be considered are turn radius, wind effect, airspeed, degree of turn, and cockpit instrumentation. An early turn, as illustrated below, is one method of adhering to airways or routes. The use of any available cockpit instrumentation, such as Distance Measuring Equipment, may be used by the pilot to lead his turn when making course changes. This is *consistent* with the intent of FAR Part 91.181, which requires pilots to operate along the centerline of an airway and along the direct course between navigational aids or fixes.

b. Turns which begin at or after fix passage may exceed airway or route boundaries. Figure 5-34[1] contains an example flight track depicting this, together with an example of an early turn.

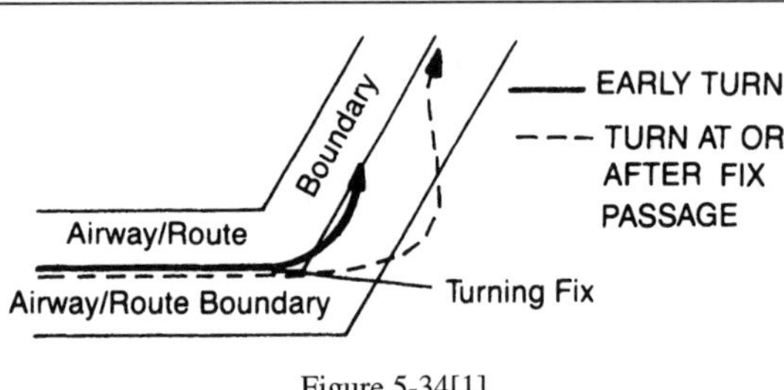

Figure 5-34[1]

c. Without such actions as leading a turn, aircraft operating in excess of 290 knots TAS can exceed the normal airway or route boundaries depending on the amount of course change required, wind direction and velocity, the character of the turn fix (DME, overhead navigation aid, or intersection), and the pilot's technique in making a course change. For example, a flight operating at 17,000 feet MSL with a TAS of 400 knots, a 25 degree bank, and a course change of more than 40 degrees would exceed the width of the airway or route; i.e., 4 nautical miles each side of centerline. However, in the airspace below 18,000 feet MSL, operations in excess of 290 knots TAS are not prevalent and the provision of additional IFR separation in all course change situations for the occasional aircraft making a turn in excess of 290 knots TAS creates an unacceptable waste of airspace and imposes a penalty upon the preponderance of traffic which operates at low speeds. Consequently, the FAA expects pilots to lead turns and take other actions they consider necessary during course changes to adhere as closely as possible to the airways or route being flown.

d. Due to the high airspeeds used at 18,000 feet MSL and above, FAA provides additional IFR separation protection for course changes made at such altitude levels.

5-35. CHANGEOVER POINTS (COP's)

a. COP's are prescribed for Federal Airways, jet routes, Area Navigation routes, or other direct routes for which an MEA is designated under FAR Part 95. The COP is a point along the route or airway segment between two adjacent navigation facilities or way points where changeover in navigation guidance should occur. At this point, the pilot should change navigation receiver frequency from the station behind the aircraft to the station ahead.

b. The COP is located midway between the navigation facilities for straight route segments, or at the intersection of radials or courses forming a dogleg in the case of dogleg route segments. When the COP is NOT located at the midway point, aeronautical charts will depict the COP location and give the mileage to the radio aids.

c. COP's are established for the purpose of preventing loss of navigation guidance, to prevent frequency interference from other facilities, and to prevent use of different facilities by different aircraft in the same airspace. Pilots are urged to observe COP's to the fullest extent.

5-36. HOLDING

a. Whenever an aircraft is cleared to a fix other than the destination airport and delay is expected, it is the responsibility of the ATC controller to issue complete holding instructions (unless the pattern is charted), an EFC time and his best estimate of any additional en route/terminal delay.

5-36a NOTE—Only those holding patterns depicted on U.S. Government or commercially produced (meeting FAA requirements) *Low/High Altitude Enroute*, and *Area* or *STAR* charts should be used.

b. If the holding pattern is charted and the controller doesn't issue complete holding instructions, the pilot is expected to hold as depicted on the appropriate chart. When the pattern is charted, the controller may omit all holding instructions except the charted holding direction and the statement *AS PUBLISHED;* e.g., *HOLD EAST AS PUBLISHED.* Controllers shall always issue complete holding instructions when pilots request them.

c. If no holding pattern is charted and holding instructions have not been issued, the pilot should ask ATC for holding instructions prior to reaching the fix. This procedure will eliminate the possibility of an aircraft entering a holding pattern other than that desired by ATC. If the pilot is unable to obtain holding instructions prior to reaching the fix (due to frequency congestion, stuck microphone, etc.), he should hold in a standard pattern on the course on which he approached the fix and request further clearance as soon as possible. In this event, the altitude/flight level of the aircraft at the clearance limit will be protected so that separation will be provided as required.

d. When an aircraft is 3 minutes or less from a clearance limit and a clearance beyond the fix has not been received, the pilot is expected to start a speed reduction so that he will cross the fix, initially, at or below the maximum holding airspeed.

e. When no delay is expected, the controller should issue a clearance beyond the fix as soon as possible and, whenever possible, at least 5 minutes before the aircraft reaches the clearance limit.

f. Pilots should report to ATC the time and altitude/flight level at which the aircraft reaches the clearance limit and report leaving the clearance limit.

5-36f NOTE—In the event of two-way communications failure, pilots are required to comply with FAR Part 91.185.

g. When holding at a VOR station, pilots should begin the turn to the outbound leg at the time of the first complete reversal of the to/from indicator.

h. Patterns at the most generally used holding fixes are depicted (charted) on U.S. Government or commercially produccd (meeting FAA requirements) Low or High Altitude En route, Area and STAR Charts. Pilots are expected to hold in the pattern depicted unless specifically advised otherwise by ATC.

i. An ATC clearance requiring an aircraft to hold at a fix where the pattern is not charted will include the following information: (See Figure 5-36[1].)

EXAMPLES OF HOLDING

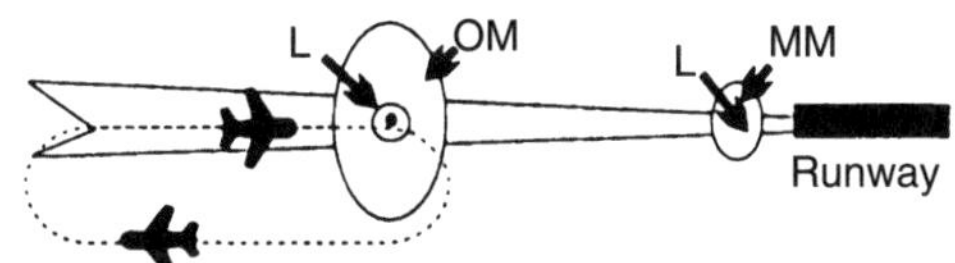

Typical procedure on an ILS outer marker

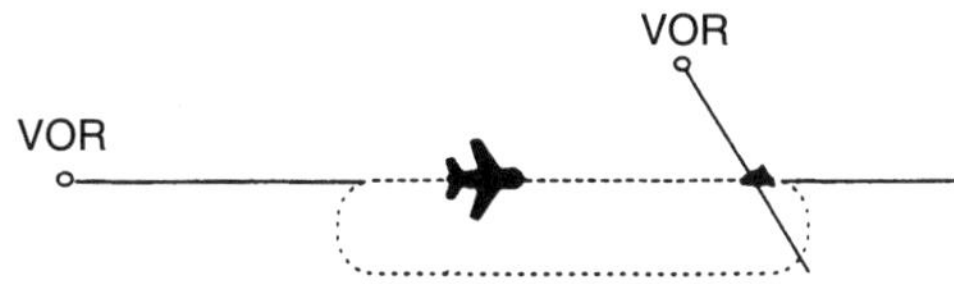

Typical procedure at intersection of VOR radials

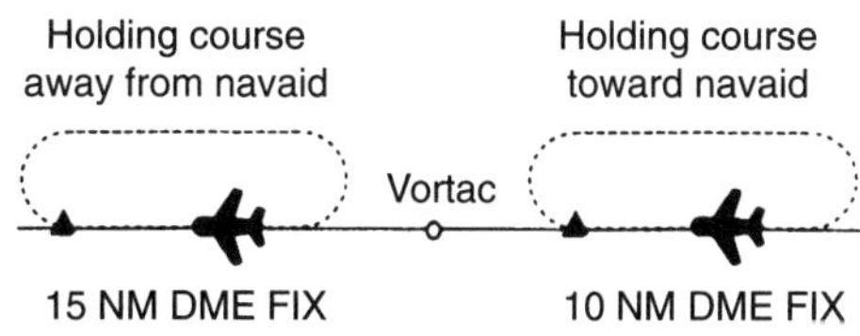

Typical procedure at DME FIX

Figure 5-36[1]

1. Direction of holding from the fix in terms of the eight cardinal compass points (i.e., N, NE, E, SE, etc.).

2. Holding fix (the fix may be omitted if included at the beginning of the transmission as the clearance limit).

3. Radial, course, bearing, airway or route on which the aircraft is to hold.

4. Leg length in miles if DME or RNAV is to be used (leg length will be specified in minutes on pilot request or if the controller considers it necessary).

5. Direction of turn if left turns are to be made, the pilot requests, or the controller considers it necessary.

6. Time to expect further clearance and any pertinent additional delay information.

j. Holding pattern airspace protection is based on the following procedures.

1. Descriptive Terms—

(a) Standard Pattern: Right turns (See Figure 5-36[2].)

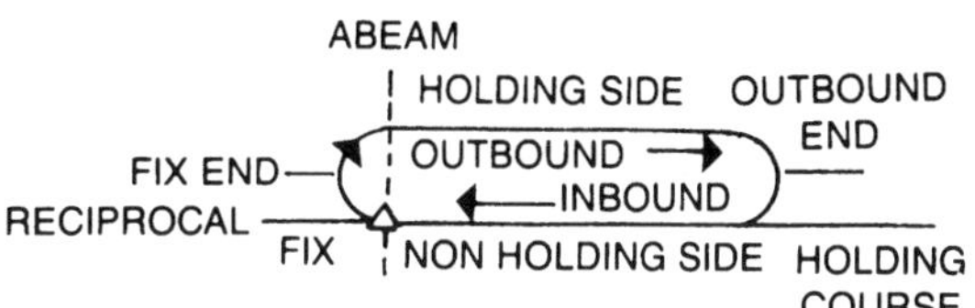

Figure 5-36[2]

(b) Nonstandard Pattern: Left turns

2. Airspeeds (maximum—See Table 5-36[1][Propeller-Driven], Table 5-36[2][Civil Turbojet], and Table 5-36[3][Military Turbojet]).

Table 5-36[1]

PROPELLER-DRIVEN	MAXIMUM AIRSPEED
1. All (including turboprop)	175K IAS

Table 5-36[2]

CIVIL TURBOJET	MAXIMUM AIRSPEED
1. MHA through 14,000 feet	230K IAS
2. Above 14,000 feet	265K IAS

Table 5-36[3]

MILITARY TURBOJET	MAXIMUM AIRSPEED
1. All —Except aircraft listed in 2, 3, and 4	230K IAS
2. USAF F-4 aircraft	280K IAS
3. B-1, F-111, and F-5	310K IAS
4. T-37	175K IAS

5-36j2 NOTE 1—Additional military exceptions may be added to Table 5-36[3].

5-36j2 NOTE 2—Holding speed depends upon weight and drag configuration.

5-36j2 NOTE 3—Civil aircraft holding at military or joint civil/military use airports should expect to operate at a maximum holding pattern airspeed of 230 knots.

3. Entry Procedures (See Figure 5-36[3])—

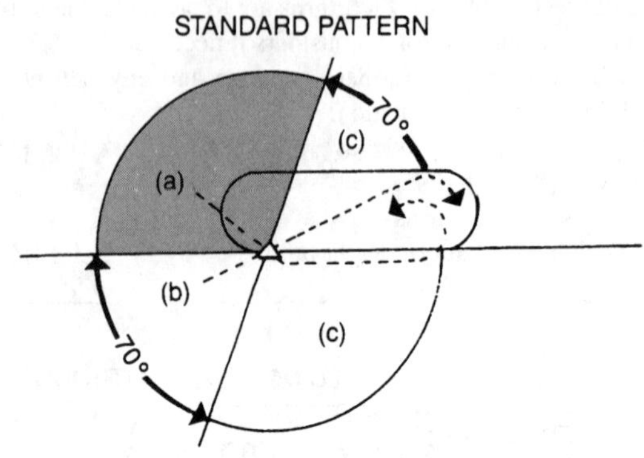

Figure 5-36[3]

(a) Parallel Procedure: When approaching the holding fix from anywhere in sector (a), the parallel entry procedure would be to turn to a heading to parallel the holding course outbound on the non-holding side for one minute, turn in the direction of the holding pattern thru more than 180 degrees, and return to the holding fix or intercept the holding course inbound.

(b) Teardrop Procedure: When approaching the holding fix from anywhere in sector (b), the teardrop entry procedure would be to fly to the fix, turn outbound to a heading for a 30 degree teardrop entry within the pattern (on the holding side) for a period of one minute, then turn in the direction of the holding pattern to intercept the inbound holding course.

(c) Direct Entry Procedure: When approaching the holding fix from anywhere in sector (c), the direct entry procedure would be to fly directly to the fix and turn to follow the holding pattern.

(d) While Other entry procedures may enable the aircraft to enter the holding pattern and remain within protected airspace, the parallel, teardrop and direct entries are the procedures for entry and holding recommended by the FAA.

4. Timing—

(a) Inbound Leg:

(1) At or below 14,000 Ft. MSL-1 minute.

(2) Above 14,000 Ft. MSL-1½ minutes.

5-36j4a2 NOTE—The *initial* outbound leg should be flown for 1 minute or 1 and ½ min. (appropriate to altitude). Timing for subsequent outbound legs should be adjusted, as necessary, to achieve proper inbound leg time. Pilots may use any navigational means available; i.e., DME, RNAV, etc., to insure the appropriate inbound leg times.

(b) Outbound leg timing begins *over/abeam* the fix,

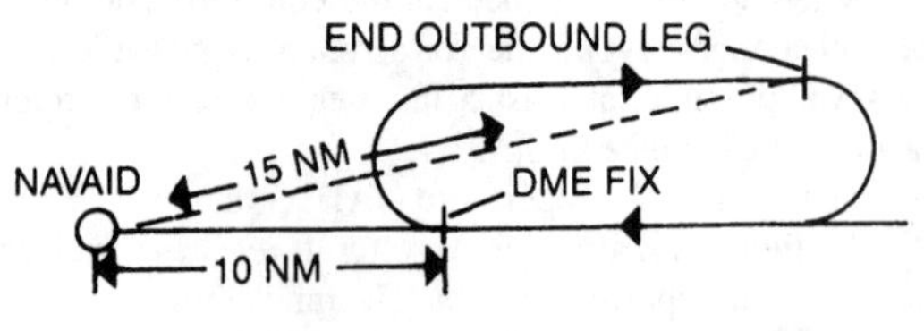

Figure 5-36[4]

whichever occurs later. If the abeam position cannot be determined, start timing when turn to outbound is completed.

NOTE—When the inbound course is *toward* the NAVAID and the fix distance is 10 NM, and the leg length is 5 NM, then the end of the outbound leg will be reached when the DME reads 15 NM.

NOTE—When the inbound course is *away* from the NAVAID and the fix distance is 28 NM, and the leg length is 8 NM, then the end of the outbound leg will be reached when the DME reads 20 NM.

5. Distance Measuring Equipment (DME)—DME holding is subject to the same entry and holding procedures except that distances (nautical miles) are used in lieu of time values. The *outbound course* of a DME holding pattern is called the outbound leg of the pattern. The length of the outbound leg will be specified by the controller. The end of the outbound leg is determined by the odometer reading. (See Figure 5-36[4] and Figure 5-36[5].)

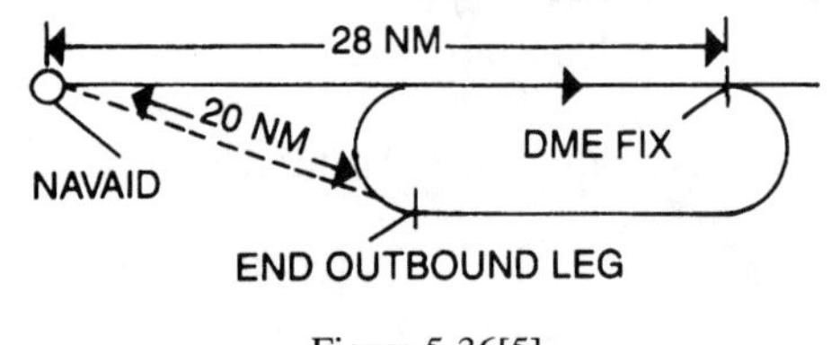

Figure 5-36[5]

6. Pilot Action—

(a) Start speed reduction when 3 minutes or less from the holding fix. Cross the holding fix, initially, at or below the maximum holding airspeed.

(b) Make all turns during entry and while holding at:

(1) 3 degrees per second, or

(2) 30 degree bank angle, or

(3) 25 degree bank provided a flight director system is used.

5-36j6b3 NOTE—Use whichever requires the least bank angle.

(c) Compensate for wind effect primarily by drift correction on the inbound and outbound legs. When outbound, triple the inbound drift correction to avoid major turning adjustments; e.g., if correcting left by 8 degrees when inbound, correct right by 24 degrees when outbound.

(d) Determine entry turn from aircraft heading upon arrival at the holding fix; ±5 degrees in heading is considered to be within allowable good operating limits for determining entry.

(e) Advise ATC immediately if any increased airspeed is necessary due to turbulence, icing, etc., or if unable to accomplish any part of the holding procedures. When such higher speeds become no longer necessary, operate according to the appropriate published holding speed and notify ATC.

5-36j6e NOTE—Airspace protection for turbulent air holding is based on a maximum of 280K IAS/Mach 0.8, whichever is lower. Considerable impact on traffic flow will result when turbulent air holding patterns are used; thus, pilot discretion will ensure their use is limited to bona fide conditions or requirements.

7. Nonstandard Holding Pattern—Fix end and outbound end turns are made to the left. Entry procedures to a nonstandard pattern are oriented in relation to the 70 degree line on the holding side just as in the standard pattern.

k. When holding at a fix and instructions are received specifying the time of departure from the fix, the pilot should adjust his flight path within the limits of the established holding pattern in order to leave the fix at the exact time specified. After departing the holding fix, normal speed is to be resumed with respect to other governing speed requirements, such as terminal area speed limits, specific ATC requests, etc. Where the fix is associated with an instrument approach and timed approaches are in effect, a procedure turn shall not be executed unless the pilot advises ATC, since aircraft holding are expected to proceed inbound on final approach directly from the holding pattern when approach clearance is received.

l. Radar surveillance of outer fix holding pattern airspace areas.

1. Whenever aircraft are holding at an outer fix, ATC will usually provide radar surveillance of the outer fix holding pattern airspace area, or any portion of it, if it is shown on the controller's radarscope.

2. The controller will attempt to detect any holding aircraft that stray outside the holding pattern airspace area and will assist any detected aircraft to return to the assigned airspace area.

5-36l2 NOTE—Many factors could prevent ATC from providing this additional service, such as workload, number of targets, precipitation, ground clutter, and radar system capability. These circumstances may make it unfeasible to maintain radar identification of aircraft to detect aircraft straying from the holding pattern. The provision of this service depends entirely upon whether the controller believes he is in a position to provide it and does not relieve a pilot of his responsibility to adhere to an accepted ATC clearance.

3. If an aircraft is established in a published holding pattern at an assigned altitude above the published minimum holding altitude and subsequently cleared for the approach, the pilot may descend to the published minimum holding altitude. The holding pattern would only be a segment of the IAP *if* it is published on the instrument procedure chart and is used in lieu of a procedure turn.

m. For those holding patterns where there are no published minimum holding altitudes, the pilot, upon receiving an approach clearance, must maintain his last assigned altitude until leaving the holding pattern and established on the inbound course. Thereafter, the published minimum altitude of the route segment being flown will apply. It is expected that the pilot will be assigned a holding altitude that will permit a normal descent on the inbound course.

5-37 thru 5-39. RESERVED

Section 4. ARRIVAL PROCEDURES

5-40. STANDARD TERMINAL ARRIVAL (STAR)

a. A STAR is an ATC coded IFR arrival route established for application to arriving IFR aircraft destined for certain airports. Its purpose is to simplify clearance delivery procedures.

b. Pilots of IFR civil aircraft destined to locations for which STARs have been published may be issued a clearance containing a STAR whenever ATC deems it appropriate. Until military STAR publications and distribution is accomplished, STARs will be issued to military pilots only when requested in the flight plan or verbally by the pilot.

c. Use of STARs requires pilot possession of at least the approved textual description. As with any ATC clearance or portion thereof, it is the responsibility of each pilot to accept or refuse an issued STAR. A pilot should notify ATC if he does not wish to use a STAR by placing "NO STAR" in the remarks section of the flight plan or by the less desirable method of verbally stating the same to ATC.

d. STAR charts are published in the *Terminal Procedures Publication (TPP)* and are available on subscription from the National Ocean Service.

5-41. LOCAL FLOW TRAFFIC MANAGEMENT PROGRAM

a. This program is a continuing effort by the FAA to enhance safety, minimize the impact of aircraft noise and conserve aviation fuel. The enhancement of safety and reduction of noise is achieved in this program by minimizing low altitude maneuvering of arriving turbojet and turboprop aircraft weighing more than 12,500 pounds and, by permitting departure aircraft to climb to higher altitudes sooner, as arrivals are operating at higher altitudes at the points where their flight paths cross. The application of these procedures also reduces exposure time between controlled aircraft and uncontrolled aircraft at the lower altitudes in and around the terminal environment. Fuel conservation is accomplished by absorbing any necessary arrival delays for aircraft included in this program operating at the higher and more fuel efficient altitudes.

b. A fuel efficient descent is basically an uninterrupted descent (except where level flight is required for speed adjustment) from cruising altitude to the point when level flight is necessary for the pilot to stabilize his final approach. The procedure for a fuel efficient descent is based on an altitude loss which is most efficient for the majority of aircraft being served. This will generally result in a descent gradient window of 250–350 feet per nautical mile.

c. When crossing altitudes and speed restrictions are issued verbally or are depicted on a chart, ATC will expect the pilot to descend first to the crossing altitude and then reduce speed. Verbal clearances for descent will normally permit an uninterrupted descent in accordance with the procedure as described in paragraph b. above. Acceptance of a charted fuel efficient descent (Runway Profile Descent) clearance requires the pilot to adhere to the altitudes, speeds, and headings depicted on the charts unless otherwise instructed by ATC. PILOTS RECEIVING A CLEARANCE FOR A FUEL EFFICIENT DESCENT ARE EXPECTED TO ADVISE ATC IF THEY DO NOT HAVE RUNWAY PROFILE DESCENT CHARTS PUBLISHED FOR THAT AIRPORT OR ARE UNABLE TO COMPLY WITH THE CLEARANCE.

5-42. APPROACH CONTROL

a. Approach control is responsible for controlling all instrument flight operating within its area of responsibility. Approach control may serve one or more airfields, and control is exercised primarily by direct pilot and controller communications. Prior to arriving at the destination radio facility, instructions will be received from ARTCC to contact approach control on a specified frequency.

b. Radar Approach Control

1. Where radar is approved for approach control service, it is used not only for radar approaches (ASR and PAR) but is also used to provide vectors in conjunction with published nonradar approaches based on radio NAVAIDs (ILS, MLS, VOR, NDB, TACAN). Radar vectors can provide course guidance and expedite traffic to the final approach course of any established IAP or to the traffic pattern for a visual approach. Approach control facilities that provide this radar service will operate in the following manner:

(a) Arriving aircraft are either cleared to an outer fix most appropriate to the route being flown with vertical separation and, if required, given holding information or, when radar handoffs are effected between the ARTCC and approach control, or between two approach control facilities, aircraft are cleared to the airport or to a fix so located that the handoff will be completed prior to the time the aircraft reaches the fix. When radar handoffs are utilized, successive arriving flights may be handed off to approach control with radar separation in lieu of vertical separation.

(b) After release to approach control, aircraft are vectored to the final approach course (ILS, MLS, VOR, ADF, etc.). Radar vectors and altitude or Flight Levels will be issued as required for spacing and separating aircraft. *Therefore, pilots must not deviate from the headings issued by approach control.* Aircraft will normally be informed when it is necessary to vector across the final approach course for spacing or other reasons. If approach course crossing is imminent and the pilot has not been informed that he will be vectored across the final approach course, he should query the controller.

(c) The pilot is not expected to turn inbound on the final approach course unless an approach clearance has been issued. This clearance will normally be issued with the final vector for interception of the final approach course, and the vector will be such as to enable the pilot to establish his aircraft on the final approach course prior to reaching the final approach fix.

(d) In the case of aircraft already inbound on the final approach course, approach clearance will be issued prior to the aircraft reaching the final approach fix. When established inbound on the final approach course, radar separation will be maintained and the pilot will be expected to complete the approach utilizing the approach aid designated in the clearance (ILS, MLS, VOR, radio beacons, etc.) as the primary means of navigation. Therefore, once established on the final approach course, pilots must not deviate from it unless a clearance to do so is received from ATC.

(e) After passing the final approach fix on final approach, aircraft are expected to continue inbound on the final approach course and complete the approach or effect the missed approach procedure published for that airport.

2. Whether aircraft are vectored to the appropriate final approach course or provide their own navigation on published routes to it, radar service is automatically terminated when the landing is completed or when instructed to change to advisory frequency at uncontrolled airports, whichever occurs first.

5-43. ADVANCE INFORMATION ON INSTRUMENT APPROACH

a. When landing at airports with approach control services and where two or more IAPs are published, pilots will be provided in advance of their arrival with the type of approach to expect or that they may be vectored for a visual approach. This information will be broadcast either by a controller or on ATIS. It will not be furnished when the visibility is three miles or better and the ceiling is at or above the highest initial approach altitude established for any low altitude LAP for the airport.

b. The purpose of this information is to aid the pilot in planning arrival actions; however, it is not an ATC clearance or commitment and is subject to change. Pilots should bear in mind that fluctuating weather, shifting winds, blocked runway, etc., are conditions which may result in changes to approach information previously received. It is important that the pilot advise ATC immediately if he is unable to execute the approach ATC advised will be used, or if he prefers another type of approach.

c. When making an IFR approach to an airport not served by a tower or FSS, after the ATC controller advises "CHANGE TO ADVISORY FREQUENCY APPROVED" you should broadcast your intentions, including the type of approach being executed, your position, and when over the final approach fix inbound (nonprecision approach) or when over the outer marker or fix used in lieu of the outer marker inbound (precision approach). Continue to monitor the appropriate frequency (UNICOM, etc.) for reports from other pilots.

5-44. INSTRUMENT APPROACH PROCEDURE CHARTS

a. FAR Part 91.175a (Instrument Approaches to Civil Airports) requires the use of SIAPs prescribed for the airport in FAR Part 97 unless otherwise authorized by the Administrator (including ATC). FAR Part 91.175g (Military Airports) requires civil pilots flying into or out of military airports to comply with the IAPs and takeoff and landing minimums prescribed by the authority having jurisdiction at those airports.

1. All IAPs (standard and special, civil and military) are based on joint civil and military criteria contained in the U.S. Standard for TERPs. The design of IAPs based on criteria contained in TERPs, takes into account the interrelationship between airports, facilities, and the surrounding environment, terrain, obstacles, noise sensitivity, etc. Appropriate altitudes, courses, headings, distances, and other limitations are specified and, once approved, the procedures are published and distributed by government and commercial cartographers as instrument approach charts.

2. Not all IAPs are published in chart form. Radar IAPs are established where requirements and facilities exist but they are printed in tabular form in appropriate U.S. Government Flight Information Publications.

3. A pilot adhering to the altitudes, flight paths, and weather minimums depicted on the IAP chart or vectors and altitudes issued by the radar controller, is assured of terrain and obstruction clearance and runway or airport alignment during approach for landing.

4. IAPs are designed to provide an IFR descent from the en route environment to a point where a safe landing can be made. They are prescribed and approved by appropriate civil or military authority to ensure a safe descent during instrument flight conditions at a specific airport. It is important that pilots understand these procedures and their use prior to attempting to fly instrument approaches.

5. TERPs criteria are provided for the following type of instrument approach procedures:

(a) Precision approaches where an electronic glide slope is provided (PAR and ILS) and,

(b) Nonprecision approaches where glide slope information is not provided (all approaches except PAR and ILS).

b. The method used to depict prescribed altitudes on instrument approach charts differs according to techniques employed by different chart publishers. Prescribed altitudes may be depicted in three different configurations: Minimum, maximum, and mandatory. The U.S. Government distributes charts produced by Defense Mapping Agency (DMA) and NOS. Altitudes are depicted on these charts in the profile view with underscore, overscore, or both to identify them as minimum, maximum, or mandatory.

1. Minimum Altitude will be depicted with the altitude value underscored. Aircraft are required to maintain altitude at or above the depicted value.

2. Maximum Altitude will be depicted with the altitude value overscored. Aircraft are required to maintain altitude at or below the depicted value.

3. Mandatory Altitude will be depicted with the altitude value both underscored and overscored. Aircraft are required to maintain altitude at the depicted value.

5-44b3 NOTE—The underscore and overscore to identify mandatory altitudes and the overscore to identify maximum altitudes are used almost exclusively by DMA for military charts. With very few exceptions, civil approach charts produced by NOS utilize only the underscore to identify minimum altitudes. Pilots are cautioned to adhere to altitudes as prescribed because, in certain instances, they may be used as the basis for vertical separation of aircraft by ATC. When a depicted altitude is specified in the ATC clearance, that altitude becomes mandatory as defined above.

c. Minimum Safe Altitudes (MSA) are published for emergency use on instrument approach procedure (IAP) charts except RNAV IAPs. The MSA is defined using NDB or VOR type facilities within 25 NM (normally) or 30 NM (maximum) of the airport. The MSA has a 25 NM (normally) or 30 NM (maximum) radius. If there is no NDB or VOR facility within 30 NM of the airport, there will be no MSA. The altitude shown provides at least 1,000 feet of clearance above the highest obstacle in the defined sector. As many as four sectors may be depicted with different altitudes for each sector displayed in rectangular boxes in the plan view of the chart. A single altitude for the entire area may be shown in the lower right portion of the plan view. Navigational course guidance is not assured at the MSA within these sectors.

d. Minimum Vectoring Altitudes (MVA) are established for use by ATC when radar ATC is exercised. MVA charts are prepared by air traffic facilities at locations where there are numerous different minimum IFR altitudes. Each MVA chart has sectors large enough to accommodate vectoring of aircraft within the sector at the MVA. Each sector boundary is at least 3 miles from the obstruction determining the MVA. To avoid a large sector with an excessively high MVA due to an isolated prominent obstruction, the obstruction may be enclosed in a buffer area whose boundaries are at least 3 miles from the obstruction. This is done to facilitate vectoring around the obstruction. (See Figure 5-44[1].)

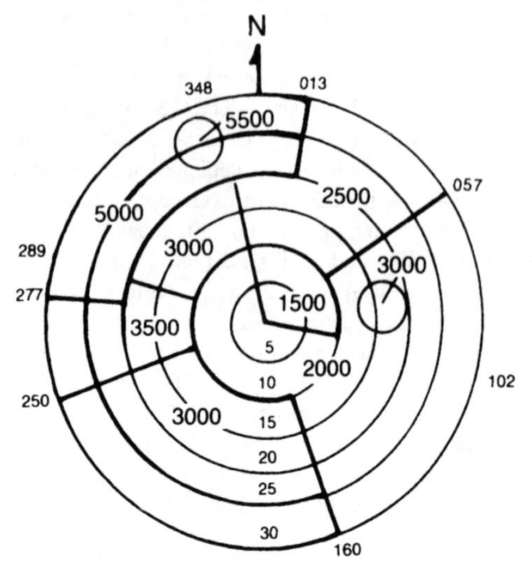

Figure 5-44[1]

1. The minimum vectoring altitude in each sector provides 1,000 feet above the highest obstacle in nonmountainous areas and 2,000 feet above the highest obstacle in designated mountainous areas. Where lower MVAs are required in designated mountainous areas to achieve compatibility with terminal routes or to permit vectoring to an IAP, 1,000 feet of obstacle clearance may be authorized with the use of Airport Surveillance Radar (ASR). The minimum vectoring altitude will provide at least 300 feet above the floor of controlled airspace.

2. Because of differences in the areas considered for MVA, and those applied to other minimum altitudes, and the ability to isolate specific obstacles, some MVAs may be lower than the nonradar Minimum en Route Altitudes (MEA), Minimum Obstruction Clearance Altitudes (MOCA) or other minimum altitudes depicted on charts for a given location. While being radar vectored, IFR altitude assignments by ATC will be at or above MVA.

e. Visual Descent Points (VDP) are incorporated in selected nonprecision approach procedures. The VDP is a defined point on the final approach course of a nonprecision straight-in approach procedure from which normal descent from the MDA to the runway touchdown point may be commenced, provided visual reference required by FAR Part 91.175(c)(3) is established. The VDP will normally be identified by DME on VOR and LOC procedures. The VDP is identified on the profile view of the approach chart by the symbol: **V**.

1. VDPs are intended to provide additional guidance where they are implemented. No special technique is required to fly a procedure with a VDP. The pilot should not descend below the MDA prior to reaching the VDP and acquiring the necessary visual reference.

2. Pilots not equipped to receive the VDP should fly the approach procedure as though no VDP had been provided.

5-45. APPROACH CLEARANCE

a. An aircraft which has been cleared to a holding fix and subsequently "cleared . . . approach" has not received new routing. Even though clearance for the approach may have been issued prior to the aircraft reaching the holding fix, ATC would expect the pilot to proceed via the holding fix (his last assigned route), and the feeder route associated with that fix (if a feeder route is published on the approach chart) to the initial approach fix (IAF) to commence the approach. *When cleared for the approach, the published off airway (feeder) routes that lead from the en route structure to the IAF are part of the approach clearance.*

b. If a feeder route to an IAF begins at a fix located along the route of flight prior to reaching the holding fix, and clearance for an approach is issued, a pilot should commence his approach via the published feeder route; i.e., the aircraft would not be expected to overfly the feeder route and return to it. The pilot is expected to commence his approach in a similar manner at the IAF, if the IAF for the procedure is located along the route of flight to the holding fix.

c. If a route of flight directly to the initial approach fix is desired, it should be so stated by the controller with phraseology to include the words "direct . . .," "proceed direct" or a similar phrase which the pilot can interpret without question. If the pilot is uncertain of his clearance, he should immediately query ATC as to what route of flight is desired.

d. The name of an instrument approach, as published, is used to identify the approach, even though a component of the approach aid, such as the glideslope on an Instrument Landing System, is inoperative or unreliable. The controller will use the name of the approach as published, but must advise the aircraft at the time an approach clearance is issued that the inoperative or unreliable approach aid component is unusable.

5-46. INSTRUMENT APPROACH PROCEDURES

a. Minimums are specified for various aircraft approach categories based upon a value 1.3 times the stalling speed of the aircraft in the landing configuration at maximum certificated gross landing weight. (See FAR Part 97.3(b)). If it is necessary, while circling-to-land, to maneuver at speeds in excess of the upper limit of the speed range for each category, due to the possibility of extending the circling maneuver beyond the area for which obstruction clearance is provided, the circling minimum for the next higher approach category should be used. For example, an aircraft which falls in Category C, but is circling to land at a speed of 141 knots or higher should use the approach category "D" minimum when circling to land.

b. When operating on an unpublished route or while being radar vectored, the pilot, when an approach clearance is received, shall, in addition to complying with the minimum altitudes for IFR operations (FAR Part 91.177), maintain his last assigned altitude unless a different altitude is assigned by ATC, or until the aircraft is established on a segment of a published route or IAP. After the aircraft is so established, published altitudes apply to descent within each succeeding route or approach segment unless a different altitude is assigned by ATC. Notwithstanding this pilot responsibility, for aircraft operating on unpublished routes or while being radar vectored, ATC will, except when conducting a radar approach, issue an IFR approach clearance only after the aircraft is established on a segment of a published route or IAP, or assign an altitude to maintain until the aircraft is established on a segment of a published route or instrument approach procedure. For this purpose, the Procedure Turn of a published IAP shall not be considered a segment of that IAP until the aircraft reaches the initial fix or navigation facility upon which the procedure turn is predicated.

EXAMPLE:
CROSS REDDING VOR AT OR ABOVE FIVE THOUSAND, CLEARED VOR RUNWAY THREE FOUR APPROACH. *or*

EXAMPLE:
FIVE MILES FROM OUTER MARKER, TURN RIGHT HEADING THREE THREE ZERO, MAINTAIN TWO THOUSAND UNTIL ESTABLISHED ON THE LOCALIZER, CLEARED ILS RUNWAY THREE SIX APPROACH.

5-46b NOTE—The altitude assigned will assure IFR obstruction clearance from the point at which the approach clearance is issued until established on a segment of a published route or IAP. If a pilot is uncertain of the meaning of his clearance, he shall immediately request clarification from ATC.

c. Several IAPs, using various navigation and approach aids may be authorized for an airport. ATC may advise that a particular approach procedure is being used, primarily to expedite traffic. (Reference—Advance Information on Instrument Approach, paragraph 5-43.) If a pilot is issued a clearance that specifies a particular approach procedure, he is expected to notify ATC immediately if he desires a different one. In this event it may be necessary for ATC to withhold clearance for the different approach until such time as traffic conditions permit. However, if the pilot is involved in an emergency situation he will be given priority. If the pilot is not familiar with the specific approach procedure, ATC should be advised and they will provide detailed information on the execution of the procedure.

d. At times ATC may not specify a particular approach procedure in the clearance, but will state "CLEARED APPROACH." Such clearance indicates that the pilot may execute any one of the authorized IAPs for that airport. This clearance does not constitute approval for the pilot to execute a contact approach or a visual approach.

e. When cleared for a specifically prescribed IAP; i.e., "cleared ILS runway one niner approach" or when "cleared approach" i.e., execution of any procedure prescribed for the airport, shall execute the entire procedure as described on the IAP Chart unless an appropriate new or revised ATC clearance is received, or the IFR flight plan is canceled.

f. Pilots planning flights to locations served by special IAPs should obtain advance approval from the owner of the procedure. Approval by the owner is necessary because special procedures are for the exclusive use of the single interest unless otherwise authorized by the owner. Additionally, some special approach procedures require certain crew qualifications training, or other special considerations in order to execute the approach. Also, some of these approach procedures are based on privately owned navigational aids. Owners of aids that are not for public use may elect to turn off the aid for whatever reason they may have; i.e., maintenance, conservation, etc. Air traffic controllers are not required to question pilots to determine if they have permission to use the procedure. Controllers presume a pilot has obtained approval and is aware of any details of the procedure if he files an IFR flight plan to that airport.

g. When executing an instrument approach and in radio contact with an FAA facility, unless in "radar contact," report passing the final approach fix inbound (nonprecision approach) or the outer marker or fix used in lieu of the outer marker inbound (precision approach).

h. If a missed approach is required, advise ATC and include the reason (unless initiated by ATC). Comply with the missed approach instructions for the instrument approach procedure being executed, unless otherwise directed by ATC. (Reference—Missed Approach, paragraph 5-56 and Missed Approach, paragraph 5-74.)

5-47. PROCEDURE TURN

a. A procedure turn is the maneuver prescribed when it is necessary to reverse direction to establish the aircraft inbound on an intermediate or final approach course. It is a required maneuver except when the symbol NoPT is shown, when RADAR VECTORING is provided, when a holding pattern is published in lieu of procedure turn, when conducting a timed approach, or when the procedure turn is not authorized. The altitude prescribed for the procedure turn is a *minimum* altitude until the aircraft is established on the inbound course. The maneuver must be completed within the distance specified in the profile view.

1. On U.S. Government charts, a barbed arrow indicates the direction or side of the outbound course on which the procedure turn is made. Headings are provided for course reversal using the 45 degree type procedure turn. However, the point at which the turn may be commenced and the type and rate of turn is left to the discretion of the pilot. Some of the options are the 45 degree procedure turn, the racetrack pattern, the tear-drop procedure turn, or the 80 degree-260 degree course reversal. Some procedure turns are specified by procedural track. These turns must be flown exactly as depicted.

2. When the approach procedure involves a procedure turn, a maximum speed of not greater than 250 knots (IAS) should be observed and the turn should be executed within the distance specified in the profile view. The normal procedure turn distance is 10 miles. This may be reduced to a minimum of 5 miles where only Category A or helicopter aircraft are to be operated or increased to as much as 15 miles to accommodate high performance aircraft.

3. A teardrop procedure or penetration turn may be specified in some procedures for a required course reversal. The teardrop procedure consists of departure from an initial approach fix on an outbound course followed by a turn toward and intercepting the inbound course at or prior to the intermediate fix or point. Its purpose is to permit an aircraft to reverse direction and lose considerable altitude within reasonably limited airspace. Where no fix is available to mark the beginning of the intermediate segment, it shall be assumed to commence at a point 10 miles prior to the final approach fix. When the facility is located on the airport, an aircraft is considered to be on final approach upon completion of the penetration turn. However, the final approach segment begins on the final approach course 10 miles from the facility.

4. A procedure turn need not be established when an approach can be made from a properly aligned holding pattern. In such cases, the holding pattern is established over an intermediate fix or a final approach fix. The holding pattern maneuver is completed when the aircraft is established on the inbound course after executing the appropriate entry. If cleared for the approach prior to returning to the holding fix, and the aircraft is at the prescribed altitude, additional circuits of the holding pattern are not necessary nor expected by ATC. If the pilot elects to make additional circuits to lose excessive altitude or to become better established on course, it is his responsibility to so advise ATC when he receives his approach clearance.

5. A procedure turn is not required when an approach can be made directly from a specified intermediate fix to the final approach fix. In such cases, the term "NoPT" is used with the appropriate course and altitude to denote that the procedure turn is not required. If a procedure turn is desired, and when cleared to do so by ATC, descent below the procedure turn altitude should not be made until the aircraft is established on the inbound course, since some NoPT altitudes may be lower than the procedure turn altitudes.

b. Limitations on Procedure Turns.

1. In the case of a radar initial approach to a final approach fix or position, or a timed approach from a holding fix, or where the procedure specifies "NoPT," no pilot may make a procedure turn unless, when he receives his final approach clearance, he so advises ATC and a clearance is received.

2. When a teardrop procedure turn is depicted and a course reversal is required, this type turn must be executed.

3. When holding pattern replaces the procedure turn, the standard entry and the holding pattern must be followed except when RADAR VECTORING is provided or when NoPT is shown on the approach course. As in the procedure turn, the descent from the minimum holding pattern altitude to the final approach fix altitude (when lower) may not commence until the aircraft is established on the inbound course.

4. The absence of the procedure turn barb in the Plan View indicates that a procedure turn is not authorized for that procedure.

5-48. TIMED APPROACHES FROM A HOLDING FIX

a. TIMED APPROACHES may be conducted when the following conditions are met:

1. A control tower is in operation at the airport where the approaches are conducted.

2. Direct communications are maintained between the pilot and the center or approach controller until the pilot is instructed to contact the tower.

3. If more than one missed approach procedure is available, none require a course reversal.

4. If only one missed approach procedure is available, the following conditions are met:

(a) Course reversal is not required; and,

(b) Reported ceiling and visibility are equal to or greater than the highest prescribed circling minimums for the IAP.

5. When cleared for the approach, pilots shall not execute a procedure turn. (FAR Part 91.175.)

b. Although the controller will not specifically state that "timed approaches are in progress," his assigning a time to depart the final approach fix inbound (nonprecision approach) or the outer marker or fix used in lieu of the outer marker inbound (precision approach) is indicative that timed approach procedures are being utilized, or in lieu of holding, he may use radar vectors to the Final Approach Course to establish a mileage interval between aircraft that will insure the appropriate time sequence between the final approach fix/outer marker or fix used in lieu of the outer marker and the airport.

c. Each pilot in an approach sequence will be given advance notice as to the time he should leave the holding point on approach to the airport. When a time to leave the holding point has been received, the pilot should adjust his flight path to leave the fix as closely as possible to the designated time. (See Figure 5-48[1].)

At 12:03 local time, in the example shown, a pilot holding, receives instructions to leave the fix inbound at 12:07. These instructions are received just as the pilot has completed turn at the outbound end of the holding pattern and is proceeding inbound towards the fix. Arriving back over the fix, the pilot notes that the time is 12:04 and that he has 3 minutes to lose in order to leave the fix at the assigned time. Since the time remaining is more than 2 minutes, the pilot plans to fly a race track pattern rather than a 360 degrees turn, which would use up 2 minutes. The runs at the ends of the race track pattern will consume approximately 2 minutes. Three minutes to go, minus 2 minutes required for turns, leaves 1 minute for level flight. Since two portions of level flight will be required to get back to the fix inbound, the pilot halves the 1 minute remaining and plans to fly level for 30 seconds outbound before starting his turn back toward the fix on final approach. If the winds were negligible at flight altitude, this procedure would bring the pilot inbound across the fix precisely at the specified time of 12:07. However, if the pilot expected a headwind on final approach, he should shorten his 30 seconds outbound course somewhat, knowing that the wind will carry him away from the fix faster while outbound and decrease his ground speed while returning to the fix. On the other hand, if the pilot knew he would have a tailwind on final approach, he should lengthen his calculated 30-second outbound heading somewhat, knowing that the wind would tend to hold him closer to the fix while outbound and increase his ground speed while returning to the fix.

Time Approach Example — a final approach procedure from a holding pattern at a final approach fix (FAF).

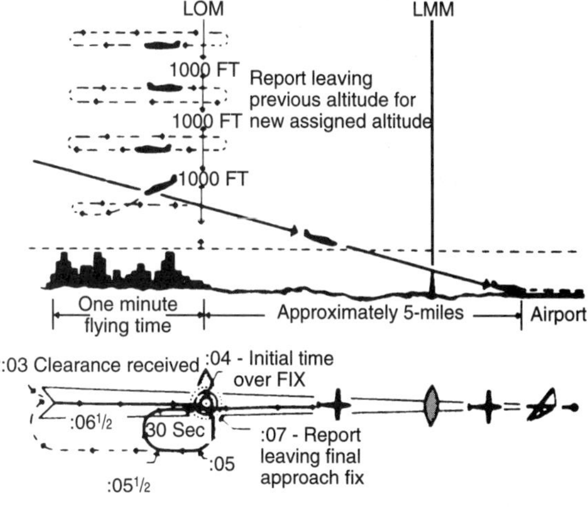

Figure 5-48[1]

5-49. RADAR APPROACHES

a. The only airborne radio equipment required for radar approaches is a functioning radio transmitter and receiver. The radar controller vectors the aircraft to align it with the runway centerline. The controller continues the vectors to keep the aircraft on course until the pilot can complete the approach and landing by visual reference to the surface. There are two types of radar approaches: Precision (PAR) and Surveillance (ASR).

b. A radar approach may be given to any aircraft upon request and may be offered to pilots of aircraft in distress or to expedite traffic, however, an ASR might not be approved unless there is an ATC operational requirement, or in an unusual or emergency situation. Acceptance of a PAR or ASR by a pilot does not waive the prescribed weather minimums for the airport or for the particular aircraft operator concerned. The decision to make a radar approach when the reported weather is below the established minimums rests with the pilot.

c. PAR and ASR minimums are published on separate pages in the NOS Terminal Procedures Publication (TPP).

1. A PRECISION APPROACH (PAR) is one in which a controller provides highly accurate navigational guidance in azimuth and elevation to a pilot. Pilots are given headings to fly, to direct them to, and keep their aircraft aligned with the extended centerline of the landing runway. They are told to anticipate glide path interception approximately 10 to 30 seconds before it occurs and when to start descent. The published Decision Height will be given only if the pilot requests it. If the aircraft is observed to deviate above or below the glide path, the pilot is given the relative amount of deviation by use of terms "slightly" or "well" and is expected to adjust his rate of descent to return to the glide path. Trend information is also issued with respect to the elevation of the aircraft and may be modified by the terms "rapidly" and "slowly"; e.g., "well above glide path, coming down rapidly." Range from touchdown is given at least once each mile. If an aircraft is observed by the controller to proceed outside of specified safety zone limits in azimuth and/or elevation and continue to operate outside these prescribed limits, the pilot will be directed to execute a missed approach or to fly a specified course unless he has the runway environment (runway, approach lights, etc.) in sight. Navigational guidance in azimuth and elevation is provided the pilot until the aircraft reaches the published Decision Height (DH). Advisory course and glidepath information is furnished by the controller until the aircraft passes over the landing threshold, at which point the pilot is advised of any deviation from the runway centerline. Radar service is automatically terminated upon completion of the approach.

2. A SURVEILLANCE APPROACH (ASR) is one in

which a controller provides navigational guidance in azimuth only. The pilot is furnished headings to fly to align his aircraft with the extended centerline of the landing runway. Since the radar information used for a surveillance approach is considerably less precise than that used for a precision approach, the accuracy of the approach will not be as great and higher minimums will apply. Guidance in elevation is not possible but the pilot will be advised when to commence descent to the Minimum Descent Altitude (MDA) or, if appropriate, to an intermediate step-down fix Minimum Crossing Altitude and subsequently to the prescribed MDA. In addition, the pilot will be advised of the location of the Missed Approach Point (MAP) prescribed for the procedure and his position each mile on final from the runway, airport or heliport or MAP, as appropriate. If requested by the pilot, recommended altitudes will be issued at each mile, based on the descent gradient established for the procedure, down to the last mile that is at or above the MDA. Normally, navigational guidance will be provided until the aircraft reaches the MAP. Controllers will terminate guidance and instruct the pilot to execute a missed approach unless at the MAP the pilot has the runway, airport or heliport in sight or, for a helicopter point-in-space approach, the prescribed visual reference with the surface is established. Also, if, at any time during the approach the controller considers that safe guidance for the remainder of the approach cannot be provided, he will terminate guidance and instruct the pilot to execute a missed approach. Similarly, guidance termination and missed approach will be effected upon pilot request and, for civil aircraft only, controllers may terminate guidance when the pilot reports the runway, airport/heliport or visual surface route (point-in-space approach) in sight or otherwise indicates that continued guidance is not required. Radar service is automatically terminated at the completion of a radar approach.

5-49c2 NOTE—The published MDA for straight-in approaches will be issued to the pilot before beginning descent. When a surveillance approach will terminate in a circle-to-land maneuver, the pilot must furnish the aircraft approach category to the controller. The controller will then provide the pilot with the appropriate MDA.

5-49c2 NOTE—ASR approaches are not available when an ATC facility is using CENRAP.

3. A NO-GYRO APPROACH is available to a pilot under radar control who experiences circumstances wherein his directional gyro or other stabilized compass is inoperative or inaccurate. When this occurs, he should so advise ATC and request a No- Gyro vector or approach. Pilots of aircraft not equipped with a directional gyro or other stabilized compass who desire radar handling may also request a No-Gyro vector or approach. The pilot should make all turns at standard rate and should execute the turn immediately upon receipt of instructions. For example, "TURN RIGHT," "STOP TURN." When a surveillance or precision approach is made, the pilot will be advised after his aircraft has been turned onto final approach to make turns at half standard rate.

5-50. RADAR MONITORING OF INSTRUMENT APPROACHES

a. PAR facilities operated by the FAA and the military services at some joint-use (civil and military) and military installations monitor aircraft on instrument approaches and issue radar advisories to the pilot when weather is below VFR minimums (1,000 and 3), at night, or when requested by a pilot. This service is provided only when the PAR Final Approach Course coincides with the final approach of the navigational aid and only during the operational hours of the PAR. The radar advisories serve only as a secondary aid since the pilot has selected the navigational aid as the primary aid for the approach.

b. Prior to starting final approach, the pilot will be advised of the frequency on which the advisories will be transmitted. If, for any reason, radar advisories cannot be furnished, the pilot will be so advised.

c. Advisory information, derived from radar observations, includes information on:

1. Passing the final approach fix inbound (non precision approach) or passing the outer marker or fix used in lieu of the outer marker inbound (precision approach).

5-50c1 NOTE—At this point, the pilot may be requested to report sighting the approach lights or the runway.

2. Trend advisories with respect to elevation and/or azimuth radar position and movement will be provided.

5-50c2 NOTE—Whenever the aircraft nears the PAR safety limit, the pilot will be advised that he is well above or below the glidepath or well left or right of course. Glidepath information is given only to those aircraft executing a precision approach, such as ILS or MLS. Altitude information is not transmitted to aircraft executing other than precision approaches because the descent portions of these approaches generally do not coincide with the depicted PAR glidepath. At locations where the MLS glidepath and PAR glidepath are not coincidental, only azimuth monitoring will be provided.

3. If, after repeated advisories, the aircraft proceeds outside the PAR safety limit or if a radical deviation is observed, the pilot will be advised to execute a missed approach unless the prescribed visual reference with the surface is established.

d. Radar service is automatically terminated upon completion of the approach.

5-51. SIMULTANEOUS ILS/MLS APPROACHES

(See Figure 5-51[1].)

a. System: An approach system permitting simultaneous ILS/MLS, or ILS and MLS approaches to airports having parallel runways separated by at least 4,300 feet between centerlines. Integral parts of a total system are ILS or MLS, radar, communications, ATC procedures, and appropriate airborne equipment. The Approach Procedure Chart permitting simultaneous approaches will contain the note "simultaneous approaches authorized Rwys 14L and

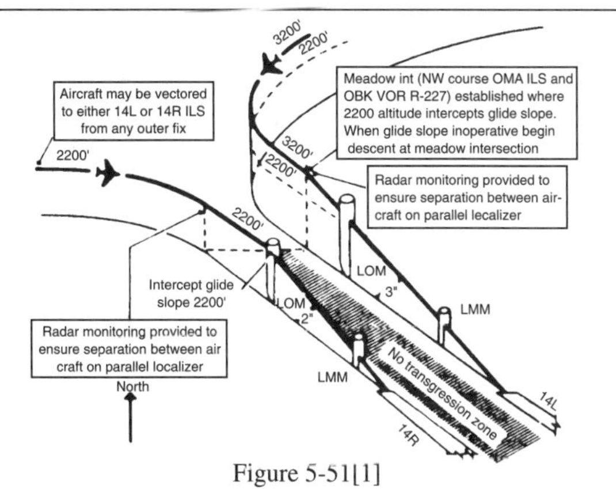

Figure 5-51[1]

14R" identifying the appropriate runways as the case may be. When advised that simultaneous ILS approaches are in progress, pilots shall advise approach control immediately of malfunctioning or inoperative receivers or if simultaneous approach is not desired.

5-51a NOTE—Simultaneous ILS/MLS Approaches are not available when CENRAP is in use.

b. Radar Monitoring: This service is provided for each ILS/MLS approach to insure prescribed lateral separation during simultaneous ILS/MLS approaches. Radar Monitoring includes instructions when an aircraft nears or exceeds the prescribed no transgression zone (an area at least 2,000 feet wide). This service will be provided as follows:

1. The monitor controller will have the capability of overriding the tower controller on the tower frequency.

2. Pilots will be advised to monitor the tower frequency to receive advisories and instructions.

3. Aircraft deviating from either final approach course to the point where the no transgression zone (an area at least 2,000 feet wide) may be penetrated will be instructed to take corrective action. If an aircraft fails to respond to such instruction, the aircraft on the adjacent final approach course may be instructed to alter course.

4. The monitor will automatically be terminated no more than one mile from the runway threshold.

5. The monitor controller will *not* advise when the monitor is terminated.

5-52. PARALLEL ILS/MLS APPROACHES

a. Parallel approaches are an ATC procedure permitting parallel ILS, MLS, or ILS and MLS approaches to airports having parallel runways separated by at least 2,500 feet between centerlines. Integral parts of a total system are ILS or MLS, radar, communications, ATC procedures, and appropriate airborne equipment.

b. A parallel approach differs from a simultaneous approach in that the minimum distance between parallel runway centerlines is reduced; there is no requirement for radar monitoring or advisories; and a staggered separation of aircraft on the adjacent localizer course is required.

c. Aircraft are afforded a minimum of two miles radar separation between successive aircraft on the adjacent localizer course and a minimum of three miles radar separation from aircraft on the same final approach course. In addition, a minimum of 1,000 feet vertical or a minimum of three miles radar separation is provided between aircraft during turn on.

d. Whenever parallel approaches are in progress, aircraft are informed that approaches to both runways are in use. In addition, the radar controller will have the interphone capability of communicating directly with the tower controller where the responsibility for radar separation is not performed by the tower controller.

5-53. SIMULTANEOUS CONVERGING INSTRUMENT APPROACHES

a. ATC may conduct instrument approaches simultaneously to converging runways; i.e., runways having an included angle from 15 to 100 degrees, at airports where a program has been specifically approved to do so.

b. The basic concept requires that dedicated, separate standard instrument approach procedures be developed for each converging runway included. Missed Approach Points must be at least 3 miles apart and missed approach procedures ensure that missed approach protected airspace does not overlap.

c. Other requirements are: radar availability, nonintersecting final approach courses, precision (ILS/MLS) approach systems on each runway and, if runways intersect, controllers must be able to apply visual separation as well as intersecting runway separation criteria. Intersecting runways also require minimums of at least 700 and 2. Straight in approaches and landings must be made.

d. Whenever simultaneous converging approaches are in progress, aircraft will be informed by the controller as soon as feasible after initial contact or via ATIS. Additionally, the radar controller will have direct communications capability with the tower controller where separation responsibility has not been delegated to the tower.

5-54. SIDE-STEP MANEUVER

a. ATC may authorize an approach procedure which serves either one of parallel runways that are separated by 1,200 feet or less followed by a straight-in landing on the adjacent runway.

b. Aircraft that will execute a side-step maneuver will be cleared for a specified approach and landing on the adjacent parallel runway. Example, "cleared ILS runway 7 left approach, side-step to runway 7 right." Pilots are expected to commence the side-step maneuver as soon as possible after the runway or runway environment is in sight.

c. Landing minimums to the adjacent runway will be higher than the minimums to the primary runway, but will normally be lower than the published circling minimums.

5-55. APPROACH AND LANDING MINIMUMS

a. Landing Minimums. The rules applicable to landing minimums are contained in FAR Part 91.175.

b. Published Approach Minimums. Approach minimums are published for different aircraft categories and consist of a minimum altitude (DH, MDA) and required visibility. These minimums are determined by applying the appropriate TERPS criteria. When a fix is incorporated in a nonprecision final segment, two sets of minimums may be published: one, for the pilot that is able to identify the fix, and a second for the pilot that cannot. Two sets of minimums may also be published when a second altimeter source is used in the procedure.

c. Obstacle Clearance. Final approach obstacle clearance is provided from the start of the final segment to the runway or Missed Approach Point, whichever occurs last. Side-step obstacle protection is provided by increasing the width of the final approach obstacle clearance area. Circling approach protected areas are defined by the tangential connection of arcs drawn from each runway end. The arc radii distance differs by aircraft approach category. Because of obstacles near the airport, a portion of the circling area may be restricted by a procedural note: e.g., "Circling NA E of RWY 17-35." Obstacle clearance is provided at the published minimums for the pilot that makes a straight-in approach, side-steps, circles, or executes the missed approach. Missed approach obstacle clearance requirements may dictate the published minimums for the approach. (See Figure 5-55[1].)

d. Straight-in Minimums. Are shown on the IAP when the final approach course is within 30 degrees of the runway alignment and a normal descent can be made from the IFR altitude shown on the IAP to the runway surface. When either the normal rate of descent or the runway alignment factor of 30 degrees is exceeded, a straight-in minimum is not published and a circling minimum applies. The fact that a straight-in minimum is not published does not preclude pilots from landing straight-in if they have the active runway in sight and have sufficient time to make a normal approach for landing. Under such conditions and when ATC has cleared them for landing on that runway, pilots are not expected to circle even though only circling minimums are published. If they desire to circle, they should advise ATC.

e. Side-Step Maneuver Minimums. Landing minimums for a side-step maneuver to the adjacent runway will normally be higher than the minimums to the primary runway.

f. Circling Minimums. In some busy terminal areas, ATC may not allow circling and circling minimums will

CIRCLING APPROACH AREA RADII

Approach Category	Radius (Miles)
A	1.3
B	1.5
C	1.7
D	2.3
E	4.5

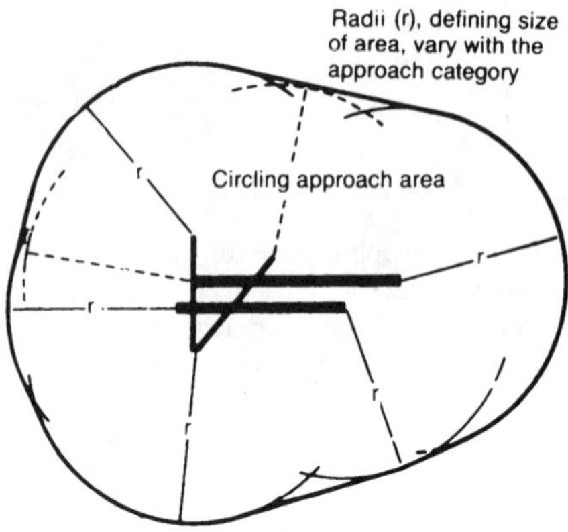

Construction of circling approach area.

Figure 5-55[1]

not be published. Published circling minimums provide obstacle clearance when pilots remain within the appropriate area of protection. Pilots should remain at or above the circling altitude until the aircraft is continuously in a position from which a descent to a landing on the intended runway can be made at a normal rate of descent using normal maneuvers. Circling may require maneuvers at low altitude, at low airspeed, and in marginal weather conditions. Pilots must use sound judgment, have an indepth knowledge of their capabilities, and fully understand the aircraft performance to determine the exact circling maneuver since weather, unique airport design, and the aircraft position, altitude, and airspeed must all be considered. The following basic rules apply:

1. Maneuver the shortest path to the base or downwind leg, as appropriate, considering existing weather conditions. There is no restriction from passing over the airport or other runways.

2. It should be recognized that circling maneuvers may be made while VFR or other flying is in progress at the airport. Standard left turns or specific instruction from the controller for maneuvering must be considered when circling to land.

3. At airports without a control tower, it may be desirable to fly over the airport to observe wind and turn indicators and other traffic which may be on the runway or flying in the vicinity of the airport.

g. Instrument Approach at a Military Field. When instrument approaches are conducted by civil aircraft at military airports, they shall be conducted in accordance with the procedures and minimums approved by the military agency having jurisdiction over the airport.

5-56. MISSED APPROACH

a. When a landing cannot be accomplished, advise ATC and, upon reaching the Missed Approach Point defined on the approach procedure chart, the pilot must comply with the missed approach instructions for the procedure being used or with an alternate missed approach procedure specified by ATC.

b. Protected obstacle clearance areas for missed approach are predicated on the assumption that the abort is initiated at the missed approach point not lower than the MDA or DH. Reasonable buffers are provided for normal maneuvers. However, no consideration is given to an abnormally early turn. Therefore, when an early missed approach is executed, pilots should, unless otherwise cleared by ATC, fly the IAP as specified on the approach plate to the missed approach point at or above the MDA or DH before executing a turning maneuver.

c. If visual reference is lost while circling-to-land from an instrument approach, the missed approach specified for that particular procedure must be followed (unless an alternate missed approach procedure is specified by ATC). To become established on the prescribed missed approach course, the pilot should make an initial climbing turn toward the landing runway and continue the turn until he is established on the missed approach course. Inasmuch as the circling maneuver may be accomplished in more than one direction, different patterns will be required to become established on the prescribed missed approach course, depending on the aircraft position at the time visual reference is lost. Adherence to the procedure will assure that an aircraft will remain within the circling and missed approach obstruction clearance areas. (See Figure 5-56[1].)

d. At locations where ATC Radar Service is provided, the pilot should conform to radar vectors when provided by ATC in lieu of the published missed approach procedure. (See Figure 5-56[2].)

e. When approach has been missed, request clearance for specific action; i.e., to alternative airport, another approach, etc.

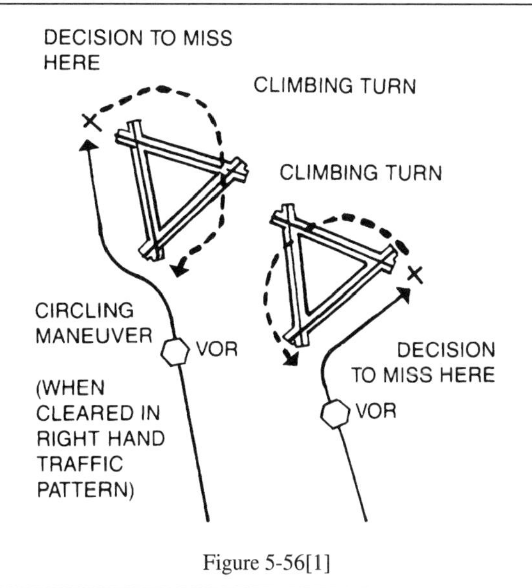

Figure 5-56[1]

MISSED APPROACH -EXAMPLE

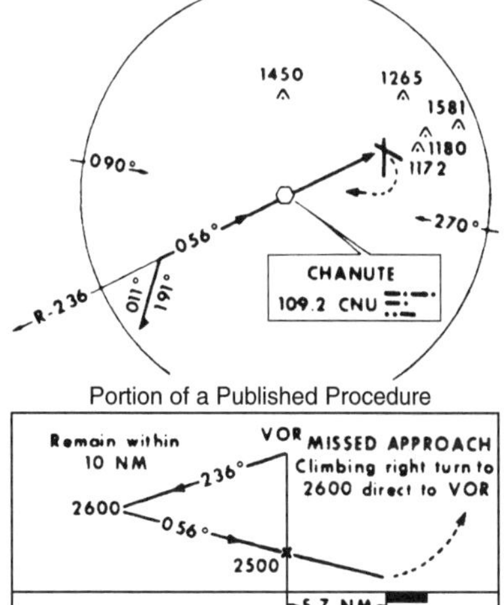

Figure 5-56[2]

5-57. VISUAL APPROACH

a. A visual approach is conducted on an IFR flight plan and authorizes a pilot to proceed visually to the airport. The pilot must have either the airport or the preceding identified aircraft in sight. This approach must be authorized and controlled by the appropriate air traffic control facility. Reported weather at the airport must have a ceiling at or above 1,000 feet and visibility 3 miles or greater. ATC may authorize this type approach when it will be operationally beneficial. Compliance with FAR 91.155 is not required.

b. OPERATING TO AN AIRPORT WITHOUT WEATHER REPORTING SERVICE: ATC will advise the pilot when weather is not available at the destination airport. ATC may initiate a visual approach provided there is a reasonable assurance that weather at the airport is a ceiling at or above 1,000 feet and visibility 3 miles or greater (e.g., area weather reports, PIREPS, etc.).

c. OPERATING TO AN AIRPORT WITH AN OPERATING CONTROL TOWER: Aircraft may be authorized to conduct a visual approach to one runway while other aircraft are conducting IFR or VFR approaches to another parallel, intersecting, or converging runway. When operating to airports with parallel runways separated by less than 2,500 feet, the succeeding aircraft must report sighting the preceding aircraft unless standard separation is being provided by ATC. When operating to parallel runways separated by at least 2,500 feet but less than 4,300 feet, controllers will clear/vector aircraft to the final at an angle not greater than 30 degrees unless radar, vertical, or visual separation is provided during the turn-on. The purpose of the 30 degree intercept angle is to reduce the potential for overshoots of the final and to preclude side-by-side operations with one or both aircraft in a belly-up configuration during the turn-on. Once the aircraft are established within 30 degrees of final, or on the final, these operations may be conducted simultaneously. When the parallel runways are separated by 4,300 feet or more, or intersecting/converging runways are in use, ATC may authorize a visual approach after advising all aircraft involved that other aircraft are conducting operations to the other runway. This may be accomplished through use of the ATIS.

d. SEPARATION RESPONSIBILITIES: If the pilot has the airport in sight but cannot see the aircraft he is following, ATC may clear the aircraft for a visual approach; however, ATC retains both separation and wake vortex separation responsibility. When visually following a preceding aircraft, acceptance of the visual approach clearance constitutes acceptance of pilot responsibility for maintaining a safe approach interval and adequate wake turbulence separation.

e. A visual approach is not an IAP and therefore has no missed approach segment. If a go around is necessary for any reason, aircraft operating at controlled airports will be issued an appropriate advisory/clearance/instruction by the tower. At uncontrolled airports, aircraft are expected to remain clear of clouds and complete a landing as soon as possible. If a landing cannot be accomplished, the aircraft is expected to remain clear of clouds and contact ATC as soon as possible for further clearance. Separation from other IFR aircraft will be maintained under these circumstances.

f. Visual approaches reduce pilot/controller workload and expedite traffic by shortening flight paths to the airport. It is the pilot's responsibility to advise ATC as soon as possible if a visual approach is not desired.

g. Authorization to conduct a visual approach is an IFR authorization and does not alter IFR flight plan cancellation responsibility. (Reference—Canceling IFR Flight Plan, paragraph 5-13.)

h. Radar service is automatically terminated, without advising the pilot, when the aircraft is instructed to change to advisory frequency.

5-58. CHARTED VISUAL FLIGHT PROCEDURES (CVFP)

a. CVFPs are charted visual approaches established at loca ions with jet operations for noise abatement purposes. The approach charts depict prominent landmarks, courses, and recommended altitudes to specific runways.

b. These procedures will be used only in a radar environment at airports with an operating control tower.

c. Most approach charts will depict some NAVID information which is for supplemental navigational guidance only.

d. Unless indicating a Class B airspace floor, all depicted altitudes are for noise abatement purposes and are recommended only. Pilots are not prohibited from flying other than recommended altitudes if operational requirements dictate.

e. When landmarks used for navigation are not visible at night, the approach will be annotated "PROCEDURE NOT AUTHORIZED AT NIGHT."

f. CVFPs usually begin within 15 flying miles from the airport.

g. Published weather minimums for CVFPs are based on minimum vectoring altitudes rather than the recommended altitudes depicted on charts.

h. CVFPs are not instrument approaches and do not have missed approach segments.

i. ATC will not issue clearances for CVFPs when the weather is less than the published minimum.

j. ATC will clear aircraft for a CVFP after the pilot reports siting a charted landmark or a preceding aircraft. If instructed to follow a preceding aircraft, pilots are responsible for maintaining a safe approach interval and wake turbulence separation.

k. Pilots should advise ATC if at any point they are unable to continue an approach or lose sight of a preceding aircraft. Missed approaches will be handled as a go-around.

5-59. CONTACT APPROACH

a. Pilots operating in accordance with an IFR flight plan, provided they are clear of clouds and have at least 1 mile flight visibility and can reasonably expect to continue to the destination airport in those conditions, may request ATC authorization for a contact approach.

b. Controllers may authorize a contact approach provided:

1. The Contact Approach is specifically requested by the pilot. ATC cannot initiate this approach.

EXAMPLE:
REQUEST CONTACT APPROACH.

2. The reported ground visibility at the destination airport is at least 1 statute mile.

3. The contact approach will be made to an airport having a standard or special instrument approach procedure.

4. Approved separation is applied between aircraft so cleared and between these aircraft and other IFR or special VFR aircraft.

EXAMPLE:
CLEARED CONTACT APPROACH (and, if required) AT OR BELOW (altitude) (Routing) IF NOT POSSIBLE (alternative procedures) AND ADVISE.

c. A Contact Approach is an approach procedure that may be used by a pilot (with prior authorization from ATC) in lieu of conducting a standard or special IAP to an airport. It is not intended for use by a pilot on an IFR flight clearance to operate to an airport not having an authorized IAP. Nor is it intended for an aircraft to conduct an instrument approach to one airport and then, when "in the clear," to discontinue that approach and proceed to another airport. In the execution of a contact approach, the pilot assumes the responsibility for obstruction clearance. If radar service is being received, it will automatically terminate when the pilot is told to contact the tower.

5-60. LANDING PRIORITY

A clearance for a specific type of approach (ILS, MLS, ADF, VOR or Straight-in Approach) to an aircraft operating on an IFR flight plan does not mean that landing priority will be given over other traffic. ATCTs handle all aircraft, regardless of the type of flight plan, on a "first-come, first-served" basis. Therefore, because of local traffic or runway in use, it may be necessary for the controller in the interest of safety, to provide a different landing sequence. In any case, a landing sequence will be issued to each aircraft as soon as possible to enable the pilot to properly adjust his flight path.

5-61. OVERHEAD APPROACH MANEUVER

a. Pilots operating in accordance with an instrument flight rules (IFR) flight plan in visual meteorological conditions (VMC) may request Air Traffic Control (ATC) authorization for an overhead maneuver. An overhead maneuver is not an instrument approach procedure. Overhead maneuver patterns are developed at airports where aircraft have an operational need to conduct the maneuver. An aircraft conducting an overhead maneuver is considered to be visual flight rules (VFR) and the IFR flight plan

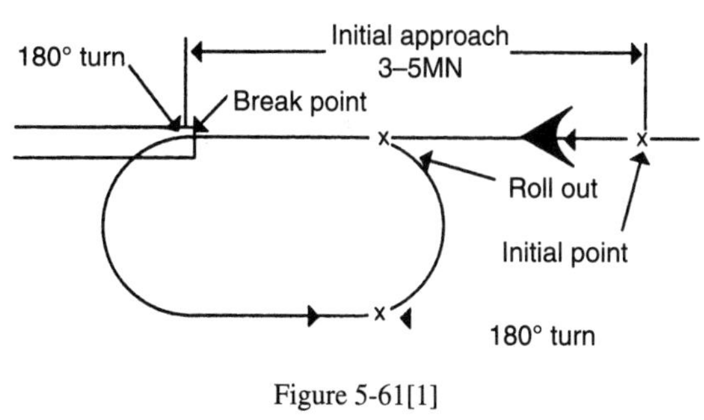

Figure 5-61[1]

is cancelled when the aircraft crosses the landing threshold on the initial approach portion of the maneuver. (See Figure 5-61[1].) The existence of a standard overhead maneuver pattern does not eliminate the possible requirement for an aircraft to conform to conventional rectangular patterns if an overhead maneuver cannot be approved. Aircraft operating to an airport without a functioning control tower must initiate cancellation of an IFR flight plan prior to executing the overhead maneuver. Cancellation of the IFR flight plan must be accomplished after crossing the landing threshold on the initial portion of the maneuver or after landing. Controllers may authorize an overhead maneuver and issue the following to arriving aircraft:

1. Pattern altitude and direction of traffic. This information may be omitted if either is standard.

PHRASEOLOGY:
PATTERN ALTITUDE (altitude). RIGHT TURNS.

2. Request for a report on initial approach.

PHRASEOLOGY:
REPORT INITIAL.

3. "Break" information and a request for the pilot to report. The "Break Point" will be specified if non-standard. Pilots may be requested to report "break" if required for traffic or other reasons.

PHRASEOLOGY:
BREAK AT (Specified point).
REPORT BREAK.

5-62 thru 5-69. RESERVED

Section 5. PILOT/CONTROLLER ROLES AND RESPONSIBILITIES

5-70. GENERAL

a. The roles and responsibilities of the pilot and controller for effective participation in the ATC system are contained in several documents. Pilot responsibilities are in the FARs and the air traffic controller's are in the Air Traffic Control Order (FAA Order 7110.65) and supplemental FAA directives. Additional and supplemental information for pilots can be found in the current Airman's Information Manual (AIM), Notices to Airmen, Advisory Circulars and aeronautical charts. Since there are many other excellent publications produced by nongovernment organizations, as well as other government organizations, with various updating cycles, questions concerning the latest or most current material can be resolved by cross-checking with the above mentioned documents.

b. The pilot in command of an aircraft is directly responsible for, and is the final authority as to the safe operation of that aircraft. In an emergency requiring immediate action, the pilot in command may deviate from any rule in the General Subpart A and Flight Rules Subpart B in accordance with FAR Part 91.3.

c. The air traffic controller is responsible to give first priority to the separation of aircraft and to the issuance of radar safety alerts, second priority to other services that are required, but do not involve separation of aircraft and third priority to additional services to the extent possible.

d. In order to maintain a safe and efficient air traffic system, it is necessary that each party fulfill his responsibilities to the fullest.

e. The responsibilities of the pilot and the controller intentionally overlap in many areas providing a degree of redundancy. Should one or the other fail in any manner, this overlapping responsibility is expected to compensate, in many cases, for failures that may affect safety.

f. The following, while not intended to be all inclusive, is a brief listing of pilot and controller responsibilities for some commonly used procedures or phases of flight. More detailed explanations are contained in other portions of this publication, the appropriate FARs, ACs and similar publications. The information provided is an overview of the principles involved and is not meant as an interpretation of the rules nor is it intended to extend or diminish responsibilities.

5-71. AIR TRAFFIC CLEARANCE

a. Pilot—

1. Acknowledges receipt and understanding of an ATC clearance.

2. Readbacks any hold short of runway instructions issued by ATC.

3. Requests clarification or amendment, as appropriate, any time a clearance is not fully understood or considered unacceptable from a safety standpoint.

4. Promptly complies with an air traffic clearance upon receipt except as necessary to cope with an emergency. Advises ATC as soon as possible and obtains an amended clearance, if deviation is necessary.

5-71a NOTE—A clearance to land means that appropriate separation on the landing runway will be ensured. A landing clearance does not relieve the pilot from compliance with any previously issued altitude crossing restriction.

b. Controller—

1. Issues appropriate clearances for the operation to be conducted, or being conducted, in accordance with established criteria.

2. Assigns altitudes in IFR clearances that are at or above the Minimum IFR Altitudes in controlled airspace.

3. Ensures acknowledgement by the pilot for issued information, clearances, or instructions.

4. Ensures that readbacks by the pilot of altitude, heading, or other items are correct. If incorrect, distorted, or incomplete, makes corrections as appropriate.

5-72. CONTACT APPROACH

a. Pilot—

1. Must request a contact approach and makes it in lieu of a standard or special instrument approach.

2. By requesting the contact approach, indicates that the flight is operating clear of clouds, has at least one mile flight visibility, and reasonably expects to continue to the destination airport in those conditions.

3. Assumes responsibility for obstruction clearance while conducting a contact approach.

4. Advises ATC immediately if unable to continue the contact approach or if encounters less than 1 mile flight visibility.

5. Is aware that if radar service is being received, it may automatically be terminated when told to contact the tower. (Reference—Pilot/Controller Glossary, Radar Service Terminated.)

b. Controller—

1. Issues clearance for a contact approach only when requested by the pilot. Does not solicit the use of this procedure.

2. Before issuing the clearance, ascertains that reported ground visibility at destination airport is at least 1 mile.

3. Provides approved separation between the aircraft cleared for a contact approach and other IFR or special VFR aircraft. When using vertical separation, does not assign a fixed altitude, but clears the aircraft at or below an altitude which is at least 1,000 feet below any IFR traffic but not below Minimum Safe Altitudes prescribed in FAR Part 91.119.

4. Issues alternative instructions if, in his judgment, weather conditions may make completion of the approach impracticable.

5-73. INSTRUMENT APPROACH

a. Pilot—

1. Be aware that the controller issues clearance for approach based only on known traffic.

2. Follows the procedure as shown on the IAP, including all restrictive notations, such as:

(a) Procedure not authorized at night;

(b) Approach not authorized when local area altimeter not available;

(c) Procedure not authorized when control tower not in operation;

(d) Procedure not authorized when glide slope not used;

(e) Straight-in minimums not authorized at night; etc.

(f) Radar required; or

(g) The circling minimums published on the instrument approach chart provide adequate obstruction clearance and the pilot should not descend below the circling altitude until the aircraft is in a position to make final descent for landing. Sound judgment and knowledge of his and the aircraft's capabilities are the criteria for a pilot to determine the exact maneuver in each instance since airport design and the aircraft position, altitude and airspeed must all be considered. (Reference— Approach and Landing Minimums, paragraph 5-55f.)

3. Upon receipt of an approach clearance while on an unpublished route or being radar vectored:

(a) Complies with the minimum altitude for IFR, and

(b) Maintains the last assigned altitude until established on a segment of a published route or IAP, at which time published altitudes apply.

b. Controller—

1. Issues an approach clearance based on known traffic.

2. Issues an IFR approach clearance only after the aircraft is established on a segment of published route or IAP, or assigns an appropriate altitude for the aircraft to maintain until so established.

5-74. MISSED APPROACH

a. Pilot—

1. Executes a missed approach when one of the following conditions exist:

(a) Arrival at the Missed Approach Point (MAP) or the Decision Height (DH) and visual reference to the runway environment is insufficient to complete the landing.

(b) Determined that a safe landing is not possible.

(c) Instructed to do so by ATC.

2. Advises ATC that a missed approach will be made. Include the reason for the missed approach unless the missed approach is initiated by ATC.

3. Complies with the missed approach instructions for the IAP being executed unless other *missed approach* instructions are specified by ATC.

4. If executing a missed approach prior to reaching the MAP or DH, flies the instrument procedure to the MAP at an altitude at or above the Minimum Descent Altitude (MDA) or DH before executing a turning maneuver.

5. Radar vectors issued by ATC when informed that a missed approach is being executed supersedes the previous missed approach procedure.

6. If making a missed approach from a radar approach, executes the missed approach procedure previously given or climbs to the altitude and flies the heading specified by the controller.

7. Following a missed approach, requests clearance for specific action; i.e., another approach, hold for improved conditions, proceed to an alternate airport, etc.

b. Controller—

1. Issues an approved alternate missed approach procedure if it is desired that the pilot execute a procedure other than as depicted on the instrument approach chart.

2. May vector a radar identified aircraft executing a missed approach when operationally advantageous to the pilot or the controller.

3. In response to the pilot's stated intentions, issues a clearance to an alternate airport, to a holding fix, or for reentry into the approach sequence, as traffic conditions permit.

5-75. RADAR VECTORS

a. Pilot—

1. Promptly complies with headings and altitudes assigned to you by the controller.

2. Questions any assigned heading or altitude believed to be incorrect.

3. If operating VFR and compliance with any radar vector or altitude would cause a violation of any FAR, advises ATC and obtains a revised clearance or instructions.

b. Controller—

1. Vectors aircraft in Class A, Class B, Class C, Class D and Class E airspace:

(a) For separation.

(b) For noise abatement.

(c) To obtain an operational advantage for the pilot or controller.

2. Vectors aircraft in Class A, Class B, Class C, Class D, Class E and Class G airspace when requested by the pilot.

3. Vectors IFR aircraft at or above Minimum Vectoring Altitudes.

4. May vector VFR aircraft, not at an ATC assigned altitude, at any altitude. In these cases, terrain separation is the pilot's responsibility.

5-76. SAFETY ALERT

a. Pilot—

1. Initiates appropriate action if a safety alert is received from ATC.

2. Be aware that this service is not always available and that many factors affect the ability of the controller to be aware of a situation in which unsafe proximity to terrain, obstructions, or another aircraft may be developing.

b. Controller—

1. Issues a safety alert if he is aware an aircraft under his control is at an altitude which, in the controller's judgment, places the aircraft in unsafe proximity to terrain, obstructions or another aircraft. Types of safety alerts are:

(a) Terrain or Obstruction Alert—Immediately issued to an aircraft under his control if he is aware the aircraft is at an altitude believed to place the aircraft in unsafe proximity to terrain or obstructions.

(b) Aircraft Conflict Alert—Immediately issued to an aircraft under his control if he is aware of an aircraft not under his control at an altitude believed to place the aircraft in unsafe proximity to each other. With the alert, he offers the pilot an alternative, if feasible.

2. Discontinues further alerts if informed by the pilot that he is taking action to correct the situation or that he has the other aircraft in sight.

5-77. SEE AND AVOID

a. Pilot—When meteorological conditions permit, regardless of type of flight plan or whether or not under control of a radar facility, the pilot is responsible to see and avoid other traffic, terrain, or obstacles.

b. Controller—

1. Provides radar traffic information to radar identified aircraft operating outside positive control airspace on a workload permitting basis.

2. Issues a safety alert to an aircraft under his control if he is aware the aircraft is at an altitude believed to place the aircraft in unsafe proximity to terrain, obstructions, or other aircraft.

5-78. SPEED ADJUSTMENTS

a. Pilot—

1. Advises ATC any time cruising airspeed varies plus or minus 5 percent or 10 knots, whichever is greater, from that given in the flight plan.

2. Complies with speed adjustments from ATC unless:

(a) The minimum or maximum safe airspeed for any particular operation is greater or less than the requested airspeed. In such cases, advises ATC.

5-78a NOTE—It is the pilot's responsibility and prerogative to refuse speed adjustments that he considers excessive or contrary to the aircraft's operating specifications.

(b) Operating at or above 10,000 feet MSL on an ATC assigned SPEED ADJUSTMENT of more than 250 knots IAS and subsequent clearance is received for descent below 10,000 feet MSL. In such cases, pilots are expected to comply with FAR Part 91.117(a).

3. When complying with speed adjustment assignments, maintains an indicated airspeed within plus or minus 10 knots or 0.02 mach number of the specified speed.

b. Controller—

1. Assigns speed adjustments to aircraft when necessary but not as a substitute for good vectoring technique.

2. Adheres to the restrictions published in the Air Traffic Control Order (FAA Order 7110.65) as to when speed adjustment procedures may be applied.

3. Avoids speed adjustments requiring alternate decreases and increases.

4. Assigns speed adjustments to a specified IAS (KNOTS)/mach number or to increase or decrease speed using increments of 10 knots or multiples thereof.

5. Advises pilots to resume normal speed when speed adjustments are no longer required.

6. Gives due consideration to aircraft capabilities to reduce speed while descending.

7. Does not assign speed adjustments to aircraft at or above FL390 without pilot consent.

5-79. TRAFFIC ADVISORIES (Traffic Information)

a. Pilot—

1. Acknowledges receipt of traffic advisories.

2. Informs controller if traffic in sight.

3. Advises ATC if a vector to avoid traffic is desired.

4. Do not expect to receive radar traffic advisories on all traffic. Some aircraft may not appear on the radar display. Be aware that the controller may be occupied with higher priority duties and unable to issue traffic information for a variety of reasons.

5. Advises controller if service not desired.

b. Controller—

1. Issues radar traffic to the maximum extent consistent with higher priority duties except in Class A airspace.

2. Provides vectors to assist aircraft to avoid observed traffic when requested by the pilot.

3. Issues traffic information to aircraft in the Class B, Class C, and Class D surface areas for sequencing purposes.

5-80. VISUAL APPROACH

a. Pilot—

1. If a visual approach is not desired, advises ATC.

2. Complies with controller's instructions for vectors toward the airport of intended landing or to a visual position behind a preceding aircraft.

3. The pilot must, at all times, have either the airport or the preceding aircraft in sight. After being cleared for a visual approach, proceed to the airport in a normal manner or follow the preceding aircraft. Remain clear of clouds while conducting a visual approach.

4. If the pilot accepts a visual approach clearance to visually follow a preceding aircraft, you are required to establish a safe landing interval behind the aircraft you were instructed to follow. You are responsible for wake turbulence separation.

5. Advise ATC immediately if the pilot is unable to continue following the preceding aircraft, cannot remain clear of clouds, or lose sight of the airport.

6. Be aware that radar service is automatically terminated, without being advised by ATC, when the pilot is instructed to change to advisory frequency.

7. Be aware that there may be other traffic in the traffic pattern and the landing sequence may differ from the traffic sequence assigned by approach control or ARTCC.

b. Controller—

1. Do not clear an aircraft for a visual approach unless reported weather at the airport is ceiling at or above 1,000 feet and visibility is 3 miles or greater. When weather is not available for the destination airport, inform the pilot and do not initiate a visual approach to that airport unless there is reasonable assurance that descent and flight to the airport can be made in VFR conditions.

2. Issue visual approach clearance when the pilot reports sighting either the airport or a preceding aircraft which is to be followed.

3. Provide separation except when visual separation is being applied by the pilot.

4. Continue flight following and traffic information until the aircraft has landed or has been instructed to change to advisory frequency.

5. Inform the pilot when the preceding aircraft is a heavy.

6. When weather is available for the destination airport, do not initiate a vector for a visual approach unless the reported ceiling at the airport is 500 feet or more above the MVA and visibility is 3 miles or more. If vectoring weather minima are not available but weather at the airport is ceiling at or above 1,000 feet and visibility of 3 miles or greater, visual approaches may still be conducted.

7. Informs the pilot conducting the visual approach of the aircraft class when pertinent traffic is known to be a heavy aircraft.

5-81. VISUAL SEPARATION

a. Pilot—

1. Acceptance of instructions to follow another aircraft or to provide visual separation from it is an acknowledgment that the pilot will maneuver his/her aircraft as necessary to avoid the other aircraft or to maintain in-trail separation.

2. If instructed by ATC to follow another aircraft or to provide visual separation from it, promptly notify the controller if you lose sight of that aircraft, are unable to maintain continued visual contact with it, or cannot accept the responsibility for your own separation for any reason.

3. The pilot also accepts responsibility for wake turbulence separation under these conditions.

b. Controller—Applies visual separation only:

1. In conjunction with visual approaches.

2. Within the terminal area when a controller has both aircraft in sight or by instructing a pilot who sees the other aircraft to maintain visual separation from it.

3. Within en route airspace when aircraft are on opposite courses and one pilot reports having seen the other aircraft and that the aircraft have passed each other.

5-82. VFR-ON-TOP

a. Pilot—

1. This clearance must be requested by the pilot on an IFR flight plan, and if approved, permits the pilot to select an altitude or Flight Level of his choice (subject to any ATC restrictions) in lieu of an assigned altitude.

5-82a1 NOTE 1—VFR-ON-TOP is not permitted in certain airspace areas, such as positive control airspace, certain restricted areas, etc. Consequently, IFR flights operating VFR-ON-TOP will avoid such airspace.

5-82a1 NOTE 2—Reference—IFR Clearance VFR-ON-TOP, paragraph 4-86, IFR Separation Standards, paragraph 4-89, Position Reporting, paragraph 5-31, and Additional Reports, paragraph 5-32.

2. By requesting a VFR-ON-TOP clearance, the pilot indicates that he is assuming the sole responsibility to be vigilant so as to see and avoid other aircraft and that he will:

(a) Fly at the appropriate VFR altitude as prescribed in FAR Part 91.159.

(b) Comply with the VFR visibility and distance from criteria in FAR Part 91.155 (Basic VFR Weather Minimums).

(c) Comply with instrument flight rules that are applicable to this flight; i.e., minimum IFR altitudes, position reporting, radio communications, course to be flown, adherence to ATC clearance, etc.

3. Should advise ATC prior to any altitude change to ensure the exchange of accurate traffic information.

b. Controller—

1. May clear an aircraft to maintain VFR-ON-TOP if the pilot of an aircraft on an IFR flight plan requests the clearance.

2. Informs the pilot of an aircraft cleared to climb to VFR-ON-TOP the reported height of the tops or that no top report is available; issues an alternate clearance if necessary; and once the aircraft reports reaching VFR-ON-TOP, reclears the aircraft to maintain VFR-ON-TOP.

3. Before issuing clearance, ascertains that the aircraft is not in or will not enter positive control airspace.

5-83. INSTRUMENT DEPARTURES

a. Pilot—

1. Prior to departure considers the type of terrain and other obstructions on or in the vicinity of the departure airport.

2. Determines if obstruction avoidance can be maintained visually or that the departure procedure should be followed.

3. Determines whether a departure procedure and/or SID is available for obstruction avoidance.

4. At airports where IAP's have not been published, hence no published departure procedure, determines what action will be necessary and takes such action that will assure a safe departure.

b. Controller—

1. At locations with airport traffic control service, when necessary, specifies direction of takeoff, turn, or initial heading to be flown after takeoff.

2. At locations without airport traffic control service but within Class E surface area when necessary to specify direction of takeoff, turn, or initial heading to be flown, obtains pilot's concurrence that the procedure will allow him to comply with local traffic patterns, terrain, and obstruction avoidance.

3. Includes established departure procedures as part of

the ATC clearance when pilot compliance is necessary to ensure separation.

5-84. MINIMUM FUEL ADVISORY

a. Pilot—

1. Advise ATC of your minimum fuel status when your fuel supply has reached a state where, upon reaching destination, you cannot accept any undue delay.

2. Be aware this is not an emergency situation, but merely an advisory that indicates an emergency situation is possible should any undue delay occur.

3. On initial contact, the term "minimum fuel" should be used after stating call sign.

EXAMPLE:

SALT LAKE APPROACH, UNITED 621, "MINIMUM FUEL"

4. Be aware a minimum fuel advisory does not imply a need for traffic priority.

5. If the remaining usable fuel supply suggests the need for traffic priority to ensure a safe landing, you should declare an emergency account low fuel and report fuel remaining in minutes. (Reference—Pilot/Controller Glossary, Fuel Remaining.)

b. Controller—

1. When an aircraft declares a state of minimum fuel, relay this information to the facility to whom control jurisdiction is transferred.

2. Be alert for any occurrence which might delay the aircraft.

5-85 thru 5-89. RESERVED

Section 6. NATIONAL SECURITY AND INTERCEPTION PROCEDURES

5-90. NATIONAL SECURITY

a. National security in the control of air traffic is governed by (FAR Part 99).

b. All aircraft entering domestic U.S. airspace from points outside must provide for identification prior to entry. To facilitate early aircraft identification of all aircraft in the vicinity of U.S. and international airspace boundaries, Air Defense Identification Zones (ADIZ) have been established. (Reference—ADIZ Boundaries and Designated Mountainous Areas, paragraph 5-94.)

c. Operational requirements for aircraft operations associated with an ADIZ are as follows:

1. Flight Plan—Except as specified in subparagraphs **d** and **e** below, an IFR or DVFR flight plan must be filed with an appropriate aeronautical facility as follows:

(a) Generally, for all operations that enter an ADIZ.

(b) For operations that will enter or exit the United States and which will operate into, within or across the Contiguous U.S. ADIZ regardless of true airspeed.

(c) The flight plan must be filed before departure except for operations associated with the Alaskan ADIZ when the airport of departure has no facility for filing a flight plan, in which case the flight plan may be filed immediately after takeoff or when within range of the aeronautical facility.

2. Two-way Radio—For the majority of operations associated with an ADIZ, an operating two-way radio is required. See FAR Part 99.1 for exceptions.

3. Transponder Requirements—Unless otherwise authorized by ATC, each aircraft conducting operations into, within, or across the Contiguous U.S. ADIZ must be equipped with an operable radar beacon transponder having altitude reporting capability (Mode C), and that transponder must be turned on and set to reply on the appropriate code or as assigned by ATC.

4. Position Reporting:

(a) For IFR flight—Normal IFR position reporting.

(b) For DVFR flights—The estimated time of ADIZ penetration must be filed with the aeronautical facility at least 15 minutes prior to penetration except for flight in the Alaskan ADIZ, in which case report prior to penetration.

(c) For inbound aircraft of foreign registry—The pilot must report to the aeronautical facility at least one hour prior to ADIZ penetration.

5. Aircraft Position Tolerances:

(a) Over land, the tolerance is within plus or minus five minutes from the estimated time over a reporting point or point of penetration and within 10 NM from the centerline of an intended track over an estimated reporting point or penetration point.

(b) Over water, the tolerance is plus or minus five minutes from the estimated time over a reporting point or point of penetration and within 20 NM from the centerline of the intended track over an estimated reporting point or point of penetration (to include the Aleutian Islands).

d. Except when applicable under FAR Part 99.7, FAR Part 99 does not apply to aircraft operations:

1. Within the 48 contiguous states and the District of Columbia, or within the State of Alaska, and remains within 10 miles of the point of departure;

2. Over any island, or within three nautical miles of the coastline of any island, in the Hawaii ADIZ; or

3. Associated with any ADIZ other than the Contiguous U.S. ADIZ, when the aircraft true airspeed is less than 180 knots.

e. Authorizations to deviate from the requirements of Part 99 may also be granted by the ARTCC, on a local basis, for some operations associated with an ADIZ.

f. An Airfiled VFR Flight Plan makes an aircraft subject to interception for positive identification when entering an ADIZ. Pilots are therefore urged to file the required DVFR flight plan either in person or by telephone prior to departure.

g. Special Security Instructions

1. During defense emergency or air defense emergency conditions, additional special security instructions may be issued in accordance with the Security Control of Air Traffic and Air Navigation Aids (SCATANA) Plan.

2. Under the provisions of the SCATANA Plan, the military will direct the action to be taken—in regard to landing, grounding, diversion, or dispersal of aircraft and the control of air navigation aids in the defense of the U.S. during emergency conditions.

3. At the time a portion or all of SCATANA is implemented, ATC facilities will broadcast appropriate instructions received from the military over available ATC frequencies. Depending on instructions received from the military, VFR flights may be directed to land at the nearest available airport, and IFR flights will be expected to proceed as directed by ATC.

4. Pilots on the ground may be required to file a flight plan and obtain an approval (through FAA) prior to conducting flight operation.

5. In view of the above, all pilots should guard an ATC or FSS frequency at all times while conducting flight operations.

5-91. INTERCEPTION PROCEDURES

a. General

1. Identification intercepts during peacetime operations are vastly different than those conducted under increased states of readiness. Unless otherwise directed by the control agency, intercepted aircraft will be identified by type only. When specific information is required (i.e., markings, serial numbers, etc.) the interceptor aircrew will respond only if the request can be conducted in a safe manner. During hours of darkness or Instrument Meteorological Conditions (IMC), identification of unknown aircraft will be by type only. The interception pattern described below is the typical peacetime method used by air interceptor aircrews. In all situations, the interceptor aircrew will use caution to avoid startling the intercepted aircrew and/or passengers.

b. Intercept phases (See Figure 5-91[1].)

1. Phase One—Approach Phase: During peacetime, intercepted aircraft will be approached from the stern. Generally two interceptor aircraft will be employed to accomplish the identification. The flight leader and his wingman will coordinate their individual positions in conjunction with the ground controlling agency. Their relationship will resemble a line abreast formation. At night or in IMC, a comfortable radar trail tactic will be used. Safe vertical separation between interceptor aircraft and unknown aircraft will be maintained at all times.

2. Phase Two—Identification Phase: The intercepted aircraft should expect to visually acquire the lead intercep-

Interception patterns for identification of intercepted aircraft (typical)

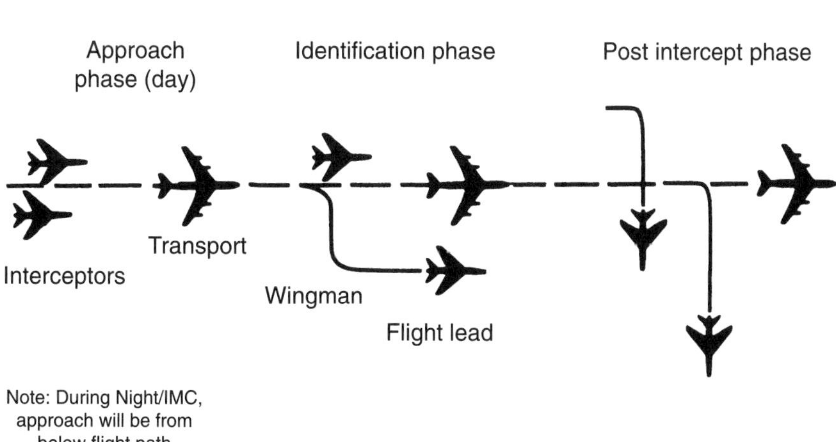

Figure 5-91[1]

tor and possibly the wingman during this phase in visual meteorological conditions (VMC). The wingman will assume a surveillance position while the flight leader approaches the unknown aircraft. Intercepted aircraft personnel may observe the use of different drag devices to allow for speed and position stabilization during this phase. The flight leader will then initiate a gentle closure toward the intercepted aircraft, stopping at a distance no closer than absolutely necessary to obtain the information needed. The interceptor aircraft will use every possible precaution to avoid startling intercepted aircrew or passengers. Additionally, the interceptor aircrews will constantly keep in mind that maneuvers considered normal to a fighter aircraft may be considered hazardous to passengers and crews of nonfighter aircraft. When interceptor aircrews know or believe that an unsafe condition exists, the identification phase will be terminated. As previously stated, during darkness or IMC identification of unknown aircraft will be by type only. Positive vertical separation will be maintained by interceptor aircraft throughout this phase.

3. Phase Three—Post Intercept Phase: Upon identification phase completion, the flight leader will turn away from the intercepted aircraft. The wingman will remain well clear and accomplish a rejoin with his leader.

c. Communication interface between interceptor aircrews and the ground controlling agency is essential to ensure successful intercept completion. Flight Safety is paramount. An aircraft which is intercepted by another aircraft shall immediately:

1. Follow the instructions given by the intercepting aircraft, interpreting and responding to the visual signals.

2. Notify, if possible, the appropriate air traffic services unit.

3. Attempt to establish radio communication with the intercepting aircraft or with the appropriate intercept control unit, by making a general call on the emergency frequency 243.0 MHz and repeating this call on the emergency frequency 121.5 MHz, if practicable, giving the identity and position of the aircraft and the nature of the flight.

4. If equipped with SSR transponder, select MODE 3/A Code 7700, unless otherwise instructed by the appropriate air traffic services unit. If any instructions received by radio from any sources conflict with those given by the intercepting aircraft by visual or radio signals, the intercepted aircraft shall request immediate clarification while continuing to comply with the instructions given by the intercepting aircraft.

5-92. LAW ENFORCEMENT OPERATIONS BY CIVIL AND MILITARY ORGANIZATIONS

a. Special law enforcement operations.

1. Special law enforcement operations include in-flight identification, surveillance, interdiction, and pursuit activities performed in accordance with official civil and/or military mission responsibilities.

2. To facilitate accomplishment of these special missions, exemptions from specified sections of the FAR have been granted to designated departments and agencies. However, it is each organization's responsibility to apprise air traffic control (ATC) of their intent to operate under an authorized exemption before initiating actual operations.

3. Additionally, some departments and agencies that perform special missions have been assigned coded identifiers to permit them to apprise ATC of ongoing mission activities and solicit special air traffic assistance.

5-93. INTERCEPTION SIGNALS

(See Table 5-93[1] and Table 5-93[2].)

5-94. ADIZ BOUNDARIES AND DESIGNATED MOUNTAINOUS AREAS

(See Figure 5-94[1].)

Table 5-93[1]
INTERCEPTION SIGNALS
Signals initiated by intercepting aircraft and responses by intercepted aircraft (as set forth in ICAO Annex 2—Appendix A, 2.1)

Series	INTERCEPTING Aircraft Signals	Meaning	INTERCEPTED Aircraft Responds	Meaning
1	DAY—Rocking wings from a position slightly above and ahead of, and normally to the left of, the intercepted aircraft and, after acknowledgement, a slow level turn, normally to the left, on to the desired heading. NIGHT—Same and, in addition, flashing navigational lights at irregular intervals. Note 1.—Meteorological conditions or terrain may require the intercepting aircraft to take up a position slightly above and ahead of, and to the right of, the intercepted aircraft and to make the subsequent turn to the right. Note 2.—If the intercepted aircraft is not able to keep pace with the intercepting aircraft, the latter is expected to fly a series of race-track patterns and to rock its wings each time it passes the intercepted aircraft.	You have been intercepted. Follow me.	AEROPLANES: DAY—Rocking wings and following. NIGHT—Same and, in addition, flashing navigational lights at irregular intervals. HELICOPTERS: DAY or NIGHT—Rocking aircraft, flashing navigational lights at irregular intervals and following.	Understood, will comply.
2	DAY or NIGHT—An abrupt break-away maneuver from the intercepted aircraft consisting of a climbing turn of 90 degrees or more without crossing the line of flight of the intercepted aircraft.	You may proceed.	AEROPLANES: DAY or NIGHT—Rocking wings. HELICOPTERS: DAY or NIGHT—Rocking aircraft.	Understood, will comply.
3	DAY—Circling aerodrome, lowering landing gear and overflying runway in direction of landing or, if the intercepted aircraft is a helicopter, overflying the helicopter landing area. NIGHT—Same and, in addition, showing steady landing lights.	Land at this aerodrome.	AEROPLANES: DAY—Lowering landing gear, following the intercepting aircraft and, if after overflying the runway landing is considered safe, proceeding to land. NIGHT—Same and, in addition, showing steady landing lights (if carried). HELICOPTERS: DAY or NIGHT—Following the intercepting aircraft and proceeding to land, showing a steady landing light (if carried).	Understood, will comply.

Table 5-93[2]
INTERCEPTION SIGNALS
Signals and Responses During Aircraft Intercept
Signals initiated by intercepted aircraft and responses by intercepting aircraft
(as set forth in ICAO Annex 2—Appendix A, 2.2)

Series	INTERCEPTED Aircraft Signals	Meaning	INTERCEPTING Aircraft Responds	Meaning
4	AEROPLANES: DAY—Raising landing gear while passing over landing runway at a height exceeding 300m (1,000 ft) but not exceeding 600m (2,000 ft) above the aerodrome level, and continuing to circle the aerodrome. NIGHT—Flashing landing lights while passing over landing runway at a height exceeding 300m (1,000 ft) but not exceeding 600m (2,000 ft) above the aerodrome level, and continuing to circle the aerodrome. If unable to flash landing lighs, flash any other lights available.	Aerodrome you have designated is inadequate.	DAY or NIGHT—If it is desired that the intercepted aircraft follow the intercepting aircraft to an alternate aerodrome, the intercepting aircraft raises its landing gear and uses the Series 1 signals prescribed for intercepting aircraft. If it is decided to release the intercepted aircraft, the intercepting aircraft uses the Series 2 signals prescribed for intercepting aircraft.	Understood, follow me. Understood, you may proceed.
5	AEROPLANES: DAY or NIGHT—Regular switching on and off of all available lights but in such a manner as to be distinct from flashing lights.	Cannot comply.	DAY or NIGHT—Use Series 2 signals prescribed for intercepting aircraft.	Understood.
6	AEROPLANES: DAY or NIGHT—Irregular flashing of all available lights. HELICOPTERS: DAY or NIGHT—Irregular flashing of all available lights.	In distress.	DAY or NIGHT—Use Series 2 signals prescribed for intercepting aircraft.	Understood.

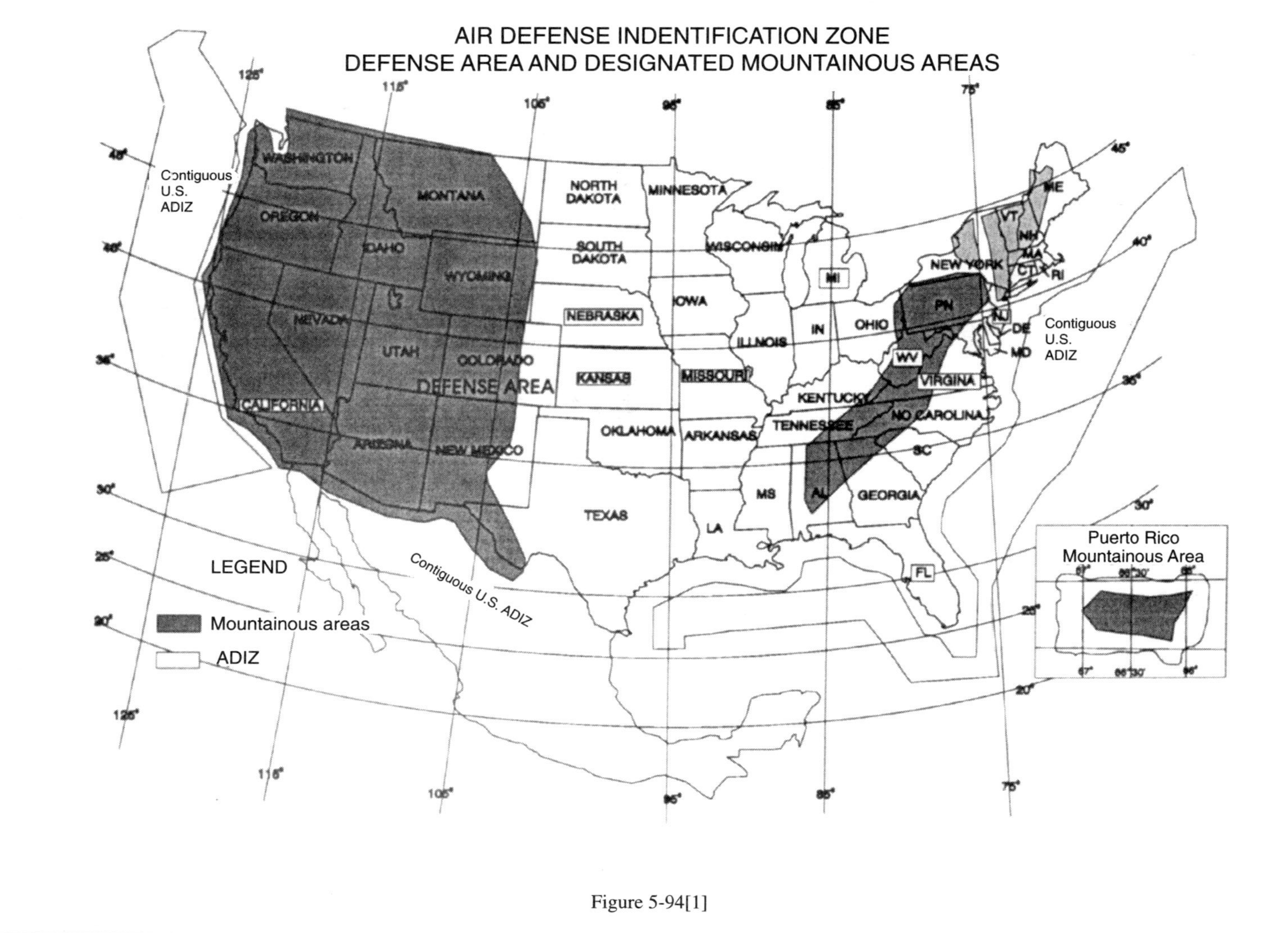
AIR DEFENSE INDENTIFICATION ZONE
DEFENSE AREA AND DESIGNATED MOUNTAINOUS AREAS
Contiguous U.S. ADIZ
Contiguous U.S. ADIZ
Contiguous U.S. ADIZ
Puerto Rico Mountainous Area
DEFENSE AREA
LEGEND
Mountainous areas
ADIZ
WASHINGTON
OREGON
CALIFORNIA
NEVADA
IDAHO
MONTANA
WYOMING
UTAH
ARIZONA
COLORADO
NEW MEXICO
NORTH DAKOTA
SOUTH DAKOTA
NEBRASKA
KANSAS
OKLAHOMA
TEXAS
MINNESOTA
IOWA
MISSOURI
ARKANSAS
LA
WISCONSIN
ILLINOIS
MS
MI
IN
KENTUCKY
TENNESSEE
AL
OHIO
GEORGIA
FL
SC
NO CAROLINA
VIRGINIA
WV
PA
NEW YORK
NJ
DE
MD
VT
NH
MA
CT
RI
ME

Figure 5-94[1]

Chapter 6. Emergency procedures
Section 1. GENERAL

6-1. PILOT RESPONSIBILITY AND AUTHORITY

a. The pilot in command of an aircraft is directly responsible for and is the final authority as to the operation of that aircraft. In an emergency requiring immediate action, the pilot in command may deviate from any rule in the FAR, Subpart A, General, and Subpart B, Flight Rules, to the extent required to meet that emergency (FAR Part 91.3(b)).

b. If the emergency authority of FAR Part 91.3.(b) is used to deviate from the provisions of an ATC clearance, the pilot in command must notify ATC as soon as possible and obtain an amended clearance.

c. Unless deviation is necessary under the emergency authority of FAR Part 91.3, pilots of IFR flights experiencing two-way radio communications failure are expected to adhere to the procedures prescribed under "IFR operations, two-way radio communications failure" (FAR Part 91.185).

6-2. EMERGENCY CONDITION—REQUEST ASSISTANCE IMMEDIATELY

a. An emergency can be either a *distress* or *urgency* condition as defined in the Pilot/Controller Glossary. Pilots do not hesitate to declare an emergency when they are faced with *distress* conditions such as fire, mechanical failure, or structural damage. However, some are reluctant to report an *urgency* condition when they encounter situations which may not be immediately perilous, but are potentially catastrophic. An aircraft is in at least an *urgency* condition the moment the pilot becomes doubtful about position, fuel endurance, weather, or any other condition that could adversely affect flight safety. This is the time to ask for help, not after the situation has developed into a *distress* condition.

b. Pilots who become apprehensive for their safety for any reason should *request assistance immediately*. Ready and willing help is available in the form of radio, radar, direction finding stations and other aircraft. Delay has caused accidents and cost lives. *Safety is not a luxury! Take action!*

6-3 thru 6-9. RESERVED

Section 2. EMERGENCY SERVICES AVAILABLE TO PILOTS

6-10. RADAR SERVICE FOR VFR AIRCRAFT IN DIFFICULTY

a. Radar equipped ATC facilities can provide radar assistance and navigation service (vectors) to VFR aircraft in difficulty when the pilot can talk with the controller, and the aircraft is within radar coverage. Pilots should clearly understand that authorization to proceed in accordance with such radar navigational assistance does not constitute authorization for the pilot to violate FARs. In effect, assistance is provided on the basis that navigational guidance information is advisory in nature, and the responsibility for flying the aircraft safely remains with the pilot.

b. Experience has shown that many pilots who are not qualified for instrument flight cannot maintain control of their aircraft when they encounter clouds or other reduced visibility conditions. In many cases, the controller will not know whether flight into instrument conditions will result from his instructions. To avoid possible hazards resulting from being vectored into IFR conditions, a pilot in difficulty should keep the controller advised of the weather conditions in which he is operating and the weather along the course ahead and observe the following:

1. If a course of action is available which will permit flight and a safe landing in VFR weather conditions, noninstrument rated pilots should choose the VFR condition rather than requesting a vector or approach that will take them into IFR weather conditions; or

2. If continued flight in VFR conditions is not possible, the noninstrument rated pilot should so advise the controller and indicating the lack of an instrument rating, declare a *distress* condition, or

3. If the pilot is instrument rated and current, and the aircraft is instrument equipped, the pilot should so indicate by requesting an IFR flight clearance. Assistance will then be provided on the basis that the aircraft can operate safely in IFR weather conditions.

6-11. TRANSPONDER EMERGENCY OPERATION

a. When a *distress* or *urgency* condition is encountered, the pilot of an aircraft with a coded radar beacon transponder, who desires to alert a ground radar facility, should squawk MODE 3/A, Code 7700/Emergency and MODE C altitude reporting and then immediately establish communications with the ATC facility.

b. Radar facilities are equipped so that Code 7700 normally triggers an alarm or special indicator at all control positions. Pilots should understand that they might not be within a radar coverage area. Therefore, they should continue squawking Code 7700 and establish radio communications as soon as possible.

6-12. DIRECTION FINDING INSTRUMENT APPROACH PROCEDURE

a. DF equipment has long been used to locate lost aircraft and to guide aircraft to areas of good weather or to airports. Now at most DF equipped airports, DF instrument approaches may be given to aircraft in a *distress* or *urgency* condition.

b. Experience has shown that most emergencies requiring DF assistance involve pilots with little flight experience. With this in mind, DF approach procedures provide maximum flight stability in the approach by using small turns, and wings-level descents. The DF specialist will give the pilot headings to fly and tell the pilot when to begin descent.

c. DF IAPs are for emergency use only and will not be used in IFR weather conditions unless the pilot has declared a *distress* or *urgency* condition.

d. To become familiar with the procedures and other benefits of DF, pilots are urged to request practice DF guidance and approaches in VFR weather conditions. DF specialists welcome the practice and will honor such requests, workload permitting.

6-13. INTERCEPT AND ESCORT

a. The concept of airborne intercept and escort is based on the Search and Rescue (SAR) aircraft establishing visual and/or electronic contact with an aircraft in difficulty, providing in-flight assistance, and escorting it to a safe landing. If bailout, crash landing or ditching becomes necessary, SAR operations can be conducted without delay. For most incidents, particularly those occurring at night and/or during instrument flight conditions, the availability of intercept and escort services will depend on the proximity of SAR units with suitable aircraft on alert for immediate dispatch. In limited circumstances, other aircraft flying in the vicinity of an aircraft in difficulty can provide these services.

b. If specifically requested by a pilot in difficulty or if a *distress* condition is declared, SAR coordinators *will* take steps to intercept and escort an aircraft. Steps *may* be initiated for intercept and escort if an *urgency* condition is declared and unusual circumstances make such action advisable.

c. It is the pilot's prerogative to refuse intercept and escort services. Escort services will normally be provided to the nearest adequate airport. Should the pilot receiving escort services continue onto another location after reaching a safe airport, or decide not to divert to the nearest safe airport, the escort aircraft is not obligated to continue and further escort is discretionary. The decision will depend on the circumstances of the individual incident.

6-14. EMERGENCY LOCATOR TRANSMITTERS

a. GENERAL. Emergency Locator Transmitters (ELTs) are required for most General Aviation airplanes (FAR Part 91.207). ELTs of various types have been developed as a means of locating downed aircraft. These electronic, battery operated transmitters emit a distinctive downward swept audio tone on 121.5 MHz and 243.0 MHz. If "armed" and when subject to crash generated forces they are designed to automatically activate and continuously emit these signals. The transmitters will operate continuously for at least 48 hours over a wide temperature range. A properly installed and maintained ELT can expedite search and rescue operations and save lives.

b. TESTING. ELTs should be tested in accordance with the manufacturer's instructions, preferably in a shielded or screened room to prevent the broadcast of signals which could trigger a false alert. When this cannot be done, aircraft operational testing is authorized on 121.5 MHz and 243.0 MHz as follows:

1. Tests should be conducted only during the first 5 minutes after any hour. If operational tests must be made outside of this time-frame, they should be coordinated with the nearest FAA Control Tower or FSS.

2. Tests should be no longer than three audible sweeps.

3. If the antenna is removable, a dummy load should be substituted during test procedures.

4. Airborne tests are not authorized.

c. FALSE ALARMS. Caution should be exercised to prevent the inadvertent activation of ELTs in the air or while they are being handled on the ground. Accidental or unauthorized activation will generate an emergency signal that cannot be distinguished from the real thing, leading to expensive and frustrating searches. A false ELT signal could also interfere with genuine emergency transmissions and hinder or prevent the timely location of crash sites. Frequent false alarms could also result in complacency and decrease the vigorous reaction that must be attached to all ELT signals. Numerous cases of inadvertent activation have occurred as a result of aerobatics, hard landings, movement by ground crews and aircraft maintenance. These false alarms can be minimized by monitoring 121.5 MHz and/or 243.0 MHz as follows:

1. In flight when a receiver is available.

2. Prior to engine shut down at the end of each flight.

3. When the ELT is handled during installation or maintenance.

4. When maintenance is being performed in the vicinity of the ELT.

5. When the aircraft is moved by a ground crew.

6. If an ELT signal is heard, turn off the ELT to determine if it is transmitting. If it has been activated, maintenance might be required before the unit is returned to the "ARMED" position.

d. IN-FLIGHT MONITORING AND REPORTING. Pilots are encouraged to monitor 121.5 MHz and/or 243.0 MHz while in-flight to assist in identifying possible emergency ELT transmissions. On receiving a signal, report the following information to the nearest air traffic facility:

1. Your position at the time the signal was first heard.

2. Your position at the time the signal was last heard.

3. Your position at maximum signal strength.

4. Your flight altitudes and frequency on which the emergency signal was heard—121.5 MHz or 243.0 MHz. If possible, positions should be given relative to a navigation aid. If the aircraft has homing equipment, provide the bearing to the emergency signal with each reported position.

6-15. FAA SPONSORED EXPLOSIVES DETECTION (DOG/HANDLER TEAM) LOCATIONS

a. At many of our major airports a program has been established by the FAA to make available explosives detection dog/handler teams. The dogs are trained by the Air Force and the overall program is run by FAA's Office of Civil Aviation Security. Local police departments are the caretakers of the dogs and are allowed to use the dogs in their normal police patrol functions. The local airport, however, has first call on the teams' services. The explosives detection teams were established so that no aircraft in flight is more than 1 hour from an airport at which it can be searched if a bomb threat is received. The following list contains those locations that presently have a team in existence. This list will be updated as more teams are established. If you desire this service, notify your company or an FAA facility.

b. Team Locations: (See Table 6-15[1].)

Table 6-15[1]

Airport Symbol	Location
ATL	Atlanta, Georgia
BHM	Birmingham, Alabama
BOS	Boston, Massachusetts
BUF	Buffalo, New York
ORD	Chicago, Illinois
CVG	Cincinnati, Ohio
COS	Colorado Springs, Colorado
DFW	Dallas, Texas
DTW	Detroit, Michigan
IAH	Houston, Texas
JAX	Jacksonville, Florida
MCI	Kansas City, Missouri
LAX	Los Angeles, California
MEM	Memphis, Tennessee
MIA	Miami, Florida
MKE	Milwaukee, Wisconsin
PHX	Phoenix, Arizona
PIT	Pittsburgh, Pennsylvania
PDX	Portland, Oregon
SLC	Salt Lake City, Utah
SAN	San Diego, California
SFO	San Francisco, California
SJU	San Juan, Puerto Rico
SEA	Seattle, Washington
STL	St. Louis, Missouri
TUS	Tucson, Arizona
TUL	Tulsa, Oklahoma

c. If due to weather or other considerations an aircraft with a suspected hidden explosive problem were to land or intended to land at an airport other than those listed in b., it is recommended that they call the FAA's Washington Operations Center (telephone 202-426-3333, if appropriate) or have an air traffic facility with which you can communicate contact the above center requesting assistance.

6-16. SEARCH AND RESCUE

a. GENERAL. SAR is a lifesaving service provided through the combined efforts of the federal agencies signatory to the National SAR Plan, and the agencies responsible for SAR within each state. Operational resources are provided by the U.S. Coast Guard, DOD components, the Civil Air Patrol, the Coast Guard Auxiliary, state, county and local law enforcement and other public safety agencies, and private volunteer organizations. Services include search for missing aircraft, survival aid, rescue, and emergency medical help for the occupants after an accident site is located.

b. NATIONAL SEARCH AND RESCUE PLAN. By federal interagency agreement, the National Search and Rescue Plan provides for the effective use of all available facilities in all types of SAR missions. These facilities include aircraft, vessels, pararescue and ground rescue teams, and emergency radio fixing. Under the Plan, the U.S. Coast Guard is responsible for the coordination of SAR in the Maritime Region, and the USAF is responsible in the Inland Region. To carry out these responsibilities, the Coast Guard and the Air Force have established Rescue Coordination Centers (RCCs) to direct SAR activities within their Regions. For aircraft emergencies, distress, and urgency, information normally will be passed to the appropriate RCC through an ARTCC or FSS.

c. COAST GUARD RESCUE COORDINATION CENTERS.

(See Table 6-16[1].)

Table 6-16[1]
Coast Guard Rescue Coordination Centers

Boston, MA 617-223-8555	Long Beach, CA 213-590-2225 310-499-5380
New York, NY 212-668-7055	San Francisco, CA 415-437-3700
Portsmouth, VA 804-398-6231	Seattle, WA 206-553-5886
Miami, FL 305-536-5611	Juneau, AK 907-463-2000
New Orleans, LA 504-589-6225	Honolulu, HI 808-541-2500
Cleveland, OH 216-522-3984	San Juan, Puerto Rico 809-729-6770
St. Louis, MO 314-262-3706	

d. AIR FORCE RESCUE COORDINATION CENTERS.

Air Force Rescue Coordination Center—48 Contiguous States. (See Table 6-16[2]).

Table 6-16[2]
Air Force Rescue Coordination Centers

Scott AFB, Illinois	Phone
Commercial	618-256-4815
WATS	800-851-3051
DSN	576-4815

Alaskan Air Command Rescue Coordination Center—Alaska. (See Table 6-16[3]).

Table 6-16[3]
Alaskan Air Command Rescue Coordination Center

Elmendorf AFB, Alaska	Phone
Commercial	907-552-5375
DSN	317-552-2426

e. JOINT RESCUE COORDINATION CENTERS

Honolulu Joint Rescue Coordination Center—Hawaii. (See Table 6-16[4]).

Table 6-16[4]
Honolulu Joint Rescue Coordination Center

HQ 14th CG District Honolulu	Phone
Commercial	808-541-2500
DSN	448-0301

f. OVERDUE AIRCRAFT.

1. ARTCCs and FSSs will alert the SAR system when information is received from any source that an aircraft is in difficulty, overdue, or missing. *A filed flight plan is the most timely and effective indicator that an aircraft is overdue.* Flight plan information is invaluable to SAR forces for search planning and executing search efforts.

2. Prior to departure on every flight, local or otherwise, someone at the departure point should be advised of your destination and route of flight if other than direct. Search efforts are often wasted and rescue is often delayed because of pilots who thoughtlessly take off without telling anyone where they are going. File a flight plan for *your* safety.

3. According to the National Search and Rescue Plan, "The life expectancy of an injured survivor decreases as much as 80 percent during the first 24 hours, while the chances of survival of uninjured survivors rapidly diminishes after the first 3 days."

4. An Air Force Review of 325 SAR missions conducted during a 23-month period revealed that "Time works against people who experience a *distress* but are not on a flight plan, since 36 hours normally pass before family concern initiates an (alert)."

g. VFR SEARCH AND RESCUE PROTECTION.

1. To receive this valuable protection, *file a VFR or DVFR Flight Plan* with an FAA FSS. For maximum protection, file only to the point of first intended landing, and refile for each leg to final destination. When a lengthy flight plan is filed, with several stops en route and an ETE to final destination, a mishap could occur on any leg, and unless other information is received, it is probable that no one would start looking for you until 30 minutes after your ETA at your final destination.

2 If you land at a location other than the intended destination, report the landing to the nearest FAA FSS and advise them of your original destination.

3. If you land en route and are delayed more than 30 minutes, report this information to the nearest FSS and give them your original destination.

4. If your ETE changes by 30 minutes or more, report a new ETA to the nearest FSS and give them your original destination. Remember that if you fail to respond within one-half hour after your ETA at final destination, a search will be started to locate you.

5. It is important that you *close your flight plan IMMEDIATELY AFTER ARRIVAL AT YOUR FINAL DESTINATION WITH THE FSS DESIGNATED WHEN YOUR FLIGHT PLAN WAS FILED. The pilot is responsible for closure of a VFR or DVFR flight plan; they are not closed automatically.* This will prevent needless search efforts.

6. The rapidity of rescue on land or water will depend on how accurately your position may be determined. If a flight plan has been followed and your position is on course, rescue will be expedited.

h. SURVIVAL EQUIPMENT.

1. For flight over uninhabited land areas, it is wise to take and know how to use survival equipment for the type of climate and terrain.

2. If a forced landing occurs at sea, chances for survival are governed by the degree of crew proficiency in emergency procedures and by the availability and effectiveness of water survival equipment.

HOW TO USE THEM

If you are forced down and are able to attract the attention of the pilot of a rescue airplane, the body signals illustrated on these pages can be used to transmit messages to him as he circles over your location. Stand in the open when you make the signals. Be sure the background, as seen from the air, is not confusing. Go through the motions slowly and repeat each signal until you are positive that the pilot understands you.

(See Figure 6-16[1][Ground-Air Visual Code for Use by Survivors].)

GROUND-AIR VISUAL CODE FOR USE BY SURVIVORS

NO.	MESSAGE	CODE SYMBOL
1	Require assistance	V
2	Require medical assistance	X
3	No or Negative	N
4	Yes or Affirmative	Y
5	Proceeding in this direction	↑

IF IN DOUBT, USE INTERNATIONAL SYMBOL **SOS**

INSTRUCTIONS

1. Lay out symbols by using strips of fabric or parachutes, pieces of wood, stones, or any available material.
2. Provide as much color contrast as possible between material used for symbols and background against which symbols are exposed.
3. Symbols should be at least 10 feet high or larger. Care should be taken to lay out symbols exactly as shown.
4. In addition to using symbols, every effort is to be made to attract attention by means of radio, flares, smoke, or other available means.
5. On snow covered ground, signals can be made by dragging, shoveling or tramping. Depressed areas forming symbols will appear black from the air.
6. Pilot should acknowledge message by rocking wings from side to side.

Figure 6-16[1]

(See Figure 6-16[2][Ground-Air Visual Code for use by Ground Search Parties].)

GROUND-AIR VISUAL CODE FOR USE BY GROUND SEARCH PARTIES

NO.	MESSAGE	CODE SYMBOL
1	Operation completed.	L L L
2	We have found all personnel.	L L (underlined)
3	We have found only some personnel.	++
4	We are not able to continue. Returning to base.	X X
5	Have divided into two groups. Each proceeding in direction indicated.	⇄
6	Information received that aircraft is in this direction.	→ →
7	Nothing found. Will continue search.	N N

Note: These visual signals have been accepted for international use and appear in Annex 12 to the Convention on International Civil Aviation.

Figure 6-16[2]

(See Figure 6-16[3][Urgent Medical Assistance].)

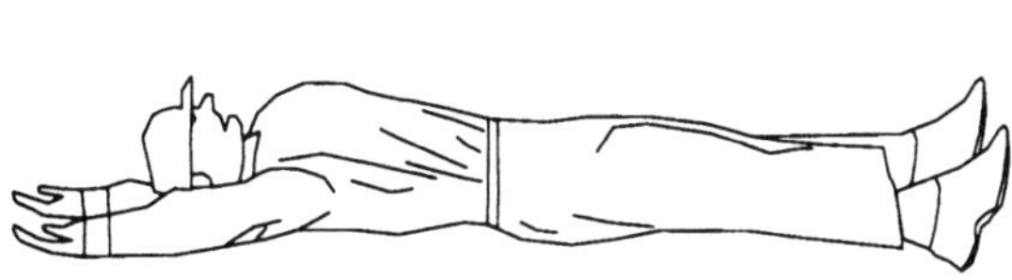

NEED MEDICAL
ASSISTANCE-URGENT

Used only when life is at stake

Figure 6-16[3]

(See Figure 6-16[4][All OK].)

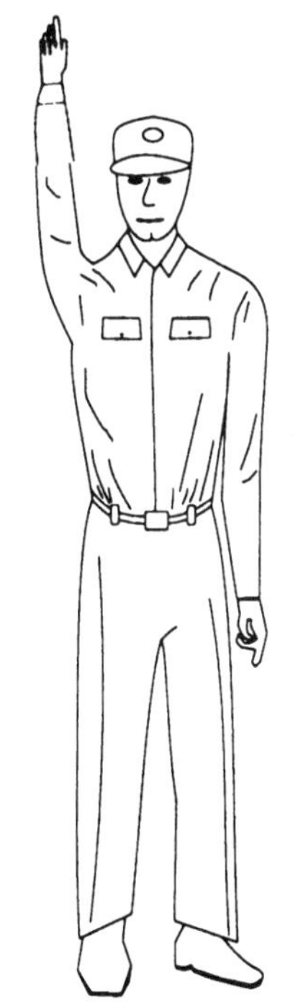

ALL OK-DO NOT WAIT

Wave one arm overhead

Figure 6-16[4]

(See Figure 6-16[5][Short Delay].)

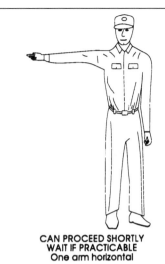

CAN PROCEED SHORTLY
WAIT IF PRACTICABLE
One arm horizontal

Figure 6-16[5]

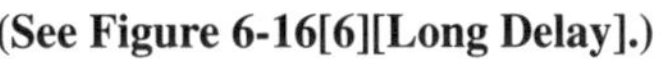

(See Figure 6-16[6][Long Delay].)

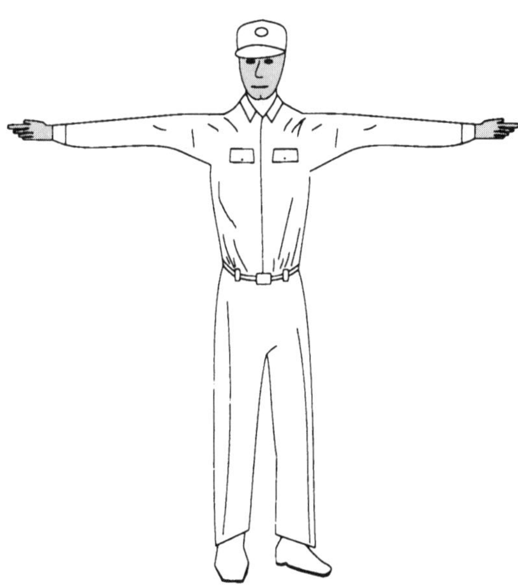

NEED MECHANICAL HELP
OR PARTS - LONG DELAY
Both arms horizontal

Figure 6-16[6]

(See Figure 6-16[7][Drop Message].)

Figure 6-16[7]

(See Figure 6-16[9][Do Not Land Here].)

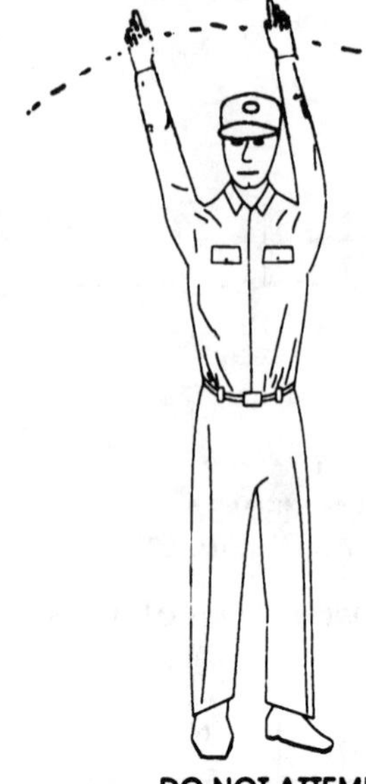

DO NOT ATTEMPT TO LAND HERE
Both arms waved across face

Figure 6-16[9]

(See Figure 6-16[8][Receiver Operates].)

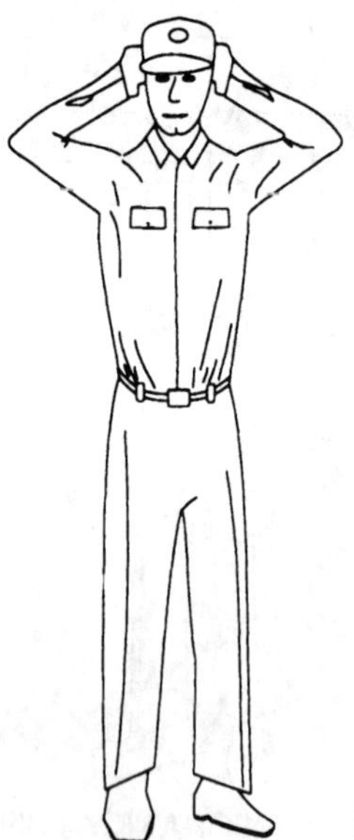

OUR RECEIVER IS OPERATING
Cup hands over ears

Figure 6-16[8]

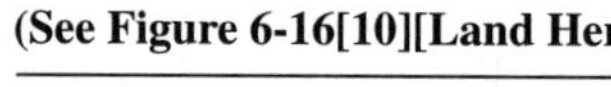

(See Figure 6-16[10][Land Here].)

LAND HERE

Both arms forward horizontally, squatting and point in direction of landing - Repeat

Figure 6-16[10]

(See Figure 6-16[11][Negative (Ground)].)

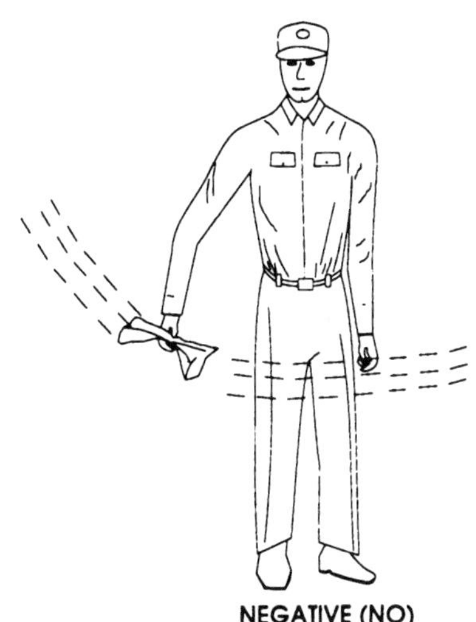

Figure 6-16[11]

(See Figure 6-16[12][Affirmative (Ground)].)

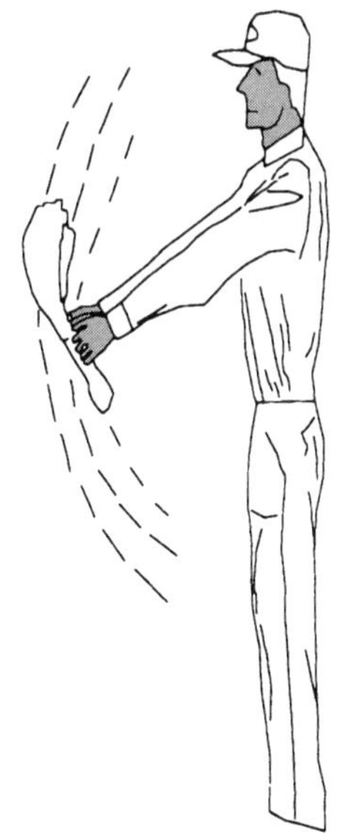

Figure 6-16[12]

(See Figure 6-16[13][Pick Us Up].)

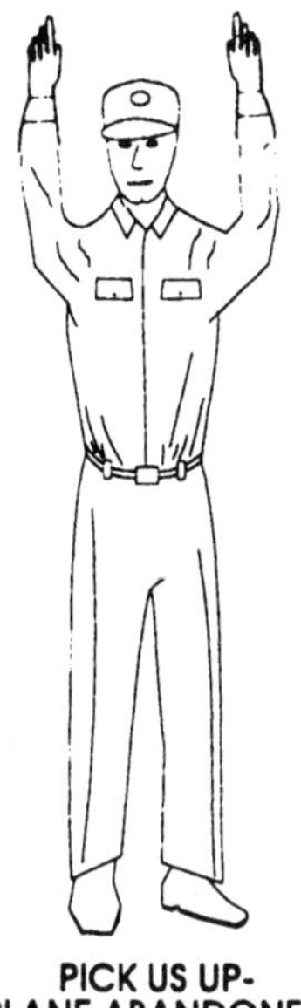

Figure 6-16[13]

(See Figure 6-16[14][Affirmative (Aircraft)].)

Affirmative reply from aircraft:

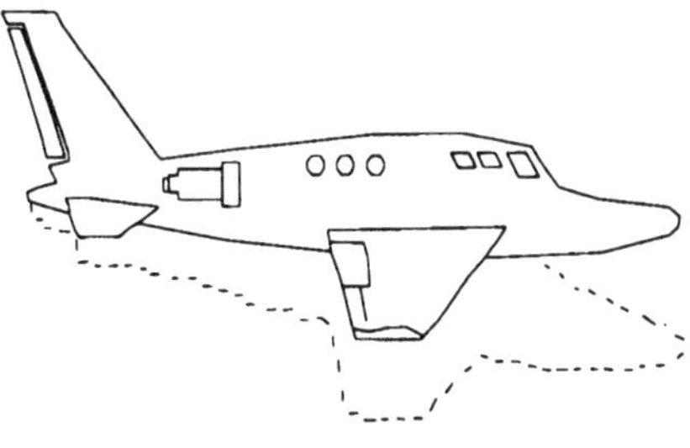

AFFIRMATIVE (YES)

Dip nose of plane several times

Figure 6-16[14]

(See Figure 6-16[15][Negative (Aircraft)].)

Negative reply from aircraft:

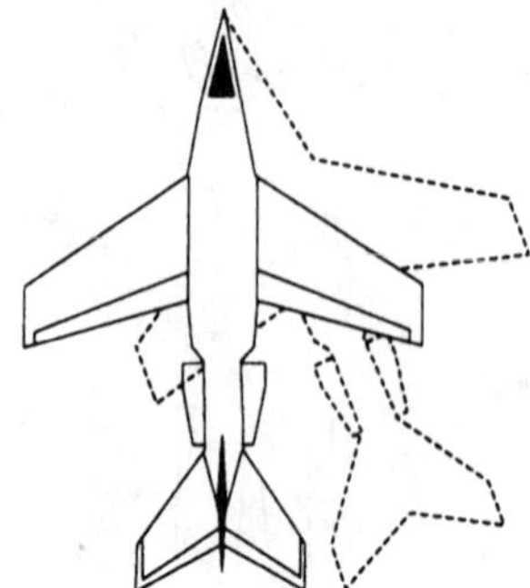

NEGATIVE (NO)

Fishtail plane

Figure 6-16[15]

(See Figure 6-16[16][Message received and understood (Aircraft)].)

Message received and understood by aircraft:
Day or moonlight - Rocking wings
Night - Green flashed from signal lamp

Figure 6-16[16]

(See Figure 6-16[17][Message received and NOT understood (Aircraft)].)

Message received and NOT understood by aircraft:
Day or moonlight - Making a complete right-hand circle
Night-Red flashes from signal lamp.

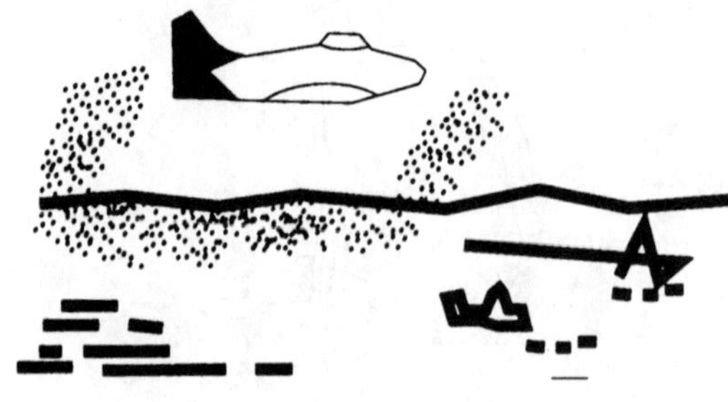

Figure 6-16[17]

i. OBSERVANCE OF DOWNED AIRCRAFT.

1. Determine if crash is marked with a yellow cross; if so, the crash has already been reported and identified.

2. If possible, determine type and number of aircraft and whether there is evidence of survivors.

3. Fix the position of the crash as accurately as possible with reference to a navigational aid. If possible provide geographic or physical description of the area to aid ground search parties.

4. Transmit the information to the nearest FAA or other appropriate radio facility.

5. If circumstances permit, orbit the scene to guide in other assisting units until their arrival or until you are relieved by another aircraft.

6. Immediately after landing, make a complete report to the nearest FAA facility, or Air Force or Coast Guard Rescue Coordination Center. The report can be made by long distance collect telephone.

6-17 thru 6-19. RESERVED

Section 3. DISTRESS AND URGENCY PROCEDURES

6-20. DISTRESS AND URGENCY COMMUNICATIONS

a. A pilot who encounters a *distress* or *urgency* condition can obtain assistance simply by contacting the air traffic facility or other agency in whose area of responsibility the aircraft is operating, stating the nature of the difficulty, pilot's intentions and assistance desires. *Distress* and *urgency* communications procedures are prescribed by the International Civil Aviation Organization (ICAO), however, and have decided advantages over the informal procedure described above.

b. *Distress* and *urgency* communications procedures discussed in the following paragraphs relate to the use of air ground voice communications.

c. The initial communication, and if considered necessary, any subsequent transmissions by an aircraft in *distress* should begin with the signal MAYDAY, preferably repeated three times. The signal PAN-PAN should be used in the same manner for an *urgency* condition.

d. *Distress* communications have absolute priority over all other communications, and the word MAYDAY commands radio silence on the frequency in use. *Urgency* communications have priority over all other communications except *distress*, and the word PAN-PAN warns other stations not to interfere with *urgency* transmissions.

e. Normally, the station addressed will be the air traffic facility or other agency providing air traffic services, on the frequency in use at the time. If the pilot is not communicating and receiving services, the station to be called will normally be the air traffic facility or other agency in whose area of responsibility the aircraft is operating, on the appropriate assigned frequency. If the station addressed does not respond, or if time or the situation dictates, the *distress* or *urgency* message may be broadcast, or a collect call may be used, addressing "Any Station (Tower)(Radio)(Radar)."

f. The station addressed should immediately acknowledge a *distress* or *urgency* message, provide assistance, coordinate and direct the activities of assisting facilities, and alert the appropriate search and rescue coordinator if warranted. Responsibility will be transferred to another station only if better handling will result.

g. All other stations, aircraft and ground, will continue to listen until it is evident that assistance is being provided. If any station becomes aware that the station being called either has not received a *distress* or *urgency* message, or cannot communicate with the aircraft in difficulty, it will attempt to contact the aircraft and provide assistance.

h. Although the frequency in use or other frequencies assigned by ATC are preferable, the following emergency frequencies can be used for distress or urgency communications, if necessary or desirable:

1. 121.5 MHz and 243.0 MHz—Both have a range generally limited to line of sight. 121.5 MHz is guarded by direction finding stations and some military and civil aircraft. 243.0 MHz is guarded by military aircraft. Both 121.5 MHz and 243.0 MHz are guarded by military towers, most civil towers, FSSs, and radar facilities. Normally ARTCC emergency frequency capability does not extend to radar coverage limits. If an ARTCC does not respond when called on 121.5 MHz or 243.0 MHz, call the nearest tower or FSS.

2. 2182 kHz—The range is generally less than 300 miles for the average aircraft installation. It can be used to request assistance from stations in the maritime service. 2182 kHz is guarded by major radio stations serving Coast Guard Rescue Coordination Centers, and Coast Guard units along the sea coasts of the U.S. and shores of the Great Lakes. The call "Coast Guard" will alert all Coast Guard Radio Stations within range. 2182 kHz is also guarded by most commercial coast stations and some ships and boats.

6-21. OBTAINING EMERGENCY ASSISTANCE

a. A pilot in any *distress* or *urgency* condition should *immediately* take the following action, not necessarily in the order listed, to obtain assistance:

1. Climb, if possible, for improved communications, and better radar and direction finding detection. However, it must be understood that unauthorized climb or descent under IFR conditions within controlled airspace is prohibited, except as permitted by FAR Part 91.3(b).

2. If equipped with a radar beacon transponder (civil) or IFF/SIF (military):

(a) Continue squawking assigned MODE A/3 discrete code/VFR code and MODE C altitude encoding when in radio contact with an air traffic facility or other agency providing air traffic services, unless instructed to do otherwise.

(b) If unable to immediately establish communications with an air traffic facility/agency, squawk MODE A/3, Code 7700/Emergency and MODE C.

3. Transmit a *distress* or *urgency* message consisting of *as many as necessary* of the following elements, preferably in the order listed:

(a) If distress, MAYDAY, MAYDAY, MAYDAY; if *urgency*, PAN-PAN, PAN-PAN, PAN-PAN.

(b) Name of station addressed.

(c) Aircraft identification and type.

(d) Nature of *distress* or *urgency*.

(e) Weather.

(f) Pilots intentions and request.

(g) Present position, and heading; or if *lost*, last known position, time, and heading since that position.

(h) altitude or Flight Level.

(i) Fuel remaining in minutes.

(j) Number of people on board.

(k) Any other useful information.

Reference—Pilot/Controller Glossary, Fuel Remaining.

b. After establishing radio contact, comply with advice and instructions received. Cooperate. Do not hesitate to ask questions or clarify instructions when you do not understand or if you cannot comply with clearance. Assist the ground station to control communications on the frequency in use. Silence interfering radio stations. Do not change frequency or change to another ground station unless absolutely necessary. If you do, advise the ground station of the new frequency and station name prior to the change, transmitting in the blind if necessary. If two-way communications cannot be established on the new frequency, return immediately to the frequency or station where two-way communications last existed.

c. When in a distress condition with bailout, crash landing or ditching imminent, take the following additional actions to assist search and rescue units:

1. Time and circumstances permitting, transmit as many as necessary of the message elements in subpargraph a(3) and any of the following that you think might be helpful

(a) ELT status.

(b) Visible landmarks.

(c) Aircraft color.

(d) Number of persons on board.

(e) Emergency equipment on board.

2. Actuate your ELT if the installation permits.

3. For bailout, and for crash landing or ditching if risk of fire is not a consideration, set your radio for continuous transmission.

4. If it becomes necessary to ditch, make every effort to ditch near a surface vessel. If time permits, an FAA facility should be able to get the position of the nearest commercial or Coast Guard vessel from a Coast Guard Rescue Coordination Center.

5. After a crash landing unless you have good reason to believe that you will not be located by search aircraft or ground teams, it is best to remain with your aircraft and prepare means for signalling search aircraft.

6-22. DITCHING PROCEDURES

a. A successful aircraft ditching is dependent on three primary factors. In order of importance they are:

1. Sea conditions and wind.

2. Type of aircraft.

3. Skill and technique of pilot.

b. Common oceanographic terminology:

1. Sea. The condition of the surface that is the result of both waves and swells.

(See Figure 6-22[1][Wind-Swell-Ditch Heading].)

WIND-SWELL-DITCH HEADING SITUATION

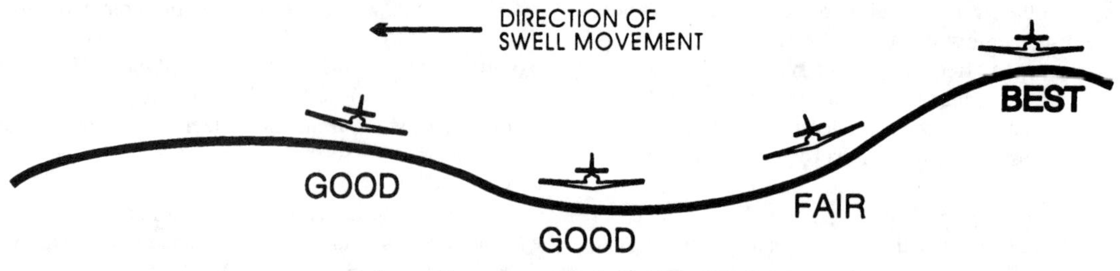

Figure 6-22[1]

(See Figure 6-22[2][Single Swell (15 knot wind)].)

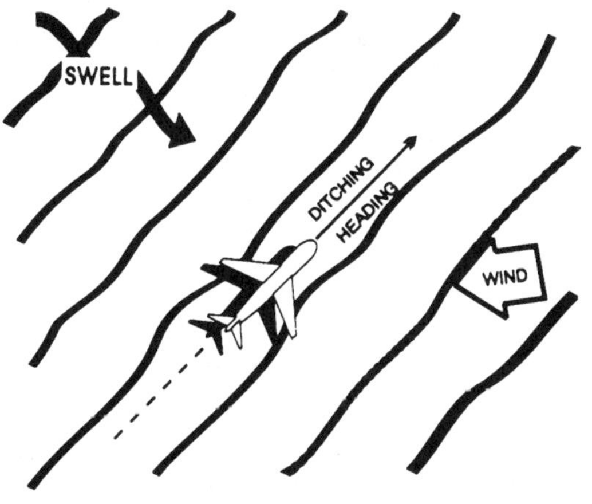

Single Swell System - Wind 15 knots

Figure 6-22[2]

(See Figure 6-22[3][Double Swell (15 knot wind)].)

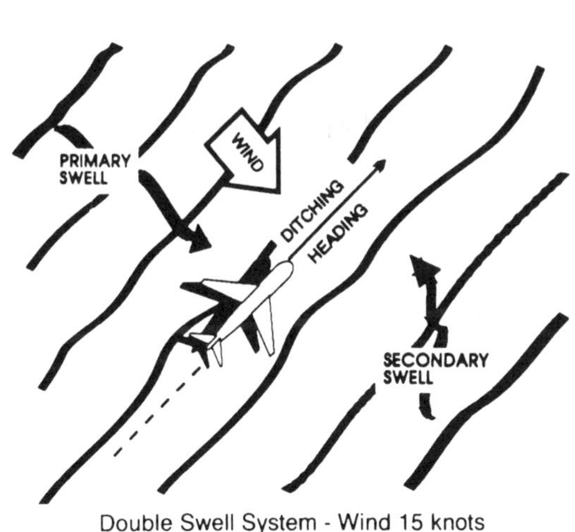

Double Swell System - Wind 15 knots

Figure 6-22[3]

(See Figure 6-22[4][Double Swell (30 knot wind)].)

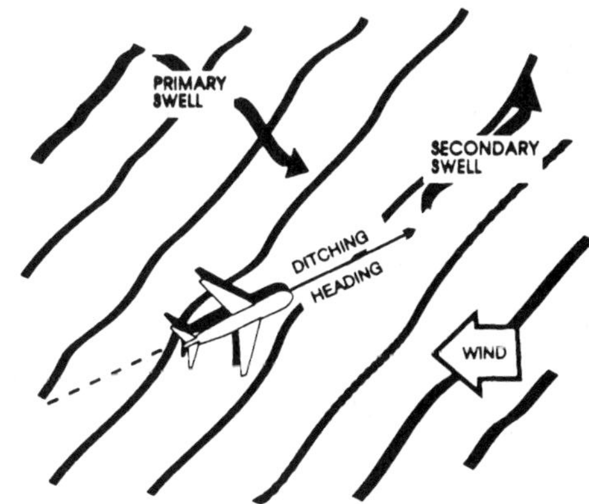

Double Swell System - Wind 30 knots

Figure 6-22[4]

(See Figure 6-22[5][(50 knot wind)].)

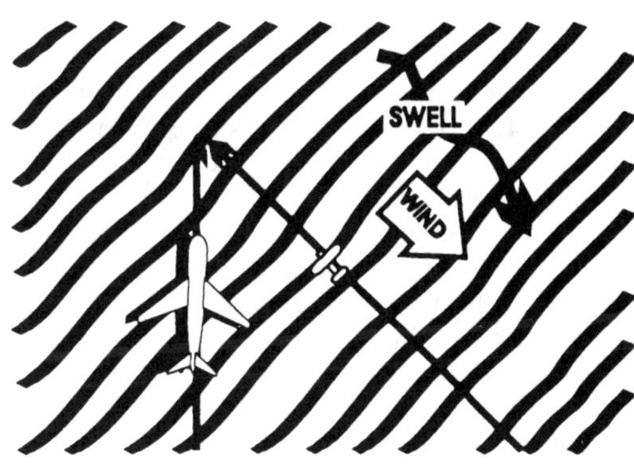

Wind - 50 knots

Aircraft with low landing speeds - land into the wind.

Aircraft with high landing speeds - choose compromise heading between wind and swell.

Both - Land on back side of swell.

Figure 6-22[5

2. Wave (or Chop). The condition of the surface caused by the local winds.

3. Swell. The condition of the surface which has been caused by a distance disturbance.

4. Swell Face. The side of the swell toward the observer. The backside is the side away from the observer. These definitions apply regardless of the direction of swell movement.

5. Primary Swell. The swell system having the greatest height from trough to crest.

6. Secondary Swells. Those swell systems of less height than the primary swell.

7. Fetch. The distance the waves have been driven by a wind blowing in a constant direction, without obstruction.

8. Swell Period. The time interval between the passage of two successive crests at the same spot in the water, measured in seconds.

9. Swell Velocity. The speed and direction of the swell with relation to a fixed reference point, measured in knots. There is little movement of water in the horizontal direction. Swells move primarily in a vertical motion, similar to the motion observed when shaking out a carpet.

10. Swell Direction. The direction *from* which a swell is moving. This direction is not necessarily the result of the wind present at the scene. The swell may be moving into or across the local wind. Swells, once set in motion, tend to maintain their original direction for as long as they continue in deep water, regardless of changes in wind direction.

11. Swell Height. The height between crest and trough, measured in feet. The vast majority of ocean swells are lower than 12 to 15 feet, and swells over 25 feet are not common at any spot on the oceans. Successive swells may differ considerably in height.

c. In order to select a good heading when ditching an aircraft, a basic evaluation of the sea is required. Selection of a good ditching heading may well minimize damage and could save your life. It can be extremely dangerous to land into the wind without regard to sea conditions; the swell system, or systems, must be taken into consideration. Remember one axiom—AVOID THE FACE OF A SWELL.

1. In ditching parallel to the swell, it makes little difference whether touchdown is on the top of the crest or in the trough. It is preferable, however, to land on the top or back side of the swell, if possible. After determining which heading (and its reciprocal) will parallel the swell, select the heading with the most into the wind component.

2. If only one swell system exists, the problem is relatively simple—even with a high, fast system. Unfortunately, most cases involve two or more swell systems running in different directions. With more than one system present, the sea presents a confused appearance. One of the most difficult situations occurs when two swell systems are at right angles. For example, if one system is eight feet high, and the other three feet, plan to land parallel to the primary system, and on the down swell of the secondary system. If both systems are of equal height, a compromise may be advisable—select an intermediate heading at 45 degrees down swell to both systems. When landing down a secondary swell, attempt to touch down on the back side, not on the face of the swell.

3. *If the swell system is formidable, it is considered advisable, in landplanes, to accept more crosswind in order to avoid landing directly into the swell.*

4. The secondary swell system is often from the same direction as the wind. Here, the landing may be made parallel to the primary system, with the wind and secondary system at an angle. There is a choice to two directions paralleling the primary system. One direction is downwind and down the secondary swell, and the other is into the wind and into the secondary swell, the choice will depend on the velocity of the wind versus the velocity and height of the secondary swell.

d. The simplest method of estimating the wind direction and velocity is to examine the windstreaks on the water. These appear as long streaks up and down wind. Some persons may have difficulty determining wind direction after seeing the streaks on the water. Whitecaps fall forward with the wind but are overrun by the waves thus producing the illusion that the foam is sliding backward. Knowing this, and by observing the direction of the streaks, the wind direction is easily determined. Wind velocity can be estimated by noting the appearance of the whitecaps, foam and wind streaks.

1. The behavior of the aircraft on making contact with the water will vary within wide limits according to the state of the sea. If landed parallel to a single swell system, the behavior of the aircraft may approximate that to be expected on a smooth sea. If landed into a heavy swell or into a confused sea, the deceleration forces may be extremely great—resulting in breaking up of the aircraft. Within certain limits, the pilot is able to minimize these forces by proper sea evaluation and selection of ditching heading.

2. When on final approach the pilot should look ahead and observe the surface of the sea. There may be shadows and whitecaps—signs of large seas. Shadows and whitecaps close together indicate short and rough seas. Touchdown in these areas is to be avoided. Select and touch down in any area (only about 500 feet is needed) where the shadows and whitecaps are not so numerous.

3. Touchdown should be at the *lowest* speed and rate of descent which permit safe handling and optimum nose up attitude on impact. Once first impact has been made, there is often little the pilot can do to control a landplane.

e. Once preditching preparations are completed, the pilot should turn to the ditching heading and commence letdown. The aircraft should be flown low over the water, and slowed down until ten knots or so above stall. At this point, additional power should be used to overcome the increased drag caused by the nose up attitude. When a smooth stretch of water appears ahead, cut power, and touch down at the best recommended speed as fully stalled as possible. By cutting power when approaching a relatively smooth area, the pilot will prevent overshooting and will touch down with less chance of planing off into a second uncontrolled landing. Most experienced seaplane pilots prefer to make contact with the water in a semi-stalled attitude, cutting power as the tail makes contact. This technique eliminates the chance of misjudging altitude with a resultant heavy drop in a fully stalled condition. Care must be taken not to drop the aircraft from too high altitude or to balloon due to excessive speed. The altitude above water depends on the aircraft. Over glassy smooth water, or at night without sufficient light, it is very easy, for even the most experienced pilots to misjudge altitude by 50 feet or more. Under such conditions, carry enough power to maintain nine to twelve degrees nose up attitude, and 10 to 20 percent over stalling speed until contact is made with the water. The proper use of power on the approach is of great importance. If power is available on one side only, a little power should be used to flatten the approach; however, the engine should not be used to such an extent that the aircraft cannot be turned against the good engines right down to the stall with a margin of rudder movement available. When near the stall, sudden application of excessive unbalanced power may result in loss of directional control. If power is available on one side only, a slightly higher than normal glide approach speed should be used. This will insure good control and some margin of speed after leveling off without excessive use of power. The use of power in ditching is so important that when it is certain that the coast cannot be reached, the pilot should, if possible, ditch before fuel is exhausted. The use of power in a night or instrument ditching is far more essential than under daylight contact conditions.

1. If no power is available, a greater than normal approach speed should be used down to the flare-out. This speed margin will allow the glide to be broken early and more gradually, thereby giving the pilot time and distance to feel for the surface—decreasing the possibility of stalling high or flying into the water. When landing parallel to a swell system, little difference is noted between landing on top of a crest or in the trough. If the wings of aircraft are trimmed to the surface of the sea rather than the horizon, there is little need to worry about a wing hitting a swell crest. The actual slope of a swell is very gradual. If forced to land into a swell, touchdown should be made just after passage of the crest. If contact is made on the face of the swell, the aircraft may be swamped or thrown violently into the air, dropping heavily into the next swell. If control surfaces remain intact, the pilot should attempt to maintain the proper nose above the horizon attitude by rapid and positive use of the controls.

f. After Touchdown: In most cases drift, caused by crosswind can be ignored; the forces acting on the aircraft after touchdown are of such magnitude that drift will be only a secondary consideration. If the aircraft is under good control, the "crab" may be kicked out with rudder just prior to touchdown. This is more important with high wing aircraft, for they are laterally unstable on the water in a crosswind and may roll to the side in ditching.

6-22f NOTE—This information has been extracted from Appendix H of the "National Search And Rescue Manual."

6-23. SPECIAL EMERGENCY (AIR PIRACY)

a. A special emergency is a condition of air piracy, or other hostile act by a person(s) aboard an aircraft, which threatens the safety of the aircraft or its passengers.

b. The pilot of an aircraft reporting a special emergency condition should:

1. If circumstances permit, apply *distress* or *urgency* radio-telephony procedures (Reference—Distress and Urgency Communications, paragraph 6-20). Include the details of the special emergency.

2. If circumstances do not permit the use of prescribed *distress* or *urgency* procedures, transmit:

(a) On the air/ground frequency in use at the time.

(b) As many as possible of the following elements spoken distinctly and in the following order:

(1) Name of the station addressed (time and circumstances permitting).

(2) The identification of the aircraft and present position.

(3) The nature of the special emergency condition and pilot intentions (circumstances permitting).

(4) If unable to provide this information, use code words and/or transponder as follows: state "TRANSPONDER SEVEN FIVE ZERO ZERO". Meaning: "I am being hijacked/forced to a new destination"; and/or use Transponder Setting MODE 3/A, Code 7500.

6-23b2b4 NOTE—Code 7500 will never be assigned by ATC without prior notification from the pilot that his aircraft is being subjected to unlawful interference. The pilot should refuse the assignment of Code 7500 in any other situation and inform the controller accordingly. Code 7500 will trigger the special emergency indicator in all radar ATC facilities.

c. Air traffic controllers will acknowledge and confirm receipt of transponder Code 7500 by asking the pilot to verify it. If the aircraft is not being subjected to unlawful interference, the pilot should respond to the query by broadcasting in the clear that he is not being subjected to unlawful interference. Upon receipt of this information, the controller will request the pilot to verify the code selection depicted in the code selector windows in the transponder control panel and change the code to the appropriate setting. If the pilot replies in the affirmative or does not reply, the controller will not ask further questions but will flight follow, respond to pilot requests and notify appropriate authorities.

d. If it is possible to do so without jeopardizing the safety of the flight, the pilot of a hijacked passenger aircraft, after departing from the cleared routing over which the aircraft was operating, will attempt to do one or more of the following things, insofar as circumstances may permit:

1. Maintain a true airspeed of no more than 400 knots, and preferably an altitude of between 10,000 and 25,000 feet.

2. Fly a course toward the destination which the hijacker has announced.

e. If these procedures result in either radio contact or air intercept, the pilot will attempt to comply with any instructions received which may direct him to an appropriate landing field.

6-24. FUEL DUMPING

a. Should it become necessary to dump fuel, the pilot should immediately advise ATC. Upon receipt of information that an aircraft will dump fuel, ATC will broadcast or cause to be broadcast immediately and every 3 minutes thereafter the following on appropriate ATC and FSS radio frequencies:

EXAMPLE:
ATTENTION ALL AIRCRAFT—FUEL DUMPING IN PROGRESS OVER—(location) AT (altitude) BY (type aircraft) (flight direction).

b. Upon receipt of such a broadcast, pilots of aircraft affected, which are not on IFR flight plans or special VFR clearances, should clear the area specified in the advisory. Aircraft on IFR flight plans or special VFR clearances will be provided specific separation by ATC. At the termination of the fuel dumping operation, pilots should advise ATC. Upon receipt of such information, ATC will issue, on the appropriate frequencies, the following:

EXAMPLE:
ATTENTION ALL AIRCRAFT—FUEL DUMPING BY—(type aircraft)—TERMINATED.

6-25 thru 6-29. RESERVED

Section 4. TWO-WAY RADIO COMMUNICATIONS FAILURE

6-30. TWO-WAY RADIO COMMUNICATIONS FAILURE

a. It is virtually impossible to provide regulations and procedures applicable to all possible situations associated with two-way radio communications failure. During two-way radio communications failure, when confronted by a situation not covered in the regulation, pilots are expected to exercise good judgment in whatever action they elect to take. Should the situation so dictate they should not be reluctant to use the emergency action contained in FAR Part 91.3(b).

b. Whether two-way communications failure constitutes an emergency depends on the circumstances, and in any event, it is a determination made by the pilot. FAR Part 91.3(b) authorizes a pilot to deviate from any rule in Subparts A and B to the extent required to meet an emergency.

c. In the event of two-way radio communications failure, ATC service will be provided on the basis that the pilot is operating in accordance with FAR Part 91.185. A pilot experiencing two-way communications failure should (unless emergency authority is exercised) comply with FAR Part 91.185 quoted below:

6-30c NOTE—Capitalization and examples added for emphasis.

1. General. Unless otherwise authorized by ATC, each pilot who has two-way radio communications failure when operating under IFR shall comply with the rules of this section.

2. VFR conditions. If the failure occurs in VFR conditions, or if VFR conditions are encountered after the failure, each pilot shall continue the flight under VFR and land as soon as practicable.

6-30c2 NOTE—This procedure also applies when two-way radio failure occurs while operating in Class A airspace. The primary objective of this provision in FAR Part 91.185 is to preclude extended IFR operation in the ATC system in VFR weather conditions. Pilots should recognize that operation under these conditions may unnecessarily as well as adversely affect other users of the airspace, since ATC may be required to reroute or delay other users in order to protect the failure aircraft. However, it is not intended that the requirement to "land as soon as practicable" be construed to mean "as soon as possible." The pilot retains his prerogative of exercising his best judgment and is not required to land at an unauthorized airport, at an airport unsuitable for the type of aircraft flown, or to land only minutes short of his destination.

3. IFR conditions. If the failure occurs in IFR conditions, or if paragraph (2) of this section cannot be complied with, each pilot shall continue the flight according to the following:

(a) Route.

(1) By the route assigned in the last ATC clearance received;

(2) If being radar vectored, by the direct route from the point of radio failure to the fix, route, or airway specified in the vector clearance;

(3) In the absence of an assigned route, by the route that ATC has advised may be expected in a further clearance; or

(4) In the absence of an assigned route or a route that ATC has advised may be expected in a further clearance by the route filed in the flight plan.

(b) Altitude. At the HIGHEST of the following altitudes or Flight Levels FOR THE ROUTE SEGMENT BEING FLOWN:

(1) The altitude or flight level assigned in the last ATC clearance received;

(2) The minimum altitude (converted, if appropriate, to minimum flight level as prescribed in FAR Part 91.121.(c)) for IFR operations; or

(3) The altitude or flight level ATC has advised may be expected in a further clearance.

6-30cb2 NOTE—The intent of the rule is that a pilot who has experienced two-way radio failure should select the appropriate altitude for the particular route segment being flown and make the necessary altitude adjustments for subsequent route segments. If the pilot received an "expect further clearance" containing a higher altitude to expect at a specified time or fix, he/she should maintain the highest of the following altitudes until that time/fix: (1) His/her last assigned altitude, or (2) The minimum altitude/flight level for IFR operations.

Upon reaching the time/fix specified, the pilot should commence his/her climb to the altitude he/she was advised to expect. If the radio failure occurs after the time/fix specified, the altitude to be expected is not applicable and the pilot should maintain an altitude consistent with 1 or 2 above.

If the pilot receives an "expect further clearance" containing lower altitude, the pilot should maintain the highest of 1 or 2 above until that time/fix specified in paragraph 6-30c(3)(c).

EXAMPLE:

A pilot experiencing two-way radio failure at an assigned altitude of 7,000 feet is cleared along a direct route which will require a climb to a minimum IFR altitude of 9,000 feet, should climb to reach 9,000 feet at the time or place where it becomes necessary (see FAR Part 91.177(b)). Later while proceeding along an airway with an MEA of 5,000 feet, the pilot would descend to *7,000 feet* (the last assigned altitude), because that altitude is *higher* than the MEA.

EXAMPLE:

A pilot experiencing two-way radio failure while being progressively descended to lower altitudes to begin an approach is assigned 2,700 feet until crossing the VOR and then cleared for the approach. The MOCA along the airway is 2,700 feet and MEA is 4,000 feet. The aircraft is within 22 NM of the VOR. The pilot should remain at 2,700 feet until crossing the VOR because that altitude is the minimum IFR altitude for the route segment being flown.

EXAMPLE:

The MEA between **A** and **B** -5,000 feet. The MEA between **B** and **C** -5,000 feet. The MEA between **C** and **D** -11,000 feet. The MEA between **D** and **E** -7,000 feet. A pilot had been cleared via **A, B, C, D,** to **E**. While flying between **A** and **B** his assigned altitude was 6,000 feet and he was told to expect a clearance to 8,000 feet at **B**. Prior to receiving the higher altitude assignment, he experienced two-way failure. The pilot would maintain 6,000 to **B**, then climb to 8,000 feet (the altitude he was advised to expect.) He would maintain 8,000 feet, then climb to 11,000 at **C**, or prior to **C** if necessary to comply with an MCA at **C**. (FAR Part 91.177(b)). Upon reaching **D**, the pilot would descend to *8,000 feet* (even though the MEA was 7,000 feet), as 8,000 was the highest of the altitude situations stated in the rule (FAR Part 91.185).

(c) Leave clearance limit.

(1) When the clearance limit is a fix from which an approach begins, commence descent or descent and approach as close as possible to the expect further clearance time if one has been received, or if one has not been received, as close as possible to the Estimated Time of Arrival (ETA) as calculated from the filed or amended (with ATC) Estimated Time en Route (ETE).

(2) If the clearance limit is not a fix from which an approach begins, leave the clearance limit at the expect further clearance time if one has been received, or if none has been received, upon arrival over the clearance limit, and proceed to a fix from which an approach begins and commence descent or descent and approach as close as possible to the estimated time of arrival as calculated from the filed or amended (with ATC) estimated time en route.

6-31. TRANSPONDER OPERATION DURING TWO-WAY COMMUNICATIONS FAILURE

a. If an aircraft with a coded radar beacon transponder experiences a loss of two-way radio capability, the pilot should adjust the transponder to reply on MODE A/3, Code 7600.

b. The pilot should understand that he may not be in an area of radar coverage.

6-32. REESTABLISHING RADIO CONTACT

a. In addition to monitoring the NAVAID voice feature, the pilot should attempt to reestablish communications by attempting contact:

1. on the previously assigned frequency, or

2. with an FSS or *ARINC.

b. If communications are established with an FSS or ARINC, the pilot should advise that radio communications on the previously assigned frequency has been lost giving the aircraft's position, altitude, last assigned frequency and then request further clearance from the controlling facility. The preceding does not preclude the use of 121.5 MHz. There is no priority on which action should be attempted first. If the capability exists, do all at the same time.

6-32b NOTE—*AERONAUTICAL RADIO/INCORPORATED (ARINC)—is a commercial communications corporation which designs, constructs, operates, leases or otherwise engages in radio activities serving the aviation community. ARINC has the capability of relaying information to/from ATC facilities throughout the country.

Chapter 7. Safety of flight

Section 1. METEOROLOGY

7-1. NATIONAL WEATHER SERVICE AVIATION PRODUCTS

a. Weather service to aviation is a joint effort of the National Weather Service (NWS), the Federal Aviation Administration (FAA), the military weather services, and other aviation oriented groups and individuals. The NWS maintains an extensive surface, upper air, and radar weather observing program; a nationwide aviation weather forecasting service; and also provides pilot briefing service. The majority of pilot weather briefings are provided by FAA personnel at Flight Service Stations (FSSs). Surface weather observations are taken by NWS, by NWS-certified FAA, contract, and supplemental observers, and by automated observing systems. (Reference—Weather Observing Programs, paragraph 7-10.)

b. Aviation forecasts are prepared by 52 Weather Service Forecast Offices (WSFO's). These offices prepare and distribute approximately 500 terminal forecasts 3 times daily for specific airports in the 50 States and the Caribbean (4 times daily in Alaska and Hawaii). These forecasts, which are amended as required, are valid for 24 hours. The last 6 hours are given in categorical outlook terms as described in Categorical Outlooks, paragraph 7-6. WSFO's also prepare a total of over 300 route forecasts and 39 synopses for Pilots Automatic Telephone Weather Answering Service (PATWAS), Transcribed Weather Broadcast (TWEB), and briefing purposes. The route forecasts that are issued during the morning and mid-day are valid for 12 hours while the evening issuance is valid for 18 hours. A centralized aviation forecast program originating from the National Aviation Weather Advisory Unit (NAWAU) in Kansas City was implemented in November 1982. In the conterminous U.S., all In-flight Advisories (SIGMETs, Convective SIGMETs, and AIRMETs) and all Area Forecasts (6 areas) are now issued by NAWAU. Area Forecasts are prepared 3 times a day in the conterminous States (4 times in Hawaii), and amended as required, while In-flight Advisories are issued only when conditions warrant. (Reference—In-Flight Weather Advisories, paragraph 7-5.) Winds aloft forecasts are provided for 176 locations in the 48 contiguous States and 21 in Alaska for flight planning purposes. (Winds aloft forecasts for Hawaii are prepared locally.) All the aviation weather forecasts are given wide distribution through the Weather Message Switching Center in Kansas City (WMSC).

c. Weather element values may be expressed by using different measurement systems depending on several factors, such as whether the weather products will be used by the general public, aviation interests, international services, or a combination of these users. Figure 7-1[1] provides conversion tables for the most used weather elements that will be encountered by pilots.

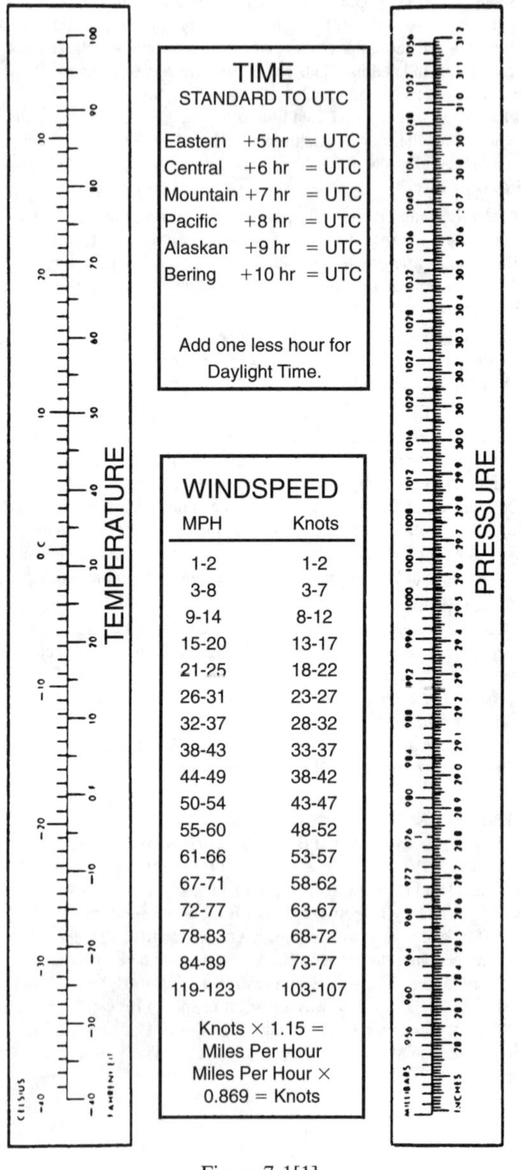

TEMPERATURE
TIME
STANDARD TO UTC
Eastern +5 hr = UTC
Central +6 hr = UTC
Mountain +7 hr = UTC
Pacific +8 hr = UTC
Alaskan +9 hr = UTC
Bering +10 hr = UTC
Add one less hour for Daylight Time.
WINDSPEED
MPH Knots
1-2 1-2
3-8 3-7
9-14 8-12
15-20 13-17
21-25 18-22
26-31 23-27
32-37 28-32
38-43 33-37
44-49 38-42
50-54 43-47
55-60 48-52
61-66 53-57
67-71 58-62
72-77 63-67
78-83 68-72
84-89 73-77
119-123 103-107
Knots × 1.15 = Miles Per Hour
Miles Per Hour × 0.869 = Knots
PRESSURE

Figure 7-1[1]

7-2. FAA WEATHER SERVICES

a. The FAA maintains a nationwide network of Flight Service Stations (FSSs) and Supplemental Weather Service Locations (SWSLs) to serve the weather needs of pilots. In addition, National Weather Service (NWS) meteorologists are assigned to most Air Route Traffic Control Centers (ARTCC's) as part of the Center Weather Service Unit (CWSU). They provide advisory service and short-term forecasts (nowcasts) to support the needs of the FAA and other users of the system.

b. The primary source of preflight weather briefings is an individual briefing obtained from a briefer at the FSS or NWS. These briefings, which are tailored to your specific flight, are available 24 hours a day through the local FSS or through the use of toll free lines (INWATS). Numbers for these services can be found in the Airport/Facility Directory under "FAA and NWS Telephone Numbers" section. They are also listed in the U.S. Government section of your local telephone directory under Department of Transportation, Federal Aviation Administration, or Department of Commerce, National Weather Service. (Reference—Preflight Briefing, paragraph 7-3 explains the types of preflight briefings available and the information contained in each.) SWSL personnel provide the weather report but not NOTAM information for the airport where they are located. NWS pilot briefers do not provide aeronautical information (NOTAM's, flow control advisories, etc.) nor do they accept flight plans.

c. Other sources of weather information are as follows:

1. The A.M. Weather telecast on the PBS television network is a jointly sponsored 15-minute weather program designed for pilots. It is broadcast Monday through Friday mornings. Check TV listings in your area for station and exact times.

2. The Transcribed Weather Broadcast (TWEB), telephone access to the TWEB (TEL-TWEB), Telephone Information Briefing Service (TIBS) (AFSS) and Pilots Automatic Telephone Weather Answering Service (PATWAS) (FSS) provide continuously updated recorded weather information for short or local flights. Separate paragraphs in this section give additional information regarding these services.

3. Weather and aeronautical information is also available from numerous private industry sources on an individual or contract pay basis. Information on how to obtain this service should be available from local pilot organizations.

4. The Direct User Access System (DUATS) can be accessed by pilots with a current medical certificate toll-free in the 48 contiguous States via personal computer. Pilots can receive alpha-numeric preflight weather data and file domestic VFR and IFR flight plans. The following are the contract DUATS vendors:

CONTEL Federal Systems
15000 Conference Center Drive
Chantilly, VA 22021-3808
Telephone—*For filing flight plans and obtaining weather briefings:* (800) 767-9989
For customer service: (800) 345-3828

Data Transformation Corporation
108-D Greentree Road
Turnerville, NJ 08012
Telephone—*For filing flight plans and obtaining weather briefings:* (800) 245-3828
For customer service: (800) 243-3828

d. In-flight weather information is available from any FSS within radio range. (Reference—Pilots Automatic Telephone Weather Answering Service (PATWAS) and Telephone Information Briefing Service (TIBS), paragraph 7-7, Transcribed Weather Broadcast (TWEB), paragraph 7-8, and In-Flight Weather Broadcasts, paragraph 7-9 for information on broadcasts.) En route Flight Advisory Service (EFAS) is provided to serve the nonroutine weather needs of pilots in flight. (Reference—En Route Flight Advisory Service (EFAS), paragraph 7-4 gives details on this service.)

7-3. PREFLIGHT BRIEFING

a. Flight Service Stations (FSSs) are the primary source for obtaining preflight briefings and in-flight weather information. In some locations, the Weather Service Office (WSO) provides preflight briefings on a limited basis. Flight Service Specialists are qualified and certificated by the NWS as Pilot Weather Briefers. They are not authorized to make original forecasts, but are authorized to translate and interpret available forecasts and reports directly into terms describing the weather conditions which you can expect along your flight route and at your destination. Available aviation weather reports and forecasts are displayed at each FSS and WSO. Some of the larger FSSs provide a separate display for pilot use. Pilots should feel free to use these self briefing displays where available, or to ask for a briefing or assistance from the specialist on duty. Three basic types of preflight briefings are available to serve your specific needs. These are: Standard Briefing, Abbreviated Briefing, and Outlook Briefing. You should specify to the briefer the type of briefing you want, along with appropriate background information. (Reference—Preflight Preparation, paragraph 5-1 for items that are required.) This will enable the briefer to tailor the information to your intended flight. The following paragraphs describe the types of briefings available and the information provided in each.

b. Standard Briefing—You should request a Standard Briefing any time you are planning a flight and you have not received a previous briefing or have not received preliminary information through mass dissemination media;

e.g., TWEB, PATWAS, VRS, etc. The briefer will automatically provide the following information in the sequence listed, except as noted, when it is applicable to your proposed flight.

1. Adverse Conditions—Significant meteorological and aeronautical information that might influence the pilot to alter the proposed flight; e.g., hazardous weather conditions, runway closures, NAVAID outages, etc.

2. VFR Flight Not Recommended—When VFR flight is proposed and sky conditions or visibilities are present or forecast, surface or aloft, that in the briefer's judgment would make flight under visual flight rules doubtful, the briefer will describe the conditions, affected locations, and use the phrase "*VFR flight is not recommended.*" This recommendation is advisory in nature. The final decision as to whether the flight can be conducted safely rests solely with the pilot.

3. Synopsis—A brief statement describing the type, location and movement of weather systems and/or air masses which might affect the proposed flight.

7-3b3 NOTE—These first 3 elements of a briefing may be combined in any order when the briefer believes it will help to more clearly describe conditions.

4. Current Conditions—Reported weather conditions applicable to the flight will be summarized from all available sources; e.g., SA's, PIREP's, RAREP's. This element will be omitted if the proposed time of departure is beyond 2 hours, unless the information is specifically requested by the pilot.

5. En Route Forecast—Forecast en route conditions for the proposed route are summarized in logical order; i.e., departure/climbout, en route, and descent.

6. Destination Forecast—The destination forecast for the planned ETA. Any significant changes within 1 hour before and after the planned arrival are included.

7. Winds Aloft—Forecast winds aloft will be summarized for the proposed route. The briefer will interpolate wind directions and speeds between levels and stations as necessary to provide expected conditions at planned altitudes.

8. Notices to Airmen (NOTAM's)—

(a) Available NOTAM (D) information pertinent to the proposed flight.

(b) NOTAM (L) information pertinent to the departure and/or local area, if available, and pertinent FDC NOTAM's within approximately 400 miles of the FSS providing the briefing.

7-3b8b NOTE 1—NOTAM information may be combined with current conditions when the briefer believes it is logical to do so.

7-3b8b NOTE 2—NOTAM (D) information and FDC NOTAM's which have been published in the Notices to Airmen publication are not included in pilot briefings unless a review of this publication is specifically requested by the pilot. For complete flight information you are urged to review the printed NOTAMs in the Notices to Airmen Publication and the Airport/Facility Directory in addition to obtaining a briefing.

9. ATC Delays—Any known ATC delays and flow control advisories which might affect the proposed flight.

10. Pilots may obtain the following from FSS briefers upon request:

(a) Information on military training routes (MTR) and military operations area (MOA) activity within the flight plan area and a 100 NM extension around the flight plan area.

7-3b10a NOTE—Pilots are encouraged to request updated information from en route FSSs.

(b) A review of the Notices to Airmen publication for pertinent NOTAM's and Special Notices.

(c) Approximate density altitude data.

(d) Information regarding such items as air traffic services and rules, customs/immigration procedures, ADIZ rules, search and rescue, etc.

(e) LORAN-C NOTAM's.

(f) Other assistance as required.

c. Abbreviated Briefing—Request an Abbreviated Briefing when you need information to supplement mass disseminated data, update a previous briefing, or when you need only one or two specific items. Provide the briefer with appropriate background information, the time you received the previous information, and/or the specific items needed. You should indicate the source of the information already received so that the briefer can limit the briefing to the information that you have not received, and/or appreciable changes in meteorological conditions since your previous briefing. To the extent possible, the briefer will provide the information in the sequence shown for a Standard Briefing. If you request only one or two specific items, the briefer will advise you if adverse conditions are present or forecast. Details on these conditions will be provided at your request.

d. Outlook Briefing—You should request an Outlook Briefing whenever your proposed time of departure is six or more hours from the time of the briefing. The briefer will provide available forecast data applicable to the proposed flight. This type of briefing is provided for planning purposes only. You should obtain a Standard or Abbreviated Briefing prior to departure in order to obtain such items as current conditions, updated forecasts, winds aloft and NOTAM's.

e. Inflight Briefing—You are encouraged to obtain your preflight briefing by telephone or in person before departure. In those cases where you need to obtain a preflight briefing or an update to a previous briefing by radio, you should contact the nearest FSS to obtain this information. After communications have been established, advise the specialist of the type briefing you require and provide appropriate background information. You will be provided information as specified in the above paragraphs, depending on the type briefing requested. In addition, the specialist will recommend shifting to the Flight Watch frequency when conditions along the intended route indicate that it would be advantageous to do so.

f. Following any briefing, feel free to ask for any information that you or the briefer may have missed. It helps to save your questions until the briefing has been completed. This way, the briefer is able to present the information in a logical sequence, and lessens the chance of important items being overlooked.

7-4. EN ROUTE FLIGHT ADVISORY SERVICE (EFAS)

a. EFAS is a service specifically designed to provide en route aircraft with timely and meaningful weather advisories pertinent to the type of flight intended, route of flight, and altitude. In conjunction with this service, EFAS is also a central collection and distribution point for pilot reported weather information. EFAS is provided by specially trained specialists in selected AFSSs/FSSs controlling multiple Remote Communications Outlets covering a large geographical area and is normally available throughout the conterminous U.S. and Puerto Rico from 6 a.m. to 10 p.m. EFAS provides communications capabilities for aircraft flying at 5,000 feet above ground level to 17,500 feet MSL on a common frequency of 122.0 MHz. Discrete EFAS frequencies have been established to ensure communications coverage from 18,000 through 45,000 MSL serving in each specific ARTCC area. These discrete frequencies may be used below 18,000 feet when coverage permits reliable communication.

7-4a NOTE—When an EFAS outlet is located in a time zone different from the zone in which the Flight Watch control station is located, the availability of service may be plus or minus one hour from the normal operating hours.

b. Contact flight watch by using the name of the Air Route Traffic Control Center facility identification serving the area of your location, followed by your aircraft identification, name of the nearest VOR to your position. The specialist needs to know this approximate location to select the most appropriate transmitter/receiver outlet for communications coverage.

EXAMPLE:
CLEVELAND FLIGHT WATCH, CESSNA ONE TWO THREE FOUR KILO, MANSFIELD V-O-R, OVER.

c. Charts depicting the location of the flight watch control stations (parent facility) and the outlets they use are contained in the Airport Facility Directories (A/FD). If you do not know in which flight watch area you are flying, initiate contact by using the words "Flight Watch," your aircraft identification, and the name of the nearest VOR. The facility will respond using the name of the flight watch facility.

EXAMPLE:
FLIGHT WATCH, CESSNA ONE TWO THREE FOUR KILO, MANSFIELD V-O-R, OVER.

d. FSSs that provide En route Flight Advisory Service are listed regionally in the Airport/Facilities Directories.

e. EFAS is not intended to be used for filing or closing flight plans, position reporting, getting complete preflight briefings, or obtaining random weather reports and forecasts. En route flight advisories are tailored to the phase of flight that begins after climb-out and ends with descent to land. Immediate destination weather and terminal forecast will be provided on request. Pilots requesting information not within the scope of flight watch will be advised of the appropriate FSS frequency to contact to obtain the information. Pilot participation is essential to the success of EFAS by providing a continuous exchange of information on weather, winds, turbulence, flight visibility, icing, etc., between pilots and flight watch specialists. Pilots are encouraged to report good weather as well as bad, and to confirm expected conditions as well as unexpected to EFAS facilities.

7-5. IN-FLIGHT WEATHER ADVISORIES

a. The NWS issues in-flight weather advisories designated as Severe Weather Forecasts Alerts (AWW), Convective SIGMETs (WST), SIGMETs (WS), Center Weather Advisories (CWA), or AIRMET's (WA). These advisories are issued individually; however, the information contained in them is also included in relevant portions of the Area Forecast (FA). When these advisories are issued subsequent to the FA, they automatically amend appropriate portions of the FA until the FA itself has been amended. In-flight advisories serve to notify en route pilots of the possibility of encountering hazardous flying conditions which may not have been forecast at the time of the preflight briefing. Whether or not the condition described is potentially hazardous to a particular flight is for the pilot and/or aircraft dispatcher in a (FAR Part 121) operation to evaluate on the basis of experience and the operational limits of the aircraft.

b. Severe Weather Forecast Alerts (AWW) are preliminary messages issued in order to alert users that a Severe Weather Bulletin (WW) is being issued. These messages define areas of possible severe thunderstorms or tornado activity. The messages are unscheduled and issued as required by the National Severe Storm Forecast Center at Kansas City, Missouri.

c. WST in the Conterminous United States: WST's concern only thunderstorms and related phenomena (tornadoes, heavy precipitation, hail, and high surface winds) over the conterminous United States and imply the associated occurrence of turbulence, icing, and convective low level wind shear. WSTs are issued for the following phenomena:

1. Tornadoes.

2. Lines of thunderstorms.

3. Embedded thunderstorms.

4. Thunderstorm areas greater than or equal to thunderstorm intensity level 4 with an area coverage of $4/10$ (40 percent) or more. (Reference—Pilot/Controller Glossary, Radar Weather Echo Intensity Levels.)

5. Hail greater than or equal to ¾ inch in diameter and/or winds gusts to 50 knots or greater.

7-5c5 NOTE—Since thunderstorms are the reason for issuing the WST, severe or greater turbulence, severe icing, and low-level wind shear (gust fronts, downbursts, microbursts, etc.) are implied and will not be specified in the advisory.

d. Convective Sigmet Bulletins—

1. Three Convective SIGMET bulletins, each covering a specified geographic area, are issued. These areas are the Eastern (E), Central (C), and Western (W) U.S. The boundaries that separate the Eastern from the Central and the Central from the Western U.S. are 87 and 107 degrees West, respectively. These bulletins are issued on a scheduled basis, hourly at 55 minutes past the hour (H+55), and as special bulletins on an unscheduled basis.

2. Each of the Convective SIGMET bulletins will be:

(a) Made up of one or more individually numbered Convective SIGMET's,

(b) Valid for two hours or until superseded by the next hourly issuance.

3. On an hourly basis, an outlook is made for each of the three Convective SIGMET regions. The outlook for a particular region is appended to the Convective SIGMET bulletin for the same region. However, it is not appended to special Convective SIGMET's. The outlook is reviewed each hour and revised when necessary. The outlook is a forecast and meteorological discussion for thunderstorm systems that are expected to require Convective SIGMET issuances during a time period 2–6 hours into the future. Furthermore, an outlook will always be made for each of the three regions, even if it is a negative statement.

e. SIGMET's within the conterminous U.S. are issued by the National Aviation Weather Advisory Unit (NAWAU) for the following hazardous weather phenomena:

1. Severe, extreme turbulence or clear air turbulence not associated with thunderstorms.

2. Severe icing not associated with thunderstorms.

3. Widespread duststorms, sandstorms, or volcanic ash lowering surface and/or in-flight visibilities to less than three miles.

4. Volcanic eruption SIGMET's are identified by an alphanumeric designator which consists of an alphabetic identifier and issuance number. The first time an advisory is issued for a phenomenon associated with a particular weather system, it will be given the next alphabetic designator in the series and will be numbered as the first for that designator. Subsequent advisories will retain the same alphabetic designator until the phenomenon ends. In the conterminous U.S., this means that a phenomenon that is assigned an alphabetic designator in one area will retain that designator as it moves within the area or into one or more other areas. Issuances for the same phenomenon will be sequentially numbered, using the same alphabetic designator until the phenomenon no longer exists. Alphabetic designators NOVEMBER through YANKEE, except SIERRA, TANGO and ZULU are used only for SIGMET's, while designators SIERRA, TANGO and ZULU are used for AIRMET's.

f. The CWA is an unscheduled in-flight, flow control, air traffic, and air crew advisory. By nature of its short lead time, the CWA is not a flight planning product. It is generally a Nowcast for conditions beginning within the next two hours. CWA's will be issued:

1. As a supplement to an existing SIGMET, Convective SIGMET, AIRMET, or Area Forecast (FA).

2. When an in-flight Advisory has not been issued but observed or expected weather conditions meet SIGMET/AIRMET criteria based on current pilot reports and reinforced by other sources of information about existing meteorological conditions.

3. When observed or developing weather conditions do not meet SIGMET, Convective SIGMET, or AIRMET criteria; e.g., in terms of intensity or area coverage, but current pilot reports or other weather information sources indicate that existing or anticipated meteorological phenomena will adversely affect the safe flow of air traffic within the Air Route Traffic Control Center (ARTCC) area of responsibility.

g. WA may be of significance to any pilot or aircraft operator and are issued for all domestic airspace. They are of particular concern to operators and pilots of aircraft sensitive to the phenomena described and to pilots without instrument ratings and are issued by the NAWAU for the following weather phenomena which are potentially hazardous to aircraft:

1. Moderate icing.

2. Moderate turbulence.

3. Sustained winds of 30 knots or more at the surface.

4. Widespread area of ceilings less than 1,000 feet and/or visibility less than three miles.

5. Extensive mountain obscurement.

7-5g5 NOTE—If the above phenomena are adequately forecast in the FA, an AIRMET will not be issued.

h. AIRMET's are issued on a scheduled basis every six hours, with unscheduled amendments issued as required. AIRMET's have fixed alphanumeric designator with ZULU for icing and freezing level data, TANGO for turbulence, strong surface winds, and windshear, and SIERRA for instrument flight rules and mountain obscuration.

i. Each CWA will have a phenomenon number (1-6) immediately following the ARTCC identifier. This number will be assigned to each meteorologically distinct condition or conditions; e.g., jet stream clear air turbulence or low IFR and icing northwest of a low pressure center, meeting CWA issuance criteria. Following the product type (CWA) a two digit issuance number will be entered starting at midnight local each day. In addition, those CWA's based on existing nonconvective SIGMET's/

AIRMET's will include the associated alphanumeric designator; e.g., ALPHA 4.

EXAMPLE:

ZKC1 CWA01/ALPHA 4

j. Each AWW is numbered sequentially beginning January 1 of each year.

EXAMPLE:

MKC AWW 161755

WW 279 SEVERE TSTM NY PA NJ

161830Z-170000Z AXIS..70 STATUTE

MILES EITHER SIDE OF LINE..

10W MSS TO 20E ABE..AVIATION

COORDS..60NM EITHER SIDE/60NW

SLK - 35W EWR..HAIL SURFACE

AND ALOFT..2 INCHES. SURFACE

WIND GUSTS..65 KNOTS. MAX TOPS

TO 540. MEAN WIND VECTOR 19020.

REPLACES WW 278..OH PA NY

Status reports are issued as needed on Severe Weather Watch Bulletins to show progress of storms and to delineate areas no longer under the threat of severe storm activity. Cancellation bulletins are issued when it becomes evident that no severe weather will develop or that storms have subsided and are no longer severe.

7-6. CATEGORICAL OUTLOOKS

a. Categorical outlook terms, describing general ceiling and visibility conditions for advanced planning purposes, are defined as follows:

1. LIFR (Low IFR)—Ceiling less than 500 feet and/or visibility less than 1 mile.

2. IFR—Ceiling 500 to less than 1,000 feet and/or visibility 1 to less than 3 miles.

3. MVFR (Marginal VFR)—Ceiling 1,000 to 3,000 feet and/or visibility 3 to 5 miles inclusive.

4. VFR—Ceiling greater than 3,000 feet and visibility greater than 5 miles; includes sky clear.

b. The cause of LIFR, IFR, or MVFR is indicated by either ceiling or visibility restrictions or both. The contraction "CIG" and/or weather and obstruction to vision symbols are used. If winds or gusts of 25 knots or greater are forecast for the outlook period, the word "WIND" is also included for all categories including VFR.

EXAMPLE:

LIFR CIG—Low IFR due to low ceiling.

EXAMPLE:

IFR F—IFR due to visibility restricted by fog.

EXAMPLE:

MVFR CIG HK—Marginal VFR due to both ceiling and visibility restricted by haze and smoke.

EXAMPLE:

IFR CIG R WIND—IFR due to both low ceiling and visibility restricted by rain; wind expected to be 25 knots or greater.

7-7. PILOTS AUTOMATIC TELEPHONE WEATHER ANSWERING SERVICE (PATWAS) AND TELEPHONE INFORMATION BRIEFING SERVICE (TIBS)

a. PATWAS is provided by nonautomated flight service stations. PATWAS is a continuous recording of meteorological and aeronautical information, which is available at selected locations by telephone. Normally, the recording contains a summary of data for an area within 50 NM of the parent station; however, at some locations route information similar to that available on the Transcribed Weather Broadcast (TWEB) is included.

b. PATWAS is not intended to substitute for specialist-provided preflight briefings. It is, however, recommended for use as a preliminary briefing, and often will be valuable in helping you to make a "go or no go" decision.

c. TIBS is provided by Automated Flight Service Stations (AFSS) and provides continuous telephone recordings of meteorological and/or aeronautical information. Specifically, TIBS provides area and/or route briefings, airspace procedures, and special announcements (if applicable) concerning aviation interests.

d. Depending on user demand, other items may be provided; i.e., surface observations, terminal forecasts, winds/temperatures aloft forecasts, etc. A TOUCH-TONE telephone is necessary to fully utilize the TIBS program.

e. Pilots are encouraged to avail themselves of this service. PATWAS and TIBS locations are found in the Airport/Facility Directory under the FSS and National Weather Service Telephone Numbers section.

7-8. TRANSCRIBED WEATHER BROADCAST (TWEB)

Equipment is provided at selected FSSs by which meteorological and aeronautical data are recorded on tapes and broadcast continuously over selected low-frequency (190-535 kHz) navigational aids (L/MF ranges or H facilities) and/or VOR's. Broadcasts are made from a series of individual tape recordings, and changes, as they occur, are transcribed onto the tapes. The information provided varies depending on the type equipment available. Generally, the broadcast contains route-oriented data with specially prepared NWS forecasts, In-flight Advisories, and winds aloft plus preselected current information, such as weather reports, NOTAM's, and special notices. In some locations, the information is broadcast over the local VOR only and is limited to such items as the hourly weather for the parent station and up to 5 immediately adjacent stations, local NOTAM information, terminal forecast (FT) for the parent station, adverse conditions extracted from In-flight Advisories, and other potentially hazardous conditions. At selected locations, telephone access to the TWEB has been provided (TEL-TWEB). Telephone numbers for this service are found in the FSS and National

Weather Service Telephone Numbers section of the Airport/Facility Directory. These broadcasts are made available primarily for preflight and in-flight planning, and as such, should not be considered as a substitute for specialist-provided preflight briefings.

7-9. IN-FLIGHT WEATHER BROADCASTS

a. Weather Advisory Broadcasts—FAA FSSs broadcast Severe Weather Forecast Alerts (AWW), Convective SIGMETs, SIGMETs, CWAs, and AIRMET's during their valid period when they pertain to the area within 150 NM of the FSS or a broadcast facility controlled by the FSS as follows:

1. Severe Weather Forecast Alerts (AWW) and Convective SIGMET—Upon receipt and at 15-minute intervals H+00, H+15, H+30, and H+45 for the first hour after issuance.

EXAMPLE:
AVIATION BROADCAST, WEATHER ADVISORY, (Severe Weather Forecast Alert or Convective SIGMET identification) (text of advisory).

2. SIGMET's, CWA's, and AIRMET's—Upon receipt and at 30-minute intervals at H+15 and H+45 for the first hour after issuance.

EXAMPLE:
AVIATION BROADCAST, WEATHER ADVISORY, (area or ARTCC identification) (SIGMET, CWA, or AIRMET identification) (text of advisory).

3. Thereafter, a summarized alert notice will be broadcast at H+15 and H+45 during the valid period of the advisories.

EXAMPLE:
AVIATION BROADCAST, WEATHER ADVISORY, A (Severe Weather Forecast Alert, Convective SIGMET, SIGMET, CWA, or AIRMET) IS CURRENT FOR (description of weather) (area affected).

4. Pilots, upon hearing the alert notice, if they have not received the advisory or are in doubt, should contact the nearest FSS and ascertain whether the advisory is pertinent to their flights.

b. ARTCC's broadcast a Severe Weather Forecast Alert (AWW), Convective SIGMET, SIGMET, or CWA alert once on all frequencies, except emergency, when any part of the area described is within 150 miles of the airspace under their jurisdiction. These broadcasts contain SIGMET or CWA (identification) and a brief description of the weather activity and general area affected.

EXAMPLE:
ATTENTION ALL AIRCRAFT, SIGMET DELTA THREE, FROM MYTON TO TUBA CITY TO MILFORD, SEVERE TURBULENCE AND SEVERE CLEAR ICING BELOW ONE ZERO THOUSAND FEET. EXPECTED TO CONTINUE BEYOND ZERO THREE ZERO ZERO ZULU.

EXAMPLE:
ATTENTION ALL AIRCRAFT, CONVECTIVE SIGMET TWO SEVEN EASTERN. FROM THE VICINITY OF ELMIRA TO PHILLIPSBURG. SCATTERED EMBEDDED THUNDERSTORMS MOVING EAST AT ONE ZERO KNOTS. A FEW INTENSE LEVEL FIVE CELLS, MAXIMUM TOPS FOUR FIVE ZERO.

EXAMPLE:
ATTENTION ALL AIRCRAFT, KANSAS CITY CENTER WEATHER ADVISORY ONE ZERO THREE. NUMEROUS REPORTS OF MODERATE TO SEVERE ICING FROM EIGHT TO NINER THOUSAND FEET IN A THREE ZERO MILE RADIUS OF ST. LOUIS. LIGHT OR NEGATIVE ICING REPORTED FROM FOUR THOUSAND TO ONE TWO THOUSAND FEET REMAINDER OF KANSAS CITY CENTER AREA.

7-9b NOTE—Terminal control facilities have the option to limit the AWW, Convective SIGMET, SIGMET, or CWA broadcast as follows: local control and approach control positions may opt to broadcast SIGMET or CWA alerts only when any part of the area described is within 50 miles of the airspace under their jurisdiction.

c. Hazardous In-flight Weather Advisory Service (HIWAS)—This is a continuous broadcast of in-flight weather advisories including summarized AWW, SIGMET's, Convective SIGMET's, CWA's, AIRMET's, and urgent PIREP's. HIWAS has been adopted as a national program and will be implemented throughout the conterminous U.S. as resources permit. In those areas where HIWAS is commissioned, ARTCC, Terminal ATC, and FSS facilities have discontinued the broadcast of in-flight advisories as described in the preceding paragraph. HIWAS is an additional source of hazardous weather information which makes these data available on a continuous basis. It is not, however, a replacement for preflight or in-flight briefings or real-time weather updates from Flight Watch (EFAS). As HIWAS is implemented in individual center areas, the commissioning will be advertised in the Notices to Airmen publication.

1. Where HIWAS has been implemented, a HIWAS alert will be broadcast on all except emergency frequencies once upon receipt by ARTCC and terminal facilities, which will include an alert announcement, frequency instruction, number, and type of advisory updated; e.g., AWW, SIGMET, Convective SIGMET, or CWA.

EXAMPLE:
ATTENTION ALL AIRCRAFT, MONITOR HIWAS OR CONTACT A FLIGHT SERVICE STATION ON FREQUENCY ONE TWO TWO POINT ZERO OR ONE TWO TWO POINT TWO FOR NEW CONVECTIVE SIGMET (identification) INFORMATION.

2. In HIWAS ARTCC areas, FSSs will broadcast a HIWAS update announcement once on all except emergency frequencies upon completion of recording an update to the HIWAS broadcast. Included in the broadcast will be the type of advisory updated; e.g., AWW, SIGMET, Convective SIGMET, CWA, etc.

EXAMPLE:
ATTENTION ALL AIRCRAFT, MONITOR HIWAS OR CONTACT FLIGHT WATCH OR FLIGHT SERVICE FOR NEW CONVECTIVE SIGMET INFORMATION.

d. Unscheduled Broadcasts—These broadcasts are made by FSSs on VOR and selected VHF frequencies upon receipt of special weather reports, PIREP's, NO-

TAM's and other information considered necessary to enhance safety and efficiency of flight. These broadcasts will be made at random times and will begin with the announcement "Aviation Broadcast" followed by identification of the data.

EXAMPLE:
AVIATION BROADCAST, SPECIAL WEATHER REPORT, (Notice to Airmen, Pilot Report, etc.) (location name twice) THREE SEVEN (past the hour) OBSERVATION...(etc.).

e. Alaskan Scheduled Broadcasts—Selected FSSs in Alaska having voice capability on radio ranges (VOR) or radio beacons (NDB) broadcast weather reports and Notice to Airmen information at 15 minutes past each hour from reporting points within approximately 150 miles from the broadcast station.

7-10. WEATHER OBSERVING PROGRAMS

a. Manual Observations—Surface weather observations are taken at more than 600 locations in the United States. With only a few exceptions, these stations are located at airport sites and most are manned by FAA or NWS personnel who manually observe, perform calculations, and enter the observation into the distribution system. The format and coding of these observations are contained in Key to Aviation Weather Observations and Forecasts, paragraph 7-27.

b. Automatic Meteorological Observing Stations (AMOS)—

1. Full parameter AMOS facilities provide data for the basic weather program at remote, unstaffed, or part-time staffed locations at approximately 90 locations in the United States. They report temperature, dew point, wind, pressure, and precipitation (liquid) amount. At staffed AMOS locations, an observer may manually add visually observed and manually calculated elements to the automatic reports. The elements manually added are sky condition, visibility, weather, obstructions to vision, and sea level pressure. The content and format of AMOS reports is the same as the manually observed reports, except the acronym "AMOS" or "RAMOS" (for Remote Automatic Meteorological Observing Station) will be the first item of the report.

2. Partial parameter AMOS stations only report some of the elements contained in the full parameter locations, nor-

Table 7-10[1]
DECODING OBSERVATIONS FROM AUTOB STATIONS

***EXAMPLE:* ENV AUTOB E25 BKN BV7 P 33/29/3606/975 PK WND 08 001**

ENCODE	DECODE	EXPLANATION
ENV	STATION IDENTIFICATION:	(Wendover, UT) Identifies report using FAA identifiers.
AUTOB	AUTOMATIC STATION IDENTIFIER	
E25 BKN	SKY & CEILING:	(Estimated 2500 ft. broken) Figures are height in 100s of feet above ground. Contraction after height is amount of sky cover. Letter preceding height indicates ceiling. WX reported if visibility is less than 2 miles and no clouds are detected. *NO CLOUDS REPORTED ABOVE 6000 FEET.*
BV7	BACKSCATTER VISIBILITY AVERAGED IN PAST MINUTE:	Reported in whole miles from 1 to 7.
P	PRECIPITATION OCCURRENCE:	(P = precipitation in past 10 minutes).
33	TEMPERATURE:	(33 degrees F.) Minus sign indicates sub-zero temperatures.
/29	DEW POINT:	(29 degrees F.) Minus sign indicates sub-zero temperatures.
/3606	WIND:	(360 degrees true at 6 knots) Direction is first two digits and is reported in tens of degrees. To decode, add a zero to first two digits. The last digits are speed; e.g., 2524 = 250 degrees at 24 knots.
/975	ALTIMETER SETTING:	(29.75 inches) The tens digit and decimal are omitted from report. To decode, prefix a 2 to code if it begins with 8 or 9. Otherwise, prefix a 3; e.g., 982 = 29.82, 017 = 30.17.
PK WIND 08	PEAK WIND SPEED:	(8 knots) Reported speed is highest detected since last hourly observation.
001	PRECIPITATION ACCUMULATION:	(0.01 inches) Amount of precipitation since last synoptic time (00, 06, 12, 1800 UTC).

NOTE: If no clouds are detected below 6,000 feet and the visibility is greater than 2 miles, the reported sky condition will be *CLR BLO 60.*

mally wind. These observations are not normally disseminated through aviation weather circuits.

c. Automatic Observing Stations (AUTOB)—There are four AUTOB's in operation. They are located at Winslow, Arizona (INW); Sandberg, California (SDB); Del Rio, Texas (DRT); and Wendover, Utah (ENV). These stations report all normal surface aviation weather elements, but cloud height and visibility are reported in a manner different from the conventional weather report. (See Table 7-10[1] for a description of these reports.)

d. Automated Weather Observing System (AWOS)—

1. Automated weather reporting systems are increasingly being installed at airports. (See Key to AWOS: Figure 7-10[1].) These systems consist of various sensors, a processor, a computer-generated voice subsystem, and a transmitter to broadcast local, minute-by-minute weather data directly to the pilot.

7-10d1 NOTE—When the barometric pressure exceeds 31.00 inches Hg., see Procedures, paragraph 7-41a2 for the altimeter setting procedures.

2. The AWOS observations will include the prefix "AWOS" to indicate that the data are derived from an automated system. Some AWOS locations will be augmented by certified observers who will provide weather and obstruction to vision information in the remarks of the report when the reported visibility is less than 3 miles. These sites, along with the hours of augmentation, are to be published in the Airport/Facility Directory. Augmentation is identified in the observation as "OBSERVER WEATHER." The AWOS wind speed, direction and gusts, temperature, dew point, and altimeter setting are exactly the same as for manual observations. The AWOS will also report density altitude when it exceeds the field elevation by more than 1,000 feet. The reported visibility is derived from a sensor near the touchdown of the primary instrument runway. The visibility sensor output is converted to a runway visibility value (RVV) equation, using a 10-minute harmonic average. The AWOS sensors have been calibrated against the FAA transmissometer standards used for runway visual range values. Since the AWOS visibility is an extrapolation of a measurement at the touchdown point of the runway, it may differ from the standard prevailing visibility. The reported sky condition/ceiling is derived from the ceilometer located next to the visibility sensor. The AWOS algorithm integrates the last 30 minutes of ceilometer data to derive cloud layers and heights. This output may also differ from the *observer* sky condition in that the AWOS is totally dependent upon the cloud advection over the sensor site.

3. These real-time systems are operationally classified into four basic levels: AWOS-A, AWOS-1, AWOS-2, and AWOS-3. AWOS-A only reports altimeter setting. AWOS-1 usually reports altimeter setting, wind data, temperature, dewpoint, and density altitude. AWOS-2 provides the information provided by AWOS-1 plus visibility. AWOS-3 provides the information provided by AWOS-2 plus cloud/ceiling data.

4. The information is transmitted over a discrete radio frequency or the voice portion of a local NAVAID. AWOS transmissions are receivable within 25 NM of the AWOS site, at or above 3,000 feet AGL. In many cases, AWOS signals may be received on the surface of the airport. The system transmits a 20 to 30 second weather message updated each minute. Pilots should monitor the designated frequency for the automated weather broadcast. A description of the broadcast is contained in subparagraph e. There is no two-way communication capability. Most AWOS sites also have a dial-up capability so that the minute-by-minute weather messages can be accessed via telephone.

5. AWOS information (system level, frequency, phone number, etc.) concerning specific locations is published, as the systems become operational, in the Airport/Facility Directory, and where applicable, on published Instrument Approach Procedures. Selected individual systems may be incorporated into nationwide data collection and dissemination networks in the future.

e. Automated Weather Observing System (AWOS) Broadcasts-Computer-generated voice is used in Automated Weather Observing Systems (AWOS) to automate the broadcast of the minute-by-minute weather observations. In addition, some systems are configured to permit the addition of an operator-generated voice message; e.g., weather remarks following the automated parameters. The phraseology used generally follows that used for other weather broadcasts. Following are explanations and examples of the exceptions.

1. *Location and Time*—The location/name and the phrase "AUTOMATED WEATHER OBSERVATION," followed by the time are announced.

(a) If the airport's specific location is included in the airport's name, the airport's name is announced.

EXAMPLES:

"BREMERTON NATIONAL AIRPORT AUTOMATED WEATHER OBSERVATION, ONE FOUR FIVE SIX ZULU;"

"RAVENSWOOD JACKSON COUNTY AIRPORT AUTOMATED WEATHER OBSERVATION, ONE FOUR FIVE SIX ZULU."

(b) If the airport's specific location *is not* included in the airport's name, the location is announced followed by the airport's name.

EXAMPLES:

"SAULT STE MARIE, CHIPPEWA COUNTY INTERNATIONAL AIRPORT AUTOMATED WEATHER OBSERVATION;"

"SANDUSKY, COWLEY FIELD AUTOMATED WEATHER OBSERVATION";

(c) The word "TEST" is added following "OBSERVATION" when the system is not in commissioned status.

EXAMPLE:

"BREMERTON NATIONAL AIRPORT AUTOMATED WEATHER OBSERVATION TEST, ONE FOUR FIVE SIX ZULU."

KEY TO AWOS (AUTOMATED WEATHER OBSERVING SYSTEM) OBSERVATIONS

LOCATION IDENTIFIER TYPE OF REPORT TIME OF REPORT STATION TYPE	SKY CONDITION AND CEILING BELOW 12,000'	VISIBILITY	TEMPERATURE / DEW POINT / WIND DIRECTION, SPEED AND CHARACTER / ALTIMETER SETTING /	<u>REMARKS</u>: **AUTOMATED** REMARKS GENERATED AUTOMATICALLY IF CONDITIONS EXIST. **AUGMENTED** REMARKS ADDED **IF** CONDITIONS EXIST **AND** CERTIFIED WEATHER OBSERVER IS ATTENDING THE SYSTEM
HTM SA 1755 AWOS	**M20 OVC**	**1V**	**36/34/2015G25/990/**	**P010/VSBY 1/2V2 WND 17V23/WEA: R — F**

LOCATION IDENTIFIER: 3 or 4 alphanumeric characters (usually the airport identifier).

TYPE OF REPORT: SA = Scheduled record (routine) observation. All observations identified as SA. Most are transmitted at 20-minute intervals (approximately 15, 35 and 55 minutes past each hour).

TIME OF REPORT: Coordinated Universal Time (UTC or Z) using 24-hour clock.

STATION TYPE: AWOS = Automated Weather Observing System site. **Note:** In the future, some systems will use " AO " designators.

SKY CONDITION AND CEILING: Sky condition contractions are for each layer in ascending order. Numbers preceding contractions are base heights in hundreds of feet above ground level (AGL).

CLR BLO 120 = No clouds below 12,000 ft.
SCT = Scattered: 0.1 to 0.5 sky cover.
BKN = Broken: 0.6 to 0.9 sky cover.
OVC = Overcast: More than 0.9 sky cover.
X = Obscured sky **—X** = Partially obscured

A letter preceding the height of a base identifies a ceiling layer and indicates how ceiling height was determined.

M = Measured **W** = Indefinite

VISIBILITY: Reported in statute miles and fractions. Visibility greater than 10 not reported. **V**= variable: see Automated Remarks

TEMPERATURE AND DEW POINT: Reported in degrees Fahrenheit.

WIND DIRECTION, SPEED & CHARACTER: Direction in tens of degrees from true north, except voice broadcast is in degrees magnetic. Speed in knots. **0000** = calm. **G** = gusts. See Automated Remarks for variable direction.

ALTIMETER SETTING: Hundredths of inches of mercury. Shown as last 3 digits only without decimal point (e.g., 30.05 inches = 005).

PRESENT WEATHER/OBSTRUCTIONS TO VISION: Reported only when observer is available. See Augmented Remarks. In the future, some systems will report precipitation, fog, and haze in the body of the observation.

AUTOMATED REMARKS: Precipitation accumulation reported in hundredth of inches (e.g., P110 = 1.10 inches; P010 = 0.10 inch). **WND V** = variable wind direction. **VSBY V** = variable visibility. **DENSITY ALTITUDE** is included in the voice broadcast when more than 1000 feet above airport elevation.

MISSING DATA: Reported as " M ".

AUGMENTED REMARKS: " WEA: " indicates manual observer data. Remarks include operationally significant weather conditions within a five mile radius of the airport (e.g., thunderstorms, precipitation, obstructions to vision when visibility is 3 miles or less, fog banks). Standard weather observation contractions are used.

DECODED REPORT: Hometown Municipal Airport, observation at 1755 UTC, AWOS report. Measured ceiling 2000 feet overcast. Visibility 1 mile variable. Temperature 36 degrees (F), dew point 34 degrees (F), wind from 200 degrees true at 15 knots gusting to 25 knots, altimeter setting 29.90 inches. Precipitation accumulation during past hour 0.10 inch. Visibility variable between 1/2 and 2 miles. Wind direction variable from 170 degrees to 230 degrees true. Observer reports light rain (R—) and fog (F).

NOTE: Refer to the *Airman's Information Manual* for more information. Refer to the *Airport/Facility Directory*, aeronautical charts, and related publications for broadcast, telephone and location data. Check *Notices to Airmen* for AWOS system status.

Figure 7-10[1]

(d) The phrase "TEMPORARILY INOPERATIVE" is added when the system is inoperative.

EXAMPLE:

"BREMERTON NATIONAL AIRPORT AUTOMATED WEATHER OBSERVING SYSTEM TEMPORARILY INOPERATIVE."

2. *Ceiling and Sky Cover—*

(a) Ceiling is announced as either "CEILING" or "INDEFINITE CEILING." The phrases "MEASURED CEILING" and "ESTIMATED CEILING" are not used. With the exception of indefinite ceilings, all automated ceiling heights are measured.

EXAMPLES:

"BREMERTON NATIONAL AIRPORT AUTOMATED WEATHER OBSERVATION, ONE FOUR FIVE SIX ZULU. CEILING TWO THOUSAND OVERCAST";

"BREMERTON NATIONAL AIRPORT AUTOMATED WEATHER OBSERVATION, ONE FOUR FIVE SIX ZULU. INDEFINITE CEILING TWO HUNDRED, SKY OBSCURED."

(b) The word "Clear" is not used in AWOS due to limitations in the height ranges of the sensors. No clouds detected is announced as "NO CLOUDS BELOW XXX" or, in newer systems as "CLEAR BELOW XXX" (where XXX is the range limit of the sensor).

EXAMPLES:

"NO CLOUDS BELOW ONE TWO THOUSAND."

"CLEAR BELOW ONE TWO THOUSAND."

(c) A sensor for determining ceiling and sky cover is not included in some AWOS. In these systems, ceiling and sky cover are not announced. "SKY CONDITION MISSING" is announced only if the system is configured with a ceilometer and the ceiling and sky cover information is not available.

3. *Visibility—*

(a) The lowest reportable visibility value in AWOS is "less than ¼." It is announced as "VISIBILITY LESS THAN ONE QUARTER."

(b) A sensor for determining visibility is not included in some AWOS. In these systems, visibility is not announced. "VISIBILITY MISSING" is announced only if the system is configured with a visibility sensor and visibility information is not available.

4. *Weather*—In the future, some AWOS's are to be configured to determine the occurrence of precipitation. However, the type and intensity may not always be determined. In these systems, the word "PRECIPITATION" will be announced if precipitation is occurring, but the type and intensity are not determined.

5. *Remarks*—If remarks are included in the observation, the word "REMARKS" is announced following the altimeter setting. Remarks are announced in the following order of priority:

(a) Automated "Remarks"

(1) Density Altitude.

(2) Variable Visibility.

(3) Variable Wind Direction.

(b) Manual Input Remarks.—Manual input remarks are prefaced with the phrase "OBSERVER WEATHER." As a general rule the manual remarks are limited to:

(1) Type and intensity of precipitation,

(2) Thunderstorms, intensity (if applicable) and direction, and

(3) Obstructions to vision when the visibility is 3 miles or less.

EXAMPLE:

"REMARKS ... DENSITY ALTITUDE, TWO THOUSAND FIVE HUNDRED ... VISIBILITY VARIABLE BETWEEN ONE AND TWO ... WIND DIRECTION VARIABLE BETWEEN TWO FOUR ZERO AND THREE ONE ZERO ... OBSERVED WEATHER ... THUNDERSTORM MODERATE RAIN SHOWERS AND FOG ... THUNDERSTORM OVERHEAD."

(c) If an automated parameter is "missing" and no manual input for that parameter is available, the parameter is announced as "MISSING." For example, a report with the dew point "missing" and no manual input available, would be announced as follows:

EXAMPLE:

"CEILING ONE THOUSAND OVERCAST ... VISIBILITY THREE ... PRECIPITATION... TEMPERATURE THREE ZERO, DEW POINT MISSING... WIND CALM ... ALTIMETER THREE ZERO ZERO ONE."

(d) "REMARKS" are announced in the following order of priority:

(1) Automated "REMARKS."

I Density Altitude

II Variable Visibility

III Variable Wind Direction

(2) Manual Input "Remarks." As a general rule, the remarks are announced in the same order as the parameters appear in the basic text of the observation; i.e., Sky Condition, Visibility, Weather and Obstructions to Vision, Temperature, Dew Point, Wind, and Altimeter Setting.

EXAMPLE:

"REMARKS ... DENSITY ALTITUDE, TWO THOUSAND FIVE HUNDRED ... VISIBILITY VARIABLE BETWEEN ONE AND TWO ... WIND DIRECTION VARIABLE BETWEEN TWO FOUR ZERO AND THREE ONE ZERO ... OBSERVER CEILING ESTIMATED TWO THOUSAND BROKEN ... OBSERVER TEMPERATURE TWO, DEW POINT MINUS FIVE."

f. Automated Surface Observation System (ASOS)—The ASOS is the primary surface weather observing system of the United States. (See Key to ASOS: Figure 7-10[2].) The program to install and operate up to 1,700 systems throughout the United States is a joint effort of the NWS, the FAA and the Department of Defense. ASOS is designed to support aviation operations and weather forecast activities. The ASOS will provide continuous minute-by-minute observations and perform the basic observing functions necessary to generate a Surface Aviation Observation (SAO) and other aviation weather information. While the automated system and the human may differ in their methods of data collection and interpretation, both produce an observation quite similar in form and content.

KEY TO ASOS (AUTOMATED SURFACE OBSERVING SYSTEM) WEATHER OBSERVATIONS

LOCATION IDENTIFIER TYPE OF REPORT TIME OF REPORT STATION TYPE	SKY CONDITION AND CEILING BELOW 12,000'	VISIBILITY, WEATHER, AND OBSTRUCTIONS TO VISION	SEA-LEVEL PRESSURE / TEMPERATURE / DEW POINT / WIND DIRECTION, SPEED AND CHARACTER / ALTIMETER SETTING /	REMARKS **AUTOMATED** REMARKS GENERATED AUTOMATICALLY IF CONDITIONS EXIST. **AUGMENTED** REMARKS ADDED IF CONDITIONS EXIST AND CERTIFIED WEATHER OBSERVER IS ATTENDING THE SYSTEM	STATUS REMARKS SYSTEM GENERATED
HTM RS 1755 AO2A	**M19V OVC**	**1R — F**	**125/36/34/2116G24/990/**	**R29LVR10V50 CIG 16V22 TWR VSBY 2 PK WND 2032/1732 PRESFR**	**ZRNO $**

LOCATION IDENTIFIER:
3 or 4 alphanumeric characters (usually airport identifier).

TYPE OF REPORT:
SA = Scheduled record (hourly) observation.
SP = Special observation indicating a significant change in one or more of the observed elements.
RS = SA that also qualifies as an SP.
USP = Urgent special observation to report tornado.

TIME OF REPORT:
Coordinated Universal Time (UTC or Z) using 24-hr clock.

STATION TYPE:
AO2 = Unattended (no observer) ASOS.
AO2A = Attended (observer present) ASOS.

SKY CONDITION AND CEILING BELOW 12,000' AGL:
Sky condition contractions are for each layer in ascending order. Numbers preceding contractions are base height in hundreds of feet above ground level (AGL).
CLR BLO 120 = Less than 0.1 sky cover below 12,000'
SCT = Scattered: 0.1 to 0.5 sky cover.
BKN = Broken: 0.6 to 0.9 sky cover.
OVC = Overcast: More than 0.9 sky cover.
A letter preceding the height of a base identifies a ceiling layer and indicates how ceiling height was determined.
M = Measured   **W** = Indefinite
E = Estimated   **X** = Obscured sky
The letter **V** is added immediately following the height of a base to indicate a variable ceiling: see Remarks.

VISIBILITY:
Reported in statute miles and fractions from **<1/4** through **10+**. **V** = variable: see Remarks.

PRESENT WEATHER:
TORNADO (when augmented).
T = Thunder (when augmented): see Status Remarks.
R = Liquid precipitation that does not freeze (e.g., rain).
P— = Light precipitation in unknown form.
ZR = Liquid precipitation that freezes on impact (e.g., freezing rain): see Status Remarks.
A = Hail (when augmented).
S = Frozen precipitation other than hail (e.g., snow).
+ = Heavy. No sign = Moderate. **—** = Light.

OBSTRUCTIONS TO VISION :
Reported only when visibility is less than 7 statute miles.
F = Fog   **H** = Haze
VOLCANIC ASH (when augmented).

SEA-LEVEL PRESSURE:
Tenths of Hectopascals (millibars). Shown as last 3 digits only without decimal point (e.g., 950 = 995.0).

TEMPERATURE AND DEW POINT:
Degrees Fahrenheit.

WIND DIRECTION, SPEED AND CHARACTER:
Direction in tens of degrees from **true** north. Voice broadcast in degrees from **magnetic**. Speed in knots. **0000** = calm. **E** = estimated. **G** = gusts. **Q** = squalls. Variable wind, peak wind, wind shift: see Remarks.

ALTIMETER SETTING:
Hundredths of inches of mercury. Shown as last 3 digits only without decimal point (e.g., 005 = 30.05 inches).

MISSING DATA: Reported as **M**.

DENSITY ALTITUDE: Included on voice broadcast only when 1000 or more feet above airport elevation.

REMARKS: Can Include:
RVR (Runway Visual Range), **VOLCANIC ASH**, **VIRGA**, **TWR VSBY** (Tower visibility), **SFC VSBY** (Surface visibility), **VSBY V** (Variable visibility), **CIG V** (Variable ceiling), **WSHFT** (Windshift), **PK WND** (Peak wind), **WND V** (Variable wind direction), **PCPN** (Precipitation amount), **PRESRR** (Pressure rising rapidly), **PRESFR** (Pressure falling rapidly), **PRJMP** (Pressure jump), **B** (Time weather began), **E** (Time weather ended).

STATUS REMARKS:
PWINO = Present weather information not available.
ZRNO = Freezing rain information not available.
TNO = Thunderstorm information not available.
$ = Maintenance check indicator.

DECODED REPORT: Hometown Municipal Airport, record special observation at 1755 UTC. ASOS with observer. Measured ceiling 1900 feet variable, overcast. Visibility 1 mile, light rain, fog. Sea-level pressure 1012.5 hectopascals, temperature 36°F, dew point 34°F, wind from 210° true at 16 knots gusting to 24 knots, altimeter 29.90 inches. Runway 29L visual range 1000 variable to 5000 feet. Ceiling 1600 variable to 2200 feet, tower visibility 2 miles, peak wind 200° true at 32 knots at 1732 UTC, pressure falling rapidly. Freezing rain information not available, maintenance check indicator.

NOTE: Refer to *ASOS Guide for Pilots* and the *Airman's Information Manual* for more information. Refer to the *Airport/Facility Directory*, aeronautical charts, and related publications for broadcast, telephone and location data. Check *Notices to Airmen* for ASOS system status.

Figure 7-10[2]

For the "objective" elements such as pressure, ambient temperature, dew point temperature, wind, and precipitation accumulation, both the automated system and the observer use a fixed location and time-averaging technique. The quantitative differences between the observer and the automated observation of these elements are negligible. For the "subjective" elements, however, observers use a fixed time, spatial averaging technique to describe the visual elements (sky condition, visibility and present weather), while the automated systems use a fixed location, time averaging technique. Although this is a fundamental change, the manual and automated techniques yield remarkably similar results within the limits of their respective capabilities.

1. *System Description:*

(a) The ASOS at each airport location consists of four main components:

(1) Individual weather sensors.

(2) Data collection package(s). (DCP)

(3) The acquisition control unit.

(4) Peripherals and displays.

(b) The ASOS sensors perform the basic function of data acquisition. They continuously sample and measure the ambient environment, derive raw sensor data and make them available to the collocated DCP.

2. *Every ASOS will contain the following basic set of sensors:*

(a) Cloud height indicator (one or possibly three).

(b) Visibility sensor (one or possibly three).

(c) Precipitation identification sensor.

(d) Freezing rain sensor.

(e) Pressure sensors (two sensors at small airports; three sensors at large airports).

(f) Ambient temperature/Dew point temperature sensor.

(g) Anemometer (wind direction and speed sensor).

(h) Rainfall accumulation sensor.

3. *The ASOS data outlets include:*

(a) Those necessary for on-site airport users.

(b) National communications networks.

(c) Computer-generated voice (available through FAA radio broadcast to pilots, and dial-in telephone line).

4. *The common ASOS reports available through these outlets include:*

(a) SAO messages which include message types; Scheduled Record Hourly (SA), Record Special (RS), and Special (SP) observations.

EXAMPLE:

ELEMENTS OF THE SURFACE AVIATION OBSERVATION (SAO) STATION IDENTIFIER/OBSERVATION/TYPE/TIME/STATION TYPE/SKY CONDITION VISIBILITY/WEATHER & OBSTRUCTIONS TO VISION/SEA-LEVEL PRESSURE/TEMPERATURE/DEW-POINT TEMPERATURE/WIND DIRECTION, SPEED & CHARACTER/ALTIMETER SETTING/REMARKS

(b) If an element in the main body of the observation (i.e., sky condition through altimeter setting) is missing, and no backup data are available, the element is encoded as M. If an element in the REMARKS section is missing, it is encoded as /.

(1) *STATION IDENTIFIER*—Three or four alphanumeric characters identifying the observation site.

(2) *OBSERVATION TYPE*—Record observations (SA) are scheduled on a routine basis; Special observations (SP) are taken whenever certain events occur; record special observations (RS) are record observations coincidental with the occurrence of certain events; urgent special observations (USP) are taken to report tornados.

(3) *TIME*—The time of the observation in coordinated universal time (UTC) using a 24-hour clock.

(4) *STATION TYPE*—Unaugmented or unedited ASOS observations are identified as AO2; ASOS observations identified as AO2A are augmented and/or edited.

(5) *SKY CONDITION*—CLR BLO 120 means no clouds detected below 12,000 feet over the ASOS cloud height indicator. SCT (scattered) means that .1 to .5 of sky is covered; BKN (broken) means .6 to .9 is covered; OVC (overcast) means all the sky is covered. The number preceding the SCT, BKN, or OVC is the height in hundreds of feet. The height of the BKN or OVC is preceded by a ceiling designator: M (measured) or E (estimated). If a V appears after the height, the ceiling is variable and a remark is included. W indicates an indefinite ceiling, the number after the W is the vertical visibility in hundreds of feet, and X indicates the sky is totally obscured.

(6) *VISIBILITY*—ASOS reports an instrumentally derived visibility value. Reportable values are: << ¼, ¼, 1/2, ¾, 1, 1¼, 1½, 1¾, 2, 2½, 3, 3½, 4, 5, 7, and 10+ statute miles. If the visibility is variable, a V is added to the average reportable value and a remark is included in the observation. When tower visibility is less than 4 miles and less than surface visibility, tower visibility will be reported in the body of the report as prevailing visibility, and surface visibility will be reported in remarks.

(7) *WEATHER AND OBSTRUCTIONS TO VISION*—Weather refers to precipitation, tornados and thunder; obstructions to vision refer to phenomena that reduce visibility but are not precipitation. ASOS reports snow (S) which is any form of frozen precipitation other than hail, rain (R), any form of liquid precipitation that does not freeze, freezing rain (ZR) any form of liquid precipitation that freezes on impact, and precipitation (P–) any light precipitation that the ASOS cannot identify as R, S or ZR. R, S, and ZR may be reported as heavy (+), moderate (no sign), or light (–). P is always light (–). Tornados (TORNADO), thunder (T), and hail (A) are augmented by an observer. Obstructions to vision are either fog (F), visible minute water droplets, or haze (H), fine suspended particles of dust, salt, combustion, or pollution products. Volcanic ash (VOLCANIC ASH) is augmented by an observer as an obstruction to vision if visibility is less than 7 miles.

(8) *SEA-LEVEL PRESSURE*—The theoretical pressure at sea-level reported in tenths of Hectopascals (millibars) using the last three digits without the decimal point.

(9) *TEMPERATURE*—Air temperature reported in degrees Fahrenheit.

(10) *DEW-POINT TEMPERATURE*—Temperature to which air must be cooled at constant pressure and water-vapor content for saturation to occur reported in degrees Fahrenheit.

(11) *WIND*—The direction reported in tens of degrees from true North from which the wind is blowing; may be estimated (E).

(12) *WIND SPEED*—Reported in whole knots; may be estimated (E). If wind is calm, wind direction and speed are reported as 0000.

(13) *WIND CHARACTER*—Gusts (G) and squalls (Q) are reported in whole knots.

(14) *ALTIMETER SETTING*—Used for setting aircraft altimeters; reported in hundredths of inches of mercury using only the units, tenths, and hundredths digits without a decimal point.

(15) *REMARKS NOTE*—Remarks underlined below are augmented and not generated automatically by ASOS.

(a) **RVR**—Runway Visual Range in hundredths of feet.

(b) **VOLCANIC ASH**—Volcanic ash is present but does not restrict visibility to less than 7 miles.

(c) **VIRGA**—Precipitation falling from clouds, but not reaching the ground, e.g., VIRGA VCNTY STN (Virga within 10 statute miles of the station when augmented by an observer).

(d) *CIG minVmax*—Variable ceiling remark, e.g., CIG 5V10 means the ceiling varies between 500 and 1,000 feet.

(e) **TWR VSBY**—Visibility reported by airport traffic control tower, e.g., TWR VSBY 1 indicates the visibility from the tower was 1 mile.

(f) *VSBY*—Visibility reported by ASOS visibility sensor.

(g) *VSBY minVmax*—Variable visibility remark, e.g., VSBY ½V2 means the visibility varies between ½ and 2 miles.

(h) *—Btt—Ett*—Time of beginning and ending of weather, e.g., RB05E20SB20E55 means that rain began at 5 minutes past the hour and ended at 20 past and snow began at 20 and ended at 55 minutes past the hour.

(i) *PCPN rrrr*—Hourly precipitation accumulation remark where rrrr is the liquid equivalent in hundredths of an inch of all precipitation since the last hourly observation, e.g., PCPN 0009 means that 9 hundredths of an inch fell since the last hourly observation; a trace is reported as PCPN 0000; PCPN M = missing.

(j) *WSHFT hhmm*—Wind shift remark with the time the wind shift began, e.g., WSHFT 1730 means the wind shift began at 1730 UTC.

(k) *WND ddVdd*—Variable wind direction remark, e.g., WND 03V12 means the wind is varying between 30 and 120 degrees.

(l) *PK WND ddff/hhmm*—Peak wind remark used whenever peak wind is over 25 knots, e.g., PK WND 2845/1715 means that a peak wind of 45 knots from 280 degrees was observed at 1715 UTC.

(m) *PRESRR*—Pressure rising rapidly remark.

(n) *PRESFR*—pressure falling rapidly remark.

(o) *PRJMP*—Pressure jump remark, e.g., PRJMP 13/1250/1312 means that a pressure jump of .13 inches of mercury began at 1250 UTC and ended at 1312 UTC.

(p) *PWINO*—Precipitation Identifier sensor is not operational.

(q) *ZRNO*—Freezing rain sensor is not operational.

(r) *TNO*—Thunderstorm information not available.

(s) *$*—Maintenance check indicator.

5. *THE AUTOMATED AND THE MANUAL SAO*—In form, the automated SAO and the manual SAO look very much alike. For instance, under the same circumstances, the ASOS would report:

(a) *RIC SA 1950 AO2 M80 BKN 5H 123/45/40/1106/013* while the observer might report:

(b) *RIC SA 1951 E80 BKN 140 OVC 6H 123/45/40/1106/013 FEW CU 20.* Notice that both SAO's contain the station ID, observation type, time, sky condition, visibility, obstructions to vision, sea-level pressure, ambient temperature, dew point temperature, wind, and altimeter setting. At first glance, the main differences are the ASOS station type (AO2) signifying an unattended observation (i.e., human oversight not provided for ASOS backup or augmentation), the 12,000 foot upper limit in ASOS for reporting cloud height, the lower visibility increment reported by the ASOS and the cloud type remark in the manual observation. Other locations operate with on-site oversight and intervention in the form of augmentation and/or backup of the SAO message. This type of station is designated as AO2A in the body of the SAO. The AO2A designation means that the observation contains augmented and/or backup data. The AO2 or AO2A designator will appear in all transmitted ASOS SAOs. Closer examination reveals that the ASOS reported ceiling is always measured (M), while the manual ceiling observation in this example is estimated (E). Clouds above 12,000 feet are not reported by the ASOS. The ASOS visibility of 5 miles, although different from the manual observation of 6 miles, is intentionally reported as 5 miles whenever the ASOS measured visibility is between 5 and 7 miles. The final difference between the human observation and the ASOS report in the above example is in the remarks. The automated system cannot determine cloud type, range, or azimuth of cloud phenomena, and, therefore, a cloud type remark is not included in the ASOS report. The following sections will examine each of the weather elements reported by the ASOS.

(1) *SKY CONDITION*—The ASOS cloud sensor is a vertically pointing laser ceilometer, and is referred to as the cloud height indicator (CHI). This CHI is used to detect the presence of clouds directly overhead up to 12,000 feet.

(2) *VISIBILITY*—ASOS sensor visibility is based on measurement of a small volume sample with extrapolation to overall areal visibility. ASOS visibility is based on the scattering of light by air molecules, precipitation, fog droplets, haze specks, or other particles suspended in the air. The visibility sensor projects a beam of light, and the light that is scattered is detected by a receiver. The amount of light scattered and then received by the sensor is converted into a visibility value.

(3) *PRESENT WEATHER/OBSTRUCTIONS TO VISION*—There are currently two ASOS present weather sensors. They are the precipitation identification (PI) sensor which discriminates between rain and snow, and the freezing rain sensor. ASOS algorithms have also been developed to evaluate multiple sensor data and infer the presence of obstructions to vision (fog or haze). The PI sensor has the capability to detect and report R–, R, R+, S–, S, S+. When the precipitation type cannot be determined (e.g., mixed rain and snow), it will report a P–. The PI sensor has a rainfall and snow detection threshold of 0.01 inch per hour. The freezing rain sensor is sensitive enough to measure accumulation rates as low as 0.01 of an inch per hour. Only one present weather phenomenon other than thunder, tornado, or hail will be reported at one time, i.e., ASOS does not report mixed precipitation. If freezing rain is reported when the PI sensor indicates no precipitation or rain, the output is freezing rain. If freezing rain is detected when the PI indicates snow, snow is reported. Present weather remarks are generated and appended to the observation. This includes weather beginning and ending times in minutes past the current hour, e.g., ZRB05E22. Obstructions to vision (OTV) are not directly measured by ASOS, but rather inferred from measurements of visibility, temperature and dew point. All OTV's are reported as either Fog (F) or Haze (H) by ASOS. OTVs are reported only when the visibility drops below 7 statute miles.

(4) *TEMPERATURE AND DEW POINT*—The ASOS temperature sensors directly measure the ambient temperature and the dew point temperature. Temperature data are stored and processed to compute various required temperature parameters such as the calendar day and monthly maximum and minimum temperatures and heating and cooling degree days.

(5) *WIND*—The rotating cup anemometer and the simple wind vane are the principal indicators of wind speed and direction. The ASOS algorithm uses a 2-minute period to obtain average wind direction and speed. Wind character (i.e., gusts and squalls) and the peak wind are obtained by comparing the average wind speed with the maximum "instantaneous" wind speed observed over a specified time interval. All wind speeds are reported to the nearest knot. Wind direction is the direction from which the wind is blowing and is reported to the nearest 10 degrees relative to true North in the SAO message and in the daily/monthly summaries. Wind direction is reported relative to magnetic north in the computer-generated voice messages. Under appropriate conditions either a gust or a squall may be reported in the SAO observation, but not both. Where there is contention, squalls are given precedence. The ASOS will report wind remarks in the SAO such as wind shift, variable wind, and peak wind. Wind data are stored and processed to compute daily peak wind and fastest 2-minute wind parameters.

(6) *PRESSURE*—The pressure measurement sensors used in ASOS consist of redundant digital pressure transducers. Because of the criticality of pressure determination, three separate and independent pressure sensors are used at towered airport locations. At other locations, two pressure sensors are used. The pressure parameters available from ASOS are:

(a) Altimeter Setting.

(b) Sea Level Pressure.

(c) Density Altitude (radio broadcast only).

(d) Pressure Altitude (not transmitted).

(e) Pressure Change/Tendency.

(f) Pressure remarks (such as Pressure Rising Rapidly [PRESSRR], Pressure Jump [PRJMP], as well as pressure change and character (5appp)

(7) *PRECIPITATION ACCUMULATION*—ASOS uses a heated tipping bucket (HTB) precipitation gauge. The HTB at high rain rates underestimates rainfall, and a correction factor is applied to 1-minute rainfall amounts to correct for this error. The gauge data are used by ASOS to compile a variety of cumulative precipitation remarks/messages. These remarks/messages include:

(a) SAO hourly precipitation (PCPN rrrr) remark.

(b) SAO 3- and 6-hourly precipitation report (6RRR/) remark.

(c) SAO 24-hour precipitation accumulation remark (7R24R24R24R24)

(d) Daily and monthly cumulative precipitation totals.

(e) SHEF messages.

6. *DATA NOT PROVIDED BY ASOS*—The reporting of certain weather phenomena is not part of the ASOS. Plans are currently underway to observe and report the occurrence of these phenomena from other sources. These alternate sources may include augmentation of the ASOS observation for specified phenomena at selected locations, separate observing networks for specific elements, complementary remote sensing technologies such as radar, satellites, and lightning detection systems, and of course, future enhancements to ASOS capabilities. The elements not currently sensed and reported as such by ASOS include:

(a) Tornado, funnel cloud, waterspout.

(b) Thunderstorms.

(c) Hail.

(d) Ice crystals.

(e) Snow pellets, snow grains, ice pellets.

(f) Drizzle, freezing drizzle.

(g) Volcanic ash.
(h) Blowing obstructions (snow, sand, dust, spray).
(i) Smoke.
(j) Snow fall.
(k) Snow depth.
(l) Water equivalent of snow on the ground.
(m) Clouds above 12,000 feet.
(n) Virga and distant precipitation.
(o) Distant clouds.
(p) Operationally significant local variations in visibility.

7. *BACKUP AND AUGMENTATION*—Backup is the process of either manually editing specific elements within the ASOS observation prior to dissemination, or providing a complete manual observation and alternate means of dissemination in case of total ASOS failure. Backup includes substituting manually observed data for "missing" or unrepresentative data to ensure that message content is correct and complete. No non-ASOS data element is added to the ASOS observation through backup. Backup may be applied to the ASOS-generated SAO messages or daily or monthly summary products. Augmentation is the process of adding information to an ASOS SAO message, or daily or monthly summary message, prior to dissemination, that is beyond the capabilities of ASOS. This information is derived by an observer. An example of augmentation would be adding the occurrence of thunder. ASOS may be augmented and backed up at the same time. Backup and augmented data are used by ASOS in the computations for both the daily and monthly summary products.

8. *MISSING DATA vs NON-EVENT DATA*:

(a) It is possible to distinguish between missing data and a nonevent, i.e., nonoccurrence, of a parameter from an ASOS SAO. Generally, if data routinely reported in the body of an ASOS SAO is missing, an M will appear in place of the missing element. An M, however, is not placed in the present weather and obstructions to vision fields of the SAO when their sensors are determined to be inoperative. This is because these elements are derived from more than one sensor and are only reported when conditions warrant.

(b) In the event that the precipitation identification sensor is not operational, the remark PWINO (Present Weather Identification Not Operational) will be placed in the remarks section of the SAO. In this case, ASOS will not be able to report R, S, or P. If the freezing rain sensor is not operational, the remark ZRNO (Freezing Rain Not Operational) will appear in the SAO's remarks. When no precipitation is occurring at the site, and if the temperature and/or dew point are not functioning, the obstruction to vision will be reported as haze if the prevailing visibility is less than seven statute miles, but greater than or equal to 4 statute miles. If the visibility is less than 4 statute miles, fog will be reported. If the temperature or dewpoint sensor, and the visibility sensor are not reporting data, it is not possible for an ASOS site to report an obstruction to vision. It is possible to distinguish missing data from nonevent occurrences in remarks by the inclusion of a slant (/) in place of the missing data.

EXAMPLE:

IAD SA 1855 AO2A CLR BLO 120 6F 101/42/41/2804 991$

(c) In this observation, ASOS was not able to automatically report temperature or dew point due to a sensor malfunction. The operator, in accordance with his/her agency's backup plan, used a sling psychrometer to observe ambient temperature and determine dew point. The operator edited the observation to enter the data manually. Note that the designation for this site was changed from AO2 to AO2A. There is, however, no indication in the body of the report that the temperature and dew point specifically were manually edited. The maintenance check indicator ($) is automatically appended to the observation by ASOS because of the inoperative temperature/dew point sensor.

EXAMPLE:

IAD SA 1555 AO2 25 SCT 7 M/M/M/3112/ 992 $

(d) In this SAO, sea-level pressure, temperature, and dew point are missing. Note that sea-level pressure is missing because of the missing temperature. M is used to indicate these missing data. The dollar sign ($) at the end of the observation is a maintenance check indicator that indicates ASOS may be in need of maintenance. This is an AO2 location where an observer is not available to back up these values.

7-11. WEATHER RADAR SERVICES

(See Figure 7-11[1] [NWS Radar Network].)

a. The National Weather Service operates a network of 56 radar sites for detecting coverage, intensity, and movement of precipitation. The network is supplemented by FAA and DOD radar sites in the western sections of the country. Another 72 local warning radar sites augment the network by operating on an as needed basis to support warning and forecast programs.

b. Scheduled radar observations are taken hourly and transmitted in alpha-numeric format on weather telecommunications circuits for flight planning purposes. Under certain conditions, special radar reports are issued in addition to the hourly transmittals. Data contained in the reports are also collected by the National Meteorological Center and used to prepare hourly national radar summary charts for dissemination on facsimile circuits.

c. All En route Flight Advisory Service facilities and many FSSs have equipment to directly access the radar displays from the individual weather radar sites. Specialists at these locations are trained to interpret the display for pilot briefing and in-flight advisory services. The Center Weather Service Units located in ARTCC's also have access to weather radar displays and provide support to all air traffic facilities within their center's area.

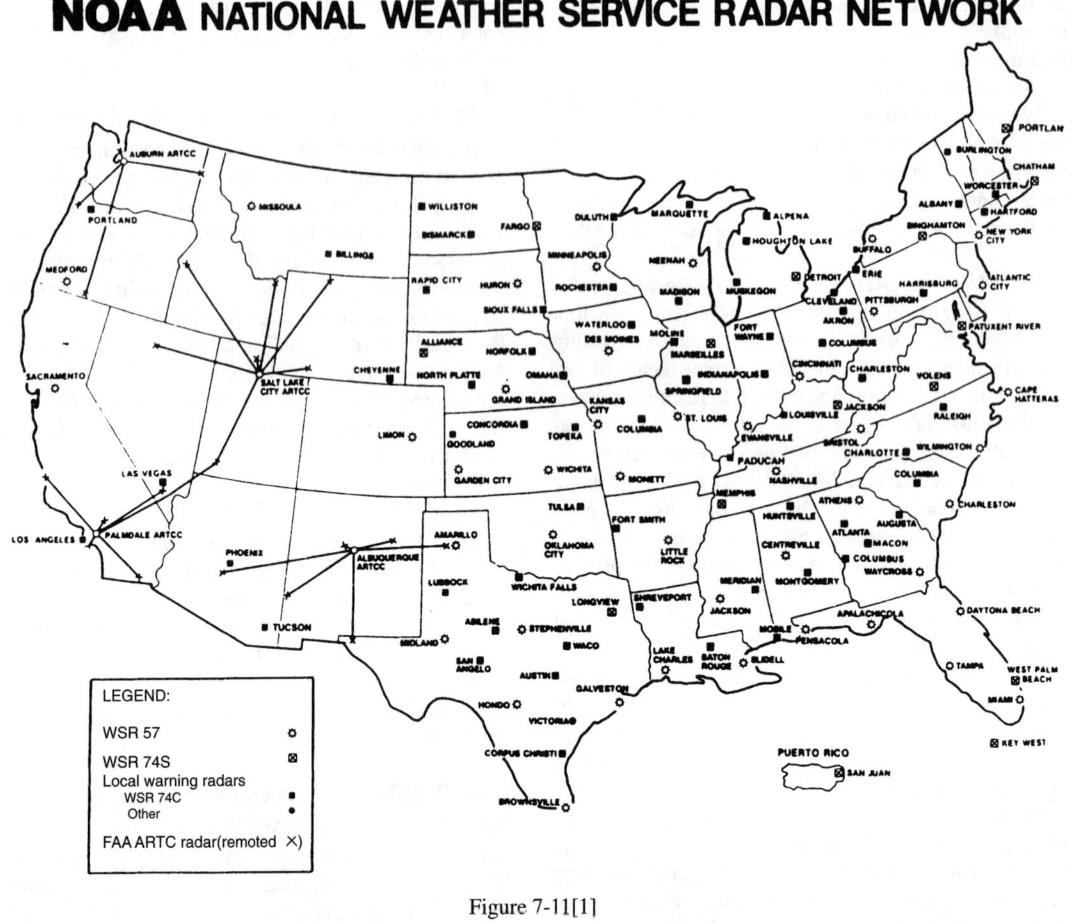

Figure 7-11[1]

d. A clear radar display (no echoes) does not mean that there is no significant weather within the coverage of the radar site. Clouds and fog are not detected by the radar. However, when echoes are present, turbulence can be implied by the intensity of the precipitation, and icing is implied by the presence of the precipitation at temperatures at or below zero degrees Celsius. Used in conjunction with other weather products, radar provides invaluable information for weather avoidance and flight planning.

e. Additional information on weather radar products and services can be found in AC 00-45, AVIATION WEATHER SERVICES. (Reference—Pilot/Controller Glossary, Radar Weather Echo Intensity Levels and Thunderstorms, paragraph 7-25.) (See A/FD charts, NWS Upper Air Observing Stations and Weather Network for the location of specific radar sites.)

7-12. ATC IN-FLIGHT WEATHER AVOIDANCE ASSISTANCE

a. ATC Radar Weather Display—

1. Areas of radar weather clutter result from rain or moisture. Radars cannot detect turbulence. The determination of the intensity of the weather displayed is based on its precipitation density. Generally, the turbulence associated with a very heavy rate of rainfall will normally be more severe than any associated with a very light rainfall rate.

2. ARTCC's are phasing in computer generated digitized radar displays to replace broadband radar display. This new system, known as Narrowband Radar, provides the controller with two distinct levels of weather intensity by assigning radar display symbols for specific precipitation densities measured by the narrowband system.

b. Weather Avoidance Assistance—

1. To the extent possible, controllers will issue pertinent information on weather or chaff areas and assist pilots in avoiding such areas when requested. Pilots should respond to a weather advisory by either acknowledging the advisory or by acknowledging the advisory and requesting an alternative course of action as follows:

(a) Request to deviate off course by stating the number of miles and the direction of the requested deviation. In this case, when the requested deviation is approved, the pilot is expected to provide his own navigation, maintain the altitude assigned by ATC and to remain within the specified mileage of his original course.

(b) Request a new route to avoid the affected area.

(c) Request a change of altitude.

(d) Request radar vectors around the affected areas.

2. For obvious reasons of safety, an IFR pilot must not deviate from the course or altitude or Flight Level without a proper ATC clearance. When weather conditions encountered are so severe that an immediate deviation is determined to be necessary and time will not permit approval by ATC, the pilot's emergency authority may be exercised.

3. When the pilot requests clearance for a route deviation or for an ATC radar vector, the controller must evaluate the air traffic picture in the affected area, and coordinate with other controllers (if ATC jurisdictional boundaries may be crossed) before replying to the request.

4. It should be remembered that the controller's primary function is to provide safe separation between aircraft. Any additional service, such as weather avoidance assistance, can only be provided to the extent that it does not derogate the primary function. It's also worth noting that the separation workload is generally greater than normal when weather disrupts the usual flow of traffic. ATC radar limitations and frequency congestion may also be a factor in limiting the controller's capability to provide additional service.

5. It is very important, therefore, that the request for deviation or radar vector be forwarded to ATC as far in advance as possible. Delay in submitting it may delay or even preclude ATC approval or require that additional restrictions be placed on the clearance. Insofar as possible the following information should be furnished to ATC when requesting clearance to detour around weather activity:

(a) Proposed point where detour will commence.

(b) Proposed route and extent of detour (direction and distance).

(c) Point where original route will be resumed.

(d) Flight conditions (IFR or VFR).

(e) Any further deviation that may become necessary as the flight progresses.

(f) Advise if the aircraft is equipped with functioning airborne radar.

6. To a large degree, the assistance that might be rendered by ATC will depend upon the weather information available to controllers. Due to the extremely transitory nature of severe weather situations, the controller's weather information may be of only limited value if based on weather observed on radar only. Frequent updates by pilots giving specific information as to the area affected, altitudes intensity and nature of the severe weather can be of considerable value. Such reports are relayed by radio or phone to other pilots and controllers and also receive widespread teletypewriter dissemination.

7. Obtaining IFR clearance or an ATC radar vector to circumnavigate severe weather can often be accommodated more readily in the en route areas away from terminals because there is usually less congestion and, therefore, offer greater freedom of action. In terminal areas, the problem is more acute because of traffic density, ATC coordination requirements, complex departure and arrival routes, adjacent airports, etc. As a consequence, controllers are less likely to be able to accommodate all requests for weather detours in a terminal area or be in a position to volunteer such routing to the pilot. Nevertheless, pilots should not hesitate to advise controllers of any observed severe weather and should specifically advise controllers if they desire circumnavigation of observed weather.

7-13. RUNWAY VISUAL RANGE (RVR)

a. RVR visibility values are measured by transmissometers mounted on towers along the runway. A full RVR system consists of:

1. Transmissometer projector and related items.

2. Transmissometer receiver (detector) and related items.

3. Analogue recorder.

4. Signal data converter and related items.

5. Remote digital or remote display programmer.

b. The transmissometer projector and receiver are mounted on towers either 250 or 500 feet apart. A known intensity of light is emitted from the projector and is measured by the receiver. Any obscuring matter such as rain, snow, dust, fog, haze or smoke reduces the light intensity arriving at the receiver. The resultant intensity measurement is then converted to an RVR value by the signal data converter. These values are displayed by readout equipment in the associated air traffic facility and updated approximately once every minute for controller issuance to pilots.

c. The signal data converter receives information on the high intensity runway edge light setting in use (step 3, 4, or 5); transmission values from the transmissometer, and the sensing of day or night conditions. From the three data sources, the system will compute appropriate RVR values. Due to variable conditions, the reported RVR values may deviate somewhat from the true observed visual range due to the slant range consideration, brief time delays between the observed RVR conditions and the time they are transmitted to the pilot, and rapidly changing visibility conditions.

d. An RVR transmissometer established on a 500 foot baseline provides digital readouts to a minimum of 1,000 feet. A system established on a 250 foot baseline provides digital readouts to a minimum of 600 feet, which are displayed in 200 foot increments to 3,000 feet and in 500 foot increments from 3,000 feet to a maximum value of 6,000 feet.

e. RVR values for Category IIIa operations extend down to 700 feet RVR; however, only 600 and 800 feet are reportable RVR increments. The 800 RVR reportable value covers a range of 701 feet to 900 feet and is therefore a valid minimum indication of Category IIIa operations.

f. Approach categories with the corresponding minimum RVR values: (See Table 7-13[1].)

Table 7-13[1]

Category	Visibility (RVR)
Nonprecision	2,400 feet
Category I	1,800 feet
Category II	1,200 feet
Category IIIa	700 feet
Category IIIb	150 feet
Category IIIc	0 feet

g. Ten minute maximum and minimum RVR values for the designated RVR runway are reported in the remarks section of the aviation weather report when the prevailing visibility is less than one mile and/or the RVR is 6,000 feet or less. ATCT's report RVR when the prevailing visibility is 1 mile or less and/or the RVR is 6,000 feet or less.

h. Details on the requirements for the operational use of RVR are contained in FAA AC 97-1, "RUNWAY VISUAL RANGE." Pilots are responsible for compliance with minimums prescribed for their class of operations in the appropriate FAR's and/or operations specifications.

7-14. REPORTING OF CLOUD HEIGHTS

a. Ceiling, by definition in the FAR's and as used in Aviation Weather Reports and Forecasts, is the height *above ground (or water) level* of the lowest layer of clouds or obscuring phenomenon that is reported as "broken," "overcast," or "obscuration" and not classified as "thin" or "partial." For example, a forecast which reads "CIGS WILL BE GENLY 1 TO 2 THSD FEET" refers to heights *above ground level.* A forecast which reads "BRKN TO OVC LYRS AT 8 TO 12 THSD MSL" states that the height is *above mean sea level.* See the Key to Aviation Weather Observations and Forecasts in paragraph 7-27 for definition of "broken," "overcast," and "obscuration."

b. Pilots usually report height values above MSL, since they determine heights by the altimeter. This is taken in account when disseminating and otherwise applying information received from pilots. ("Ceiling" heights are always above ground level.) In reports disseminated as PIREP's, height references are given the same as received from pilots, that is, above MSL. In the following example of an hourly observation, a pilot report of the heights of the bases and tops of an overcast layer in the terminal area is converted by the reporting station to reflect AGL to report the base of the overcast layer (E12). The pilot's reported top of the overcast layer (23) is not converted to AGL and is shown in the remarks section (last item) of the weather reporting (top ovc 23)

EXAMPLE:

E12 OVC 2FK 132/49/47/0000/002/TOP OVC 23

c. In aviation forecasts (Terminal, Area, or In-flight Advisories) ceilings are denoted by the prefix "C" when used with sky cover symbols as in "LWRG to C5 OVC-1TRW," or by the contraction "CIG" before, or the contraction "AGL" after, the forecast cloud height value. When the cloud base is given in height above MSL, it is so indicated by the contraction "MSL" or "ASL" following the height value. The heights of clouds tops, freezing level, icing, and turbulence are always given in heights above ASL or MSL.

7-15. REPORTING PREVAILING VISIBILITY

a. Surface (horizontal) visibility is reported in weather observations in terms of statute miles and increments thereof; e.g., 1/16, 1/8, 1/4, 1/2, 3/4, 1, 1 1/4, etc. Visibility is determined through the ability to see and identify preselected and prominent objects at a known distance from the usual point of observation. Visibilities which are determined to be less than 7 miles, identify the obscuring atmospheric condition; e.g., fog, haze, smoke, etc., or combinations thereof.

b. Prevailing visibility is the greatest visibility equalled or exceeded throughout at least half of the horizon circle, which need not be continuous. Segments of the horizon circle which may have a significantly lower visibility may be reported in the remarks section of the weather report; i.e., the southeastern quadrant of the horizon circle may be determined to be 2 miles in fog while the remaining quadrants are determined to be 3 miles in fog.

c. When the prevailing visibility at the usual point of observation, or at the tower level, is less than 4 miles, certificated tower personnel will take visibility observations in addition to those taken at the usual point of observation. The lower of these two values will be used as the prevailing visibility for aircraft operations.

7-16. ESTIMATING INTENSITY OF PRECIPITATION

a. Light-Scattered drops or flakes that do not completely wet or cover an exposed surface, regardless of duration, to 0.10 inch per hour; maximum 0.01 inch in 6 minutes.

b. Moderate-0.11 inch to 0.30 inch per hour; more than 0.01 inch to 0.03 inch in 6 minutes.

c. Heavy—More than 0.30 inch per hour; more than 0.03 inch in 6 minutes.

7-17. ESTIMATING INTENSITY OF DRIZZLE

a. Light—Scattered drops that do not completely wet surface, regardless of duration, to 0.01 inch per hour.

b. Moderate—More than 0.01 inch to 0.02 inch per hour.

c. Heavy—More than 0.02 inch per hour.

7-18. ESTIMATING INTENSITY OF SNOW

a. Light—Visibility ⅝ statute mile or more.

b. Moderate—Visibility less than ⅝ statute mile but not less than 5/16 statute mile.

c. Heavy—Visibility less than 5/16 statute mile.

7-19. PILOT WEATHER REPORTS (PIREP'S)

a. FAA air traffic facilities are required to solicit PIREP's when the following conditions are reported or forecast: Ceilings at or below 5,000 feet; Visibility at or below 5 miles; Thunderstorms and related phenomena; Icing of light degree or greater; Turbulence of moderate degree or greater; and Windshear.

b. Pilots are urged to cooperate and promptly volunteer reports of these conditions and other atmospheric data such as: Cloud bases, tops and layers; Flight visibility; Precipitation; Visibility restrictions such as haze, smoke and dust; Wind at altitude; and Temperature aloft.

c. PIREP's should be given to the ground facility with which communications are established; i.e., EFAS, FSS, ARTCC, or terminal ATC. One of the primary duties of EFAS facilities, radio call "FLIGHT WATCH," is to serve as a collection point for the exchange of PIREP's with en route aircraft.

d. If pilots are not able to make PIREP's by radio, reporting upon landing of the in-flight conditions encountered to the nearest FSS or Weather Service Office will be helpful. Some of the uses made of the reports are:

1. The ATCT uses the reports to expedite the flow of air traffic in the vicinity of the field and for hazardous weather avoidance procedures.

2. The FSS uses the reports to brief other pilots, to provide in-flight advisories, and weather avoidance information to en route aircraft.

3. The ARTCC uses the reports to expedite the flow of en route traffic, to determine most favorable altitudes, and to issue hazardous weather information within the center's area.

4. The NWS uses the reports to verify or amend conditions contained in aviation forecast and advisories. In some cases, pilot reports of hazardous conditions are the triggering mechanism for the issuance of advisories. They also use the reports for pilot weather briefings.

5. The NWS, other government organizations, the military, and private industry groups use PIREP's for research activities in the study of meteorological phenomena.

6. All air traffic facilities and the NWS forward the reports received from pilots into the weather distribution system to assure the information is made available to all pilots and other interested parties.

e. The FAA, NWS, and other organizations that enter PIREP's into the weather reporting system use the format listed in Table 7-19[1]. Items 1 through 6 are included in all transmitted PIREP's along with one or more of items 7 through 13. Although the PIREP should be as complete and concise as possible, pilots should not be overly concerned with strict format or phraseology. The important thing is that the information is relayed so other pilots may

Table 7-19[1]

PIREP ELEMENTS	PIREP CODE	CONTENTS
1. 3-letter station identifier	XXX	Nearest weather reporting location to the reported phenomenon
2. Report type	UA or UUA	Routine or Urgent PIREP
3. Location	/OV	In relation to a VOR
4. Time	/TM	Coordinated Universal Time
5. Altitude	/FL	Essential for turbulence and icing reports
6. Type Aircraft	/TP	Essential for turbulence and icing reports
7. Sky cover	/SK	Cloud height and coverage (scattered, broken, or overcast)
8. Weather	/WX	Flight visibility, precipitation, restrictions to visibility, etc.
9. Temperature	/TA	Degrees Celsius
10. Wind	/WV	Direction in degrees true and speed in knots
11. Turbulence	/TB	See paragraph 7-21
12. Icing	/IC	See paragraph 7-20
13. Remarks	/RM	For reporting elements not included or to clarify previously reported items

benefit from your observation. If a portion of the report needs clarification, the ground station will request the information. Completed PIREP's will be transmitted to weather circuits as in the following examples:

CMH UA /OV APE 230010/TM 1516/FL085/TP BE80/SK BKN 065/WX FV03 H K/TA 20/TB LGT

Translation: One zero miles southwest of Appleton VOR; Time 1516 UTC; altitude eight thousand five hundred; aircraft type BE80; top of the broken cloud layer is six thousand five hundred; flight visibility 3 miles with haze and smoke; air Temperature 20 degrees Celsius; light turbulence.

CRW UV /OV BKW 360015—CRW/TM 1815/FL120//TP BE99/SK OVC/WX R/TA -08 /WV 290030/TB LGT-MDT/IC LGT RIME/RM MDT MXD ICG DURGC ROA NWBND FL080-100 1750

Translation: From 15 miles north of Beckley VOR to Charleston VOR; time 1815 UTC; altitude 12,000 feet; type aircraft, BE-99; in clouds; rain; temperature -08 Celsius; wind 290 degrees true at 30 knots; light to moderate turbulence; light rime icing; encountered moderate mixed icing during climb northwestbound from Roanoke, VA between 8000 and 10,000 feet at 1750 UTC.

7-20. PIREP'S RELATING TO AIRFRAME ICING

a. The effects of ice on aircraft are cumulative—thrust is reduced, drag increases, lift lessens, and weight increases. The results are an increase in stall speed and a deterioration of aircraft performance. In extreme cases, 2 to 3 inches of ice can form on the leading edge of the airfoil in less than 5 minutes. It takes but ½ inch of ice to reduce the lifting power of some aircraft by 50 percent and increases the frictional drag by an equal percentage.

b. A pilot can expect icing when flying in visible precipitation, such as rain or cloud droplets, and the temperature is 0 degrees Celsius or colder. When icing is detected, a pilot should do one of two things (particularly if the aircraft is not equipped with deicing equipment), he should get out of the area of precipitation or go to an altitude where the temperature is above freezing. This "warmer" altitude may not always be a lower altitude. Proper preflight action includes obtaining information on the freezing level and the above-freezing levels in precipitation areas. Report icing to ATC/FSS, and if operating IFR, request new routing or altitude if icing will be a hazard. Be sure to give the type of aircraft to ATC when reporting icing. The following describes how to report icing conditions.

1. Trace—Ice becomes perceptible. Rate of accumulation is slightly greater than the rate of sublimation. It is not hazardous even though deicing/anti-icing equipment is not utilized unless encountered for an extended period of time (over 1 hour).

2. Light—The rate of accumulation may create a problem if flight is prolonged in this environment (over 1 hour). Occasional use of deicing/anti-icing equipment removes/prevents accumulation. It does not present a problem if the deicing/anti-icing equipment is used.

3. Moderate—The rate of accumulation is such that even short encounters become potentially hazardous and use of deicing/anti-icing equipment or flight diversion is necessary.

4. Severe—The rate of accumulation is such that deicing/anti-icing equipment fails to reduce or control the hazard. Immediate flight diversion is necessary.

EXAMPLE:

Pilot Report: Give Aircraft Identification, Location, Time (UTC), Intensity of Type, Altitude/FL, Aircraft Type, IAS, and Outside Air Temperature.

7-20b4 NOTE 1—Rime Ice: rough, milky, opaque ice formed by the instantaneous freezing of small supercooled water droplets

7-20b4 NOTE 2—Clear Ice: a glossy, clear, or translucent ice formed by the relatively slow freezing of large supercooled water droplets.

7-20b4 NOTE 3—The Outside Air Temperature (OAT) should be requested by the FSS/ATC if not included in the PIREP.

7-21. PIREP'S RELATING TO TURBULENCE

a. When encountering turbulence, pilots are urgently requested to report such conditions to ATC as soon as practicable. PIREP's relating to turbulence should state:

1. Aircraft location.

2. Time of occurrence in UTC.

3. Turbulence intensity.

4. Whether the turbulence occurred in or near clouds.

5. Aircraft altitude or Flight Level.

6. Type of Aircraft.

7. Duration of turbulence.

EXAMPLE:

OVER OMAHA, 1232Z, MODERATE TURBULENCE IN CLOUDS AT FLIGHT LEVEL THREE ONE ZERO, BOEING 707.

EXAMPLE:

FROM FIVE ZERO MILES SOUTH OF ALBUQUERQUE TO THREE ZERO MILES NORTH OF PHOENIX, 1250Z, OCCASIONAL MODERATE CHOP AT FLIGHT LEVEL THREE THREE ZERO, DC8.

b. Duration and classification of intensity should be made using Table 7-21[1].

7-22. WIND SHEAR PIREP'S

a. Because unexpected changes in wind speed and direction can be hazardous to aircraft operations at low altitudes on approach to and departing from airports, pilots are urged to promptly volunteer reports to controllers of wind shear conditions they encounter. An advance warning of this information will assist other pilots in avoiding or coping with a wind shear on approach or departure.

b. When describing conditions, use of the terms "negative" or "positive" wind shear should be avoided. PIREP's

Table 7-21[1]

TURBULENCE REPORTING CRITERIA TABLE

Intensity	**Aircraft Reaction**	**Reaction Inside Aircraft**	**Reporting Term-Definition**
Light	Turbulence that momentarily causes slight, erratic changes in altitude and/or attitude (pitch, roll, yaw). Report as **Light Turbulence;**[1] *or* Turbulence that causes slight, rapid and somewhat rhythmic bumpiness without appreciable changes in altitude or attitude. Report as **Light Chop**.	Occupants may feel a slight strain against seatbelts or shoulder straps. Unsecured objects may be displaced slightly. Food service may be conducted and little or no difficulty is encountered in walking.	Occasional—Less than ⅓ of the time. Intermittent—⅓ to ⅔. Continuous—More than ⅔.
Moderate	Turbulence that is similar to Light Turbulence but of greater intensity. Changes in altitude and/or attitude occur but the aircraft remains in positive control at all times. It usually causes variations in indicated airspeed. Report as **Moderate Turbulence;**[1] *or* Turbulence that is similar to Light Chop but of greater intensity. It causes rapid bumps or jolts without appreciable changes in aircraft altitude or attitude. Report as **Moderate Chop.**[1]	Occupants feel definite strains against seatbelts or shoulder straps. Unsecured objects are dislodged. Food service and walking are difficult.	**NOTE** 1. Pilots should report location(s), time (UTC), intensity, whether in or near clouds, altitude, type of aircraft and, when applicable, duration of turbulence. 2. Duration may be based on time between two locations or over a single location. All locations should be readily identifiable.
Severe	Turbulence that causes large, abrupt changes in altitude and/or attitude. It usually causes large variations in indicated airspeed. Aircraft may be momentarily out of control. Report as **Severe Turbulence.**[1]	Occupants are forced violently against seatbelts or shoulder straps. Unsecured objects are tossed about. Food service and walking are impossible.	**EXAMPLES:** a. Over Omaha. 1232Z, Moderate Turbulence, in cloud, Flight Level 310, B707. b. From 50 miles south of Albuquerque to 30 miles north of Phoenix, 1210Z to 1250Z, occasional Moderate Chop, Flight Level 330, DC8.
Extreme	Turbulence in which the aircraft is violently tossed about and is practically impossible to control. It may cause structural damage. Report as **Extreme Turbulence.**[1]		

[1] High level turbulence (normally above 15,000 feet ASL) not associated with cumuliform cloudiness, including thunderstorms, should be reported as CAT (clear air turbulence) preceded by the appropriate intensity, or light or moderate chop.

of "*negative* wind shear on final," intended to describe loss of airspeed and lift, have been interpreted to mean that *no* wind shear was encountered. The recommended method for wind shear reporting is to state the loss or gain of airspeed and the altitudes at which it was encountered.

EXAMPLE:
DENVER TOWER, CESSNA 1234 ENCOUNTERED WIND SHEAR, LOSS OF 20 KNOTS AT 400 FEET.

EXAMPLE:
TULSA TOWER, AMERICAN 721 ENCOUNTERED WIND SHEAR ON FINAL, GAINED 25 KNOTS BETWEEN 600 AND 400 FEET FOLLOWED BY LOSS OF 40 KNOTS BETWEEN 400 FEET AND SURFACE.

1. Pilots who are not able to report wind shear in these specific terms are encouraged to make reports in terms of the effect upon their aircraft.

EXAMPLE:
MIAMI TOWER, GULFSTREAM 403 CHARLIE ENCOUNTERED AN ABRUPT WIND SHEAR AT 800 FEET ON FINAL, MAX THRUST REQUIRED.

2. Pilots using Inertial Navigation Systems (INS) should report the wind and altitude both above and below the shear level.

7-23. CLEAR AIR TURBULENCE (CAT) PIREP'S

CAT has become a very serious operational factor to flight operations at all levels and especially to jet traffic flying in excess of 15,000 feet. The best available information on this phenomenon must come from pilots via the PIREP reporting procedures. All pilots encountering CAT conditions are urgently requested to report *time, location,* and *intensity* (light, moderate, severe, or extreme) of the element to the FAA facility with which they are maintaining radio contact. If time and conditions permit, elements should be reported according to the standards for other PIREP's and position reports. (See Pireps Relating to Turbulence, paragraph 7-21).

7-24. MICROBURSTS

a. Relatively recent meteorological studies have confirmed the existence of microburst phenomenon. Microbursts are small-scale intense downdrafts which, on reaching the surface, spread outward in all directions from the downdraft center. This causes the presence of both vertical and horizontal wind shears that can be extremely hazardous to all types and categories of aircraft, especially at low altitudes. Due to their small size, short life-span, and the fact that they can occur over areas without surface precipitation, microbursts are not easily detectable using conventional weather radar or wind shear alert systems.

b. Parent clouds producing microburst activity can be any of the low or middle layer convective cloud types. Note, however, that microbursts commonly occur within the heavy rain portion of thunderstorms, and in much weaker, benign appearing convective cells that have little or no precipitation reaching the ground.

c. The life cycle of a microburst as it descends in a convective rain shaft is seen in Figure 7-24[1]. An important consideration for pilots is the fact that the microburst intensifies for about 5 minutes after it strikes the ground.

d. Characteristics of microbursts include:

1. Size—The microburst downdraft is typically less than 1 mile in diameter as it descends from the cloud base to about 1,000-3,000 feet above the ground. In the transition zone near the ground, the downdraft changes to a horizontal outflow that can extend to approximately 2½ miles in diameter.

2 Intensity—The downdrafts can be as strong as 6,000 feet per minute. Horizontal winds near the surface can be as strong as 45 knots resulting in a 90 knot shear (headwind to tailwind change for a traversing aircraft) across the microburst. These strong horizontal winds occur within a few hundred feet of the ground.

3. Visual Signs—Microbursts can be found almost any-

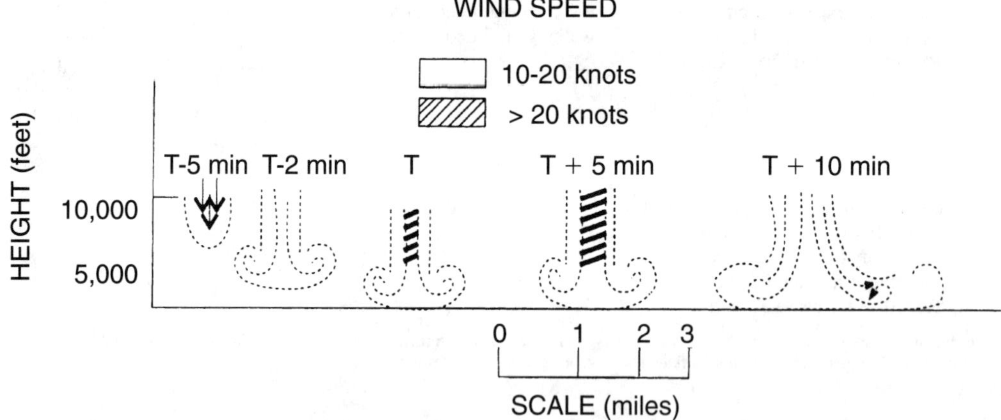

Vertical cross section of the evolution of a microburst wind field. T is the time of initial divergence at the surface. The shading refers to the vector wind speeds. Figure adapted from Wilson et al., 1984, Microburst Wind Structure and Evaluation of Doppler Radar for Wind Shear Detection, DOT/FAA Report No. DOT/FAA/PM-84/29, National Technical Information Service, Springfield, VA 37 pp.

Figure 7-24[1]

where that there is convective activity. They may be embedded in heavy rain associated with a thunderstorm or in light rain in benign appearing virga. When there is little or no precipitation at the surface accompanying the microburst, a ring of blowing dust may be the only visual clue of its existence.

4. Duration—An individual microburst will seldom last longer than 15 minutes from the time it strikes the ground until dissipation. The horizontal winds continue to increase during the first 5 minutes with the maximum intensity winds lasting approximately 2–4 minutes. Sometimes microbursts are concentrated into a line structure, and under these conditions, activity may continue for as long as an hour. Once microburst activity starts, multiple microbursts in the same general area are not uncommon and should be expected.

e. Microburst wind shear may create a severe hazard for aircraft within 1,000 feet of the ground, particularly during the approach to landing and landing and take-off phases. The impact of a microburst on aircraft which have the unfortunate experience of penetrating one is characterized in Figure 7-24[2]. The aircraft may encounter a headwind (performance increasing) followed by a downdraft and tailwind (both performance decreasing), possibly resulting in terrain impact.

f. Pilots should heed wind shear PIREP's, as a previous pilot's encounter with a microburst may be the only indication received. However, since the wind shear intensifies rapidly in its early stages, a PIREP may not indicate the current severity of a microburst. Flight in the vicinity of suspected or reported microburst activity should always be avoided. Should a pilot encounter one, a wind shear PIREP should be made at once.

7-25. THUNDERSTORMS

a. Turbulence, hail, rain, snow, lightning, sustained updrafts and downdrafts, icing conditions—all are present in

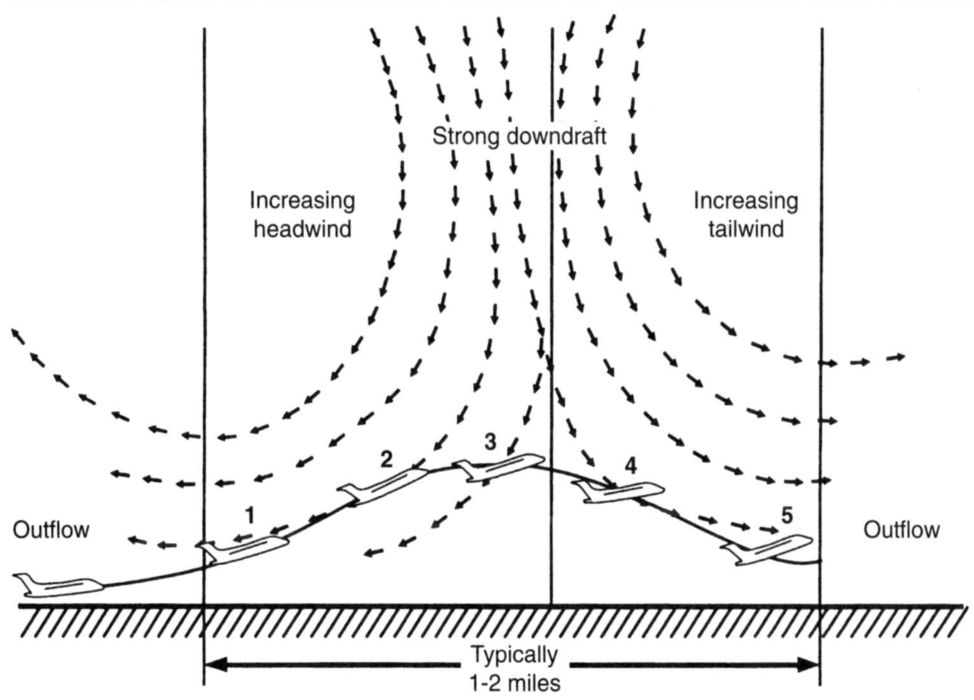

A microburst encounter during takeoff. The airplane first encounters a headwind and experiences increasing performance (1), this followed in short succession by a decreasing headwind component (2), a downdraft (3), and finally a strong tailwind (4), where 2 through 5 all result in decreasing performance of the airplane. Position (5) represents an extreme situation just prior to impact. Figure courtesy of Walter Frost, FWG Associates, Inc., Tullahoma, Tennessee.

Figure 7-24[2]

thunderstorms. While there is some evidence that maximum turbulence exists at the middle level of a thunderstorm, recent studies show little variation of turbulence intensity with altitude.

b. There is no useful correlation between the external visual appearance of thunderstorms and the severity or amount of turbulence or hail within them. The visible thunderstorm cloud is only a portion of a turbulent system whose updrafts and downdrafts often extend far beyond the visible storm cloud. Severe turbulence can be expected up to 20 miles from severe thunderstorms. This distance decreases to about 10 miles in less severe storms.

c. Weather radar, airborne or ground based, will normally reflect the areas of moderate to heavy precipitation (radar does not detect turbulence). The frequency and severity of turbulence generally increases with the radar reflectivity which is closely associated with the areas of highest liquid water content of the storm. NO FLIGHT PATH THROUGH AN AREA OF STRONG OR VERY STRONG RADAR ECHOES SEPARATED BY 20–30 MILES OR LESS MAY BE CONSIDERED FREE OF SEVERE TURBULENCE.

d. Turbulence beneath a thunderstorm should not be minimized. This is especially true when the relative humidity is low in any layer between the surface and 15,000 feet. Then the lower altitudes may be characterized by strong out-flowing winds and severe turbulence.

e. The probability of lightning strikes occurring to aircraft is greatest when operating at altitudes where temperatures are between minus 5 degrees Celsius and plus 5 degrees Celsius. Lightning can strike aircraft flying in the clear in the vicinity of a thunderstorm.

f. The NWS recognizes only two classes of intensities of thunderstorms as applied to aviation surface weather observations:

1. T—Thunderstorm; and

2. T+—Severe thunderstorm.

g. NWS radar systems are able to objectively determine radar weather echo intensity levels by use of Video Integrator Processor (VIP) equipment. These thunderstorm intensity levels are on a scale of one to six. (Reference—Pilot/Controller Glossary, Radar Weather Echo Intensity Levels.)

EXAMPLE:

Alert provided by an ATC facility to an aircraft:

(Aircraft identification) LEVEL FIVE INTENSE WEATHER ECHO BETWEEN TEN O'CLOCK AND TWO O'CLOCK, ONE ZERO MILES, MOVING EAST AT TWO ZERO KNOTS, TOPS FLIGHT LEVEL THREE NINE ZERO.

EXAMPLE:

Alert provided by an FSS:

(Aircraft identification) LEVEL FIVE INTENSE WEATHER ECHO, TWO ZERO MILES WEST OF ATLANTA V-O-R, TWO FIVE MILES WIDE, MOVING EAST AT TWO ZERO KNOTS, TOPS FLIGHT LEVEL THREE NINE ZERO.

7-26. THUNDERSTORM FLYING

a. Above all, remember this: never regard any thunderstorm "lightly" even when radar observers report the echoes are of light intensity. Avoiding thunderstorms is the best policy. Following are some Do's and Don'ts of thunderstorm avoidance:

1. Don't land or take off in the face of an approaching thunderstorm. A sudden gust front of low level turbulence could cause loss of control.

2. Don't attempt to fly under a thunderstorm even if you can see through to the other side. Turbulence and wind shear under the storm could be disastrous.

3. Don't fly without airborne radar into a cloud mass containing scattered embedded thunderstorms. Scattered thunderstorms not embedded usually can be visually circumnavigated.

4. Don't trust the visual appearance to be a reliable indicator of the turbulence inside a thunderstorm.

5. Do avoid by at least 20 miles any thunderstorm identified as severe or giving an intense radar echo. This is especially true under the anvil of a large cumulonimbus.

6. Do clear the top of a known or suspected severe thunderstorm by at least 1,000 feet altitude for each 10 knots of wind speed at the cloud top. This should exceed the altitude capability of most aircraft.

7. Do circumnavigate the entire area if the area has 6/10 thunderstorm coverage.

8. Do remember that vivid and frequent lightning indicates the probability of a severe thunderstorm.

9. Do regard as extremely hazardous any thunderstorm with tops 35,000 feet or higher whether the top is visually sighted or determined by radar.

b. If you cannot avoid penetrating a thunderstorm, following are some Do's *before* entering the storm:

1. Tighten your safety belt, put on your shoulder harness if you have one and secure all loose objects.

2. Plan and hold your course to take you through the storm in a minimum time.

3. To avoid the most critical icing, establish a penetration altitude below the freezing level or above the level of minus 15 degrees Celsius.

4. Verify that pitot heat is on and turn on carburetor heat or jet engine anti-ice. Icing can be rapid at any altitude and cause almost instantaneous power failure and/or loss of airspeed indication.

5. Establish power settings for turbulence penetration airspeed recommended in your aircraft manual.

6. Turn up cockpit lights to highest intensity to lessen temporary blindness from lightning.

7. If using automatic pilot, disengage altitude hold mode and speed hold mode. The automatic altitude and speed controls will increase maneuvers of the aircraft thus increasing structural stress.

8. If using airborne radar, tilt the antenna up and down occasionally. This will permit you to detect other thunderstorm activity at altitudes other than the one being flown.

c. Following are some Do's and Don'ts *during* the thunderstorm penetration:

1. Do keep your eyes on your instruments. Looking outside the cockpit can increase danger of temporary blindness from lightning.

2. Don't change power settings; maintain settings for the recommended turbulence penetration airspeed.

3. Do maintain constant attitude; let the aircraft "ride the waves." Maneuvers in trying to maintain constant altitude increase stress on the aircraft.

4. Don't turn back once you are in the thunderstorm. A straight course through the storm most likely will get you out of the hazards most quickly. In addition, turning maneuvers increase stress on the aircraft.

7-27. KEY TO AVIATION WEATHER OBSERVATIONS AND FORECASTS

KEY TO AVIATION WEATHER FORECASTS

TERMINAL FORECASTS (FT) contain information for specific airports. They are issued 3 times a day, amended as needed, and are valid for a 24 hour period. The last six hours of each period is covered by a categorical forecast indicating that VFR, MVFR, IFR, or LIFR conditions are expected. Terminal forecasts are written in the following form:

Airport Identifier: 3 or 4 alphanumeric characters.
Date And Valid Time Period Of Forecast: Z or UTC.
Message Type: RTD (Delayed), COR (Corrected), or AMD (Amended).
Ceiling: Identified by the letter "C" prefix.
Cloud Heights: In hundreds of feet above the airport (AGL)
Cloud Amount: CLR (Clear), SCT (Scattered), BKN (Broken), OVC (Overcast), or X (Obscured).
Visibility: In statute miles (6+ indicates unrestricted).
Weather & Obstruction To Visibility: Standard weather / visibility symbols used.
Surface Wind: In tens of degrees and knots. Omitted when less than 6 knots. Gusts indicated by a G followed by maximum speed.
Ceiling And Visibility Categories:

Category	Ceiling (feet)		Visibility (miles)
LIFR	less than 500	and / or	less than 1
IFR	500 to 1000	and / or	1 to 3
MVFR	1000 thru 3000	and / or	3 thru 5
VFR	more than 3000	and	more than 5

Example Of Terminal Forecast:

DCA 221010 10 SCT C18 BKN 5SW— 3415G25 OCNL C8 X 1/2 SW.
12Z C50 BKN 3312G22.
04Z MVFR CIG..

Decoded Example: Washington National Airport for the 22nd of the month valid from 10Z to 10Z. Scattered clouds at 1000 feet, ceiling 1800 feet broken, visibility 5 miles in light snow showers, surface wind 340 degrees at 15 knots, gusts to 25 knots. Occasional ceiling 800 feet, sky obscured, visibility one-half mile in moderate snow showers. By 12Z becoming ceiling 5000 feet broken, surface wind 330 degrees at 12 knots, gusts to 22 knots. The categorical outlook for the last 6 hours beginning at 04Z calls for marginal VFR conditions due to ceiling.

AREA FORECASTS (FA) provide an 18-hour synopsis of expected weather patterns; a 12-hour forecast of VFR cloud cover, weather and visibility; and a 6-hour categorical outlook. FAs are prepared 3 times a day (4 times a day in Alaska and Hawaii) and are supplemented and updated by SIGMETs, AIRMETs, and by FA amendments. Heights in the FA are above mean sea level (MSL) unless stated as above ground level (AGL). Ceilings (CIG) are always AGL.

WIND and TEMPERATURE ALOFT FORECASTS (FD) are 6, 12, and 24-hour forecasts of wind direction, speed, and temperatures for selected altitudes to 53,000 feet MSL at specified locations. Direction is relative to true north rounded to the nearest 10 degrees. Speed is in knots. Temperatures aloft (in degrees Celsius) are included with wind data for all but the 3000-foot MSL level and those levels within 2500 feet of the ground. Temperatures above 24,000 feet MSL are always negative. Winds at other locations and intermediate altitudes can be obtained by interpolation.

Example of Winds Aloft Forecast:

FT	3000	6000	9000	etc.
ACY	2925	2833 + 02	2930 — 03	etc.

Decoded example: For Atlantic City, N.J., at 6000 feet MSL wind from 280 degrees true at 33 knots, temperature 2 degrees Celsius.

IN-FLIGHT ADVISORIES of potentially hazardous weather and include SIGMETs, CONVECTIVE SIGMETs, AIRMETs, and Center Weather Advisories (CWA). **SIGMETs** warn of hazardous conditions of importance to all aircraft i.e., severe icing or turbulence, duststorms, sandstorms, and volcanic ash. **AIRMETs** warn of less severe conditions which may be hazardous to some aircraft or pilots. SIGMETs are issued as needed, AIRMET bulletins are issued routinely and supplement the Area Forecast (FA). **CONVECTIVE SIGMETs** are issued hourly for thunderstorms in the conterminous U.S. **Center Weather Advisories**, issued as needed, are detailed advisories of conditions which meet or approach SIGMET or AIRMET criteria.

Example Of Sigmet:

SIGMET OSCAR 2 VALID UNTIL 052100
KS NE
FROM PWE TO OSW TO LBL TO PWE
SVR TURBC BLO 60 XPCD DUE TO STG NWLY FLOW BHD CDFNT.
CONDS CONTG BYD 2100Z.

Decoded Example: SIGMET OSCAR 2 is valid until 2100Z on the 5th day of the month. For Kansas and Nebraska from Pawnee City VORTAC to Oswego VORTAC to Liberal VORTAC to Pawnee City VORTAC. Severe turbulence below 6000 feet expected due to strong northwesterly flow behind a coldfront. Conditions continuing beyond 2100Z.

TRANSCRIBED WEATHER BROADCASTs (TWEB) are continuous broadcasts of recorded NOTAM and weather information prepared for a 50-nautical mile wide zone along a route and for selected terminal areas. TWEBs are broadcast over selected NDB and VOR facilities and generally contain a weather synopsis, in-flight advisories, route and/or local vicinity forecasts, Winds Aloft Forecasts, current weather reports, NOTAMs, and special notices.

U.S. DEPARTMENT OF COMMERCE — NATIONAL OCEANIC AND ATMOSPHERIC ADMINISTRATION — NATIONAL WEATHER SERVICE 4/14/93

KEY TO MANUAL AVIATION WEATHER OBSERVATIONS

LOCATION IDENTIFIER, TYPE AND TIME OF REPORT	SKY CONDITION AND CEILING	VISIBILITY, WEATHER, AND OBSTRUCTIONS TO VISION	SEA-LEVEL PRESSURE	TEMPERATURE AND DEW POINT	WIND DIRECTION, SPEED AND CHARACTER	ALTIMETER SETTING	REMARKS AND CODED DATA
MCI SA 0758	**15 SCT M15 OVC**	**1R — F**	**132**	**/58/56**	**/1807**	**/993/**	**R01VR20V40**

LOCATION IDENTIFIER:
3 or 4 alphanumeric characters (airport identifier).

TYPE OF REPORT:
SA = Scheduled record (hourly) observation.
SP = Special observation indicating a significant change in one or more of the observed elements.
RS = SA that also qualifies as an SP.
USP = Urgent special observation (tornado).

TIME OF REPORT:
Coordinated Universal Time (UTC or Z) using 24-hour clock. Example: 2255 = 10:55 pm.

SKY CONDITION AND CEILING:
Sky condition contractions are for each layer in ascending order. Numbers preceding contractions are base height in hundreds of feet above ground level (AGL). Sky condition contractions are: (— = Thin).
CLR = Clear: Less than 0.1 sky cover.
SCT = Scattered: 0.1 to 0.5 sky cover.
BKN = Broken: 0.6 to 0.9 sky cover.
OVC = Overcast: More than 0.9 sky cover.
— X = Partially obscured. 0.9 or less of sky hidden by precipitation or obstruction to vision (cloud bases at the surface).
X = Obscured: entire sky hidden.
A letter preceding height of a layer identifies a ceiling and indicates how ceiling was obtained.
E = Estimated. **M** = Measured.
W = Vertical visibility into obscured sky.
V following height = variable ceiling.

VISIBILITY:
Reported in **statute** miles and fractions.
V = Variable.

WEATHER & OBSTRUCTIONS TO VISION:

A	Hail	**GF**	Ground fog
D	Dust	**ZR**	Freezing rain
R	Rain	**BD**	Blowing dust
F	Fog	**SP**	Snow pellets
S	Snow	**BN**	Blowing sand
H	Haze	**SW**	Snow showers
K	Smoke	**BS**	Blowing snow
IF	Ice fog	**T**	Thunderstorm
L	Drizzle	**T+**	Severe thunderstorm
IP	Ice pellets	**RW**	Rain showers
IC	Ice crystals	**IPW**	Ice pellet showers
SG	Snow grains	**ZL**	Freezing drizzle

— = Light. (no sign) = Moderate. **+** = Heavy.

SEA-LEVEL PRESSURE:
Pressure in hectopascals (millibars): Shown as 3 digits. Leading 9 or 10 and decimal point is omitted. Examples: 150 = 1015.0 950 = 995.0

TEMPERATURE AND DEW POINT:
Reported in Degrees Fahrenheit (°F).

WIND DIRECTION SPEED & CHARACTER:
Direction in tens of degrees from **true** north, speed in knots. **0000** = calm. **G** = gusty. **Q** = squall. Peak speed of gusts in the past ten minutes follows G or Q. **WSHFT** in Remarks = windshift occurred at time indicated. Example: 3627G40 = 360° at 27 peak gusts 40 knots.

ALTIMETER SETTING:
Actual altimeter setting with first digit omitted. Examples: 005 = 30.05" 992 = 29.92"

RUNWAY VISUAL RANGE (RVR):
RVR is reported for some stations. Value(s) during 10 minutes prior to observation are given in hundreds of feet. Runway number precedes RVR report. **V** = Variable.

DECODED REPORT:
Kansas City Int'l Airport: Record observation completed at 0758Z. 1500 feet scattered clouds, measured ceiling 2500 feet overcast, visibility 1 mile, light rain, fog, sea level pressure 1013.2 hectopascals, temperature 58° F, dewpoint 56° F, wind 180°, 7 knots, altimeter setting 29.93". Runway 01 visual range varying from 2000 to 4000 feet in the past 10 minutes.

PILOT REPORTS (PIREPS):
A PIREP describes actual in-flight conditions. Pilots are encouraged to provide PIREPS to an FAA facility. **Example:** UA /OV FRR 275045 /TM 1745 /FL330 /TP B727 /SK 185 BKN 220 280 BKN 310 /TA-53 /WV 290120 /TB LGT-MDT CAT ABV 310. **Decoded:** Pilot report, Front Royal VORTAC, 275 radial 45nm, at 1745Z, flight level 330; Boeing 727; cloud base 18500 broken, tops 22000, second layer 28000 broken, tops 31000; air temperature minus 53 degrees Celsius; wind 290 degrees 120 knots; light to moderate clear air turbulence above 31000.

NOAA/PA 93055

U.S. DEPARTMENT OF COMMERCE — NATIONAL OCEANIC AND ATMOSPHERIC ADMINISTRATION — NATIONAL WEATHER SERVICE 4/14/93

Figure 7-27[1]

7-28. AUTOB DECODING KEY

Table 7-28[1]

DECODING OBSERVATIONS FROM AUTOB STATIONS

***EXAMPLE:* ENV AUTOB E25 BKN BV7 P 33/29/3606/975 PK WND 08 001**

ENCODE	DECODE	EXPLANATION
ENV	STATION IDENTIFICATION:	(Wendover, UT) Identifies report using FAA identifiers.
AUTOB	AUTOMATIC STATION IDENTIFIER	
E25 BKN	SKY & CEILIING:	(Estimated 2500 ft. broken) Figures are height in 100s of feet above ground. Contraction after height in amount of sky cover. Letter preceding height indicates ceiling. WX reported if visibility is less than 2 miles and no clouds are detected. *NO CLOUDS REPORTED ABOVE 6000 FEET.*
BV7	BACKSCATTER VISIBILITY AVERAGED IN PAST MINUTE:	Reported in whole miles from 1 to 7.
P	PRECIPITATION OCCURRENCE:	(P=precipitation in past 10 minutes).
33	TEMPERATURE:	(33 degrees F.) Minus sign indicates sub-zero temperatures.
/29	DEW POINT:	(29 degrees F.) Minus sign indicates sub-zero temperatures.
/3606	WIND:	(360 degrees true at 6 knots) Direction is first two digits and is reported in tens of degrees. To decode, add a zero to first two digits. The last digits are speed; e.g., 2524 = 250 degrees at 24 knots.
/975	ALTIMETER SETTING:	(29.75 inches) The tens digit and decimal are omitted from report. To decode, prefix a 2 to code if it begins with 8 or 9. Otherwise, prefix a 3; e.g., 982 = 29.82, 017 = 30.17.
PK WND 08	PEAK WIND SPEED:	(8 knots) Reported speed is highest detected since last hourly observation.
001	PRECIPITATION ACCUMULATION:	(0.01 inch) Amount of precipitation since last synoptic time (00, 06, 12, 1800 UTC).

NOTE: If no clouds are detected below 6,000 feet and the visibility is greater than 2 miles, the reported sky condition will be *CLR BLO 60.*

7-29. INTERNATIONAL CIVIL AVIATION ORGANIZATION (ICAO) TERMINAL FORECAST (TAF)

Terminal forecasts for international locations and domestic military locations are available to the Flight Service Station specialist, via their weather computer. Domestic military locations are available to the pilot, via the Direct User Access Terminal (DUAT), but are in an international alphanumeric code. They are scheduled four times daily, for 24-hour periods, beginning at 0000Z, 0600Z, 1200Z and 1800Z.

a. Format. The TAF is a series of groups made up of digits and letters. An individual group is identified by its position in the sequence, by its alphanumeric coding or by a numerical indicator. Listed below are a few contractions used in the TAF. Some of the contractions are followed by time entries indicated by "tt" or "tttt" or by probability, "pp."

b. Significant weather change indicators.

GRADU tttt—A gradual change occurring during a period in excess of one-half hour. "tttt" are the beginning and ending times of the expected change to the nearest hour; i.e., "GRADU 1213" means the transition will occur between 1200Z and 1300Z.

RAPID tt—A rapid change occurring in one-half hour or less. "tt" is the time to the nearest hour of the change; i.e., "RAPID 23" means the change will occur about 2300Z.

Variability terms—indicate that short time period variations from prevailing conditions are expected with the total occurrence of these variations less than ½ of the time period during which they are called for.

TEMPO tttt—Temporary changes from prevailing conditions of less than one hour duration in each instance. There may be more than one (1) instance for a specified time period. "tttt" are the earliest and latest times during which the temporary changes are expected; i.e., "TEMPO

0107" means the temporary changes may occur between 0100Z and 0700Z.

INTER tttt—Changes from prevailing conditions are expected to occur frequently and briefly. "tttt" are the earliest and latest times the brief changes are expected; i.e., "INTER 1518" means that the brief, but frequent, changes may occur between 1500Z and 1800Z. INTER has shorter and more frequent changes than TEMPO.

c. Probability.

PROB pp—Probability of conditions occurring. "pp" is the probability in percent; i.e., "PROB 20" means a 10 or 20% probability of the conditions occurring. "PROB 40" means a 30 to 50% inclusive probability.

d. Cloud and weather terms.

CAVOK—No clouds below 5,000 feet or below the highest minimum sector altitude whichever is greater, and no cumulonimbus. Visibility 6 miles or greater. No precipitation, thunderstorms, shallow fog or low drifting snow.

WX NIL—No significant weather (no precipitation, thunderstorms or obstructions to vision).

SKC—Sky clear.

e. Following is a St. Louis MO forecast in TAF code.
EXAMPLE:

KSTL 1212 33025/35 0800 71SN 9//05 INTER 1215 0000 75XXSN 9//000 GRADU 1516 33020 4800 38BLSN 7SC030 PROB 40 85SNSH GRADU 2122 33015 9999 WX NIL 3SC030 RAPID 00 VRB05 9999 SKC GRADU 0304 24015/25 CAVOK 0

1. The forecast is broken down into the elements lettered "a" to "l" to aid in the discussion. Not included in the example but explained at the end are three optional forecast groups for "m" icing, "n" turbulence and "o" temperature. (See Table 7-29[1].)

Table 7-29[1]

KSTL	1212	33025/35
a.	b.	c.
0800	71SN	9//005
d.	e.	f.

INTER 1215 0000 75XXSN 9//000
g.

GRADU 1516 33020 4800 38BLSN 7SC030
h.

PROB 40 85SNSH
i.

GRADU 2122 33015 9999 WX NIL 3SC030
j.

RAPID 00 VRB05 9999 SKC
k.

GRADU 0304 24015125 CAVOK 0
l.

(a) *Station identifier.* The TAF code uses ICAO 4-letter station identifiers. In the contiguous 48 states the 3-letter identifier is prefixed with a "K"; i.e., the 3-letter identifier for Seattle is SEA while the ICAO identifier is KSEA. Elsewhere, the first two letters of the ICAO identifier tell what region the station is in. "MB" means Panama/Canal Zone (MBHO is Howard AFB); "MI" means Virgin Islands (MISX is St. Croix); "TJ" is Puerto Rico (TJSJ is San Juan); "PA" is Alaska (PACD is Cold Bay); "PH" is Hawaii (PHTO is Hilo).

(b) *Valid time.* Valid time of the forecast follows station identifier. "1212" means a 24-hour forecast valid from 1200Z until 1200Z the following day.

(c) *Wind.* Wind is forecast usually by a 5-digit group giving degrees in 3 digits and speed in 2 digits. When wind is expected to be 100 knots or more, the group is 6 digits with speed given in 3 digits. When speed is gusty or variable, peak speed is separated from average speed with a slash. For example, in the KSTL TAF, "33025/35" means wind 330 degrees, average speed 25 knots, peak speed 35 knots. A group "160115/130" means wind 160 degrees, 115 knots, peak speed 130 knots. "00000" means calm; "VRB" followed by speed indicates direction variable; i.e., "VRB10" means wind direction variable at 10 knots.

(d) *Visibility.* Visibility is in meters. "0800" means 800 meters converted from table to ½ mile. (See Table 7-29[2] for converting meters to miles and fractions.)

Table 7-29[2]
Visibility conversion TAF code to miles

Meters	Miles	Meters	Miles	Meters	Miles
0000	0	1200	3/4	3000	1 7/8
0100	1/16	1400	7/8	3200	2
0200	1/8	1600	1	3600	2 1/4
0300	3/16	1800	1	4000	2 1/2
0400	1/4	2000	1 1/4	4800	3
0500	5/16	2200	1 3/8	6000	4
0600	3/8	2400	1 1/2	8000	5
0800	1/2	2600	1 5/8	9000	6
1000	5/8	2800	1 3/4	9999	>6

(e) *Significant weather.* Significant weather is decoded using Table 7-29[3]. Groups in the table are numbered sequentially. Each number is followed by an acronym suggestive of the weather; you can soon learn to read most of the acronyms without reference to the table. Examples: "177TS," thunderstorm; "18SQ," squall; "31SA," sandstorm; "60RA," rain; "85SNSH," snow shower. "XX" freezing rain. In the KSTL forecast, "71SN" means light snow. The TAF encodes only the single most significant type of weather; the U.S. domestic FT permits encoding of multiple weather types. (See Table 7-29[4] to convert weather from FT to TAF.)

Table 7-29[3]

TAF weather codes

Code	Simple Definition	Detailed Definition
04FU	Smoke	Visibility reduced by smoke. No visibility restriction.
05HZ	Dust haze	Visibility reduced by haze. No visibility restriction.
06HZ	Dust haze	Visibility reduced by dust suspended in the air but wind not strong enough to be adding more dust. No visibility restriction.* * While this may seem to be contradictory, it means that while visibility is restricted, the amount of the restriction is not limited.
07SA	Duststorm, sandstorm rising dust or sand	Visibility reduced by dust suspended in the air and wind strong enough to be adding sand more dust. No well developed dust devils, duststorm or sandstorm. Visibility 6 miles or less.
08PO	Dust devil	Basically the same as 07SA but with well developed dust devils. Visibility 6 miles or less.
10BR	Mist	Fog, ground fog or ice fog with visibility ⅝ to 6 miles.
11MIFG	Shallow fog	Patchy shallow fog (less than 6 feet deep and coverage less than half) with visibility in the fog less than ⅝ mile.
12MIFG	Shallow fog	Shallow fog (less than 6 feet deep with more or less continuous coverage) with visibility in the fog less than ⅝ miles.
17TS	Thunderstorm	Thunderstorm at the station but with no precipitation.
18SQ	Squall	No precipitation. A sudden increase of at least 15 knots in average wind speed, sustained at 20 knots or more for at least one (1) minute.
19FC	Funnel cloud	Used to forecast a tornado, funnel cloud or waterspout at or near the station. Also not easy to forecast and likely to be overshadowed by some other more violent weather such as thunderstorms.
30SA	Duststorm, sandstorm rising dust or sand	Duststorm or sandstorm, visibility 5/16 to less than ⅝ mile, increasing in intensity.
31SA		Basically the same as 30SA but with no change in intensity.
32SA		Basically the same as 30SA but increasing in intensity
33XXSA	Heavy duststorm or sandstorm	Severe duststorm or sandstorm, visibility, less than 5/16 mile, decreasing in intensity.
34XXSA		Basically the same as 33XXSA but with no change in intensity.
35XXSA		Basically the same as 33XXSA but increasing in intensity.
36DRSN	Low drifting snow	Low drifting snow (less than 6 feet) with visibility in drifting snow less than 5/16 miles.
37DRS		Low drifting snow (less than 6 feet) with visibility in drifting snow less than 5/16 miles.
38BLSN	Blowing snow	Blowing snow (more than 6 feet) with visibility 5/16 to 6 miles.
39BLSN		Blowing snow (more than 6 feet) with visibility 5/16 to 6 miles.
40BCFG	Fog Patches	Distant fog (not at station).
41BCFG		Patchy fog at the station, visibility in the fog patches less than ⅝ of a mile.
42FG		Fog at the station, visibility less than ⅝ mile, sky visible, fog thinning.
43FG		Fog at the station, visibility less than ⅝ mile, sky not visible, fog thinning.
44FG		Fog at the station, less than ⅝ mile, sky visible, no change in intensity.

Code	Simple Definition	Detailed Definition
45FG		Fog at the station, visibility less than ⅝ mile, sky not visible, no change in intensity.
46FG		Fog at the station, visibility less than ⅝ mile, sky visible, fog thickening.
47FG		Fog at the station, visibility less than ⅝ mile, sky not visible, fog thickening.
		NOTE: In code figures 40 through 47, "fog" includes both fog and ice fog. See FMH No. 1 (Surface Observations) for definitions of precipitation intensities.
48FZFG	Freezing	Fog depositing rime ice, visibility fogless than ⅝ mile, sky visible.
49FZFG		Fog depositing rime ice, visibility less than ⅝ mile, sky not visible.
50DZ	Drizzle	Light intermittent drizzle.
51DZ		Light continuous drizzle.
52DZ		Moderate intermittent drizzle.
53DZ		Moderate continuous drizzle.
54XXDZ	Heavy drizzle	Heavy intermittent drizzle.
55XXDZ		Heavy continuous drizzle.
56XXDZ	Freezing drizzle	Light freezing drizzle.
57XXFZDZ	Heavy freezing drizzle	Moderate or heavy freezing drizzle.
58RA		Mixed rain and drizzle, light.
59RA		Mixed rain and drizzle, moderate or heavy.
60RA	Rain	Light intermittent rain.
61RA		Light continuous rain.
62RA		Moderate intermittent rain.
63RA		Moderate continuous rain.
64XXRA	Heavy rain	Heavy intermittent rain.
65XXRA		Heavy continuous rain.
66FZRA	Freezing rain	Freezing rain or mixed freezing rain and freezing drizzle, light.
67XXFZRA	Heavy freezing rain	Freezing rain or mixed freezing rain and freezing drizzle, moderate or heavy.
68RASN	Rain and snow	Mixed rain and snow or drizzle and snow, light.
69XXRASN	Heavy rain and snow	Mixed rain and snow or drizzle and snow, snow, moderate or heavy.
70SN		Light intermittent snow.
71SN		Light continuous snow.
72SN	Snow	Moderate intermittent snow.
73SN		Moderate continuous snow.
74XXSN	Heavy snow	Heavy intermittent snow.
75XXSN		Heavy continuous snow.
77SG	Snow grains	Snow grains, and intensity. May be accompanied by fog or ice fog.
79PE	Ice pellets	Ice pellets, any intensity. May be mixed with some other precipitation.
80RASH	Showers	Light rain showers.
81XXSH	Heavy	Moderate or heavy rain showers.

Table 7-29[3] *Continued*

Code	Simple Definition	Detailed Definition
82XXSH	Showers	Violent rain showers (more than 1 inch per hour or 0.1 inch in 6 minutes).
83RASN	Showers of rain and	Mixed rain showers and snow showers. Intensity of both showers is light.
84XXRASN	Heavy showers of rain and snow	Mixed rain showers and snow showers. Intensity of either shower is moderate or heavy.
85SNSH	Snow showers	Light snow showers.
86XXSNSH	Heavy snow showers	Moderate or heavy snow showers.
87GR		Light ice pellet showers. There may also be rain or mixed rain or snow.
88GR	Soft hail	Moderate or heavy ice pellet showers. There may also be rain or mixed rain and snow.
89GR	Hail	Hail, not associated with a Thunderstorm. There may also be rain or mixed rain and snow.
90XXGR	Heavy hail	Moderate or heavy hail, not associated with a thunderstorm. There may also be rain or mixed rain or snow.
91RA	Rain	Light rain or light rain shower at the time of the forecast and thunderstorms during the preceding hour, but not at the time of the forecast.
92XXRA	Heavy rain	Basically the same as 91RA but the intensity of the rain or rain shower is moderate or heavy.
93GR	Hail	Basically the same as 91RA, but the precipitation is light snow or snow showers, or light mixed rain and snow or rain showers and snow showers, or light ice pellets or ice pellet showers.
94XXGR	Heavy hail	Basically the same as 93GR but the intensity of any precipitation is moderate or heavy.
95TS	Thunderstorm	Thunderstorm with rain or snow, or a mixture of rain and snow, but no hail, ice pellets or snow pellets.
96TSGR	Thunderstorm with hail	Thunderstorm with hail, ice pellets or snow pellets. There may also be rain or snow, or mixed rain and snow.
97XXTS	Heavy thunderstorm	Severe thunderstorm with rain or snow, or a mixture of rain and snow, but no hail, ice pellets or snow pellets.
98TSSA	Thunderstorm with duststorm or sandstorm	Thunderstorm with duststorm or sandstorm. There may also be some form of precipitation with the thunderstorm.
99XXTSGR	Heavy thunderstorm with hail	Basically the same as 97XXTS but in addition to everything else there is hail.

(f) *Clouds*. A cloud group is a 6-character group. The first digit is coverage in octas (eighths) of the individual cloud layer only. The summation of cloud layer to determine total sky cover from a ground observer's point of view is NOT used. (See Table 7-29[5].) The two letters identify cloud type as shown in the same table. The last three digits are cloud height in hundreds of feet above ground level (AGL). In the KSTL TAF, "9//005" means sky obscured (9), clouds not observed (//), vertical visibility 500 feet (005). The TAF may include as many cloud groups as necessary to describe expected sky condition.

(g) Expected variation from prevailing conditions. Variations from prevailing conditions are identified by the contractions INTER and TEMPO as defined earlier. In the KSTL TAF, "INTER 1215 0000 75XXSN 9//000" means intermittently from 1200Z to 1500Z (1215) visibility zero meters (0000) or zero miles, heavy snow (75XXSN), sky obscured, clouds not observed, vertical visibility zero (9//000).

(h), (i), (j), (k), and (l) An expected change in prevailing conditions is indicated by the contraction GRADU and RAPID as defined earlier. In the KSTL TAF, "GRADU 1516 33020 4800 38BLSN 7SC030" means a gradual change between 1500Z and 1600Z to wind 330 degrees at 20 knots, visibility 4,800 meters or 3 miles (See Table 7-29[2]), blowing snow, ⅞ stratocumulus (See Table 7-29[5]) at 3000 feet AGL. "PROB 40 85SNSH" means there is a 30 to 50% probability that light snow showers will occur between

Table 7-29[4]
Converting significant weather from U.S. terms to ICAO terms.

Precipitation & Intensity

US	TAF Code	Light	Moderate	Heavy
A	89GR			
BD or BN (vsby 5/16 to ½ mi)	31SA			
BD or BN (vsby 0 to ¼ mi)	34XXSA			
BS (vsby 6 mi or less)	38BLSN			
D (vsby 6 mi or less)	06HZ			
GF (vsby ½ mi or less)	44FG			
H (vsby 6 mi or less)	05HZ			
F or IF (vsby ½ mi or less)	45FG			
F, GF or IF (vsby ⅝ to 6 mi)	10BR			
IP	79PE			
IPW	87GR			
K	04FU			
L		51DZ	53DZ	55DZ
R		61RA	63RA	64RA
RS		68RASN	68RASN	69XXRASN
RW		80RASH	80RASH	81XXSH
RWSW		83RASN	83RASN	84XXRASN
S		71SN	73SN	75XXSN
SG	77SG			
SP	87GR			
SW		85SNSH	85SNSH	86XXSNSH
ZL		56FZDZ	56FZDZ	57XXFZDZ
ZR		66FZRA	67XXFZRA	67XXFZRA
TRW– or TRW	95TS			INTER 81XXSH
TRW+	95TS			INTER 82XXSH*
TRW–A or TRWA	96TSGR			
T–RW	97XXTS			
T–RW+	97XXTS 97XXTS			INTER 81XXSH INTER 82XXSH*
T–RWA	99XXTSGR			
T–RW+A	99XXTSGR 99XXTSGR			INTER 81XXSH INTER 82XXSH*

*INTER 82XXSH is to be encoded in a TAF only when a violent rainshower (at least 1 inch of rain per hour or 0.10 inch in 6 minutes) is forecast.

NOTE: Conversions from TAF to FT will not be exact in some cases due to a lack of a one to one relationship.

1600Z and 2100Z. "GRADU 2122 33015 9999 WX NIL 3SC030" means a gradual change between 2100Z and 2200Z to wind 330 degrees at 15 knots, visibility 10 kilometers or more (more than 6 miles), no significant weather, ⅜ stratocumulus at 3000 feet. "RAPID 00 VRB05 9999 SKC" means a rapid change about 0000Z to wind direction variable at 5 knots, visibility more than 6 miles, sky clear. "GRADU 0304 24015/25 CAVOK 0" means a gradual change between 0300Z and 0400Z to wind 240 degrees at 15 knots, peak gust to 25 knots with CAVOK conditions. 0 means end of message.

(m) *Icing*. An icing group may be included. It is a 6-digit group. The first digit is always a 6, identifying it as an icing group. The second digit is the type of ice accretion from Table 7-29[6]. The next three digits are height of the base of the icing layer in hundreds of feet (AGL). The last

Table 7-29-[5]
TAF CLOUD CODE

Code	Cloud amount	Cloud Type
0	0 (clear)	CI Cirrus
1	1 octa or less but not zero	CC Cirrocumulus
2	2 octas	CS Cirrostratus
3	3 octas	AC Altocumulus
4	4 octas	AS Altostratus
5	5 octas	NS Nimbostratus
6	6 octas	SC Stratocumulus
7	7 octas or more but not 8 octas	CU Cumulus
8	8 octas (overcast)	CB Cumulonimbus
9	Sky obscured or cloud	// cloud not visible amount not estimated due to darkness of obscuring phenomena

Table 7-29[6]
TAF ICING

Figure Code	Amount of ice accretion
0	No icing
1	Light icing
2	Light icing in cloud
3	Light icing in precipitation
4	Moderate icing
5	Moderate icing in cloud
6	Moderate icing in precipitation
7	Severe icing
8	Severe icing in cloud
9	Severe icing in precipitation

digit is the thickness of the layer in thousands of feet. For example, let's decode the group "680304." The "6" indicates an icing forecast; the "8" indicates severe icing in cloud; "030" says the base of the icing is at 3,000 feet (AGL); and "4" specifies a layer 4,000 feet thick.

(n) *Turbulence*. A turbulence group also may be included. It also is a 6-digit group coded the same as the icing group except a "5" identifies the group as a turbulence forecast. Type of turbulence is from Table 7-29[7]. For example, decoding the group "590359," the "5" identifies a turbulence forecast; the "9" specifies frequent severe turbulence in cloud (See Table 7-29[7]); "035" says the base of the turbulent layer is 3,500 feet (AGL); the "9" specifies that the turbulence layer is 9,000 feet thick. When either an icing layer or a turbulent layer is expected to be more than 9,000 feet thick, multiple groups are used. The top specified in one group is coincident with the base in the following group. Let's assume the forecaster expects frequent turbulence from the surface to 45,000 feet with the most hazardous turbulence at mid-levels. This could be encoded "530005 550509 592309 553209 554104." While you most likely will never see such a complex coding with this many groups, the flexible TAF code permits it.

Table 7-29[7]
TURBULENCE

Figure Code	Turbulence
0	None
1	Light turbulence
2	Moderate turbulence in clear air, infrequent
3	Moderate turbulence in clear air, frequent
4	Moderate turbulence in cloud, infrequent
5	Moderate turbulence in cloud, frequent
6	Severe turbulence in clear air, infrequent
7	Severe turbulence in clear air, frequent
8	Severe turbulence in cloud, infrequent
9	Severe turbulence in cloud, frequent

(o) *Temperature*. A temperature code is seldom included in a terminal forecast. However, it may be included if critical to aviation. It may be used to alert the pilot to high density altitude or possible frost when on the ground. The temperature group is identified by the digit "0." The next two (2) digits are the time to the nearest whole hour (UTC) to which the forecast temperature applies. The last two (2) digits are temperature in degrees Celsius. A minus temperature is preceded by the letter "M."

EXAMPLE:
"02137" means temperature at 2100Z is expected to be 37 degrees Celsius (about 99 degrees F); "012M02" means temperature at 1200Z is expected to be minus 2 degrees Celsius. A forecast may include more than one temperature group.

7-30 thru 7-39. RESERVED

Section 2. ALTIMETER SETTING PROCEDURES

7-40. GENERAL

a. The accuracy of aircraft altimeters is subject to the following factors:

1. Nonstandard temperatures of the atmosphere.

2. Nonstandard atmospheric pressure.

3. Aircraft static pressure systems (position error), and

4. Instrument error.

b. EXTREME CAUTION SHOULD BE EXERCISED WHEN FLYING IN PROXIMITY TO OBSTRUCTIONS OR TERRAIN IN LOW TEMPERATURES AND PRESSURES. This is especially true in extremely cold temperatures that cause a large differential between the Standard

Day temperature and actual temperature. This circumstance can cause serious errors that result in the aircraft being significantly lower than the indicated altitude.

7-40b NOTE—Standard temperature at sea level is 15 degrees Celsius (59 degrees Fahrenheit). The temperature gradient from sea level is minus 2 degrees Celsius (3.6 degrees Fahrenheit) per 1,000 feet. Pilots should apply corrections for static pressure systems and/or instruments, if appreciable errors exist.

c. The adoption of a standard altimeter setting at the higher altitudes eliminates station barometer errors, some altimeter instrument errors, and errors caused by altimeter settings derived from different geographical sources.

7-41. PROCEDURES

The cruising altitude or flight level of aircraft shall be maintained by reference to an altimeter which shall be set, when operating:

a. Below 18,000 feet MSL:

1. When the barometric pressure is 31.00 inches Hg. or less—to the current reported altimeter setting of a station along the route and within 100 NM of the aircraft, or if there is no station within this area, the current reported altimeter setting of an appropriate available station. When an aircraft is en route on an instrument flight plan, air traffic controllers will furnish this information to the pilot at least once while the aircraft is in the controllers' area of jurisdiction. In the case of an aircraft not equipped with a radio, set to the elevation of the departure airport or use an appropriate altimeter setting available prior to departure.

2. When the barometric pressure exceeds 31.00 inches Hg.—the following procedures will be placed in effect by NOTAM defining the geographic area affected:

(a) For all aircraft—Set 31.00 inches for en route operations below 18,000 feet MSL. Maintain this setting until beyond the affected area or until reaching final approach segment. At the beginning of the final approach segment, the current altimeter setting will be set, if possible. If not possible, 31.00 inches will remain set throughout the approach. Aircraft on departure or missed approach will set 31.00 inches prior to reaching any mandatory/crossing altitude or 1,500 feet AGL, whichever is lower. (Air traffic control will issue actual altimeter settings and advise pilots to set 31.00 inches in their altimeters for en route operations below 18,000 feet MSL in affected areas.)

(b) During preflight, barometric altimeters shall be checked for normal operation to the extent possible.

(c) For aircraft with the capability of setting the current altimeter setting and operating into airports with the capability of measuring the current altimeter setting, no additional restrictions apply.

(d) For aircraft operating VFR, there are no additional restrictions, however, extra diligence in flight planning and in operating in these conditions is essential.

(e) Airports unable to accurately measure barometric pressures above 31.00 inches of Hg. will report the barometric pressure as "missing" or "in excess of 31.00 inches of Hg." Flight operations to and from those airports are restricted to VFR weather conditions.

(f) For aircraft operating IFR and unable to set the current altimeter setting, the following restrictions apply:

(1) To determine the suitability of departure alternate airports, destination airports, and destination alternate airports, increase ceiling requirements by 100 feet and visibility requirements by ¼ statute mile for each ⅒ of an inch of Hg., or any portion thereof, over 31.00 inches. These adjusted values are then applied in accordance with the requirements of the applicable operating regulations and operations specifications.

EXAMPLE:

Destination altimeter is 31.28 inches, ILS DH 250 feet (200–½). When flight planning, add 300–¾ to the weather requirements which would become 500–1¼.

(2) On approach, 31.00 inches will remain set. Decision Height or minimum descent altitude shall be deemed to have been reached when the published altitude is displayed on the altimeter.

7-41a1f2 NOTE—Although visibility is normally the limiting factor on an approach, pilots should be aware that when reaching DH the aircraft will be higher than indicated. Using the example above the aircraft would be approximately 300 feet higher.

(3) These restrictions do not apply to authorized Category II and III ILS operations nor do they apply to certificate holders using approved QFE altimetry systems.

(g) The FAA Regional Flight Standards Division Manager of the Affected area is authorized to approve temporary waivers to permit emergency resupply or emergency medical service operation.

b. At or above 18,000 feet MSL—to 29.92 inches of mercury (standard setting). The lowest usable flight level is determined by the atmospheric pressure in the area of operation as shown in Table 7-41[1].

Table 7-41[1]

Altimeter Setting (Current Reported)	Lowest Usable Flight Level
29.92 or higher	180
29.91 to 29.42	185
29.41 to 28.92	190
28.91 to 28.42	195
28.41 to 27.92	200

c. Where the minimum altitude, as prescribed in FAR Part 91.159 and FAR Part 91.177, is above 18,000 feet MSL, the lowest usable flight level shall be the flight level equivalent of the minimum altitude plus the number of feet specified in Table 7-41[2].

Table 7-41[2]

Altimeter Setting	Correction Factor
29.92 or higher	none
29.91 to 29.42	500 Feet
29.41 to 28.92	1000 Feet
28.91 to 28.42	1500 Feet
28.41 to 27.92	2000 Feet
27.91 to 27.42	2500 Feet

EXAMPLE:
The minimum safe altitude of a route is 19,000 feet MSL and the altimeter setting is reported between 29.92 and 29.42 inches of mercury, the lowest usable flight level will be 195, which is the flight level equivalent of 19,500 feet MSL (minimum altitude plus 500 feet).

7-42. ALTIMETER ERRORS

a. Most pressure altimeters are subject to mechanical, elastic, temperature, and installation errors. (Detailed information regarding the use of pressure altimeters is found in the Instrument Flying Handbook, Chapter IV.) Although manufacturing and installation specifications, as well as the periodic test and inspections required by regulations (FAR Part 43, Appendix E), act to reduce these errors, any scale error may be observed in the following manner:

1. Set the current reported altimeter setting on the altimeter setting scale.

2. Altimeter should now read field elevation if you are located on the same reference level used to establish the altimeter setting.

3. Note the variation between the known field elevation and the altimeter indication. If this variation is in the order of plus or minus 75 feet, the accuracy of the altimeter is questionable and the problem should be referred to an appropriately rated repair station for evaluation and possible correction.

b. Once in flight, it is very important to obtain frequently current altimeter settings en route. If you do not reset your altimeter when flying *from* an area of high pressure *into* an area of low pressure, *your aircraft will be closer to the surface than your altimeter indicates.* An inch error in the altimeter setting equals 1,000 feet of altitude. To quote an old saying: **"GOING FROM A HIGH TO A LOW, LOOK OUT BELOW."**

c. Temperature also has an effect on the accuracy of altimeters and your altitude. The crucial values to consider are standard temperature versus the ambient (at altitude) temperature. It is this "difference" that causes the error in indicated altitude. When the air is warmer than standard, you are higher than your altimeter indicates. Subsequently, when the air is colder than standard you are lower than indicated. It is the magnitude of this "difference" that determines the magnitude of the error. When flying into a cooler air mass while maintaining a constant indicated altitude, you are losing true altitude. However, flying into a cooler air mass does not necessarily mean you will be lower than indicated if the *difference* is still on the plus side. For example, while flying at 10,000 feet (where **STANDARD** temperature is –5 degrees Celsius (C)), the outside air temperature cools from +5 degrees C to 0 degrees C, the temperature error will nevertheless cause the aircraft to be **HIGHER** than indicated. It is the extreme "cold" difference that normally would be of concern to the pilot. Also, when flying in cold conditions over mountainous country, the pilot should exercise caution in flight planning both in regard to route and altitude to ensure adequate enroute terrain clearance.

d. The possible results of the above situations is obvious, particularly if operating at the minimum altitude or when conducting an instrument approach. If the altimeter is in error, you may still be on instruments when reaching the minimum altitude (as indicated on the altimeter), whereas you might have been in the clear and able to complete the approach if the altimeter setting was correct.

7-43. HIGH BAROMETRIC PRESSURE

a. Cold, dry air masses may produce barometric pressures in excess of 31.00 inches of Mercury, and many altimeters do not have an accurate means of being adjusted for settings of these levels. When the altimeter cannot be set to the higher pressure setting, the aircraft actual altitude will be higher than the altimeter indicates. (Reference—Altimeter Errors, paragraph 7-42.)

b. When the barometric pressure exceeds 31.00 inches, air traffic controllers will issue the actual altimeter setting, and:

1. En Route/Arrivals—Advise pilots to remain set on 31.00 inches until reaching the final approach segment.

2. Departures—Advise pilots to set 31.00 inches prior to reaching any mandatory/crossing altitude or 1,500 feet, whichever is lower.

c. The altimeter error caused by the high pressure will be in the opposite direction to the error caused by the cold temperature.

7-44. LOW BAROMETRIC PRESSURE

When abnormally low barometric pressure conditions occur (below 28.00), flight operations by aircraft unable to set the actual altimeter setting are not recommended.

7-44 NOTE—The true altitude of the aircraft is lower than the indicated altitude if the pilot is unable to set the actual altimeter setting.

7-45 thru 7-49. RESERVED

Section 3. WAKE TURBULENCE

7-50. GENERAL

a. Every aircraft generates a wake while in flight. Initially, when pilots encountered this wake in flight, the disturbance was attributed to "prop wash." It is known, however, that this disturbance is caused by a pair of counter rotating vortices trailing from the wing tips. The vortices from larger aircraft pose problems to encountering aircraft. For instance, the wake of these aircraft can impose rolling moments exceeding the roll-control authority of the encountering aircraft. Further, turbulence generated within the vortices can damage aircraft components and equipment if encountered at close range. The pilot must learn to envision the location of the vortex wake generated by larger (transport category) aircraft and adjust the flight path accordingly.

b. During ground operations and during takeoff, jet engine blast (thrust stream turbulence) can cause damage and upsets if encountered at close range. Exhaust velocity versus distance studies at various thrust levels have shown a need for light aircraft to maintain an adequate separation behind large turbojet aircraft. Pilots of larger aircraft should be particularly careful to consider the effects of their "jet blast" on other aircraft, vehicles, and maintenance equipment during ground operations.

7-51. VORTEX GENERATION

Lift is generated by the creation of a pressure differential over the wing surface. The lowest pressure occurs over the upper wing surface and the highest pressure under the wing. This pressure differential triggers the roll up of the airflow aft of the wing resulting in swirling air masses trailing downstream of the wing tips. After the roll up is completed, the wake consists of two counter rotating cylindrical vortices. (See Figure 7-51[1].) Most of the energy is within a few feet of the center of each vortex, but pilots should avoid a region within about 100 feet of the vortex core.

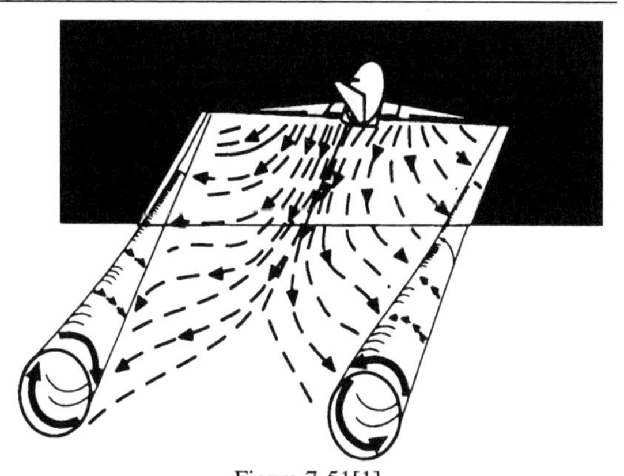

Figure 7-51[1]

7-52. VORTEX STRENGTH

a. The strength of the vortex is governed by the weight, speed, and shape of the wing of the generating aircraft. The vortex characteristics of any given aircraft can also be changed by extension of flaps or other wing configuring devices as well as by change in speed. However, as the basic factor is weight, the vortex strength increases proportionately. Peak vortex tangential speeds exceeding 300 feet per second have been recorded. The greatest vortex strength occurs when the generating aircraft is HEAVY, CLEAN, and SLOW.

b. INDUCED ROLL

1. In rare instances a wake encounter could cause inflight structural damage of catastrophic proportions. However, the usual hazard is associated with induced rolling moments which can exceed the roll-control authority of the encountering aircraft. In flight experiments, aircraft have been intentionally flown directly up trailing vortex cores of larger aircraft. It was shown that the capability of an aircraft to counteract the roll imposed by the wake vortex primarily depends on the wingspan and counter-control responsiveness of the encountering aircraft.

2. Counter control is usually effective and induced roll minimal in cases where the wingspan and ailerons of the encountering aircraft extend beyond the rotational flow field of the vortex. It is more difficult for aircraft with short wingspan (relative to the generating aircraft) to counter the imposed roll induced by vortex flow. Pilots of short span aircraft, even of the high performance type, must be especially alert to vortex encounters. (See Figure 7-52[1].)

3. The wake of larger aircraft requires the respect of all pilots.

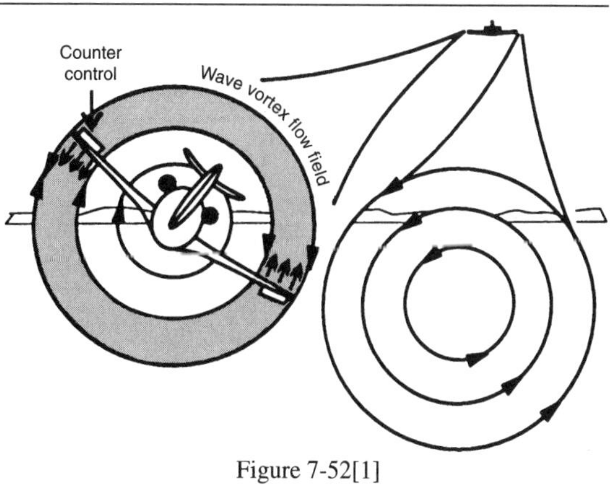

Figure 7-52[1]

7-53. VORTEX BEHAVIOR

a. Trailing vortices have certain behavioral characteristics which can help a pilot visualize the wake location and thereby take avoidance precautions.

1. Vortices are generated from the moment aircraft leave the ground, since trailing vortices are a by-product of wing lift. Prior to takeoff or touchdown pilots should note the rotation or touchdown point of the preceding aircraft. (See Figure 7-53[1][Wake Begins/Ends].)

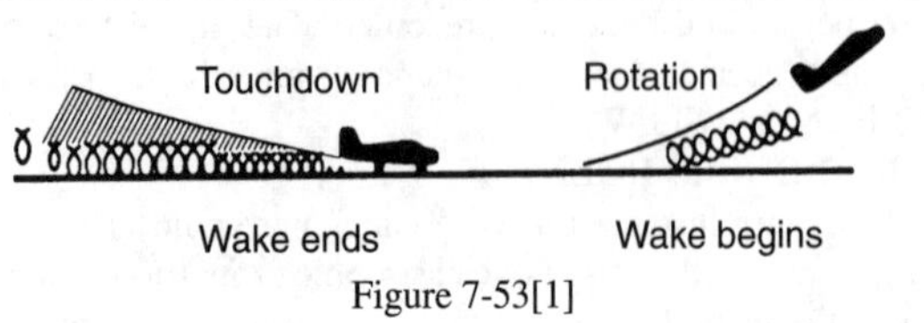

Figure 7-53[1]

2. The vortex circulation is outward, upward and around the wing tips when viewed from either ahead or behind the aircraft. Tests with large aircraft have shown that the vortices remain spaced a bit less than a wingspan apart, drifting with the wind, at altitudes greater than a wingspan from the ground. In view of this, if persistent vortex turbulence is encountered, a slight change of altitude and lateral position (preferably upwind) will provide a flight path clear of the turbulence.

3. Flight tests have shown that the vortices from larger (transport category) aircraft sink at a rate of several hundred feet per minute, slowing their descent and diminishing in strength with time and distance behind the generating aircraft. Atmospheric turbulence hastens breakup. Pilots should fly at or above the preceding aircraft's flight path, altering course as necessary to avoid the area behind and below the generating aircraft. (See Figure 7-53[2][Vortex Flow Field].) However vertical separation of 1,000 feet may be considered safe.

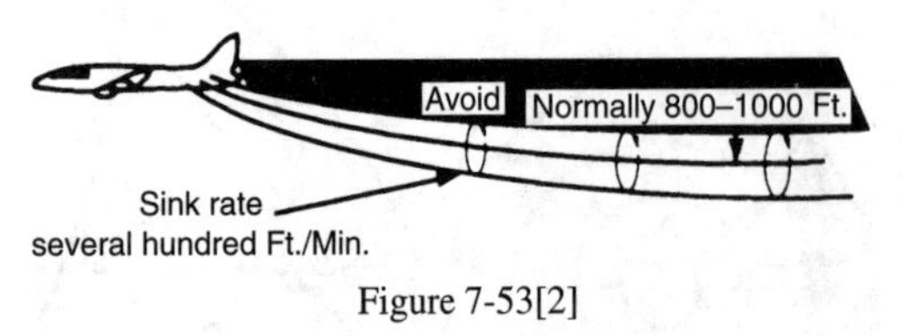

Figure 7-53[2]

4. When the vortices of larger aircraft sink close to the ground (within 100 to 200 feet), they tend to move laterally over the ground at a speed of 2 or 3 knots. (See Figure 7-53[3][Vortex Sink Rate].)

b. A crosswind will decrease the lateral movement of the upwind vortex and increase the movement of the downwind vortex. Thus a light wind with a cross runway component of 1 to 5 knots could result in the upwind vortex remaining in the touchdown zone for a period of time and hasten the drift of the downwind vortex toward another runway. (See Figure 7-53[4][Vortex Movement in Ground Effect (No Wind)].) Similarly, a tailwind condition can move the vortices of the preceding aircraft forward into the touchdown zone. THE LIGHT QUARTERING TAILWIND REQUIRES MAXIMUM CAUTION. Pilots should be alert to large aircraft upwind from their approach and takeoff flight paths. (See Figure 7-53[5][Vortex Movement in Ground Effect (Wind)].)

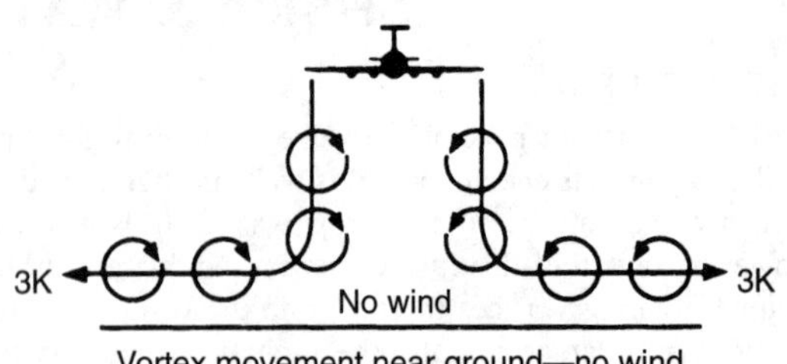

Figure 7-53[3]

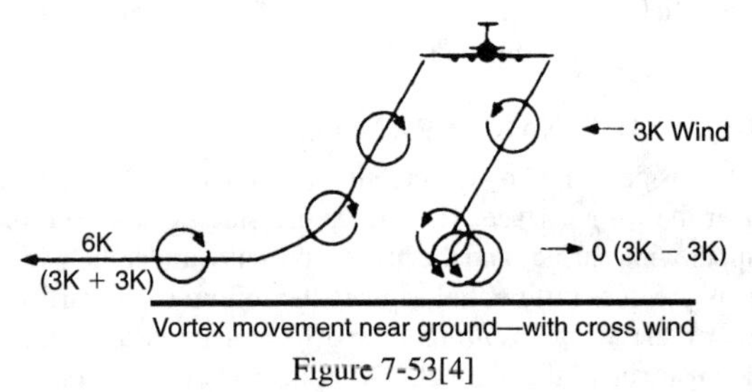

Figure 7-53[4]

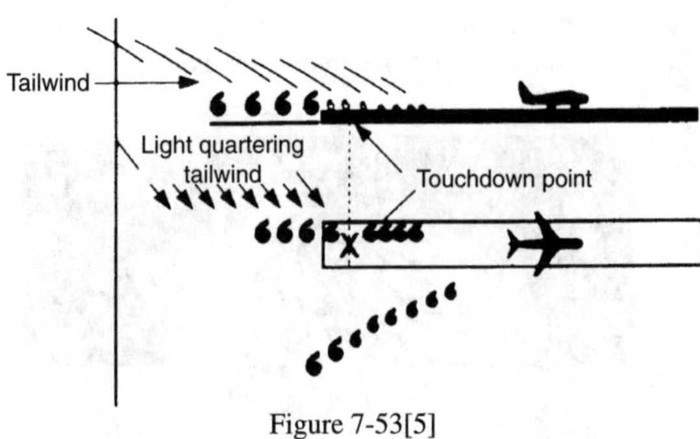

Figure 7-53[5]

7-54. OPERATIONS PROBLEM AREAS

a. A wake encounter can be catastrophic. In 1972 at Fort Worth a DC-9 got too close to a DC-10 (two miles back), rolled, caught a wingtip, and cartwheeled coming to rest in an inverted position on the runway. All aboard were killed. Serious and even fatal GA accidents induced by wake vortices are not uncommon. However, a wake encounter is not necessarily hazardous. It can be one or more jolts with varying severity depending upon the direction of the encounter, weight of the generating aircraft, size of the encountering aircraft, distance from the generating aircraft, and point of vortex encounter. The probability of induced roll increases when the encountering aircraft's heading is generally aligned with the flight path of the generating aircraft.

b. AVOID THE AREA BELOW AND BEHIND THE GENERATING AIRCRAFT, ESPECIALLY AT LOW ALTITUDE WHERE EVEN A MOMENTARY WAKE ENCOUNTER COULD BE HAZARDOUS. This is not easy to do. Some accidents have occurred even though the pilot of the trailing aircraft had carefully noted that the aircraft in front was at a considerably lower altitude. Unfortunately, this does not ensure that the flight path of the lead aircraft will be below that of the trailing aircraft.

c. Pilots should be particularly alert in calm wind conditions and situations where the vortices could:

1. Remain in the touchdown area.

2. Drift from aircraft operating on a nearby runway.

3. Sink into the takeoff or landing path from a crossing runway.

4. Sink into the traffic pattern from other airport operations.

5. Sink into the flight path of VFR aircraft operating on the hemispheric altitude 500 feet below.

d. Pilots of all aircraft should visualize the location of the vortex trail behind larger aircraft and use proper vortex avoidance procedures to achieve safe operation. It is equally important that pilots of larger aircraft plan or adjust their flight paths to minimize vortex exposure to other aircraft.

7-55. VORTEX AVOIDANCE PROCEDURES

a. Under certain conditions, airport traffic controllers apply procedures for separating IFR aircraft. The controllers will also provide to VFR aircraft, with whom they are in communication and which in the tower's opinion may be adversely affected by wake turbulence from a larger aircraft, the position, altitude and direction of flight of larger aircraft followed by the phrase "CAUTION—WAKE TURBULENCE." WHETHER OR NOT A WARNING HAS BEEN GIVEN, HOWEVER, THE PILOT IS EXPECTED TO ADJUST HIS OR HER OPERATIONS AND FLIGHT PATH AS NECESSARY TO PRECLUDE SERIOUS WAKE ENCOUNTERS.

b. The following vortex avoidance procedures are recommended for the various situations:

1. Landing behind a larger aircraft—same runway: Stay at or above the larger aircraft's final approach flight path—note its touchdown point—land beyond it.

2. Landing behind a larger aircraft—when parallel runway is closer than 2,500 feet: Consider possible drift to your runway. Stay at or above the larger aircraft's final approach flight path—note its touchdown point.

3. Landing behind a larger aircraft—crossing runway: Cross above the larger aircraft's flight path.

4. Landing behind a departing larger aircraft—same runway: Note the larger aircraft's rotation point—land well prior to rotation point.

5. Landing behind a departing larger aircraft—crossing runway: Note the larger aircraft's rotation point—if past the intersection—continue the approach—land prior to the intersection. If larger aircraft rotates prior to the intersection, avoid flight below the larger aircraft's flight path. Abandon the approach unless a landing is ensured well before reaching the intersection.

6. Departing behind a larger aircraft: Note the larger aircraft's rotation point—rotate prior to larger aircraft's rotation point—continue climb above the larger aircraft's climb path until turning clear of his wake. Avoid subsequent headings which will cross below and behind a larger aircraft. Be alert for any critical takeoff situation which could lead to a vortex encounter.

7. Intersection takeoffs—same runway: Be alert to adjacent larger aircraft operations, particularly upwind of your runway. If intersection takeoff clearance is received, avoid subsequent heading which will cross below a larger aircraft's path.

8. Departing or landing after a larger aircraft executing a low approach, missed approach or touch-and-go landing: Because vortices settle and move laterally near the ground, the vortex hazard may exist along the runway and in your flight path after a larger aircraft has executed a low approach, missed approach or a touch-and-go landing, particularly in light quartering wind conditions. You should ensure that an interval of at least 2 minutes has elapsed before your takeoff or landing.

9. En route VFR (thousand-foot altitude plus 500 feet): Avoid flight below and behind a large aircraft's path. If a larger aircraft is observed above on the same track (meeting or overtaking) adjust your position laterally, preferably upwind.

7-56. HELICOPTERS

In a slow hover taxi or stationary hover near the surface, helicopter main rotor(s) generate downwash producing high velocity outwash vortices to a distance approximately three times the diameter of the rotor. When rotor downwash hits the surface, the resulting outwash vortices have behavioral characteristics similar to wing tip vortices produced by fixed wing aircraft. However, the vortex circulation is outward, upward, around, and away from the main rotor(s) in all directions. Pilots of small aircraft should avoid operating within three rotor diameters of any helicopter in a slow hover taxi or stationary hover. In forward flight, departing or landing helicopters produce a pair of strong, high-speed trailing vortices similar to wing tip vortices of larger fixed wing aircraft. Pilots of small aircraft should use caution when operating behind or crossing behind landing and departing helicopters.

7-57. PILOT RESPONSIBILITY

a. Government and industry groups are making concerted efforts to minimize or eliminate the hazards of trailing vortices. However, the flight disciplines necessary to ensure vortex avoidance during VFR operations must be exercised by the pilot. Vortex visualization and avoidance procedures should be exercised by the pilot using the same degree of concern as in collision avoidance.

b. Wake turbulence may be encountered by aircraft in flight as well as when operating on the airport movement area. (Reference—Pilot/Controller Glossary, Wake Turbulence).

c. Pilots are reminded that in operations conducted behind all aircraft, acceptance of instructions from ATC in the following situations is an acknowledgment that the pilot will ensure safe takeoff and landing intervals and accepts the responsibility of providing his own wake turbulence separation.

1. Traffic information,

2. Instructions to follow an aircraft, and

3. The acceptance of a visual approach clearance.

d. For operations conducted behind **heavy** aircraft, ATC will specify the word "**heavy**" when this information is known. Pilots of **heavy** aircraft should always use the word "**heavy**" in radio communications.

7-58. AIR TRAFFIC WAKE TURBULENCE SEPARATIONS

a. Because of the possible effects of wake turbulence, controllers are required to apply no less than specified minimum separation for aircraft operating behind a **heavy** jet and, in certain instances, behind large **nonheavy** aircraft.

1. Separation is applied to aircraft operating directly behind a **heavy** jet at the same altitude or less than 1,000 feet below:

(a) Heavy jet behind **heavy** jet—4 miles.

(b) Small/large aircraft behind heavy jet—5 miles.

2. Also, separation, measured at the time the preceding aircraft is over the landing threshold, is provided to small aircraft:

(a) Small aircraft landing behind **heavy** jet—6 miles.

(b) Small aircraft landing behind **large** aircraft—4 miles.

7-58a2b NOTE—See Pilot/Controller Glossary, Aircraft Classes.

3. Additionally, appropriate time or distance intervals are provided to departing aircraft:

(a) Two minutes or the appropriate 4 or 5 mile radar separation when takeoff behind a **heavy** jet will be:

—from the same threshold

—on a crossing runway and projected flight paths will cross

—from the threshold of a parallel runway when staggered ahead of that of the adjacent runway by less than 500 feet and when the runways are separated by less than 2,500 feet.

7-58a3a NOTE—Pilots, after considering possible wake-turbulence effects, may specifically request waiver of the 2-minute interval by stating, "request waiver of 2-minute interval" or a similar statement. Controllers may acknowledge this statement as pilot acceptance of responsibility for wake turbulence separation and, if traffic permits, issue takeoff clearance.

b. A 3-minute interval will be provided when a **small** aircraft will take off:

1. From an intersection on the same runway (same or opposite direction) behind a departing **large** aircraft,

2. In the opposite direction on the same runway behind a large aircraft takeoff or low/missed approach.

7-58b2 NOTE—This 3-minute interval may be waived upon specific pilot request.

c. A 3-minute interval will be provided for all aircraft taking off when the operations are as described in b(1) and (2) above, the preceding aircraft is a **heavy** jet, and the operations are on either the same runway or parallel runways separated by less than 2,500 feet. Controllers may not reduce or waive this interval.

d. Pilots may request additional separation; i.e., 2 minutes instead of 4 or 5 miles for wake turbulence avoidance. This request should be made as soon as practical on ground control and at least before taxiing onto the runway.

7-58d NOTE—FAR Part 91.3(a) states: "The pilot in command of an aircraft is directly responsible for and is the final authority as to the operation of that aircraft."

e. Controllers may anticipate separation and need not withhold a takeoff clearance for an aircraft departing behind a **large/heavy** aircraft if there is reasonable assurance the required separation will exist when the departing aircraft starts takeoff roll.

7-59 thru 7-69. RESERVED

Section 4. BIRD HAZARDS AND FLIGHT OVER NATIONAL REFUGES, PARKS, AND FORESTS

7-70. MIGRATORY BIRD ACTIVITY

a. Bird strike risk increases because of bird migration during the months of March through April, and August through November.

b. The altitudes of migrating birds vary with winds aloft, weather fronts, terrain elevations, cloud conditions, and other environmental variables. While over 90 percent of the reported bird strikes occur at or below 3,000 feet AGL, strikes at higher altitudes are common during migration. Ducks and geese are frequently observed up to 7,000 feet AGL and pilots are cautioned to minimize en route flying at lower altitudes during migration.

c. Considered the greatest potential hazard to aircraft because of their size, abundance, or habit of flying in dense flocks are gulls, waterfowl, vultures, hawks, owls, egrets, blackbirds, and starlings. Four major migratory flyways exist in the United States. The Atlantic flyway parallels the Atlantic Coast. The Mississippi Flyway stretches from Canada through the Great Lakes and follows the Mississippi River. The Central Flyway represents a broad area east of the Rockies, stretching from Canada through Central America. The Pacific Flyway follows the west coast and overflies major parts of Washington, Oregon, and California. There are also numerous smaller flyways which cross these major north-south migratory routes.

7-71. REDUCING BIRD STRIKE RISKS

a. The most serious strikes are those involving ingestion into an engine (turboprops and turbine jet engines) or windshield strikes. These strikes can result in emergency situations requiring prompt action by the pilot.

b. Engine ingestions may result in sudden loss of power or engine failure. Review engine out procedures, especially when operating from airports with known bird hazards or when operating near high bird concentrations.

c. Windshield strikes have resulted in pilots experiencing confusion, disorientation, loss of communications, and aircraft control problems. Pilots are encouraged to review their emergency procedures before flying in these areas.

d. When encountering birds en route, climb to avoid collision, because birds in flocks generally distribute themselves downward, with lead birds being at the highest altitude.

e. Avoid overflight of known areas of bird concentration and flying at low altitudes during bird migration. Charted wildlife refuges and other natural areas contain unusually high local concentration of birds which may create a hazard to aircraft.

7-72. REPORTING BIRD STRIKES

Pilots are urged to report any bird or other wildlife strike using FAA Form 5200-7, Bird Strike Incident/Ingestion Report. (Form available on pages 243 and 244.) Additional forms are available at any FSS, General Aviation District Office, Air Carrier District Office, or at an FAA Regional Office. The data derived from these reports is used to develop standards to cope with this potential hazard to aircraft and for documentation of necessary habitat control on airports.

7-73. REPORTING BIRD AND OTHER WILDLIFE ACTIVITIES

If you observe birds or other animals on or near the runway, request airport management to disperse the wildlife before taking off. Also contact the nearest FAA ARTCC, FSS, or tower (including non-Federal towers) regarding large flocks of birds and report the:

1. Geographic location

2. Bird type (geese, ducks, gulls, etc.)

3. Approximate numbers

4. Altitude

5. Direction of bird flight path

7-74. PILOT ADVISORIES ON BIRD AND OTHER WILDLIFE HAZARDS

Many airports advise pilots of other wildlife hazards caused by large animals on the runway through the Airport/Facility Directory and the NOTAM system. Collisions of landing and departing aircraft and animals on the runway are increasing and are not limited to rural airports. These accidents have also occurred at several major airports. Pilots should exercise extreme caution when warned of the presence of wildlife on and in the vicinity of airports. If you observe deer or other large animals in close proximity to movement areas, advise the FSS, tower, or airport management.

7-75. FLIGHTS OVER CHARTED U.S. WILDLIFE REFUGES, PARKS, AND FOREST SERVICE AREAS

a. The landing of aircraft is prohibited on lands or waters administered by the National Park Service, U.S. Fish and Wildlife Service, or U.S. Forest Service without authorization from the respective agency. Exceptions, including:

1. when forced to land due to an emergency beyond the control of the operator,

2. at officially designated landing sites, or

3. an approved official business of the Federal Government.

b. All aircraft are requested to maintain a minimum altitude of 2,000 feet above the *surface* of the following: National Parks, Monuments, Seashores, Lakeshores, Recreation Areas and Scenic Riverways administered by the National Park Service, National Wildlife Refuges, Big Game Refuges, Game Ranges and Wildlife Ranges administered by the U.S. Fish and Wildlife Service, and Wilderness and Primitive areas administered by the U.S. Forest Service.

7-75b NOTE—FAA Advisory Circular 91-36, Visual Flight Rules (VFR) Flight Near Noise-Sensitive Areas, defines the *surface* of a National Park Area (including Parks, Forests, Primitive Areas, Wilderness Areas, Recreational Areas, National Seashores, National Monuments, National Lakeshores, and National Wildlife Refuge and Range Areas) as: the highest terrain within 2,000 feet laterally of the route of flight, *or* the upper-most rim of a canyon or valley.

c. Federal statutes prohibit certain types of flight activity and/or provide altitude restrictions over *designated* U.S. Wildlife Refuges, Parks, and Forest Service Areas. These designated areas, for example: Boundary Waters Canoe Wilderness Areas, Minnesota; Haleakala National Park, Hawaii; Yosemite National Park, California; are charted on Sectional Charts.

d. Federal regulations also prohibit airdrops by parachute or other means of persons, cargo, or objects from aircraft on lands administered by the three agencies without authorization from the respective agency. Exceptions include:

1. emergencies involving the safety of human life, or

2. threat of serious property loss.

7-76 thru 7-79. RESERVED.

Section 5. POTENTIAL FLIGHT HAZARDS

7-80. ACCIDENT CAUSE FACTORS

a. The 10 most frequent cause factors for General Aviation Accidents that involve the pilot-in-command are:

1. Inadequate preflight preparation and/or planning.

2. Failure to obtain and/or maintain flying speed.

3. Failure to maintain direction control.

4. Improper level off.

5. Failure to see and avoid objects or obstructions.

6. Mismanagement of fuel.

7. Improper in-flight decisions or planning.

8. Misjudgment of distance and speed.

9. Selection of unsuitable terrain.

10. Improper operation of flight controls.

b. This list remains relatively stable and points out the need for continued refresher training to establish a higher level of flight proficiency for all pilots. A part of the FAA's continuing effort to promote increased aviation safety is the General Aviation Accident Prevention Program. For information on Accident Prevention activities contact your nearest General Aviation or Flight Standards District Office.

c. ALERTNESS—Be alert at all times, especially when the weather is good. Most pilots pay attention to business when they are operating in full IFR weather conditions, but strangely, air collisions almost invariably have occurred under ideal weather conditions. Unlimited visibility appears to encourage a sense of security which is not at all justified. Considerable information of value may be obtained by listening to advisories being issued in the terminal area, even though controller workload may prevent a pilot from obtaining individual service.

d. GIVING WAY—If you think another aircraft is too close to you, give way instead of waiting for the other pilot to respect the right-of-way to which you may be entitled. It is a lot safer to pursue the right-of-way angle after you have completed your flight.

7-81. VFR IN CONGESTED AREAS

A high percentage of near midair collisions occur below 8,000 feet AGL and within 30 miles of an airport. When operating VFR in these highly congested areas, whether you intend to land at an airport within the area or are just flying through, it is recommended that extra vigilance be maintained and that you monitor an appropriate control frequency. Normally the appropriate frequency is an approach control frequency. By such monitoring action you can "get the picture" of the traffic in your area. When the approach controller has radar, radar traffic advisories may be given to VFR pilots upon request. (Reference—Radar Traffic Information Service, paragraph 4-14.)

7-82. OBSTRUCTIONS TO FLIGHT

a. GENERAL

Many structures exist that could significantly affect the safety of your flight when operating below 500 feet above ground level (AGL), and particularly below 200 feet AGL. While FAR Part 91.119 allows flight below 500 AGL when over sparsely populated areas or open water, such operations are very dangerous. At and below 200 feet AGL there are numerous power lines, antenna towers, etc., that are not marked and lighted as obstructions and therefore may not be seen in time to avoid a collision. Notices to Airmen (NOTAMs) are issued on those lighted structures experiencing temporary light outages. However, some time may pass before the FAA is notified of these outages, and the NOTAM issued, thus pilot vigilance is imperative.

b. Antenna Towers

Extreme caution should be exercised when flying less than 2,000 feet above ground level (AGL) because of numerous skeletal structures, such as radio and television antenna towers, that exceed 1,000 feet AGL with some extending higher than 2,000 feet AGL. Most skeletal structures are supported by guy wires which are very difficult to see in good weather and can be invisible at dusk or during periods of reduced visibility. These wires can extend about 1,500 feet horizontally from a structure; therefore, all skeletal structures should be avoided horizontally by at least 2,000 feet. Additionally, new towers may not be on your current chart because the information was not received prior to the printing of the chart.

c. Overhead Wires

Overhead transmission and utility lines often span approaches to runways, natural flyways such as lakes, rivers, gorges, and canyons, and cross other landmarks pilots frequently follow such as highways, railroad tracks, etc. As with antenna towers, these high voltage/power lines or the supporting structures of these lines may not always be readily visible and the wires may be virtually impossible to see under certain conditions. In some locations, the supporting structures of overhead transmission lines are equipped with unique sequence flashing white strobe light systems to indicate that there are wires between the structures. However, many power lines do not require notice to the FAA and, therefore, are not marked and/or lighted. Many of those that do require notice do not exceed 200 feet AGL or meet the Obstruction Standard of FAR Part 77 and, therefore, are not marked and/or lighted. All pilots are cautioned to remain extremely vigilant for these power lines or their supporting structures when following natural flyways or during the approach and landing phase. This is particularly important for seaplane and/or float equipped aircraft when landing on, or departing from, unfamiliar lakes or rivers.

d. Other Objects/Structures

There are other objects or structures that could adversely affect your flight such as construction cranes near an airport, newly constructed buildings, new towers, etc. Many of these structures do not meet charting requirements or may not yet be charted because of the charting cycle. Some structures do not require obstruction marking and/or lighting and some may not be marked and lighted even though the FAA recommended it.

7-83. AVOID FLIGHT BENEATH UNMANNED BALLOONS

a. The majority of unmanned free balloons currently being operated have, extending below them, either a suspension device to which the payload or instrument package is attached, or a trailing wire antenna, or both. In many instances these balloon subsystems may be invisible to the pilot until his aircraft is close to the balloon, thereby creating a potentially dangerous situation. Therefore, good judgment on the part of the pilot dictates that aircraft should remain well clear of all unmanned free balloons and flight below them should be avoided at all times.

b. Pilots are urged to report any unmanned free balloons sighted to the nearest FAA ground facility with which communication is established. Such information will assist FAA ATC facilities to identify and flight follow unmanned free balloons operating in the airspace.

7-84. MOUNTAIN FLYING

a. Your first experience of flying over mountainous terrain (particularly if most of your flight time has been over the flatlands of the midwest) could be a never-to-be-forgotten nightmare if proper planning is not done and if you are not aware of the potential hazards awaiting. Those familiar section lines are not present in the mountains; those flat, level fields for forced landings are practically nonexistent; abrupt changes in wind direction and velocity occur; severe updrafts and downdrafts are common, particularly near or above abrupt changes of terrain such as cliffs or rugged areas; even the clouds look different and can build up with startling rapidity. Mountain flying need not be hazardous if you follow the recommendations below:

b. File a flight plan. Plan your route to avoid topography which would prevent a safe forced landing. The route should be over populated areas and well known mountain passes. Sufficient altitude should be maintained to permit gliding to a safe landing in the event of engine failure.

c. Don't fly a light aircraft when the winds aloft, at your proposed altitude, exceed 35 miles per hour. Expect the winds to be of much greater velocity over mountain passes than reported a few miles from them. Approach mountain passes with as much altitude as possible. Downdrafts of from 1,500 to 2,000 feet per minute are not uncommon on the leeward side.

d. Don't fly near or above abrupt changes in terrain. Severe turbulence can be expected, especially in high wind conditions.

e. Some canyons run into a dead end. Don't fly so far up a canyon that you get trapped. ALWAYS BE ABLE TO MAKE A 180 DEGREE TURN!

f. VFR flight operations may be conducted at night in mountainous terrain with the application of sound judgment and common sense. Proper pre-flight planning, giving ample consideration to winds and weather, knowledge of the terrain and pilot experience in mountain flying are prerequisites for safety of flight. Continuous visual contact with the surface and obstructions is a major concern and flight operations under an overcast or in the vicinity of clouds should be approached with extreme caution.

g. When landing at a high altitude field, the same indicated airspeed should be used as at low elevation fields. *Remember:* that due to the less dense air at altitude, this same indicated airspeed actually results in higher true airspeed, a faster landing speed, and more important, a longer landing distance. During gusty wind conditions which often prevail at high altitude fields, a power approach and power landing is recommended. Additionally, due to the faster groundspeed, your takeoff distance will increase considerably over that required at low altitudes.

h. Effects of Density Altitude. Performance figures in the aircraft owner's handbook for length of takeoff run, horsepower, rate of climb, etc., are generally based on standard atmosphere conditions (59 degrees Fahrenheit (15 degrees Celsius), pressure 29.92 inches of mercury) at sea level. However, inexperienced pilots, as well as experienced pilots, may run into trouble when they encounter an altogether different set of conditions. This is particularly true in hot weather and at higher elevations. Aircraft operations at altitudes above sea level and at higher than standard temperatures are commonplace in mountainous areas. Such operations quite often result in a drastic reduction of aircraft performance capabilities because of the changing air density. Density altitude is a measure of air density. It is not to be confused with pressure altitude, true altitude or absolute altitude. It is not to be used as a height reference, but as a determining criteria in the performance capability of an aircraft. Air density decreases with altitude. As air density decreases, density altitude increases. The further effects of high temperature and high humidity are cumulative, resulting in an increasing high density altitude condition. High density altitude reduces all aircraft performance parameters. To the pilot, this means that the normal horsepower output is reduced, propeller efficiency is reduced and a higher true airspeed is required to sustain the aircraft throughout its operating parameters. It means an increase in runway length requirements for takeoff and landings, and decreased rate of climb. An average small airplane, for example, requiring 1,000 feet for takeoff at sea level under standard atmospheric conditions will require a takeoff run of approximately 2,000 feet at an operational altitude of 5,000 feet.

7-84h NOTE—A turbo-charged aircraft engine provides some slight advantage in that it provides sea level horsepower up to a specified altitude above sea level.

1. Density Altitude Advisories—at airports with elevations of 2,000 feet and higher, control towers and FSSs will broadcast the advisory "Check Density Altitude" when the temperature reaches a predetermined level. These advisories will be broadcast on appropriate tower frequencies or, where available, ATIS. FSSs will broadcast these advisories as a part of Local Airport Advisory, and on TWEB.

2. These advisories are provided by air traffic facilities, as a reminder to pilots that high temperatures and high field elevations will cause significant changes in aircraft characteristics. The pilot retains the responsibility to compute density altitude, when appropriate, as a part of preflight duties.

7-84h2 NOTE—All FSSs will compute the current density altitude upon request.

i. Mountain Wave. Many pilots go all their lives without understanding what a mountain wave is. Quite a few have lost their lives because of this lack of understanding. One need not be a licensed meteorologist to understand the mountain wave phenomenon.

1. Mountain waves occur when air is being blown over a mountain range or even the ridge of a sharp bluff area. As the air hits the upwind side of the range, it starts to climb, thus creating what is generally a smooth updraft which turns into a turbulent downdraft as the air passes the crest of the ridge. From this point, for many miles downwind, there will be a series of downdrafts and updrafts. Satellite photos of the Rockies have shown mountain waves extending as far as 700 miles downwind of the range. Along the east coast area, such photos of the Appalachian chain have picked up the mountain wave phenomenon over a hundred miles eastward. All it takes to form a mountain wave is wind blowing across the range at 15 knots or better at an intersection angle of not less than 30 degrees.

2. Pilots from flatland areas should understand a few things about mountain waves in order to stay out of trouble. When approaching a mountain range from the upwind side (generally the west), there will usually be a smooth updraft; therefore, it is not quite as dangerous an area as the lee of the range. From the leeward side, it is always a good idea to add an extra thousand feet or so of altitude because downdrafts can exceed the climb capability of the aircraft. Never expect an updraft when approaching a mountain chain from the leeward. Always be prepared to cope with a downdraft and turbulence.

3. When approaching a mountain ridge from the downwind side, it is recommended that the ridge be approached at approximately a 45 degree angle to the horizontal direction of the ridge. This permits a safer retreat from the ridge with less stress on the aircraft should severe turbulence and downdraft be experienced. If severe turbulence is encountered, simultaneously reduce power and adjust pitch until aircraft approaches maneuvering speed, then adjust power and trim to maintain maneuvering speed and fly away from the turbulent area.

7-85. SEAPLANE SAFETY

a. Acquiring a seaplane class rating affords access to many areas not available to landplane pilots. Adding a seaplane class rating to your pilot certificate can be relatively uncomplicated and inexpensive. However, more effort is required to become a safe, efficient, competent "bush" pilot. The natural hazards of the backwoods have given way to modern man-made hazards. Except for the far north, the available bodies of water are no longer the exclusive domain of the airman. Seaplane pilots must be vigilant for hazards such as electric power lines, power, sail and rowboats, rafts, mooring lines, water skiers, swimmers, etc.

b. Seaplane pilots must have a thorough understanding of the right-of-way rules as they apply to aircraft versus other vessels. Seaplane pilots are expected to know and adhere to both the United States Coast Guard's (USCG) Inland Navigation Rules and FAR Part 91.115, Right of Way Rules; Water Operations. The navigation rules of the road are a set of collision avoidance rules as they apply to aircraft on the water. A seaplane is considered a vessel when on the water for the purposes of these collision avoidance rules. In general, a seaplane on the water shall keep well clear of all vessels and avoid impeding their navigation. The FAR requires, in part, that aircraft operating on the water "...shall, insofar as possible, keep clear of all vessels and avoid impeding their navigation and shall give way to any vessel or other aircraft that is given the right of way...." This means that a seaplane should avoid boats and commercial shipping when on the water. If on a collision course, the seaplane should slow, stop, or maneuver to the right, away from the bow of the oncoming vessel. Also, while on the surface with an engine running, an aircraft must give way to all non-powered vessels. Since a seaplane in the water may not be as maneuverable as one in the air, the aircraft on the water has right-of-way over one in the air, and one taking off has right-of-way over one landing. A seaplane is exempt from the USCG safety equipment requirements, including the requirements for Personal Floatation Devices (PFD). Requiring seaplanes on the water to comply with USCG equipment requirements in addition to the FAA equipment requirements would be an unnecessary burden on seaplane owners and operators.

c. Unless they are under Federal jurisdiction, navigable bodies of water are under the jurisdiction of the state, or in a few cases, privately owned. Unless they are specifically restricted, aircraft have as much right to operate on these bodies of water as other vessels. To avoid problems, check with Federal or local officials in advance of operating on unfamiliar waters. In addition to the agencies listed in Table 7-85[1], the nearest Flight Standards District Office can usually offer some practical suggestions as well as regulatory information. If you land on a restricted body of water because of an inflight emergency, or in ignorance of the restrictions you have violated, report as quickly as practical to the nearest local official having jurisdiction and explain your situation.

d. When operating over or into remote areas, appropriate attention should be given to survival gear. Minimum kits are recommended for summer and winter, and are required by law for flight into sparsely settled areas of Canada and Alaska. Alaska State Department of Transportation and Canadian Ministry of Transport officials can provide specific information on survival gear requirements. The kit should be assembled in one container and be easily reachable and preferably floatable.

e. The FAA recommends that each seaplane owner or operator provide flotation gear for occupants any time a seaplane operates on or near water. FAR Section 91.205(b)(11) requires approved flotation gear for aircraft operated for hire over water and beyond power-off gliding distance from shore. FAA-approved gear differs from that required for navigable waterways under USCG rules. FAA-approved life vests are inflatable designs as compared to the USCG's non-inflatable PFD's that may consist of solid, bulky material. Such USCG PFD's are impractical for seaplanes and other aircraft because they may block passage through the relatively narrow exits available to pilots and passengers. Life vests approved under Technical Standard Order (TSO) C13E contain fully inflatable compartments. The wearer inflates the compartments (AFTER exiting the aircraft) primarily by independent CO_2 cartridges, with an oral inflation tube as a backup. The flotation gear also contains a water-activated, self-illuminating signal light. The fact that pilots and passengers can easily don and wear inflatable life vests (when not inflated) provides maximum effectiveness and allows for unrestricted movement. It is imperative that passengers are briefed on the location and proper use of available PFD's prior to leaving the dock.

Table 7-85[1]
AUTHORITY TO CONSULT FOR USE OF A BODY OF WATER

Location	Authority	Contact
Wilderness Area	U.S. Department of Agriculture, Forest Service	local forest ranger
National Forest	USDA Forest Service	local forest ranger
National Park	U.S. Department of the Interior, National Park Service	local park ranger
Indian Reservation	USDI, Bureau of Indian Affairs	local Bureau office
State Park	State government or state forestry or park service	local state aviation office for further information
Canadian National and Provincial Parks	Supervised and restricted on an individual basis from province to province and by different departments of the Canadian government; consult Canadian Flight Information Manual and/or Water Aerodrome Supplement	Park Superintendent in an emergency

f. The FAA recommends that seaplane owners and operators obtain Advisory Circular (AC) 91-69, Seaplane Safety, free from the U.S. Department of Transportation,

Utilization and Storage Section, M-443.2, Washington, DC 20590. The Navigation Rules are available from the Government Printing Office for $8 and can be ordered using Mastercard or Visa at (202) 783-3238.

7-86. FLIGHT OPERATIONS IN VOLCANIC ASH

a. Severe volcanic eruptions which send ash into the upper atmosphere occur somewhere around the world several times each year. Flying into a volcanic ash cloud can be exceedingly dangerous. A B747-200 lost all four engines after such an encounter and a B747-400 had the same nearly catastrophic experience. Piston-powered aircraft are less likely to lose power but severe damage is almost certain to ensue after an encounter with a volcanic ash cloud which is only a few hours old.

b. Most important is to avoid any encounter with volcanic ash. The ash plume may not be visible, especially in instrument conditions or at night; and even if visible, it is difficult to distinguish visually between an ash cloud and an ordinary weather cloud. Volcanic ash clouds are **not** displayed on airborne or ATC radar. The pilot must rely on reports from air traffic controllers and other pilots to determine the location of the ash cloud and use that information to remain well clear of the area. Every attempt should be made to remain on the upwind side of the volcano.

c. It is recommended that pilots encountering an ash cloud should immediately reduce thrust to idle (altitude permitting), and reverse course in order to escape from the cloud. Ash clouds may extend for hundreds of miles and pilots should not attempt to fly through or climb out of the cloud. In addition, the following procedures are recommended:

1. Disengage the autothrottle if engaged. This will prevent the autothrottle from increasing engine thrust;

2. Turn on continuous ignition;

3. Turn on all accessory airbleeds including all air conditioning packs, nacelles, and wing anti-ice. This will provide an additional engine stall margin by reducing engine pressure.

d. The following has been reported by flightcrews who have experienced encounters with volcanic dust clouds:

1. Smoke or dust appearing in the cockpit;

2. An acrid odor similar to electrical smoke;

3. Multiple engine malfunctions, such as compressor stalls, increasing EGT, torching from tailpipe, and flameouts;

4. At night, St. Elmo's fire or other static discharges accompanied by a bright orange glow in the engine inlets;

5. A fire warning in the forward cargo area.

e. It may become necessary to shut down and then restart engines to prevent exceeding EGT limits. Volcanic ash may block the pitot system and result in unreliable airspeed indications.

f. If you see a volcanic eruption and have not been previously notified of it, you may have been the first person to observe it. In this case, immediately contact ATC and alert them to the existence of the eruption. If possible, use the Volcanic Activity Reporting form (VAR) on page 245. Items 1 through 8 of the VAR should be transmitted immediately. The information requested in items 9 through 16 should be passed after landing. If a VAR form is not immediately available, relay enough information to identify the position and nature of the volcanic activity. Do not become unnecessarily alarmed if there is merely steam or very low-level eruptions of ash.

g. When landing at airports where volcanic ash has been deposited on the runway be aware that even a thin layer of dry ash can be detrimental to braking action. Wet ash on the runway may also reduce effectiveness of braking. It is recommended that reverse thrust be limited to minimum practical to reduce the possibility of reduced visibility and engine ingestion of airborne ash.

h. When departing from airports where volcanic ash have been deposited it is recommended that pilots avoid operating in visible airborne ash. Allow ash to settle before initiating takeoff roll. It is also recommended that flap extension be delayed until initiating the before takeoff checklist and that a rolling takeoff be executed to avoid blowing ash back into the air.

7-87. EMERGENCY AIRBORNE INSPECTION OF OTHER AIRCRAFT

a. Providing airborne assistance to another aircraft may involve flying in very close proximity to that aircraft. Most pilots receive little, if any, formal training or instruction in this type of flying activity. Close proximity flying without sufficient time to plan (i.e., in an emergency situation), coupled with the stress involved in a perceived emergency can be hazardous.

b. The pilot in the best position to assess the situation should take the responsibility of coordinating the airborne intercept and inspection, and take into account the unique flight characteristics and differences of the category(s) of aircraft involved.

c. Some of the safety considerations are:

1. Area, direction and speed of the intercept;

2. Aerodynamic effects (i.e., rotorcraft downwash) which may also affect;

3. Minimum safe separation distances;

4. Communications requirements, lost communications procedures, coordination with ATC;

5. Suitability of diverting the distressed aircraft to the nearest safe airport; and

6. Emergency actions to terminate the intercept.

d. Close proximity, in-flight inspection of another aircraft is uniquely hazardous. The pilot in command of the aircraft experiencing the problem/emergency must not relinquish his/her control of the situation and jeopardize the safety of his/her aircraft. The maneuver must be accomplished with minimum risk to both aircraft.

7-88 thru 7-89. RESERVED

Section 6. SAFETY, ACCIDENT, AND HAZARD REPORTS

7-90. AVIATION SAFETY REPORTING PROGRAM

a. The FAA has established a voluntary Aviation Safety Reporting Program designed to stimulate the free and unrestricted flow of information concerning deficiencies and discrepancies in the aviation system. This is a positive program intended to ensure the safest possible system by identifying and correcting unsafe conditions before they lead to accidents. The primary objective of the program is to obtain information to evaluate and enhance the safety and efficiency of the present system.

b. This program applies primarily to that part of the system involving the safety of aircraft operations, including departure, en route, approach and landing operations and procedures, ATC procedures, pilot/controller communications, the aircraft movement area of the airport, and near midair collisions. Pilots, air traffic controllers, and all other members of the aviation community and the general public are asked to file written reports of any discrepancy or deficiency noted in these areas.

c. The report should give the date, time, location, persons and aircraft involved (if applicable), nature of the event, and all pertinent details.

d. To ensure receipt of this information, the program provides for the waiver of certain disciplinary actions against persons, including pilots and air traffic controllers, who file timely written reports concerning potentially unsafe incidents. To be considered timely, reports must be delivered or postmarked within 10 days of the incident unless that period is extended for good cause. Reporting forms are available at FAA facilities.

e. The FAA utilizes the National Aeronautics and Space Administration (NASA) to act as an independent third party to receive and analyze reports submitted under the program. This program is described in AC 00-46.

7-91. AIRCRAFT ACCIDENT AND INCIDENT REPORTING

a. Occurrences Requiring Notification—The operator of an aircraft shall immediately, and by the most expeditious means available, notify the nearest National Transportation Safety Board (NTSB) Field Office when:

1. An aircraft accident or any of the following listed incidents occur:

(a) Flight control system malfunction or failure;

(b) Inability of any required flight crew member to perform his normal flight duties as a result of injury or illness;

(c) Failure of structural components of a turbine engine excluding compressor and turbine blades and vanes.

(d) In-flight fire;

(e) Aircraft collide in flight.

(f) Damage to property, other than the aircraft, estimated to exceed $25,000 for repair (including materials and labor) or fair market value in the event of total loss, whichever is less.

(g) For large multi-engine aircraft (more than 12,500 pounds maximum certificated takeoff weight):

(1) In-flight failure of electrical systems which requires the sustained use of an emergency bus powered by a back-up source such as a battery, auxiliary power unit, or air-driven generator to retain flight control or essential instruments;

(2) In-flight failure of hydraulic systems that results in sustained reliance on the sole remaining hydraulic or mechanical system for movement of flight control surfaces;

(3) Sustained loss of the power or thrust produced by two or more engines; and

(4) An evacuation of aircraft in which an emergency egress system is utilized.

2. An aircraft is overdue and is believed to have been involved in an accident.

b. Manner of Notification—

1. The most expeditious method of notification to the NTSB by the operator will be determined by the circumstances existing at that time. The NTSB has advised that any of the following would be considered examples of the type of notification that would be acceptable:

(a) Direct telephone notification.

(b) Telegraphic notification.

(c) Notification to the FAA who would in turn notify the NTSB by direct communication; i.e., dispatch or telephone.

c. Items To Be Notified—The notification required above shall contain the following information, if available:

1. Type, nationality, and registration marks of the aircraft;

2. Name of owner and operator of the aircraft;

3. Name of the pilot-in-command;

4. Date and time of the accident, or incident;

5. Last point of departure, and point of intended landing of the aircraft;

6. Position of the aircraft with reference to some easily defined geographical point;

7. Number of persons aboard, number killed, and number seriously injured;

8. Nature of the accident, or incident, the weather, and the extent of damage to the aircraft so far as is known; and

9. A description of any explosives, radioactive materials, or other dangerous articles carried.

d. Followup Reports—

1. The operator shall file a report on NTSB Form 6120.1 or 6120.2, available from NTSB Field Offices or from the NTSB, Washington, DC, 20594.

(a) Within 10 days after an accident;

(b) When, after 7 days, an overdue aircraft is still missing;

(c) A report on an incident for which notification is required as described in subparagraph a(l) shall be filed only as requested by an authorized representative of the NTSB.

2. Each crewmember, if physically able at the time the report is submitted, shall attach a statement setting forth the facts, conditions and circumstances relating to the accident or incident as they appear to him. If the crewmember is incapacitated, he shall submit the statement as soon as he is physically able.

e. Where To File the Reports—

1. The operator of an aircraft shall file with the NTSB Field Office nearest the accident or incident any report required by this section.

2. The NTSB Field Offices are listed under U.S. Government in the telephone directories in the following cities: Anchorage, AK; Atlanta, GA; Chicago, IL; Denver, CO; Forth Worth, TX; Kansas City, MO; Los Angeles, CA; Miami, FL; New York, NY; Seattle, WA.

7-92. NEAR MIDAIR COLLISION REPORTING

a. Purpose and Data Uses—The primary purpose of the Near Midair Collision (NMAC) Reporting Program is to provide information for use in enhancing the safety and efficiency of the National Airspace System. Data obtained from NMAC reports are used by the FAA to improve the quality of FAA services to users and to develop programs, policies, and procedures aimed at the reduction of NMAC occurrences. All NMAC reports are thoroughly investigated by Flight Standards Facilities in coordination with Air Traffic Facilities. Data from these investigations are transmitted to FAA Headquarters in Washington, DC, where they are compiled and analyzed, and where safety programs and recommendations are developed.

b. Definition—A near midair collision is defined as an incident associated with the operation of an aircraft in which a possibility of collision occurs as a result of proximity of less than 500 feet to another aircraft, or a report is received from a pilot or a flight crew member stating that a collision hazard existed between two or more aircraft.

c. Reporting Responsibility—It is the responsibility of the pilot and/or flight crew to determine whether a near midair collision did actually occur and, if so, to initiate an NMAC report. Be specific, as ATC will not interpret a casual remark to mean that an NMAC is being reported. The pilot should state "I wish to report a near midair collision."

d. Where To File Reports—Pilots and/or flight crew members involved in NMAC occurrences are urged to report each incident immediately:

1. By radio or telephone to the nearest FAA ATC facility or FSS.

2. In writing, in lieu of the above, to the nearest Air Carrier District Office (ACDO), General Aviation District Office (GADO), or Flight Standards District Office (FSDO).

e. Items To Be Reported—

1. Date and Time (UTC) of incident.

2. Location of incident and altitude.

3. Identification and type of reporting aircraft, aircrew destination, name and home base of pilot.

4. Identification and type of other aircraft, aircrew destination, name and home base of pilot.

5. Type of flight plans; station altimeter setting used.

6. Detailed weather conditions at altitude or Flight Level.

7. Approximate courses of both aircraft: indicate if one or both aircraft were climbing or descending.

8. Reported separation in distance at first sighting, proximity at closest point horizontally and vertically, and length of time in sight prior to evasive action.

9. Degree of evasive action taken, if any (from both aircraft, if possible).

10. Injuries, if any.

f. Investigation—The district office responsible for the investigation and reporting of NMAC's will be:

1. The Air Carrier or Flight Standards District Office in whose area the incident occurred when an air carrier aircraft is involved.

2. The General Aviation or Flight Standards District Office in whose area the incident occurred in all other cases.

g. Existing radar, communication, and weather data will be examined in the conduct of the investigation. When possible, all cockpit crew members will be interviewed regarding factors involving the NMAC incident. Air Traffic controllers will be interviewed in cases where one or more of the involved aircraft was provided ATC service. Both flight and ATC procedures will be evaluated. When the investigation reveals a violation of an FAA regulation, enforcement action will be pursued.

Form Approved
OMB No. 2120-0045

BIRD STRIKE INCIDENT/INGESTION REPORT
Other Wildlife Species May Be Described Here
Operation Cost and Engine Damage Information

1. Name of Operator	2. Aircraft Make/Model	3. Engine Make/Model
4. Aircraft Registration	5. Date of Incident (DD, MM, YY)	6. Local Time of Incident ☐ Dawn ☐ Dusk ☐ Day ☐ Night
7. Aerodrome Name	8. Runway Used	9. Location if En Route (Nearest Town/Reference and State)
10. Height (AGL) *feet*	11. Speed (IAS) *knots*	

12. Phase of Flight
- ☐ A. Parked
- ☐ B. Taxi
- ☐ C. Take-off
- ☐ D. Climb
- ☐ E. En Route
- ☐ F. Descent
- ☐ G. Approach
- ☐ H. Landing Roll

13. Part(s) of Aircraft Struck or Damaged

	Struck	Damaged		Struck	Damaged
A. Radome	☐	☐	H. Propeller	☐	☐
B. Windshield	☐	☐	I. Wing/Rotor	☐	☐
C. Nose	☐	☐	J. Fuselage	☐	☐
D. Engine No. 1	☐	☐	K. Landing Gear	☐	☐
E. Engine No. 2	☐	☐	L. Tail	☐	☐
F. Engine No. 3	☐	☐	M. Lights	☐	☐
G. Engine No. 4	☐	☐	N. Other *(specify)*	☐	☐

14. Effect on Flight
- ☐ None
- ☐ Aborted Take-Off
- ☐ Precautionary Landing
- ☐ Engines Shut Down
- ☐ Other *(specify)*

15. Sky Condition
- ☐ No Cloud
- ☐ Some Cloud
- ☐ Overcast

16. Precipitation
- ☐ Fog
- ☐ Rain
- ☐ Snow

17. Bird Species

18. Number of birds seen and/or struck

Number of Birds	Seen	Struck
1	☐	☐
2-10	☐	☐
11-100	☐	☐
more than 100	☐	☐

19. Size of Bird(s)
- ☐ Small
- ☐ Medium
- ☐ Large

20. Pilot Warned of Birds
☐ Yes ☐ No

21. Remarks *(describe damage, injuries and other pertinent information).*

ENGINE DAMAGE COST INFORMATION

22. Aircraft time out of service: hours	23. Estimated cost of repairs or replacement ($ U.S. in thousands): $	24. Estimated other cost ($ U.S. thousands) (e.g. loss of revenue, fuel, hotels): $

Reported by *(Optional)*	Title	Date

FAA Form 5200-7 (2-90) Supersedes Previous Edition

Continued on Reverse

BIRD STRIKE INCIDENT/INGESTION REPORT (Continued)					
SPECIAL INFORMATION ON ENGINE DAMAGE STRIKES					
Reason for failure/shutdown	Engine 1	Engine 2	Engine 3	Engine 4	Comments
Unconditional Failure	☐	☐	☐	☐	
Fire	☐	☐	☐	☐	
Shutdown — Vibration	☐	☐	☐	☐	
Shutdown — Temperature	☐	☐	☐	☐	
Shutdown — Fire warning	☐	☐	☐	☐	
Shutdown — Other *(specify)*	☐	☐	☐	☐	
Shutdown — Unknown	☐	☐	☐	☐	
	☐	☐	☐	☐	
Estimated percentage of thrust loss*					
Estimated number of birds ingested					

*These may be difficult to determine but even estimates are useful.

Agency Display Of Estimated Burden For Bird Strike Incident/Ingestion Report

The public report burden for this collection of information is estimated to average 5 minutes per response.

If you wish to comment on the accuracy of the estimate or make suggestions for reducing this burden, please direct your comments to OMB and the FAA at the following addresses:

Office of Management and Budget	**— and —**	**U.S. Department of Transportation**
Paperwork Reduction Project 2120-0045		Federal Aviation Administration
Washington, D.C. 20503		Program Support Branch, ARP-11
		800 Independence Avenue, S.W.
		Washington, D.C. 20591

FAA Form 5200-7 (2-90)

US Department
of Transportation
FEDERAL AVIATION ADMINISTRATION
800 Independence Ave. S.W.
Washington, D. C. 20591

NO POSTAGE NECESSARY IF MAILED IN THE UNITED STATES

BUSINESS REPLY MAIL

FIRST CLASS PERMIT NO 12438 WASHINGTON, D. C.

POSTAGE WILL BE PAID BY THE FEDERAL AVIATION ADMINISTRATION

FEDERAL AVIATION ADMINISTRATION
OFFICE OF AIRPORT SAFETY AND STANDARDS, AAS-310
800 INDEPENDENCE AVENUE, S.W.
WASHINGTON, D.C. 20591

VOLCANIC ACTIVITY REPORTING FORM (VAR)

Date ________________

SECTION 1 - Transmit to ATC via radio

{PRIVATE }**1.** Aircraft Identification	
2. Position	
3. Time (UTC)	
4. Flight level or altitude	
5. Position/location of volcanic activity or ash cloud	
6. Air temperature	
7. Wind	
8. Supplementary Information (Brief description of activity including vertical and lateral extent of the ash cloud, horizontal movement, rate of growth, etc., as available.)	

SECTION 2 - Complete and forward as directed

Mark the appropriate box(s)			
{PRIVATE }**9.** *Density of ash cloud*	☐ wispy	☐ moderately dense	☐ very dense
10. *Color of ash*	☐ white ☐ black	☐ light gray	☐ dark gray
11. *Eruption*	☐ continuous	☐ intermittent	☐ not visible
12. *Position of activity*	☐ summit ☐ multiple	☐ side ☐ not observed	☐ single
13. *Other observed features of eruption*	☐ lightning ☐ ash fallout	☐ glow ☐ mushroom cloud	☐ large rocks ☐ none
14. *Effect on aircraft*	☐ communications ☐ pitot static ☐ none	☐ navigation system ☐ windscreen	☐ engines ☐ other windows
15. *Other effects*	☐ turbulence ☐ ash deposits	☐ St. Elmo's fire	☐ fumes
16. *Other information deemed useful*			

Forward completed form via mail to:
Global Volcanism Program
NHB-119
Smithsonian Institution
Washington, DC 20560

Or Fax to:
Global Volcanism Program
(202) 357-2476

A2

Chapter 8. Medical facts for pilots
Section 1. FITNESS FOR FLIGHT

8-1. FITNESS FOR FLIGHT

a. Medical Certification—

1. All pilots except those flying gliders and free air balloons must possess valid medical certificates in order to exercise the privileges of their airman certificates. The periodic medical examinations required for medical certification are conducted by designated Aviation Medical Examiners, who are physicians with a special interest in aviation safety and training in aviation medicine.

2. The standards for medical certification are contained in FAR Part 67. Pilots who have a history of certain medical conditions described in these standards are mandatorily disqualified from flying. These medical conditions include a personality disorder manifested by overt acts, a psychosis, alcoholism, drug dependence, epilepsy, an unexplained disturbance of consciousness, myocardial infarction, angina pectoris and diabetes requiring medication for its control. Other medical conditions may be temporarily disqualifying, such as acute infections, anemia, and peptic ulcer. Pilots who do not meet medical standards may still be qualified under special issuance provisions or the exemption process. This may require that either additional medical information be provided or practical flight tests be conducted.

3. Student pilots should visit an Aviation Medical Examiner as soon as possible in their flight training in order to avoid unnecessary training expenses should they not meet the medical standards. For the same reason, the student pilot who plans to enter commercial aviation should apply for the highest class of medical certificate that might be necessary in the pilot's career.

Caution: The FAR's prohibit a pilot who possesses a current medical certificate from performing crewmember duties while the pilot has a known medical condition or increase of a known medical condition that would make the pilot unable to meet the standards for the medical certificate.

b. Illness—

1. Even a minor illness suffered in day-to-day living can seriously degrade performance of many piloting tasks vital to safe flight. Illness can produce fever and distracting symptoms that can impair judgment, memory, alertness, and the ability to make calculations. Although symptoms from an illness may be under adequate control with a medication, the medication itself may decrease pilot performance.

2. The safest rule is not to fly while suffering from any illness. If this rule is considered too stringent for a particular illness, the pilot should contact an Aviation Medical Examiner for advice.

c. Medication—

1. Pilot performance can be seriously degraded by both prescribed and over-the-counter medications, as well as by the medical conditions for which they are taken. Many medications, such as tranquilizers, sedatives, strong pain relievers, and cough- suppressant preparations, have primary effects that may impair judgment, memory, alertness, coordination, vision, and the ability to make calculations. Others, such as antihistamines, blood pressure drugs, muscle relaxants, and agents to control diarrhea and motion sickness, have side effects that may impair the same critical functions. Any medication that depresses the nervous system, such as a sedative, tranquilizer or antihistamine, can make a pilot much more susceptible to hypoxia.

2. The FAR's prohibit pilots from performing crewmember duties while using any medication that affects the faculties in any way contrary to safety. The safest rule is not to fly as a crewmember while taking any medication, unless approved to do so by the FAA.

d. Alcohol—

1. Extensive research has provided a number of facts about the hazards of alcohol consumption and flying. As little as one ounce of liquor, one bottle of beer or four ounces of wine can impair flying skills, with the alcohol consumed in these drinks being detectable in the breath and blood for at least 3 hours. Even after the body completely destroys a moderate amount of alcohol, a pilot can still be severely impaired for many hours by hangover. There is simply no way of increasing the destruction of alcohol or alleviating a hangover. Alcohol also renders a pilot much more susceptible to disorientation and hypoxia.

2. A consistently high alcohol related fatal aircraft accident rate serves to emphasize that alcohol and flying are a potentially lethal combination. The FAR's prohibit pilots from performing crewmember duties within 8 hours after drinking any alcoholic beverage or while under the influence of alcohol. However, due to the slow destruction of alcohol, a pilot may still be under influence 8 hours after drinking a moderate amount of alcohol. Therefore, an excellent rule is to allow at least 12 to 24 hours between "bottle and throttle," depending on the amount of alcoholic beverage consumed.

e. Fatigue—

1. Fatigue continues to be one of the most treacherous hazards to flight safety, as it may not be apparent to a pilot until serious errors are made. Fatigue is best described as either acute (short-term) or chronic (long-term).

2. A normal occurrence of everyday living, acute fatigue is the tiredness felt after long periods of physical and mental strain, including strenuous muscular effort, immobility, heavy mental workload, strong emotional pressure, monotony, and lack of sleep. Consequently, coordination and alertness, so vital to safe pilot performance, can be reduced. Acute fatigue is prevented by adequate rest and sleep, as well as by regular exercise and proper nutrition.

3. Chronic fatigue occurs when there is not enough time for full recovery between episodes of acute fatigue. Performance continues to fall off, and judgment becomes impaired so that unwarranted risks may be taken. Recovery from chronic fatigue requires a prolonged period of rest.

f. Stress—

1. Stress from the pressures of everyday living can impair pilot performance, often in very subtle ways. Difficulties, particularly at work, can occupy thought processes enough to markedly decrease alertness. Distraction can so interfere with judgment that unwarranted risks are taken, such as flying into deteriorating weather conditions to keep on schedule. Stress and fatigue (see above) can be an extremely hazardous combination.

2. Most pilots do not leave stress "on the ground." Therefore, when more than usual difficulties are being experienced, a pilot should consider delaying flight until these difficulties are satisfactorily resolved.

g. Emotion—

Certain emotionally upsetting events, including a serious argument, death of a family member, separation or divorce, loss of job, and financial catastrophe, can render a pilot unable to fly an aircraft safely. The emotions of anger, depression, and anxiety from such events not only decrease alertness but also may lead to taking risks that border on self-destruction. Any pilot who experiences an emotionally upsetting event should not fly until satisfactorily recovered from it.

h. Personal Checklist—Aircraft accident statistics show that pilots should be conducting preflight checklists on themselves as well as their aircraft for pilot impairment contributes to many more accidents than failures of aircraft systems. A personal checklist, which includes all of the categories of pilot impairment as discussed in this section, that can be easily committed to memory is being distributed by the FAA in the form of a wallet-sized card.

PERSONAL CHECKLIST

I'm physically and mentally safe to fly—not being impaired by:

Illness,
Medication,
Stress,
Alcohol,
Fatigue,
Emotion.

8-2. EFFECTS OF ALTITUDE

a. Hypoxia—

1. Hypoxia is a state of oxygen deficiency in the body sufficient to impair functions of the brain and other organs. Hypoxia from exposure to altitude is due only to the reduced barometric pressures encountered at altitude, for the concentration of oxygen in the atmosphere remains about 21 percent from the ground out to space.

2. Although a deterioration in night vision occurs at a cabin pressure altitude as low as 5,000 feet, other significant effects of altitude hypoxia usually do not occur in the normal healthy pilot below 12,000 feet. From 12,000 to 15,000 feet of altitude, judgment, memory, alertness, coordination and ability to make calculations are impaired, and headache, drowsiness, dizziness and either a sense of well-being (euphoria) or belligerence occur. The effects appear following increasingly shorter periods of exposure to increasing altitude. In fact, pilot performance can seriously deteriorate within 15 minutes at 15,000 feet.

3. At cabin pressure altitudes above 15,000 feet, the periphery of the visual field grays out to a point where only central vision remains (tunnel vision). A blue coloration (cyanosis) of the fingernails and lips develops. The ability to take corrective and protective action is lost in 20 to 30 minutes at 18,000 feet and 5 to 12 minutes at 20,000 feet, followed soon thereafter by unconsciousness.

4. The altitude at which significant effects of hypoxia occur can be lowered by a number of factors. Carbon monoxide inhaled in smoking or from exhaust fumes, lowered hemoglobin (anemia), and certain medications can reduce the oxygen-carrying capacity of the blood to the degree that the amount of oxygen provided to body tissues will already be equivalent to the oxygen provided to the tissues when exposed to a cabin pressure altitude of several thousand feet. Small amounts of alcohol and low doses of certain drugs, such as antihistamines, tranquilizers, sedatives and analgesics can, through their depressant action, render the brain much more susceptible to hypoxia. Extreme heat and cold, fever, and anxiety increase the body's demand for oxygen, and hence its susceptibility to hypoxia.

5. The effects of hypoxia are usually quite difficult to recognize, especially when they occur gradually. Since symptoms of hypoxia do not vary in an individual, the ability to recognize hypoxia can be greatly improved by experiencing and witnessing the effects of hypoxia during an altitude chamber "flight." The FAA provides this opportunity through aviation physiology training, which is conducted at the FAA Civil Aeromedical Institute and at many military facilities across the United States, to attend the Physiological Training Program at the Civil Aeromedical Institute, Mike Monroney Aeronautical Center, Oklahoma City, OK, contact by telephone (405) 680-4837, or by writing Airmen Education Branch, AAM-420, CAMI,

Mike Monroney Aeronautical Center, P.O. Box 25082, Oklahoma City, OK 73125.

8-2a5 NOTE—To attend the Physiological Training Program at one of the military installations having the training capability, an application form and a fee must be submitted. Full particulars about location, fees, scheduling procedures, course content, individual requirements, etc. are contained in the Physiological Training Application, form number AC-3150-7, which is obtained by contacting the Accident Prevention Specialist or the Office Forms Manager in the nearest FAA office.

6. Hypoxia is prevented by heeding factors that reduce tolerance to altitude, by enriching the inspired air with oxygen from an appropriate oxygen system, and by maintaining a comfortable, safe cabin pressure altitude. For optimum protection, pilots are encouraged to use supplemental oxygen above 10,000 feet during the day, and above 5,000 feet at night. The FAR's require that at the minimum, flight crew be provided with and use supplemental oxygen after 30 minutes of exposure to cabin pressure altitudes between 12,500 and 14,000 feet and immediately on exposure to cabin pressure altitudes above 14,000 feet. Every occupant of the aircraft must be provided with supplemental oxygen at cabin pressure altitudes above 15,000 feet.

b. Ear Block—

1. As the aircraft cabin pressure decreases during ascent, the expanding air in the middle ear pushes the eustachian tube open, and by escaping down it to the nasal passages, equalizes in pressure with the cabin pressure. But during descent, the pilot must periodically open the eustachian tube to equalize pressure. This can be accomplished by swallowing, yawning, tensing muscles in the throat, or if these do not work, by a combination of closing the mouth, pinching the nose closed, and attempting to blow through the nostrils (Valsalva maneuver).

2. Either an upper respiratory infection, such as a cold or sore throat, or a nasal allergic condition can produce enough congestion around the eustachian tube to make equalization difficult. Consequently, the difference in pressure between the middle ear and aircraft cabin can build up to a level that will hold the eustachian tube closed, making equalization difficult if not impossible. The problem is commonly referred to as an "ear block."

3. An ear block produces severe ear pain and loss of hearing that can last from several hours to several days. Rupture of the ear drum can occur in flight or after landing. Fluid can accumulate in the middle ear and become infected.

4. An ear block is prevented by not flying with an upper respiratory infection or nasal allergic condition. Adequate protection is usually not provided by decongestant sprays or drops to reduce congestion around the eustachian tubes. Oral decongestants have side effects that can significantly impair pilot performance.

5. f an ear block does not clear shortly after landing, a physician should be consulted.

c. Sinus Block—

1. During ascent and descent, air pressure in the sinuses equalizes with the aircraft cabin pressure through small openings that connect the sinuses to the nasal passages. Either an upper respiratory infection, such as a cold or sinusitis, or a nasal allergic condition can produce enough congestion around an opening to slow equalization, and as the difference in pressure between the sinus and cabin mounts, eventually plug the opening. This "sinus block" occurs most frequently during descent.

2. A sinus block can occur in the frontal sinuses, located above each eyebrow, or in the maxillary sinuses, located in each upper cheek. It will usually produce excruciating pain over the sinus area. A maxillary sinus block can also make the upper teeth ache. Bloody mucus may discharge from the nasal passages.

3. A sinus block is prevented by not flying with an upper respiratory infection or nasal allergic condition. Adequate protection is usually not provided by decongestant sprays or drops to reduce congestion around the sinus openings. Oral decongestants have side effects that can impair pilot performance.

4. If a sinus block does not clear shortly after landing, a physician should be consulted.

d. Decompression Sickness After Scuba Diving—

1. A pilot or passenger who intends to fly after scuba diving should allow the body sufficient time to rid itself of excess nitrogen absorbed during diving. If not, decompression sickness due to evolved gas can occur during exposure to low altitude and create a serious in-flight emergency.

2. The recommended waiting time before going to flight altitudes of up to 8,000 feet is at least 12 hours after diving which has not required controlled ascent (nondecompression stop diving), and at least 24 hours after diving which has required controlled ascent (decompression stop diving). The waiting time before going to flight altitudes above 8,000 feet should be at least 24 hours after any SCUBA dive. These recommended altitudes are actual flight altitudes above mean sea level (AMSL) and not pressurized cabin altitudes. This takes into consideration the risk of decompression of the aircraft during flight.

8-3. HYPERVENTILATION IN FLIGHT

a. Hyperventilation, or an abnormal increase in the volume of air breathed in and out of the lungs, can occur subconsciously when a stressful situation is encountered in flight. As hyperventilation "blows off" excessive carbon dioxide from the body, a pilot can experience symptoms of lightheadedness, suffocation, drowsiness, tingling in the extremities, and coolness and react to them with even greater hyperventilation. Incapacitation can eventually result from incoordination, disorientation, and painful muscle spasms. Finally, unconsciousness can occur.

b. The symptoms of hyperventilation subside within a few minutes after the rate and depth of breathing are consciously brought back under control. The buildup of carbon dioxide in the body can be hastened by controlled breathing in and out of a paper bag held over the nose and mouth.

c. Early symptoms of hyperventilation and hypoxia are similar. Moreover, hyperventilation and hypoxia can occur at the same time. Therefore, if a pilot is using an oxygen system when symptoms are experienced, the oxygen regulator should immediately be set to deliver 100 percent oxygen, and then the system checked to assure that it has been functioning effectively before giving attention to rate and depth of breathing.

8-4. CARBON MONOXIDE POISONING IN FLIGHT

a. Carbon monoxide is a colorless, odorless, and tasteless gas contained in exhaust fumes. When breathed even in minute quantities over a period of time, it can significantly reduce the ability of the blood to carry oxygen. Consequently, effects of hypoxia occur.

b. Most heaters in light aircraft work by air flowing over the manifold. Use of these heaters while exhaust fumes are escaping through manifold cracks and seals is responsible every year for several nonfatal and fatal aircraft accidents from carbon monoxide poisoning.

c. A pilot who detects the odor of exhaust or experiences symptoms of headache, drowsiness, or dizziness while using the heater should suspect carbon monoxide poisoning, and immediately shut off the heater and open air vents. If symptoms are severe or continue after landing, medical treatment should be sought.

8-5. ILLUSIONS IN FLIGHT

a. Introduction—Many different illusions can be experienced in flight. Some can lead to spatial disorientation. Others can lead to landing errors. Illusions rank among the most common factors cited as contributing to fatal aircraft accidents.

b. Illusions Leading to Spatial Disorientation—

1. Various complex motions and forces and certain visual scenes encountered in flight can create illusions of motion and position. Spatial disorientation from these illusions can be prevented only by visual reference to reliable, fixed points on the ground or to flight instruments.

2. The leans—An abrupt correction of a banked attitude, which has been entered too slowly to stimulate the motion sensing system in the inner ear, can create the illusion of banking in the opposite direction. The disoriented pilot will roll the aircraft back into its original dangerous attitude, or if level flight is maintained, will feel compelled to lean in the perceived vertical plane until this illusion subsides.

(a) Coriolis illusion—An abrupt head movement in a prolonged constant-rate turn that has ceased stimulating the motion sensing system can create the illusion of rotation or movement in an entirely different axis. The disoriented pilot will maneuver the aircraft into a dangerous attitude in an attempt to stop rotation. This most overwhelming of all illusions in flight may be prevented by not making sudden, extreme head movements, particularly while making prolonged constant-rate turns under IFR conditions.

(b) Graveyard spin—A proper recovery from a spin that has ceased stimulating the motion sensing system can create the illusion of spinning in the opposite direction. The disoriented pilot will return the aircraft to its original spin.

(c) Graveyard spiral—An observed loss of altitude during a coordinated constant-rate turn that has ceased stimulating the motion sensing system can create the illusion of being in a descent with the wings level. The disoriented pilot will pull back on the controls, tightening the spiral and increasing the loss of altitude.

(d) Somatogravic illusion—A rapid acceleration during takeoff can create the illusion of being in a nose up attitude. The disoriented pilot will push the aircraft into a nose low, or dive attitude. A rapid deceleration by a quick reduction of the throttles can have the opposite effect, with the disoriented pilot pulling the aircraft into a nose up, or stall attitude.

(e) Inversion illusion—An abrupt change from climb to straight and level flight can create the illusion of tumbling backwards. The disoriented pilot will push the aircraft abruptly into a nose low attitude, possibly intensifying this illusion.

(f) Elevator illusion—An abrupt upward vertical acceleration, usually by an updraft, can create the illusion of being in a climb. The disoriented pilot will push the aircraft into a nose low attitude. An abrupt downward vertical acceleration, usually by a downdraft, has the opposite effect, with the disoriented pilot pulling the aircraft into a nose up attitude.

(g) False horizon-Sloping cloud formations, an obscured horizon, a dark scene spread with ground lights and stars, and certain geometric patterns of ground light can create illusions of not being aligned correctly with the actual horizon. The disoriented pilot will place the aircraft in a dangerous attitude.

(h) Autokinesis—In the dark, a static light will appear to move about when stared at for many seconds. The disoriented pilot will lose control of the aircraft in attempting to align it with the light.

3. Illusions Leading to Landing Errors—

(a) Various surface features and atmospheric conditions encountered in landing can create illusions of incorrect height above and distance from the runway threshold. Landing errors from these illusions can be prevented by anticipating them during approaches, aerial visual inspection of unfamiliar airports before landing, using electronic glide slope or VASI systems when available, and maintaining optimum proficiency in landing procedures.

(b) Runway width illusion—A narrower-than-usual runway can create the illusion that the aircraft is at a higher altitude than it actually is. The pilot who does not recognize this illusion will fly a lower approach, with the risk of striking objects along the approach path or landing short. A wider-than-usual runway can have the opposite effect, with the risk of leveling out high and landing hard or overshooting the runway.

(c) Runway and terrain slopes illusion—An upsloping runway, upsloping terrain, or both, can create the illusion that the aircraft is at a higher altitude than it actually is. The pilot who does not recognize this illusion will fly a lower approach. A downsloping runway, downsloping approach terrain, or both, can have the opposite effect.

(d) Featureless terrain illusion—An absence of ground features, as when landing over water, darkened areas, and terrain made featureless by snow, can create the illusion that the aircraft is at a higher altitude than it actually is. The pilot who does not recognize this illusion will fly a lower approach

(e) Atmospheric illusions—Rain on the windscreen can create the illusion of greater height, and atmospheric haze the illusion of being at a greater distance from the runway. The pilot who does not recognize these illusions will fly a lower approach. Penetration of fog can create the illusion of pitching up. The pilot who does not recognize this illusion will steepen the approach, often quite abruptly.

(f) Ground lighting illusions—Lights along a straight path, such as a road, and even lights on moving trains can be mistaken for runway and approach lights. Bright runway and approach lighting systems, especially where few lights illuminate the surrounding terrain, may create the illusion of less distance to the runway. The pilot who does not recognize this illusion will fly a higher approach. Conversely, the pilot overflying terrain which has few lights to provide height cues may make a lower than normal approach.

8-6. VISION IN FLIGHT

a. Introduction—Of the body senses, vision is the most important for safe flight. Major factors that determine how effectively vision can be used are the level of illumination and the technique of scanning the sky for other aircraft.

b. Vision Under Dim and Bright Illumination—

1. Under conditions of dim illumination, small print and colors on aeronautical charts and aircraft instruments become unreadable unless adequate cockpit lighting is available. Moreover, another aircraft must be much closer to be seen unless its navigation lights are on.

2. In darkness, vision becomes more sensitive to light, a process called dark adaptation. Although exposure to total darkness for at least 30 minutes is required for complete dark adaptation, a pilot can achieve a moderate degree of dark adaptation within 20 minutes under dim red cockpit lighting. Since red light severely distorts colors, especially on aeronautical charts, and can cause serious difficulty in focusing the eyes on objects inside the aircraft, its use is advisable only where optimum outside night vision capability is necessary. Even so, white cockpit lighting must be available when needed for map and instrument reading, especially under IFR conditions. Dark adaptation is impaired by exposure to cabin pressure altitudes above 5,000 feet, carbon monoxide inhaled in smoking and from exhaust fumes, deficiency of Vitamin A in the diet, and by prolonged exposure to bright sunlight. Since any degree of dark adaptation is lost within a few seconds of viewing a bright light, a pilot should close one eye when using a light to preserve some degree of night vision.

3. Excessive illumination, especially from light reflected off the canopy, surfaces inside the aircraft, clouds, water, snow, and desert terrain, can produce glare, with uncomfortable squinting, watering of the eyes, and even temporary blindness. Sunglasses for protection from glare should absorb at least 85 percent of visible light (15 percent transmittance) and all colors equally (neutral transmittance), with negligible image distortion from refractive and prismatic errors.

c. Scanning for Other Aircraft—

1. Scanning the sky for other aircraft is a key factor in collision avoidance. It should be used continuously by the pilot and copilot (or right seat passenger) to cover all areas of the sky visible from the cockpit. Although pilots must meet specific visual acuity requirements, the ability to read an eye chart does not ensure that one will be able to efficiently spot other aircraft. Pilots must develop an effective scanning technique which maximizes one's visual capabilities. The probability of spotting a potential collision threat obviously increases with the time spent looking outside the cockpit. Thus, one must use timesharing techniques to efficiently scan the surrounding airspace while monitoring instruments as well.

2. While the eyes can observe an approximate 200 degree arc of the horizon at one glance, only a very small center area called the fovea, in the rear of the eye, has the ability to send clear, sharply focused messages to the brain. All other visual information that is not processed directly through the fovea will be of less detail. An aircraft at a distance of 7 miles which appears in sharp focus within the foveal center of vision would have to be as close as 7/10 of a mile in order to be recognized if it were outside of foveal vision. Because the eyes can focus only on this narrow viewing area, effective scanning is accomplished with a series of short, regularly spaced eye movements that bring successive areas of the sky into the central visual field. Each movement should not exceed 10 degrees, and each area should be observed for at least 1 second to enable detection. Although horizontal back-and-forth eye movements seem preferred by most pilots, each pilot should develop a scanning pattern that is most comfortable and then adhere to it to assure optimum scanning.

3. Studies show that the time a pilot spends on visual tasks inside the cabin should represent no more that 1/4 to 1/3 of the scan time outside, or no more than 4 to 5 seconds on the instrument panel for every 16 seconds outside. Since the brain is already trained to process sight information that is presented from left to right, one may find it easier to start scanning over the left shoulder and proceed across the windshield to the right.

4. Pilots should realize that their eyes may require several seconds to refocus when switching views between items in the cockpit and distant objects. The eyes will also tire more quickly when forced to adjust to distances immediately after close-up focus, as required for scanning the instrument panel. Eye fatigue can be reduced by looking from the instrument panel to the left wing past the wing tip to the center of the first scan quadrant when beginning the exterior scan. After having scanned from left to right, allow the eyes to return to the cabin along the right wing from its tip inward. Once back inside, one should automatically commence the panel scan.

5. Effective scanning also helps avoid "empty-field myopia." This condition usually occurs when flying above the clouds or in a haze layer that provides nothing specific to focus on outside the aircraft. This causes the eyes to relax and seek a comfortable focal distance which may range from 10 to 30 feet. For the pilot, this means looking without seeing, which is dangerous.

8-7. AEROBATIC FLIGHT

a. Airmen planning to engage in aerobatics should be aware of the physiological stresses associated with accelerative forces during aerobatic maneuvers. Many prospective aerobatic trainees enthusiastically enter aerobatic instruction but find their first experiences with G forces to be unanticipated and very uncomfortable. To minimize or avoid potential adverse effects, the aerobatic instructor and trainee must have a basic understanding of the physiology of G force adaptation.

b. Forces experienced with a rapid push-over maneuver result in the blood and body organs being displaced toward the head. Depending on forces involved and individual tolerance, the airman may experience discomfort, headache, "red out," and even unconsciousness.

c. Forces experienced with a rapid pull-up maneuver result in the blood and body organ displacement toward the lower part of the body away from the head. Since the brain requires continuous blood circulation for an adequate oxygen supply, there is a physiologic limit to the time the pilot can tolerate higher forces before losing consciousness. As the blood circulation to the brain decreases as a result of forces involved, the airman will experience "narrowing" of visual fields, "gray-out," "black-out," and unconsciousness. Even a brief loss of consciousness in a maneuver can lead to improper control movement causing structural failure of the aircraft or collision with another object or terrain.

d. In steep turns, the centrifugal forces tend to push the pilot into the seat, thereby resulting in blood and body organ displacement toward the lower part of the body as in the case of rapid pull-up maneuvers and with the same physiologic effects and symptoms.

e. Physiologically, humans progressively adapt to imposed strains and stress, and with practice, any maneuver will have decreasing effect. Tolerance to G forces is dependent on human physiology and the individual pilot. These factors include the skeletal anatomy, the cardiovascular architecture, the nervous system, the quality of the blood, the general physical state, and experience and recency of exposure. The airman should consult an Aviation Medical Examiner prior to aerobatic training and be aware that poor physical condition can reduce tolerance to accelerative forces.

f. The above information provides the airman a brief summary of the physiologic effects of G forces. It does not address methods of "counteracting" these effects. There are numerous references on the subject of G forces during aerobatics available to the airman. Among these are "G Effects on the Pilot During Aerobatics," FAA-AM-72-28, and "G Incapacitation in Aerobatic Pilots: A Flight Hazard" FAA-AM-82-13. These are available from the National Technical Information Service, Springfield, Virginia 22161. (Reference-FAA AC 91-61, "A Hazard in Aerobatics: Effects of G-Forces on Pilots".)

8-8. JUDGMENT ASPECTS OF COLLISION AVOIDANCE

a. Introduction—The most important aspects of vision and the techniques to scan for other aircraft are described in Vision in Flight, paragraph 8-6. Pilots should also be familiar with the following information to reduce the possibility of mid-air collisions.

b. Determining Relative Altitude—Use the horizon as a reference point. If the other aircraft is above the horizon, it is probably on a higher flight path. If the aircraft appears to be below the horizon, it is probably flying at a lower altitude.

c. Taking Appropriate Action—Pilots should be familiar with rules on right-of-way, so if an aircraft is on an obvious collision course, one can take immediately evasive action, preferably in compliance with applicable Federal Aviation Regulations.

d. Consider Multiple Threats—The decision to climb, descend, or turn is a matter of personal judgment, but one should anticipate that the other pilot may also be making a quick maneuver. Watch the other aircraft during the maneuver and begin your scanning again immediately since there may be other aircraft in the area.

e. Collision Course Targets—Any aircraft that appears to have no relative motion and stays in one scan quadrant is likely to be on a collision course. Also, if a target shows no lateral or vertical motion, but increases in size, take evasive action.

f. Recognize High Hazard Areas—

1. Airways, especially near VOR's, and Class B, Class C, Class D, and Class E surface areas are places where aircraft tend to cluster.

2. Remember, most collisions occur during days when the weather is good. Being in a "radar environment" still requires vigilance to avoid collisions.

g. Cockpit Management—Studying maps, checklists, and manuals before flight, with other proper preflight planning; e.g., noting necessary radio frequencies and organizing cockpit materials, can reduce the amount of time required to look at these items during flight, permitting more scan time.

h. Windshield Conditions—Dirty or bug-smeared windshields can greatly reduce the ability of pilots to see other aircraft. Keep a clean windshield.

i. Visibility Conditions—Smoke, haze, dust, rain, and flying towards the sun can also greatly reduce the ability to detect targets.

j. Visual Obstructions in the Cockpit—

1. Pilots need to move their heads to see around blind spots caused by fixed aircraft structures, such as door posts, wings, etc. It will be necessary at times to maneuver the aircraft; e.g., lift a wing, to facilitate seeing.

2. Pilots must insure curtains and other cockpit objects; e.g., maps on glare shield, are removed and stowed during flight.

k. Lights On—

1. Day or night, use of exterior lights can greatly increase the conspicuity of any aircraft.

2. Keep interior lights low at night.

l. ATC support—ATC facilities often provide radar traffic advisories on a workload-permitting basis. Flight through Class C and Class D airspace requires communication with ATC. Use this support whenever possible or when required.

Chapter 9. Aeronautical charts and related publications

Section 1. TYPES OF CHARTS AVAILABLE

9-1. GENERAL

Aeronautical charts for the U.S., its territories, and possessions are produced by the National Ocean Service (NOS), a part of the Department of Commerce, from information furnished by the FAA.

9-2. OBTAINING CIVIL AERONAUTICAL CHARTS

a. Enroute Aeronautical Charts, Terminal Procedure Publication Charts, Regional Airport/Facility Directories, and other publications described in this Chapter are available upon subscription and one time sales from:

NOAA Distribution Branch (N/CG33)
National Ocean Service
Riverdale, Maryland 20737 - 1199
Telephone: (301) 436-6990

b. Charts may also be purchased directly from authorized NOS chart agents who are located worldwide. A listing of these chart agents may be found in the back of the NOS' Aeronautical Charts and Related Products free catalog. Many fixed base operators are NOS chart agents.

9-3. A FEW OF THE CHARTS AND PRODUCTS THAT ARE AVAILABLE

Sectional and VFR Terminal Area Charts
World Aeronautical Charts (U.S.)
Enroute Low, High, and Alaska
Oceanic Planning Charts
Terminal Procedures Publication (TPP)
Alaska Terminal Publication
Helicopter Route Charts
Airport Facility Directory
Supplement Alaska & Chart Supplement Pacific

9-4. GENERAL DESCRIPTION OF EACH CHART SERIES

a. Sectional and VFR Terminal Area Charts: (See Figure 9-4[1].)

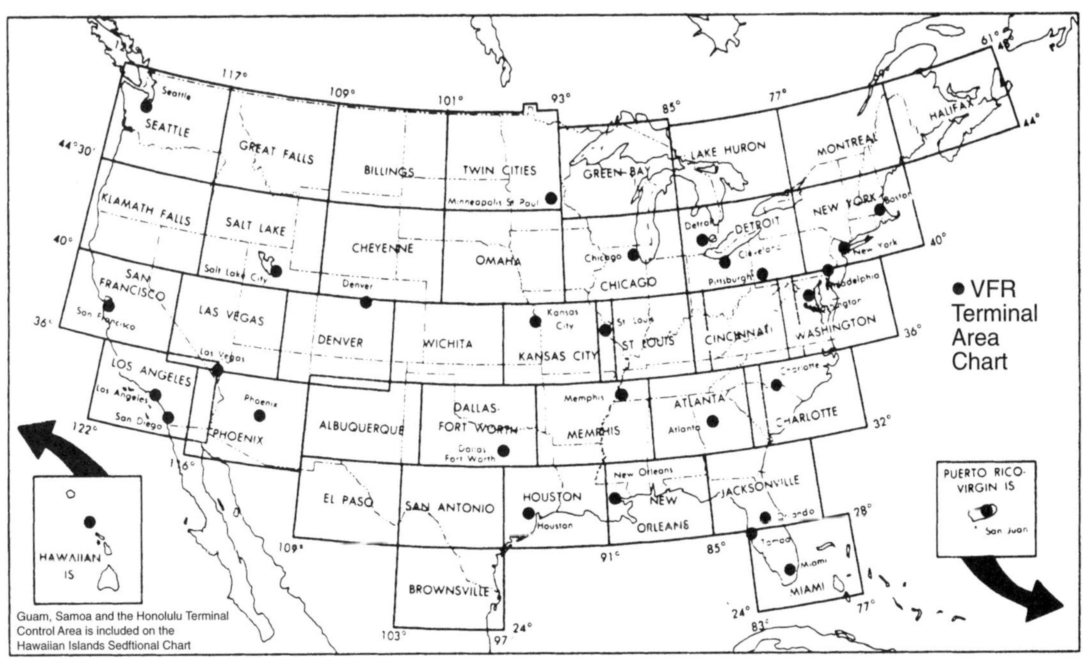

Sectional and VFR Terminal Area Charts for the Conterminous United States, Hawaiian Islands, Puerto Rico and the Virgin Islands

Figure 9-4[1]

1. These charts are designed for visual navigation of slow and medium speed aircraft. They are produced to the following scales:

Sectional Charts—1:500,000 (1 in = 6.86 NM)

VFR Terminal Area Charts—1:250,000 (1 in = 3.43 NM)

2. Topographic information features the portrayal of relief and a judicious selection of visual check points for VFR Right. VFR Terminal Area Charts include populated places, drainage, roads, railroads, and other distinctive landmarks. Aeronautical information includes visual and radio aids to navigation, airports, controlled airspace, restricted areas, obstructions, and related data. These charts also depict the airspace designated as "Class B airspace," which provides for the control or segregation of all aircraft operating within that airspace. The Puerto Rico—Virgin Islands Terminal Area Chart contains basically the same information as that shown on Sectional and Terminal Area Charts. It includes the Gulf of Mexico and Caribbean Planning Chart on the reverse side (See Planning Charts). Charts are revised semiannually, except for several Alaskan Sectionals and the Puerto Rico—Virgin Islands Terminal Area which are revised annually.

b. World Aeronautical Charts—These charts are designed to provide a standard series of aeronautical charts, covering land areas of the world, at a size and scale convenient for navigation by moderate speed aircraft. They are produced at a scale of 1:1,000,000 (1 in = 13.7 NM). Topographic information includes cities and towns, principal roads, railroads, distinctive landmarks, drainage, and relief. The latter is shown by spot elevations, contours, and gradient tints. Aeronautical information includes visual and radio aids to navigation, airports, airways, restricted areas, obstructions, and other pertinent data. These charts are revised annually except several Alaskan charts and the Mexican/Caribbean charts which are revised every 2 years.

c. Enroute Low Altitude Charts—These charts are designed to provide aeronautical information for en route navigation under Instrument Flight Rules (IFR) in the low altitude stratum. The series also includes Enroute Area Charts, which furnish terminal data at a large scale in congested areas and are included with the subscription to the series. Information includes the portrayal of L/MF and VHF airways, limits of controlled airspace, position, identification and frequencies of radio aids, selected airports, minimum en route altitudes and obstruction clearance altitudes, airway distances, reporting points, special use airspace areas, military training routes, and related information. Charts are printed back to back and are revised every 56 days effective with the date of airspace changes. An Enroute Change Notice may be issued as required. (See Figure 9-4[2].)

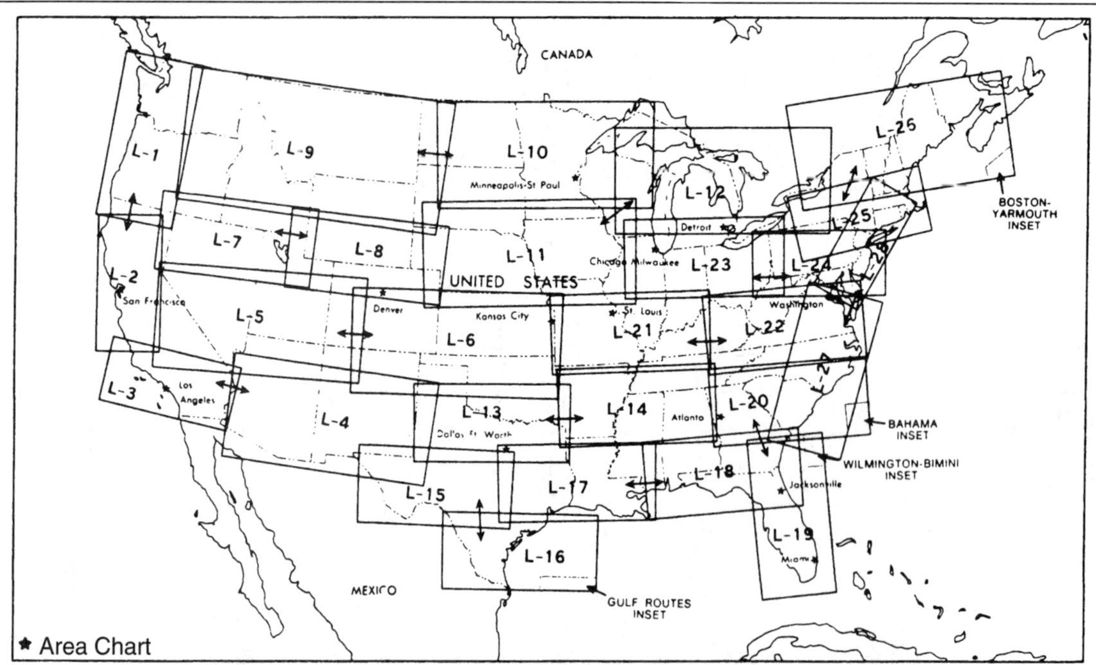

Enroute Low Altitude Instrument Charts for the Conterminous U.S. (Includes Area Charts)

Figure 9-4[2]

d. Enroute High Altitude Charts—These charts are designed to provide aeronautical information for en route navigation under IFR in the high altitude stratum. Information includes the portrayal of jet routes, position, identification and frequencies of radio aids, selected airports, distances, time zones, special use airspace areas, and related information. Charts are revised every 56 days effective with the date of airspace changes. An Enroute Change Notice may be issued as required. (See Figure 9-4[3].)

e. Alaska Enroute Charts (Low and High)—These charts are produced in a low altitude series and a high altitude series with the purpose and makeup identical to Enroute Low and High Altitude Charts described above. Charts are revised every 56 days effective with the date of airspace changes. An Enroute Change Notice may be issued as required. (See Figures 9-4[4] and 9-4[5].)

f. Charted VFR Flyway Planning Chart—These charts are designed to identify flight paths clear of the major controlled traffic flows. The program is intended to provide charts showing multiple VFR routings through high density traffic areas which may be used as an alternative to flight within Class B airspace. Ground references are provided as guides for improved visual navigation. These charts are not intended to discourage VFR operations within the Class B airspace, but are designed for information and planning purposes. They are produced at a scale of 1:250,000 (1 in = 3.43 NM). These charts are revised semi-annually and are published on the back of the existing VFR Terminal Area Charts.

g. Planning Charts

1. Gulf of Mexico and Caribbean Planning Chart—This chart is designated for preflight planning for VFR flights. It is produced at a scale of 1:6,270,551 (1 in = 86 NM). This chart is on the reverse of the Puerto Rico-Virgin Islands Terminal Area Chart. Information includes mileage between Airports of Entry, a selection of special use airspace areas, and a Directory of Airports with their available facilities and servicing. (See Figure 9-4[6].)

2. North Atlantic Route Chart—This five-color chart is designed for use by air traffic controllers in monitoring transatlantic flights and by FAA planners. Oceanic control areas, coastal navigation aids, major coastal airports, and oceanic reporting points are depicted. Geographic coordinates for NAVAID's and reporting points are included. The chart may be used for pre- and in-flight planning. This chart is revised each 24 weeks. The chart available in two sizes, full size (58 by 41 inches) scale: 1:5,500,000; half size (29 by 20½ inches) scale: 1:11,000,000.

3. North Pacific Oceanic Route Chart—This chart series, like the North Atlantic Route Chart series, is designed for FAA air traffic controllers' use in monitoring transoceanic air traffic. Charts are available in two scales: one 1:12,000,000 composite small scale planning chart, which covers the entire North Pacific, and four 1:7,000,000 Area Charts. They are revised every 56 days. The charts are available unfolded (flat only) and contain established intercontinental air routes including all reporting points with geographic positions. (See Figure 9-4[7].)

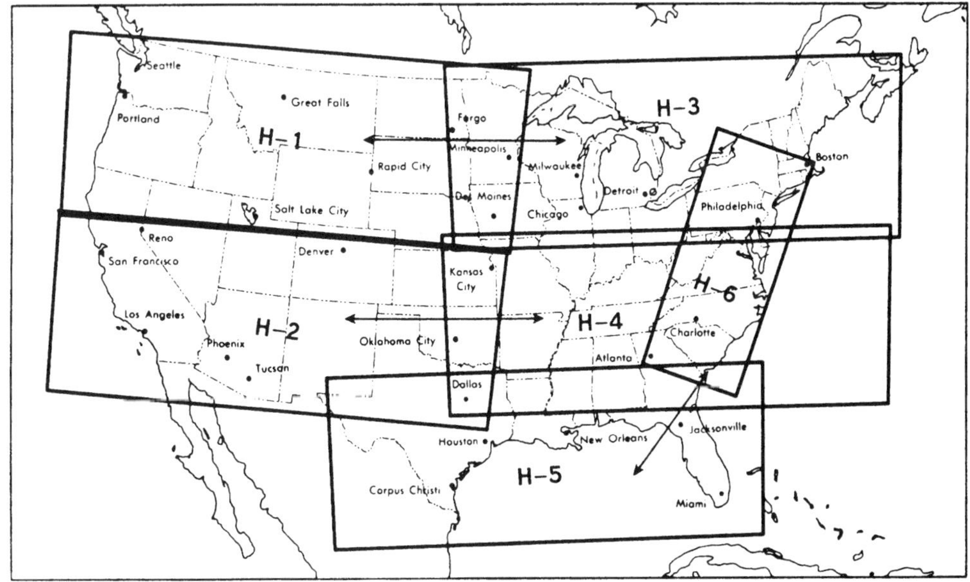

Enroute High Altitude Charts for the Conterminous U.S.

Figure 9-4[3]

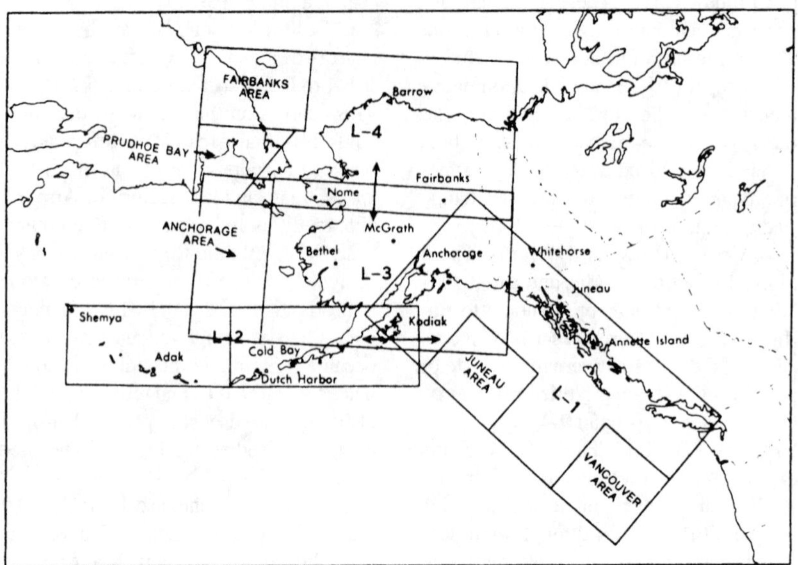

Alaska Enroute Low Altitude Chart

Figure 9-4[4]

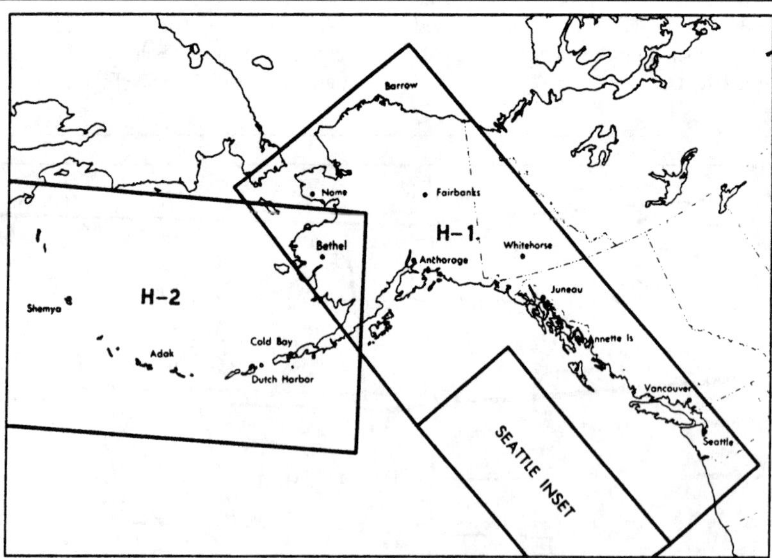

Alaska Enroute High Altitude Chart

Figure 9-4[5]

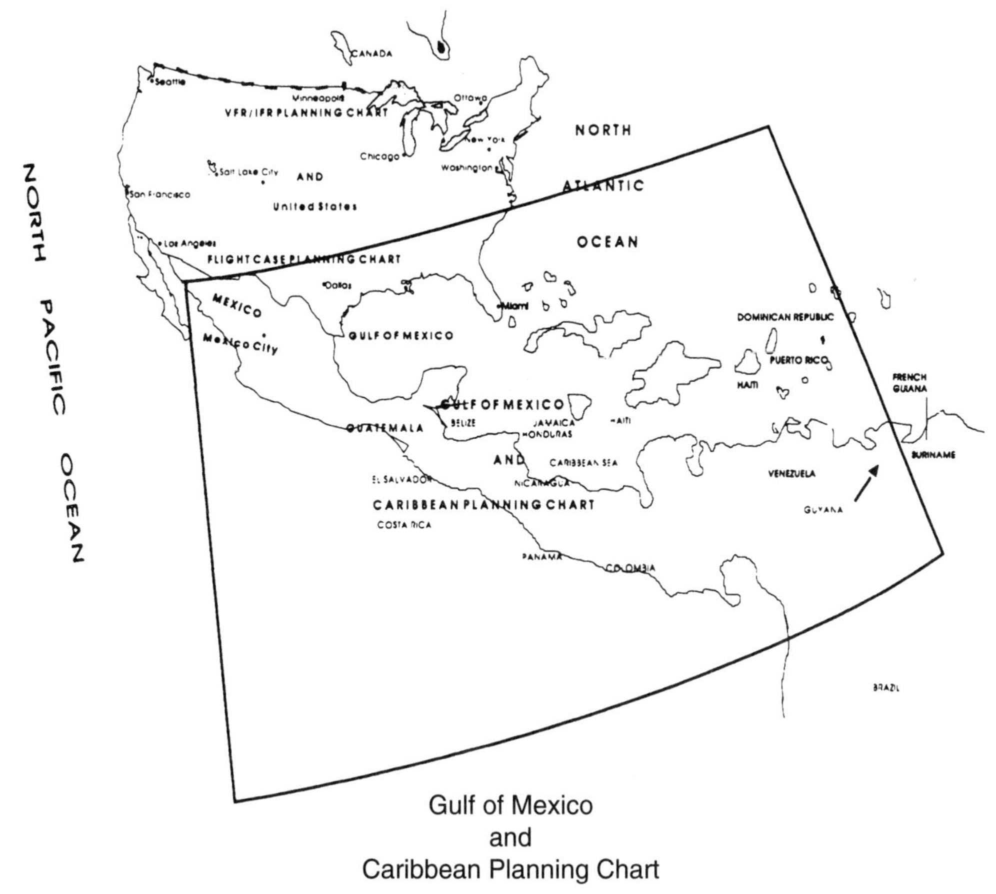

Gulf of Mexico
and
Caribbean Planning Chart

Figure 9-4[6]

h. Terminal Procedures Publication (TPP)—This publication contains charts depicting Instrument Approach Procedures (IAP), Standard Terminal Arrivals (STAR), and Standard Instrument Departures (SID).

1. Instrument Approach Procedure (IAP) Charts—IAP charts portray the aeronautical data which is required to execute instrument approaches to airports. Each chart depicts the IAP, all related navigation data, communications information, and an airport sketch. Each procedure is designated for use with a specific electronic navigational aid, such as ILS, VOR, NDB, RNAV, etc. Airport Diagram Charts, where published, are included. (See Figure 9-4[8] for U.S. Terminal Publication Volumes.)

2. Standard Instrument Departure (SID) Charts—These charts are designed to expedite clearance delivery and to facilitate transition between takeoff and en route operations. They furnish pilots departure routing clearance information in graphic and textual form.

3. Standard Terminal Arrival (STAR) Charts—These charts are designed to expedite ATC arrival procedures and to facilitate transition between en route and instrument approach operations. They present to the pilot preplanned IFR ATC arrival procedures in graphic and textual form. Each STAR procedure is presented as a separate chart and may serve a single airport or more than one airport in a given geographic location.

These charts are published in 16 bound volumes covering the conterminous U.S. and the Puerto Rico—Virgin Islands. Each volume is superseded by a new volume each 56 days. Changes to procedures occurring between the 56-day publication cycle is reflected in a Change Notice volume, issued on the 28-day midcycle. These changes are in the form of a new chart. The publication of a new 56-day volume incorporates all the changes and replaces the preceding volume and the change notice. The volumes are $5\frac{3}{8} \times 8\frac{1}{4}$ inches and are bound on the top edge.

i. Alaska Terminal Publication

1. This publication contains charts depicting all terminal

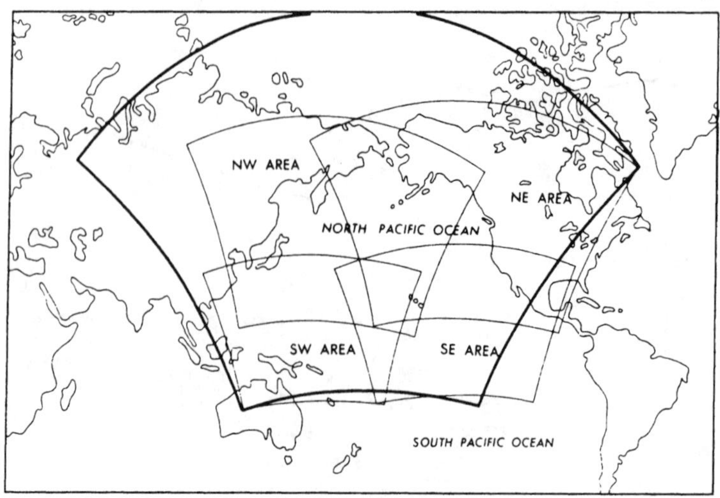

North Pacific Route Charts

Figure 9-4[7]

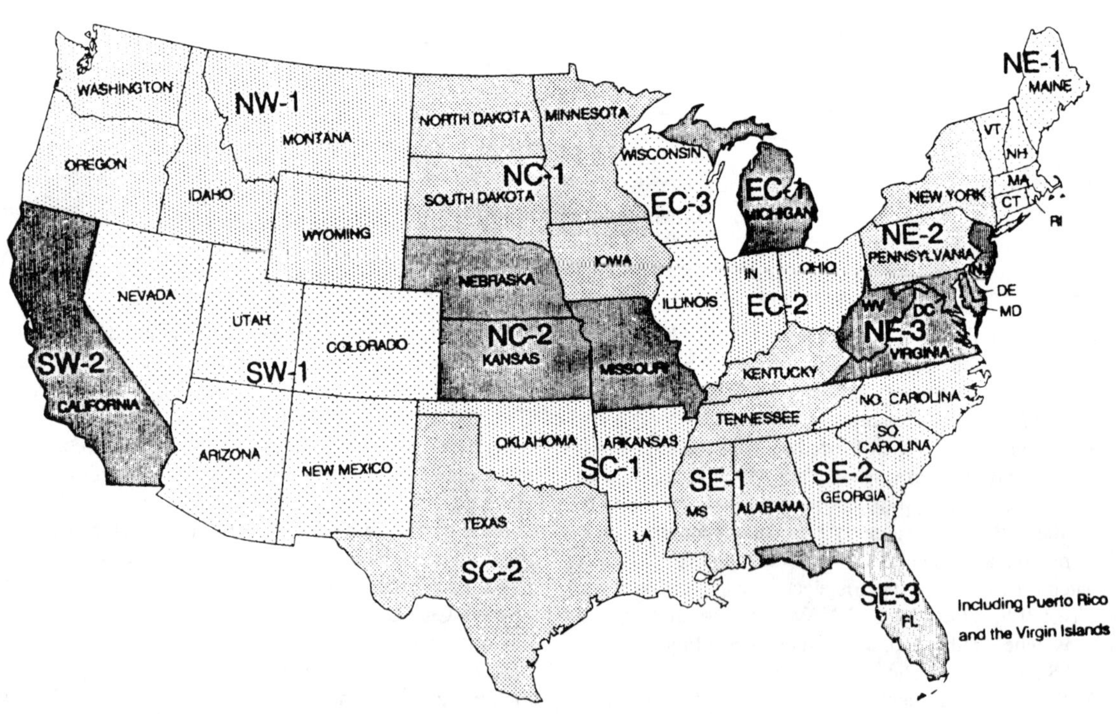

U.S. Terminal Publication Volumes

Figure 9-4[8]

flight procedures in the State of Alaska for civil and military aviation. They are:

(a) Instrument Approach Procedure (IAP) Charts.

(b) Standard Instrument Departure (SID) Charts.

(c) Standard Terminal Arrival (STAR) Charts.

(d) Airport Diagram Charts.

(e) Radar Minimums.

2. All supplementary supporting data; i.e., IFR Takeoff and Departure Procedures, IFR Alternate Minimums, Rate of Descent Table, Inoperative Components Table, etc., is also included.

3. The Alaska Terminal is published in a bound book, 5⅜ inches by 8¼ inches. The publication is issued every 56 days with provisions for an as required "Terminal Change" on the 28-day midpoint.

j. Helicopter Route Charts

1. Prepared under the auspices of the FAA Helicopter Route Chart Program, these charts enhance helicopter operator access into, egress from, and operation within selected high density traffic areas. The scale is 1:125,000; however, some include smaller scale insets. Graphic information includes urban tint, principal roads, pictorial symbols, and spot elevations. Aeronautical information includes routes, operating zones, altitudes or flight ceilings/bases, heliports, helipads, NAVAID's, special use airspace, selected obstacles, ATC and traffic advisory radio communications frequencies, Class B surface area tint, and other important flight aids. These charts are revised when significant aeronautical information changes and/or safety-related events occur. Historically, new editions are published about every 2 years. See the "Dates of Latest Editions" for current editions.

2. Air traffic facility managers are responsible for determining the need for new chart development or existing chart revision. Therefore, requests for new charts or revisions to existing charts should be directed to these managers. Guidance pertinent to mandatory chart features and managerial evaluation of requests is contained in FAA Order 7210.3, Facility Operation and Administration.

9-5. RELATED PUBLICATIONS

a. The Airport/Facility Directory—This directory is issued in seven volumes with each volume covering a specific geographic area of the conterminous U.S., including Puerto Rico and the U.S. Virgin Islands. The directory is 5⅜ × 8¼ inches and is bound on the side. Each volume is reissued in its entirety each 56 days. Each volume is indexed alphabetically by state, airport, navigational aid, and ATC facility for the area of coverage. All pertinent information concerning each entry is included. (See Figure 9-5[1].)

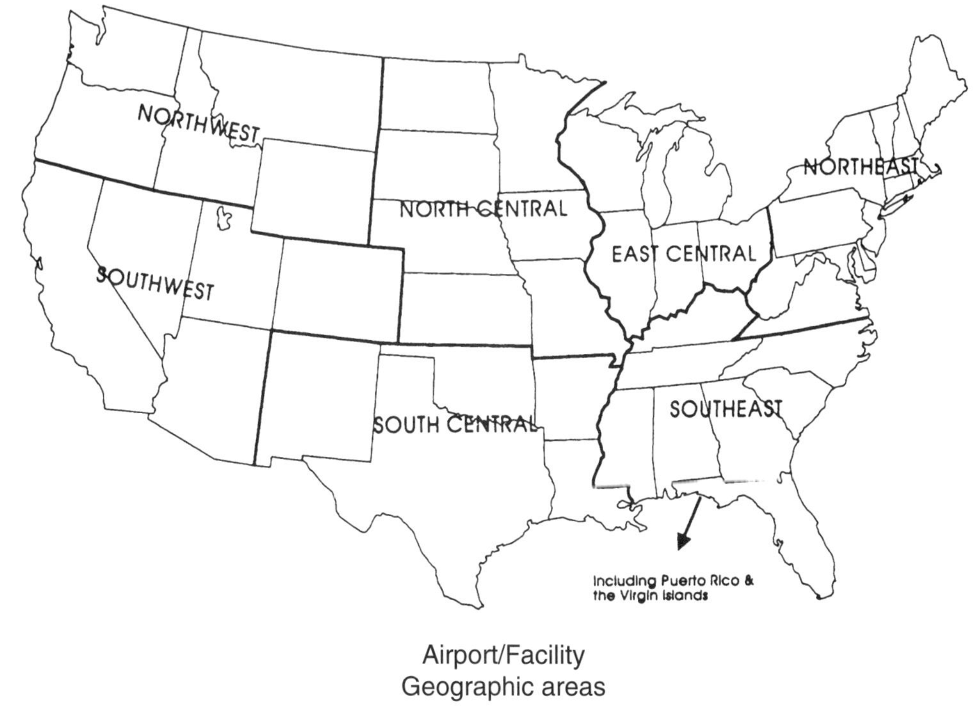

Airport/Facility
Geographic areas

Figure 9-5[1]

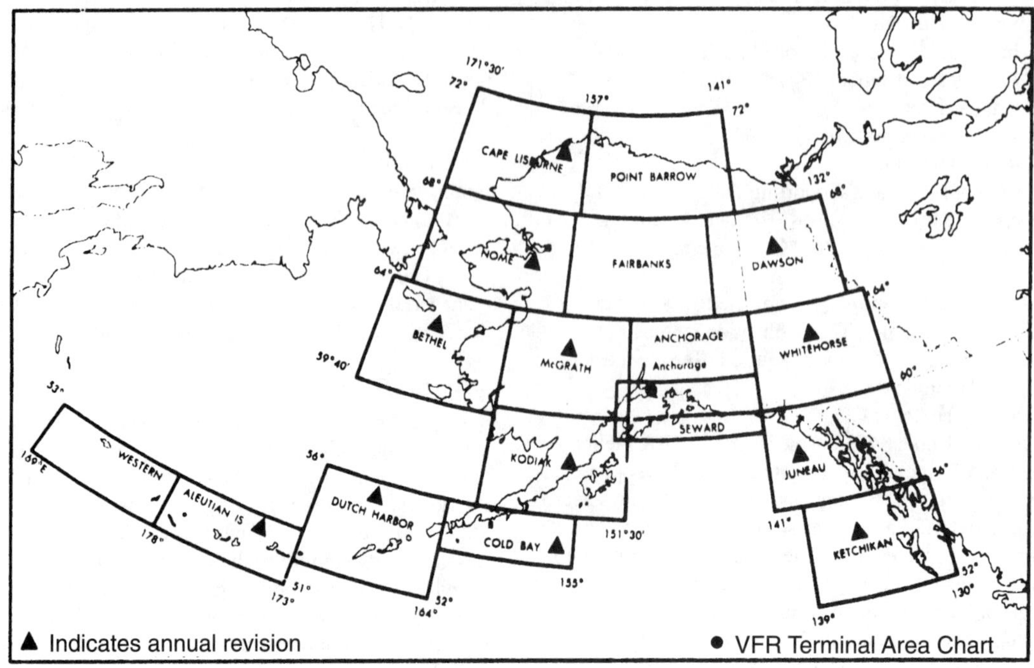

Sectional and VFR Terminal Area Charts for Alaska

Figure 9-5[2]

b. Alaska Supplement—This supplement is a joint Civil and Military Flight Information Publication (FLIP), published and distributed every 56 days by the NOS. It is designed for use with the Flight Information Publication En route Charts, Alaska Terminal, WAC and Sectional Aeronautical Charts. (See Figure 9-5[2] and Figure 9-5[3].) This Supplement contains an Airport/Facility Directory of all airports shown on En route Charts, and those requested by appropriate agencies, communications data, navigational facilities, special notices and procedures applicable to the area of chart coverage.

c. Pacific Supplement—This Chart Supplement is a Civil Flight Information Publication, published and distributed every 56 days by the NOS. It is designed for use with the Flight Information En route Publication Charts and the Sectional Aeronautical Chart covering the State of Hawaii and that area of the Pacific served by U.S. facilities. An Amendment Notice is published 4 weeks after each issue of the Supplement. This chart Supplement contains an Airport/Facility Directory of all airports open to the public, and those requested by appropriate agencies, communications data, navigational facilities, special notices and procedures applicable to the Pacific area.

d. Digital Aeronautical Chart Supplement (DACS). The DACS is a subset of the data NOAA provides to FAA controllers every 56 days. It reflects digitally exactly what is shown on the En Route and Air Traffic Controller Charts. The DACS is designed to assist with flight planning and should not be considered a substitute for a chart. The supplement comes in either a 3.5" or 5.25" diskette compressed format. The supplement is divided into nine individual sections. They are:

Section 1—High Altitude Airways, Conterminous U.S.

Section 2—Low Altitude Airways, Conterminous U.S.

Section 3—Selected Instrument Approach Procedure NAVAID and FIX Data

Section 4—Military Training Routes

Section 5—Alaska, Hawaii, Puerto Rico, Bahama and Selected Oceanic Routes

Section 6—STAR's, Standard Terminal Arrivals and Profile Descent Procedures

Section 7—SID's, Standard Instrument Departures

Section 8—Preferred IFR Routes (Low and High Altitudes)

Section 9—Air Route and Airport Surveillance Radar Facilities.

9-5d NOTE—Section 3 has a Change Notice that will be issued at the mid-28 day point and contains changes that occurred after the 56 day publication.

e. NOAA Aeronautical Chart User's Guide—This guide is designed to be used as a teaching aid, reference document, and an introduction to the wealth of information pro-

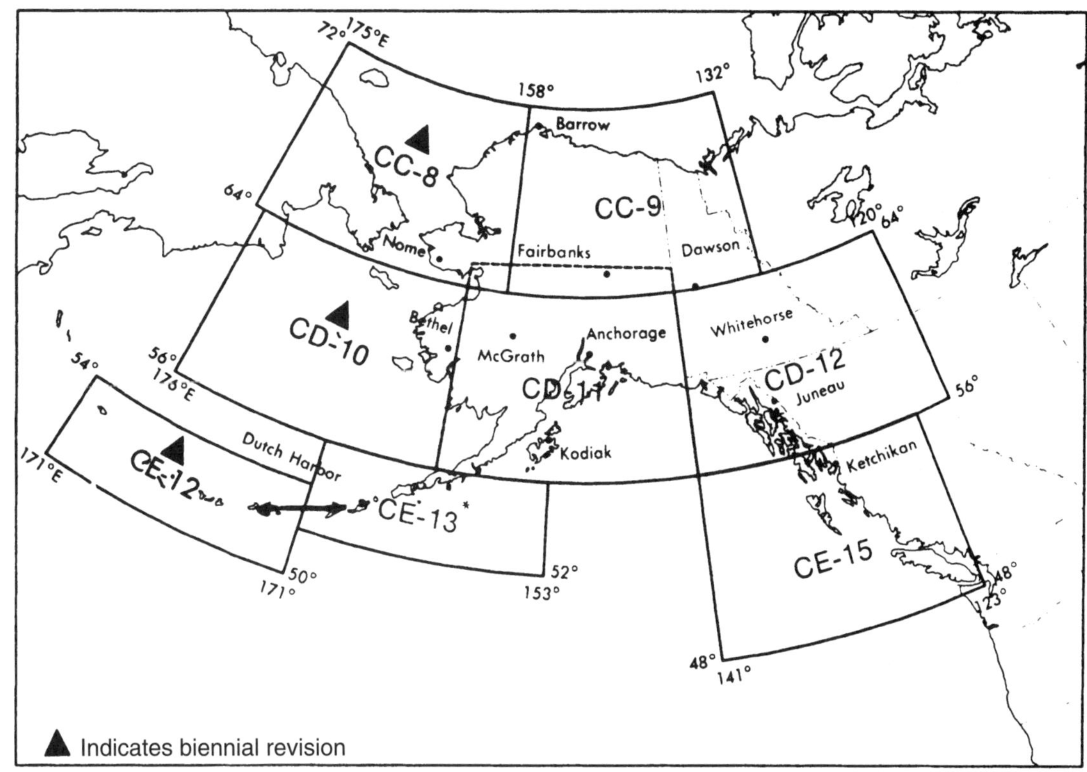

World Aeronautical Charts for Alaska

Figure 9-5[3]

vided on NOAA's aeronautical charts and publications. The guide includes discussion of IFR chart terms and symbols.

f. Airport Obstruction Charts (OC)—The Airport Obstruction Chart is a 1:12,000 scale graphic depicting FAR Part 77 surfaces, a representation of objects that penetrate these surfaces, aircraft movement and apron areas, navigational aids, prominent airport buildings, and a selection of roads and other planimetric detail in the airport vicinity. Also included are tabulations of runway and other operational data.

g. Defense Mapping Agency Aerospace Center (DMAAC) Publications—Defense Mapping Agency Aeronautical Charts and Products are available prepaid from:

National Ocean Service
NOAA Distribution Branch (N/CG33)
Riverdale, Maryland 20737
Telephone: (301) 436-6990

1. Pilotage Charts (PC/TPC)—Scale 1:500,000 used for detail preflight planning and mission analysis. Emphasis in design is on ground features significant in visual and radar, low-level high speed navigation.

2. Jet Navigation Charts (JNC-A)—Scale 1:3,000,000. Designed to provide greater coverage than the 1:2,000,000 scale Jet Navigation Charts described below. Uses include preflight planning and en route navigation by long-range jet aircraft with dead reckoning, radar, celestial and grid navigation capabilities.

3. LORAN Navigation & Consol LORAN Navigation Charts (LJC/CJC)—Scale 1:2,000,000. Used for preflight planning and in-flight navigation on long-range flights in the Polar areas and adjacent regions utilizing LORAN and CONSOL navigation aids.

4. Continental Entry Chart (CEC)—Scale 1:2,000,000. Used for CONSOLAN and LORAN navigation for entry into the U.S. when a high degree of accuracy is required to comply with Air Defense identification and reporting procedures. Also suitable as a basic dead reckoning sheet and for celestial navigation.

5. Aerospace Planning Chart (ASC)—Scale 1:9,000,000 and 1:18,000,000. Six charts at each scale and with various projections, cover the world. Charts are useful for general planning, briefings, and studies.

6. Air Distance/Geography Chart (GH-2, 2a)—Scales 1:25,000,000 and 1:50,000,000. This chart shows great circle distances between major airports. It also shows major cities, international boundaries, shaded relief and gradient tints.

7. LORAN C Navigation Chart (LCC)—Scale of 1:3,000,000. Primarily designed for preflight and in-flight long-range navigation where LORAN-C is used as the basic navigation aid.

8. DOD Weather Plotting Chart (WPC)—Various scales. Designed as non-navigational outline charts which depict locations and identifications of meteorological observing stations. Primarily used to forecast and monitor weather and atmospheric conditions throughout the world.

9. Flight Information Publications (FLIP)—These include En route Low Altitude and High Altitude Charts, En route Supplements, Terminal (Instrument Approach Charts), and other informational publications for various areas of the world.

9-5g9 NOTE—FLIP. Terminal publications do not necessarily include all instrument approach procedures for all airports. They include only those required for military operations.

10. World Aeronautical (WAC) and Operational Navigation Charts (ONC)—The Operational Navigation Charts (ONC) have the same purpose and contain essentially the same information as the WAC series except the terrain is portrayed by shaded relief as well as contours. The ONC series is replacing the WAC series and the WAC's will be available only where the ONC's have not been issued. ONC's are 42 × 57½ inches, WAC's are 22 × 30 inches. These charts are revised on a regular schedule. (See Figure 9-5[3] and Figure 9-5[4].)

11. Jet Navigation Charts—These charts are designed to fulfill the requirements for long-range high altitude, high speed navigation. They are produced at a scale of 1:2,000,000 (1 in = 27.4 NM). Topographic features include large cities, roads, railroads, drainage and relief. The latter is indicated by contours, spot elevations and gradient tints. All aeronautical information necessary to conform to the purpose of the chart is shown. This includes restricted areas, L/MF and VOR ranges, radio beacons and a selection of standard broadcasting stations and airports. The runway patterns of the airports are shown to exaggerated scale in order that they may be readily identified as visual landmarks. Universal Jet Navigation Charts are used as

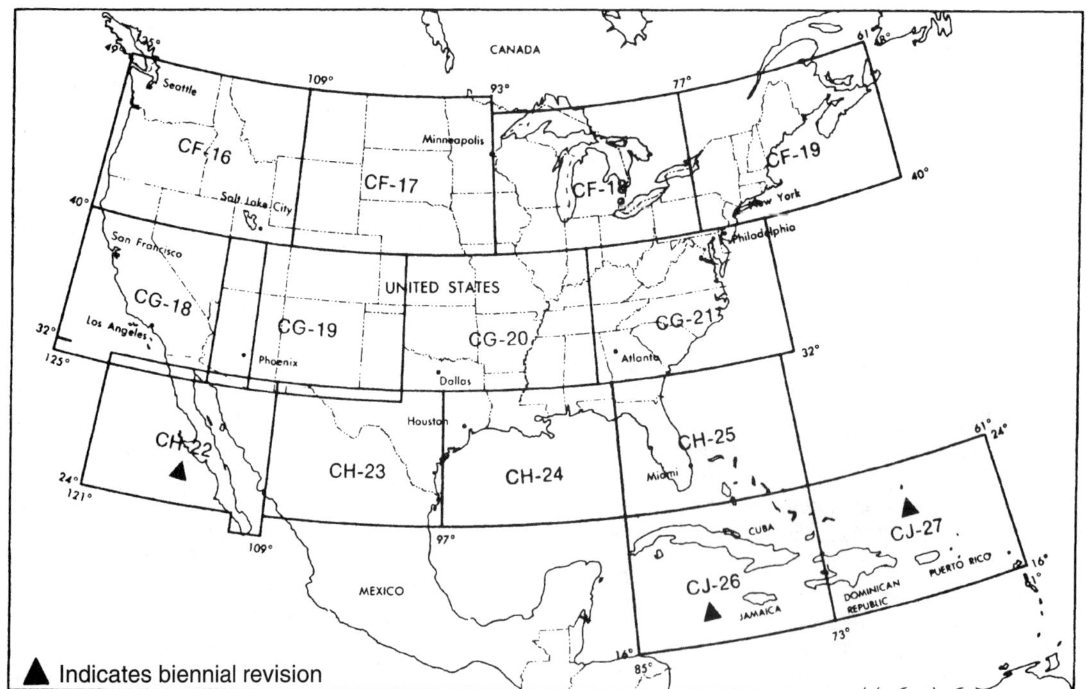

World Aeronautical Charts for the Conterminous United States, Mexico, and the Caribbean Areas

Figure 9-5[4]

plotting charts in the training and practice of celestial and dead reckoning navigation. They may also be used for grid navigational training.

12. Global Navigational Charts—These charts are designed to fulfill the requirements for aeronautical planning, operations over long distances, and en route navigation in long-range, high altitude, high speed aircraft. They are produced at a scale of 1:5,000,000 (1 in = 68.58 NM). Global Navigation Charts (GNC) are 42 by 57½ inches. They show principal cities, towns and drainage, primary roads and railroads, prominent culture and shadient relief augmented with tints and spot elevations. Aeronautical data includes radio aids to navigation, aerodrome and restricted areas. Charts 1 and 26 have a polar navigation grid and charts 2 and 6 have sub-polar navigation grids. Global LORAN Navigation Charts (GLC's) are the same size and scale and cover the same area as the GNC charts. They contain major cities only, coast lines, major lakes and rivers, and land tint. No relief or vegetation. Aeronautical data includes radio aids to navigation and LORAN lines of position.

9-6. AUXILIARY CHARTS

a. Military Training Routes—

1. Charts and Booklet: The Defense Mapping Agency Aerospace Center (DMAAC) publishes a narrative description in booklet form and charts depicting the IFR and VFR Training Routes.

2. The charts and booklet are published every 56 days. Both the charts and narrative route description booklet are available to the general public as a brochure by single copy or annual subscription.

3. Subscription and single-copy requests should be for the "DOD Area Planning AP/1B, Military Training Routes." (Reference—Military Training Routes (MTR), paragraph 3-41.)

9-6a3 NOTE—The DOD provides these booklets and charts to each FSS for use in preflight pilot briefings. Pilots should review this information to acquaint themselves with those routes that are located along their route of flight and in the vicinity of the airports from which they operate.

PILOT/CONTROLLER GLOSSARY

This Glossary was compiled to promote a common understanding of the terms used in the Air Traffic Control system. It includes those terms which are intended for pilot/controller communications. Those terms most frequently used in pilot/controller communications are printed in ***bold italics***. The definitions are primarily defined in an operational sense applicable to both users and operators of the National Airspace System. Use of the Glossary will preclude any misunderstandings concerning the system's design, function, and purpose.

Because of the international nature of flying, terms used in the *Lexicon*, published by the International Civil Aviation Organization (ICAO), are included when they differ from FAA definitions. These terms are followed by "[ICAO]." For the reader's convenience, there are also cross references to related terms in other parts of the Glossary and to other documents, such as the Federal Aviation Regulations (FAR's) and the Airman's Information Manual (AIM).

This Glossary will be revised, as necessary, to maintain a common understanding of the system.

AAI (See ARRIVAL AIRCRAFT INTERVAL.)

AAR (See AIRPORT ACCEPTANCE RATE.)

ABBREVIATED IFR FLIGHT PLANS An authorization by ATC requiring pilots to submit only that information needed for the purpose of ATC. It includes only a small portion of the usual IFR flight plan information. In certain instances, this may be only aircraft identification, location, and pilot request. Other information may be requested if needed by ATC for separation/control purposes. It is frequently used by aircraft which are airborne and desire an instrument approach or by aircraft which are on the ground and desire a climb to VFR-on-top.
(See VFR-ON-TOP.) (Refer to AIM.)

ABEAM An aircraft is "abeam" a fix, point, or object when that fix, point, or object is approximately 90 degrees to the right or left of the aircraft track. Abeam indicates a general position rather than a precise point.

ABORT To terminate a preplanned aircraft maneuver; e.g., an aborted takeoff.

ACC [ICAO] (See AREA CONTROL CENTER.)

ACCELERATE-STOP DISTANCE AVAILABLE The runway plus stopway length declared available and suitable for the acceleration and deceleration of an airplane aborting a takeoff.

ACCELERATE-STOP DISTANCE AVAILABLE [ICAO] The length of the take-off run available plus the length of the stopway if provided.

ACDO (See AIR CARRIER DISTRICT OFFICE.)

ACKNOWLEDGE Let me know that you have received my message.
(See ICAO term ACKNOWLEDGE.)

ACKNOWLEDGE [ICAO] Let me know that you have received and understood this message.

ACLS (See AUTOMATIC CARRIER LANDING SYSTEM.)

ACLT (See ACTUAL CALCULATED LANDING TIME.)

ACROBATIC FLIGHT An intentional maneuver involving an abrupt change in an aircraft's attitude, an abnormal attitude, or abnormal acceleration not necessary for normal flight.
(Refer to Part 91.) (See ICAO term ACROBATIC FLIGHT.)

ACROBATIC FLIGHT [ICAO] Manoeuvres intentionally performed by an aircraft involving an abrupt change in its attitude, an abnormal attitude, or an abnormal variation in speed.

ACTIVE RUNWAY (See RUNWAY IN USE/ACTIVE RUNWAY/DUTY RUNWAY.)

ACTUAL CALCULATED LANDING TIME ACLT is a flight's frozen calculated landing time. An actual time determined at freeze calculated landing time (FCLT) or meter list display interval (MLDI) for the adapted vertex for each arrival aircraft based upon runway configuration, airport acceptance rate, airport arrival delay period, and other metered arrival aircraft. This time is either the vertex time of arrival (VTA) of the aircraft or the tentative calculated landing time (TCLT)/ACLT of the previous aircraft plus the arrival aircraft interval (AAI), whichever is later. This time will not be updated in response to the aircraft's progress.

ADDITIONAL SERVICES Advisory information provided by ATC which includes but is not limited to the following:

1. Traffic advisories.
2. Vectors, when requested by the pilot, to assist aircraft receiving traffic advisories to avoid observed traffic.
3. Altitude deviation information of 300 feet or more from an assigned altitude as observed on a verified (reading correctly) automatic altitude readout (Mode C).
4. Advisories that traffic is no longer a factor.
5. Weather and chaff information.
6. Weather assistance.
7. Bird activity information.
8. Holding pattern surveillance. Additional services are

provided to the extent possible contingent only upon the controller's capability to fit them into the performance of higher priority duties and on the basis of limitations of the radar, volume of traffic, frequency congestion, and controller workload. The controller has complete discretion for determining if he is able to provide or continue to provide a service in a particular case. The controller's reason not to provide or continue to provide a service in a particular case is not subject to question by the pilot and need not be made known to him.
(See TRAFFIC ADVISORIES.) (Refer to AIM.)

ADF (See AUTOMATIC DIRECTION FINDER.)

ADIZ (See AIR DEFENSE IDENTIFICATION ZONE.)

ADLY (See ARRIVAL DELAY.)

ADMINISTRATOR The Federal Aviation Administrator or any person to whom he has delegated his authority in the matter concerned.

ADVISE INTENTIONS Tell me what you plan to do.

ADVISORY Advice and information provided to assist pilots in the safe conduct of flight and aircraft movement.
(See ADVISORY SERVICE.)

ADVISORY FREQUENCY The appropriate frequency to be used for Airport Advisory Service.
(See LOCAL AIRPORT ADVISORY.) (See UNICOM.) (Refer to ADVISORY CIRCULAR NO. 90-42.) (Refer to AIM.)

ADVISORY SERVICE Advice and information provided by a facility to assist pilots in the safe conduct of flight and aircraft movement.
(See LOCAL AIRPORT ADVISORY.) (See TRAFFIC ADVISORIES.) (See SAFETY ALERT.) (See ADDITIONAL SERVICES.) (See RADAR ADVISORY.) (See EN ROUTE FLIGHT ADVISORY SERVICE.) (Refer to AIM.)

AERIAL REFUELING A procedure used by the military to transfer fuel from one aircraft to another during flight.
(Refer to VFR/IFR Wall Planning Charts.)

AERODROME A defined area on land or water (including any buildings, installations and equipment) intended to be used either wholly or in part for the arrival, departure, and movement of aircraft.

AERODROME BEACON [ICAO] Aeronautical beacon used to indicate the location of an aerodrome from the air.

AERODROME CONTROL SERVICE [ICAO] Air traffic control service for aerodrome traffic.

AERODROME CONTROL TOWER [ICAO] A unit established to provide air traffic control service to aerodrome traffic.

AERODROME ELEVATION [ICAO] The elevation of the highest point of the landing area.

AERODROME TRAFFIC CIRCUIT [ICAO] The specified path to be flown by aircraft operating in the vicinity of an aerodrome.

AERONAUTICAL BEACON A visual NAVAID displaying flashes of white and/or colored light to indicate the location of an airport, a heliport, a landmark, a certain point of a Federal airway in mountainous terrain, or an obstruction.
(See AIRPORT ROTATING BEACON.) (Refer to AIM.)

AERONAUTICAL CHART A map used in air navigation containing all or part of the following: Topographic features, hazards and obstructions, navigation aids, navigation routes, designated airspace, and airports. Commonly used aeronautical charts are:

1. Sectional Aeronautical Charts (1:500,000). Designed for visual navigation of slow or medium speed aircraft. Topographic information on these charts features the portrayal of relief and a judicious selection of visual check points for VFR flight. Aeronautical information includes visual and radio aids to navigation, airports, controlled airspace, restricted areas, obstructions, and related data.
2. VFR Terminal Area Charts (1:250,000). Depict Class B airspace which provides for the control or segregation of all the aircraft within Class B airspace. The chart depicts topographic information and aeronautical information which includes visual and radio aids to navigation, airports, controlled airspace, restricted areas, obstructions, and related data.
3. World Aeronautical Charts (WAC) (1:1,000,000). Provide a standard series of aeronautical charts covering land areas of the world at a size and scale convenient for navigation by moderate speed aircraft. Topographic information includes cities and towns, principal roads, railroads, distinctive landmarks, drainage, and relief. Aeronautical information includes visual and radio aids to navigation, airports, airways, restricted areas, obstructions, and other pertinent data.
4. En Route Low Altitude Charts. Provide aeronautical information for en route instrument navigation (IFR) in the low altitude stratum. Information includes the portrayal of airways, limits of controlled airspace, position identification and frequencies of radio aids, selected airports, minimum en route and minimum obstruction clearance altitudes, airway distances, reporting points, restricted areas, and related data. Area charts, which are a part of this series, furnish terminal data at a larger scale in congested areas.
5. En Route High Altitude Charts. Provide aeronautical information for en route instrument navigation (IFR) in the high altitude stratum. Information includes the portrayal of jet routes, identification and frequencies of radio aids, selected airports, distances, time zones, special use airspace, and related information.
6. Instrument Approach Procedures (IAP) Charts. Portray the aeronautical data which is required to execute an instrument approach to an airport. These charts depict the procedures, including all related data, and the airport dia-

gram. Each procedure is designated for use with a specific type of electronic navigation system including NDB, TACAN, VOR, ILS/MLS, and RNAV. These charts are identified by the type of navigational aid(s) which provide final approach guidance.

7. Standard Instrument Departure (SID) Charts. Designed to expedite clearance delivery and to facilitate transition between takeoff and en route operations. Each SID procedure is presented as a separate chart and may serve a single airport or more than one airport in a given geographical location.

8. Standard Terminal Arrival (STAR) Charts. Designed to expedite air traffic control arrival procedures and to facilitate transition between en route and instrument approach operations. Each STAR procedure is presented as a separate chart and may serve a single airport or more than one airport in a given geographical location.

9. Airport Taxi Charts. Designed to expedite the efficient and safe flow of ground traffic at an airport. These charts are identified by the official airport name; e.g., Washington National Airport.

(See ICAO term AERONAUTICAL CHART.)

AERONAUTICAL CHART [ICAO] A representation of a portion of the earth, its culture and relief, specifically designated to meet the requirements of air navigation.

AERONAUTICAL INFORMATION PUBLICATION [AIP] [ICAO] A publication issued by or with the authority of a State and containing aeronautical information of a lasting character essential to air navigation.

A/FD (See AIRPORT/FACILITY DIRECTORY.)

AFFIRMATIVE Yes.

AIM (See AIRMAN'S INFORMATION MANUAL.)

AIP [ICAO] See AERONAUTICAL INFORMATION PUBLICATION.)

AIRBORNE DELAY Amount of delay to be encountered in airborne holding.

AIR CARRIER DISTRICT OFFICE An FAA field office serving an assigned geographical area, staffed with Flight Standards personnel serving the aviation industry and the general public on matters related to the certification and operation of scheduled air carriers and other large aircraft operations.

AIRCRAFT Device(s) that are used or intended to be used for flight in the air, and when used in air traffic control terminology, may include the flight crew.

(See ICAO term AIRCRAFT.)

AIRCRAFT [ICAO] Any machine that can derive support in the atmosphere from the reactions of the air other than the reactions of the air against the earth's surface.

AIRCRAFT APPROACH CATEGORY A grouping of aircraft based on a speed of 1.3 times the stall speed in the landing configuration at maximum gross landing weight. An aircraft shall fit in only one category. If it is necessary to maneuver at speeds in excess of the upper limit of a speed range for a category, the minimums for the next higher category should be used. For example, an aircraft which falls in Category A, but is circling to land at a speed in excess of 91 knots, should use the approach Category B minimums when circling to land. The categories are as follows:

1. Category A. Speed less than 91 knots.
2. Category B. Speed 91 knots or more but less 121 knots.
3. Category C. Speed 121 knots or more but less than 141 knots.
4. Category D. Speed 141 knots or more but less than 166 knots.
5. Category E. Speed 166 knots or more.

(Refer to Part 1.) (Refer to Part 97.)

AIRCRAFT CLASSES For the purposes of Wake Turbulence Separation Minima, ATC classifies aircraft as Heavy, Large, and Small as follows:

1. Heavy. Aircraft capable of takeoff weights of 300,000 pounds or more whether or not they are operating at this weight during a particular phase of flight.
2. Large. Aircraft of more than 12,500 pounds, maximum certificated takeoff weight, up to 300,000 pounds.
3. Small. Aircraft of 12,500 pounds or less maximum certificated takeoff weight.

(Refer to AIM.)

AIRCRAFT SITUATION DISPLAY ASD is a computer system that receives radar track data from all 20 CONUS ARTCC's, organizes this data into a mosaic display, and presents it on a computer screen. The display allows the traffic management coordinator multiple methods of selection and high-lighting of individual aircraft or groups of aircraft. The user has the option of superimposing these aircraft positions over any number of background displays. These background options include ARTCC boundaries, any stratum of en route sector boundaries, fixes, airways, military and other special use airspace, airports, and geopolitical boundaries. By using the ASD, a coordinator can monitor any number of traffic situations or the entire systemwide traffic flows.

AIRCRAFT SURGE LAUNCH AND RECOVERY Procedures used at USAF bases to provide increased launch and recovery rates in instrument flight rules conditions. ASLAR is based on:

a. Reduced separation between aircraft which is based on time or distance. Standard arrival separation applies between participants including multiple flights until the DRAG point. The DRAG point is a published location on an ASLAR approach where aircraft landing second in a formation slows to a predetermined airspeed. The DRAG point is the reference point at which MARSA applies as expanding elements effect separation within a flight or between subsequent participating flights.

b. ASLAR procedures shall be covered in a Letter of Agreement between the responsible USAF military ATC facility and the concerned Federal Aviation Administration facility. Initial Approach Fix spacing requirements are normally addressed as a minimum.

AIR DEFENSE EMERGENCY A military emergency condition declared by a designated authority. This condition exists when an attack upon the continental U.S., Alaska, Canada, or U.S. installations in Greenland by hostile aircraft or missiles is considered probable, is imminent, or is taking place.

(Refer to AIM.)

AIR DEFENSE IDENTIFICATION ZONE The area of airspace over land or water, extending upward from the surface, within which the ready identification, the location, and the control of aircraft are required in the interest of national security.

1. Domestic Air Defense Identification Zone. An ADIZ within the United States along an international boundary of the United States.

2. Coastal Air Defense Identification Zone. An ADIZ over the coastal waters of the United States.

3. Distant Early Warning Identification Zone (DEWIZ). An ADIZ over the coastal waters of the State of Alaska.

ADIZ locations and operating and flight plan requirements for civil aircraft operations are specified in Part 99.

(Refer to AIM.)

AIRMAN'S INFORMATION MANUAL A primary FAA publication whose purpose is to instruct airmen about operating in the National Airspace System of the U.S. It provides basic flight information, ATC Procedures and general instructional information concerning health, medical facts, factors affecting flight safety, accident and hazard reporting, and types of aeronautical charts and their use.

AIRMAN'S METEOROLOGICAL INFORMATION (See AIRMET.)

AIRMET In-flight weather advisories issued only to amend the area forecast concerning weather phenomena which are of operational interest to all aircraft and potentially hazardous to aircraft having limited capability because of lack of equipment, instrumentation, or pilot qualifications. AIRMET's concern weather of less severity than that covered by SIGMET's or Convective SIGMET's. AIRMET's cover moderate icing, moderate turbulence, sustained winds of 30 knots or more at the surface, widespread areas of ceilings less than 1,000 feet and/or visibility less than 3 miles, and extensive mountain obscurement.

(See AWW.) (See SIGMET.) (See CONVECTIVE SIGMET.) (See CWA.) (Refer to AIM.)

AIR NAVIGATION FACILITY Any facility used in, available for use in, or designed for use in, aid of air navigation, including landing areas, lights, any apparatus or equipment for disseminating weather information, for signaling, for radio-directional finding, or for radio or other electrical communication, and any other structure or mechanism having a similar purpose for guiding or controlling flight in the air or the landing and take-off of aircraft.

(See NAVIGATIONAL AID.)

AIRPORT An area on land or water that is used or intended to be used for the landing and takeoff of aircraft and includes its buildings and facilities, if any.

AIRPORT ACCEPTANCE RATE A dynamic input parameter specifying the number of arriving aircraft which an airport or airspace can accept from the ARTCC per hour. The AAR is used to calculate the desired interval between successive arrival aircraft.

AIRPORT ADVISORY AREA The area within ten miles of an airport without a control tower or where the tower is not in operation, and on which a Flight Service Station is located.

(See LOCAL AIRPORT ADVISORY.) (Refer to AIM.)

AIRPORT ELEVATION The highest point of an airport's usable runways measured in feet from mean sea level.

(See TOUCHDOWN ZONE ELEVATION.) (See ICAO term AERODROME ELEVATION.)

AIRPORT/FACILITY DIRECTORY A publication designed primarily as a pilot's operational manual containing all airports, seaplane bases, and heliports open to the public including communications data, navigational facilities, and certain special notices and procedures. This publication is issued in seven volumes according to geographical area.

AIRPORT INFORMATION AID (See AIRPORT INFORMATION DESK.)

AIRPORT INFORMATION DESK An airport unmanned facility designed for pilot self-service briefing, flight planning, and filing of flight plans.

(Refer to AIM.)

AIRPORT LIGHTING Various lighting aids that may be installed on an airport. Types of airport lighting include:

1. Approach Light System (ALS). An airport lighting facility which provides visual guidance to landing aircraft by radiating light beams in a directional pattern by which the pilot aligns the aircraft with the extended centerline of the runway on his final approach for landing. Condenser-Discharge Sequential Flashing Lights/Sequenced Flashing Lights may be installed in conjunction with the ALS at some airports. Types of Approach Light Systems are:

a. ALSF-1. Approach Light System with Sequenced Flashing Lights in ILS Cat-I configuration.

b. ALSF-2. Approach Light System with Sequenced

Flashing Lights in ILS Cat-II configuration. The ALSF-2 may operate as an SSALR when weather conditions permit.

c. SSALF. Simplified Short Approach Light System with Sequenced Flashing Lights.

d. SSALR. Simplified Short Approach Light System with Runway Alignment Indicator Lights.

e. MALSF. Medium Intensity Approach Light System with Sequenced Flashing Lights.

f. MALSR. Medium Intensity Approach Light System with Runway Alignment Indicator Lights.

g. LDIN. Lead-in-light system: Consists of one or more series of flashing lights installed at or near ground level that provides positive visual guidance along an approach path, either curving or straight, where special problems exist with hazardous terrain, obstructions, or noise abatement procedures.

h. RAIL. Runway Alignment Indicator Lights (Sequenced Flashing Lights which are installed only in combination with other light systems.)

i. ODALS. Omnidirectional Approach Lighting System consists of seven omnidirectional flashing lights located in the approach area of a nonprecision runway. Five lights are located on the runway centerline extended with the first light located 300 feet from the threshold and extending at equal intervals up to 1,500 feet from the threshold. The other two lights are located, one on each side of the runway threshold, at a lateral distance of 40 feet from the runway edge, or 75 feet from the runway edge when installed on a runway equipped with a VASI.

(Refer to Order 6850.2.)

2. Runway Lights/Runway Edge Lights. Lights having a prescribed angle of emission used to define the lateral limits of a runway. Runway lights are uniformly spaced at intervals of approximately 200 feet, and the intensity may be controlled or preset.

3. Touchdown Zone Lighting. Two rows of tranverse light bars located symmetrically about the runway centerline normally at 100 foot intervals. The basic system extends 3,000 feet along the runway.

4. Runway Centerline Lighting. Flush centerline lights spaced at 50-foot intervals beginning 75 feet from the landing threshold and extending to within 75 feet of the opposite end of the runway.

5. Threshold Lights. Fixed green lights arranged symmetrically left and right of the runway centerline, identifying the runway threshold.

6. Runway End Identifier Lights (REIL). Two synchronized flashing lights, one on each side of the runway threshold, which provide rapid and positive identification of the approach end of a particular runway.

7. Visual Approach Slope Indicator (VASI). An airport lighting facility providing vertical visual approach slope guidance to aircraft during approach to landing by radiating a directional pattern of high intensity red and white focused light beams which indicate to the pilot that he is "on path" if he sees red/white, "above path" if white/white, and "below path" if red/red. Some airports serving large aircraft have three-bar VASIs which provide two visual glide paths to the same runway.

8. Boundary Lights. Lights defining the perimeter of an airport or landing area.

(Refer to AIM.)

AIRPORT MARKING AIDS Markings used on runway and taxiway surfaces to identify a specific runway, a runway threshold, a centerline, a hold line, etc. A runway should be marked in accordance with its present usage such as:

1. Visual.
2. Nonprecision instrument.
3. Precision instrument.

(Refer to AIM.)

AIRPORT RESERVATION OFFICE Office responsible for monitoring the operation of the high density rule. Receives and processes requests for IFR operations at high density traffic airports.

AIRPORT ROTATING BEACON A visual NAVAID operated at many airports. At civil airports, alternating white and green flashes indicate the location of the airport. At military airports, the beacons flash alternately white and green, but are differentiated from civil beacons by dualpeaked (two quick) white flashes between the green flashes.

(See SPECIAL VFR OPERATIONS.) (See INSTRUMENT FLIGHT RULES.) (Refer to AIM.) (See ICAO term AERODROME BEACON.)

AIRPORT SURFACE DETECTION EQUIPMENT Radar equipment specifically designed to detect all principal features on the surface of an airport, including aircraft and vehicular traffic, and to present the entire image on a radar indicator console in the control tower. Used to augment visual observation by tower personnel of aircraft and/or vehicular movements on runways and taxiways.

AIRPORT SURVEILLANCE RADAR Approach control radar used to detect and display an aircraft's position in the terminal area. ASR provides range and azimuth information but does not provide elevation data. Coverage of the ASR can extend up to 60 miles.

AIRPORT TAXI CHARTS (See AERONAUTICAL CHART.)

AIRPORT TRAFFIC CONTROL SERVICE A service provided by a control tower for aircraft operating on the movement area and in the vicinity of an airport.

(See MOVEMENT AREA.) (See TOWER.) (See ICAO term AERODROME CONTROL SERVICE.)

AIRPORT TRAFFIC CONTROL TOWER (See TOWER.)

AIR ROUTE SURVEILLANCE RADAR Air route traffic control center (ARTCC) radar used primarily to detect and display an aircraft's position while en route between terminal areas. The ARSR enables controllers to provide radar air traffic control service when aircraft are within the ARSR coverage. In some instances, ARSR may enable an ARTCC to provide terminal radar services similar to but usually more limited than those provided by a radar approach control.

AIR ROUTE TRAFFIC CONTROL CENTER A facility established to provide air traffic control service to aircraft operating on IFR flight plans within controlled airspace and principally during the en route phase of flight. When equipment capabilities and controller workload permit, certain advisory/assistance services may be provided to VFR aircraft.

(See NAS STAGE A.) (See EN ROUTE AIR TRAFFIC CONTROL SERVICES.) (Refer to AIM.)

AIRSPACE HIERARCHY Within the airspace classes, there is a hierarchy and, in the event of an overlap of airspace: Class A preempts Class B, Class B preempts Class C, Class C preempts Class D, Class D preempts Class E, and Class E preempts Class G.

AIRSPEED The speed of an aircraft relative to its surrounding air mass. The unqualified term "airspeed" means one of the following:

1. Indicated Airspeed. The speed shown on the aircraft airspeed indicator. This is the speed used in pilot/controller communications under the general term "airspeed."

(Refer to Part 1.)

2. True Airspeed. The airspeed of an aircraft relative to undisturbed air. Used primarily in flight planning and en route portion of flight. When used in pilot/controller communications, it is referred to as "true airspeed" and not shortened to "airspeed."

AIRSTART The starting of an aircraft engine while the aircraft is airborne, preceded by engine shutdown during training flights or by actual engine failure.

AIR TAXI Used to describe a helicopter/VTOL aircraft movement conducted above the surface but normally not above 100 feet AGL. The aircraft may proceed either via hover taxi or flight at speeds more than 20 knots. The pilot is solely responsible for selecting a safe airspeed/altitude for the operation being conducted.

(See HOVER TAXI.) (Refer to AIM.)

AIR TRAFFIC Aircraft operating in the air or on an airport surface, exclusive of loading ramps and parking areas.

(See ICAO term AIR TRAFFIC.)

AIR TRAFFIC [ICAO] All aircraft in flight or operating on the manoeuvring area of an aerodrome.

AIR TRAFFIC CLEARANCE An authorization by air traffic control, for the purpose of preventing collision between known aircraft, for an aircraft to proceed under specified traffic conditions within controlled airspace. The pilot-in-command of an aircraft may not deviate from the provisions of a visual flight rules (VFR) or instrument flight rules (IFR) air traffic clearance unless an amended clearance has been obtained. Additionally, the pilot may request a different clearance from that which has been issued by air traffic control (ATC) if information available to the pilot makes another course of action more practicable or if aircraft equipment limitations or company procedures forbid compliance with the clearance issued. Pilots may also request clarification or amendment, as appropriate, any time a clearance is not fully understood, or considered unacceptable because of safety of flight. Controllers should, in such instances and to the extent of operational practicality and safety, honor the pilot's request. Part 91.3(a) states: "The pilot-in-command of an aircraft is directly responsible for, and is the final authority as to, the operation of that aircraft." THE PILOT IS RESPONSIBLE TO REQUEST AN AMENDED CLEARANCE if ATC issues a clearance that would cause a pilot to deviate from a rule or regulation, or in the pilot's opinion, would place the aircraft in jeopardy.

(See ATC INSTRUCTIONS.) (See ICAO term AIR TRAFFIC CONTROL CLEARANCE.)

AIR TRAFFIC CONTROL A service operated by appropriate authority to promote the safe, orderly and expeditious flow of air traffic.

(See ICAO term AIR TRAFFIC CONTROL SERVICE.)

AIR TRAFFIC CONTROL CLEARANCE [ICAO] Authorization for an aircraft to proceed under conditions specified by an air traffic control unit.

Note 1: For convenience, the term air traffic control clearance is frequently abbreviated to clearance when used in appropriate contexts.

Note 2: The abbreviated term clearance may be prefixed by the words taxi, takeoff, departure, en route, approach or landing to indicate the particular portion of flight to which the air traffic control clearance relates.

AIR TRAFFIC CONTROL SERVICE (See AIR TRAFFIC CONTROL.)

AIR TRAFFIC CONTROL SERVICE [ICAO] A service provided for the purpose of:

1. Preventing collisions:

a. Between aircraft; and

b. On the maneuvering area between aircraft and obstructions; and

2. Expediting and maintaining an orderly flow of air traffic.

AIR TRAFFIC CONTROL SPECIALIST A person authorized to provide air traffic control service.

(See AIR TRAFFIC CONTROL.) (See FLIGHT SERVICE STATION.) (See ICAO term CONTROLLER.)

AIR TRAFFIC CONTROL SYSTEM COMMAND CENTER An Air Traffic Operations Service facility consisting of four operational units.

1. Central Flow Control Function (CFCF). Responsible for coordination and approval of all major intercenter flow control restrictions on a system basis in order to obtain maximum utilization of the airspace.

(See QUOTA FLOW CONTROL.)

2. Central Altitude Reservation Function (CARF). Responsible for coordinating, planning, and approving special user requirements under the Altitude Reservation (ALTRV) concept.

(See ALTITUDE RESERVATION.)

3. Airport Reservation Office (ARO). Responsible for approving IFR flights at designated high density traffic airports (John F. Kennedy, LaGuardia, O'Hare, and Washington National) during specified hours.

(Refer to Part 93 and AIRPORT/FACILITY DIRECTORY.)

4. ATC Contingency Command Post. A facility which enables the FAA to manage the ATC system when significant portions of the system's capabilities have been lost or are threatened.

AIR TRAFFIC SERVICE A generic term meaning:

1. Flight Information Service:
2. Alerting Service:
3. Air Traffic Advisory Service:
4. Air Traffic Control Service:

a. Area Control Service,
b. Approach Control Service, or
c. Airport Control Service.

AIRWAY A Class E airspace area established in the form of a corridor, the centerline of which is defined by radio navigational aids.

(See FEDERAL AIRWAYS.) (Refer to Part 71.) (Refer to AIM.) (See ICAO term AIRWAY.)

AIRWAY [ICAO] A control area or portion thereof established in the form of corridor equipped with radio navigational aids.

AIRWAY BEACON Used to mark airway segments in remote mountain areas. The light flashes Morse Code to identify the beacon site.

(Refer to AIM.)

AIT (See AUTOMATED INFORMATION TRANSFER.)

ALERFA (Alert Phase) [ICAO] A situation wherein apprehension exists as to the safety of an aircraft and its occupants.

ALERT AREA (See SPECIAL USE AIRSPACE.)

ALERT NOTICE A request originated by a flight service station (FSS) or an air route traffic control center (ARTCC) for an extensive communication search for overdue, unreported, or missing aircraft.

ALERTING SERVICE A service provided to notify appropriate organizations regarding aircraft in need of search and rescue aid and assist such organizations as required.

ALNOT (See ALERT NOTICE.)

ALPHANUMERIC DISPLAY Letters and numerals used to show identification, altitude, beacon code, and other information concerning a target on a radar display.

(See AUTOMATED RADAR TERMINAL SYSTEMS.) (See NAS STAGE A.)

ALTERNATE AERODROME [ICAO] An aerodrome to which an aircraft may proceed when it becomes either impossible or inadvisable to proceed to or to land at the aerodrome of intended landing.

Note: The aerodrome from which a flight departs may also be an en-route or a destination alternate aerodrome for the flight.

ALTERNATE AIRPORT An airport at which an aircraft may land if a landing at the intended airport becomes inadvisable.

(See FAA term ICAO term ALTERNATE AERODROME.)

ALTIMETER SETTING The barometric pressure reading used to adjust a pressure altimeter for variations in existing atmospheric pressure or to the standard altimeter setting (29.92).

(Refer to Part 91.) (Refer to AIM.)

ALTITUDE The height of a level, point, or object measured in feet Above Ground Level (AGL) or from Mean Sea Level (MSL).

(See FLIGHT LEVEL.)

1. MSL Altitude. Altitude expressed in feet measured from mean sea level.

2. AGL Altitude. Altitude expressed in feet measured above ground level.

3. Indicated Altitude. The altitude as shown by an altimeter. On a pressure or barometric altimeter it is altitude as shown uncorrected for instrument error and uncompensated for variation from standard atmospheric conditions.

(See ICAO term ALTITUDE.)

ALTITUDE [ICAO] The vertical distance of a level, a point or an object considered as a point, measured from mean sea level (MSL).

ALTITUDE READOUT An aircraft's altitude, transmitted via the Mode C transponder feature, that is visually displayed in 100-foot increments on a radar scope having readout capability.

(See AUTOMATED RADAR TERMINAL SYSTEMS.) (See NAS STAGE A.) (See ALPHANUMERIC DISPLAY.) (Refer to AIM.)

ALTITUDE RESERVATION Airspace utilization under prescribed conditions normally employed for the mass movement of aircraft or other special user requirements which cannot otherwise be accomplished. ALTRVs are approved by the appropriate FAA facility.

(See AIR TRAFFIC CONTROL SYSTEM COMMAND CENTER.)

ALTITUDE RESTRICTION An altitude or altitudes, stated in the order flown, which are to be maintained until reaching a specific point or time. Altitude restrictions may be issued by ATC due to traffic, terrain, or other airspace considerations.

ALTITUDE RESTRICTIONS ARE CANCELED Adherence to previously imposed altitude restrictions is no longer required during a climb or descent.

ALTRV (See ALTITUDE RESERVATION.)

AMVER (See AUTOMATED MUTUAL-ASSISTANCE VESSEL RESCUE SYSTEM.)

APPROACH CLEARANCE Authorization by ATC for a pilot to conduct an instrument approach. The type of instrument approach for which a clearance and other pertinent information is provided in the approach clearance when required.

(See INSTRUMENT APPROACH PROCEDURE.) (See CLEARED APPROACH.) (Refer to AIM and Part 91.)

APPROACH CONTROL FACILITY A terminal ATC facility that provides approach control service in a terminal area.

(See APPROACH CONTROL SERVICE.) (See RADAR APPROACH CONTROL FACILITY.)

APPROACH CONTROL SERVICE Air traffic control service provided by an approach control facility for arriving and departing VFR/IFR aircraft and, on occasion, en route aircraft. At some airports not served by an approach control facility, the ARTCC provides limited approach control service.

(Refer to AIM.) (See ICAO term APPROACH CONTROL SERVICE.)

APPROACH CONTROL SERVICE [ICAO] Air traffic control service for arriving or departing controlled flights.

APPROACH GATE An imaginary point used within ATC as a basis for vectoring aircraft to the final approach course. The gate will be established along the final approach course 1 mile from the outer marker (or the fix used in lieu of the outer marker) on the side away from the airport for precision approaches and 1 mile from the final approach fix on the side away from the airport for nonprecision approaches. In either case when measured along the final approach course, the gate will be no closer than 5 miles from the landing threshold.

APPROACH LIGHT SYSTEM (See AIRPORT LIGHTING.)

APPROACH SEQUENCE The order in which aircraft are positioned while on approach or awaiting approach clearance.

(See LANDING SEQUENCE.) (See ICAO term APPROACH SEQUENCE.)

APPROACH SEQUENCE [ICAO] The order in which two or more aircraft are cleared to approach to land at the aerodrome.

APPROACH SPEED The recommended speed contained in aircraft manuals used by pilots when making an approach to landing. This speed will vary for different segments of an approach as well as for aircraft weight and configuration.

APPROPRIATE ATS AUTHORITY [ICAO] The relevant authority designated by the State responsible for providing air traffic services in the airspace concerned. In the United States, the "appropriate ATS authority" is the Director, Office of Air Traffic System Management, ATM-1.

APPROPRIATE AUTHORITY

1. Regarding flight over the high seas: the relevant authority is the State of Registry.
2. Regarding flight over other than the high seas: the relevant authority is the State having sovereignty over the territory being overflown.

APPROPRIATE OBSTACLE CLEARANCE MINIMUM ALTITUDE Any of the following:

(See MINIMUM IFR ALTITUDE MIA.) (See MINIMUM EN ROUTE ALTITUDE MEA.) (See MINIMUM OBSTRUCTION CLEARANCE ALTITUDE MOCA.) (See MINIMUM VECTORING ALTITUDE MVA.)

APPROPRIATE TERRAIN CLEARANCE MINIMUM ALTITUDE Any of the following:

(See MINIMUM IFR ALTITUDE MIA.) (See MINIMUM EN ROUTE ALTITUDE MEA.) (See MINIMUM OBSTRUCTION CLEARANCE ALTITUDE MOCA.) (See MINIMUM VECTORING ALTITUDE MVA.)

APRON A defined area on an airport or heliport intended to accommodate aircraft for purposes of loading or unloading passengers or cargo, refueling, parking, or maintenance. With regard to seaplanes, a ramp is used for access to the apron from the water.

(See ICAO term APRON.)

APRON [ICAO] A defined area, on a land aerodrome, intended to accommodate aircraft for purposes of loading or unloading passengers, mail or cargo, refuelling, parking or maintenance.

ARC The track over the ground of an aircraft flying at a constant distance from a navigational aid by reference to distance measuring equipment (DME).

AREA CONTROL CENTER [ICAO] An ICAO term for an air traffic control facility primarily responsible for ATC services being provided IFR aircraft during the en route phase of flight. The U.S. equivalent facility is an air route traffic control center (ARTCC).

AREA NAVIGATION A method of navigation that permits aircraft operation on any desired course within the coverage of station-referenced navigation signals or within the limits of a self-contained system capability. Random area navigation routes are direct routes, based on area navigation capability, between waypoints defined in terms of latitude/longitude coordinates, degree/distance fixes, or offsets from published or established routes/airways at a specified distance and direction. The major types of equipment are:

1. VORTAC referenced or Course Line Computer (CLC) systems, which account for the greatest number of RNAV units in use. To function, the CLC must be within the service range of a VORTAC.

2. OMEGA/VLF, although two separate systems, can be considered as one operationally. A long-range navigation system based upon Very Low Frequency radio signals transmitted from a total of 17 stations worldwide.

3. Inertial (INS) systems, which are totally self-contained and require no information from external references. They provide aircraft position and navigation information in response to signals resulting from inertial effects on components within the system.

4. MLS Area Navigation (MLS/RNAV), which provides area navigation with reference to an MLS ground facility.

5. LORAN-C is a long-range radio navigation system that uses ground waves transmitted at low frequency to provide user position information at ranges of up to 600 to 1,200 nautical miles at both en route and approach altitudes. The usable signal coverage areas are determined by the signal-to-noise ratio, the envelope-to-cycle difference, and the geometric relationship between the positions of the user and the transmitting stations.

6. GPS is a space based radio positioning, navigation, and time-transfer system. The system provides highly accurate position and velocity information, and precise time, on a continuous global basis, to an unlimited number of properly equipped users. The system is unaffected by weather, and provides a worldwide common grid reference system.

(See ICAO term AREA NAVIGATION.)

AREA NAVIGATION [ICAO] A method of navigation which permits aircraft operation on any desired flight path within the coverage of station-referenced navigation aids or within the limits of the capability of self-contained aids, or a combination of these.

ARINC An acronym for Aeronautical Radio, Inc., a corporation largely owned by a group of airlines. ARINC is licensed by the FCC as an aeronautical station and contracted by the FAA to provide communications support for air traffic control and meteorological services in portions of international airspace.

ARMY AVIATION FLIGHT INFORMATION BULLETIN A bulletin that provides air operation data covering Army, National Guard, and Army Reserve aviation activities.

ARO (See AIRPORT RESERVATION OFFICE.)

ARRESTING SYSTEM A safety device consisting of two major components, namely, engaging or catching devices and energy absorption devices for the purpose of arresting both tailhook and/or nontailhook-equipped aircraft. It is used to prevent aircraft from overrunning runways when the aircraft cannot be stopped after landing or during aborted takeoff. Arresting systems have various names; e.g., arresting gear, hook device, wire barrier cable.

(See ABORT.) (Refer to AIM.)

ARRIVAL AIRCRAFT INTERVAL An internally generated program in hundredths of minutes based upon the AAR. AAI is the desired optimum interval between successive arrival aircraft over the vertex.

ARRIVAL CENTER The ARTCC having jurisdiction for the impacted airport.

ARRIVAL DELAY A parameter which specifies a period of time in which no aircraft will be metered for arrival at the specified airport.

ARRIVAL SECTOR An operational control sector containing one or more meter fixes.

ARRIVAL SECTOR ADVISORY LIST An ordered list of data on arrivals displayed at the PVD of the sector which controls the meter fix.

ARRIVAL SEQUENCING PROGRAM The automated program designed to assist in sequencing aircraft destined for the same airport.

ARRIVAL TIME The time an aircraft touches down on arrival.

ARSR (See AIR ROUTE SURVEILLANCE RADAR.)

ARTCC (See AIR ROUTE TRAFFIC CONTROL CENTER.)

ARTS (See AUTOMATED RADAR TERMINAL SYSTEMS.)

ASD (See AIRCRAFT SITUATION DISPLAY.)

ASDA (See ACCELERATE-STOP DISTANCE AVAILABLE.)

ASDA [ICAO] See ICAO Term ACCELERATE-STOP DISTANCE AVAILABLE.)

ASDE (See AIRPORT SURFACE DETECTION EQUIPMENT.)

ASLAR (See AIRCRAFT SURGE LAUNCH AND RECOVERY.)

ASP (See ARRIVAL SEQUENCING PROGRAM.)

ASR (See AIRPORT SURVEILLANCE RADAR.)

ASR APPROACH (See SURVEILLANCE APPROACH.)

ATC (See AIR TRAFFIC CONTROL.)

ATCAA (See ATC ASSIGNED AIRSPACE.)

ATC ADVISES Used to prefix a message of noncontrol information when it is relayed to an aircraft by other than an air traffic controller.
(See ADVISORY.)

ATC ASSIGNED AIRSPACE Airspace of defined vertical/lateral limits, assigned by ATC, for the purpose of providing air traffic segregation between the specified activities being conducted within the assigned airspace and other IFR air traffic.
(See SPECIAL USE AIRSPACE.)

ATC CLEARANCE (See AIR TRAFFIC CLEARANCE.)

ATC CLEARS Used to prefix an ATC clearance when it is relayed to an aircraft by other than an air traffic controller.

ATC INSTRUCTIONS Directives issued by air traffic control for the purpose of requiring a pilot to take specific actions; e.g., "Turn left heading two five zero," "Go around," "Clear the runway."
(Refer to Part 91.)

ATCRBS (See RADAR.)

ATC REQUESTS Used to prefix an ATC request when it is relayed to an aircraft by other than an air traffic controller.

ATCSCC (See AIR TRAFFIC CONTROL SYSTEM COMMAND CENTER.)

ATCSCC DELAY FACTOR The amount of delay calculated to be assigned prior to departure.

ATCT (See TOWER.)

ATIS (See AUTOMATIC TERMINAL INFORMATION SERVICE.)

ATIS [ICAO] (See ICAO Term AUTOMATIC TERMINAL INFORMATION SERVICE.)

ATS Route [ICAO] A specified route designed for channelling the flow of traffic as necessary for the provision of air traffic services.

Note: The term "ATS Route" is used to mean variously, airway, advisory route, controlled or uncontrolled route, arrival or departure, etc.

AUTOLAND APPROACH An autoland approach is a precision instrument approach to touchdown and, in some cases, through the landing rollout. An autoland approach is performed by the aircraft autopilot which is receiving position information and/or steering commands from onboard navigation equipment.
(See COUPLED APPROACH.)

Note: Autoland and coupled approaches are flown in VFR and IFR. It is common for carriers to require their crews to fly coupled approaches and autoland approaches (if certified) when the weather conditions are less than approximately 4,000 RVR.

AUTOMATED INFORMATION TRANSFER A precoordinated process, specifically defined in facility directives, during which a transfer of altitude control and/or radar identification is accomplished without verbal coordination between controllers using information communicated in a full data block.

AUTOMATED MUTUAL-ASSISTANCE VESSEL RESCUE SYSTEM A facility which can deliver, in a matter of minutes, a surface picture (SURPIC) of vessels in the area of a potential or actual search and rescue incident, including their predicted positions and their characteristics.
(See Order 7110.65, paragraph 10-73, In-Flight Contingencies.)

AUTOMATED RADAR TERMINAL SYSTEMS The generic term for the ultimate in functional capability afforded by several automation systems. Each differs in functional capabilities and equipment. ARTS plus a suffix roman numeral denotes a specific system. A following letter indicates a major modification to that system. In general, an ARTS displays for the terminal controller aircraft identification, flight plan data, other flight associated information; e.g., altitude, speed, and aircraft position symbols in conjunction with his radar presentation. Normal radar coexists with the alphanumeric display. In addition to enhancing visualization of the air traffic situation, ARTS facilitate intra/inter-facility transfer and coordination of flight information. These capabilities are enabled by specially designed computers and subsystems tailored to the radar and communications equipments and operational requirements of each automated facility. Modular design permits adoption of improvements in computer software and electronic technologies as they become available while retaining the characteristics unique to each system.

1. ARTS II. A programmable nontracking, computer-aided display subsystem capable of modular expansion. ARTS II systems provide a level of automated air traffic control capability at terminals having low to medium activity. Flight identification and altitude may be associated with the display of secondary radar targets. The system has the capability of communicating with ARTCC's and other ARTS II, IIA, III, and IIIA facilities.

2. ARTS IIA. A programmable radar-tracking computer subsystem capable of modular expansion. The ARTS IIA detects, tracks, and predicts secondary radar targets. The targets are displayed by means of computer-generated symbols, ground speed, and flight plan data. Although it does not track primary radar targets, they are displayed coincident with the secondary radar as well as the symbols and alphanumerics. The system has the capability of communicating with ARTCC's and other ARTS II, IIA, III, and IIIA facilities.

3. ARTS III. The Beacon Tracking Level of the modular programmable automated radar terminal system in use at medium to high activity terminals. ARTS III detects, tracks, and predicts secondary radar-derived aircraft targets. These are displayed by means of computer-generated symbols and alphanumeric characters depicting flight identification, aircraft altitude, ground speed, and flight plan data. Although it does not track primary targets, they are displayed coincident with the secondary radar as well as the symbols and alphanumerics. The system has the capability of communicating with ARTCC's and other ARTS III facilities.

4. ARTS IIIA. The Radar Tracking and Beacon Tracking Level (RT&BTL) of the modular, programmable automated radar terminal system. ARTS IIIA detects, tracks, and predicts primary as well as secondary radar-derived aircraft targets. This more sophisticated computer-driven system upgrades the existing ARTS III system by providing improved tracking, continuous data recording, and fail-soft capabilities.

AUTOMATIC ALTITUDE REPORT (See ALTITUDE READOUT.)

AUTOMATIC ALTITUDE REPORTING That function of a transponder which responds to Mode C interrogations by transmitting the aircraft's altitude in 100-foot increments.

AUTOMATIC CARRIER LANDING SYSTEM U.S. Navy final approach equipment consisting of precision tracking radar coupled to a computer data link to provide continuous information to the aircraft, monitoring capability to the pilot, and a backup approach system.

AUTOMATIC DIRECTION FINDER An aircraft radio navigation system which senses and indicates the direction to an L/MF nondirectional radio beacon (NDB) ground transmitter. Direction is indicated to the pilot as a magnetic bearing or as a relative bearing to the longitudinal axis of the aircraft depending on the type of indicator installed in the aircraft. In certain applications, such as military, ADF operations may be based on airborne and ground transmitters in the VHF/UHF frequency spectrum.

(See BEARING.) (See NONDIRECTIONAL BEACON.)

AUTOMATIC TERMINAL INFORMATION SERVICE The continuous broadcast of recorded noncontrol information in selected terminal areas. Its purpose is to improve controller effectiveness and to relieve frequency congestion by automating the repetitive transmission of essential but routine information; e.g., "Los Angeles information Alfa. One three zero zero Coordinated Universal Time. Weather, measured ceiling two thousand overcast, visibility three, haze, smoke, temperature seven one, dew point five seven, wind two five zero at five, altimeter two niner niner six. I-L-S Runway Two Five Left approach in use, Runway Two Five Right closed, advise you have Alfa."

(Refer to AIM.) (See ICAO term AUTOMATIC TERMINAL INFORMATION SERVICE.)

AUTOMATIC TERMINAL INFORMATION SERVICE [ICAO] The provision of current, routine information to arriving and departing aircraft by means of continuous and repetitive broadcasts throughout the day or a specified portion of the day.

AUTOROTATION A rotorcraft flight condition in which the lifting rotor is driven entirely by action of the air when the rotorcraft is in motion.

1. Autorotative Landing/Touchdown Autorotation. Used by a pilot to indicate that he will be landing without applying power to the rotor.

2. Low Level Autorotation. Commences at an altitude well below the traffic pattern, usually below 100 feet AGL and is used primarily for tactical military training.

3. 180 degrees Autorotation. Initiated from a downwind heading and is commenced well inside the normal traffic pattern. "Go around" may not be possible during the latter part of this maneuver.

AVIATION WEATHER SERVICE A service provided by the National Weather Service (NWS) and FAA which collects and disseminates pertinent weather information for pilots, aircraft operators, and ATC. Available aviation weather reports and forecasts are displayed at each NWS office and FAA FSS.

(See EN ROUTE FLIGHT ADVISORY SERVICE.) (See TRANSCRIBED WEATHER BROADCAST.) (See WEATHER ADVISORY.) (See PILOTS AUTOMATIC TELEPHONE WEATHER ANSWERING SERVICE.) (Refer to AIM.)

AWW (See SEVERE WEATHER FORECAST ALERTS.)

AZIMUTH (MLS) A magnetic bearing extending from an MLS navigation facility.

Note: azimuth bearings are described as magnetic and are referred to as "azimuth" in radio telephone communications.

BASE LEG (See TRAFFIC PATTERN.)

BEACON (See RADAR.)

(See NONDIRECTIONAL BEACON.) (See MARKER BEACON.) (See AIRPORT ROTATING BEACON.) (See AERONAUTICAL BEACON.) (See AIRWAY BEACON.)

BEARING The horizontal direction to or from any point, usually measured clockwise from true north, magnetic north, or some other reference point through 360 degrees.

(See NONDIRECTIONAL BEACON.)

BELOW MINIMUMS Weather conditions below the minimums prescribed by regulation for the particular action involved; e.g., landing minimums, takeoff minimums.

BLAST FENCE A barrier that is used to divert or dissipate jet or propeller blast.

BLIND SPEED The rate of departure or closing of a target relative to the radar antenna at which cancellation of the primary radar target by moving target indicator (MTI)

circuits in the radar equipment causes a reduction or complete loss of signal.
(See ICAO term BLIND VELOCITY.)

BLIND SPOT An area from which radio transmissions and/or radar echoes cannot be received. The term is also used to describe portions of the airport not visible from the control tower.

BLIND TRANSMISSION (See TRANSMITTING IN THE BLIND.)

BLIND VELOCITY [ICAO] The radial velocity of a moving target such that the target is not seen on primary radars fitted with certain forms of fixed echo suppression.

BLIND ZONE (See BLIND SPOT.)

BLOCKED Phraseology used to indicate that a radio transmission has been distorted or interrupted due to multiple simultaneous radio transmissions.

BOUNDARY LIGHTS (See AIRPORT LIGHTING.)

BRAKING ACTION (GOOD, FAIR, POOR, OR NIL) A report of conditions on the airport movement area providing a pilot with a degree/quality of braking that he might expect. Braking action is reported in terms of good, fair, poor, or nil.
(See RUNWAY CONDITION READING.)

BRAKING ACTION ADVISORIES When tower controllers have received runway braking action reports which include the terms "poor" or "nil," or whenever weather conditions are conducive to deteriorating or rapidly changing runway braking conditions, the tower will include on the ATIS broadcast the statement, "BRAKING ACTION ADVISORIES ARE IN EFFECT." During the time Braking Action Advisories are in effect, ATC will issue the latest braking action report for the runway in use to each arriving and departing aircraft. Pilots should be prepared for deteriorating braking conditions and should request current runway condition information if not volunteered by controllers. Pilots should also be prepared to provide a descriptive runway condition report to controllers after landing.

BROADCAST Transmission of information for which an acknowledgement is not expected.
(See ICAO term BROADCAST.)

BROADCAST [ICAO] A transmission of information relating to air navigation that is not addressed to a specific station or stations.

CALCULATED LANDING TIME A term that may be used in place of tentative or actual calculated landing time, whichever applies.

CALL UP Initial voice contact between a facility and an aircraft, using the identification of the unit being called and the unit initiating the call.
(Refer to AIM.)

CALL FOR RELEASE Wherein the overlying ARTCC requires a terminal facility to initiate verbal coordination to secure ARTCC approval for release of a departure into the en route environment.

CANADIAN MINIMUM NAVIGATION PERFORMANCE SPECIFICATION AIRSPACE That portion of Canadian domestic airspace within which MNPS separation may be applied.

CARDINAL ALTITUDES "Odd" or "Even" thousand-foot altitudes or flight levels; e.g., 5,000, 6,000, 7,000, FL 250, FL 260, FL 270.
(See ALTITUDE.) (See FLIGHT LEVEL.)

CARDINAL FLIGHT LEVELS (See CARDINAL ALTITUDES.)

CAT (See CLEAR-AIR TURBULENCE.)

CDT PROGRAMS (See CONTROLLED DEPARTURE TIME PROGRAMS.)

CEILING The heights above the earth's surface of the lowest layer of clouds or obscuring phenomena that is reported as "broken," "overcast," or "obscuration," and not classified as "thin" or "partial."
(See ICAO term CEILING.)

CEILING [ICAO] The height above the ground or water of the base of the lowest layer of cloud below 6,000 metres (20,000 feet) covering more than half the sky.

CENRAP (See CENTER RADAR ARTS PRESENTATION/PROCESSING.)

CENRAP-PLUS (See CENTER RADAR ARTS PRESENTATION/PROCESSING-PLUS.)

CENTER (See AIR ROUTE TRAFFIC CONTROL CENTER.)

CENTER'S AREA The specified airspace within which an air route traffic control center (ARTCC) provides air traffic control and advisory service.
(See AIR ROUTE TRAFFIC CONTROL CENTER.) (Refer to AIM.)

CENTER RADAR ARTS PRESENTATION/PROCESSING A computer program developed to provide a back-up system for airport surveillance radar in the event of a failure or malfunction. The program uses air route traffic control center radar for the processing and presentation of data on the ARTS IIA or IIIA displays.

CENTER RADAR ARTS PRESENTATION/PROCESSING-PLUS A computer program developed to provide a back-up system for airport surveillance radar in the event of a terminal secondary radar system failure. The program uses a combination of Air Route Traffic Control Center Radar and terminal airport surveillance radar primary targets displayed simultaneously for the processing and presentation of data on the ARTS IIA or IIIA displays.

CENTER WEATHER ADVISORY
An unscheduled weather advisory issued by Center Weather Service Unit meteorologists for ATC use to alert pilots of existing or anticipated adverse weather conditions within the next 2 hours. A CWA may modify or redefine a SIGMET.
(See AWW.) (See SIGMET.) (See CONVECTIVE SIGMET.) (See AIRMET.) (Refer to AIM.)

CENTRAL EAST PACIFIC An organized route system between the U.S. West Coast and Hawaii.

CEP (See CENTRAL EAST PACIFIC.)

CERAP (See COMBINED CENTER-RAPCON.)

CFR (See CALL FOR RELEASE.)

CHAFF Thin, narrow metallic reflectors of various lengths and frequency responses, used to reflect radar energy. These reflectors when dropped from aircraft and allowed to drift downward result in large targets on the radar display.

CHARTED VFR FLYWAYS Charted VFR Flyways are flight paths recommended for use to bypass areas heavily traversed by large turbine-powered aircraft. Pilot compliance with recommended flyways and associated altitudes is strictly voluntary. VFR Flyway Planning charts are published on the back of existing VFR Terminal Area charts.

CHARTED VISUAL FLIGHT PROCEDURE APPROACH An approach conducted on an instrument flight rules (IFR) flight plan which authorizes the pilot of an aircraft under radar control to proceed visually and clear of clouds to the airport via visual landmarks and other information depicted on a charted visual flight procedure. This approach must be authorized and under the control of the appropriate air traffic control facility. Weather minimums required are depicted on the chart.

CHASE An aircraft flown in proximity to another aircraft normally to observe its performance during training or testing.

CHASE AIRCRAFT (See CHASE.)

CIRCLE-TO-LAND MANEUVER A maneuver initiated by the pilot to align the aircraft with a runway for landing when a straight-in landing from an instrument approach is not possible or is not desirable. This maneuver is made only after ATC authorization has been obtained and the pilot has established required visual reference to the airport.
(See CIRCLE TO RUNWAY.) (See LANDING MINIMUMS.) (Refer to AIM.)

CIRCLE TO RUNWAY (RUNWAY NUMBER) Used by ATC to inform the pilot that he must circle to land because the runway in use is other than the runway aligned with the instrument approach procedure. When the direction of the circling maneuver in relation to the airport/runway is required, the controller will state the direction (eight cardinal compass points) and specify a left or right downwind or base leg as appropriate; e.g., "Cleared VOR Runway Three Six Approach circle to Runway Two Two," or "Circle northwest of the airport for a right downwind to Runway Two Two."
(See CIRCLE-TO-LAND MANEUVER.) (See LANDING MINIMUMS.) (Refer to AIM.)

CIRCLING APPROACH (See CIRCLE-TO-LAND MANEUVER.)

CIRCLING MANEUVER (See CIRCLE-TO-LAND MANEUVER.)

CIRCLING MINIMA (See LANDING MINIMUMS.)

CLASS A AIRSPACE (See CONTROLLED AIRSPACE.)

CLASS B AIRSPACE (See CONTROLLED AIRSPACE.)

CLASS C AIRSPACE (See CONTROLLED AIRSPACE.)

CLASS D AIRSPACE (See CONTROLLED AIRSPACE.)

CLASS E AIRSPACE (See CONTROLLED AIRSPACE.)

CLASS G AIRSPACE That airspace not designated as Class A, B, C, D or E.

CLEAR-AIR TURBULENCE Turbulence encountered in air where no clouds are present. This term is commonly applied to high-level turbulence associated with wind shear. CAT is often encountered in the vicinity of the jet stream.
(See WIND SHEAR.) (See JET STREAM.)

CLEAR OF THE RUNWAY
1. A taxiing aircraft, which is approaching a runway, is clear of the runway when all parts of the aircraft are held short of the applicable holding position marking.
2. A pilot or controller may consider an aircraft, which is exiting or crossing a runway, to be clear of the runway when all parts of the aircraft are beyond the runway edge and there is no ATC restriction to its continued movement beyond the applicable holding position marking.
3. Pilots and controllers shall exercise good judgment to ensure that adequate separation exists between all aircraft on runways and taxiways at airports with inadequate runway edge lines or holding position markings.

CLEARANCE (See AIR TRAFFIC CLEARANCE.)

CLEARANCE LIMIT The fix, point, or location to which an aircraft is cleared when issued an air traffic clearance.
(See ICAO term CLEARANCE LIMIT.)

CLEARANCE LIMIT [ICAO] The point of which an aircraft is granted an air traffic control clearance.

CLEARANCE VOID IF NOT OFF BY (TIME) Used by ATC to advise an aircraft that the departure clearance is automatically canceled if takeoff is not made prior to a specified time. The pilot must obtain a new clearance or cancel his IFR flight plan if not off by the specified time.
(See ICAO term CLEARANCE VOID TIME.)

CLEARANCE VOID TIME [ICAO] A time specified by an air traffic control unit at which a clearance ceases to be valid unless the aircraft concerned has already taken action to comply therewith.

CLEARED AS FILED Means the aircraft is cleared to proceed in accordance with the route of flight filed in the flight plan. This clearance does not include the altitude, SID, or SID Transition.
(See REQUEST FULL ROUTE CLEARANCE.) (Refer to AIM.)

CLEARED (Type Of) APPROACH ATC authorization for an aircraft to execute a specific instrument approach procedure to an airport; e.g., "Cleared ILS Runway Three Six Approach."
(See INSTRUMENT APPROACH PROCEDURE.) (See APPROACH CLEARANCE.) (Refer to AIM.) (Refer to Part 91.)

CLEARED APPROACH ATC authorization for an aircraft to execute any standard or special instrument approach procedure for that airport. Normally, an aircraft will be cleared for a specific instrument approach procedure.
(See INSTRUMENT APPROACH PROCEDURE.) (See CLEARED (TYPE OF) APPROACH.) (Refer to AIM.) (Refer to Part 91.)

CLEARED FOR TAKEOFF ATC authorization for an aircraft to depart. It is predicated on known traffic and known physical airport conditions.

CLEARED FOR THE OPTION ATC authorization for an aircraft to make a touch-and-go, low approach, missed approach, stop and go, or full stop landing at the discretion of the pilot. It is normally used in training so that an instructor can evaluate a student's performance under changing situations.
(See OPTION APPROACH.) (Refer to AIM.)

CLEARED THROUGH ATC authorization for an aircraft to make intermediate stops at specified airports without refiling a flight plan while en route to the clearance limit.

CLEARED TO LAND ATC authorization for an aircraft to land. It is predicated on known traffic and known physical airport conditions.

CLEARWAY An area beyond the takeoff runway under the control of airport authorities within which terrain or fixed obstacles may not extend above specified limits. These areas may be required for certain turbine-powered operations and the size and upward slope of the clearway will differ depending on when the aircraft was certificated.
(Refer to Part 1.)

CLIMBOUT That portion of flight operation between takeoff and the initial cruising altitude.

CLIMB TO VFR ATC authorization for an aircraft to climb to VFR conditions within Class B, C, D, and E surface areas when the only weather limitation is restricted visibility. The aircraft must remain clear of clouds while climbing to VFR.
(See SPECIAL VFR.) (Refer to AIM.)

CLOSED RUNWAY A runway that is unusable for aircraft operations. Only the airport management/military operations office can close a runway.

CLOSED TRAFFIC Successive operations involving takeoffs and landings or low approaches where the aircraft does not exit the traffic pattern.

CLT (See CALCULATED LANDING TIME.)

CLUTTER In radar operations, clutter refers to the reception and visual display of radar returns caused by precipitation, chaff, terrain, numerous aircraft targets, or other phenomena. Such returns may limit or preclude ATC from providing services based on radar.
(See GROUND CLUTTER.) (See CHAFF.) (See PRECIPITATION.) (See TARGET.) (See ICAO term RADAR CLUTTER.)

CMNPS (See CANADIAN MINIMUM NAVIGATION PERFORMANCE SPECIFICATION AIRSPACE.)

COASTAL FIX A navigation aid or intersection where an aircraft transitions between the domestic route structure and the oceanic route structure.

CODES The number assigned to a particular multiple pulse reply signal transmitted by a transponder.
(See DISCRETE CODE.)

COMBINED CENTER-RAPCON An air traffic facility which combines the functions of an ARTCC and a radar approach control facility.
(See AIR ROUTE TRAFFIC CONTROL CENTER.) (See RADAR APPROACH CONTROL FACILITY.)

COMMON POINT A *significant point* over which two or more aircraft will report passing or have reported passing before proceeding on the same or diverging tracks. To establish/maintain longitudinal separation, a controller may determine a common point not originally in the aircraft's flight plan and then clear the aircraft to fly over the point. See significant point.

COMMON PORTION (See COMMON ROUTE.)

COMMON ROUTE That segment of a North American Route between the inland navigation facility and the coastal fix.

COMMON TRAFFIC ADVISORY FREQUENCY (CTAF) A frequency designed for the purpose of carrying out airport advisory practices while operating to or from an airport without an operating control tower. The CTAF may be a UNICOM, Multicom, FSS, or tower frequency and is identified in appropriate aeronautical publications.

COMPASS LOCATOR A low power, low or medium frequency (L/MF) radio beacon installed at the site of the outer or middle marker of an instrument landing system (ILS). It can be used for navigation at distances of approximately 15 miles or as authorized in the approach procedure.

1. Outer Compass Locator (LOM). A compass locator installed at the site of the outer marker of an instrument landing system.

(See OUTER MARKER.)

2. Middle Compass Locator (LMM). A compass locator installed at the site of the middle marker of an instrument landing system.

(See MIDDLE MARKER.) (See ICAO term LOCATOR.)

COMPASS ROSE A circle, graduated in degrees, printed on some charts or marked on the ground at an airport. It is used as a reference to either true or magnetic direction.

COMPOSITE FLIGHT PLAN A flight plan which specifies VFR operation for one portion of flight and IFR for another portion. It is used primarily in military operations.

(Refer to AIM.)

COMPOSITE ROUTE SYSTEM An organized oceanic route structure, incorporating reduced lateral spacing between routes, in which composite separation is authorized.

COMPOSITE SEPARATION A method of separating aircraft in a composite route system where, by management of route and altitude assignments, a combination of half the lateral minimum specified for the area concerned and half the vertical minimum is applied.

COMPULSORY REPORTING POINTS Reporting points which must be reported to ATC. They are designated on aeronautical charts by solid triangles or filed in a flight plan as fixes selected to define direct routes. These points are geographical locations which are defined by navigation aids/fixes. Pilots should discontinue position reporting over compulsory reporting points when informed by ATC that their aircraft is in "radar contact."

CONFLICT ALERT A function of certain air traffic control automated systems designed to alert radar controllers to existing or pending situations between tracked targets (known IFR or VFR aircraft) that require his immediate attention/action.

(See MODE C INTRUDER ALERT.)

CONFLICT RESOLUTION The resolution of potential conflictions between aircraft that are radar identified and in communication with ATC by ensuring that radar targets do not touch. Pertinent traffic advisories shall be issued when this procedure is applied.

Note: This procedure shall not be provided utilizing mosaic radar systems.

CONSOLAN A low frequency, long-distance NAVAID used principally for transoceanic navigations.

CONTACT

1. Establish communication with (followed by the name of the facility and, if appropriate, the frequency to be used).

2. A flight condition wherein the pilot ascertains the attitude of his aircraft and navigates by visual reference to the surface.

(See CONTACT APPROACH.) (See RADAR CONTACT.)

CONTACT APPROACH An approach wherein an aircraft on an IFR flight plan, having an air traffic control authorization, operating clear of clouds with at least 1 mile flight visibility and a reasonable expectation of continuing to the destination airport in those conditions, may deviate from the instrument approach procedure and proceed to the destination airport by visual reference to the surface. This approach will only be authorized when requested by the pilot and the reported ground visibility at the destination airport is at least 1 statute mile.

(Refer to AIM.)

CONTERMINOUS U.S. The 48 adjoining States and the District of Columbia.

CONTINENTAL UNITED STATES The 49 States located on the continent of North America and the District of Columbia.

CONTROL AREA [ICAO] A controlled airspace extending upwards from a specified limit above the earth.

CONTROLLED AIRSPACE An airspace of defined dimensions within which air traffic control service is provided to IFR flights and to VFR flights in accordance with the airspace classification.

Note 1: Controlled airspace is a generic term that covers Class A, Class B, Class C, Class D, and Class E airspace.

Note 2: Controlled airspace is also that airspace within which all aircraft operators are subject to certain pilot qualifications, operating rules, and equipment requirements in Part 91 (for specific operating requirements, please refer to Part 91). For IFR operations in any class of controlled airspace, a pilot must file an IFR flight plan and receive an appropriate ATC clearance. Each Class B, Class C, and Class D airspace area designated for an airport contains at least one primary airport around which the airspace is designated (for specific designations and descriptions of the airspace classes, please refer to Part 71).

Controlled airspace in the United States is designated as follows:

1. CLASS A: Generally, that airspace from 18,000 feet MSL up to and including FL600, including the airspace overlying the waters within 12 nautical miles of the coast of the 48 contiguous States and Alaska. Unless otherwise authorized, all persons must operate their aircraft under IFR.

2. CLASS B: Generally, that airspace from the surface to 10,000 feet MSL surrounding the nation's busiest airports in terms of airport operations or passenger enplanements. The configuration of each Class B airspace area is individually tailored and consists of a surface area and two or more layers (some Class B airspaces areas resemble upside-

down wedding cakes), and is designed to contain all published instrument procedures once an aircraft enters the airspace. An ATC clearance is required for all aircraft to operate in the area, and all aircraft that are so cleared receive separation services within the airspace. The cloud clearance requirement for VFR operations is "clear of clouds."

3. CLASS C: Generally, that airspace from the surface to 4,000 feet above the airport elevation (charted in MSL) surrounding those airports that have an operational control tower, are serviced by a radar approach control, and that have a certain number of IFR operations or passenger enplanements. Although the configuration of each Class C area is individually tailored, the airspace usually consists of a surface area with a 5 nautical mile (NM) radius, an outer circle with a 10 nm radius that extends from 1,200 feet to 4,000 feet above the airport elevation and an outer area. Each person must establish two-way radio communications with the ATC facility providing air traffic services prior to entering the airspace and thereafter maintain those communications while within the airspace. VFR aircraft are only separated from IFR aircraft within the airspace. *(See OUTER AREA.)*

4. CLASS D: Generally, that airspace from the surface to 2,500 feet above the airport elevation (charted in MSL) surrounding those airports that have an operational control tower. The configuration of each Class D airspace area is individually tailored and when instrument procedures are published, the airspace will normally be designed to contain the procedures. Arrival extensions for instrument approach procedures may be Class D or Class E airspace. Unless otherwise authorized, each person must establish two-way radio communications with the ATC facility providing air traffic services prior to entering the airspace and thereafter maintain those communications while in the airspace. No separation services are provided to VFR aircraft.

5. CLASS E: Generally, if the airspace is not Class A, Class B, Class C, or Class D, and it is controlled airspace, it is Class E airspace. Class E airspace extends upward from either the surface or a designated altitude to the overlying or adjacent controlled airspace. When designated as a surface area, the airspace will be configured to contain all instrument procedures. Also in this class are Federal airways, airspace beginning at either 700 or 1,200 feet AGL used to transition to/from the terminal or enroute environment, enroute domestic, and offshore airspace areas designated below 18,000 feet MSL. Unless designated at a lower altitude, Class E airspace begins at 14,500 MSL over the United States, including that airspace overlying the waters within 12 nautical miles of the coast of the 48 contiguous States and Alaska, up to, but not including 18,000 feet MSL, and the airspace above FL 600.

CONTROLLED AIRSPACE [ICAO] An airspace of defined dimensions within which air traffic control service is provided to IFR flights and to VFR flights in accordance with the airspace classification. *Note*-Controlled airspace is a generic term which covers ATS airspace Classes A, B, C, D, and E.

CONTROLLED DEPARTURE TIME PROGRAMS These programs are the flow control process whereby aircraft are held on the ground at the departure airport when delays are projected to occur in either the en route system or the terminal of intended landing. The purpose of these programs is to reduce congestion in the air traffic system or to limit the duration of airborne holding in the arrival center or terminal area. A CDT is a specific departure slot shown on the flight plan as an expected departure clearance time (EDCT).

CONTROLLED TIME OF ARRIVAL The original estimated time of arrival adjusted by the ATCSCC ground delay factor.

CONTROLLER (See AIR TRAFFIC CONTROL SPECIALIST.)

CONTROLLER [ICAO] A person authorized to provide air traffic control services.

CONTROL SECTOR An airspace area of defined horizontal and vertical dimensions for which a controller or group of controllers has air traffic control responsibility, normally within an air route traffic control center or an approach control facility. Sectors are established based on predominant traffic flows, altitude strata, and controller workload. Pilot-communications during operations within a sector are normally maintained on discrete frequencies assigned to the sector.
(See DISCRETE FREQUENCY.)

CONTROL SLASH A radar beacon slash representing the actual position of the associated aircraft. Normally, the control slash is the one closest to the interrogating radar beacon site. When ARTCC radar is operating in narrowband (digitized) mode, the control slash is converted to a target symbol.

CONVECTIVE SIGMET A weather advisory concerning convective weather significant to the safety of all aircraft. Convective SIGMET's are issued for tornadoes, lines of thunderstorms, embedded thunderstorms of any intensity level, areas of thunderstorms greater than or equal to VIP level 4 with an area coverage of $\frac{4}{10}$ (40%) or more, and hail ¾ inch or greater.
(See AWW.) (See SIGMET.) (See CWA.) (See AIRMET.) (Refer to AIM.)

CONVECTIVE SIGNIFICANT METEOROLOGICAL INFORMATION (See CONVECTIVE SIGMET.)

COORDINATES The intersection of lines of reference, usually expressed in degrees/minutes/seconds of latitude and longitude, used to determine position or location.

COORDINATION FIX The fix in relation to which facilities will handoff, transfer control of an aircraft, or coordinate flight progress data. For terminal facilities, it may also serve as a clearance for arriving aircraft.

COPTER (See HELICOPTER.)

CORRECTION An error has been made in the transmission and the correct version follows.

COUPLED APPROACH A coupled approach is an instrument approach performed by the aircraft autopilot which is receiving position information and/or steering commands from onboard navigation equipment. In general, coupled nonprecision approaches must be discontinued and flown manually at altitudes lower than 50 feet below the minimum descent altitude, and coupled precision approaches must be flown manually below 50 feet ALG.

(See AUTOLAND APPROACH.)

Note: Coupled and autoland approaches are flown in VFR and IFR. It is common for carriers to require their crews to fly coupled approaches and autoland approaches (if certified) when the weather conditions are less than approximately 4,000 RVR.

COURSE

1. The intended direction of flight in the horizontal plane measured in degrees from north.
2. The ILS localizer signal pattern usually specified as the front course or the back course.
3. The intended track along a straight, curved, or segmented MLS path.

(See BEARING.) (See RADIAL.) (See INSTRUMENT LANDING SYSTEM.) (See MICROWAVE LANDING SYSTEM.)

CPL [ICAO] (See CURRENT FLIGHT PLAN.)

CRITICAL ENGINE The engine which, upon failure, would most adversely affect the performance or handling qualities of an aircraft.

CROSS (FIX) AT (ALTITUDE) Used by ATC when a specific altitude restriction at a specified fix is required.

CROSS (FIX) AT OR ABOVE (ALTITUDE) Used by ATC when an altitude restriction at a specified fix is required. It does not prohibit the aircraft from crossing the fix at a higher altitude than specified; however, the higher altitude may not be one that will violate a succeeding altitude restriction or altitude assignment.

(See ALTITUDE RESTRICTION.) (Refer to AIM.)

CROSS (FIX) AT OR BELOW (ALTITUDE) Used by ATC when a maximum crossing altitude at a specific fix is required. It does not prohibit the aircraft from crossing the fix at a lower altitude; however, it must be at or above the minimum IFR altitude.

(See MINIMUM IFR ALTITUDES.) (See ALTITUDE RESTRICTION.) (Refer to Part 91.)

CROSSWIND

1. When used concerning the traffic pattern, the word means "crosswind leg."

(See TRAFFIC PATTERN.)

2. When used concerning wind conditions, the word means a wind not parallel to the runway or the path of an aircraft.

(See CROSSWIND COMPONENT.)

CROSSWIND COMPONENT The wind component measured in knots at 90 degrees to the longitudinal axis of the runway.

CRUISE Used in an ATC clearance to authorize a pilot to conduct flight at any altitude from the minimum IFR altitude up to and including the altitude specified in the clearance. The pilot may level off at any intermediate altitude within this block of airspace. Climb/descent within the block is to be made at the discretion of the pilot. However, once the pilot starts descent and verbally reports leaving an altitude in the block, he may not return to that altitude without additional ATC clearance. Further, it is approval for the pilot to proceed to and make an approach at destination airport and can be used in conjunction with:

1. An airport clearance limit at locations with a standard/special instrument approach procedure. The FAR's require that if an instrument letdown to an airport is necessary, the pilot shall make the letdown in accordance with a standard/special instrument approach procedure for that airport, or
2. An airport clearance limit at locations that are within/below/outside controlled airspace and without a standard/special instrument approach procedure. Such a clearance is NOT AUTHORIZATION for the pilot to descend under IFR conditions below the applicable minimum IFR altitude nor does it imply that ATC is exercising control over aircraft in Class G airspace; however, it provides a means for the aircraft to proceed to destination airport, descend, and land in accordance with applicable FAR's governing VFR flight operations. Also, this provides search and rescue protection until such time as the IFR flight plan is closed.

(See INSTRUMENT APPROACH PROCEDURE.)

CRUISING ALTITUDE An altitude or flight level maintained during en route level flight. This is a constant altitude and should not be confused with a cruise clearance.

(See ALTITUDE.) (See ICAO term CRUISING LEVEL.)

CRUISING LEVEL [ICAO] A level maintained during a significant portion of a flight.

CRUISE CLIMB A climb technique employed by aircraft, usually at a constant power setting, resulting in an increase of altitude as the aircraft weight decreases.

CRUISING LEVEL (See CRUISING ALTITUDE.)

CT MESSAGE An EDCT time generated by the ATCSCC to regulate traffic at arrival airports. Normally, a CT message is automatically transferred from the Traffic Management System computer to the NAS en route computer and appears as an EDCT. In the event of a communication failure between the TMS and the NAS, the CT message can be manually entered by the TMC at the en route facility.

CTA (See CONTROLLED TIME OF ARRIVAL.)

CTA (See CONTROL AREA [ICAO].)

CTAF (See COMMON TRAFFIC ADVISORY FREQUENCY.)

CURRENT FLIGHT PLAN [ICAO] The flight plan, including changes, if any, brought about by subsequent clearances.

CVFP APPROACH (See CHARTED VISUAL FLIGHT PROCEDURE APPROACH.)

CWA (See CENTER WEATHER ADVISORY.)
(See WEATHER ADVISORY.)

DA [ICAO] (See ICAO Term DECISION ALTITUDE/ DECISION HEIGHT.)

DAIR (See DIRECT ALTITUDE AND IDENTITY READOUT.)

DANGER AREA [ICAO] An airspace of defined dimensions within which activities dangerous to the flight of aircraft may exist at specified times.

Note: The term "Danger Area" is not used in reference to areas within the United States or any of its possessions or territories.

DATA BLOCK (See ALPHANUMERIC DISPLAY.)

DEAD RECKONING Dead reckoning, as applied to flying, is the navigation of an airplane solely by means of computations based on airspeed, course, heading, wind direction, and speed, groundspeed, and elapsed time.

DECISION ALTITUDE/DECISION HEIGHT [ICAO] A specified altitude or height (A/H) in the precision approach at which a missed approach must be initiated if the required visual reference to continue the approach has not been established.

Note 1: Decision altitude [DA] is referenced to mean sea level [MSL] and decision height [DH] is referenced to the threshold elevation.

Note 2: The required visual reference means that section of the visual aids or of the approach area which should have been in view for sufficient time for the pilot to have made an assessment of the aircraft position and rate of change of position, in relation to the desired flight path.

DECISION HEIGHT With respect to the operation of aircraft, means the height at which a decision must be made during an ILS, MLS, or PAR instrument approach to either continue the approach or to execute a missed approach.
(See ICAO term DECISION ALTITUDE/DECISION HEIGHT.)

DECODER The device used to decipher signals received from ATCRBS transponders to effect their display as select codes.
(See CODES.) (See RADAR.)

DEFENSE VISUAL FLIGHT RULES Rules applicable to flights within an ADIZ conducted under the visual flight rules in Part 91.
(See AIR DEFENSE IDENTIFICATION ZONE.) (Refer to Part 91.) (Refer to Part 99.)

DELAY INDEFINITE (REASON IF KNOWN) EXPECT FURTHER CLEARANCE (TIME) Used by ATC to inform a pilot when an accurate estimate of the delay time and the reason for the delay cannot immediately be determined; e.g., a disabled aircraft on the runway, terminal or center area saturation, weather below landing minimums, etc.
(See EXPECT FURTHER CLEARANCE (TIME).)

DELAY TIME The amount of time that the arrival must lose to cross the meter fix at the assigned meter fix time. This is the difference between ACLT and VTA.

DEPARTURE CENTER The ARTCC having jurisdiction for the airspace that generates a flight to the impacted airport.

DEPARTURE CONTROL A function of an approach control facility providing air traffic control service for departing IFR and, under certain conditions, VFR aircraft.
(See APPROACH CONTROL FACILITY.) (Refer to AIM.)

DEPARTURE SEQUENCING PROGRAM A program designed to assist in achieving a specified interval over a common point for departures.

DEPARTURE TIME The time an aircraft becomes airborne.

DESCENT SPEED ADJUSTMENTS Speed deceleration calculations made to determine an accurate VTA. These calculations start at the transition point and use arrival speed segments to the vertex.

DETRESFA (DISTRESS PHASE) [ICAO] The code word used to designate an emergency phase wherein there is reasonable certainty that an aircraft and its occupants are threatened by grave and imminent danger or require immediate assistance.

DEVIATIONS

1. A departure from a current clearance, such as an off course maneuver to avoid weather or turbulence.
2. Where specifically authorized in the FAR's and requested by the pilot, ATC may permit pilots to deviate from certain regulations.

(Refer to AIM.)

DF (See DIRECTION FINDER.)

DF APPROACH PROCEDURE Used under emergency conditions where another instrument approach procedure cannot be executed. DF guidance for an instrument approach is given by ATC facilities with DF capability.

(See DF GUIDANCE.) (See DIRECTION FINDER.) (Refer to AIM.)

DF FIX The geographical location of an aircraft obtained by one or more direction finders.

(See DIRECTION FINDER.)

DF GUIDANCE Headings provided to aircraft by facilities equipped with direction finding equipment. These headings, if followed, will lead the aircraft to a predetermined point such as the DF station or an airport. DF guidance is given to aircraft in distress or to other aircraft which request the service. Practice DF guidance is provided when workload permits.

(See DIRECTION FINDER.) (See DF FIX.) (Refer to AIM.)

DF STEER (See DF GUIDANCE.)

DH (See DECISION HEIGHT.)

DH [ICAO] (See ICAO Term DECISION ALTITUDE/DECISION HEIGHT.)

DIRECT Straight line flight between two navigational aids, fixes, points, or any combination thereof. When used by pilots in describing off-airway routes, points defining direct route segments become compulsory reporting points unless the aircraft is under radar contact.

DIRECT ALTITUDE AND IDENTITY READOUT The DAIR System is a modification to the AN/TPX-42 Interrogator System. The Navy has two adaptations of the DAIR System-Carrier Air Traffic Control Direct Altitude and Identification Readout System for Aircraft Carriers and Radar Air Traffic Control Facility Direct Altitude and Identity Readout System for land-based terminal operations. The DAIR detects, tracks, and predicts secondary radar aircraft targets. Targets are displayed by means of computer-generated symbols and alphanumeric characters depicting flight identification, altitude, ground speed, and flight plan data. The DAIR System is capable of interfacing with ARTCC's.

DIRECTION FINDER A radio receiver equipped with a directional sensing antenna used to take bearings on a radio transmitter. Specialized radio direction finders are used in aircraft as air navigation aids. Others are ground-based, primarily to obtain a "fix" on a pilot requesting orientation assistance or to locate downed aircraft. A location "fix" is established by the intersection of two or more bearing lines plotted on a navigational chart using either two separately located Direction Finders to obtain a fix on an aircraft or by a pilot plotting the bearing indications of his DF on two separately located ground-based transmitters, both of which can be identified on his chart. UDF's receive signals in the ultra high frequency radio broadcast band; VDF's in the very high frequency band; and UVDF's in both bands. ATC provides DF service at those air traffic control towers and flight service stations listed in the Airport/Facility Directory and the DOD FLIP IFR En Route Supplement.

(See DF GUIDANCE.) (See DF FIX.)

DISCRETE BEACON CODE (See DISCRETE CODE).

DISCRETE CODE As used in the Air Traffic Control Radar Beacon System (ATCRBS), any one of the 4096 selectable Mode 3/A aircraft transponder codes except those ending in zero zero; e.g., discrete codes: 0010, 1201, 2317, 7777; nondiscrete codes: 0100, 1200, 7700. Nondiscrete codes are normally reserved for radar facilities that are not equipped with discrete decoding capability and for other purposes such as emergencies (7700), VFR aircraft (1200), etc.

(See RADAR.) (Refer to AIM.)

DISCRETE FREQUENCY A separate radio frequency for use in direct pilot-controller communications in air traffic control which reduces frequency congestion by controlling the number of aircraft operating on a particular frequency at one time. Discrete frequencies are normally designated for each control sector in en route/terminal ATC facilities. Discrete frequencies are listed in the Airport/Facility Directory and the DOD FLIP IFR En Route Supplement.

(See CONTROL SECTOR.)

DISPLACED THRESHOLD A threshold that is located at a point on the runway other than the designated beginning of the runway.

(See THRESHOLD.) (Refer to AIM.)

DISTANCE MEASURING EQUIPMENT Equipment (airborne and ground) used to measure, in nautical miles, the slant range distance of an aircraft from the DME navigational aid.

(See TACAN.) (See VORTAC.) (See MICROWAVE LANDING SYSTEM.)

DISTRESS A condition of being threatened by serious and/or imminent danger and of requiring immediate assistance.

DIVE BRAKES (See SPEED BRAKES.)

DIVERSE VECTOR AREA In a radar environment, that area in which a prescribed departure route is not required as the only suitable route to avoid obstacles. The area in which random radar vectors below the MVA/MIA, established in accordance with the TERPS criteria for diverse departures obstacles and terrain avoidance, may be issued to departing aircraft.

DME (See DISTANCE MEASURING EQUIPMENT.)

DME FIX A geographical position determined by reference to a navigational aid which provides distance and azimuth information. It is defined by a specific distance in nautical miles and a radial, azimuth, or course (i.e., localizer) in degrees magnetic from that aid.

(See DISTANCE MEASURING EQUIPMENT.) (See FIX.) (See MICROWAVE LANDING SYSTEM.)

DME SEPARATION Spacing of aircraft in terms of distances (nautical miles) determined by reference to distance measuring equipment (DME).
(See DISTANCE MEASURING EQUIPMENT.)

DOD FLIP Department of Defense Flight Information Publications used for flight planning, en route, and terminal operations. FLIP is produced by the Defense Mapping Agency for world-wide use. United States Government Flight Information Publications (en route charts and instrument approach procedure charts) are incorporated in DOD FLIP for use in the National Airspace System (NAS).

DOMESTIC AIRSPACE Airspace which overlies the continental land mass of the United States plus Hawaii and U.S. possessions. Domestic airspace extends to 12 miles offshore.

DOWNBURST A strong downdraft which induces an outburst of damaging winds on or near the ground. Damaging winds, either straight or curved, are highly divergent. The sizes of downbursts vary from ½ mile or less to more than 10 miles. An intense downburst often causes widespread damage. Damaging winds, lasting 5 to 30 minutes, could reach speeds as high as 120 knots.

DOWNWIND LEG (See TRAFFIC PATTERN.)

DRAG CHUTE A parachute device installed on certain aircraft which is deployed on landing roll to assist in deceleration of the aircraft.

DSP (See DEPARTURE SEQUENCING PROGRAM.)

DT (See DELAY TIME.)

DUE REGARD A phase of flight wherein an aircraft commander of a State-operated aircraft assumes responsibility to separate his aircraft from all other aircraft.
(See also Chapter 1, Word Meanings.)

DUTY RUNWAY (See RUNWAY IN USE/ACTIVE RUNWAY/DUTY RUNWAY.)

DVA (See DIVERSE VECTOR AREA.)

DVFR (See DEFENSE VISUAL FLIGHT RULES.)

DVFR FLIGHT PLAN A flight plan filed for a VFR aircraft which intends to operate in airspace within which the ready identification, location, and control of aircraft are required in the interest of national security.

DYNAMIC Continuous review, evaluation, and change to meet demands.

DYNAMIC RESTRICTIONS Those restrictions imposed by the local facility on an "as needed" basis to manage unpredictable fluctuations in traffic demands.

EARTS (See EN ROUTE AUTOMATED RADAR TRACKING SYSTEM.)

EDCT (See EXPECTED DEPARTURE CLEARANCE TIME.)

EFC (See EXPECT FURTHER CLEARANCE (TIME).)

ELT (See EMERGENCY LOCATOR TRANSMITTER.)

EMERGENCY A *distress* or an *urgency* condition.

EMERGENCY LOCATOR TRANSMITTER A radio transmitter attached to the aircraft structure which operates from its own power source on 121.5 MHz and 243.0 MHz. It aids in locating downed aircraft by radiating a downward sweeping audio tone, 2–4 times per second. It is designed to function without human action after an accident.
(Refer to Part 91.3.) (Refer to AIM.)

EMERGENCY SAFE ALTITUDE (See MINIMUM SAFE ALTITUDE.)

E-MSAW (See EN ROUTE MINIMUM SAFE ALTITUDE WARNING.)

ENTRY POINT The point at which an aircraft transitions from an offshore control area to oceanic airspace.

ENGINEERED PERFORMANCE STANDARDS A mathematically derived runway capacity standard. EPS's are calculated for each airport on an individual basis and reflect that airport's aircraft mix, operating procedures, runway layout, and specific weather conditions. EPS's do not give consideration to staffing, experience levels, equipment outages, and in-trail restrictions as does the AAR.

EN ROUTE AIR TRAFFIC CONTROL SERVICES Air traffic control service provided aircraft on IFR flight plans, generally by centers, when these aircraft are operating between departure and destination terminal areas. When equipment, capabilities, and controller workload permit, certain advisory/assistance services may be provided to VFR aircraft.
(See NAS STAGE A.) (See AIR ROUTE TRAFFIC CONTROL CENTER.) (Refer to AIM.)

EN ROUTE AUTOMATED RADAR TRACKING SYSTEM An automated radar and radar beacon tracking system. Its functional capabilities and design are essentially the same as the terminal ARTS IIIA system except for the EARTS capability of employing both short-range (ASR) and long-range (ARSR) radars, use of full digital radar displays, and fail-safe design.
(See AUTOMATED RADAR TERMINAL SYSTEMS.)

EN ROUTE CHARTS (See AERONAUTICAL CHART.)

EN ROUTE DESCENT Descent from the en route cruising altitude which takes place along the route of flight.

EN ROUTE FLIGHT ADVISORY SERVICE A service specifically designed to provide, upon pilot request, timely weather information pertinent to his type of flight, in-

tended route of flight, and altitude. The FSS's providing this service are listed in the Airport/Facility Directory.
(See FLIGHT WATCH.) (Refer to AIM.)

EN ROUTE HIGH ALTITUDE CHARTS (See AERONAUTICAL CHART.)

EN ROUTE LOW ALTITUDE CHARTS (See AERONAUTICAL CHART.)

EN ROUTE MINIMUM SAFE ALTITUDE WARNING A function of the NAS Stage A en route computer that aids the controller by alerting him when a tracked aircraft is below or predicted by the computer to go below a predetermined minimum IFR altitude (MIA).

EN ROUTE SPACING PROGRAM A program designed to assist the exit sector in achieving the required in-trail spacing.

EPS (See ENGINEERED PERFORMANCE STANDARDS.)

ESP (See EN ROUTE SPACING PROGRAM.)

ESTIMATED ELAPSED TIME [ICAO] The estimated time required to proceed from one significant point to another. (See ICAO Term TOTAL ESTIMATED ELAPSED TIME).

ESTIMATED OFF-BLOCK TIME [ICAO] The estimated time at which the aircraft will commence movement associated with departure.

ESTIMATED TIME OF ARRIVAL The time the flight is estimated to arrive at the gate (scheduled operators) or the actual runway on times for nonscheduled operators.

ESTIMATED TIME EN ROUTE The estimated flying time from departure point to destination (lift-off to touchdown).

ETA (See ESTIMATED TIME OF ARRIVAL.)

ETE (See ESTIMATED TIME EN ROUTE.)

EXECUTE MISSED APPROACH Instructions issued to a pilot making an instrument approach which means continue inbound to the missed approach point and execute the missed approach procedure as described on the Instrument Approach Procedure Chart or as previously assigned by ATC. The pilot may climb immediately to the altitude specified in the missed approach procedure upon making a missed approach. No turns should be initiated prior to reaching the missed approach point. When conducting an ASR or PAR approach, execute the assigned missed approach procedure immediately upon receiving instructions to "execute missed approach."
(Refer to AIM.)

EXPECT (ALTITUDE) AT (TIME) or (FIX) Used under certain conditions to provide a pilot with an altitude to be used in the event of two-way communications failure. It also provides altitude information to assist the pilot in planning.
(Refer to AIM.)

EXPECTED DEPARTURE CLEARANCE TIME The runway release time assigned to an aircraft in a controlled departure time program and shown on the flight progress strip as an EDCT.

EXPECT FURTHER CLEARANCE (TIME) The time a pilot can expect to receive clearance beyond a clearance limit.

EXPECT FURTHER CLEARANCE VIA (AIRWAYS, ROUTES OR FIXES) Used to inform a pilot of the routing he can expect if any part of the route beyond a short range clearance limit differs from that filed.

EXPEDITE Used by ATC when prompt compliance is required to avoid the development of an imminent situation.

FAF (See FINAL APPROACH FIX.)

FA MESSAGE The data entered into the ARTCC computers that activates delay processing for an impacted airport. The FA data includes the delay factor for flight plans that have not been assigned delays under CT message processing. The delay factor appears on flight progress strips in the form of an EDCT (e.g., EDCT 1820). FA processing assigns delays in 15-minute time blocks. FA's control numbers of aircraft within a specified time block but do not spread aircraft out evenly throughout the time block.

FAP (See FINAL APPROACH POINT.)

FAST FILE A system whereby a pilot files a flight plan via telephone that is tape recorded and then transcribed for transmission to the appropriate air traffic facility. Locations having a fast file capability are contained in the Airport/Facility Directory.
(Refer to AIM.)

FCLT (See FREEZE CALCULATED LANDING TIME.)

FEATHERED PROPELLER A propeller whose blades have been rotated so that the leading and trailing edges are nearly parallel with the aircraft flight path to stop or minimize drag and engine rotation. Normally used to indicate shutdown of a reciprocating or turboprop engine due to malfunction.

FEDERAL AIRWAYS (See LOW ALTITUDE AIRWAY STRUCTURE.)

FEEDER FIX The fix depicted on Instrument Approach Procedure Charts which establishes the starting point of the feeder route.

FEEDER ROUTE A route depicted on instrument approach procedure charts to designate routes for aircraft to

proceed from the en route structure to the initial approach fix (IAF).
(See INSTRUMENT APPROACH PROCEDURE.)

FERRY FLIGHT A flight for the purpose of:
1. Returning an aircraft to base.
2. Delivering an aircraft from one location to another.
3. Moving an aircraft to and from a maintenance base.
Ferry flights, under certain conditions, may be conducted under terms of a special flight permit.

FIELD ELEVATION (See AIRPORT ELEVATION.)

FILED Normally used in conjunction with flight plans, meaning a flight plan has been submitted to ATC.

FILED EN ROUTE DELAY Any of the following preplanned delays at points/areas along the route of flight which require special flight plan filing and handling techniques.
1. Terminal Area Delay. A delay within a terminal area for touch-and-go, low approach, or other terminal area activity.
2. Special Use Airspace Delay. A delay within a Military Operating Area, Restricted Area, Warning Area, or ATC Assigned Airspace.
3. Aerial Refueling Delay. A delay within an Aerial Refueling Track or Anchor.

FILED FLIGHT PLAN The flight plan as filed with an ATS unit by the pilot or his designated representative without any subsequent changes or clearances.

FINAL Commonly used to mean that an aircraft is on the final approach course or is aligned with a landing area.
(See FINAL APPROACH COURSE.) (See FINAL APPROACH-IFR). (See TRAFFIC PATTERN.) (See SEGMENTS OF AN INSTRUMENT APPROACH PROCEDURE.)

FINAL APPROACH [ICAO] That part of an instrument approach procedure which commences at the specified final approach fix or point, or where such a fix or point is not specified,
a) At the end of the last procedure turn, base turn or inbound turn of a racetrack procedure, if specified; or
b) At the point of interception of the last track specified in the approach procedure; and ends at a point in the vicinity of an aerodrome from which:
1) A landing can be made; or
2) A missed approach procedure is initiated.

FINAL APPROACH COURSE A published MLS course, a straight line extension of a localizer, a final approach radial/bearing, or a runway centerline all without regard to distance.
(See FINAL APPROACH-IFR.) (See TRAFFIC PATTERN.)

FINAL APPROACH FIX The fix from which the final approach (IFR) to an airport is executed and which identifies the beginning of the final approach segment. It is designated on Government charts by the Maltese Cross symbol for nonprecision approaches and the lightning bolt symbol for precision approaches; or when ATC directs a lower-than-published Glideslope/path Intercept Altitude, it is the resultant actual point of the glideslope/path intercept.
(See FINAL APPROACH POINT.) (See GLIDESLOPE INTERCEPT ALTITUDE.) (See SEGMENTS OF AN INSTRUMENT APPROACH PROCEDURE.)

FINAL APPROACH-IFR The flight path of an aircraft which is inbound to an airport on a final instrument approach course, beginning at the final approach fix or point and extending to the airport or the point where a circle-to-land maneuver or a missed approach is executed.
(See SEGMENTS OF AN INSTRUMENT APPROACH PROCEDURE.) (See FINAL APPROACH FIX.) (See FINAL APPROACH COURSE.) (See FINAL APPROACH POINT.) (See ICAO term FINAL APPROACH.)

FINAL APPROACH POINT The point, applicable only to a nonprecision approach with no depicted FAF (such as an on-airport VOR), where the aircraft is established inbound on the final approach course from the procedure turn and where the final approach descent may be commenced. The FAP serves as the FAF and identifies the beginning of the final approach segment.
(See FINAL APPROACH FIX.) (See SEGMENTS OF AN INSTRUMENT APPROACH PROCEDURE.)

FINAL APPROACH SEGMENT (See SEGMENTS OF AN INSTRUMENT APPROACH PROCEDURE.)

FINAL APPROACH SEGMENT [ICAO] That segment of an instrument approach procedure in which alignment and descent for landing are accomplished.

FINAL APPROACH-VFR (See TRAFFIC PATTERN.)

FINAL CONTROLLER The controller providing information and final approach guidance during PAR and ASR approaches utilizing radar equipment.
(See RADAR APPROACH.)

FINAL MONITOR AID A high resolution color display that is equipped with the controller alert system hardware/software which is used in the precision runway monitor (PRM) system. The display includes alert algorithms providing the target predictors, a color change alert when a target penetrates or is predicted to penetrate the no transgression zone (NTZ), a color change alert if the aircraft transponder becomes inoperative, synthesized voice alerts, digital mapping, and like features contained in the PRM system.
(See RADAR APPROACH.)

FIR (See FLIGHT INFORMATION REGION.)

FIRST TIER CENTER The ARTCC immediately adjacent to the impacted center.

FIX A geographical position determined by visual reference to the surface, by reference to one or more radio

NAVAIDs, by celestial plotting, or by another navigational device.

FIX BALANCING A process whereby aircraft are evenly distributed over several available arrival fixes reducing delays and controller workload.

FLAG A warning device incorporated in certain airborne navigation and flight instruments indicating that:

1. Instruments are inoperative or otherwise not operating satisfactorily, or

2. Signal strength or quality of the received signal falls below acceptable values.

FLAG ALARM (See FLAG.)

FLAMEOUT Unintended loss of combustion in turbine engines resulting in the loss of engine power.

FLIGHT CHECK A call-sign prefix used by FAA aircraft engaged in flight inspection/certification of navigational aids and flight procedures. The word "recorded" may be added as a suffix; e.g., "Flight Check 320 recorded" to indicate that an automated flight inspection is in progress in terminal areas.

(See FLIGHT INSPECTION.) (Refer to AIM.)

FLIGHT FOLLOWING (See TRAFFIC ADVISORIES.)

FLIGHT INFORMATION REGION An airspace of defined dimensions within which Flight Information Service and Alerting Service are provided.

1. Flight Information Service. A service provided for the purpose of giving advice and information useful for the safe and efficient conduct of flights.

2. Alerting Service. A service provided to notify appropriate organizations regarding aircraft in need of search and rescue aid and to assist such organizations as required.

FLIGHT INFORMATION SERVICE A service provided for the purpose of giving advice and information useful for the safe and efficient conduct of flights.

FLIGHT INSPECTION Inflight investigation and evaluation of a navigational aid to determine whether it meets established tolerances.

(See NAVIGATIONAL AID.) (See FLIGHT CHECK.)

FLIGHT LEVEL A level of constant atmospheric pressure related to a reference datum of 29.92 inches of mercury. Each is stated in three digits that represent hundreds of feet. For example, flight level 250 represents a barometric altimeter indication of 25,000 feet; flight level 255, an indication of 25,500 feet.

(See ICAO term FLIGHT LEVEL.)

FLIGHT LEVEL [ICAO] A surface of constant atmospheric pressure which is related to a specific pressure datum, 1013.2 hPa (1013.2 mb), and is separated from other such surfaces by specific pressure intervals.

Note 1: A pressure type altimeter calibrated in accordance with the standard atmosphere:

a) When set to a QNH altimeter setting, will indicate altitude;

b) When set to a QFE altimeter setting, will indicate height above the QFE reference datum; and

c) When set to a pressure of 1013.2 hPa (1013.2 mb), may be used to indicate flight levels.

Note 2: The terms 'height' and 'altitude,' used in Note 1 above, indicate altimetric rather than geometric heights and altitudes.

FLIGHT LINE A term used to describe the precise movement of a civil photogrammetric aircraft along a predetermined course(s) at a predetermined altitude during the actual photographic run.

FLIGHT MANAGEMENT SYSTEMS A computer system that uses a large data base to allow routes to be preprogrammed and fed into the system by means of a data loader. The system is constantly updated with respect to position accuracy by reference to conventional navigation aids. The sophisticated program and its associated data base insures that the most appropriate aids are automatically selected during the information update cycle.

FLIGHT MANAGEMENT SYSTEM PROCEDURE An arrival, departure, or approach procedure developed for use by aircraft with a slant (/) G equipment suffix.

FLIGHT PATH A line, course, or track along which an aircraft is flying or intended to be flown.

(See TRACK.) (See COURSE.)

FLIGHT PLAN Specified information relating to the intended flight of an aircraft that is filed orally or in writing with an FSS or an ATC facility.

(See FAST FILE.) (See FILED.) (Refer to AIM.)

FLIGHT PLAN AREA The geographical area assigned by regional air traffic divisions to a flight service station for the purpose of search and rescue for VFR aircraft, issuance of notams, pilot briefing, in-flight services, broadcast, emergency services, flight data processing, international operations, and aviation weather services. Three letter identifiers are assigned to every flight service station and are annotated in AFD's and Order 7350.6 as tie-in-facilities.

(See FAST FILE.) (See FILED.) (Refer to AIM.)

FLIGHT RECORDER A general term applied to any instrument or device that records information about the performance of an aircraft in flight or about conditions encountered in flight. Flight recorders may make records of airspeed, outside air temperature, vertical acceleration, engine RPM, manifold pressure, and other pertinent variables for a given flight.

(See ICAO term FLIGHT RECORDER.)

FLIGHT RECORDER [ICAO] Any type of recorder installed in the aircraft for the purpose of complementing accident/incident investigation.

Note: See Annex 6 Part I, for specifications relating to flight recorders.

FLIGHT SERVICE STATION Air traffic facilities which provide pilot briefing, en route communications and VFR search and rescue services, assist lost aircraft and aircraft in emergency situations, relay ATC clearances, originate Notices to Airmen, broadcast aviation weather and NAS information, receive and process IFR flight plans, and monitor NAVAID's. In addition, at selected locations, FSS's provide Enroute Flight Advisory Service (Flight Watch), take weather observations, issue airport advisories, and advise Customs and Immigration of transborder flights.
(Refer to AIM.)

FLIGHT STANDARDS DISTRICT OFFICE
An FAA field office serving an assigned geographical area and staffed with Flight Standards personnel who serve the aviation industry and the general public on matters relating to the certification and operation of air carrier and general aviation aircraft. Activities include general surveillance of operational safety, certification of airmen and aircraft, accident prevention, investigation, enforcement, etc.

FLIGHT TEST A flight for the purpose of:

1. Investigating the operation/flight characteristics of an aircraft or aircraft component.
2. Evaluating an applicant for a pilot certificate or rating.

FLIGHT VISIBILITY (See VISIBILITY.)

FLIGHT WATCH A shortened term for use in air-ground contacts to identify the flight service station providing En Route Flight Advisory Service; e.g., "Oakland Flight Watch."
(See EN ROUTE FLIGHT ADVISORY SERVICE.)

FLIP (See DOD FLIP.)

FLOW CONTROL Measures designed to adjust the flow of traffic into a given airspace, along a given route, or bound for a given aerodrome (airport) so as to ensure the most effective utilization of the airspace.
(See QUOTA FLOW CONTROL.) (Refer to AIRPORT/FACILITY DIRECTORY.)

FLY HEADING (DEGREES) Informs the pilot of the heading he should fly. The pilot may have to turn to, or continue on, a specific compass direction in order to comply with the instructions. The pilot is expected to turn in the shorter direction to the heading unless otherwise instructed by ATC.

FMA (See FINAL MONITOR AID.)

FMS (See FLIGHT MANAGEMENT SYSTEM.)

FMSP (See FLIGHT MANAGEMENT SYSTEM PROCEDURE.)

FORMATION FLIGHT More than one aircraft which, by prior arrangement between the pilots, operate as a single aircraft with regard to navigation and position reporting. Separation between aircraft within the formation is the responsibility of the flight leader and the pilots of the other aircraft in the flight. This includes transition periods when aircraft within the formation are maneuvering to attain separation from each other to effect individual control and during join-up and breakaway.

1. A standard formation is one in which a proximity of no more than 1 mile laterally or longitudinally and within 100 feet vertically from the flight leader is maintained by each wingman.
2. Nonstandard formations are those operating under any of the following conditions:
   a. When the flight leader has requested and ATC has approved other than standard formation dimensions.
   b. When operating within an authorized altitude reservation (ALTRV) or under the provisions of a letter of agreement.
   c. When the operations are conducted in airspace specifically designed for a special activity.

(See ALTITUDE RESERVATION.) (Refer to Part 91.)

FRC (See REQUEST FULL ROUTE CLEARANCE.)

FREEZE/FROZEN Terms used in referring to arrivals which have been assigned ACLTs and to the lists in which they are displayed.

FREEZE CALCULATED LANDING TIME A dynamic parameter number of minutes prior to the meter fix calculated time of arrival for each aircraft when the TCLT is frozen and becomes an ACLT (i.e., the VTA is updated and consequently the TCLT is modified as appropriate until FCLT minutes prior to meter fix calculated time of arrival, at which time updating is suspended and an ACLT and a frozen meter fix crossing time (MFT) is assigned).

FREEZE SPEED PARAMETER A speed adapted for each aircraft to determine fast and slow aircraft. Fast aircraft freeze on parameter FCLT and slow aircraft freeze on parameter MLDI.

FSDO (See FLIGHT STANDARDS DISTRICT OFFICE.)

FSPD (See FREEZE SPEED PARAMETER.)

FSS (See FLIGHT SERVICE STATION.)

FUEL DUMPING Airborne release of usable fuel. This does not include the dropping of fuel tanks.
(See JETTISONING OF EXTERNAL STORES.)

FUEL REMAINING A phrase used by either pilots or controllers when relating to the fuel remaining on board until actual fuel exhaustion. When transmitting such information in response to either a controller question or pilot initiated cautionary advisory to air traffic control, pilots will state the APPROXIMATE NUMBER OF MINUTES the flight can continue with the fuel remaining. All reserve fuel SHOULD BE INCLUDED in the time stated, as should an allowance for established fuel gauge system error.

FUEL SIPHONING Unintentional release of fuel caused by overflow, puncture, loose cap, etc.

FUEL VENTING (See FUEL SIPHONING.)

GADO (See GENERAL AVIATION DISTRICT OFFICE.)

GATE HOLD PROCEDURES Procedures at selected airports to hold aircraft at the gate or other ground location whenever departure delays exceed or are anticipated to exceed 15 minutes. The sequence for departure will be maintained in accordance with initial call-up unless modified by flow control restrictions. Pilots should monitor the ground control/clearance delivery frequency for engine start/taxi advisories or new proposed start/taxi time if the delay changes.
(See FLOW CONTROL.)

GCA (See GROUND CONTROLLED APPROACH.)

GENERAL AVIATION That portion of civil aviation which encompasses all facets of aviation except air carriers holding a certificate of public convenience and necessity from the Civil Aeronautics Board and large aircraft commercial operators.
(See ICAO term GENERAL AVIATION.)

GENERAL AVIATION [ICAO] All civil aviation operations other than scheduled air services and nonscheduled air transport operations for remuneration or hire.

GENERAL AVIATION DISTRICT OFFICE An FAA field office serving a designated geographical area and staffed with Flight Standards personnel who have the responsibility for serving the aviation industry and the general public on all matters relating to the certification and operation of general aviation aircraft.

GEO MAP The digitized map markings associated with the ASR-9 Radar System.

GLIDEPATH (See GLIDESLOPE.)

GLIDEPATH INTERCEPT ALTITUDE (See GLIDESLOPE INTERCEPT ALTITUDE.)

GLIDESLOPE Provides vertical guidance for aircraft during approach and landing. The glideslope/glidepath is based on the following:

1. Electronic components emitting signals which provide vertical guidance by reference to airborne instruments during instrument approaches such as ILS/ MLS, or
2. Visual ground aids, such as VASI, which provide vertical guidance for a VFR approach or for the visual portion of an instrument approach and landing.
3. PAR. Used by ATC to inform an aircraft making a PAR approach of its vertical position (elevation) relative to the descent profile.

(See ICAO term GLIDEPATH.)

GLIDEPATH [ICAO] A descent profile determined for vertical guidance during a final approach.

GLIDESLOPE INTERCEPT ALTITUDE The minimum altitude to intercept the glideslope/path on a precision approach. The intersection of the published intercept altitude with the glideslope/path, designated on Government charts by the lightning bolt symbol, is the precision FAF; however, when ATC directs a lower altitude, the resultant lower intercept position is then the FAF.
(See FINAL APPROACH FIX.) (See SEGMENTS OF AN INSTRUMENT APPROACH PROCEDURE.)

GLOBAL POSITIONING SYSTEM A space-base radio positioning, navigation, and time-transfer system. The system provides highly accurate position and velocity information, and precise time, on a continuous global basis, to an unlimited number of properly equipped users. The system will be unaffected by weather, and will provide a worldwide common grid reference system. The GPS concept is predicated upon accurate and continuous knowledge of the spatial position of each satellite in the system with respect to time and distance from a transmitting satellite to the user. The GPS receiver automatically selects appropriate signals from the satellites in view and translates these into a three-dimensional position, velocity, and time. System accuracy for civil users is normally 100 meters horizontally.

GO AHEAD Proceed with your message. Not to be used for any other purpose.

GO AROUND Instructions for a pilot to abandon his approach to landing. Additional instructions may follow. Unless otherwise advised by ATC, a VFR aircraft or an aircraft conducting visual approach should overfly the runway while climbing to traffic pattern altitude and enter the traffic pattern via the crosswind leg. A pilot on an IFR flight plan making an instrument approach should execute the published missed approach procedure or proceed as instructed by ATC; e.g., "Go around" (additional instructions if required).
(See LOW APPROACH.) (See MISSED APPROACH.)

GPS (See Global Positioning System.)

GROUND CLUTTER A pattern produced on the radar scope by ground returns which may degrade other radar returns in the affected area. The effect of ground clutter is minimized by the use of moving target indicator (MTI) circuits in the radar equipment resulting in a radar presentation which displays only targets which are in motion.
(See CLUTTER.)

GROUND CONTROLLED APPROACH A radar approach system operated from the ground by air traffic con-

trol personnel transmitting instructions to the pilot by radio. The approach may be conducted with surveillance radar (ASR) only or with both surveillance and precision approach radar (PAR). Usage of the term "GCA" by pilots is discouraged except when referring to a GCA facility. Pilots should specifically request a "PAR" approach when a precision radar approach is desired or request an "ASR" or "surveillance" approach when a nonprecision radar approach is desired.
(See RADAR APPROACH.)

GROUND DELAY The amount of delay attributed to ATC, encountered prior to departure, usually associated with a CDT program.

GROUND SPEED The speed of an aircraft relative to the surface of the earth.

GROUND STOP Normally, the last initiative to be utilized; this method mandates that the terminal facility will not allow any departures to enter the ARTCC airspace until further notified.

GROUND VISIBILITY (See VISIBILITY.)

HAA (See HEIGHT ABOVE AIRPORT.)

HAL (See HEIGHT ABOVE LANDING.)

HANDOFF An action taken to transfer the radar identification of an aircraft from one controller to another if the aircraft will enter the receiving controller's airspace and radio communications with the aircraft will be transferred.

HAT (See HEIGHT ABOVE TOUCHDOWN.)

HAVE NUMBERS Used by pilots to inform ATC that they have received runway, wind, and altimeter information only.

HAZARDOUS INFLIGHT WEATHER ADVISORY SERVICE Continuous recorded hazardous inflight weather forecasts broadcasted to airborne pilots over selected VOR outlets defined as an HIWAS BROADCAST AREA.

HAZARDOUS WEATHER INFORMATION Summary of significant meteorological information (SIGMET/WS), convective significant meteorological information (convective SIGMET/WST), urgent pilot weather reports (urgent PIREP/UUA), center weather advisories (CWA), airmen's meteorological information (AIRMET/WA) and any other weather such as isolated thunderstorms that are rapidly developing and increasing in intensity, or low ceilings and visibilities that are becoming widespread which is considered significant and are not included in a current hazardous weather advisory.

HEAVY (AIRCRAFT) (See AIRCRAFT CLASSES.)

HEIGHT ABOVE AIRPORT The height of the Minimum Descent Altitude above the published airport elevation. This is published in conjunction with circling minimums.
(See MINIMUM DESCENT ALTITUDE.)

HEIGHT ABOVE LANDING The height above a designated helicopter landing area used for helicopter instrument approach procedures.
(Refer to Part 97.)

HEIGHT ABOVE TOUCHDOWN The height of the Decision Height or Minimum Descent Altitude above the highest runway elevation in the touchdown zone (first 3,000 feet of the runway). HAT is published on instrument approach charts in conjunction with all straight-in minimums.
(See DECISION HEIGHT.) (See MINIMUM DESCENT ALTITUDE.)

HELICOPTER Rotorcraft that, for its horizontal motion, depends principally on its engine-driven rotors.
(See ICAO term HELICOPTER.)

HELICOPTER [ICAO] A heavier-than-air aircraft supported in flight chiefly by the reactions of the air on one or more power-driven rotors on substantially vertical axes.

HELIPAD A small, designated area, usually with a prepared surface, on a heliport, airport, landing/takeoff area, apron/ramp, or movement area used for takeoff, landing, or parking of helicopters.

HELIPORT An area of land, water, or structure used or intended to be used for the landing and takeoff of helicopters and includes its buildings and facilities if any.

HERTZ The standard radio equivalent of frequency in cycles per second of an electromagnetic wave. Kilohertz (kHz) is a frequency of one thousand cycles per second. Megahertz (MHz) is a frequency of one million cycles per second.

HF (See HIGH FREQUENCY.)

HF COMMUNICATIONS (See HIGH FREQUENCY COMMUNICATIONS.)

HIGH FREQUENCY The frequency band between 3 and 30 MHz.
(See HIGH FREQUENCY COMMUNICATIONS.)

HIGH FREQUENCY COMMUNICATIONS High radio frequencies (HF) between 3 and 30 MHz used for air-to-ground voice communication in overseas operations.

HIGH SPEED EXIT (See HIGH SPEED TAXIWAY.)

HIGH SPEED TAXIWAY A long radius taxiway designed and provided with lighting or marking to define the path of aircraft, traveling at high speed (up to 60 knots), from the runway center to a point on the center of a taxiway. Also referred to as long radius exit or turn-off taxiway. The high speed taxiway is designed to expedite aircraft turning off the runway after landing, thus reducing runway occupancy time.

HIGH SPEED TURNOFF (See HIGH SPEED TAXIWAY.)

HIWAS (See HAZARDOUS INFLIGHT WEATHER ADVISORY SERVICE.)

HIWAS AREA (See HAZARDOUS INFLIGHT WEATHER ADVISORY SERVICE).

HIWAS BROADCAST AREA A geographical area of responsibility including one or more HIWAS outlet areas assigned to an AFSS/FSS for hazardous weather advisory broadcasting.

HIWAS OUTLET AREA An area defined as a 150 NM radius of a HIWAS outlet, expanded as necessary to provide coverage.

HOLDING PROCEDURE (See HOLD PROCEDURE.)

HOLD PROCEDURE A predetermined maneuver which keeps aircraft within a specified airspace while awaiting further clearance from air traffic control. Also used during ground operations to keep aircraft within a specified area or at a specified point while awaiting further clearance from air traffic control.
(See HOLDING FIX.) (Refer to AIM.)

HOLDING FIX A specified fix identifiable to a pilot by NAVAID's or visual reference to the ground used as a reference point in establishing and maintaining the position of an aircraft while holding.
(See FIX.) (See VISUAL HOLDING.) (Refer to AIM.)

HOLDING POINT [ICAO] A specified location, identified by visual or other means, in the vicinity of which the position of an aircraft in flight is maintained in accordance with air traffic control clearances.

HOLD FOR RELEASE Used by ATC to delay an aircraft for traffic management reasons; i.e., weather, traffic volume, etc. Hold for release instructions (including departure delay information) are used to inform a pilot or a controller (either directly or through an authorized relay) that an IFR departure clearance is not valid until a release time or additional instructions have been received.
(See ICAO term HOLDING POINT.)

HOMING Flight toward a NAVAID, without correcting for wind, by adjusting the aircraft heading to maintain a relative bearing of zero degrees.
(See BEARING.) (See ICAO term HOMING.)

HOMING [ICAO] The procedure of using the direction-finding equipment of one radio station with the emission of another radio station, where at least one of the stations is mobile, and whereby the mobile station proceeds continuously towards the other station.

HOVER CHECK Used to describe when a helicopter/VTOL aircraft requires a stabilized hover to conduct a performance/power check prior to hover taxi, air taxi, or takeoff. Altitude of the hover will vary based on the purpose of the check.

HOVER TAXI Used to describe a helicopter/VTOL aircraft movement conducted above the surface and in ground effect at airspeeds less than approximately 20 knots. The actual height may vary, and some helicopters may require hover taxi above 25 feet AGL to reduce ground effect turbulence or provide clearance for cargo slingloads.
(See AIR TAXI.) (See HOVER CHECK.) (Refer to AIM.)

HOW DO YOU HEAR ME? A question relating to the quality of the transmission or to determine how well the transmission is being received.

HZ (See HERTZ.)

IAF (See INITIAL APPROACH FIX.)

IAP (See INSTRUMENT APPROACH PROCEDURE.)

ICAO (See INTERNATIONAL CIVIL AVIATION ORGANIZATION.)

ICAO [ICAO] (See ICAO Term INTERNATIONAL CIVIL AVIATION ORGANIZATION.)

IDENT A request for a pilot to activate the aircraft transponder identification feature. This will help the controller to confirm an aircraft identity or to identify an aircraft.
(Refer to AIM.)

IDENT FEATURE The special feature in the Air Traffic Control Radar Beacon System (ATCRBS) equipment. It is used to immediately distinguish one displayed beacon target from other beacon targets.
(See IDENT.)

IF (See INTERMEDIATE FIX.)

IFIM (See INTERNATIONAL FLIGHT INFORMATION MANUAL.)

IF NO TRANSMISSION RECEIVED FOR (TIME) Used by ATC in radar approaches to prefix procedures which should be followed by the pilot in event of lost communications.
(See LOST COMMUNICATIONS.)

IFR (See INSTRUMENT FLIGHT RULES.)

IFR AIRCRAFT An aircraft conducting flight in accordance with instrument flight rules.

IFR CONDITIONS Weather conditions below the minimum for flight under visual flight rules.
(See INSTRUMENT METEOROLOGICAL CONDITIONS.)

IFR DEPARTURE PROCEDURE (See IFR TAKEOFF MINIMUMS AND DEPARTURE PROCEDURES.)
(Refer to AIM.)

IFR FLIGHT (See IFR AIRCRAFT.)

IFR LANDING MINIMUMS (See LANDING MINIMUMS).

IFR MILITARY TRAINING ROUTES (IR) Routes used by the Department of Defense and associated Reserve and Air Guard units for the purpose of conducting low-altitude navigation and tactical training in both IFR and VFR weather conditions below 10,000 feet MSL at airspeeds in excess of 250 knots IAS.

IFR TAKEOFF MINIMUMS AND DEPARTURE PROCEDURES Federal Aviation Regulations, Part 91, prescribes standard takeoff rules for certain civil users. At some airports, obstructions or other factors require the establishment of nonstandard takeoff minimums, departure procedures, or both to assist pilots in avoiding obstacles during climb to the minimum en route altitude. Those airports are listed in NOS/DOD Instrument Approach Charts (IAP's) under a section entitled "IFR Takeoff Minimums and Departure Procedures." The NOS/DOD IAP chart legend illustrates the symbol used to alert the pilot to nonstandard takeoff minimums and departure procedures. When departing IFR from such airports or from any airports where there are no departure procedures, SID's, or ATC facilities available, pilots should advise ATC of any departure limitations. Controllers may query a pilot to determine acceptable departure directions, turns, or headings after takeoff. Pilots should be familiar with the departure procedures and must assure that their aircraft can meet or exceed any specified climb gradients.

ILS (See INSTRUMENT LANDING SYSTEM.)

ILS CATEGORIES 1. ILS Category I. An ILS approach procedure which provides for approach to a height above touchdown of not less than 200 feet and with runway visual range of not less than 1,800 feet. 2. ILS Category II. An ILS approach procedure which provides for approach to a height above touchdown of not less than 100 feet and with runway visual range of not less than 1,200 feet. 3. ILS Category III:

a. IIIA. An ILS approach procedure which provides for approach without a decision height minimum and with runway visual range of not less than 700 feet.

b. IIIB. An ILS approach procedure which provides for approach without a decision height minimum and with runway visual range of not less than 150 feet.

c. IIIC. An ILS approach procedure which provides for approach without a decision height minimum and without runway visual range minimum.

IM (See INNER MARKER.)

IMC (See INSTRUMENT METEOROLOGICAL CONDITIONS.)

IMMEDIATELY Used by ATC when such action compliance is required to avoid an imminent situation.

INCERFA (Uncertainty Phase) [ICAO] A situation wherein uncertainty exists as to the safety of an aircraft and its occupants.

INCREASE SPEED TO SPEED) (See SPEED ADJUSTMENT.)

INERTIAL NAVIGATION SYSTEM An RNAV system which is a form of self-contained navigation.
(See Area Navigation/RNAV.)

INFLIGHT REFUELING (See AERIAL REFUELING.)

INFLIGHT WEATHER ADVISORY (See WEATHER ADVISORY.)

INFORMATION REQUEST A request originated by an FSS for information concerning an overdue VFR aircraft.

INITIAL APPROACH FIX The fixes depicted on instrument approach procedure charts that identify the beginning of the initial approach segment(s).
(See FIX.) (See SEGMENTS OF AN INSTRUMENT APPROACH PROCEDURE.)

INITIAL APPROACH SEGMENT (See SEGMENTS OF AN INSTRUMENT APPROACH PROCEDURE.)

INITIAL APPROACH SEGMENT [ICAO] That segment of an instrument approach procedure between the initial approach fix and the intermediate approach fix or, where applicable, the final approach fix or point.

INLAND NAVIGATION FACILITY A navigation aid on a North American Route at which the common route and/or the noncommon route begins or ends.

INNER MARKER A marker beacon used with an ILS (CAT II) precision approach located between the middle marker and the end of the ILS runway, transmitting a radiation pattern keyed at six dots per second and indicating to the pilot, both aurally and visually, that he is at the designated decision height (DH), normally 100 feet above the touchdown zone elevation, on the ILS CAT II approach. It also marks progress during a CAT III approach.
(See INSTRUMENT LANDING SYSTEM.) (Refer to AIM.)

INNER MARKER BEACON (See INNER MARKER.)

INREQ (See INFORMATION REQUEST.)

INS (See INERTIAL NAVIGATION SYSTEM.)

INSTRUMENT APPROACH (See INSTRUMENT APPROACH PROCEDURE.)

INSTRUMENT APPROACH PROCEDURE A series of predetermined maneuvers for the orderly transfer of an aircraft under instrument flight conditions from the beginning of the initial approach to a landing or to a point from which a landing may be made visually. It is prescribed and approved for a specific airport by competent authority.

(See SEGMENTS OF AN INSTRUMENT APPROACH PROCEDURE.) (Refer to Part 91.) (See AIM.)

1. U.S. civil standard instrument approach procedures are approved by the FAA as prescribed under Part 97 and are available for public use.

2. U.S. military standard instrument approach procedures are approved and published by the Department of Defense.

3. Special instrument approach procedures are approved by the FAA for individual operators but are not published in Part 97 for public use.

(See ICAO term INSTRUMENT APPROACH PROCEDURE.)

INSTRUMENT APPROACH PROCEDURE [ICAO] A series of predetermined maneuvers by reference to flight instruments with specified protection from obstacles from the initial approach fix, or where applicable, from the beginning of a defined arrival route to a point from which a landing can be completed and thereafter, if a landing is not completed, to a position at which holding or en route obstacle clearance criteria apply.

INSTRUMENT APPROACH PROCEDURES CHARTS (See AERONAUTICAL CHART.)

INSTRUMENT FLIGHT RULES Rules governing the procedures for conducting instrument flight. Also a term used by pilots and controllers to indicate type of flight plan.

(See VISUAL FLIGHT RULES.) (See INSTRUMENT METEOROLOGICAL CONDITIONS.) (See VISUAL METEOROLOGICAL CONDITIONS.) (Refer to AIM.) (See ICAO term INSTRUMENT FLIGHT RULES.)

INSTRUMENT FLIGHT RULES [ICAO] A set of rules governing the conduct of flight under instrument meteorological conditions.

INSTRUMENT LANDING SYSTEM A precision instrument approach system which normally consists of the following electronic components and visual aids:

1. Localizer.
(See LOCALIZER.)

2. Glideslope.
(See GLIDESLOPE.)

3. Outer Marker.
(See OUTER MARKER.)

4. Middle Marker.
(See MIDDLE MARKER.)

5. Approach Lights.
(See AIRPORT LIGHTING.)
(Refer to Part 91.) (See AIM.)

INSTRUMENT METEOROLOGICAL CONDITIONS Meteorological conditions expressed in terms of visibility, distance from cloud, and ceiling less than the minima specified for visual meteorological conditions.

(See VISUAL METEOROLOGICAL CONDITIONS.) (See INSTRUMENT FLIGHT RULES.) (See VISUAL FLIGHT RULES.)

INSTRUMENT RUNWAY A runway equipped with electronic and visual navigation aids for which a precision or nonprecision approach procedure having straight-in landing minimums has been approved.

(See ICAO term INSTRUMENT RUNWAY.)

INSTRUMENT RUNWAY [ICAO] One of the following types of runways intended for the operation of aircraft using instrument approach procedures:

a) Nonprecision Approach Runway An instrument runway served by visual aids and a nonvisual aid providing at least directional guidance adequate for a straight-in approach.

b) Precision Approach Runway, Category I An instrument runway served by ILS and visual aids intended for operations down to 60 m (200 feet) decision height and down to an RVR of the order of 800 m.

c) Precision Approach Runway, Category II An instrument runway served by ILS and visual aids intended for operations down to 30 m (100 feet) decision height and down to an RVR of the order of 400 m.

d) Precision Approach Runway, Category III An instrument runway served by ILS to and along the surface of the runway and:

A. Intended for operations down to an RVR of the order of 200 m (no decision height being applicable) using visual aids during the final phase of landing;

B. Intended for operations down to an RVR of the order of 50 m (no decision height being applicable) using visual aids for taxiing;

C. Intended for operations without reliance on visual reference for landing or taxiing.

Note 1: See Annex 10 Volume I, Part I Chapter 3, for related ILS specifications.

Note 2: Visual aids need not necessarily be matched to the scale of nonvisual aids provided. The criterion for the selection of visual aids is the conditions in which operations are intended to be conducted.

INTERMEDIATE APPROACH SEGMENT (See SEGMENTS OF AN INSTRUMENT APPROACH PROCEDURE).

INTERMEDIATE APPROACH SEGMENT [ICAO] That segment of an instrument approach procedure between either the intermediate approach fix and the final approach fix or point, or between the end of a reversal, race track or dead reckoning track procedure and the final approach fix or point, as appropriate.

INTERMEDIATE FIX The fix that identifies the beginning of the intermediate approach segment of an instrument approach procedure. The fix is not normally identified on the instrument approach chart as an intermediate fix (IF).

(See SEGMENTS OF AN INSTRUMENT APPROACH PROCEDURE.)

INTERMEDIATE LANDING On the rare occasion that

this option is requested, it should be approved. The departure center, however, must advise the ATCSCC so that the appropriate delay is carried over and assigned at the intermediate airport. An intermediate landing airport within the arrival center will not be accepted without coordination with and the approval of the ATCSCC.

INTERNATIONAL AIRPORT Relating to international flight, it means:

1. An airport of entry which has been designated by the Secretary of Treasury or Commissioner of Customs as an international airport for customs service.

2. A landing rights airport at which specific permission to land must be obtained from customs authorities in advance of contemplated use.

3. Airports designated under the Convention on International Civil Aviation as an airport for use by international commercial air transport and/or international general aviation.

(Refer to AIRPORT/FACILITY DIRECTORY.) (Refer to IFIM.)
(See ICAO term INTERNATIONAL AIRPORT.)

INTERNATIONAL AIRPORT [ICAO] Any airport designated by the Contracting State in whose territory it is situated as an airport of entry and departure for international air traffic, where the formalities incident to customs, immigration, public health, animal and plant quarantine and similar procedures are carried out.

INTERNATIONAL CIVIL AVIATION ORGANIZATION [ICAO] A specialized agency of the United Nations whose objective is to develop the principles and techniques of international air navigation and to foster planning and development of international civil air transport.

ICAO Regions include:

AFI African-Indian Ocean Region
CAR Caribbean Region
EUR European Region
MID/ASIA Middle East/Asia Region
NAM North American Region
NAT North Atlantic Region
PAC Pacific Region
SAM South American Region

INTERNATIONAL FLIGHT INFORMATION MANUAL A publication designed primarily as a pilot's preflight planning guide for flights into foreign airspace and for flights returning to the U.S. from foreign locations

INTERROGATOR The ground-based surveillance radar beacon transmitter-receiver, which normally scans in synchronism with a primary radar, transmitting discrete radio signals which repetitiously request all transponders on the mode being used to reply. The replies received are mixed with the primary radar returns and displayed on the same plan position indicator (radar scope). Also, applied to the airborne element of the TACAN/DME system.

(See TRANSPONDER.) (Refer to AIM.)

INTERSECTING RUNWAYS Two or more runways which cross or meet within their lengths.

(See INTERSECTION.)

INTERSECTION

1. A point defined by any combination of courses, radials, or bearings of two or more navigational aids.

2. Used to describe the point where two runways, a runway and a taxiway, or two taxiways cross or meet.

INTERSECTION DEPARTURE A departure from any runway intersection except the end of the runway.

(See INTERSECTION.)

INTERSECTION TAKEOFF (See INTERSECTION DEPARTURE.)

IR (See IFR MILITARY TRAINING ROUTES.)

I SAY AGAIN The message will be repeated.

JAMMING Electronic or mechanical interference which may disrupt the display of aircraft on radar or the transmission/reception of radio communications/navigation.

JET BLAST Jet engine exhaust (thrust stream turbulence).

(See WAKE TURBULENCE.)

JET ROUTE A route designed to serve aircraft operations from 18,000 feet MSL up to and including flight level 450. The routes are referred to as "J" routes with numbering to identify the designated route; e.g., J105.

(See Class A airspace.) (Refer to Part 71.)

JET STREAM A migrating stream of high-speed winds present at high altitudes.

JETTISONING OF EXTERNAL STORES Airborne release of external stores; e.g., tiptanks, ordnance.

(See FUEL DUMPING.) (Refer to Part 91.)

JOINT USE RESTRICTED AREA (See RESTRICTED AREA.)

KNOWN TRAFFIC With respect to ATC clearances, means aircraft whose altitude, position, and intentions are known to ATC.

LAA (See LOCAL AIRPORT ADVISORY.)

LAAS (See LOW ALTITUDE ALERT SYSTEM.)

LANDING AREA Any locality either on land, water, or structures, including airports/heliports and intermediate landing fields, which is used, or intended to be used, for the landing and takeoff of aircraft whether or not facilities are provided for the shelter, servicing, or for receiving or discharging passengers or cargo.

(See ICAO term LANDING AREA.)

LANDING AREA [ICAO] That part of a movement area intended for the landing or takeoff of aircraft.

LANDING DIRECTION INDICATOR A device which

visually indicates the direction in which landings and takeoffs should be made.

(See TETRAHEDRON.) (Refer to AIM.)

LANDING DISTANCE AVAILABLE [ICAO] The length of runway which is declared available and suitable for the ground run of an aeroplane landing.

LANDING MINIMUMS The minimum visibility prescribed for landing a civil aircraft while using an instrument approach procedure. The minimum applies with other limitations set forth in Part 91 with respect to the Minimum Descent Altitude (MDA) or Decision Height (DH) prescribed in the instrument approach procedures as follows:

1. Straight-in landing minimums. A statement of MDA and visibility, or DH and visibility, required for a straight-in landing on a specified runway, or
2. Circling minimums. A statement of MDA and visibility required for the circle-to-land maneuver.

Descent below the established MDA or DH is not authorized during an approach unless the aircraft is in a position from which a normal approach to the runway of intended landing can be made and adequate visual reference to required visual cues is maintained.

(See STRAIGHT-IN LANDING.) (See CIRCLE-TO-LAND MANEUVER.) (See DECISION HEIGHT.) (See MINIMUM DESCENT ALTITUDE.) (See VISIBILITY.) (See INSTRUMENT APPROACH PROCEDURE.) (Refer to Part 91.)

LANDING ROLL The distance from the point of touchdown to the point where the aircraft can be brought to a stop or exit the runway.

LANDING SEQUENCE The order in which aircraft are positioned for landing.

(See APPROACH SEQUENCE.)

LAST ASSIGNED ALTITUDE The last altitude/flight level assigned by ATC and acknowledged by the pilot.

(See MAINTAIN.) (Refer to Part 91.)

LATERAL SEPARATION The lateral spacing of aircraft at the same altitude by requiring operation on different routes or in different geographical locations.

(See SEPARATION.)

LDA (See LOCALIZER TYPE DIRECTIONAL AID.)

LDA [ICAO] (See ICAO Term LANDING DISTANCE AVAILABLE.)

LF (See LOW FREQUENCY.)

LIGHTED AIRPORT An airport where runway and obstruction lighting is available.

(See AIRPORT LIGHTING.) (Refer to AIM.)

LIGHT GUN A handheld directional light signaling device which emits a brilliant narrow beam of white, green, or red light as selected by the tower controller. The color and type of light transmitted can be used to approve or disapprove anticipated pilot actions where radio communication is not available. The light gun is used for controlling traffic operating in the vicinity of the airport and on the airport movement area.

(Refer to AIM.)

LOCALIZER The component of an ILS which provides course guidance to the runway.

(See INSTRUMENT LANDING SYSTEM.) (Refer to AIM.) (See ICAO term LOCALIZER COURSE.)

LOCALIZER COURSE [ICAO] The locus of points, in any given horizontal plane, at which the DDM (difference in depth of modulation) is zero.

LOCALIZER TYPE DIRECTIONAL AID A NAVAID used for nonprecision instrument approaches with utility and accuracy comparable to a localizer but which is not a part of a complete ILS and is not aligned with the runway.

(Refer to AIM.)

LOCALIZER USABLE DISTANCE The maximum distance from the localizer transmitter at a specified altitude, as verified by flight inspection, at which reliable course information is continuously received.

(Refer to AIM.)

LOCAL AIRPORT ADVISORY [LAA] A service provided by flight service stations or the military at airports not serviced by an operating control tower. This service consists of providing information to arriving and departing aircraft concerning wind direction and speed, favored runway, altimeter setting, pertinent known traffic, pertinent known field conditions, airport taxi routes and traffic patterns, and authorized instrument approach procedures. This information is advisory in nature and does not constitute an ATC clearance.

(See AIRPORT ADVISORY AREA.)

LOCAL TRAFFIC Aircraft operating in the traffic pattern or within sight of the tower, or aircraft known to be departing or arriving from flight in local practice areas, or aircraft executing practice instrument approaches at the airport.

(See TRAFFIC PATTERN.)

LOCATOR [ICAO] An LM/MF NDB used as an aid to final approach.

Note: A locator usually has an average radius of rated coverage of between 18.5 and 46.3 km (10 and 25 NM).

LONGITUDINAL SEPARATION The longitudinal spacing of aircraft at the same altitude by a minimum distance expressed in units of time or miles.

(See SEPARATION.) (Refer to AIM.)

LONG RANGE NAVIGATION (See LORAN.)

LORAN An electronic navigational system by which hyperbolic lines of position are determined by measuring the difference in the time of reception of synchronized pulse signals from two fixed transmitters. Loran A oper-

ates in the 1750–1950 kHz frequency band. Loran C and D operate in the 100–110 kHz frequency band.
(Refer to AIM.)

LOST COMMUNICATIONS Loss of the ability to communicate by radio. Aircraft are sometimes referred to as NORDO (No Radio). Standard pilot procedures are specified in Part 91. Radar controllers issue procedures for pilots to follow in the event of lost communications during a radar approach when weather reports indicate that an aircraft will likely encounter IFR weather conditions during the approach.
(Refer to Part 91.) (See AIM.)

LOW ALTITUDE AIRWAY STRUCTURE The network of airways serving aircraft operations up to but not including 18,000 feet MSL.
(See AIRWAY.) (Refer to AIM.)

LOW ALTITUDE ALERT, CHECK YOUR ALTITUDE IMMEDIATELY (See SAFETY ALERT.)

LOW ALTITUDE ALERT SYSTEM An automated function of the TPX-42 that alerts the controller when a Mode C transponder-equipped aircraft on an IFR flight plan is below a predetermined minimum safe altitude. If requested by the pilot, LAAS monitoring is also available to VFR Mode C transponder-equipped aircraft.

LOW APPROACH An approach over an airport or runway following an instrument approach or a VFR approach including the go-around maneuver where the pilot intentionally does not make contact with the runway.
(Refer to AIM.)

LOW FREQUENCY The frequency band between 30 and 300 kHz.
(Refer to AIM.)

MAA (See MAXIMUM AUTHORIZED ALTITUDE.)

MACH NUMBER The ratio of true airspeed to the speed of sound; e.g., MACH .82, MACH 1.6.
(See AIRSPEED.)

MACH TECHNIQUE [ICAO] Describes a control technique used by air traffic control whereby turbojet aircraft operating successively along suitable routes are cleared to maintain appropriate MACH numbers for a relevant portion of the en route phase of flight. The principal objective is to achieve improved utilization of the airspace and to ensure that separation between successive aircraft does not decrease below the established minima.

MAINTAIN

1. Concerning altitude/flight level, the term means to remain at the altitude/flight level specified. The phrase "climb and" or "descend and" normally precedes "maintain" and the altitude assignment; e.g., "descend and maintain 5,000."

2. Concerning other ATC instructions, the term is used in its literal sense; e.g., maintain VFR.

MAKE SHORT APPROACH Used by ATC to inform a pilot to alter his traffic pattern so as to make a short final approach.
(See TRAFFIC PATTERN.)

MANDATORY ALTITUDE An altitude depicted on an instrument Approach Procedure Chart requiring the aircraft to maintain altitude at the depicted value.

MAP (See MISSED APPROACH POINT.)

MARKER BEACON An electronic navigation facility transmitting a 75 MHz vertical fan or boneshaped radiation pattern. Marker beacons are identified by their modulation frequency and keying code, and when received by compatible airborne equipment, indicate to the pilot, both aurally and visually, that he is passing over the facility.
(See OUTER MARKER.) (See MIDDLE MARKER.) (See INNER MARKER.) (Refer to AIM.)

MARSA (See MILITARY AUTHORITY ASSUMES RESPONSIBILITY FOR SEPARATION OF AIRCRAFT.)

MAXIMUM AUTHORIZED ALTITUDE A published altitude representing the maximum usable altitude or flight level for an airspace structure or route segment. It is the highest altitude on a Federal airway, jet route, area navigation low or high route, or other direct route for which an MEA is designated in Part 95 at which adequate reception of navigation aid signals is assured.

MAYDAY The international radiotelephony distress signal. When repeated three times, it indicates imminent and grave danger and that immediate assistance is requested.
(See PAN-PAN.) (Refer to AIM.)

MCA (See MINIMUM CROSSING ALTITUDE.)

MDA (See MINIMUM DESCENT ALTITUDE.)

MEA (See MINIMUM EN ROUTE IFR ALTITUDE.)

METEOROLOGICAL IMPACT STATEMENT An unscheduled planning forecast describing conditions expected to begin within 4 to 12 hours which may impact the flow of air traffic in a specific center's (ARTCC) area.

METER FIX TIME/SLOT TIME A calculated time to depart the meter fix in order to cross the vertex at the ACLT. This time reflects descent speed adjustment and any applicable time that must be absorbed prior to crossing the meter fix.

METER LIST DISPLAY INTERVAL A dynamic parameter which controls the number of minutes prior to the flight plan calculated time of arrival at the meter fix for each aircraft, at which time the TCLT is frozen and becomes an ACLT; i.e., the VTA is updated and consequently the TCLT modified as appropriate until frozen at which time updating is suspended and an ACLT is assigned. When frozen, the flight entry is inserted into the arrival sector's meter list for display on the sector PVD. MLDI is used if filed true airspeed is less than or equal to freeze speed parameters (FSPD).

METERING   A method of time-regulating arrival traffic flow into a terminal area so as not to exceed a predetermined terminal acceptance rate.

METERING AIRPORTS   Airports adapted for metering and for which optimum flight paths are defined. A maximum of 15 airports may be adapted.

METERING FIX   A fix along an established route from over which aircraft will be metered prior to entering terminal airspace. Normally, this fix should be established at a distance from the airport which will facilitate a profile descent 10,000 feet above airport elevation [AAE] or above.

METERING POSITION(S)   Adapted PVD's and associated "D" positions eligible for display of a metering position list. A maximum of four PVD's may be adapted.

METERING POSITION LIST   An ordered list of data on arrivals for a selected metering airport displayed on a metering position PVD.

MFT   (See METER FIX TIME/SLOT TIME.)

MHA   (See MINIMUM HOLDING ALTITUDE.)

MIA   (See MINIMUM IFR ALTITUDES.)

MICROBURST   A small downburst with outbursts of damaging winds extending 2.5 miles or less. In spite of its small horizontal scale, an intense microburst could induce wind speeds as high as 150 knots.
(Refer to AIM.)

MICROWAVE LANDING SYSTEM   A precision instrument approach system operating in the microwave spectrum which normally consists of the following components:

1. Azimuth Station.
2. Elevation Station.
3. Precision Distance Measuring Equipment.

(See MLS CATEGORIES.)

MIDDLE COMPASS LOCATOR   (See COMPASS LOCATOR.)

MIDDLE MARKER   A marker beacon that defines a point along the glideslope of an ILS normally located at or near the point of decision height (ILS Category I). It is keyed to transmit alternate dots and dashes, with the alternate dots and dashes keyed at the rate of 95 dot/dash combinations per minute on a 1300 Hz tone, which is received aurally and visually by compatible airborne equipment.
(See MARKER BEACON.) (See INSTRUMENT LANDING SYSTEM.) (Refer to AIM.)

MID RVR   (See VISIBILITY.)

MILES-IN-TRAIL   A specified distance between aircraft, normally, in the same stratum associated with the same destination or route of flight.

MILITARY AUTHORITY ASSUMES RESPONSIBILITY FOR SEPARATION OF AIRCRAFT   A condition whereby the military services involved assume responsibility for separation between participating military aircraft in the ATC system. It is used only for required IFR operations which are specified in letters of agreement or other appropriate FAA or military documents.

MILITARY OPERATIONS AREA   (See SPECIAL USE AIRSPACE.)

MILITARY TRAINING ROUTES   Airspace of defined vertical and lateral dimensions established for the conduct of military flight training at airspeeds in excess of 250 knots IAS.
(See IFR MILITARY TRAINING ROUTES.) (See VFR MILITARY TRAINING ROUTES.)

MINIMA   (See MINIMUMS.)

MINIMUM CROSSING ALTITUDE   The lowest altitude at certain fixes at which an aircraft must cross when proceeding in the direction of a higher minimum en route IFR altitude (MEA).
(See MINIMUM EN ROUTE IFR ALTITUDE.)

MINIMUM DESCENT ALTITUDE   The lowest altitude, expressed in feet above mean sea level, to which descent is authorized on final approach or during circle-to-land maneuvering in execution of a standard instrument approach procedure where no electronic glideslope is provided.
(See NONPRECISION APPROACH PROCEDURE.)

MINIMUM EN ROUTE IFR ALTITUDE   The lowest published altitude between radio fixes which assures acceptable navigational signal coverage and meets obstacle clearance requirements between those fixes. The MEA prescribed for a Federal airway or segment thereof, area navigation low or high route, or other direct route applies to the entire width of the airway, segment, or route between the radio fixes defining the airway, segment, or route.
(Refer to Part 91.) (Refer to Part 95.) (Refer to AIM.)

MINIMUM FUEL   Indicates that an aircraft's fuel supply has reached a state where, upon reaching the destination, it can accept little or no delay. This is not an emergency situation but merely indicates an emergency situation is possible should any undue delay occur.
(Refer to AIM.)

MINIMUM HOLDING ALTITUDE   The lowest altitude prescribed for a holding pattern which assures navigational signal coverage, communications, and meets obstacle clearance requirements.

MINIMUM IFR ALTITUDES   Minimum altitudes for IFR operations as prescribed in Part 91. These altitudes are published on aeronautical charts and prescribed in Part 95 for airways and routes, and in Part 97 for standard instrument approach procedures. If no applicable minimum altitude is prescribed in FAR 95 or FAR 97, the following minimum IFR altitude applies:

1. In designated mountainous areas, 2,000 feet above the highest obstacle within a horizontal distance of 4 nautical miles from the course to be flown; or

2. Other than mountainous areas, 1,000 feet above the highest obstacle within a horizontal distance of 4 nautical miles from the course to be flown; or

3. As otherwise authorized by the Administrator or assigned by ATC.

(See MINIMUM EN ROUTE IFR ALTITUDE.) (See MINIMUM OBSTRUCTION CLEARANCE ALTITUDE.) (See MINIMUM CROSSING ALTITUDE.) (See MINIMUM SAFE ALTITUDE.) (See MINIMUM VECTORING ALTITUDE.) (Refer to Part 91.)

MINIMUM NAVIGATION PERFORMANCE SPECIFICATION A set of standards which require aircraft to have a minimum navigation performance capability in order to operate in MNPS designated airspace. In addition, aircraft must be certified by their State of Registry for MNPS operation.

MINIMUM NAVIGATION PERFORMANCE SPECIFICATIONS AIRSPACE Designated airspace in which MNPS procedures are applied between MNPS certified and equipped aircraft. Under certain conditions, non-MNPS aircraft can operate in MNPSA. However, standard oceanic separation minima is provided between the non-MNPS aircraft and other traffic. Currently, the only designated MNPSA is described as follows:

1. Between FL 275 and FL 400;

2. Between latitudes 27° N and the North Pole;

3. In the east, the eastern boundaries of the CTA's Santa Maria Oceanic, Shanwick Oceanic, and Reykjavik;

4. In the west, the western boundaries of CTA's Reykjavik and Gander Oceanic and New York Oceanic excluding the area west of 60° W and south of 38° 30'N.

MINIMUM OBSTRUCTION CLEARANCE ALTITUDE The lowest published altitude in effect between radio fixes on VOR airways, off-airway routes, or route segments which meets obstacle clearance requirements for the entire route segment and which assures acceptable navigational signal coverage only within 25 statute (22 nautical) miles of a VOR.

(Refer to Part 91.) (Refer to Part 95.)

MINIMUM RECEPTION ALTITUDE The lowest altitude at which an intersection can be determined.

(Refer to Part 95.)

MINIMUM SAFE ALTITUDE

1. The minimum altitude specified in Part 91 for various aircraft operations.

2. Altitudes depicted on approach charts which provide at least 1,000 feet of obstacle clearance for emergency use within a specified distance from the navigation facility upon which a procedure is predicated. These altitudes will be identified as Minimum Sector Altitudes or Emergency Safe Altitudes and are established as follows:

a. Minimum Sector Altitudes. Altitudes depicted on approach charts which provide at least 1,000 feet of obstacle clearance within a 25-mile radius of the navigation facility upon which the procedure is predicated. Sectors depicted on approach charts must be at least 90 degrees in scope. These altitudes are for emergency use only and do not necessarily assure acceptable navigational signal coverage.

(See ICAO term MINIMUM SECTOR ALTITUDE.)

b. Emergency Safe Altitudes. Altitudes depicted on approach charts which provide at least 1,000 feet of obstacle clearance in nonmountainous areas and 2,000 feet of obstacle clearance in designated mountainous areas within a 100-mile radius of the navigation facility upon which the procedure is predicated and normally used only in military procedures. These altitudes are identified on published procedures as "Emergency Safe Altitudes."

MINIMUM SAFE ALTITUDE WARNING A function of the ARTS III computer that aids the controller by alerting him when a tracked Mode C-equipped aircraft is below or is predicted by the computer to go below a predetermined minimum safe altitude.

(Refer to AIM.)

MINIMUM SECTOR ALTITUDE [ICAO] The lowest altitude which may be used under emergency conditions which will provide a minimum clearance of 300 m (1,000 feet) above all obstacles located in an area contained within a sector of a circle of 46 km (25 NM) radius centered on a radio aid to navigation.

MINIMUMS Weather condition requirements established for a particular operation or type of operation; e.g., IFR takeoff or landing, alternate airport for IFR flight plans, VFR flight, etc.

(See LANDING MINIMUMS.) (See IFR TAKEOFF MINIMUMS AND DEPARTURE PROCEDURES.) (See VFR CONDITIONS.) (See IFR CONDITIONS.) (Refer to Part 91.) (Refer to AIM.)

MINIMUM VECTORING ALTITUDE The lowest MSL altitude at which an IFR aircraft will be vectored by a radar controller, except as otherwise authorized for radar approaches, departures, and missed approaches. The altitude meets IFR obstacle clearance criteria. It may be lower than the published MEA along an airway or J-route segment. It may be utilized for radar vectoring only upon the controller's determination that an adequate radar return is being received from the aircraft being controlled. Charts depicting minimum vectoring altitudes are normally available only to the controllers and not to pilots.

(Refer to AIM.)

MINUTES-IN-TRAIL A specified interval between aircraft expressed in time. This method would more likely be utilized regardless of altitude.

MIS (See METEOROLOGICAL IMPACT STATEMENT.)

MISSED APPROACH

1. A maneuver conducted by a pilot when an instrument approach cannot be completed to a landing. The route of flight and altitude are shown on instrument approach procedure charts. A pilot executing a missed approach prior to the Missed Approach Point (MAP) must continue along the final approach to the MAP. The pilot may climb immediately to the altitude specified in the missed approach procedure.

2. A term used by the pilot to inform ATC that he is executing the missed approach.

3. At locations where ATC radar service is provided, the pilot should conform to radar vectors when provided by ATC in lieu of the published missed approach procedure.

(See MISSED APPROACH POINT.) (Refer to AIM.)

MISSED APPROACH POINT A point prescribed in each instrument approach procedure at which a missed approach procedure shall be executed if the required visual reference does not exist.

(See MISSED APPROACH.) (See SEGMENTS OF AN INSTRUMENT APPROACH PROCEDURE.)

MISSED APPROACH PROCEDURE [ICAO] The procedure to be followed if the approach cannot be continued.

MISSED APPROACH SEGMENT (See SEGMENTS OF AN INSTRUMENT APPROACH PROCEDURE.)

MLDI (See METER LIST DISPLAY INTERVAL.)

MLS (See MICROWAVE LANDING SYSTEM.)

MLS CATEGORIES

1. MLS Category I. An MLS approach procedure which provides for an approach to a height above touchdown of not less than 200 feet and a runway visual range of not less than 1,800 feet.

2. MLS Category II. Undefined until data gathering/analysis completion.

3. MLS Category III. Undefined until data gathering/analysis completion.

MM (See MIDDLE MARKER.)

MNPS (See MINIMUM PERFORMANCE SPECIFICATION.)

MNPSA (See MINIMUM PERFORMANCE SPECIFICATIONS AIRSPACE.)

MOA (See MILITARY OPERATIONS AREA.)

MOCA (See MINIMUM OBSTRUCTION CLEARANCE ALTITUDE.)

MODE The letter or number assigned to a specific pulse spacing of radio signals transmitted or received by ground interrogator or airborne transponder components of the Air Traffic Control Radar Beacon System (ATCRBS). Mode A (military Mode 3) and Mode C (altitude reporting) are used in air traffic control.

(See TRANSPONDER.) (See INTERROGATOR.) (See RADAR.) (Refer to AIM.) (See ICAO term MODE.)

MODE (SSR MODE) [ICAO] The letter or number assigned to a specific pulse spacing of the interrogation signals transmitted by an interrogator. There are 4 modes, A, B, C, and D specified in Annex 10, corresponding to four different interrogation pulse spacings.

MODE C INTRUDER ALERT A function of certain air traffic control automated systems designed to alert radar controllers to existing or pending situations between a tracked target (known IFR or VFR aircraft) and an untracked target (unknown IFR or VFR aircraft) that requires immediate attention/action.

(See CONFLICT ALERT.)

MONITOR (When used with communication transfer) listen on a specific frequency and stand by for instructions. Under normal circumstances do not establish communications.

MOVEMENT AREA The runways, taxiways, and other areas of an airport/heliport which are utilized for taxiing/hover taxiing, air taxiing, takeoff, and landing of aircraft, exclusive of loading ramps and parking areas. At those airports/heliports with a tower, specific approval for entry onto the movement area must be obtained from ATC.

(See ICAO term MOVEMENT AREA.)

MOVEMENT AREA [ICAO] That part of an aerodrome to be used for the takeoff, landing and taxiing of aircraft, consisting of the manoeuvring area and the apron(s).

MOVING TARGET INDICATOR An electronic device which will permit radar scope presentation only from targets which are in motion. A partial remedy for ground clutter.

MRA (See MINIMUM RECEPTION ALTITUDE.)

MSA (See MINIMUM SAFE ALTITUDE.)

MSAW (See MINIMUM SAFE ALTITUDE WARNING.)

MTI (See MOVING TARGET INDICATOR.)

MTR (See MILITARY TRAINING ROUTES.)

MULTICOM A mobile service not open to public correspondence used to provide communications essential to conduct the activities being performed by or directed from private aircraft.

MULTIPLE RUNWAYS The utilization of a dedicated arrival runway(s) for departures and a dedicated departure runway(s) for arrivals when feasible to reduce delays and enhance capacity.

MVA (See MINIMUM VECTORING ALTITUDE.)

NAS (See NATIONAL AIRSPACE SYSTEM.)

NAS STAGE A The en route ATC system's radar, computers and computer programs, controller plan view displays (PVDs/Radar Scopes), input/output devices, and the related communications equipment which are integrated to form the heart of the automated IFR air traffic control system, This equipment performs Flight Data Processing (FDP) and Radar Data Processing (RDP). It interfaces with automated terminal systems and is used in the control of en route IFR aircraft.
(Refer to AIM.)

NATIONAL AIRSPACE SYSTEM The common network of U.S. airspace; air navigation facilities, equipment and services, airports or landing areas; aeronautical charts, information and services; rules, regulations and procedures, technical information, and manpower and material. Included are system components shared jointly with the military.

NATIONAL BEACON CODE ALLOCATION PLAN AIRSPACE Airspace over United States territory located within the North American continent between Canada and Mexico, including adjacent territorial waters outward to about boundaries of oceanic control areas (CTA)/Flight Information Regions (FIR).
(See FLIGHT INFORMATION REGION.)

NATIONAL FLIGHT DATA CENTER
A facility in Washington D.C., established by FAA to operate a central aeronautical information service for the collection, validation, and dissemination of aeronautical data in support of the activities of government, industry, and the aviation community. The information is published in the National Flight Data Digest.
(See NATIONAL FLIGHT DATA DIGEST.)

NATIONAL FLIGHT DATA DIGEST A daily (except weekends and Federal holidays) publication of flight information appropriate to aeronautical charts, aeronautical publications, Notices to Airmen, or other media serving the purpose of providing operational flight data essential to safe and efficient aircraft operations.

NATIONAL SEARCH AND RESCUE PLAN An interagency agreement which provides for the effective utilization of all available facilities in all types of search and rescue missions.

NAVAID (See NAVIGATIONAL AID.)

NAVAID CLASSES VOR, VORTAC, and TACAN aids are classed according to their operational use. The three classes of NAVAID's are:

T—Terminal.

L—Low altitude.

H—High altitude.

The normal service range for T, L, and H class aids is found in the FAAAIMTOC. Certain operational requirements make it necessary to use some of these aids at greater service ranges than specified. Extended range is made possible through flight inspection determinations. Some aids also have lesser service range due to location, terrain, frequency protection, etc. Restrictions to service range are listed in Airport/Facility Directory.

NAVIGABLE AIRSPACE Airspace at and above the minimum flight altitudes prescribed in the FAR's including airspace needed for safe takeoff and landing.
(Refer to Part 91.)

NAVIGATIONAL AID Any visual or electronic device airborne or on the surface which provides point-to-point guidance information or position data to aircraft in flight.
(See AIR NAVIGATION FACILITY.)

NBCAP AIRSPACE (See NATIONAL BEACON CODE ALLOCATION PLAN AIRSPACE.)

NDB (See NONDIRECTIONAL BEACON.)

NEGATIVE "No," or "permission not granted," or "that is not correct."

NEGATIVE CONTACT Used by pilots to inform ATC that:

1. Previously issued traffic is not in sight. It may be followed by the pilot's request for the controller to provide assistance in avoiding the traffic.
2. They were unable to contact ATC on a particular frequency.

NFDC (See NATIONAL FLIGHT DATA CENTER.)

NFDD (See NATIONAL FLIGHT DATA DIGEST.)

NIGHT The time between the end of evening civil twilight and the beginning of morning civil twilight, as published in the American Air Almanac, converted to local time.
(See ICAO term NIGHT.)

NIGHT [ICAO] The hours between the end of evening civil twilight and the beginning of morning civil twilight or such other period between sunset and sunrise as may be specified by the appropriate authority.

Note: Civil twilight ends in the evening when the centre of the sun's disk is 6 degrees below the horizon and begins in the morning when the centre of the sun's disk is 6 degrees below the horizon.

NO GYRO APPROACH A radar approach/vector provided in case of a malfunctioning gyro-compass or directional gyro. Instead of providing the pilot with headings to be flown, the controller observes the radar track and issues control instructions "turn right/left" or "stop turn" as appropriate.
(Refer to AIM.)

NO GYRO VECTOR (See NO GYRO APPROACH.)

NONAPPROACH CONTROL TOWER Authorizes aircraft to land or take-off at the airport controlled by the tower or to transit the Class D airspace. The primary function of a nonapproach control tower is the sequencing of aircraft in the traffic pattern and on the landing area. Nonapproach control towers also separate aircraft operating under instrument flight rules clearances from approach controls and centers. They provide ground control services to aircraft, vehicles, personnel, and equipment on the airport movement area.

NONCOMMON ROUTE/PORTION That segment of a North American Route between the inland navigation facility and a designated North American terminal.

NONCOMPOSITE SEPARATION Separation in accordance with minima other than the composite separation minimum specified for the area concerned.

NONDIRECTIONAL BEACON An L/MF or UHF radio beacon transmitting nondirectional signals whereby the pilot of an aircraft equipped with direction finding equipment can determine his bearing to or from the radio beacon and "home" on or track to or from the station. When the radio beacon is installed in conjunction with the Instrument Landing System marker, it is normally called a Compass Locator.
(See COMPASS LOCATOR.) (See AUTOMATIC DIRECTION FINDER.)

NONMOVEMENT AREAS Taxiways and apron (ramp) areas not under the control of air traffic.

NONPRECISION APPROACH (See NONPRECISION APPROACH PROCEDURE).

NONPRECISION APPROACH PROCEDURE A standard instrument approach procedure in which no electronic glideslope is provided; e.g., VOR, TACAN, NDB, LOC, ASR, LDA, or SDF approaches.

NONRADAR Precedes other terms and generally means without the use of radar, such as:

1. Nonradar Approach. Used to describe instrument approaches for which course guidance on final approach is not provided by ground-based precision or surveillance radar. Radar vectors to the final approach course may or may not be provided by ATC. Examples of nonradar approaches are VOR, NDB, TACAN, and ILS/MLS approaches.
(See FINAL APPROACH-IFR.) (See FINAL APPROACH COURSE.) (See RADAR APPROACH.) (See INSTRUMENT APPROACH PROCEDURE.)

2. Nonradar Approach Control. An ATC facility providing approach control service without the use of radar.
(See APPROACH CONTROL FACILITY.) (See APPROACH CONTROL SERVICE.)

3. Nonradar Arrival. An aircraft arriving at an airport without radar service or at an airport served by a radar facility and radar contact has not been established or has been terminated due to a lack of radar service to the airport.
(See RADAR ARRIVAL.) (See RADAR SERVICE.)

4. Nonradar Route. A flight path or route over which the pilot is performing his own navigation. The pilot may be receiving radar separation, radar monitoring, or other ATC services while on a nonradar route.
(See RADAR ROUTE.)

5. Nonradar Separation. The spacing of aircraft in accordance with established minima without the use of radar; e.g., vertical, lateral, or longitudinal separation.
(See RADAR SEPARATION.) (See ICAO term NONRADAR SEPARATION.)

NONRADAR SEPARATION [ICAO]
The separation used when aircraft position information is derived from sources other than radar.

NOPAC (See NORTH PACIFIC.)

NORDO (See LOST COMMUNICATIONS.)

NORTH AMERICAN ROUTE A numerically coded route preplanned over existing airway and route systems to and from specific coastal fixes serving the North Atlantic. North American Routes consist of the following:

1. Common Route/Portion. That segment of a North American Route between the inland navigation facility and the coastal fix.

2. NonCommon Route/Portion. That segment of a North American Route between the inland navigation facility and a designated North American terminal.

3. Inland Navigation Facility. A navigation aid on a North American Route at which the common route and/or the noncommon route begins or ends.

4. Coastal Fix. A navigation aid or intersection where an aircraft transitions between the domestic route structure and the oceanic route structure.

NORTH MARK A beacon data block sent by the host computer to be displayed by the ARTS on a 360 degree bearing at a locally selected radar azimuth and distance. The North Mark is used to ensure correct range/azimuth orientation during periods of CENRAP.

NORTH PACIFIC An organized route system between the Alaskan west coast and Japan.

NOTAM (See NOTICE TO AIRMEN.)

NOTICE TO AIRMEN A notice containing information (not known sufficiently in advance to publicize by other means) concerning the establishment, condition, or change in any component (facility, service, or procedure of, or hazard in the National Airspace System) the timely knowledge of which is essential to personnel concerned with flight operations.

1. NOTAM(D). A NOTAM given (in addition to local dissemination) distant dissemination beyond the area of responsibility of the Flight Service Station. These NOTAM's will be stored and available until canceled.

2. NOTAM(L). A NOTAM given local dissemination by voice and other means, such as telautograph and telephone, to satisfy local user requirements.

3. FDC NOTAM. A NOTAM regulatory in nature, transmitted by USNOF and given system wide dissemination.
(See ICAO term NOTAM.)

NOTAM [ICAO] A notice containing information concerning the establishment, condition or change in any aeronautical facility, service, procedure or hazard, the timely knowledge of which is essential to personnel concerned with flight operations.

Class I Distribution—Distribution by means of telecommunication.

Class II Distribution—Distribution by means other than telecommunications.

NOTICES TO AIRMEN PUBLICATION A publication issued every 14 days, designed primarily for the pilot, which contains current NOTAM information considered essential to the safety of flight as well as supplemental data to other aeronautical publications. The contraction NTAP is used in NOTAM text.

(See NOTICE TO AIRMEN.)

NTAP (See NOTICES TO AIRMEN PUBLICATION.)

NUMEROUS TARGETS VICINITY (LOCATION) A traffic advisory issued by ATC to advise pilots that targets on the radar scope are too numerous to issue individually.

(See TRAFFIC ADVISORIES.)

OALT (See OPERATIONAL ACCEPTABLE LEVEL OF TRAFFIC.)

OBSTACLE An existing object, object of natural growth, or terrain at a fixed geographical location or which may be expected at a fixed location within a prescribed area with reference to which vertical clearance is or must be provided during flight operation.

OBSTACLE FREE ZONE The OFZ is a three dimensional volume of airspace which protects for the transition of aircraft to and from the runway. The OFZ clearing standard precludes taxiing and parked airplanes and object penetrations, except for frangible NAVAID locations that are fixed by function. Additionally, vehicles, equipment, and personnel may be authorized by air traffic control to enter the area using the provisions of Order 7110.65, Air Traffic Control, paragraph 3-5. The runway OFZ and when applicable, the inner-approach OFZ, and the inner-transitional OFZ, comprise the OFZ.

1. Runway OFZ. The runway OFZ is a defined volume of airspace centered above the runway. The runway OFZ is the airspace above a surface whose elevation at any point is the same as the elevation of the nearest point on the runway centerline. The runway OFZ extends 200 feet beyond each end of the runway. The width is as follows:

(a) For runways serving large airplanes, the greater of:

(1) 400 feet, or

(2) 180 feet, plus the wingspan of the most demanding airplane, plus 20 feet per 1,000 feet of airport elevation.

(b) For runways serving only small airplanes:

(1) 300 feet for precision instrument runways.

(2) 250 feet for other runways serving small airplanes with approach speeds of 50 knots, or more.

(3) 120 feet for other runways serving small airplanes with approach speeds of less than 50 knots.

2. Inner-approach OFZ. The inner-approach OFZ is a defined volume of airspace centered on the approach area. The inner-approach OFZ applies only to runways with an approach lighting system. The inner-approach OFZ begins 200 feet from the runway threshold at the same elevation as the runway threshold and extends 200 feet beyond the last light unit in the approach lighting system. The width of the inner-approach OFZ is the same as the runway OFZ and rises at a slope of 50 (horizontal) to 1 (vertical) from the beginning.

3. Inner-transitional OFZ. The inner transitional surface OFZ is a defined volume of airspace along the sides of the runway and inner-approach OFZ and applies only to precision instrument runways. The inner-transitional surface OFZ slopes 3 (horizontal) to 1 (vertical) out from the edges of the runway OFZ and inner-approach OFZ to a height of 150 feet above the established airport elevation.

(Refer to AC 150/5300-13, Chapter 3 and FAA Order 7110.65 paragraph 3-5.)

OBSTRUCTION Any object/obstacle exceeding the obstruction standards specified by Part 77, Subpart C.

OBSTRUCTION LIGHT A light or one of a group of lights, usually red or white, frequently mounted on a surface structure or natural terrain to warn pilots of the presence of an obstruction.

OCEANIC AIRSPACE Airspace over the oceans of the world, considered international airspace, where oceanic separation and procedures per the International Civil Aviation Organization are applied. Responsibility for the provisions of air traffic control service in this airspace is delegated to various countries, based generally upon geographic proximity and the availability of the required resources.

OCEANIC DISPLAY AND PLANNING SYSTEM An automated digital display system which provides flight data processing, conflict probe, and situation display for oceanic air traffic control.

OCEANIC NAVIGATIONAL ERROR REPORT A report filed when an aircraft exiting oceanic airspace has been observed by radar to be off course. ONER reporting parameters and procedures are contained in Order 7110.82, Monitoring of Navigational Performance In Oceanic Areas.

OCEANIC PUBLISHED ROUTE A route established in international airspace and charted or described in flight information publications, such as Route Charts, DOD Enroute Charts, Chart Supplements, NOTAM's, and Track Messages.

OCEANIC TRANSITION ROUTE An ATS route established for the purpose of transitioning aircraft to/from an organized track system.

ODAPS (See OCEANIC DISPLAY AND PLANNING SYSTEM.)

OFF COURSE A term used to describe a situation where an aircraft has reported a position fix or is observed on radar at a point not on the ATC-approved route of flight.

OFFSHORE CONTROL AREA That portion of airspace between the U.S. 12-mile limit and the oceanic CTA/FIR boundary within which air traffic control is exercised. These areas are established to permit the application of domestic procedures in the provision of air traffic control services. Offshore control area is generally synonymous with Federal Aviation Regulations, Part 71, Subpart E, "Control Areas and Control Area Extensions."

OFF-ROUTE VECTOR A vector by ATC which takes an aircraft off a previously assigned route. Altitudes assigned by ATC during such vectors provide required obstacle clearance.

OFFSET PARALLEL RUNWAYS Staggered runways having centerlines which are parallel.

OFT (See OUTER FIX TIME.)

OM (See OUTER MARKER.)

OMEGA An RNAV system designed for long-range navigation based upon ground-based electronic navigational aid signals.

ONER (See OCEANIC NAVIGATIONAL ERROR REPORT.)

OPERATIONAL (See DUE REGARD.)

ON COURSE

1. Used to indicate that an aircraft is established on the route centerline.

2. Used by ATC to advise a pilot making a radar approach that his aircraft is lined up on the final approach course.

(See ON-COURSE INDICATION.)

ON-COURSE INDICATION An indication on an instrument, which provides the pilot a visual means of determining that the aircraft is located on the centerline of a given navigational track, or an indication on a radar scope that an aircraft is on a given track.

OPERATIONAL ACCEPTABLE LEVEL OF TRAFFIC An air traffic activity level associated with the designed capacity for a sector or airport. The OALT considers dynamic changes in staffing, personnel experience levels, equipment outages, operational configurations, weather, traffic complexity, aircraft performance mixtures, transitioning flights, adjacent airspace, handoff/point-out responsibilities, and other factors that may affect an air traffic operational position or system element. The OALT is normally considered to be the total number of aircraft that any air traffic functional position can accommodate for a defined period of time under a given set of circumstances.

OPPOSITE DIRECTION AIRCRAFT Aircraft are operating in opposite directions when:

1. They are following the same track in reciprocal directions; or

2. Their tracks are parallel and the aircraft are flying in reciprocal directions; or

3. Their tracks intersect at an angle of more than 135°.

OPTION APPROACH An approach requested and conducted by a pilot which will result in either a touch-and-go, missed approach, low approach, stop-and-go, or full stop landing.

(See CLEARED FOR THE OPTION.) (Refer to AIM.)

ORGANIZED TRACK SYSTEM A movable system of oceanic tracks that traverses the North Atlantic between Europe and North America the physical position of which is determined twice daily taking the best advantage of the winds aloft.

ORGANIZED TRACK SYSTEM A series of ATS routes which are fixed and charted; i.e., CEP, NOPAC, or flexible and described by NOTAM; i.e., NAT TRACK MESSAGE.

OTR (See OCEANIC TRANSITION ROUTE.)

OTS (See ORGANIZED TRACK SYSTEM.)

OUT The conversation is ended and no response is expected.

OUTER AREA (associated with Class C airspace) Non-regulatory airspace surrounding designated Class C airspace airports wherein ATC provides radar vectoring and sequencing on a full-time basis for all IFR and participating VFR aircraft. The service provided in the outer area is called Class C service which includes: IFR/IFR-standard IFR separation; IFR/VFR-traffic advisories and conflict resolution; and VFR/VFR-traffic advisories and, as appropriate, safety alerts. The normal radius will be 20 nautical miles with some variations based on site-specific requirements. The outer area extends outward from the primary Class C airspace airport and extends from the lower limits of radar/radio coverage up to the ceiling of the approach control's delegated airspace excluding the Class C charted area and other airspace as appropriate.

(See CONTROLLED AIRSPACE.) (See CONFLICT RESOLUTION.)

OUTER COMPASS LOCATOR (See COMPASS LOCATOR).

OUTER FIX A general term used within ATC to describe fixes in the terminal area, other than the final approach fix. Aircraft are normally cleared to these fixes by an Air Route Traffic Control Center or an Approach Control Facility. Aircraft are normally cleared from these fixes to the final approach fix or final approach course.

OUTER FIX An adapted fix along the converted route of flight, prior to the meter fix, for which crossing times are calculated and displayed in the metering position list.

OUTER FIX TIME A calculated time to depart the outer fix in order to cross the vertex at the ACLT. The time reflects descent speed adjustments and any applicable delay time that must be absorbed prior to crossing the meter fix.

OUTER MARKER A marker beacon at or near the glideslope intercept altitude of an ILS approach. It is keyed to transmit two dashes per second on a 400 Hz tone, which is received aurally and visually by compatible airborne equipment. The OM is normally located four to seven miles from the runway threshold on the extended centerline of the runway.

(See MARKER BEACON.) (See INSTRUMENT LANDING SYSTEM.) (Refer to AIM.)

OVER My transmission is ended; I expect a response.

OVERHEAD MANEUVER A series of predetermined maneuvers prescribed for aircraft (often in formation) for entry into the visual flight rules (VFR) traffic pattern and to proceed to a landing. An overhead maneuver is not an instrument flight rules (IFR) approach procedure. An aircraft executing an overhead maneuver is considered VFR and the IFR flight plan is canceled when the aircraft reaches the "initial point" on the initial approach portion of the maneuver. The pattern usually specifies the following:

1. The radio contact required of the pilot.
2. The speed to be maintained.
3. An initial approach 3 to 5 miles in length.
4. An elliptical pattern consisting of two 180 degree turns.
5. A break point at which the first 180 degree turn is started.
6. The direction of turns.
7. Altitude (at least 500 feet above the conventional pattern).
8. A "Roll-out" on final approach not less than ¼ mile from the landing threshold and not less than 300 feet above the ground.

OVERLYING CENTER The ARTCC facility that is responsible for arrival/departure operations at a specific terminal.

P TIME (See PROPOSED DEPARTURE TIME.)

PAN-PAN The international radio-telephony urgency signal. When repeated three times, indicates uncertainty or alert followed by the nature of the urgency.

(See MAYDAY.) (Refer to AIM.)

PAR (See PRECISION APPROACH RADAR.)

PAR [ICAO] (See ICAO Term PRECISION APPROACH RADAR).

PARALLEL ILS APPROACHES Approaches to parallel runways by IFR aircraft which, when established inbound toward the airport on the adjacent final approach courses, are radar-separated by at least 2 miles.

(See FINAL APPROACH COURSE.) (See SIMULTANEOUS ILS APPROACHES.)

PARALLEL MLS APPROACHES (See PARALLEL ILS APPROACHES).

PARALLEL OFFSET ROUTE A parallel track to the left or right of the designated or established airway/route. Normally associated with Area Navigation (RNAV) operations.

(See AREA NAVIGATION.)

PARALLEL RUNWAYS Two or more runways at the same airport whose centerlines are parallel. In addition to runway number, parallel runways are designated as L (left) and R (right) or, if three parallel runways exist, L (left), C (center), and R (right).

PATWAS (See PILOTS AUTOMATIC TELEPHONE WEATHER ANSWERING SERVICE.)

PBCT (See PROPOSED BOUNDARY CROSSING TIME.)

PERMANENT ECHO Radar signals reflected from fixed objects on the earth's surface; e.g., buildings, towers, terrain. Permanent echoes are distinguished from "ground clutter" by being definable locations rather than large areas. Under certain conditions they may be used to check radar alignment.

PHOTO RECONNAISSANCE Military activity that requires locating individual photo targets and navigating to the targets at a preplanned angle and altitude. The activity normally requires a lateral route width of 16 NM and altitude range of 1,500 feet to 10,000 feet AGL.

PIDP (See PROGRAMMABLE INDICATOR DATA PROCESSOR.)

PILOT BRIEFING A service provided by the FSS to assist pilots in flight planning. Briefing items may include weather information, NOTAMs, military activities, flow control information, and other items as requested.

(Refer to AIM.)

PILOT IN COMMAND The pilot responsible for the operation and safety of an aircraft during flight time.

(Refer to Part 91.)

PILOTS AUTOMATIC TELEPHONE WEATHER ANSWERING SERVICE A continuous telephone recording containing current and forecast weather information for pilots.

(See FLIGHT SERVICE STATION.) (Refer to AIM.)

PILOT'S DISCRETION When used in conjunction with altitude assignments, means that ATC has offered the pilot the option of starting climb or descent whenever he wishes and conducting the climb or descent at any rate he wishes.

He may temporarily level off at any intermediate altitude. However, once he has vacated an altitude, he may not return to that altitude.

PILOT WEATHER REPORT   A report of meteorological phenomena encountered by aircraft in flight.
(Refer to AIM.)

PIREP   (See PILOT WEATHER REPORT.)

POINT OUT   (See RADAR POINT OUT.)

POLAR TRACK STRUCTURE   A system of organized routes between Iceland and Alaska which overlie Canadian MNPS Airspace.

POSITION REPORT   A report over a known location as transmitted by an aircraft to ATC.
(Refer to AIM.)

POSITION SYMBOL   A computer-generated indication shown on a radar display to indicate the mode of tracking.

POSITIVE CONTROL   The separation of all air traffic within designated airspace by air traffic control.

PRACTICE INSTRUMENT APPROACH
An instrument approach procedure conducted by a VFR or an IFR aircraft for the purpose of pilot training or proficiency demonstrations.

PREARRANGED COORDINATION   A standardized procedure which permits an air traffic controller to enter the airspace assigned to another air traffic controller without verbal coordination. The procedures are defined in a facility directive which ensures standard separation between aircraft.

PRECIPITATION   Any or all forms of water particles (rain, sleet, hail, or snow) that fall from the atmosphere and reach the surface.

PRECISION APPROACH   (See PRECISION APPROACH PROCEDURE.)

PRECISION APPROACH PROCEDURE   A standard instrument approach procedure in which an electronic glideslope/glidepath is provided; e.g., ILS/MLS and PAR.
(See INSTRUMENT LANDING SYSTEM.) (See MICROWAVE LANDING SYSTEM.) (See PRECISION APPROACH RADAR.)

PRECISION APPROACH RADAR   Radar equipment in some ATC facilities operated by the FAA and/or the military services at joint-use civil/military locations and separate military installations to detect and display azimuth, elevation, and range of aircraft on the final approach course to a runway. This equipment may be used to monitor certain nonradar approaches, but is primarily used to conduct a precision instrument approach (PAR) wherein the controller issues guidance instructions to the pilot based on the aircraft's position in relation to the final approach course (azimuth), the glidepath (elevation), and the distance (range) from the touchdown point on the runway as displayed on the radar scope.
(See GLIDEPATH.) (See PAR.) (Refer to AIM.) The abbreviation "PAR" is also used to denote preferential arrival routes in ARTCC computers.
(See PREFERENTIAL ROUTES.) (See ICAO term PRECISION APPROACH RADAR.)

PRECISION APPROACH RADAR [ICAO]
Primary radar equipment used to determine the position of an aircraft during final approach, in terms of lateral and vertical deviations relative to a nominal approach path, and in range relative to touchdown.
Note: Precision approach radars are designed to enable pilots of aircraft to be given guidance by radio communication during the final stages of the approach to land.

PRECISION RUNWAY MONITOR   Provides air traffic controllers with high precision secondary surveillance data for aircraft on final approach to closely spaced parallel runways. High resolution color monitoring displays (FMA) are required to present surveillance track data to controllers along with detailed maps depicting approaches and no transgression zone.

PREFERENTIAL ROUTES   Preferential routes (PDR's, PAR's, and PDAR's) are adapted in ARTCC computers to accomplish inter/intrafacility controller coordination and to assure that flight data is posted at the proper control positions. Locations having a need for these specific inbound and outbound routes normally publish such routes in local facility bulletins, and their use by pilots minimizes flight plan route amendments. When the workload or traffic situation permits, controllers normally provide radar vectors or assign requested routes to minimize circuitous routing. Preferential routes are usually confined to one ARTCC's area and are referred to by the following names or acronyms:

1. Preferential Departure Route (PDR). A specific departure route from an airport or terminal area to an en route point where there is no further need for flow control. It may be included in a Standard Instrument Departure (SID) or a Preferred IFR Route.

2. Preferential Arrival Route (PAR). A specific arrival route from an appropriate en route point to an airport or terminal area. It may be included in a Standard Terminal Arrival (STAR) or a Preferred IFR Route. The abbreviation "PAR" is used primarily within the ARTCC and should not be confused with the abbreviation for Precision Approach Radar.

3. Preferential Departure and Arrival Route (PDAR). A route between two terminals which are within or immediately adjacent to one ARTCC's area. PDAR's are not synonymous with Preferred IFR Routes but may be listed as such as they do accomplish essentially the same purpose.
(See PREFERRED IFR ROUTES.) (See NAS STAGE A.)

PREFERRED IFR ROUTES   Routes established between busier airports to increase system efficiency and capacity.

They normally extend through one or more ARTCC areas and are designed to achieve balanced traffic flows among high density terminals. IFR clearances are issued on the basis of these routes except when severe weather avoidance procedures or other factors dictate otherwise. Preferred IFR Routes are listed in the Airport/Facility Directory. If a flight is planned to or from an area having such routes but the departure or arrival point is not listed in the Airport/Facility Directory, pilots may use that part of a Preferred IFR Route which is appropriate for the departure or arrival point that is listed. Preferred IFR Routes are correlated with SID's and STAR's and may be defined by airways, jet routes, direct routes between NAVAID's, Waypoints, NAVAID radials/ DME, or any combinations thereof.

(See STANDARD INSTRUMENT DEPARTURE.) (See STANDARD TERMINAL ARRIVAL.) (See PREFERENTIAL ROUTES.) (See CENTER'S AREA.) (Refer to AIRPORT/FACILITY DIRECTORY.) (Refer to NOTICES TO AIRMEN PUBLICATION.)

PREFLIGHT PILOT BRIEFING (See PILOT BRIEFING.)

PREVAILING VISIBILITY (See VISIBILITY.)

PRM (See PRECISION RUNWAY MONITOR.)

PROCEDURE TURN The maneuver prescribed when it is necessary to reverse direction to establish an aircraft on the intermediate approach segment or final approach course. The outbound course, direction of turn, distance within which the turn must be completed, and minimum altitude are specified in the procedure. However, unless otherwise restricted, the point at which the turn may be commenced and the type and rate of turn are left to the discretion of the pilot.

(See ICAO term PROCEDURE TURN.)

PROCEDURE TURN [ICAO] A manoeuver in which a turn is made away from a designated track followed by a turn in the opposite direction to permit the aircraft to intercept and proceed along the reciprocal of the designated track.

Note 1: Procedure turns are designated "left" or "right" according to the direction of the initial turn.

Note 2: Procedure turns may be designated as being made either in level flight or while descending, according to the circumstances of each individual approach procedure.

PROCEDURE TURN INBOUND That point of a procedure turn maneuver where course reversal has been completed and an aircraft is established inbound on the intermediate approach segment or final approach course. A report of "procedure turn inbound" is normally used by ATC as a position report for separation purposes.

(See FINAL APPROACH COURSE.) (See PROCEDURE TURN.) (See SEGMENTS OF AN INSTRUMENT APPROACH PROCEDURE.)

PROFILE DESCENT An uninterrupted descent (except where level flight is required for speed adjustment; e.g., 250 knots at 10,000 feet MSL) from cruising altitude/level to interception of a glideslope or to a minimum altitude specified for the initial or intermediate approach segment of a nonprecision instrument approach. The profile descent normally terminates at the approach gate or where the glideslope or other appropriate minimum altitude is intercepted.

PROGRAMMABLE INDICATOR DATA PROCESSOR The PIDP is a modification to the AN/TPX-42 interrogator system currently installed in fixed RAPCON's. The PIDP detects, tracks, and predicts secondary radar aircraft targets. These are displayed by means of computer-generated symbols and alphanumeric characters depicting flight identification, aircraft altitude, ground speed, and flight plan data. Although primary radar targets are not tracked, they are displayed coincident with the secondary radar targets as well as with the other symbols and alphanumerics. The system has the capability of interfacing with ARTCC's.

PROGRESS REPORT (See POSITION REPORT.)

PROGRESSIVE TAXI Precise taxi instructions given to a pilot unfamiliar with the airport or issued in stages as the aircraft proceeds along the taxi route.

PROHIBITED AREA (See SPECIAL USE AIRSPACE.)

(See ICAO term PROHIBITED AREA.)

PROHIBITED AREA [ICAO] An airspace of defined dimensions, above the land areas or territorial waters of a State, within which the flight of aircraft is prohibited.

PROPOSED BOUNDARY CROSSING TIME Each center has a PBCT parameter for each internal airport. Proposed internal flight plans are transmitted to the adjacent center if the flight time along the proposed route from the departure airport to the center boundary is less than or equal to the value of PBCT or if airport adaptation specifies transmission regardless of PBCT.

PROPOSED DEPARTURE TIME The time a scheduled flight will depart the gate (scheduled operators) or the actual runway off time for nonscheduled operators. For EDCT purposes, the ATCSCC adjusts the "P" time for scheduled operators to reflect the runway off times.

PROTECTED AIRSPACE The airspace on either side of an oceanic route/track that is equal to one-half the lateral separation minimum except where reduction of protected airspace has been authorized.

PT (See PROCEDURE TURN.)

PTS (See POLAR TRACK STRUCTURE.)

PUBLISHED ROUTE A route for which an IFR altitude has been established and published; e.g., Federal Airways, Jet Routes, Area Navigation Routes, Specified Direct Routes.

QUEUING (See STAGING/QUEUING.)

LOW (See QUOTA FLOW CONTROL.)

QNE The barometric pressure used for the standard altimeter setting (29.92 inches Hg.).

QNH The barometric pressure as reported by a particular station.

QUADRANT A quarter part of a circle, centered on a NAVAID, oriented clockwise from magnetic north as follows: NE quadrant 000-089, SE quadrant 090-179, SW quadrant 180-269, NW quadrant 270-359.

QUICK LOOK A feature of NAS Stage A and ARTS which provides the controller the capability to display full data blocks of tracked aircraft from other control positions.

QUOTA FLOW CONTROL A flow control procedure by which the Central Flow Control Function (CFCF) restricts traffic to the ARTC Center area having an impacted airport, thereby avoiding sector/area saturation.
(See AIR TRAFFIC CONTROL SYSTEM COMMAND CENTER.) (Refer to AIRPORT/FACILITY DIRECTORY.)

RADAR A device which, by measuring the time interval between transmission and reception of radio pulses and correlating the angular orientation of the radiated antenna beam or beams in azimuth and/or elevation, provides information on range, azimuth, and/or elevation of objects in the path of the transmitted pulses.

1. Primary Radar. A radar system in which a minute portion of a radio pulse transmitted from a site is reflected by an object and then received back at that site for processing and display at an air traffic control facility.

2. Secondary Radar/Radar Beacon (ATCRBS). A radar system in which the object to be detected is fitted with cooperative equipment in the form of a radio receiver/transmitter (transponder). Radar pulses transmitted from the searching transmitter/receiver (interrogator) site are received in the cooperative equipment and used to trigger a distinctive transmission from the transponder. This reply transmission, rather than a reflected signal, is then received back at the transmitter/receiver site for processing and display at an air traffic control facility.
(See TRANSPONDER.) (See INTERROGATOR.) (Refer to AIM.) (See ICAO term RADAR.) (See ICAO term PRIMARY RADAR.) (See ICAO term SECONDARY RADAR.)

RADAR [ICAO] A radio detection device which provides information on range, azimuth and/or elevation of objects.

Primary Radar.—Radar system which uses reflected radio signals.

Secondary Radar.—Radar system wherein a radio signal transmitted from a radar station initiates the transmission of a radio signal from another station.

RADAR ADVISORY The provision of advice and information based on radar observations.
(See ADVISORY SERVICE.)

RADAR ALTIMETER (See RADIO ALTIMETER.)

RADAR APPROACH An instrument approach procedure which utilizes Precision Approach Radar (PAR) or Airport Surveillance Radar (ASR).
(See SURVEILLANCE APPROACH.) (See AIRPORT SURVEILLANCE RADAR.) (See PRECISION APPROACH RADAR.) (See INSTRUMENT APPROACH PROCEDURE.) (Refer to AIM.) (See ICAO term RADAR APPROACH.)

RADAR APPROACH [ICAO] An approach, executed by an aircraft, under the direction of a radar controller.

RADAR APPROACH CONTROL FACILITY A terminal ATC facility that uses radar and nonradar capabilities to provide approach control services to aircraft arriving, departing, or transiting airspace controlled by the facility.
(See APPROACH CONTROL SERVICE.)
Provides radar ATC services to aircraft operating in the vicinity of one or more civil and/or military airports in a terminal area. The facility may provide services of a ground controlled approach (GCA); i.e., ASR and PAR approaches. A radar approach control facility may be operated by FAA, USAF, US Army, USN, USMC, or jointly by FAA and a military service. Specific facility nomenclatures are used for administrative purposes only and are related to the physical location of the facility and the operating service generally as follows:

Army Radar Approach Control (ARAC) (Army).
Radar Air Traffic Control Facility (RATCF) (Navy/FAA).
Radar Approach Control (RAPCON) (Air Force/FAA).
Terminal Radar Approach Control (TRACON) (FAA).
Tower/Airport Traffic Control Tower (ATCT) (FAA).
(Only those towers delegated approach control authority).

RADAR ARRIVAL An aircraft arriving at an airport served by a radar facility and in radar contact with the facility.
(See NONRADAR.)

RADAR BEACON (See RADAR.)

RADAR CONTACT

1. Used by ATC to inform an aircraft that it is identified on the radar display and radar flight following will be provided until radar identification is terminated. Radar service may also be provided within the limits of necessity and capability. When a pilot is informed of "radar contact," he automatically discontinues reporting over compulsory reporting points.
(See RADAR FLIGHT FOLLOWING.) (See RADAR CONTACT LOST.) (See RADAR SERVICE.) (See RADAR SERVICE TERMINATED.) (Refer to AIM.)

2. The term used to inform the controller that the aircraft is identified and approval is granted for the aircraft to enter the receiving controller's airspace.
(See ICAO term RADAR CONTACT.)

RADAR CONTACT LOST Used by ATC to inform a pi-

lot that radar data used to determine the aircraft's position is no longer being received, or is no longer reliable and radar service is no longer being provided. The loss may be attributed to several factors including the aircraft merging with weather or ground clutter, the aircraft operating below radar line of sight coverage, the aircraft entering an area of poor radar return, failure of the aircraft transponder, or failure of the ground radar equipment.
(See CLUTTER.) (See RADAR CONTACT.)

RADAR CLUTTER [ICAO] The visual indication on a radar display of unwanted signals.

RADAR CONTACT [ICAO] The situation which exists when the radar blip or radar position symbol of a particular aircraft is seen and identified on a radar display.

RADAR ENVIRONMENT An area in which radar service may be provided.
(See RADAR CONTACT.) (See RADAR SERVICE.) (See ADDITIONAL SERVICES.) (See TRAFFIC ADVISORIES.)

RADAR FLIGHT FOLLOWING The observation of the progress of radar identified aircraft, whose primary navigation is being provided by the pilot, wherein the controller retains and correlates the aircraft identity with the appropriate target or target symbol displayed on the radar scope.
(See RADAR CONTACT.) (See RADAR SERVICE.) (Refer to AIM.)

RADAR IDENTIFICATION The process of ascertaining that an observed radar target is the radar return from a particular aircraft.
(See RADAR CONTACT.) (See RADAR SERVICE.) (See ICAO term RADAR IDENTIFICATION.)

RADAR IDENTIFICATION [ICAO] The process of correlating a particular radar blip or radar position symbol with a specific aircraft.

RADAR IDENTIFIED AIRCRAFT An aircraft, the position of which has been correlated with an observed target or symbol on the radar display.
(See RADAR CONTACT.) (See RADAR CONTACT LOST.)

RADAR MONITORING (See RADAR SERVICE.)

RADAR NAVIGATIONAL GUIDANCE (See RADAR SERVICE.)

RADAR POINT OUT An action taken by a controller to transfer the radar identification of an aircraft to another controller if the aircraft will or may enter the airspace or protected airspace of another controller and radio communications will not be transferred.

RADAR REQUIRED A term displayed on charts and approach plates and included in FDC Notams to alert pilots that segments of either an instrument approach procedure or a route are not navigable because of either the absence or unusability of a NAVAID. The pilot can expect to be provided radar navigational guidance while transiting segments labeled with this term.
(See RADAR ROUTE.) (See RADAR SERVICE.)

RADAR ROUTE A flight path or route over which an aircraft is vectored. Navigational guidance and altitude assignments are provided by ATC.
(See FLIGHT PATH.) (See ROUTE.)

RADAR SEPARATION (See RADAR SERVICE.)

RADAR SERVICE A term which encompasses one or more of the following services based on the use of radar which can be provided by a controller to a pilot of a radar identified aircraft.

1. Radar Monitoring. The radar flight-following of aircraft, whose primary navigation is being performed by the pilot, to observe and note deviations from its authorized flight path, airway, or route. When being applied specifically to radar monitoring of instrument approaches; i.e., with precision approach radar (PAR) or radar monitoring of simultaneous ILS/MLS approaches, it includes advice and instructions whenever an aircraft nears or exceeds the prescribed PAR safety limit or simultaneous ILS/MLS no transgression zone.
(See ADDITIONAL SERVICES.) (See TRAFFIC ADVISORIES.)

2. Radar Navigational Guidance. Vectoring aircraft to provide course guidance.

3. Radar Separation. Radar spacing of aircraft in accordance with established minima.
(See ICAO term RADAR SERVICE.)

RADAR SERVICE [ICAO] Term used to indicate a service provided directly by means of radar.

Radar Monitoring.—The use of radar for the purpose of providing aircraft with information and advice relative to significant deviations from nominal flight path.

Radar Separation.—The separation used when aircraft position information is derived from radar sources.

RADAR SERVICE TERMINATED Used by ATC to inform a pilot that he will no longer be provided any of the services that could be received while in radar contact. Radar service is automatically terminated, and the pilot is not advised in the following cases:

1. An aircraft cancels its IFR flight plan, except within Class B airspace, Class C airspace, a TRSA, or where Basic Radar service is provided.

2. An aircraft conducting an instrument, visual, or contact approach has landed or has been instructed to change to advisory frequency.

3. An arriving VFR aircraft, receiving radar service to a tower-controlled airport within Class B airspace, Class C airspace, a TRSA, or where sequencing service is provided, has landed; or to all other airports, is instructed to change to tower or advisory frequency.

4. An aircraft completes a radar approach.

RADAR SURVEILLANCE The radar observation of a given geographical area for the purpose of performing some radar function.

RADAR TRAFFIC ADVISORIES Advisories issued to alert pilots to known or observed radar traffic which may affect the intended route of flight of their aircraft.
(See TRAFFIC ADVISORIES.)

RADAR TRAFFIC INFORMATION SERVICE (See TRAFFIC ADVISORIES.)

RADAR VECTORING [ICAO] Provision of navigational guidance to aircraft in the form of specific headings, based on the use of radar.

RADAR WEATHER ECHO INTENSITY LEVELS Existing radar systems cannot detect turbulence. However, there is a direct correlation between the degree of turbulence and other weather features associated with thunderstorms and the radar weather echo intensity. The National Weather Service has categorized radar weather echo intensity for precipitation into six levels. These levels are sometimes expressed during communications as "VIP LEVEL" 1 through 6 (derived from the component of the radar that produces the information-Video Integrator and Processor). The following list gives the "VIP LEVELS" in relation to the precipitation intensity within a thunderstorm:

Level 1. WEAK
Level 2. MODERATE
Level 3. STRONG
Level 4. VERY STRONG
Level 5. INTENSE
Level 6. EXTREME
(See AC00-45.)

RADIAL A magnetic bearing extending from a VOR/VORTAC/TACAN navigation facility.

RADIO

1. A device used for communication.

2. Used to refer to a flight service station; e.g., "Seattle Radio" is used to call Seattle FSS.

RADIO ALTIMETER Aircraft equipment which makes use of the reflection of radio waves from the ground to determine the height of the aircraft above the surface.

RADIO BEACON (See NONDIRECTIONAL BEACON.)

RADIO DETECTION AND RANGING (See RADAR.)

RADIO MAGNETIC INDICATOR An aircraft navigational instrument coupled with a gyro compass or similar compass that indicates the direction of a selected NAVAID and indicates bearing with respect to the heading of the aircraft.

RAMP (See APRON.)

RANDOM ALTITUDE An altitude inappropriate for direction of flight and/or not in accordance with paragraph 4-60.

RANDOM ROUTE Any route not established or charted/published or not otherwise available to all users.

RC (See ROAD RECONNAISSANCE.)

RCAG (See REMOTE COMMUNICATIONS AIR/GROUND FACILITY.)

RCC (See RESCUE COORDINATION CENTER.)

RCO (See REMOTE COMMUNICATIONS OUTLET.)

RCR (See RUNWAY CONDITION READING.)

READ BACK Repeat my message back to me.

RECEIVING CONTROLLER A controller/facility receiving control of an aircraft from another controller/facility.

RECEIVING FACILITY (See RECEIVING CONTROLLER.)

REDUCE SPEED TO (SPEED) (See SPEED ADJUSTMENT.)

REIL (See RUNWAY END IDENTIFIER LIGHTS.)

RELEASE TIME A departure time restriction issued to a pilot by ATC (either directly or through an authorized relay) when necessary to separate a departing aircraft from other traffic.
(See ICAO term RELEASE TIME.)

RELEASE TIME [ICAO] Time prior to which an aircraft should be given further clearance or prior to which it should not proceed in case of radio failure.

REMOTE COMMUNICATIONS AIR/GROUND FACILITY An unmanned VHF/UHF transmitter/receiver facility which is used to expand ARTCC air/ground communications coverage and to facilitate direct contact between pilots and controllers. RCAG facilities are sometimes not equipped with emergency frequencies 121.5 MHz and 243.0 MHz.
(Refer to AIM.)

REMOTE COMMUNICATIONS OUTLET
An unmanned communications facility remotely controlled by air traffic personnel. RCO's serve FSS's. RTR's serve terminal ATC facilities. An RCO or RTR may be UHF or VHF and will extend the communication range of the air traffic facility. There are several classes of RCO's and RTR's. The class is determined by the number of transmitters or receivers. Classes A through G are used primarily for air/ground purposes. RCO and RTR class O facilities are nonprotected outlets subject to undetected and prolonged outages. RCO (O's) and RTR (O's) were established for the express purpose of providing ground-to-ground communications between air traffic control specialists and pilots lo-

cated at a satellite airport for delivering en route clearances, issuing departure authorizations, and acknowledging instrument flight rules cancellations or departure/landing times. As a secondary function, they may be used for advisory purposes whenever the aircraft is below the coverage of the primary air/ground frequency.

REMOTE TRANSMITTER/RECEIVER (See REMOTE COMMUNICATIONS OUTLET.)

REPORT Used to instruct pilots to advise ATC of specified information; e.g., "Report passing Hamilton VOR."

REPORTING POINT A geographical location in relation to which the position of an aircraft is reported.
(See COMPULSORY REPORTING POINTS.) (Refer to AIM.) (See ICAO term REPORTING POINT.)

REPORTING POINT [ICAO] A specified geographical location in relation to which the position of an aircraft can be reported.

REQUEST FULL ROUTE CLEARANCE Used by pilots to request that the entire route of flight be read verbatim in an ATC clearance. Such request should be made to preclude receiving an ATC clearance based on the original filed flight plan when a filed IFR flight plan has been revised by the pilot, company, or operations prior to departure.

RESCUE COORDINATION CENTER A search and rescue (SAR) facility equipped and manned to coordinate and control SAR operations in an area designated by the SAR plan. The U.S. Coast Guard and the U.S. Air Force have responsibility for the operation of RCC's.
(See ICAO term RESCUE CO-ORDINATION CENTRE.)

RESCUE CO-ORDINATION CENTRE [ICAO] A unit responsible for promoting efficient organization of search and rescue service and for coordinating the conduct of search and rescue operations within a search and rescue region.

RESTRICTED AREA (See SPECIAL USE AIRSPACE.)
(See ICAO term RESTRICTED AREA.)

RESTRICTED AREA [ICAO] An airspace of defined dimensions, above the land areas or territorial waters of a State, within which the flight of aircraft is restricted in accordance with certain specified conditions.

RESUME OWN NAVIGATION Used by ATC to advise a pilot to resume his own navigational responsibility. It is issued after completion of a radar vector or when radar contact is lost while the aircraft is being radar vectored.
(See RADAR CONTACT LOST.) (See RADAR SERVICE TERMINATED.)

RMI (See RADIO MAGNETIC INDICATOR.)

RNAV (See AREA NAVIGATION.)

RNAV [ICAO] (See ICAO Term AREA NAVIGATION.)

RNAV APPROACH An instrument approach procedure which relies on aircraft area navigation equipment for navigational guidance.
(See INSTRUMENT APPROACH PROCEDURE). (See AREA NAVIGATION.)

ROAD RECONNAISSANCE Military activity requiring navigation along roads, railroads, and rivers. Reconnaissance route/route segments are seldom along a straight line and normally require a lateral route width of 10 NM to 30 NM and an altitude range of 500 feet to 10,000 feet AGL.

ROGER I have received all of your last transmission. It should not be used to answer a question requiring a yes or a no answer.
(See AFFIRMATIVE.) (See NEGATIVE.)

ROLLOUT RVR (See VISIBILITY.)

ROUTE A defined path, consisting of one or more courses in a horizontal plane, which aircraft traverse over the surface of the earth.
(See AIRWAY.) (See JET ROUTE.) (See PUBLISHED ROUTE.) (See UNPUBLISHED ROUTE.)

ROUTE SEGMENT As used in Air Traffic Control, a part of a route that can be defined by two navigational fixes, two NAVAID's, or a fix and a NAVAID.
(See FIX.) (See ROUTE.) (See ICAO term ROUTE SEGMENT.)

ROUTE SEGMENT [ICAO] A portion of a route to be flown, as defined by two consecutive significant points specified in a flight plan.

RSA (See RUNWAY SAFETY AREA.)

RTR (See REMOTE TRANSMITTER/RECEIVER.)

RUNWAY A defined rectangular area on a land airport prepared for the landing and takeoff run of aircraft along its length. Runways are normally numbered in relation to their magnetic direction rounded off to the nearest 10 degrees; e.g., Runway 01, Runway 25.
(See PARALLEL RUNWAYS.) (See ICAO term RUNWAY.)

RUNWAY [ICAO] A defined rectangular area on a land aerodrome prepared for the landing and takeoff of aircraft.

RUNWAY CENTERLINE LIGHTING (See AIRPORT LIGHTING).

RUNWAY CONDITION READING Numerical decelerometer readings relayed by air traffic controllers at USAF and certain civil bases for use by the pilot in determining runway braking action. These readings are routinely relayed only to USAF and Air National Guard Aircraft.
(See BRAKING ACTION.)

RUNWAY END IDENTIFIER LIGHTS (See AIRPORT LIGHTING.)

RUNWAY GRADIENT The average slope, measured in percent, between two ends or points on a runway. Runway gradient is depicted on Government aerodrome sketches when total runway gradient exceeds 0.3%.

RUNWAY HEADING The magnetic direction that corresponds with the runway centerline extended, not the painted runway number. When cleared to "fly or maintain runway heading," pilots are expected to fly or maintain the heading that corresponds with the extended centerline of the departure runway. Drift correction shall not be applied; e.g., Runway 4, actual magnetic heading of the runway centerline 044, fly 044.

RUNWAY IN USE/ACTIVE RUNWAY/DUTY RUNWAY Any runway or runways currently being used for takeoff or landing. When multiple runways are used, they are all considered active runways. In the metering sense, a selectable adapted item which specifies the landing runway configuration or direction of traffic flow. The adapted optimum flight plan from each transition fix to the vertex is determined by the runway configuration for arrival metering processing purposes.

RUNWAY LIGHTS (See AIRPORT LIGHTING.)

RUNWAY MARKINGS (See AIRPORT MARKING AIDS.)

RUNWAY OVERRUN In military aviation exclusively, a stabilized or paved area beyond the end of a runway, of the same width as the runway plus shoulders, centered on the extended runway centerline.

RUNWAY PROFILE DESCENT An instrument flight rules (IFR) air traffic control arrival procedure to a runway published for pilot use in graphic and/or textual form and may be associated with a STAR. Runway Profile Descents provide routing and may depict crossing altitudes, speed restrictions, and headings to be flown from the en route structure to the point where the pilot will receive clearance for and execute an instrument approach procedure. A Runway Profile Descent may apply to more than one runway if so stated on the chart.

(Refer to AIM.)

RUNWAY SAFETY AREA A defined surface surrounding the runway prepared, or suitable, for reducing the risk of damage to airplanes in the event of an undershoot, overshoot, or excursion from the runway. The dimensions of the RSA vary and can be determined by using the criteria contained within Advisory Circular 150/5300-13, Chapter 3. Figure 3-1 in Advisory Circular 150/5300-13 depicts the RSA. The design standards dictate that the RSA shall be:

1. Cleared, graded, and have no potentially hazardous ruts, humps, depressions, or other surface variations;
2. Drained by grading or storm sewers to prevent water accumulation;
3. Capable, under dry conditions, of supporting snow removal equipment, aircraft rescue and firefighting equipment, and the occasional passage of aircraft without causing structural damage to the aircraft; and,
4. Free of objects, except for objects that need to be located in the runway safety area because of their function. These objects shall be constructed on low impact resistant supports (frangible mounted structures) to the lowest practical height with the frangible point no higher than 3 inches above grade.

(Refer to AC 150/5300-13, Chap. 3.)

RUNWAY USE PROGRAM A noise abatement runway selection plan designed to enhance noise abatement efforts with regard to airport communities for arriving and departing aircraft. These plans are developed into runway use programs and apply to all turbojet aircraft 12,500 pounds or heavier; turbojet aircraft less than 12,500 pounds are included only if the airport proprietor determines that the aircraft creates a noise problem. Runway use programs are coordinated with FAA offices, and safety criteria used in these programs are developed by the Office of Flight Operations. Runway use programs are administered by the Air Traffic Service as "Formal" or "Informal" programs.

1. Formal Runway Use Program. An approved noise abatement program which is defined and acknowledged in a Letter of Understanding between Flight Operations, Air Traffic Service, the airport proprietor, and the users. Once established, participation in the program is mandatory for aircraft operators and pilots as provided for in Part 91.129.
2. Informal Runway Use Program. An approved noise abatement program which does not require a Letter of Understanding, and participation in the program is voluntary for aircraft operators/pilots.

RUNWAY VISIBILITY VALUE (See VISIBILITY.)

RUNWAY VISUAL RANGE (See VISIBILITY.)

SAFETY ALERT A safety alert issued by ATC to aircraft under their control if ATC is aware the aircraft is at an altitude which, in the controller's judgment, places the aircraft in unsafe proximity to terrain, obstructions, or other aircraft. The controller may discontinue the issuance of further alerts if the pilot advises he is taking action to correct the situation or has the other aircraft in sight.

1. Terrain/Obstruction Alert. A safety alert issued by ATC to aircraft under their control if ATC is aware the aircraft is at an altitude which, in the controller's judgment, places the aircraft in unsafe proximity to terrain/obstructions; e.g., "Low Altitude Alert, check your altitude immediately."
2. Aircraft Conflict Alert. A safety alert issued by ATC to aircraft under their control if ATC is aware of an aircraft that is not under their control at an altitude which, in the controller's judgment, places both aircraft in unsafe proximity to each other. With the alert, ATC will offer the pilot an alternate course of action when feasible; e.g., "Traffic Alert, advise you turn right heading zero niner zero or climb to eight thousand immediately."

The issuance of a safety alert is contingent upon the capability of the controller to have an awareness of an unsafe condition. The course of action provided will be predicated on other traffic under ATC control. Once the alert is issued, it is solely the pilot's prerogative to determine what course of action, if any, he will take.

SAIL BACK A maneuver during high wind conditions (usually with power off where float plane movement is controlled by water rudders/opening and closing cabin doors.

SAME DIRECTION AIRCRAFT Aircraft are operating in the same direction when:

1. They are following the same track in the same direction; or
2. Their tracks are parallel and the aircraft are flying in the same direction; or
3. Their tracks intersect at an angle of less than 45 degrees.

SAR (See SEARCH AND RESCUE.)

SAY AGAIN Used to request a repeat of the last transmission. Usually specifies transmission or portion thereof not understood or received; e.g., "Say again all after ABRAM VOR."

SAY ALTITUDE Used by ATC to ascertain an aircraft's specific altitude/flight level. When the aircraft is climbing or descending, the pilot should state the indicated altitude rounded to the nearest 100 feet.

SAY HEADING Used by ATC to request an aircraft heading. The pilot should state the actual heading of the aircraft.

SDF (See SIMPLIFIED DIRECTIONAL FACILITY.)

SEA LANE A designated portion of water outlined by visual surface markers for and intended to be used by aircraft designed to operate on water.

SEARCH AND RESCUE A service which seeks missing aircraft and assists those found to be in need of assistance. It is a cooperative effort using the facilities and services of available Federal, state and local agencies. The U.S. Coast Guard is responsible for coordination of search and rescue for the Maritime Region, and the U.S. Air Force is responsible for search and rescue for the Inland Region. Information pertinent to search and rescue should be passed through any air traffic facility or be transmitted directly to the Rescue Coordination Center by telephone.

(See FLIGHT SERVICE STATION.) (See RESCUE COORDINATION CENTER.) (Refer to AIM.)

SEARCH AND RESCUE FACILITY A facility responsible for maintaining and operating a search and rescue (SAR) service to render aid to persons and property in distress. It is any SAR unit, station, NET, or other operational activity which can be usefully employed during an SAR Mission; e.g., a Civil Air Patrol Wing, or a Coast Guard Station.

(See SEARCH AND RESCUE.)

SECTIONAL AERONAUTICAL CHARTS (See AERONAUTICAL CHART.)

SECTOR LIST DROP INTERVAL A parameter number of minutes after the meter fix time when arrival aircraft will be deleted from the arrival sector list.

SEE AND AVOID When weather conditions permit, pilots operating IFR and VFR are required to observe and maneuver to avoid other aircraft. Right-of-way rules are contained in Part 91.

SEGMENTED CIRCLE A system of visual indicators designed to provide traffic pattern information at airports without operating control towers.

(Refer to AIM.)

SEGMENTS OF AN INSTRUMENT APPROACH PROCEDURE An instrument approach procedure may have as many as four separate segments depending on how the approach procedure is structured.

1. Initial Approach. The segment between the initial approach fix and the intermediate fix or the point where the aircraft is established on the intermediate course or final approach course.

(See ICAO term INITIAL APPROACH SEGMENT.)

2. Intermediate Approach. The segment between the intermediate fix or point and the final approach fix.

(See ICAO term INTERMEDIATE APPROACH SEGMENT.)

3. Final Approach. The segment between the final approach fix or point and the runway, airport, or missed approach point.

(See ICAO term FINAL APPROACH SEGMENT.)

4. Missed Approach. The segment between the missed approach point or the point of arrival at decision height and the missed approach fix at the prescribed altitude.

(Refer to Part 97.) (See ICAO term MISSED APPROACH PROCEDURE.)

SELECTED GROUND DELAYS A traffic management procedure whereby selected flights are issued ground delays to better regulate traffic flows over a particular fix or area.

SEPARATION In air traffic control, the spacing of aircraft to achieve their safe and orderly movement in flight and while landing and taking off.

(See SEPARATION MINIMA). (See ICAO term SEPARATION.)

SEPARATION [ICAO] Spacing between aircraft, levels or tracks.

SEPARATION MINIMA The minimum longitudinal, lateral, or vertical distances by which aircraft are spaced through the application of air traffic control procedures.

(See SEPARATION.)

SERVICE A generic term that designates functions or assistance available from or rendered by air traffic control. For example, Class C service would denote the ATC services provided within a Class C airspace area.

SEVERE WEATHER AVOIDANCE PLAN An approved plan to minimize the effect of severe weather on traffic flows in impacted terminal and/or ARTCC areas. SWAP is normally implemented to provide the least disruption to the ATC system when flight through portions of airspace is difficult or impossible due to severe weather.

SEVERE WEATHER FORECAST ALERTS Preliminary messages issued in order to alert users that a Severe Weather Watch Bulletin (WW) is being issued. These messages define areas of possible severe thunderstorms or tornado activity. The messages are unscheduled and issued as required by the National Severe Storm Forecast Center at Kansas City, Missouri.
(See SIGMET.) (See CONVECTIVE SIGMET.) (See CWA.) (See AIRMET.)

SFA (See SINGLE FREQUENCY APPROACH.)

SFO (See SIMULATED FLAMEOUT.)

SHF (See SUPER HIGH FREQUENCY.)

SHORT RANGE CLEARANCE A clearance issued to a departing IFR flight which authorizes IFR flight to a specific fix short of the destination while air traffic control facilities are coordinating and obtaining the complete clearance.

SHORT TAKEOFF AND LANDING AIRCRAFT An aircraft which, at some weight within its approved operating weight, is capable of operating from a STOL runway in compliance with the applicable STOL characteristics, airworthiness, operations, noise, and pollution standards.
(See VERTICAL TAKEOFF AND LANDING AIRCRAFT.)

SIAP (See STANDARD INSTRUMENT APPROACH PROCEDURE.)

SIDESTEP MANEUVER A visual maneuver accomplished by a pilot at the completion of an instrument approach to permit a straight-in landing on a parallel runway not more than 1,200 feet to either side of the runway to which the instrument approach was conducted.
(Refer to AIM.)

SIGMET A weather advisory issued concerning weather significant to the safety of all aircraft. SIGMET advisories cover severe and extreme turbulence, severe icing, and widespread dust or sandstorms that reduce visibility to less than 3 miles.
(See AWW.) (See CONVECTIVE SIGMET.) (See CWA.) (See AIRMET.) (Refer to AIM.) (See ICAO term SIGMET INFORMATION.)

SIGMET INFORMATION [ICAO] Information issued by a meteorological watch office concerning the occurrence or expected occurrence of specified en-route weather phenomena which may affect the safety of aircraft operations.

SIGNIFICANT METEOROLOGICAL INFORMATION (See SIGMET.)

SIGNIFICANT POINT A point, whether a named intersection, a NAVAID, a fix derived from a NAVAID(s), or geographical coordinate expressed in degrees of latitude and longitude, which is established for the purpose of providing separation, as a reporting point, or to delineate a route of flight.

SIMPLIFIED DIRECTIONAL FACILITY A NAVAID used for nonprecision instrument approaches. The final approach course is similar to that of an ILS localizer except that the SDF course may be offset from the runway, generally not more than 3 degrees, and the course may be wider than the localizer, resulting in a lower degree of accuracy.
(Refer to AIM.)

SIMULATED FLAMEOUT A practice approach by a jet aircraft (normally military) at idle thrust to a runway. The approach may start at a relatively high altitude over a runway (high key) and may continue on a relatively high and wide downwind leg with a high rate of descent and a continuous turn to final. It terminates in a landing or low approach. The purpose of this approach is to simulate a flameout.
(See FLAMEOUT.)

SIMULTANEOUS ILS APPROACHES An approach system permitting simultaneous ILS/MLS approaches to airports having parallel runways separated by at least 4,300 feet between centerlines. Integral parts of a total system are ILS/MLS, radar, communications, ATC procedures, and appropriate airborne equipment.
(See PARALLEL RUNWAYS.) (Refer to AIM.)

SIMULTANEOUS MLS APPROACHES (See SIMULTANEOUS ILS APPROACHES.)

SIMULTANEOUS OPERATIONS ON INTERSECTING RUNWAYS Operations which include simultaneous takeoffs and landings and/or simultaneous landings when a landing aircraft is able and is instructed by the controller to hold short of the intersecting runway or designated hold short point. Pilots are expected to promptly inform the controller if the hold short clearance cannot be accepted.
(See PARALLEL RUNWAYS.) (Refer to AIM.)

SINGLE DIRECTION ROUTES Preferred IFR Routes which are sometimes depicted on high altitude en route charts and which are normally flown in one direction only.
(See PREFERRED IFR ROUTES.) (Refer to AIRPORT/FACILITY DIRECTORY.)

SINGLE FREQUENCY APPROACH A service provided under a letter of agreement to military single-piloted turbojet aircraft which permits use of a single UHF frequency during approach for landing. Pilots will not normally be required to change frequency from the beginning of the approach to touchdown except that pilots conducting an en route descent are required to change frequency when control is transferred from the air route traffic control center to the terminal facility. The abbreviation "SFA" in the DOD FLIP IFR Supplement under "Communications" indicates this service is available at an aerodrome.

SINGLE-PILOTED AIRCRAFT A military turbojet aircraft possessing one set of flight controls, tandem cockpits, or two sets of flight controls but operated by one pilot is considered single-piloted by ATC when determining the appropriate air traffic service to be applied.
(See SINGLE FREQUENCY APPROACH.)

SLASH A radar beacon reply displayed as an elongated target.

SLDI (See SECTOR LIST DROP INTERVAL.)

SLOT TIME (See METER FIX TIME/SLOT TIME.)

SLOW TAXI To taxi a float plane at low power or low RPM.

SN (See SYSTEM STRATEGIC NAVIGATION.)

SPEAK SLOWER Used in verbal communications as a request to reduce speech rate.

SPECIAL EMERGENCY A condition of air piracy or other hostile act by a person(s) aboard an aircraft which threatens the safety of the aircraft or its passengers.

SPECIAL INSTRUMENT APPROACH PROCEDURE (See INSTRUMENT APPROACH PROCEDURE.)

SPECIAL USE AIRSPACE Airspace of defined dimensions identified by an area on the surface of the earth wherein activities must be confined because of their nature and/or wherein limitations may be imposed upon aircraft operations that are not a part of those activities. Types of special use airspace are:

1. Alert Area. Airspace which may contain a high volume of pilot training activities or an unusual type of aerial activity, neither of which is hazardous to aircraft. Alert Areas are depicted on aeronautical charts for the information of nonparticipating pilots. All activities within an Alert Area are conducted in accordance with Federal Aviation Regulations, and pilots of participating aircraft as well as pilots transiting the area are equally responsible for collision avoidance.

2. Controlled Firing Area. Airspace wherein activities are conducted under conditions so controlled as to eliminate hazards to nonparticipating aircraft and to ensure the safety of persons and property on the ground.

3. Military Operations Area (MOA). An MOA is an airspace assignment of defined vertical and lateral dimensions established outside Class A airspace to separate/segregate certain military activities from IFR traffic and to identify for VFR traffic where these activities are conducted.
(Refer to AIM.)

4. Prohibited Area. Designated airspace within which the flight of aircraft is prohibited.
(Refer to En Route Charts, AIM.)

5. Restricted Area. Airspace designated under Part 73, within which the flight of aircraft, while not wholly prohibited, is subject to restriction. Most restricted areas are designated joint use and IFR/VFR operations in the area may be authorized by the controlling ATC facility when it is not being utilized by the using agency. Restricted areas are depicted on en route charts. Where joint use is authorized, the name of the ATC controlling facility is also shown.
(Refer to Part 73.) (Refer to AIM.)

6. Warning Area. Airspace which may contain hazards to nonparticipating aircraft in international airspace.

SPECIAL VFR CONDITIONS Meteorological conditions that are less than those required for basic VFR flight in Class B, C, D, or E surface areas and in which some aircraft are permitted flight under visual flight rules.
(See SPECIAL VFR OPERATIONS.) (Refer to Part 91.)

SPECIAL VFR FLIGHT [ICAO] A VFR flight cleared by air traffic control to operate within Class B, C, D, and E surface areas in meteorological conditions below VMC.

SPECIAL VFR OPERATIONS Aircraft operating in accordance with clearances within Class B, C, D, and E surface areas in weather conditions less than the basic VFR weather minima. Such operations must be requested by the pilot and approved by ATC.
(See SPECIAL VFR CONDITIONS.) (See ICAO term SPECIAL VFR FLIGHT.)

SPEED (See AIRSPEED.)
(See GROUND SPEED.)

SPEED ADJUSTMENT An ATC procedure used to request pilots to adjust aircraft speed to a specific value for the purpose of providing desired spacing. Pilots are expected to maintain a speed of plus or minus 10 knots or 0.02 mach number of the specified speed.

Examples of speed adjustments are:

1. "Increase/reduce speed to mach point (number)."

2. "Increase/reduce speed to (speed in knots)" or "Increase/reduce speed (number of knots) knots."

SPEED BRAKES Moveable aerodynamic devices on aircraft that reduce airspeed during descent and landing.

SPEED SEGMENTS Portions of the arrival route between the transition point and the vertex along the optimum flight path for which speeds and altitudes are specified. There is one set of arrival speed segments adapted from each transition point to each vertex. Each set may contain up to six segments.

SQUAWK (Mode, Code, Function) Activate specific modes/codes/functions on the aircraft transponder; e.g., "Squawk three/alpha, two one zero five, low."
(See TRANSPONDER.)

STAGING/QUEUING The placement, integration, and segregation of departure aircraft in designated movement areas of an airport by departure fix, EDCT, and/or restriction.

STANDARD INSTRUMENT APPROACH PROCEDURE (See INSTRUMENT APPROACH PROCEDURE.)

STANDARD INSTRUMENT DEPARTURE
A preplanned instrument flight rule (IFR) air traffic control departure procedure printed for pilot use in graphic and/or textual form. SID's provide transition from the terminal to the appropriate en route structure.

(See IFR TAKEOFF MINIMUMS AND DEPARTURE PROCEDURES.) (Refer to AIM.)

STANDARD INSTRUMENT DEPARTURE CHARTS (See AERONAUTICAL CHART.)

STANDARD RATE TURN A turn of three degrees per second.

STANDARD TERMINAL ARRIVAL A preplanned instrument flight rule (IFR) air traffic control arrival procedure published for pilot use in graphic and/or textual form. STAR's provide transition from the en route structure to an outer fix or an instrument approach fix/arrival waypoint in the terminal area.

STANDARD TERMINAL ARRIVAL CHARTS (See AERONAUTICAL CHART.)

STAND BY Means the controller or pilot must pause for a few seconds, usually to attend to other duties of a higher priority. Also means to wait as in "stand by for clearance." The caller should reestablish contact if a delay is lengthy. "Stand by" is not an approval or denial.

STAR (See STANDARD TERMINAL ARRIVAL.)

STATE AIRCRAFT Aircraft used in military, customs and police service, in the exclusive service of any government, or of any political subdivision, thereof including the government of any state, territory, or possession of the United States or the District of Columbia, but not including any government-owned aircraft engaged in carrying persons or property for commercial purposes.

STATIC RESTRICTIONS Those restrictions that are usually not subject to change, fixed, in place, and/or published.

STATIONARY RESERVATIONS Altitude reservations which encompass activities in a fixed area. Stationary reservations may include activities, such as special tests of weapons systems or equipment, certain U.S. Navy carrier, fleet, and anti-submarine operations, rocket, missile and drone operations, and certain aerial refueling or similar operations.

STEPDOWN FIX A fix permitting additional descent within a segment of an instrument approach procedure by identifying a point at which a controlling obstacle has been safely overflown.

STEP TAXI To taxi a float plane at full power or high RPM.

STEP TURN A maneuver used to put a float plane in a planing configuration prior to entering an active sea lane for takeoff. The STEP TURN maneuver should only be used upon pilot request.

STEREO ROUTE A routinely used route of flight established by users and ARTCC's identified by a coded name; e.g., ALPHA 2. These routes minimize flight plan handling and communications.

STOL AIRCRAFT (See SHORT TAKEOFF AND LANDING AIRCRAFT.)

STOP ALTITUDE SQUAWK Used by ATC to inform an aircraft to turn-off the automatic altitude reporting feature of its transponder. It is issued when the verbally reported altitude varies 300 feet or more from the automatic altitude report.

(See ALTITUDE READOUT.) (See TRANSPONDER.)

STOP AND GO A procedure wherein an aircraft will land, make a complete stop on the runway, and then commence a takeoff from that point.

(See LOW APPROACH.) (See OPTION APPROACH.)

STOP BURST (See STOP STREAM.)

STOP BUZZER (See STOP STREAM.)

STOPOVER FLIGHT PLAN A flight plan format which permits in a single submission the filing of a sequence of flight plans through interim full-stop destinations to a final destination.

STOP SQUAWK (Mode or Code) Used by ATC to tell the pilot to turn specified functions of the aircraft transponder off.

(See STOP ALTITUDE SQUAWK.) (See TRANSPONDER.)

STOP STREAM Used by ATC to request a pilot to suspend electronic countermeasure activity.

(See JAMMING.)

STOPWAY An area beyond the takeoff runway no less wide than the runway and centered upon the extended centerline of the runway, able to support the airplane during an aborted takeoff, without causing structural damage to the airplane, and designated by the airport authorities for use in decelerating the airplane during an aborted takeoff.

STRAIGHT-IN APPROACH-IFR An instrument approach wherein final approach is begun without first having executed a procedure turn, not necessarily completed with a straight-in landing or made to straight-in landing minimums.

(See STRAIGHT-IN LANDING.) (See LANDING MINIMUMS.) (See STRAIGHT-IN APPROACH-VFR.)

STRAIGHT-IN APPROACH-VFR Entry into the traffic pattern by interception of the extended runway centerline (final approach course) without executing any other portion of the traffic pattern.
(See TRAFFIC PATTERN.)

STRAIGHT-IN LANDING A landing made on a runway aligned within 30° of the final approach course following completion of an instrument approach.
(See STRAIGHT-IN APPROACH-IFR.)

STRAIGHT-IN LANDING MINIMUMS (See LANDING MINIMUMS.)

STRAIGHT-IN MINIMUMS (See STRAIGHT-IN LANDING MINIMUMS.)

SUBSTITUTIONS Users are permitted to exchange CTA's. Normally, the airline dispatcher will contact the ATCSCC with this request. The ATCSCC shall forward approved substitutions to the TMU's who will notify the appropriate terminals. Permissible swapping must not change the traffic load for any given hour of an EQF program.

SUBSTITUTE ROUTE A route assigned to pilots when any part of an airway or route is unusable because of NAVAID status. These routes consist of:

1. Substitute routes which are shown on U.S. Government charts.
2. Routes defined by ATC as specific NAVAID radials or courses.
3. Routes defined by ATC as direct to or between NAVAID's.

SUNSET AND SUNRISE The mean solar times of sunset and sunrise as published in the Nautical Almanac, converted to local standard time for the locality concerned. Within Alaska, the end of evening civil twilight and the beginning of morning civil twilight, as defined for each locality.

SUPER HIGH FREQUENCY The frequency band between 3 and 30 gigahertz (GHz). The elevation and azimuth stations of the microwave landing system operate from 5031 MHz to 5091 MHz in this spectrum.

SUPPLEMENTAL WEATHER SERVICE LOCATION Airport facilities staffed with contract personnel who take weather observations and provide current local weather to pilots via telephone or radio. (All other services are provided by the parent FSS.)

SUPPS Refers to ICAO Document 7030 Regional Supplementary Procedures. SUPPS contain procedures for each ICAO Region which are unique to that Region and are not covered in the worldwide provisions identified in the ICAO Air Navigation Plan. Procedures contained in chapter 8 are based in part on those published in SUPPS.

SURFACE AREA The airspace contained by the lateral boundary of the Class B, C, D, or E airspace designated for an airport that begins at the surface and extends upward.

SURPIC A description of surface vessels in the area of a Search and Rescue incident including their predicted positions and their characteristics.
(See Paragraph 10-73, In-Flight Contingencies.)

SURVEILLANCE APPROACH An instrument approach wherein the air traffic controller issues instructions, for pilot compliance, based on aircraft position in relation to the final approach course (azimuth), and the distance (range) from the end of the runway as displayed on the controller's radar scope. The controller will provide recommended altitudes on final approach if requested by the pilot.
(Refer to AIM.)

SWAP (See SEVERE WEATHER AVOIDANCE PLAN.)

SWSL (See SUPPLEMENTAL WEATHER SERVICE LOCATION.)

SYSTEM STRATEGIC NAVIGATION Military activity accomplished by navigating along a preplanned route using internal aircraft systems to maintain a desired track. This activity normally requires a lateral route width of 10 NM and altitude range of 1,000 feet to 6,000 feet AGL with some route segments that permit terrain following.

TACAN (See TACTICAL AIR NAVIGATION.)

TACAN-ONLY AIRCRAFT An aircraft, normally military, possessing TACAN with DME but no VOR navigational system capability. Clearances must specify TACAN or VORTAC fixes and approaches.

TACTICAL AIR NAVIGATION An ultra-high frequency electronic rho-theta air navigation aid which provides suitably equipped aircraft a continuous indication of bearing and distance to the TACAN station.
(See VORTAC.) (Refer to AIM.)

TAILWIND Any wind more than 90 degrees to the longitudinal axis of the runway. The magnetic direction of the runway shall be used as the basis for determining the longitudinal axis.

TAKEOFF AREA (See LANDING AREA.)

TAKE-OFF DISTANCE AVAILABLE [ICAO] The length of the take-off run available plus the length of the clearway, if provided.

TAKE-OFF RUN AVAILABLE [ICAO] The length of runway declared available and suitable for the ground run of an aeroplane take-off.

TARGET The indication shown on a radar display resulting from a primary radar return or a radar beacon reply.
(See RADAR.) (See TARGET SYMBOL.) (See ICAO term TARGET.)

TARGET [ICAO] In radar:

1. Generally, any discrete object which reflects or retransmits energy back to the radar equipment.
2. Specifically, an object of radar search or surveillance.

TARGET RESOLUTION A process to ensure that correlated radar targets do not touch. Target resolution shall be applied as follows:

1. Between the edges of two primary targets or the edges of the ASR-9 primary target symbol.
2. Between the end of the beacon control slash and the edge of a primary target.
3. Between the ends of two beacon control slashes.— MANDATORY TRAFFIC ADVISORIES AND SAFETY ALERTS SHALL BE ISSUED WHEN THIS PROCEDURE IS USED.

Note: This procedure shall not be provided utilizing mosaic radar systems.

TARGET SYMBOL A computer-generated indication shown on a radar display resulting from a primary radar return or a radar beacon reply.

TAXI The movement of an airplane under its own power on the surface of an airport (Part 135.100-Note). Also, it describes the surface movement of helicopters equipped with wheels.

(See AIR TAXI.) (See HOVER TAXI.) (Refer to AIM.) (Refer to Part 135.100.)

TAXI INTO POSITION AND HOLD Used by ATC to inform a pilot to taxi onto the departure runway in takeoff position and hold. It is not authorization for takeoff. It is used when takeoff clearance cannot immediately be issued because of traffic or other reasons.

(See CLEARED FOR TAKEOFF.)

TAXI PATTERNS Patterns established to illustrate the desired flow of ground traffic for the different runways or airport areas available for use.

TCAS (See TRAFFIC ALERT AND COLLISION AVOIDANCE SYSTEM.)

TCH (See THRESHOLD CROSSING HEIGHT.)

TCLT (See TENTATIVE CALCULATED LANDING TIME.)

TDZE (See TOUCHDOWN ZONE ELEVATION.)

TELEPHONE INFORMATION BRIEFING SERVICE A continuous telephone recording of meteorological and/or aeronautical information.

(Refer to AIM.)

TENTATIVE CALCULATED LANDING TIME A projected time calculated for adapted vertex for each arrival aircraft based upon runway configuration, airport acceptance rate, airport arrival delay period, and other metered arrival aircraft. This time is either the VTA of the aircraft or the TCLT/ACLT of the previous aircraft plus the AAI, whichever is later. This time will be updated in response to an aircraft's progress and its current relationship to other arrivals.

TERMINAL AREA A general term used to describe airspace in which approach control service or airport traffic control service is provided.

TERMINAL AREA FACILITY A facility providing air traffic control service for arriving and departing IFR, VFR, Special VFR, and on occasion en route aircraft.

(See APPROACH CONTROL FACILITY.) (See TOWER.)

TERMINAL VFR RADAR SERVICE A national program instituted to extend the terminal radar services provided instrument flight rules (IFR) aircraft to visual flight rules (VFR) aircraft. The program is divided into four types service referred to as basic radar service, terminal radar service area (TRSA) service, Class B service and Class C service. The type of service provided at a particular location is contained in the Airport/Facility Directory.

1. Basic Radar Service: These services are provided for VFR aircraft by all commissioned terminal radar facilities. Basic radar service includes safety alerts, traffic advisories, limited radar vectoring when requested by the pilot, and sequencing at locations where procedures have been established for this purpose and/or when covered by a letter of agreement. The purpose of this service is to adjust the flow of arriving IFR and VFR aircraft into the traffic pattern in a safe and orderly manner and to provide traffic advisories to departing VFR aircraft.
2. TRSA Service: This service provides, in addition to basic radar service, sequencing of all IFR and participating VFR aircraft to the primary airport and separation between all participating VFR aircraft. The purpose of this service is to provide separation between all participating VFR aircraft and all IFR aircraft operating within the area defined as a TRSA.
3. Class C Service: This service provides, in addition to basic radar service, approved separation between IFR and VFR aircraft, and sequencing of VFR aircraft, and sequencing of VFR arrivals to the primary airport.
4. Class B Service: This service provides, in addition to basic radar service, approved separation of aircraft based on IFR, VFR, and/or weight, and sequencing of VFR arrivals to the primary airport(s).

(See CONTROLLED AIRSPACE.) (See TERMINAL RADAR SERVICE AREA.) (Refer to AIM.) (Refer to AIRPORT/FACILITY DIRECTORY.)

TERMINAL RADAR SERVICE AREA Airspace surrounding designated airports wherein ATC provides radar vectoring, sequencing, and separation on a full-time basis for all IFR and participating VFR aircraft. Service provided in a TRSA is called Stage III Service. The AIM contains an explanation of TRSA. TRSA's are depicted on

VFR aeronautical charts. Pilot participation is urged but is not mandatory.
(See TERMINAL RADAR PROGRAM.) (Refer to AIM.) (Refer to AIRPORT/FACILITY DIRECTORY.)

TERMINAL-VERY HIGH FREQUENCY OMNIDIRECTIONAL RANGE STATION A very high frequency terminal omnirange station located on or near an airport and used as an approach aid.
(See NAVIGATIONAL AID.) (See VOR.)

TERRAIN FOLLOWING The flight of a military aircraft maintaining a constant AGL altitude above the terrain or the highest obstruction. The altitude of the aircraft will constantly change with the varying terrain and/or obstruction.

TETRAHEDRON A device normally located on uncontrolled airports and used as a landing direction indicator. The small end of a tetrahedron points in the direction of landing. At controlled airports, the tetrahedron, if installed, should be disregarded because tower instructions supersede the indicator.
(See SEGMENTED CIRCLE.) (Refer to AIM.)

TF (See TERRAIN FOLLOWING.)

THAT IS CORRECT The understanding you have is right.

360 OVERHEAD (See OVERHEAD APPROACH.)

THRESHOLD The beginning of that portion of the runway usable for landing.
(See AIRPORT LIGHTING.) (See DISPLACED THRESHOLD.)

THRESHOLD CROSSING HEIGHT
The theoretical height above the runway threshold at which the aircraft's glideslope antenna would be if the aircraft maintains the trajectory established by the mean ILS glideslope or MLS glidepath.
(See GLIDESLOPE.) (See THRESHOLD.)

THRESHOLD LIGHTS (See AIRPORT LIGHTING.)

TIBS (See TELEPHONE INFORMATION BRIEFING SERVICE.)

TIME GROUP Four digits representing the hour and minutes from the 24-hour clock. Time groups without time zone indicators are understood to be UTC (Coordinated Universal Time); e.g., "0205." The term "Zulu" is used when ATC procedures require a reference to UTC. A time zone designator is used to indicate local time; e.g., "0205M." The end and the beginning of the day are shown by "2400" and "0000," respectively.

TMPA (See TRAFFIC MANAGEMENT PROGRAM ALERT.)

TMU (See TRAFFIC MANAGEMENT UNIT.)

TODA [ICAO] (See ICAO Term TAKE-OFF DISTANCE AVAILABLE.)

TORA [ICAO] (See ICAO Term TAKE-OFF RUN AVAILABLE.)

TORCHING The burning of fuel at the end of an exhaust pipe or stack of a reciprocating aircraft engine, the result of an excessive richness in the fuel air mixture.

TOTAL ESTIMATED ELAPSED TIME [ICAO] For IFR flights, the estimated time required from take-off to arrive over that designated point, defined by reference to navigation aids, from which it is intended that an instrument approach procedure will be commenced, or, if no navigation aid is associated with the destination aerodrome, to arrive over the destination aerodrome. For VFR flights, the estimated time required from takeoff to arrive over the destination aerodrome.
(See ESTIMATED ELAPSED TIME.)

TOUCH-AND-GO An operation by an aircraft that lands and departs on a runway without stopping or exiting the runway.

TOUCH-AND-GO LANDING (See TOUCH-AND-GO.)

TOUCHDOWN
1. The point at which an aircraft first makes contact with the landing surface.
2. Concerning a precision radar approach (PAR), it is the point where the glide path intercepts the landing surface.
(See ICAO term TOUCHDOWN.)

TOUCHDOWN [ICAO] The point where the nominal glide path intercepts the runway.
Note: Touchdown as defined above is only a datum and is not necessarily the actual point at which the aircraft will touch the runway.

TOUCHDOWN RVR (See VISIBILITY.)

TOUCHDOWN ZONE The first 3,000 feet of the runway beginning at the threshold. The area is used for determination of Touchdown Zone Elevation in the development of straight-in landing minimums for instrument approaches.
(See ICAO term TOUCHDOWN ZONE).

TOUCHDOWN ZONE [ICAO] The portion of a runway, beyond the threshold, where it is intended landing aircraft first contact the runway.

TOUCHDOWN ZONE ELEVATION The highest elevation in the first 3,000 feet of the landing surface. TDZE is indicated on the instrument approach procedure chart when straight-in landing minimums are authorized.
(See TOUCHDOWN ZONE.)

TOUCHDOWN ZONE LIGHTING (See AIRPORT LIGHTING.)

TOWER A terminal facility that uses air/ground communications, visual signaling, and other devices to provide ATC services to aircraft operating in the vicinity of an airport or on the movement area. Authorizes aircraft to land

or take off at the airport controlled by the tower or to transit the Class D airspace area regardless of flight plan or weather conditions (IFR or VFR). A tower may also provide approach control services (radar or nonradar).
(See AIRPORT TRAFFIC CONTROL SERVICE.) (See APPROACH CONTROL FACILITY.) (See APPROACH CONTROL SERVICE.) (See MOVEMENT AREA.) (See TOWER EN ROUTE CONTROL SERVICE.) (Refer to AIM.) (See ICAO term AERODROME CONTROL TOWER.)

TOWER EN ROUTE CONTROL SERVICE The control of IFR en route traffic within delegated airspace between two or more adjacent approach control facilities. This service is designed to expedite traffic and reduce control and pilot communication requirements.

TOWER TO TOWER (See TOWER EN ROUTE CONTROL SERVICE.)

TPX-42 A numeric beacon decoder equipment/system. It is designed to be added to terminal radar systems for beacon decoding. It provides rapid target identification, reinforcement of the primary radar target, and altitude information from Mode C.
(See AUTOMATED RADAR TERMINAL SYSTEMS.) (See TRANSPONDER.)

TRACK The actual flight path of an aircraft over the surface of the earth.
(See COURSE.) (See ROUTE.) (See FLIGHT PATH.) (See ICAO term TRACK.)

TRACK [ICAO] The projection on the earth's surface of the path of an aircraft, the direction of which path at any point is usually expressed in degrees from North (True, Magnetic, or Grid).

TRAFFIC

1. A term used by a controller to transfer radar identification of an aircraft to another controller for the purpose of coordinating separation action. Traffic is normally issued:

(a) in response to a handoff or point out,

(b) in anticipation of a handoff or point out,

or

(c) in conjunction with a request for control of an aircraft.

2. A term used by ATC to refer to one or more aircraft.

TRAFFIC ADVISORIES Advisories issued to alert pilots to other known or observed air traffic which may be in such proximity to the position or intended route of flight of their aircraft to warrant their attention. Such advisories may be based on:

1. Visual observation.

2. Observation of radar identified and nonidentified aircraft targets on an ATC radar display, or

3. Verbal reports from pilots or other facilities.

The word "traffic" followed by additional information, if known, is used to provide such advisories; e.g., "Traffic, 2 o'clock, one zero miles, southbound, eight thousand."

Traffic advisory service will be provided to the extent possible depending on higher priority duties of the controller or other limitations; e.g., radar limitations, volume of traffic, frequency congestion, or controller workload. Radar/nonradar traffic advisories do not relieve the pilot of his responsibility to see and avoid other aircraft. Pilots are cautioned that there are many times when the controller is not able to give traffic advisories concerning all traffic in the aircraft's proximity; in other words, when a pilot requests or is receiving traffic advisories, he should not assume that all traffic will be issued.
(Refer to AIM.)

(Identification), TRAFFIC ALERT. ADVISE YOU TURN LEFT/RIGHT (specific heading if appropriate), AND/OR CLIMB/DESCEND (specific altitude if appropriate) IMMEDIATELY. (See SAFETY ALERT.)

TRAFFIC ALERT AND COLLISION AVOIDANCE SYSTEM An airborne collision avoidance system based on radar beacon signals which operates independent of ground-based equipment. TCAS-I generates traffic advisories only. TCAS-II generates traffic advisories, and resolution (collision avoidance) advisories in the vertical plane.

TRAFFIC INFORMATION (See TRAFFIC ADVISORIES.)

TRAFFIC IN SIGHT Used by pilots to inform a controller that previously issued traffic is in sight.
(See NEGATIVE CONTACT.) (See TRAFFIC ADVISORIES.)

TRAFFIC MANAGEMENT PROGRAM ALERT A term used in a Notice to Airmen (NOTAM) issued in conjunction with a special traffic management program to alert pilots to the existence of the program and to refer them to either the Notices to Airmen publication or a special traffic management program advisory message for program details. The contraction TMPA is used in NOTAM text.

TRAFFIC MANAGEMENT UNIT
The entity in ARTCC's and designated terminals responsible for direct involvement in the active management of facility traffic. Usually under the direct supervision of an assistant manager for traffic management.

TRAFFIC NO FACTOR Indicates that the traffic described in a previously issued traffic advisory is no factor.

TRAFFIC NO LONGER OBSERVED Indicates that the traffic described in a previously issued traffic advisory is no longer depicted on radar, but may still be a factor.

TRAFFIC PATTERN The traffic flow that is prescribed for aircraft landing at, taxiing on, or taking off from an airport. The components of a typical traffic pattern are upwind leg, crosswind leg, downwind leg, base leg, and final approach.

1. Upwind Leg. A flight path parallel to the landing runway in the direction of landing.

2. Crosswind Leg. A flight path at right angles to the landing runway off its upwind end.

3. Downwind Leg. A flight path parallel to the landing runway in the direction opposite to landing. The downwind leg normally extends between the crosswind leg and the base leg.

4. Base Leg. A flight path at right angles to the landing runway off its approach end. The base leg normally extends from the downwind leg to the intersection of the extended runway centerline.

5. Final Approach. A flight path in the direction of landing along the extended runway centerline. The final approach normally extends from the base leg to the runway. An aircraft making a straight-in approach VFR is also considered to be on final approach.

(See STRAIGHT-IN APPROACH-VFR.) (See TAXI PATTERNS.) (Refer to AIM.) (Refer to Part 91.) (See ICAO term AERODROME TRAFFIC CIRCUIT.)

TRANSCRIBED WEATHER BROADCAST A continuous recording of meteorological and aeronautical information that is broadcast on L/MF and VOR facilities for pilots.

(Refer to AIM.)

TRANSFER OF CONTROL That action whereby the responsibility for the separation of an aircraft is transferred from one controller to another.

(See ICAO term TRANSFER OF CONTROL.)

TRANSFER OF CONTROL [ICAO] Transfer of responsibility for providing air traffic control service.

TRANSFERRING CONTROLLER A controller/facility transferring control of an aircraft to another controller/facility.

(See ICAO term TRANSFERRING UNIT/CONTROLLER.)

TRANSFERRING FACILITY (See TRANSFERRING CONTROLLER.)

TRANSFERRING UNIT/CONTROLLER [ICAO] Air traffic control unit/air traffic controller in the process of transferring the responsibility for providing air traffic control service to an aircraft to the next air traffic control unit/air traffic controller along the route of flight.

Note: See definition of *accepting unit/controller*.

TRANSITION

1. The general term that describes the change from one phase of flight or flight condition to another; e.g., transition from en route flight to the approach or transition from instrument flight to visual flight.

2. A published procedure (SID Transition) used to connect the basic SID to one of several en route airways/jet routes, or a published procedure (STAR Transition) used to connect one of several en route airways/jet routes to the basic STAR.

(Refer to SID/STAR Charts.)

TRANSITIONAL AIRSPACE That portion of controlled airspace wherein aircraft change from one phase of flight or flight condition to another.

TRANSITION POINT A point at an adapted number of miles from the vertex at which an arrival aircraft would normally commence descent from its en route altitude. This is the first fix adapted on the arrival speed segments.

TRANSMISSOMETER An apparatus used to determine visibility by measuring the transmission of light through the atmosphere. It is the measurement source for determining runway visual range (RVR) and runway visibility value (RVV).

(See VISIBILITY.)

TRANSMITTING IN THE BLIND A transmission from one station to other stations in circumstances where two-way communication cannot be established, but where it is believed that the called stations may be able to receive the transmission.

TRANSPONDER The airborne radar beacon receiver/transmitter portion of the Air Traffic Control Radar Beacon System (ATCRBS) which automatically receives radio signals from interrogators on the ground, and selectively replies with a specific reply pulse or pulse group only to those interrogations being received on the mode to which it is set to respond.

(See INTERROGATOR.) (Refer to AIM.) (See ICAO term TRANSPONDER.)

TRANSPONDER [ICAO] A receiver/transmitter which will generate a reply signal upon proper interrogation; the interrogation and reply being on different frequencies.

TRANSPONDER CODES (See CODES.)

TRSA (See TERMINAL RADAR SERVICE AREA.)

TURBOJET AIRCRAFT An aircraft having a jet engine in which the energy of the jet operates a turbine which in turn operates the air compressor.

TURBOPROP AIRCRAFT An aircraft having a jet engine in which the energy of the jet operates a turbine which drives the propeller.

TWEB (See TRANSCRIBED WEATHER BROADCAST.)

TVOR (See TERMINAL-VERY HIGH FREQUENCY OMNIDIRECTIONAL RANGE STATION.)

TWO-WAY RADIO COMMUNICATIONS FAILURE (See LOST COMMUNICATIONS.)

UDF (See DIRECTION FINDER.)

UHF (See ULTRAHIGH FREQUENCY.)

ULTRAHIGH FREQUENCY The frequency band between 300 and 3,000 MHz. The bank of radio frequencies used for military air/ground voice communications. In some instances this may go as low as 225 MHz and still be referred to as UHF.

ULTRALIGHT VEHICLE An aeronautical vehicle operated for sport or recreational purposes which does not require FAA registration, an airworthiness certificate, nor pilot certification. They are primarily single occupant vehicles, although some two-place vehicles are authorized for training purposes. Operation of an ultralight vehicle in certain airspace requires authorization from ATC.
(See Part 103.)

UNABLE Indicates inability to comply with a specific instruction, request, or clearance.

UNDER THE HOOD Indicates that the pilot is using a hood to restrict visibility outside the cockpit while simulating instrument flight. An appropriately rated pilot is required in the other control seat while this operation is being conducted.
(Refer to Part 91.)

UNICOM A nongovernment communication facility which may provide airport information at certain airports. Locations and frequencies of UNICOMs are shown on aeronautical charts and publications.
(See AIRPORT/FACILITY DIRECTORY.) (Refer to AIM.)

UNPUBLISHED ROUTE A route for which no minimum altitude is published or charted for pilot use. It may include a direct route between NAVAIDs, a radial, a radar vector, or a final approach course beyond the segments of an instrument approach procedure.
(See PUBLISHED ROUTE.) (See ROUTE.)

UPWIND LEG (See TRAFFIC PATTERN.)

URGENCY A condition of being concerned about safety and of requiring timely but not immediate assistance; a potential *distress* condition.
(See ICAO term URGENCY.)

URGENCY [ICAO] A condition concerning the safety of an aircraft or other vehicle, or of person on board or in sight, but which does not require immediate assistance.

USAFIB (See ARMY AVIATION FLIGHT INFORMATION BULLETIN.)

UVDF (See DIRECTION FINDER.)

VASI (See VISUAL APPROACH SLOPE INDICATOR.)

VDF (See DIRECTION FINDER.)

VDP (See VISUAL DESCENT POINT.)

VECTOR A heading issued to an aircraft to provide navigational guidance by radar.
(See ICAO term RADAR VECTORING.)

VERIFY Request confirmation of information; e.g., "verify assigned altitude."

VERIFY SPECIFIC DIRECTION OF TAKEOFF (OR TURNS AFTER TAKEOFF) Used by ATC to ascertain an aircraft's direction of takeoff and/or direction of turn after takeoff. It is normally used for IFR departures from an airport not having a control tower. When direct communication with the pilot is not possible, the request and information may be relayed through an FSS, dispatcher, or by other means.
(See IFR TAKEOFF MINIMUMS AND DEPARTURE PROCEDURES.)

VERTEX The last fix adapted on the arrival speed segments. Normally, it will be the outer marker of the runway in use. However, it may be the actual threshold or other suitable common point on the approach path for the particular runway configuration.

VERTEX TIME OF ARRIVAL A calculated time of aircraft arrival over the adapted vertex for the runway configuration in use. The time is calculated via the optimum flight path using adapted speed segments.

VERTICAL SEPARATION Separation established by assignment of different altitudes or flight levels.
(See SEPARATION.) (See ICAO term VERTICAL SEPARATION.)

VERTICAL SEPARATION [ICAO] Separation between aircraft expressed in units of vertical distance.

VERTICAL TAKEOFF AND LANDING AIRCRAFT Aircraft capable of vertical climbs and/or descents and of using very short runways or small areas for takeoff and landings. These aircraft include, but are not limited to, helicopters.
(See SHORT TAKEOFF AND LANDING AIRCRAFT.)

VERY HIGH FREQUENCY The frequency band between 30 and 300 MHz. Portions of this band, 108 to 118 MHz, are used for certain NAVAIDS; 118 to 136 MHz are used for civil air/ground voice communications. Other frequencies in this band are used for purposes not related to air traffic control.

VERY HIGH FREQUENCY OMNIDIRECTIONAL RANGE STATION (See VOR.)

VERY LOW FREQUENCY The frequency band between 3 and 30 kHz.

VFR (See VISUAL FLIGHT RULES.)

VFR AIRCRAFT An aircraft conducting flight in accordance with visual flight rules.
(See VISUAL FLIGHT RULES.)

VFR CONDITIONS Weather conditions equal to or better than the minimum for flight under visual flight rules. The term may be used as an ATC clearance/instruction only when:

1. An IFR aircraft requests a climb/descent in VFR conditions.

2. The clearance will result in noise abatement benefits where part of the IFR departure route does not conform to

an FAA approved noise abatement route or altitude.

3. A pilot has requested a practice instrument approach and is not on an IFR flight plan. All pilots receiving this authorization must comply with the VFR visibility and distance from cloud criteria in Part 91. Use of the term does not relieve controllers of their responsibility to separate aircraft in Class B and Class C airspace or TRSA's as required by FAA Order 7110.65. When used as an ATC clearance/instruction, the term may be abbreviated "VFR"; e.g., "MAINTAIN VFR," "CLIMB/DESCEND VFR," etc.

VFR FLIGHT (See VFR AIRCRAFT.)

VFR MILITARY TRAINING ROUTES Routes used by the Department of Defense and associated Reserve and Air Guard units for the purpose of conducting low-altitude navigation and tactical training under VFR below 10,000 feet MSL at airspeeds in excess of 250 knots IAS.

VFR NOT RECOMMENDED An advisory provided by a flight service station to a pilot during a preflight or inflight weather briefing that flight under visual flight rules is not recommended. To be given when the current and/or forecast weather conditions are at or below VFR minimums. It does not abrogate the pilot's authority to make his own decision.

VFR-ON-TOP ATC authorization for an IFR aircraft to operate in VFR conditions at any appropriate VFR altitude (as specified in FAR and as restricted by ATC). A pilot receiving this authorization must comply with the VFR visibility, distance from cloud criteria, and the minimum IFR altitudes specified in Part 91. The use of this term does not relieve controllers of their responsibility to separate aircraft in Class B and Class C airspace or TRSA's as required by FAA Order 7110.65.

VFR TERMINAL AREA CHARTS (See AERONAUTICAL CHART.)

VHF (See VERY HIGH FREQUENCY.)

VHF OMNIDIRECTIONAL RANGE/TACTICAL AIR NAVIGATION (See VORTAC.)

VIDEO MAP An electronically displayed map on the radar display that may depict data such as airports, heliports, runway centerline extensions, hospital emergency landing areas, NAVAID's and fixes, reporting points, airway/route centerlines, boundaries, handoff points, special use tracks, obstructions, prominent geographic features, map alignment indicators, range accuracy marks, minimum vectoring altitudes.

VISIBILITY The ability, as determined by atmospheric conditions and expressed in units of distance, to see and identify prominent unlighted objects by day and prominent lighted objects by night. Visibility is reported as statute miles, hundreds of feet or meters.

(Refer to Part 91.) (See AIM.)

1. Flight Visibility. The average forward horizontal distance, from the cockpit of an aircraft in flight, at which prominent unlighted objects may be seen and identified by day and prominent lighted objects may be seen and identified by night.

2. Ground Visibility. Prevailing horizontal visibility near the earth's surface as reported by the United States National Weather Service or an accredited observer.

3. Prevailing Visibility. The greatest horizontal visibility equaled or exceeded throughout at least half the horizon circle which need not necessarily be continuous.

4. Runway Visibility Value (RVV). The visibility determined for a particular runway by a transmissometer. A meter provides a continuous indication of the visibility (reported in miles or fractions of miles) for the runway. RVV is used in lieu of prevailing visibility in determining minimums for a particular runway.

5. Runway Visual Range (RVR). An instrumentally derived value, based on standard calibrations, that represents the horizontal distance a pilot will see down the runway from the approach end. It is based on the sighting of either high intensity runway lights or on the visual contrast of other targets whichever yields the greater visual range. RVR, in contrast to prevailing or runway visibility, is based on what a pilot in a moving aircraft should see looking down the runway. RVR is horizontal visual range, not slant visual range. It is based on the measurement of a transmissometer made near the touchdown point of the instrument runway and is reported in hundreds of feet. RVR is used in lieu of RVV and/or prevailing visibility in determining minimums for a particular runway.

a. Touchdown RVR. The RVR visibility readout values obtained from RVR equipment serving the runway touchdown zone.

b. Mid-RVR. The RVR readout values obtained from RVR equipment located midfield of the runway.

c. Rollout RVR. The RVR readout values obtained from RVR equipment located nearest the rollout end of the runway.

(See ICAO term VISIBILITY.) (See ICAO term FLIGHT VISIBILITY.) (See ICAO term GROUND VISIBILITY.) (See ICAO term RUNWAY VISUAL RANGE.)

VISIBILITY [ICAO] The ability, as determined by atmospheric conditions and expressed in units of distance, to see and identify prominent unlighted objects by day and prominent lighted objects by night.

Flight Visibility.—The visibility forward from the cockpit of an aircraft in flight.

Ground Visibility.—The visibility at an aerodrome as reported by an accredited observer.

Runway Visual Range [RVR].—The range over which the pilot of an aircraft on the centre line of a runway can see the runway surface markings or the lights delineating the runway or identifying its centre line.

VISUAL APPROACH An approach conducted on an instrument flight rules (IFR) flight plan which authorizes the pilot to proceed visually and clear of clouds to the airport. The pilot must, at all times, have either the airport or the preceding aircraft in sight. the approach must be authorized and under the control of the appropriate air traffic control facility. Reported weather at the airport must be ceiling at or above 1,000 feet and visibility of 3 miles or greater.

(See ICAO term VISUAL APPROACH.)

VISUAL APPROACH [ICAO] An approach by an IFR flight when either part or all of an instrument approach procedure is not completed and the approach is executed in visual reference to terrain.

VISUAL APPROACH SLOPE INDICATOR (See AIRPORT LIGHTING.)

VISUAL DESCENT POINT A defined point on the final approach course of a nonprecision straight-in approach procedure from which normal descent from the MDA to the runway touchdown point may be commenced, provided the approach threshold of that runway, or approach lights, or other markings identifiable with the approach end of that runway are clearly visible to the pilot.

VISUAL FLIGHT RULES Rules that govern the procedures for conducting flight under visual conditions. The term "VFR" is also used in the United States to indicate weather conditions that are equal to or greater than minimum VFR requirements. In addition, it is used by pilots and controllers to indicate type of flight plan.

(See INSTRUMENT FLIGHT RULES.) (See INSTRUMENT METEOROLOGICAL CONDITIONS.) (See VISUAL METEOROLOGICAL CONDITIONS.) (Refer to Part 91.) (Refer to AIM.)

VISUAL HOLDING The holding of aircraft at selected, prominent geographical fixes which can be easily recognized from the air.

(See HOLDING FIX.)

VISUAL METEOROLOGICAL CONDITIONS Meteorological conditions expressed in terms of visibility, distance from cloud, and ceiling equal to or better than specified minima.

(See INSTRUMENT FLIGHT RULES.) (See INSTRUMENT METEOROLOGICAL CONDITIONS.) (See VISUAL FLIGHT RULES.)

VISUAL SEPARATION A means employed by ATC to separate aircraft in terminal areas. There are two ways to effect this separation:

1. The tower controller sees the aircraft involved and issues instructions, as necessary, to ensure that the aircraft avoid each other.

2. A pilot sees the other aircraft involved and upon instructions from the controller provides his own separation by maneuvering his aircraft as necessary to avoid it. This may involve following another aircraft or keeping it in sight until it is no longer a factor.

(See and Avoid.) (Refer to Part 91.)

VLF (See VERY LOW FREQUENCY.)

VMC (See VISUAL METEOROLOGICAL CONDITIONS.)

VOICE SWITCHING AND CONTROL SYSTEM The VSCS is a computer controlled switching system that provides air traffic controllers with all voice circuits (air to ground and ground to air) necessary for air traffic control.

(See VOICE SWITCHING AND CONTROL SYSTEM.) (Refer to AIM.)

VOR A ground-based electronic navigation aid transmitting very high frequency navigation signals, 360 degrees in azimuth, oriented from magnetic north. Used as the basis for navigation in the National Airspace System. The VOR periodically identifies itself by Morse Code and may have an additional voice identification feature. Voice features may be used by ATC or FSS for transmitting instructions/information to pilots.

(See NAVIGATIONAL AID.) (Refer to AIM.)

VORTAC A navigation aid providing VOR azimuth, TACAN azimuth, and TACAN distance measuring equipment (DME) at one site.

(See DISTANCE MEASURING EQUIPMENT.) (See NAVIGATIONAL AID.) (See TACAN.) (See VOR.) (Refer to AIM.)

VORTICES Circular patterns of air created by the movement of an airfoil through the air when generating lift. As an airfoil moves through the atmosphere in sustained flight, an area of low pressure is created above it. The air flowing from the high pressure area to the low pressure area around and about the tips of the airfoil tends to roll up into two rapidly rotating vortices, cylindrical in shape. These vortices are the most predominant parts of aircraft wake turbulence and their rotational force is dependent upon the wing loading, gross weight, and speed of the generating aircraft. The vortices from medium to heavy aircraft can be of extremely high velocity and hazardous to smaller aircraft.

(See AIRCRAFT CLASSES.) (See WAKE TURBULENCE.) (Refer to AIM.)

VOR TEST SIGNAL (See VOT.)

VOT A ground facility which emits a test signal to check VOR receiver accuracy. Some VOT's are available to the user while airborne, and others are limited to ground use only.

(Refer to Part 91.) (See AIM.) (See AIRPORT/FACILITY DIRECTORY.)

VR (See VFR MILITARY TRAINING ROUTES.)

VTA (See VERTEX TIME OF ARRIVAL.)

VSCS (See VOICE SWITCHING AND CONTROL SYSTEM.)

VTOL AIRCRAFT (See VERTICAL TAKEOFF AND LANDING AIRCRAFT.)

WA (See AIRMET.)
(See WEATHER ADVISORY.)

WAKE TURBULENCE Phenomena resulting from the passage of an aircraft through the atmosphere. The term includes vortices, thrust stream turbulence, jet blast, jet wash, propeller wash, and rotor wash both on the ground and in the air.
(See AIRCRAFT CLASSES.) (See JET BLAST.) (See VORTICES.) (Refer to AIM.)

WARNING AREA (See SPECIAL USE AIRSPACE.)

WAYPOINT A predetermined geographical position used for route/instrument approach definition, or progress reporting purposes, that is defined relative to a VORTAC station or in terms of latitude/longitude coordinates.

WEATHER ADVISORY In aviation weather forecast practice, an expression of hazardous weather conditions not predicted in the area forecast, as they affect the operation of air traffic and as prepared by the NWS.
(See SIGMET.) (See AIRMET.)

WHEN ABLE When used in conjunction with ATC instructions, gives the pilot the latitude to delay compliance until a condition or event has been reconciled. Unlike "pilot discretion," when instructions are prefaced "when able," the pilot is expected to seek the first opportunity to comply. Once a maneuver has been initiated, the pilot is expected to continue until the specifications of the instructions have been met. "When able," should not be used when expeditious compliance is required.

WILCO I have received your message, understand it, and will comply with it.

WIND SHEAR A change in wind speed and/or wind direction in a short distance resulting in a tearing or shearing effect. It can exist in a horizontal or vertical direction and occasionally in both.

WING TIP VORTICES (See VORTICES.)

WORDS TWICE

1. As a request: "Communication is difficult. Please say every phrase twice."
2. As information: "Since communications are difficult, every phrase in this message will be spoken twice."

WORLD AERONAUTICAL CHARTS (See AERONAUTICAL CHART.)

WS (See SIGMET.)
(See WEATHER ADVISORY.)

WST (See CONVECTIVE SIGMET.)
(See WEATHER ADVISORY.)

FAA's AIM Index

References are to paragraph numbers

A/FD 4-9, 5-3, 7-4, 7-11
ABBREVIATED DEPARTURE CLEARANCE 5-22
ABNORMAL ATMOSPHERIC PHENOMENA 1-30
ABNORMAL BREATHING 8-3
ABNORMAL FLIGHT ATTITUDES 4-19
ABNORMAL 1-30, 4-19, 5-56, 7-44, 8-3
ABNORMALLY EARLY TURN 5-56
ABNORMALLY LOW BAROMETRIC PRESSURE 7-44
ACCIDENT CAUSE FACTORS 7-80, 8-1, 8-4, 8-5
ACCIDENT CAUSES 6-2
ACCIDENT INVESTIGATION 4-4
ACCIDENT POTENTIAL 4-50
ACCIDENT PREVENTION PROGRAM 7-80
ACCIDENT PREVENTION SPECIALIST 8-2
ACCIDENT PREVENTION 4-4
ACCIDENT SCENE 4-66
ACCIDENT SITE LOCATION 6-16
ACCIDENT 3-42, 4-4, 4-33, 4-50, 4-66, 6-2, 6-14, 6-16, 7-54, 7-74, 7-80, 7-90, 7-91, 7-92, 8-1, 8-2, 8-4, 8-5
ACCIDENTAL ELT ACTIVATION 6-14
ACROBATIC FLIGHT 3-34, 8-7
ACROBATICS 3-34, 6-14, 8-7
ADDITIONAL REPORTS 5-32
ADHERENCE TO CLEARANCE 4-88, 5-71, 7-12
ADIZ 4-19, 5-4, 5-5, 5-90, 5-94, 7-3
ADVISORY CIRCULAR 5-70
ADVISORY FREQUENCY 4-10, 5-42, 5-43, 5-57, 5-80
ADVISORY INFORMATION 4-57, 4-58, 5-49, 6-10
ADVISORY 1-10, 1-12, 1-17, 1-30, 3-34, 3-40, 3-44, 4-3, 4-5, 4-8, 4-9, 4-10, 4-11, 4-12, 4-14, 4-16, 4-17, 4-19, 4-20, 4-32, 4-35, 4-43, 4-51, 4-52, 4-56, 4-61, 4-64, 4-70, 4-80, 4-89, 5-13, 5-42, 5-43, 5-49, 5-50, 5-51, 5-52, 5-57, 5-70, 5-79, 5- 80, 5-84, 6-10, 6-24, 7-1, 7-2, 7-3, 7-4, 7-5, 7-8, 7-9, 7-11, 7-12, 7-14, 7-19, 7-74, 7-80, 7-81, 7-84, 8-8, 9-4
AERONAUTICAL ADVISORY STATIONS 4-9, 4-11, 4-12
AERONAUTICAL CHART 5-1, 9-1, 9-2, 9-3, 9-5, 9-6
AERONAUTICAL LIGHT BEACON 2-8, 2-20, 2-21, 2-22
AERONAUTICAL MULTICOM SERVICE 4-9, 4-11
AERONAUTICAL TIME 4-41
AIR AMBULANCE FLIGHTS 4-33
AIR CARRIER AIRCRAFT 7-92
AIR DEFENSE EMERGENCY 5-90
AIR DEFENSE IDENTIFICATION ZONE 4-19, 5-4, 5-5, 5-90, 5-94, 7-3
AIR ROUTE SURVEILLANCE RADAR 1-32
AIR ROUTE TRAFFIC CONTROL CENTER 1-30, 3-42, 4-1, 4-4, 4-19, 4-67, 4-70, 4-90, 5-7, 5-22, 5-30, 5-42, 6-16, 6-20, 7-9, 7-19, 7-73
AIR TAXI 4-66
AIR TO AIR COMMUNICATIONS 4-11
AIR TRAFFIC CONTROL CLEARANCE 4-66, 4-80, 4-81, 4-82, 4-83, 4-84, 4-85, 4-86, 4-87, 5-71
AIR TRAFFIC CONTROL RADAR BEACON SYSTEM 1-31
AIR TRAFFIC FACILITIES 4-1, 4-2, 4-3, 4-6, 4-7
AIRBORNE VOR CHECK POINTS 1-4
AIRCRAFT ACCIDENT REPORTING 7-91
AIRCRAFT APPROACH CATEGORY 1-10, 7-13
AIRCRAFT ARRESTING DEVICE 2-43
AIRCRAFT CONFLICT ADVISORY 5-79
AIRCRAFT CONFLICT ALERT 5-76
AIRCRAFT IDENTIFICATION 4-33, 4-80, 5-1
AIRCRAFT INCIDENT REPORTING 7-91
AIRCRAFT INTERCEPTION 5-91, 5-93
AIRCRAFT LIGHTS 4-72
AIRCRAFT TRAFFIC CONTROL LIGHT SIGNALS 4-62
AIRCRAFT TRANSPONDER EQUIPMENT 5-7
AIRCRAFT TYPE SUFFIX 5-7
AIRCRAFT 1-10, 2-43, 4-33, 4-34, 4-62, 4-72, 4-80, 5-1, 5-7, 5-76, 5-79, 5-91, 5-93, 6-21, 6-22, 7-13, 7-91, 7-92
AIRFRAME ICING REPORTING CRITERIA 7-20
AIRFRAME ICING 7-20
AIRMAN'S METEOROLOGICAL INFORMATION 7-5
AIRMEN'S METEOROLOGICAL INFORMATION 7-1, 7-9
AIRMET 7-1, 7-5, 7-9
AIRPORT ADVISORY AREA 3-40
AIRPORT ADVISORY FREQUENCY 4-5
AIRPORT BEACON 2-8
AIRPORT DESTINATION SIGNS 2-40
AIRPORT DIRECTION SIGNS 2-39
AIRPORT INFORMATION SIGNS 2-41
AIRPORT LIGHT CONTROL 2-6
AIRPORT LOCATION SIGNS 2-38
AIRPORT MANAGEMENT RESPONSIBILITIES 4-58
AIRPORT MANDATORY INSTRUCTION SIGNS 2-37
AIRPORT MARKING AIDS AND SIGNS 2-30
AIRPORT PAVEMENT MARKINGS 2-31
AIRPORT RADAR SERVICE AREA 3-46
AIRPORT RESERVATIONS OFFICE 4-21
AIRPORT SIGNS 2-36
AIRPORT SURVEILLANCE RADAR 1-32, 5-42, 5-44, 5-49
AIRPORT TRAFFIC CONTROL TOWER 4-2
AIRPORT/FACILITY DIRECTORY 1-1, 1-4, 1-8, 1-15, 1-19, 1-30, 2-7, 3-44, 4-9, 4-13, 4-17, 4-43, 4-56, 5-3, 5-7, 5-20, 7-4, 7-7, 7-10, 7-11, 7-71, 7-74, 9-5
AIRPORT 1-32, 2-6, 2-8, 2-42, 2-43, 3-40, 3-47, 4-2, 4-5, 4-8, 4-9, 4-43, 4-51, 4-52, 4-61, 4-70, 5-1, 5-4, 5-7, 5-8, 5-13, 5-43, 5-44, 5-55, 5-74, 7-41, 7-84
AIRSPACE PARAMETERS 3-1, 3-2
AIRSPACE 3-1, 3-2, 3-30, 3-31, 3-32, 3-33, 3-34, 3-35, 3-36, 5-9, 5-33
AIRWAY BEACON 2-21
AIRWAY COURSE CHANGES 5-34
AIRWAY DEPICTION ON FLIGHT PLANS 5-7
AIRWAY POSITION REPORTING 5-31
AIRWAY SYSTEM EXPLAINED 5-33
AIRWAY 2-21, 5-7, 5-31, 5-33, 5-34
ALASKA SUPPLEMENT 4-9
ALASKA TERMINAL PUBLICATION 4-9
ALCOHOL USE 8-1
ALERT AREA 3-35
ALERT NOTICE 7-9
ALNOT 7-9
ALS 1-10, 2-1, 2-2, 2-6, 2-7, 5-49, 5-50, 8-5
ALTERNATE AIRPORT 5-4, 5-7, 5-8, 5-74, 7-41
ALTIMETER ACCURACY 7-40, 7-42, 7-43, 7-44
ALTIMETER ERRORS 7-40, 7-42, 7-43
ALTIMETER SETTING PROCEDURES 7-10, 7-40, 7-41, 7-42, 7-43

ALTIMETER SETTING 1-12, 4-8, 4-9, 4-13, 4-19, 7-10, 7-28, 7-40, 7-41, 7-42, 7-44, 7-92
ALTIMETER 1-12, 4-8, 4-9, 4-13, 4-19, 7-10, 7-28, 7-40, 7-41, 7-42, 7-43, 7-44, 7-92
ALTITUDE ASSIGNMENT 4-82, 4-85
ALTITUDE CHANGE 4-82, 5-32
ALTITUDE DEVIATION 4-80
ALTITUDE HYPOXIA 8-2
ALTITUDE RESERVATION 3-32
ALTITUDE RESTRICTION 4-88
ALTITUDE VERIFICATION 4-14, 5-30
ALTITUDE 1-3, 1-4, 1-7, 1-8, 1-10, 1-33, 3-32, 3-42, 3-44, 4-9, 4-14, 4-15, 4-17, 4-18, 4-19, 4-20, 4-31, 4-32, 4-38, 4-52, 4-53, 4-80, 4-82, 4-84, 4-85, 4-86, 4-87, 4-88, 4-89, 5-1, 5-2, 5-4, 5-7, 5-8, 5-10, 5-13, 5-22, 5-25, 5-30, 5-31, 5-32, 5-33, 5-36, 5-41, 5-42, 5-43, 5-44, 5-46, 5-47, 5-49, 5-55, 5-58, 5-71, 5-72, 5-73, 5-74, 5-75, 5-76, 5-77, 5-82, 6-14, 6-22, 6-23, 6-30, 7-10, 7-12, 7-20, 7-22, 7-26, 7-40, 7-41, 7-42, 7-43, 7-44, 7-53, 7-55, 7-58, 7-70, 7-71, 7-75, 7-84, 8-2, 8-5, 8-6, 8-8, 9-4
AMBULANCE FLIGHTS 4-33
AMENDED CLEARANCE 4-80, 4-83, 4-88, 5-71, 6-1
AMENDED FLIGHT PLAN 5-10, 5-11, 5-22
APPROACH AND LANDING MINIMUMS 5-55
APPROACH AZIMUTH GUIDANCE 1-12
APPROACH CLEARANCE 4-33, 4-70, 4-90, 5-42, 5-45, 5-46, 5-47, 5-60, 5-73, 5-80
APPROACH CONTROL 4-8, 4-17, 4-18, 5-42
APPROACH ILLUSIONS 8-5
APPROACH LIGHT SYSTEM 1-10, 2-1, 2-2, 2-6, 2-7, 5-49, 5-50, 8-5
APPROACH MINIMUMS 5-55
APPROACH PROCEDURES CHART 4-9, 5-44
APPROACH SEQUENCE 4-17, 5-48, 5-74
APPROACH 1-10, 1-12, 2-1, 2-2, 2-6, 2-7, 4-8, 4-9, 4-17, 4-18, 4-51, 4-60, 4-61, 4-70, 4-71, 4-90, 5-32, 5-42, 5-43, 5-44, 5-46, 5-48, 5-49, 5-50, 5-51, 5-52, 5-55, 5-56, 5-57, 5-59, 5-72, 5-74, 5-77, 5-80, 5-81, 8-5
AREA NAVIGATION CAPABILITY 5-7
AREA NAVIGATION CLEARANCE 5-7
AREA NAVIGATION EQUIPMENT 5-7
AREA NAVIGATION ROUTE 5-7, 5-33
AREA NAVIGATION 5-7, 5-33
ARO 4-21
ARRIVAL PROCEDURES 4-51, 4-52, 5-40, 5-41, 5-42, 5-43, 5-44, 5-45, 5-46, 5-47, 5-48, 5-49, 5-58, 5-59, 5-60, 5-72, 5-73, 5-74, 5-75, 5-79, 5-80, 6-30
ARRIVAL SPEED ADJUSTMENTS 4-90
ARRIVAL 4-13, 4-51, 4-52, 4-90, 5-7, 5-9, 5-22, 5-40, 5-41, 5-42, 5-43, 5-44, 5-45, 5-46, 5-47, 5-48, 5-49, 5-50, 5-51, 5-52, 5-53, 5-54, 5-55, 5-56, 5-57, 5-58, 5-59, 5-60, 5-72, 5-73, 5-74, 5-75, 5-79, 5-80, 6-30, 7-43
ARTCC 1-30, 3-42, 4-4, 4-19, 4-67, 4-70, 4-90, 5-7, 5-22, 5-30, 5-42, 6-16, 6-20, 7-9, 7-19, 7-73
ARTS 1-31, 4-15
ASOS 7-10
ASR 1-32, 5-42, 5-44, 5-49
ASSIGNED ALTITUDE 5-22, 5-30, 5-32, 5-36, 7-12
ATC CLEARANCE 4-17, 4-33, 4-55, 4-66, 4-80, 4-81, 4-82, 4-83, 4-84, 4-85, 4-86, 4-87, 5-6, 5-7, 5-71, 5-82, 6-1, 7-12
ATIS BROADCAST 4-8, 4-13, 4-17, 4-43, 4-57, 5-43
ATIS CODE 4-8, 4-13
ATIS 4-8, 4-9, 4-13, 4-17, 4-31, 4-32, 4-43, 4-57, 4-63, 5-43, 5-53, 5-57, 7-84
AUTOB CODE 7-10, 7-28
AUTOB 7-10, 7-28
AUTOMATED OBSERVING SYSTEMS 7-1
AUTOMATED RADAR TERMINAL SYSTEMS 1-31, 4-15
AUTOMATED SURFACE OBSERVING SYSTEM 7-10
AUTOMATED WEATHER OBSERVING SYSTEM BROADCASTS 7-10
AUTOMATED WEATHER OBSERVING SYSTEM 4-9, 7-10
AUTOMATIC ALTITUDE REPORTING 4-14, 4-19
AUTOMATIC METEOROLOGICAL OBSERVING STATIONS 7-10
AUTOMATIC OBSERVING STATIONS 7-10, 7-28
AUTOMATIC TERMINAL INFORMATION SERVICE 4-8, 4-9, 4-13, 4-17, 4-31, 4-32, 4-43, 4-57, 4-63, 5-43, 5-53, 5-57, 7-84
AVIATION SAFETY REPORTING PROGRAM 7-90
AVIATION WEATHER FORECAST 7-1, 7-27
AWOS 4-9, 7-10

B

BACK AZIMUTH GUIDANCE 1-12
BACK COURSE MARKER 1-10
BALLOONS 4-11, 4-19, 7-83, 8-1
BASE LEG 4-51
BASIC VFR WEATHER MINIMUMS 3-3
BEACON CODE 1-2, 1-10
BIRD ACTIVITY 7-71
BIRD ADVISORY 7-74
BIRD HAZARDS 7-71
BIRD INCIDENTS 7-72
BIRD MIGRATION 7-70
BIRD STRIKES 7-72
BRAKING ACTION ADVISORIES 4-57
BRAKING ACTION REPORTS 4-57
BRAKING ACTION 4-57
BREAK POINT 5-61
BRIEFING 3-41, 4-3, 4-82, 5-1, 5-2, 5-3, 5-4, 7-1, 7-2, 7-3, 7-7, 7-8, 7-9, 7-11
BROADCAST 1-10, 1-18, 4-8, 4-9, 4-13, 4-17, 4-43, 4-57, 5-43, 5-90, 6-24, 7-1, 7-2, 7-7, 7-8, 7-9, 7-10, 7-84

C

CABIN PRESSURE ALTITUDE 8-2, 8-6
CALL SIGN 4-30, 4-32, 4-33, 4-35, 4-73, 4-80, 4-85, 5-4, 5-7, 5-22
CANADIAN AIRSPACE 5-33
CANCEL INSTRUMENT FLIGHT RULES 5-13
CANCEL VISUAL FLIGHT RULES 5-12
CARBON MONOXIDE POISONING 8-4
CATEGORICAL OUTLOOKS 7-1, 7-6
CENRAP 1-32, 3-12, 3-13, 4-8, 4-17, 4-89, 5-49, 5-51
CENTER RADAR ARTS PRESENTATION/PROCESSING 1-32
CENTER WEATHER ADVISORY 7-9
CHANGEOVER POINTS 5-35
CHARTED VISUAL FLIGHT PROCEDURES 5-58
CHARTED VISUAL FLIGHT RULES FLYWAY PLANNING CHART 9-4
CHARTS 3-46
CIRCLING MINIMUMS 5-55
CLASS A AIRSPACE 3-3, 3-4, 3-10, 3-11, 3-15, 3-20, 4-14, 4-19, 5-75, 5-79, 6-30
CLASS B AIRSPACE 2-8, 3-3, 3-10, 3-12, 3-13, 3-15, 3-20, 3-34, 3-42, 3-45, 3-47, 4-17, 4-19, 4-51, 4-61, 4-65, 4-67, 4-70, 5-7, 5-13, 5-58, 5-75, 5-79, 8-8, 9-4
CLASS B SURFACE AREA 4-84, 4-88, 4-90
CLASS C AIRSPACE 2-8, 3-3, 3-10, 3-13, 3-15, 3-20, 3-34, 3-42, 3-45, 4-17, 4-19, 4-51, 4-61, 4-65, 4-67, 4-70, 5-7, 5-13, 5-75, 5-79, 8-8

CLASS C SURFACE AREA 4-84, 4-88, 4-90
CLASS D AIRSPACE 2-8, 3-3, 3-10, 3-14, 3-15, 3-20, 3-34, 3-42, 3-45, 3-46, 4-17, 4-19, 4-42, 4-51, 4-61, 4-65, 4-67, 5-7, 5-75, 5-79, 8-8
CLASS D SURFACE AREA 4-84, 4-88, 4-90
CLASS E AIRSPACE 2-8, 3-3, 3-10, 3-14, 3-15, 3-20, 3-34, 3-46, 4-17, 4-65, 5-7, 5-75, 5-83, 8-8
CLASS E SURFACE AREA 4-84
CLASS G AIRSPACE 3-3, 3-20, 3-21, 3-22, 4-19, 5-75
CLEAR AIR TURBULENCE 7-21, 7-23
CLEARANCE DELAY 5-20
CLEARANCE DELIVERY 4-63, 4-64, 5-20
CLEARANCE DEVIATION 4-88
CLEARANCE INTERPRETATION 5-45
CLEARANCE ITEMS 4-82
CLEARANCE LIMITS 4-82, 4-86, 5-6, 5-36, 6-30
CLEARANCE PREFIX 4-81
CLEARANCE READBACK 4-85
CLEARANCE RECORD 4-85, 5-71
CLEARANCE RELAY 4-81
CLEARANCE RESTRICTION 4-88
CLEARANCE VOID TIMES 5-23
CLEARANCE 1-10, 1-12, 2-2, 2-8, 3-32, 4-17, 4-33, 4-55, 4-59, 4-60, 4-63, 4-64, 4-66, 4-67, 4-70, 4-80, 4-81, 4-82, 4-83, 4-84, 4-85, 4-86, 4-87, 4-88, 4-90, 4-91, 4-92, 5-2, 5-6, 5-7, 5-20, 5-21, 5-22, 5-23, 5-25, 5-36, 5-41, 5-42, 5-44, 5-45, 5-46, 5-47, 5-55, 5-56, 5-59, 5-60, 5-71, 5-72, 5-73, 5-75, 5-80, 5-82, 5-83, 6-1, 6-30, 7-12, 7-55, 7-57, 7-58, 7-82, 7-83
CLEARED AS FILED 5-22
CLEARED FOR THE OPTION APPROACH 4-71
CLOUD CEILING 7-10, 7-14
CODE 7500 4-19, 6-23
CODE 7600 4-19, 6-31
CODE 7700 4-19, 5-91, 6-11, 6-21
CODE BEACONS 2-21
CODE WORDS 6-23
CODE 1-2, 1-3, 1-6, 1-7, 1-10, 1-11, 1-13, 1-14, 1-16, 1-31, 2-21, 4-8, 4-13, 4-14, 4-17, 4-19, 4-36, 5-4, 5-7, 5-13, 5-22, 5-24, 5-31, 5-90, 5-91, 6-11, 6-16, 6-21, 6-23, 6-31, 7-10, 7-19, 7-28
COLLISION AVOIDANCE 8-6, 8-8
COMMON TRAFFIC ADVISORY FREQUENCY 3-44, 4-9, 4-10
COMMUNICATION FAILURE 4-82, 6-30
COMMUNICATION PROCEDURES 4-9, 4-11, 4-13, 4-30, 4-32, 4-33, 4-34, 4-35, 4-36, 5-91, 6-1, 6-20, 6-21, 7-3, 7-57
COMMUNICATION 1-12, 1-16, 3-42, 4-4, 4-9, 4-11, 4-13, 4-20, 4-30, 4-31, 4-32, 4-33, 4-34, 4-35, 4-36, 4-37, 4-38, 4-39, 4-40, 4-41, 4-42, 4-43, 4-51, 4-63, 4-67, 4-81, 4-82, 4-85, 4-90, 5-21, 5-30, 5-48, 5-91, 6-1, 6-11, 6-20, 6-21, 6-30, 6-32, 7-3, 7-57, 7-81, 7-84, 8-8
COMPASS LOCATOR 1-2, 1-8, 1-10
COMPLIANCE WITH FREQUENCY CHANGES 4-32
COMPOSITE FLIGHT PLAN 5-6
COMPULSORY REPORTING POINTS 5-31, 5-32
COMPULSORY VISUAL FLIGHT RULES FLIGHT 5-5
CONFLICT ADVISORY 5-79
CONFLICT ALERT 4-15, 5-76, 5-77
CONTACT APPROACH 5-59, 5-72
CONTROLLED AIRSPACE 3-10
CONTROLLED FIRING AREA 3-36
CONTROLLER LIMITATIONS 1-30
CONTROLLER RESPONSIBILITIES 3-10, 3-12, 3-13, 3-14, 3-32, 4-13, 4-14, 4-15, 4-16, 4-17, 4-30, 4-32, 4-33, 4-51, 4-55, 4-57, 4-58, 4-59, 4-63, 4-64, 4-66, 4-67, 4-70, 4-82, 4-83, 4-84, 4-88, 4-92, 4-94, 5-1, 5-7, 5-13, 5-22, 5-23, 5-24, 5-25, 5-31, 5-33, 5-36, 5-42, 5-45, 5-46, 5-48, 5-49, 5-52, 5-53, 5-57, 5-58, 5-59, 5-60, 5-61, 5-70, 5-71, 5-72, 5-73, 5-74, 5-75, 5-76, 5-77, 5-78, 5-79, 5-80, 5-81, 5-82, 5-83, 5-84, 6-10, 6-23, 6-24, 7-12, 7-41, 7-43, 7-55, 7-58, 7-90
CONTROLLER 1-30, 1-31, 4-13, 4-14, 4-15, 4-16, 4-17, 4-30, 4-32, 4-33, 4-51, 4-54, 4-55, 4-57, 4-59, 4-60, 4-63, 4-64, 4-66, 4-67, 4-70, 4-82, 4-83, 4-84, 4-88, 4-92, 5-1, 5-7, 5-13, 5-22, 5-23, 5-24, 5-25, 5-31, 5-33, 5-36, 5-42, 5-45, 5-46, 5-48, 5-49, 5-52, 5-53, 5-57, 5-58, 5-59, 5-60, 5-70, 5-71, 5-72, 5-73, 5-74, 5-75, 5-76, 5-77, 5-78, 5-79, 5-80, 5-81, 5-82, 5-83, 5-84, 6-10, 6-23, 6-24, 7-12, 7-41, 7-43, 7-55, 7-58, 7-90
CONVECTIVE SIGMET 7-5
COURSE LIGHTS 2-21
CROSS WIND LEG 4-51
CROSSING ALTITUDE 4-88
CRUISE CLEARANCE 4-82
CRUISING ALTITUDE 7-41
CRUISING ALTITUDES AND FLIGHT LEVELS 4-38
CTAF 2-7, 3-44
CUSTOMS NOTIFICATION 5-9

D

DATA COMMUNICATIONS 1-12
DECISION HEIGHT 1-10, 5-49, 5-74
DECLARED DISTANCES 4-55
DECOMPRESSION SICKNESS 8-2
DEFENSE VISUAL FLIGHT RULES FLIGHT PLAN 5-5, 5-90, 6-16
DEFINITION OF CLEARANCE 4-80, 5-71
DENSITY ALTITUDE ADVISORY 7-84
DENSITY ALTITUDE BROADCAST 7-84
DENSITY ALTITUDE 7-10, 7-84
DEPARTURE CLEARANCE 4-59, 4-63, 4-64, 4-67, 4-91, 5-20, 5-22, 5-83
DEPARTURE CONTROL 5-24
DEPARTURE DELAYS 4-64
DEPARTURE MINIMUMS 5-25, 5-83
DEPARTURE OBSTRUCTION CLEARANCE 5-25, 5-83
DEPARTURE PROCEDURES 4-59, 4-82, 4-88, 5-24, 5-25, 5-83
DEPARTURE RESTRICTIONS 5-23
DEPARTURE SEPARATION 4-59, 7-58
DEPARTURE TIME CHANGE 5-11
DESCENT CLEARANCE 4-88, 4-90, 5-41
DESIGNATED MOUNTAINOUS AREAS 5-94
DESIGNATED RUNWAY 4-55
DIRECT ROUTES 5-31
DIRECT USER ACCESS TERMINAL (DUAT) 7-29
DIRECT USER ACCESS TERMINAL SYSTEM 5-1
DIRECTION FINDER SERVICES 1-19, 6-2, 6-12, 6-21
DISASTER AREA 3-42
DISTANCE MEASURING EQUIPMENT 1-7, 1-8, 1-10
DISTANT EARLY WARNING IDENTIFICATION ZONE 5-90
DITCHING PROCEDURES 6-22
DOG/HANDLER TEAMS 6-15
DOPPLER RADAR 1-21
DOWNDRAFTS 7-24
DOWNED AIRCRAFT 6-21, 6-22
DOWNWIND LEG 4-51
DRUG AND SEDATIVE USE 8-1, 8-2
DUAT 7-29
DUATS 5-1

E

EAR BLOCK 8-2
EFFECTS OF ALTITUDE 8-2
EFIS 5-7
ELECTRONIC FLIGHT INSTRUMENT SYSTEM 5-7
EMERGENCY AIRBORNE INSPECTION OF OTHER AIRCRAFT 7-87
EMERGENCY COMMUNICATIONS 4-43, 6-20, 6-21, 6-32
EMERGENCY FREQUENCY 6-21

EMERGENCY INTERCEPT AND ESCORT 6-13
EMERGENCY LOCATOR TRANSMITTER 6-14
EMERGENCY PROCEDURES 6-2, 6-10, 6-11, 6-12, 6-13, 6-14, 6-15, 6-16, 6-20, 6-21, 6-22, 6-23, 6-24, 6-30, 6-31, 6-32
EMERGENCY RADIO TRANSMISSIONS 6-20, 6-21
EMERGENCY TRANSPONDER OPERATION 4-19, 6-11, 6-21, 6-23, 6-31
EMERGENCY VISUAL FLIGHT RULES NAVIGATIONAL ASSISTANCE 4-16, 6-2, 6-12
EN ROUTE CLEARANCE 4-88
EN ROUTE COMMUNICATIONS 4-43, 5-30, 6-32
EN ROUTE FLIGHT ADVISORY SERVICE 4-3, 4-43, 7-4
EN ROUTE HOLDING 4-82, 5-32, 5-36, 6-30
EN ROUTE OBSTRUCTION CLEARANCE 5-2, 7-82, 7-83
ENGINE STARTUP ADVISORY 4-64
EQUIPMENT CODE 5-7
EXPECT FURTHER CLEARANCE 5-36

F

FAN MARKER 1-9
FAST FILE 5-1
FAVORED RUNWAY 4-55
FINAL APPROACH 4-51, 5-32, 5-42, 5-43
FLIGHT CHECK 4-73
FLIGHT INFORMATION PUBLICATION 3-41
FLIGHT INSPECTION 4-73
FLIGHT LIMITATIONS 3-43
FLIGHT MANAGEMENT SYSTEM 1-22, 5-7
FLIGHT OPERATIONS IN VOLCANIC ASH 7-86
FLIGHT PLAN CHANGE 5-10, 5-11
FLIGHT PLAN FORM 5-1
FLIGHT PLAN FORMAT 4-18, 5-4, 5-7, 5-33
FLIGHT PLAN 4-18, 5-1, 5-4, 5-5, 5-6, 5-7, 5-10, 5-11, 5-22, 5-33, 5-90
FLIGHT RESTRICTION 3-43
FLIGHT SERVICE STATION 4-3
FLIGHT STANDARDS DISTRICT OFFICE 7-80
FLIGHTS OVER NATIONAL PARKS 7-75
FLIP 3-41
FLOW CONTROL ADVISORY 7-2, 7-3
FMS 1-22, 5-7
FOREIGN AIRSPACE 5-9, 5-33
FORMAL RUNWAY USE PROGRAM 4-55
FORMATION FLYING 7-87
FREQUENCY CHANGES 4-32, 5-30
FRICTION MEASUREMENT DEVICES 4-58
FUEL DUMPING BROADCAST 6-24
FUEL DUMPING 6-24
FULL APPROACH 4-70
FURTHER CLEARANCE 6-30

G

GATE HOLD PROCEDURES 4-64
GENERAL DIMENSIONS OF AIRSPACE SEGMENTS 3-2
GLIDESLOPE INFORMATION 1-10, 1-13
GLOBAL POSITIONING SYSTEM 1-23
GO-AROUND 5-57
GPS 1-23
GROUND CONTROL 4-63, 5-20
GROUND MOVEMENTS 4-66
GROUND STATION IDENTIFICATION 4-35
GROUND VEHICLE OPERATIONS 4-10, 4-62
GROUND VOR CHECK POINTS 1-4

H

HAZARDOUS AREA REPORTING SERVICE 4-20
HAZARDOUS IN-FLIGHT WEATHER ADVISORY SERVICE 4-43, 7-9
HELICOPTER OPERATIONS 4-42, 4-66
HELICOPTER VISUAL FLIGHT RULES FLIGHTS 4-66
HELIPAD 4-66
HELIPORT 4-66
HIGH ALTITUDE DESTINATIONS 5-8
HIGH BAROMETRIC PRESSURE 7-43
HIGH DENSITY TRAFFIC AIRPORT 4-21, 5-1
HIGH DENSITY TRAFFIC AREAS 9-4
HIGH INTENSITY RUNWAY LIGHTS 2-4
HIGH PERFORMANCE CLIMB 4-88
HIJACK PROCEDURES 6-23
HIWAS BROADCAST 4-43
HIWAS 4-43
HOLD FOR RELEASE 5-23
HOLDING FIX 4-82, 5-32, 5-36, 5-45, 5-48, 6-30
HOLDING INSTRUCTIONS 5-36, 6-30
HOLDING MINIMUMS 5-36
HOLDING PATTERN PROCEDURES 4-82, 5-36, 5-47, 5-48, 6-30
HOLDING POSITIONS MARKINGS 2-34
HOLDING SPEED 5-36
HOT AIR BALLOONS 4-11
HOVER TAXI 4-66
HYPOXIA 8-2

I

ICAO 7-29
IDENTIFICATION CODE 1-2, 1-3, 1-6, 1-7, 1-10, 1-11, 1-13, 1-14, 1-16, 5-4, 5-7
IFR CLEARANCE 3-32, 4-80
IFR REQUIREMENTS 3-22
IFR RESERVATIONS 4-21
IFR 5-61
IN-FLIGHT WEATHER ADVISORY 7-5
IN-FLIGHT WEATHER AVOIDANCE 7-12
IN-FLIGHT WEATHER BROADCASTS 7-9
IN-RUNWAY LIGHTING 2-5
INCIDENT REPORT 7-72
INERTIAL NAVIGATION SYSTEM ACCURACY 1-20
INERTIAL NAVIGATION SYSTEM 1-20
INFORMAL RUNWAY USE PROGRAM 4-55
INITIAL APPROACH ALTITUDE 5-43
INITIAL CONTACT 4-32, 4-34, 4-51, 5-20
INNER MARKER 1-10
INOPERATIVE COMPONENT TABLE 1-10
INS 1-20
INSTRUMENT APPROACH 5-46
INSTRUMENT FLIGHT RULES ALTITUDES 4-38
INSTRUMENT FLIGHT RULES FLIGHT LEVELS 4-38
INSTRUMENT FLIGHT RULES FLIGHT PLAN 4-86, 5-7, 5-90
INSTRUMENT FLIGHT RULES RADAR VECTOR 5-33, 5-42, 5-44, 5-49, 5-75, 7-12
INSTRUMENT FLIGHT RULES SEPARATION STANDARDS 4-89, 5-34
INSTRUMENT LANDING SYSTEM ALTITUDES 1-10
INSTRUMENT LANDING SYSTEM APPROACH 5-51, 5-52
INSTRUMENT LANDING SYSTEM COURSE DISTORTION 1-10
INSTRUMENT LANDING SYSTEM CRITICAL AREA 1-10
INSTRUMENT LANDING SYSTEM DISTANCE MEASURING EQUIPMENT 1-7, 1-10
INSTRUMENT LANDING SYSTEM FREQUENCY 1-10
INSTRUMENT LANDING SYSTEM MARKER 1-2, 1-9, 1-10
INSTRUMENT LANDING SYSTEM 1-10
INTERCEPTION PROCEDURES 5-91, 5-93
INTERCHANGE AIRCRAFT 4-34
INTERNATIONAL CIVIL AVIATION ORGANIZATION 7-29

INTERSECTION DEPARTURE 4-59
INTERSECTION TAKEOFF CLEARANCE 7-55
INTERSECTION TAKEOFF 4-59

J

JET ENGINE BLAST 7-50
JET ROUTES 5-7, 5-33
JOINT USE AIRPORT 2-43
JUDGING ALTITUDE 6-22

K

KEY TO AVIATION WEATHER OBSERVATIONS AND FORECASTS 7-14

L

LAA 3-40, 4-9, 4-43, 4-58, 4-70, 7-84
LANDING CLEARANCE 2-8, 4-60
LANDING DIRECTION INDICATOR 4-52
LANDING ERROR ILLUSIONS 8-5
LANDING MINIMUMS 5-55
LANDING PRIORITY 5-60
LANDING STRIP INDICATOR 4-52
LANDING 4-69
LAW ENFORCEMENT OPERATIONS BY CIVIL AND MILITARY ORGANIZATIONS 5-92
LEAD-IN LIGHTS 2-1
LEASED AIRCRAFT 4-34
LIGHT SIGNALS 4-62
LOCAL AIRPORT ADVISORY 3-40, 4-9, 4-43, 4-58, 4-70, 7-84
LOCAL CONTROL 4-63
LOCAL FLOW TRAFFIC MANAGEMENT PROGRAM 5-41
LOCALIZER CRITICAL AREA 1-10
LOCALIZER DIRECTIONAL AID 1-10
LOCALIZER DISTANCE MEASURING EQUIPMENT 1-7
LOCALIZER TRANSMITTER 1-10, 1-13
LORAN NAVIGATION SYSTEM 1-17
LOW ALTITUDE ALERT 4-15
LOW APPROACH 4-61
LOW BAROMETRIC PRESSURE 7-40
LOW INTENSITY RUNWAY LIGHTS 2-4
LOW LEVEL WIND SHEAR ALERT SYSTEM 4-56

M

MACH NUMBER 4-90, 5-78
MAGNETIC BEARING 4-39
MAGNETIC COURSE 4-39
MAGNETIC HEADING 4-39
MANDATORY ALTITUDE 5-44
MARKER BEACON 1-9, 1-10, 1-13
MAXIMUM ALTITUDE 5-44
MEA 5-25
MEDICAL REQUIREMENTS 8-1
MEDICAL WAIVER 8-1
MEDIUM INTENSITY RUNWAY LIGHTS 2-4
MICROBURSTS 7-24
MICROWAVE LANDING SYSTEM 1-12, 1-13
MIDDLE MARKER 1-10
MIGRATORY BIRD ACTIVITY 7-70
MIGRATORY FLYWAYS 7-71
MILITARY AIRPORT BEACON 2-8
MILITARY AIRPORT 2-8, 2-43, 5-44, 5-55
MILITARY MINIMUMS 5-55
MILITARY OPERATIONS AREA 3-34
MILITARY TRAFFIC PATTERN ALTITUDE 4-53
MILITARY TRAFFIC PATTERNS 4-53
MILITARY TRAINING ROUTES 3-41
MINIMUM ALTITUDE 5-44, 7-41, 7-75
MINIMUM DESCENT ALTITUDE 5-8
MINIMUM ENROUTE ALTITUDE 5-25
MINIMUM FUEL ADVISORY 5-84
MINIMUM FUEL 5-84
MINIMUM HOLDING ALTITUDE 5-36
MINIMUM SAFE ALTITUDE WARNING 4-15
MINIMUM SAFE ALTITUDE 5-44, 7-41
MINIMUM VECTORING ALTITUDE 5-44
MISSED APPROACH 5-32, 5-46, 5-48, 5-49, 5-56, 5-74
MISSED COMMUNICATIONS 1-16
MOA 3-34
MODE 3/A MONITORING 4-19
MODE C VEIL AIRPORTS 3-47
MONITORING RADIO FREQUENCIES 4-4, 4-30, 5-1
MORSE CODE 4-36
MOUNTAIN FLYING 5-8, 7-84
MOUNTAIN WAVE 7-84
MOUNTAINOUS TERRAIN 5-8
MOVEMENT AREAS 4-66
MTR 3-41
MULTICOM 4-11

N

NATIONAL AIRSPACE SYSTEM 4-3
NATIONAL PARK SERVICE 7-75
NATIONAL SECURITY 5-5
NATIONAL TRANSPORTATION SAFETY BOARD 7-91
NATIONAL WEATHER SERVICE 7-1, 7-2
NAVAID ALTITUDE AND DISTANCE LIMITATIONS 1-3, 1-7, 1-8
NAVAID CERTIFICATION 4-73
NAVAID CLASSES 1-8
NAVAID IDENTIFICATION 1-2, 1-6, 1-11, 1-13, 1-14
NAVAID IDENTIFIER REMOVAL 1-14
NAVAID INTERFERENCE 1-2
NAVAID MAINTENANCE 1-14
NAVAID SERVICE VOLUME 1-8
NAVAID USER REPORT 1-16
NAVAID VOICE 1-15
NEAR MIDAIR COLLISION 7-92
NO-GYRO APPROACH 5-49
NOISE ABATEMENT 5-24, 5-31, 5-41
NONCOMPULSORY REPORTING POINTS 5-31, 5-32
NONDIRECTIONAL BEACON 1-2, 1-8
NONPRECISION APPROACHES 1-17, 5-44
NOTAM 5-3, 7-82
NOTICE TO AIRMEN 1-8, 1-17, 1-18, 3-42, 3-43, 4-3, 4-8, 4-58, 5-1, 5-2, 5-3, 5-9, 5-70, 7-3, 7-8, 7-9
NUCLEAR ACCIDENT 3-42
NUMBERS USAGE 4-37

O

OBSTACLE CLEARANCE 1-10, 1-12, 2-2, 4-17, 5-44, 5-55, 5-56, 5-59, 5-72, 5-73
OBSTRUCTION ALERT 4-15, 5-76
OBSTRUCTION LIGHTS 2-22
OBSTRUCTIONS CLEARANCE 4-87
OBSTRUCTIONS TO FLIGHT 7-82
OMEGA NAVIGATION SYSTEM 1-18
OMEGA STATUS BROADCAST 1-18
OPERATION LIGHTS ON 4-72
OPERATION RAINCHECK 4-7
OPERATION TAKE-OFF 4-7
OTHER MARKINGS 2-35
OUTER MARKER 1-10
OVERDUE AIRCRAFT 7-91
OVERHEAD APPROACH MANEUVER 5-61

P

PARACHUTE JUMP AIRCRAFT OPERATIONS 3-44
PARACHUTE JUMP ALTITUDES 3-44
PARACHUTE JUMP OPERATIONS 3-44, 4-9
PARALLEL INSTRUMENT LANDING SYSTEM APPROACHES 5-52
PERSONAL CHECKLIST 8-1
PHONETIC ALPHABET 4-36
PHYSIOLOGICAL DIFFICULTIES 8-2, 8-3, 8-4, 8-5, 8-6
PHYSIOLOGICAL STRESSES 8-7
PHYSIOLOGICAL TRAINING PROGRAM 8-2
PILOT AUTOMATIC TELEPHONE WEATHER ANSWERING SERVICE 7-7
PILOT BRIEFING 3-41, 4-3, 4-82, 5-1, 5-4, 7-1, 7-2, 7-3, 7-7, 7-11
PILOT BROADCAST 4-9
PILOT CLEARANCE RESPONSIBILITY 4-66, 4-82, 4-83, 4-84, 4-85, 4-88, 4-90, 4-92, 5-71, 5-73, 5-75, 7-12
PILOT CONTROL OF AIRPORT LIGHTING 2-7, 4-10

PILOT MANUEVERS IN THE TRAFFIC PATTERN 4-54
PILOT RESPONSIBILITIES 3-10, 3-11, 3-12, 3-13, 3-14, 3-15, 3-32, 3-35, 3-41, 3-42, 3-43, 3-44, 3-46, 3-47, 4-17, 4-20, 4-30, 4-32, 4-33, 4-34, 4-43, 4-50, 4-51, 4-54, 4-55, 4-57, 4-58, 4-59, 4-60, 4-61, 4-65, 4-66, 4-67, 4-68, 4-69, 4-70, 4-80, 4- 82, 4-84, 4-85, 4-86, 4-87, 4-88, 4-90, 4-91, 4-92, 4-94, 5-4, 5-6, 5-7, 5-9, 5-10, 5-12, 5-13, 5-21, 5-22, 5-23, 5-25, 5-30, 5-31, 5-32, 5-33, 5-34, 5-36, 5-41, 5-42, 5-46, 5-47, 5-49, 5-51, 5-55, 5-57, 5-58, 5-59, 5-61, 5-70, 5-71, 5-72, 5-73, 5-74, 5-75, 5-76, 5-77, 5-78, 5-79, 5-80, 5-81, 5-82, 5-83, 5-84, 6-1, 6-2, 6-16, 6-20, 6-21, 6-23, 6-30, 7-12, 7-13, 7-50, 7-52, 7-55, 7-82, 7-84, 7-87, 7-90, 8-1, 8-6, 8-8
PILOT TRAFFIC ADVISORY PRACTICES 4-9, 4-52, 4-61, 5-79
PILOT WEATHER REPORT 5-32, 7-14, 7-19, 7-20, 7-21, 7-22, 7-23, 7-24
PILOT/CONTROLLER COMMUNICATIONS 4-30
PILOT/CONTROLLER GLOSSARY 4-30
PILOT 2-7, 3-42, 4-3, 4-9, 4-10, 4-20, 4-30, 4-32, 4-33, 4-34, 4-43, 4-50, 4-51, 4-52, 4-54, 4-55, 4-57, 4-59, 4-60, 4-61, 4-65, 4-66, 4-67, 4-68, 4-69, 4-70, 4-80, 4-82, 4-83, 4-84, 4-85, 4-86, 4-87, 4-88, 4-90, 4-91, 4-92, 5-1, 5-4, 5-6, 5-7, 5-9, 5-10, 5-12, 5-13, 5-21, 5-22, 5-23, 5-25, 5-30, 5-31, 5-32, 5-34, 5-36, 5-41, 5-42, 5-46, 5-47, 5-49, 5-51, 5-55, 5-57, 5-58, 5-59, 5-70, 5-71, 5-72, 5-73, 5-74, 5-75, 5-76, 5-77, 5-78, 5-79, 5-80, 5-81, 5-82, 5-83, 5-84, 6-1, 6-2, 6-16, 6-20, 6-21, 6-23, 6-30, 7-1, 7-2, 7-3, 7-7, 7-11, 7-12, 7-13, 7-14, 7-19, 7-20, 7-21, 7-22, 7-23, 7-24, 7-50, 7-52, 7-55, 7-82, 7-84, 7-90, 8-1, 8-6, 8-8
PIREP CODE 7-19
PIREP 7-19, 7-20, 7-21, 7-22, 7-23, 7-24
POSITION REPORTING 5-31, 5-32
POSITION UPDATE ALIGNMENT 1-20
PRACTICE DIRECTION FINDER PROCEDURES 6-12
PRACTICE INSTRUMENT APPROACH 4-70
PRECIPITATION INTENSITIES 7-16, 7-17, 7-18
PRECISION APPROACH PATH INDICATOR 2-2
PRECISION APPROACH RADAR PROCEDURES 1-33, 5-44, 5-49, 5-50, 5-51, 5-75
PRECISION DISTANCE MEASURING EQUIPMENT 1-12
PREFLIGHT BRIEFING 5-1, 5-2, 5-3, 7-2, 7-3, 7-7, 7-8, 7-9
PRETAXI CLEARANCE PROCEDURES 5-20
PROCEDURE TURN 5-36, 5-47
PROHIBITED AREA 3-31
PROHIBITIONS 3-43
PUBLISHED VFR ROUTES 3-45
PULSATING VISUAL APPROACH SLOPE INDICATOR 2-2

R

RADAR ADVISORY SERVICE 4-14, 4-17, 4-20, 4-51, 5-13, 5-79
RADAR ADVISORY 5-50, 5-52
RADAR ALERT SERVICE 5-76
RADAR APPLICATION 1-30
RADAR APPROACH MONITORING 1-33
RADAR APPROACH 5-49
RADAR BEACON PHRASEOLOGY 4-19
RADAR BEACON 1-31, 4-19
RADAR EQUIPMENT PERFORMANCE 1-30
RADAR IDENTIFICATION 5-30
RADAR NAVIGATION SYSTEM 1-17
RADAR SERVICE LIMITATIONS 1-30
RADAR SERVICE TERMINATED 5-72
RADAR TRAFFIC ADVISORIES 4-51
RADAR WEATHER ECHO INTENSITY LEVELS 7-25
RADIO COMMUNICATIONS TECHNIQUE 5-91
RADIO COMMUNICATIONS TECHNIQUE 4-31, 4-36, 4-37, 4-38, 4-39, 4-40, 4-41, 4-42, 4-43, 4-63, 4-67, 4-81, 4-85, 4-90, 5-21, 5-30, 5-48, 6-1, 6-21, 6-32, 7-3, 7-57
RADIO CONTACT PROCEDURE 4-32, 4-34, 4-51, 6-32
RADIO FAILURE PROCEDURES 4-42, 5-30, 6-30, 6-31, 6-32
RADIO FREQUENCY DESIGNATIONS 4-11
RADIO PROCEDURES 4-11
RADIO RANGE 4-31
RANGE GUIDANCE 1-12
RE-ESTABLISHING RADIO CONTACT 6-32
RECORDING OPERATIONAL COMMUNICATIONS 4-4
REDUCING BIRD STRIKE RISKS 7-71
RELATIVE ALTITUDE 8-8
RELEASE TIMES 5-23
REPORTING CLOUD HEIGHTS 7-10, 7-14
REPORTING POINTS 4-82, 5-7, 5-10, 5-31, 5-32, 5-33, 5-90, 7-9, 9-4
REPORTING WILDLIFE HAZARDS 7-73, 7-74
RESCUE COORDINATION CENTER 6-16
RESPIRATORY DIFFICULTIES 8-2
RESTRICTED AIRSPACE 3-32
RESTRICTED ALTITUDE 3-42, 4-82, 4-85
RESTRICTED AREA 3-32
RIGHT OF WAY 4-52
ROTATING BEACON 2-8
ROTOR DOWNWASH 4-66
ROUTE COURSE CHANGES 5-34
RUNWAY CENTERLINE LIGHTS 2-5
RUNWAY CROSSING 4-67
RUNWAY DISTANCE REMAINING MARKERS 2-9
RUNWAY DISTANCE REMAINING SIGNS 2-42
RUNWAY EDGE LIGHTS 2-4
RUNWAY END IDENTIFIER LIGHTS 2-3
RUNWAY EXITING 4-69
RUNWAY FRICTION REPORTS 4-58
RUNWAY HOLDING PROCEDURES 4-60, 4-67, 4-91
RUNWAY MARKINGS 2-32, 4-55
RUNWAY PROFILE DESCENT 5-41
RUNWAY REMAINING LIGHTS 2-5
RUNWAY USE PROGRAM 4-55
RUNWAY VISIBILITY 7-13

S

SAFETY ALERT 4-15, 5-76
SAFETY ALERTS 4-17
SCATANA BROADCAST 5-90
SCHEDULED BROADCAST 7-9
SCUBA DIVING 8-2
SEAPLANE SAFETY 7-85
SEARCH AND RESCUE FACILITIES 6-16
SEARCH AND RESCUE PROCEDURES 5-12, 5-23, 6-13, 6-16, 6-20
SEGMENTED CIRCLE 4-52
SELF-ANNOUNCE 4-9
SEQUENCED FLASHING LIGHT SYSTEM 2-1
SEVERE WEATHER AVOIDANCE 7-12
SEVERE WEATHER FORECAST ALERTS 7-5, 7-9
SFAR 3-47
SIDE STEP MANEUVER 5-54, 5-55
SIDE STEP MINIMUMS 5-55
SIMPLIFIED DIRECTIONAL FACILITY 1-11
SIMULATED INSTRUMENT FLIGHT RULES FLIGHT 5-2
SIMULTANEOUS CONVERGING INSTRUMENT APPROACHES 5-53
SIMULTANEOUS DEPARTURES 4-60
SIMULTANEOUS INSTRUMENT LANDING SYSTEM APPROACH 4-60, 5-51, 5-52
SPACE-BASED RADIO POSITIONING 1-23

SPATIAL DISORIENTATION 8-5
SPECIAL FEDERAL AVIATION REGULATION 3-47
SPECIAL LAW ENFORCEMENT OPERATIONS 5-92
SPECIAL MEDICAL EVALUATION 8-1
SPECIAL TRAFFIC MANAGEMENT PROGRAMS 4-21
SPECIAL USE AIRSPACE 3-30, 3-31, 3-32, 3-33, 3-34, 3-35, 3-36
SPECIAL VFR CLEARANCE 4-84
SPECIAL VISUAL FLIGHT RULES CLEARANCE 4-84
SPEED ADJUSTMENTS 4-90, 5-41, 5-46, 5-78
SPEED MEASUREMENTS 4-40, 4-90, 5-78
STAGE I SERVICE 4-17
STAGE II SERVICE 4-17
STAGE III SERVICE 4-17
STANDARD ALTIMETER SETTING 7-41
STANDARD INSTRUMENT DEPARTURE 4-9, 5-22, 5-25
STANDARD TERMINAL ARRIVAL 5-22, 5-40
STRAIGHT-IN MINIMUMS 5-55
STUDENT PILOT RADIO IDENTIFICATION 4-33
SUPPLEMENTAL WEATHER SERVICE LOCATION 5-3
SUPPLEMENTAL WEATHER SERVICE LOCATIONS 4-3, 4-43, 5-1, 7-2
SURVEILLANCE APPROACH 5-49

T

TAC 3-45
TACTICAL AIR NAVIGATION 1-5, 1-6, 1-7, 1-8
TAF 7-29
TAKEOFF CLEARANCE 4-59, 4-66, 4-67, 7-58
TAXI CLEARANCE 4-67,5-21
TAXI PROCEDURES 4-63, 4-66, 4-67, 4-68, 4-69, 4-72, 5-20, 5-21
TAXIWAY LIGHTS 2-9
TAXIWAY MARKINGS 2-33
TAXIWAY TURNOFF LIGHTS 2-5
TCAS 4-80, 4-94, 5-7
TEL-TWEB 7-2, 7-8
TELEPHONE INFORMATION BRIEFING SERVICE 7-7
TEMPERATURE INVERSIONS 1-30
TEMPORARY FLIGHT RESTRICTION 3-42
TERMINAL AREA VFR ROUTES 3-45
TERMINAL AREA 5-81
TERMINAL FORECAST (TAF) 7-29
TERMINAL INSTRUMENT APPROACH PROCEDURES 1-10, 5-44, 5-55
TERMINAL INSTRUMENT PROCEDURES 1-10, 5-44
TERMINAL PROCEDURES PUBLICATION 1-12, 5-40, 9-3, 9-4
TERMINAL RADAR SERVICE AREA 3-46, 4-17, 5-13
TERRAIN ALERT 4-15
TERRAIN CLEARANCE 1-10, 4-17, 4-87, 5-44
TETRAHEDRON 4-52
THRESHOLD CROSSING HEIGHT 1-10
THUNDERSTORMS 7-25, 7-26
TIMED APPROACH 5-48
TOLL FREE LINES 7-2
TOUCHDOWN ZONE LIGHTING 2-5
TOWER EN ROUTE CONTROL 4-18
TOWER RADAR DISPLAY 4-51
TPX-42 SYSTEM 4-15
TRAFFIC ADVISORY FREQUENCY 9-4
TRAFFIC ADVISORY 1-30, 3-34, 3-44, 4-8, 4-10, 4-14, 4-17, 4-19, 4-32, 4-51, 4-61, 4-89, 5-79, 7-81, 8-8
TRAFFIC ALERT AND COLLISION AVOIDANCE SYSTEM 4-80, 4-94
TRAFFIC CONTROL LIGHT SIGNALS 4-42, 4-62, 4-67
TRAFFIC PATTERN ALTITUDE 4-53
TRAFFIC PATTERN INDICATOR 4-52
TRAFFIC PATTERN 4-51, 4-52, 4-53, 4-54
TRANSCRIBED WEATHER BROADCAST 1-10, 7-1, 7-8
TRANSPONDER CODE 1-31, 4-14, 4-17, 4-19, 5-7, 5-13, 5-22, 5-24, 5-31, 5-90, 5-91, 6-11, 6-21, 6-23, 6-31
TRANSPONDER MODE C 4-19
TRANSPONDER OPERATION 1-31, 4-19, 5-7, 5-24, 5-31, 5-90
TREND ADVISORY 5-50
TRICOLOR VISUAL APPROACH SLOPE INDICATOR 2-2
TRSA 3-46
TRUE BEARING 4-39
TRUE COURSE 4-39
TRUE HEADING 4-39
TURBULENCE REPORTING CRITERIA 7-21
TWEB BROADCAST 7-8
TWO WAY RADIO COMMUNICATIONS FAILURE 6-30
TWO WAY RADIO 5-90

U

U. S. FISH AND WILDLIFE SERVICE 7-75
UNICOM 4-11
UNMANNED BALLOONS 7-83
UNSAFE CONDITIONS 7-90
UNSCHEDULED WEATHER BROADCAST 7-9
UPWIND LEG 4-51
USE OF AIRCRAFT LIGHTS 4-72

V

VERY HIGH FREQUENCY OMNI-RANGE IDENTIFICATION 1-3
VERY HIGH FREQUENCY OMNI-RANGE RECEIVER CHECK 1-4
VERY HIGH FREQUENCY OMNI-RANGE RECEIVER TOLERANCE 1-3, 1-4
VERY HIGH FREQUENCY OMNI-RANGE TEST ALTITUDES 1-4
VERY HIGH FREQUENCY OMNI-RANGE TEST 1-4
VERY HIGH FREQUENCY OMNI-RANGE WITH DISTANCE MEASURING EQUIPMENT 1-5, 1-7
VERY HIGH FREQUENCY OMNI-RANGE WITH TACTICAL AIR NAVIGATION 1-5, 1-6, 1-7, 1-8
VERY HIGH FREQUENCY OMNI-RANGE 1-3, 1-6, 1-7, 1-8, 5-33
VERY LOW FREQUENCY NAVIGATION SYSTEM 1-18
VFR CORRIDORS 3-45
VFR CRUISING ALTITUDES AND FLIGHT LEVELS 3-4
VFR FLYWAYS 3-45
VFR REQUIREMENTS 3-21
VFR TERMINAL AREA CHARTS 3-45
VFR 5-61
VICTOR AIRWAYS 5-33
VISUAL APPROACH CLEARANCE 5-80, 7-57
VISUAL APPROACH LIGHT SYSTEM 2-2
VISUAL APPROACH SLOPE INDICATORS 2-2
VISUAL APPROACH 5-57, 5-77, 5-80, 5-81
VISUAL CLEARING PROCEDURES 4-93
VISUAL CODE 6-16
VISUAL DESCENT GUIDANCE 2-2
VISUAL DESCENT POINT 5-44
VISUAL FLIGHT RULES (VFR) RESERVATIONS 4-21
VISUAL FLIGHT RULES ALTITUDES 4-38
VISUAL FLIGHT RULES COMMUNICATIONS 4-43, 4-51, 6-21, 7-81, 7-84
VISUAL FLIGHT RULES FLIGHT FOLLOWING 4-20
VISUAL FLIGHT RULES FLIGHT PLAN 5-4
VISUAL FLIGHT RULES HELICOPTER OPERATIONS 4-66
VISUAL FLIGHT RULES ON TOP 4-86, 4-89, 5-31, 5-32, 7-5
VISUAL FLIGHT RULES OVERWATER FLIGHT FOLLOWING 4-20
VISUAL FLIGHT RULES RADAR NAVIGATION ASSISTANCE 4-16, 4-51

VISUAL FLIGHT RULES 7-84
VISUAL GLIDE PATH 2-2
VISUAL GLIDESLOPE INDICATORS 2-2
VISUAL METEOROLOGICAL CONDITIONS 5-61
VISUAL SEPARATION 4-92, 5-81
VOLCANIC DUST 7-86

W

WAKE TURBULENCE AVOIDANCE PROCEDURES 7-55, 7-58
WAKE TURBULENCE 4-59, 4-92, 7-50, 7-51, 7-52, 7-53, 7-54, 7-55, 7-56, 7-57, 7-58
WAKE VORTICES INDUCED ACCIDENTS 7-54
WARNING AREA 3-33
WAYPOINT 5-33
WEATHER ADVISORY 7-1, 7-2, 7-5, 7-11, 7-12, 7-14, 7-19
WEATHER BRIEFING 5-1, 5-2, 5-3, 5-4, 7-1, 7-2, 7-3, 7-7, 7-11
WEATHER BROADCAST 7-2, 7-7, 7-9, 7-10
WEATHER RADAR 7-11, 7-12
WEATHER REPORTS 7-1, 7-14, 7-15, 7-16, 7-17, 7-18, 7-27
WEATHER SERVICES 7-2
WILDLIFE ACCIDENTS 7-74
WILDLIFE ADVISORY 7-74
WILDLIFE REFUGE AREA 7-73
WIND DIRECTION INDICATOR 4-52
WIND DIRECTION 4-39, 4-55
WIND SHEAR 4-56, 7-22, 7-24

Z

Z-MARKER 1-9

Federal Aviation Regulations Part

Federal Aviation Regulations	Part
Definitions and abbreviations	1
Maintenance	43
Airman certification	61
Medical certification	67
Airspace	71
Special use airspace	73
Operating and flight rules	91
Instrument approach procedures	97
Security control of air traffic	99
Ultralight vehicles	103
Parachute jumping	105
Drug testing & alcohol prevention	121
Air taxi and commercial operators	135
Pilot schools	141
Ground instructors	143
NTSB accident/incident notification	830

Part 1 – Definitions and abbreviations

Source: Docket No. 1150 (27 FR 4588, 5/15/62)

§ 1.1 General definitions.

As used in subchapters A through K of this chapter unless the context requires otherwise:

"Administrator" means the Federal Aviation Administrator or any person to whom he has delegated his authority in the matter concerned.

"Aerodynamic coefficients" means nondimensional coefficients for aerodynamic forces and moments.

"Air carrier" means a person who undertakes directly by lease, or other arrangement, to engage in air transportation.

"Air commerce" means interstate, overseas, or foreign air commerce or the transportation of mail by aircraft or any operation or navigation of aircraft within the limits of any Federal airway or any operation or navigation of aircraft when directly affects, or which may endanger safety in, interstate, overseas, or foreign air commerce.

"Aircraft" means a device that is used or intended to be used for flight in the air.

"Aircraft engine" means an engine that is used or intended to be used for propelling aircraft. It includes turbosuperchargers, appurtenances, and accessories necessary for its functioning, but does not include propellers.

"Airframe" means the fuselage, booms, nacelles, cowlings, fairings, airfoil surfaces (including rotors but excluding propellers and rotating airfoils of engines), and landing gear of an aircraft and their accessories and controls.

"Airplane" means an engine-driven fixed-wing aircraft heavier than air, that is supported in flight by the dynamic reaction of the air against its wings.

"Airport" means an area of land or water that is used or intended to be used for the landing and takeoff of aircraft, and includes its buildings and facilities, if any.

"Airship" means an engine-driven lighter-than-air aircraft that can be steered.

"Air traffic" means aircraft operating in the air or on an airport surface, exclusive of loading ramps and parking areas.

"Air traffic clearance" means an authorization by air traffic control, for the purpose of preventing collision between known aircraft, for an aircraft to proceed under specified traffic conditions within controlled airspace.

"Air traffic control" means a service operated by appropriate authority to promote the safe, orderly, and expeditious flow of air traffic.

"Air transportation" means interstate, overseas, or foreign air transportation or the transportation of mail by aircraft.

"Alternate airport" means an airport at which an aircraft may land if a landing at the intended airport becomes inadvisable.

"Altitude engine" means a reciprocating aircraft engine having a rated takeoff power that is producible from sea level to an established higher altitude.

"Appliance" means any instrument, mechanism, equipment, part, apparatus, appurtenance, or accessory, including communications equipment, that is used or intended to be used in operating or controlling an aircraft in flight, is installed in or attached to the aircraft, and is not part of an airframe, engine, or propeller.

"Approved," unless used with reference to another person, means approved by the Administrator.

"Area navigation (RNAV)" means a method of navigation that permits aircraft operations on any desired course within the coverage of station-referenced navigation signals or within the limits of self-contained system capability.

"Area navigation low route" means an area navigation route within the airspace extending upward from 1,200 feet above the surface of the earth to, but not including, 18,000 feet MSL.

"Area navigation high route" means an area navigation route within the airspace extending upward from, and including, 18,000 feet MSL to flight level 450.

"Armed Forces" means the Army, Navy, Air Force, Marine Corps, and Coast Guard, including their regular and reserve components and members serving without component status.

"Autorotation" means a rotorcraft flight condition in which the lifting rotor is driven entirely by action of the air when the rotorcraft is in motion.

"Auxiliary rotor" means a rotor that serves either to counteract the effect of the main rotor torque on a rotorcraft or to maneuver the rotorcraft about one or more of its three principal axes.

"Balloon" means a lighter-than-air aircraft that is not engine driven.

"Brake horsepower" means the power delivered at the propeller shaft (main drive or main output) of an aircraft engine.

"Calibrated airspeed" means indicated airspeed of an aircraft, corrected for position and instrument error. Calibrated airspeed is equal to true airspeed in standard atmosphere at sea level.

"Canard" means the forward wing of a canard configuration and may be a fixed, movable, or variable geometry surface, with or without control surfaces.

"Canard configuration" means a configuration in which the span of the forward wing is substantially less than that of the main wing.

"Category":

(1) As used with respect to the certification, ratings, privileges, and limitations of airmen, means a broad classification of aircraft. Examples include: airplane; rotorcraft; glider; and lighter-than-air; and

(2) As used with respect to the certification of aircraft, means a grouping of aircraft based upon intended use of operating limitations. Examples include: transport, normal, utility, acrobatic, limited, restricted, and provisional.

"Category A," with respect to transport category rotorcraft, means multiengine rotorcraft designed with engine and system isolation features specified in Part 29 and utilizing scheduled takeoff and landing operations under a critical engine failure concept which assures adequate designated surface area and adequate performance capability for continued safe flight in the event of engine failure.

"Category B," with respect to transport category rotorcraft, means single-engine or multiengine rotorcraft which do not fully meet all Category A standards. Category B rotorcraft have no guaranteed stay-up ability in the event of engine failure and unscheduled landing is assumed.

"Category II operations," with respect to the operation of aircraft, means a straight-in ILS approach to the runway of an airport under a Category II ILS instrument approach procedure issued by the Administrator or other appropriate authority.

"Category III operations," with respect to the operation of aircraft, means an ILS approach to, and landing on, the runway of an airport using a Category III ILS instrument approach procedure issued by the Administrator or other appropriate authority.

"Ceiling" means the height above the earth's surface of the lowest layer of clouds or obscuring phenomena that is reported as "broken," "overcast," or "obscuration," and not classified as "thin" or "partial."

"Civil aircraft" means aircraft other than public aircraft.

"Class":

(1) As used with respect to the certification, ratings, privileges, and limitations of airmen, means a classification of aircraft within a category having similar operating characteristics. Examples include: single engine; multiengine; land; water; gyroplane, helicopter; airship; and free balloon; and

(2) As used with respect to the certification of aircraft, means a broad grouping of aircraft having similar characteristics of propulsion, flight, or landing. Examples include: airplane; rotorcraft; glider; balloon; landplane; and seaplane.

"Clearway" means:

(1) For turbine engine powered airplanes certificated after August 29, 1959, an area beyond the runway, not less than 500 feet wide, centrally located about the extended centerline of the runway, and under the control of the airport authorities. The clearway is expressed in terms of a clearway plane, extending from the end of the runway with an upward slope not exceeding 1.25 percent, above which no object nor any terrain protrudes. However, threshold lights may protrude above the plane if their height above the runway is 26 inches or less and if they are located to each side of the runway.

(2) For turbine engine powered airplanes certificated after September 30, 1958, but before August 30, 1959, an area beyond the takeoff runway extending no less than 300 feet on either side of the extended centerline of the runway, at an elevation no higher than the elevation of the end of the runway, clear of all fixed obstacles, and under the control of the airport authorities.

"Climbout speed," with respect to rotorcraft, means a referenced airspeed which results in a flight path clear of the height-velocity envelope during initial climbout.

"Commercial operator" means a person who, for compensation or hire, engages in the carriage by aircraft in air commerce of persons or property, other than as an air carrier or foreign air carrier or under the authority of Part 375 of this Title. Where it is doubtful that an operation is for "compensation or hire," the test applied is whether the carriage by air is merely incidental to the person's other business or is, in itself, a major enterprise for profit.

["Controlled airspace" means an airspace of defined dimensions within which air traffic control service is provided to IFR flights and to VFR flights in accordance with the airspace classification.

[Note—Controlled airspace is a generic term that covers Class A Class B, Class C, Class D, and Class E airspace.]

[Effective 9/16/93 per Amdt. 1-38, Eff. 9/16/93.]

"Crewmember" means a person assigned to perform duty in an aircraft during flight time.

"Critical altitude" means the maximum altitude at which, in standard atmosphere, it is possible to maintain, at a specified rotational speed, a specified power or a specified manifold pressure. Unless otherwise stated, the critical altitude is the maximum altitude at which it is possible to maintain, at the maximum continuous rotational speed, one of the following:

(1) The maximum continuous power, in the case of engines for which this power rating is the same at sea level and at the rated altitude.

(2) The maximum continuous rated manifold pressure, in the case of engines the maximum continuous power of which, is governed by a constant manifold pressure.

"Critical engine" means the engine whose failure would most adversely affect the performance or handling qualities of an aircraft.

"Decision height," with respect to the operation of aircraft, means the height at which a decision must be made, during an ILS or PAR instrument approach, to either continue the approach or to execute a missed approach.

"Equivalent airspeed" means the calibrated airspeed of an aircraft corrected for adiabatic compressible flow for the particular altitude. Equivalent airspeed is equal to calibrated airspeed in standard atmosphere at sea level.

"Extended over-water operation" means—

(1) With respect to aircraft other than helicopters, and operation over water at a horizontal distance of more than 50 nautical miles from the nearest shoreline; and

(2) With respect to helicopters, an operation over water at a horizontal distance of more than 50 nautical miles from the nearest shoreline and more than 50 nautical miles from an off-shore heliport structure.

"External load" means a load that is carried, or extends, outside of the aircraft fuselage.

"External-load attaching means" means the structural components used to attach an external load to an aircraft, including external-load containers, the backup structure at the attachment points, and any quick-release device used to jettison the external load.

"Fireproof"—

(1) With respect to materials and parts used to confine fire in a designated fire zone, means the capacity to withstand at least as well as steel in dimensions appropriate for the purpose for which they are used, the heat produced when there is a severe fire of extended duration in that zone; and

(2) With respect to other materials and parts, means the capacity to withstand the heat associated with fire at least as well as steel in dimensions appropriate for the purpose for which they are used.

"Fire resistant"—

(1) With respect to sheet or structural members means the capacity to withstand the heat associated with fire at least as well as aluminum alloy in dimensions appropriate for the purpose for which they are used; and

(2) With respect to fluid-carrying lines, fluid system parts, wiring, air ducts, fittings, and powerplant controls, means the capacity to perform the intended functions under the heat and other conditions likely to occur when there is a fire at the place concerned.

"Flame resistant" means not susceptible to combustion to the point of propagating a flame, beyond safe limits, after the ignition source is removed.

"Flammable," with respect to a fluid or gas, means susceptible to igniting readily or to exploding.

"Flap extended speed" means the highest speed permissible with wing flaps in a prescribed extended position.

"Flash resistant" means not susceptible to burning violently when ignited.

"Flight crewmember" means a pilot, flight engineer, or flight navigator assigned to duty in an aircraft during flight time.

"Flight level" means a level of constant atmospheric pressure related to a reference datum of 29.92 inches of mercury. Each is stated in three digits that represent hundreds of feet. For example, flight level 250 represents a barometric altimeter indication of 25,000 feet; flight level 255, an indication of 25,500 feet.

"Flight plan" means specified information, relating to the intended flight of an aircraft, that is filed orally or in writing with air traffic control.

"Flight time" means the time from the moment the aircraft first moves under its own power for the purpose of flight until the moment it comes to rest at the next point of landing ("Block-to-block" time).

"Flight visibility" means the average forward horizontal distance, from the cockpit of an aircraft in flight, at which prominent unlighted objects may be seen and identified by day and prominent lighted objects may be seen and identified by night.

"Foreign air carrier" means any person other than a citizen of the United States, who undertakes directly, by lease or other arrangement, to engage in air transportation.

"Foreign air commerce" means the carriage by aircraft of persons or property for compensation or hire, or the carriage of mail by aircraft, or the operation or navigation of aircraft in the conduct or furtherance of a business or vocation, in commerce between a place in the United States and any place outside thereof; whether such commerce moves wholly by aircraft or partly by aircraft and partly by other forms of transportation.

"Foreign air transportation" means the carriage by aircraft of persons or property as a common carrier for compensation or hire, or the carriage of mail by aircraft, in commerce between a place in the United States and any place outside of the United States, whether that commerce moves wholly by aircraft or partly by aircraft and partly by, other forms of transportation.

"Forward wing" means a forward lifting surface of a canard configuration or tandem-wing configuration airplane. The surface may be a fixed, movable, or variable geometry surface, with or without control surfaces.

"Glider" means a heavier-than-air aircraft, that is supported in flight by the dynamic reaction of the air against its lifting surfaces and whose free flight does not depend principally on an engine.

"Ground visibility" means prevailing horizontal visibility near the earth's surface as reported by the United States National Weather Service or an accredited observer.

"Gyrodyne" means a rotorcraft whose rotors are normally engine-driven for takeoff, hovering, and landing, and for forward flight through part of its speed range, and whose means of propulsion, consisting usually of conventional propellers, is independent of the rotor system.

"Gyroplane" means a rotorcraft whose rotors are not engine-driven except for initial starting, but are made to rotate by action of the air when the rotorcraft is moving; and whose means of propulsion, consisting usually of conventional propellers, is independent of the rotor system.

"Helicopter" means a rotorcraft that, for its horizontal motion, depends principally on its engine-driven rotors.

"Heliport" means an area of land, water, or structure used or intended to be used for the landing and takeoff of helicopters.

"Idle thrust" means the jet thrust obtained with the engine power control lever set at the stop for the least thrust position at which it can be placed.

"IFR conditions" means weather conditions below the minimum for flight under visual flight rules.

"IFR over-the-top," with respect to the operation of aircraft, means the operation of an aircraft over-the-top on an IFR flight plan when cleared by air traffic control to maintain "VFR conditions" or "VFR conditions on top".

"Indicated airspeed" means the speed of an aircraft as shown on its pitot static airspeed indicator calibrated to reflect standard atmosphere adiabatic compressible flow, at sea level uncorrected for airspeed system errors.

"Instrument" means a device using an internal mechanism to show visually or aurally the attitude, altitude, or operation of an aircraft or aircraft part. It includes electronic devices for automatically controlling an aircraft in flight.

"Interstate air commerce" means the carriage by aircraft of persons or property for compensation or hire, or the carriage of mail by aircraft, or the operation or navigation of aircraft in the conduct or furtherance of a business or vocation, in commerce between a place in any State of the United States, or the District of Columbia, and a place in any other State of the United States, or the District of Columbia; or between places in the same State of the United States through the airspace over any place outside thereof, or between places in the same territory or possession of the United States, or the District of Columbia.

"Interstate air transportation" means the carriage by aircraft of persons or property as a common carrier for compensation or hire, or the carriage of mail by aircraft in commerce:

(1) Between a place in a State or the District of Columbia and another place in another State or the District of Columbia;

(2) Between places in the same State through the airspace over any place outside that State; or

(3) Between places in the same possession of the United States;

whether that commerce moves wholly by aircraft or partly by aircraft and partly by other forms of transportation.

"Intrastate air transportation" means the carriage of persons or property as a common carrier for compensation or hire, by turbojet-powered aircraft capable of carrying thirty or more persons, wholly within the same State of the United States.

"Kite" means a framework, covered with paper, cloth, metal, or other material, intended to be flown at the end of a rope or cable, and having as its only support the force of the wind moving past its surfaces.

"Landing gear extended speed" means the maximum speed at which an aircraft can be safely flown with the landing gear extended.

"Landing gear operating speed" means the maximum speed at which the landing gear can be safely extended or retracted.

"Large aircraft" means aircraft of more than 12,500 pounds, maximum certificated takeoff weight.

"Lighter-than-air aircraft" means aircraft that can rise and remain suspended by using contained gas weighing less than the air that is displaced by the gas.

"Load factor" means the ratio of a specified load to the total weight of the aircraft. The specified load is expressed in terms of any of the following: aerodynamic forces, inertia forces, or ground or water reactions.

"Mach number" means the ratio of true airspeed to the speed of sound.

"Main rotor" means the rotor that supplies the principal lift to a rotorcraft.

"Maintenance" means inspection, overhaul, repair, preservation, and the replacement of parts, but excludes preventive maintenance.

"Major alteration" means an alteration not listed in the aircraft, aircraft engine, or propeller specifications—

(1) That might appreciably affect weight, balance, structural strength, performance, powerplant operation, flight characteristics, or other qualities affecting airworthiness; or

(2) That is not done according to accepted practices or cannot be done by elementary operations.

"Major repair" means a repair—

(1) That, if improperly done, might appreciably affect weight, balance, structural strength, performance, powerplant operation, flight characteristics, or other qualities affecting airworthiness; or

(2) That is not done according to accepted practices or cannot be done by elementary operations.

"Manifold pressure" means absolute pressure as measured at the appropriate point in the induction system and usually expressed in inches of mercury.

"Medical certificate" means acceptable evidence of physical fitness on a form prescribed by the Administrator.

"Minimum descent altitude" means the lowest altitude, expressed in feet above mean sea level, to which descent is authorized on final approach or during circle-to-land maneuvering in execution of a standard instrument approach procedure, where no electronic glide slope is provided.

"Minor alteration" means an alteration other than a major alteration.

"Minor repair" means a repair other than a major repair.

"Navigable airspace" means airspace at and above the minimum flight altitudes prescribed by or under this chapter, including airspace needed for safe takeoff and landing.

"Night" means the time between the end of evening civil twilight and the beginning of morning civil twilight, as published in the American Air Almanac, converted to local time.

"Nonprecision approach procedure" means a standard instrument approach procedure in which no electronic glide slope is provided.

"Operate," with respect to aircraft, means use, cause to

use or authorize to use aircraft, for the purpose (except as provided in § 91.13 of this chapter) of air navigation including the piloting of aircraft, with or without the right of legal control (as owner, lessee, or otherwise).

"Operational control," with respect to a flight, means the exercise of authority over initiating, conducting, or terminating a flight.

"Overseas air commerce" means the carriage by aircraft of persons or property for compensation or hire, or the carriage of mail by aircraft, or the operation or navigation of aircraft in the conduct or furtherance of a business or vocation, in commerce between a place in any State of the United States, or the District of Columbia, and any place in a territory or possession of the United States; or between a place in a territory or possession of the United States, and a place in any other territory or possession of the United States.

"Overseas air transportation" means the carriage by aircraft of persons or property as a common carrier or compensation or hire, or the carriage of mail by aircraft, in commerce—

(1) Between a place in a State or the District of Columbia and a place in a possession of the United States; or

(2) Between a place in a possession of the United States and a place in another possession of the United States; whether that commerce moves wholly by aircraft or partly by aircraft and partly by other forms of transportation.

"Over-the-top" means above the layer of clouds or other obscuring phenomena forming the ceiling.

"Parachute" means a device used or intended to be used to retard the fall of a body or object through the air.

"Person" means an individual, firm, partnership, corporation, company, association, joint-stock association, or governmental entity. It includes a trustee, receiver, assignee, or similar representative of any of them.

"Pilotage" means navigation by visual reference to landmarks.

"Pilot in command" means the pilot responsible for the operation and safety of an aircraft during flight time.

"Pitch setting" means the propeller blade setting as determined by the blade angle measured in a manner, and at a radius, specified by the instruction manual for the propeller.

"Positive control" means control of all air traffic, within designated airspace, by air traffic control.

"Precision approach procedure" means a standard instrument approach procedure in which an electronic glide slope is provided, such as ILS and PAR.

"Preventive maintenance" means simple or minor preservation operations and the replacement of small standard parts not involving complex assembly operations.

"Prohibited area" means designated airspace within which the flight of aircraft is prohibited.

"Propeller" means a device for propelling an aircraft that has blades on an engine-driven shaft and that, when rotated, produces by its action on the air, a thrust approximately perpendicular to its plane of rotation. It includes control components normally supplied by its manufacturer, but does not include main and auxiliary rotors or rotating airfoils of engines.

"Public aircraft" means aircraft used only in the service of a government, or a political subdivision. It does not include any government-owned aircraft engaged in carrying persons or property for commercial purposes.

"Rated continuous OEI power," with respect to rotorcraft turbine engines, means the approved brake horsepower developed under static conditions at specified altitudes and temperatures within the operating limitations established for the engine under Part 33 of this chapter, and limited in use to the time required to complete the flight after the failure of one engine of a multiengine rotorcraft.

"Rated maximum continuous augmented thrust," with respect to turbojet engine type certification, means the approved jet thrust that is developed statically or in flight, in standard atmosphere at a specified altitude, with fluid injection or with the burning of fuel in a separate combustion chamber, within the engine operating limitations established under Part 33 of this chapter, and approved for unrestricted periods of use.

"Rated maximum continuous power," with respect to reciprocating, turbopropeller, and turboshaft engines, means the approved brake horsepower that is developed statically or in flight, in standard atmosphere at a specified altitude, within the engine operating limitations established under Part 33, and approved for unrestricted periods of use.

"Rated maximum continuous thrust," with respect to turbojet engine type certification, means the approved jet thrust that is developed statically or in flight, in standard atmosphere at a specified altitude, without fluid injection and without the burning of fuel in a separate combustion chamber, within the engine operating limitations established under Part 33 of this chapter, and approved for unrestricted periods of use.

"Rated takeoff augmented thrust," with respect to turbojet engine type certification, means the approved jet thrust that is developed statically under standard sea level conditions, with fluid injection or with the burning of fuel in a separate combustion chamber, within the engine operating limitations established under Part 33 of this chapter, and limited in use to periods of not over 5 minutes for takeoff operation.

"Rated takeoff power," with respect to reciprocating, turbopropeller, and turboshaft engine type certification, means the approved brake horsepower that is developed statically under standard sea level conditions, within the engine operating limitations established under Part 33, and limited in use to periods of not over 5 minutes for takeoff operation.

"Rated takeoff thrust," with respect to turbojet engine type certification, means the approved jet thrust that is developed statically under standard sea level conditions, without fluid injection and without the burning of fuel in a separate combustion chamber, within the engine operating limitations established under Part 33 of this chapter, and limited in use to periods of not over 5 minutes for takeoff operation.

"Rated 30-minute OEI power," with respect to rotorcraft turbine engines, means the approved brake horsepower developed under static conditions at specified altitudes and temperatures within the operating limitations established for the engine under Part 33 of this chapter, and limited in use to a period of not more than 30 minutes after the failure of one engine of a multiengine rotorcraft.

"Rated 2½-minute OEI power," with respect to rotorcraft turbine engines, means the approved brake horsepower developed under static conditions at specified altitudes and temperatures within the operating limitations established for the engine under Part 33 of this chapter, and limited in use to a period of not more than 2½ minutes after the failure of one engine of a multiengine rotorcraft.

"Rating" means a statement that, as a part of a certificate, sets forth special conditions, privileges, or limitations.

"Reporting point" means a geographical location in relation to which the position of an aircraft is reported.

"Restricted area" means airspace designated under Part 73 of this chapter within which the flight of aircraft, while not wholly prohibited, is subject to restriction.

"RNAV way point (W/P)" means a predetermined geographical position used for route or instrument approach definition or progress reporting purposes that is defined relative to a VORTAC station position.

"Rocket" means an aircraft propelled by ejected expanding gases generated in the engine from self-contained propellants and not dependent on the intake of outside substances. It includes any part which becomes separated during the operation.

"Rotorcraft" means a heavier-than-air aircraft that depends principally for its support in flight on the lift generated by one or more rotors.

"Rotorcraft-load combination" means the combination of a rotorcraft and an external-load, including the external-load attaching means. Rotorcraft-load combinations are designated as Class A, Class B, Class C, and Class D, as follows:

(1) "Class A rotorcraft-load combination" means one in which the external load cannot move freely, cannot be jettisoned, and does not extend below the landing gear.

(2) "Class B rotorcraft-load combination" means one in which the external load is jettisonable and is lifted free of land or water during the rotorcraft operation.

(3) "Class C rotorcraft-load combination" means one in which the external load is jettisonable and remains in contact with land or water during the rotorcraft operation.

(4) "Class D rotorcraft-load combination" means one in which the external-load is other than a Class A, B, or C and has been specifically approved by the Administrator for that operation.

"Route segment" means a part of a route. Each end of that part is identified by—

(1) a continental or insular geographical location; or

(2) a point at which a definite radio fix can be established.

"Sea level engine" means a reciprocating aircraft engine having a rated takeoff power that is producible only at sea level.

"Second in command" means a pilot who is designated to be second in command of an aircraft during flight time.

"Show," unless the context otherwise requires, means to show to the satisfaction of the Administrator.

"Small aircraft" means aircraft of 12,500 pounds or less, maximum certificated takeoff weight.

[Special VFR conditions mean meteorological conditions that are less than those required for basic VFR flight in controlled airspace and in which some aircraft are permitted flight under visual flight rules.]

[Special VFR operations means aircraft operating in accordance with clearances within controlled airspace in meteorological conditions less than the basic VFR weather minima. Such operations must be requested by the pilot and approved by ATC.]

[Effective 9/16/93 per Amdt. 1-38, Eff. 9/16/93.)]

"Standard atmosphere" means the atmosphere defined in U.S. Standard Atmosphere, 1962 (Geo-potential altitude tables).

"Stopway" means an area beyond the takeoff runway, no less wide than the runway and centered upon the extended centerline of the runway, able to support the airplane during an aborted takeoff, without causing structural damage to the airplane, and designated by the airport authorities for use in decelerating the airplane during an aborted takeoff.

"Takeoff power"—

(1) With respect to reciprocating engines, means the brake horsepower that is developed under standard sea level conditions, and under the maximum conditions of crankshaft rotational speed and engine manifold pressure approved for the normal takeoff, and limited in continuous use to the period of time shown in the approved engine specification; and

(2) With respect to turbine engines, means the brake horsepower that is developed under static conditions at a specified altitude and atmospheric temperature, and under the maximum conditions of rotorshaft rotational speed and gas temperature approved for the normal take off, and limited in continuous use to the period of time shown in the approved engine specification.

"Takeoff safety speed" means a referenced airspeed obtained after lift-off at which the required one-engine-inoperative climb performance can be achieved.

"Takeoff thrust," with respect to turbine engines, means the jet thrust that is developed under static conditions at a specific altitude and atmospheric temperature under the maximum conditions of rotorshaft rotational speed and gas temperature approved for the normal takeoff, and limited in continuous use to the period of time shown in the approved engine specification.

"Tandem wing configuration" means a configuration having two wings of similar span, mounted in tandem.

"TCAS I" means a TCAS that utilizes interrogations of, and replies from, airborne radar beacon Transponders and provides traffic advisories to the pilot.

"TCAS II" means a TCAS that utilizes interrogations of, and replies from airborne radar beacon transponders and provides traffic advisories and resolution advisories in the vertical plane.

"TCAS III" means a TCAS that utilizes interrogation of, and replies from, airborne radar beacon transponders and provides traffic advisories and resolution advisories in the vertical and horizontal planes to the pilot.

"Time in service," with respect to maintenance time records, means the time from the moment an aircraft leaves the surface of the earth until it touches it at the next point of landing.

"True airspeed" means the airspeed of an aircraft relative to undisturbed air. True airspeed is equal to equivalent airspeed multiplied by $(\rho 0/\rho)^{1/2}$.

"Traffic pattern" means the traffic flow that is prescribed for aircraft landing at, taxiing on, or taking off from, an airport.

"Type":

(1) As used with respect to the certification, ratings, privileges, and limitations of airmen, means a specific make and basic model of aircraft, including modifications thereto that do not change its handling or flight characteristics. Examples include: DC-7, 1049, and F-27; and

(2) As used with respect to the certification of aircraft, means those aircraft which are similar in design. Examples include: DC-7 and DC-7C; 1049G and 1049H; and F-27 and F-27F.

(3) As used with respect to the certification of aircraft engines means those engines which are similar in design. For example, JT8D and JT8D-7 are engines of the same type, and JT9D-3A and JT9D-7 are engines of the same type.

"United States," in a geographical sense, means (1) the States, the District of Columbia, Puerto Rico, and the possessions, including the territorial waters, and (2) the airspace of those areas.

"United States air carrier" means a citizen of the United States who undertakes directly by lease, or other arrangement, to engage in air transportation.

"VFR over-the-top," with respect to the operation of aircraft, means the operation of an aircraft over-the-top under VFR when it is not being operated on an IFR flight plan.

"Winglet or tip fin" means an out-of-plane surface extending from a lifting surface. The surface may or may not have control surfaces.

(Amdt. 1-9, Eff. 9/26/65); (Amdt. 1-10, Eff. 3/29/66); (Amdt. 1-12, Eff. 8/7/67); (Amdt. 1-13, Eff. 8/3/67); (Amdt. 1-14, Eff. 11/18/67); (Amdt. 1-16, Eff. 5/8/70); (Amdt. 1-17, Eff. 6/25/70); (Amdt. 1-19, Eff. 9/18/70); (Amdt. 1-20, Eff. 2/4/71); (Amdt. 1-21, Eff. 7/20/71); (Amdt. 1-22, Eff. 4/14/72); (Amdt. 1-23, Eff. 10/31/74); (Amdt. 1-24, Eff. 3/15/75); (Amdt. 1-25, Eff. 12/9/76); (Amdt. 1-26, Eff. 5/2/77); (Amdt. 1-29, Eff. 3/1/78); (Amdt. 1-30, Eff. 5/8/81); (Amdt. 1-31, Eff. 3/2/83); (Amdt. 1-32, Eff. 12/6/84); (Amdt. 1-33, Eff. 1/6/87); (Amdt. 1-34, Eff. 10/3/88); (Amdt. 1-35, Eff. 2/9/89); (Amdt. 1-36, Eff. 8/18/90); (Amdt. 1-37, Eff. 2/4/91); [(Amdt. 1-38, Eff. 9/16/93)]

§ 1.2 Abbreviations and symbols.

In Subchapters A through K of this chapter:

"AGL" means above ground level.

"ALS" means approach light system.

"ASR" means airport surveillance radar.

"ATC" means air traffic control.

"CAS" means calibrated airspeed.

"CAT II" means Category II.

"CONSOL or CONSOLAN" means a kind of low or medium frequency long range navigational aid.

"DH" means decision height.

"DME" means distance measuring equipment compatible with TACAN.

"EAS" means equivalent airspeed.

"FAA" means Federal Aviation Administration.

"FM" means fan marker.

"GS" means glide slope.

"HIRL" means high-intensity runway light system.

"IAS" means indicated airspeed.

"ICAO" means International Civil Aviation Organization.

"IFR" means instrument flight rules.

"ILS" means instrument landing system.

"IM" means ILS inner marker.

"INT" means intersection.

"LDA" means localizer-type directional aid.

"LFR" means low-frequency radio range.

"LMM" means compass locator at middle marker.

"LOC" means ILS localizer.

"LOM" means compass locator at outer marker.

"M" means mach number.

"MAA" means maximum authorized IFR altitude.

"MALS" means medium intensity approach light system.

"MALSR" means medium intensity approach light system with runway alignment indicator lights.

"MCA" means minimum crossing altitude.

"MDA" means minimum descent altitude.

"MEA" means minimum en route IFR altitude.

"MM" means ILS middle marker.

"MOCA" means minimum obstruction clearance altitude.

"MRA" means minimum reception altitude.

"MSL" means mean sea level.

"NDB(ADF)" means nondirectional beacon (automatic direction finder).

"NOPT" means no procedure turn required.

"OEI" means one engine inoperative.

"OM" means ILS outer marker.

"PAR" means precision approach radar.

"RAIL" means runway alignment indicator light system.

"RBN" means radio beacon.

"RCLM" means runway centerline marking.

"RCLS" means runway centerline light system.

"REIL" means runway end identification lights.

"RR" means low or medium frequency radio range station.

"RVR" means runway visual range as measured in the touchdown zone area.

"SALS" means short approach light system.

"SSALS" means simplified short approach light system.

"SSALSR" means simplified short approach light system with runway alignment indicator lights.

"TACAN" means ultra-high frequency tactical air navigational aid.

"TAS" means true airspeed.

"TCAS" means a traffic alert and collision avoidance system.

"TDZL" means touchdown zone lights.

"TVOR" means very high frequency terminal omnirange station.

V_A means design maneuvering speed.

V_B means design speed for maximum gust intensity.

V_C means design cruising speed.

V_D means design diving speed.

V_{DF}/M_{DF} means demonstrated flight diving speed.

V_F means design flap speed.

V_{FC}/M_{FC} means maximum speed for stability characteristics.

V_{FE} means maximum extended speed.

V_H means maximum speed in level flight with maximum continuous power.

V_{LE} means maximum landing gear extended speed.

V_{LO} means maximum landing gear operating speed.

V_{LOF} means lift-off speed.

V_{MC} means minimum control speed with the critical engine inoperative.

V_{MO}/M_{MO} means maximum operating limit speed.

V_{MU} means minimum unstick speed.

V_{NE} means never-exceed speed.

V_{NO} means maximum structural cruising speed.

V_R means rotation speed.

V_S means the stalling speed or the minimum steady flight speed at which the airplane is controllable.

V_{SO} means the stalling speed or the minimum steady flight speed in the landing configuration.

V_{S1} means the stalling speed or the minimum steady flight speed obtained in a specific configuration.

V_{TOSS} means takeoff safety speed for Category A rotorcraft.

V_X means speed for best angle of climb.

V_Y means speed for best rate of climb.

V_1 means takeoff decision speed (formerly denoted as critical engine failure speed).

V_2 means takeoff safety speed.

$V_{2\ MIN}$ means minimum takeoff safety speed.

"VFR" means visual flight rules.

"VHF" means very high frequency.

"VOR" means very high frequency omnirange station.

"VORTAC" means collocated VOR and TACAN.

(Amdt. 1-10, Eff. 3/29/66); (Amdt. 1-14, Eff. 11/18/67); (Amdt. 1-15, Eff. 2/25/68); (Amdt. 1-18, Eff. 8/18/70); (Amdt. 1-27, Eff. 9/14/77); (Amdt. 1-28, Eff. 9/21/77); (Amdt. 1-29, Eff. 3/1/78); (Amdt. 1-32, Eff. 12/6/84); (Amdt. 1-34, Eff. 10/3/88); (Amdt. 1-35, Eff. 2/9/89)

§ 1.3 Rules of construction.

(a) In Subchapters A through K of this chapter, unless the context requires otherwise:

(1) Words importing the singular include the plural;

(2) Words importing the plural include the singular; and

(3) Words importing the masculine gender include the feminine.

(b) In Subchapters A through K of this chapter, the word:

(1) "Shall" is used in an imperative sense;

(2) "May" is used in a permissive sense to state authority or permission to do the act prescribed, and the words "no person may . . ." or "a person may not . . ." mean that no person is required, authorized, or permitted to do the act prescribed; and

(3) "Includes" means "includes but is not limited to."

(Amdt. 1-10, Eff. 3/29/66)

Part 43 – Maintenance, preventive maintenance, rebuilding, and alteration

43.1 Applicability
43.2 Records of overhaul and rebuilding
43.3 Persons authorized to perform maintenance, preventive maintenance, rebuilding, and alterations
43.5 Approval for return to service after maintenance, preventive maintenance, rebuilding, or alteration
43.7 Persons authorized to approve aircraft, airframes, aircraft engines, propellers, appliances, or component parts for return to service after maintenance, preventive maintenance, rebuilding, or alteration
43.9 Content, form, and disposition of maintenance, preventive maintenance, rebuilding, and alteration records (except inspections performed in accordance with Part 91, Part 123, Part 125, § 135.411(a)(1), and § 135.419 of this chapter)
43.11 Content, form, and disposition of records for inspections conducted under Parts 91 and 125 and §§ 135.411(a)(1) and 135.419 of this chapter
43.12 Maintenance records: Falsification, reproduction, or alteration
43.13 Performance rules (general)
43.15 Additional performance rules for inspections
43.16 Airworthiness limitations
43.17 [Maintenance, preventive maintenance, and alterations performed on U.S. aeronautical products by certain Canadian persons]

Appendix A—Major alterations, major repairs, and preventive maintenance

Appendix B—Recording of major repairs and major alterations

Appendix C—[Reserved]

Appendix D—Scope and detail of items (as applicable to the particular aircraft) to be included in annual and 100-hour inspections

Appendix E—Altimeter system test and inspection

Appendix F—ATC transponder tests and inspections

Source: Docket No. 1993 (29 FR 5451, 4/23/64) effective 7/6/64 unless otherwise noted.

§ 43.1 Applicability.

(a) Except as provided in paragraph (b) of this section, this part prescribes rules governing the maintenance, preventive maintenance, rebuilding, and alteration of any—

(1) Aircraft having a U.S. airworthiness certificate;

(2) Foreign-registered civil aircraft used in common carriage or carriage of mail under the provisions of Part 121, 127, or 135 of this chapter; and

(3) Airframe, aircraft engines, propellers, appliances, and component parts of such aircraft.

(b) This part does not apply to any aircraft for which an experimental airworthiness certificate has been issued, unless a different kind of airworthiness certificate had previously been issued for that aircraft.

(Amdt. 43-23, Eff. 10/15/82)

§ 43.2 Records of overhaul and rebuilding.

(a) No person may describe in any required maintenance entry or form an aircraft, airframe, aircraft engine, propeller, appliance, or component part as being overhauled unless—

(1) Using methods, techniques, and practices acceptable to the Administrator, it has been disassembled, cleaned, inspected, repaired as necessary, and reassembled; and

(2) It has been tested in accordance with approved standards and technical data, or in accordance with current standards and technical data acceptable to the Administrator, which have been developed and documented by the holder of the type certificate, supplemental type certificate, or a material, part, process, or appliance approval under § 21.305 of this chapter.

(b) No person may describe in any required maintenance entry or form an aircraft, airframe, aircraft engine, propeller, appliance, or component part as being rebuilt unless it has been disassembled, cleaned, inspected, repaired as necessary, reassembled, and tested to the same tolerances and limits as a new item, using either new parts or used parts that either conform to new part tolerances and limits or to approved oversized or undersized dimensions.

Docket No. 21071 (47 FR 41076, 9/16/82)
(Amdt. 43-23, Eff. 10/15/82)

§ 43.3 Persons authorized to perform maintenance, preventive maintenance, rebuilding, and alterations.

(a) Except as provided in this section and § 43.17, no person may maintain, rebuild, alter, or perform preventive maintenance on an aircraft, airframe, aircraft engine, propeller, appliance, or component part to which this part applies. Those items, the performance of which is a major alteration, a major repair, or preventive maintenance, are listed in Appendix A.

(b) The holder of a mechanic certificate may perform maintenance, preventive maintenance, and alterations as provided in Part 65 of this chapter.

(c) The holder of a repairman certificate may perform maintenance and preventive maintenance as provided in Part 65 of this chapter.

(d) A person working under the supervision of a holder of a mechanic or repairman certificate may perform the maintenance, preventive maintenance, and alterations that his supervisor is authorized to perform, if the supervisor personally observes the work being done to the extent necessary to ensure that it is being done properly and if the supervisor is readily available, in person, for consultation. However, this paragraph does not authorize the performance of any inspection required by Part 91 or Part 125 of this chapter or any inspection performed after a major repair or alteration.

(e) The holder of a repair station certificate may perform maintenance, preventive maintenance, and alterations as provided in Part 145 of this chapter.

(f) The holder of an air carrier operating certificate or an operating certificate issued under Part 121, 127, or 135, may perform maintenance, preventive maintenance, and alterations as provided in Part 121, 127, or 135.

(g) The holder of a pilot certificate issued under Part 61 may perform preventive maintenance on any aircraft owned or operated by that pilot which is not used under Part 121, 127, 129, or 135.

(h) Notwithstanding the provisions of paragraph (g) of this section, the Administrator may approve a certificate holder under Part 135 of this chapter, operating rotorcraft in a remote area, to allow a pilot to perform specific preventive maintenance items provided—

(1) The items of preventive maintenance are a result of a known or suspected mechanical difficulty or malfunction that occurred en route to or in a remote area;

(2) The pilot has satisfactorily completed an approved training program and is authorized in writing by the certificate holder for each item of preventive maintenance that the pilot is authorized to perform;

(3) There is no certificated mechanic available to perform preventive maintenance;

(4) The certificate holder has procedures to evaluate the accomplishment of a preventive maintenance item that requires a decision concerning the airworthiness of the rotorcraft; and

(5) The items of preventive maintenance authorized by this section are those listed in paragraph (c) of Appendix A of this part.

(i) A manufacturer may—

(1) Rebuild or alter any aircraft, aircraft engine, propeller, or appliance manufactured by him under a type or production certificate;

(2) Rebuild or alter any appliance or part of aircraft, aircraft engines, propellers, or appliances manufactured by him under a Technical Standard Order Authorization, an FAA-Parts Manufacturer Approval, or Product and Process Specification issued by the Administrator; and

(3) Perform any inspection required by Part 91 or Part 125 of this chapter on aircraft it manufacturers, while currently operating under a production certificate or under a currently approved production inspection system for such aircraft.

(Amdt. 43-3, Eff. 4/2/66); (Amdt. 43-4, Eff. 10/1/66); (Amdt. 43-12, Eff. 11/15/69); (Amdt. 43-23, Eff. 10/15/82); (Amdt. 43-25, Eff. 1/6/87)

§ 43.5 Approval for return to service after maintenance, preventive maintenance, rebuilding, or alteration.

No person may approve for return to service any aircraft, airframe, aircraft engine, propeller, or appliance, that has undergone maintenance, preventive maintenance, rebuilding, or alteration unless—

(a) The maintenance record entry required by § 43.9 or § 43.11, as appropriate, has been made;

(b) The repair or alteration form authorized by or furnished by the Administrator has been executed in a manner prescribed by the Administrator; and

(c) If a repair or an alteration results in any change in the aircraft operating limitations or flight data contained in the approved aircraft flight manual, those operating limitations or flight data are appropriately revised and set forth as prescribed in § 91.9 of this chapter.

(Amdt. 43-23, Eff. 10/15/82); (Amdt. 43-31, Eff. 8/18/90)

§ 43.7 Persons authorized to approve aircraft, airframes, aircraft engines, propellers, appliances, or component parts for return to service after maintenance, preventive maintenance, rebuilding, or alteration.

(a) Except as provided in this section and § 43.17, no person, other than the Administrator, may approve an aircraft, airframe, aircraft engine, propeller, appliance, or component part for return to service after it has undergone maintenance, preventive maintenance, rebuilding, or alteration.

(b) The holder of a mechanic certificate or an inspection authorization may approve an aircraft, airframe, aircraft engine, propeller, appliance, or component part for return to service as provided in Part 65 of this chapter.

(c) The holder of a repair station certificate may approve an aircraft, airframe, aircraft engine, propeller, appliance, or component part for return to service as provided in Part 145 of this chapter.

(d) A manufacturer may approve for return to service any aircraft, airframe, aircraft engine, propeller, appliance, or component part which that manufacturer has worked on under § 43.3(h). However, except for minor alterations, the work must have been done in accordance with technical data approved by the Administrator.

(e) The holder of an air carrier operating certificate or an operating certificate issued under Part 121, 127, or 135, may approve an aircraft, airframe, aircraft engine, propeller, appliance, or component part for return to service as provided in Part 121, 127, or 135 of this chapter, as applicable.

(f) A person holding at least a private pilot certificate may approve an aircraft for return to service after performing preventive maintenance under the provisions of § 43.3(g).

(Amdt. 43-6, Eff. 7/6/66); (Amdt. 43-12, Eff. 11/15/69); (Amdt. 43-23, Eff. 10/15/82)

§ 43.9 Content, form, and disposition of maintenance, preventive maintenance, rebuilding, and alteration records (except inspections performed in accordance with Part 91, Part 123, Part 125, § 135.411(a)(1), and § 135.419 of this chapter).

(a) *Maintenance record entries.* Except as provided in paragraphs (b) and (c) of this section, each person who maintains, performs preventive maintenance, rebuilds, or alters an aircraft, airframe, aircraft engine, propeller, appliance, or component part shall make an entry in the maintenance record of that equipment containing the following information:

(1) A description (or reference to data acceptable to the Administrator) of work performed.

(2) The date of completion of the work performed.

(3) The name of the person performing the work if other than the person specified in paragraph (a)(4) of this section.

(4) If the work performed on the aircraft, airframe, aircraft engine, propeller, appliance, or component part has been performed satisfactorily, the signature, certificate number, and kind of certificate held by the person approving the work. The signature constitutes the approval for return to service only for the work performed.

In addition to the entry required by this paragraph, major repairs and major alterations shall be entered on a form, and the form disposed of, in the manner prescribed in Appendix B, by the person performing the work.

(b) Each holder of an air carrier operating certificate or an operating certificate issued under Part 121, 127, or 135, that is required by its approved operations specifications to provide for a continuous airworthiness maintenance program, shall make a record of the maintenance, preventive maintenance, rebuilding, and alteration, on aircraft, airframes, aircraft engines, propellers, appliances, or component parts which it operates in accordance with the applicable provisions of Part 121, 127, or 135 of this chapter, as appropriate.

(c) This section does not apply to persons performing inspections in accordance with Part 91, 123, 125, § 135.411(a)(1), or § 135.419 of this chapter.

(Amdt. 43-1, Eff. 4/1/65); (Amdt. 43-3, Eff. 4/2/66); (Amdt. 43-11, Eff. 10/16/69); (Amdt. 43-15, Eff. 10/23/72); (Amdt. 43-16, Eff. 9/8/72); (Amdt. 43-23, Eff. 10/15/82)

§ 43.11 Content, form, and disposition of records for inspections conducted under Parts 91 and 125 and §§ 135.411(a)(1) and 135.419 of this chapter.

(a) *Maintenance record entries.* The person approving or disapproving for return to service an aircraft, airframe, aircraft engine, propeller, appliance, or component part after any inspection performed in accordance with Part 91, 123, 125, § 135.411(a)(1), or § 135.419 shall make an entry in the maintenance record of that equipment containing the following information:

(1) The type of inspection and a brief description of the extent of the inspection.

(2) The date of the inspection and aircraft total time in service.

(3) The signature, the certificate number, and kind of certificate held by the person approving or disapproving for return to service the aircraft, airframe, aircraft engine, propeller, appliance, component part, or portions thereof.

(4) Except for progressive inspections, if the aircraft is found to be airworthy and approved for return to service, the following or a similarly worded statement—"I certify that this aircraft has been inspected in accordance with (insert type) inspection and was determined to be in airworthy condition."

(5) Except for progressive inspections, if the aircraft is not approved for return to service because of needed maintenance, noncompliance with applicable specifications, airworthiness directives, or other approved data, the following or a similarly worded statement—"I certify that

this aircraft has been inspected in accordance with (insert type) inspection and a list of discrepancies and unairworthy items dated (date) has been provided for the aircraft owner or operator."

(6) For progressive inspections, the following or a similarly worded statement—"I certify that in accordance with a progressive inspection program, a routine inspection of (identify whether aircraft or components) and a detailed inspection of (identify components) were performed and the (aircraft or components) are (approved or disapproved) for return to service." If disapproved, the entry will further state "and a list of discrepancies and unairworthy items dated (date) has been provided to the aircraft owner or operator."

(7) If an inspection is conducted under an inspection program provided for in Part 91, 123, 125, or § 135.411(a)(1), the entry must identify the inspection program, that part of the inspection program accomplished, and contain a statement that the inspection was performed in accordance with the inspections and procedures for that particular program.

(b) *Listing of discrepancies and placards.* If the person performing any inspection required by Part 91 or 125 or § 135.411(a)(1) of this chapter finds that the aircraft is unairworthy or does not meet the applicable type certificate data, airworthiness directives, or other approved data upon which its airworthiness depends, that person must give the owner or lessee a signed and dated list of those discrepancies. For those items permitted to be inoperative under § 91.30(d)(2), that person shall place a placard, that meets the aircraft's airworthiness certification regulations, on each inoperative instrument and the cockpit control of each item of inoperative equipment, marking it "Inoperative," and shall add the items to the signed and dated list of discrepancies given to the owner or lessee.

(Amdt. 43-3, Eff. 4/2/66); (Amdt. 43-13, Eff. 6/15/70 (Amdt. 43-16, Eff. 9/8/72); (Amdt. 43-23, Eff. 10/15/82); (Amdt. 43-30, Eff. 12/13/88)

§ 43.12 Maintenance records: Falsification, reproduction, or alteration.

(a) No person may make or cause to be made:

(1) Any fraudulent or intentionally false entry in any record or report that is required to be made, kept, or used to show compliance with any requirement under this part;

(2) Any reproduction, for fraudulent purpose, of any record or report under this part; or

(3) Any alteration, for fraudulent purpose, of any record or report under this part.

(b) The commission by any person of an act prohibited under paragraph (a) of this section is a basis for suspending or revoking the applicable airman, operator, or production certificate, Technical Standard Order Authorization, FAA-Parts Manufacturer Approval, or Product and Process Specification issued by the Administrator and held by that person.

Docket No. 16383 (43 FR 22636)

(Amdt. 43-19, Eff. 6/26/78); (Amdt. 43-23, Eff. 10/15/82)

§ 43.13 Performance rules (general).

(a) Each person performing maintenance, alteration, or preventive maintenance on an aircraft, engine, propeller, or appliance shall use the methods, techniques, and practices prescribed in the current manufacturer's maintenance manual or Instructions for Continued Airworthiness prepared by its manufacturer, or other methods, techniques, and practices acceptable to the Administrator, except as noted in § 43.16. He shall use the tools, equipment, and test apparatus necessary to assure completion of the work in accordance with accepted industry practices. If special equipment or test apparatus is recommended by the manufacturer involved, he must use that equipment or apparatus or its equivalent acceptable to the Administrator.

(b) Each person maintaining or altering, or performing preventive maintenance, shall do that work in such a manner and use materials of such a quality, that the condition of the aircraft, airframe, aircraft engine, propeller, or appliance worked on will be at least equal to its original or properly altered condition (with regard to aerodynamic function, structural strength, resistance to vibration and deterioration, and other qualities affecting airworthiness).

(c) *Special provisions for holders of air carrier operating certificates and operating certificates issued under the provisions of Part 121, 127, or 135 and Part 129 operators holding operations specifications.* Unless otherwise notified by the administrator, the methods, techniques, and practices contained in the maintenance manual or the maintenance part of the manual of the holder of an air carrier operating certificate or an operating certificate under Part 121, 127, or 135 and Part 129 operators holding operations specifications (that is required by its operating specifications to provide a continuous airworthiness maintenance and inspection program) constitute acceptable means of compliance with this section.

(Amdt. 43-15, Eff. 10/23/72); (Amdt. 43-20, Eff. 10/14/80); (Amdt. 43-23, Eff. 10/15/82); (Amdt. 43-28, Eff. 8/25/87)

§ 43.15 Additional performance rules for inspections.

(a) *General.* Each person performing an inspection required by Part 91, 123, 125, or 135 of this chapter, shall—

(1) Perform the inspection so as to determine whether the aircraft, or portion(s) thereof under inspection, meets all applicable airworthiness requirements; and

(2) If the inspection is one provided for in Part 123, 125, 135, or § 91.409(e) of this chapter, perform the inspection in accordance with the instructions and procedures set forth in the inspection program for the aircraft being inspected.

(b) *Rotorcraft.* Each person performing an inspection required by Part 91 on a rotorcraft shall inspect the following systems in accordance with the maintenance manual or Instructions for Continued Airworthiness of the manufacturer concerned:

(1) The drive shafts or similar systems.

(2) The main rotor transmission gear box for obvious defects.

(3) The main rotor and center section (or the equivalent area).

(4) The auxiliary rotor on helicopters.

(c) *Annual and 100-hour inspections.* (1) Each person performing an annual or 100-hour inspection shall use a checklist while performing the inspection. The checklist may be of the person's own design, one provided by the manufacturer of the equipment being inspected or one obtained from another source. This checklist must include the scope and detail of the items contained in Appendix D to this part and paragraph (b) of this section.

(2) Each person approving a reciprocating-engine-powered aircraft for return to service after an annual or 100-hour inspection shall, before that approval, run the aircraft engine or engines to determine satisfactory performance in accordance with the manufacturer's recommendations of—

(i) Power output (static and idle r.p.m.);

(ii) Magnetos;

(iii) Fuel and oil pressure; and

(iv) Cylinder and oil temperature.

(3) Each person approving a turbine-engine-powered aircraft for return to service after an annual, 100-hour, or progressive inspection shall, before that approval, run the aircraft engine or engines to determine satisfactory performance in accordance with the manufacturer's recommendations.

(d) *Progressive inspection.* (1) Each person performing a progressive inspection shall, at the start of a progressive inspection system, inspect the aircraft completely. After this initial inspection, routine and detailed inspections must be conducted as prescribed in the progressive inspection schedule. Routine inspections consist of visual examination or check of the appliances, the aircraft, and its components and systems, insofar as practicable without disassembly. Detailed inspections consist of a thorough examination of the appliances, the aircraft, and its components and systems, with such disassembly as is necessary. For the purposes of this subparagraph, the overhaul of a component or system is considered to be a detailed inspection.

(2) If the aircraft is away from the station where inspections are normally conducted, an appropriately rated mechanic, a certificated repair station, or the manufacturer of the aircraft may perform inspections in accordance with the procedures and using the forms of the person who would otherwise perform the inspection.

(Amdt. 43-3, Eff. 4/2/66); (Amdt. 43-8, Eff. 10/14/68); (Amdt. 43-21, Eff. 2/1/81); (Amdt. 43-22, Eff. 4/1/81); (Amdt. 43-22A, Eff. 2/3/81); (Amdt. 43-23, Eff. 10/15/82); (Amdt. 43-25, Eff. 1/6/87); (Amdt. 43-31, Eff. 8/18/90)

§ 43.16 Airworthiness limitations.

Each person performing an inspection or other maintenance specified in an Airworthiness Limitations section of a manufacturer's maintenance manual or Instructions for Continued Airworthiness shall perform the inspection or other maintenance in accordance with that section, or in accordance with operations specifications approved by the Administrator under Parts 121, 123, 127, or 135, or an inspection program approved under § 91.409(e).

Docket No. 8444 (33 FR 14104, 9/18/68)

(Amdt. 43-9, Eff. 10/17/68); (Amdt. 43-20, Eff. 10/14/80); (Amdt. 43-23, Eff. 10/15/82); (Amdt. 43-31, Eff. 8/18/90)

§ 43.17 [Maintenance, preventive maintenance, and alterations performed on U.S. aeronautical products by certain Canadian persons.

[(a) *Definitions.* For purposes of this section:

[Aeronautical product means any civil aircraft or airframe, aircraft engine, propeller, appliance, component, or part to be installed thereon.

[Canadian aeronautical product means any civil aircraft or airframe, aircraft engine, propeller, or appliance under airworthiness regulation by the Canadian Department of Transport, or component or part to be installed thereon.

[U.S. aeronautical product means any civil aircraft or airframe, aircraft engine, propeller, or appliance under airworthiness regulation by the FAA or component or part to be installed thereon.

[(b) *Applicability.* This section does not apply to any U.S. aeronautical products maintained or altered under any bilateral agreement made between Canada and any other than the United States.

[(c) *Authorized persons.*

[(1) A person holding a valid Canadian Department of Transport license (Aircraft Maintenance Engineer) and appropriate ratings may, with respect to a U.S.-registered aircraft located in Canada, perform maintenance, preventive maintenance, and alterations in accordance with the requirements of paragraph (d) of this section and approve the affected aircraft for return to service in accordance with the requirements of paragraph (e) of this section.

[(2) A company (Approved Maintenance Organization) (AMO) whose system of quality control for the maintenance, alteration, and inspection of aeronautical products has been approved by the Canadian Department of Transport, or a person who is an authorized employee performing work for such a company may, with respect to a U.S.-registered aircraft located in Canada or other U.S. aeronautical products transported to Canada from the United States, perform maintenance, preventive maintenance, and alterations in accordance with the requirements of paragraph (d) of this section and approve the affected products for return to service in accordance with the requirements of paragraph (e) of this section.

[(d) *Performance requirements.* A person authorized in paragraph (c) of this section may perform maintenance (including any inspection required by § 91.409 of this chapter, except an annual inspection), preventive maintenance, and alterations, provided:

[(1) The person performing the work is authorized by the Canadian Department of Transport to perform the same type of work respect to Canadian aeronautical products;

[(2) The work is performed in accordance with §§ 43.13, 43.15, 43.16 of this chapter, as applicable;

[(3) The work is performed such that the affected product complies with the applicable requirements of Part 36 of this chapter; and

[(4) The work is recorded in accordance with §§ 43.2(a), 43.9, and 43.11 of this chapter, as applicable.

[(e) *Approval requirements.*

[(1) To return an affected product to service, a person authorized in paragraph (c) of this section must approve (certify) maintenance, preventive maintenance, and alterations performed under this section, except that an Aircraft Maintenance Engineer may not approve a major repair or major alteration.

[(2) An AMO whose system of quality control for the maintenance, preventive maintenance, alteration, and inspection of aeronautical products has been approved by the Canadian Department of Transport, or an authorized employee performing work for such an AMO, may approve (certify) a major repair or major alteration performed under this section if the work was performed in accordance with technical data approved by the Administrator.

[(f) No person may operate in air commerce an aircraft, airframe, aircraft engine, propeller, or appliance on which maintenance, preventive maintenance, or alteration has been performed under this section unless it has been approved for return to service by a person authorized in this section.]

Docket No. 6992 (31 FR 5948, 4/19/66)

(Amdt. 43-5, Eff. 5/2/66); (Amdt. 43-10, Eff. 11/29/68); (Amdt. 43-23, Eff. 10/15/82); (Amdt. 43-31, Eff. 8/18/90); [(Amdt. 43-33, Eff. 2/10/92)]

Appendix A—Major alterations, major repairs, and preventive maintenance

(a) *Major alterations*—(1) *Airframe major alterations.* Alterations of the following parts and alterations of the following types, when not listed in the aircraft specifications issued by the FAA, are airframe major alterations:

(i) Wings.

(ii) Tail surfaces.

(iii) Fuselage.

(iv) Engine mounts.

(v) Control system.

(vi) Landing gear.

(vii) Hull or floats.

(viii) Elements of an airframe including spars, ribs, fittings, shock absorbers, bracing, cowling, fairings, and balance weights.

(ix) Hydraulic and electrical actuating system of components.

(x) Rotor blades.

(xi) Changes to the empty weight or empty balance which result in an increase in the maximum certificated weight or center of gravity limits of the aircraft.

(xii) Changes to the basic design of the fuel, oil, cooling, heating, cabin pressurization, electrical, hydraulic, de-icing, or exhaust systems.

(xiii) Changes to the wing or to fixed or movable control surfaces which affect flutter and vibration characteristics.

(2) *Powerplant major alterations.* The following alterations of a powerplant when not listed in the engine specifications issued by the FAA, are powerplant major alterations.

(i) Conversion of an aircraft engine from one approved model to another, involving any changes in compression ratio, propeller reduction gear, impeller gear ratios or the substitution of major engine parts which requires extensive rework and testing of the engine.

(ii) Changes to the engine by replacing aircraft engine structural parts with parts not supplied by the original manufacturer or parts not specifically approved by the Administrator.

(iii) Installation of an accessory which is not approved for the engine.

(iv) Removal of accessories that are listed as required equipment on the aircraft or engine specification.

(v) Installation of structural parts other than the type of parts approved for the installation.

(vi) Conversions of any sort for the purpose of using fuel of a rating or grade other than that listed in the engine specifications.

(3) *Propeller major alterations.* The following alterations of a propeller when not authorized in the propeller specifications issued by the FAA are propeller major alterations:

(i) Changes in blade design.

(ii) Changes in hub design.

(iii) Changes in the governor or control design.

(iv) Installation of a propeller governor or feathering system.

(v) Installation of propeller de-icing system.

(vi) Installation of parts not approved for the propeller.

(4) *Appliance major alterations.* Alterations of the basic design not made in accordance with recommendations of the appliance manufacturer or in accordance with an FAA Airworthiness Directive are appliance major alterations. In

addition, changes in the basic design of radio communication and navigation equipment approved under type certification or a Technical Standard Order that have an effect on frequency stability, noise level, sensitivity, selectivity, distortion, spurious radiation, AVC characteristics, or ability to meet environmental test conditions and other changes that have an effect on the performance of the equipment are also major alterations.

(b) *Major repairs—(1) Airframe major repairs.* Repairs to the following parts of an airframe and repairs of the following types, involving the strengthening, reinforcing, splicing, and manufacturing of primary structural members or their replacement, when replacement is by fabrication such as riveting or welding, are airframe major repairs.

(i) Box beams.

(ii) Monocoque or semimonocoque wings or control surfaces.

(iii) Wing stringers or chord members.

(iv) Spars.

(v) Spar flanges.

(vi) Members of truss-type beams.

(vii) Thin sheet webs of beams.

(viii) Keel and chine members of boat hulls or floats.

(ix) Corrugated sheet compression members which act as flange material of wings or tail surfaces.

(x) Wing main ribs and compression members.

(xi) Wing or tail surface brace struts.

(xii) Engine mounts.

(xiii) Fuselage longerons.

(xiv) Members of the side truss, horizontal truss, or bulkheads.

(xv) Main seat support braces and brackets.

(xvi) Landing gear brace struts.

(xvii) Axles.

(xviii) Wheels.

(xix) Skis, and ski pedestals.

(xx) Parts of the control system such as control columns, pedals, shafts, brackets, or horns.

(xxi) Repairs involving the substitution of material.

(xxii) The repair of damaged areas in metal or plywood stressed covering exceeding six inches in any direction.

(xxiii) The repair of portions of skin sheets by making additional seams.

(xxiv) The splicing of skin sheets.

(xxv) The repair of three or more adjacent wing or control surface ribs or the leading edge of wings and control surfaces, between such adjacent ribs.

(xxvi) Repair of fabric covering involving an area greater than that required to repair two adjacent ribs.

(xxvii) Replacement of fabric on fabric covered parts such as wings, fuselages, stabilizers, and control surfaces.

(xxviii) Repairing, including rebottoming, of removable or integral fuel tanks and oil tanks.

(2) *Powerplant major repairs.* Repairs of the following parts of an engine and repairs of the following types, are powerplant major repairs:

(i) Separation or disassembly of a crankcase or crankshaft of a reciprocating engine equipped with an integral supercharger.

(ii) Separation or disassembly of a crankcase or crankshaft of a reciprocating engine equipped with other than spur-type propeller reduction gearing.

(iii) Special repairs to structural engine parts by welding, plating, metalizing, or other methods.

(3) *Propeller major repairs.* Repairs of the following types to a propeller are propeller major repairs:

(i) Any repairs to, or straightening of steel blades.

(ii) Repairing or machining of steel hubs.

(iii) Shortening of blades.

(iv) Retipping of wood propellers.

(v) Replacement of outer laminations on fixed pitch wood propellers.

(vi) Repairing elongated bolt holes in the hub of fixed pitch wood propellers.

(vii) Inlay work on wood blades.

(viii) Repairs to composition blades.

(ix) Replacement of tip fabric.

(x) Replacement of plastic covering.

(xi) Repair of propeller governors.

(xii) Overhaul of controllable pitch propellers.

(xiii) Repairs to deep dents, cuts, scars, nicks, etc., and straightening of aluminum blades.

(xiv) The repair or replacement of internal elements of blades.

(4) *Appliance major repairs.* Repairs of the following types to appliances are appliance major repairs:

(i) Calibration and repair of instruments.

(ii) Calibration of radio equipment.

(iii) Rewinding the field coil of an electrical accessory.

(iv) Complete disassembly of complex hydraulic power valves.

(v) Overhaul of pressure type carburetors, and pressure type fuel, oil and hydraulic pumps.

(c) *Preventive maintenance.* Preventive maintenance is limited to the following work, provided it does not involve complex assembly operations: (1) Removal, installation, and repair of landing gear tires.

(2) Replacing elastic shock absorber cords on landing gear.

(3) Servicing landing gear shock struts by adding oil, air, or both.

(4) Servicing landing gear wheel bearings, such as cleaning and greasing.

(5) Replacing defective safety wiring or cotter keys.

(6) Lubrication not requiring disassembly other than removal of nonstructural items such as cover plates, cowlings, and fairings.

(7) Making simple fabric patches not requiring rib stitching or the removal of structural parts or control sur-

faces. In the case of balloons, the making of small fabric repairs to envelopes (as defined in, and in accordance with, the balloon manufacturers' instructions) not requiring load tape repair or replacement.

(8) Replenishing hydraulic fluid in the hydraulic reservoir.

(9) Refinishing decorative coating of fuselage, balloon baskets, wings tail group surfaces (excluding balanced control surfaces), fairings, cowlings, landing gear, cabin, or cockpit interior when removal or disassembly of any primary structure or operating system is not required.

(10) Applying preservative or protective material to components where no disassembly of any primary structure or operating system is involved and where such coating is not prohibited or is not contrary to good practices.

(11) Repairing upholstery and decorative furnishings of the cabin, cockpit, or balloon basket interior when the repairing does not require disassembly of any primary structure or operating system or interfere with an operating system or affect the primary structure of the aircraft.

(12) Making small simple repairs to fairings, nonstructural cover plates, cowlings, and small patches and reinforcements not changing the contour so as to interfere with proper air flow.

(13) Replacing side windows where that work does not interfere with the structure or any operating system such as controls, electrical equipment, etc.

(14) Replacing safety belts.

(15) Replacing seats or seat parts with replacement parts approved for the aircraft, not involving disassembly of any primary structure or operating system.

(16) Troubleshooting and repairing broken circuits in landing light wiring circuits.

(17) Replacing bulbs, reflectors, and lenses of position and landing lights.

(18) Replacing wheels and skis where no weight and balance computation is involved.

(19) Replacing any cowling not requiring removal of the propeller or disconnection of flight controls.

(20) Replacing or cleaning spark plugs and setting of spark plug gap clearance.

(21) Replacing any hose connection except hydraulic connections.

(22) Replacing prefabricated fuel lines.

(23) Cleaning or replacing fuel and oil strainers or filter elements.

(24) Replacing and servicing batteries.

(25) Cleaning of balloon burner pilot and main nozzles in accordance with the balloon manufacturer's instructions.

(26) Replacement or adjustment of nonstructural standard fasteners incidental to operations.

(27) The interchange of balloon baskets and burners on envelopes when the basket or burner is designated as interchangeable in the balloon type certificate data and the baskets and burners are specifically designed for quick removal and installation.

(28) The installations of anti-misfueling devices to reduce the diameter of fuel tank filler openings provided the specific device has been made a part of the aircraft type certificate data by the aircraft manufacturer, the aircraft manufacturer has provided FAA-approved instructions for installation of the specific device, and installation does not involve the disassembly of the existing tank filler opening.

(29) Removing, checking, and replacing magnetic chip detectors.

[(30) The inspection and maintenance tasks prescribed and specifically identified as preventive maintenance in a primary category aircraft type certificate or supplemental type certificate holder's approved special inspection and preventive maintenance program when accomplished on a primary category aircraft provided:

[(i) They are performed by the holder of at least a private pilot certificate issued under part 61 who is the registered owner (including co-owners) of the affected aircraft and who holds a certificate of competency for the affected aircraft (1) issued by a school approved under § 147.21(f) of this chapter; (2) issued by the holder of the production certificate for that primary category aircraft that has a special training program approved under § 21.24 of this subchapter; or (3) issued by another entity that has a course approved by the Administrator; and

[(ii) The inspections and maintenance tasks are performed in accordance with instructions contained in the special inspection and preventive maintenance program approved as part of the aircraft's type design or supplemental type design.]

(Amdt. 43-14, Eff. 8/18/72); (Amdt. 43-23, Eff. 10/15/82); (Amdt. 43-24, Eff. 11/7/84); (Amdt. 43-25, Eff. 1/6/87); (Amdt. 43-27, Eff. 6/5/87); [(Admt. 43-34, Eff. 12/31/92)]

Appendix B—Recording of major repairs and major alterations

(a) Except as provided in paragraphs (b), (c), and (d) of this appendix, each person performing a major repair or major alteration shall—

(1) Execute FAA Form 337 at least in duplicate;

(2) Give a signed copy of that form to the aircraft owner; and

(3) Forward a copy of that form to the local Flight Standards District Office within 48 hours after the aircraft, airframe, aircraft engine, propeller, or appliance is approved for return to service.

(b) For major repairs made in accordance with a manual or specifications acceptable to the Administrator, a certificated repair station may, in place of the requirements of paragraph (a)—

(1) Use the customer's work order upon which the repair is recorded;

(2) Give the aircraft owner a signed copy of the work order and retain a duplicate copy for at least two years from the date of approval for return to service of the aircraft, airframe, aircraft engine, propeller, or appliance;

(3) Give the aircraft owner a maintenance release signed by an authorized representative of the repair station and incorporating the following information:

(i) Identity of the aircraft, airframe, aircraft engine, propeller or appliance.

(ii) If an aircraft, the make, model, serial number, nationality and registration marks, and location of the repaired area.

(iii) If an airframe, aircraft engine, propeller, or appliance, give the manufacturer's name, name of the part, model, and serial numbers (if any); and

(4) Include the following or a similarly worded statement—

"The aircraft, airframe, aircraft engine, propeller, or appliance identified above was repaired and inspected in accordance with current Regulations of the Federal Aviation Agency and is approved for return to service.

Pertinent details of the repair are on file at this repair station under Order No.________,

No.________ Date________________

Signed ______________________________________ for
(signature of authorized representative)

____________________ ____________________
(repair station name) (certificate number)

____."
(address)

(c) For a major repair or major alteration made by a person authorized in § 43.17, the person who performs the major repair or major alteration and the person authorized by § 43.17 to approve that work shall execute a FAA Form 337 at least in duplicate. A completed copy of that form shall be—

(1) Given to the aircraft owner; and

(2) Forwarded to the Federal Aviation Administration, Aircraft Registration Branch, Post Office Box 25082, Oklahoma City, Okla. 73125, within 48 hours after the work is inspected.

(d) For extended-range fuel tanks installed within the passenger compartment or a baggage compartment, the person who performs the work and the person authorized to approve the work by § 43.7 of this part shall execute an FAA Form 337 in at least triplicate. One (1) copy of the FAA Form 337 shall be placed on board the aircraft as specified in § 91.417 of this chapter. The remaining forms shall be distributed as required by paragraph (a) (2) and (3) or (c) (1) and (2) of this paragraph as appropriate.

(Amdt. 43-10, Eff. 11/29/68); (Amdt. 43-29, Eff. 12/8/87); (Amdt. 43-31, Eff. 8/18/90)

Appendix C—[Reserved]

(Amdt. 43-3, Eff. 4/2/66); (Amdt. 43-13, Eff. 6/15/70)

Appendix D—Scope and detail of items (as applicable to the particular aircraft) to be included in annual and 100-hour inspections

(a) Each person performing an annual or 100-hour inspection shall, before that inspection, remove or open all necessary inspection plates, access doors, fairing, and cowling. He shall thoroughly clean the aircraft and aircraft engine.

(b) Each person performing an annual or 100-hour inspection shall inspect (where applicable) the following components of the fuselage and hull group:

(1) Fabric and skin—for deterioration, distortion, other evidence of failure, and defective or insecure attachment of fittings.

(2) Systems and component—for improper installation, apparent defects, and unsatisfactory operation.

(3) Envelope, gas bags, ballast tanks, and related parts—for poor condition.

(c) Each person performing an annual or 100-hour inspection shall inspect (where applicable) the following components of the cabin and cockpit group:

(1) Generally—for uncleanliness and loose equipment that might foul the controls.

(2) Seats and safety belts—for poor condition and apparent defects.

(3) Windows and windshields—for deterioration and breakage.

(4) Instruments—for poor condition, mounting, marking, and (where practicable) improper operation.

(5) Flight and engine controls—for improper installation and improper operation.

(6) Batteries—for improper installation and improper charge.

(7) All systems—for improper installation, poor general condition, apparent and obvious defects, and insecurity of attachment.

(d) Each person performing an annual or 100-hour inspection shall inspect (where applicable) components of the engine and nacelle group as follows:

(1) Engine section—for visual evidence of excessive oil, fuel, or hydraulic leaks, and sources of such leaks.

(2) Studs and nuts—for improper torquing and obvious defects.

(3) Internal engine—for cylinder compression and for metal particles or foreign matter on screens and sump drain plugs. If there is weak cylinder compression, for improper internal condition and improper internal tolerances.

(4) Engine mount—for cracks, looseness of mounting, and looseness of engine to mount.

(5) Flexible vibration dampeners—for poor condition and deterioration.

(6) Engine controls—for defects, improper travel, and improper safetying.

(7) Lines, hoses, and clamps—for leaks, improper condition and looseness.

(8) Exhaust stacks—for cracks, defects, and improper attachment.

(9) Accessories—for apparent defects in security of mounting.

(10) All systems—for improper installation, poor general condition, defects, and insecure attachment.

(11) Cowling—for cracks, and defects.

(e) Each person performing an annual or 100-hour inspection shall inspect (where applicable) the following components of the landing gear group:

(1) All units—for poor condition and insecurity of attachment.

(2) Shock absorbing devices—for improper oleo fluid level.

(3) Linkages, trusses, and members—for undue or excessive wear fatigue, and distortion.

(4) Retracting and locking mechanism—for improper operation.

(5) Hydraulic lines—for leakage.

(6) Electrical system—for chafing and improper operation of switches.

(7) Wheels—for cracks, defects, and condition of bearings.

(8) Tires—for wear and cuts.

(9) Brakes—for improper adjustment.

(10) Floats and skis—for insecure attachment and obvious or apparent defects.

(f) Each person performing an annual or 100-hour inspection shall inspect (where applicable) all components of the wing and center section assembly for poor general condition, fabric or skin deterioration, distortion, evidence of failure, and insecurity of attachment.

(g) Each person performing an annual or 100-hour inspection shall inspect (where applicable) all components and systems that make up the complete empennage assembly for poor general condition, fabric or skin deterioration, distortion, evidence of failure, insecure attachment, improper component installation, and improper component operation.

(h) Each person performing an annual or 100-hour inspection shall inspect (where applicable) the following components of the propeller group:

(1) Propeller assembly—for cracks, nicks, binds, and oil leakage.

(2) Bolts—for improper torquing and lack of safetying.

(3) Anti-icing devices—for improper operations and obvious defects.

(4) Control mechanisms—for improper operation, insecure mounting, and restricted travel.

(i) Each person performing an annual or 100-hour inspection shall inspect (where applicable) the following components of the radio group:

(1) Radio and electronic equipment—for improper installation and insecure mounting.

(2) Wiring and conduits—for improper routing, insecure mounting, and obvious defects.

(3) Bonding and shielding—for improper installation and poor condition.

(4) Antenna including trailing antenna—for poor condition, insecure mounting, and improper operation.

(j) Each person performing an annual or 100-hour inspection shall inspect (where applicable) each installed miscellaneous item that is not otherwise covered by this listing for improper installation and improper operation.

(Amdt. 43-3, Eff. 4/2/66)

Appendix E—Altimeter system test and inspection

Each person performing the altimeter system tests and inspections required by § 91.411 shall comply with the following:

(a) Static pressure system:

(1) Ensure freedom from entrapped moisture and restrictions.

(2) Determine that leakage is within the tolerances established in § 23.1325 or § 25.1325, whichever is applicable.

(3) Determine that the static port heater, if installed, is operative.

(4) Ensure that no alterations or deformations of the airframe surface have been made that would affect the relationship between air pressure in the static pressure system and true ambient static air pressure for any flight condition.

(b) Altimeter:

(1) Test by an appropriately rated repair facility in accordance with the following subparagraphs. Unless otherwise specified, each test for performance may be conducted with the instrument subjected to vibration. When tests are conducted with the temperature substantially different from ambient temperature of approximately 25 degrees C., allowance shall be made for the variation from the specified condition.

(i) *Scale error.* With the barometric pressure scale at 29.92 inches of mercury, the altimeter shall be subjected successively to pressures corresponding to the altitude specified in Table I up to the maximum normally expected operating altitude of the airplane in which the altimeter is to be installed. The reduction in pressure shall be made at a rate not in excess of 20,000 feet per minute to within approximately 2,000 feet of the test point. The test point shall be approached at a rate compatible with the test equipment. The altimeter shall be kept at the pressure corresponding to each test point for at least 1 minute, but not more than 10 minutes, before a reading is taken. The error at all test points must not exceed the tolerances specified in Table I.

(ii) *Hysteresis.* The hysteresis test shall begin not more than 15 minutes after the altimeter's initial exposure to the pressure corresponding to the upper limit of the scale error test prescribed in subparagraph (i); and while the altimeter is at this pressure, the hysteresis test shall commence. Pressure shall be increased at a rate simulating a descent in altitude at the rate of 5,000 to 20,000 feet per minute until within 3,000 feet of the first test point (50 percent of maximum altitude). The test point shall then be approached at a rate of approximately 3,000 feet per minute. The altimeter shall be kept at this pressure for at least 5 minutes, but not more than 15 minutes, before the test reading is taken. After the reading has been taken, the pressure shall be increased further, in the same manner as before, until the pressure corresponding to the second test point (40 percent of maximum altitude) is reached. The altimeter shall be kept at this pressure for at least 1 minute, but not more than 10 minutes, before the test reading is taken. After the reading has been taken, the pressure shall be increased further, in the same manner as before, until atmospheric pressure is reached. The reading of the altimeter at either of the two test points shall not differ by more than the tolerance specified in Table II from the reading of the altimeter for the corresponding altitude recorded during the scale error test prescribed in paragraph (b)(i).

(iii) *After effect.* Not more than 5 minutes after the completion of the hysteresis test prescribed in paragraph (b)(ii), the reading of the altimeter (corrected for any change in atmospheric pressure) shall not differ from the original atmospheric pressure reading by more than the tolerance specified in Table II.

(iv) *Friction.* The altimeter shall be subjected to a steady rate of decrease of pressure approximating 750 feet per minute. At each altitude listed in Table III, the change in reading of the pointers after vibration shall not exceed the corresponding tolerance listed in Table III.

(v) *Case leak.* The leakage of the altimeter case, when the pressure within it corresponds to an altitude of 18,000 feet, shall not change the altimeter reading by more than the tolerance shown in Table II during an interval of 1 minute.

(vi) *Barometric scale error.* At constant atmospheric pressure, the barometric pressure scale shall be set at each of the pressures (falling within its range of adjustment) that are listed in Table IV, and shall cause the pointer to indicate the equivalent altitude difference shown in Table IV with a tolerance of 25 feet.

(2) Altimeters which are the air data computer type with associated computing systems, or which incorporate air data correction internally, may be tested in a manner and to specifications developed by the manufacturer which are acceptable to the Administrator.

(c) Automatic Pressure Altitude Reporting Equipment and ATC Transponder System Integration Test. The test must be conducted by an appropriately rated person under the conditions specified in paragraph (a). Measure the automatic pressure altitude at the output of the installed ATC transponder when interrogated on Mode C at a sufficient number of test points to ensure that the altitude reporting equipment, altimeters, and ATC transponders perform their intended functions as installed in the aircraft. The difference between the automatic reporting output and the altitude displayed at the altimeter shall not exceed 125 feet.

(d) Records: Comply with the provisions of § 43.9 of this chapter as to content, form, and disposition of the records. The person performing the altimeter tests shall record on the altimeter the date and maximum altitude to which the altimeter has been tested and the persons approving the airplane for return to service shall enter that data in the airplane log or other permanent record.

TABLE I

Altitude	Equivalent pressure (inches of mercury)	Tolerance ± (feet)
—1,000	31.018	20
0	29.921	20
500	29.385	20
1,000	28.856	20
1,500	28.335	25
2,000	27.821	30
3,000	26.817	30
4,000	25.842	35
6,000	23.978	40
8,000	22.225	60
10,000	20.577	80
12,000	19.029	90
14,000	17.577	100
16,000	16.216	110
18,000	14.942	120
20,000	13.750	130
22,000	12.636	140
25,000	11.104	155
30,000	8.885	180
35,000	7.041	205
40,000	5.538	230
45,000	4.355	255
50,000	3.425	280

TABLE II—TEST TOLERANCES

Test	Tolerance (feet)
Case Leak Test	±100
Hysteresis Test:	
First Test Point (50 percent of maximum altitude)	75
Second Test Point (40 percent of maximum altitude)	75
After Effect Test	30

TABLE III—FRICTION

Altitude (feet)	Tolerance (feet)
1,000	±70
2,000	70
3,000	70
5,000	70
10,000	80
15,000	90
20,000	100
25,000	120
30,000	140
35,000	160
40,000	180
50,000	250

TABLE IV—PRESSURE-ALTITUDE DIFFERENCE

Pressure (inches of Hg)	Altitude difference (feet)
28.10	–1,727
28.50	–1,340
29.00	–863
29.50	–392
29.92	0
30.50	+531
30.90	+893
30.99	+974

(Amdt. 43-2, Eff. 7/29/65); (Amdt. 43-7, Eff. 8/1/67); (Amdt. 43-19, Eff. 6/26/78); (Amdt. 43-23, Eff. 10/15/82); (Amdt. 43-31, Eff. 8/18/90)

Appendix F—ATC transponder tests and inspections

The ATC transponder tests required by § 91.413 of this chapter may be conducted using a bench check or portable test equipment and must meet the requirements prescribed in paragraphs (a) through (j) of this appendix. If portable test equipment with appropriate coupling to the aircraft antenna system is used, operate the test equipment for ATCRBS transponders at a nominal rate of 235 interrogations per second to avoid possible ATCRBS interference. Operate the test equipment at a nominal rate of 50 Mode S interrogations per second for Mode S. An additional 3 dB loss is allowed to compensate for antenna coupling errors during receiver sensitivity measurements conducted in accordance with paragraph (c)(1) when using portable test equipment.

(a) Radio Reply Frequency:

(1) For all classes of ATCRBS transponders, interrogate the transponder and verify that the reply frequency is 1090±3 Megahertz (MHz).

(2) For classes 1B, 2B, and 3B Mode S transponders, interrogate the transponder and verify that the reply frequency is 1090±3 MHz.

(3) For classes 1B, 2B, and 3B Mode S transponders that incorporate the optional 1090±1 MHz reply frequency, interrogate the transponder and verify that the reply frequency is correct.

(4) For classes 1A, 2A, 3A, and 4 Mode S transponders, interrogate the transponder and verify that the reply frequency is 1090±1 MHz.

(b) Suppression: When Classes 1B and 2B ATCRBS Transponders, or Classes 1B, 2B, and 3B Mode S transponders are interrogated Mode 3/A at an interrogation rate between 230 and 1,000 interrogations per second; or when Classes 1A and 2A ATCRBS Transponders, or Classes 1B, 2A, 3A, and 4 Mode S transponders are interrogated at a rate between 230 and 1,200 Mode 3/A interrogations per second:

(1) Verify that the transponder does not respond to more than 1 percent of ATCRBS interrogations when the amplitude of P_2 pulse is equal to the P_1 pulse.

(2) Verify that the transponder replies to at least 90 percent of ATCRBS interrogations when the amplitude of the P_2 pulse is 9 dB less than the P_1 pulse. If the test is conducted with a radiated test signal, the interrogation rate shall be 235±5 interrogations per second unless a higher rate has been approved for the test equipment used at that location.

(c) Receiver Sensitivity:

(1) Verify that for any class of ATCRBS Transponder, the receiver minimum triggering level (MTL) of the system is –73±4 dbm, or that for any class of Mode S transponder the receiver MTL for Mode S format (P6 type) interrogations is –74±3 dbm by use of a test set either:

(i) Connected to the antenna end of the transmission line;

(ii) Connected to the antenna terminal of the transponder with a correction for transmission line loss; or

(iii) Utilized radiated signal.

(2) Verify that the difference in Mode 3/A and Mode C receiver sensitivity does not exceed 1 db for either any class of ATCRBS transponder or any class of Mode S transponder.

(d) Radio Frequency (RF) Peak Output Power:

(1) Verify that the transponder RF output power is within specifications for the class of transponder. Use the same conditions as described in (c)(1) (i), (ii), and (iii) above.

(i) For Class 1A and 2A ATCRBS transponders, verify that the minimum RF peak output power is at least 21.0 dbw (125 watts).

(ii) For Class 1B and 2B ATCRBS Transponders, verify that the minimum RF peak output power is at least 18.5 dbw (70 watts).

(iii) For Class 1A, 2A, 3A, and 4 and those Class 1B, 2B, and 3B Mode S transponders that include the optional high RF peak output power, verify that the minimum RF peak output power is at least 21.0 dbw (125 watts).

(iv) For Classes 1B, 2B, and 3B Mode S transponders, verify that the minimum RF peak output power is at least 18.5 dbw (70 watts).

(v) For any class of ATCRBS or any class of Mode S transponders, verify that the maximum RF peak output power does not exceed 27.0 dbw (500 watts).

Note: The tests in (e) through (j) apply only to Mode S transponders.

(e) Mode S Diversity Transmission Channel Isolation: For any class of Mode S transponder that incorporates diversity operation, verify that the RF peak output power transmitted from the selected antenna exceeds the power transmitted from the nonselected antenna by at least 20 db.

(f) Mode S Address: Interrogate the Mode S transponder and verify that it replies only to its assigned address. Use the correct address and at least two incorrect addresses. The interrogations should be made at a nominal rate of 50 interrogations per second.

(g) Mode S Formats: Interrogate the Mode S transponder with uplink formats (UF) for which it is equipped and verify that the replies are made in the correct format. Use the surveillance formats UF = 4 and 5. Verify that the altitude reported in the replies to UF = 4 are the same as that reported in a valid ATCRBS Mode C reply. Verify that the identity reported in the replies to UF = 5 are the same as that reported in a valid ATCRBS Mode 3/A reply. If the transponder is so equipped, use the communication formats UF = 20, 21, and 24.

(h) Mode S All-Call Interrogations: Interrogate the Mode S transponder with the Mode S only all-call format UF = 11, and the ATCRBS/Mode S all-call formats (1.6 microsecond P_4 pulse) and verify that the correct address and capability are reported in the replies (downlink format DF = 11).

(i) ATCRBS-Only All-Call Interrogation: Interrogate the Mode S transponder with the ATCRBS-only all-call interrogation (0.8 microsecond P_4 pulse) and verify that no reply is generated.

(j) Squitter: Verify that the Mode S transponder generates a correct squitter approximately once per second.

(k) Records: Comply with the provisions of § 43.9 of this chapter as to content, form, and disposition of the records.

(Amdt. 43-17, Eff. 1/26/73); (Amdt. 43-18, Eff. 12/31/73); (Amdt. 43-19, Eff. 6/26/78); (Amdt. 43-26, Eff. 4/6/87); (Amdt. 43-31, Eff. 8/18/90)

Part 61 – Certification: pilots and flight instructors

Subpart A—General

61.1 Applicability
61.2 Certification of foreign pilots and flight instructors
61.3 Requirements for certificates, ratings, and authorizations
61.5 Certificates and ratings issued under this part
61.7 Obsolete certificates and ratings
61.9 Exchange of obsolete certificates and ratings for current certificates and ratings
61.11 Expired pilot certificates and reissuance
61.13 Application and qualification
61.14 Refusal to submit to drug [or alcohol] test
61.15 Offenses involving alcohol or drugs
61.16 Refusal to submit to an alcohol test or to furnish test results
61.17 Temporary certificate
61.19 Duration of pilot and flight instructor certificates
61.21 Duration of Category II pilot authorization
61.23 Duration of medical certificates
61.25 Change of name
61.27 Voluntary surrender or exchange of certificate
61.29 Replacement of lost or destroyed certificate
61.31 General limitations
61.33 Tests: General procedure
61.35 Written test: Prerequisites and passing grades
61.37 Written tests: Cheating or other unauthorized conduct
61.39 Prerequisites for flight tests
61.41 Flight instruction received from flight instructors not certificated by FAA
61.43 Flight tests: General procedures
61.45 Flight tests: Required aircraft and equipment
61.47 Flight tests: Status of FAA inspectors and other authorized flight examiners
61.49 Retesting after failure
61.51 Pilot logbooks
61.53 Operations during medical deficiency
61.55 Second-in-command qualifications
61.56 Flight review
61.57 Recent flight experience: Pilot in command
61.58 Pilot-in-command proficiency check: Operation of aircraft requiring more than one required pilot
61.59 Falsification, reproduction or alteration of applications, certificates, logbooks, reports, or records
61.60 Change of address

Subpart B—Aircraft ratings and special certificates

61.61 Applicability
61.63 Additional aircraft ratings (other than airline transport pilot)
61.65 Instrument rating requirements
61.67 Category II pilot authorization requirements
61.69 Glider towing: Experience and instruction requirements
61.71 Graduates of certificated [pilot] schools: Special rules
61.73 Military pilots or former military pilots: Special rules
61.75 Pilot certificate issued on basis of a foreign pilot license
61.77 Special purpose pilot certificate: Operation of U.S.-registered civil airplanes leased by a person not a U.S. citizen

Subpart C—Student and recreational pilots

61.81 Applicability
61.83 Eligibility requirements: Student pilots
61.85 Application
61.87 Solo flight requirements for student pilots
61.89 General limitations
61.91 Aircraft limitations: Pilot in command
61.93 Cross-country flight requirements (for student and recreational pilots seeking private pilot certification)
61.95 Operations in airspace and at airports located within Class B
61.96 Eligibility requirements: Recreational pilots
61.97 Aeronautical knowledge
61.98 Flight proficiency
61.99 Airplane rating: Aeronautical experience
61.100 Rotorcraft rating: Aeronautical experience
61.101 Recreational pilot privileges and limitations

Subpart D—Private pilots

61.102 Applicability
61.103 Eligibility requirements: General
61.105 Aeronautical knowledge
61.107 Flight proficiency
61.109 Airplane rating: Aeronautical experience
61.111 Cross-country flights: Pilots based on small islands
61.113 Rotorcraft rating: Aeronautical experience
61.115 Glider rating: Aeronautical experience
61.117 Lighter-than-air rating: Aeronautical experience

61.118 Private pilot privileges and limitations: Pilot in command
61.119 Free balloon rating: Limitations
61.120 Private pilot privileges and limitations: Second in command of aircraft requiring more than one required pilot

Subpart E—Commercial pilots

61.121 Applicability
61.123 Eligibility requirements: General
61.125 Aeronautical knowledge
61.127 Flight proficiency
61.129 Airplane rating: Aeronautical experience
61.131 Rotorcraft ratings: Aeronautical experience
61.133 Glider rating: Aeronautical experience
61.135 Airship rating: Aeronautical experience
61.137 Free balloon rating: Aeronautical experience
61.139 Commercial pilot privileges and limitations: General
61.141 Airship and free balloon ratings: Limitations

Subpart F—Airline transport pilots

61.151 Eligibility requirements: General
61.153 Airplane rating: Aeronautical knowledge
61.155 Airplane rating: Aeronautical experience
61.157 Airplane rating: Aeronautical skill
61.159 Rotorcraft rating: Aeronautical knowledge
61.161 Rotorcraft rating: Aeronautical experience
61.163 Rotorcraft rating: Aeronautical skill
61.165 Additional category ratings
61.167 Tests
61.169 Instruction in air transportation service
61.171 General privileges and limitations

Subpart G—Flight instructors

61.181 Applicability
61.183 Eligibility requirements: General
61.185 Aeronautical knowledge
61.187 Flight proficiency
61.189 Flight instructor records
61.191 Additional flight instructor ratings
61.193 Flight instructor authorizations
61.195 Flight instructor limitations
61.197 Renewal of flight instructor certificates
61.199 Expired flight instructor certificates and ratings
61.201 Conversion to new system of instructor ratings

Appendix A—Practical Test Requirements for Airplane Airline Transport Pilot Certificates and Associated Class and Type Ratings

Appendix B—Practical Test Requirements for Rotorcraft Airline Transport Pilot Certificates with a Helicopter Class Rating and Associated Type Ratings

Special Federal Aviation Regulation

Special Federal Aviation Regulation No. 58 *(Rule)*

Subpart A—General

Source: Docket No. 11802, Amdt. 61-60, (38 FR 3161, 2/1/73) Eff. 11/1/73, for each subpart, unless otherwise noted.

§ 61.1 Applicability.

(a) This part prescribes the requirements for issuing pilot and flight instructor certificates and ratings, the conditions under which those certificates and ratings are necessary, and the privileges and limitations of those certificates and ratings.

(b) [Except as provided in § 61.71, an applicant for a certificate or rating must meet the requirements of this part.]

(Amdt. 61-14, Eff. 4/3/65); (Amdt. 61-63, Eff. 11/1/74); [(Amdt. 61-90, Eff. 4/15/91)]

§ 61.2 Certification of foreign pilots and flight instructors.

A person who is neither a United States citizen nor a resident alien is issued a certificate under this part (other than under § 61.75 or § 61.77), outside the United States, only when the Administrator finds that the pilot certificate is needed for the operation of a U.S.-registered civil aircraft or finds that the flight instructor certificate is needed for the training of students who are citizens of the United States.

(Amdt. 61-72, Eff. 10/18/82)

§ 61.3 Requirements for certificates, ratings, and authorizations.

(a) *Pilot certificate.* No person may act as pilot in command or in any other capacity as a required pilot flight crewmember of a civil aircraft of United States registry unless he has in his personal possession a current pilot certificate issued to him under this part. However, when the aircraft is operated within a foreign country a current pilot license issued by the country in which the aircraft is operated may be used.

(b) *Pilot certificate: foreign aircraft.* No person may, within the United States, act as pilot in command or in any other capacity as a required pilot flight crewmember of a civil aircraft of foreign registry unless he has in his personal possession a current pilot certificate issued to him under this part, or a pilot license issued to him or validated for him by the country in which the aircraft is registered.

(c) *Medical certificate.* Except for free balloon pilots piloting balloons and glider pilots piloting gliders, no person may act as pilot in command or in any other capacity as a required pilot flight crewmember of an aircraft under a certificate issued to him under this part, unless he has in his personal possession an appropriate current medical certificate issued under Part 67 of this chapter. However, when the aircraft is operated within a foreign country with a current pilot license issued by that country, evidence of current medical qualification for that license, issued by that country, may be used. In the case of a pilot certificate issued on the basis of a foreign pilot license under § 61.75, evidence of current medical qualification accepted for the issue of that license is used in place of a medical certificate.

(d) *Flight instructor certificate.* Except for lighter-than-air flight instruction in lighter-than-air aircraft, and for instruction in air transportation service given by the holder of an Airline Transport Pilot Certificate under § 61.169, no person other than the holder of a flight instructor certificate issued by the Administrator with an appropriate rating on that certificate may—

(1) Give any of the flight instruction required to qualify for a solo flight, solo cross-country flight, or for the issue of a pilot or flight instructor certificate or rating;

(2) Endorse a pilot logbook to show that he has given any flight instruction; or

(3) Endorse a student pilot certificate or logbook for solo operating privileges.

(e) *Instrument rating.* No person may act as pilot in command of a civil aircraft under instrument flight rules, or in weather conditions less than the minimums prescribed for VFR flight unless—

(1) In the case of an airplane, he holds an instrument rating or an airline transport pilot certificate with an airplane category rating on it;

(2) In the case of a helicopter, he holds a helicopter instrument rating or an airline transport pilot certificate with a rotorcraft category and helicopter class rating not limited to VFR;

(3) In the case of a glider, he holds an instrument rating (airplane) or an airline transport pilot certificate with an airplane category rating; or

(4) In the case of an airship, he holds a commercial pilot certificate with lighter-than-air category and airship class ratings.

(f) *Category II pilot authorization.*

(1) No person may act as pilot in command of a civil aircraft in a Category II operation unless he holds a current Category II pilot authorization for that type aircraft or, in the case of a civil aircraft of foreign registry, he is authorized by the country of registry to act as pilot in command of that aircraft in Category II operations.

(2) No person may act as second in command of a civil aircraft in a Category II operation unless he holds a current appropriate instrument rating or an appropriate airline transport pilot certificate or, in the case of a civil aircraft of foreign registry, he is authorized by the country of registry to act as second in command of that aircraft in Category II operations. This paragraph does not apply to operations conducted by the holder of a certificate issued under Parts 121 and 135 of this chapter.

(g) *Category A aircraft pilot authorization.* The Administrator may issue a certificate of authorization to the pilot of a small aircraft identified as a Category A aircraft in § 97.3(b) (1) of this chapter to use that aircraft in a Category II operation, if he finds that the proposed operation can be safely conducted under the terms of the certificate. Such authorization does not permit operation of the aircraft carrying persons or property for compensation or hire.

(h) *Inspection of certificate.* Each person who holds a pilot certificate, flight instructor certificate, medical certificate, authorization, or license required by this part shall present it for inspection upon the request of the Administrator, an authorized representative of the National Transportation Safety Board, or any Federal, State, or local law enforcement officer.

(Amdt. 61-18, Eff. 9/26/65); (Amdt. 61-22, Eff. 9/19/66); (Amdt. 61-24, Eff. 10/17/66); (Amdt. 61-25, Eff. 11/19/66); (Amdt. 61-32, Eff. 8/7/67); (Amdt. 61-40, Eff. 5/10/68); (Amdt. 61-47, Eff. 4/30/70); (Amdt. 61-53, Eff. 5/9/71); (Amdt. 61-77, Eff. 1/6/87)

§ 61.5 Certificates and ratings issued under this part.

(a) The following certificates are issued under this part:

(1) Pilot certificates:

(i) Student pilot.

(ii) Recreational pilot.

(iii) Private pilot.

(iv) Commercial pilot.

(v) Airline transport pilot.

(2) Flight instructor certificates.

(b) The following ratings are placed on pilot certificates (other than student pilot) where applicable:

(1) Aircraft category ratings:

(i) Airplane.

(ii) Rotorcraft.

(iii) Glider.
(iv) Lighter-than-air.
(2) Airplane class ratings:
(i) Single-engine land.
(ii) Multiengine land.
(iii) Single-engine sea.
(iv) Multiengine sea.
(3) Rotorcraft class ratings:
(i) Helicopter.
(ii) Gyroplane.
(4) Lighter-than-air class ratings:
(i) Airship.
(ii) Free balloon.

(5) Aircraft type ratings are listed in Advisory Circular 61-1 entitled "Aircraft Type Ratings." This list includes ratings for the following:

(i) Large aircraft, other than lighter-than-air.
(ii) Small turbojet-powered airplanes.
(iii) Small helicopters for operations requiring an airline transport pilot certificate.
(iv) Other aircraft type ratings specified by the Administrator through aircraft type certificate procedures.

(6) Instrument ratings (on private and commercial pilot certificates only):

(i) Instrument—airplanes.
(ii) Instrument—helicopter.

(c) The following ratings are placed on flight instructor certificates where applicable:

(1) Aircraft category ratings:
(i) Airplane.
(ii) Rotorcraft.
(iii) Glider.
(2) Airplane class ratings:
(i) Single-engine.
(ii) Multiengine.
(3) Rotorcraft class ratings:
(i) Helicopter.
(ii) Gyroplane.
(4) Instrument ratings:
(i) Instrument—airplane.
(ii) Instrument—helicopter.

(Amdt. 61-14, Eff. 4/3/65); (Amdt. 61-25, Eff. 11/19/66); (Amdt. 61-32, Eff. 8/7/67); (Amdt. 61-64, Eff. 12/22/76); (Amdt. 61-82, Eff. 8/31/89)

§ 61.7 Obsolete certificates and ratings.

(a) The holder of a free balloon pilot certificate issued before November 1, 1973, may not exercise the privileges of that certificate.

(b) The holder of a pilot certificate that bears any of the following category ratings without an associated class rating, may not exercise the privileges of that category rating:

(1) Rotorcraft.
(2) Lighter-than-air.
(3) Helicopter.
(4) Autogiro.

§ 61.9 Exchange of obsolete certificates and ratings for current certificates and ratings.

(a) The holder of an unexpired free balloon pilot certificate, or an unexpired pilot certificate with an obsolete category rating listed in § 61.7(b) of this part may exchange that certificate for a certificate with the following applicable category and class rating, without a further showing of competency, until October 31, 1975. After that date, a free balloon pilot certificate or certificate with an obsolete rating expires.

(b) *Private or commercial pilot certificate with rotorcraft category rating.* The holder of a private or commercial pilot certificate with a rotorcraft category rating is issued that certificate with a rotorcraft category rating, and a helicopter or gyroplane class rating, depending upon whether a helicopter or a gyroplane is used to qualify for the rotorcraft category rating.

(c) *Private or commercial pilot certificate with helicopter or autogiro category rating.* The holder of a private or commercial pilot certificate with a helicopter or autogiro category rating is issued that certificate with a rotorcraft category rating and a helicopter class rating (in the case of a helicopter category rating), or a gyroplane class rating (in the case of an autogiro rating).

(d) *Airline transport pilot certificate with helicopter or autogiro category rating.* The holder of an airline transport pilot certificate with a helicopter or autogiro category rating is issued that certificate with a rotorcraft category rating (limited to VFR) and a helicopter class and type rating (in the case of a helicopter category rating) or a gyroplane class rating (in the case of an autogiro category rating).

(e) *Airline transport pilot certificate with a rotorcraft category rating (without a class rating).* The holder of an airline transport pilot certificate with a rotorcraft category rating (without a class rating) is issued that certificate with a rotorcraft category rating limited to VFR, and a helicopter and type rating or a gyroplane class rating, depending upon whether a helicopter or gyroplane is used to qualify for the rotorcraft category rating.

(f) *Free balloon pilot certificate.* The holder of a free balloon pilot certificate is issued a commercial pilot certificate with a lighter-than-air category rating and a free balloon class rating. However, a free balloon class rating may be issued with the limitations provided in § 61.141.

(g) *Lighter-than-air pilot certificate or pilot certificate with lighter-than-air category (without a class rating).*

(1) In the case of an application made before November 1, 1975, the holder of a lighter-than-air pilot certificate or a pilot certificate with a lighter-than-air category rating (without a class rating) is issued a private or commercial pilot certificate, as appropriate, with a lighter-than-air cat-

egory rating and airship and free balloon class ratings.

(2) In the case of an application made after October 31, 1975, the holder of a lighter-than-air pilot certificate with an airship rating issued prior to November 1, 1973, may be issued a free balloon class rating upon passing the appropriate flight test in a free balloon.

(Amdt. 61–14, Eff. 4/3/65); (Amdt. 61-18, Eff. 9/26/65); (Amdt. 61-25, Eff. 11/19/66); (Amdt. 61-33, Eff. 6/15/67); (Amdt. 61-38, Eff. 12/21/67); (Amdt. 61-64, Eff. 12/22/76)

§ 61.11 Expired pilot certificates and reissuance.

(a) No person who holds an expired pilot certificate or rating may exercise the privileges of that pilot certificate, or rating.

(b) Except as provided, the following certificates and ratings have expired and are not reissued:

(1) An airline transport pilot certificate issued before May 1, 1949, or containing a horsepower rating. However, an airline transport pilot certificate bearing an expiration date and issued after April 30, 1949, may be reissued without an expiration date if it does not contain a horsepower rating.

(2) A private or commercial pilot certificate, or a lighter-than-air or free balloon pilot certificate, issued before July 1, 1945. However, each of those certificates issued after June 30, 1945, and bearing an expiration date, may be reissued without an expiration date.

(c) A private or commercial pilot certificate or a special purpose pilot certificate, issued on the basis of a foreign pilot license, expires on the expiration date stated thereon. A certificate without an expiration date is issued to the holder of the expired certificate only if he meets the requirements of § 61.75 of this part for the issue of a pilot certificate based on a foreign pilot license.

§ 61.13 Application and qualification.

(a) An application for a certificate and rating or for an additional rating under this part is made on a form and in a manner prescribed by the Administrator. Each person who is neither a United States citizen nor a resident alien must show evidence that the fee prescribed by Appendix A of Part 187 of this chapter has been paid if that person—

(1) Applies for a student pilot certificate to be issued outside the United States; or

(2) Applies for a written or practical test to be administered outside the United States for any certificate or rating issued under this part.

(b) An applicant who meets the requirements of this part is entitled to an appropriate pilot certificate with aircraft ratings. Additional aircraft category, class, type and other ratings, for which the applicant is qualified, are added to his certificate. However, the Administrator may refuse to issue certificates to persons who are not citizens of the United States and who do not reside in the United States.

(c) An applicant who cannot comply with all of the flight proficiency requirements prescribed by this part because the aircraft used by him for his flight training or flight test is characteristically incapable of performing a required pilot operation, but who meets all other requirements for the certificate or rating sought, is issued the certificate or rating with appropriate limitations.

(d) An applicant for a pilot certificate who holds a medical certificate under § 67.19 of this chapter with special limitations on it, but who meets all other requirements for that pilot certificate, is issued a pilot certificate containing such operating limitations as the Administrator determines are necessary because of the applicant's medical deficiency.

(e) A Category II pilot authorization is issued as a part of the applicant's instrument rating or airline transport pilot certificate. Upon original issue the authorization contains a limitation for Category II operations of 1,600 feet RVR and a 150 foot decision height. This limitation is removed when the holder shows that since the beginning of the sixth preceding month he has made three Category II ILS approaches to a landing under actual or simulated instrument conditions with a 150 foot decision height.

(f) Unless authorized by the Administrator—

(1) A person whose pilot certificate is suspended may not apply for any pilot or flight instructor certificate or rating during the period of suspension; and

(2) A person whose flight instructor certificate only is suspended may not apply for any rating to be added to that certificate during the period of suspension.

(g) Unless the order of revocation provides otherwise—

(1) A person whose pilot certificate is revoked may not apply for any pilot or flight instructor certificate or rating for one year after the date of revocation; and

(2) A person whose flight instructor certificate only is revoked may not apply for any flight instructor certificate for one year after the date of revocation.

(Amdt. 61-4, Eff. 9/16/63); (Amdt. 61-25, Eff. 11/19/66); (Amdt. 61-51, Eff. 9/4/70); (Amdt. 61-52, Eff. 4/12/71); (Amdt. 61-72, Eff. 10/18/82)

§ 61.14 Refusal to submit to drug [or alcohol] test.

[(a) This section applies to an employee who performs a function listed in Appendix I or Appendix J to Part 121 of this chapter directly or by contract for a Part 121 certificate holder, a Part 135 certificate holder, or an operator as defined in § 135.1(c) of this chapter.

(b) Refusal by the holder of a certificate issued under this part to take a drug test required under the provisions of Appendix I to Part 121 or an alcohol test required under the provisions of Appendix J to Part 121 is grounds for—

(1) Denial of an application for any certificate or rating issued under this part for a period of up to 1 year after the date of such refusal; and

(2) Suspension or revocation of any certificate or rating issued under this part.]

Docket No. 25148 (53 FR 47056) Eff. 11/21/88; (Amdt. 61-81, Eff. 12/21/88); (Amdt. 61-83, Eff. 4/11/89); [(Amdt. 61-94, Eff. 3/17/94)]

§ 61.15 Offenses involving alcohol or drugs.

(a) A conviction for the violation of any Federal or state statute relating to the growing, processing, manufacture, sale, disposition, possession, transportation, or importation of narcotic drugs, marijuana, or depressant or stimulant drugs or substances is grounds for—

(1) Denial of an application for any certificate or rating issued under this part for a period of up to 1 year after the date of final conviction; or

(2) Suspension or revocation of any certificate or rating issued under this part.

(b) The commission of an act prohibited by § 91.17(a) or § 91.19(a) of this chapter is grounds for—

(1) Denial of an application for a certificate or rating issued under this part for a period of up to 1 year after the date of that act; or

(2) Suspension or revocation of any certificate or rating issued under this part.

(c) For the purposes of paragraphs (d) and (e) of this section, a motor vehicle action means—

(1) A conviction after November 29, 1990, for the violation of any Federal or state statute relating to the operation of a motor vehicle while intoxicated by alcohol or a drug, while impaired by alcohol or a drug, or while under the influence of alcohol or a drug;

(2) The cancellation, suspension, or revocation of a license to operate a motor vehicle by a state after November 29, 1990, for a cause related to the operation of a motor vehicle while intoxicated by alcohol or a drug, while impaired by alcohol or a drug, or while under the influence of alcohol or a drug; or

(3) The denial after November 29, 1990, of an application for a license to operate a motor vehicle by a state for a cause related to the operation of a motor vehicle while intoxicated by alcohol or a drug, while impaired by alcohol or a drug, or while under the influence of alcohol or a drug.

(d) Except in the case of a motor vehicle action that results from the same incident or arises out of the same factual circumstances, a motor vehicle action occurring within 3 years of a previous motor vehicle action is grounds for—

(1) Denial of an application for any certificate or rating issued under this part for a period of up to 1 year after the date of the last motor vehicle action; or

(2) Suspension or revocation of any certificate or rating issued under this part.

(e) Each person holding a certificate issued under this part shall provide a written report of each motor vehicle action to the FAA, Civil Aviation Security Division (AAC-700), P.O. Box 25810, Oklahoma City, OK 73125, not later than 60 days after the motor vehicle action. The report must include—

(1) The person's name, address, date of birth, and airman certificate number;

(2) The type of violation that resulted in the conviction or the administrative action;

(3) The date of the conviction or administrative action;

(4) The state that holds the record of conviction or administrative action; and

(5) A statement of whether the motor vehicle action resulted from the same incident or arose out of the same factual circumstances related to a previously-reported motor vehicle action.

(f) Failure to comply with paragraph (e) of this section is grounds for—

(1) Denial of an application for any certificate or rating issued under this part for a period of up to 1 year after the date of the motor vehicle action; or

(2) Suspension or revocation of any certificate or rating issued under this part.

Docket No. 21956 (50 FR 15379) Eff. 4/17/85; (Amdt. 61-1, Eff. 11/1/62); (Amdt. 61-14, Eff. 4/3/65); (Amdt. 61-20, Eff. 12/16/65); (Amdt. 61-24, Eff. 10/17/66); (Amdt. 61-40, Eff. 5/10/68); (Amdt. 61-44, Eff. 11/22/69); (Amdt. 61-74, Eff. 6/17/85); (Amdt. 61-84, Eff. 8/18/90); (Amdt. 61-87, Eff. 11/29/90)

§ 61.16 Refusal to submit to an alcohol test or to furnish test results.

A refusal to submit to a test to indicate the percentage by weight of alcohol in the blood, when requested by a law enforcement officer in accordance with § 91.17(c) of this chapter, or a refusal to furnish or authorize the release of the test results requested by the Administrator in accordance with § 91.17(c) or (d) of this chapter, is grounds for—

(a) Denial of an application for any certificate or rating issued under this part for a period of up to 1 year after the date of that refusal; or

(b) Suspension or revocation of any certificate or rating issued under this part.

Docket No. 21956 (51 FR 1229) Eff. 1/9/86; (Amdt. 61-20, Eff. 12/16/65); (Amdt. 61-40, Eff. 5/10/68); (Amdt. 61-44, Eff. 11/22/69); (Amdt. 61-74, Eff. 6/17/85); (Amdt. 61-76, Eff. 4/9/86); (Amdt. 61-84, Eff. 8/18/90)

§ 61.17 Temporary certificate.

(a) A temporary pilot or flight instructor certificate, or a rating, effective for a period of not more than 120 days, is issued to a qualified applicant pending a review of his qualifications and the issuance of a permanent certificate or rating by the Administrator. The permanent certificate or rating is issued to an applicant found qualified and a denial thereof is issued to an applicant found not qualified.

(b) A temporary certificate issued under paragraph (a) of this section expires—

(1) At the end of the expiration date stated thereon; or

(2) Upon receipt by the applicant of—

(i) The certificate or rating sought; or

(ii) Notice that the certificate or rating sought is denied.

(Amdt. 61-1, Eff. 11/1/62); (Amdt. 61-5, Eff. 10/22/63); (Amdt. 61-14, Eff. 4/3/65); (Amdt. 61-15, Eff. 4/4/65); (Amdt. 61-18, Eff. 9/26/65); (Amdt. 61-20, Eff. 12/16/65); (Amdt. 61-49, Eff. 6/1/70); (Amdt. 61-66, Eff. 6/26/78)

§ 61.19 Duration of pilot and flight instructor certificates.

(a) *General.* The holder of a certificate with an expiration date may not, after that date, exercise the privileges of that certificate.

(b) *Student pilot certificate.* A student pilot certificate expires at the end of the 24th month after the month in which it is issued.

(c) *Other pilot certificates.* Any pilot certificate (other than a student pilot certificate) issued under this part is issued without a specific expiration date. However, the holder of a pilot certificate issued on the basis of a foreign pilot license may exercise the privileges of that certificate only while the foreign pilot license on which that certificate is based is effective.

(d) *Flight instructor certificate.* A flight instructor certificate—

(1) Is effective only while the holder has a current pilot certificate and a medical certificate appropriate to the pilot privileges being exercised; and

(2) Expires at the end of the 24th month after the month in which it was last issued or renewed.

(e) *Surrender, suspension, or revocation.* Any pilot certificate or flight instructor certificate issued under this part ceases to be effective if it is surrendered, suspended, or revoked.

(f) *Return of certificate.* The holder of any certificate issued under this part that is suspended or revoked shall, upon the Administrator's request, return it to the Administrator.

(Amdt. 61-11, Eff. 12/4/64)

§ 61.21 Duration of Category II pilot authorization.

A Category II pilot authorization expires at the end of the sixth month after it was last issued or renewed. Upon passing a practical test it is renewed for each type aircraft for which an authorization is held. However, an authorization for any particular type aircraft for which an authorization is held will not be renewed to extend beyond the end of the twelfth month after the practical test was passed in that type aircraft. If the holder of the authorization passes the practical test for a renewal in the month before the authorization expires, he is considered to have passed it during the month the authorization expired.

(Amdt. 61-16, Eff. 5/10/65); (Amdt. 61-18, Eff. 9/26/65); (Amdt. 61-21, Eff. 6/26/66); (Amdt. 61-29, Eff. 4/23/67); (Amdt. 61-50, Eff. 9/11/70); (Amdt. 61-77, Eff. 1/6/87)

§ 61.23 Duration of medical certificates.

(a) A first-class medical certificate expires at the end of the last day of—

(1) The sixth month after the month of the date of examination shown on the certificate, for operations requiring an airline transport pilot certificate;

(2) The 12th month after the month of the date of examination shown on the certificate, for operations requiring only a commercial pilot certificate.

(3) The 24th month after the month of the date of examination shown on the certificate, for operations requiring only a private, recreational, or student pilot certificate.

(b) A second-class medical certificate expires at the end of the last day of—

(1) The 12th month after the month of the date of examination shown on the certificate, for operations requiring a commercial pilot certificate, or an air traffic control tower operator certificate; and

(2) The 24th month after the month of the date of examination shown on the certificate, for operations requiring only a private, recreational, or student pilot certificate.

(c) A third-class medical certificate expires at the end of the last day of the 24th month after the month of the date of examination shown on the certificate, for operations requiring a private, recreational, or student pilot certificate.

(Amdt. 61-14, Eff. 4/3/65); (Amdt. 61-68, Eff. 4/23/80); (Amdt. 61-68A, Eff. 5/7/80); (Amdt. 61-82, Eff. 8/31/89)

§ 61.25 Change of name.

An application for the change of a name on a certificate issued under this part must be accompanied by the applicant's current certificate and a copy of the marriage license, court order, or other document verifying the change. The documents are returned to the applicant after inspection.

(Amdt. 61-14, Eff. 4/3/65)

§ 61.27 Voluntary surrender or exchange of certificate.

The holder of a certificate issued under this part may voluntarily surrender it for cancellation, or for the issue of a certificate of lower grade, or another certificate with specific ratings deleted. If he so requests, he must include the following signed statement or its equivalent:

"This request is made for my own reasons, with full knowledge that my (insert name of certificate or rating, as appropriate) may not be reissued to me unless I again pass the tests prescribed for its issue."

(Amdt. 61-1, Eff. 11/1/62); (Amdt. 61-14, Eff. 4/3/65); (Amdt. 61-17, Eff. 5/13/65); (Amdt. 61-18, Eff. 9/26/65); (Amdt. 61-19, Eff. 9/1/65); (Amdt. 61-25, Eff. 11/19/66); (Amdt. 61-27, Eff. 4/15/67); (Amdt. 61-29, Eff. 4/23/67)

§ 61.29 Replacement of lost or destroyed certificate.

(a) An application for the replacement of a lost or destroyed airman certificate issued under this part is made by letter to the Department of Transportation, Federal Aviation Administration, Airman Certification Branch, P.O. Box 25082, Oklahoma City, Oklahoma 73125. The letter must—

(1) State the name of the person to whom the certificate was issued, the permanent mailing address (including zip code), social security number (if any), date and place of birth of the certificate holder, and any available information regarding the grade, number, and date of issue of the certificate, and the ratings on it; and

(2) Be accompanied by a check or money order for $2.00, payable to the Federal Aviation Administration.

(b) An application for the replacement of a lost or destroyed medical certificate is made by letter to the Department of Transportation, Federal Aviation Administration, Aeromedical Certification Branch, P.O. Box 25082, Oklahoma City, Oklahoma 73125, accompanied by a check or money order for $2.00.

(c) A person who has lost a certificate issued under this part, or a medical certificate issued under Part 67 of this chapter, or both, may obtain a telegram from the FAA confirming that it was issued. The telegram may be carried as a certificate for a period not to exceed 60 days pending his receipt of a duplicate certificate under paragraph (a) or (b) of this section, unless he has been notified that the certificate has been suspended or revoked. The request for such a telegram may be made by letter or prepaid telegram, including the date upon which a duplicate certificate was previously requested, if a request had been made, and a money order for the cost of the duplicate certificate. The request for a telegraphic certificate is sent to the office listed in paragraph (a) or (b) of this section, as appropriate. However, a request for both airman and medical certificates at the same time must be sent to the office prescribed in paragraph (a) of this section.

(Amdt. 61-23, Eff. 7/6/66); (Amdt. 61-25, Eff. 11/19/66)

§ 61.31 General limitations.

(a) *Type ratings required.* A person may not act as pilot in command of any of the following aircraft unless he holds a type rating for that aircraft:

(1) A large aircraft (except lighter-than-air).

(2) A helicopter, for operations requiring an airline transport pilot certificate.

(3) A turbojet powered airplane.

(4) Other aircraft specified by the Administrator through aircraft type certificate procedures.

(b) *Authorization in lieu of a type rating.*

(1) In lieu of a type rating required under paragraphs (a) (1), (3), and (4) of this section, an aircraft may be operated under an authorization issued by the Administrator, for a flight or series of flights within the United States, if—

(i) The particular operation for which the authorization is requested involves a ferry flight, a practice or training flight, a flight test for a pilot type rating, or a test flight of an aircraft, for a period that does not exceed 60 days;

(ii) The applicant shows that compliance with paragraph (a) of this section is impracticable for the particular operation; and

(iii) The Administrator finds that an equivalent level of safety may be achieved through operating limitations on the authorization.

(2) Aircraft operated under an authorization issued under this paragraph—

(i) May not be operated for compensation or hire; and

(ii) May carry only flight crewmembers necessary for the flight.

(3) An authorization issued under this paragraph may be reissued for an additional 60-day period for the same operation if the applicant shows that he was prevented from carrying out the purpose of the particular operation before his authorization expired.

The prohibition of paragraph (b)(2)(i) of this section does not prohibit compensation for the use of an aircraft by a pilot solely to prepare for or take a flight test for a type rating.

(c) *Category and class rating: carrying another person or operating for compensation or hire.* Unless he holds a category and class rating for that aircraft, a person may not act as pilot in command of an aircraft that is carrying another person or is operated for compensation or hire. In addition, he may not act as pilot in command of that aircraft for compensation or hire.

(d) *Category and class rating: other operations.* No person may act as pilot in command of an aircraft in solo flight in operations not subject to paragraph (c) of this section, unless he meets at least one of the following:

(1) He holds a category and class rating appropriate to that aircraft.

(2) He has received flight instruction in the pilot operations required by this part, appropriate to the category and class of aircraft for first solo, given to him by a certificated flight instructor who found him competent to solo that category and class of aircraft and has so endorsed his pilot logbook.

(3) He has soloed and logged pilot-in-command time in that category and class of aircraft before November 1, 1973.

(e) *High performance airplanes.* A person holding a private or commercial pilot certificate may not act as pilot in command of an airplane that has more than 200 horsepower, or that has a retractable landing gear, flaps, and a controllable propeller, unless he has received flight instruction from an authorized flight instructor who has certified in his logbook that he is competent to pilot an airplane that has more than 200 horsepower, or that has a retractable landing gear, flaps, and a controllable propeller, as the case may be. However, this instruction is not required if he has logged flight time as pilot in command in high performance airplanes before November 1, 1973.

[(f) *High altitude airplanes.*

[(1) Except as provided in paragraph (f)(2) of this section, no person may act as pilot in command of a pressurized airplane that has a service ceiling or maximum operating altitude, whichever is lower, above 25,000 feet MSL unless that person has completed the ground and flight training specified in paragraphs (f)(1)(i) and (ii) of this section and has received a logbook or training record endorsement from an authorized instructor certifying satisfactory completion of the training. The training shall consist of:

[(i) Ground training that includes instruction on high altitude aerodynamics and meteorology; respiration; effects, symptoms, and causes of hypoxia and any other high altitude sicknesses; duration of consciousness without supplemental oxygen; effects of prolonged usage of supplemental oxygen; causes and effects of gas expansion and gas bubble formations; preventive measures for eliminating gas expansion, gas bubble formations, and high altitude sicknesses; physical phenomena and incidents of decompression; and any other physiological aspects of high altitude flight; and

[(ii) Fight training in an airplane, or in a simulator that meets the requirements of § 121.407 of this chapter, and which is representative of an airplane as described in paragraph (f)(1) of this section. This training shall include normal cruise flight operations while operating above 25,000 feet MSL; the proper emergency procedures for simulated rapid decompression without actually depressurizing the airplane; and emergency descent procedures;

[(2) The training required in paragraph (f)(1) of this section is not required if a person can document accomplishment of any of the following in an airplane, or in a simulator that meets the requirements of § 121.407 of this section, and that is representative of an airplane described in paragraph (f)(1) of this section:

[(i) Served as pilot in command prior to April 15, 1991;

[(ii) Completed a pilot proficiency check for a pilot certificate or rating conducted by the FAA prior to April 15, 1991;

[(iii) Completed an official pilot-in-command check by the military services of the United States; or

[(iv) Completed a pilot-in-command proficiency check under Parts 121, 125, or 135 conducted by the FAA or by an approved pilot check airman.

[(g) *Tailwheel Airplanes.* No person may act as pilot in command of a tailwheel airplane unless that pilot has received flight instruction from an authorized flight instructor who has found the pilot competent to operate a tailwheel airplane and has made a one time endorsement so stating in the pilot's logbook. The endorsement must certify that the pilot is competent in normal and crosswind takeoffs and landings, wheel landings unless the manufacturer has recommended against such landings, and go-around procedures. This endorsement is not required if a pilot has logged flight time as pilot in command of tailwheel airplanes prior to April 15, 1991.]

[(h)] *Exception.* This section does not require a class rating for gliders, or category and class ratings for aircraft that are not type certificated as airplanes, rotorcraft, or lighter-than-air aircraft. In addition, the rating limitations of this section do not apply to—

(1) The holder of a student pilot certificate;

(2) The holder of a recreational pilot certificate when operating under the provisions of § 61.101 (f), (g), and (h).

(3) The holder of a pilot certificate when operating an aircraft under the authority of an experimental or provisional type certificate;

(4) An applicant when taking a flight test given by the Administrator; or

(5) The holder of a pilot certificate with a lighter-than-air category rating when operating a hot air balloon without an airborne heater.

(Amdt. 61-1, Eff. 11/1/62); (Amdt. 61-5, Eff. 10/22/63); (Amdt. 61-14, Eff. 4/3/65); (Amdt. 61-24, Eff. 10/17/66); (Amdt. 61-35, Eff. 7/20/67); (Amdt. 61-36, Eff. 11/5/67); (Amdt. 61-44, Eff. 11/22/69); (Amdt. 61-82, Eff. 8/31/89); [(Amdt. 61-90, Eff. 4/15/91)]

§ 61.33 Tests: General procedure.

Tests prescribed by or under this part are given at times and places, and by persons, designated by the Administrator.

(Amdt. 61-1, Eff. 11/1/62); (Amdt. 61-23, Eff. 7/6/66); (Amdt. 61-33, Eff. 6/15/67)

§ 61.35 Written test: Prerequisites and passing grades.

(a) An applicant for a written test must—

(1) Show that he has satisfactorily completed the ground instruction or home study course required by this part for the certificate or rating sought;

(2) Present as personal identification an airman certificate, driver's license, or other official document; and

(3) Present a birth certificate or other official document showing that he meets the age requirement prescribed in this part for the certificate sought not later than 2 years from the date of application for the test.

(b) The minimum passing grade is specified by the Administrator on each written test sheet or booklet furnished to the applicant.

This section does not apply to the written test for an airline transport pilot certificate or a rating associated with that certificate.

(Amdt. 61-9, Eff. 7/27/64); (Amdt. 61-24, Eff. 10/17/66); (Amdt. 61-44, Eff. 11/22/69)

§ 61.37 Written tests: Cheating or other unauthorized conduct.

(a) Except as authorized by the Administrator, no person may—

(1) Copy, or intentionally remove, a written test under this part;

(2) Give to another, or receive from another, any part or copy of that test;

(3) Give help on that test to, or receive help on that test from, any person during the period that test is being given;

(4) Take any part of that test in behalf of another person;

(5) Use any material or aid during the period that test is being given; or

(6) Intentionally cause, assist, or participate in any act prohibited by this paragraph.

(b) No person whom the Administrator finds to have committed an act prohibited by paragraph (a) of this section is eligible for any airman or ground instructor certificate or rating, or to take any test therefor, under this chapter for a period of one year after the date of that act. In addition, the commission of that act is a basis for suspending or revoking any airman or ground instructor certificate or rating held by that person.

(Amdt. 61-24, Eff. 10/17/66); (Amdt. 61-28, Eff. 5/17/67); (Amdt. 61-32, Eff. 8/7/67)

§ 61.39 Prerequisites for flight tests.

(a) To be eligible for a flight test for a certificate, or an aircraft or instrument rating issued under this part, the applicant must—

(1) Have passed any required written test since the beginning of the 24th month before the month in which he takes the flight test;

(2) Have the applicable instruction and aeronautical experience prescribed in this part;

(3) Hold a current medical certificate appropriate to the certificate he seeks or, in the case of a rating to be added to his pilot certificate, at least a third-class medical certificate issued since the beginning of the 24th month before the month in which he takes the flight test;

(4) Except for a flight test for an airline transport pilot certificate, meet the age requirement for the issuance of the certificate or rating he seeks; and

(5) Have a written statement from an appropriately certificated flight instructor certifying that he has given the applicant flight instruction in preparation for the flight test within 60 days preceding the date of application, and finds him competent to pass the test and to have satisfactory knowledge of the subject areas in which he is shown to be deficient by his FAA airman written test report. However, an applicant need not have this written statement if he—

(i) Holds a foreign pilot license issued by a contracting State to the Convention on International Civil Aviation that authorizes at least the pilot privileges of the airman certificate sought by him;

(ii) Is applying for a type rating only, or a class rating with an associated type rating; or

(iii) Is applying for an airline transport pilot certificate or an additional aircraft rating on that certificate.

(b) Notwithstanding the provisions of paragraph (a)(1) of this section, an applicant for an airline transport pilot certificate or rating may take the flight test for that certificate or rating if—

(1) The applicant—

(i) Within the period ending 24 calendar months after the month in which the applicant passed the first of any required written tests, was employed as a flight crewmember by a U.S. air carrier or commercial operator operating either under Part 121 or as a commuter air carrier under Part 135 (as defined in Part 298 of this title) and is employed by such a certificate holder at the time of the flight test;

(ii) Has completed initial training, and, if appropriate, transition or upgrade training; and

(iii) Meets the recurrent training requirements of the applicable part; or

(2) Within the period ending 24 calendar months after the month in which the applicant passed the first of any required written tests, the applicant participated as a pilot in a pilot training program of a U.S. scheduled military air transportation service and is currently participating in that program.

(Amdt. 61-14, Eff. 4/3/65); (Amdt. 61-18, Eff. 9/26/65); (Amdt. 61-20, Eff. 12/16/65); (Amdt. 61-25, Eff. 11/19/66); (Amdt. 61-40, Eff. 5/10/68); (Amdt. 61-44, Eff. 11/22/69); (Amdt. 61-71, Eff. 4/28/82)

§ 61.41 Flight instruction received from flight instructors not certificated by FAA.

Flight instruction may be credited toward the requirements for a pilot certificate or rating issued under this part if it is received from—

(a) An Armed Force of either the United States or a foreign contracting State to the Convention on International Civil Aviation in a program for training military pilots; or

(b) A flight instructor who is authorized to give that flight instruction by the licensing authority of a foreign contracting State to the Convention on International Civil Aviation and the flight instruction is given outside the United States.

(Amdt. 61-1, Eff. 11/1/62)

§ 61.43 Flight tests: General procedures.

(a) The ability of an applicant for a private or commercial pilot certificate, or for an aircraft or instrument rating on that certificate to perform the required pilot operations is based on the following:

(1) Executing procedures and maneuvers within the aircraft's performance capabilities and limitations, including use of the aircraft's systems.

(2) Executing emergency procedures and maneuvers appropriate to the aircraft.

(3) Piloting the aircraft with smoothness and accuracy.

(4) Exercising judgment.

(5) Applying his aeronautical knowledge.

(6) Showing that he is the master of the aircraft, with the successful outcome of a procedure or maneuver never seriously in doubt.

(b) If the applicant fails any of the required pilot operations in accordance with the applicable provisions of paragraph (a) of this section, the applicant fails the flight test. The applicant is not eligible for the certificate or rating sought until he passes any pilot operations he has failed.

(c) The examiner or the applicant may discontinue the test at any time when the failure of a required pilot operation makes the applicant ineligible for the certificate or rating sought. If the test is discontinued the applicant is entitled to credit for only those entire pilot operations that he has successfully performed.

(Amdt. 61-1, Eff. 11/1/62); (Amdt. 61-14, Eff. 4/3/65)

§ 61.45 Flight tests: Required aircraft and equipment.

(a) *General.* An applicant for a certificate or rating under this part must furnish, for each flight test that he is required to take, an appropriate aircraft of United States registry that has a current standard or limited airworthiness certificate. However, the applicant may, at the discretion of the inspector or examiner conducting the test, furnish an aircraft of U.S. registry that has a current airworthiness certificate other than standard or limited, an aircraft of foreign registry that is properly certificated by the country of registry, or a military aircraft in an operational status if its use is allowed by an appropriate military authority.

(b) *Required equipment (other than controls).* Aircraft furnished for a flight test must have—

(1) The equipment for each pilot operation required for the flight test;

(2) No prescribed operating limitations that prohibit its use in any pilot operation required on the test;

(3) Pilot seats with adequate visibility for each pilot to operate the aircraft safely, except as provided in paragraph (d) of this section; and

(4) Cockpit and outside visibility adequate to evaluate the performance of the applicant, where an additional jump seat is provided for the examiner.

(c) *Required controls.* An aircraft (other than lighter-than-air) furnished under paragraph (a) of this section for any pilot flight test must have engine power controls and flight controls that are easily reached and operable in a normal manner by both pilots, unless after considering all the factors, the examiner determines that the flight test can be conducted safely without them. However, an aircraft having other controls such as nosewheel steering, brakes, switches, fuel selectors, and engine air flow controls that are not easily reached and operable in a normal manner by both pilots may be used, if more than one pilot is required under its airworthiness certificate, or if the examiner determines that the flight can be conducted safely.

(d) *Simulated instrument flight equipment.* An applicant for any flight test involving flight maneuvers solely by reference to instruments must furnish equipment satisfactory to the examiner that excludes the visual reference of the applicant outside of the aircraft.

(e) *Aircraft with single controls.* At the discretion of the examiner, an aircraft furnished under paragraph (a) of this section for a flight test may, in the cases listed herein, have a single set of controls. In such case, the examiner determines the competence of the applicant by observation from the ground or from another aircraft.

(1) A flight test for addition of a class or type rating, not involving demonstration of instrument skills, to a private or commercial pilot certificate.

(2) A flight test in a single-place gyroplane for—

(i) A private pilot certificate with a rotorcraft category rating and gyroplane class rating, in which case the certificate bears the limitation "rotorcraft single-place gyroplane only;" or

(ii) Addition of a rotorcraft category rating and gyroplane class rating to a pilot certificate, in which case a certificate higher than a private pilot certificate bears the limitation "rotorcraft single-place gyroplane, private pilot privileges, only."

The limitations prescribed by this subparagraph may be removed if the holder of the certificate passes the appropriate flight test in a gyroplane with two pilot stations or otherwise passes the appropriate flight test for a rotorcraft category rating.

(Amdt. 61-4, Eff. 9/16/63)

§ 61.47 Flight tests: Status of FAA i nspectors and other authorized flight examiners.

An FAA inspector or other authorized flight examiner conducts the flight test of an applicant for a pilot certificate or rating for the purpose of observing the applicant's ability to perform satisfactorily the procedures and maneuvers on the flight test. The inspector or other examiner is not pilot in command of the aircraft during the flight test unless he acts in that capacity for the flight, or portion of the flight, by prior arrangement with the applicant or other person who would otherwise act as pilot in command of the flight, or portion of the flight. Notwithstanding the type of aircraft used during a flight test, the applicant and the inspector or other examiner are not, with respect to each other (or other occupants authorized by the inspector or other examiner), subject to the requirements or limitations for the carriage of passengers specified in this chapter.

(Amdt. 61-1, Eff. 11/1/62), (Amdt. 61-6, Eff. 9/7/64); (Amdt. 61-20, Eff. 12/16/65); (Amdt. 61-24, Eff.

10/17/66); (Amdt. 61-32, Eff. 8/7/67); (Amdt. 61-48, Eff. 5/7/70)

§ 61.49 Retesting after failure.

[(a) An applicant for a written or practical test who fails that test may not apply for retesting until 30 days after the date the test was failed. However, in the case of a first failure, the applicant may apply for retesting before the 30 days have expired provided the applicant presents a logbook or training record endorsement from an authorized instructor who has given the applicant remedial instruction and finds the applicant competent to pass the test.

[(b) An applicant for a flight instructor certificate with an airplane category rating, or for a flight instructor certificate with a glider category rating, who has failed the practical test due to deficiencies of knowledge or skill relating to stall awareness, spin entry, spins, or spin recovery techniques must, during the retest, satisfactorily demonstrate both knowledge and skill in these areas in an aircraft of the appropriate category that is certificated for spins.]

(Amdt. 61-14, Eff. 4/3/65); (Amdt. 61-25, Eff. 11/19/66); [(Amdt. 61-90, Eff. 4/15/91)]

§ 61.51 Pilot logbooks.

(a) The aeronautical training and experience used to meet the requirements for a certificate or rating, or the recent flight experience requirements of this part must be shown by a reliable record. The logging of other flight time is not required.

(b) *Logbook entries.* Each pilot shall enter the following information for each flight or lesson logged:

(1) *General.*

(i) Date.

(ii) Total time of flight.

(iii) Place, or points of departure and arrival.

(iv) Type and identification of aircraft.

(2) *Type of pilot experience or training.*

(i) Pilot in command or solo.

(ii) Second in command.

(iii) Flight instruction received from an authorized flight instructor.

(iv) Instrument flight instruction from an authorized flight instructor.

(v) Pilot ground trainer instruction.

(vi) Participating crew (lighter-than-air).

(vii) Other pilot time.

(3) *Conditions of flight.*

(i) Day or night.

(ii) Actual instrument.

(iii) Simulated instrument conditions.

(c) *Logging of pilot time.*

(1) *Solo flight time.* A pilot may log as solo flight time only that flight time when he is the sole occupant of the aircraft. However, a student pilot may also log as solo flight time that time during which he acts as the pilot in command of an airship requiring more than one flight crewmember.

(2) *Pilot-in-command flight time.*

(i) A recreational, private, or commercial pilot may log pilot-in-command time only that flight time during which that pilot is the sole manipulator of the controls of an aircraft for which the pilot is rated, or when the pilot is the sole occupant of the aircraft, or, except for a recreational pilot, when acting as pilot in command of an aircraft on which more than one pilot is required under the type certification of the aircraft or the regulations under which the flight is conducted.

(ii) An airline transport pilot may log as pilot in command time all of the flight time during which he acts as pilot-in-command.

(iii) A certificated flight instructor may log as pilot-in-command time all flight time during which he acts as a flight instructor.

(3) *Second-in-command flight time.* A pilot may log as second in command time all flight time during which he acts as second-in-command of an aircraft on which more than one pilot is required under the type certification of the aircraft, or the regulations under which the flight is conducted.

(4) *Instrument flight time.* A pilot may log as instrument flight time only that time during which he operates the aircraft solely by reference to instruments, under actual or simulated instrument flight conditions. Each entry must include the place and type of each instrument approach completed, and the name of the safety pilot for each simulated instrument flight. An instrument flight instructor may log as instrument time that time during which he acts as instrument flight instructor in actual instrument weather conditions.

(5) *Instruction time.* All time logged as flight instruction, instrument flight instruction, pilot ground trainer instruction, or ground instruction time must be certified by the appropriately rated and certificated instructor from whom it was received.

(d) *Presentation of logbook.*

(1) A pilot must present his logbook (or other record required by this section) for inspection upon reasonable request by the Administrator, an authorized representative of the National Transportation Safety Board, or any State or local law enforcement officer.

(2) A student pilot must carry his logbook (or other record required by this section) with him on all solo cross-country flights, as evidence of the required instructor clearances and endorsements.

(3) A recreational pilot must carry his or her logbook that has the required instructor endorsement on all solo flights—

(i) In excess of 50 nautical miles from an airport at which instruction as received;

(ii) In airspace in which communication with air traffic control is required;

(iii) Between sunset and sunrises; and

(iv) In an aircraft for which the pilot is not rated.

(Amdt. 61-51, Eff. 9/4/70); (Amdt. 61-82, Eff. 8/31/89)

§ 61.53 Operations during medical deficiency.

No person may act as pilot in command, or in any other capacity as a required pilot flight crewmember while he has a known medical deficiency, or increase of a known medical deficiency, that would make him unable to meet the requirements for his current medical certificate.

§ 61.55 Second-in-command qualifications.

(a) Except as provided in paragraph (d) of this section, no person may serve as second in command of an aircraft type certificated for more than one required pilot flight crewmember unless that person holds—

(1) At least a current private pilot certificate with appropriate category and class ratings; and

(2) An appropriate instrument rating in the case of flight under IFR.

(b) Except as provided in paragraph (d) of this section, no person may serve as second in command of an aircraft type certificated for more than one required pilot flight crewmember unless, since the beginning of the 12th calendar month before the month in which the pilot serves, the pilot has, with respect to that type of aircraft—

(1) Become familiar with all information concerning the aircraft's powerplant, major components and systems, major appliances, performance and limitations, standard and emergency operating procedures, and the contents of the approved aircraft flight manual or approved flight manual material, placards, and markings.

(2) Except as provided in paragraph (e) of this section, performed and logged—

(i) Three takeoffs and three landings to a full stop in the aircraft as the sole manipulator of the flight controls; and

(ii) Engine-out procedures and maneuvering with an engine out while executing the duties of a pilot in command. For airplanes, this requirement may be satisfied in a simulator acceptable to the Administrator.

For the purpose of meeting the requirements of paragraph (b)(2) of this section, a person may act as second in command of a flight under day VFR or day IFR, if no persons or property, other than as necessary for the operation, are carried.

(c) If a pilot complies with the requirements in paragraph (5) of this section in the calendar month before, or the calendar month after, the month in which compliance with those requirements is due, he is considered to have complied with them in the month they are due.

(d) This section does not apply to a pilot who—

(1) Meets the pilot in command proficiency check requirements of Part 121, 125, 127, or 135 of this chapter;

(2) Is designated as the second in command of an aircraft operated under the provisions of Part 121, 125, 127, or 135 of this chapter; or

(3) Is designated as the second in command of an aircraft for the purpose of receiving flight training required by this section and no passengers or cargo are carried on that aircraft.

(e) The holder of a commercial or airline transport pilot certificate with appropriate category and class ratings need not meet the requirements of paragraph (b)(2) of this section for the conduct of ferry flights, aircraft flight tests, or airborne equipment evaluation, if no persons or property other than as necessary for the operation are carried.

(Amdt. 61-64, Eff. 12/22/76); (Amdt. 61-65, Eff. 5/9/77); (Amdt. 61-77, Eff. 1/6/87)

§ 61.56 Flight review.

(a) [A flight review consists of a minimum of 1 hour of flight instruction and 1 hour of ground instruction. The review must include—

[(1) A review of the current general operating and flight rules of Part 91 of this chapter; and

[(2) A review of those maneuvers and procedures which, at the discretion of the person giving the review, are necessary for the pilot to demonstrate the safe exercise of the privileges of the pilot certificate.

(b) [Glider pilots may substitute a minimum of three instructional flights in a glider, each of which includes a 360-degree turn, in lieu of the 1 hour of flight instruction required in paragraph (a) of this section.

(c) [Except as provided in paragraphs (d) and (e) of this section, no person may act as pilot in command of an aircraft unless, since the beginning of the 24th calendar month before the month in which that pilot acts as pilot in command, that person has—

[(1) Accomplished a flight review given in an aircraft for which that pilot is rated by an appropriately rated instructor certificated under this part or other person designated by the Administrator; and

[(2) A logbook endorsed by the person who gave the review certifying that the person has satisfactorily completed the review.

(d) [A person who has, within the period specified in paragraph (c) of this section, satisfactorily completed a pilot proficiency check conducted by the FAA, an approved pilot check airman, or a U.S. Armed Force, for a pilot certificate, rating, or operating privilege, need not accomplish the flight review required by this section.

(e) [A person who has, within the period specified in paragraph (c) of this section, satisfactorily completed one or more phases of an FAA-sponsored pilot proficiency award program need not accomplish the flight review required by this section.

(f) [A person who holds a current flight instructor certificate who has, within the period specified in paragraph (c) of this section, satisfactorily completed a renewal of a flight instructor certificate under the provisions on § 61.197 (c), need not accomplish the 1 hour of ground instruction specified in subparagraph (a)(1) of this section.

(g) [The requirements of this section may be accomplished in combination with the requirements of § 61.57 and other applicable recency requirements at the discretion of the instructor.]

Docket No. 24695 (54 FR 13037) Eff. 3/29/89; (Amdt. 61-82, Eff. 8/31/89); (Amdt. 61-86, Eff. 10/5/89); (Amdt. 61-89, Eff. 12/5/90); (Amdt. 61-90, Eff. 4/15/91); (Amdt. 61-91, Eff. 9/5/91); [(Amdt. 61-93, Eff. 8/31/93)]

§ 61.57 Recent flight experience: Pilot in command.

(a) (Reserved)

(b) (Reserved)

(c) *General experience.* No person may act as pilot in command of an aircraft carrying passengers, nor of an aircraft certificated for more than one required pilot flight crewmember, unless within the preceding 90 days, he has made three takeoffs and three landings as the sole manipulator of the flight controls in an aircraft of the same category and class and, if a type rating is required, of the same type. If the aircraft is a tailwheel airplane, the landings must have been made to a full stop in a tailwheel airplane. For the purpose of meeting the requirements of the paragraph, a person may act as pilot in command of a flight under day VFR or day IFR if no persons or property other than as necessary for his compliance thereunder, are carried. This paragraph does not apply to operations requiring an airline transport pilot certificate, or to operations conducted under Part 135 of this chapter.

(d) *Night experience.* No person may act as pilot in command of an aircraft carrying passengers during the period beginning 1 hour after sunset and ending 1 hour before sunrise (as published in the American Air Almanac) unless, within the preceding 90 days, he has made at least three takeoffs and three landings to a full stop during that period in the category and class of aircraft to be used. This paragraph does not apply to operations requiring an airline transport pilot certificate.

(e) *Instrument.*

(1) *Recent IFR experience.* No pilot may act as pilot in command under IFR, nor in weather conditions less than the minimums prescribed for VFR, unless he has, within the past 6 calender months—

(i) In the case of an aircraft other than a glider, logged at least 6 hours of instrument time under actual or simulated IFR conditions, at least 3 of which were in flight in the category of aircraft involved, including at least 6 instrument approaches, or passed an instrument competency check in the category of aircraft involved.

(ii) In the case of a glider, logged at least 3 hours of instrument time, at least half of which were in a glider or an airplane. If a passenger is carried in the glider, at least 3 hours of instrument flight time must have been in gliders.

(2) *Instrument competency check.* A pilot who does not meet the recent instrument experience requirements of paragraph (e)(1) of this section during the prescribed time or 6 calendar months thereafter may not serve as pilot in command under IFR, nor in weather conditions less than the minimums prescribed for VFR, until he passes an instrument competency check in the category of aircraft involved, given by an FAA inspector, a member of an Armed Force in the United States authorized to conduct flight tests, an FAA-approved check pilot, or a certificated instrument flight instructor. The Administrator may authorize the conduct of part or all of this check in a pilot ground trainer equipped for instruments or an aircraft simulator.

(Amdt. 61-77, Eff. 1/6/87); (Amdt. 61-82, Eff. 8/31/89)

§ 61.58 Pilot-in-command proficiency check: Operation of aircraft requiring more than one required pilot.

(a) [Except as provided in paragraph (e) of this section, no person may act as pilot in command of an aircraft that is type certificated for more than one required pilot crewmember unless the proficiency checks or flight checks prescribed in paragraphs (5) and (c) of this section are satisfactorily completed.]

(b) Since the beginning of the 12th calendar month before the month in which a person acts as pilot in command of an aircraft that is type certificated for more than one required pilot crewmember he must have completed one of the following:

(1) For an airplane—a proficiency or flight check in either an airplane that is type certificated for more than one required pilot crewmember, or in an approved simulator or other training device, given to him by an FAA inspector or designated pilot examiner and consisting of those maneuvers and procedures set forth in Appendix F of Part 121 of this chapter which may be performed in a simulator or training device.

(2) For other aircraft—a proficiency or flight check in an aircraft that is type certificated for more than one required pilot crewmember given to him by an FAA inspector or designated pilot examiner which includes those maneuvers and procedures required for the original issuance of a type rating for the aircraft used in the check.

(3) A pilot in command proficiency check given to him in accordance with the provisions for that check under Part 121, 123 or 135 of this chapter. However, in the case of a person acting as pilot in command of a helicopter he may complete a proficiency check given to him in accordance with Part 127 of this chapter.

(4) A flight test required for an aircraft type rating.

(5) An initial or periodic flight check for the purpose of the issuance of a pilot examiner or check airman designation.

(6) A military proficiency check required for pilot in command and instrument privileges in an aircraft which the military requires to be operated by more than one pilot.

(c) Except as provided in paragraph (d) of this section, since the beginning of the 24th calendar month before the month in which a person acts as pilot in command of an aircraft that is type certificated for more than one required pilot crewmember he must have completed one of the following proficiency or flight checks in the particular type aircraft in which he is to serve as pilot in command:

(1) A proficiency check or flight check given to him by an FAA inspector or a designated pilot examiner which includes the maneuvers, procedures, and standards required for the original issuance of a type rating for the aircraft used in the check.

(2) A pilot-in-command proficiency check given to him in accordance with the provisions for that check under Part 121, 123 or 135 of this chapter. However, in the case of a person acting as pilot in command of a helicopter he may complete a proficiency check given to him in accordance with Part 127 of this chapter.

(3) A flight test required for an aircraft type rating.

(4) An initial or periodic flight check for the purpose of the issuance of a pilot examiner or check airman designation.

(5) A military proficiency check required for pilot in command and instrument privileges in an aircraft which the military requires to be operated by more than one pilot.

(d) For airplanes, the maneuvers and procedures required for the checks and test prescribed in paragraphs (c)(1), (2), (4), and (5) of this section, and paragraph (c)(3) of this section in the case of type ratings obtained in conjunction with a Part 121 of this chapter training program may be performed in a simulator or training device if—

(1) The maneuver or procedure can be performed in a simulator or training device as set forth in Appendix F to Part 121 of this chapter; and

(2) The simulator or training device is one that is approved for the particular maneuver or procedure.

(e) This section does not apply to persons conducting operations subject to Parts 121, 127, 133, 135, and 137 of this chapter or to persons maintaining continuing qualification under an Advanced Qualification Program approved under SFAR 58.

(f) For the purpose of meeting the proficiency check requirements of paragraphs (b) and (c) of this section, a person may act as pilot in command of a flight under day VFR or day IFR if no persons or property, other than as necessary for his compliance thereunder, are carried.

(g) If a pilot takes the proficiency check required by paragraph (a) of this section in the calendar month before, or the calendar month after, the month in which it is due, he is considered to have taken it in the month it is due.

(Amdt. 61-88, Eff. 10/2/90); [(Amdt. 61-90, Eff. 4/15/91)]

§ 61.59 Falsification, reproduction or alteration of applications, certificates, logbooks, reports, or records.

(a) No person may make or cause to be made—

(1) Any fraudulent or intentionally false statement on any application for a certificate, rating, or duplicate thereof, issued under this part;

(2) Any fraudulent or intentionally false entry in any logbook, record, or report that is required to be kept, made, or used, to show compliance with any requirement for the issuance, or exercise of the privileges, or any certificate or rating under this part;

(3) Any reproduction, for fraudulent purpose, of any certificate or rating under this part; or

(4) Any alteration of any certificate or rating under this part.

(b) The commission by any person of an act prohibited under paragraph (a) of this section is a basis for suspending or revoking any airman or ground instructor certificate or rating held by that person.

(Amdt. 61-82, Eff. 8/31/89)

§ 61.60 Change of address.

The holder of a pilot or flight instructor certificate who has made a change in his permanent mailing address may not after 30 days from the date he moved, exercise the privileges of his certificate unless he has notified in writing the Department of Transportation, Federal Aviation Administration, Airman Certification Branch, Box 25082, Oklahoma City, OK 73125, of his new address.

Subpart B—Aircraft ratings and special certificates

§ 61.61 Applicability.

This subpart prescribes the requirements for the issuance of additional aircraft ratings after a pilot or instructor certificate is issued, and the requirements and limitations for special pilot certificates and ratings issued by the Administrator.

(Amdt. 61-26, Eff. 5/1/67)

§ 61.63 Additional aircraft ratings (other than airline transport pilot).

(a) *General.* To be eligible for an aircraft rating after his certificate is issued to him an applicant must meet the requirements of paragraphs (5) through (d) of this section, as appropriate to the rating sought.

(b) *Category rating.* An applicant for a category rating

to be added on his pilot certificate must meet the requirements of this part for the issue of the pilot certificate appropriate to the privileges for which the category rating is sought. However, the holder of a category rating for powered aircraft is not required to take a written test for the addition of a category rating on his pilot certificate.

(c) *Class rating.* An applicant for an aircraft class rating to be added on his pilot certificate must—

(1) Present a logbook record certified by an authorized flight instructor showing that the applicant has received flight instruction in the class of aircraft for which a rating is sought and has been found competent in the pilot operations appropriate to the pilot certificate to which his category rating applies; and

(2) Pass a flight test appropriate to his pilot certificate and applicable to the aircraft category and class rating sought.

A person who holds a lighter-than-air category rating with a free balloon class rating, who seeks an airship class rating, must meet the requirements of paragraph (b) of this section as though seeking a lighter-than-air category rating.

(d) *Type rating.* An applicant for a type rating to be added on his pilot certificate must meet the following requirements:

(1) He must hold, or concurrently obtain, an instrument rating appropriate to the aircraft for which a type rating is sought.

(2) He must pass a flight test showing competence in pilot operations appropriate to the pilot certificate he holds and to the type rating sought.

(3) He must pass a flight test showing competence in pilot operations under instrument flight rules in an aircraft of the type for which the type rating is sought or, in the case of a single pilot station airplane, meet the requirements of paragraph (d)(3)(i) or (ii) of this section, whichever is applicable.

(i) [The applicant must have met the requirements of this paragraph in a multiengine airplane for which a type rating is required.]

(ii) If he does not meet the requirements of paragraph (d)(3)(i) of this section and he seeks a type rating for a single engine airplane, he must meet the requirements of this subparagraph in either a single or multiengine airplane, and have the recent instrument experience set forth in § 61.57(e), when he applies for the flight test under paragraph (d)(2) of this section.

(4) An applicant who does not meet the requirements of paragraphs (d)(1) and (3) of this section may obtain a type rating limited to "VFR only." Upon meeting these instrument requirements or the requirements of § 61.73(e)(2), the "VFR only" limitation may be removed for the particular type of aircraft in which competence is shown.

(5) When an instrument rating is issued to the holder of one or more type ratings, the type ratings on the amended certificate bear the limitation described in paragraph (d)(4) of this section for each airplane type rating for which he has not shown his instrument competency under this paragraph.

[(6) On and after April 15, 1991, an applicant for a type rating to be added to a pilot certificate must—

[(i) Have completed ground and flight training on the maneuvers and procedures of Appendix A of this part that is appropriate to the airplane for which a type rating is sought, and received an endorsement from an authorized instructor in the person's logbook or training records certifying satisfactory completion of the training; or

[(ii) For a pilot employee of a Part 121 or Part 135 certificate holder, have completed the certificate holder's approved ground and flight training that is appropriate to the airplane for which a type rating is sought.]

(Amdt. 61-2, Eff. 5/2/63); (Amdt. 61-10, Eff. 6/12/64); (Amdt. 61-12, Eff. 2/26/65); (Amdt. 61-18, Eff. 9/26/65); (Amdt. 61-23, Eff. 7/6/66); (Amdt. 61-44, Eff. 11/22/69); [(Amdt. 61-90, Eff. 4/15/91)]

§ 61.65 Instrument rating requirements.

(a) *General.* To be eligible for an instrument rating (airplane) or an instrument rating (helicopter), an applicant must—

(1) Hold at least a current private pilot certificate with an aircraft rating appropriate to the instrument rating sought;

(2) Be able to read, speak, and understand the English language; and

(3) Comply with the applicable requirements of this section.

(b) *Ground instruction.* An applicant for the written test for an instrument rating must have received ground instruction, or have logged home study in at least the following areas of aeronautical knowledge appropriate to the rating sought.

(1) The regulations of this chapter that apply to flight under IFR conditions, the Airman's Information Manual, and the IFR air traffic system and procedures;

(2) Dead reckoning appropriate to IFR navigation, IFR navigation by radio aids using the VOR, ADF, and ILS systems, and the use of IFR charts and instrument approach plates;

(3) The procurement and use of aviation weather reports and forecasts, and the elements of forecasting weather trends on the basis of that information and personal observation of weather conditions; and

(4) The safe and efficient operation of airplanes or helicopters, as appropriate, under instrument weather conditions.

(c) *Flight instruction and skill—airplanes.* An applicant for the flight test for an instrument rating (airplane) must present a logbook record certified by an authorized flight instructor showing that he has received instrument flight instruction in an airplane in the following pilot operations, and has been found competent in each of them:

(1) Control and accurate maneuvering of an airplane solely by reference to instruments.

(2) IFR navigation by the use of the VOR and ADF systems, including compliance with air traffic control instructions and procedures.

(3) Instrument approaches to published minimums using the VOR, ADF, and ILS system (instruction in the use of the ADF and ILS may be received in an instrument ground trainer and instruction in the use of the ILS glide slope may be received in an airborne ILS simulator).

(4) Cross-country flying in simulated or actual IFR conditions, on Federal airways or as routed by ATC, including one such trip of at least 250 nautical miles, including VOR, ADF, and ILS approaches at different airports.

(5) Simulated emergencies, including the recovery from unusual attitudes, equipment or instrument malfunctions, loss of communications, and engine-out emergencies if a multiengine airplane is used, and missed approach procedure.

(d) *Instrument instruction and skill—(helicopter).* An applicant for the flight test for an instrument rating (helicopter) must present a logbook record certified to by an authorized flight instructor showing that he has received instrument flight instruction in a helicopter in the following pilot operations, and has been found competent in each of them:

(1) The control and accurate maneuvering of a helicopter solely by reference to instruments.

(2) IFR navigation by the use of the VOR and ADF systems, including compliance with air traffic control instructions and procedures.

(3) Instrument approaches to published minimums using the VOR, ADF, and ILS systems (instruction in the use of the ADF and ILS may be received in an instrument ground trainer, and instruction in the use of the ILS glide slope may be received in an airborne ILS simulator).

(4) Cross-country flying under simulated or actual IFR conditions, on Federal airways or as routed by ATC, including one flight of at least 100 nautical miles, including VOR, ADF, and ILS approaches at different airports.

(5) Simulated IFR emergencies, including equipment malfunctions, missed approach procedures, and deviations to unplanned alternates.

(e) *Flight experience.* An applicant for an instrument rating must have at least the following flight time as a pilot:

(1) A total of 125 hours of pilot flight time, of which 50 hours are as pilot in command in cross-country flight in a powered aircraft with other than a student pilot certificate. Each cross-country flight must have a landing at a point more than 50 nautical miles from the original departure point.

(2) 40 hours of simulated or actual instrument time, of which not more than 20 hours may be instrument instruction by an authorized instructor in an instrument ground trainer acceptable to the Administrator.

(3) 15 hours of instrument flight instruction by an authorized flight instructor, including at least 5 hours in an airplane or a helicopter, as appropriate.

(f) *Written test.* An applicant for an instrument rating must pass a written test appropriate to the instrument rating sought on the subjects in which ground instruction is required by paragraph (b) of this section.

(g) *Practical test.* An applicant for an instrument rating must pass a flight test in an airplane or a helicopter, as appropriate. The test must include instrument flight procedures selected by the inspector or examiner conducting the test to determine the applicant's ability to perform competently the IFR operations on which instruction is required by paragraph (c) or (d) of this section.

(Amdt. 61-12, Eff. 2/26/65); (Amdt. 61-18, Eff. 9/26/65); (Amdt. 61-70, Eff. 1/25/82); (Amdt. 61-75, Eff. 6/7/85)

§ 61.67 Category II pilot authorization requirements.

(a) *General.* An applicant for a Category II pilot authorization must hold—

(1) A pilot certificate with an instrument rating or an aircraft transport pilot certificate; and

(2) A type rating for the aircraft type if the authorization is requested for a large aircraft or a small turbojet aircraft.

(b) *Experience requirements.* Except for the holder of an airline transport pilot certificate, an applicant for a Category II authorization must have at least—

(1) 50 hours of night flight time under VFR conditions as pilot in command;

(2) 75 hours of instrument time under actual or simulated conditions that may include 25 hours in a synthetic trainer; and

(3) 250 hours of cross-country flight time as pilot in command.

Night flight and instrument flight time used to meet the requirements of paragraphs (b)(1) and (2) of this section may also be used to meet the requirements of paragraph (b)(3) of this section.

(c) *Practical test required.*

(1) The practical test must be passed by—

(i) An applicant for issue or renewal of an authorization; and

(ii) An applicant for the addition of another type aircraft to his authorization.

(2) To be eligible for the practical test an applicant must meet the requirements of paragraph (a) of this section and, if he has not passed a practical test since the beginning of the twelfth month before the test, he must meet the following recent experience requirements:

(i) The requirements of § 61.57(e).

(ii) At least six ILS approaches since the beginning of the sixth month before the test. These approaches must be under actual or simulated instrument flight conditions down to the minimum landing altitude for the ILS approach in the type airplane in which the flight test is to be conducted. However, the approaches need not be conducted down to the decision heights authorized for Category II operations. At least three of these approaches must have been conducted manually, without the use of an approach coupler.

The flight time acquired in meeting the requirements of paragraph (c)(2)(ii) of this section may be used to meet the requirements of paragraph (c)(2)(i) of this section.

(d) *Practical test procedures.* The practical test consists of two phases:

(1) *Phase I—oral operational test.* The applicant must demonstrate his knowledge of the following:

(i) Required landing distance.

(ii) Recognition of the decision height.

(iii) Missed approach procedures and techniques utilizing computed or fixed attitude guidance displays.

(iv) RVR, its use and limitations.

(v) Use of visual clues, their availability or limitations, and altitude at which they are normally discernible at reduced RVR readings.

(vi) Procedures and techniques related to transition from nonvisual to visual flight during a final approach under reduced RVR.

(vii) Effects of vertical and horizontal wind shear.

(viii) Characteristics and limitations of the ILS and runway lighting system.

(ix) Characteristics and limitations of the flight director system, auto approach coupler (including split axis type if equipped), auto throttle system (if equipped), and other required Category II equipment.

(x) Assigned duties of the second in command during Category II approaches.

(xi) Instrument and equipment failure warning systems.

(2) Phase II—flight test. The flight test must be taken in an aircraft that meets the requirements of Part 91 of this chapter for Category II operations. The test consists of at least two ILS approaches to 100 feet including at least one landing and one missed approach. All approaches must be made with the approved flight control guidance system. However, if an approved automatic approach coupler is installed, at least one approach must be made manually. In the case of a multiengine aircraft that has performance capability to execute a missed approach with an engine out, the missed approach must be executed with one engine set in idle or zero thrust position before reaching the middle marker. The required flight maneuvers must be performed solely by reference to instruments and in coordination with a second in command who holds a class rating and, in the case of a large aircraft or a small turbojet aircraft, a type rating for that aircraft.

(Amdt. 61-18, Eff. 9/26/65); (Amdt. 61-77, Eff. 1/6/87)

§ 61.69 Glider towing: Experience and instruction requirements.

No person may act as pilot in command of an aircraft towing a glider unless he meets the following requirements:

(a) He holds a current pilot certificate (other than a student or recreational pilot certificate) issued under this part.

(b) He has an endorsement in his pilot logbook from a person authorized to give flight instruction in gliders, certifying that he has received ground and flight instruction in gliders and is familiar with the techniques and procedures essential to the safe towing of gliders, including airspeed limitations, emergency procedures, signals used, and maximum angles of bank.

(c) He has made and entered in his pilot logbook—

(1) At least three flights as sole manipulator of the controls of an aircraft towing a glider while accompanied by a pilot who has met the requirements of this section and made and logged at least 10 flights as pilot in command of an aircraft towing a glider; or

(2) At least three flights as sole manipulator of the controls of an aircraft simulating glider towing flight procedures (while accompanied by a pilot who meets the requirements of this section), and at least three flights as pilot or observer in a glider being towed by an aircraft.

However, any person who, before May 17, 1967, made, and entered in his pilot logbook, 10 or more flights as pilot in command of an aircraft towing a glider in accordance with a certificate of waiver need not comply with paragraphs (c)(1) and (2) of this section.

(d) If he holds only a private pilot certificate he must have had, and entered in his pilot logbook at least—

(1) 100 hours of pilot flight time in powered aircraft; or

(2) 200 total hours of pilot flight time in powered or other aircraft.

(e) Within the preceding 12 months he has—

(1) Made at least 3 actual or simulated glider tows while accompanied by a qualified pilot who meets the requirements of this section; or

(2) Made at least 3 flights as pilot in command of a glider towed by an aircraft.

(Amdt. 61-18, Eff. 9/26/65); (Amdt. 61-64, Eff. 12/22/76); (Amdt. 61-82, Eff. 8/31/89)

§ 61.71 Graduates of certificated [pilot] schools: Special rules.

(a) A graduate of a flying school that is certificated under Part 141 of this chapter is considered to meet the applicable aeronautical experience requirements of this part if he presents an appropriate graduation certificate within 60 days after the date he is graduated. However, if he applies for a flight test for an instrument rating he must hold a commercial pilot certificate, or hold a private pilot certificate and meet the requirements of § 61.65(e)(1) and § 61.123 (except paragraphs (d) and (e) thereof). In addition, if he applies for a flight instructor certificate he must hold a commercial pilot certificate.

(b) [An applicant for a certificate or rating under this part is considered to meet the aeronautical knowledge and skill requirements, or both, applicable to that certificate or rating if the applicant applies within 90 days after graduation from an appropriate course given by a pilot school that is certificated under Part 141 of this chapter and is authorized to test applicants on aeronautical knowledge or skill, or both.]

(Amdt. 61-14, Eff. 4/3/65); (Amdt. 61-23, Eff. 7/6/66); (Amdt. 61-40, Eff. 5/10/68); (Amdt. 61-44, Eff. 11/22/69); (Amdt. 61-63, Eff. 11/1/74); [(Amdt. 61-90, Eff. 4/15/91)]

§ 61.73 Military pilots or former military pilots: Special rules.

(a) *General.* A rated military pilot or former rated military pilot who applies for a private or commercial pilot certificate, or an aircraft or instrument rating, is entitled to that certificate with appropriate ratings or to the addition of a rating on the pilot certificate he holds, if he meets the applicable requirements of this section. This section does not apply to a military pilot or former military pilot who has been removed from flying status for lack of proficiency or because of disciplinary action involving aircraft operation.

(b) *Military pilots on active flying status within 12 months.* A rated military pilot or former rated military pilot who has been on active flying status within the 12 months before he applies must pass a written test on the parts of this chapter relating to pilot privileges and limitations, air traffic and general operating rules, he must present documents showing that he meets the requirements of paragraph (d) of this section for at least one aircraft rating, and that he is, or was at any time since the beginning of the twelfth month before the month in which he applies—

(1) A rated military pilot on active flying status in an armed force of the United States; or

(2) A rated military pilot of an armed force of a foreign contracting State to the Convention on International Civil Aviation, assigned to pilot duties (other than flight training) with an armed force of the United States who holds, at the time he applies, a current civil pilot license issued by that foreign State authorizing at least the privileges of the pilot certificate he seeks.

(c) *Military pilots not on active flying status within previous 12 months.* A rated military pilot or former military pilot who has not been on active flying status within the 12 months before he applies must pass the appropriate written and flight tests prescribed in this part for the certificate or rating he seeks. In addition, he must show that he holds an FAA medical certificate appropriate to the pilot certificate he seeks and present documents showing that he was, before the beginning of the twelfth month before the month in which he applies, a rated military pilot as prescribed by either subparagraph (1) or (2) of paragraph (b) of this section.

(d) *Aircraft ratings: other than airplane category and type.* An applicant for a category, class, or type rating (other than airplane category and type rating) to be added on the pilot certificate he holds, or for which he has applied, is issued that rating if he presents documentary evidence showing one of the following:

(1) That he has passed an official United States military checkout as pilot in command of aircraft of the category, class, or type for which he seeks a rating since the beginning of the twelfth month before the month in which he applies.

(2) That he has had at least 10 hours of flight time serving as pilot in command of aircraft of the category, class, or type for which he seeks a rating since the beginning of the twelfth month before the month in which he applies and previously has had an official United States military checkout as pilot in command of that aircraft.

(3) That he has met the requirements of subparagraph (1) or (2) of paragraph (b) of this section, has had an official United States military checkout in the category of aircraft for which he seeks a rating, and that he passes an FAA flight test appropriate to that category and the class or type rating he seeks. To be eligible for that flight test, he must have a written statement from an authorized flight instructor, made not more than 60 days before he applies for the flight test, certifying that he is competent to pass the test.

A type rating is issued only for aircraft types that the Administrator has certificated for civil operations. Any rating placed on an airline transport pilot certificate is limited to commercial pilot privileges.

(e) *Airplane category and type ratings.*

(1) An applicant for a commercial pilot certificate with an airplane category rating, or an applicant for the addition of an airplane category rating on his commercial pilot certificate, must hold an airplane instrument rating, or his certificate is endorsed with the following limitation: "not valid for the carriage of passengers or property for hire in airplanes on cross-country flights of more than 50 nautical miles, or at night."

(2) An applicant for a private or commercial pilot certificate with an airplane type rating, or for the addition of an airplane type rating on his private or commercial pilot certificate who holds an instrument rating (airplane), must present documentary evidence showing that he has demonstrated instrument competency in the type of airplane for which the type rating is sought, or his certificate is endorsed with the following limitation: "VFR only."

(f) *Instrument rating.* An applicant for an airplane instrument rating or a helicopter instrument rating to be added on the pilot certificate he holds, or for which he has applied, is entitled to that rating if he has, within the 12 months preceding the month in which he applies, satisfactorily accomplished an instrument flight check of a U.S. Armed Force in an aircraft of the category for which he seeks the instrument rating and is authorized to conduct IFR flights on Federal airways. A helicopter instrument rating added on an airline transport pilot certificate is limited to commercial pilot privileges.

(g) *Evidentiary documents.* The following documents are satisfactory evidence for the purposes indicated:

(1) To show that the applicant is a member of the armed forces, an official identification card issued to the applicant by an armed force may be used.

(2) To show the applicant's discharge or release from an armed force, or his former membership therein, an original or a copy of a certificate of discharge or release may be used.

(3) To show current or previous status as a rated military pilot on flying status with a U.S. Armed Force, one of the following may be used:

(i) An official U.S. Armed Force order to flight duty as a military pilot.

(ii) An official U.S. Armed Force form or logbook showing military pilot status.

(iii) An official order showing that the applicant graduated from a United States military pilot school and is rated as a military pilot.

(4) To show flight time in military aircraft as a member of a U.S. Armed Force, an appropriate U.S. Armed Force form or summary of it, or a certified United States military logbook may be used.

(5) To show pilot-in-command status, an official U.S. Armed Force record of a military checkout as pilot in command, may be used.

(6) To show instrument pilot qualification, a current instrument card issued by a U.S. Armed Force, or an official record of the satisfactory completion of an instrument flight check within the 12 months preceding the month of the application may be used. However, a Tactical (Pink) instrument card issued by the U.S. Army is not acceptable.

(Amdt. 61-18, Amdt. 9/26/65); (Amdt. 61-37, Eff. 11/5/67)

§ 61.75 Pilot certificate issued on basis of a foreign pilot license.

(a) *Purpose.* The holder of a current private, commercial, senior commercial, or airline transport pilot license issued by a foreign contracting State to the Convention on International Civil Aviation may apply for a pilot certificate under this section authorizing him to act as a pilot of a civil aircraft of U.S. registry.

(b) *Certificate issued.* A pilot certificate is issued to an applicant under this section, specifying the number and State of issuance of the foreign pilot license on which it is based. An applicant who holds a foreign private pilot license is issued a private pilot certificate, and an applicant who holds a foreign commercial, senior commercial, or airline transport pilot license is issued a commercial pilot certificate, if—

(1) He meets the requirements of this section;

(2) His foreign pilot license does not contain an endorsement that he has not met all of the standards of ICAO for that license; and

(3) He does not hold a U.S. pilot certificate of private pilot grade or higher.

(c) *Limitation on licenses used as basis for U.S. certificate.* Only one foreign pilot license may be used as a basis for issuing a pilot certificate under this section.

(d) *Aircraft ratings issued.* Aircraft ratings listed on the applicant's foreign pilot license, in addition to any issued after testing under the provisions of this part, are placed on the applicant's pilot certificate.

(e) *Instrument rating issued.* An instrument rating is issued to an applicant if—

(1) His foreign pilot license authorizes instrument privileges; and

(2) Within 24 months preceding the month in which he makes application for a certificate, he passed a test on the instrument flight rules in Subpart B of Part 91 of this chapter, including the related procedures for the operation of the aircraft under instrument flight rules.

(f) *Medical standards and certification.* An applicant must submit evidence that he currently meets the medical standards for the foreign pilot license on which the application for a certificate under this section is based. A current medical certificate issued under Part 67 of this chapter is accepted as evidence that the applicant meets those standards. However, a medical certificate issued under Part 67 of this chapter is not evidence that the applicant meets those standards outside the United States, unless the State that issued the applicant's foreign pilot license also accepts that medical certificate as evidence of meeting the medical standards for his foreign pilot license.

(g) *Limitations placed on pilot certificate.*

(1) If the applicant cannot read, speak, and understand the English language, the Administrator places any limitation on the certificate that he considers necessary for safety.

(2) A certificate issued under this section is not valid for agricultural aircraft operations, or the operation of an aircraft in which persons or property are carried for compensation or hire. This limitation is also placed on the certificate.

(h) *Operating privileges and limitations.* The holder of a pilot certificate issued under this section may act as a pilot of a civil aircraft of U.S. registry in accordance with the pilot privileges authorized by the foreign pilot license on which that certificate is based, subject to the limitations of this part and any additional limitations placed on his certificate by the Administrator. He is subject to these limitations while he is acting as a pilot of the aircraft within or outside the United States. However, he may not act as pilot in command, or in any other capacity as a required pilot flight crewmember, of a civil aircraft of U.S. registry that is carrying persons or property for compensation or hire.

(i) *Flight instructor certificate.* A pilot certificate issued under this section does not satisfy any of the requirements of this part for the issuance of a flight instructor certificate.

§ 61.77 Special purpose pilot certificate: operation of U.S.-registered civil airplanes leased by a person not a U.S. citizen.

(a) *General.* The holder of a current foreign pilot certificate or license issued by a foreign contracting State to the Convention on International Civil Aviation, who meets the requirements of this section, may hold a special purpose pilot certificate authorizing the holder to perform pilot duties on a civil airplane of U.S. registry, leased to a person not a citizen of the United States, carrying persons or property for compensation or hire. Special purpose pilot certificates are issued under this section only for airplane types that can have a maximum passenger seating configuration, excluding any flight crewmember seat, of more than 30 seats or a maximum payload capacity (as defined in § 135.2(e) of this chapter) of more than 7,500 pounds.

(b) *Eligibility.* To be eligible for the issuance or renewal of a certificate under this section, an applicant or a representative of the applicant must present the following to the Administrator:

(1) A current foreign pilot certificate or license, issued by the aeronautical authority of a foreign contracting State to the Convention on International Civil Aviation, or a facsimile acceptable to the Administrator. The certificate or license must authorize the applicant to perform the pilot duties to be authorized by a certificate issued under this section on the same airplane type as the leased airplane.

(2) A current certification by the lessee of the airplane—

(i) Stating that the applicant is employed by the lessee;

(ii) Specifying the airplane type on which the applicant will perform pilot duties; and

(iii) Stating that the applicant has received ground and flight instruction which qualifies the applicant to perform the duties to be assigned on the airplane.

(3) Documentation showing that the applicant has not reached the age of 60 and that the applicant currently meets the medical standards for the foreign pilot certificate or license required by paragraph (b)(1) of this section, except that a U.S. medical certificate issued under Part 67 of this chapter is not evidence that the applicant meets those standards unless the State which issued the applicant's foreign pilot certificate or license accepts a U.S. medical certificate as evidence of medical fitness for a pilot certificate or license.

(c) *Privileges.* The holder of a special purpose pilot certificate issued under this section may exercise the same privileges as those shown on the certificate or license specified in paragraph (b)(1), subject to the limitations specified in this section. The certificate holder is not subject to the requirements of §§ 61.55, 61.57, and 61.58 of this part.

(d) *Limitations.* Each certificate issued under this section is subject to the following limitations:

(1) It is valid only—

(i) For flights between foreign countries or for flights in foreign air commerce;

(ii) While it and the foreign pilot certificate or license required by paragraph (b) (1) of this section are in the certificate holder's personal possession and are current;

(iii) While the certificate holder is employed by the person to whom the airplane described in the certification required by paragraph (b)(2) of this section is leased;

(iv) While the certificate holder is performing pilot duties on the U.S.-registered civil airplane described in the certification required by paragraph (b)(2) of this section;

(v) While the medical documentation required by paragraph (b)(3) of this section is in the certificate holder's personal possession and is currently valid; and

(vi) While the certificate holder is under 60 years of age.

(2) Each certificate issued under this section contains the following:

(i) The name of the person to whom the U.S.-registered civil aircraft is leased.

(ii) The type of aircraft.

(iii) The limitation: "Issued under, and subject to, § 61.77 of the Federal Aviation Regulations."

(iv) The limitation: "Subject to the privileges and limitations shown on the holder's foreign pilot certificate or license."

(3) Any additional limitations placed on the certificate which the Administrator considers necessary.

(e) *Termination.* Each special purpose pilot certificate issued under this section terminates—

(1) When the lease agreement for the airplane described in the certification required by paragraph (b)(2) of this section terminates;

(2) When the foreign pilot certificate or license, or the medical documentation, required by paragraph (b) of this section is suspended, revoked, or no longer valid;

(3) When the certificate holder reaches the age of 60; or

(4) After 24 months after the month in which the special purpose pilot certificate was issued.

(f) *Surrender of certificate.* The certificate holder shall surrender the special purpose pilot certificate to the Administrator within 7 days after the date it terminates.

(g) *Renewal.* The certificate holder may have the certificate renewed by complying with the requirements of paragraph (b) of this section at the time of application for renewal.

Docket No. 19300 (45 FR 5671) Eff. 1/24/80; (Amdt. 61-67, Eff. 2/25/80)

Subpart C—Student and recreational pilots

§ 61.81 Applicability.

This subpart prescribes the requirements for the issuance of student pilot certificates and recreational pilot certificates and ratings, the conditions under which those certificates and ratings are necessary, and the general operating rules and limitations for the holders of those certificates and ratings.

Docket No. 24695 (54 FR 13038) Eff. 3/29/89; (Amdt. 61-14, Eff. 4/3/65); (Amdt. 61-82, Eff. 8/31/89)

§ 61.83 Eligibility requirements: Student pilots.

To be eligible for a student pilot certificate, a person must—

(a) Be at least 16 years of age, or at least 14 years of age for a student pilot certificate limited to the operation of a glider or free balloon;

(b) Be able to read, speak, and understand English language, or have such operating limitations placed on his pilot certificate as are necessary for the safe operation of aircraft, to be removed when he shows that he can read, speak, and understand the English language; and

(c) Hold at least a current third-class medical certificate issued under Part 67 of this chapter, or, in the case of glider or free balloon operations, certify that he has no known medical defect that makes him unable to pilot a glider or a free balloon.

(Amdt. 61-3, Eff. 5/17/63); (Amdt. 61-64, Eff. 12/22/76); (Amdt. 61-82, Eff. 8/31/89)

§ 61.85 Application.

An application for a student pilot certificate is made on a form and in a manner provided by the Administrator and is submitted to—

(a) A designated aviation medical examiner when applying for an FAA medical certificate in the United States; or

(b) An FAA operations inspector or designated pilot examiner, accompanied by a current FAA medical certificate, or in the case of an application for a glider or free balloon pilot certificate it may be accompanied by a certification by applicant that he has no known medical defect that makes him unable to pilot a glider or free balloon.

(Amdt. 61-2, Eff. 5/2/63); (Amdt. 61-29, Eff. 4/23/67); (Amdt. 61-41, Eff. 4/11/68); (Amdt. 61-72, Eff. 10/18/82)

§ 61.87 Solo flight requirements for student pilots.

(a) *General.* A student pilot may not operate an aircraft in solo flight unless that student meets the requirements of this section. The term "solo flight," as used in this subpart, means that flight time during which a student pilot is the sole occupant of the aircraft, or that flight time during which the student acts as pilot in command of an airship requiring more than one flight crewmember.

(b) *Aeronautical knowledge.* A student pilot must have demonstrated satisfactory knowledge to an authorized instructor, of the appropriate portions of Parts 61 and 91 of the Federal Aviation Regulations that are applicable to student pilots. This demonstration must include the satisfactory completion of a written examination to be administered and graded by the instructor who endorses the student's pilot certificate for solo flight. The written examination must include questions on the applicable regulations and the flight characteristics and operational limitations for the make and model aircraft to be flown.

(c) *Pre-solo flight training.* Prior to being authorized to conduct a solo flight, a student pilot must have received and logged instruction in at least the applicable maneuvers and procedures listed in paragraphs (d) through (j) of this section for the make and model of aircraft to be flown in solo flight, and must have demonstrated proficiency to an acceptable performance level as judged by the instructor who endorses the student's pilot certificate.

(d) For all aircraft (as appropriate to the aircraft to be flown in solo flight), the student pilot must have received pre-solo flight training in—

(1) Flight preparation procedures, including preflight inspections, powerplant operation, and aircraft systems;

(2) Taxiing or surface operations, including runups;

(3) Takeoffs and landings, including normal and crosswind;

(4) Straight and level flight, shallow, medium, and steep banked turns in both directions;

(5) Climbs and climbing turns;

(6) Airport traffic patterns including entry and departure procedures, and collision and wake turbulence avoidance;

(7) Descents with and without turns using high and low drag configurations;

(8) Flight at various airspeeds from cruising to minimum controllable airspeed;

(9) Emergency procedures and equipment malfunctions; and

(10) Ground reference maneuvers.

(e) For airplanes, in addition to the maneuvers and procedures in paragraph (d) of this section, the student pilot must have received pre-solo flight training in—

(1) Approaches to the landing area with engine power at idle and with partial power;

(2) Slips to a landing;

(3) Go-arounds from final approach and from the landing flare in various flight configurations including turns;

(4) Forced landing procedures initiated on takeoff, during initial climb, cruise, descent, and in the landing pattern; and

(5) Stall entries from various flight attitudes and power combinations with recovery initiated at the first indication of a stall, and recovery from a full stall.

(f) For rotorcraft (other than single-place gyroplanes), in addition to the maneuvers and procedures in paragraph (d) of this section and as allowed by the aircraft's performance and maneuver limitations, the student pilot must have received pre-solo flight training in—

(1) Approaches to the landing area;

(2) Hovering turns and air taxiing (for helicopters only) and ground maneuvers;

(3) Go-arounds from landing hover and from final approach;

(4) Simulated emergency procedures, including autorotational descents with a power recovery or running landing in gyroplanes, a power recovery to a hover in a single engine helicopter, or approaches to a hover or landing with one engine inoperative in multiengine helicopters; and

(5) Rapid decelerations (helicopters only).

(g) For single-place gyroplanes, in addition to the appropriate maneuvers and procedures in paragraph (d) of this section, the student pilot must have received pre-solo flight-training in—

(1) Simulated emergency procedures, including autorotational descents with a power recovery or a running landing;

(2) At least three successful flights in gyroplanes under the observation of a qualified instructor, and

(3) For nonpowered single-place gyroplanes only, at least three successful flights in a gyroplane towed from the ground under the observation of the flight instructor who endorses the student's pilot certificate.

(h) For gliders, in addition to the appropriate maneuvers and procedures in paragraph (d) of this section, the student pilot must have received pre-solo flight training in—

(1) Preflight inspection of towline rigging, review of signals, and release procedures to be used;

(2) Aerotows, ground tows, or self-launch;

(3) Principles of glider disassembly and assembly;

(4) Stall entries from various flight attitudes with recovery initiated at the first indication of a stall, and recovery from a full stall;

(5) Straight glides, turns, and spirals;

(6) Slips to a landing;

(7) Procedures and techniques for thermalling in convergence lift or ridge lift as appropriate to the training area; and

(8) Emergency operations including towline break procedures.

(i) In airships, in addition to the appropriate maneuvers and procedures in paragraph (d) of this section, the student pilot must have received pre-solo flight training in—

(1) Rigging, ballasting, controlling pressure in the ballonets, and superheating; and

(2) Landings with positive and with negative static balance.

(j) In free balloons, in addition to the appropriate maneuvers and procedures in paragraph (d) of this section, the student pilot must have received pre-solo flight training in—

(1) Operation of hot air or gas source, ballast, valves, and rip panels, as appropriate;

(2) Emergency use of rip panel (may be simulated);

(3) The effects of wind on climb and approach angles; and

(4) Obstruction detection and avoidance techniques.

(k) The instruction required by this section must be given by an authorized flight instructor who is certificated—

(1) In the category and class of airplanes, for airplanes;

(2) Except as provided in paragraph (k)(3) of this section, in helicopters or gyroplanes, as appropriate, for rotorcraft;

(3) In airplanes or gyroplanes, for single-place gyroplanes; and

(4) In gliders for gliders.

(l) The holder of a commercial pilot certificate with a lighter-than-air category rating may give the instruction required by this section in—

(1) Airships, if that commercial pilot holds an airship class rating; and

(2) Free balloons, if that commercial pilot holds a free balloon class rating.

(m) Flight instructor endorsements. No student pilot may operate an aircraft in solo flight unless that student's pilot certificate and logbook have been endorsed for the specific make and model aircraft to be flown by an authorized flight instructor certificated under this part, and the student's logbook has been endorsed, within the 90 days prior to that student operating in solo flight, by an authorized flight instructor certificated under this part who has flown with the student. No flight instructor may authorize solo flight without endorsing the student's logbook. The instructor's endorsement must certify that the instructor—

(1) Has given the student instruction in the make and model aircraft in which the solo flight is to be made;

(2) Finds that the student has met the flight training requirements of this section; and

(3) Finds that the student is competent to make a safe solo flight in that aircraft.

(n) Notwithstanding the requirements of paragraphs (a) through (m) of this section, each student pilot, whose student pilot certificate and logbook are endorsed for solo flight by an authorized flight instructor on or before August 30, 1989, may operate an aircraft in solo flight until the 90th day after the date on which the logbook was endorsed for solo flight.

Docket No. 24695 (54 FR 13038) Eff. 3/29/89; (Amdt. 61-64, Eff. 12/22/76); (Amdt. 61-77, Eff. 1/6/87); (Amdt. 61-82, Eff. 8/31/89); (Amdt. 61-86, Eff. 10/5/89)

§ 61.89 General limitations.

(a) A student pilot may not act as pilot in command of an aircraft—

(1) That is carrying a passenger;

(2) That is carrying property for compensation or hire;

(3) For compensation or hire;

(4) In furtherance of a business;

(5) On an international flight, except that a student pilot may make solo training flights from Haines, Gustavus, or Juneau, Alaska, to White Horse, Yukon, Canada, and return, over the province of British Columbia;

(6) With a flight or surface visibility of less than 3 statute miles during daylight hours or 5 statute miles at night;

(7) When the flight cannot be made with visual reference to the surface; or

(8) In a manner contrary to any limitations placed in the pilot's logbook by the instructor.

(b) A student pilot may not act as a required pilot flight crewmember on any aircraft for which more than one pilot is required by the type certificate of the aircraft or regulations under which the flight is conducted, except when receiving flight instruction from an authorized flight instructor on board an airship and no person other than a required flight crewmember is carried on the aircraft.

(Amdt. 61-29, Eff. 4/23/67); (Amdt. 61-82, Eff. 8/31/89)

§ 61.91 Aircraft limitations: Pilot in command.

A student pilot may not serve as pilot in command of any airship requiring more than one flight crewmember unless he has met the pertinent requirements prescribed in § 61.87.

(Amdt. 61-3, Eff. 5/17/63)

§ 61.93 Cross-country flight requirements (for student and recreational pilots seeking private pilot certification).

(a) *General.* No student pilot may operate an aircraft in solo cross-country flight, nor may that student, except in an emergency, make a solo flight landing at any point other than the airport of takeoff, unless the student has met the requirements of this section. The term cross-country flight, as used in this section, means a flight beyond a radius of 25 nautical miles from the point of departure.

(b) Notwithstanding paragraph (a) of this section, an authorized flight instructor, certificated under this part, may permit the student to practice solo takeoffs and landings at another airport within 25 nautical miles from the airport at which the student receives instruction if the flight instructor—

(1) Determines that the student pilot is competent and proficient to make those landings and takeoffs;

(2) Has flown with that student prior to authorizing those takeoffs and landings; and

(3) Endorses the student pilot's logbook with an authorization to make those landings and takeoffs.

(c) *Flight training.* A student pilot, in addition to the presolo flight training maneuvers and procedures required by § 61.87(c), must have received and logged instruction from an authorized flight instructor in the appropriate pilot maneuvers and procedures of this section. Additionally, a student pilot must have demonstrated an acceptable standard of performance, as judged by the authorized flight instructor certificated under this part, who endorses the student's pilot certificate in the appropriate pilot maneuvers and procedures of this section.

(1) For all aircraft—

(i) The use of aeronautical charts for VFR navigation using pilotage and dead reckoning with the aid of a magnetic compass;

(ii) Aircraft cross-country performance and procurement and analysis of aeronautical weather reports and forecasts, including recognition of critical weather situations and estimating visibility while in flight;

(iii) Cross-country emergency conditions including lost procedures, adverse weather conditions, and simulated precautionary off-airport approaches and landing procedures;

(iv) Traffic pattern procedures, including normal area arrival and departure, collision avoidance, and wake turbulence precautions;

(v) Recognition of operational problems associated with the different terrain features in the geographical area in which the cross-country flight is to be flown; and

(vi) Proper operation of the instruments and equipment installed in the aircraft to be flown.

(2) For airplanes, in addition to paragraph (c)(1) of this section—

(i) Short and soft field takeoff, approach, and landing procedures, including crosswind takeoffs and landings;

(ii) Takeoffs at best angle and rate of climb;

(iii) Control and maneuvering solely by reference to flight instruments including straight and level flight, turns, descents, climbs, and the use of radio aids and radar directives;

(iv) The use of radios for VFR navigation and for two-way communication; and

(v) For those student pilots seeking night flying privileges, night flying procedures including takeoffs, landings, go-arounds, and VFR navigation.

(3) For rotorcraft, in addition to paragraph (c)(1) of this section and as appropriate to the aircraft being flown—

(i) High altitude takeoff and landing procedures;

(ii) Steep and shallow approaches to a landing hover;

(iii) Rapid decelerations (helicopters only); and

(iv) The use of radios for VFR navigation and two-way communication.

(4) For gliders, in addition to the appropriate maneuvers and procedures in paragraph (c)(1) of this section—

(i) Landings accomplished without the use of the altimeter from at least 2,000 feet above the surface;

(ii) Recognition of weather conditions and conditions favorable for cross-country soaring; and

(iii) The use of radios for two-way radio communications.

(5) For airships, in addition to the appropriate maneuvers and procedures in paragraph (c)(1) of this section—

(i) Control of gas pressure with regard to superheating and altitude; and

(ii) Control of the airship solely by reference to flight instruments.

(6) For free balloons, the appropriate maneuvers and procedures in paragraph (c)(1) of this section.

(d) No student pilot may operate an aircraft in solo cross-country flight, unless—

(1) The instructor is an authorized instructor certificated under this part and the student's certificate has been endorsed by the instructor attesting that the student has received the instruction and demonstrated an acceptable level of competency and proficiency in the maneuvers and procedures of this section for the category of aircraft to be flown; and

(2) The instructor has endorsed the student's logbook—

(i) For each solo cross-country flight, after reviewing the student's preflight planning and preparation, attesting that the student is prepared to make the flight safely under the known circumstances and subject to any conditions listed in the logbook by the instructor; and

(ii) For repeated specific solo cross-country flights that are not greater than 50 nautical miles from the point of departure, after giving that student flight instruction in both directions over the route, including takeoffs and landings at the airports to be used, and has specified the conditions for which the flights can be made.

Docket No. 24695 (54 FR 13039) Eff. 3/29/89; (Amdt. 61-29, Eff. 4/23/67); (Amdt. 61-44, Eff. 11/22/69); (Amdt. 61-80, Eff. 1/12/89); (Amdt. 61-82, Eff. 8/31/89)

§ 61.95 Operations in airspace and at airports located within Class B.

(a) [A student pilot may not operate an aircraft on a solo flight in Class B airspace unless—

(1) [The pilot has received both ground and flight instruction from an authorized instructor on that Class B airspace area and the flight instruction was received in the specific Class B airspace area for which solo flight is authorized;

(2) [The logbook of that pilot has been endorsed within the preceding 90 days for conducting solo flight in that Class B airspace area by the instructor who gave the flight training; and

(3) [The logbook endorsement specifies that the pilot has received the required ground and flight instruction and has been found competent to conduct solo flight in that specific Class B airspace area.

(b) [Pursuant to § 91.131(b), a student pilot may not operate an aircraft on a solo flight to, from, or at an airport located within Class B airspace unless—]

(1) That student pilot has received both ground and flight instruction from an authorized instructor to operate at that airport and the flight and ground instruction has been received at the specific airport for which the solo flight is authorized;

(2) The logbook of that student pilot has been endorsed within the preceding 90 days for conducting solo flight at that specific airport by the instructor who gave the flight training; and

(3) The logbook endorsement specifies that the student pilot has received the required ground and flight instruction and has been found competent to conduct solo flight operations at that specific airport.

Docket No. 25304 (53 FR 40322) Eff. 10/14/88; (Amdt. 61-58, Eff. 8/7/72); (Amdt. 61-80, Eff. 1/12/89); [(Amdt. 61-92, Eff. 9/16/93)]

§ 61.96 Eligibility requirements: Recreational pilots.

To be eligible for a recreational pilot certificate, a person must—

(a) Be at least 17 years of age;

(b) Be able to read, speak, and understand the English language, or have such operating limitations placed on the pilot certificate as are necessary for the safe operation of aircraft, to be removed when the recreational pilot shows the ability to read, speak, and understand the English language;

(c) Hold at least a current third-class medical certificate issued under Part 67 of this chapter;

(d) Pass a written test on the subject areas on which instruction or home study is required by § 61.97;

(e) Pass an oral and flight test on maneuvers and procedures selected by an FAA inspector or designated pilot examiner to determine the applicant's competency in the appropriate flight operations listed in § 61.98; and

(f) Comply with the sections of this part that apply to the rating sought.

Docket No. 24695 (54 FR 13040) Eff. 3/29/89; (Amdt. 61-82, Eff. 8/31/89)

§ 61.97 Aeronautical knowledge.

An applicant for a recreational pilot certificate must have logged ground instruction from an authorized instructor, or must present evidence showing satisfactory completion of a course of instruction or home study in at least the following areas of aeronautical knowledge appropriate to the category and class of aircraft for which a rating is sought:

(a) The Federal Aviation Regulations applicable to recreational pilot privileges, limitations, and flight operations, the accident reporting requirements of the National Transportation Safety Board, and the use of the applicable portions of the "Airman's Information Manual" and the FAA advisory circulars;

(b) The use of aeronautical charts for VFR navigation using pilotage with the aid of a magnetic compass;

(c) The recognition of critical weather situations from the ground and in flight and the procurement and use of aeronautical weather reports and forecasts;

(d) The safe and efficient operation of aircraft including collision and wake turbulence avoidance;

(e) The effects of density altitude on takeoff and climb performance;

(f) Weight and balance computations;

(g) Principles of aerodynamics, powerplants, and aircraft systems; and

(h) Stall awareness, spin entry, spins, and spin recovery techniques.

Docket No. 24695 (54 FR 13040) Eff. 3/29/89; (Amdt. 61-14, Eff. 4/3/65); (Amdt. 61-82, Eff. 8/31/89); [(Amdt. 61-90, Eff. 4/15/91)]

§ 61.98 Flight proficiency.

The applicant for a recreational pilot certificate must have logged instruction from an authorized flight instructor in at least the pilot operations listed in this section. In addition, the applicant's logbook must contain an endorsement by an authorized flight instructor who has found the applicant competent to perform each of those operations safely as a recreational pilot.

(a) *In airplanes.*

(1) Preflight operations, including weight and balance determination, line inspection, airplane servicing, powerplant operations, and aircraft systems;

(2) Airport and traffic pattern operations, collision and wake turbulence avoidance;

(3) Flight maneuvering by reference to ground objects;

(4) Pilotage with the aid of magnetic compass;

(5) [Flight at slow airspeeds with realistic distractions and the recognition of and recovery from stalls entered from straight flight and from turns;]

(6) Emergency operations, including simulated aircraft and equipment malfunctions;

(7) Maximum performance takeoffs and landings; and

(8) Normal and crosswind takeoffs and landings.

(b) *In helicopters.*

(1) Preflight operations including weight and balance determination, line inspection, helicopter servicing, powerplant operations, and aircraft systems;

(2) Airport and traffic pattern operations, collision and wake turbulence avoidance;

(3) Hovering, air taxiing, and maneuvering by reference to ground objects;

(4) Pilotage with the aid of magnetic compass;

(5) High altitude takeoffs and roll-on landings, and rapid decelerations; and

(6) Emergency operations, including autorotative descents.

(c) *In gyroplanes.*

(1) Preflight operations including weight and balance determination, line inspection, gyroplane servicing, powerplant operations, and aircraft systems;

(2) Airport and traffic pattern operations, collision and wake turbulence avoidance;

(3) Flight maneuvering by reference to ground objects;

(4) Pilotage with the aid of a magnetic compass;

(5) Maneuvering at critically slow air speeds, and the recognition of and recovery from high rates of descent at low airspeeds; and

(6) Emergency procedures, including maximum performance takeoffs and landings.

Docket No. 24695 (54 FR 13040) Eff. 3/29/89; (Amdt. 61-82, Eff. 8/31/89); [(Amdt. 61-90, Eff. 4/15/91)]

§ 61.99 Airplane rating: Aeronautical experience.

(a) An applicant for a recreational pilot certificate with an airplane rating must have had at least a total of 30 hours of flight instruction and solo flight time which must include the following:

(1) Fifteen hours of flight instruction from an authorized flight instructor, including at least—

(i) Except as provided for in paragraph (b), 2 hours outside of the vicinity of the airport at which instruction is given, including at least three landings at another airport that is located more than 25 nautical miles from the airport of departure; and

(ii) Two hours in airplanes in preparation for the recreational pilot flight test within the 60-day period before the test.

(2) Fifteen hours of solo flight time in airplanes.

(b) Pilots based on small islands.

(1) An applicant who is located on an island from which the flight required in § 61.99(a)(1)(i) cannot be accomplished without flying over water more than 10 nautical miles from the nearest shoreline need not comply with § 61.99(a)(1)(i). However, if other airports that permit civil operations are available to which a flight may be made without flying over water more than 10 nautical miles from the nearest shoreline, the applicant must show completion of a dual flight between those two airports which must include three landings at the other airport.

(2) The pilot certificate issued to a person under paragraph (b)(1) of this section contains an endorsement with the following limitation which may subsequently be amended to include another island if the applicant complies with paragraph (b)(1) of this section with respect to that island: Passenger carrying prohibited in flights more than 10 nautical miles from (appropriate island).

(3) The holder of a recreational pilot certificate with an endorsement described in paragraph (b)(2) of this section is entitled to removal of the endorsement if the holder presents satisfactory evidence of compliance with the applicable flight requirements of § 61.93(c) to an FAA inspector or designated pilot examiner.

Docket No. 24695 (54 FR 13041) Eff. 3/29/89; (Amdt. 61-14, Eff. 4/3/65); (Amdt. 61-23, Eff. 7/6/66); (Amdt. 61-82, Eff. 8/31/89)

§ 61.100 Rotorcraft rating: Aeronautical experience.

An applicant for a recreational pilot certificate with a rotorcraft category rating must have at least the following aeronautical experience:

(a) For a helicopter rating, an applicant must have a minimum of 30 hours of flight instruction and solo flight time in aircraft, which must include the following:

(1) Fifteen hours of flight instruction from an authorized flight instructor including at least—

(i) Two hours of flight instruction in helicopters from an authorized flight instructor outside the vicinity of the airport at which instruction is given, including at least three landings at another airport that is located more than 25 nautical miles from the airport of departure; and

(ii) Two hours of flight instruction in preparation for the flight test within the 60-day period preceding the test.

(2) Fifteen hours of solo time in helicopters including—

(i) A takeoff and landing at an airport that serves both airplanes and helicopters; and

(ii) A flight with a landing at a point other than an airport.

(b) For a gyroplane rating, an applicant must have a minimum of 30 hours of flight instruction and solo flight time in aircraft, which must include the following:

(1) Fifteen hours of flight instruction from an authorized flight instructor including at least—

(i) Two hours of flight instruction in gyroplanes from an authorized flight instructor outside the vicinity of the airport at which instruction is given, including at least three landings at another airport that is located more than 25 nautical miles from the airport of departure; and

(ii) Two hours of flight instruction in preparation for the flight test within the 60-day period preceding the test.

(2) Ten hours of solo flight time in a gyroplane, including flights with takeoffs and landings at paved and unpaved airports.

Docket No. 24695 (54 FR 13041) Eff. 3/29/89; (Amdt, 61-82, Eff. 8/31/89)

§61.101 Recreational pilot privileges and limitations.

(a) A recreational pilot may—

(1) Carry not more than one passenger; and

(2) Share the operating expenses of the flight with the passenger.

(3) Act as pilot in command of an aircraft only when—

(i) The flight is within 50 nautical miles of an airport at which the pilot has received ground and flight instruction from an authorized instructor certificated under this part;

(ii) The flight lands at an airport within 50 nautical miles of the departure airport; and

(iii) The pilot carries, in that pilot's personal possession, a logbook that has been endorsed by the instructor attesting to the instruction required by paragraph (a)(3)(i) of this section.

(b) Except as provided in paragraphs (f) and (g) of this section, a recreational pilot may not act as pilot in command of an aircraft—

(1) That is certificated—

(i) For more than four occupants;

(ii) With more than one powerplant;

(iii) With a powerplant of more than 180 horsepower; or

(iv) With retractable landing gear.

(2) That is classified as a glider, airship, or balloon;

(3) That is carrying a passenger or property for compensation or hire;

(4) For compensation or hire;

(5) In furtherance of a business;

(6) Between sunset and sunrise;

(7) In airspace in which communication with air traffic control is required;

(8) At an altitude of more than 10,000 feet MSL or 2,000 feet AGL, whichever is higher;

(9) When the flight or surface visibility is less than 3 statute miles;

(10) Without visual reference to the surface;

(11) On a flight outside the United States;

(12) To demonstrate that aircraft in flight to a prospective buyer;

(13) That is used in a passenger-carrying airlift and sponsored by a charitable organization; and

(14) That is towing any object.

(c) A recreational pilot may not act as a required pilot flight crewmember on any aircraft for which more than one pilot is required by the type certificate of the aircraft or the regulations under which the flight is conducted, except when receiving flight instruction from an authorized flight instructor on board an airship and no person other than a required flight crewmember is carried on the aircraft.

(d) A recreational pilot who has logged fewer than 400 flight hours and who has not logged pilot-in-command time in an aircraft within the preceding 180 days may not act as pilot in command of an aircraft until the pilot has received flight instruction from an authorized flight instructor who certifies in the pilot's logbook that the pilot is competent to act as pilot in command of the aircraft. This requirement can be met in combination with the requirements of §§ 61.56 and 61.57 at the discretion of the instructor.

(e) The recreational pilot certificate issued under this subpart carries the notation "Holder does not meet ICAO requirements."

(f) For the purpose of obtaining additional certificates or ratings, while under the supervision of, an authorized flight instructor, a recreational pilot may fly as sole occupant of an aircraft—

(1) For which the pilot does not hold an appropriate category or class rating;

(2) Within airspace that requires communication with air traffic control; or

(3) Between sunset and sunrise, provided the flight or surface visibility is at least 5 statute miles.

(g) In order to fly solo as provided in paragraph (f) of this section, the recreational pilot must meet the appropriate aeronautical knowledge and flight training requirements of § 61.87 for that aircraft. When operating an aircraft under the conditions specified in paragraph (f) of this section, the recreational pilot shall carry the logbook that has been endorsed for each flight by an authorized flight instructor who—

(1) Has given the recreational pilot instruction in the make and model of aircraft in which the solo flight is to be made;

(2) Has found that the recreational pilot has met the applicable requirements of § 61.87; and

(3) Has found that the recreational pilot is competent to make solo flights in accordance with the logbook endorsement.

(h) Notwithstanding paragraph 61.101(a)(3), a recreational pilot may, for the purpose of obtaining an additional certificate or rating, while under the supervision of an authorized flight instructor, act as pilot in command of an aircraft on a flight in excess of 50 nautical miles from an airport at which flight instruction is received if the pilot meets the flight training requirements of § 61.93 and in that pilot's personal possession is the logbook that has been endorsed by an authorized instructor attesting that:

(1) The recreational pilot has received instruction in solo cross-country flight and the training described in § 61.93 applicable to the aircraft to be operated, and is competent to make solo cross-country flights in the make and model of aircraft to be flown; and

(2) The instructor has reviewed the student's preflight planning and preparation for the specific solo cross-country flight and that the recreational pilot is prepared to make the flight safely under the known circumstances and subject to any conditions listed in the logbook by the instructor.

Docket No. 24695 (54 FR 13041) Eff. 3/29/89; (Amdt. 61-8, Eff. 5/4/64); (Amdt. 61-20, Eff. 12/16/65); (Amdt. 61-40, Eff. 5/10/68); (Amdt. 61-58, Eff. 8/7/72); (Amdt. 61-82, Eff. 8/31/89)

Subpart D—Private pilots

§ 61.102 Applicability.

This subpart prescribes the requirements for the issuance of private pilot certificates and ratings, the conditions under which those certificates and ratings are necessary, and the general operating rules for the holders of those certificates and ratings.

(Amdt. 61-82, Eff. 8/31/89)

§ 61.103 Eligibility requirements: General.

To be eligible for a private pilot certificate, a person must—

(a) Be at least 17 years of age, except that a private pilot certificate with a free balloon or a glider rating only may be issued to a qualified applicant who is at least 16 years of age;

(b) Be able to read, speak, and understand the English language, or have such operating limitations placed on his pilot certificate as are necessary for the safe operation of aircraft, to be removed when he shows that he can read, speak, and understand the English language;

(c) Hold at least a current third class medical certificate issued under Part 67 of this chapter, or, in the case of a glider or free balloon rating, certify that he has no known medical defect that makes him unable to pilot a glider or free balloon, as appropriate;

(d) Pass a written test on the subject areas on which instruction or home study is required by § 61.105;

(e) Pass an oral and flight test on procedures and maneuvers selected by an FAA inspector or examiner to determine the applicant's competency in the flight operations on which instruction is required by the flight proficiency provisions of § 61.107; and

(f) Comply with the sections of this part that apply to the rating he seeks.

§ 61.105 Aeronautical knowledge.

An applicant for a private pilot certificate must have logged ground instruction from an authorized instructor, or must present evidence showing that he has satisfactorily completed a course of instruction or home study in at least the following areas of aeronautical knowledge appropriate to the category of aircraft for which a rating is sought.

(a) *Airplanes and rotorcraft.*

(1) The accident reporting requirements of the National Transportation Safety Board and the Federal Aviation Regulations applicable to private pilot privileges, limitations, and flight operations for airplanes or rotorcraft, as appropriate, the use of the "Airman's Information Manual," and FAA advisory circulars;

(2) VFR navigation, using pilotage, dead reckoning, and radio aids;

(3) The recognition of critical weather situations from the ground and in flight, the procurement and use of aeronautical weather reports and forecasts;

(4) [The safe and efficient operation of airplanes or rotorcraft, as appropriate, including high-density airport operations, collision avoidance precautions, and radio communication procedures;

(5) [Basic aerodynamics and the principles of flight which apply to airplanes or rotorcraft, as appropriate; and

[(6) Stall awareness, spin entry, spins, and spin recovery techniques for airplanes.]

(b) *Gliders.*

(1) The accident reporting requirements of the National Transportation Safety Board and the Federal Aviation Regulations applicable to glider pilot privileges, limitations, and flight operations;

(2) Glider navigation, including the use of aeronautical charts and the magnetic compass;

(3) [Recognition of weather situations of concern to the glider pilot, and the procurement and use of aeronautical weather reports and forecasts;

(4) [The safe and efficient operation of gliders, including ground and/or aero tow procedures as appropriate, signals, and safety precautions; and

[(5) Stall awareness, spin entry, spins, and spin recovery techniques for gliders.]

(c) *Airships.*

(1) The Federal Aviation Regulations applicable to private lighter-than-air pilot privileges, limitations, and airship flight operations;

(2) Airship navigation, including pilotage, dead reckoning, and the use of radio aids;

(3) The recognition of weather conditions of concern to the airship pilot, and the procurement and use of aeronautical weather reports and forecasts; and

(4) Airship operations, including free ballooning, the effects of superheating, and positive and negative lift.

(d) *Free balloons.*

(1) The Federal Aviation Regulations applicable to private free balloon pilot privileges, limitations, and flight operations;

(2) The use of aeronautical charts and the magnetic compass for free balloon navigation;

(3) The recognition of weather conditions of concern to the free balloon pilot, and the procurement and use of aeronautical weather reports and forecasts appropriate to free balloon operations; and

(4) Operating principles and procedures of free balloons, including gas and hot air inflation systems.

(Amdt. 61-77, Eff. 1/6/87); [(Amdt. 61-90, Eff. 4/15/91)]

§ 61.107 Flight proficiency.

The applicant for a private pilot certificate must have logged instruction from an authorized flight instructor in at least the following pilot operations. In addition, his logbook must contain an endorsement by an authorized flight instructor who has found him competent to perform each of those operations safely as a private pilot.

(a) *In airplanes.*

(1) Preflight operations, including weight and balance determination, line inspection, and airplane servicing;

(2) Airport and traffic pattern operations, including operations at controlled airports, radio communications, and collision avoidance precautions;

(3) Flight maneuvering by reference to ground objects;

(4) [Flight at slow airspeeds with realistic distractions, and the recognition of and recovery from stalls entered from straight flight and from turns;]

(5) Normal and crosswind takeoffs and landings;

(6) Control and maneuvering an airplane solely by reference to instruments, including descents and climbs using radio aids or radar directives;

(7) Cross-country flying, using pilotage, dead reckoning, and radio aids, including one 2-hour flight;

(8) Maximum performance takeoffs and landings;

(9) Night flying, including takeoffs, landings, and VFR navigation; and

(10) Emergency operations, including simulated aircraft and equipment malfunctions.

(b) *In helicopters.*

(1) Preflight operations, including the line inspection and servicing of helicopters;

(2) Hovering, air taxiing, and maneuvering by ground references;

(3) Airport and traffic pattern operations, including collision avoidance precautions;

(4) Cross-country flying, using pilotage, dead reckoning, and radio aids, including one 1-hour flight;

(5) Operations in confined areas and on pinnacles, rapid decelerations, landings on slopes, high-altitude takeoffs, and run-on landings;

(6) Night flying, including takeoffs, landings, and VFR navigation; and

(7) Simulated emergency procedures, including aircraft and equipment malfunctions, approaches to a hover or landing with an engine inoperative in a multiengine helicopter, or autorotational descents with a power recovery to a hover in single-engine helicopters.

(c) *In gyroplanes.*

(1) Preflight operations, including the line inspection and servicing of gyroplanes;

(2) Flight maneuvering by ground references;

(3) Maneuvering at critically slow airspeeds, and the recognition of and recovery from high rates of descent at low airspeeds;

(4) Airport and traffic pattern operations, including collision avoidance precautions and radio communication procedures;

(5) Cross-country flying by pilotage, dead reckoning, and the use of radio aids; and

(6) Emergency procedures, including maximum performance takeoffs and landings.

(d) *In gliders.*

(1) Preflight operations including the installation of wings and tail surfaces specifically designed for quick removal and installation by pilots, and line inspection;

(2) Ground (auto or winch) tow or aero tow (the applicant's certificate is limited to the kind of tow selected);

(3) Precision maneuvering, including steep turns and spirals in both directions;

(4) The correct use of critical sailplane performance speeds;

(5) [Flight at slow airspeeds with realistic distractions, and the recognition of and recovery from stalls entered from straight flight and from turns; and]

(6) Accuracy approaches and landings with the nose of the glider stopping short of and within 200 feet of a line or mark.

(e) *In airships.*

(1) Ground handling, mooring, rigging, and preflight operations;

(2) Takeoffs and landing with static lift, and with negative and positive lift, and the use of two-way radio;

(3) Straight and level flight, climbs, turns, and descents;

(4) Precision flight maneuvering;

(5) Navigation, using pilotage, dead reckoning, and radio aids; and

(6) Simulated emergencies, including equipment malfunction, the valving of gas, and the loss of power on one engine.

(f) *In free balloons.*

(1) Rigging and tethering, including the installation of baskets and burners specifically designed for quick removal or installation by a pilot; and the interchange of baskets or burners, when provided for in the type certificate data, classified as preventive maintenance, and subject to the recording requirements of § 43.9 of this chapter;

(2) Operation of burner, if airborne heater used;

(3) Ascents and descents;

(4) Landing; and

(5) Emergencies, including the use of the rip cord (may be simulated).

(Amdt. 61-77, 1/6/87); (Amdt. 61-79, Eff. 6/5/87); [(Amdt. 61-90, Eff. 4/15/91)]

§ 61.109 Airplane rating: Aeronautical experience.

An applicant for a private pilot certificate with an airplane rating must have had at least a total of 40 hours of flight instruction and solo flight time which must include the following:

(a) Twenty hours of flight instruction from an authorized flight instructor, including at least—

(1) Three hours of cross-country;

(2) Three hours at night, including 10 takeoffs and landings for applicants seeking night flying privileges; and

(3) Three hours in airplanes in preparation for the private pilot flight test within 60 days prior to that test. An applicant who does not meet the night flying requirement in paragraph (a)(2) of this section is issued a private pilot certificate bearing the limitation "Night flying prohibited." This limitation may be removed if the holder of the certificate shows that he has met the requirements of paragraph (a)(2) of this section.

(b) Twenty hours of solo flight time, including at least:

(1) Ten hours in airplanes.

(2) Ten hours of cross-country flights, each flight with a landing at a point more than 50 nautical miles from the original departure point. One flight must be of at least 300 nautical miles with landings at a minimum of three points, one of which is at least 100 nautical miles from the original departure point.

(3) Three solo takeoffs and landings to a full stop at an airport with an operating control tower.

(Amdt. 61-73, Eff. 11/15/82)

§ 61.111 Cross-country flights: Pilots based on small islands.

(a) An applicant who shows that he is located on an island from which the required flights cannot be accomplished without flying over water more than 10 nautical miles from the nearest shoreline need not comply with paragraph (b)(2) of § 61.109. However, if other airports that permit civil operations are available to which a flight may be made without flying over water more than 10 nautical miles from the nearest shoreline, he must show that he has completed two round trip solo flights between those two airports that are farthest apart, including a landing at each airport on both flights.

(b) The pilot certificate issued to a person under paragraph (a) of this section contains an endorsement with the following limitation which may be subsequently amended to include another island if the applicant complies with paragraph (a) of this section with respect to that island:

"Passenger carrying prohibited on flights more than 10 nautical miles from (appropriate island)."

(c) If an applicant for a private pilot certificate under paragraph (a) of this section does not have at least 3 hours of solo cross-country flight time, including a round trip flight to an airport at least 50 nautical miles from the place of departure with at least two full stop landings at different points along the route, his pilot certificate is also endorsed as follows:

"Holder does not meet the cross-country flight requirements of ICAO."

(d) The holder of a private pilot certificate with an endorsement described in paragraph (b) or (c) of this section, is entitled to a removal of the endorsement, if he presents satisfactory evidence to an FAA inspector or designated pilot examiner that he has complied with the applicable solo cross-country flight requirements and has passed a practical test on cross-country flying.

(Amdt. 61-14, Eff. 4/3/65); (Amdt. 61-29, Eff. 4/23/67)

§ 61.113 Rotorcraft rating: Aeronautical experience.

An applicant for a private pilot certificate with a rotorcraft category rating must have at least the following aeronautical experience:

(a) [*Helicopter class rating.* A total of 40 hours of flight instruction and solo flight time in aircraft, including at least—]

(1) 20 hours of flight instruction from an authorized flight instructor, 15 hours of which must be in a helicopter, including—

(i) 3 hours of cross-country flying in helicopters;

(ii) 3 hours of night flying in helicopters, including 10 takeoffs and landings, each of which must be separated by an en route phase of flight;

(iii) 3 hours in helicopters in preparation for the private pilot flight test within 60 days before that test; and

(iv) A flight in a helicopter with a landing at a point other than an airport; and

(2) 20 hours of solo flight time, 15 hours of which must be in a helicopter, including at least—

(i) 3 hours of cross-country flying in helicopters, including one flight with a landing at three or more points, each of which must be more than 25 nautical miles from each of the other two points; and

(ii) Three takeoffs and landings in helicopters at an airport with an operating control tower, each of which must be separated by an en route phase of flight.

(b) [*Gyroplane class rating.* A total of 40 hours of flight instruction and solo flight time in aircraft, including at least—]

(1) 20 hours of flight instruction from an authorized flight instructor, 15 hours of which must be in a gyroplane, including—

(i) 3 hours of cross-country flying in gyroplanes;

(ii) 3 hours of night flying in gyroplanes, including 10 takeoffs and landings; and

(iii) 3 hours in gyroplanes in preparation for the private pilot flight test within 60 days before that test; and

(2) 20 hours of solo flight time, 10 hours of which must be in a gyroplane, including—

(i) 3 hours of cross-country flying in gyroplanes, including one flight with a landing at three or more points, each of which must be more than 25 nautical miles from each of the other two points; and

(ii) Three takeoffs and landings in gyroplanes at an airport with an operating control tower.

(c) [An applicant who does not meet the night flying requirement in paragraph (a)(1)(ii) or (b)(1)(ii) of this section is issued a private pilot certificate bearing the limitation "night flying prohibited." This limitation may be removed if the holder of the certificate demonstrates compliance with the requirements of paragraph (a)(1)(ii) or (b)(1)(ii) of this section, as appropriate.]

Docket No. 24550 (51 FR 40704) Eff. 11/7/86; (Amdt. 61-14, Eff. 4/3/65); (Amdt. 61-40, Eff. 5/10/68); (Amdt. 61-54, Eff. 7/20/71); (Amdt. 61-77, Eff. 1/6/87); (Amdt. 61-78, Eff. 2/16/87); [(Amdt. 61-90, Eff. 4/15/91)]

§ 61.115 Glider rating: Aeronautical experience.

An applicant for a private pilot certificate with a glider rating must have logged at least one of the following:

(a) Seventy solo glider flights, including 20 flights during which 360° turns were made.

(b) Seven hours of solo flight in gliders, including 35 glider flights launched by ground tows, or 20 glider flights launched by aero tows.

(c) Forty hours of flight time in gliders and single-engine airplanes, including 10 solo glider flights during which 360° turns were made.

(Amdt. 61-1, Eff. 11/1/62); (Amdt. 61-29, Eff. 4/23/67)

§ 61.117 Lighter-than-air rating: Aeronautical experience.

An applicant for a private pilot certificate with a lighter-than-air category rating must have at least the aeronautical experience prescribed in paragraph (a) or (b) of this section, appropriate to the rating sought.

(a) *Airships.*

A total of 50 hours of flight time as pilot with at least 25 hours in airships, which must include 5 hours of solo flight time in airships, or time performing the functions of pilot in command of an airship for which more than one pilot is required.

(b) *Free balloons.*

(1) If a gas balloon or a hot air balloon with an airborne heater is used, a total of 10 hours in free balloons with at least 6 flights under the supervision of a person holding a commercial pilot certificate with a free balloon rating. These flights must include:

(i) Two flights, each of at least one hour's duration, if a gas balloon is used, or of 30 minutes' duration, if a hot air balloon with an airborne heater is used;

(ii) One ascent under control to 5,000 feet above the point of takeoff, if a gas balloon is used, or 3,000 feet above the point of takeoff, if a hot air balloon with an airborne heater is used; and

(iii) One solo flight in a free balloon.

(2) If a hot air balloon without an airborne heater is used, six flights in a free balloon under the supervision of a commercial balloon pilot, including at least one solo flight.

§ 61.118 Private pilot privileges and limitations: Pilot in command.

Except as provided in paragraphs (a) through (d) of this section, a private pilot may not act as pilot in command of an aircraft that is carrying passengers or property for compensation or hire; nor may he, for compensation or hire, act as pilot in command of an aircraft.

(a) A private pilot may, for compensation or hire, act as pilot in command of an aircraft in connection with any business or employment if the flight is only incidental to that business or employment and the aircraft does not carry passengers or property for compensation or hire.

For the purpose of paragraph (d) of this section, a "charitable organization" means an organization listed in Publication No. 78 of the Department of the Treasury called the "Cumulative List of Organization" described in section 170(c) of the Internal Revenue Code of 1954, as amended from time to time by published supplemental lists.

(b) A private pilot may share the operating expenses of a flight with his passengers.

(c) A private pilot who is an aircraft salesman and who has at least 200 hours of logged flight time may demonstrate an aircraft in flight to a prospective buyer.

(d) A private pilot may act as pilot in command of an aircraft used in a passenger-carrying airlift sponsored by a charitable organization, and for which the passengers make a donation to the organization, if—

(1) The sponsor of the airlift notifies the FAA Flight Standards District Office having jurisdiction over the area concerned, at least 7 days before the flight, and furnishes any essential information that the office requests;

(2) The flight is conducted from a public airport adequate for the aircraft used, or from another airport that has been approved for the operation by an FAA inspector;

(3) He has logged at least 200 hours of flight time;

(4) No acrobatic or formation flights are conducted;

(5) Each aircraft used is certificated in the standard category and complies with the 100-hour inspection requirement of 91.409 of this chapter; and

(6) The flight is made under VFR during the day.

(Amdt. 61-84, Eff. 8/18/90); (Amdt. 61-85, Eff. 10/25/89)

§ 61.119 Free balloon rating: Limitations.

(a) If the applicant for a free balloon rating takes his flight test in a hot air balloon with an airborne heater, his pilot certificate contains an endorsement restricting the exercise of the privilege of that rating to hot air balloons with airborne heaters. The restrictions may be deleted when the holder of the certificate obtains the pilot experience required for a rating on a gas balloon.

(b) If the applicant for a free balloon rating takes his flight test in a hot air balloon without an airborne heater, his pilot certificate contains an endorsement restricting the exercise of the privileges of that rating to hot air balloons without airborne heaters. The restriction may be deleted when the holder of the certificate obtains the pilot experience and passes the tests required for a rating on a free balloon with an airborne heater or a gas balloon.

(Amdt. 61-1, Eff. 11/1/62); (Amdt. 61-29, Eff. 4/23/67)

§ 61.120 Private pilot privileges and limitations: Second in command of aircraft requiring more than one required pilot.

Except as provided in paragraphs (a) through (d) of § 61.118 a private pilot may not, for compensation or hire, act as second in command of an aircraft that is type certificated for more than one required pilot, nor may he act as second in command of such an aircraft that is carrying passengers or property for compensation or hire.

Subpart E—Commercial pilots

§ 61.121 Applicability.

This subpart prescribes the requirements for the issuance of commercial pilot certificates and ratings, the conditions under which those certificates and ratings are necessary, and the limitations upon these certificates and ratings.

(Amdt. 61-3, Eff. 5/17/63)

§ 61.123 Eligibility requirements: General

To be eligible for a commercial pilot certificate, a person must—

(a) Be at least 18 years of age;

(b) Be able to read, speak, and understand the English language, or have such operating limitations placed on his pilot certificate as are necessary for safety, to be removed when he shows that he can read, speak, and understand the English language;

(c) Hold at least a valid second-class medical certificate issued under Part 67 of this chapter, or, in the case of a glider or free balloon rating, certify that he has no known medical deficiency that makes him unable to pilot a glider or a free balloon, as appropriate;

(d) Pass a written examination appropriate to the aircraft rating sought on the subjects in which ground instruction is required by § 61.125;

(e) Pass an oral and flight test appropriate to the rating he seeks, covering items selected by the inspector or examiner from those on which training is required by § 61.127; and

(f) Comply with the provision of this subpart which apply to the rating he seeks.

(Amdt. 61-29, Eff. 4/23/67); (Amdt. 61-44, Eff. 11/22/69); (Amdt. 61-64, Eff. 12/22/76)

§ 61.125 Aeronautical knowledge.

An applicant for a commercial pilot certificate must have logged ground instruction from an authorized instructor, or must present evidence showing that he has satisfactorily completed a course of instruction or home study, in at least the following areas of aeronautical knowledge appropriate to the category of aircraft for which a rating is sought.

(a) *Airplanes.*

(1) The regulations of this chapter governing the operations, privileges, and limitations of a commercial pilot, and the accident reporting requirements of the National Transportation Safety Board.

(2) [Basic aerodynamics and the principles of flight which apply to airplanes;

(3) [Airplane operations, including the use of flaps, retractable landing gears, controllable propellers, high altitude operation with and without pressurization, loading and balance computations, and the significance and use of airplane performance speeds; and

[(4) Stall awareness, spin entry, spins, and spin recovery techniques for airplanes.]

(b) *Rotorcraft.*

(1) The regulations of this chapter which apply to the operations, privileges, and limitations of a commercial rotorcraft pilot, and the accident reporting requirements of the National Transportation Safety Board;

(2) Meteorology, including the characteristics of air masses and fronts, elements of weather forecasting, and the procurement and use of aeronautical weather reports and forecasts;

(3) The use of aeronautical charts and the magnetic compass for pilotage and dead reckoning, and the use of radio aids for VFR navigation;

(4) The safe and efficient operation of helicopters or gyroplanes, as appropriate to the rating sought; and

(5) Basic aerodynamics and principles of flight which apply to rotocraft and the significance and use of performance charts.

(c) *Gliders.*

(1) The regulations of this chapter pertinent to commercial glider pilot operations, privileges, and limitations, and the accident reporting requirements of the National Transportation Safety Board;

(2) Glider navigation, including the use of aeronautical charts and the magnetic compass, and radio orientation;

(3) [The recognition of weather situations of concern to the glider pilot from the ground and in flight, and the procurement and use of aeronautical weather reports and forecasts;

(4) [The safe and efficient operation of gliders, including ground and/or aero tow procedures as appropriate, signals, critical glider performance speeds, and safety precautions; and

[(5) Stall awareness, spin entry, spins, and spin recovery techniques for gliders.]

(d) *Airships.*

(1) The regulations of this chapter pertinent to airship operations, VFR and IFR, including the privileges and limitations of a commercial airship pilot;

(2) Airship navigation, including pilotage, dead reckoning, and the use of radio aids for VFR and IFR navigation, and IFR approaches;

(3) The use and limitations of the required flight instruments;

(4) ATC procedures for VFR and IFR operations, and the use of IFR charts and approach plates;

(5) Meteorology, including the characteristics of air masses and fronts, and the procurement and use of aeronautical weather reports and forecasts;

(6) Airship ground and flight instruction procedures; and

(7) Airship operating procedures and emergency operations, including free ballooning procedures.

(e) *Free balloons.*

(1) The regulations of this chapter pertinent to commercial free balloon piloting privileges,

(2) The use of aeronautical charts and the magnetic compass for free balloon navigation;

(3) The recognition of weather conditions significant to free balloon flight operations, and the procurement and use of aeronautical weather reports and forecasts appropriate to free ballooning;

(4) Free balloon flight and ground instruction procedures; and

(5) Operating principles and procedures for free balloons, including emergency procedures such as crowd control and protection, high wind and water landings, and operations in proximity to buildings and power lines.

(Amdt. 61-58, Eff. 8/7/72); (Amdt. 61-77, Eff. 1/6/87); [(Amdt. 61-90, Eff. 4/15/91)]

§ 61.127 Flight proficiency.

The applicant for a commercial pilot certificate must have logged Instruction from an authorized flight instructor in at least the following pilot operations. In addition, his logbook must contain an endorsement by an authorized flight instructor who has given him the instruction certifying that he has found the applicant prepared to perform each of those operations competently as a commercial pilot.

(a) *Airplanes.*

(1) Preflight duties, including load and balance determination, line inspection, and aircraft servicing;

(2) [Flight at slow airspeeds with realistic distractions, and the recognition of and recovery from stalls entered from straight flight and from turns;]

(3) Normal and crosswind takeoffs and landings, using precision approaches, flaps, power as appropriate, and specified approach speeds;

(4) Maximum performance takeoffs and landings, climbs, and descents;

(5) Operation of an airplane equipped with a retractable landing gear, flaps, and controllable propeller(s), including normal and emergency operations; and

(6) Emergency procedures, such as coping with power loss or equipment malfunctions, fire in flight, collision avoidance precautions, and engine-out procedures if a multiengine airplane is used.

(b) *Helicopters.*

(1) Preflight duties, including line inspection and helicopter servicing;

(2) Straight and level flight, climbs, turns, and descents;

(3) Air taxiing, hovering, and maneuvering by ground references;

(4) Normal and crosswind takeoffs and landings;

(5) Recognition of and recovery from imminent flight at critical rapid descent with power (settling with power);

(6) Airport and traffic pattern operations, including collision avoidance precautions and radio communications;

(7) Cross-country flight operations;

(8) Operations in confined areas and on pinnacles, rapid decelerations, landing on slopes, high-altitude takeoffs, and run-on landings; and

(9) Simulated emergency procedures, including failure of an engine or other component or system, and approaches to a hover or landing with one engine inoperative in multiengine helicopters, or autorotational descents with a power recovery to a hover in single-engine helicopters.

(c) *Gyroplanes.*

(1) Preflight operations, including line inspection and gyroplane servicing;

(2) Straight and level flight, turns, climbs, and descents;

(3) Flight maneuvering by ground references;

(4) Maneuvering at critically slow airspeeds, and the recognition of and recovery from high rates of descent at slow airspeeds;

(5) Normal and crosswind takeoffs and landings;

(6) Airport and traffic pattern operations, including collision avoidance precautions and radio communications;

(7) Cross-country flight operations; and

(8) Emergency procedures, such as power failures, equipment malfunctions, maximum performance takeoffs and landings and simulated liftoffs at low airspeed and high angles of attack.

(d) *Gliders.*

(1) Preflight duties, including glider assembly and preflight inspection;

(2) Glider launches by ground (auto or winch) or by aero tows (the applicant's certificate is limited to the kind of tow selected);

(3) Precision maneuvering, including straight glides, turns to headings, steep turns, and spirals in both directions;

(4) [The correct use of glider's performance speeds, flight at slow airspeeds with realistic distractions, and the recognition of and recovery from stalls entered from straight flight and from turns; and]

(5) Accuracy approaches and landings, with the nose of the glider coming to rest short of and within 100 feet of a line or mark.

(e) *Airships.*

(1) Ground handling, mooring, and preflight operations;

(2) Straight and level flight, turns, climbs, and descents, under VFR and simulated IFR conditions;

(3) Takeoffs and landings with positive and with negative static lift;

(4) Turns and figure eights;

(5) Precision turns to headings under simulated IFR conditions;

(6) Preparing and filing IFR flight plans, and complying with IFR clearances;

(7) IFR radio navigation and instrument approach procedures;

(8) Cross-country flight operations, using pilotage, dead reckoning, and radio aids; and

(9) Emergency operations, including engine-out operations, free ballooning an airship, and ripcord procedures (may be simulated).

(f) *Free balloons.*

(1) Assembly of basket and burner to the envelope, and rigging, inflating, and tethering of a free balloon;

(2) Ground and flight crew briefing;

(3) Ascents;

(4) Descents;

(5) Landings;

(6) Operation of airborne heater, if balloon is so equipped; and

(7) Emergency operations, including the use of the ripcord (may be simulated), and recovery from a terminal velocity descent if a balloon with an airborne heater is used.

(Amdt. 61-14, Eff. 4/3/65); (Amdt. 61-77, Eff. 1/6/87); (Amdt. 61-79, Eff. 6/5/87); [(Amdt. 61-90, Eff. 4/15/91)]

§ 61.129 Airplane rating: Aeronautical experience.

(a) *General.* An applicant for a commercial pilot certificate with an airplane rating must hold a private pilot certificate with an airplane rating. If he does not hold that certificate and rating he must meet the flight experience requirements for a private pilot certificate and airplane rating and pass the applicable written and practical test prescribed in Subpart D of this part. In addition, the applicant must hold an instrument rating (airplane), or the commercial pilot certificate that is issued is endorsed with a limitation prohibiting the carriage of passengers for hire in airplanes on cross-country flights of more than 50 nautical miles, or at night.

(b) *Flight time as pilot.* An applicant for a commercial pilot certificate with an airplane rating must have a total of at least 250 hours of flight time as pilot, which may include not more than 50 hours of instruction from an authorized instructor in a ground trainer acceptable to the Administrator. The total flight time as pilot must include—

(1) 100 hours in powered aircraft, including at least—

(i) 50 hours in airplanes, and

(ii) 10 hours of flight instruction and practice given by an authorized flight instructor in an airplane having a retractable landing gear, flaps and a controllable pitch propeller; and

(2) 50 hours of flight instruction given by an authorized flight instructor, including—

(i) 10 hours of instrument instruction, of which at least 5 hours must be in flight in airplanes, and

(ii) 10 hours of instruction in preparation for the commercial pilot flight test; and

(3) 100 hours of pilot in command time, including at least:

(i) 50 hours in airplanes.

(ii) 50 hours of cross-country flights, each flight with a landing at a point more than 50 nautical miles from the original departure point. One flight must have landings at a minimum of three points, one of which is at least 150 nautical miles from the original departure point if the flight is conducted in Hawaii, or at least 250 nautical miles from the original departure point if it is conducted elsewhere.

(iii) 5 hours of night flying including at least 10 takeoffs and landings as sole manipulator of the controls.

(Amdt. 61-23, Eff. 7/6/66); (Amdt. 61-73, Eff. 11/15/82)

§ 61.131 Rotorcraft ratings: Aeronautical experience.

An applicant for a commercial pilot certificate with a rotorcraft category rating must have at least the following aeronautical experience as a pilot:

(a) [*Helicopter class rating.* A total of 150 hours of flight time, including at least 100 hours in powered aircraft, 50 hours of which must be in a helicopter, including at least—]

(1) 40 hours of flight instruction from an authorized flight instructor, including 15 hours of which must be in a helicopter, including—

(i) 3 hours of cross-country flying in helicopters;

(ii) 3 hours of night flying in helicopters, including 10 takeoffs and landings, each of which must be separated by an en route phase of flight;

(iii) 3 hours in helicopters preparing for the commercial pilot flight test within 60 days before that test; and

(iv) Takeoffs and landings at three points other than airports; and

(2) 100 hours of pilot-in-command flight time, 35 hours of which must be in a helicopter, including at least—

(i) 10 hours of cross-country flying in helicopters, including one flight with a landing at three or more points, each of which must be more than 50 nautical miles from each of the other two points; and

(ii) Three takeoffs and landings in helicopters, each of which must be separated by an en route phase of flight, at an airport with an operating control tower.

(b) [*Gyroplane class rating.* A total of 150 hours of flight time in aircraft, including at least 100 hours in powered aircraft, 25 hours of which must be in a gyroplane, including at least—]

(1) 40 hours of flight instruction from an authorized flight instructor, 10 hours of which must be in a gyroplane, including at least—

(i) 3 hours of cross-country flying in gyroplanes;

(ii) 3 hours of night flying in gyroplanes, including 10 takeoffs and landings; and

(iii) 3 hours in gyroplanes preparing for the commercial pilot flight test within 60 days before that test; and

(2) 100 hours of pilot-in-command flight time, 15 hours of which must be in a gyroplane, including at least—

(i) 10 hours of cross-country flying in gyroplanes, including one flight with a landing at three or more points, each of which is more than 50 nautical miles from each of the other two points; and

(ii) Three takeoffs and landings in gyroplanes at an airport with an operating control tower.

Docket No. 24550 (51 FR 40704) Eff. 11/7/86; (Amdt. 61-14, Eff. 4/3/65); (Amdt. 61-18, Eff. 9/26/65); (Amdt. 61-20, Eff. 12/16/65); (Amdt. 61-40, Eff. 5/10/68); (Amdt. 61-58, Eff. 8/7/72); (Amdt. 61-77, Eff. 1/6/87); (Amdt. 61-78, Eff. 2/16/87); [(Amdt. 61-90, Eff. 4/15/91)]

§ 61.133 Glider rating: Aeronautical experience.

An applicant for a commercial pilot certificate with a glider rating must meet either of the following aeronautical experience requirements:

(a) A total of at least 25 hours of pilot time in aircraft, including 20 hours in gliders, and a total of 100 glider flights as pilot in command, including 25 flights during which 360° turns were made; or

(b) A total of 200 hours of pilot time in heavier-than-air aircraft, including 20 glider flights as pilot in command during which 360° turns were made.

§ 61.135 Airship rating: Aeronautical experience.

An applicant for a commercial pilot certificate with an airship rating must have a total of at least 200 hours of flight time as pilot, including—

(a) 50 hours of flight time as pilot in airships;

(b) 30 hours of flight time, performing the duties of pilot in command in airships, including—

(1) 10 hours of cross-country flight; and

(2) 10 hours of night flight; and

(c) 40 hours of instrument time, of which at least 20 hours must be in flight with 10 hours of that flight time in airships.

§ 61.137 Free balloon rating: Aeronautical experience.

An applicant for a commercial pilot certificate with a free balloon rating must have the following flight time as pilot:

(a) If a gas balloon or a hot air balloon with an airborne heater is used, a total of at least 35 hours of flight time as pilot, including—

(1) 20 hours in free balloons; and

(2) 10 flights in free balloons, including—

(i) Six flights under the supervision of a commercial free balloon pilot;

(ii) Two solo flights;

(iii) Two flights of at least 2 hours duration if a gas balloon is used, or at least 1 hour duration if a hot air balloon with an airborne heater is used; and

(iv) One ascent under control to more than 10,000 feet above the takeoff point if a gas balloon is used or 5,000 feet above the takeoff point if a hot air balloon with an airborne heater is used.

(b) If a hot air balloon without an airborne heater is used, ten flights in free balloons, including—

(1) Six flights under the supervision of a commercial free balloon pilot; and

(2) Two solo flights.

§ 61.139 Commercial pilot privileges and limitations: General.

The holder of a commercial pilot certificate may:

(a) Act as pilot in command of an aircraft carrying persons or property for compensation or hire;

(b) Act as pilot in command of an aircraft for compensation or hire; and

(c) Give flight instruction in an airship if he holds a lighter-than-air category and an airship class rating, or in a free balloon if he holds a free balloon class rating.

§ 61.141 Airship and free balloon ratings: Limitations.

(a) If the applicant for a free balloon class rating takes his flight test in a hot air balloon without an airborne heater, his pilot certificate contains an endorsement restricting the exercise of the privileges of that rating to hot air balloons without airborne heaters. The restriction may be deleted when the holder of the certificate obtains the pilot experience and passes the test required for a rating on a free balloon with an airborne heater or a gas balloon.

(b) If the applicant for a free balloon class rating takes his flight test in a hot air balloon with an airborne heater, his pilot certificate contains an endorsement restricting the exercise of the privileges of that rating to hot air balloons with airborne heaters. The restriction may be deleted when the holder of the certificate obtains the pilot experience required for a rating on a gas balloon.

(Amdt. 61-21, Eff. 6/26/66)

Subpart F—Airline transport pilots

§ 61.151 Eligibility requirements: General.

To be eligible for an airline transport pilot certificate, a person must—

(a) Be at least 23 years of age;

(b) Be of good moral character;

(c) Be able to read, write, and understand the English language and speak it without accent or impediment of speech that would interfere with two-way radio conversation;

(d) Be a high school graduate, or its equivalent in the Administrator's opinion, based on the applicant's general experience and aeronautical experience, knowledge, and skill;

(e) Have a first-class medical certificate issued under Part 67 of this chapter within the 6 months before the date he applies; and

(f) Comply with the sections of this part that apply to the rating he seeks.

Docket No. 1179 (27 FR 7965) Eff. 8/10/62; (Amdt. 61-24, Eff. 10/17/66)

§ 61.153 Airplane rating: Aeronautical knowledge.

An applicant for an airline transport pilot certificate with an airplane rating must, after meeting the requirements of § 61.151 (except paragraph (a) thereof) and § 61.155, pass a written test on—

(a) The sections of this part relating to airline transport pilots and Part 121, Subpart C of Part 65, and §§ 91.1, 91.3, 91.5, 91.11, 91.13, 91.103, 91.105, 91.189, 91.193, 91.703, and Subpart B of Part 91 of this chapter, and so much of Parts 21 and 25 of this chapter as relate to the operations of air carrier aircraft;

(b) The fundamentals of air navigation and use of formulas, instruments, and other navigational aids, both in aircraft and on the ground, that are necessary for navigating aircraft by instruments;

(c) The general system of weather collection and dissemination;

(d) Weather maps, weather forecasting, and weather sequence abbreviations, symbols, and nomenclature;

(e) Elementary meteorology, including knowledge of cyclones as associated with fronts;

(f) Cloud forms;

(g) National Weather Service Federal Meteorological Handbook No. 1, as amended;

(h) Weather conditions, including icing conditions and upper-air winds, that affect aeronautical activities;

(i) Air navigation facilities used on Federal airways, including rotating beacons, course lights, radio ranges, and radio marker beacons;

(j) Information from airplane weather observations and meteorological data reported from observations made by pilots on air carrier flights;

(k) The influence of terrain on meteorological conditions and developments, and their relation to air carrier flight operations;

(l) Radio communication procedure in aircraft operations; and

(m) Basic principles of loading and weight distribution and their effect on flight characteristics.

Docket No. 1179 (27 FR 7965), Eff. 8/10/62; (Amdt. 61-24, Eff. 10/17/66); (Amdt. 61-64, Eff. 12/22/76); (Amdt. 61-84, Eff. 8/18/90)

§ 61.155 Airplane rating: Aeronautical experience.

(a) An applicant for an airline transport pilot certificate with an airplane rating must hold a commercial pilot certificate or a foreign airline transport pilot or commercial pilot license without limitations, issued by a member state of ICAO, or he must be a pilot in an Armed Force of the United States whose military experience qualifies him for a commercial pilot certificate under § 61.73.

(b) An applicant must have had—

(1) At least 250 hours of flight time as pilot in command of an airplane, or as copilot of an airplane performing the duties and functions of a pilot in command under the supervision of a pilot in command, or any combination thereof, at least 100 hours of which were cross-country time and 25 hours of which were night flight time; and

(2) At least 1,500 hours of flight time as a pilot, including at least—

(i) 500 hours of cross-country flight time;

(ii) 100 hours of night flight time; and

(iii) 75 hours of actual or simulated instrument time, at least 50 hours of which were in actual flight.

Flight time used to meet the requirements of paragraph (b)(1) of this section may also be used to meet the requirements of paragraph (b)(2) of this section. Also, an applicant who has made at least 20 night takeoffs and landings to a full stop may substitute one additional night takeoff and landing to a full stop for each hour of night flight time required by paragraph (b)(2)(ii) of this section. However, not more than 25 hours of night flight time may be credited in this manner.

(c) If an applicant with less than 150 hours of pilot-in-command time otherwise meets the requirements of paragraph (b)(1) of this section, his certificate will be endorsed "Holder does not meet the pilot-in-command flight experience requirements of ICAO," as prescribed by Article 39 of the "Convention on International Civil Aviation." Whenever he presents satisfactory written evidence that he has accumulated the 150 hours of pilot-in-command time, he is entitled to a new certificate without the endorsement.

(d) A commercial pilot may credit the following flight time toward the 1,500 hours total flight time requirement of paragraph (b)(2) of this section:

(1) All second-in-command time acquired in airplanes required to have more than one pilot by their approved Aircraft Flight Manuals or airworthiness certificates; and

(2) Flight engineer time acquired in airplanes required to have a flight engineer by their approved Aircraft Flight Manuals, while participating at the same time in an approved pilot training program approved under Part 121 of this chapter.

However, the applicant may not credit under paragraph (d)(2) of this section more than 1 hour for each 3 hours of flight engineer flight time so acquired, nor more than a total of 500 hours.

(e) If an applicant who credits second-in-command or flight engineer time under paragraph (d) of this section toward the 1500 hours total flight time requirements of paragraph (b)(2) of this section—

(1) Does not have at least 1200 hours of flight time as a pilot including no more than 50 percent of his second-in-command time and none of his flight engineer time; but

(2) Otherwise meets the requirements of paragraph (b)(2) of this section, his certificate will be endorsed "Holder does not meet the pilot flight experience requirements of ICAO," as prescribed by Article 39 of the "Convention on International Civil Aviation." Whenever he presents satisfactory evidence that he has accumulated 1200 hours of flight time as a pilot including no more than 50 percent of his second-in-command time and none of his flight engineer time, he is entitled to a new certificate without the endorsement.

(f) [Reserved]

Docket No. 1179, (27 FR 7965) Eff. 8/10/62; (Amdt. 61-20, Eff. 12/16/65); (Amdt. 61-24, Eff. 10/17/66); (Amdt. 61-31, Eff. 4/15/67); (Amdt. 61-64, Eff. 12/22/76); (Amdt. 61-71, Eff. 4/28/82)

§ 61.157 Airplane rating: Aeronautical skill.

(a) An applicant for an airline transport pilot certificate with a single-engine or multiengine class rating or an additional type rating must pass a practical test that includes the items set forth in Appendix A of this part. The FAA inspector or designated examiner may modify any required maneuver where necessary for the reasonable and safe operation of the airplane being used and, unless specifically prohibited in Appendix A, may combine any required maneuvers and may permit their performance in any convenient sequence.

(b) Whenever an applicant for an airline transport pilot certificate does not already have an instrument rating he shall, as part of the oral part of the practical test, comply with § 61.65(g), and, as part of the flight part, perform each additional maneuver required by § 61.65(g) that is appropriate to the airplane type and not required in Appendix A of this part.

(c) Unless the Administrator requires certain or all maneuvers to be performed, the person giving a flight test for an airline transport pilot certificate or additional airplane class or type rating may, in his discretion, waive any of the maneuvers for which a specific waiver authority is contained in Appendix A of this part if a pilot being checked—

(1) Is employed as a pilot by a Part 121 certificate holder; and

(2) Within the preceding six calendar months, has successfully completed that certificate holder's approved training program for the airplane type involved.

(d) The items specified in paragraph (a) of this section may be performed in the airplane simulator or other training device specified in Appendix A to this part for the particular item if—

(1) The airplane simulator or other training device meets the requirements of § 121.407 of this chapter; and

(2) In the case of the items preceded by an asterisk (*) in Appendix A, the applicant has successfully completed the training set forth in § 121.424(d) of this chapter.

However, the FAA inspector or designated examiner may require Items 11(d), V(f), or V(g) of Appendix A to this part to be performed in the airplane if he determines that action is necessary to determine the applicant's competence with respect to that maneuver.

(e) An approved simulator may be used instead of the airplane to satisfy the in-flight requirements of Appendix A of this part, if the simulator—

(1) Is approved under § 121.407 of this chapter and meets the appropriate simulator requirements of Appendix H of Part 121; and

(2) Is used as part of an approved program that meets the training requirements of § 121.424(a) and (c) and Appendix H of Part 121 of this chapter.

[(f) On and after April 15, 1991, an applicant for a type rating to be added to an airline transport pilot certificate, or for issuance of an airline transport pilot certificate in an airplane requiring a type rating, must—

[(1) Have completed ground and flight training on the maneuvers and procedures of Appendix A of this part that is appropriate to the airplane for which a type rating is sought and received an endorsement from an authorized instructor in the person's logbook or training records certifying satisfactory completion of the training; or

[(2) For a pilot employee of a Part 121 or Part 135 certificate holder, have completed ground and flight training that is appropriate to the airplane for which a type rating is sought and is approved under Parts 121 and 135.]

(Amdt. 61-29, Eff. 4/23/67); (Amdt. 61-64, Eff. 12/22/76); (Amdt. 61-69, Eff. 7/30/80); [(Amdt. 61-90, Eff. 4/15/91)]

§ 61.159 Rotorcraft rating: Aeronautical knowledge.

An applicant for an airline transport pilot certificate with a rotorcraft category and a helicopter class rating must pass a written test on—

(a) So much of this chapter as relates to air carrier rotorcraft operations;

(b) Rotorcraft design, components, systems and performance limitations;

(c) Basic principles of loading and weight distribution and their effect on rotorcraft flight characteristics;

(d) Air traffic control systems and procedures relating to rotorcraft;

(e) Procedures for operating rotorcraft in potentially hazardous meteorological conditions;

(f) Flight theory as applicable to rotorcraft; and

(g) The items listed under paragraphs (b) through (m) of § 61.153.

Docket No. 24550 (51 FR 40705) Eff. 11/7/86; (Amdt. 61-1, Eff. 11/1/62); (Amdt. 61-20, Eff. 12/16/65); (Amdt. 61-64, Eff. 12/22/76); (Amdt. 61-77, Eff. 1/6/87)

§ 61.161 Rotorcraft rating: Aeronautical experience.

(a) An applicant for an airline transport pilot certificate with a rotorcraft category and helicopter class rating must hold a commercial pilot certificate, or a foreign airline transport pilot or commercial pilot certificate with a rotorcraft category and helicopter class rating issued by a member of ICAO, or be a pilot in an armed force of the United States whose military experience qualifies that pilot for the issuance of a commercial pilot certificate under § 61.73.

(b) An applicant must have had at least 1,200 hours of flight time as a pilot, including at least—

(1) 500 hours of cross-country flight time;

(2) 100 hours at night, including at least 15 hours are in helicopters;

(3) 200 hours in helicopters, including at least 75 hours as pilot in command, or as second in command performing the duties and functions of a pilot in command under the supervision of a pilot in command, or any combination thereof; and

(4) 75 hours of instrument time under actual or simulated instrument conditions of which at least 50 hours were completed in flight with at least 25 hours in helicopters as pilot in command, or as second in command performing the duties of a pilot in command under the supervision of a pilot in command, or any combination thereof.

Docket No. 24550 (51 FR 40705) Eff. 11/7/86; (Amdt. 61-77, Eff. 1/6/87)

§ 61.163 Rotorcraft rating: Aeronautical skill.

(a) An applicant for an airline transport pilot certificate with a rotorcraft category and helicopter class rating, or additional aircraft rating, must pass a practical test on those maneuvers set forth in Appendix B of this part in a helicopter. The FAA inspector or designated examiner may modify or waive any maneuver where necessary for the reasonable and safe operation of the rotorcraft being used and may combine any maneuvers and permit their performance in any convenient sequence to determine the applicant's competency.

(b) Whenever an applicant for an airline transport pilot certificate with a rotorcraft category and helicopter class rating does not already have an instrument rating, the applicant shall, as part of the practical test, comply with § 61.65(g).

Docket No. 24550 (51 FR 40704) Eff. 11/7/86; (Amdt. 61-77, Eff. 1/6/87)

§ 61.165 Additional category ratings.

(a) *Rotorcraft category with a helicopter class rating.* The holder of an airline transport pilot certificate (airplane category) who applies for a rotorcraft category with a helicopter class rating must meet the applicable requirements of §§ 61.159 and 61.161, and 61.163 and—

(1) Have at least 100 hours, including at least 15 hours at night, of rotorcraft flight time as pilot in command or as second in command performing the duties and functions of a pilot in command under the supervision of a pilot in command who holds an airline transport pilot certificate with an appropriate rotorcraft rating, or any combination thereof; or

(2) Complete a training program conducted by a certificated air carrier or other approved agency requiring at least 75 hours of rotorcraft flight time as pilot in command, second in command, or as flight instruction from an appropriately rated FAA certificated flight instructor or an airline transport pilot, or any combination thereof, including at least 15 hours of night flight time.

(b) *Airplane rating.* The holder of an airline transport pilot certificate (rotorcraft category) who applies for an airplane category must comply with §§ 61.153, 61.155 (except § 61.155(b)(1)), and 61.157 and—

(1) Have at least 100 hours, including at least 15 hours at night, of airplane flight time as pilot in command or as second in command performing the duties and functions of a pilot in command under the supervision of a pilot in command who holds an airline transport pilot certificate with an appropriate airplane rating, or any combination thereof, or

(2) Complete a training program conducted by a certificated air carrier or other approved agency requiring at least 75 hours of airplane flight time as pilot in command, second in command, or as flight instruction from an appropriately rated FAA certificated flight instructor or an airline transport pilot, or any combination thereof, including at least 15 hours of night flight time.

Docket No. 1179 (27 FR 7965) Eff. 8/10/62; (Amdt. 61-20, Eff. 12/16/65); (Amdt. 61-64, Eff. 12/22/76); (Amdt. 61-77, Eff. 1/6/87)

§ 61.167 Tests.

(a) Each applicant for an airline transport pilot certificate must pass each practical and theoretical test to the satisfaction of the Administrator. The minimum passing grade in each subject is 70 percent. Each flight maneuver is graded separately. Other tests are graded as a whole.

(b) Information collected incidentally to such a test shall be treated as a confidential matter by the persons giving the test and by employees of the FAA.

Docket No. 1179 (27 FR 7965) Eff. 8/10/62

§ 61.169 Instruction in air transportation service.

An airline transport pilot may instruct other pilots in air transportation service in aircraft of the category, class, and type for which he is rated. However, he may not instruct for more than 8 hours in one day nor more than 36 hours in any 7-day period. He may instruct under this section only in aircraft with functioning dual controls. Unless he has a flight instructor certificate, an airline transport pilot may instruct only as provided in this section.

Docket No. 1179 (27 FR 7965) Eff. 8/10/62

§ 61.171 General privileges and limitations.

An airline transport pilot has the privileges of a commercial pilot with an instrument rating. The holder of a commercial pilot certificate who qualifies for an airline transport pilot certificate retains the ratings on his commercial pilot certificate, but he may exercise only the privileges of a commercial pilot with respect to them.

Docket No. 1179 (27 FR 7965) Eff. 8/10/62; (Amdt. 61-1, Eff. 11/1/62); (Amdt. 61-3, Eff. 5/17/63)

Subpart G—Flight instructors

§ 61.181 Applicability.

This subpart prescribes the requirements for the issuance of flight instructor certificates and ratings, the conditions under which those certificates and ratings are necessary, and the limitations upon these certificates and ratings.

(Amdt. 61-11, Eff. 12/4/64)

§ 61.183 Eligibility requirements: General.

To be eligible for a flight instructor certificate a person must—

(a) Be at least 18 years of age;

(b) Read, write, and converse fluently in English;

(c) Hold—

(1) A commercial or airline transport pilot certificate with an aircraft rating appropriate to the flight instructor rating sought, and

(2) An instrument rating, if the person is applying for an airplane or an instrument instructor rating;

(d) Pass a written test on the subjects in which ground instruction is required by § 61.185; and

(e) [Pass a practical test on all items in which instruction is required by § 61.187 and, in the case of an applicant for a flight instructor-airplane or flight instructor-glider rating, present a logbook endorsement from an appropriately certificated and rated flight instructor who has provided the applicant with spin entry, spin, and spin recovery training in an aircraft of the appropriate category that is certificated for spins, and has found that applicant competent and proficient in those training areas. Except in the case of a retest after a failure for the deficiencies stated in § 61.49(b), the person conducting the practical test may either accept the spin training logbook endorsement or require demonstration of the spin entry, spin, and spin recovery maneuver on the flight portion of the practical test.]

[(Amdt. 61-90, Eff. 4/15/91)]

§ 61.185 Aeronautical knowledge.

(a) Present evidence showing that he has satisfactorily completed a course of instruction in at least the following subjects:

(1) The learning process.

(2) Elements of effective teaching.

(3) Student evaluation, quizzing, and testing.

(4) Course development.

(5) Lesson planning.

(6) Classroom instructing techniques.

(b) Have logged ground instruction from an authorized ground or flight instructor in all of the subjects in which ground instruction is required for a private and commercial pilot certificate, and for an instrument rating, if an airplane or instrument instructor rating is sought.

§ 61.187 Flight proficiency.

(a) An applicant for a flight instructor certificate must have received flight instruction, appropriate to the instructor rating sought in the subjects listed in this paragraph by a person authorized in paragraph (b) of this section. In addition, his logbook must contain an endorsement by the person who has given him the instruction certifying that he has found the applicant competent to pass a practical on the following subjects:

(1) Preparation and conduct of lesson plans for students with varying backgrounds and levels of experience and ability.

(2) The evaluation of student flight performance.

(3) Effective preflight and postflight instruction.

(4) Flight instructor responsibilities and certifying procedures.

(5) Effective analysis and correction of common student pilot flight errors.

(6) [Performance and analysis of standard flight training procedures and maneuvers appropriate to the flight instructor rating sought. For flight instructor-airplane and flight instructor-glider applicants, this shall include the satisfactory demonstration of stall awareness, spin entry, spins, and spin recovery techniques in an aircraft of the appropriate category that is certificated for spins.]

(b) The flight instruction required by paragraph (a) of this section must be given by a person who has held a flight instructor certificate during the 24 months immediately preceding the date the instruction is given, who meets the general requirements for a flight instructor certificate prescribed in § 61.183, and who has given at least 200 hours of flight instruction, or 80 hours in the case of glider instruction, as a certificate flight instructor.

[(Amdt. 61-90, Eff. 4/15/91)]

§ 61.189 Flight instructor records.

(a) Each certificated flight instructor shall sign the logbook of each person to whom he has given flight or ground instruction and specify in that book the amount of the time and the date on which it was given. In addition, he shall maintain a record in his flight instructor logbook, or in a separate document containing the following:

(1) The name of each person whose logbook or student pilot certificate he has endorsed for solo flight privileges. The record must include the type and date of each endorsement.

(2) The name of each person for whom he has signed a certification for a written, flight, or practical test, including the kind of test, date of his certification, and the result of the test.

(b) The record required by this section shall be retained by flight instructor separately or in his logbook for at least 3 years.

§ 61.191 Additional flight instructor ratings.

The holder of a flight instructor certificate who applies for an additional rating on that certificate must—

(a) Hold an effective pilot certificate with ratings appropriate to the flight instructor rating sought.

(b) Have had at least 15 hours as pilot in command in the category and class of aircraft appropriate to the rating sought; and

(c) Pass the written and practical test prescribed in this subpart for the issuance of a flight instructor certificate with the rating sought.

§ 61.193 Flight instructor authorizations.

(a) The holder of a flight instructor certificate is authorized, within the limitations of that person's flight instructor certificate and ratings, to give the—

(1) Flight instruction required by this part for a pilot certificate or rating;

(2) Ground instruction or a home study course required by this part for a pilot certificate and rating;

(3) Ground and flight instruction required by this subpart for a flight instructor certificate and rating, if that person meets the requirements prescribed in § 61.187(b);

(4) Flight instruction required for an initial solo or cross-country flight;

(5) Flight review required in § 61.56 in a manner acceptable to the Administrator;

(6) Instrument competency check required in § 61.57 (e)(2);

(7) Pilot-in-command flight instruction required under § 61.101 (d); and

(8) Ground and flight instruction required by this part

for the issuance of the endorsements specified in paragraph (b) of this section.

(b) The holder of a flight instructor certificate is authorized within the limitations of that person's flight instructor certificate and rating, to endorse—

(1) In accordance with §§ 61.87(m) and 61.93(c) and (d), the pilot certificate of a student pilot the flight instructor has instructed authorizing the student to conduct solo or solo cross-country flights, or to act as pilot in command of an airship requiring more than one flight crewmember;

(2) In accordance with §§ 61.87(m) and 61.93(b) and (d), the logbook of a student pilot the flight instructor has instructed, authorizing single or repeated solo flights;

(3) In accordance with § 61.93(d), the logbook of a student pilot whose preparation and preflight planning for a solo cross-country flight the flight instructor has reviewed and found adequate for a safe flight under the conditions the flight instructor has listed in the logbook;

(4) [In accordance with § 61.95, the logbook of a student pilot the flight instructor has instructed authorizing solo flights in a Class B airspace area or at an airport within a Class B airspace area;]

(5) The logbook of a pilot or another flight instructor who has been trained by the person described in paragraph (b) of this section, certifying that the pilot or other flight instructor is prepared for an operating privilege, a written test, or practical test required by this part;

(6) In accordance with §§ 61.57(e)(2) and 61.101(d), the logbook of a pilot the flight instructor has instructed authorizing the pilot to act as pilot in command;

(7) [Reserved]; and

(8) In accordance with § 61.101(g) and (h), the logbook of a recreational pilot the flight instructor has instructed authorizing solo flight.

(Amdt. 61-80, Eff. 1/12/89); (Amdt. 61-82, Eff. 8/31/89); (Amdt. 61-90, Eff. 4/15/91); [(Amdt. 61-92, Eff. 9/16/93)]

§ 61.195 Flight instructor limitations.

The holder of a flight instructor certificate is subject to the following limitations:

(a) *Hours of instruction.* He may not conduct more than eight hours of flight instruction in any period of 24 consecutive hours.

(b) *Ratings.* Flight instruction may not be conducted in any aircraft for which the flight instructor does not hold a category, class, and if appropriate, a type rating, on the flight instructor's pilot and flight instructor certificates.

(c) *Endorsement of student pilot certificate.* He may not endorse a student pilot certificate for initial solo or solo cross-country flight privileges, unless he has given that student pilot flight instruction required by this part for the endorsement, and considers that the student is prepared to conduct the flight safely with the aircraft involved.

(d) *Logbook endorsement.* He may not endorse a student pilot's logbook—

(1) For solo flight unless he has given that student flight instruction and found that student pilot prepared for solo flight in the type of aircraft involved;

(2) For a cross-country flight, unless he has reviewed the student's flight preparation, planning, equipment, and proposed procedures and found them to be adequate for the flight proposed under existing circumstances; or

(3) [For solo flight in a Class B airspace area or at an airport within a Class B airspace area unless the flight instructor has given that student ground and flight instruction and has found that student prepared and competent to conduct the operations authorized.]

(e) *Solo flights.* He may not authorize any student pilot to make a solo flight unless he possesses a valid student pilot certificate endorsed for solo in the make and model aircraft to be flown. In addition, he may not authorize any student pilot to make a solo cross-country flight unless he possesses a valid student pilot certificate endorsed for solo cross-country flight in the category of aircraft to be flown.

(f) *Instruction in multiengine airplane or helicopter.* He may not give flight instruction required for the issuance of a certificate or a category, or class rating, in a multiengine airplane or a helicopter, unless he has at least five hours of experience as pilot in command in the make and model of that airplane or helicopter, as the case may be.

(g) *Recreational pilot endorsements.* The flight instructor may not endorse a recreational pilot's logbook unless the instructor has given that pilot the ground and flight instruction required under this part for the endorsement and found that pilot competent to pilot the aircraft safely.

(Amdt. 61-80, Eff. 1/12/89); (Amdt. 61-82, Eff. 8/31/89); (Amdt. 61-90, Eff. 4/15/91); [(Amdt. 61-92, Eff. 9/16/93)]

§ 61.197 Renewal of flight instructor certificates.

The holder of a flight instructor certificate may have his certificate renewed for an additional period of 24 months if he passes the practical test for a flight instructor certificate and the rating involved, or those portions of that test that the Administrator considers necessary to determine his competency as a flight instructor. His certificate may be renewed without taking the practical test if—

(a) His record of instruction shows that he is a competent flight instructor;

(b) He has a satisfactory record as a company check pilot, chief flight instructor, pilot in command of an aircraft operated under Part 121 of this chapter, or other activity involving the regular evaluation of pilots, and passes any oral test that may be necessary to determine that instructor's knowledge of current pilot training and certification requirements and standards; or

[(c) He or she has successfully completed, within 90 days before the application for the renewal of his or her certificate, an approved flight instructor refresher course consisting of ground or flight instruction, or both.]

[(Amdt. 61-95, Eff. 4/13/94)]

§ 61.199 Expired flight instructor certificates and ratings.

(a) *Flight instructor certificates.* The holder of an expired flight instructor certificate may exchange that certificate for a new certificate by passing the practical test prescribed in § 61.187.

(b) *Flight instructor ratings.* A flight instructor rating or a limited flight instructor rating on a pilot certificate is no longer valid and may not be exchanged for a similar rating or a flight instructor certificate. The holder of either of those ratings is issued a flight instructor certificate only if he passes the written and practical test prescribed in this subpart for the issue of that certificate.

§ 61.201 Conversion to new system of instructor ratings.

General. The holder of a flight instructor certificate that does not bear any of the new class or instrument ratings listed in § 61.5(c)(2), (3), or (4) for a flight instructor certificate, may not exercise the privileges of that certificate. The holder of a flight instructor certificate with a glider rating need not convert that rating to a new class rating to exercise the privileges of that certificate and rating.

[(b) through (g) removed]

(Amdt. 61-90, Eff. 4/15/91); [(OST Docket 48146, Notice No. 92-28; 57 FR 60725, effective 12/22/92)]

Appendix A to Part 61
Practical test requirements for airplane airline transport pilot certificates and associated class and type ratings

Throughout the maneuvers prescribed in this appendix, good judgment commensurate with a high level of safety must be demonstrated. In determining whether such judgment has been shown, the FAA inspector or designated examiner who conducts the check considers adherence to approved procedures, actions based on analysis of situations for which there is no prescribed procedure or recommended practice, and qualities of prudence and care in selecting a course of action.

Each maneuver or procedure must be performed inflight except to the extent that certain maneuvers or procedures may be performed in an airplane simulator with a visual system (visual simulator) or an airplane simulator without a visual system (non-visual simulator) or may be waived as indicated by an X in the appropriate columns. A maneuver authorized to be performed in a non-visual simulator may be performed in a visual simulator, and a maneuver authorized to be performed in a training device may be performed in a non-visual or a visual simulator.

An asterisk (*) preceding a maneuver or procedure indicates that the maneuver or procedure may be performed in an airplane simulator or other training device as indicated, provided the applicant has successfully completed the training set forth in § 121.424(d) of this chapter.

When a maneuver or procedure is preceded by this symbol (#), it indicates that the FAA Inspector or designated examiner may require the maneuver or procedure to be performed in the airplane if he determines such action is necessary to determine the applicant's competence with respect to that maneuver.

An X and asterisk (X*) indicates that a particular condition is specified in connection with the maneuver, procedure, or waiver provisions.

Maneuvers/Procedures	*Required* Simulated Instrument Conditions	*Required* Inflight	*Permitted* Visual Simulator	*Permitted* Non-Visual Simulator	*Permitted* Training Device	Waiver Provisions of §61.157(c)
The procedures and maneuvers set forth in this appendix must be performed in a manner that satisfactorily demonstrates knowledge and skill with respect to—	....	....	....	....	....	
(1) The airplane, its systems and components;	....	....	....	....	....	
(2) Proper control of airspeed, configuration, direction, altitude, and attitude in accordance with procedures and limitations contained in the approved Airplane Flight Manual, check lists, or other approved material appropriate to the airplane type; and	....	....	....	....	....	
(3) Compliance with approved en route, instrument approach, missed approach, ATC, or other applicable procedures.	....	....	....	....	....	
1. Preflight.	....	....	....	....	....	
(a) Equipment examination (oral). As part of the practical test the equipment examination must be closely coordinated with, and related to, the flight maneuvers portion but may not be given during the flight maneuvers portion. Nothwithstanding §61.21 the equipment examination may be given to an applicant who has completed a ground school that is part of an approved training program under Federal Aviation Regulations Part 121 for the airplane type involved and who is recommended by his instructor. The equipment examination must be repeated if the flight maneuvers portion is not satisfactorily completed within 60 days. The equipment examination must cover—	....	....	....	....	X	
(1) Subjects requiring a practical knowledge of the airplane, its powerplants, systems, components, operational, and performance factors;	....	....	....	....	....	
(2) Normal, abnormal, and emergency procedures, and the operations and limitations relating thereto; and	....	....	....	....	....	
(3) The appropriate provisions of the approved Airplane Flight Manual.	....	....	....	....	....	
(b) Preflight Inspection. The pilot must—	....	....	....	....	....	
(1) Conduct an actual visual inspection of the exterior and interior of the airplane, locating each item and explaining briefly the purpose of inspecting it; and	....	....	....	....	X	X*
(2) Demonstrate the use of the prestart check list, appropriate control system checks, starting procedures, radio and electronic equipment checks, and the selection of proper navigation and communications radio facilities and frequencies prior to flight.	....	....	....	X	....	

Maneuvers/Procedures	*Required* Simulated Instrument Conditions	*Required* Inflight	*Permitted* Visual Simulator	*Permitted* Non-Visual Simulator	*Permitted* Training Device	Waiver Provisions of §61.157(c)
If a flight engineer is a required crewmember for the particular type airplane, the actual visual inspection may either be waived or it may be replaced by using an approved pictorial means that realistically portrays the location and detail of inspection items.	....	....	....	....	....	
(c) Taxiing. This maneuver includes taxiing, sailing, or docking procedures in compliance with instructions issued by the appropriate traffic control authority or by the FAA inspector or designated examiner.	....	X	....	....	....	
(d) Powerplant checks. As appropriate to the airplane type.	....	....	....	....	X	
II. Takeoffs.						
(a) Normal. One normal takeoff which, for the purpose of this maneuver, begins when the airplane is taxied into position on the runway to be used.	....	X	....	....	....	
*(b) Instrument. One takeoff with instrument conditions simulated at or before reaching an altitude of 100 feet above the airport elevation.	X	....	X	....	....	
(c) Cross wind. One cross wind takeoff, if practicable under the existing meteorological, airport, and traffic conditions.	....	X*	....	....	....	
*#(d) Powerplant failure. One takeoff with a simulated failure of the most critical powerplant—	....	....	X	....	....	
(1) At a point after V_1, and before V_2 that in the judgment of the person conducting the check is appropriate to the airplane type under the prevailing conditions; or	....	....	....	....	....	
(2) At a point as close as possible after V1 when V_1, and V2 or V1, and Vr are identical; or	....	....	....	....	....	
(3) At the appropriate speed for non-transport category airplanes.	....	....	....	....	....	
For additional type rating in an airplane group with engines mounted in similar positions or from wing-mounted engines to aft fuselage-mounted engines this maneuver may be performed in a non-visual simulator.	....	....	....	....	....	
(e) Rejected. a rejected takeoff performed in an airplane during a normal takeoff run after reaching a reasonable speed determined by giving due consideration to aircraft characteristics, runway length, surface conditions, wind direction and velocity, brake heat energy, and any other pertinent factors that may adversely affect safety or the airplane.	....	....	....	X	....	X*
III. Instrument Procedures.	....	....	....	....	....	
*(a) Area departure and area arrival. During each of these maneuvers the applicant must—	X	....	....	X	....	X*
(1) Adhere to actual or simulated ATC clearances (including assigned radials); and	....	....	....	....	....	

Maneuvers/Procedures	Required: Simulated Instrument Conditions	Required: Inflight	Permitted: Visual Simulator	Permitted: Non-Visual Simulator	Permitted: Training Device	Waiver Provisions of §61.157(c)
(2) Properly use available navigation facilities. Either area arrival or area departure, but not both, may be waived under §61.157(c).	....	....	....	....	....	
*(b) Holding. This maneuver includes entering, maintaining, and leaving holding patterns. It may be performed under either area departure or area arrival.	X	....	....	X	....	X*
(c) ILS and other instrument approaches. There must be the following:	....	....	....	....	....	
*(1) At least one normal ILS approach	X	....	X	....	....	
#(2) At least one manually controlled ILS approach with a simulated failure of one powerplant. The simulated failure should occur before initiating the final approach course and must continue to touchdown or through the missed approach procedure.	X	....	X	....	....	
However, either the normal ILS approach or the manually controlled ILS approach must be performed in flight.	....	....	....	....	....	
(3) At least one nonprecision approach procedure that is representative of the nonprecision approach procedures that the applicant is likely to use.	X	....	X	....	....	
(4) Demonstration of at least one nonprecision approach procedure on a letdown aid other than the approach procedure performed under sub-paragraph (3) of this paragraph that the applicant is likely to use. If performed in a synthetic instrument trainer, the procedures must be observed by the FAA inspector or designated examiner, or if the applicant has completed an approved training course under Part 121 of this chapter for the airplane type involved, the procedures may be observed by a person qualified to act as an instructor or check airman under that approved training program.	X	....	....	....	X	
Each instrument approach must be performed according to any procedures and limitations approved for the approach facility used. The instrument approach begins when the airplane is over the initial approach fix for the approach procedure being used (or turned over to the final approach controller in the case of GCA approach) and ends when the airplane touches down on the runway or when transition to a missed approach configuration is completed. Instrument conditions need not be simulated below 100 feet above touchdown zone elevation.	....	....	....	....	....	
(d) Circling approaches. At least one circling approach must be made under the following conditions:	....	....	X	....	....	X*
(1) The portion of the circling approach to the authorized minimum circling	X	....	....	....	....	

Maneuvers/Procedures	Required: Simulated Instrument Conditions	Required: Inflight	Permitted: Visual Simulator	Permitted: Non-Visual Simulator	Permitted: Training Device	Waiver Provisions of §61.157(c)
approach altitude must be made under simulated instrument conditions.						
(2) The approach must be made to the authorized minimum circling approach altitude followed by a change in heading and the necessary maneuvering (by visual reference) to maintain a flight path that permits a normal landing on a runway at least 90 degrees from the final approach course of the simulated instrument portion of the approach.	....	....	....	....	....	
(3) The circling approach must be performed without excessive maneuvering, and without exceeding the normal operating limits of the airplane. The angle of bank should not exceed 30 degrees.	....	....	....	....	....	
When the maneuver is performed in an airplane, it may be waived as provided in §61.157(c) if local conditions beyond the control of the pilot prohibit the maneuver or prevent it from being performed as required.	....	....	....	....	....	
The circling approach maneuver is not required for a pilot employed by a certificate holder subject to the operating rules of Part 121 of this chapter, if the certificate holder's manual prohibits a circling approach in weather conditions below 1000-3 (ceiling and visibility).	....	....	....	....	....	
*(e) Missed approaches. Each applicant must perform at least two missed approaches, with at least one missed approach from an ILS approach. A complete approved missed approach procedure must be accomplished at least once and, at the discretion of the FAA inspector or designated examiner, a simulated powerplant failure may be required during any of the missed approaches. These maneuvers may be performed either independently or in conjunction with maneuvers required under Sections III or V of this appendix. At least one must be performed inflight.	X	X*	X*	....	....	
IV. Inflight Maneuvers.	....	....	....	....	....	
*(a) Steep turns. At least one steep turn in each direction must be performed. Each steep turn must involve a bank angle of 45 degrees with a heading change of at least 180 degrees but not more than 360 degrees.	X	....	....	X	....	X
*(b) Approaches to stalls. For the purpose of this maneuver the required approach to a stall is reached when there is a perceptible buffet or other response to the initial stall entry. Except as provided below, there must be at least three approaches to stalls as follows:	X	....	....	X	....	X*
(1) One must be in the takeoff configuration (except where the airplane uses only a zero-flap takeoff configuration).	....	....	....	....	....	

Maneuvers/Procedures	*Required* Simulated Instrument Conditions	*Required* Inflight	*Permitted* Visual Simulator	*Permitted* Non-Visual Simulator	*Permitted* Training Device	Waiver Provisions of §61.157(c)
(2) One in a clean configuration.	....	....	....	....	....	
(3) One in a landing configuration.	....	....	....	....	....	
At the discretion of the FAA inspector or designated examiner, one approach to a stall must be performed in one of the above configurations while in a turn with a bank angle between 15 and 30 degrees. Two out of the three approaches required by this paragraph may be waived as provided in §61.157(c).	....	....	....	....	....	
*(c) Specific flight characteristics. Recovery from specific flight characteristics that are peculiar to the airplane type.	....	....	....	X	....	X
(d) Powerplant failures. In addition to the specific requirements for maneuvers with simulated powerplant failures, the FAA inspector or designated examiner may require a simulated powerplant failure at any time during the check.	....	X	....	....	....	
V. Landings and Approaches to Landings. Notwithstanding the authorizations for combining of maneuvers and for waiver of maneuvers, at least three actual landings (one to a full stop) must be made. These landings must include the types listed below but more than one type can be combined where appropriate:	....	....	....	....	....	
(a) Normal landing	....	X*	....	....	....	
#(b) Landing in sequence from an ILS instrument approach except that if circumstances beyond the control of the pilot prevent an actual landing, the person conducting the check may accept an approach to a point where in his judgement a landing to a full stop could have been made. In addition, where a simulator approved for the landing maneuver out of an ILS approach is used, the approach may be continued through the landing and credit given for one of the three landings required by this section.	....	....	X*	....	....	
(c) Cross wind landing, if practical under existing meteorological, airport, and traffic conditions.	....	X*	....	....	....	
#(d) Maneuvering to a landing with simulated powerplant failure, as follows:	....	....	X*	....	....	
(1) In the case of 3-engine airplanes, maneuvering to a landing with an approved procedure that approximates the loss of two powerplants (center and one outboard-engine); or	....	....	....	....	....	
(2) In the case of other multiengine airplanes, maneuvering to a landing with a simulated failure of 50 percent of available powerplants, with the simulated loss of power on one side of the airplane. However, before January 1, 1975, in the	....	....	....	....	....	

Maneuvers/Procedures	*Required* Simulated Instrument Conditions	*Required* Inflight	*Permitted* Visual Simulator	*Permitted* Non-Visual Simulator	*Permitted* Training Device	Waiver Provisions of §61.157(c)
case of a four engine turbojet powered airplane, maneuvering to a landing with a simulated failure of the most critical powerplant may be substituted therefore, if a flight instructor in an approved training program under Part 121 of this chapter certifies to the Administrator that he has observed the applicant satisfactorily perform a landing in that type airplane with a simulated failure of 50 percent of the available powerplants. The substitute maneuver may not be used if the Administrator determines that training in the two-engine out landing maneuver provided in the training program is unsatisfactory.						
If an applicant performs this maneuver in a visual simulator, he must, in addition, maneuver in flight to a landing with a simulated failure of the most critical powerplant.	....	....	....	....	....	
#(e) Except as provided in paragraph (f), landing under simulated circling approach conditions, except that if circumstances beyond the control of the pilot prevent a landing, the person conducting the check may accept an approach to a point where in his judgment a landing to a full stop could have been made.	....	....	X*	....	....	
The circling approach maneuver is not required for a pilot employed by a certificate holder subject to the operating rules of Part 121 of this chapter, if the certificate holder's manual prohibits a circling approach in weather conditions below 1000-3 (ceiling and visibility).	....	....	....	....	....	
*#(f) A rejected landing, including a normal missed approach procedure, that is rejected approximately 50 feet over the runway and approximately over the runway threshold. This maneuver may be combined with instrument, circling, or missed approach procedures, but instrument conditions need not be simulated below 100 feet above the runway.	X*	....	X*	....	....	
#(g) A zero-flap visual approach to a point where in the judgment of the person conducting the check, a landing to a full stop on the appropriate runway could be made. This maneuver is not required for a particular airplane type if the Administrator has determined that the probability of flap extension failure on that type is extremely remote due to system design. In making its determination, the administrator determines whether checking on slats only and partial flap approaches is necessary.	....	....	X*	....	....	
(h) For a single powerplant rating only, unless the applicant holds a commercial pilot certificate, he must accomplish accuracy	....	X	....	....	....	

Maneuvers/Procedures	Required: Simulated Instrument Conditions	Required: Inflight	Permitted: Visual Simulator	Permitted: Non-Visual Simulator	Permitted: Training Device	Waiver Provisions of §61.157(c)
approaches and spot landings that include a series of three landings from an altitude of 1000 feet or less, with the engine throttled and 180 degrees change in direction. The airplane must touch the ground in a normal landing attitude beyond and within 200 feet from a designated line. At least one landing must be from a forward slip. One hundred eighty degree approaches using two 90 degree turns with a straight base leg are preferred although circular approaches are acceptable.						
VI. Normal and Abnormal Procedures. Each applicant must demonstrate the proper use of as many of the systems and devices listed below as the FAA inspector or designated examiner finds are necessary to determine that the person being checked has a practical knowledge of the use of the systems and devices appropriate to the aircraft type:	....	....	....	....	....	
(a) Anti-icing and de-icing systems	....	....	....	X	....	
(b) Auto-pilot systems	....	....	....	X	....	
(c) Automatic or other approach aid systems	....	....	....	X	....	
(d) Stall warning devices, stall avoidance devices, and stability augmentation devices	....	....	....	X	....	
(e) Airborne radar devices	....	....	....	X	....	
(f) Any other systems, devices, or aids available	....	....	....	X	....	
(g) Hydraulic and electrical system failures and malfunctions.	....	....	....	X	....	
(h) Landing gear and flap systems failures or malfunctions.	....	....	....	X	....	
(i) Failure of navigation or communications equipment.	....	....	....	X	....	
VII. Emergency Procedures. Each applicant must demonstrate the proper emergency procedures for as many of the emergency situations listed below as the FAA inspector or designated examiner finds are necessary to determine that the person being checked has an adequate knowledge of, and ability to perform, such procedures:	....	....	....	....	....	
(a) Fire inflight	....	....	....	X	....	
(b) Smoke control	....	....	....	X	....	
(c) Rapid decompression	....	....	....	X	....	
(d) Emergency descent	....	....	....	X	....	
(e) Any other emergency procedures outlined in the appropriate approved airplane flight manual.	....	....	....	X	....	

(Amdt. 61-34, Eff. 5/19/67); (Amdt. 61-39, Eff. 12/1/67); (Amdt. 61-42, Eff. 5/1/68); (Amdt. 61-45, Eff. 2/2/70); (Amdt. 61-56, Eff. 5/26/72); (Amdt. 61-57, Eff. 6/27/72); (Amdt. 61-62, Eff. 12/19/73); (Amdt. 61-77, Eff. 1/6/87)

Appendix B to Part 61—Practical test requirements for rotorcraft airline transport pilot certificates with a helicopter class rating and associated type ratings

Docket No. 24550 (51 FR 40705) effective 11/7/86.

Throughout the maneuvers prescribed in this appendix, good judgment commensurate with a high level of safety must be demonstrated. In determining whether such judgment has been shown, the FAA inspector or designated pilot examiner who conducts the check considers adherence to approved procedures, actions based on analysis of situations for which there is no prescribed procedure or recommended practice, and qualities of prudence and care in selecting a course of action. The successful outcome of a procedure or maneuver will never be in doubt.

Maneuvers/Procedures

The maneuvers and procedures in this appendix must be performed in a manner that satisfactorily demonstrates knowledge and skill with respect to—

(1) The helicopter, its systems and components;

(2) Proper control of airspeed, direction, altitude, and attitude in accordance with procedures and limitations contained in the approved Rotorcraft Flight Manual, checklists, or other approved material appropriate to the rotorcraft type; and

(3) Compliance with approved en route, instrument approach, missed approach, ATC, and other applicable procedures.

I. Preflight

(a) *Equipment examination (oral).* The equipment examination must be repeated if the flight maneuvers portion is not satisfactorily completed within 60 days. The equipment examination must cover—

(1) Subjects requiring a practical knowledge of the helicopter, its powerplants, systems, components, and operational and performance factors;

(2) Normal, abnormal, and emergency procedures and related operations and limitations; and

(3) The appropriate provisions of the approved helicopter Flight Manual or manual material.

(b) *Preflight inspection.* The pilot must—

(1) Conduct an actual visual inspection of the exterior and interior of the helicopter, locating each item and explaining briefly the purpose of inspecting it; and

(2) Demonstrate the use of the prestart checklist, appropriate control system checks, starting procedures, radio and electronic equipment checks, and the selection of proper navigation and communications radio facilities and frequencies before flight.

(c) *Taxiing.* The maneuver includes ground taxiing, hover taxiing (including performance checks), and docking procedures, as appropriate, in compliance with instructions issued by ATC, the FAA inspector, or the designated pilot examiner.

(d) *Powerplant checks.* As appropriate to the helicopter type in accordance with the Rotorcraft Flight Manual procedures.

II. Takeoffs

(a) *Normal.* One normal takeoff from a nstabilized hover which begins when the helicopter is taxied into position for takeoff.

(b) *Instrument.* One takeoff with instrument conditions simulated at or before reaching 100 feet above airport elevation.

(c) *Crosswind.* One crosswind takeoff from a stabilized hover, if practical under the existing meteorological, airport, and traffic conditions.

(d) *Powerplant failure.*

(1) For single-engine rotorcraft, one normal takeoff with simulated powerplant failure.

(2) For multiengine rotorcraft, one normal takeoff with simulated failure of one engine—

(i) At an appropriate airspeed that would allow continued climb performance in forward flight; or

(ii) At an appropriate airspeed that is 50 percent of normal cruise speed, if there is no published single-engine climb airspeed for that type of helicopter.

(e) *Rejected.* One normal takeoff that is rejected after simulated engine failure at a reasonable airspeed, determined by giving due consideration to the helicopter's characteristics, length of landing area, surface conditions, wind direction and velocity, and any other pertinent factors that may adversely affect safety.

III. Instrument Procedures

(a) *Area departure and arrival.* During each of these maneuvers, the applicant must—

(1) Adhere to actual or simulated ATC clearance (including assigned bearings or radials); and

(2) Properly use available navigation facilities.

(b) *Holding.* This maneuver includes entering, maintaining, and leaving holding patterns.

(c) *ILS and other instrument approaches.* The instrument approach begins when the helicopter is over the initial approach fix for the approach procedures being used (or turned over to the final controller in case of a surveillance or precision radar approach) and ends when the helicopter terminates at a hover or touches down or where transition to a missed approach is completed. The following approaches must be performed:

(1) At least on normal ILS approach.

(2) For multiengine rotorcraft, at least one manually controlled ILS approach with a simulated failure of one powerplant. The simulated engine failure should occur be-

fore initiating the final approach course and continue to a hover to touchdown or through the missed approach procedure.

(3) At least one nonprecision approach procedure that is representative of the nonprecision approach procedure that the applicant is likely to use.

(4) At least one nonprecision approach procedure on a letdown aid other than the approach procedure performed under subparagraph (3) of this paragraph that the applicant is likely to use.

(d) *Circling approaches.* At least one circling approach must be made under the following conditions:

(1) The portion of the circling approach to the authorized minimum circling approach altitude must be made under simulated instrument conditions.

(2) The approach must be made to the authorized minimum circling approach altitude followed by a change in heading and the necessary maneuvering (by visual reference) to maintain a flight path that permits a normal landing on a runway at least 90 degrees from the final approach course of the simulated instrument portion of the approach.

(3) The circling approach must be performed without excessive maneuvering and without exceeding the normal operating limits of the rotorcraft. The angle of bank should not exceed 30 degrees.

(e) *Missed approaches.* Each applicant must perform at least two missed approaches with at least one missed approach from the ILS approach. At the discretion of the FAA inspector or designated examiner, a simulated powerplant failure may be required during any of the missed approaches. The maneuvers may be performed either independently or in conjunction with maneuvers required under section III or V of the Appendix. At least one must be performed in flight.

IV. In-flight Maneuvers

(a) *Steep turns.* At least one steep turn in each direction must be performed. Each steep turn must involve a bank angle of 30 degrees with a heading change of at least 180 degrees but not more than 360 degrees.

(b) *Settling with power.* Demonstrate recognition of and recovery from imminent flight at critical/rapid descent with power. For the purpose of this maneuver, settling with power is reached when a perceptive buffet or other indications of imminent settling with power have been induced.

(c) *Powerplant failure.* In addition to the specific requirements for maneuvers with simulated powerplant failures, the FAA inspector or designated examiner may require a simulated powerplant failure at any time during the check.

(d) Recovery from unusual attitudes.

V. Approaches and Landings

(a) *Normal.* One normal approach to a stabilized hover or to the ground must be performed.

(b) *Instrument.* One approach to a hover or to a landing in sequence from an ILS instrument approach.

(c) *Crosswind.* One crosswind approach to a hover or to the ground, if practical under the existing meteorological, airport, or traffic conditions.

(d) *Powerplant failure.* For a multiengine rotorcraft, maneuvering to a landing with simulated powerplant failure of one engine.

(e) *Rejected.* Rejected landing, include a normal missed approach procedure at approximately 50 feet above the runway. This maneuver may be combined with instrument or missed approach procedures, but instrument conditions need not be simulated below 100 feet above the runway or landing area.

(f) *Autorotative landings.* Autorotative landings in a single-engine helicopter. The applicant may be required to accomplish at least one autorotative approach and landing from any phase of flight as specified by the FAA inspector or designated examiner.

VI. Normal and Abnormal Procedures

Each applicant must demonstrate the proper use of as many systems and devices listed below as the FAA inspector or designated examiner finds are necessary to determine that the applicant has a practical knowledge of the use of the systems and devices appropriate to helicopter type:

(a) Anti-icing or deicing system.

(b) Autopilot or other stability augmentation devices.

(c) Airborne radar devices.

(d) Hydraulic and electrical systems failures or malfunctions.

(e) Landing gear failures or malfunctions.

(f) Failure of navigation or communications equipment.

(g) Any other system appropriate to the helicopter as outlined in the approved Rotorcraft Flight Manual.

VII. Emergency Procedures

Each applicant must demonstrate the proper emergency procedures for as many of the emergency situations listed below as the FAA inspector or designated examiner finds are necessary to determine that the applicant has adequate knowledge of, and ability to perform, such procedures:

(a) Fire or smoke control in flight.

(b) Ditching.

(c) Evacuation.

(d) Operations of emergency equipment.

(e) Emergency descent.

(f) Any other emergency procedure outline in the approved Rotorcraft Flight Manual.

Docket No. 24550 (51 FR 40705) Eff. 11/7/86; (Amdt. 61-77, Eff. 1/6/87)

Special Federal Aviation Regulation 58.
Advanced Qualification Program

(Published in 55 FR 40262, October 2, 1990)

(SFAR 58, Eff. 10/2/90)

SUMMARY: This Special Federal Aviation Regulation (SFAR) establishes a voluntary, alternative method for the training, evaluation, certification, and qualification requirements of flight crewmembers, flight attendants, aircraft dispatchers, instructors, evaluators and other operations personnel subject to the training and qualification requirements of 14 CFR Parts 121 and 135. The FAA has developed this alternative method in response to recommendations made by representatives from the government, airlines, aircrew professional organizations, and airline industry organizations. The SFAR is designed to improve aircrew performance and allows certificate holders that are subject to the training requirements of Parts 121 and 135 to develop innovative training programs that incorporate the most recent advances in training methods and techniques.

[1. *Purpose and Eligibility.*

(a) This Special Federal Aviation Regulation provides for approval of an alternate method (known as "Advanced Qualification Program" or "AQP") for qualifying, training, certifying, and otherwise ensuring competency of crewmembers, aircraft dispatchers, other operations personnel, instructors, and evaluators who are required to be trained or qualified under Parts 121 and 135 of the FAR or under this SFAR.

(b) A certificate holder is eligible under this Special Federal Aviation Regulation if the certificate holder is required to have an approved training program under 121.401 or § 135.341 of the FAR, or elects to have an approved training program under § 135.341.

(c) A certificate holder obtains approval of each proposed curriculum under this AQP as specified in section 10 of this SFAR.

(d) A curriculum approved under the AQP may include elements of present Part 121 and Part 135 training programs. Each curriculum must specify the make, model, and series aircraft (or variant) and each crewmember position or other positions to be covered by that curriculum. Positions to be covered by the AQP must include all flight crewmember positions, instructors, and evaluators and may include other positions, such as flight attendants, aircraft dispatchers, and other operations personnel.

(e) Each certificate holder that obtains approval of an AQP under this SFAR shall comply with all of the requirements of that program.

2. *Definitions.*

As used in this SFAR:

"Curriculum" means a portion of an Advanced Qualification Program that covers one of three program areas: (1) indoctrination, (2) qualification, or (3) continuing qualification. A qualification or continuing qualification curriculum addresses the required training and qualification activities for a specific make, model, and series aircraft (or variant) and for a specific duty position.

"Evaluator" means a person who has satisfactorily completed training and evaluation that qualifies that person to evaluate the performance of crewmembers, instructors, other evaluators, aircraft dispatchers, and other operations personnel.

"Facility" means the physical environment required for training and qualification (e.g., buildings, classrooms).

"Training center" means an independent organization that provides training under contract or other arrangement to certificate holders. A training center may be a certificate holder that provides training to another certificate holder, an aircraft manufacturer that provides training to certificate holders, or any non-certificate holder that provides training to a certificate holder.

"Variant" means a specifically configured aircraft for which the FAA has identified training and qualification requirements that are significantly different from those applicable to other aircraft of the same make, model, and series.

3. *Required Curriculums.*

Each AQP must have separate curriculums for indoctrination, qualification, and continuing qualification as specified in §§ 4, 5, and 6 of this SFAR.

4. *Indoctrination Curriculums.*

Each indoctrination curriculum must include the following:

(a) For newly hired persons being trained under an AQP: Company policies and operating practices and general operational knowledge.

(b) For newly hired flight crewmembers and aircraft dispatchers: General aeronautical knowledge.]

[(c) For instructors: The fundamental principles of the teaching and learning process; methods and theories of instruction; and the knowledge necessary to use aircraft, flight training devices, flight simulators, and other training equipment in advanced qualification curriculums.

(d) For evaluators: Evaluation requirements specified in each approved curriculum; methods of evaluating crewmembers and aircraft dispatchers and other operations personnel; and policies and practices used to conduct the kinds of evaluations particular to an advanced qualification curriculum (e.g., proficiency and online).

5. *Qualification Curriculums.*

Each qualification curriculum must include the following:

(a) The certificate holder's planned hours of training, evaluation, and supervised operating experience.

(b) A list of and text describing the training, qualification, and certification activities, as applicable for specific positions subject to the AQP, as follows:

(1) *Crewmembers, aircraft dispatchers, and other operations personnel.* Training, evaluation, and certification activities which are aircraft- and equipment-specific to qualify a person for a particular duty position on, or duties related to the operation of a specific make, model, and series aircraft (or variant); a list of and text describing the knowledge requirements, subject materials, job skills, and each maneuver and procedure to be trained and evaluated;

the practical test requirements in addition to or in place of the requirements of Parts 61, 63, and 65; and a list of and text describing supervised operating experience.

(2) *Instructors.* Training and evaluation to qualify a person to impart instruction on how to operate, or on how to ensure the safe operation of a particular make, model, and series aircraft (or variant).

(3) *Evaluators.* Training, evaluation, and certification activities that are aircraft and equipment specific to qualify a person to evaluate the performance of persons who operate or who ensure the safe operation of, a particular make, model, and series aircraft (or variant).

6. *Continuing Qualification Curriculums.*

Continuing qualification curriculums must comply with the following requirements:

(a) *General.* A continuing qualification curriculum must be based on—

(1) A continuing qualification cycle that ensures that during each cycle each person qualified under an AQP, including instructors and evaluators, will receive a balanced mix of training and evaluation on all events and subjects necessary to ensure that each person maintains the minimum proficiency level of knowledge, skills, and attitudes required for original qualification; and

(2) If applicable, flight crewmember or aircraft dispatcher recency of experience requirements.

(b) *Continuing Qualification Cycle Content.* Each continuing qualification cycle must include at least the following:

(1) *Evaluation period.* An evaluation period during which each person qualified under an AQP must receive at least one training session and a proficiency evaluation at a training facility. The number and frequency of training sessions must be approved by the Administrator.

A training session, including any proficiency evaluation completed at that session, that occurs any time during the two calendar months before the last date for completion of an evaluation period can be considered by the certificate holder to be completed in the last calendar month.

(2) *Training.* Continuing qualification must include training in all events and major subjects required for original qualification, as follows:

(i) For pilots in command, seconds in command, flight engineers, and instructors and evaluators: Ground training including a general review of knowledge and skills covered in qualification training, updated information on newly developed procedures, and safety information.]

[(ii) For crewmembers, aircraft dispatchers, instructors, evaluators, and other operation personnel who conduct their duties in flight: Proficiency training in an aircraft, flight training device, or flight simulator on normal, abnormal, and emergency flight procedures and maneuvers.

(iii) For instructors and evaluators who are limited to conducting their duties in flight simulators and flight training devices: Proficiency training in a flight training device and/or flight simulator regarding operation of this training equipment and in operational flight procedures and maneuvers (normal, abnormal, and emergency).

(3) *Evaluations.* Continuing qualification must include evaluation in all events and major subjects required for original qualification, and online evaluations for pilots in command and other eligible flight crewmembers. Each person qualified under an AQP must successfully complete a proficiency evaluation and, if applicable, an online evaluation during each evaluation period. An individual's proficiency evaluation may be accomplished over several training sessions if a certificate holder provides more than one training session in an evaluation period. The following evaluation requirements apply:

(i) Proficiency evaluations as follows:

(A) For pilots in command, seconds in command, and flight engineers: A proficiency evaluation, portions of which may be conducted in an aircraft, flight simulator, or flight training device as approved in the certificate holder's curriculum which must be completed during each evaluation period.

(B) For any other persons covered by an AQP a means to evaluate their proficiency in the performance of their duties in their assigned tasks in an operational setting.

(ii) Online evaluations as follows:

(A) For pilots in command: An online evaluation conducted in an aircraft during actual flight operations under Part 121 or Part 135 or during operationally (line) oriented flights, such as ferry flights or proving flights. An online evaluation in an aircraft must be completed in the calendar month that includes the midpoint of the evaluation period. An online evaluation that is satisfactorily completed in the calendar month before or the calendar month after the calendar month in which it becomes due is considered to have been completed during the calendar month it became due. However, in no case is an online evaluation under this paragraph required more often than once during an evaluation period.

(B) During the online evaluations required under paragraph (b)(3)(ii)(A) of this section, each person performing duties as a pilot in command, second in command, or flight engineer for that flight, must be individually evaluated to determine whether he or she—(1) Remains adequately trained and currently proficient with respect to the particular aircraft, crew position, and type of operation in which he or she serves; and (2) Has sufficient knowledge and skills to operate effectively as part of a crew.

(4) *Recency of experience.* For pilots in command and seconds in command, and, if the certificate holder elects, flight engineers and aircraft dispatchers, approved recency of experience requirements.

(c) *Duration periods.* Initially the continuing qualification cycle approved for an AQP may not exceed 26 calendar months and the evaluation period may not exceed 13 calendar months. Thereafter, upon demonstration by a certificate holder that an extension is warranted, the Administrator may approve extensions of the continuing qualification cycle and the evaluation period in increments not exceeding 3 calendar months. However, a continuing qualification cycle may not exceed 39 calendar months and an evaluation period may not exceed 26 calendar months.

(d) *Requalification.* Each continuing qualification curriculum must include a curriculum segment that covers the requirements for requalifying a crewmember, aircraft dispatcher, or other operations personnel who has not maintained continuing qualification.

7. *Other Requirements.* In addition to the requirements of sections 4, 5, and 6, each AQP qualification and continuing qualification curriculum must include the following requirements:]

[(a) Approved Cockpit Resource Management (CRM) Training applicable to each position for which training is provided under an AQP.

(b) Approved training on and evaluation of skills and proficiency of each person being trained under an AQP to use their cockpit resource management skills and their technical (piloting or other) skills in an actual or simulated operations scenario. For flight crewmembers this training and evaluation must be conducted in an approved flight training device or flight simulator.

(c) Data collection procedures that will ensure that the certificate holder provides information from its crewmembers, instructors, and evaluators that will enable the FAA to determine whether the training and evaluations are working to accomplish the overall objectives of the curriculum.

8. *Certification.*

A person enrolled in an AQP is eligible to receive a commercial or airline transport pilot, flight engineer, or aircraft dispatcher certificate or appropriate rating based on the successful completion of training and evaluation events accomplished under that program if the following requirements are met:

(a) Training and evaluation of required knowledge and skills under the AQP must meet minimum certification and rating criteria established by the Administrator in Parts 61, 63, or 65. The Administrator may accept substitutes for the practical test requirements of Parts 61, 63, or 65, as applicable.

(b) The applicant satisfactorily completes the appropriate qualification curriculum.

(c) The applicant shows competence in required technical knowledge and skills (e.g., piloting) and cockpit resource management knowledge and skills in scenarios that test both types of knowledge and skills together.

(d) The applicant is otherwise eligible under the applicable requirements of Part 61, 63, or 65.

9. *Training Devices and Simulators.*

(a) *Qualification and approval of flight training devices and flight simulators.*

(1) Any training device or simulator that will be used in an AQP for one of the following purposes must be evaluated by the Administrator for assignment of a flight training device or flight simulator qualification level:

(i) Required evaluation of individual or crew proficiency.

(ii) Training activities that determine if an individual or crew is ready for a proficiency evaluation.

(iii) Activities used to meet recency of experience requirements.

(iv) Line Operational Simulations (LOS).

(2) To be eligible to request evaluation for a qualification level of a flight training device or flight simulator an applicant must—

(i) Hold an operating certificate; or

(ii) Be a training center that has applied for authorization to the Administrator or has been authorized by the Administrator to conduct training or qualification under an AQP.

(3) Each flight training device or flight simulator to be used by a certificate holder or training center for any of the purposes set forth in paragraph (a)(1) of this section must—

(i) Be, or have been, evaluated against a set of criteria established by the Administrator for a particular qualification level of simulation;

(ii) Be approved for its intended use in a specified AQP; and

(iii) Be part of a flight simulator or flight training device continuing qualification program approved by the Administrator.]

[(b) *Approval of other Training Equipment.*

(1) Any training device that is intended to be used in an AQP for purposes other than those set forth in paragraph (a)(1) of this section must be approved by the Administrator for its intended use.

(2) An applicant for approval of a training device under this paragraph must identify the device by its nomenclature and describe its intended use.

(3) Each training device approved for use in an AQP must be part of a continuing program to provide for its serviceability and fitness to perform its intended function as approved by the Administrator.

10. *Approval of Advanced Qualification Program.*

(a) *Approval Process.* Each applicant for approval of an AQP curriculum under this SFAR shall apply for approval of that curriculum. Application for approval is made to the certificate holder's FAA Flight Standards District Office.

(b) *Approval Criteria.* An application for approval of an AQP curriculum will be approved if the program meets the following requirements:

(1) It must be submitted in a form and manner acceptable to the Administrator.

(2) It must meet all of the requirements of this SFAR.

(3) It must indicate specifically the requirements of Parts 61, 63, 65, 121 or 135, as applicable, that would be re-

placed by an AQP curriculum. If a requirement of Parts 61, 63, 65, 121, or 135 is replaced by an AQP curriculum, the certificate holder must show how the AQP curriculum provides an equivalent level of safety for each requirement that is replaced. Each applicable requirement of Parts 61, 63, 65, 121 or 135 that is not specifically addressed in an AQP curriculum continues to apply to the certificate holder.

(c) *Application and Transition.* Each certificate holder that applies for one or more advanced qualification curriculums or for a revision to a previously approved curriculum must comply with § 121.405 or § 135.325, as applicable, and must include as part of its application a proposed transition plan (containing a calendar of events) for moving from its present approved training to the advanced qualification training.

(d) *Advanced Qualification Program Revisions or Rescissions of Approval.* If after a certificate holder begins operations under an AQP, the Administrator finds that the certificate holder is not meeting the provisions of its approved AQP, the Administrator may require the certificate holder to make revisions in accordance with § 121.405 or § 135.325, as applicable, or to submit and obtain approval for a plan (containing a schedule of events) that the certificate holder must comply with and use to transition to an approved Part 121 or Part 135 training program, as appropriate.

11. *Approval of Training, Qualification, or Evaluation by a Person who Provides Training by Arrangement.*

(a) A certificate holder under Part 121 or Part 135 may arrange to have AQP required training, qualification, or evaluation functions performed by another person (a "training center") if the following requirements are met:

(1) The training center's training and qualification curriculums, curriculum segments, or portions of curriculum segments must be provisionally approved by the Administrator. A training center may apply for provisional approval independently or in conjunction with a certificate holder's application for AQP approval. Application for provisional approval must be made to the FAA's Flight Standards District Office that has responsibility for the training center.

(2) The specific use of provisionally approved curriculums, curriculum segments, or portions of curriculum segments in a certificate holder's AQP must be approved by the Administrator as set forth in Section 10 of this SFAR.]

[(b) An applicant for provisional approval of a curriculum, curriculum segment, or portion of a curriculum segment under this paragraph must show that the following requirements are met:

(1) The applicant must have a curriculum for the qualification and continuing qualification of each instructor or evaluator employed by the applicant.

(2) The applicant's facilities must be found by the Administrator to be adequate for any planned training, qualification, or evaluation for a Part 121 or Part 135 certificate holder.

(3) Except for indoctrination curriculums, the curriculum, curriculum segment, or portion of a curriculum segment must identify the specific make, model, and series aircraft (or variant) and crewmember or other positions for which it is designed.

(c) A certificate holder who wants approval to use a training center's provisionally approved curriculum, curriculum segment, or portion of a curriculum segment in its AQP, must show that the following requirements are met:

(1) Each instructor or evaluator used by the training center must meet all of the qualification and continuing qualification requirements that apply to employees of the certificate holder that has arranged for the training, including knowledge of the certificate holder's operations.

(2) Each provisionally approved curriculum, curriculum segment, or portion of a curriculum segment must be approved by the Administrator for use in the certificate holder's AQP. The Administrator will either provide approval or require modifications to ensure that each curriculum, curriculum segment, or portion of a curriculum segment is applicable to the certificate holder's AQP.

12. *Recordkeeping Requirements.*

Each certificate holder and each training center holding AQP provisional approval shall show that it will establish and maintain records in sufficient detail to establish the training, qualification, and certification of each person qualified under an AQP in accordance with the training, qualification, and certification requirements of this SFAR.

13. *Expiration.*

This Special Federal Aviation Regulation terminates on October 2, 1995, unless sooner terminated.]

Part 67 – Medical standards and certification

Subpart A—General

67.1 Applicability
[67.3 Access to the National Driver Register]
67.11 Issue
67.12 Certification of foreign airmen
67.13 First-class medical certificate
67.15 Second-class medical certificate
67.17 Third-class medical certificate
67.19 Special issue of medical certificates
67.20 Applications, certificates, logbooks, reports, and records: Falsification, reproduction or alteration

Subpart B—Certification procedures

67.21 Applicability
67.23 Medical examinations: Who may give
67.25 Delegation of authority
67.27 Denial of medical certificate
67.29 Medical certificates by senior flight surgeons of Armed Forces
67.31 Medical records

Subpart A—General

§ 67.1 Applicability.

This subpart prescribes the medical standards for issuing medical certificates for airmen.

[§ 67.3 Access to the National Driver Register.

At the time of application for a certificate issued under this part, each person who applies for a medical certificate shall execute an express consent form authorizing the Administrator to request the chief driver licensing official of any state designated by the Administrator to transmit information contained in the National Driver Register about the person to the Administrator. The Administrator shall make information received from the National Driver Register, if any, available on request to the person for review and written comment.]

§ 67.11 Issue.

Except as provided in § 67.12, an applicant who meets the medical standards prescribed in this Part, based on medical examination and evaluation of his history and condition, is entitled to an appropriate medical certificate.

§ 67.12 Certification of foreign airmen.

A person who is neither a United States citizen nor a resident alien is issued a certificate under this part, outside the United States, only when the Administrator finds that the certificate is needed for operation of a U.S.-registered civil aircraft.

§ 67.13 First-class medical certificate.

(a) To be eligible for a first-class medical certificate, an applicant must meet the requirements of paragraphs (b) through (f) of this section.

(b) *Eye* :

(1) Distant visual acuity of 20/20 or better in each eye separately, without correction; or of at least 20/100 in each eye separately corrected to 20/20 or better with corrective lenses (glasses or contact lenses), in which case the applicant may be qualified only on the condition that he wears those corrective lenses while exercising the privileges of his airman certificate.

(2) Near vision of at least v = 1.00 at 18 inches with each eye separately, with or without corrective glasses.

(3) Normal color vision.

(4) Normal fields of vision.

(5) No acute or chronic pathological condition of either eye or adenexae that might interfere with its proper function, might progress to that degree, or might be aggravated by flying.

(6) Bifoveal fixation and vergencephoria relationship sufficient to prevent a break in fusion under conditions that may reasonably occur in performing airman duties.

Tests for the factors named in subparagraph (6) of this paragraph are not required except for applicants found to have more than one prism diopter of hyperphoria, six prism diopters of esophoria, or six prism diopters of exophoria. If these values are exceeded, the Federal Air Surgeon may require the applicant to be examined by a qualified eye specialist to determine if there is bifoveal fixation and adequate vergencephoria relationship. However, if the applicant is otherwise qualified, he is entitled to a medical certificate pending the results of the examination.

(c) *Ear, nose, throat, and equilibrium:*

(1) Ability to—

(i) Hear the whispered voice at a distance of at least 20 feet with each ear separately; or

(ii) Demonstrate a hearing acuity of at least 50 percent of normal in each ear throughout the effective speech and radio range as shown by a standard audiometer.

(2) No acute or chronic disease of the middle or internal ear.

(3) No disease of the mastoid.

(4) No unhealed (unclosed) perforation of the eardrum.

(5) No disease or malformation of the nose or throat that might interfere with, or be aggravated by, flying.

(6) No disturbance in equilibrium.

(d) *Mental and neurologic:*

(1) Mental.

(i) No established medical history or clinical diagnosis of any of the following:

(a) A personality disorder that is severe enough to have repeatedly manifested itself by overt acts.

(b) A psychosis.

[*(c)* Alcoholism, unless there is established clinical evidence, satisfactory to the Federal Air Surgeon, of recovery, including sustained total abstinence from alcohol for not less than the preceding 2 years. As used in this section, "alcoholism" means a condition in which a person's intake of alcohol is great enough to damage physical health or personal or social functioning, or when alcohol has become a prerequisite to normal functioning.]

(d) Drug dependence. As used in this section, "drug dependence" means a condition in which a person is addicted to or dependent on drugs other than alcohol, tobacco, or ordinary caffeine-containing beverages, as evidenced by habitual use or a clear sense of need for the drug.

(ii) No other personality disorder, neurosis, or mental condition that the Federal Air Surgeon finds—

(a) Makes the applicant unable to safely perform the duties or exercise the privileges of the airman certificate that he holds or for which he is applying; or

(b) May reasonably be expected, within two years after the finding, to make him unable to perform those duties or exercise those privileges;

and the findings are based on the case history and appropriate qualified, medical judgement relating to the condition involved.

(2) Neurologic.

(i) No established medical history or clinical diagnosis of either of the following:

(a) Epilepsy.

(b) A disturbance of consciousness without satisfactory medical explanation of the cause.

(ii) No other convulsive disorder, disturbance of consciousness, or neurologic condition that the Federal Air Surgeon finds—

(a) Makes the applicant unable to safely perform the duties or exercise the privileges of the airman certificate that he holds or for which he is applying; or

(b) May reasonably be expected, within two years after the finding, to make him unable to perform those duties or exercise those privileges;

and the findings are based on the case history and appropriate, qualified, medical judgement relating to the condition involved.

(e) *Cardiovascular:*

(1) No established medical history or clinical diagnosis of—

[(i) Myocardial infarction;

[(ii) Angina pectoris; or

[(iii) Coronary heart disease that has required treatment or, if untreated, that has been symptomatic or clinically significant.]

(2) If the applicant has passed his thirty-fifth birthday but not his fortieth, he must, on the first examination after his thirty-fifth birthday, show an absence of myocardial infarction on electrocardiographic examination.

(3) If the applicant has passed his fortieth birthday, he must annually show an absence of myocardial infarction on electrocardiographic examination.

(4) Unless the adjusted maximum readings apply, the applicant's reclining blood pressure may not be more than the maximum reading for his age group in the following table:

Age Group	Maximum readings (reclining blood pressure in mm)		Adjusted maximum readings (reclining blood pressure in mm)[1]	
	Systolic	Diastolic	Systolic	Diastolic
20–29	140	88	—	—
30–39	145	92	155	98
40–49	155	96	165	100
50 and over	160	98	170	100

[1]For an applicant at least 30 years of age whose reclining blood pressure is more than the maximum reading for his age group and whose cardiac and kidney conditions, after complete cardiovascular examination, are found to be normal.

(5) If the applicant is at least 40 years of age, he must show a degree of circulatory efficiency that is compatible with the safe operation of aircraft at high altitudes.

An electrocardiogram, made according to acceptable standards and techniques within the 90 days before an examination for a first-class certificate, is accepted at the time of the physical examination as meeting the requirements of subparagraphs (2) and (3) of this paragraph.

(f) *General medical condition:*

(1) No established medical history or clinical diagnosis of diabetes mellitus that requires insulin or any other hypoglycemic drug for control.

(2) No other organic, functional, or structural disease, defect, or limitation that the Federal Air Surgeon finds—

(i) Makes the applicant unable to safely perform the duties or exercise the privileges of the airman certificate that he holds or for which he is applying; or

(ii) May reasonably be expected, within two years after the finding, to make him unable to perform those duties or exercise those privileges;

and the findings are based on the case history and appropriate, qualified, medical judgment relating to the condition involved.

[(g) An applicant who does not meet the provisions of paragraphs (b) through (f) of this section may apply for the discretionary issuance of a certificate under § 67.19.]

§ 67.15 Second-class medical certificate.

(a) To be eligible for a second-class medical certificate, an applicant must meet the requirements of paragraphs (b) through (f) of this section.

(b) *Eye:*

(1) Distant visual acuity of 20/20 or better in each eye separately, without correction; or of at least 20/100 in each eye separately corrected to 20/20 or better with corrective lenses (glasses or contact lenses), in which case the applicant may be qualified only on the condition that he wears those corrective lenses while exercising the privileges of his airman certificate.

(2) Enough accommodation to pass a test prescribed by the Administrator based primarily on ability to read official aeronautical maps.

(3) Normal fields of vision.

(4) No pathology of the eye.

(5) Ability to distinguish aviation signal red, aviation signal green, and white.

(6) Bifoveal fixation and vergencephoria relationship sufficient to prevent a break in fusion under conditions that may reasonably occur in performing airman duties.

Tests for the factors named in subparagraph (6) of this paragraph are not required except for applicants found to have more than one prism diopter of hyperphoria, six prism diopters of esophoria, or six prism diopters of exophoria. If these values are exceeded, the Federal Air Surgeon may require the applicant to be examined by a qualified eye specialist to determine if there is bifoveal fixation and adequate vergencephoria relationship. However, if the applicant is otherwise qualified, he is entitled to a medical certificate pending the results of the examination.

(c) *Ear, nose, throat, and equilibrium:*

(1) Ability to hear the whispered voice at 8 feet with each ear separately.

(2) No acute or chronic disease of the middle or internal ear.

(3) No disease of the mastoid.

(4) No unhealed (unclosed) perforation of the eardrum.

(5) No disease or malformation of the nose or throat that might interfere with, or be aggravated by, flying.

(6) No disturbance in equilibrium.

(d) *Mental and neurologic:*

(1) Mental.

(i) No established medical history or clinical diagnosis of any of the following:

(a) A personality disorder that is severe enough to have repeatedly manifested itself by overt acts.

(b) A psychosis.

[*(c)* Alcoholism, unless there is established clinical evidence, satisfactory to the Federal Air Surgeon, of recovery, including sustained total abstinence from alcohol for not less than the preceding 2 years. As used in this section, "alcoholism" means a condition in which a person's intake of alcohol is great enough to damage physical health or personal or social functioning, or when alcohol has become a prerequisite to normal functioning.]

(d) Drug dependence. As used in this section, "drug dependence" means a condition in which a person is addicted to or dependent on drugs other than alcohol, tobacco, or ordinary caffeine-containing beverages, as evidenced by habitual use or a clear sense of need for the drug.

(ii) No other personality disorder, neurosis, or mental condition that the Federal Air Surgeon finds—

(a) Makes the applicant unable to safely perform the duties or exercise the privileges of the airman certificate that he holds or for which he is applying; or

(b) May reasonably be expected, within two years after the finding, to make him unable to perform those duties or exercise those privileges;

and the findings are based on the case history and appropriate, qualified, medical judgment relating to the condition involved.

(2) Neurologic.

(i) No established medical history or clinical diagnosis of either of the following:

(a) Epilepsy.

(b) A disturbance of consciousness without satisfactory medical explanation of the cause.

(ii) No other convulsive disorder, disturbance of consciousness, or neurologic condition that the Federal Air Surgeon finds—

(a) Makes the applicant unable to safely perform the duties or exercise the privileges of the airman certificate that he holds or for which he is applying; or

(b) May reasonably be expected, within two years after the finding, to make him unable to perform those duties or exercise those privileges;

and the findings are based on the case history and appropriate, qualified, medical judgment relating to the condition involved.

(e) *Cardiovascular:*

(1) No established medical history or clinical diagnosis of—

[(i) Myocardial infarction;

[(ii) Angina pectoris; or

[(iii) Coronary heart disease that has required treatment or, if untreated, that has been symptomatic or clinically significant.]

(f) *General medical condition:*

(1) No established medical history or clinical diagnosis of diabetes mellitus that requires insulin or any other hypoglycemic drug for control.

(2) No other organic, functional, or structural disease, defect, or limitation that the Federal Air Surgeon finds—

(i) Makes the applicant unable to safely perform the duties or exercise the privileges of the airman certificate that he holds or for which he is applying; or

(ii) May reasonably be expected, within 2 years after the finding to make him unable to perform those duties or exercise those privileges;

and the findings are based on the case history and appropriate, qualified, medical judgment relating to the condition involved.

[(g) An applicant who does not meet the provisions of paragraphs (b) through (f) of this section may apply for the discretionary issuance of a certificate under § 67.19.]

§ 67.17 Third-class medical certificate.

(a) To be eligible for a third-class medical certificate, an applicant must meet the requirements of paragraphs (b) through (f) of this section.

(b) *Eye:*

(1) Distant visual acuity of 20/50 or better in each eye separately, without correction; or if the vision in either or both eyes is poorer than 20/50 and is corrected to 20/30 or better in each eye with corrective lenses (glasses or contact lenses), the applicant may be qualified on the condition that he wears those corrective lenses while exercising the privileges of his airman certificate.

(2) No serious pathology of the eye.

(3) Ability to distinguish aviation signal red, aviation signal green, and white.

(c) *Ears, nose, throat, and equilibrium:*

(1) Ability to hear the whispered voice at 3 feet.

(2) No acute or chronic disease of the internal ear.

(3) No disease or malformation of the nose or throat that might interfere with, or be aggravated by, flying.

(4) No disturbance in equilibrium.

(d) *Mental and neurologic:*

(1) Mental.

(i) No established medical history or clinical diagnosis of any of the following:

(a) A personality disorder that is severe enough to have repeatedly manifested itself by overt acts.

(b) A psychosis.

[*(c)* Alcoholism, unless there is established clinical evidence, satisfactory to the Federal Air Surgeon, of recovery, including sustained total abstinence from alcohol for not less than the preceding 2 years. As used in this section, "alcoholism" means a condition in which a person's intake of alcohol is great enough to damage physical health or personal or social functioning, or when alcohol has become a prerequisite to normal functioning.]

(d) Drug dependence. As used in this section, "drug dependence" means a condition in which a person is addicted to or dependent on drugs other than alcohol, tobacco, or ordinary caffeine-containing beverages, as evidenced by habitual use or a clear sense of need for the drug.

(ii) No other personality disorder, neurosis, or mental condition that the Federal Air Surgeon finds—

(a) Makes the applicant unable to safely perform the duties or exercise the privileges of the airman certificate that he holds or for which he is applying; or

(b) May reasonably be expected, within two years after the finding, to make him unable to perform those duties or exercise those privileges;

and the findings are based on the case history and appropriate, qualified, medical judgment relating to the condition involved.

(2) Neurologic.

(i) No established medical history or clinical diagnosis of either of the following:

(a) Epilepsy

(b) A disturbance of consciousness without satisfactory medical explanation of the cause.

(ii) No other convulsive disorder, disturbance of consciousness, or neurologic condition that the Federal Air Surgeon finds—

(a) Makes the applicant unable to safely perform the duties or exercise the privileges of the airman certificate that he holds or for which he is applying; or

(b) May reasonably be expected, within two years after the finding, to make him unable to perform those duties or exercise those privileges;

and the findings are based on the case history and appropriate, qualified, medical judgment relating to the condition involved.

(e) *Cardiovascular:*

(1) No established medical history or clinical diagnosis of—

[(i) Myocardial infarction;

[(ii) Angina pectoris; or

[(iii) Coronary heart disease that has required treatment or, if untreated, that has been symptomatic or clinically significant.]

(f) *General medical condition:*

(1) No established medical history or clinical diagnosis of diabetes mellitus that requires insulin or any other hypoglycemic drug for control;

(2) No other organic, functional or structural disease, defect, or limitation that the Federal Air Surgeon finds—

(i) Makes the applicant unable to safely perform the duties or exercise the privileges of the airman certificate that he holds or for which he is applying; or

(ii) May reasonably be expected, within 2 years after the finding, to make him unable to perform those duties or exercise those privileges;

and the findings are based on the case history and appropriate, qualified, medical judgment relating to the condition involved.

[(g) An applicant who does not meet the provisions of paragraphs (b) through (f) of this section may apply for the discretionary issuance of a certificate under § 67.19.]

§ 67.19 Special issue of medical certificates.

(a) At the discretion of the Federal Air Surgeon, a medical certificate may be issued to an applicant who does not meet the applicable provisions of § 67.13, 67.15, or 67.17 if the applicant shows to the satisfaction of the Federal Air Surgeon that the duties authorized by the class of medical certificate applied for can be performed without endangering air commerce during the period in which the certificate would be in force. The Federal Air Surgeon may authorize a special medical flight test, practical test, or medical evaluation for this purpose.

(b) The Federal Air Surgeon may consider the applicant's operational experience and any medical facts that may affect the ability of the applicant to perform airman duties including:

(1) The combined effect on the applicant of failure to meet more than one requirement of this part; and

(2) The prognosis derived from professional consideration of all available information regarding the airman.

(c) In determining whether the special issuance of a third-class medical certificate should be made to an applicant, the Federal Air Surgeon considers the freedom of an airman, exercising the privileges of a private pilot certificate, to accept reasonable risks to his or her person and property that are not acceptable in the exercise of commercial or airline transport privileges, and, at the same time, considers the need to protect the public safety of persons and property in other aircraft and on the ground.

(d) In issuing a medical certificate under this section, the Federal Air Surgeon may do any or all of the following:

(1) Limit the duration of the certificate.

(2) Condition the continued effect of the certificate on the results of subsequent medical tests, examinations, or evaluations.

(3) Impose any operational limitation on the certificate needed for safety.

(4) Condition the continued effect of a second- or third-class medical certificate on compliance with a statement of functional limitations issued to the applicant in coordination with the [Director, Flight Standards Service] or the Director's designee.

(e) An applicant who has been issued a medical certificate under this section based on a special medical flight or practical test need not take the test again during later physical examinations unless the Federal Air Surgeon determines that the physical deficiency has become enough more pronounced to require another special medical flight or practical test.

(f) The authority of the Federal Air Surgeon under this section is also exercised by the [Manager], Aeromedical Certification Branch, Civil Aeromedical Institute, and each Regional Flight Surgeon.

§ 67.20 Applications, certificates, logbooks, reports, and records: Falsification, reproduction, or alteration.

(a) No person may make or cause to be made—

(1) Any fraudulent or intentionally false statement on any application for a medical certificate under this part;

(2) Any fraudulent or intentionally false entry in any logbook, record, or report that is required to be kept, made, or used, to show compliance with any requirement for any medical certificate under this part;

(3) Any reproduction, for fraudulent purpose, of any medical certificate under this part; or

(4) Any alteration of any medical certificate under this part.

(b) The commission by any person of an act prohibited under paragraph (a) of this section is a basis for suspending or revoking any airman, ground instructor, or medical certificate or rating held by that person.

Subpart B—Certification procedures

§ 67.21 Applicability.

This subpart prescribes the general procedures that apply to the issue of medical certificates for airmen.

§ 67.23 Medical examinations: Who may give.

(a) *First class.* Any aviation medical examiner who is specifically designated for the purpose may give the examination for the first class certificate. Any interested person may obtain a list of these aviation medical examiners, in any area, from the FAA [Regional Administrator] of the region in which the area is located.

(b) *Second class and third class.* Any aviation medical examiner may give the examination for the second or third class certificate. Any interested person may obtain a list of aviation medical examiners, in any area, from the FAA [Regional Administrator] of the region in which the area is located.

§ 67.25 Delegation of authority.

(a) The authority of the Administrator, under section 602 of the Federal Aviation Act of 1958 (49 U.S.C. 1422), to issue or deny medical certificates is delegated to the

Federal Air Surgeon, to the extent necessary to—

(1) Examine applicants for and holders of medical certificates for compliance with applicable medical standards; and

(2) Issue, renew, or deny medical certificates to applicants and holders based upon compliance or noncompliance with applicable medical standards.

Subject to limitations in this chapter, the authority delegated in paragraphs (1) and (2) of this section is also delegated to aviation medical examiners and to authorized representatives of the Federal Air Surgeon within the FAA.

(b) The authority of the Administrator, under subsection 314(b) of the Federal Aviation Act of 1958 (49 U.S.C. 1355(b)), to reconsider the action of an aviation medical examiner is delegated to the Federal Air Surgeon, the [Chief, Aeromedical Certification Division] , and each Regional Flight Surgeon. Where the applicant does not meet the standards of § 67.13(d) (1) (ii), (d) (2) (ii), or (f) (2), § 67.15(d) (1) (ii), (d) (2) (ii), or (f) (2), or § 67.17(d) (1) (ii), (d) (2) (ii), or (f) (2), any action taken under this paragraph other than by the Federal Air Surgeon is subject to reconsideration by the Federal Air Surgeon. A certificate issued by an aviation medical examiner is considered to be affirmed as issued unless an FAA official named in this paragraph on his own initiative reverses that issuance within 60 days after the date of issuance. However, if within 60 days after the date of issuance that official requests the certificate holder to submit additional medical information, he may on his own initiative reverse the issuance within 60 days after he receives the requested information.

(c) The authority of the Administrator, under section 609 of the Federal Aviation Act of 1958 (49 U.S.C. 1429), to reexamine any civil airman, to the extent necessary to determine an airman's qualification to continue to hold an airman medical certificate, is delegated to the Federal Air Surgeon and his authorized representatives within the FAA.

§ 67.27 Denial of medical certificate.

(a) Any person who is denied a medical certificate by an aviation medical examiner may, within 30 days after the date of the denial, apply in writing and in duplicate to the Federal Air Surgeon, Attention: [Manager, Aeromedical Certification Division, AAM-300], Federal Aviation Administration, P.O. Box 25082, Oklahoma City, Okla. 73125, for reconsideration of that denial. If he does not apply for reconsideration during the 30 day period after the date of the denial, he is considered to have withdrawn his application for a medical certificate.

(b) The denial of a medical certificate—

(1) By an aviation medical examiner is not a denial by the Administrator under section 602 of the Federal Aviation Act of 1958 (49 U.S.C. 1422);

(2) By the Federal Air Surgeon is considered to be a denial by the Administrator under that section of the Act; and

(3) By the [Manager, Aeromedical Certification Division, AAM-300], or a Regional Flight Surgeon is considered to be a denial by the Administrator under the Act except where the applicant does not meet the standards of § 67.13 (d) (1) (ii), (d) (2) (ii), or (f) (2), § 67.15(d) (1) (ii), (d) (2) (ii), (f) (2), or § 67.17(d) (1) (ii), (d) (2) (ii), or (f) (2).

(c) Any action taken under § 67.25(b) that wholly or partly reverses the issue of a medical certificate by an aviation medical examiner is the denial of a medical certificate under paragraph (b) of this section.

(d) If the issue of a medical certificate is wholly or partly reversed upon reconsideration by the Federal Air Surgeon, the [Manager, Aeromedical Certification Division, AAM-300], or a Regional Flight Surgeon, the person holding that certificate shall surrender it, upon request of the FAA.

§ 67.29 Medical certificates by senior flight surgeons of Armed Forces.

(a) The FAA has designated senior flight surgeons of the Armed Forces on specified military posts, stations, and facilities, as aviation medical examiners.

(b) An aviation medical examiner described in paragraph (a) of this section may give physical examinations to applicants for FAA medical certificates who are on active duty or who are, under Department of Defense medical programs, eligible for FAA medical certification as civil airmen. In addition, such an examiner may issue or deny an appropriate FAA medical certificate in accordance with the regulations of this chapter and the policies of the FAA.

(c) Any interested person may obtain a list of the military posts, stations and facilities at which a senior flight surgeon has been designated as an aviation medical examiner, from the Surgeon General of the Armed Force concerned or from the [Manager, Aeromedical Certification Division, AAM-300], Department of Transportation, Federal Aviation Administration, Civil Aeromedical Institute, P. O. Box 25082, Oklahoma City, Oklahoma 73125.

§ 67.31 Medical records.

Whenever the Administrator finds that additional medical information or history is necessary to determine whether an applicant for or the holder of a medical certificate meets the medical standards for it, he requests that person to furnish that information or authorize any clinic, hospital, doctor, or other person to release to the Administrator any available information or records concerning that history. If the applicant, or holder, refuses to provide the requested medical information or history or to authorize the release so requested, the Administrator may suspend, modify, or revoke any medical certificate that he holds or may, in the case of an applicant, refuse to issue a medical certificate to him.

Part 71 – Designation of Class A, Class B, Class C, Class D, and Class E airspace areas; airways; routes; and reporting points

Subpart A—General; Class A airspace

71.1 [Applicability
71.3 [Reserved]
71.5 Reporting points
71.7 Bearings, radial, and mileages
71.9 Overlapping airspace designations
71.31 Class A airspace
71.33 Class A airspace areas

Subpart B—Class B airspace

71.41 Class B airspace

Subpart C—Class C airspace

71.51 Class C airspace

Subpart D—Class D airspace

71.61 Class D airspace

Subpart E—Class E airspace

71.71 Class E airspace
71.73 Classification of Federal airways
71.75 Extent of Federal airways
71.77 [Removed and Reserved]
71.79 Designation of VOR Federal airways

Subpart F—[Reserved]

Subpart G—[Reserved]

Subpart H—Reporting points

71.901 Applicability

Subpart A—General; Class A airspace

Source: Docket No. 24456, (56 FR 65638), December 17, 1991.

§ 71.1 [Applicability.

[The complete listing for all Class A, Class B, Class C, Class D, and Class E airspace areas and for all reporting points can be found in FAA Order 7400.9B, Airspace Designations and Reporting Points, dated July 18, 1994. This incorporation by reference was approved by the Director of the Federal Register in accordance with 5 U.S.C. 552(a) and 1 CFR part 51. The approval to incorporate by reference FAA Order 7400.9B is effective September 16, 1994, through September 15, 1995. During the incorporation by reference period, proposed changes to the listings of Class A, Class B, Class C, Class D, and Class E airspace areas and to reporting points will be published in full text as proposed rule documents in the *Federal Register*. Amendments to the listings of Class A, Class B, Class C, Class D, and Class E airspace areas and to reporting points will be published in full text as final rules in the *Federal Register*. Periodically, the final rule amendments will be integrated into a revised edition of the order and submitted to the Director of the Federal Register for approval for incorporation by reference in this section. Copies of FAA Order 7400.9B may be obtained from the Document Inspection Facility, APA-220, Federal Aviation Administration, 800 Independence Avenue, SW., Washington, DC 20591, (202) 267-3485. Copies of FAA Order 7400.9B may be inspected in Docket No. 27855 at the Federal Aviation Administration, Office of the Chief Counsel, AGC-200, Room 915G, 800 Independence Avenue, SW., Washington, D.C. weekdays between 8:30 a.m. and 5:00 pm., or at the Office of the Federal Register, 800 North Capitol Street, NW., Suite 700, Washington, DC. This section is effective September 16, 1994, through September 15, 1995.]

(Amdt. 71-16, Eff. 9/16/93); (Amdt. 71-19, Eff. 9/16/93); [(Amdt. 71-20, Eff. 9/16/93)]

§ 71.3 [Reserved]

§ 71.5 Reporting points.

The reporting points listed in subpart H of [FAA Order 7400.9B] (incorporated by reference, see § 71.1) consist of geographic locations at which the position of an aircraft must be reported in accordance with part 91 of this chapter. [(Amdt. 71-20, Eff. 9/16/93)]

§ 71.7 Bearings, radials, and mileages.

All bearings and radials in this part are true and are applied from point of origin and all mileages in this part are stated as nautical miles.

§ 71.9 Overlapping airspace designations.

(a) When overlapping airspace designations apply to the same airspace, the operating rules associated with the more restrictive airspace designation apply.

(b) For the purpose of this section—

(1) Class A airspace is more restrictive than Class B, Class C, Class D, Class E, or Class G airspace;

(2) Class B airspace is more restrictive than Class C, Class D, Class E, or Class G airspace;

(3) Class C airspace is more restrictive than Class D, Class E, or Class G airspace;

(4) Class D airspace is more restrictive than Class E or Class G airspace; and

(5) Class E is more restrictive than Class G airspace.

§ 71.31 Class A airspace.

The airspace descriptions contained in § 71.33 of this part and the routes contained in subpart A of [FAA Order 7400.9B] (incorporated by reference, see § 71.1) are designated as Class A airspace within which all pilots and aircraft are subject to the rating requirements, operating rules, and equipment requirements of part 91 of this chapter. [(Amdt. 71-20, Eff. 9/16/93)]

§ 71.33 Class A airspace areas.

(a) That airspace of the United States, including that airspace overlying the waters within 12 nautical miles of the coast of the 48 contiguous states, from 18,000 feet MSL to and including FL600 excluding the states of Alaska and Hawaii, Santa Barbara Island, Farallon Island, and the airspace south of latitude 25°04'00" North.

(b) That airspace of the State of Alaska, including that airspace overlying the waters within 12 nautical miles of the coast, from 18,000 feet MSL to and including FL600 but not including the airspace less than 1,500 feet above the surface of the earth and the Alaska Peninsula west of longitude 160°00'00" West.

[(c) The airspace areas listed as offshore airspace areas in subpart A of FAA Order 7400.9B (incorporated by reference, see § 71.1) that are designated in international airspace within areas of domestic radio navigational signal or ATC radar coverage, and within which domestic ATC procedures are applied.]

[(Amdt. 71-19, Eff. 9/16/93)]

Subpart B—Class B airspace

§ 71.41 Class B airspace.

The Class B airspace areas listed in subpart B of [FAA Order 7400.9B] (incorporated by reference, see § 71.1) consist of specified airspace within which all aircraft operators are subject to the minimum pilot qualification requirements, operating rules, and aircraft equipment requirements of part 91 of this chapter. Each Class B airspace area designated for an airport in subpart B of [FAA Order 7400.9B] (incorporated by reference, see § 71.1) contains at least one primary airport around which the airspace is designated.

[(Amdt. 71-20, Eff. 9/16/93)]

Subpart C—Class C airspace

§ 71.51 Class C airspace.

The Class C airspace areas listed in subpart C of [FAA Order 7400.9B] (incorporated by reference, see § 71.1) consist of specified airspace within which all aircraft operators are subject to operating rules and equipment requirements specified in part 91 of this chapter. Each Class C airspace area designated for an airport in subpart C of [FAA Order 7400.9B] (incorporated by reference, see § 71.1) contains at least one primary airport around which the airspace is designated. [(Amdt. 71-20, Eff. 9/16/93)]

Subpart D—Class D airspace

§ 71.61 Class D airspace.

The Class D airspace areas listed in subpart D of [FAA Order 7400.9B] (incorporated by reference, see § 71.1) consist of specified airspace within which all aircraft operators are subject to operating rules and equipment requirements specified in part 91 of this chapter. Each Class D airspace area designated for an airport in subpart D of [FAA Order 7400.9B] (incorporated by reference, see § 71.1) contains at least one primary airport around which the airspace is designated. [(Amdt. 71-20, Eff. 9/16/93)]

Subpart E—Class E airspace

§ 71.71 Class E airspace.

Class E Airspace consists of:

(a) [The airspace of the United States, including that airspace overlying the waters within 12 nautical miles of the coast of the 48 contiguous states and Alaska, extending upward from 14,500 feet MSL up to, but not including 18,000 feet MSL, and the airspace above FL600, excluding—*]

(1) The Alaska peninsula west of longitude 160°00'00"W.; and

(2) The airspace below 1,500 feet above the surface of the earth.

(b) The airspace areas designated for an airport in subpart E of [FAA Order 7400.9B] (incorporated by reference, see § 71.1) within which all aircraft operators are subject to the operating rules specified in part 91 of this chapter.

(c) The airspace areas listed as domestic airspace areas in subpart E of [FAA Order 7400.9B] (incorporated by reference, see § 71.1) which extend upward from 700 feet or more above the surface of the earth when designated in conjunction with an airport for which an approved instrument approach procedure has been prescribed, or from 1,200 feet or more above the surface of the earth for the purpose of transitioning to or from the terminal or en route environment. When such areas are designated in conjunction with airways or routes, the extent of such designation has the lateral extent identical to that of a Federal airway and extends upward from 1,200 feet or higher. Unless otherwise specified, the airspace areas in the paragraph extend upward from 1,200 feet or higher above the surface to, but not including, 14,500 feet MSL.

(d) The Federal airways described in subpart E of [FAA Order 7400.9B] (incorporated by reference, see § 71.1).

(e) The airspace areas listed as en route domestic airspace areas in subpart E of [FAA Order 7400.9B] (incorporated by reference, see § 71.1). Unless otherwise specified, each airspace area has a lateral extent identical to that of a Federal airway and extends upward from 1,200 feet above the surface of the earth to the overlying or adjacent controlled airspace.

(f) The airspace areas listed as offshore airspace areas in subpart E of [FAA Order 7400.9B] (incorporated by reference, see § 71.1) which are designated in international airspace within areas of domestic radio navigational signal or ATC radar coverage, and within which domestic ATC procedures are applied. Unless otherwise specified, each airspace area extends upward from a specified altitude up to, but not including, 18,000 feet MSL. (Amdt. 71-16, Eff. 9/16/93); (Amdt. 71-19, Eff. 9/16/93); [(Amdt. 71-20, Eff. 9/16/93)]; [*(Amdt. 71-21, Eff. 9/16/93)]

§ 71.73 Classification of Federal airways.

Federal airways are classified as follows:

(a) Colored Federal airways:

(1) Green Federal airways.

(2) Amber Federal airways.

(3) Red Federal airways.

(4) Blue Federal airways.

(b) VOR Federal airways.

§ 71.75 Extent of Federal airways.

(a) Each Federal airway is based on a center line that extends from one navigational aid or intersection to another navigational aid (or through several navigational aids or intersections) specified for that airway.

(b) Unless otherwise specified:

(1) Each Federal airway includes the airspace within parallel boundary lines 4 miles each side of the center line. Where an airway changes direction, it includes that airspace enclosed by extending the boundary lines of the airway segments until they meet.

(2) Where the changeover point for an airway segment is more than 51 miles from either of the navigational aids defining that segment, and—

(i) The changeover point is midway between the navigational aids, the airway includes the airspace between lines diverging at angles of 4.5° from the center line at each navigational aid and extending until they intersect opposite the changeover point; or

(ii) The changeover point is not midway between the navigational aids, the airway includes the airspace between lines diverging at angles of 4.5° from the center line at the navigational aid more distant from the changeover point, and extending until they intersect with the bisector of the angle of the center lines at the changeover point; and between lines connecting these points of intersection and the navigational aid nearer to the changeover point.

(3) Where an airway terminates at a point or intersection more than 51 miles from the closest associated navigational aid, it includes the additional airspace within lines diverging at angles of 4.5° from the center line extending from the associated navigational aid to a line perpendicular to the center line at the termination point.

(4) Where an airway terminates, it includes the airspace within a circle centered at the specified navigational aid or intersection having a diameter equal to the airway width at that point. However, an airway does not extend into an oceanic control area.

(c) Unless otherwise specified—

(1) Each Federal airway includes that airspace extending upward from 1,200 feet above the surface of the earth to, but not including, 18,000 feet MSL, except that Federal airways for Hawaii have no upper limits. Variations of the lower limits of an airway are expressed in digits representing hundreds of feet above the surface or MSL and, unless

otherwise specified, apply to the segment of an airway between adjoining navigational aids or intersections; and

(2) The airspace of a Federal airway, within the lateral limits of a Class E airspace area with a lower floor, has a floor coincident with the floor of that area.

(d) A Federal airway does not include the airspace of a prohibited area.

§ 71.77 [Removed and Reserved.]

(Amdt. 71-19, Eff. 9/16/93)]

§ 71.79 Designation of VOR Federal airways.

Unless otherwise specified the place names appearing in the descriptions of airspace areas in subpart E of [FAA Order 7400.9B] (incorporated by reference, see § 71.1) designated as VOR Federal airways indicate VOR or VORTAC navigational facilities identified by those names. [(Amdt. 71-20, Eff. 9/16/93)]

Subpart F—[Reserved]

Subpart G—[Reserved]

Subpart H—Reporting points

§ 71.901 Applicability.

Unless otherwise designated:

(a) Each reporting point listed in subpart H of [FAA Order 7400.9B] (incorporated by reference, see § 71.1) applies to all directions of flight. In any case where a geographic location is designated as a reporting point for less than all airways passing through that point, or for a particular direction of flight along an airway only, it is so indicated by including the airways or direction of flight in the designation of geographical location.

(b) Place names appearing in the reporting point descriptions indicate VOR or VORTAC facilities identified by those names.

[(Amdt. 71-20, Eff. 9/16/93)]

Part 73 – Special use airspace

Subpart A—General

73.1 Applicability
73.3 Special use airspace
73.5 Bearings; radials; miles

Subpart B—Restricted areas

73.11 Applicability
73.13 Restrictions
73.15 Using agency
73.17 Controlling agency
73.19 Reports by using agency
73.21 through 73.72

Subpart C—Prohibited areas

73.81 Applicability
73.83 Restrictions
73.85 Using agency
73.87 through 73.99

Special Federal Aviation Regulation

Special Federal Aviation Regulation No. 53 *(Rule)*

Subpart A—General

§ 73.1 Applicability.

The airspace that is described in Subpart B and Subpart C of this Part is designated as special use airspace. This Part prescribes the requirements for the use of that airspace.

§ 73.3 Special use airspace.

(a) Special use airspace consists of airspace of defined dimensions identified by an area on the surface of the earth wherein activities must be confined because of their nature, or wherein limitations are imposed upon aircraft operations that are not a part of those activities, or both.

(b) The vertical limits of special use airspace are measured by designated altitude floors and ceilings expressed as flight levels or as feet above mean sea level. Unless otherwise specified, the word "to" (an altitude or flight level) means "to and including" (that altitude or flight level).

(c) The horizontal limits of special use airspace are measured by boundaries described by geographic coordinates or other appropriate references that clearly define their perimeter.

(d) The period of time during which a designation of special use airspace is in effect is stated in the designation.

§ 73.5 Bearings; radials; miles.

(a) All bearings and radials in this Part are true from point of origin.

(b) Unless otherwise specified, all mileages in this Part as stated as statute miles.

Subpart B—Restricted areas

§ 73.11 Applicability.

This subpart designates restricted areas and prescribes limitations on the operation of aircraft within them.

§ 73.13 Restrictions.

No person may operate an aircraft within a restricted area between the designated altitudes and during the time of designation, unless he has the advance permission of—

(a) The using agency described in § 73.15; or

(b) The controlling agency described in § 73.17.

§ 73.15 Using agency.

(a) For the purposes of this subpart, the following are using agencies:

(1) The agency, organization, or military command whose activity within a restricted area necessitated the area being so designated.

[(2) [Reserved]]

(b) Upon the request of the FAA, the using agency shall execute a letter establishing procedures for joint use of a restricted area by the using agency and the controlling agency, under which the using agency would notify the controlling agency whenever the controlling agency may grant permission for transit through the restricted area in accordance with the terms of the letter.

(c) The using agency shall—

(1) Schedule activities within the restricted area;

(2) Authorize transit through, or flight within, the restricted area as feasible; and

(3) Contain within the restricted area all activities conducted therein in accordance with the purpose for which it was designated.

§ 73.17 Controlling agency.

For the purposes of this Part, the controlling agency is the FAA facility that may authorize transit through or flight within a restricted area in accordance with a joint-use letter issued under § 73.15.

§ 73.19 Reports by using agency.

(a) Each using agency shall prepare a report on the use of each restricted area assigned thereto during any part of the preceding 12-month period ended September 30, and transmit it by the following January 31 of each year to the [Manager], Air Traffic Division in the regional office of the Federal Aviation Administration having jurisdiction over the area in which the restricted area is located, with a copy to the Director, Office of Air Traffice System Management, Federal Aviation Administration, Washington, D.C. 20591.

(b) In the report under this section the using agency shall:

(1) State the name and number of the restricted area as published in this part, and the period covered by the report.

(2) State the activities (including average daily number of operations if appropriate) conducted in the area, and any other pertinent information concerning current and future electronic monitoring devices.

(3) State the number of hours daily, the days of the week, and the number of weeks during the year that the area was used.

(4) For restricted areas having a joint-use designation, also state the number of hours daily, the days of the week, and the number of weeks during the year that the restricted area was released to the controlling agency for public use.

(5) State the mean sea level altitudes or flight levels (whichever is appropriate) used in aircraft operations and the maximum and average ordinate of surface firing (expressed in feet, mean sea level altitude) used on a daily, weekly, and yearly basis.

(6) Include a chart of the area (of optional scale and design) depicting, if used, aircraft operating areas, flight patterns, ordnance delivery areas, surface firing points, and target, fan, and impact areas. After once submitting an appropriate chart, subsequent annual charts are not required unless there is a change in the area, activity or altitude (or flight levels) used, which might alter the depiction of the activities originally reported. If no change is to be submitted, a statement indicating "no change" shall be included in the report.

(7) Include any other information not otherwise required under this part which is considered pertinent to activities carried on in the restricted area.

(c) If it is determined that the information submitted under paragraph (b) of this section is not sufficient to evaluate the nature and extent of the use of a restricted area, the FAA may request the using agency to submit supplementary reports. Within 60 days after receiving a request for additional information, the using agency shall submit such information as the Director, Office of Air Traffice System Management considers appropriate. Supplementary reports must be sent to the FAA officials designated in paragraph (a) of this section.

§§ 73.21 through 73.72

[Redesignations.] [§§ 608.21 through 608.72 of the Regulations of the Administrator are hereby redesignated as §§ 73.21 through 73.72, respectively.]*

Subpart C—Prohibited areas

§ 73.81 Applicability.

This subpart designates prohibited areas and prescribes limitations on the operation of aircraft therein.

§ 73.83 Restrictions.

No person may operate an aircraft within a prohibited area unless authorization has been granted by the using agency.

§ 73.85 Using agency.

For the purpose of this subpart, the using agency is the agency, organization or military command that established the requirements for the prohibited area.

§ 73.87 through 73.99*

These sections are reserved for descriptions of designated prohibited areas.

Ch. 4 (Amdt. 73-5, Eff. 10/25/89)

Special Federal Aviation Regulation No. 53. Establishment of warning areas in the airspace overlying the waters between 3 and 12 nautical miles from the United States coast

[**1. Applicability.** This rule establishes warning areas in the same location as nonregulatory warning areas previ-

*The airspace descriptions in this part and their subsequent changes are published in the Federal Register. Due to their complexity and length, they will not be included in this publication of Part 73.

ously designated over international waters. This special regulation does not affect the validity of any nonregulatory warning area which is designated over international waters beyond 12 nautical miles from the coast of the United States. This special regulation expires on January 15, 1996.]

2. [Definition-Warning area. A warning area established under this special rule is airspace of defined dimensions, extending from 3 to 12 nautical miles from the coast of the United States, that contains activity which may be hazardous to nonparticipating aircraft. The purpose of such warning areas is to warn nonparticipating pilots of the potential danger. Part 91 is applicable within the airspace designated under this special rule.]

3. Participating aircraft. Each person conducting an aircraft operation within a warning area designated under this special rule and operating with the approval of the using agency may deviate from the rules of Part 91, Subpart B, to the extent that the rules are not compatible with approved operations.

4. Nonparticipating aircraft. Nonparticipating pilots, while not excluded from the warning areas established by this SFAR, are on notice that military activity, which may be hazardous to nonparticipating aircraft, is conducted in these areas.

Ch. 6 (SFAR 53-2, Eff. 12/27/90)

Part 91 – General operating and flight rules

Subpart A—General

91.1 Applicability
91.3 Responsibility and authority of the pilot in command
91.5 Pilot in command of aircraft requiring more than one required pilot
91.7 Civil aircraft airworthiness
91.9 Civil aircraft flight manual, marking, and placard requirements
91.11 Prohibition against interference with crewmembers
91.13 Careless or reckless operation
91.15 Dropping objects
91.17 Alcohol or drugs
91.19 Carriage of narcotic drugs, marijuana, and depressant or stimulant drugs or substances
91.21 Portable electronic devices
91.23 Truth-in-leasing clause requirement in leases and conditional sales contracts
91.25 Aviation Safety Reporting Program: Prohibition against use of reports for enforcement purposes
91.27–91.99 [Reserved]

Subpart B—Flight rules general

91.101 Applicability
91.103 Preflight action
91.105 Flight crewmembers at stations
91.107 [Use of safety belts, shoulder harnesses, and child restraint systems]
91.109 Flight instruction; Simulated instrument flight and certain flight tests
91.111 Operating near other aircraft
91.113 Right-of-way rules: Except water operations
91.115 Right-of-way rules: Water operations
91.117 Aircraft speed
91.119 Minimum safe altitudes: General
91.121 Altimeter settings
91.123 Compliance with ATC clearances and instructions
91.125 ATC light signals
91.126 Operating on or in the vicinity of an airport in Class G airspace.
91.127 Operating on or in the vicinity of an airport in Class E airspace
91.129 Operations in Class D airspace
91.130 Operations in Class C airspace
91.131 Operations in Class B airspace
91.133 Restricted and prohibited areas
91.135 Operations in Class A airspace
91.137 Temporary flight restrictions
[91.138 Temporary flight restrictions in national disaster areas in the State of Hawaii]
91.139 Emergency air traffic rules
91.141 Flight restrictions in the proximity of the Presidential and other parties
91.143 Flight limitation in the proximity of space flight operations
91.144 Temporary restriction on flight operations during abnormally high barometric pressure conditions
91.145–91.149 [Reserved]

Visual flight rules

91.151 Fuel requirements for flight in VFR conditions
91.153 VFR flight plan: Information required
91.155 Basic VFR weather minimums
[91.157 Special VFR weather minimums]
91.159 VFR cruising altitude or flight level
91.161–91.165 [Reserved]

Instrument flight rules

91.167 Fuel requirements for flight in IFR conditions
91.169 IFR flight plan: Information required
91.171 VOR equipment check for IFR operations
91.173 ATC clearance and flight plan required
91.175 Takeoff and landing under IFR
91.177 Minimum altitudes for IFR operations
91.179 IFR cruising altitude or flight level
91.181 Course to be flown
91.183 IFR radio communications
91.185 IFR operations: Two-way radio communications failure
91.187 Operation under IFR in controlled airspace: Malfunction reports
91.189 Category II and III operations: General operating rules
91.191 Category II manual
91.193 Certificate of authorization for certain Category II operations
91.195–91.199 [Reserved]

Subpart C—Equipment, instrument, and certificate requirements

91.201 [Reserved]
91.203 Civil aircraft: Certifications required
91.205 Powered civil aircraft with standard category U.S. airworthiness certificates: Instrument and equipment requirements

91.207 Emergency locator transmitters
91.209 Aircraft lights
91.211 Supplemental oxygen
91.213 Inoperative instruments and equipment
91.215 ATC transponder and altitude reporting equipment and use
91.217 Data correspondence between automatically reported pressure altitude data and the pilot's altitude reference
91.219 Altitude alerting system or device: Turbojet-powered civil airplanes
91.221 Traffic alert and collision avoidance system equipment and use
91.223–91.299 [Reserved]

Subpart D—Special flight operations

91.301 [Reserved]
91.303 Aerobatic flight
91.305 Flight test areas
91.307 Parachutes and parachuting
91.309 Towing: Gliders
91.311 Towing: Other than under § 91.309
91.313 Restricted category civil aircraft: Operating limitations
91.315 Limited category civil aircraft: Operating limitations
91.317 Provisionally certificated civil aircraft: Operating limitations
91.319 Aircraft having experimental certificates: Operating limitations
91.321 Carriage of candidates in Federal elections
91.323 Increased maximum certificated weights for certain airplanes operated in Alaska
91.325 Primary category aircraft: Operating limitations
91.327–91.399 [Reserved]

Subpart E—Maintenance, preventive maintenance, and alterations

91.401 Applicability
91.403 General
91.405 Maintenance required.
91.407 Operation after maintenance, preventive maintenance, rebuilding, or alteration
91.409 Inspections
91.411 Altimeter system and altitude reporting equipment tests and inspections
91.413 ATC transponder tests and inspections
91.415 Changes to aircraft inspection programs
91.417 Maintenance records
91.419 Transfer of maintenance records
91.421 Rebuilt engine maintenance records
91.423–91.499 [Reserved]

Subpart F—Large and turbine-powered multiengine airplanes

91.501 Applicability
91.503 Flying equipment and operating information
91.505 Familiarity with operating limitations and emergency equipment
91.507 Equipment requirements: Over-the-top or night VFR operations
91.509 Survival equipment for overwater operations
91.511 Radio equipment for overwater operations
91.513 Emergency equipment
91.515 Flight altitude rules
91.517 Passenger information
91.519 Passenger briefing
91.521 Shoulder harness
91.523 Carry-on baggage
91.525 Carriage of cargo
91.527 Operating in icing conditions
91.529 Flight engineer requirements
91.531 Second in command requirements
91.533 Flight attendant requirements
[91.535 Stowage of food, beverage, and passenger service equipment during aircraft movement on the surface, takeoff, and landing]
91.537–91.599 [Reserved]

Subpart G—Additional equipment and operating requirements for large and transport category aircraft

91.601 Applicability
91.603 Aural speed warning device
91.605 Transport category civil airplane weight limitations
91.607 Emergency exits for airplanes carrying passengers for hire
91.609 Flight recorders and cockpit voice recorders
91.611 Authorization for ferry flight with one engine inoperative
91.613 Materials for compartment interiors
91.615–91.699 [Reserved]

Subpart H—Foreign aircraft operations and operations of U.S.-registered civil aircraft outside of the United States

91.701 Applicability
91.703 Operations of civil aircraft of U.S. registry outside of the United States
91.705 Operations within the North Atlantic Minimum Navigation Performance Specifications airspace
91.707 Flights between Mexico or Canada and the United States
91.709 Operations to Cuba
91.711 Special rules for foreign civil aircraft

91.713 Operation of civil aircraft of Cuban registry
91.715 Special flight authorizations for foreign civil aircraft
91.717–91.799 [Reserved]

Subpart I—Operating noise limits

91.801 Applicability: Relation to Part 36
91.803 Part 125 operators: Designation of applicable regulations
91.805 Final compliance: Subsonic airplanes
91.807 Phased compliance under Parts 121, 125, and 135: Subsonic airplanes
91.809 Replacement airplanes
91.811 Service to small communities exemption: Two-engine, subsonic airplanes
91.813 Compliance plans and status: U.S. operations of subsonic airplanes.
91.815 Agricultural and fire fighting airplanes: Noise operating limitations.
91.817 Civil aircraft sonic boom
91.819 Civil supersonic airplanes that do not comply with Part 36
91.821 Civil supersonic airplanes: Noise limits
91.823–91.849 [Reserved]
91.851 Definitions
91.853 Final compliance: Civil subsonic airplanes
91.855 Entry and nonaddition rule
91.857 Airplanes imported to points outside the contiguous United States
91.859 Modification to meet Stage 3 noise levels
91.861 Base level
91.863 Transfers of Stage 2 airplanes with base level
91.865 Phased compliance for operators with base level
91.867 Phased compliance for new entrants
91.869 Carry-forward compliance
91.871 Waivers from interim compliance requirements
91.873 Waivers from final compliance
91.875 Annual progress reports
91.877–91.899 [Reserved]

Subpart J—Waivers

91.901 [Reserved]
91.903 Policy and procedures
91.905 List of rules subject to waivers
91.907–91.999 [Reserved]

Appendix A

Category II operations: Manual, instruments, equipment, and maintenance

Appendix B

Authorizations to exceed Mach 1 (section 91.817)

Appendix C

Operations in the North Atlantic (NAT) Minimum Navigation Performance Specifications (MNPS) airspace

Appendix D

Airports/locations: special operating restrictions

Appendix E

Airplane flight recorder specifications

Appendix F

Helicopter flight recorder specifications

Special Federal Aviation Regulations

Special Federal Aviation Regulation No. 50-2. Special flight rules in the vicinity of the Grand Canyon National Park, AZ *(Rule)*
Special Federal Aviation Regulation No. 51-1. Special flight rules in the vicinity of Los Angeles International Airport *(Rule)*
Special Federal Aviation Regulation No. 60. Air Traffic Control System emergency operation *(Rule)*
Special Federal Aviation Regulation No. 62. Suspension of certain aircraft operations from the transponder with automatic pressure altitude reporting capability requirement *(Rule)*
Special Federal Aviation Regulation No. 64. Special flight authorizations for noise restricted aircraft *(Rule)*
Special Federal Aviation Regulation No. 66. Prohibition against certain flights between the United States and Yugoslavia *(Rule)*
Special Federal Aviation Regulation No. 67. Prohibition against certain flights within the territory and airspace of Afghanistan *(Rule)*
Special Federal Aviation Regulation No. 68. Prohibition against certain flights within the territory and airspace of Yemen *(Rule)*
Special Federal Aviation Regulation No. 69. Prohibition against certain flights between the United States and Haiti *(Rule)*

Subpart A—General

§ 91.1 Applicability.

(a) Except as provided in paragraph (b) of this section and § 91.703, this part prescribes rules governing the operation of aircraft (other than moored balloons, kites, unmanned rockets, and unmanned free balloons, which are governed by part 101 of this chapter, and ultralight vehicles operated in accordance with part 103 of this chapter) within the United States, including the waters within 3 nautical miles of the U.S. coast.

(b) Each person operating an aircraft in the airspace overlying the waters between 3 and 12 nautical miles from the coast of the United States shall comply with §§ 91.1 through 91.21; §§ 91.101 through 91.143; §§ 91.151 through 91.159; §§ 91.167 through 91.193; § 91.203; § 91.205; §§ 91.209 through 91.217; § 91.221; §§ 91.303 through 91.319; § 91.323; § 91.605; § 91.609; §§ 91.703 through 91.715; and § 91.903.

§ 91.3 Responsibility and authority of the pilot in command.

(a) The pilot in command of an aircraft is directly responsible for, and is the final authority as to, the operation of that aircraft.

(b) In an in-flight emergency requiring immediate action, the pilot in command may deviate from any rule of this part to the extent required to meet that emergency.

(c) Each pilot in command who deviates from a rule under paragraph (b) of this section shall, upon the request of the Administrator, send a written report of that deviation to the Administrator.

(Approved by the Office of Management and Budget under OMB control number 2120-0005).

§ 91.5 Pilot in command of aircraft requiring more than one required pilot.

No person may operate an aircraft that is type certificated for more than one required pilot flight crewmember unless the pilot in command meets the requirements of § 61.58 of this chapter.

§ 91.7 Civil aircraft airworthiness.

(a) No person may operate a civil aircraft unless it is in an airworthy condition.

(b) The pilot in command of a civil aircraft is responsible for determining whether that aircraft is in condition for safe flight. The pilot in command shall discontinue the flight when unairworthy mechanical, electrical, or structural conditions occur.

§ 91.9 Civil aircraft flight manual, marking, and placard requirements.

(a) Except as provided in paragraph (d) of this section, no person may operate a civil aircraft without complying with the operating limitations specified in the approved Airplane or Rotorcraft Flight Manual, markings, and placards, or as otherwise prescribed by the certificating authority of the country of registry.

(b) No person may operate a U.S.-registered civil aircraft—

(1) For which an Airplane or Rotorcraft Flight Manual is required by § 21.5 of this chapter unless there is available in the aircraft a current, approved Airplane or Rotorcraft Flight Manual or the manual provided for in § 121.141(b); and

(2) For which an Airplane or Rotorcraft Flight Manual is not required by § 21.5 of this chapter, unless there is available in the aircraft a current approved Airplane or Rotorcraft Flight Manual, approved manual material, markings, and placards, or any combination thereof.

(c) No person may operate a U.S.-registered civil aircraft unless that aircraft is identified in accordance with part 45 of this chapter.

(d) Any person taking off or landing a helicopter certificated under part 29 of this chapter at a heliport constructed over water may make such momentary flight as is necessary for takeoff or landing through the prohibited range of the limiting height-speed envelope established for the helicopter if that flight through the prohibited range takes place over water on which a safe ditching can be accomplished and if the helicopter is amphibious or is equipped with floats or other emergency flotation gear adequate to accomplish a safe emergency ditching on open water.

§ 91.11 Prohibition against interference with crewmembers.

No person may assault, threaten, intimidate, or interfere with a crewmember in the performance of the crewmember's duties aboard an aircraft being operated.

§ 91.13 Careless or reckless operation.

(a) *Aircraft operations for the purpose of air navigation.* No person may operate an aircraft in a careless or reckless manner so as to endanger the life or property of another.

(b) *Aircraft operations other than for the purpose of air navigation.* No person may operate an aircraft, other than for the purpose of air navigation, on any part of the surface of an airport used by aircraft for air commerce (including areas used by those aircraft for receiving or discharging persons or cargo), in a careless or reckless manner so as to endanger the life or property of another.

§ 91.15 Dropping objects.

No pilot in command of a civil aircraft may allow any object to be dropped from that aircraft in flight that creates a hazard to persons or property. However, this section does not prohibit the dropping of any object if reasonable precautions are taken to avoid injury or damage to persons or property.

§ 91.17 Alcohol or drugs.

(a) No person may act or attempt to act as a crewmember of a civil aircraft—

(1) Within 8 hours after the consumption of any alcoholic beverage;

(2) While under the influence of alcohol;

(3) While using any drug that affects the person's faculties in any way contrary to safety; or

(4) While having .04 percent by weight or more alcohol in the blood.

(b) Except in an emergency, no pilot of a civil aircraft may allow a person who appears to be intoxicated or who demonstrates by manner or physical indications that the individual is under the influence of drugs (except a medical patient under proper care) to be carried in that aircraft.

(c) A crewmember shall do the following:

(1) On request of a law enforcement officer, submit to a test to indicate the percentage by weight of alcohol in the blood, when—

(i) The law enforcement officer is authorized under State or local law to conduct the test or to have the test conducted; and

(ii) The law enforcement officer is requesting submission to the test to investigate a suspected violation of State or local law governing the same or substantially similar conduct prohibited by paragraph (a)(1), (a)(2), or (a)(4) of this section.

(2) Whenever the Administrator has a reasonable basis to believe that a person may have violated paragraph (a)(1), (a)(2), or (a)(4) of this section, that person shall, upon request by the Administrator, furnish the Administrator, or authorize any clinic, hospital, doctor, or other person to release to the Administrator, the results of each test taken within 4 hours after acting or attempting to act as a crewmember that indicates percentage by weight of alcohol in the blood.

(d) Whenever the Administrator has a reasonable basis to believe that a person may have violated paragraph (a)(3) of this section, that person shall, upon request by the Administrator, furnish the Administrator, or authorize any clinic, hospital, doctor, or other person to release to the Administrator, the results of each test taken within 4 hours after acting or attempting to act as a crewmember that indicates the presence of any drugs in the body.

(e) Any test information obtained by the Administrator under paragraph (c) or (d) of this section may be evaluated in determining a person's qualifications for any airman certificate or possible violations of this chapter and may be used as evidence in any legal proceeding under section 602, 609, or 901 of the Federal Aviation Act of 1958.

§ 91.19 Carriage of narcotic drugs, marijuana, and depressant or stimulant drugs or substances.

(a) Except as provided in paragraph (b) of this section, no person may operate a civil aircraft within the United States with knowledge that narcotic drugs, marijuana, and depressant or stimulant drugs or substances as defined in Federal or State statutes are carried in the aircraft.

(b) Paragraph (a) of this section does not apply to any carriage of narcotic drugs, marijuana, and depressant or stimulant drugs or substances authorized by or under any Federal or State statute or by any Federal or State agency.

§ 91.21 Portable electronic devices.

(a) Except as provided in paragraph (b) of this section, no person may operate, nor may any operator or pilot in command of an aircraft allow the operation of, any portable electronic device on any of the following U.S.-registered civil aircraft:

(1) Aircraft operated by a holder of an air carrier operating certificate or an operating certificate; or

(2) Any other aircraft while it is operated under IFR.

(b) Paragraph (a) of this section does not apply to—

(1) Portable voice recorders;

(2) Hearing aids;

(3) Heart pacemakers;

(4) Electric shavers; or

(5) Any other portable electronic device that the operator of the aircraft has determined will not cause interference with the navigation or communication system of the aircraft on which it is to be used.

(c) In the case of an aircraft operated by a holder of an air carrier operating certificate or an operating certificate, the determination required by paragraph (b)(5) of this section shall be made by that operator of the aircraft on which the particular device is to be used. In the case of other aircraft, the determination may be made by the pilot in command or other operator of the aircraft.

§ 91.23 Truth-in-leasing clause requirement in leases and conditional sales contracts.

(a) Except as provided in paragraph (b) of this section, the parties to a lease or contract of conditional sale involving a U.S.-registered large civil aircraft and entered into after January 2, 1973, shall execute a written lease or contract and include therein a written truth-in-leasing clause as a concluding paragraph in large print, immediately preceding the space for the signature of the parties, which contains the following with respect to each such aircraft:

(1) Identification of the Federal Aviation Regulations under which the aircraft has been maintained and inspected

during the 12 months preceding the execution of the lease or contract of conditional sale, and certification by the parties thereto regarding the aircraft's status of compliance with applicable maintenance and inspection requirements in this part for the operation to be conducted under the lease or contract of conditional sale.

(2) The name and address (printed or typed) and the signature of the person responsible for operational control of the aircraft under the lease or contract of conditional sale, and certification that each person understands that person's responsibilities for compliance with applicable Federal Aviation Regulations.

(3) A statement that an explanation of factors bearing on operational control and pertinent Federal Aviation Regulations can be obtained from the nearest FAA Flight Standards district office.

(b) The requirements of paragraph (a) of this section do not apply—

(1) To a lease or contract of conditional sale when—

(i) The party to whom the aircraft is furnished is a foreign air carrier or certificate holder under part 121, 125, 127, 135, or 141 of this chapter, or

(ii) The party furnishing the aircraft is a foreign air carrier, certificate holder under part 121, 125, 127, or of this chapter, or a certificate holder under part 135 of this chapter having appropriate authority to engage in air taxi operations with large aircraft.

(2) To a contract of conditional sale, when the aircraft involved has not been registered anywhere prior to the execution of the contract, except as a new aircraft under a dealer's aircraft registration certificate issued in accordance with § 47.61 of this chapter.

(c) No person may operate a large civil aircraft of U.S. registry that is subject to a lease or contract of conditional sale to which paragraph (a) of this section applies, unless—

(1) The lessee or conditional buyer, or the registered owner if the lessee is not a citizen of the United States, has mailed a copy of the lease or contract that complies with the requirements of paragraph (a) of this section, within 24 hours of its execution, to the Aircraft Registration Branch, Attn: Technical Section, P.O. Box 25724, Oklahoma City, Oklahoma 73125;

(2) A copy of the lease or contract that complies with the requirements of paragraph (a) of this section is carried in the aircraft. The copy of the lease or contract shall be made available for review upon request by the Administrator, and

(3) The lessee or conditional buyer, or the registered owner if the lessee is not a citizen of the United States, has notified by telephone or in person the FAA Flight Standards district office nearest the airport where the flight will originate. Unless otherwise authorized by that office, the notification shall be given at least 48 hours before takeoff in the case of the first flight of that aircraft under that lease or contract and inform the FAA of—

(i) The location of the airport of departure;

(ii) The departure time; and

(iii) The registration number of the aircraft involved.

(d) The copy of the lease or contract furnished to the FAA under paragraph (c) of this section is commercial or financial information obtained from a person. It is, therefore, privileged and confidential and will not be made available by the FAA for public inspection or copying under 5 U.S.C. 552(b)(4) unless recorded with the FAA under part 49 of this chapter.

(e) For the purpose of this section, a lease means any agreement by a person to furnish an aircraft to another person for compensation or hire, whether with or without flight crewmembers, other than an agreement for the sale of an aircraft and a contract of conditional sale under section 101 of the Federal Aviation Act of 1958. The person furnishing the aircraft is referred to as the lessor, and the person to whom it is furnished the lessee.

(Approved by the Office of Management and Budget under OMB control number 2120-0005)

(Amdt. 91-212, Eff. 8/18/90)

§ 91.25 Aviation Safety Reporting Program: Prohibition against use of reports for enforcement purposes.

The Administrator of the FAA will not use reports submitted to the National Aeronautics and Space Administration under the Aviation Safety Reporting Program (or information derived therefrom) in any enforcement action except information concerning accidents or criminal offenses which are wholly excluded from the Program.

§§ 91.27–91.99 [Reserved]

[Amdt. 91–240, Corrected, Eff. 5/12/94)]

Subpart B—Flight rules general

§ 91.101 Applicability.

This subpart prescribes flight rules governing the operation of aircraft within the United States and within 12 nautical miles from the coast of the United States.

§ 91.103 Preflight action.

Each pilot in command shall, before beginning a flight, become familiar with all available information concerning that flight. This information must include—

(a) For a flight under IFR or a flight not in the vicinity of an airport, weather reports and forecasts, fuel requirements, alternatives available if the planned flight cannot be completed, and any known traffic delays of which the pilot in command has been advised by ATC;

(b) For any flight, runway lengths at airports of intended use, and the following takeoff and landing distance information:

(1) For civil aircraft for which an approved Airplane or Rotorcraft Flight Manual containing takeoff and landing distance data is required, the takeoff and landing distance data contained therein; and

(2) For civil aircraft other than those specified in paragraph (b)(1) of this section, other reliable information appropriate to the aircraft, relating to aircraft performance under expected values of airport elevation and runway slope, aircraft gross weight, and wind and temperature.

§ 91.105 Flight crewmembers at stations.

(a) During takeoff and landing, and while en route, each required flight crewmember shall—

(1) Be at the crewmember station unless the absence is necessary to perform duties in connection with the operation of the aircraft or in connection with physiological needs; and

(2) Keep the safety belt fastened while at the crewmember station.

(b) [Each required flight crewmember of a U.S.-registered civil aircraft shall, during takeoff and landing, keep his or her shoulder harness fastened while at his or her assigned duty station. This paragraph does not apply if—]

(1) The seat at the crewmember's station is not equipped with a shoulder harness; or

(2) The crewmember would be unable to perform required duties with the shoulder harness fastened.

[(Admt. 91-231, Eff. 10/15/92)]

§ 91.107 [Use of safety belts, shoulder harnesses, and child restraint systems.]

[(a) Unless otherwise authorized by the Administrator—

[(1) No pilot may take off a U.S.-registered civil aircraft (except a free balloon that incorporates a basket or gondola, or an airship type certificated before November 2, 1987) unless the pilot in command of that aircraft ensures that each person on board is briefed on how to fasten and unfasten that person's safety belt and, if installed, shoulder harness.

[(2) No pilot may cause to be moved on the surface, take off, or land a U.S.-registered civil aircraft (except a free balloon that incorporates a basket or gondola, or an airship type certificated before November 2, 1987) unless the pilot in command of that aircraft ensures that each person on board has been notified to fasten his or her safety belt and, if installed, his or her shoulder harness.

[(3) Except as provided in this paragraph, each person on board a U.S.-registered civil aircraft (except a free balloon that incorporates a basket or gondola or an airship type certificated before November 2, 1987) must occupy an approved seat or berth with a safety belt and, if installed, shoulder harness, properly secured about him or her during movement on the surface, takeoff, and landing. For seaplane and float equipped rotorcraft operations during movement on the surface, the person pushing off the seaplane or rotorcraft from the dock and the person mooring the seaplane or rotorcraft at the dock are excepted from preceding seating and safety belt requirements. Notwithstanding the preceding requirements of this paragraph, a person may:

[(i) Be held by an adult who is occupying a seat or berth if that person has not reached his or her second birthday;

[(ii) Use the floor of the aircraft as a seat, provided that the person is on board for the purpose of engaging in sport parachuting; or

[(iii) Notwithstanding any other requirement of this chapter, occupy an approved child restraint system furnished by the operator or one of the persons described in paragraph (a)(3)(iii)(A) of this section provided that:

[(A) The child is accompanied by a parent, guardian, or attendant designated by the child's parent or guardian to attend to the safety of the child during the flight;

[(B) The approved child restraint system bears one or more labels as follows:

[(1) Seats manufactured to U.S. standards between January 1, 1981, and February 25, 1985, must bear the label: "This child restraint system conforms to all applicable Federal motor vehicle safety standards." Vest- and harness-type child restraint systems manufactured before February 26, 1985, bearing such a label are not approved for the purposes of this section;

[(2) Seats manufactured to U.S. standards on or after February 26, 1985, must bear two labels:

[(i) "This child restraint system conforms to all applicable Federal motor vehicle safety standards"; and

[(ii) THIS RESTRAINT IS CERTIFIED FOR USE IN MOTOR VEHICLES AND AIRCRAFT" in red lettering;

[(3) Seats that do not qualify under paragraphs (a)(3)(iii)
(B)(1) and (a)(3)(iii)(B)(2) of this section must bear either a label showing approval of a foreign government or a label showing that the seat was manufactured under the standards of the United Nations; and

[(C) The operator complies with the following requirements:

[(1) The restraint system must be properly secured to an approved forward-facing seat or berth;

[(2) The child must be properly secured in the restraint system and must not exceed the specified weight limit for the restraint system; and

[(3) The restraint system must bear the appropriate label(s).

[(b) Unless otherwise stated, this section does not apply to operations conducted under Part 121, 125, or 135 of this chapter. Paragraph (a)(3) of this section does not apply to persons subject to § 91.105.]

[(Admt. 91-231, Eff. 10/15/92)]

§ 91.109 Flight instruction; Simulated instrument flight and certain flight tests.

(a) No person may operate a civil aircraft (except a manned free balloon) that is being used for flight instruction unless that aircraft has fully functioning dual controls. However, instrument flight instruction may be given in a single-engine airplane equipped with a single, functioning throwover control wheel in place of fixed, dual controls of the elevator and ailerons when—

(1) The instructor has determined that the flight can be conducted safely; and

(2) The person manipulating the controls has at least a private pilot certificate with appropriate category and class ratings.

(b) No person may operate a civil aircraft in simulated instrument flight unless—

(1) The other control seat is occupied by a safety pilot who possesses at least a private pilot certificate with category and class ratings appropriate to the aircraft being flown.

(2) The safety pilot has adequate vision forward and to each side of the aircraft, or a competent observer in the aircraft adequately supplements the vision of the safety pilot; and

(3) Except in the case of lighter-than-air aircraft, that aircraft is equipped with fully functioning dual controls. However, simulated instrument flight may be conducted in a single-engine airplane, equipped with a single, functioning, throwover control wheel, in place of fixed, dual controls of the elevator and ailerons, when—

(i) The safety pilot has determined that the flight can be conducted safely; and

(ii) The person manipulating the controls has at least a private pilot certificate with appropriate category and class ratings.

(c) No person may operate a civil aircraft that is being used for a flight test for an airline transport pilot certificate or a class or type rating on that certificate, or for a part 121 proficiency flight test, unless the pilot seated at the controls, other than the pilot being checked, is fully qualified to act as pilot in command of the aircraft.

§ 91.111 Operating near other aircraft.

(a) No person may operate an aircraft so close to another aircraft as to create a collision hazard.

(b) No person may operate an aircraft in formation flight except by arrangement with the pilot in command of each aircraft in the formation.

(c) No person may operate an aircraft, carrying passengers for hire, in formation flight.

§ 91.113 Right-of-way rules: Except water operations.

(a) *Inapplicability.* This section does not apply to the operation of an aircraft on water.

(b) *General.* When weather conditions permit, regardless of whether an operation is conducted under instrument flight rules or visual flight rules, vigilance shall be maintained by each person operating an aircraft so as to see and avoid other aircraft. When a rule of this section gives another aircraft the right-of-way, the pilot shall give way to that aircraft and may not pass over, under, or ahead of it unless well clear.

(c) *In distress.* An aircraft in distress has the right-of-way over all other air traffic.

(d) *Converging.* When aircraft of the same category are converging at approximately the same altitude (except head-on, or nearly so), the aircraft to the other's right has the right-of-way. If the aircraft are of different categories—

(1) A balloon has the right-of-way over any other category of aircraft;

(2) A glider has the right-of-way over an airship, airplane, or rotorcraft; and

(3) An airship has the right-of-way over an airplane or rotorcraft.

However, an aircraft towing or refueling other aircraft has the right-of-way over all other engine-driven aircraft.

(e) *Approaching head-on.* When aircraft are approaching each other head-on, or nearly so, each pilot of each aircraft shall alter course to the right.

(f) *Overtaking.* Each aircraft that is being overtaken has the right-of-way and each pilot of an overtaking aircraft shall alter course to the right to pass well clear.

(g) *Landing.* Aircraft, while on final approach to land or while landing, have the right-of-way over other aircraft in flight or operating on the surface, except that they shall not take advantage of this rule to force an aircraft off the runway surface which has already landed and is attempting to make way for an aircraft on final approach. When two or more aircraft are approaching an airport for the purpose of landing, the aircraft at the lower altitude has the right-of-way, but it shall not take advantage of this rule to cut in front of another which is on final approach to land or to overtake that aircraft.

§ 91.115 Right-of-way rules: Water operations.

(a) *General.* Each person operating an aircraft on the water shall, insofar as possible, keep clear of all vessels and avoid impeding their navigation, and shall give way to any vessel or other aircraft that is given the right-of-way by any rule of this section.

(b) *Crossing.* When aircraft, or an aircraft and a vessel, are on crossing courses, the aircraft or vessel to the other's right has the right-of-way.

(c) *Approaching head-on.* When aircraft, or an aircraft and a vessel, are approaching head-on, or nearly so, each shall alter its course to the right to keep well clear.

(d) *Overtaking.* Each aircraft or vessel that is being overtaken has the right-of-way, and the one overtaking shall alter course to keep well clear.

(e) *Special circumstances.* When aircraft, or an aircraft and a vessel, approach so as to involve risk of collision, each aircraft or vessel shall proceed with careful regard to existing circumstances, including the limitations of the respective craft.

§ 91.117 Aircraft speed.

[(a) Unless otherwise authorized by the Administrator, no person may operate an aircraft below 10,000 feet MSL at an indicated airspeed of more than 250 knots (288 m.p.h.).]

[(b) Unless otherwise authorized or required by ATC, no person may operate an aircraft at or below 2,500 feet above the surface within 4 nautical miles of the primary airport of a Class C or Class D airspace area at an indicated airspeed of more than 200 knots (230 mph.). This paragraph (b) does not apply to any operations within a Class B airspace area. Such operations shall comply with paragraph (a) of this section.]

(c) No person may operate an aircraft in the airspace underlying a Class B airspace area designated for an airport or in a VFR corridor designated through such a Class B airspace area, at an indicated airspeed of more than 200 knots (230 mph).

(d) If the minimum safe airspeed for any particular operation is greater than the maximum speed prescribed in this section, the aircraft may be operated at that minimum speed.

(Amdt. 91-219, Eff. 8/24/90); (Amdt. 91-227, Eff. 9/16/93); (Amdt. 91-227, Corrected, Eff. 9/16/93); [***(Amdt. 91-233, Eff. 9/16/93***]

§ 91.119 Minimum safe altitudes: General.

Except when necessary for takeoff or landing, no person may operate an aircraft below the following altitudes:

(a) *Anywhere.* An altitude allowing, if a power unit fails, an emergency landing without undue hazard to persons or property on the surface.

(b) *Over congested areas.* Over any congested area of a city, town, or settlement, or over any open air assembly of persons, an altitude of 1,000 feet above the highest obstacle within a horizontal radius of 2,000 feet of the aircraft.

(c) *Over other than congested areas.* An altitude of 500 feet above the surface, except over open water or sparsely populated areas. In those cases, the aircraft may not be operated closer than 500 feet to any person, vessel, vehicle, or structure.

(d) *Helicopters.* Helicopters may be operated at less than the minimums prescribed in paragraph (b) or (c) of this section if the operation is conducted without hazard to persons or property on the surface. In addition, each person operating a helicopter shall comply with any routes or altitudes specifically prescribed for helicopters by the Administrator.

§ 91.121 Altimeter settings.

(a) Each person operating an aircraft shall maintain the cruising altitude or flight level of that aircraft, as the case may be, by reference to an altimeter that is set, when operating—

(1) Below 18,000 feet MSL, to—

(i) The current reported altimeter setting of a station along the route and within 100 nautical miles of the aircraft;

(ii) If there is no station within the area prescribed in paragraph (a)(1)(i) of this section, the current reported altimeter setting of an appropriate available station; or

(iii) In the case of an aircraft not equipped with a radio, the elevation of the departure airport or an appropriate altimeter setting available before departure; or

(2) At or above 18,000 feet MSL, to 29.92" Hg.

(b) The lowest usable flight level is determined by the atmospheric pressure in the area of operation as shown in the following table:

Current altimeter setting	Lowest usable flight level
29.92" (or higher)	180
29.91" through 29.42"	185
29.41" through 28.92"	190
28.91" through 28.42"	195
28.41" through 27.92"	200
27.91" through 27.42"	205
27.41" through 26.92"	210

(c) To convert minimum altitude prescribed under § 91.119 and § 91.177 to the minimum flight level, the pilot shall take the flight level equivalent of the minimum altitude in feet and add the appropriate number of feet specified below, according to the current reported altimeter setting:

Current altimeter setting	Adjustment factor
29.92" (or higher)	None
29.91" through 29.42"	500
29.41" through 28.92"	1,000
28.91" through 28.42"	1,500
28.41" through 27.92"	2,000
27.91" through 27.42"	2,500
27.41" through 26.92"	3,000

§ 91.123 Compliance with ATC clearances and instructions.

[(a) When an ATC clearance has been obtained, a pilot in command may not deviate from that clearance, except in an emergency, unless that pilot obtains an amended clearance. However, except in Class A airspace, this paragraph does not prohibit that pilot from canceling an IFR flight plan if the operation is being conducted in VFR weather conditions. When a pilot is uncertain of an ATC clearance, that pilot must immediately request clarification from ATC.]

(b) Except in an emergency, no person may operate an aircraft contrary to an ATC instruction in an area in which air traffic control is exercised.

(c) Each pilot in command who, in an emergency, deviates from an ATC clearance or instruction shall notify ATC of that deviation as soon as possible.

(d) Each pilot in command who (though not deviating from a rule of this subpart) is given priority by ATC in an emergency, shall submit a detailed report of that emergency within 48 hours to the manager of that ATC facility, if requested by ATC.

(e) Unless otherwise authorized by ATC, no person operating an aircraft may operate that aircraft according to any clearance or instruction that has been issued to the pilot of another aircraft for radar air traffic control purposes.

(Approved by the Office of Management and Budget under OMB control number 2120-0005).

[(Amdt. 91-227, Eff. 9/16/93)]

§ 91.125 ATC light signals.

ATC light signals have the meaning shown in the following table:

Color and type of signal	Meaning with respect to aircraft on the surface	Meaning with respect to aircraft in flight
Steady green	Cleared for takeoff.	Cleared to land.
Flashing green	Cleared to taxi.	Return for landing (to be followed by steady green at proper time).
Steady red	Stop.	Give way to other aircraft and continue circling.
Flashing red	Taxi clear of runway in use.	Airport unsafe—do not land.
Flashing white	Return to starting point on airport.	Not applicable.
Alternating red and green	Exercise extreme caution.	Exercise extreme caution.

[§ 91.126 Operating on or in the vicinity of an airport in Class G airspace.

[(a) *General.* Unless otherwise authorized or required, each person operating an aircraft on or in the vicinity of an airport in a Class G airspace area must comply with the requirements of this section.

(b) *Direction of turns.* [(When approaching to land at an airport without an operating control tower in Class G airspace—]

(1) Each pilot of an airplane must make all turns of that airplane to the left unless the airport displays approved light signals or visual markings indicating that turns should be made to the right, in which case the pilot must make all turns to the right; and

(2) Each pilot of a helicopter must avoid the flow of fixed-wing aircraft.

(c) *Flap settings.* Except when necessary for training or certification, the pilot in command of a civil turbojet-powered aircraft must use, as a final flap setting, the minimum certificated landing flap setting set forth in the approved performance information in the Airplane Flight Manual for the applicable conditions. However, each pilot in command has the final authority and responsibility for the safe operation of the pilot's airplane, and may use a different flap setting for that airplane if the pilot determines that it is necessary in the interest of safety.]

[(d) *Communications with control towers.* Unless otherwise authorized or required by ATC, no person may operate an aircraft to, from, through, or on an airport having an operational control tower unless two-way radio communications are maintained between that aircraft and the control tower. Communications must be established prior to 4 nautical miles from the airport, up to and including 2,500 feet AGL. However, if the aircraft radio fails in flight, the pilot in command may operate that aircraft and land if weather conditions are at or above basic VFR weather minimums, visual contact with the tower is maintained, and a clearance to land is received. If the aircraft radio fails while in flight under IFR, the pilot must comply with § 91.185.]

(Amdt. 91-227, Eff. 9/16/93); [(Admt. 91–239, Eff. 3/11/94)]

§ 91.127 Operating on or in the vicinity of an airport in Class E airspace.

(a) Unless otherwise required by Part 93 of this chapter or unless otherwise authorized or required by the ATC facility having jurisdiction over the Class E airspace area, each person operating an aircraft on or in the vicinity of an airport in a Class E airspace area must comply with the requirements of § 91.126.

(b) *Departures.* Each pilot of an aircraft must comply with any traffic patterns established for that airport in Part 93 of this chapter.

[(c) *Communications with control towers.* Unless otherwise authorized or required by ATC, no person may operate an aircraft to, from, through, or on an airport having an operational control tower unless two-way radio communications are maintained between that aircraft and the control tower. Communications must be established prior to 4 nautical miles from the airport, up to and including 2,500 feet AGL. However, if the aircraft radio fails in flight, the pilot in command may operate that aircraft and land if weather conditions are at or above basic VFR weather minimums, visual contact with the tower is maintained, and a clearance to land is received. If the aircraft radio fails while in flight under IFR, the pilot must comply with §91.185.]

(Amdt. 91-227, Eff. 9/16/93); (Amdt. 91-239, Eff. 3/11/94)]

§ 91.129 Operations in Class D airspace.

(a) *General.* Unless otherwise authorized or required by the ATC facility having jurisdiction over the Class D airspace area, each person operating an aircraft in Class D airspace must comply with the applicable provisions of this section. In addition, each person must comply with §§ 91.126 and 91.127. For the purpose of this section, the primary airport is the airport for which the Class D airspace area is designated. A satellite airport is any other airport within the Class D airspace area.

(b) *Deviations.* An operator may deviate from any provision of this section under the provisions of an ATC authorization issued by the ATC facility having jurisdiction over the airspace concerned. ATC may authorize a deviation on a continuing basis or for an individual flight, as appropriate.

(c) *Communications.* Each person operating an aircraft in Class D airspace must meet the following two-way radio communications requirements:

(1) Arrival or through flight. Each person must establish two-way radio communications with the ATC facility (including foreign ATC in the case of foreign airspace designated in the United States) providing air traffic services prior to entering that airspace and thereafter maintain those communications while within that airspace.

(2) Departing flight. Each person—

(i) From the primary airport or satellite airport with an operating control tower must establish and maintain two-way radio communications with the control tower, and thereafter as instructed by ATC while operating in the Class D airspace area; or

(ii) From a satellite airport without an operating control tower, must establish and maintain two-way radio communications with the ATC facility having jurisdiction over the Class D airspace area as soon as practicable after departing.

(d) *Communications failure.* Each person who operates an aircraft in a Class D airspace area must maintain two-way radio communications with the ATC facility having jurisdiction over that area.

(1) If the aircraft radio fails in flight under IFR, the pilot must comply with § 91.185 of the part.

(2) If the aircraft radio fails in flight under VFR, the pilot in command may operate that aircraft and land if—

(i) Weather conditions are at or above basic VFR weather minimums;

(ii) Visual contact with the tower is maintained; and

(iii) A clearance to land is received.

[(e) *Minimum Altitudes.* When operating to an airport in Class D airspace, each pilot of—

(1) [A large or turbine-powered airplane shall, unless otherwise required by the applicable distance from cloud criteria, enter the traffic pattern at an altitude of at least 1,500 feet above the elevation of the airport and maintain at least 1,500 feet until further descent is required for a safe landing;

(2) [A large or turbine-powered airplane approaching to land on a runway served by an instrument landing system (ILS), if the airplane is ILS equipped, shall fly that airplane at an altitude at or above the glide slope between the outer marker (or point of interception of glide slope, if compliance with the applicable distance from cloud criteria requires interception closer in) and the middle marker; and

(3) [An airplane approaching to land on a runway served by a visual approach slope indicator shall maintain an altitude at or above the glide slope until a lower altitude is necessary for a safe landing.]

Paragraphs (e)(2) and (e)(3) of this section do not prohibit normal bracketing maneuvers above or below the glide slope that are conducted for the purpose of remaining on the glide slope.

(f) *Approaches.* Except when conducting a circling approach under Part 97 of this chapter or unless otherwise required by ATC, each pilot must—

(1) Circle the airport to the left, if operating an airplane; or

(2) Avoid the flow of fixed-wing aircraft, if operating a helicopter.

(g) *Departures.* No person may operate an aircraft departing from an airport except in compliance with the following:

(1) Each pilot must comply with any departure procedures established for that airport by the FAA.

(2) Unless otherwise required by the prescribed departure procedure for that airport or the applicable distance from clouds criteria, each pilot of a turbine-powered airplane and each pilot of a large airplane must climb to an altitude of 1,500 feet above the surface as rapidly as practicable.

(h) *Noise abatement.* Where a formal runway use program has been established by the FAA, each pilot of a large or turbine-powered airplane assigned a noise abatement runway by ATC must use that runway. However, consistent with the final authority of the pilot in command concerning the safe operation of the aircraft as prescribed in § 91.3(a), ATC may assign a different runway if requested by the pilot in the interest of safety.

(i) *Takeoff, landing, taxi clearance.* No person may, at any airport with an operating control tower, operate an aircraft on a runway or taxiway, or take off or land an aircraft, unless an appropriate clearance is received from ATC. A clearance to "taxi to" the takeoff runway assigned to the aircraft is not a clearance to cross that assigned takeoff runway, or to taxi on that runway at any point, but is a clearance to cross other runways that intersect the taxi route to that assigned takeoff runway. A clearance to "taxi to" any point other than an assigned takeoff runway is clearance to cross all runways that intersect the taxi route to that point.

(Amdt. 91-227, Eff. 9/16/93); (Amdt. 91-234, Eff. 9/16/93)]

§ 91.130 Operations in Class C airspace.

(a) *General.* Unless otherwise authorized by ATC, each aircraft operation in Class C airspace must be conducted in compliance with this section and § 91.129. For the purpose of this section, the primary airport is the airport for which the Class C airspace area is designated. A satellite airport is any other airport within the Class C airspace area.

(b) *Traffic patterns.* No person may take off or land an aircraft at a satellite airport within a Class C airspace area except in compliance with FAA arrival and departure traffic patterns.

(c) *Communications.* Each person operating an aircraft in Class C airspace must meet the following two-way radio communications requirements:

(1) Arrival or through flight. Each person must establish two-way radio communications with the ATC facility (including foreign ATC in the case of foreign airspace designated in the United States) providing air traffic services prior to entering that airspace and thereafter maintain those communications while within that airspace.

(2) Departing flight. Each person—

(i) From the primary airport or satellite airport with an operating control tower must establish and maintain two-way radio communications with the control tower, and thereafter as instructed by ATC while operating in the Class C airspace area; or

(ii) From a satellite airport without an operating control tower, must establish and maintain two-way radio communications with the ATC facility having jurisdiction over the Class C airspace area as soon as practicable after departing.

(d) *Equipment requirements.* Unless otherwise authorized by the ATC having jurisdiction over the Class C airspace area, no person may operate an aircraft within a Class C airspace area designated for an airport unless that aircraft is equipped with the applicable equipment specified in § 91.215.

[(e) *Deviations.* An operator may deviate from any provision of this section under the provisions of an ATC authorization issued by the ATC facility having jurisdiction over the airspace concerned. ATC may authorize a deviation on a continuing basis or for an individual flight, as appropriate.]

(Amdt. 91-215, Eff. 8/18/90); (Amdt. 91-227, Eff. 9/16/93); (Amdt. 91-232, Eff. 9/16/93; [(Amdt. 91-239, Eff. 3/11/94)]

[§ 91.131 Operations in Class B airspace.

[(a) *Operating rules.* No person may operate an aircraft within a Class B airspace area except in compliance with § 91.129 and the following rules:

[(1) The operator must receive an ATC clearance from the ATC facility having jurisdiction for that area before operating an aircraft in that area.

[(2) Unless otherwise authorized by ATC, each person operating a large turbine engine-powered airplane to or from a primary airport for which a Class B airspace area is designated must operate at or above the designated floors of the Class B airspace area while within the lateral limits of that area.

[(3) Any person conducting pilot training operations at an airport within a Class B airspace area must comply with any procedures established by ATC for such operations in that area.

[(b) *Pilot requirements.*

[(1) No person may take off or land a civil aircraft at an airport within a Class B airspace area or operate a civil aircraft within a Class B airspace area unless—

[(i) The pilot in command holds at least a private pilot certificate; or

[(ii) The aircraft is operated by a student pilot or recreational pilot who seeks private pilot certification and has met the requirements of § 61.95 of this chapter.

[(2) Notwithstanding the provisions of paragraph (b)(1)(ii) of this section, no person may take off or land a civil aircraft at those airports listed in Section 4 of Appendix D of this part unless the pilot in command holds at least a private pilot certificate.

[(c) *Communications and navigation equipment requirements.* Unless otherwise authorized by ATC, no person may operate an aircraft within a Class B airspace area unless that aircraft is equipped with—

[(1) For IFR operation. An operable VOR or TACAN receiver; and

[(2) For all operations. An operable two-way radio capable of communications with ATC on appropriate frequencies for that Class B airspace area.

[(d) *Transponder requirements.* No person may operate an aircraft in a Class B airspace area unless the aircraft is equipped with the applicable operating transponder and automatic altitude reporting equipment specified in paragraph (a) of § 91.215, except as provided in paragraph (d) of that section.]

(Amdt. 91-214, Eff. 8/18/90); (Amdt. 91-216, Eff. 8/18/90); [(Amdt. 91-227, Eff. 9/16/93)]

§ 91.133 Restricted and prohibited areas.

(a) No person may operate an aircraft within a restricted area (designated in Part 73) contrary to the restrictions imposed, or within a prohibited area, unless that person has the permission of the using or controlling agency, as appropriate.

(b) Each person conducting, within a restricted area, an aircraft operation (approved by the using agency) that creates the same hazards as the operations for which the restricted area was designated may deviate from the rules of this subpart that are not compatible with the operation of the aircraft.

[§ 91.135 Operations in Class A airspace.

[Except as provided in paragraph (d) of this section, each person operating an aircraft in Class A airspace must conduct that operation under instrument flight rules (IFR) and in compliance with the following:

[(a) *Clearance*. Operations may be conducted only under an ATC clearance received prior to entering the airspace.

[(b) *Communications*. Unless otherwise authorized by ATC, each aircraft operating in Class A airspace must be equipped with a two-way radio capable of communicating with ATC on a frequency assigned by ATC. Each pilot must maintain two-way radio communications with ATC while operating in Class A airspace.

[(c) *Transponder requirement.* Unless otherwise authorized by ATC, no person may operate an aircraft within Class A airspace unless that aircraft is equipped with the applicable equipment specified in § 91.215.

[(d) *ATC authorizations*. An operator may deviate from any provision of this section under the provisions of an ATC authorization issued by the ATC facility having jurisdiction of the airspace concerned. In the case of an inoperative transponder, ATC may immediately approve an operation within a Class A airspace area allowing flight to continue, if desired, to the airport of ultimate destination, including any intermediate stops, or to proceed to a place where suitable repairs can be made, or both. Requests for deviation from any provision of this section must be submitted in writing, at least 4 days before the proposed operation. ATC may authorize a deviation on a continuing basis or for an individual flight.]

[(Amdt. 91-227, Eff. 9/16/93)]

§ 91.137 Temporary flight restrictions.

(a) The Administrator will issue a Notice to Airmen (NOTAM) designating an area within which temporary flight restrictions apply and specifying the hazard or condition requiring their imposition, whenever he determines it is necessary in order to—

(1) Protect persons and property on the surface or in the air from a hazard associated with an incident on the surface;

(2) Provide a safe environment for the operation of disaster relief aircraft; or

(3) Prevent an unsafe congestion of sightseeing and other aircraft above an incident or event which may generate a high degree of public interest.

The Notice to Airmen will specify the hazard or condition that requires the imposition of temporary flight restrictions.

(b) When a NOTAM has been issued under paragraph (a)(1) of this section, no person may operate an aircraft within the designated area unless that aircraft is participating in the hazard relief activities and is being operated under the direction of the official in charge of on scene emergency response activities.

(c) When a NOTAM has been issued under paragraph (a)(2) of this section, no person may operate an aircraft within the designated area unless at least one of the following conditions are met:

(1) The aircraft is participating in hazard relief activities and is being operated under the direction of the official in charge of on scene emergency response activities.

(2) The aircraft is carrying law enforcement officials.

(3) The aircraft is operating under the ATC approved IFR flight plan.

(4) The operation is conducted directly to or from an airport within the area, or is necessitated by the impracticability of VFR flight above or around the area due to weather, or terrain; notification is given to the Flight Service Station (FSS) or ATC facility specified in the NOTAM to receive advisories concerning disaster relief aircraft operations; and the operation does not hamper or endanger relief activities and is not conducted for the purpose of observing the disaster.

(5) The aircraft is carrying properly accredited news representatives, and, prior to entering the area, a flight plan is filed with the appropriate FAA or ATC facility specified in the Notice to Airmen and the operation is conducted above the altitude used by the disaster relief aircraft, unless otherwise authorized by the official in charge of on scene emergency response activities.

(d) When a NOTAM has been issued under paragraph (a)(3) of this section, no person may operate an aircraft within the designated area unless at least one of the following conditions is met:

(1) The operation is conducted directly to or from an airport within the area, or is necessitated by the impracticability of VFR flight above or around the area due to weather or terrain, and the operation is not conducted for the purpose of observing the incident or event.

(2) The aircraft is operating under an ATC approved IFR flight plan.

(3) The aircraft is carrying incident or event personnel, or law enforcement officials.

(4) The aircraft is carrying properly accredited news representatives and, prior to entering that area, a flight plan is filed with the appropriate FSS or ATC facility specified in the NOTAM.

(e) Flight plans filed and notifications made with an FSS or ATC facility under this section shall include the following information:

(1) Aircraft identification, type and color.

(2) Radio communications frequencies to be used.

(3) Proposed times of entry of, and exit from, the designated area.

(4) Name of news media or organization and purpose of flight.

(5) Any other information requested by ATC.

[§ 91.138 Temporary flight restrictions in national disaster areas in the State of Hawaii.

[(a) When the Administrator has determined, pursuant to a request and justification provided by the Governor of the State of Hawaii, or the Governor's designee, that an inhabited area within a declared national disaster area in the State of Hawaii is in need of protection for humanitarian reasons, the Administrator will issue a Notice to Airmen (NOTAM) designating an area within which temporary flight restrictions apply. The Administrator will designate the extent and duration of the temporary flight restrictions necessary to provide for the protection of persons and property on the surface.

[(b) When a NOTAM has been issued in accordance with this section, no person may operate an aircraft within the designated airspace unless:

[(1) That person has obtained authorization from the official in charge of associated emergency or disaster relief response activities, and is operating the aircraft under the conditions of that authorization;

[(2) The aircraft is carrying law enforcement officials;

[(3) The aircraft is carrying persons involved in an emergency or a legitimate scientific purpose;

[(4) The aircraft is carrying properly accredited newspersons, and that prior to entering the area, a flight plan is filed with the appropriate FAA or ATC facility specified in the NOTAM and the operation is conducted in compliance with the conditions and restrictions established by the official in charge of on-scene emergency response activities; or

[(5) The aircraft is operating in accordance with an ATC clearance or instruction.

[(c) A NOTAM issued under this section is effective for 90 days or until the national disaster designation is terminated, whichever comes first, unless terminated by notice or extended by the Administrator at the request of the Governor of the State of Hawaii or the Governor's designee.]

[(Amdt. 91-222, Eff. 5/20/91)]

§ 91.139 Emergency air traffic rules.

(a) This section prescribes a process for utilizing Notices to Airmen (NOTAMs) to advise of the issuance and operations under emergency air traffic rules and regulations and designates the official who is authorized to issue NOTAMs on behalf of the Administrator in certain matters under this section.

(b) Whenever the Administrator determines that an emergency condition exists, or will exist, relating to the FAA's ability to operate the air traffic control system and during which normal flight operations under this chapter cannot be conducted consistent with the required levels of safety and efficiency—

(1) The Administrator issues an immediately effective air traffic rule or regulation in response to that emergency condition; and

(2) The Administrator or the Associate Administrator for Air Traffic may utilize the NOTAM system to provide notification of the issuance of the rule or regulation.

Those NOTAMs communicate information concerning the rules and regulations that govern flight operations, the use of navigation facilities, and designation of that airspace in which the rules and regulations apply.

(c) When a NOTAM has been issued under this section, no person may operate an aircraft, or other device governed by the regulation concerned, within the designated airspace except in accordance with the authorizations, terms, and conditions prescribed in the regulation covered by the NOTAM.

§ 91.141 Flight restrictions in the proximity of the Presidential and other parties.

No person may operate an aircraft over or in the vicinity of any area to be visited or traveled by the President, the Vice President, or other public figures contrary to the restrictions established by the Administrator and published in a Notice to Airmen (NOTAM).

§ 91.143 Flight limitation in the proximity of space flight operations.

No person may operate any aircraft of U.S. registry, or pilot any aircraft under the authority of an airman certificate issued by the Federal Aviation Administration within areas designated in a Notice to Airmen (NOTAM) for space flight operations except when authorized by ATC, or operated under the control of the Department of Defense Manager for Space Transportation System Contingency Support Operations.

§ 91.144 Temporary restriction on flight operations during abnormally high barometric pressure conditions.

(a) *Special flight restrictions*. When any information indicates that barometric pressure on the route of flight currently exceeds or will exceed 31 inches of mercury, no person may operate an aircraft or initiate a flight contrary to the requirements established by the Administrator and published in a Notice to Airmen issued under this section.

(b) *Waivers*. The Administrator is authorized to waive any restriction issued under paragraph (a) of this section to permit emergency supply, transport, or medical services to be delivered to isolated communities, where the operation can be conducted with an acceptable level of safety.]

[(Amdt. 91-240, Eff. 5/12/94)]

§§ 91.145 – 91.149 [Reserved]

Visual flight rules (Subpart B—Flight rules)

§ 91.151 Fuel requirements for flight in VFR conditions.

(a) No person may begin a flight in an airplane under VFR conditions unless (considering wind and forecast weather conditions) there is enough fuel to fly to the first point of intended landing and, assuming normal cruising speed—

(1) During the day, to fly after that for at least 30 minutes; or

(2) At night, to fly after that for at least 45 minutes.

(b) No person may begin a flight in a rotorcraft under VFR conditions unless (considering wind and forecast weather conditions) there is enough fuel to fly to the first point of intended landing and, assuming normal cruising speed, to fly after that for at least 20 minutes.

§ 91.153 VFR flight plan: Information required.

(a) *Information required.* Unless otherwise authorized by ATC, each person filing a VFR flight plan shall include in it the following information:

(1) The aircraft identification number and, if necessary, its radio call sign.

(2) The type of the aircraft or, in the case of a formation flight, the type of each aircraft and the number of aircraft in the formation.

(3) The full name and address of the pilot in command or, in the case of a formation flight, the formation commander.

(4) The point and proposed time of departure.

(5) The proposed route, cruising altitude (or flight level), and true airspeed at that altitude.

(6) The point of first intended landing and the estimated elapsed time until over that point.

(7) The amount of fuel on board (in hours).

(8) The number of persons in the aircraft, except where that information is otherwise readily available to the FAA.

(9) Any other information the pilot in command or ATC believes is necessary for ATC purposes.

(b) *Cancellation.* When a flight plan has been activated, the pilot in command, upon canceling or completing the flight under the flight plan, shall notify an FAA Flight Service Station or ATC facility.

§ 91.155 Basic VFR weather minimums.

(a) Except as provided in paragraph (b) of this section and § 91.157, no person may operate an aircraft under VFR when the flight visibility is less, or at a distance from clouds that is less, than that prescribed for the corresponding altitude and class of airspace in the table on page 439:

(b) *Class G Airspace.* Notwithstanding the provisions of paragraph (a) of this section, the following operations may be conducted in Class G airspace below 1,200 feet above the surface:

(1) Helicopter. A helicopter may be operated clear of clouds if operated at a speed that allows the pilot adequate opportunity to see any air traffic or obstruction in time to avoid a collision.

(2) Airplane. When the visibility is less than 3 statute miles but not less than 1 statute mile during night hours, an airplane may be operated clear of clouds if operated in an airport traffic pattern within one-half mile of the runway.

(c) [Except as provided in § 91.157, no person may operate an aircraft beneath the ceiling under VFR within the lateral boundaries of controlled airspace designated to the surface for an airport when the ceiling is less than 1,000 feet.]

(d) Except as provided in § 91.157 of this part, no person may take off or land an aircraft, or enter the traffic pattern of an airport, under VFR, within the lateral boundaries of the surface areas of Class B, Class C, Class D, or Class E airspace designated for an airport—

(1) Unless ground visibility at that airport is at least 3 statute miles; or

(2) If ground visibility in not reported at that airport, unless flight visibility during landing or takeoff, or while operating in the traffic pattern is at least 3 statute miles.

(e) For the purpose of this section, an aircraft operating at the base altitude of a Class E airspace area is considered to be within the airspace directly below that area.

(Amdt. 91-213, Eff. 8/18/90); (Amdt. 91-224, Eff. 9/23/91); (Amdt. 91-227, Eff. 9/16/93); [(Amdt. 91-235, Eff. 10/5/93)]

[§ 91.157 Special VFR weather minimums.

[(a) Except as provided in Appendix D, Section 3, of this part, special VFR operations may be conducted under the weather minimums and requirements of this section, instead of those contained in § 91.155, below 10,000 feet MSL within the airspace contained by the upward extension of the lateral boundaries of the controlled airspace designated to the surface for an airport.

[(b) Special VFR operations may only be conducted—

[(1) With an ATC clearance;

[(2) Clear of clouds;

[(3) Except for helicopters, when flight visibility is at least 1 statute mile; and

[(4) Except for helicopters, between sunrise and sunset (or in Alaska, when the sun is 6° or more above the horizon) unless—

[(i) The person being granted the ATC clearance meets the applicable requirements for instrument flight under part 61 of this chapter; and

[(ii) The aircraft is equipped as required in § 91.205(d).

Airspace	Flight visibility	Distance from clouds
Class A	Not Applicable	Not Applicable.
Class B	3 statute miles	Clear of Clouds.
Class C	3 statute miles	500 feet below. 1,000 feet above. 2,000 feet horizontal.
Class D	3 statute miles	500 feet below. 1,000 feet above. 2,000 feet horizontal.
Class E		
Less than 10,000 feet MSL	3 statute miles	500 feet below. 1,000 feet above. 2,000 feet horizontal.
At or above 10,000 feet MSL	5 statute miles	1,000 feet below. 1,000 feet above. 1 statute mile horizontal.
Class G		
1,200 feet or less above the surface (regardless of MSL altitude)		
Day, except as provided in §91.155(b)	1 statute mile	Clear of clouds.
Night, except as provided in §91.155(b)	3 statute miles	500 feet below. 1,000 feet above. 2,000 feet horizontal.
More than 1,200 feet above the surface but less than 10,000 feet MSL		
Day	1 statute mile	500 feet below. 1,000 feet above. 2,000 feet horizontal.
Night	3 statute miles	500 feet below. 1,000 feet above. 2,000 feet horizontal.
More than 1,200 feet above the surface and at or above 10,000 feet MSL	5 statute miles	1,000 feet below. 1,000 feet above. 1 statute mile horizontal.

[(c) No person may take off or land an aircraft (other than a helicopter) under special VFR—

[(1) Unless ground visibility is at least 1 statute mile; or

[(2) If ground visibility is not reported, unless flight visibility is at least 1 statute mile.]

(Amdt. 91-227, Eff. 9/16/93); [(Amdt. 91-235, Eff. 10/5/93)]

§ 91.159 VFR cruising altitude or flight level.

Except while holding in a holding pattern of 2 minutes or less, or while turning, each person operating an aircraft under VFR in level cruising flight more than 3,000 feet above the surface shall maintain the appropriate altitude or flight level prescribed below, unless otherwise authorized by ATC:

(a) When operating below 18,000 feet MSL and—

(1) On a magnetic course of zero degrees through 179 degrees, any odd thousand foot MSL altitude +500 feet (such as 3,500, 5,500, or 7,500); or

(2) On a magnetic course of 180 degrees through 359 degrees, any even thousand foot MSL altitude +500 feet (such as 4,500, 6,500, or 8,500).

(b) When operating above 18,000 feet MSL to flight level 290 (inclusive) and—

(1) On a magnetic course of zero degrees through 179

degrees, any odd flight level +500 feet (such as 195, 215, or 235); or

(2) On a magnetic course of 180 degrees through 359 degrees, any even flight level +500 feet (such as 185, 205, or 225).

(c) When operating above flight level 290 and—

(1) On a magnetic course of zero degrees through 179 degrees, any flight level, at 4,000-foot intervals, beginning at and including flight level 300 (such as flight level 300, 340, or 380); or

(2) On a magnetic course of 180 degrees through 359 degrees, any flight level, at 4,000-foot intervals, beginning at and including flight level 320 (such as flight level 320, 360, or 400).

§§ 91.161–91.165 [Reserved]

Instrument flight rules (Subpart B—Flight rules)

§ 91.167 Fuel requirements for flight in IFR conditions.

(a) Except as provided in paragraph (b) of this section, no person may operate a civil aircraft in IFR conditions unless it carries enough fuel (considering weather reports and forecasts and weather conditions) to—

(1) Complete the flight to the first airport of intended landing;

(2) Fly from that airport to the alternate airport; and

(3) Fly after that for 45 minutes at normal cruising speed or, for helicopters, fly after that for 30 minutes at normal cruising speed.

(b) Paragraph (a)(2) of this section does not apply if—

(1) Part 97 of this chapter prescribes a standard instrument approach procedure for the first airport of intended landing; and

(2) For at least 1 hour before and 1 hour after the estimated time of arrival at the airport, the weather reports or forecasts or any combination of them indicate—

(i) The ceiling will be at least 2,000 feet above the airport elevation; and

(ii) Visibility will be at least 3 statute miles.

§ 91.169 IFR flight plan: Information required.

(a) *Information required.* Unless otherwise authorized by ATC, each person filing an IFR flight plan shall include in it the following information:

(1) Information required under § 91.153(a).

(2) An alternate airport, except as provided in paragraph (b) of this section.

(b) *Exceptions to applicability of paragraph (a)(2) of this section.* Paragraph (a)(2) of this section does not apply if Part 97 of this chapter prescribes a standard instrument approach procedure for the first airport of intended landing and, for at least 1 hour before and 1 hour after the estimated time of arrival, the weather reports or forecasts, or any combination of them, indicate—

(1) The ceiling will be at least 2,000 feet above the airport elevation; and

(2) The visibility will be at least 3 statute miles.

(c) *IFR alternate airport weather minimums.* Unless otherwise authorized by the Administrator, no person may include an alternate airport in an IFR flight plan unless current weather forecasts indicate that, at the estimated time of arrival at the alternate airport, the ceiling and visibility at that airport will be at or above the following alternate airport weather minimums:

(1) If an instrument approach procedure has been published in Part 97 of this chapter for that airport, the alternate airport minimums specified in that procedure or, if none are so specified, the following minimums:

(i) Precision approach procedure: Ceiling 600 feet and visibility 2 statute miles.

(ii) Nonprecision approach procedure: Ceiling 800 feet and visibility 2 statute miles.

(2) If no instrument approach procedure has been published in Part 97 of this chapter for that airport, the ceiling and visibility minimums are those allowing descent from the MEA, approach, and landing under basic VFR.

(d) *Cancellation.* When a flight plan has been activated, the pilot in command, upon canceling or completing the flight under the flight plan, shall notify an FAA Flight Service Station or ATC facility.

§ 91.171 VOR equipment check for IFR operations.

(a) No person may operate a civil aircraft under IFR using the VOR system of radio navigation unless the VOR equipment of that aircraft—

(1) Is maintained, checked, and inspected under an approved procedure; or

(2) Has been operationally checked within the preceding 30 days, and was found to be within the limits of the permissible indicated bearing error set forth in paragraph (b) or (c) of this section.

(b) Except as provided in paragraph (c) of this section, each person conducting a VOR check under paragraph (a)(2) of this section shall—

(1) Use, at the airport of intended departure, an FAA-operated or approved test signal or a test signal radiated by a certificated and appropriately rated radio repair station or, outside the United States, a test signal operated or approved by an appropriate authority to check the VOR equipment (the maximum permissible indicated bearing error is plus or minus 4 degrees); or

(2) Use, at the airport of intended departure, a point on the airport surface designated as a VOR system checkpoint by the Administrator, or, outside the United States, by an appropriate authority (the maximum permissible bearing error is plus or minus 4 degrees);

(3) If neither a test signal nor a designated checkpoint on the surface is available, use an airborne checkpoint designated by the Administrator or, outside the United States, by an appropriate authority (the maximum permissible bearing error is plus or minus 6 degrees); or

(4) If no check signal or point is available, while in flight—

(i) Select a VOR radial that lies along the centerline of an established VOR airway;

(ii) Select a prominent ground point along the selected radial preferably more than 20 nautical miles from the VOR ground facility and maneuver the aircraft directly over the point at a reasonably low altitude; and

(iii) Note the VOR bearing indicated by the receiver when over the ground point (the maximum permissible variation between the published radial and the indicated bearing is 6 degrees).

(c) If dual system VOR (units independent of each other except for the antenna) is installed in the aircraft, the person checking the equipment may check one system against the other in place of the check procedures specified in paragraph (b) of this section. Both systems shall be tuned to the same VOR ground facility and note the indicated bearings to that station. The maximum permissible variation between the two indicated bearings is 4 degrees.

(d) Each person making the VOR operational check, as specified in paragraph (b) or (c) of this section, shall enter the date, place, bearing error, and sign the aircraft log or other record. In addition, if a test signal radiated by a repair station, as specified in paragraph (b)(1) of this section, is used, an entry must be made in the aircraft log or other record by the repair station certificate holder or the certificate holder's representative certifying to the bearing transmitted by the repair station for the check and the date of transmission.

(Approved by the Office of Management and Budget under OMB control number 2120-0005).

§ 91.173 ATC clearance and flight plan required.

No person may operate an aircraft in controlled airspace under IFR unless that person has—

(a) Filed an IFR flight plan; and

(b) Received an appropriate ATC clearance.

§ 91.175 Takeoff and landing under IFR.

(a) *Instrument approaches to civil airports.* Unless otherwise authorized by the Administrator, when an instrument letdown to a civil airport is necessary, each person operating an aircraft, except a military aircraft of the United States, shall use a standard instrument approach procedure prescribed for the airport in part 97 of this chapter.

(b) *Authorized DH or MDA.* For the purpose of this section, when the approach procedure being used provides for and requires the use of a DH or MDA, the authorized DH or MDA is the highest of the following:

(1) The DH or MDA prescribed by the approach procedure.

(2) The DH or MDA prescribed for the pilot in command.

(3) The DH or MDA for which the aircraft is equipped.

(c) *Operation below DH or MDA.* Where a DH or MDA is applicable, no pilot may operate an aircraft, except a military aircraft of the United States, at any airport below the authorized MDA or continue an approach below the authorized DH unless—

(1) The aircraft is continuously in a position from which a descent to a landing on the intended runway can be made at a normal rate of descent using normal maneuvers, and for operations conducted under Part 121 or Part 135 unless that descent rate will allow touchdown to occur within the touchdown zone of the runway of intended landing;

(2) The flight visibility is not less than the visibility prescribed in the standard instrument approach being used; and

(3) Except for a Category II or Category III approach where any necessary visual reference requirements are specified by the Administrator, at least one of the following visual references for the intended runway is distinctly visible and identifiable to the pilot:

(i) The approach light system, except that the pilot may not descend below 100 feet above the touchdown zone elevation using the approach lights as a reference unless the red terminating bars or the red side row bars are also distinctly visible and identifiable.

(ii) The threshold.

(iii) The threshold markings.

(iv) The threshold lights.

(v) The runway end identifier lights.

(vi) The visual approach slope indicator.

(vii) The touchdown zone or touchdown zone markings.

(viii) The touchdown zone lights.

(ix) The runway or runway markings.

(x) The runway lights.

(d) *Landing.* No pilot operating an aircraft, except a military aircraft of the United States, may land that aircraft when the flight visibility is less than the visibility prescribed in the standard instrument approach procedure being used.

(e) *Missed approach procedures.* Each pilot operating an aircraft, except a military aircraft of the United States, shall immediately execute an appropriate missed approach procedure when either of the following conditions exist:

(1) Whenever the requirements of paragraph (c) of this section are not met at either of the following times:

(i) When the aircraft is being operated below MDA; or

(ii) Upon arrival at the missed approach point, including a DH where a DH is specified and its use is required, and at any time after that until touchdown.

(2) Whenever an identifiable part of the airport is not distinctly visible to the pilot during a circling maneuver at or above MDA, unless the inability to see an identifiable part of the airport results only from a normal bank of the aircraft during the circling approach.

(f) *Civil airport takeoff minimums.* Unless otherwise authorized by the Administrator, no pilot operating an aircraft under Parts 121, 125, 127, 129, or 135 of this chapter may take off from a civil airport under IFR unless weather conditions are at or above the weather minimum for IFR takeoff prescribed for that airport under Part 97 of this chapter. If takeoff minimums are not prescribed under Part 97 of this chapter for a particular airport, the following minimums apply to takeoffs under IFR for aircraft operating under those parts:

(1) For aircraft, other than helicopters, having two engines or less—1 statute mile visibility.

(2) For aircraft having more than two engines—½ statute mile visibility.

(3) For helicopters—½ statute mile visibility.

(g) *Military airports.* Unless otherwise prescribed by the Administrator, each person operating a civil aircraft under IFR into or out of a military airport shall comply with the instrument approach procedures and the takeoff and landing minimum prescribed by the military authority having jurisdiction of that airport.

(h) *Comparable values of RVR and ground visibility.*

(1) Except for Category II or Category III minimums, if RVR minimums for takeoff or landing are prescribed in an instrument approach procedure, but RVR is not reported for the runway of intended operation, the RVR minimum shall be converted to ground visibility in accordance with the table in paragraph (h)(2) of this section and shall be the visibility minimum for takeoff or landing on that runway.

(2) RVR Table:

RVR (feet)	Visibility (statute miles)
1,600	¼
2,400	½
3,200	⅝
4,000	¾
4,500	⅞
5,000	1
6,000	1¼

(i) *Operations on unpublished routes and use of radar in instrument approach procedures.* When radar is approved at certain locations for ATC purposes, it may be used not only for surveillance and precision radar approaches, as applicable, but also may be used in conjunction with instrument approach procedures predicated on other types of radio navigational aids. Radar vectors may be authorized to provide course guidance through the segments of an approach to the final course or fix. When operating on an unpublished route or while being radar vectored, the pilot, when an approach clearance is received, shall, in addition to complying with § 91.177, maintain the last altitude assigned to that pilot until the aircraft is established on a segment of a published route or instrument approach procedure unless a different altitude is assigned by ATC. After the aircraft is so established, published altitudes apply to descent within each succeeding route or approach segment unless a different altitude is assigned by ATC. Upon reaching the final approach course or fix, the pilot may either complete the instrument approach in accordance with a procedure approved for the facility or continue a surveillance or precision radar approach to a landing.

(j) *Limitation on procedure turns.* In the case of a radar vector to a final approach course or fix, a timed approach from a holding fix, or an approach for which the procedure specifies "No PT," no pilot may make a procedure turn unless cleared to do so by ATC.

(k) *ILS components.* The basic ground components of an ILS are the localizer, glide slope, outer marker, middle marker, and, when installed for use with Category II or Category III instrument approach procedures, an inner marker. A compass locator or precision radar may be substituted for the outer or middle marker. DME, VOR, or nondirectional beacon fixes authorized in the standard instrument approach procedure or surveillance radar may be substituted for the outer marker. Applicability of, and substitution for, the inner marker for Category II or III approaches is determined by the appropriate Part 97 approach procedure, letter of authorization, or operations specification pertinent to the operations.

§ 91.177 Minimum altitudes for IFR operations.

(a) *Operation of aircraft at minimum attitudes.* Except when necessary for takeoff or landing, no person may operate an aircraft under IFR below—

(1) The applicable minimum altitudes prescribed in Parts 95 and 97 of this chapter; or

(2) If no applicable minimum altitude is prescribed in those parts—

(i) In the case of operations over an area designated as a mountainous area in Part 95, an altitude of 2,000 feet above the highest obstacle within a horizontal distance of 4 nautical miles from the course to be flown; or

(ii) In any other case, an altitude of 1,000 feet above the highest obstacle within a horizontal distance of 4 nautical miles from the course to be flown.

However, if both a MEA and a MOCA are prescribed for a particular route or route segment, a person may operate an aircraft below the MEA down to, but not below, the MOCA, when within 22 nautical miles of the VOR concerned (based on the pilot's reasonable estimate of that distance).

(b) *Climb.* Climb to a higher minimum IFR altitude shall begin immediately after passing the point beyond which that minimum altitude applies, except that when ground obstructions intervene, the point beyond which that higher minimum altitude applies shall be crossed at or above the applicable MCA.

§ 91.179 IFR cruising altitude or flight level.

(a) *In controlled airspace.* Each person operating an aircraft under IFR in level cruising flight in controlled airspace shall maintain the altitude or flight level assigned that aircraft by ATC. However, if the ATC clearance assigns "VFR conditions on-top," that person shall maintain an altitude or flight level as prescribed by § 91.159.

(b) *In uncontrolled airspace.* Except while in a holding pattern of 2 minutes or less or while turning, each person operating an aircraft under IFR in level cruising flight in uncontrolled airspace shall maintain an appropriate altitude as follows:

(1) When operating below 18,000 feet MSL and—

(i) On a magnetic course of zero degrees through 179 degrees, any odd thousand foot MSL altitude (such as 3,000, 5,000, or 7,000); or

(ii) On a magnetic course of 180 degrees through 359 degrees, any even thousand foot MSL altitude (such as 2,000, 4,000, or 6,000).

(2) When operating at or above 18,000 feet MSL but below flight level 290, and—

(i) On a magnetic course of zero degrees through 179 degrees, any odd flight level (such as 190, 210, or 230); or

(ii) On a magnetic course of 180 degrees through 359 degrees, any even flight level (such as 180, 200, or 220).

(3) When operating at flight level 290 and above, and—

(i) On a magnetic course of zero degrees through 179 degrees, any flight level, at 4,000-foot intervals, beginning at and including flight level 290 (such as flight level 290, 330, or 370); or

(ii) On a magnetic course of 180 degrees through 359 degrees, any flight level, at 4,000-foot intervals, beginning at and including flight level 310 (such as flight level 310, 350, or 390).

§ 91.181 Course to be flown.

Unless otherwise authorized by ATC, no person may operate an aircraft within controlled airspace under IFR except as follows:

(a) On a Federal airway, along the centerline of that airway.

(b) On any other route, along the direct course between the navigational aids or fixes defining that route. However, this section does not prohibit maneuvering the aircraft to pass well clear of other air traffic or the maneuvering of the aircraft in VFR conditions to clear the intended flight path both before and during climb or descent.

§ 91.183 IFR radio communications.

The pilot in command of each aircraft operated under IFR in controlled airspace shall have a continuous watch maintained on the appropriate frequency and shall report by radio as soon as possible—

(a) The time and altitude of passing each designated reporting point, or the reporting points specified by ATC, except that while the aircraft is under radar control, only the passing of those reporting points specifically requested by ATC need be reported;

(b) Any unforecast weather conditions encountered; and

(c) Any other information relating to the safety of flight.

§ 91.185 IFR operations: Two-way radio communications failure.

(a) *General.* Unless otherwise authorized by ATC, each pilot who has two-way radio communications failure when operating under IFR shall comply with the rules of this section.

(b) *VFR conditions.* If the failure occurs in VFR conditions, or if VFR conditions are encountered after the failure, each pilot shall continue the flight under VFR and land as soon as practicable.

(c) *IFR conditions.* If the failure occurs in IFR conditions, or if paragraph (b) of this section cannot be complied with, each pilot shall continue the flight according to the following:

(1) *Route.*

(i) By the route assigned in the last ATC clearance received;

(ii) If being radar vectored, by the direct route from the point of radio failure to the fix, route, or airway specified in the vector clearance;

(iii) In the absence of an assigned route, by the route that ATC has advised may be expected in a further clearance; or

(iv) In the absence of an assigned route or a route that ATC has advised may be expected in a further clearance, by the route filed in the flight plan.

(2) *Altitude.* At the highest of the following altitudes or flight levels for the route segment being flown:

(i) The altitude or flight level assigned in the last ATC clearance received;

(ii) The minimum altitude (converted, if appropriate, to minimum flight level as prescribed in § 91.121(c)) for IFR operations; or

(iii) The altitude or flight level ATC has advised may be expected in a further clearance.

(3) *Leave clearance limit.*

(i) When the clearance limit is a fix from which an approach begins, commence descent or descent and approach as close as possible to the expect-further-clearance time if one has been received, or if one has not been received, as close as possible to the estimated time of arrival as calculated from the filed or amended (with ATC) estimated time en route.

(ii) If the clearance limit is not a fix from which an approach begins, leave the clearance limit at the expect-further-clearance time if one has been received, or if none has been received, upon arrival over the clearance limit, and proceed to a fix from which an approach begins and commence descent or descent and approach as close as possible to the estimated time of arrival as calculated from the filed or amended (with ATC) estimated time en route.

§ 91.187 Operation under IFR in controlled airspace: Malfunction reports.

(a) The pilot in command of each aircraft operated in controlled airspace under IFR shall report as soon as practical to ATC any malfunctions of navigational, approach, or communication equipment occurring in flight.

(b) In each report required by paragraph (a) of this section, the pilot in command shall include the—

(1) Aircraft identification;

(2) Equipment affected;

(3) Degree to which the capability of the pilot to operate under IFR in the ATC system is impaired; and

(4) Nature and extent of assistance desired from ATC.

§ 91.189 Category II and III operations: General operating rules.

(a) No person may operate a civil aircraft in a Category II or III operation unless—

(1) The flight crew of the aircraft consists of a pilot in command and a second in command who hold the appropriate authorizations and ratings prescribed in § 61.3 of this chapter;

(2) Each flight crewmember has adequate knowledge of, and familiarity with, the aircraft and the procedures to be used; and

(3) The instrument panel in front of the pilot who is controlling the aircraft has appropriate instrumentation for the type of flight control guidance system that is being used.

(b) Unless otherwise authorized by the Administrator, no person may operate a civil aircraft in a Category II or Category III operation unless each ground component required for that operation and the related airborne equipment is installed and operating.

(c) *Authorized DH.* For the purpose of this section, when the approach procedure being used provides for and requires the use of a DH, the authorized DH is the highest of the following:

(1) The DH prescribed by the approach procedure.

(2) The DH prescribed for the pilot in command.

(3) The DH for which the aircraft is equipped.

(d) Unless otherwise authorized by the Administrator, no pilot operating an aircraft in a Category II or Category III approach that provides and requires use of a DH may continue the approach below the authorized decision height unless the following conditions are met:

(1) The aircraft is in a position from which a descent to a landing on the intended runway can be made at a normal rate of descent using normal maneuvers, and where that descent rate will allow touchdown to occur within the touchdown zone of the runway of intended landing.

(2) At least one of the following visual references for the intended runway is distinctly visible and identifiable to the pilot:

(i) The approach light system, except that the pilot may not descend below 100 feet above the touchdown zone elevation using the approach lights as a reference unless the red terminating bars or the red side row bars are also distinctly visible and identifiable.

(ii) The threshold.

(iii) The threshold markings.

(iv) The threshold lights.

(v) The touchdown zone or touchdown zone markings.

(vi) The touchdown zone lights.

(e) Unless otherwise authorized by the Administrator, each pilot operating an aircraft shall immediately execute an appropriate missed approach whenever, prior to touchdown, the requirements of paragraph (d) of this section are not met.

(f) No person operating an aircraft using a Category III approach without decision height may land that aircraft except in accordance with the provisions of the letter of authorization issued by the Administrator.

(g) Paragraphs (a) through (f) of this section do not apply to operations conducted by the holders of certificates issued under Part 121, 125, 129, or 135 of this chapter. No person may operate a civil aircraft in a Category II or Category III operation conducted by the holder of a certificate issued under Part 121, 125, 129, or 135 of this chapter unless the operation is conducted in accordance with that certificate holder's operations specifications.

§ 91.191 Category II manual.

(a) No person may operate a civil aircraft of United States registry in a Category II operation unless—

(1) There is available in the aircraft a current, approved Category II manual for that aircraft;

(2) The operation is conducted in accordance with the procedures, instructions, and limitations in that manual; and

(3) The instruments and equipment listed in the manual that are required for a particular Category II operation

have been inspected and maintained in accordance with the maintenance program contained in that manual.

(b) Each operator shall keep a current copy of the approved manual at its principal base of operations and shall make it available for inspection upon request of the Administrator.

(c) This section does not apply to operations conducted by the holder of a certificate issued under Part 121 of this chapter.

(Approved by the Office of Management and Budget under OMB control number 2120-0005).

§ 91.193 Certificate of authorization for certain Category II operations.

The Administrator may issue a certificate of authorization authorizing deviations from the requirements of § 91.189, § 91.191, and § 91.205(f) for the operation of small aircraft identified as Category A aircraft in § 97.3 of this chapter in Category II operations if the Administrator finds that the proposed operation can be safely conducted under the terms of the certificate. Such authorization does not permit operation of the aircraft carrying persons or property for compensation or hire.

§§ 91.195–91.199 [Reserved]

Subpart C—Equipment, instrument, and certificate requirements

§ 91.201[Reserved]

§ 91.203 Civil aircraft: Certifications required.

(a) Except as provided in § 91.715, no person may operate a civil aircraft unless it has within it the following:

(1) An appropriate and current airworthiness certificate. Each U.S. airworthiness certificate used to comply with this subparagraph (except a special flight permit, a copy of the applicable operations specifications issued under § 21.197(c) of this chapter, appropriate sections of the air carrier manual required by Parts 121 and 135 of this chapter containing that portion of the operations specifications issued under § 21.197(c), or an authorization under § 91.611) must have on it the registration number assigned to the aircraft under Part 47 of this chapter. However, the airworthiness certificate need not have on it an assigned special identification number before 10 days after that number is first affixed to the aircraft. A revised airworthiness certificate having on it an assigned special identification number, that has been affixed to an aircraft, may only be obtained upon application to an FAA Flight Standards district office.

(2) An effective U.S. registration certificate issued to its owner or, for operation within the United States, the second duplicate copy (pink) of the Aircraft Registration Application as provided for in § 47.31(b), or a registration certificate issued under the laws of a foreign country.

(b) No person may operate a civil aircraft unless the airworthiness certificate required by paragraph (a) of this section or a special flight authorization issued under § 91.715 is displayed at the cabin or cockpit entrance so that it is legible to passengers or crew.

(c) No person may operate an aircraft with a fuel tank installed within the passenger compartment or a baggage compartment unless the installation was accomplished pursuant to Part 43 of this chapter, and a copy of FAA Form 337 authorizing that installation is on board the aircraft.

[(d) No person may operate a civil airplane (domestic or foreign) into or out of an airport in the United States unless it complies with the fuel venting and exhaust emissions requirements of Part 34 of this chapter.]

[(Amdt. 91-218, Eff. 9/10/90)]

§ 91.205 Powered civil aircraft with standard category U.S. airworthiness certificates: Instrument and equipment requirements.

(a) *General.* Except as provided in paragraphs (c)(3) and (e) of this section, no person may operate a powered civil aircraft with a standard category U.S. airworthiness certificate in any operation described in paragraphs (b) through (f) of this section unless that aircraft contains the instruments and equipment specified in those paragraphs (or FAA-approved equivalents) for that type of operation, and those instruments and items of equipment are in operable condition.

(b) *Visual-flight rules (day).* For VFR flight during the day, the following instruments and equipment are required:

(1) Airspeed indicator.

(2) Altimeter.

(3) Magnetic direction indicator.

(4) Tachometer for each engine.

(5) Oil pressure gauge for each engine using pressure system.

(6) Temperature gauge for each liquid-cooled engine.

(7) Oil temperature gauge for each air-cooled engine.

(8) Manifold pressure gauge for each altitude engine.

(9) Fuel gauge indicating the quantity of fuel in each tank.

(10) Landing gear position indicator, if the aircraft has a retractable landing gear.

(11) If the aircraft is operated for hire over water and beyond power-off gliding distance from shore, approved flotation gear readily available to each occupant and at least one pyrotechnic signaling device. As used in this section, "shore" means that area of the land adjacent to the water which is above the high water mark and excludes land areas which are intermittently under water.

(12) [An approved safety belt with an approved metal-to-metal latching device for each occupant 2 years of age or older.]

(13) For small civil airplanes manufactured after July 18, 1978, an approved shoulder harness for each front seat. The shoulder harness must be designed to protect the occupant from serious head injury when the occupant experiences the ultimate inertia forces specified in § 23.561(b)(2) of this chapter. Each shoulder harness installed at a flight crewmember station must permit the crewmember, when seated and with the safety belt and shoulder harness fastened, to perform all functions necessary for flight operations. For purposes of this paragraph—

(i) The date of manufacture of an airplane is the date the inspection acceptance records reflect that the airplane is complete and meets the FAA-approved type design data; and

(ii) A front seat is a seat located at a flight crewmember station or any seat located alongside such a seat.

(14) An emergency locator transmitter, if required by § 91.207.

(15) For normal, utility, and acrobatic category airplanes with a seating configuration, excluding pilot seats, of 9 or less, manufactured after December 12, 1986, a shoulder harness for—

(i) Each front seat that meets the requirements of § 23.785(g) and (h) of this chapter in effect on December 12, 1985;

(ii) Each additional seat that meets the requirements of § 23.785(g) of this chapter in effect on December 12, 1985.

(16) For rotorcraft manufactured after September 16, 1992, a shoulder harness for each seat that meets the requirements of § 27.2 or § 29.2 of this chapter in effect on September 16, 1991.

(c) *Visual flight rules (night).* For VFR flight at night, the following instruments and equipment are required:

(1) Instruments and equipment specified in paragraph (b) of this section.

(2) Approved position lights.

(3) An approved aviation red or aviation white anticollision light system on all U.S.-registered civil aircraft. Anticollision light systems initially installed after August 11, 1971, on aircraft for which a type certificate was issued or applied for before August 11, 1971, must at least meet the anticollision light standards of Part 23, 25, 27, or 29 of this chapter, as applicable, that were in effect on August 10, 1971, except that the color may be either aviation red or aviation white. In the event of failure of any light of the anticollision light system, operations with the aircraft may be continued to a stop where repairs or replacement can be made.

(4) If the aircraft is operated for hire, one electric landing light.

(5) An adequate source of electrical energy for all installed electrical and radio equipment.

(6) One spare set of fuses, or three spare fuses of each kind required, that are accessible to the pilot in flight.

(d) *Instrument flight rules.* For IFR flight, the following instruments and equipment are required:

(1) Instruments and equipment specified in paragraph (b) of this section, and, for night flight, instruments and equipment specified in paragraph (c) of this section.

(2) Two-way radio communications system and navigational equipment appropriate to the ground facilities to be used.

(3) Gyroscopic rate-of-turn indicator, except on the following aircraft:

(i) Airplanes with a third attitude instrument system usable through flight attitudes of 360 degrees of pitch and roll and installed in accordance with the instrument requirements prescribed in § 121.305(j) of this chapter; and:

(ii) Rotorcraft with a third attitude instrument system usable through flight attitudes of ±80 degrees of pitch and ±120 degrees of roll and installed in accordance with § 29.1303(g) of this chapter.

(4) Slip-skid indicator.

(5) Sensitive altimeter adjustable for barometric pressure.

(6) A clock displaying hours, minutes, and seconds with a sweep-second pointer or digital presentation.

(7) Generator or alternator of adequate capacity.

(8) Gyroscopic pitch and bank indicator (artificial horizon).

(9) Gyroscopic direction indicator (directional gyro or equivalent).

(e) *Flight at and above 24,000 ft. MSL (FL 240).* If VOR navigational equipment is required under paragraph (d)(2) of this section, no person may operate a U.S.-registered civil aircraft within the 50 states and the District of Columbia at or above FL 240 unless that aircraft is equipped with approved distance measuring equipment (DME). When DME required by this paragraph fails at and above FL 240, the pilot in command of the aircraft shall notify ATC immediately, and then may continue operations at and above FL 240 to the next airport of intended landing at which repairs or replacement of the equipment can be made.

(f) *Category II operations.* For Category II operations the instruments and equipment specified in paragraph (d) of this section and Appendix A to this part are required. This paragraph does not apply to operations conducted by the holder of a certificate issued under Part 121 of this chapter.

(Amdt. 91-220, Eff. 11/26/90); (Amdt. 91-223, Eff. 9/16/91); [(Amdt. 91-231, Eff. 10/15/92)]

§ 91.207 Emergency locator transmitters.

(a) Except as provided in paragraphs (e) and (f) of this section, no person may operate a U.S.-registered civil airplane unless—

(1) There is attached to the airplane an approved automatic type emergency locator transmitter that is in operable condition for the following operations, except that after June 21, 1995, an emergency locator transmitter that meets the requirements of TSO-C91 may not be used for new installations:

(i) Those operations governed by the supplemental air carrier and commercial operator rules of Parts 121 and 125;

(ii) Charter flights governed by the domestic and flag air carrier rules of Part 121 of this chapter; and

(iii) Operations governed by Part 135 of this chapter; or

(2) For operations other than those specified in paragraph (a)(1) of this section, there must be attached to the airplane an approved personal type or an approved automatic type emergency locator transmitter that is in operable condition, except that after June 21, 1995, an emergency locator transmitter that meets the requirements of TSO-C91 may not be used for new installations.

(b) Each emergency locator transmitter required by paragraph (a) of this section must be attached to the airplane in such a manner that the probability of damage to the transmitter in the event of crash impact is minimized. Fixed and deployable automatic type transmitters must be attached to the airplane as far aft as practicable.

(c) Batteries used in the emergency locator transmitters required by paragraphs (a) and (b) of this section must be replaced (or recharged, if the batteries are rechargeable)—

(1) When the transmitter has been in use for more than 1 cumulative hour; or

(2) When 50 percent of their useful life (or, for rechargeable batteries, 50 percent of their useful life of charge) has expired, as established by the transmitter manufacturer under its approval.

The new expiration date for replacing (or recharging) the battery must be legibly marked on the outside of the transmitter and entered in the aircraft maintenance record. Paragraph (c)(2) of this section does not apply to batteries (such as water-activated batteries) that are essentially unaffected during probable storage intervals.

(d) Each emergency locator transmitter required by paragraph (a) of this section must be inspected within 12 calendar months after the last inspection for—

(1) Proper installation;

(2) Battery corrosion;

(3) Operation of the controls and crash sensor; and

(4) The presence of a sufficient signal radiated from its antenna.

(e) Notwithstanding paragraph (a) of this section, a person may—

(1) Ferry a newly acquired airplane from the place where possession of it was taken to a place where the emergency locator transmitter is to be installed; and

(2) Ferry an airplane with an inoperative emergency locator transmitter from a place where repairs or replacements cannot be made to a place where they can be made.

No person other than required crewmembers may be carried aboard an airplane being ferried under paragraph (e) of this section.

(f) Paragraph (a) of this section does not apply to—

(1) Turbojet-powered aircraft;

(2) Aircraft while engaged in scheduled flights by scheduled air carriers;

(3) Aircraft while engaged in training operations conducted entirely within a 50-nautical mile radius of the airport from which such local flight operations began;

(4) Aircraft while engaged in flight operations incident to design and testing;

(5) New aircraft while engaged in flight operations incident to their manufacture, preparation, and delivery;

(6) Aircraft while engaged in flight operations incident to the aerial application of chemicals and other substances for agricultural purposes;

(7) Aircraft certificated by the Administrator for research and development purposes;

(8) Aircraft while used for showing compliance with regulations, crew training, exhibition, air racing, or market surveys;

(9) Aircraft equipped to carry not more than one person; and

(10) An aircraft during any period for which the transmitter has been temporarily removed for inspection, repair, modification, or replacement, subject to the following:

(i) No person may operate the aircraft unless the aircraft records contain an entry which includes the date of initial removal, the make, model, serial number, and reason for removing the transmitter, and a placard located in view of the pilot to show "ELT not installed."

(ii) No person may operate the aircraft more than 90 days after the ELT is initially removed from the aircraft.

(Amdt. 91-242, Eff. 6/21/94)

§ 91.209 Aircraft lights.

No person may, during the period from sunset to sunrise (or, in Alaska, during the period a prominent unlighted object cannot be seen from a distance of 3 statute miles or the sun is more than 6 degrees below the horizon)—

(a) Operate an aircraft unless it has lighted position lights;

(b) Park or move an aircraft in, or in dangerous proximity to, a night flight operations area of an airport unless the aircraft—

(1) Is clearly illuminated;

(2) Has lighted position lights; or

(3) Is in an area which is marked by obstruction lights;

(c) Anchor an aircraft unless the aircraft—

(1) Has lighted anchor lights; or

(2) Is in an area where anchor lights are not required on vessels; or

(d) Operate an aircraft, required by § 91.205(c)(3) to be equipped with an anticollision light system, unless it has approved and lighted aviation red or aviation white anticollision lights. However, the anticollision lights need not be lighted when the pilot in command determines that, because of operating conditions, it would be in the interest of safety to turn the lights off.

§ 91.211 Supplemental oxygen.

(a) *General.* No person may operate a civil aircraft of U.S. registry—

(1) At cabin pressure altitudes above 12,500 feet (MSL) up to and including 14,000 feet (MSL) unless the required minimum flight crew is provided with and uses supplemental oxygen for that part of the flight at those altitudes that is of more than 30 minutes duration;

(2) At cabin pressure altitudes above 14,000 feet (MSL) unless the required minimum flight crew is provided with and uses supplemental oxygen during the entire flight time at those altitudes; and

(3) At cabin pressure altitudes above 15,000 feet (MSL) unless each occupant of the aircraft is provided with supplemental oxygen.

(b) *Pressurized cabin aircraft.*

(1) No person may operate a civil aircraft of U.S. registry with a pressurized cabin—

(i) At flight altitudes above flight level 250 unless at least a 10-minute supply of supplemental oxygen, in addition to any oxygen required to satisfy paragraph (a) of this section, is available for each occupant of the aircraft for use in the event that a descent is necessitated by loss of cabin pressurization; and

(ii) At flight altitudes above flight level 350 unless one pilot at the controls of the airplane is wearing and using an oxygen mask that is secured and sealed and that either supplies oxygen at all times or automatically supplies oxygen whenever the cabin pressure altitude of the airplane exceeds 14,000 feet (MSL), except that the one pilot need not wear and use an oxygen mask while at or below flight level 410 if there are two pilots at the controls and each pilot has a quick-donning type of oxygen mask that can be placed on the face with one hand from the ready position within 5 seconds, supplying oxygen and properly secured and sealed.

(2) Notwithstanding paragraph (b)(1)(ii) of this section, if for any reason at any time it is necessary for one pilot to leave the controls of the aircraft when operating at flight altitudes above flight level 350, the remaining pilot at the controls shall put on and use an oxygen mask until the other pilot has returned to that crewmember's station.

§ 91.213 Inoperative instruments and equipment.

(a) Except as provided in paragraph (d) of this section, no person may take off an aircraft with inoperative instruments or equipment installed unless the following conditions are met:

(1) An approved Minimum Equipment List exists for that aircraft.

(2) The aircraft has within it a letter of authorization, issued by the FAA Flight Standards district office having jurisdiction over the area in which the operator is located, authorizing operation of the aircraft under the Minimum Equipment List. The letter of authorization may be obtained by written request of the airworthiness certificate holder. The Minimum Equipment List and the letter of authorization constitute a supplemental type certificate for the aircraft.

(3) The approved Minimum Equipment List must—

(i) Be prepared in accordance with the limitations specified in paragraph (b) of this section; and

(ii) Provide for the operation of the aircraft with the instruments and equipment in an inoperable condition,

(4) The aircraft records available to the pilot must include an entry describing the inoperable instruments and equipment.

(5) The aircraft is operated under all applicable conditions and limitations contained in the Minimum Equipment List and the letter authorizing the use of the list.

(b) The following instruments and equipment may not be included in a Minimum Equipment List:

(1) Instruments and equipment that are either specifically or otherwise required by the airworthiness requirements under which the aircraft is type certificated and which are essential for safe operations under all operating conditions.

(2) Instruments and equipment required by an airworthiness directive to be in operable condition unless the airworthiness directive provides otherwise.

(3) Instruments and equipment required for specific operations by this part.

(c) A person authorized to use an approved Minimum Equipment List issued for a specific aircraft under Part 121, 125, or 135 of this chapter shall use that Minimum Equipment List in connection with operations conducted with that aircraft under this part without additional approval requirements.

(d) Except for operations conducted in accordance with paragraph (a) or (c) of this section, a person may take off an aircraft in operations conducted under this part with inoperative instruments and equipment without an approved Minimum Equipment List provided—

(1) The flight operation is conducted in a—

(i) Rotorcraft, nonturbine-powered airplane, glider, or lighter-than-air aircraft for which a master Minimum Equipment List has not been developed; or

(ii) Small rotorcraft, nonturbine-powered small airplane, glider, or lighter-than-air aircraft for which a Master Minimum Equipment List has been developed; and

(2) The inoperative instruments and equipment are not—

(i) Part of the VFR-day type certification instruments and equipment prescribed in the applicable airworthiness regulations under which the aircraft was type certificated;

(ii) Indicated as required on the aircraft's equipment list, or on the Kinds of Operations Equipment List for the kind of flight operation being conducted;

(iii) Required by § 91.205 or any other rule of this part for the specific kind of flight operation being conducted; or

(iv) Required to be operational by an airworthiness directive; and

(3) The inoperative instruments and equipment are—

(i) Removed from the aircraft, the cockpit control placarded, and the maintenance recorded in accordance with § 43.9 of this chapter; or

(ii) Deactivated and placarded "Inoperative." If deactivation of the inoperative instrument or equipment involves maintenance, it must be accomplished and recorded in accordance with Part 43 of this chapter; and

(4) A determination is made by a pilot, who is certificated and appropriately rated under Part 61 of this chapter, or by a person, who is certificated and appropriately rated to perform maintenance on the aircraft, that the inoperative instrument or equipment does not constitute a hazard to the aircraft. An aircraft with inoperative instruments or equipment as provided in paragraph (d) of this section is considered to be in a properly altered condition acceptable to the Administrator.

(e) Notwithstanding any other provision of this section, an aircraft with inoperable instruments or equipment may be operated under a special flight permit issued in accordance with § 21.197 and § 21.199 of this chapter.

§ 91.215 ATC transponder and altitude reporting equipment and use.

(a) *All airspace: U.S.-registered civil aircraft.* [For operations not conducted under part 121, 127 or 135 of this chapter, ATC transponder equipment installed must meet the performance and environmental requirements of any class of TSO-C74b (Mode A) or any class of TSO-C74c (Mode A with altitude reporting capability) as appropriate, or the appropriate class of TSO-C112 (Mode S).]

(b) *All airspace.* Unless otherwise authorized or directed by ATC, no person may operate an aircraft in the airspace described in paragraphs (b)(1) through (b)(5) of this section, unless that aircraft is equipped with an operable coded radar beacon transponder having either Mode 3/A 4096 code capability, replying to Mode 3/A interrogations with the code specified by ATC, or a Mode S capability, replying to Mode 3/A interrogations with the code specified by ATC and intermode and Mode S interrogations in accordance with the applicable provisions specified in TSO C-112, and that aircraft is equipped with automatic pressure altitude reporting equipment having a Mode C capability that automatically replies to Mode C interrogations by transmitting pressure altitude information in 100-foot increments. This requirement applies—

(1) All aircraft. In Class A, Class B, and Class C airspace areas;

(2) All aircraft. In all airspace within 30 nautical miles of an airport listed in Appendix D, Section 1 of this part from the surface upward to 10,000 feet MSL;

(3) Notwithstanding paragraph (b)(2) of this section, any aircraft which was not originally certificated with an engine-driven electrical system or which has not subsequently been certified with such a system installed, balloon or glider may conduct operations in the airspace within 30 nautical miles of an airport listed in Appendix D, Section 1 of this part provided such operations are conducted—

(i) Outside any Class A, Class B, or Class C airspace area; and

(ii) Below the altitude of the ceiling of a Class B or Class C airspace area designated for an airport or 10,000 feet MSL, whichever is lower; and

(4) All aircraft in all airspace above the ceiling and within the lateral boundaries of a Class B or Class C airspace area designated for an airport upward to 10,000 feet MSL; and

(5) All aircraft except any aircraft which was not originally certificated with an engine-driven electrical system or which has not subsequently been certified with such a system installed, balloon, or glider—

(i) In all airspace of the 48 contiguous states and the District of Columbia at and above 10,000 feet MSL, excluding the airspace at and below 2,500 feet above the surface; and

(ii) In the airspace from the surface to 10,000 feet MSL within a 10-nautical-mile radius of any airport listed in Appendix D, Section 2 of this part, excluding the airspace below 1,200 feet outside of the lateral boundaries of the surface area of the airspace designated for that airport.

(c) *Transponder-on operation.* While in the airspace as specified in paragraph (b) of this section or in all controlled airspace, each person operating an aircraft equipped with an operable ATC transponder maintained in accordance with § 91.413 of this part shall operate the transponder, including Mode C equipment if installed, and shall reply on the appropriate code or as assigned by ATC.

(d) *ATC authorized deviations.* Requests for ATC authorized deviations must be made to the ATC facility having jurisdiction over the concerned airspace within the time periods specified as follows:

(1) For operation of an aircraft with an operating transponder but without operating automatic pressure altitude reporting equipment having a Mode C capability, the request may be made at any time.

(2) For operation of an aircraft with an inoperative transponder to the airport of ultimate destination, including any intermediate stops, or to proceed to a place where suitable repairs can be made or both, the request may be made at any time.

(3) For operation of an aircraft that is not equipped with a transponder, the request must be made at least one hour before the proposed operation.

(Amdt. 91-221, Eff. 1/4/91); (Amdt. 91-227, Eff. 12/12/91); [(***Amdt. 91-229, Eff. 7/30/92***)]

§ 91.217 Data correspondence between automatically reported pressure altitude data and the pilot's altitude reference.

No person may operate any automatic pressure altitude reporting equipment associated with a radar beacon transponder—

(a) When deactivation of that equipment is directed by ATC;

(b) Unless, as installed, that equipment was tested and calibrated to transmit altitude data corresponding within 125 feet (on a 95 percent probability basis) of the indicated or calibrated datum of the altimeter normally used to maintain flight altitude, with that altimeter referenced to 29.92" of mercury for altitudes from sea level to the maximum operating altitude of the aircraft; or

(c) Unless the altimeters and digitizers in that equipment meet the standards of TSO C10b and TSO C88, respectively.

§ 91.219 Altitude alerting system or device: Turbojet-powered civil airplanes.

(a) Except as provided in paragraph (d) of this section, no person may operate a turbojet-powered U.S.-registered civil airplane unless that airplane is equipped with an approved altitude alerting system or device that is in operable condition and meets the requirements of paragraph (b) of this section.

(b) Each altitude alerting system or device required by paragraph (a) of this section must be able to—

(1) Alert the pilot—

(i) Upon approaching a preselected altitude in either ascent or descent, by a sequence of both aural and visual signals in sufficient time to establish level flight at that preselected altitude; or

(ii) Upon approaching a preselected altitude in either ascent or descent, by a sequence of visual signals in sufficient time to establish level flight at that preselected altitude, and when deviating above and below that preselected altitude, by an aural signal;

(2) Provide the required signals from sea level to the highest operating altitude approved for the airplane in which it is installed;

(3) Preselect altitudes in increments that are commensurate with the altitudes at which the aircraft is operated;

(4) Be tested without special equipment to determine proper operation of the alerting signals; and

(5) Accept necessary barometric pressure settings if the system or device operates on barometric pressure.

However, for operation below 3,000 feet AGL, the system or device need only provide one signal, either visual or aural, to comply with this paragraph. A radio altimeter may be included to provide the signal if the operator has an approved procedure for its use to determine DH or MDA, as appropriate.

(c) Each operator to which this section applies must establish and assign procedures for the use of the altitude alerting system or device and each flight crewmember must comply with those procedures assigned to him.

(d) Paragraph (a) of this section does not apply to any operation of an airplane that has an experimental certificate or to the operation of any airplane for the following purposes:

(1) Ferrying a newly acquired airplane from the place where possession of it was taken to a place where the altitude alerting system or device is to be installed.

(2) Continuing a flight as originally planned, if the altitude alerting system or device becomes inoperative after the airplane has taken off; however, the flight may not depart from a place where repair or replacement can be made.

(3) Ferrying an airplane with any inoperative altitude alerting system or device from a place where repairs or replacements cannot be made to a place where it can be made.

(4) Conducting an airworthiness flight test of the airplane.

(5) Ferrying an airplane to a place outside the United States for the purpose of registering it in a foreign country.

(6) Conducting a sales demonstration of the operation of the airplane.

(7) Training foreign flight crews in the operation of the airplane before ferrying it to a place outside the United States for the purpose of registering it in a foreign country.

§ 91.221 Traffic alert and collision avoidance system equipment and use.

(a) *All airspace: U.S.-registered civil aircraft.* Any traffic alert and collision avoidance system installed in a U.S.-registered civil aircraft must be approved by the Administrator.

(b) *Traffic alert and collision avoidance system, operation required.* Each person operating an aircraft equipped with an operable traffic alert and collision avoidance system shall have that system on and operating.

§§ 91.223–91.299 [Reserved]

Subpart D—Special flight operations

§ 91.301 [Reserved]

§ 91.303 Aerobatic flight.

No person may operate an aircraft in aerobatic flight—

(a) Over any congested area of a city, town, or settlement;

(b) Over an open air assembly of persons;

(c) [***Within the lateral boundaries of the surface areas of Class B, Class C, Class D, or Class E airspace designated for an airport;***]

(d) [***Within 4 nautical miles of the center line of any Federal airway;***]

(e) [***Below an altitude of 1,500 feet above the surface; or***]

[***(f) When flight visibility is less than 3 statute miles. For the purposes of this section, aerobatic flight means an intentional maneuver involving an abrupt change in an aircraft's attitude, an abnormal attitude, or abnormal acceleration, not necessary for normal flight.***]

For the purposes of this section, aerobatic flight means an intentional maneuver involving an abrupt change in an aircraft's attitude, an abnormal attitude, or abnormal acceleration, not necessary for normal flight.

[***(Amdt. 91-227, Eff. 9/16/93)***]

§ 91.305 Flight test areas.

No person may flight test an aircraft except over open water, or sparsely populated areas, having light air traffic.

§ 91.307 Parachutes and parachuting.

(a) No pilot of a civil aircraft may allow a parachute that is available for emergency use to be carried in that aircraft unless it is an approved type and—

(1) If a chair type (canopy in back), it has been packed by a certificated and appropriately rated parachute rigger within the preceding 120 days; or

(2) If any other type, it has been packed by a certificated and appropriately rated parachute rigger—

(i) Within the preceding 120 days, if its canopy, shrouds, and harness are composed exclusively of nylon, rayon, or other similar synthetic fiber or materials that are substantially resistant to damage from mold, mildew, or other fungi and other rotting agents propagated in a moist environment; or

(ii) Within the preceding 60 days, if any part of the parachute is composed of silk, pongee, or other natural fiber, or materials not specified in paragraph (a)(2)(i) of this section.

(b) Except in an emergency, no pilot in command may allow, and no person may make, a parachute jump from an aircraft within the United States except in accordance with Part 105.

(c) Unless each occupant of the aircraft is wearing an approved parachute, no pilot of a civil aircraft carrying any person (other than a crewmember) may execute any intentional maneuver that exceeds—

(1) A bank of 60 degrees relative to the horizon; or

(2) A nose-up or nose-down attitude of 30 degrees relative to the horizon.

(d) Paragraph (c) of this section does not apply to—

(1) Flight tests for pilot certification or rating; or

(2) Spins and other flight maneuvers required by the regulations for any certificate or rating when given by—

(i) A certificated flight instructor; or

(ii) An airline transport pilot instructing in accordance with 61.169 of this chapter.

(e) For the purposes of this section, "approved parachute" means—

(1) A parachute manufactured under a type certificate or a technical standard order (C-23 series); or

(2) A personnel-carrying military parachute identified by an NAF, AAF, or AN drawing number, an AAF order number, or any other military designation or specification number.

§ 91.309 Towing: Gliders.

(a) No person may operate a civil aircraft towing glider unless—

(1) The pilot in command of the towing aircraft is qualified under § 61.69 of this chapter;

(2) The towing aircraft is equipped with a tow-hitch of a kind, and installed in a manner, that is approved by the Administrator;

(3) The towline used has breaking strength not less than 80 percent of the maximum certificated operating weight of the glider and not more than twice this operating weight. However, the towline used may have a breaking strength more than twice the maximum certificated operating weight of the glider if—

(i) A safety link is installed at the point of attachment of the towline to the glider with a breaking strength not less than 80 percent of the maximum certificated operating weight of the glider and not greater than twice this operating weight.

(ii) A safety link is installed at the point of attachment of the towline to the towing aircraft with a breaking strength greater, but not more than 25 percent greater, than that of the safety link at the towed glider end of the towline and not greater than twice the maximum certificated operating weight of the glider;

(4) [***Before conducting any towing operation within the lateral boundaries of the surface areas of Class B, Class C, Class D, or Class E airspace designated for an airport, or before making each towing flight within such***

controlled airspace if required by ATC, the pilot in command notifies the control tower. If a control tower does not exist or is not in operation, the pilot in command must notify the FAA flight service station serving that controlled airspace before conducting any towing operations in that airspace; and]

(5) The pilots of the towing aircraft and the glider have agreed upon a general course of action, including takeoff and release signals, airspeeds, and emergency procedures for each pilot.

(b) No pilot of a civil aircraft may intentionally release a towline, after release of a glider, in a manner that endangers the life or property of another.

[*(Amdt. 91-227, Eff. 9/16/93)*]

§ 91.311 Towing: Other than under § 91.309

No pilot of a civil aircraft may tow anything with that aircraft (other than under § 91.309) except in accordance with the terms of a certificate of waiver issued by the Administrator.

§ 91.313 Restricted category civil aircraft: Operating limitations.

(a) No person may operate a restricted category civil aircraft—

(1) For other than the special purpose for which it is certificated; or

(2) In an operation other than one necessary to accomplish the work activity directly associated with that special purpose.

(b) For the purpose of paragraph (a) of this section, operating a restricted category civil aircraft to provide flight crewmember training in a special purpose operation for which the aircraft is certificated is considered to be an operation for that special purpose.

(c) No person may operate a restricted category civil aircraft carrying persons or property for compensation or hire. For the purposes of this paragraph, a special purpose operation involving the carriage of persons or material necessary to accomplish that operation, such as crop dusting, seeding, spraying, and banner towing (including the carrying of required persons or material to the location of that operation), and operation for the purpose of providing flight crewmember training in a special purpose operation, are not considered to be the carriage of persons or property for compensation or hire.

(d) No person may be carried on a restricted category civil aircraft unless that person—

(1) Is a flight crewmember;

(2) Is a flight crewmember trainee;

(3) Performs an essential function in connection with a special purpose operation for which the aircraft is certificated; or

(4) Is necessary to accomplish the work activity directly associated with that special purpose.

(e) Except when operating in accordance with the terms and conditions of a certificate of waiver or special operating limitations issued by the Administrator, no person may operate a restricted category civil aircraft within the United States—

(1) Over a densely populated area;

(2) In a congested airway; or

(3) Near a busy airport where passenger transport operations are conducted.

(f) This section does not apply to nonpassenger-carrying civil rotorcraft external-load operations conducted under Part 133 of this chapter.

(g) No person may operate a small restricted-category civil airplane manufactured after July 18, 1978, unless an approved shoulder harness is installed for each front seat. The shoulder harness must be designed to protect each occupant from serious head injury when the occupant experiences the ultimate inertia forces specified in § 23.561 (b)(2) of this chapter. The shoulder harness installation at each flight crewmember station must permit the crewmember, when seated and with the safety belt and shoulder harness fastened, to perform all functions necessary for flight operation. For purposes of this paragraph—

(1) The date of manufacture of an airplane is the date the inspection acceptance records reflect that the airplane is complete and meets the FAA-approved type design data; and

(2) A front seat is a seat located at a flight crewmember station or any seat located alongside such a seat.

§ 91.315 Limited category civil aircraft: Operating limitations.

No person may operate a limited category civil aircraft carrying persons or property for compensation or hire.

§ 91.317 Provisionally certificated civil aircraft: Operating limitations.

(a) No person may operate a provisionally certificated civil aircraft unless that person is eligible for a provisional airworthiness certificate under § 21.213 of this chapter.

(b) No person may operate a provisionally certificated civil aircraft outside the United States unless that person has specific authority to do so from the Administrator and each foreign country involved.

(c) Unless otherwise authorized by the Director, Flight Standards Service, no person may operate a provisionally certificated civil aircraft in air transportation.

(d) Unless otherwise authorized by the Administrator, no person may operate a provisionally certificated civil aircraft except—

(1) In direct conjunction with the type or supplemental type certification of that aircraft;

(2) For training flight crews, including simulated air carrier operations;

(3) Demonstration flight by the manufacturer for prospective purchasers;

(4) Market surveys by the manufacturer;

(5) Flight checking of instruments, accessories, and equipment that do not affect the basic airworthiness of the aircraft; or

(6) Service testing of the aircraft.

(e) Each person operating a provisionally certificated civil aircraft shall operate within the prescribed limitations displayed in the aircraft or set forth in the provisional aircraft flight manual or other appropriate document. However, when operating in direct conjunction with the type or supplemental type certification of the aircraft, that person shall operate under the experimental aircraft limitations of § 21.191 of this chapter and when flight testing, shall operate under the requirement of § 91.305 of this part.

(f) Each person operating a provisionally certificated civil aircraft shall establish approved procedures for—

(1) The use and guidance of flight and ground personnel in operating under this section; and

(2) Operating in and out of airports where takeoffs or approaches over populated areas are necessary. No person may operate that aircraft except in compliance with the approved procedures.

(g) Each person operating a provisionally certificated civil aircraft shall ensure that each flight crewmember is properly certificated and has adequate knowledge of, and familiarity with, the aircraft and procedures to be used by that crewmember.

(h) Each person operating a provisionally certificated civil aircraft shall maintain it as required by applicable regulations and as may be specially prescribed by the Administrator.

(i) Whenever the manufacturer, or the Administrator, determines that a change in design, construction, or operation is necessary to ensure safe operation, no person may operate a provisionally certificated civil aircraft until that change has been made and approved. Section § 21.99 of this chapter applies to operations under this section.

(j) Each person operating a provisionally certificated civil aircraft—

(1) May carry in that aircraft only persons who have a proper interest in the operations allowed by this section or who are specifically authorized by both the manufacturer and the Administrator; and

(2) Shall advise each person carried that the aircraft is provisionally certificated.

(k) The Administrator may prescribe additional limitations or procedures that the Administrator considers necessary, including limitations on the number of persons who may be carried in the aircraft.

(Approved by the Office of Management and Budget under OMB control number 2120-0005)

(Amdt. 91-212, Eff. 8/18/90)

§ 91.319 Aircraft having experimental certificates: Operating limitations.

(a) No person may operate an aircraft that has an experimental certificate—

(1) For other than the purpose for which the certificate was issued; or

(2) Carrying persons or property for compensation or hire.

(b) No person may operate an aircraft that has an experimental certificate outside of an area assigned by the Administrator until it is shown that—

(1) The aircraft is controllable throughout its normal range of speeds and throughout all the maneuvers to be executed; and

(2) The aircraft has no hazardous operating characteristics or design features.

(c) Unless otherwise authorized by the Administrator in special operating limitations, no person may operate an aircraft that has an experimental certificate over a densely populated area or in a congested airway. The Administrator may issue special operating limitations for particular aircraft to permit takeoffs and landings to be conducted over a densely populated area or in a congested airway, in accordance with terms and conditions specified in the authorization in the interest of safety in air commerce.

(d) Each person operating an aircraft that has an experimental certificate shall—

(1) Advise each person carried of the experimental nature of the aircraft;

(2) Operate under VFR, day only, unless otherwise specifically authorized by the Administrator; and

(3) Notify the control tower of the experimental nature of the aircraft when operating the aircraft into or out of airports with operating control towers.

(e) The Administrator may prescribe additional limitations that the Administrator considers necessary, including limitations on the persons that may be carried in the aircraft.

(Approved by the Office of Management and Budget under OMB control number 2120-0005).

§ 91.321 Carriage of candidates in Federal elections.

(a) An aircraft operator, other than one operating an aircraft under the rules of Part 121, 125, or 135 of this chapter, may receive payment for the carriage of a candidate in a Federal election, an agent of the candidate, or a person traveling on behalf of the candidate, if—

(1) That operator's primary business is not as an air carrier or commercial operator;

(2) The carriage is conducted under the rules of this Part 91; and

(3) The payment for the carriage is required, and does not exceed the amount required to be paid, by regulations of the Federal Election Commission (11 CFR et seq.).

(b) For the purposes of this section, the terms "candidate" and "election" have the same meaning as that set forth in the regulations of the Federal Election Commission.

§ 91.323 Increased maximum certificated weights for certain airplanes operated in Alaska.

(a) Notwithstanding any other provision of the Federal Aviation Regulations, the Administrator will approve, as provided in this section, an increase in the maximum certificated weight of an airplane type certificated under Aeronautics Bulletin No. 7 A of the U.S. Department of Commerce dated January 1, 1931, as amended, or under the normal category of part 4a of the former Civil Air Regulations (14 CFR Part 4a, 1964 ed.) if that airplane is operated in the State of Alaska by—

(1) An air taxi operator or other air carrier; or

(2) The U.S. Department of Interior in conducting its game and fish law enforcement activities or its management, fire detection, and fire suppression activities concerning public lands.

(b) The maximum certificated weight approved under this section may not exceed—

(1) 12,500 pounds;

(2) 115 percent of the maximum weight listed in the FAA aircraft specifications;

(3) The weight at which the airplane meets the positive maneuvering load factor requirement for the normal category specified in § 23.337 of this chapter; or

(4) The weight at which the airplane meets the climb performance requirements under which it was type certificated,

(c) In determining the maximum certificated weight, the Administrator considers the structural soundness of the airplane and the terrain to be traversed.

(d) The maximum certificated weight determined under this section is added to the airplane's operation limitations and is identified as the maximum weight authorized for operations within the State of Alaska.

[§ 91.325 Primary category aircraft: Operating limitations.

[(a) No person may operate a primary category aircraft carrying persons or property for compensation or hire.

[(b) No person may operate a primary category aircraft that is maintained by the pilot-owner under an approved special inspection and maintenance program except—

[(1) The pilot-owner; or

[(2) A designee of the pilot-owner, provided that the pilot-owner does not receive compensation for the use of the aircraft.]

[(Amdt. 91-230, Eff. 12/31/92)]

§§ 91.327–91.399 [Reserved]

Subpart E—Maintenance, preventive maintenance, and alterations

§ 91.401 Applicability.

(a) This subpart prescribes rules governing the maintenance, preventive maintenance, and alterations of U.S.-registered civil aircraft operating within or outside of the United States.

(b) Sections 91.405, 91.409, 91.411, 91.417, and 91.419 of this subpart do not apply to an aircraft maintained in accordance with a continuous airworthiness maintenance program as provided in Part 121, 127, 129, or § 135.411(a) (2) of this chapter.

(c) Sections 91.405 and 91.409 of this part do not apply to an airplane inspected in accordance with Part 125 of this chapter.

§ 91.403 General.

(a) The owner or operator of an aircraft is primarily responsible for maintaining that aircraft in an airworthy condition, including compliance with Part 39 of this chapter.

(b) No person may perform maintenance, preventive maintenance, or alterations on an aircraft other than as prescribed in this subpart and other applicable regulations, including Part 43 of this chapter.

(c) No person may operate an aircraft for which a manufacturer's maintenance manual or instructions for continued airworthiness has been issued that contains an airworthiness limitations section unless the mandatory replacement times, inspection intervals, and related procedures specified in that section or alternative inspection intervals and related procedures set forth in an operations specification approved by the Administrator under Part 121, 127, or 135 of this chapter or in accordance with an inspection program approved under § 91.409(e) have been compiled with.

§ 91.405 Maintenance required.

Each owner or operator of an aircraft—

(a) Shall have that aircraft inspected as prescribed in Subpart E of this part and shall between required inspections, except as provided in paragraph (c) of this section, have discrepancies repaired as prescribed in Part 43 of this chapter.

(b) Shall ensure that maintenance personnel make appropriate entries in the aircraft maintenance records indicating the aircraft has been approved for return to service;

(c) Shall have any inoperative instrument or item of equipment, permitted to be inoperative by § 91.213(d) (2) of this part, repaired, replaced, removed, or inspected at the next required inspection; and

(d) When listed discrepancies include inoperative instruments or equipment, shall ensure that a placard has been installed as required by § 43.11 of this chapter.

§ 91.407 Operation after maintenance, preventive maintenance, rebuilding, or alteration.

(a) No person may operate any aircraft that has undergone maintenance, preventive maintenance, rebuilding, or alteration unless—

(1) It has been approved for return to service by a person authorized under § 43.7 of this chapter; and

(2) The maintenance record entry required by § 43.9 or § 43.11, as applicable, of this chapter has been made.

(b) No person may carry any person (other than crewmembers) in an aircraft that has been maintained, rebuilt, or altered in a manner that may have appreciably changed its flight characteristics or substantially affected its operation in flight until an appropriately rated pilot with at least a private pilot certificate flies the aircraft, makes an operational check of the maintenance performed or alteration made, and logs the flight in the aircraft records.

(c) The aircraft does not have to be flown as required by paragraph (b) of this section, if, prior to flight, ground tests, inspections, or both show conclusively that the maintenance, preventive maintenance, rebuilding, or alteration has not appreciably changed the flight characteristics or substantially affected the flight operation of the aircraft.

(Approved by the Office of Management and Budget under OMB control number 2120-0005)

§ 91.409 Inspections.

(a) Except as provided in paragraph (c) of this section, no person may operate an aircraft unless within the preceding 12 calendar months, it has had—

(1) An annual inspection in accordance with Part 43 of this chapter and has been approved for return to service by a person authorized by § 43.7 of this chapter, or

(2) An inspection for the issue of an airworthiness certificate in accordance with Part 21 of this chapter.

No inspection performed under paragraph (b) of the section may be substituted for any inspection required by this paragraph unless it is performed by a person authorized to perform annual inspections and is entered as an "annual" inspection in the required maintenance records.

(b) Except as provided in paragraph (c) of this section, no person may operate an aircraft carrying any person (other than a crewmember) for hire, and no person may give flight instruction for hire in an aircraft which that person provides, unless within the preceding 100 hours of time in service the aircraft has received an annual or 100-hour inspection and been approved for return to service in accordance with Part 43 of this chapter or has received an inspection for the issuance of an airworthiness certificate in accordance with Part 21 of this chapter. The 100-hour limitation may be exceeded by not more than 10 hours while en route to reach a place where the inspection can be done. The excess time used to reach a place where the inspection can be done must be included in computing the next 100 hours of time in service.

(c) Paragraphs (a) and (b) of this section do not apply to—

(1) An aircraft that carries a special flight permit, a current experimental certificate, or a provisional airworthiness certificate:

(2) An aircraft inspected in accordance with an approved aircraft inspection program under Part 125, 127, or 135 of this chapter and so identified by the registration number in the operations specifications of the certificate holder having the approved inspection program;

(3) An aircraft subject to the requirements of paragraph (d) or (e) of this section; or

(4) Turbine-powered rotorcraft when the operator elects to inspect that rotorcraft in accordance with paragraph (e) of this section.

(d) *Progressive inspection.* Each registered owner or operator of an aircraft desiring to use a progressive inspection program must submit a written request to the FAA Flight Standards district office having jurisdiction over the area in which the applicant is located, and shall provide—

(1) A certificated mechanic holding an inspection authorization, a certificated airframe repair station, or the manufacturer of the aircraft to supervise or conduct the progressive inspection;

(2) A current inspection procedures manual available and readily understandable to pilot and maintenance personnel containing, in detail—

(i) An explanation of the progressive inspection, including the continuity of inspection responsibility, the making of reports, and the keeping of records and technical reference material;

(ii) An inspection schedule, specifying the intervals in hours or days when routine and detailed inspections will be performed and including instructions for exceeding an inspection interval by not more than 10 hours while en route and for changing an inspection interval because of service experience;

(iii) Sample routine and detailed inspection forms and instructions for their use; and

(iv) Sample reports and records and instructions for their use;

(3) Enough housing and equipment for necessary disassembly and proper inspection of the aircraft; and

(4) Appropriate current technical information for the aircraft.

The frequency and detail of the progressive inspection shall provide for the complete inspection of the aircraft within each 12 calendar months and be consistent with the manufacturer's recommendations, field service experience, and the kind of operation in which the aircraft is engaged. The progressive inspection schedule must ensure that the aircraft, at all times, will be airworthy and will conform to all applicable FAA aircraft specifications, type certificate data sheets, airworthiness directives, and other approved data. If the progressive inspection is discontinued, the owner or operator shall immediately notify the local FAA Flight Standards district office, in writing, of the discontinuance. After the discontinuance, the first annual inspection under § 91.409(a) (1) is due within 12 calendar months after the last complete inspection of the aircraft under the progressive inspection. The 100-hour inspection under § 91.409(b) is due within 100 hours after that complete inspection. A complete inspection of the aircraft, for the purpose of determining when the annual and 100-hour inspections are due, requires a detailed inspection of the aircraft and all its components in accordance with the progressive inspection. A routine inspection of the aircraft and a detailed inspection of several components is not considered to be a complete inspection.

(e) *Large airplanes (to which Part 125 is not applicable), turbojet multiengine airplanes, turbopropeller-powered multiengine airplanes, and turbine-powered rotorcraft.* No person may operate a large airplane, turbojet multiengine airplane, or turbopropeller-powered multiengine airplane, or turbine-powered rotorcraft unless the replacement times for life-limited parts specified in the aircraft specifications, type data sheets, or other documents approved by the Administrator are complied with and the airplane or turbine-powered rotorcraft, including the airframe, engines, propellers, rotors, appliances, survival equipment, and emergency equipment, is inspected in accordance with an inspection program selected under the provisions of paragraph (f) of this section, except that, the owner or operator of a turbine-powered rotorcraft may elect to use the inspection provisions of § 91.409(a), (b), (c), or (d) in lieu of an inspection option of § 91.409(f).

(f) *Selection of inspection program under paragraph (e) of this section.* The registered owner or operator of each airplane or turbine-powered rotorcraft described in paragraph (e) of this section must select, identify in the aircraft maintenance records, and use one of the following programs for the inspection of the aircraft:

(1) A continuous airworthiness inspection program that is part of a continuous airworthiness maintenance program currently in use by a person holding an air carrier operating certificate or an operating certificate issued under Part 121, 127, or 135 of this chapter and operating that make and model aircraft under Part 121 of this chapter or operating that make and model under Part 135 of this chapter and maintaining it under § 135.411(a) (2) of this chapter.

(2) An approved aircraft inspection program approved under § 135.419 of this chapter and currently in use by a person holding an operating certificate issued under Part 135 of this chapter.

(3) A current inspection program recommended by the manufacturer.

(4) Any other inspection program established by the registered owner or operator of that airplane or turbine-powered rotorcraft and approved by the Administrator under paragraph (g) of this section. However, the Administrator may require revision of this inspection program in accordance with the provisions of § 91.415.

Each operator shall include in the selected program the name and address of the person responsible for scheduling the inspections required by the program and make a copy of that program available to the person performing inspections on the aircraft and, upon request, to the Administrator.

(g) *Inspection program approved under paragraph (e), of this section.* Each operator of an airplane or turbine-powered rotorcraft desiring to establish or change an approved inspection program under paragraph (f) (4) of this section must submit the program for approval to the local FAA Flight Standards district office having jurisdiction over the area in which the aircraft is based. The program must be in writing and include at least the following information:

(1) Instructions and procedures for the conduct of inspections for the particular make and model airplane or turbine-powered rotorcraft, including necessary tests and checks. The instructions and procedures must set forth in detail the parts and areas of the airframe, engines, propellers, rotors, and appliances, including survival and emergency equipment required to be inspected.

(2) A schedule for performing the inspections that must be performed under the program expressed in terms of the time in service, calendar time, number of system operations, or any combination of these.

(h) *Changes from one inspection program to another.* When an operator changes from one inspection program under paragraph (f) of this section to another, the time in service, calendar times, or cycles of operation accumulated under the previous program must be applied in determining inspection due times under the new program.

(Approved by the Office of Management and Budget under OMB control number 2120-0005)

§ 91.411 Altimeter system and altitude reporting equipment tests and inspections.

(a) No person may operate an airplane, or helicopter, in controlled airspace under IFR unless—

(1) Within the preceding 24 calendar months, each static pressure system, each altimeter instrument, and each automatic pressure altitude reporting system has been tested and inspected and found to comply with Appendix E of Part 43 of this chapter;

(2) Except for the use of system drain and alternate static pressure valves, following any opening and closing of the static pressure system, that system has been tested and inspected and found to comply with paragraph (a), Appendices E and F, of Part 43 of this chapter; and

(3) Following installation or maintenance on the automatic pressure altitude reporting system of the ATC transponder where data correspondence error could be introduced, the integrated system has been tested, inspected, and found to comply with paragraph (c), Appendix E, of Part 43 of this chapter.

(b) The tests required by paragraph (a) of this section must be conducted by—

(1) The manufacturer of the airplane, or helicopter, on which the tests and inspections are to be performed;

(2) A certificated repair station properly equipped to perform those functions and holding—

(i) An instrument rating, Class I;

(ii) A limited instrument rating appropriate to the make and model of appliance to be tested;

(iii) A limited rating appropriate to the test to be performed;

(iv) An airframe rating appropriate to the airplane, or helicopter, to be tested; or

(v) A limited rating for a manufacturer issued for the appliance in accordance with § 145.101(b) (4) of this chapter; or

(3) A certificated mechanic with an airframe rating (static pressure system tests and inspections only).

(c) Altimeter and altitude reporting equipment approved under Technical Standard Orders are considered to be tested and inspected as of the date of their manufacture.

(d) No person may operate an airplane, or helicopter, in controlled airspace under IFR at an altitude above the maximum altitude at which all altimeters and the automatic altitude reporting system of that airplane, or helicopter, have been tested.

§ 91.413 ATC transponder tests and inspections.

(a) No person may use an ATC transponder that is specified in § 91.215(a), § 121.345(c), § 127.123(b) or § 135.143(c) of this chapter unless, within the preceding 24 calendar months, the ATC transponder has been tested and inspected and found to comply with Appendix F of Part 43 of this chapter; and

(b) Following any installation or maintenance on an ATC transponder where data correspondence error could be introduced, the integrated system has been tested, inspected, and found to comply with paragraph (c), Appendix E, of Part 43 of this chapter.

(c) The tests and inspections specified in this section must be conducted by—

(1) A certificated repair station properly equipped to perform those functions and holding—

(i) A radio rating, Class III;

(ii) A limited radio rating appropriate to the make and model transponder to be tested;

(iii) A limited rating appropriate to the test to be performed;

(iv) A limited rating for a manufacturer issued for the transponder in accordance with § 145.101(b) (4) of this chapter; or

(2) A holder of a continuous airworthiness maintenance program as provided in Part 121, 127, or § 135.411(a) (2) of this chapter; or

(3) The manufacturer of the aircraft on which the transponder to be tested is installed, if the transponder was installed by that manufacturer.

§ 91.415 Changes to aircraft inspection programs.

(a) Whenever the Administrator finds that revisions to an approved aircraft inspection program under § 91.409(f) (4) are necessary for the continued adequacy of the program, the owner or operator shall, after notification by the Administrator, make any changes in the program found to be necessary by the Administrator.

(b) The owner or operator may petition the Administrator to reconsider the notice to make any changes in a program in accordance with paragraph (a) of this section.

(c) The petition must be filed with the FAA Flight Standards district office which requested the change to the program within 30 days after the certificate holder receives the notice.

(d) Except in the case of an emergency requiring immediate action in the interest of safety, the filing of the petition stays the notice pending a decision by the Administrator.

§ 91.417 Maintenance records.

(a) Except for work performed in accordance with § 91.411 and 91.413, each registered owner or operator shall keep the following records for the periods specified in paragraph (b) of this section:

(1) Records of the maintenance, preventive maintenance, and alteration and records of the 100-hour, annual, progressive, and other required or approved inspections, as appropriate, for each aircraft (including the airframe) and each engine, propeller, rotor, and appliance of an aircraft. The records must include—

(i) A description (or reference to data acceptable to the Administrator) of the work performed; and

(ii) The date of completion of the work performed; and

(iii) The signature, and certificate number of the person approving the aircraft for return to service.

(2) Records containing the following information:

(i) The total time in service of the airframe, each engine, each propeller, and each rotor.

(ii) The current status of life-limited parts of each airframe, engine, propeller, rotor, and appliance.

(iii) The time since last overhaul of all items installed on the aircraft which are required to be overhauled on a specified time basis.

(iv) The current inspection status of the aircraft, including the time since the last inspection required by the inspection program under which the aircraft and its appliances are maintained.

(v) The current status of applicable airworthiness directives (AD) including, for each, the method of compliance, the AD number, and revision date. If the AD involves recurring action, the time and date when the next action is required.

(vi) Copies of the forms prescribed by § 43.9(a) of this chapter for each major alteration to the airframe and currently installed engines, rotors, propellers, and appliances.

(b) The owner or operator shall retain the following records for the periods prescribed:

(1) The records specified in paragraph (a) (1) of this section shall be retained until the work is repeated or superseded by other work or for 1 year after the work is performed.

(2) The records specified in paragraph (a) (2) of this section shall be retained and transferred with the aircraft at the time the aircraft is sold.

(3) A list of defects furnished to a registered owner or operator under § 43.11 of this chapter shall be retained until the defects are repaired and the aircraft is approved for return to service.

(c) The owner or operator shall make all maintenance records required to be kept by this section available for inspection by the Administrator or any authorized representative of the National Transportation Safety Board (NTSB). In addition, the owner or operator shall present the Form 337 described in paragraph (d) of this section for inspection upon request of any law enforcement officer.

(d) When a fuel tank is installed within the passenger compartment or a baggage compartment pursuant to Part 43 of this chapter, a copy of the FAA Form 337 shall be kept on board the modified aircraft by the owner or operator.

(Approved by the Office of Management and Budget under OMB control number 2120-0005)

§ 91.419 Transfer of maintenance records.

Any owner or operator who sells a U.S.-registered aircraft shall transfer to the purchaser, at the time of sale, the following records of that aircraft, in plain language form or in coded form at the election of the purchaser, if the coded form provides for the preservation and retrieval of information in a manner acceptable to the Administrator:

(a) The records specified in § 91.417(a) (2).

(b) The records specified in § 91.417(a) (1) which are not included in the records covered by paragraph (a) of this section, except that the purchaser may permit the seller to keep physical custody of such records. However, custody of records by the seller does not relieve the purchaser of the responsibility under § 91.417(c) to make the records available for inspection by the Administrator or any authorized representative of the National Transportation Safety Board (NTSB).

§ 91.421 Rebuilt engine maintenance records.

(a) The owner or operator may use a new maintenance record, without previous operating history, for an aircraft engine rebuilt by the manufacturer or by an agency approved by the manufacturer.

(b) Each manufacturer or agency that grants zero time to an engine rebuilt by it shall enter in the new record—

(1) A signed statement of the date the engine was rebuilt;

(2) Each change made as required by airworthiness directives; and

(3) Each change made in compliance with manufacturer's service bulletins, if the entry is specifically requested in that bulletin.

(c) For the purposes of this section, a rebuilt engine is a used engine that has been completely disassembled, inspected, repaired as necessary, reassembled, tested, and approved in the same manner and to the same tolerances and limits as a new engine with either new or used parts. However, all parts used in it must conform to the production drawing tolerances and limits for new parts or be of approved oversized or undersized dimensions for a new engine.

§ 91.423–91.499 [Reserved]

Subpart F—Large and turbine-powered multiengine airplanes

§ 91.501 Applicability.

(a) This subpart prescribes operating rules, in addition to those prescribed in other subparts of this part, governing the operation of large and of turbojet-powered multiengine civil airplanes of U.S. registry. The operating rules in this subpart do not apply to those airplanes when they are required to be operated under Parts 121, 125, 129, 135, and 137 of this chapter. (Section § 91.409 prescribes an inspection program for large and for turbine-powered (turbojet and turboprop) multiengine airplanes of U.S. registry when they are operated under this part or Part 129 or 137.)

(b) Operations that may be conducted under the rules in this subpart instead of those in Parts 121, 129, 135, and 137 of this chapter when common carriage is not involved, include—

(1) Ferry or training flights;

(2) Aerial work operations such as aerial photography or survey, or pipeline patrol, but not including fire fighting operations;

(3) Flights for the demonstration of an airplane to

prospective customers when no charge is made except for those specified in paragraph (d) of this section;

(4) Flights conducted by the operator of an airplane for his personal transportation, or the transportation of his guests when no charge, assessment, or fee is made for the transportation;

(5) Carriage of officials, employees, guests, and property of a company on an airplane operated by that company, or the parent or a subsidiary of the company or a subsidiary of the parent, when the carriage is within the scope of, and incidental to, the business of the company (other than transportation by air) and no charge, assessment or fee is made for the carriage in excess of the cost of owning, operating, and maintaining the airplane, except that no charge of any kind may be made for the carriage of a guest of a company, when the carriage is not within the scope of, and incidental to, the business of that company;

(6) The carriage of company officials, employees, and guests of the company on an airplane operated under a time sharing, interchange, or joint ownership agreement as defined in paragraph (c) of this section;

(7) The carriage of property (other than mail) on an airplane operated by a person in the furtherance of a business or employment (other than transportation by air) when the carriage is within the scope of, and incidental to, that business or employment and no charge, assessment, or fee is made for the carriage other than those specified in paragraph (d) of this section;

(8) The carriage on an airplane of an athletic team, sports group, choral group, or similar group having a common purpose or objective when there is no charge, assessment, or fee of any kind made by any person for that carriage; and

(9) The carriage of persons on an airplane operated by a person in the furtherance of a business other than transportation by air for the purpose of selling them land, goods, or property, including franchises or distributorships, when the carriage is within the scope of, and incidental to, that business and no charge, assessment, or fee is made for that carriage.

(c) As used in this section—

(1) A "time sharing agreement" means an arrangement whereby a person leases his airplane with flight crew to another person, and no charge is made for the flights conducted under that arrangement other than those specified in paragraph (d) of this section;

(2) An "interchange agreement" means an arrangement whereby a person leases his airplane to another person in exchange for equal time, when needed, on the other person's airplane, and no charge, assessment, or fee is made, except that a charge may be made not to exceed the difference between the cost of owning, operating, and maintaining the two airplanes;

(3) A "joint ownership agreement" means an arrangement whereby one of the registered joint owners of an airplane employs and furnishes the flight crew for that airplane and each of the registered joint owners pays a share of the charge specified in the agreement.

(d) The following may be charged, as expenses of a specific flight, for transportation as authorized by paragraphs (b)(3) and (7) and (c)(1) of this section:

(1) Fuel, oil, lubricants, and other additives.

(2) Travel expenses of the crew, including food, lodging, and ground transportation.

(3) Hangar and tie-down costs away from the aircraft's base of operation.

(4) Insurance obtained for the specific flight.

(5) Landing fees, airport taxes, and similar assessments.

(6) Customs, foreign permit, and similar fees directly related to the flight.

(7) In flight food and beverages.

(8) Passenger ground transportation.

(9) Flight planning and weather contract services.

(10) An additional charge equal to 100 percent of the expenses listed in paragraph (d)(1) of this section.

§ 91.503 Flying equipment and operating information.

(a) The pilot in command of an airplane shall ensure that the following flying equipment and aeronautical charts and data, in current and appropriate form, are accessible for each flight at the pilot station of the airplane:

(1) A flashlight having at least two size "D" cells, or the equivalent, that is in good working order.

(2) A cockpit checklist containing the procedures required by paragraph (b) of this section.

(3) Pertinent aeronautical charts.

(4) For IFR, VFR over-the-top, or night operations, each pertinent navigational en route, terminal area, and approach and letdown chart.

(5) In the case of multiengine airplanes, one-engine inoperative climb performance data.

(b) Each cockpit checklist must contain the following procedures and shall be used by the flight crewmembers when operating the airplane:

(1) Before starting engines.

(2) Before takeoff.

(3) Cruise.

(4) Before landing.

(5) After landing.

(6) Stopping engines.

(7) Emergencics.

(c) Each emergency cockpit checklist procedure required by paragraph (b)(7) of this section must contain the following procedures, as appropriate:

(1) Emergency operation of fuel, hydraulic, electrical, and mechanical systems.

(2) Emergency operation of instruments and controls.

(3) Engine inoperative procedures.

(4) Any other procedures necessary for safety.

(d) The equipment, charts, and data prescribed in this

section shall be used by the pilot in command and other members of the flight crew, when pertinent.

§ 91.505 Familiarity with operating limitations and emergency equipment.

(a) Each pilot in command of an airplane shall, before beginning a flight, become familiar with the Airplane Flight Manual for that airplane, if one is required, and with any placards, listings, instrument markings, or any combination thereof, containing each operating limitation prescribed for that airplane by the Administrator, including those specified in § 91.9(b).

(b) Each required member of the crew shall, before beginning a flight, become familiar with the emergency equipment installed on the airplane to which that crewmember is assigned and with the procedures to be followed for the use of that equipment in an emergency situation.

§ 91.507 Equipment requirements: Over-the-top or night VFR operations.

No person may operate an airplane over-the-top or at night under VFR unless that airplane is equipped with the instruments and equipment required for IFR operations under § 91.205(d) and one electric landing light for night operations. Each required instrument and item of equipment must be in operable condition.

§ 91.509 Survival equipment for overwater operations.

(a) No person may take off an airplane for a flight over water more than 50 nautical miles from the nearest shore unless that airplane is equipped with a life preserver or an approved flotation means for each occupant of the airplane.

(b) No person may take off an airplane for a flight over water more than 30 minutes flying time or 100 nautical miles from the nearest shore unless it has on board the following survival equipment:

(1) A life preserver, equipped with an approved survivor locator light, for each occupant of the airplane.

(2) Enough liferafts (each equipped with an approved survival locator light) of a rated capacity and buoyancy to accommodate the occupants of the airplane.

(3) At least one pyrotechnic signaling device for each liferaft.

(4) One self-buoyant, water-resistant, portable emergency radio signaling device that is capable of transmission on the appropriate emergency frequency or frequencies and not dependent upon the airplane power supply.

(5) A lifeline stored in accordance with § 25.1411(g) of this chapter.

(c) The required liferafts, life preservers, and signaling devices must be installed in conspicuously marked locations and easily accessible in the event of a ditching without appreciable time for preparatory procedures.

(d) A survival kit, appropriately equipped for the route to be flown, must be attached to each required liferaft.

(e) As used in this section, the term shore means that area of the land adjacent to the water which is above the high water mark and excludes land areas which are intermittently under water.

§ 91.511 Radio equipment for overwater operations.

(a) Except as provided in paragraphs (c) and (d) of this section, no person may take off an airplane for a flight over water more than 30 minutes flying time or 100 nautical miles from the nearest shore unless it has at least the following operable equipment:

(1) Radio communication equipment appropriate to the facilities to be used and able to transmit to, and receive from, any place on the route, at least one surface facility:

(i) Two transmitters.

(ii) Two microphones.

(iii) Two headsets or one headset and one speaker.

(iv) Two independent receivers.

(2) Appropriate electronic navigational equipment consisting of at least two independent electronic navigation units capable of providing the pilot with the information necessary to navigate the airplane within the airspace assigned by air traffic control. However, a receiver that can receive both communications and required navigational signals may be used in place of a separate communications receiver and a separate navigational signal receiver or unit.

(b) For the purposes of paragraphs (a)(1)(iv) and (a)(2) of this section, a receiver or electronic navigation unit is independent if the function of any part of it does not depend on the functioning of any part of another receiver or electronic navigation unit.

(c) Notwithstanding the provisions of paragraph (a) of this section, a person may operate an airplane on which no passengers are carried from a place where repairs or replacement cannot be made to a place where they can be made, if not more than one of each of the dual items of radio communication and navigational equipment specified in paragraphs (a)(1)(i) through (iv) and (a)(2) of this section malfunctions or becomes inoperative,

(d) Notwithstanding the provisions of paragraph (a) of this section, when both VHF and HF communications equipment are required for the route and the airplane has two VHF transmitters and two VHF receivers for communications, only one HF transmitter and one HF receiver is required for communications.

(e) As used in this section, the term "shore" means that area of the land adjacent to the water which is above the high-water mark and excludes land areas which are intermittently under water.

§ 91.513 Emergency equipment.

(a) No person may operate an airplane unless it is equipped with the emergency equipment listed in this section.

(b) Each item of equipment—

(1) Must be inspected in accordance with § 91.409 to ensure its continued serviceability and immediate readiness for its intended purposes;

(2) Must be readily accessible to the crew;

(3) Must clearly indicate its method of operation; and

(4) When carried in a compartment or container, must have that compartment or container marked as to contents and date of last inspection.

(c) Hand fire extinguishers must be provided for use in crew, passenger, and cargo compartments in accordance with the following:

(1) The type and quantity of extinguishing agent must be suitable for the kinds of fires likely to occur in the compartment where the extinguisher is intended to be used.

(2) At least one hand fire extinguisher must be provided and located on or near the flight deck in a place that is readily accessible to the flight crew.

(3) At least one hand fire extinguisher must be conveniently located in the passenger compartment of each airplane accommodating more than six but less than 31 passengers, and at least two hand fire extinguishers must be conveniently located in the passenger compartment of each airplane accommodating more than 30 passengers.

(4) Hand fire extinguishers must be installed and secured in such a manner that they will not interfere with the safe operation of the airplane or adversely affect the safety of the crew and passengers. They must be readily accessible and, unless the locations of the fire extinguishers are obvious, their stowage provisions must be properly identified.

(d) First aid kits for treatment of injuries likely to occur in flight or in minor accidents must be provided.

(e) Each airplane accommodating more than 19 passengers must be equipped with a crash axe.

(f) Each passenger-carrying airplane must have a portable battery-powered megaphone or megaphones readily accessible to the crewmembers assigned to direct emergency evacuation, installed as follows:

(1) One megaphone on each airplane with a seating capacity of more than 60 but less than 100 passengers, at the most rearward location in the passenger cabin where it would be readily accessible to a normal flight attendant seat. However, the Administrator may grant a deviation from the requirements of this subparagraph if the Administrator finds that a different location would be more useful for evacuation of persons during an emergency.

(2) On each airplane with a seating capacity of 100 or more passengers, one megaphone installed at the forward end and one installed at the most rearward location where it would be readily accessible to a normal flight attendant seat.

§ 91.515 Flight altitude rules.

(a) Notwithstanding § 91.119, and except as provided in paragraph (b) of this section, no person may operate an airplane under VFR at less than—

(1) One thousand feet above the surface, or 1,000 feet from any mountain, hill, or other obstruction to flight, for day operations; and

(2) The altitudes prescribed in § 91.177, for night operations.

(b) This section does not apply—

(1) During takeoff or landing;

(2) When a different altitude is authorized by a waiver to this section under subpart J of this part; or

(3) When a flight is conducted under the special VFR weather minimums of § 91.157 with an appropriate clearance from ATC.

§ 91.517 [Passenger information.

[(a) Except as provided in paragraph (b) of this section, no person may operate an airplane carrying passengers unless it is equipped with signs that are visible to passengers and flight attendants to notify them when smoking is prohibited and when safety belts must be fastened. The signs must be so constructed that the crew can turn them on and off. They must be turned on during airplane movement on the surface, for each takeoff, for each landing, and when otherwise considered to be necessary by the pilot in command.

[(b) The pilot in command of an airplane that is not required, in accordance with applicable aircraft and equipment requirements of this chapter, to be equipped as provided in paragraph (a) of this section shall ensure that the passengers are notified orally each time that it is necessary to fasten their safety belts and when smoking is prohibited.

[(c) If passenger information signs are installed, no passenger or crewmember may smoke while any "no smoking" sign is lighted nor may any passenger or crewmember smoke in any lavatory.

[(d) Each passenger required by § 91.107(a)(3) to occupy a seat or berth shall fasten his or her safety belt about him or her and keep it fastened while any "fasten seat belt" sign is lighted.

[(e) Each passenger shall comply with instructions given him or her by crewmembers regarding compliance with paragraphs (b), (c), and (d) of this section.]

[(Amdt. 91-231, Eff. 10/15/92)]

§ 91.519 Passenger briefing.

(a) Before each takeoff the pilot in command of an airplane carrying passengers shall ensure that all passengers have been orally briefed on—

(1) [Smoking: Each passenger shall be briefed on when, where, and under what condition smoking is prohibited. This briefing shall include a statement, as appropriate, that

the Federal Aviation Regulations require passenger compliance with lighted passenger information signs and no smoking placards, prohibit smoking in lavatories, and require compliance with crewmember instructions with regard to these items;

(2) [Use of safety belts and shoulder harnesses: Each passenger shall be briefed on when, where, and under what conditions it is necessary to have his or her safety belt and, if installed, his or her shoulder harness fastened about him or her. This briefing shall include a statement, as appropriate, that Federal Aviation Regulations require passenger compliance with the lighted passenger sign and/or crewmember instructions with regard to these items;]

(3) Location and means for opening the passenger entry door and emergency exits;

(4) Location of survival equipment;

(5) Ditching procedures and the use of flotation equipment required under § 91.509 for a flight over water; and

(6) The normal and emergency use of oxygen equipment installed on the airplane.

(b) The oral briefing required by paragraph (a) of this section shall be given by the pilot in command or a member of the crew, but need not be given when the pilot in command determines that the passengers are familiar with the contents of the briefing. It may be supplemented by printed cards for the use of each passenger containing—

(1) A diagram of, and methods of operating, the emergency exits; and

(2) Other instructions necessary for use of emergency equipment.

(c) Each card used under paragraph (b) must be carried in convenient locations on the airplane for the use of each passenger and must contain information that is pertinent only to the type and model airplane on which it is used.

[(Amdt. 91-231, Eff. 10/15/92)]

§ 91.521 Shoulder harness.

(a) No person may operate a transport category airplane that was type certificated after January 1, 1958, unless it is equipped at each seat at a flight deck station with a combined safety belt and shoulder harness that meets the applicable requirements specified in § 25.785 of this chapter, except that—

(1) Shoulder harnesses and combined safety belt and shoulder harnesses that were approved and installed before March 6, 1980, may continue to be used; and

(2) Safety belt and shoulder harness restraint systems may be designed to the inertia load factors established under the certification basis of the airplane.

(b) No person may operate a transport category airplane unless it is equipped at each required flight attendant seat in the passenger compartment with a combined safety belt and shoulder harness that meets the applicable requirements specified in § 25.785 of this chapter, except that—

(1) Shoulder harnesses and combined safety belt and shoulder harnesses that were approved and installed before March 6, 1980, may continue to be used; and

(2) Safety belt and shoulder harness restraint systems may be designed to the inertia load factors established under the certification basis of the airplane.

§ 91.523 Carry-on baggage.

No pilot in command of an airplane having a seating capacity of more than 19 passengers may permit a passenger to stow baggage aboard that airplane except—

(a) In a suitable baggage or cargo storage compartment, or as provided in § 91.525; or

(b) Under a passenger seat in such a way that it will not slide forward under crash impacts severe enough to induce the ultimate inertia forces specified in § 25.561(b)(3) of this chapter, or the requirements of the regulations under which the airplane was type certificated. Restraining devices must also limit sideward motion of under-seat baggage and be designed to withstand crash impacts severe enough to induce sideward forces specified in § 25.561(b)(3) of this chapter.

§ 91.525 Carriage of cargo.

(a) No pilot in command may permit cargo to be carried in any airplane unless—

(1) It is carried in an approved cargo rack, bin, or compartment installed in the airplane;

(2) It is secured by means approved by the Administrator; or

(3) It is carried in accordance with each of the following:

(i) It is properly secured by a safety belt or other tiedown having enough strength to eliminate the possibility of shifting under all normally anticipated flight and ground conditions.

(ii) It is packaged or covered to avoid possible injury to passengers.

(iii) It does not impose any load on seats or on the floor structure that exceeds the load limitation for those components.

(iv) It is not located in a position that restricts the access to or use of any required emergency or regular exit, or the use of the aisle between the crew and the passenger compartment.

(v) It is not carried directly above seated passengers.

(b) When cargo is carried in cargo compartments that are designed to require the physical entry of a crewmember to extinguish any fire that may occur during flight, the cargo must be loaded so as to allow a crewmember to effectively reach all parts of the compartment with the contents of a hand fire extinguisher.

§ 91.527 Operating in icing conditions.

(a) No pilot may take off an airplane that has—

(1) Frost, snow, or ice adhering to any propeller, windshield, or powerplant installation or to an airspeed, altimeter, rate of climb, or flight attitude instrument system;

(2) Snow or ice adhering to the wings or stabilizing or control surfaces; or

(3) Any frost adhering to the wings or stabilizing or control surfaces, unless that frost has been polished to make it smooth.

(b) Except for an airplane that has ice protection provisions that meet the requirements in section 34 of Special Federal Aviation Regulation No. 23, or those for transport category airplane type certification, no pilot may fly—

(1) Under IFR into known or forecast moderate icing conditions; or

(2) Under VFR into known light or moderate icing conditions unless the aircraft has functioning de-icing or anti-icing equipment protecting each propeller, windshield, wing, stabilizing or control surface, and each airspeed, altimeter, rate of climb, or flight attitude instrument system.

(c) Except for an airplane that has ice protection provisions that meet the requirements in section 34 of Special Federal Aviation Regulation No. 23, or those for transport category airplane type certification, no pilot may fly an airplane into known or forecast severe icing conditions.

(d) If current weather reports and briefing information relied upon by the pilot in command indicate that the forecast icing conditions that would otherwise prohibit the flight will not be encountered during the flight because of changed weather conditions since the forecast, the restrictions in paragraphs (b) and (c) of this section based on forecast conditions do not apply.

§ 91.529 Flight engineer requirements.

(a) No person may operate the following airplanes without a flight crewmember holding a current flight engineer certificate:

(1) An airplane for which a type certificate was issued before January 2, 1964, having a maximum certificated takeoff weight of more than 80,000 pounds.

(2) An airplane type certificated after January 1, 1964, for which a flight engineer is required by the type certification requirements.

(b) No person may serve as a required flight engineer on an airplane unless, within the preceding 6 calendar months, that person has had at least 50 hours of flight time as a flight engineer on that type airplane or has been checked by the Administrator on that type airplane and is found to be familiar and competent with all essential current information and operating procedures.

§ 91.531 Second in command requirements.

(a) Except as provided in paragraph (b) of this section, no person may operate the following airplanes without a pilot who is designated as second in command of that airplane:

(1) A large airplane, except that a person may operate an airplane certificated under SFAR 41 without a pilot who is designated as second in command if that airplane is certificated for operation with one pilot.

(2) A turbojet-powered multiengine airplane for which two pilots are required under the type certification requirements for that airplane.

(3) A commuter category airplane, except that a person may operate a commuter category airplane notwithstanding paragraph (a)(1) of this section, that has a passenger seating configuration, excluding pilot seats, of nine or less without a pilot who is designated as second in command if that airplane is type certificated for operations with one pilot.

(b) The Administrator may issue a letter of authorization for the operation of an airplane without compliance with the requirements of paragraph (a) of this section if that airplane is designed for and type certificated with only one pilot station. The authorization contains any conditions that the Administrator finds necessary for safe operation.

(c) No person may designate a pilot to serve as second in command, nor may any pilot serve as second in command, of an airplane required under this section to have two pilots unless that pilot meets the qualifications for second in command prescribed in § 61.55 of this chapter.

§ 91.533 Flight attendant requirements.

(a) No person may operate an airplane unless at least the following number of flight attendants are on board the airplane:

(1) For airplanes having more than 19 but less than 51 passengers on board, one flight attendant.

(2) For airplanes having more than 50 but less than 101 passengers on board, two flight attendants.

(3) For airplanes having more than 100 passengers on board, two flight attendants plus one additional flight attendant for each unit (or part of a unit) of 50 passengers above 100.

(b) No person may serve as a flight attendant on an airplane when required by paragraph (a) of this section unless that person has demonstrated to the pilot in command familiarity with the necessary functions to be performed in an emergency or a situation requiring emergency evacuation and is capable of using the emergency equipment installed on that airplane.

[§ 91.535 Stowage of food, beverage, and passenger service equipment during aircraft movement on the surface, takeoff, and landing.

[(a) No operator may move an aircraft on the surface, take off, or land when any food, beverage, or tableware furnished by the operator is located at any passenger seat.

[(b) No operator may move an aircraft on the surface, take off, or land unless each food and beverage tray and seat back tray table is secured in its stowed position.

[(c) No operator may permit an aircraft to move on the surface, take off, or land unless each passenger serving cart is secured in its stowed position.

[(d) No operator may permit an aircraft to move on the surface, take off, or land unless each movie screen that extends into the aisle is stowed.

[(e) Each passenger shall comply with instructions given by a crewmember with regard to compliance with this section.]

[(Amdt. 91-231, Eff. 10/15/92)]

§§ 91.537–91.599 [Reserved]

Subpart G—Additional equipment and operating requirements for large and transport category aircraft

§ 91.601 Applicability.

This subpart applies to operation of large and transport category U.S.-registered civil aircraft.

§ 91.603 Aural speed warning device.

No person may operate a transport category airplane in air commerce unless that airplane is equipped with an aural speed warning device that complies with § 25.1303(c)(1).

§ 91.605 Transport category civil airplane weight limitations.

(a) No person may take off any transport category airplane (other than a turbine-engine-powered airplane certificated after September 30, 1958) unless—

(1) The takeoff weight does not exceed the authorized maximum takeoff weight for the elevation of the airport of takeoff;

(2) The elevation of the airport of takeoff is within the altitude range for which maximum takeoff weights have been determined;

(3) Normal consumption of fuel and oil in flight to the airport of intended landing will leave a weight on arrival not in excess of the authorized maximum landing weight for the elevation of that airport; and

(4) The elevations of the airport of intended landing and of all specified alternate airports are within the altitude range for which the maximum landing weights have been determined.

(b) No person may operate a turbine-engine-powered transport category airplane certificated after September 30, 1958, contrary to the Airplane Flight Manual, or take off that airplane unless—

(1) The takeoff weight does not exceed the takeoff weight specified in the Airplane Flight Manual for the elevation of the airport and for the ambient temperature existing at the time of takeoff;

(2) Normal consumption of fuel and oil in flight to the airport of intended landing and to the alternate airports will leave a weight on arrival not in excess of the landing weight specified in the Airplane Flight Manual for the elevation of each of the airports involved and for the ambient temperatures expected at the time of landing;

(3) The takeoff weight does not exceed the weight shown in the Airplane Flight Manual to correspond with the minimum distances required for takeoff considering the elevation of the airport, the runway to be used, the effective runway gradient, and the ambient temperature and wind component existing at the time of takeoff; and

(4) Where the takeoff distance includes a clearway, the clearway distance is not greater than one-half of—

(i) The takeoff run, in the case of airplanes certificated after September 30, 1958, and before August 30, 1959; or

(ii) The runway length, in the case of airplanes certificated after August 29, 1959.

(c) No person may take off a turbine-engine-powered transport category airplane certificated after August 29, 1959, unless, in addition to the requirements of paragraph (b) of this section—

(1) The accelerate-stop distance is no greater than the length of the runway plus the length of the stopway (if present); and

(2) The takeoff distance is no greater than the length of the runway plus the length of the clearway (if present); and

(3) The takeoff run is no greater than the length of the runway.

§ 91.607 Emergency exits for airplanes carrying passengers for hire.

(a) Notwithstanding any other provision of this chapter, no person may operate a large airplane (type certificated under the Civil Air Regulations effective before April 9, 1957) in passenger-carrying operations for hire, with more than the number of occupants—

(1) Allowed under Civil Air Regulations 4b.362 (a), (b), and (c) as in effect on December 20, 1951; or

(2) Approved under Special Civil Air Regulations SR 387, SR 389, SR 389A, or SR 389B, or under this section as in effect.

However, an airplane type listed in the following table may be operated with up to the listed number of occupants (including crewmembers) and the corresponding number of exits (including emergency exits and doors) approved for the emergency exit of passengers or with an occupant-exit configuration approved under paragraph (b) or (c) of this section.

Airplane type	Max. number of occupants including all crewmembers	Corresponding number of exits authorized for passenger use
B-307	61	4
B-377	96	9
C-46	67	4
CV-240	53	6
CV-340 and CV-440.	53	6
DC-3	35	4
DC-3 (Super)	39	5
DC-4	86	5
DC-6	87	7
DC-6B	112	11
L-18	17	3
L-049, L-649, L-749.	87	7
L-1049	96	9
M-202	53	6
M-404	53	7
Viscount 700 Series.	53	7

(b) Occupants in addition to those authorized under paragraph (a) of this section may be carried as follows:

(1) For each additional floor-level exit at least 24 inches wide by 48 inches high, with an unobstructed 20-inch-wide access aisleway between the exit and the main passenger aisle, 12 additional occupants.

(2) For each additional window exit located over a wing that meets the requirements of the airworthiness standards under which the airplane was type certificated or that is large enough to inscribe an ellipse 19 × 26 inches, eight additional occupants.

(3) For each additional window exit that is not located over a wing but that otherwise complies with paragraph (b)(2) of this section, five additional occupants.

(4) For each airplane having a ratio (as computed from the table in paragraph (a) of this section) of maximum number of occupants to number of exits greater than 14:1, and for each airplane that does not have at least one full-size, door-type exit in the side of the fuselage in the rear part of the cabin, the first additional exit must be a floor-level exit that complies with paragraph (b)(1) of this section and must be located in the rear part of the cabin on the opposite side of the fuselage from the main entrance door. However, no person may operate an airplane under this section carrying more than 115 occupants unless there is such an exit on each side of the fuselage in the rear part of the cabin.

(c) No person may eliminate any approved exit except in accordance with the following:

(1) The previously authorized maximum number of occupants must be reduced by the same number of additional occupants authorized for that exit under this section.

(2) Exits must be eliminated in accordance with the following priority schedule: First, non-over-wing window exits; second, over-wing window exits; third, floor-level exits located in the forward part of the cabin; and fourth, floor-level exits located in the rear of the cabin.

(3) At least one exit must be retained on each side of the fuselage regardless of the number of occupants.

(4) No person may remove any exit that would result in a ratio of maximum number of occupants to approved exits greater than 14:1.

(d) This section does not relieve any person operating under Part 121 of this chapter from complying with § 121.291.

§ 91.609 Flight recorders and cockpit voice recorders.

(a) No holder of an air carrier operating certificate or an operating certificate may conduct any operation under this part with an aircraft listed in the holder's operations specifications or current list of aircraft used in air transportation unless that aircraft complies with any applicable flight recorder and cockpit voice recorder requirements of the part under which its certificate is issued except that the operator may—

(1) Ferry an aircraft with an inoperative flight recorder or cockpit voice recorder from a place where repair or replacement cannot be made to a place where they can be made;

(2) Continue a flight as originally planned, if the flight recorder or cockpit voice recorder becomes inoperative after the aircraft has taken off;

(3) Conduct an airworthiness flight test during which the flight recorder or cockpit voice recorder is turned off to test it or to test any communications or electrical equipment installed in the aircraft; or

(4) Ferry a newly acquired aircraft from the place where possession of it is taken to a place where the flight recorder or cockpit voice recorder is to be installed.

(b) [Notwithstanding paragraphs (c) and (e) of this section, an operator other than the holder of an air carrier or a commercial operator certificate may—

[(1) Ferry an aircraft with an inoperative flight recorder or cockpit voice recorder from a place where repair or replacement cannot be made to a place where they can be made;

[(2) Continue a flight as originally planned if the flight recorder or cockpit voice recorder becomes inoperative after the aircraft has taken off;

[(3) Conduct an airworthiness flight test during which the flight recorder or cockpit voice recorder is turned off to test it or to test any communications or electrical equipment installed in the aircraft;

[(4) Ferry a newly acquired aircraft from a place where possession of it was taken to a place where the flight recorder or cockpit voice recorder is to be installed; or

[(5) Operate an aircraft:

[(i) For not more than 15 days while the flight recorder and/or cockpit voice recorder is inoperative and/or removed for repair provided that the aircraft maintenance records contain an entry that indicates the date of failure, and a placard is located in view of the pilot to show that the flight recorder or cockpit voice recorder is inoperative.

[(ii) For not more than an additional 15 days, provided that the requirements in paragraph (b)(5)(i) are met and that a certificated pilot, or a certificated person authorized to return an aircraft to service under § 43.7 of this chapter, certifies in the aircraft maintenance records that additional time is required to complete repairs or obtain a replacement unit.]

(c) No person may operate a U.S. civil registered, multiengine, turbine-powered airplane or rotorcraft having a passenger seating configuration, excluding any pilot seats of 10 or more that has been manufactured after October 11, 1991, unless it is equipped with one or more approved flight recorders that utilize a digital method of recording and storing data and a method of readily retrieving that data from the storage medium, that are capable of recording the data specified in Appendix E to this part, for an airplane, or Appendix F to this part, for a rotorcraft, of this part within the range, accuracy, and recording interval specified, and that are capable of retaining no less than 8 hours of aircraft operation.

(d) Whenever a flight recorder, required by this section, is installed, it must be operated continuously from the instant the airplane begins the takeoff roll or the rotorcraft begins lift-off until the airplane has completed the landing roll or the rotorcraft has landed at its destination.

(e) Unless otherwise authorized by the Administrator, after October 11, 1991, no person may operate a U.S. civil registered multiengine, turbine-powered airplane or rotorcraft having a passenger seating configuration of six passengers or more and for which two pilots are required by type certification or operating rule unless it is equipped with an approved cockpit voice recorder that:

(1) Is installed in compliance with § 23.1457(a) (1) and (2), (b), (c), (d), (e), (f), and (g); § 25.1457(a) (1) and (2), (b), (c), (d), (e), (f), and (g); 27.1457(a) (1) and (2), (b), (c), (d), (e), (f), and (g); or § 29.1457(a) (1) and (2), (b), (c), (d), (e), (f), and (g) of this chapter, as applicable; and

(2) Is operated continuously from the use of the checklist before the flight to completion of the final checklist at the end of the flight.

(f) In complying with this section, an approved cockpit voice recorder having an erasure feature may be used, so that at any time during the operation of the recorder, information recorded more than 15 minutes earlier may be erased or otherwise obliterated.

(g) In the event of an accident or occurrence requiring immediate notification to the National Transportation Safety Board under Part 830 of its regulations that results in the termination of the flight, any operator who has installed approved flight recorders and approved cockpit voice recorders shall keep the recorded information for at least 60 days or, if requested by the Administrator or the Board, for a longer period. Information obtained from the record is used to assist in determining the cause of accidents or occurrences in connection with the investigation under Part 830. The Administrator does not use the cockpit voice recorder record in any civil penalty or certificate action.

(Amdt. 91-226, Eff. 10/11/91); [(Amdt. 91-228, Eff. 5/5/92)]

§ 91.611 Authorization for ferry flight with one engine inoperative.

(a) *General.* The holder of an air carrier operating certificate or an operating certificate issued under Part 125 may conduct a ferry flight of a four-engine airplane or a turbine-engine-powered airplane equipped with three engines, with one engine inoperative, to a base for the purpose of repairing that engine subject to the following:

(1) The airplane model has been test flown and found satisfactory for safe flight in accordance with paragraph (b) or (c) of this section, as appropriate. However, each operator who before November 19, 1966, has shown that a model of airplane with an engine inoperative is satisfactory for safe flight by a test flight conducted in accordance with performance data contained in the applicable Airplane Flight Manual under paragraph (a)(2) of this section need not repeat the test flight for that model.

(2) The approved Airplane Flight Manual contains the following performance data and the flight is conducted in accordance with that data:

(i) Maximum weight.

(ii) Center of gravity limits.

(iii) Configuration of the inoperative propeller (if applicable).

(iv) Runway length for takeoff (including temperature accountability).

(v) Altitude range.

(vi) Certificate limitations.

(vii) Ranges of operational limits.

(viii) Performance information.

(ix) Operating procedures.

(3) The operator has FAA approved procedures for the safe operation of the airplane, including specific requirements for—

(i) Limiting the operating weight on any ferry flight to the minimum necessary for the flight plus the necessary reserve fuel load;

(ii) A limitation that takeoffs must be made from dry runways unless, based on a showing of actual operating takeoff techniques on wet runways with one engine inoperative, takeoffs with full controllability from wet runways have been approved for the specific model aircraft and included in the Airplane Flight Manual;

(iii) Operations from airports where the runways may require a takeoff or approach over populated areas; and

(iv) Inspection procedures for determining the operating condition of the operative engines.

(4) No person may take off an airplane under this section if—

(i) The initial climb is over thickly populated areas; or

(ii) Weather conditions at the takeoff or destination airport are less than those required for VFR flight.

(5) Persons other than required flight crewmembers shall not be carried during the flight.

(6) No person may use a flight crewmember for flight under this section unless that crewmember is thoroughly familiar with the operating procedures for one-engine inoperative ferry flight contained in the certificate holder's manual and the limitations and performance information in the Airplane Flight Manual.

(b) *Flight tests: reciprocating-engine-powered airplanes.* The airplane performance of a reciprocating-engine-powered airplane with one engine inoperative must be determined by flight test as follows:

(1) A speed not less than 1.3 V_{S1} must be chosen at which the airplane may be controlled satisfactorily in a climb with the critical engine inoperative (with its propeller removed or in a configuration desired by the operator and with all other engines operating at the maximum power determined in paragraph (b)(3) of this section.

(2) The distance required to accelerate to the speed listed in paragraph (b)(1) of this section and to climb to 50 feet must be determined with—

(i) The landing gear extended;

(ii) The critical engine inoperative and its propeller removed or in a configuration desired by the operator; and

(iii) The other engines operating at not more than maximum power established under paragraph (b)(3) of this section.

(3) The takeoff, flight and landing procedures, such as the approximate trim settings, method of power application, maximum power, and speed must be established.

(4) The performance must be determined at a maximum weight not greater than the weight that allows a rate of climb of at least 400 feet per minute in the en route configuration set forth in § 25.67(d) of this chapter in effect on January 31, 1977, at an altitude of 5,000 feet.

(5) The performance must be determined using temperature accountability for the takeoff field length, computed in accordance with § 25.61 of this chapter in effect on January 31, 1977.

(c) *Flight tests: Turbine-engine-powered airplanes.* The airplane performance of a turbine-engine-powered airplane with one engine inoperative must be determined by flight tests, including at least three takeoff tests, in accordance with the following:

(1) Takeoff speeds V_R and V_2, not less than the corresponding speeds under which the airplane was type certificated under § 25.107 of this chapter, must be chosen at which the airplane may be controlled satisfactorily with the critical engine inoperative (with its propeller removed or in a configuration desired by the operator, if applicable) and with all other engines operating at not more than the power selected for type certification as set forth in § 25.101 of this chapter.

(2) The minimum takeoff field length must be the horizontal distance required to accelerate and climb to the 35-foot height at V_2 speed (including any additional speed increment obtained in the tests) multiplied by 115 percent and determined with—

(i) The landing gear extended;

(ii) The critical engine inoperative and its propeller removed or in a configuration desired by the operator (if applicable); and

(iii) The other engine operating at not more than the power selected for type certification as set forth in § 25.101 of this chapter.

(3) The takeoff, flight, and landing procedures such as the approximate trim setting, method of power application, maximum power, and speed must be established. The airplane must be satisfactorily controllable during the entire takeoff run when operated according to these procedures.

(4) The performance must be determined at a maximum weight not greater than the weight determined under § 25.121(c) of this chapter but with—

(i) The actual steady gradient of the final takeoff climb requirement not less than 1.2 percent at the end of the takeoff path with two critical engines inoperative; and

(ii) The climb speed not less than the two-engine inoperative trim speed for the actual steady gradient of the final takeoff climb prescribed by paragraph (c)(4)(i) of this section.

(5) The airplane must be satisfactorily controllable in a climb with two critical engines inoperative. Climb performance may be shown by calculations based on, and equal in accuracy to, the results of testing.

(6) The performance must be determined using temperature accountability for takeoff distance and final takeoff climb computed in accordance with § 25.101 of this chapter.

For the purpose of paragraphs (c)(4) and (5) of this section, "two critical engines" means two adjacent engines on one side of an airplane with four engines, and the center engine and one outboard engine on an airplane with three engines.

§ 91.613 Materials for compartment interiors.

No person may operate an airplane that conforms to an amended or supplemental type certificate issued in accordance with SFAR No. 41 for a maximum certificated takeoff weight in excess of 12,500 pounds unless within 1 year after issuance of the initial airworthiness certificate under that SFAR the airplane meets the compartment interior requirements set forth in § 25.853(a), (b), (b-1), (b-2), and (b-3) of this chapter in effect on September 26, 1978.

§§ 91.615–91.699 [Reserved]

Subpart H—Foreign aircraft operations and operations of U.S.-registered civil aircraft outside of the United States

§ 91.701 Applicability.

This subpart applies to the operations of civil aircraft of U.S. registry outside of the United States and the operations of foreign civil aircraft within the United States.

§ 91.703 Operations of civil aircraft of U.S. registry outside of the United States.

(a) Each person operating a civil aircraft of U.S. registry outside of the United States shall—

(1) When over the high seas, comply with annex 2 (Rules of the Air) to the Convention on International Civil Aviation and with §§ 91.117(c), 91.130, and 91.131;

(2) When within a foreign country, comply with the regulations relating to the flight and maneuver of aircraft there in force;

(3) Except for §§ 91.307(b), 91.309, 91.323, and 91.711, comply with this part so far as it is not inconsistent with applicable regulations of the foreign country where the aircraft is operated or annex 2 of the Convention on International Civil Aviation; and

(4) When over the North Atlantic within airspace designated as Minimum Navigation Performance Specifications airspace, comply with § 91.705.

(b) Annex 2 to the Convention on International Civil Aviation, Eighth Edition—July 1986, with amendments through Amendment 28 effective November 1987, to which reference is made in this part, is incorporated into this part and made a part hereof as provided in 5 U.S.C. 552 and pursuant to 1 CFR Part 51. Annex 2 (including a complete historic file of changes thereto) is available for public inspection at the Rules Docket, AGC-10, Federal Aviation Administration, 800 Independence Avenue SW., Washington, DC 20591. In addition, Annex 2 may be purchased from the International Civil Aviation Organization (Attention: Distribution Officer), P.O. Box 400, Succursale, Place de L'Aviation Internationale, 1000 Sherbrooke Street West, Montreal, Quebec, Canada H3A 2R2.

[*(Amdt. 91-227, Eff. 9/16/93)*]

§ 91.705 Operations within the North Atlantic Minimum Navigation Performance Specifications airspace.

No person may operate a civil aircraft of U.S. registry in North Atlantic (NAT) airspace designated as Minimum Navigation Performance Specifications (MNPS) airspace unless—

(a) The aircraft has approved navigation performance capability which complies with the requirements of Appendix C of this part; and

(b) The operator is authorized by the Administrator to perform such operations.

(c) The Administrator authorizes deviations from the requirements of this section in accordance with section 3 of Appendix C to this part.

§ 91.707 Flights between Mexico or Canada and the United States.

Unless otherwise authorized by ATC, no person may operate a civil aircraft between Mexico or Canada and the United States without filing an IFR or VFR flight plan, as appropriate.

§ 91.709 Operations to Cuba.

No person may operate a civil aircraft from the United States to Cuba unless—

(a) Departure is from an international airport of entry designated in § 6.13 of the Air Commerce Regulations of the Bureau of Customs (19 CFR 6.13); and

(b) In the case of departure from any of the 48 contiguous States or the District of Columbia, the pilot in command of the aircraft has filed—

(1) A DVFR or IFR flight plan as prescribed in §§ 99.11 or 99.13 of this chapter; and

(2) A written statement, within 1 hour before departure, with the Office of Immigration and Naturalization Service at the airport of departure, containing—

(i) All information in the flight plan;

(ii) The name of each occupant of the aircraft;

(iii) The number of occupants of the aircraft; and

(iv) A description of the cargo, if any.

This section does not apply to the operation of aircraft by a scheduled air carrier over routes authorized in operations specifications issued by the Administrator.

(Approved by the Office of Management and Budget under OMB control number 2120-0005).

§ 91.711 Special rules for foreign civil aircraft.

(a) *General.* In addition to the other applicable regulations of this part, each person operating a foreign civil aircraft within the United States shall comply with this section.

(b) *VFR.* No person may conduct VFR operations which require two-way radio communications under this part unless at least one crewmember of that aircraft is able to conduct two-way radio communications in the English language and is on duty during that operation.

(c) *IFR.* No person may operate a foreign civil aircraft under IFR unless— (1) That aircraft is equipped with—

(i) **[*Radio equipment allowing two-way radio communication with ATC when it is operated in controlled airspace; and*]**

(ii) Radio navigational equipment appropriate to the navigational facilities to be used;

(2) Each person piloting the aircraft—

(i) Holds a current United States instrument rating or is authorized by his foreign airman certificate to pilot under IFR; and

(ii) Is thoroughly familiar with the United States en route, holding, and letdown procedures; and

(3) At least one crewmember of that aircraft is able to conduct two-way radiotelephone communications in the English language and that crewmember is on duty while the aircraft is approaching, operating within, or leaving the United States.

(d) *Over water.* Each person operating a foreign civil aircraft over water off the shores of the United States shall give flight notification or file a flight plan in accordance with the Supplementary Procedures for the ICAO region concerned.

(e) *Flight at and above FL 240.* If VOR navigational equipment is required under paragraph (c)(1)(ii) of this section, no person may operate a foreign civil aircraft within the 50 States and the District of Columbia at or above FL 240, unless the aircraft is equipped with distance measuring equipment (DME) capable of receiving and indicating distance information from the VORTAC facilities to be used. When DME required by this paragraph fails at and above FL 240, the pilot in command of the aircraft shall notify ATC immediately and may then continue operations at and above FL 240 to the next airport of intended landing at which repairs or replacement of the equipment can be made. However, paragraph (e) of this section does not apply to foreign civil aircraft that are not equipped with DME when operated for the following purposes and if ATC is notified prior to each takeoff:

(1) Ferry flights to and from a place in the United States where repairs or alterations are to be made.

(2) Ferry flights to a new country of registry.

(3) Flight of a new aircraft of U.S. manufacture for the purpose of—

(i) Flight testing the aircraft;

(ii) Training foreign flight crews in the operation of the aircraft; or

(iii) Ferrying the aircraft for export delivery outside the United States.

(4) Ferry, demonstration, and test flight of an aircraft brought to the United States for the purpose of demonstration or testing the whole or any part thereof.

[*(Amdt. 91-227, Eff. 9/16/93)*]

§ 91.713 Operation of civil aircraft of Cuban registry.

No person may operate a civil aircraft of Cuban registry except in controlled airspace and in accordance with air traffic clearance or air traffic control instructions that may require use of specific airways or routes and landings at specific airports.

§ 91.715 Special flight authorizations for foreign civil aircraft.

(a) Foreign civil aircraft may be operated without airworthiness certificates required under § 91.203 if a special flight authorization for that operation is issued under this section. Application for a special flight authorization must be made to the [Flight Standards Division Manager or Aircraft Certification Directorate Manager] of the FAA region in which the applicant is located or to the region within which the U.S. point of entry is located. However, in the case of an aircraft to be operated in the U.S. for the purpose of demonstration at an airshow, the application may be made to the [Flight Standards Division Manager or Aircraft Certification Directorate Manager] of the FAA region in which the airshow is located.

(b) The Administrator may issue a special flight authorization for a foreign civil aircraft subject to any conditions and limitations that the Administrator considers necessary for safe operation in the U.S. airspace.

(c) No person may operate a foreign civil aircraft under a special flight authorization unless that operation also complies with Part 375 of the Special Regulations of the Department of Transportation (14 CFR Part 375).

(Approved by the Office of Management and Budget under OMB control number 2120-0005).

(Amdt. 91-212, Eff. 8/18/90)

§§ 91.717–91.799 [Reserved]

Subpart I—Operating noise limits

§ 91.801 Applicability: Relation to Part 36.

(a) This subpart prescribes operating noise limits and related requirements that apply, as follows, to the operation of civil aircraft in the United States.

(1) Sections 91.803, 91.805, 91.807, 91.809, and 91.811 apply to civil subsonic turbojet airplanes with maximum weights of more than 75,000 pounds and—

(i) If U.S. registered, that have standard airworthiness certificates; or

(ii) If foreign registered, that would be required by this chapter to have a U.S. standard airworthiness certificate in order to conduct the operations intended for the airplane were it registered in the United States. Those sections apply to operations to or from airports in the United States under this part and Parts 121, 125, 129, and 135 of this chapter.

(2) Section 91.813 applies to U.S. operators of civil subsonic turbojet airplanes covered by this subpart. This sec-

tion applies to operators operating to or from airports in the United States under this part and Parts 121, 125, and 135, but not to those operating under Part 129 of this chapter.

(3) Sections 91.803, 91.819, and 91.821 apply to U.S.-registered civil supersonic airplanes having standard airworthiness certificates and to foreign-registered civil supersonic airplanes that, if registered in the United States, would be required by this chapter to have U.S. standard airworthiness certificates in order to conduct the operations intended for the airplane. Those sections apply to operations under this part and under Parts 121, 125, 129, and 135 of this chapter.

(b) Unless otherwise specified, as used in this subpart "Part 36" refers to 14 CFR Part 36, including the noise levels under Appendix C of that part, notwithstanding the provisions of that part excepting certain airplanes from the specified noise requirements. For purposes of this subpart, the various stages of noise levels, the terms used to describe airplanes with respect to those levels, and the terms "subsonic airplane" and "supersonic airplane" have the meanings specified under Part 36 of this chapter. For purposes of this subpart, for subsonic airplanes operated in foreign air commerce in the United States, the Administrator may accept compliance with the noise requirements under annex 16 of the International Civil Aviation Organization when those requirements have been shown to be substantially compatible with, and achieve results equivalent to those achievable under, Part 36 for that airplane. Determinations made under these provisions are subject to the limitations of § 36.5 of this chapter as if those noise levels were Part 36 noise levels.

(c) Sections 91.851 through 91.875 of this subpart prescribe operating noise limits and related requirements that apply to any civil subsonic turbojet airplane with a maximum certificated weight of more than 75,000 pounds operating to or from an airport in the 48 contiguous United States and the District of Columbia under this part, Part 121, 125, 129, or 135 of this chapter on and after September 25, 1991).

(Amdt. 91-225, Eff. 9/25/91)

§ 91.803 Part 125 operators: Designation of applicable regulations.

For airplanes covered by this subpart and operated under Part 125 of this chapter, the following regulations apply as specified:

(a) For each airplane operation to which requirements prescribed under this subpart applied before November 29, 1980, those requirements of this subpart continue to apply.

(b) For each subsonic airplane operation to which requirements prescribed under this subpart did not apply before November 29, 1980, because the airplane was not operated in the United States under this part or Part 121, 129, or 135 of this chapter, the requirements prescribed under §§ 91.805, 91.809, 91.811, and 91.813 of this subpart apply.

(c) For each supersonic airplane operation to which requirements prescribed under this subpart did not apply before November 29, 1980, because the airplane was not operated in the United States under this part or Part 121, 129, or 135 of this chapter, the requirements of §§ 91.819 and 91.821 of this subpart apply.

(d) For each airplane required to operate under Part 125 for which a deviation under that part is approved to operate, in whole or in part, under this part or Part 121, 129, or 135 of this chapter, notwithstanding the approval, the requirements prescribed under paragraphs (a), (b), and (c) of this section continue to apply.

§ 91.805 Final compliance: Subsonic airplanes.

Except as provided in §§ 91.809 and 91.811, on and after January 1, 1985, no person may operate to or from an airport in the United States any subsonic airplane covered by this subpart unless that airplane has been shown to comply with Stage 2 or Stage 3 noise levels under Part 36 of this chapter.

§ 91.807 Phased compliance under Parts 121, 125, and 135: Subsonic airplanes.

(a) *General.* Each person operating airplanes under Part 121, 125, or 135 of this chapter, as prescribed under § 91.803 of this subpart, regardless of the state of registry of the airplane, shall comply with this section with respect to subsonic airplanes covered by this subpart.

(b) *Compliance schedules.* Except for airplanes shown to be operated in foreign air commerce under paragraph (c) of this section or covered by an exemption (including those issued under § 91.811), airplanes operated by U.S. operators in air commerce in the United States must be shown to comply with Stage 2 or Stage 3 noise levels under Part 36 of this chapter, in accordance with the following schedule, or they may not be operated to or from airports in the United States:

(1) By January 1, 1981—

(i) At least one quarter of the airplanes that have four engines with no bypass ratio or with a bypass ratio less than two; and

(ii) At least half of the airplanes powered by engines with any other bypass ratio or by another number of engines.

(2) By January 1, 1983—

(i) At least one-half of the airplanes that have four engines with no bypass ratio or with a bypass ratio less than two; and

(ii) All airplanes powered by engines with any other bypass ratio or by another number of engines.

(c) *Apportionment of airplanes.* For purposes of paragraph (b) of this section, a person operating airplanes engaged in domestic and foreign air commerce in the United States may elect not to comply with the phased schedule with respect to that portion of the airplanes operated by that person shown, under an approved method of apportionment, to be engaged in foreign air commerce in the United States.

§ 91.809 Replacement airplanes.

A Stage 1 airplane may be operated after the otherwise applicable compliance dates prescribed under §§ 91.805 and 91.807 if, under an approved plan, a replacement airplane has been ordered by the operator under a binding contract as follows:

(a) For replacement of an airplane powered by two engines, until January 1, 1986, but not after the date specified in the plan, if the contract is entered into by January 1, 1983, and specifies delivery before January 1, 1986, of a replacement airplane which has been shown to comply with Stage 3 noise levels under Part 36 of this chapter.

(b) For replacement of an airplane powered by three engines, until January 1, 1985, but not after the date specified in the plan, if the contract is entered into by January 1, 1983, and specifies delivery before January 1, 1985, of a replacement airplane which has been shown to comply with Stage 3 noise levels under Part 36 of this chapter.

(c) For replacement of any other airplane, until January 1, 1985, but not after the date specified in the plan, if the contract specifies delivery before January 1, 1985, of a replacement airplane which—

(1) Has been shown to comply with Stage 2 or Stage 3 noise levels under Part 36 of this chapter prior to issuance of an original standard airworthiness certificate; or

(2) Has been shown to comply with Stage 3 noise levels under Part 36 of this chapter prior to issuance of a standard airworthiness certificate other than original issue.

(d) Each operator of a Stage 1 airplane for which approval of a replacement plan is requested under this section shall submit to the Director, Office of Environment and Energy, an application constituting the proposed replacement plan (or revised plan) that contains the information specified under this paragraph and which is certified (under penalty of 18 U.S.C. 1001) as true and correct. Each application for approval must provide information corresponding to that specified in the contract, upon which the FAA may rely in considering its approval, as follows:

(1) Name and address of the applicant.

(2) Aircraft type and model and registration number for each airplane to be replaced under the plan.

(3) Aircraft type and model of each replacement airplane.

(4) Scheduled dates of delivery and introduction into service of each replacement airplane.

(5) Names and addresses of the parties to the contract and any other persons who may effectively cancel the contract or otherwise control the performance of any party.

(6) Information specifying the anticipated disposition of the airplanes to be replaced.

(7) A statement that the contract represents a legally enforceable, mutual agreement for delivery of an eligible replacement airplane.

(8) Any other information or documentation requested by the Director, Office of Environment and Energy, reasonably necessary to determine whether the plan should be approved.

§ 91.811 Service to small communities exemption: Two-engine, subsonic airplanes.

(a) A Stage 1 airplane powered by two engines may be operated after the compliance dates prescribed under §§ 91.805, 91.807, and 91.809 when, with respect to that airplane, the Administrator issues an exemption to the operator from the noise level requirements under this subpart. Each exemption issued under this section terminates on the earliest of the following dates:

(1) For an exempted airplane sold, or otherwise disposed of, to another person on or after January 1, 1983, on the date of delivery to that person.

(2) For an exempted airplane with a seating configuration of 100 passenger seats or less, on January 1, 1988.

(3) For an exempted airplane with a seating configuration of more than 100 passenger seats, on January 1, 1985.

(b) For the purpose of this section, the seating configuration of an airplane is governed by that shown to exist on December 1, 1979, or an earlier date established for that airplane by the Administrator.

§ 91.813 Compliance plans and status: U.S. operations of subsonic airplanes.

(a) Each U.S. operator of a civil subsonic airplane covered by this subpart (regardless of the state of registry) shall submit to the Director, Office of Environment and Energy, in accordance with this section, the operator's current compliance status and plan for achieving and maintaining compliance with the applicable noise level requirements of this subpart. If appropriate, an operator may substitute for the required plan a notice, certified as true (under penalty of 18 U.S.C. 1001) by that operator, that no change in the plan or status of any airplane affected by the plan has occurred since the date of the plan most recently submitted under this section.

(b) Each compliance plan, including each revised plan, must contain the information specified under paragraph (c) of this section for each airplane covered by this section that is operated by the operator. Unless otherwise approved by the Administrator, compliance plans must provide the required plan and status information as it exists on the date 30 days before the date specified for submission of the plan. Plans must be certified by the operator as true and complete (under penalty of 18 U.S.C. 1001) and be submitted for each airplane covered by this section on or before 90 days after initially commencing operation of airplanes covered by this section, whichever is later, and thereafter—

(1) Thirty days after any change in the operator's fleet or compliance planning decisions that has a separate or cumulative effect on 10 percent or more of the airplanes in either class of airplanes covered by § 91.807(b); and

(2) Thirty days after each compliance date applicable to that airplane under this subpart, and annually thereafter through 1985, or until any later date for that airplane prescribed under this subpart, on the anniversary of that submission date, to show continuous compliance with this subpart.

(c) Each compliance plan submitted under this section must identify the operator and include information regarding the compliance plan and status for each airplane covered by the plan as follows:

(1) Name and address of the airplane operator.

(2) Name and telephone number of the person designated by the operator to be responsible for the preparation of the compliance plan and its submission.

(3) The total number of airplanes covered by this section and in each of the following classes and subclasses:

(i) For airplanes engaged in domestic air commerce—

(A) Airplanes powered by four turbojet engines with no bypass ratio or with a bypass ratio less than two;

(B) Airplanes powered by engines with any other bypass ratio or by another number of engines; and

(C) Airplanes covered by an exemption issued under § 91.811 of this subpart.

(ii) For airplanes engaged in foreign air commerce under an approved apportionment plan—

(A) Airplanes powered by four turbojet engines with no bypass ratio or with a bypass ratio less than two;

(B) Airplanes powered by engines with any other bypass ratio or by another number of engines; and

(C) Airplanes covered by an exemption issued under § 91.811 of this subpart.

(4) For each airplane covered by this section—

(i) Aircraft type and model;

(ii) Aircraft registration number;

(iii) Aircraft manufacturer serial number;

(iv) Aircraft powerplant make and model;

(v) Aircraft year of manufacture;

(vi) Whether Part 36 noise level compliance has been shown; "Yes/No";

(vii) The appropriate code prescribed under paragraph (c)(5) of this section which indicates the acoustical technology installed, or to be installed, on the airplane;

(viii) For airplanes on which acoustical technology has been or will be applied, following the appropriate code entry, the actual or scheduled month and year of installation on the airplane;

(ix) For DC-8 and B-707 airplanes operated in domestic U.S. air commerce which have been or will be retired from service in the United States without replacement between January 24, 1977, and January 1, 1985, the appropriate code prescribed under paragraph (c)(5) of this section followed by the actual or scheduled month and year of retirement of the airplane from service;

(x) For DC-8 and B-707 airplanes operated in foreign air commerce in the United States which have been or will be retired from service in the United States without replacement between April 14, 1980, and January 1, 1985, the appropriate code prescribed under paragraph (c)(5) of this section followed by the actual or scheduled month and year of retirement of the airplane from service;

(xi) For airplanes covered by an approved replacement plan under § 91.807(c) of this subpart, the appropriate code prescribed under paragraph (c)(5) of this section followed by the scheduled month and year for replacement of the airplane;

(xii) For airplanes designated as "engaged in foreign commerce" in accordance with an approved method of apportionment under § 91.807(c) of this subpart, the appropriate code prescribed under paragraph (c)(5) of this section;

(xiii) For airplanes covered by an exemption issued to the operator granting relief from noise level requirements of this subpart, the appropriate code prescribed under paragraph (c)(5) of this section followed by the actual or scheduled month and year of expiration of the exemption and the appropriate code and applicable dates which indicate the compliance strategy planned or implemented for the airplane;

(xiv) For all airplanes covered by this section, the number of spare shipsets of acoustical components needed for continuous compliance and the number available on demand to the operator in support of those airplanes; and

(xv) For airplanes for which none of the other codes prescribed under paragraph (c)(5) of this section describes either the technology applied or to be applied to the airplane in accordance with the certification requirements under Parts 21 and 36 of this chapter, or the compliance strategy or methodology following the code "OTH," enter the date of any certificate action and attach an addendum to the plan explaining the nature and the extent of the certificated technology, strategy, or methodology employed, with reference to the type certificate documentation.

(5) Table of Acoustical Technology/Strategy Codes

Code	Airplane type/model	Certificate technology
A	B-707-120B; B-707-320B/C; B-720B	Quiet nacelles + 1-ring.
B	B-727-100	Double wall fan duct treatment.
C	B-727-200	Double wall fan duct treatment (pre-January 1977 installations and amended type certificate).

(Table of codes continued.)

Code	Airplane type/model	Certificate technology
D	B-727-200; B-737-100; B-737-200	Quiet nacelles + double wall fan duct treatment.
E	B-747-100 (pre-December 1971); B-747-200 (pre-December 1971)	Fixed lip inlets + and bullet with treatment + fan duct treatment areas.
F	DC-8	New extended inlet and bullet with treatment + fan duct treatment areas.
G	DC-9	P 36 sound absorbing material treatment kit.
H	BAC-111-200	Silencer kit (BAC Acoustic Report 522).
I	BAC-111-400	Silencer kit (BAC Acoustic Report 598).
J	B-707; DC-8	Reengined with high bypass ratio turbojet engines + quiet nacelles (if certificated under stage 3 noise level requirements)

REP—For airplanes covered by an approved replacement plan under § 91.807(c) of this subpart.

EFC—For airplanes designated as "engaged in foreign commerce" in accordance with an approved method of apportionment under § 91.811 of this subpart.

RET—For DC-8 and B-707 airplanes operated in domestic U.S. air commerce and retired from service in the United States without replacement between January 24, 1977, and January 1, 1985.

RFC—For DC-8 and B-707 airplanes operated by U.S. operators in foreign air commerce in the United States and retired from service in the United States without replacement between April 14, 1980, and January 1, 1985.

EXD—For airplanes exempted from showing compliance with the noise level requirements of this subpart.

OTH—For airplanes for which no other prescribed code describes either the certificated technology applied or to be applied to the airplane, or the compliance strategy or methodology. (An addendum must explain the nature and extent of technology, strategy, or methodology and reference the type certificate documentation.)

§ 91.815 Agricultural and fire fighting airplanes: Noise operating limitations.

(a) This section applies to propeller-driven, small airplanes having standard airworthiness certificates that are designed for "agricultural aircraft operations" (as defined in 137.3 of this chapter, as effective on January 1, 1966) or for dispensing fire fighting materials.

(b) If the Airplane Flight Manual, or other approved manual material information, markings, or placards for the airplane indicate that the airplane has not been shown to comply with the noise limits under Part 36 of this chapter, no person may operate that airplane, except—

(1) To the extent necessary to accomplish the work activity directly associated with the purpose for which it is designed;

(2) To provide flight crewmember training in the special purpose operation for which the airplane is designed; and

(3) To conduct "nondispensing aerial work operations" in accordance with the requirements under § 137.29(c) of this chapter.

§ 91.817 Civil aircraft sonic boom.

(a) No person may operate a civil aircraft in the United States at a true flight Mach number greater than 1 except in compliance with conditions and limitations in an authorization to exceed Mach 1 issued to the operator under Appendix B of this part.

(b) In addition, no person may operate a civil aircraft for which the maximum operating limit speed M_{M0} exceeds a Mach number of 1, to or from an airport in the United States, unless—

(1) Information available to the flight crew includes flight limitations that ensure that flights entering or leaving the United States will not cause a sonic boom to reach the surface within the United States; and

(2) The operator complies with the flight limitations prescribed in paragraph (b)(1) of this section or complies with conditions and limitations in an authorization to exceed Mach 1 issued under Appendix B of this part.

(Approved by the Office of Management and Budget under OMB control number 2120-0005).

§ 91.819 Civil supersonic airplanes that do not comply with Part 36.

(a) *Applicability.* This section applies to civil supersonic airplanes that have not been shown to comply with the Stage 2 noise limits of Part 36 in effect on October 13, 1977, using applicable trade-off provisions, and that are operated in the United States, after July 31, 1978.

(b) *Airport use.* Except in an emergency, the following apply to each person who operates a civil supersonic airplane to or from an airport in the United States:

(1) Regardless of whether a type design change approval is applied for under Part 21 of this chapter, no person may land or take off an airplane covered by this section for which the type design is changed, after July 31, 1978, in a manner constituting an "acoustical change" under § 21.93 unless the acoustical change requirements of Part 36 are complied with.

(2) No flight may be scheduled, or otherwise planned, for takeoff or landing after 10 p.m. and before 7 a.m. local time.

§ 91.821 Civil supersonic airplanes: Noise limits.

Except for Concorde airplanes having flight time before January 1, 1980, no person may operate in the United States, a civil supersonic airplane that does not comply with Stage 2 noise limits of Part 36 in effect on October 13, 1977, using applicable trade-off provisions.

§§ 91.823–91.849 [Reserved]

[§ 91.851 Definitions.

[For the purposes of §§ 91.851 through 91.875 of this subpart:

[*Contiguous United States* means the area encompassed by the 48 contiguous United States and the District of Columbia.

[*Fleet* means those civil subsonic turbojet airplanes with a maximum certificated weight of more than 75,000 pounds that are listed on an operator's operations specifications as eligible for operation in the contiguous United States.

[*Import* means a change in ownership of an airplane from a non-U.S. person to a U.S. person when the airplane is brought into the United States for operation.

[*Operations specifications* means an enumeration of airplanes by type, model, series, and serial number operated by the operator or foreign air carrier on a given day, regardless of how or whether such airplanes are formally listed or designated by the operator.

[*Owner* means any person that has indicia of ownership sufficient to register the airplane in the United States pursuant to Part 47 of this chapter.

[*New entrant* means an air carrier or foreign air carrier that, on or before November 5, 1990, did not conduct operations under Part 121, 125, 129, or 135 of this chapter using an airplane covered by this subpart to or from any airport in the contiguous United States, but that initiates such operation after that date.

[*Stage 2 noise levels* means the requirements for Stage 2 noise levels as defined in Part 36 of this chapter in effect on November 5, 1990.

[*Stage 3 noise levels* means the requirements for Stage 3 noise levels as defined in Part 36 of this chapter in effect on November 5, 1990.

[*Stage 2 airplane* means a civil subsonic turbojet airplane with a maximum certificated weight of 75,000 pounds or more that complies with Stage 2 noise levels as defined in Part 36 of this chapter.

[*Stage 3 airplane* means a civil subsonic turbojet airplane with a maximum certificated weight of 75,000 pounds or more that complies with Stage 3 noise levels as defined in Part 36 of this chapter.]

[(Amdt. 91-225, Eff. 9/25/91)]

§ 91.853 Final compliance: Civil subsonic airplanes.

[Except as provided in § 91.873, after December 31, 1999, no person shall operate to or from any airport in the contiguous United States any airplane subject to § 91.801(c) of this subpart, unless that airplane has been shown to comply with Stage 3 noise levels.]

[(Amdt. 91-225, Eff. 9/25/91)]

[§ 91.855 Entry and nonaddition rule.

[No person may operate any airplane subject to § 91.801(c) of this subpart to or from an airport in the contiguous United States unless one or more of the following apply:

[(a) The airplane complies with Stage 3 noise levels.

[(b) The airplane complies with Stage 2 noise levels and was owned by a U.S. person on and since November 5, 1990. Stage 2 airplanes that meet these criteria and are leased to foreign airlines are also subject to the return provisions of paragraph (e) of this section.

[(c) The airplane complies with Stage 2 noise levels, is owned by a non-U.S. person, and is the subject of a binding lease to a U.S. person effective before and on September 19, 1991. Any such airplane may be operated for the term of the lease in effect on that date, and any extensions thereof provided for in that lease.

[(d) The airplane complies with Stage 2 noise levels and is operated by a foreign air carrier.

[(e) The airplane complies with Stage 2 noise levels and is operated by a foreign operator other than for the purpose of foreign air commerce.

[(f) The airplane complies with Stage 2 noise levels and—

[(1) On November 5, 1990, was owned by:

[(i) A corporation, trust, or partnership organized under the laws of the United States or any State (including individual States, territories, possessions, and the District of Columbia);

[(ii) An individual who is a citizen of the United States; or

[(iii) An entity owned or controlled by a corporation, trust, partnership, or individual described in paragraph (g)(1)(i) or (ii) of this section; and

[(2) Enters into the United States not later than 6 months after the expiration of a lease agreement (including any extensions thereof) between an owner described in paragraph (f)(1) of this section and a foreign airline.

[(g) The airplane complies with Stage 2 noise levels and was purchased by the importer under a written contract executed before November 5, 1990.

[(h) Any Stage 2 airplane described in this section is eligible for operation in the contiguous United States only as provided under § 91.865 or § 91.867.]

[(Amdt. 91-225, Eff. 9/25/91)]

[§ 91.857 Airplanes imported to points outside the contiguous United States.

[An operator of a Stage 2 airplane that was imported into a noncontiguous State, territory, or possession of the United States on or after November 5, 1990, shall:

[(a) Include in its operations specifications a statement that such airplane may not be used to provide air transportation to or from any airport in the contiguous United States.

[(b) Obtain a special flight authorization to operate that airplane into the contiguous United States for the purpose of maintenance. The special flight authorization must include a statement indicating that this regulation is the basis for the authorization.

[(Amdt. 91-225, Eff. 9/25/91)]

[§ 91.859 Modification to meet Stage 3 noise levels.

[For an airplane subject to § 91.801(c) of this subpart and otherwise prohibited from operation to or from an airport in the contiguous United States by § 91.855, any person may apply for a special flight authorization for that airplane to operate in the contiguous United States for the purpose of obtaining modifications to meet Stage 3 noise levels.]

[(Amdt. 91-225, Eff. 9/25/91)]

[§ 91.861 Base level.

[(a) U.S. Operators. The base level of a U.S. operator is equal to the number of owned or leased Stage 2 airplanes subject to § 91.801(c) of this subpart that were listed on that operator's operations specifications for operations to or from airports in the contiguous United States on any one day selected by the operator during the period January 1, 1990 through July 1, 1991, plus or minus adjustments made pursuant to paragraphs (a)(1) and (2).

[(1) The base level of a U.S. operator shall be increased by a number equal to the total of the following—

[(i) The number of Stage 2 airplanes returned to service in the United States pursuant to § 91.855(f);

[(ii) The number of Stage 2 airplanes purchased pursuant to § 91.855(g); and

[(iii) Any U.S. operator base level acquired with a Stage 2 airplane transferred from another person under § 91.863.

[(2) The base level of a U.S. operator shall be decreased by the amount of U.S. operator base level transferred with the corresponding number of Stage 2 airplanes to another person under § 91.863.

[(b) Foreign air carriers. The base level of a foreign air carrier is equal to the number of owned or leased Stage 2 airplanes that were listed on that carrier's U.S. operations specifications on any one day during the period January 1, 1990, through July 1, 1991, plus or minus any adjustments to the base levels made pursuant to paragraphs (b)(1) and (2).

[(1) The base level of a foreign air carrier shall be increased by the amount of foreign air carrier base level acquired with a Stage 2 airplane from another person under § 91.863.

[(2) The base level of a foreign air carrier shall be decreased by the amount of foreign air carrier base level transferred with a Stage 2 airplane to another person under § 91.863.

[(c) New entrants do not have a base level.]

[(Amdt. 91-225, Eff. 9/25/91)]

[§ 91.863 Transfers of Stage 2 airplanes with base level.

[(a) Stage 2 airplanes may be transferred with or without the corresponding amount of base level. Base level may not be transferred without the corresponding number of Stage 2 airplanes.

[(b) No portion of a U.S. operator's base level established under § 91.861(a) may be used for operations by a foreign air carrier. No portion of a foreign air carrier's base level established under § 91.861(b) may be used for operations by a U.S. operator.

[(c) Whenever a transfer of Stage 2 airplanes with base level occurs, the transferring and acquiring parties shall, within 10 days, jointly submit written notification of the transfer to the FAA, Office of Environment and Energy. Such notification shall state:

[(1) The names of the transferring and acquiring parties;

[(2) The name, address, and telephone number of the individual responsible for submitting the notification on behalf of the transferring and acquiring parties;

[(3) The total number of Stage 2 airplanes transferred, listed by airplane type, model, series, and serial number;

[(4) The corresponding amount of base level transferred and whether it is U.S. operator or foreign air carrier base level; and

[(5) The effective date of the transaction.

[(d) If, taken as a whole, a transaction or series of transactions made pursuant to this section does not produce an increase or decrease in the number of Stage 2 airplanes for either the acquiring or transferring operator, such transaction or series of transactions may not be used to establish compliance with the requirements of § 91.865.]

[(Amdt. 91-225, Eff. 9/25/91)]

[§ 91.865 Phased compliance for operators with base level.

[Except as provided in paragraph (a) of this section, each operator that operates an airplane under Part 91, 121, 125, 129, or 135 of this chapter, regardless of the national registry of the airplane, shall comply with paragraph (b) or (d) of this section at each interim compliance date with regard to its subsonic airplane fleet covered by § 91.801(c) of this subpart.

[(a) This section does not apply to new entrants covered

by § 91.867 or to foreign operators not engaged in foreign air commerce.

[(b) Each operator that chooses to comply with this paragraph pursuant to any interim compliance requirement shall reduce the number of Stage 2 airplanes it operates that are eligible for operation in the contiguous United States to a maximum of:

[(1) After December 31, 1994, 75 percent of the base level held by the operator;

[(2) After December 31, 1996, 50 percent of the base level held by the operator;

[(3) After December 31, 1998, 25 percent of the base level held by the operator.

[(c) Except as provided under § 91.871, the number of Stage 2 airplanes that must be reduced at each compliance date contained in paragraph (b) of this section shall be determined by reference to the amount of base level held by the operator on that compliance date, as calculated under § 91.861.

[(d) Each operator that chooses to comply with this paragraph pursuant to any interim compliance requirement shall operate a fleet that consists of:

[(1) After December 31, 1994, not less than 55 percent Stage 3 airplanes;

[(2) After December 31, 1996, not less than 65 percent Stage 3 airplanes;

[(3) After December 31, 1998, not less than 75 percent Stage 3 airplanes.

[(e) Calculations resulting in fractions may be rounded to permit the continued operation of the next whole number of Stage 2 airplanes.]

[(Amdt. 91-225, Eff. 9/25/91)]

[§ 91.867 Phased compliance for new entrants.

[(a) New entrant U.S. air carriers.

[(1) A new entrant initiating operations under Part 121, 125, or 135 of this chapter on or before December 31, 1994, may initiate service without regard to the percentage of its fleet composed of Stage 3 airplanes.

[(2) After December 31, 1994, at least 25 percent of the fleet of a new entrant must comply with Stage 3 noise levels.

[(3) After December 31, 1996, at least 50 percent of the fleet of a new entrant must comply with Stage 3 noise levels.

[(4) After December 31, 1998, at least 75 percent of the fleet of a new entrant must comply with Stage 3 noise levels.

[(b) New entrant foreign air carriers.

[(1) A new entrant foreign air carrier initiating Part 129 operations on or before December 31, 1994, may initiate service without regard to the percentage of its fleet composed of Stage 3 airplanes.

[(2) After December 31, 1994, at least 25 percent of the fleet on U.S. operations specifications of a new entrant foreign air carrier must comply with Stage 3 noise levels.

[(3) After December 31, 1996, at least 50 percent of the fleet on U.S. operations specifications of a new entrant foreign air carrier must comply with Stage 3 noise levels.

[(4) After December 31, 1998, at least 75 percent of the fleet on U.S. operations specifications of a new entrant foreign air carrier must comply with Stage 3 noise levels.

[(c) Calculations resulting in fractions may be rounded to permit the continued operation of the next whole number of Stage 2 airplanes.]

[(Amdt. 91-225, Eff. 9/25/91)]

[§ 91.869 Carry-forward compliance.

[(a) Any operator that exceeds the requirements of paragraph (b) of § 91.865 of this part on or before December 31, 1994, or on or before December 31, 1996, may claim a credit that may be applied at a subsequent interim compliance date.

[(b) Any operator that eliminates or modifies more Stage 2 airplanes pursuant to § 91.865(b) than required as of December 31, 1994, or December 31, 1996, may count the number of additional Stage 2 airplanes reduced as a credit toward—

[(1) The number of Stage 2 airplanes that it would otherwise be required to reduce following a subsequent interim compliance date specified in § 91.865(b); or

[(2) The number of Stage 3 airplanes it would otherwise be required to operate in its fleet following a subsequent interim compliance date to meet the percentage requirements specified in § 91.865(d).]

[Amdt. 91-225, Eff. 9/25/91)]

[§ 91.871 Waivers from interim compliance requirements.

[(a) Any U.S. operator or foreign air carrier subject to the requirements of § 91.865 or § 91.867 of this subpart may request a waiver from any individual compliance requirement.

[(b) Applications must be filed with the Secretary of Transportation at least 120 days prior to the compliance date from which the waiver is requested.

[(c) Applicants must show that a grant of waiver would be in the public interest, and must include in its application its plans and activities for modifying its fleet, including evidence of good faith efforts to comply with the requirements of § 91.865 or § 91.867. The application should contain all information the applicant considers relevant, including, as appropriate, the following:

[(1) The applicant's balance sheet and cash flow positions;

[(2) The composition of the applicant's current fleet; and

[(3) The applicant's delivery position with respect to new airplanes or noise-abatement equipment.

[(d) Waivers will be granted only upon a showing by the applicant that compliance with the requirements of

§ 91.865 or § 91.867 at a particular interim compliance date is financially onerous, physically impossible, or technologically infeasible, or that it would have an adverse effect on competition or on service to small communities.

[(e) The conditions of any waiver granted under this section shall be determined by the circumstances presented in the application, but in no case may the term extend beyond the next interim compliance date.

[(f) A summary of any request for a waiver under this section will be published in the *Federal Register*, and public comment will be invited. Unless the Secretary finds that circumstances require otherwise, the public comment period will be at least 14 days.]

[(Amdt. 91-225, Eff. 9/25/91)]

[§ 91.873 Waivers from final compliance.

[(a) A U.S. air carrier may apply for a waiver from the prohibition contained in § 91.853 for its remaining Stage 2 airplanes, provided that, by July 1, 1999, at least 85 percent of the airplanes used by the carrier to provide service to or from an airport in the contiguous United States will comply with the Stage 3 noise levels.

[(b) An application for the waiver described in paragraph (a) of this section must be filed with the Secretary of Transportation no later than January 1, 1999. Such application must include a plan with firm orders for replacing or modifying all airplanes to comply with Stage 3 noise levels at the earliest practicable time.

[(c) To be eligible to apply for the waiver under this section, a new entrant U.S. air carrier must initiate service no later than January 1, 1999, and must comply fully with all provisions of this section.

[(d) The Secretary may grant a waiver under this section if the Secretary finds that granting such waiver is in the public interest. In making such a finding, the Secretary shall include consideration of the effect of granting such a waiver on competition in the air carrier industry and the effect on small community air service, and any other information submitted by the applicant that the Secretary considers relevant.

[(e) The term of any waiver granted under this section shall be determined by the circumstances presented in the application, but in no case will the waiver permit the operation of any Stage 2 airplane covered by this subchapter in the contiguous United States after December 31, 2003.

[(f) A summary of any request for a waiver under this section will be published in the *Federal Register*, and public comment will be invited. Unless the Secretary finds that circumstances require otherwise, the public comment period will be at least 14 days.]

(Amdt. 91-225, Eff. 9/25/91)]

[§ 91.875 Annual progress reports.

[(a) Each operator subject to § 91.865 or § 91.867 of this chapter shall submit an annual report to the FAA, Office of Environment and Energy, on the progress it has made toward complying with the requirements of that section. Such reports shall be submitted no later than 45 days after the end of a calendar year. All progress reports must provide the information through the end of the calendar year, be certified by the operator as true and complete (under penalty of 18 U.S.C. 1001), and include the following information:

[(1) The name and address of the operator;

[(2) The name, title, and telephone number of the person designated by the operator to be responsible for ensuring the accuracy of the information in the report;

[(3) The operator's progress during the reporting period toward compliance with the requirements of § 91.853, § 91.865 or § 91.867. For airplanes on U.S. operations specifications, each operator shall identify the airplanes by type, model, series, and serial number.

[(i) Each Stage 2 airplane added or removed from operation or U.S. operations specifications (grouped separately by those airplanes acquired with and without base level);

[(ii) Each Stage 2 airplane modified to Stage 3 noise levels (identifying the manufacturer and model of noise abatement retrofit equipment);

[(iii) Each Stage 3 airplane on U.S. operations specifications as of the last day of the reporting period; and

[(iv) For each Stage 2 airplane transferred or acquired, the name and address of the recipient or transferor; and if base level was transferred, the person to or from whom base level was transferred or acquired pursuant to Section 91.863 along with the effective date of each base level transaction, and the type of base level transferred or acquired.

[(b) Each operator subject to § 91.865 or § 91.867 of this chapter shall submit an initial progress report covering the period from January 1, 1990, through December 31, 1991, and provide:

[(1) For each operator subject to § 91.865:

[(i) The date used to establish its base level pursuant to § 91.861 (a); and

[(ii) a list of those Stage 2 airplanes (by type, model, series, and serial number) in its base level, including adjustments made pursuant to § 91.861 after the date its base level was established.

[(2) For each U.S. operator:

[(i) A plan to meet the compliance schedules in § 91.865 or § 91.867 and the final compliance date of § 91.853, including the schedule for delivery of replacement Stage 3 airplanes or the installation of noise abatement retrofit equipment; and

[(ii) A separate list (by type, model, series, and serial number) of those airplanes included in the operator's base level, pursuant to § 91.861 (a)(1)(i) and (ii), under the categories "returned" or "purchased," along with the date each was added to its operations specifications.

[(c) Each operator subject to § 91.865 or § 91.867 of this chapter shall submit subsequent annual progress reports covering the calendar year preceding the report and including any changes in the information provided in paragraphs (a) and (b) of this section; including the use of any carry-forward credits pursuant to § 91.869.

[(d) An operator may request, in any report, that specific planning data be considered proprietary.

[(e) If an operator's actions during any reporting period cause it to achieve compliance with § 91.853, the report should include a statement to that effect. Further progress reports are not required unless there is any change in the information reported pursuant to (a) of this section.

[(f) For each U.S. operator subject to § 91.865, progress reports submitted for calendar years 1994, 1996, and 1998, shall also state how the operator achieved compliance with the requirements of that section, i.e.—

[(1) By reducing the number of Stage 2 airplanes in its fleet to no more than the maximum permitted percentage of its base level under § 91.865(b), or

[(2) By operating a fleet that consists of at least the minimum required percentage of Stage 3 airplanes required under § 91.865(d).]

[(Amdt. 91-225, Eff. 9/25/91)]

§§ 91.877–91.899 [Reserved]

Subpart J—Waivers

§ 91.901 [Reserved]

§ 91.903 Policy and procedures.

(a) The Administrator may issue a certificate of waiver authorizing the operation of aircraft in deviation from any rule listed in this subpart if the Administrator finds that the proposed operation can be safely conducted under the terms of that certificate of waiver.

(b) An application for a certificate of waiver under this part is made on a form and in a manner prescribed by the Administrator and may be submitted to any FAA office.

(c) A certificate of waiver is effective as specified in that certificate of waiver.

§ 91.905 List of rules subject to waivers.

Sec	
91.107	Use of safety belts.
91.111	Operating near other aircraft.
91.113	Right-of-way rules: Except water operations.
91.115	Right-of-way rules: Water operations.
91.117	Aircraft speed.
91.119	Minimum safe altitudes: General.
91.121	Altimeter settings.
91.123	Compliance with ATC clearances and instructions.
91.125	ATC light signals.
[*91.126*	***Operating on or in the vicinity of an airport in Class G airspace.***]
[*91.127*	***Operating on or in the vicinity of an airport in Class E airspace.***]
[*91.129*	[***Operations in Class D airspace.***]
[*91.130*	***Operations in Class C airspace.***]
[*91.131*	***Operations in Class B airspace.***]
91.133	Restricted and prohibited areas.
[*91.135*	***Operations in Class A airspace.***]
91.137	Temporary flight restrictions.
91.141	Flight restrictions in the proximity of the Presidential and other parties.
91.143	Flight limitation in the proximity of space flight operations.
91.153	VFR flight plan: Information required.
91.155	Basic VFR weather minimums.
91.157	Special VFR weather minimums.
91.159	VFR cruising altitude or flight level.
91.169	IFR flight plan: Information required.
91.173	ATC clearance and flight plan required.
91.175	Takeoff and landing under IFR.
91.177	Minimum altitudes for IFR operations.
91.179	IFR cruising altitude or flight level.
91.181	Course to be flown.
91.183	IFR radio communications.
91.185	IFR operations: Two-way radio communications failure.
91.187	Operation under IFR in controlled airspace: Malfunction reports.
91.209	Aircraft lights.
91.303	Aerobatic flights.
91.305	Flight test areas.
91.311	Towing: Other than under §91.309
91.313(e)	Restricted category civil aircraft: Operating limitations.
91.515	Flight altitude rules.
91.705	Operations within the North Atlantic Minimum Navigation Performance Specifications Airspace.
91.707	Flights between Mexico or Canada and the United States.
91.713	Operation of civil aircraft of Cuban registry.

[*(Amdt. 91-227, Eff. 9/16/93)*]

§§ 91.907–91.999 [Reserved]

Appendix A
Category II operations: Manual, instruments, equipment, and maintenance

1. Category II manual

(a) *Application for approval.* An applicant for approval of a Category II manual or an amendment to an approved Category II manual must submit the proposed manual or amendment to the Flight Standards District Office having jurisdiction of the area in which the applicant is located. If the application requests an evaluation program, it must include the following:

(1) The location of the aircraft and the place where the demonstrations are to be conducted; and

(2) The date the demonstrations are to commence (at least 10 days after filing the application).

(b) *Contents.* Each Category II manual must contain:

(1) The registration number, make, and model of the aircraft to which it applies;

(2) A maintenance program as specified in Section 4 of this appendix; and

(3) The procedures and instructions related to recognition of decision height, use of runway visual range information, approach monitoring, the decision region (the region between the middle marker and the decision height), the maximum permissible deviations of the basic ILS indicator within the decision region, a missed approach, use of airborne low approach equipment, minimum altitude for the use of the autopilot, instrument and equipment failure warning systems, instrument failure, and other procedures, instructions, and limitations that may be found necessary by the Administrator.

2. Required instruments and equipment

The instruments and equipment listed in this section must be installed in each aircraft operated in a Category II operation. This section does not require duplication of instruments and equipment required by § 91.205 or any other provisions of this chapter.

(a) *Group I.*

(1) Two localizer and glide slope receiving systems. Each system must provide a basic ILS display and each side of the instrument panel must have a basic ILS display. However, a single localizer antenna and a single glide slope antenna may be used.

(2) A communications system that does not affect the operation of at least one of the ILS systems.

(3) A marker beacon receiver that provides distinctive aural and visual indications of the outer and the middle markers.

(4) Two gyroscopic pitch and bank indicating systems.

(5) Two gyroscopic direction indicating systems.

(6) Two airspeed indicators.

(7) Two sensitive altimeters adjustable for barometric pressure, each having a placarded correction for altimeter scale error and for the wheel height of the aircraft. After June 26, 1979, two sensitive altimeters adjustable for barometric pressure, having markings at 20-foot intervals and each having a placarded correction for altimeter scale error and for the wheel height of the aircraft.

(8) Two vertical speed indicators.

(9) A flight control guidance system that consists of either an automatic approach coupler or a flight director system. A flight director system must display computed information as steering command in relation to an ILS localizer and, on the same instrument, either computed information as pitch command in relation to an ILS glide slope or basic ILS glide slope information. An automatic approach coupler must provide at least automatic steering in relation to an ILS localizer. The flight control guidance system may be operated from one of the receiving systems required by subparagraph (1) of this paragraph.

(10) For Category II operations with decision heights below 150 feet either a marker beacon receiver providing aural and visual indications of the inner marker or a radio altimeter.

(b) *Group II.*

(1) Warning systems for immediate detection by the pilot of system faults in items (1), (4), (5), and (9) of Group I and, if installed for use in Category III operations, the radio altimeter and autothrottle system.

(2) Dual controls.

(3) An externally vented static pressure system with an alternate static pressure source.

(4) A windshield wiper or equivalent means of providing adequate cockpit visibility for a safe visual transition by either pilot to touchdown and rollout.

(5) A heat source for each airspeed system pitot tube installed or an equivalent means of preventing malfunctioning due to icing of the pitot system.

3. Instruments and equipment approval

(a) *General.* The instruments and equipment required by Section 2 of this appendix must be approved as provided in this section before being used in Category II operations. Before presenting an aircraft for approval of the instruments and equipment, it must be shown that since the beginning of the 12th calendar month before the date of submission—

(1) The ILS localizer and glide slope equipment were bench checked according to the manufacturer's instructions and found to meet those standards specified in RTCA Paper 23 63/DO 117 dated March 14, 1963, "Standard Ad-

justment Criteria for Airborne Localizer and Glide Slope Receivers," which may be obtained from the RTCA Secretariat, 1425 K St., NW., Washington, DC 20005.

(2) The altimeters and the static pressure systems were tested and inspected in accordance with Appendix E to Part 43 of this chapter; and

(3) All other instruments and items of equipment specified in Section 2(a) of this appendix that are listed in the proposed maintenance program were bench checked and found to meet the manufacturer's specifications.

(b) *Flight control guidance system.* All components of the flight control guidance system must be approved as installed by the evaluation program specified in paragraph (e) of this section if they have not been approved for Category III operations under applicable type or supplemental type certification procedures. In addition, subsequent changes to make, model, or design of the components must be approved under this paragraph. Related systems or devices, such as the autothrottle and computed missed approach guidance system, must be approved in the same manner if they are to be used for Category II operations.

(c) *Radio altimeter.* A radio altimeter must meet the performance criteria of this paragraph for original approval and after each subsequent alteration.

(1) It must display to the flight crew clearly and positively the wheel height of the main landing gear above the terrain.

(2) It must display wheel height above the terrain to an accuracy of plus or minus 5 feet or 5 percent, whichever is greater, under the following conditions:

(i) Pitch angles of zero to plus or minus 5 degrees about the mean approach attitude.

(ii) Roll angles of zero to 20 degrees in either direction.

(iii) Forward velocities from minimum approach speed up to 200 knots.

(iv) Sink rates from zero to 15 feet per second at altitudes from 100 to 200 feet.

(3) Over level ground, it must track the actual altitude of the aircraft without significant lag or oscillation.

(4) With the aircraft at an altitude of 200 feet or less, any abrupt change in terrain representing no more than 10 percent of the aircraft's altitude must not cause the altimeter to unlock, and indicator response to such changes must not exceed 0.1 second and, in addition, if the system unlocks for greater changes, it must reacquire the signal in less than 1 second.

(5) Systems that contain a push-to-test feature must test the entire system (with or without an antenna) at a simulated altitude of less than 500 feet.

(6) The system must provide to the flight crew a positive failure warning display any time there is a loss of power or an absence of ground return signals within the designed range of operating altitudes.

(d) *Other instruments and equipment.* All other instruments and items of equipment required by § 2 of this appendix must be capable of performing as necessary for Category II operations. Approval is also required after each subsequent alteration to these instruments and items of equipment.

(e) *Evaluation program*—

(1) *Application.* Approval by evaluation is requested as a part of the application for approval of the Category II manual.

(2) *Demonstrations.* Unless otherwise authorized by the Administrator, the evaluation program for each aircraft requires the demonstrations specified in this paragraph. At least 50 ILS approaches must be flown with at least five approaches on each of three different ILS facilities and no more than one half of the total approaches on any one ILS facility. All approaches shall be flown under simulated instrument conditions to a 100-foot decision height and 90 percent of the total approaches made must be successful. A successful approach is one in which—

(i) At the 100-foot decision height, the indicated airspeed and heading are satisfactory for a normal flare and landing (speed must be plus or minus 5 knots of programmed airspeed, but may not be less than computed threshold speed if autothrottles are used);

(ii) The aircraft at the 100-foot decision height, is positioned so that the cockpit is within, and tracking so as to remain within, the lateral confines of the runway extended;

(iii) Deviation from glide slope after leaving the outer marker does not exceed 50 percent of full-scale deflection as displayed on the ILS indicator;

(iv) No unusual roughness or excessive attitude changes occur after leaving the middle marker; and

(v) In the case of an aircraft equipped with an approach coupler, the aircraft is sufficiently in trim when the approach coupler is disconnected at the decision height to allow for the continuation of a normal approach and landing.

(3) *Records.* During the evaluation program the following information must be maintained by the applicant for the aircraft with respect to each approach and made available to the Administrator upon request:

(i) Each deficiency in airborne instruments and equipment that prevented the initiation of an approach.

(ii) The reasons for discontinuing an approach, including the altitude above the runway at which it was discontinued.

(iii) Speed control at the 100-foot decision height if auto throttles are used.

(iv) Trim condition of the aircraft upon disconnecting the auto coupler with respect to continuation to flare and landing.

(v) Position of the aircraft at the middle marker and at the decision height indicated both on a diagram of the basic ILS display and a diagram of the runway extended to the middle marker. Estimated touchdown point must be indicated on the runway diagram.

(vi) Compatibility of flight director with the auto coupler, if applicable.

(vii) Quality of overall system performance.

(4) *Evaluation.* A final evaluation of the flight control guidance system is made upon successful completion of the demonstrations. If no hazardous tendencies have been displayed or are otherwise known to exist, the system is approved as installed.

4. Maintenance program

(a) Each maintenance program must contain the following:

(1) A list of each instrument and item of equipment specified in § 2 of this appendix that is installed in the aircraft and approved for Category II operations, including the make and model of those specified in § 2(a).

(2) A schedule that provides for the performance of inspections under subparagraph (5) of this paragraph within 3 calendar months after the date of the previous inspection. The inspection must be performed by a person authorized by part 43 of this chapter, except that each alternate inspection may be replaced by a functional flight check. This functional flight check must be performed by a pilot holding a Category II pilot authorization for the type aircraft checked.

(3) A schedule that provides for the performance of bench checks for each listed instrument and item of equipment that is specified in section 2(a) within 12 calendar months after the date of the previous bench check.

(4) A schedule that provides for the performance of a test and inspection of each static pressure system in accordance with Appendix E to Part 43 of this chapter within 12 calendar months after the date of the previous test and inspection.

(5) The procedures for the performance of the periodic inspections and functional flight checks to determine the ability of each listed instrument and item of equipment specified in section 2(a) of this appendix to perform as approved for Category II operations including a procedure for recording functional flight checks.

(6) A procedure for assuring that the pilot is informed of all defects in listed instruments and items of equipment.

(7) A procedure for assuring that the condition of each listed instrument and item of equipment upon which maintenance is performed is at least equal to its Category II approval condition before it is returned to service for Category II operations.

(8) A procedure for an entry in the maintenance records required by § 43.9 of this chapter that shows the date, airport, and reasons for each discontinued Category II operation because of a malfunction of a listed instrument or item of equipment.

(b) *Bench check.* A bench check required by this section must comply with this paragraph.

(1) It must be performed by a certificated repair station holding one of the following ratings as appropriate to the equipment checked:

(i) An instrument rating.

(ii) A radio rating.

(iii) A rating issued under Subpart D of Part 145 of this chapter.

(2) It must consist of removal of an instrument or item of equipment and performance of the following:

(i) A visual inspection for cleanliness, impending failure, and the need for lubrication, repair, or replacement of parts;

(ii) Correction of items found by that visual inspection; and

(iii) Calibration to at least the manufacturer's specifications unless otherwise specified in the approved Category II manual for the aircraft in which the instrument or item of equipment is installed.

(c) *Extensions.* After the completion of one maintenance cycle of 12 calendar months, a request to extend the period for checks, tests, and inspections is approved if it is shown that the performance of particular equipment justifies the requested extension.

Appendix B
Authorizations to exceed Mach 1 (section 91.817)

Section 1. Application

(a) An applicant for an authorization to exceed Mach 1 must apply in a form and manner prescribed by the Administrator and must comply with this appendix.

(b) In addition, each application for an authorization to exceed Mach 1 covered by section 2(a) of this appendix must contain all information requested by the Administrator necessary to assist him in determining whether the designation of a particular test area or issuance of a particular authorization is a "major Federal action significantly affecting the quality of the human environment" within the meaning of the National Environmental Policy Act of 1969 (42 U.S.C. 4321 et seq.), and to assist him in complying with that act and with related Executive Orders, guidelines, and orders prior to such action.

(c) In addition, each application for an authorization to exceed Mach 1 covered by section 2(a) of this appendix must contain—

(1) Information showing that operation at a speed greater than Mach 1 is necessary to accomplish one or more of the purposes specified in section 2(a) of this appendix, including a showing that the purpose of the test cannot be safely or properly accomplished by overocean testing;

(2) A description of the test area proposed by the applicant, including an environmental analysis of that area meeting the requirements of paragraph (b) of this section; and

(3) Conditions and limitations that will ensure that no measurable sonic boom overpressure will reach the surface outside of the designated test area.

(d) An application is denied if the Administrator finds that such action is necessary to protect or enhance the environment.

Section 2. Issuance

(a) For a flight in a designated test area, an authorization to exceed Mach 1 may be issued when the Administrator has taken the environmental protective actions specified in section 1(b) of this appendix and the applicant shows one or more of the following:

(1) The flight is necessary to show compliance with airworthiness requirements.

(2) The flight is necessary to determine the sonic boom characteristics of the airplane or to establish means of reducing or eliminating the effects of sonic boom.

(3) The flight is necessary to demonstrate the conditions and limitations under which speeds greater than a true flight Mach number of 1 will not cause a measurable sonic boom overpressure to reach the surface.

(b) For a flight outside of a designated test area, an authorization to exceed Mach 1 may be issued if the applicant shows conservatively under paragraph (a)(3) of this section that—

(1) The flight will not cause a measurable sonic boom overpressure to reach the surface when the aircraft is operated under conditions and limitations demonstrated under paragraph (a)(3) of this section; and

(2) Those conditions and limitations represent all foreseeable operating conditions.

Section 3. Duration

(a) An authorization to exceed Mach 1 is effective until it expires or is surrendered, or until it is suspended or terminated by the Administrator. Such an authorization may be amended or suspended by the Administrator at any time if the Administrator finds that such action is necessary to protect the environment. Within 30 days of notification of amendment, the holder of the authorization must request reconsideration or the amendment becomes final. Within 30 days of notification of suspension, the holder of the authorization must request reconsideration or the authorization is automatically terminated. If reconsideration is requested within the 30-day period, the amendment or suspension continues until the holder shows why the authorization should not be amended or terminated. Upon such showing, the Administrator may terminate or amend the authorization if the Administrator finds that such action is necessary to protect the environment, or he may reinstate the authorization without amendment if he finds that termination or amendment is not necessary to protect the environment.

(b) Findings and actions by the Administrator under this section do not affect any certificate issued under Title VI of the Federal Aviation Act of 1958.

Appendix C
Operations in the North Atlantic (NAT) Minimum Navigation Performance Specifications (MNPS) airspace

Section 1. NAT MNPS airspace is that volume of airspace between FL 275 and FL 400 extending between latitude 27 degrees north and the North Pole, bounded in the east by the eastern boundaries of control areas Santa Maria Oceanic, Shanwick Oceanic, and Reykjavik Oceanic and in the west by the western boundary of Reykjavik Oceanic Control Area, the western boundary of Gander Oceanic Control Area, and the western boundary of New York Oceanic Control Area, excluding the area west of 60 degrees west and south of 38 degrees 30 minutes north.

Section 2. The navigation performance capability required for aircraft to be operated in the airspace defined in section 1 of this appendix is as follows:

(a) The standard deviation of lateral track errors shall be less than 6.3 NM (11.7 Km). Standard deviation is a statistical measure of data about a mean value. The mean is zero nautical miles. The overall form of data is such that the plus and minus 1 standard deviation about the mean encompasses approximately 68 percent of the data and plus or minus 2 deviations encompasses approximately 95 percent.

(b) The proportion of the total flight time spent by aircraft 30 NM (55.6 Km) or more off the cleared track shall be less than 5.3×10^{-4} (less than 1 hour in 1,887 flight hours).

(c) The proportion of the total flight time spent by aircraft between 50 NM and 70 NM (92.6 Km and 129.6 Km) off the cleared track shall be less than 13×10^{-5} (less than 1 hour in 7,693 flight hours.)

Section 3. Air traffic control (ATC) may authorize an aircraft operator to deviate from the requirements of § 91.705 for a specific flight if, at the time of flight plan filing for that flight, ATC determines that the aircraft may be provided appropriate separation and that the flight will not interfere with, or impose a burden upon, the operations of other aircraft which meet the requirements of § 91.705.

Appendix D
Airports/locations: Special operating restrictions

Section 1. Locations at which the requirements of § 91.215(b)(2) apply.

The requirements of § 91.215(b)(2) apply below 10,000 feet above the surface within 30-nautical-mile radius of each location in the following list:

Atlanta, GA (The William B. Hartsfield Atlanta International Airport)

Baltimore, MD (Baltimore Washington International Airport)

Boston, MA (General Edward Lawrence Logan International Airport)

Chantilly, VA (Washington Dulles International Airport)

Charlotte, NC (Charlotte/Douglas International Airport)

Chicago, IL (Chicago-O'Hare International Airport)

Cleveland, OH (Cleveland-Hopkins International Airport)

Dallas, TX (Dallas/Fort Worth Regional Airport)

Denver, CO (Stapleton International Airport)

Denver, CO ([Denver]* International Airport)

Detroit, MI (Metropolitan Wayne County Airport)

Honolulu, HI (Honolulu International Airport)

Houston, TX (Houston Intercontinental Airport)

Kansas City, KS (Mid-Continent International Airport)

Las Vegas, NV (McCarran International Airport)

Los Angeles, CA (Los Angeles International Airport)

Memphis, TN (Memphis International Airport)

Miami, FL (Miami International Airport)

Minneapolis, MN (Minneapolis-St. Paul International Airport)

Newark, NJ (Newark International Airport)

New Orleans, LA (New Orleans International Airport-Moisant Field)

New York, NY (John F. Kennedy International Airport)

New York, NY (LaGuardia Airport)

Orlando, FL (Orlando International Airport)

Philadelphia, PA (Philadelphia International Airport)

Phoenix, AZ (Phoenix Sky Harbor International Airport)

Pittsburgh, PA (Greater Pittsburgh International Airport)

St. Louis, MO (Lambert-St. Louis International Airport)

Salt Lake City, UT (Salt Lake City International Airport)

San Diego, CA (San Diego International Airport)

San Francisco, CA (San Francisco International Airport)

Seattle, WA (Seattle-Tacoma International Airport)

Tampa, FL (Tampa International Airport)

Washington, DC (Washington National Airport)

Section 2. Airports at which the requirements of § 91.215(b)(5)(ii) apply.

The requirements of § 91.215(b)(5)(ii) apply to operations in the vicinity of each of the following airports:

Billings, MT (Logan International Airport)

Section 3. Locations at which fixed-wing Special VFR operations are prohibited.

The Special VFR weather minimums of § 91.157 do not apply to the following airports:

Atlanta, GA (The William B. Hartsfield Atlanta International Airport)

Baltimore, MD (Baltimore/Washington International Airport)

Boston, MA (General Edward Lawrence Logan International Airport)

Buffalo, NY (Greater Buffalo International Airport)

Chicago, IL (Chicago-O'Hare International Airport)

Cleveland, OH (Cleveland-Hopkins International Airport)

Columbus, OH (Port Columbus International Airport)

Covington, KY (Greater Cincinnati International Airport)

Dallas, TX (Dallas/Fort Worth Regional Airport)

Dallas, TX (Love Field)

Denver, CO (Stapleton International Airport)

Denver, CO ([Denver]* International Airport)

Detroit, MI (Metropolitan Wayne County Airport)

Honolulu, HI (Honolulu International Airport)

Houston, TX (Houston Intercontinental Airport)

Indianapolis, IN (Indianapolis International Airport)

Los Angeles, CA (Los Angeles International Airport)

Louisville, KY (Standiford Field)

Memphis, TN (Memphis International Airport)

Miami, FL (Miami International Airport)

Minneapolis, MN (Minneapolis-St. Paul International Airport)

Newark, NJ (Newark International Airport)

New York, NY (John F. Kennedy International Airport)

New York, NY (LaGuardia Airport)

New Orleans, LA (New Orleans International Airport-Moisant Field)

Philadelphia, PA (Philadelphia International Airport)

Pittsburgh, PA (Greater Pittsburgh International Airport)

Portland, OR (Portland International Airport)

San Francisco, CA (San Francisco International Airport)

Seattle, WA (Seattle-Tacoma International Airport)

St. Louis, MO (Lambert-St. Louis International Airport)

Tampa, FL (Tampa International Airport)

Washington, DC (Washington National Airport)

Section 4. Locations at which solo student pilot activity is not permitted.

Pursuant to § 91.131(b)(2), solo student pilot operations are not permitted at any of the following airports:

Atlanta, GA (The William B. Hartsfield Atlanta International Airport)

Boston, MA (General Edward Lawrence Logan International Airport)

Chicago, IL (Chicago-O'Hare International Airport)

Dallas, TX (Dallas/Fort Worth Regional Airport)

Los Angeles, CA (Los Angeles International Airport)
Miami, FL (Miami International Airport)
Newark, NJ (Newark International Airport)
New York, NY (John F. Kennedy International Airport)
New York, NY (LaGuardia Airport)
San Francisco, CA (San Francisco International Airport)
Washington, DC (Washington National Airport)
Andrews Air Force Base, MD)]

(Amdt. 91-217, Eff. 7/23/90); (Amdt. 91-227, Eff. 9/16/93); (Amdt. 91-235, Eff. 10/5/93); [Amdt. 91-236 & 91-237, Eff. 3/9/94 and Amdt. 91-238, Eff. 5/15/94)]; [*(***Amdt. 91-241 delays effective date of name change indefinitely.***)]

Appendix E to Part 91—Airplane flight recorder specifications

Parameters	Range	Installed system[1] minimum accuracy (to recovered data)	Sampling interval (per second)	Resolution[4] read out
Relative Time (from recorded on prior to takeoff).	8 hr minimum	±0.125% per hour	1	1 sec.
Indicated Airspeed.	Vso to VD (KIAS)	±5% or ±10 kts., whichever is greater. Resolution 2 kts. below 175 KIAS	1	1%[3]
Altitude.	–1,000 ft. to max cert. alt. of A/C	±100 to ±700 ft. (see Table 1, TSO C51-a)	1	25 to 150 ft.
Magnetic Heading.	360°	±5°	1	1½
Vertical Acceleration.	–3g to +6g	±0.2g in addition to ±0.3g maximum datum	4 (or 1 per second where peaks, ref. to 1g are recorded)	0.03g
Longitudinal Acceleration.	±1.0g	±1.5% max. range excluding datum error of ±5%	2	0.01g
Pitch Attitude.	100% of usable	±2°	1	0.8°
Roll Attitude.	±60° or 100% of usable range, whichever is greater	±2°	1	0.8°
Stabilizer Trim Position, OR.	Full Range	±3% unless higher uniquely required	1	1%[3]
Pitch Control Position.				
Engine Power, Each Engine:	Full Range	±3% unless higher uniquely required	1	1%[3]
Fan or N1 Speed or EPR or Cockpit Indications. Used for Aircraft Certification OR.	Maximum Range	±5%	1	1%[3]
Prop. Speed and Torque (Sample once/sec as close together as practicable).			1 (prop Speed) 1 (torque)	1%[3] 1%[3]

Parameters	Range	Installed system[1] minimum accuracy (to recovered data)	Sampling interval (per second)	Resolution[4] read out
Altitude Rate[2] (need depends on altitude resolution).	±8,000 fpm	±10%. Resolution 250 fpm below 12,000 ft. indicated	1	250 fpm below 12,000
Angle of Attack[2] (need depends on altitude resolution).	–20° to 40° or 100% of usable range	±2°	1	0.8%[3]
Radio Transmitter Keying (Discrete).	On/Off		1	
TE Flaps (Discrete or Analog).	Each discrete position (U, D, T/O, AAP) OR		1	
LE Flaps (Discrete or Analog).	Analog 0–100% range	±3°	1	1%[3]
	Each discrete position (U, D, T/O, AAP) OR		1	
Thrust Reverser, Each Engine (Discrete).	Analog 0–100% range Stowed or full reverse	±3°	1	1%[3]
Spoiler/Speedbrake (Discrete).	Stowed or out		1	
Autopilot Engaged (Discrete).	Engaged or Disengaged		1	

[1] When data sources are aircraft instruments (except altimeters) of acceptable quality to fly the aircraft the recording system excluding these sensors (but including all other characteristics of the recording system) shall contribute no more than half of the values in this column.
[2] If data from the altitude encoding altimeter (100 ft. resolution) is used, then either one of these parameters should also be recorded. If however, altitude is recorded at a minimum resolution of 25 feet, then these two parameters can be omitted.
[3] Percent of full range.
[4] This column applies to aircraft manufactured after October 11, 1991.

Appendix F to Part 91—Helicopter flight recorder specifications

Parameters	Range	Installed system[1] minimum accuracy (to recovered data)	Sampling interval (per second)	Resolution[3] read out
Relative Time (from recorded on prior to takeoff).	4 hr minimum	±0.125% per hour	1	1 sec.
Indicated Airspeed.	VM in to VD (KIAS) (minimum airspeed signal attainable with installed pilot-static system)	±5% or ±10 kts., whichever is greater	1	1 kt.
Altitude.	–1,000 ft. to 20,000 ft. pressure altitude	±100 to ±700 ft. (see Table 1, TSO C51-a)	1	25 to 150 ft.

Parameters	Range	Installed system[1] minimum accuracy (to recovered data)	Sampling interval (per second)	Resolution[3] read out
Magnetic Heading.	360°	±5°	1	1°
Vertical Acceleration.	–3g to +6g	±0.2g in addition to ±0.3g maximum datum	4 (or 1 per second where peaks, ref. to 1g are recorded)	0.05g
Longitudinal Acceleration.	±1.0g	±1.5% max. range excluding datum error of ±5%	2	0.03g
Pitch Attitude.	100% of usable range	±2°	1	0.8°
Roll Attitude.	±60 or 100% of usable range, whichever is greater	±2°	1	0.8°
Altitude Rate.	±8,000 fpm	±10%. Resolution 250 fpm below 12,000 ft. indicated	1	250 fpm below 12,000
Engine Power, Each Engine				
Main Rotor Speed.	Maximum Range	±5%	1	1%[2]
Free or Power Turbine.	Maximum Range	±5%	1	1%[2]
Engine Torque.	Maximum Range	±5%	1	1%[2]
Flight Control Hydraulic Pressure				
Primary (Discrete).	High/Low		1	
Secondary—if applicable (Discrete).	High/Low		1	
Radio Transmitter Keying (Discrete).	On/Off		1	
Autopilot Engaged (Discrete).	Engaged or Disengaged		1	
SAS Status-Engaged (Discrete).	Engaged or Disengaged		1	
SAS Fault Status (Discrete).	Fault/OK		1	
Flight Controls				
Collective.	Full range	±3%	2	1%[2]
Pedal Position.	Full range	±3%	2	1%[2]
Lat. Cyclic.	Full range	±3%	2	1%[2]

Parameters	Range	Installed system[1] minimum accuracy (to recovered data)	Sampling interval (per second)	Resolution[3] read out
Long. Cyclic.	Full range	±3%	2	1%[2]
Controllable Stabilator Position.	Full range	±3%	2	1%[2]

[1] When data sources are aircraft instruments (except altimeters) of acceptable quality to fly the aircraft the recording system excluding these sensors (but including all other characteristics of the recording system) shall contribute no more than half of the values in this column.
[2] Percent of full range.
[3] This column applies to aircraft manufactured after October 11, 1991.

Special Federal Aviation Regulations

Special Federal Aviation Regulation No. 50-2. Special flight rules in the vicinity of the Grand Canyon National Park, AZ.

Section 1. *Applicability.* This rule prescribes special operating rules for all persons operating aircraft in the following airspace, designated as the Grand Canyon National Park Special Flight Rules Area:

That airspace extending upward from the surface up to but not including 14,500 feet MSL within an area bounded by a line beginning at Lat. 36°09'30" N., Long. 114°03'00" W.; northeast to Lat. 36°14'00" N., Long. 113°09'50" W.; thence northeast along the boundary of the Grand Canyon National Park to 36°22'55" N., Long. 112°52'00" W.; to Lat. 36°30'30" N., Long. 112°36'15" W. to Lat. 36°21'30" N., Long. 112°00'00" W. to Lat. 36°35'30" N., Long. 111°53'10" W. to Lat. 36°53'00" N., Long. 111°36'45" W. to Lat. 36°53'00" N., Long. 111°33'00" W.; to Lat. 36°19'00"N., Long. 111°50'50" W.; to Lat. 36°17'00" N., Long. 111°42'00" W.; to Lat. 35°59'30"N., Long. 111°42'00" W.; to Lat. 35°57'30" N., Long. 112°03'55" W.; thence counterclockwise via the 5 statute mile radius of the Grand Canyon Airport airport reference point (Lat. 35°57'09" N., Long. 112°08'47" W.) to Lat. 35°57'30" N., Long. 112°14'00" W.; to Lat. 35°57'30" N., Long. 113°11'00" W.; to Lat. 35°42'30" N., Long. 113°11'00" W.; to 35°38'30" N; Long. 113°27'30" W; thence counterclockwise via the 5 statute mile radius of the Peach Springs VORTAC to Lat. 35°41'20" N., Long. 113°36'00" W; to Lat. 35°55'25" N., Long. 113°49'10" W.; to Lat. 35°57'45" N., 113°45'20" W.; thence northwest along the park boundary to Lat. 36°02'20" N., Long. 113°50'15" W; to 36°00'10" N., Long. 113°53'45" W.; thence to the point of beginning.

Section 2. *Definitions.* For the purposes of this special regulation:

"Flight Standards District Office" means the FAA Flight Standards District Office with jurisdiction for the geographical area containing the Grand Canyon.

"Park" means the Grand Canyon National Park.

"Special Flight Rules Area" means the Grand Canyon National Park Special Flight Rules Area.

Section 3. *Aircraft operations: general.* Except in an emergency, no person may operate an aircraft in the Special Flight Rules Area under VFR on or after September 22, 1988, or under IFR on or after April 6, 1989, unless the operation—

(a) Is conducted in accordance with the following procedures:

NOTE: THE FOLLOWING PROCEDURES DO NOT RELIEVE THE PILOT FROM SEE-AND-AVOID RESPONSIBILITY OR COMPLIANCE WITH FAR [91.119].

(1) Unless necessary to maintain a safe distance from other aircraft or terrain—

(i) remain clear of the areas described in Section 4; and

(ii) remain at or above the following altitudes in each sector of the canyon:

Eastern section from Lees Ferry to North Canyon and North Canyon to Boundary Ridge: as prescribed in Section 5.

Boundary Ridge to Supai Point (Yumtheska Point): 10,000 feet MSL.

Supai Point to Diamond Creek: 9,000 feet MSL.

Western section from Diamond Creek to the Grand Wash Cliffs: 8,000 feet MSL.

(2) Proceed through the four flight corridors described in Section 4 at the following altitudes unless otherwise authorized in writing by the Flight Standards District Office:

Northbound: 11,500 or 13,500 feet MSL

Southbound: 10,500 or 12,500 feet MSL

(b) Is authorized in writing by the Flight Standard District Office and is conducted in compliance with the conditions contained in that authorization. Normally authorization will be granted for operation in the area described in Section 4 or below the altitudes listed in Section 5 only for operations of aircraft necessary for law enforcement, firefighting, emergency medical treatment/evacuation of persons in the vicinity of the Park; for support of Park maintenance or activities; or for aerial access to and maintenance of other property located within the Special Flight Rules Area. Authorization may be issued on a continuing basis.

(c)(1) Prior to November 1, 1988, is conducted in accordance with a specific authorization to operate in that airspace incorporated in the operator's Part 135 operations specifications in accordance with the provisions of SFAR 50-1, notwithstanding the provisions of Sections 4 and 5; and

(2) On or after November 1, 1988, is conducted in accordance with a specific authorization to operate in that airspace incorporated in the operator's Part 135 operations specifications and approved by the Flight Standards District Office in accordance with the provisions of SFAR 50-2.

(d) Is a search and rescue mission directed by the U.S. Air Force Rescue Coordination Center.

(e) Is conducted within 3 nautical miles of Whitmore Airstrip, Pearce Ferry Airstrip, North Rim Airstrip, Cliff Dwellers Airstrip, or Marble Canyon Airstrip at an altitude less than 3,000 feet above airport elevation, for the purpose of landing at or taking off from that facility.

(f) Is conducted under an IFR clearance and the pilot is acting in accordance with ATC instructions. An IFR flight plan may not be filed on a route or at an altitude that would require operation in an area described in Section 4.

Section 4. *Flight-Free zones.* Except in an emergency or if otherwise necessary for safety of flight, or unless otherwise authorized by the Flight Standards District Office for a purpose listed in Section 3(5), no person may operate an aircraft in the Special Flight Rules Area within the following areas:

(a) *Desert View Flight-Free Zone.* Within an area bounded by a line beginning at Lat. 35°59'30" N., Long. 111°46'20" W.; to 35°59'30" N., Long 111°52'45" W.; to Lat. 36°04'50" N., Long 111°52'00" W.; to Lat. 36°06'00" N., Long. 111°46'20" W.; to the point of origin; but not including the airspace at and above 10,500 feet MSL within 1 mile of the western boundary of the zone. The area between the Desert View and Bright Angel Flight-Free Zones is designated the "Zuni Point Corridor."

(b) Bright Angel Flight-Free Zone. [Within an area bounded by a line beginning at Lat. 35°59'30" N., Long 111°55'30" W.; to Lat. 35°59'30" N., Long 112°04'00" W.; thence counterclockwise via the 5-statute mile radius of the Grand Canyon Airport point (Lat. 35°57'09" N., Long 112°08'47" W.) to Lat. 36°01'30" N., Long. 112°11'00" W.; to Lat. 36°06'15" N., Long. 112°12'50" W.; to Lat. 36°14'40" N., Long. 112°08'50" W.; to Lat. 36°14'40" N., Long. 111°57'30" W.; to Lat. 36°12'30" N., Long. 111°53'50" W.; to the point of origin; but not including the airspace at and above 10,500 feet MSL within 1 mile of the eastern boundary between the southern boundary and Lat. 36°04'50" N. or the airspace at and above 10,500 feet MSL within 2 miles of the northwest boundary. The area bounded by the Bright Angel and Shinumo Flight-Free Zones is designated the "Dragon Corridor."]

(c) Shinumo Flight-Free Zone. [Within an area bounded by a line beginning at Lat. 36°04'00" N., Long. 112°16'40" W.; northwest along the park boundary to a point at Lat. 36°11'45" N., Long. 112°32'15" W.; to Lat. 36°21'15" N., Long. 112°20'20" W.; east along the park boundary to Lat. 36°21'15" N., Long. 112°13'55" W.; to Lat. 36°14'40" N., Long. 112°11'25" W.; to the point of origin. The area between the Thunder River/Toroweap and Shinumo Flight Free Zones is designated the "Fossil Canyon Corridor."]

(d) Toroweap/Thunder River Flight-Free Zone. Within an area bounded by line beginning at Lat. 36°22'45" N., Long. 112°20'35" W.; thence northeast along the boundary of the Grand Canyon National Park to Lat. 36°15'00" N., Long. 113°03'15" W.; to Lat. 36°15'00" N., Long. 113°07'10" W.; to Lat. 36°10'30" N., Long 113°07'10" W.; thence east along the Colorado River to the confluence of Havasu Canyon (Lat. 36°18'40" N., Long. 112°45'45" W.;) including that area within a 1.5-nautical-mile radius of Toroweap Overlook (Lat. 36°12'45" N., Long. 113°03'30" W.); to the point of origin; but not including the following airspace designated as the "Tuckup Corridor": at or above 10,500 feet MSL within 2 nautical miles either side of a line extending between Lat. 36°22'55" N., Long. 112°48'50" W. and Lat. 36°17'10" N. Long. 112°48'50" W.; to the point of origin.

Section 5. *Minimum flight altitudes.* Except in an emergency or if otherwise necessary for safety of flight, or unless otherwise authorized by the Flight Standards District Office for a purpose listed in Section 3(b), no person may operate an aircraft in the Special Flight Rules Area at an altitude lower than the following:

(a) Eastern section from Lees Ferry to North Canyon: 5,000 feet MSL.

(b) Eastern section from North Canyon to Boundary Ridge: 6,000 feet MSL.

(c) Boundary Ridge to Supai (Yumtheska) Point: 7,500 feet MSL.

(d) Supai Point to Diamond Creek: 6,500 feet MSL.

(e) Western section from Diamond Creek to the Grand Wash Cliffs: 5,000 feet MSL.

Section 6. *Commercial sightseeing flights.*

(a) Notwithstanding the provisions of Federal Aviation Regulations § 135.1(5)(2), nonstop sightseeing flights that begin and end at the same airport, are conducted within a 25-statute-mile radius of that airport, and operate in or through the Special Flight Rules Area during any portion of the flight are governed by the provisions of Part 135.

(b) No person holding or required to hold an operating certificate under Part 135 may operate an aircraft in the Special Flight Rules Area except as authorized by operations specifications issued under that part.

Section 7. *Minimum terrain clearance.* Except in an emergency, when necessary for takeoff or landing, or unless authorized by the Flight Standards District Office for a purpose listed in Section 3(b), no person may operate an

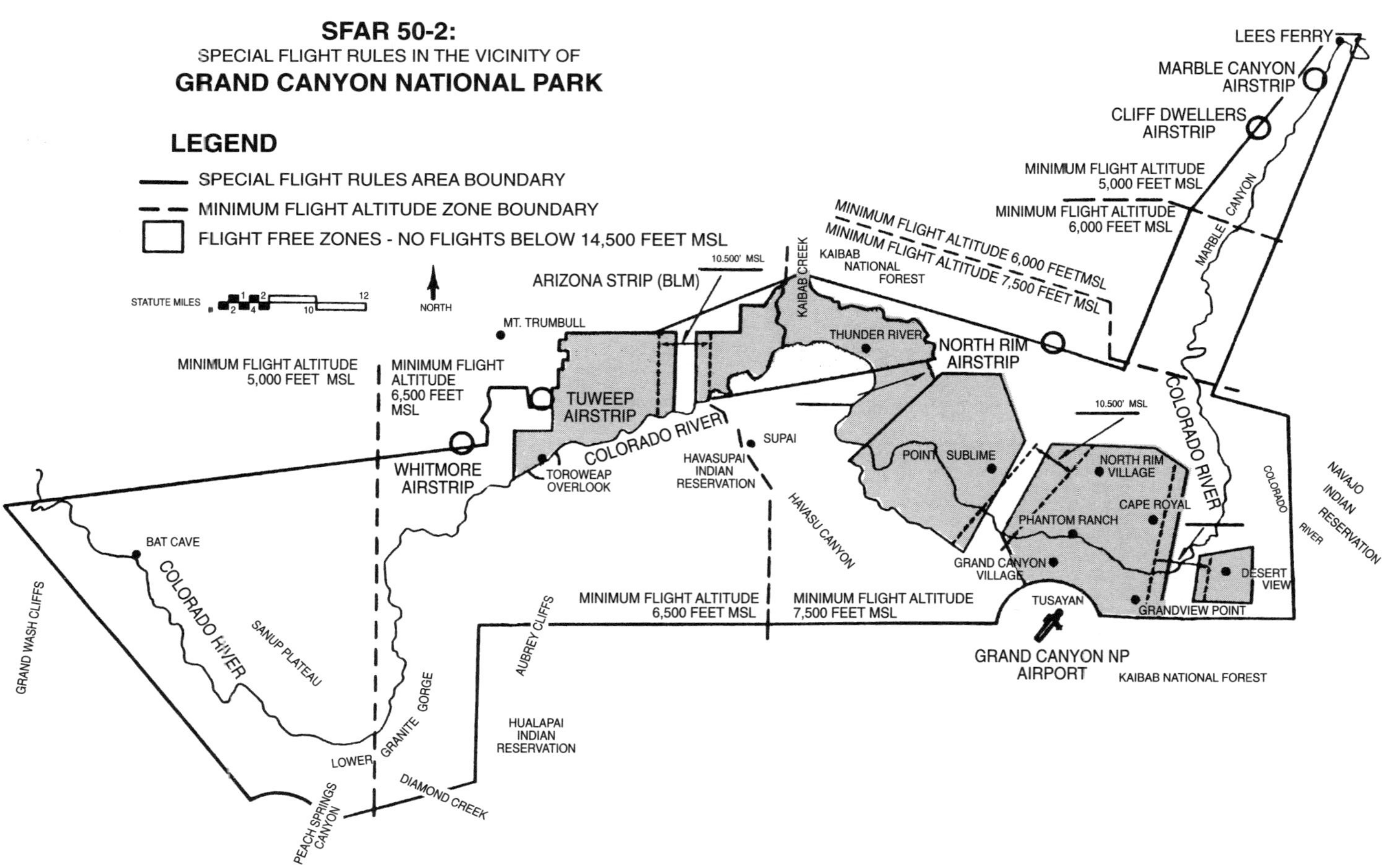
SFAR 50-2:
SPECIAL FLIGHT RULES IN THE VICINITY OF
GRAND CANYON NATIONAL PARK
LEGEND
SPECIAL FLIGHT RULES AREA BOUNDARY
MINIMUM FLIGHT ALTITUDE ZONE BOUNDARY
FLIGHT FREE ZONES - NO FLIGHTS BELOW 14,500 FEET MSL
STATUTE MILES
0 1 2 4 10 12
NORTH
LEES FERRY
MARBLE CANYON AIRSTRIP
CLIFF DWELLERS AIRSTRIP
MINIMUM FLIGHT ALTITUDE 5,000 FEET MSL
MINIMUM FLIGHT ALTITUDE 6,000 FEET MSL
MARBLE CANYON
MINIMUM FLIGHT ALTITUDE 6,000 FEETMSL
MINIMUM FLIGHT ALTITUDE 7,500 FEET MSL
KAIBAB NATIONAL FOREST
KAIBAB CREEK
10,500' MSL
ARIZONA STRIP (BLM)
MT. TRUMBULL
THUNDER RIVER
NORTH RIM AIRSTRIP
MINIMUM FLIGHT ALTITUDE 5,000 FEET MSL
MINIMUM FLIGHT ALTITUDE 6,500 FEET MSL
TUWEEP AIRSTRIP
COLORADO RIVER
SUPAI
WHITMORE AIRSTRIP
TOROWEAP OVERLOOK
HAVASUPAI INDIAN RESERVATION
HAVASU CANYON
POINT SUBLIME
10,500' MSL
NORTH RIM VILLAGE
CAPE ROYAL
PHANTOM RANCH
GRAND CANYON VILLAGE
COLORADO RIVER
COLORADO RIVER
NAVAJO INDIAN RESERVATION
DESERT VIEW
TUSAYAN
GRANDVIEW POINT
BAT CAVE
COLORADO RIVER
GRAND WASH CLIFFS
SANUP PLATEAU
MINIMUM FLIGHT ALTITUDE 6,500 FEET MSL
MINIMUM FLIGHT ALTITUDE 7,500 FEET MSL
AUBREY CLIFFS
GRAND CANYON NP AIRPORT
KAIBAB NATIONAL FOREST
LOWER GRANITE GORGE
HUALAPAI INDIAN RESERVATION
DIAMOND CREEK
PEACH SPRINGS CANYON

aircraft within 500 feet of any terrain or structure located between the north and south rims of the Grand Canyon.

Section 8. Communications. Except when in contact with the Grand Canyon National Park Airport Traffic Control Tower during arrival or departure or on a search and rescue mission directed by the U.S. Air Force Rescue Coordination Center, no person may operate an aircraft in the Special Flight Rules Area unless he monitors the appropriate frequency continuously while in that airspace.

Section 9. *Termination date.* This Special Federal Aviation Regulation expires on June 15, [1995].

Authority. 49 U.S.C. 1303, 1348, 1354(a), 1421, and 1422; 16 U.S.C. 228g; P.L. 100-91, August 18, 1987; 49 U.S.C. 106(g) (Revised Pub. L. 97-449, January 12, 1983).]

Special Federal Aviation Regulation No. 51-1. Special flight rules in the vicinity of Los Angeles International Airport.

Section 1. Applicability. [This rule establishes a special operating area for persons operating aircraft under visual flight rules (VFR) in the following airspace of the Los Angeles Class B airspace area designated as the Los Angeles Special Flight Rules Area:]

That part of Area A of the Los Angeles TCA between 3,500 feet above mean sea level (MSL) and 4,500 feet MSL, inclusive, bounded on the north by Ballona Creek, on the east by the San Diego Freeway; on the south by Imperial Highway and on the west by the Pacific Ocean shoreline.

Section 2. *Aircraft operations, general.* Unless otherwise authorized by the Administrator, no person may operate an aircraft in the airspace described in Section 1 unless the operation is conducted under the following rules:

(a) [The flight must be conducted under VFR and only when operation may be conducted in compliance with § 91.155(a)].

(b) [The aircraft must be equipped as specified in § 91.215(b) replying on Code 1201 prior to entering and while operating in this area.]

Intentionally
left
blank.

(c) The pilot shall have a current Los Angeles Terminal Area Chart in the aircraft.

(d) The pilot shall operate on the Santa Monica very high frequency omni-directional radio range (VOR) 132 degree radial.

(e) Operations in a southeasterly direction shall be in level flight at 3,500 feet MSL.

(f) Operations in a northwesterly direction shall be in level flight at 4,500 feet MSL.

(g) Indicated airspeed shall not exceed 140 knots.

(h) Anticollision lights and aircraft position/navigation lights shall be on. Use of landing lights is recommended.

(i) Turbojet aircraft are prohibited from VFR operations in this area.

Section 3. [Notwithstanding the provisions of § 91.131 (a), an air traffic control authorization is not required in the Los Angeles Special Flight Rules Area for operations in compliance with Section 2 of this SFAR. All other provisions of § 91.131 apply to operate in the Special Flight Rules Area.]

Authority: 49 U.S.C. app. 1303, 1348, 1354(a), 1421, and 1422; 49 U.S.C. 106(g).

Special Federal Aviation Regulation No. 60. Air Traffic Control System emergency operation.

1. Each person shall, before conducting any operation under the Federal Aviation Regulations (14 CFR Chapter 1), be familiar with all available information concerning that operation, including Notices to Airmen issued under § 91.139 and, when activated, the provisions of the National Air Traffic Reduced Complement Operations Plan available for inspection at operating air traffic facilities and Regional air traffic division offices, and the General Aviation Reservation Program. No operator may change the designated airport of intended operation for any flight contained in the October 1, 1990, OAG.

2. Notwithstanding any provision of the Federal Aviation Regulations to the contrary, no person may operate an aircraft in the Air Traffic Control System:

a. Contrary to any restriction, prohibition, procedure or other action taken by the Director of the Office of Air Traffic Systems Management (Director) pursuant to Paragraph 3 of this regulation and announced in a Notice to Airmen pursuant to § 91.139 of the Federal Aviation Regulations.

b. When the National Air Traffic Reduced Complement Operations Plan is activated pursuant to Paragraph 4 of this regulation, except in accordance with the pertinent provisions of the National Air Traffic Reduced Complement Operations Plan.

3. Prior to or in connection with the implementation of the RCOP, and as conditions warrant, the Director is authorized to:

a. Restrict, prohibit, or permit VFR and/or IFR operations at any airport, Class B airspace area, Class C airspace area, or other class of controlled airspace.

b. Give priority at any airport to flights that are of military necessity, or are medical emergency flights, Presidential flights, and flights transporting critical Government employees.

c. Implement, at any airport, traffic management procedures, that may include reduction of flight operations. Reduction of flight operations will be accomplished, to the extent practical, on a pro rata basis among and between air carrier, commercial operator, and general aviation operations. Flights cancelled under this SFAR at a high density traffic airport will be considered to have been operated for purposes of Part 93 of the Federal Aviation Regulations.

4. The Director may activate the National Air Traffic Reduced Complement Operations Plan at any time he finds that it is necessary for the safety and efficiency of the National Airspace System. Upon activation of the RCOP and notwithstanding any provision of the FAR to the contrary, the Director is authorized to suspend or modify any airspace designation.

5. Notice of restrictions, prohibitions, procedures and other actions taken by the Director under this regulation with respect to the operation of the Air Traffic Control system will be announced in Notices to Airmen issued pursuant to § 91.139 of the Federal Aviation Regulations.

6. The Director may delegate his authority under this regulation to the extent he considers necessary for the safe and efficient operation of the National Air Traffic Control System.

APPENDIX 1
Key Airports

ATL Atlanta Hartsfield International
BNA Nashville International
BOS Boston Logan International
CLT Charlotte-Douglas International
CVG Greater Cincinnati International
DAL Dallas-Love
DAY Cox-Dayton International
DCA Washington National
DEN Stapleton International
DFW Dallas-Fort Worth International
DTW Detroit Metropolitan Wayne County
EWR Newark International
FLL Ft. Lauderdale-Hollywood International
HOU William B. Hobby
IAD Washington-Dulles International
IAH Houston Intercontinental
IND Indianapolis International
JFK John F. Kennedy International
LAX Los Angeles International
LGA La Guardia
MCO Orlando International
MDW Chicago Midway
MEM Memphis International
MIA Miami International
ORD Chicago-O'Hare International
PPI Palm Beach International
PHL Philadelphia International
PHX Phoenix Sky Harbor International
PIT Greater Pittsburgh International
RDU Raleigh-Durham International
SAN San Diego Lindbergh International
SEA Seattle-Tacoma International
SFO San Francisco International
SJC San Jose International
SLC Salt Lake City International
STL Lambert-St. Louis International

Special Federal Aviation Regulation No. 62.
Suspension of certain aircraft operations from the transponder with automatic pressure altitude reporting capability requirement.

Section 1. For purposes of this SFAR:

(a) The airspace within 30 nautical miles of Class B airspace area primary airport, from the surface upward to 10,000 feet MSL, excluding the airspace designated as a Class B airspace area is referred to as the Mode C veil.

(b) Effective until [three years after the date of the publication of the final rule], *the transponder with automatic altitude reporting capability requirements of § 91.215(b)(2) do not apply to the operation of an aircraft:

(1) in the airspace at or below the specified altitude and within a 2-nautical-mile radius, or, if directed by ATC, within a 5-nautical mile radius, of an airport listed in Section 2 of this SFAR; and

(2) in the airspace at or below the specified altitude along the most direct and expeditious routing, or on a routing directed by ATC, between an airport listed in Section 2 of this SFAR and the outer boundary of the Mode C veil airspace overlying that airport, consistent with established traffic patterns, noise abatement procedures, and safety.

Section 2. Effective until [three years after the date of the publication of the final rule], *airports at which the provisions of § 91.215(b)(2) do not apply.

*(Amendments were pending at deadline.)

(1) Airports within a 30-nautical-mile radius of The William B. Hartsfield Atlanta International Airport.

Airport Name	Arpt ID	Alt. (AGL)
Air Acres Airport, Woodstock, GA	5GA4	1,500
B & L Strip Airport, Hollonville, GA	GA29	1,500
Camfield Airport, McDonough, GA	GA36	1,500
Cobb County-McCollum Field Airport, Marietta, GA	RYY	1,500

Airport Name	Arpt ID	Alt. (AGL)
Covington Municipal Airport, Covington, GA	9A1	1,500
Diamond R Ranch Airport, Villa Rica, GA	3GA5	1,500
Dresden Airport, Newnan, GA	GA79	1,500
Eagles Landing Airport Williamson, GA	5GA3	1,500
Fagundes Field Airport, Haralson, GA	6GA1	1,500
Gable Branch Airport, Haralson, GA	5GA0	1,500
Georgia Lite Flite Ultralight Airport, Acworth, GA	31GA	1,500
Griffin-Spalding County Airport, Griffin, GA	6A2	1,500
Howard Private Airport, Jackson, GA	GA02	1,500
Newnan Coweta County Airport, Newnan, GA	CCO	1,500
Peach State Airport, Williamson, GA	3GA7	1,500
Poole Farm Airport, Oxford, GA	2GA1	1,500
Powers Airport, Hollonville, GA	GA31	1,500
S & S Landing Strip Airport, Griffin, GA	8GA6	1,500
Shade Tree Airport, Hollonville, GA	GA73	1,500

(2) Airports within a 30-nautical-mile radius of the General Edward Lawrence Logan International Airport.

Airport Name	Arpt ID	Alt. (AGL)
Berlin Landing Area Airport, Berlin, MA	MA19	2,500
Hopedale Industrial Park Airport, Hopedale, MA	IB6	2,500
Larson's SPB, Tyngsboro, MA	MA74	2,500
Moore AAF, Ayer/Fort Devens, MA	AYE	2,500
New England Gliderport, Salem, NH	NH29	2,500
Plum Island Airport, Newburyport, MA	2B2	2,500
Plymouth Municipal Airport, Plymouth, MA	PYM	2,500
Taunton Municipal Airport, Taunton, MA	TAN	2,500
Unknown Field Airport, Southborough, MA	1MA5	2,500

(3) Airports within a 30-nautical-mile radius of the Charlotte/Douglas International Airport.

Airport Name	Arpt ID	Alt. (AGL)
Arant Airport, Wingate, NC	1NC6	2,500
Bradley Outernational Airport, China Grove, NC	NC29	2,500
Chester Municipal Airport, Chester, SC	9A6	2,500
China Grove Airport, China Grove, NC	76A	2,500
Goodnight's Airport, Kannapolis, NC	2NC8	2,500
Knapp Airport, Marshville, NC	3NC4	2,500
Lake Norman Airport, Mooresville, NC	14A	2,500
Lancaster County Airport, Lancaster, SC	LKR	2,500
Little Mountain Airport, Denver, NC	66A	2,500
Long Island Airport, Long Island, NC	NC26	2,500
Miller Airport, Mooresville, NC	8A2	2,500
U S Heliport, Wingate, NC	NC56	2,500
Unity Aerodrome Airport, Lancaster, SC	SC76	2,500
Wilhelm Airport, Kannapolis, NC	6NC2	2,500

(4) Airports within a 30-nautical-mile radius of the Chicago O'Hare International Airport.

Airport Name	Arpt ID	Alt. (AGL)
Aurora Municipal Airport, Chicago/Aurora, IL	ARR	1,200
Donald Alfred Gade Airport, Antioch, IL	IL11	1,200
Dr. Joseph W. Esser Airport, Hampshire, IL	7IL6	1,200
Flying M. Farm Airport, Aurora, IL	IL20	1,200
Fox Lake SPB, Fox Lake, IL	IS03	1,200
Graham SPB, Crystal Lake, IL	IS79	1,200
Herbert C. Mass Airport, Zion, IL	IL02	1,200
Landings Condominium Airport, Romeoville, IL	C49	1,200
Lewis University Airport, Romeoville, IL	LOT	1,200

Airport Name	Arpt ID	Alt. (AGL)
McHenry Farms Airport, McHenry, IL	44IL	1,200
Olson Airport, Plato Center, IL	LL53	1,200
Redeker Airport, Milford, IL	IL85	1,200
Reid RLA Airport, Gilberts, IL	6IL6	1,200
Shamrock Beef Cattle Farm Airport, McHenry, IL	49LL	1,200
Sky Soaring Airport, Union, IL	55LL	1,200
Waukegan Regional Airport, Waukegan, IL	UGN	1,200
Wormley Airport, Oswego, IL	85LL	1,200

(5) Airports within a 30-nautical-mile radius of the Cleveland-Hopkins International Airport.

Airport Name	Arpt ID	Alt. (AGL)
Akron Fulton International Airport, Akron, OH	AKR	1,300
Bucks Airport, Newbury, OH	400H	1,300
Derecsky Airport, Auburn Center, OH	60I0	1,300
Hannum Airport, Streetsboro, OH	690H	1,300
Kent State University Airport, Kent, OH	1G3	1,300
Lost Nation Airport, Willoughby, OH	LNN	1,300
Mills Airport, Mantua, OH	OH06	1,300
Portage County Airport, Ravenna, OH	29G	1,300
Stoney's Airport, Ravenna, OH	0I32	1,300
Wadsworth Municipal Airport, Wadsworth, OH	3G3	1,300

(6) Airports within a 30-nautical-mile radius of the Dallas/Fort Worth International Airport.

Airport Name	Arpt ID	Alt. (AGL)
Beggs Ranch/Aledo Airport, Aledo, TX	TX15	1,800
Belcher Airport, Sanger, TX	TA25	1,800
Bird Dog Field Airport, Krum, TX	TA48	1,800
Boe-Wrinkle Airport, Azle, TX	28TS	1,800
Flying V Airport, Sanger, TX	71XS	1,800
Graham Ranch Airport, Celina, TX	TX44	1,800
Haire Airport, Bolivar, TX	TX33	1,800
Hartlee Field Airport, Denton, TX	1F3	1,800
Hawkin's Ranch Strip Airport, Rhome, TX	TA02	1,800
Horseshoe Lake Airport, Sanger, TX	TE24	1,800
Ironhead Airport, Sanger, TX	T58	1,800
Kezer Air Ranch Airport, Springtown, TX	61F	1,800
Lane Field Airport, Sanger, TX	58F	1,800
Log Cabin Airport, Aledo, TX	TX16	1,800
Lone Star Airpark Airport, Denton, TX	T32	1,800
Rhome Meadows Airport, Rhome, TX	TS72	1,800
Richards Airport, Krum, TX	TA47	1,800
Tallows Field Airport, Celina, TX	79TS	1,800
Triple S Airport, Aledo, TX	42XS	1,800
Warshun Ranch Airport, Denton, TX	4TA1	1,800
Windy Hill Airport, Denton, TX	46XS	1,800
Aero Country Airport, McKinney, TX	TX05	1,400
Bailey Airport, Midlothian, TX	7TX8	1,400
Bransom Farnt Airport, Burleson, TX	TX42	1,400
Carroll Air Park Airport, De Soto, TX	F66	1,400
Carroll Lake-View Airport, Venus, TX	70TS	1,400
Eagle's Nest Estates Airport, Ovilla, TX	2T36	1,400
Flying B Ranch Airport, Ovilla, TX	TS71	1,400
Lancaster Airport, Lancaster, TX	LNC	1,400
Lewis Farm Airport, Lucas, TX	6TX1	1,400
Markum Ranch Airport, Fort Worth, TX	TX79	1,400
McKinney Municipal Airport, McKinney, TX	TK1	1,400
O'Brien Airpark Airport, Waxahachie, TX	F25	1,400
Phil L. Hudson Municipal Airport, Mesquite, TX	HQZ	1,400

Airport Name	Arpt ID	Alt. (AGL)
Plover Heliport, Crowley, TX	82Q	1,400
Venus Airport, Venus, TX	75TS	1,400

(7) Airports within a 30-nautical-mile radius of the Stapleton International Airport.

Airport Name	Arpt ID	Alt. (AGL)
Athanasiou Valley Airport, Blackhawk, CO	C007	1,200
Boulder Municipal Airport, Boulder, CO	1V5	1,200
Bowen Farms No. 2 Airport, Strasburg, CO	3C05	1,200
Carrera Airpark Airport, Mead, CO	93CO	1,200
Cartwheel Airport, Mead, CO	OC08	1,200
Colorado Antique Field Airport, Niwot, CO	8C07	1,200
Comanche Airfield Airport, Strasburg, CO	3C06	1,200
Comanche Livestock Airport, Strasburg, CO	59CO	1,200
Flying J Ranch Airport, Evergreen, CO	27CO	1,200
Frederick-Firestone Air Strip Airport, Frederick, CO	C058	1,200]
Frontier Airstrip Airport, Mead, CO	84CO	1,200
Hoy Airstrip Airport, Bennett, CO	76CO	1,200
J & S Airport, Bennett, CO	CD14	1,200
Kugel-Strong Airport, Platteville, CO	27V	1,200
Land Airport, Keenesburg, CO	C082	1,200
Lindys Airpark Airport, Hudson, CO	7C03	1,200
Marshdale STOL, Evergreen, CO	C052	1,200
Meyer Ranch Airport, Conifer, CO	5C06	1,200
Parkland Airport, Erie, CO	7C00	1,200
Pine View Airport, Elizabeth, CO	02V	1,200
Platte Valley Airport, Hudson, CO	18V	1,200
Rancho De Aereo Airport, Mead, CO	05CO	1,200
Spickard Farm Airport, Byers, CO	5CO4	1,200
Vance Brand Airport, Longmont, CO	2V2	1,200
Yoder Airstrip Airport, Bennett, CO	CD09	1,200

(8) Airports within a 30-nautical-mile radius of the Detroit Metropolitan Wayne County Airport.

Airport Name	Arpt ID	Alt. (AGL)
Al Meyers Airport, Tecumseh, MI	3TE	1,400
Brighton Airport, Brighton, MI	45G	1,400
Cackleberry Airport, Dexter, MI	2MI9	1,400
Erie Aerodome Airport, Erie, MI	05MI	1,400
Ham-A-Lot Field Airport, Petersburg, MI	MI48	1,400
Merillat Airport, Tecumseh, MI	34G	1,400
Rossettie Airport, Manchester, MI	75G	1,400
Tecumseh Products Airport, Tecumseh, MI	0D2	1,400

(9) Airports within a 30-nautical-mile radius of the Honolulu International Airport.

Airport Name	Arpt ID	Alt. (AGL)
Dillingham Airfield Airport, Mokuleia, HI	HDH	2,500

(10) Airports within a 30-nautical-mile radius of the Houston Intercontinental Airport.

Airport Name	Arpt ID	Alt. (AGL)
Ainsworth Airport, Cleveland, TX	0T6	1,200
Biggin Hill Airport, Hockley, TX	0TA3	1,200
Cleveland Municipal Airport, Cleveland, TX	6R3	1,200
Fay Ranch Airport, Cedar Lane, TX	0T2	1,200
Freeman Property Airport, Katy, TX	61T	1,200
Gum Island Airport, Dayton, TX	3T6	1,200
Harbican Airpark Airport, Katy, TX	9XS9	1,200
Harold Freeman Farm, Airport, Katy, TX	8XS1	1,200
Hoffpauir Airport, Katy, TX	59T	1,200
Horn-Katy Hawk International Airport, Katy, TX	57T	1,200
Houston-Hull Airport, Houston, TX	SGR	1,200
Houston-Southwest Airport, Houston, TX	AXH	1,200

Airport Name	Arpt ID	Alt. (AGL)
King Air Airport, Katy, TX	55T	1,200
Lake Bay Gall Airport, Cleveland, TX	OT5	1,200
Lake Bonanza Airport, Montgomery, TX	33TA	1,200
R W J Airpark Airport, Baytown, TX	54TX	1,200
Westheimer Air Park Airport, Houston, TX	5TA4	1,200

(11) Airports within a 30-nautical-mile radius of the Kansas City International Airport.

Airport Name	Arpt ID	Alt. (AGL)
Amelia Earhart Airport, Atchison, KS	K59	1,000
Booze Island Airport, St. Joseph, MO	64MO	1,000
Cedar Air Park Airport, Olathe, KS	51K	1,000
D'Field Airport, McLouth, KS	KS90	1,000
Dorei Airport, McLouth, KS	K69	1,000
East Kansas City Airport, Grain Valley, MO	3GV	1,000
Excelsior Springs Memorial Airport, Excelsior Springs, MO	3EX	1,000
Flying T Airport, Oskaloosa, KS	7KS0	1,000
Hermon Farm Airport, Gardner, KS	KS59	1,000
Hillside Airport, Stilwell, KS	63K	1,000
Independence Memorial Airport, Independence, MO	3IP	1,000
Johnson County Executive Airport, Olathe, KS	OJC	1,000
Johnson County Industrial Airport, Olathe, KS	IXD	1,000
Kimray Airport, Plattsburg, MO	7M07	1,000
Lawrence Municipal Airport, Lawrence, KS	LWC	1,000
Martins Airport, Lawson, MO	21MO	1,000
Mayes Homestead Airport, Polo, MO	37MO	1,000
McComas-Lee's Summit Municipal Airport, Lee's Summit, MO	K84	1,000
Mission Road Airport, Stilwell, KS	64K	1,000
Northwood Airport, Holt, MO	2M02	1,000
Plattsburg Airpark Airport, Plattsburg, MO	M028	1,000
Richards-Gebaur Airport, Kansas City, MO	GVW	1,000
Rosecrans Memorial Airport, St. Joseph, MO	STJ	1,000
Runway Ranch Airport, Kansas City, MO	2MO9	1,000
Sheller's Airport, Tonganoxie, KS	11KS	1,000
Shomin Airport, Oskaloosa, KS	0KS1	1,000
Stonehenge Airport, Williamstown, KS	71KS	1,000
Threshing Bee Airport, McLouth, KS	41K	1,000

(12) Airports within a 30-nautical-mile radius of the McCarran International Airport.

Airport Name	Arpt ID	Alt. (AGL)
Sky Ranch Estates Airport, Sandy Valley, NV	3L2	2,500

(13) Airports within a 30-nautical-mile radius of the Memphis International Airport.

Airport Name	Arpt ID	Alt. (AGL)
Bernard Manor Airport, Earle, AR	65M	2,500
Holly Springs-Marshall County Airport, Holly Springs, MS	M41	2,500
McNeely Airport, Earle, AR	M63	2,500
Price Field Airport, Joiner, AR	8OM	2,500
Tucker Field Airport, Hughes, AR	78M	2,500
Tunica Airport, Tunica, MS	30M	2,500
Tunica Municipal Airport, Tunica, MS	M97	2,500

(14) Airports within a 30-nautical-mile radius of the Minneapolis St. Paul International Wold-Chamberlain Airport.

Airport Name	Arpt ID	Alt. (AGL)
Belle Plaine Airport, Belle Plaine, MN	7Y7	1,200
Carleton Airport, Stanton, MN	SYN	1,200
Empire Farm Strip Airport, Bongards, MN	MN15	1,200
Flying M Ranch Airport, Roberts, WI	78WI	1,200

Airport Name	Arpt ID	Alt. (AGL)
Johnson Airport, Rockford, MN	MY86	1,200
River Falls Airport, River Falls, WI	Y53	1,200
Rusmar Farms Airport, Roberts, WI	WS41	1,200
Waldref SPB, Forest Lake, MN	9Y6	1,200
Ziermann Airport, Mayer, MN	MN71	1,200

(15) Airports within 30-nautical-mile radius of the New Orleans International/Moisant Field Airport.

Airport Name	Arpt ID	Alt. (AGL)
Bollinger SPB, Larose, LA	L38	1,500
Clovelly Airport, Cut Off, LA	LA09	1,500

(16) Airports within a 30-nautical-mile radius of the John F. Kennedy International Airport, the La Guardia Airport, and the Newark International Airport.

Airport Name	Arpt ID	Alt. (AGL)
Allaire Airport, Belmar/Farmingdale, NJ	BLM	2,000
Cuddihy Landing Strip Airport, Freehold, NJ	NJ60	2,000
Ekdahl Airport, Freehold, NJ	NJ59	2,000
Fla-Net Airport, Netcong, NJ	0NJ5	2,000
Forrestal Airport, Princeton, NJ	N21	2,000
Greenwood Lake Airport, West Milford, NJ	4N1	2,000
Greenwood Lake SPB, West Milford, NJ	6NJ7	2,000
Lance Airport, Whitehouse Station, NJ	6NJ8	2,000
Mar Bar L Farms, Englishtown, NJ	NJ46	2,000
Peekskill SPB, Peekskill, NY	7N2	2,000
Peters Airport, Somelle, NJ	4NJ8	2,000
Princeton Airport, Princeton/Rocky Hill, NJ	39N	2,000
Solberg-Hunterdon Airport, Readington, NJ	N51	2,000

(17) Airports within a 30-nautical-mile radius of the Orlando International Airport.

Airport Name	Arpt ID	Alt. (AGL)
Arthur Dunn Air Park Airport, Titusville, FL	X21	1,400
Space Center Executive Airport, Titusville, FL	TIX	1,400

(18) Airports within a 30-nautical-mile radius of the Philadelphia International Airport.

Airport Name	Arpt ID	Alt. (AGL)
Ginns Airport, West Grove, PA	78N	1,000
Hammonton Municipal Airport, Hammonton, NJ	N81	1,000
Li Calzi Airport, Bridgeton, NJ	N50	1,000
New London Airport, New London, PA	N01	1,000
Wide Sky Airpark Airport, Bridgeton, NJ	N39	1,000

(19) Airports within a 30-nautical-mile radius of the Phoenix Sky Harbor International Airport.

Airport Name	Arpt ID	Alt. (AGL)
Ak Chin Community Airfield Airport, Maricopa, AZ	E31	2,500
Boulais Ranch Airport, Maricopa, AZ	9E7	2,500
Estrella Sailport, Maricopa, AZ	E68	2,500
Hidden Valley Ranch Airport, Maricopa, AZ	AZ17	2,500
Millar Airport, Maricopa, AZ	2AZ4	2,500
Pleasant Valley Airport, New River, AZ	AZ05	2,500
Serene Field Airport, Maricopa, AZ	AZ31	2,500
Sky Ranch Carefree Airport, Carefree, AZ	E18	2,500
Sycamore Creek Airport, Fountain Hills, AZ	0AS0	2,500
University of Arizona, Maricopa Agricultural Center Airport, Maricopa, AZ	3AZ2	2,500

(20) Airports within a 30-nautical-mile radius of the Lambert/St. Louis International Airport.

Airport Name	Arpt ID	Alt. (AGL)
Blackhawk Airport, Old Monroe, MO	6M00	1,000
Lebert Flying L Airport, Lebanon, IL	3H5	1,000
Shafer Metro East Airport, St. Jacob, IL	3K6	1,000
Sloan's Airport, Elsberry, MO	OM08	1,000
Wentzville Airport, Wentzville, MO	M050	1,000
Woodliff Airpark Airport, Foristell, MO	98MO	1,000

(21) Airports within a 30-nautical-mile radius of the Salt Lake City International Airport.

Airport Name	Arpt ID	Alt. (AGL)
Bolinder Field-Tooele Valley Airport, Tooele, UT	TVY	2,500
Cedar Valley Airport, Cedar Fort, UT	UT10	2,500
Morgan County Airport, Morgan, UT	42U	2,500
Tooele Municipal Airport, Tooele, UT	U26	2,500

(22) Airports within a 30-nautical-mile radius of the Seattle Tacoma International Airport.

Airport Name	Arpt ID	Alt. (AGL)
Firstair Field Airport, Monroe, WA	WA38	1,500
Gower Field Airport, Olympia, WA	6WAZ	1,500
Harvey Field Airport, Snohomish, WA	S43	1,500

(23) Airports within a 30-nautical-mile radius of the Tampa International Airport.

Airport Name	Arpt ID	Alt. (AGL)
Hernando County Airport, Brooksville, FL	BKV	1,500
Lakeland Municipal Airport, Lakeland, FL	LAL	1,500
Zephyrhills Municipal Airport, Zephyrhills, FL	ZPH	1,500

(24) [Effective upon the establishment of the Washington Tri-Area Class B airspace area: Airports within a 30-nautical-mile radius of the Washington National Airport, Andrews Air Force Base Airport, Baltimore Washington International Airport, and Dulles International Airport.

Airport Name	Arpt ID	Alt. (AGL)
Albrecht Airstrip Airport, Long Green, MD	MD48	2,000
Armacost Farms Airport, Hampstead, MD	MD38	2,000
Barnes Airport, Lisbon, MD	MD47	2,000
Bay Bridge Airport, Stevensville, MD	W29	2,000
Carroll County Airport, Westminster, MD	W54	2,000
Castle Marina Airport, Chester, MD	OW6	2,000
Clearview Airpark Airport, Westminster, MD	2W2	2,000
Davis Airport, Laytonsville, MD	W50	2,000
Fallston Airport, Fallston, MD	W42	2,000
Faux-Burhans Airport, Frederick, MD	3MD0	2,000
Forest Hill Airport, Forest Hill, MD	MD31	2,000
Fort Detrick Helipad Heliport, Fort Detrick (Frederick), MD	MD32	2,000
Frederick Municipal Airport, Frederick, MD	FDK	2,000
Fremont Airport, Kemptown, MD	MD41	2,000
Good Neighbor Farm Airport, Unionville, MD	MD74	2,000
Happy Landings Farm Airport, Unionville, MD	MD73	2,000
Harris Airport, Still Pond, MD	MD69	2,000
Hybarc Farm Airport, Chestertown, MD	MD19	2,000
Kennersley Airport, Church Hill, MD	MD23	2,000
Kentmorr Airpark Airport, Stevensville, MD	3W3	2,000
Montgomery County Airpark Airport, Gaithersburg, MD	GAI	2,000
Phillips AAF, Aberdeen, MD	APG	2,000
Pond View Private Airport, Chestertown, MD	0MD4	2,000
Reservoir Airport, Finksburg, MD	1W8	2,000
Scheeler Field Airport, Chestertown, MD	0W7	2,000
Stolcrest STOL, Urbana, MD	MD75	2,000
Tinsley Airstrip Airport, Butler, MD	MD17	2,000

Airport Name	Arpt ID	Alt. (AGL)
Walters Airport, Mount Airy, MD	0MD6	2,000
Waredaca Farm Airport, Brookeville, MD	MD16	2,000
Weide AAF, Edgewood Arsenal, MD	EDG	2,000
Woodbine Gliderport, Woodbine, MD	MD78	2,000
Wright Field Airport, Chestertown, MD	MD11	2,000
Aviacres Airport, Warrenton, VA	3VA2	1,500
Birch Hollow Airport, Hillsboro, VA	W60	1,500
Flying Circus Aerodrome Airport, Warrenton, VA	3VA3	1,500
Fox Acres Airport, Warrenton, VA	15VA	1,500
Hartwood Airport, Somerville, VA	8W8	1,500
Horse Feathers Airport, Midland, VA	53VA	1,500
Krens Farm Airport, Hillsboro, VA	14VA	1,500
Scott Airpark Airport, Lovettsville, VA	VA61	1,500
The Grass Patch Airport, Lovettsville, VA	VA62	1,500
Walnut Hill Airport, Calverton, VA	58VA	1,500
Warrenton Air Park Airport, Warrenton, VA	9W0	1,500
Warrenton-Fauquier Airport, Warrenton, VA	W66	1,500
Whitman Strip Airport, Manassas, VA	0V5	1,500
Aqua-Land/Cliffton Skypark Airport, Newburg, MD	2W8	1,000
Buds Ferry Airport, Indian Head, MD	MD39	1,000
Burgess Field Airport, Riverside, MD	3W1	1,000
Chimney View Airport, Fredericksburg, VA	5VA5	1,000
Holly Springs Farni Airport, Nanjemoy, MD	MD55	1,000
Lanseair Farms Airport, La Plata, MD	MD97	1,000
Nyce Airport, Mount Victoria, MD	MD84	1,000
Parks Airpark Airport, Nanjemoy, MD	MD54	1,000
Pilots Cue Airport, Tompkinsville, MD	MD06	1,000
Quantico MCAF, Quantico, VA	NYG	1,000
Stewart Airport, St. Michaels, MD	MD64	1,000
U.S. Naval Weapons Center, Dahlgren Lab Airport, Dahlgren, VA	NDY	1,000

SFAR No. 64. Special flight authorizations for noise restricted aircraft

[1.] Contrary provisions of Part 91, Subpart I notwithstanding, an operator of a civil subsonic turbojet airplane with maximum weight of more than 75,000 pounds may conduct an approved limited nonrevenue operation of that airplane to or from a U.S. airport when such operation has been authorized by the FAA under paragraph [2] of this SFAR; and

(a) The operator complies with all conditions and limitations established by this SFAR and the authorization;

(b) A copy of the authorization is carried aboard the airplane during all operations to or from a U.S. airport;

(c) The airplane carries an appropriate airworthiness certificate issued by the country of registration and meets the registration and identification requirements of that country; and

(d) Whenever the application is for operation to a location at which FAA-approved noise abatement retrofit equipment is to be installed to make the aircraft comply with Stage 2 or Stage 3 noise levels as defined in Part 36 of this chapter, the applicant must have a valid contract for such equipment.

[2.] Authorization for the operation of a Stage 1 or Stage 2 civil turbojet airplane to or from a U.S. airport may be issued by the FAA for the following purposes:

Stage 1 Airplanes

(a) For a Stage 1 airplane owned by a U.S. owner/applicant on and since November 4, 1990:

(i) obtaining modifications necessary to meet Stage 2 noise levels as defined in Part 36 of this chapter;

(ii) obtaining modifications necessary to meet Stage 3 noise levels as defined in Part 36 of this chapter;

(iii) Scrapping the airplane, as deemed necessary by the FAA, to obtain spare parts to support U.S. programs for the national defense or safety.

(b) For a Stage 1 airplane owned by a non-U.S. owner/applicant:

(i) obtaining modifications necessary to meet Stage 2 noise levels as defined in Part 36 of this chapter;

(ii) obtaining modifications necessary to meet Stage 3 noise levels as defined in Part 36 of this chapter; or

(iii) Scrapping the airplane, as deemed necessary by the FAA, to obtain spare parts to support U.S. programs for the national defense or safety.

(c) For a Stage 1 airplane purchased by a U.S. owner/applicant on or after November 5, 1990:

(i) obtaining modifications necessary to meet Stage 2 noise levels as defined in Part 36 of this chapter, provided that the airplane does not subsequently operate in the contiguous United States;

(ii) obtaining modifications necessary to meet Stage 3 noise levels as defined in Part 36 of this chapter; or

(iii) Scrapping the airplane, as deemed necessary by the FAA, to obtain spare parts to support U.S. programs for the national defense or safety.

Stage 2 Airplanes

(d) For a Stage 2 airplane purchased by a U.S. owner/applicant on or after November 5, 1990, obtaining modification to meet Stage 3 noise levels as defined in part 36 of this chapter.

(e) For Stage 2 airplanes that were U.S.-owned on and since November 4, 1990, and that have been removed from service to achieve compliance with section 91.865 or section 91.867 of this part:

(i) obtaining modifications to meet Stage 3 noise levels as defined in Part 36 of this chapter;

(ii) Prior to January 1, 2000, exporting an airplane, including flying the airplane to or from any airport in the contiguous United States necessary for the exportation of that airplane; or

(iii) Prior to January 1, 2000, operating the airplane as deemed necessary by the FAA for the sale, lease, storage, or scrapping of the airplane.

[3] An application for a special flight authorization under this Special Federal Aviation Regulation shall be submitted to the FAA, Director of the Office of Environment and Energy, received no less than five days prior to the requested flight, and include the following:

(a) The applicant's name and telephone number,

(b) The name of the airplane operator;

(c) The make, model, registration number, and serial number of the airplane;

(d) The reason why such authorization is necessary;

(e) The purpose of the flight;

(f) Each U.S. airport at which the flight will be operated and the number of takeoffs and landings at each;

(g) The approximate dates of the flights;

(h) The number of people on board the airplane and the function of each person;

(i) Whether a special flight permit under FAR Part 21.199 or a special flight authorization under FAR Part 91.715 is required for the flight;

(j) A copy of the contract for noise abatement retrofit equipment, if appropriate; and

(k) Any other information or documentation requested by the Director, Office of Environment and Energy, as necessary to determine whether the application should be approved.

[4.] The Special Federal Aviation Regulation terminates on December 31, 1999, unless sooner rescinded or superseded.

Special Federal Aviation Regulation No. 66.
Prohibition against certain flights between the United States and Yugoslavia.

1. *Applicability.* Except as provided in paragraphs 3 and 4 of this Special Federal Aviation Regulation, this rule applies to all aircraft operations originating from, destined to land in, or overflying the territory of the United States.

2. *Special flight restrictions.* Except as provided in paragraph 3 of this SFAR—

(a) No person shall operate an aircraft or initiate a flight from any point in the United States to any point in the Federal Republic of Yugoslavia (Serbia and Montenegro) (hereinafter "Yugoslavia"), or to any intermediate destination on a flight the ultimate destination of which is in Yugoslavia or which includes a landing at any point in Yugoslavia in its intended itinerary;

(b) No person shall operate an aircraft to any point in the United States from any point in Yugoslavia, or from any intermediate point of departure on a flight the origin of which is in Yugoslavia, or which includes a departure from any point in Yugoslavia in its intended itinerary; and

(c) No person shall operate an aircraft over the territory of the United States if that aircraft's flight itinerary includes any landing at or departure from any point in Yugoslavia.

3. *Permitted operations.* This SFAR shall not prohibit the takeoff or landing of an aircraft, the initiation of a flight, or the overflight of United States territory by an aircraft authorized to conduct such operations by the United States Government in consultation with the United Nations Security Council Committee established by UN Security Council Resolution 757 (1992).

4. *Emergency situations.* In an emergency that requires immediate decision and action for the safety of the flight, the pilot in command of an aircraft may deviate from this SFAR to the extent required by that emergency. Any deviation required by an emergency shall be reported to the Air Traffic Control Facility having jurisdiction as soon as possible.

5. *Expiration.* This special Federal Aviation Regulation expires (pending publication of updated rule in the *Federal Register*).

Special Federal Aviation Regulation No. 67.
Prohibition against certain flights within the territory and airspace of Afghanistan.

1. *Applicability.* This rule applies to all U.S. air carriers and commercial operators, all persons exercising the privileges of an airman certificate issued by the FAA, and all operators using aircraft registered in the United States except where the operator of such aircraft is a foreign air carrier.

2. *Flight prohibitions.* Except as provided in paragraph 3 and 4 of this SFAR, no person described in paragraph 1 may conduct flight operations within the territory and airspace of Afghanistan.

3. *Permitted operations.* This SFAR does not prohibit persons described in paragraph 1 from conducting flight

operations within the territory and airspace of Afghanistan where such operations are authorized either by exemption issued by the Administrator or by another agency of the United States Government with the approval of the FAA.

4. *Emergency situations.* In an emergency that requires immediate decision and action for the safety of the flight, the pilot in command of an aircraft may deviate from this SFAR to the extent required by that emergency. Except for U.S. air carriers and commercial operators that are subject to the requirements of 14 CFR 121.557, 121.559, or 135.19, each person who deviates from this rule shall, within ten (10) days of the deviation, excluding Saturdays, Sundays, and Federal holidays, submit to the nearest FAA Flight Standards District Office a complete report of the operations of the aircraft involved in the deviation, including a description of the deviation and the reasons therefor.

5. *Expiration.* This Special Federal Aviation Regulation expires May 10, 1995.

Special Federal Aviation Regulation No. 68. Prohibition against certain flights within the territory and airspace of Yemen.

1. *Applicability.* This rule applies to all U.S. air carriers and commercial operators, all persons exercising the privileges of an airman certificate issued by the FAA, and all operators using aircraft registered in the United States except where the operator of such aircraft is a foreign air carrier.

2. *Flight prohibition.* Except as provided in paragraph 3 and 4 of this SFAR, no person described in paragraph 1 may conduct flight operations within the territory and airspace of Yemen.

3. *Permitted operations.* This SFAR does not prohibit persons described in paragraph 1 from conducting flight operations within the territory and airspace of Yemen where such operations are authorized either by exemption issued by the Administrator or by another agency of the United States Government with the approval of the FAA.

4. *Emergency situations.* In an emergency that requires immediate decision and action for the safety of the flight, the pilot in command of an aircraft may deviate from this SFAR to the extent required by that emergency. Except for U.S. air carriers and commercial operators that are subject to the requirements of 14 CFR. 121.557, 121.559, or 135.19, each person who deviates from this rule shall, within ten (10) days of the deviation, excluding Saturdays, Sundays, and Federal holidays, submit to the nearest FAA Flight Standards District Office a complete report of the operation of the aircraft involved in the deviation, including a description of the deviation and the reason therefor.

5. *Expiration.* This special Federal Aviation Regulation expires May 10, 1995.

Special Federal Aviation Regulation No. 69. Prohibition against certain flights between the United States and Haiti.

1. *Applicability.* This rule applies to all aircraft operations originating from, landing in, or overflying the territory of the United States.

2. *Special flight restrictions.* Except as provided in paragraph 3 and 4 of this SFAR—

(a) No person may operate an aircraft or initiate a flight from any point in the United States to any point in Haiti, a flight having any intermediate or ultimate destination in Haiti, or a flight that includes a landing at any point in Haiti in its itinerary;

(b) No person may operate an aircraft to any point in the United States from any point in Haiti, from any intermediate point of departure on a flight the origin of which is in Haiti, or which includes a departure from any point in Haiti in its itinerary; or

(c) No person may operate an aircraft over the territory of the United States if that aircraft's flight itinerary includes any landing at or departure from any point in Haiti.

3. *Permitted operations.* This SFAR does not prohibit the takeoff or landing of an aircraft, the initiation of a flight, or the overflight of United States territory by any civil aircraft;

(a) Operated by a U.S. or foreign air carrier to conduct scheduled passenger-carrying operations between the United States and Haiti; or

(b) Authorized to conduct such operations either by the Administrator or by another agency of the United States Government with the approval of the FAA.

4. *Emergency situations.* In an emergency that requires immediate decision and action for the safety of the flight, the pilot in command of an aircraft may deviate from this SFAR to the extent required by that emergency. Except for U.S. air carriers and commercial operators that are subject to the requirements of 14 CFR 121.557, 121.559. or 135.19, each person who deviates from this rule shall, within ten (10) days of the deviation, excluding Saturdays, Sundays, and Federal holidays, submit to the nearest FAA Flight Standards District Office a complete report of the operations of the aircraft involved in the deviation, including a description of the deviation and the reasons therefor.

5. *Expiration.* This Special Federal Aviation Regulation expires May 13, 1995.

Part 97 – Standard instrument approach procedures

Subpart A—General

97.1 Applicability
97.3 Symbols and terms used in procedures
97.5 Bearings; courses; headings; radials; miles

Subpart B—Procedures

97.10 General
97.11 Low or medium frequency range, automatic direction finding, and very high frequency omnirange procedures
97.13 Terminal very high frequency omnirange procedures
97.15 Very high frequency omnirange-distance measuring equipment procedures
97.17 Instrument landing system procedures
97.19 Radar procedures

Subpart C—TERPS procedures

97.20 General
97.21 Low or medium frequency range (L/MF) procedures
97.23 Very high frequency omnirange (VOR) and very high frequency distance measuring equipment (VOR/DME) procedures
97.25 Localizer (LOC) and localizer-type directional aid (LDA) procedures
97.27 Nondirectional beacon (automatic direction finder) (NDB(ADF)) procedures
97.29 Instrument landing system (ILS) procedures
97.31 Precision approach radar (PAR) and airport surveillance radar (ASR) procedures
97.33 Area navigation (RNAV) procedures
97.35 Helicopter procedures

Subpart A—General

§ 97.1 Applicability.

This Part prescribes standard instrument approach procedures for instrument letdown to airports in the United States and the weather minimums that apply to takeoffs and landings under IFR at those airports.

§ 97.3 Symbols and terms used in procedures.

As used in the standard terminal instrument procedures prescribed in this part—

(a) "A" means alternate airport weather minimum.

[(b) "Aircraft approach category" means a grouping of aircraft based on a speed of 1.3 V_{so} (at maximum certificated landing weight). V_{so} and the maximum certificated landing weight are, those values as established for the aircraft by the certificating authority of the country of registry. The categories are as follows:

[(1) Category A: Speed less than 91 knots.

[(2) Category B: Speed 91 knots or more but less than 121 knots.

[(3) Category C: Speed 121 knots or more but less than 141 knots.

[(4) Category D: Speed 141 knots or more but less than 166 knots.

[(5) Category E: Speed 166 knots or more.]

(c) Approach procedure segments for which altitudes (all altitudes prescribed are minimum altitudes unless otherwise specified) of courses, or both, are prescribed in procedures, are as follows:

(1) "Initial approach" is the segment between the initial approach fix and the intermediate fix or the point where the aircraft is established on the intermediate course or final approach course.

(2) "Initial approach altitude" means the altitude (or altitudes, in High Altitude Procedures) prescribed for the initial approach segment of an instrument approach.

(3) "Intermediate approach" is the segment between the intermediate fix or point and the final approach fix.

(4) "Final approach" is the segment between the final approach fix or point and the runway, airport, or missed-approach point.

(5) "Missed approach" is the segment between the missed-approach point, or point of arrival at decision height, and the missed-approach fix at the prescribed altitude.

(d) "C" means circling landing minimum, a statement of ceiling and visibility values, or minimum descent altitude and visibility, required for the circle-to-land maneuver.

(d-1) "Copter procedures" means helicopter procedures, with applicable minimums as prescribed in § 97.35 of this Part. Helicopters may also use other procedures prescribed in Subpart C of this Part and may use the Category A minimum descent altitude (MDA) or decision height (DH). The required visibility minimum may be reduced to one-half the published visibility minimum for Category A aircraft, but in no case may it be reduced to less than one-quarter mile or 1,200 feet RVR.

(e) "Ceiling minimum" means the minimum ceiling, expressed in feet above the surface of the airport, required for takeoff or required for designating an airport as an alternate airport.

(f) "d" means day.

(g) "FAF" means final approach fix.

(h) "HAA" means height above airport.

(h-1) "HAL" means height above a designated helicopter landing area used for helicopter instrument approach procedures.

(i) "HAT" means height above touchdown.

(j) "MAP" means missed approach point.

(k) "More than 65 knots" means an aircraft that has a stalling speed of more than 65 knots (as established in an approved flight manual) at maximum certificated landing weight with full flaps, landing gear extended, and power off.

(l) "MSA" means minimum safe altitude, an emergency altitude expressed in feet above mean sea level, which provides 1,000 feet clearance over all obstructions in that sector within 25 miles of the facility on which the procedure is based (LOM in ILS procedures).

(m) "n" means night.

(n) "NA" means not authorized.

(o) "NOPT" means no procedure turn required (altitude prescribed applies only if procedure turn is not executed).

(o-1) "Point in space approach" means a helicopter instrument approach procedure to a missed approach point that is more than 2,600 feet from an associated helicopter landing area.

(p) "Procedure turn" means the maneuver prescribed when it is necessary to reverse direction to establish the aircraft on an intermediate or final approach course. The outbound course, direction of turn, distance within which the turn must be completed, and minimum altitude are specified in the procedure. However, the point at which the turn may be commenced, and the type and rate of turn, is left to the discretion of the pilot.

(q) "RA" means radio altimeter setting height.

(r) "RVV" means runway visibility value.

(s) "S" means straight-in landing minimum, a statement of ceiling and visibility, minimum descent altitude and visibility, or decision height and visibility, required for a straight-in landing on a specified runway. The number appearing with the "S" indicates the runway to which the minimum applies. If a straight-in minimum is not prescribed in the procedure, the circling minimum specified applies to a straight-in landing.

(t) "Shuttle" means a shuttle, or race-track-type, pattern with 2-minute legs prescribed in lieu of a procedure turn.

(u) "65 knots or less" means an aircraft that has a stalling speed of 65 knots or less (as established in an approved flight manual) at maximum certificated landing weight with full flaps, landing gear extended, and power off.

(v) "T" means takeoff minimum.

(w) "TDZ" means touchdown zone.

(x) "Visibility minimum" means the minimum visibility specified for approach, or landing, or takeoff, expressed in statute miles, or in feet where RVR is reported.

§ 97.5 Bearings; courses; headings; radials; miles.

(a) All bearings, courses, headings, and radials in this Part are magnetic.

(b) RVR values are stated in feet. Other visibility values are stated in statute miles. All other mileages are stated in nautical miles.

Subpart B—Procedures

§ 97.10 General.

This subpart prescribes standard instrument approach procedures other than those based on the criteria contained in the U.S. Standard for Terminal Instrument Approach Procedures (TERPS). Standard instrument approach procedures adopted by the FAA and described on FAA Form 3139 are incorporated into this Part and made a part hereof as provided in 5 U.S.C. 522 (a) (1) and pursuant to 1 CFR Part 20. The incorporated standard instrument approach procedures are available for examination at the Rules Docket and at the National Flight Data Center, Federal Aviation Administration, 800 Independence Avenue, S.W., Washington, D.C. 20590. Copies of SIAPs adopted in a particular FAA Region are also available for examination at the Headquarters of that Region. Moreover, copies of SIAPs originating in a particular Flight Inspection District Office are available for examination at that Office. Based on the information contained on FAA Form 3139, standard instrument approach procedures are portrayed on charts prepared for the use of pilots by the United States Coast and Geodetic Survey and other publishers of aeronautical charts.

§ 97.11 Low or medium frequency range, automatic direction finding, and very high frequency omnirange procedures.

Section 609.100 of the Regulations of the Administrator is hereby designated as § 97.11.

§ 97.13 Terminal very high frequency omnirange procedures.

Section 609.200 of the Regulations of the Administrator is hereby designated as § 97.13.

§ 97.15 Very high frequency omnirange-distance measuring equipment procedures.

Section 609.300 of the Regulations of the Administrator is hereby designated as § 97.15.

§ 97.17 Instrument landing system procedures.

Section 609.400 of the Regulations of the Administrator is hereby designated as § 97.17.

§ 97.19 Radar procedures.

Section 609.500 of the Regulations of the Administrator is hereby designated as § 97.19.

Subpart C—TERPS procedures

§ 97.20 General.

This subpart prescribes standard instrument approach procedures based on the criteria contained in the U.S. Standard for Terminal Instrument Approach Procedures (TERPS). The standard instrument approach procedures adopted by the FAA and described on FAA Forms 8260-3, 8260-4, or 8260-5 are incorporated into this Part and made a part hereof as provided in 5 U.S.C. 552(a) (1) and pursuant to 1 CFR Part 20. The incorporated standard instrument approach procedures are available for examination at the Rules Docket and at the National Flight Data Center, Federal Aviation Administration, 800 Independence Avenue, S.W., Washington, D.C. 20590. Copies of SIAPs adopted in a particular FAA Region are also available for examination at the headquarters of that Region. Moreover, copies of SIAPs originating in a particular Flight Inspection District Office are available for examination at that Office. Based on the information contained on FAA Forms 8260-3, 8260-4, and 8260-5, standard instrument approach procedures are portrayed on charts prepared for the use of pilots by the U.S. Coast and Geodetic Survey and other publishers of aeronautical charts.

§ 97.21 Low or medium frequency range (L/MF) procedures.

§ 97.23 Very high frequency omnirange (VOR) and very high frequency distance measuring equipment (VOR/DME) procedures.

§ 97.25 Localizer (LOC) and localizer-type directional aid (LDA) procedures.

§ 97.27 Nondirectional beacon (automatic direction finder) (NDB(ADF)) procedures.

§ 97.29 Instrument landing system (ILS) procedures.

§ 97.31 Precision approach radar (PAR) and airport surveillance radar (ASR) procedures.

§ 97.33 Area navigation (RNAV) procedures.

§ 97.35 Helicopter procedures.

NOTE: The procedures set forth in § 97.35 are not carried in the Code of Federal Regulations. For Federal Register citations affecting these procedures see List of CFR Sections Affected.

Part 99 – Security control of air traffic

Subpart A—General

99.1 Applicability
99.3 General
99.5 Emergency situations
99.7 Special security instructions
99.9 Radio requirements
99.11 ADIZ flight plan requirements
99.12 Transponder-on requirements
99.15 Arrival or completion notice
99.17 Position reports; aircraft operating in or penetrating an ADIZ; IFR
99.19 Position reports; aircraft operating in or penetrating an ADIZ; DVFR
99.21 Position reports; aircraft entering the U.S. through an ADIZ; U.S. aircraft
99.23 Position reports; aircraft entering the U.S. through an ADIZ; foreign aircraft
99.27 Deviation from flight plans and ATC clearances and instructions
99.29 Radio failure; DVFR
99.31 Radio failure; IFR

Subpart B—Designated air defense identification zones

99.41 General
99.42 Contiguous U.S. ADIZ
99.43 Alaska ADIZ
99.45 Guam ADIZ
99.47 Hawaii ADIZ
99.49 Defense area

Subpart A—General

§ 99.1 Applicability.

(a) This subpart prescribes rules for operating civil aircraft in a defense area, or into, within, or out of the United States through an Air Defense Identification Zone (ADIZ), designated in Subpart B.

(b) [Except for §§ 99.7 and 99.12, this subpart does not apply to the operation of any aircraft—

[(1) Within the 48 contiguous States and the District of Columbia, or within the State of Alaska, on a flight which remains within 10 nautical miles of the point of departure;

[(2) Operating at true airspeed of less than 180 knots in the Hawaii ADIZ or over any island, or within 3 nautical miles or the coastline of any island, in the Hawaii ADIZ;

[(3) Operating at true airspeed of less than 180 knots in the Alaska ADIZ while the pilot maintains a continuous listening watch on the appropriate frequency; or

[(4) Operating at true airspeed of less than 180 knots in the Guam ADIZ.]

(c) Except as provided in § 99.7, the radio and position reporting requirements of this subpart do not apply to the operation of an aircraft within the 48 contiguous States and the District of Columbia, or within the State of Alaska, if that aircraft does not have two-way radio and is operated in accordance with a filed DVFR flight plan containing the time and point of ADIZ penetration and that aircraft departs within 5 minutes of the estimated departure time contained in the flight plan.

(d) An FAA ATC center may exempt the following operations from this subpart (except § 99.7), on a local basis only, with the concurrence of the military commanders concerned:

(1) Aircraft operations that are conducted wholly within the boundaries of an ADIZ and are not currently significant to the air defense system.

(2) Aircraft operations conducted in accordance with special procedures prescribed by the military authorities concerned.

§ 99.3 General.

(a) Air Defense Identification Zone (ADIZ) is an area of airspace over land or water in which the ready identification, location, and control of civil aircraft is required in the interest of national security.

(b) Unless designated as an ADIZ, a Defense Area is any airspace of the United States in which the control of aircraft is required for reasons of national security.

(c) For the purposes of this part, a Defense Visual Flight Rules (DVFR) flight is a flight within an ADIZ conducted under the visual flight rules in Part 91.

§ 99.5 Emergency situations.

In an emergency that requires immediate decision and action for the safety of the flight, the pilot in command of an aircraft may deviate from the rules in this part to the extent required by that emergency. He shall report the reasons for the deviation to the communications facility where flight plans or position reports are normally filed (referred to in this part as "an appropriate aeronautical facility") as soon as possible.

§ 99.7 Special security instructions.

Each person operating an aircraft in an ADIZ or Defense Area shall, in addition to the applicable operating rules of this part, comply with special security instructions issued by the Administrator in the interest of national security and that are consistent with appropriate agreements between the FAA and the Department of Defense.

§ 99.9 Radio requirements.

Except as provided in § 99.1(c), no person may operate an aircraft in an ADIZ unless the aircraft has a functioning two-way radio.

§ 99.11 ADIZ flight plan requirements.

(a) Unless otherwise authorized by ATC, no person may operate an aircraft into, within, or across an ADIZ unless that person has filed a flight plan with an appropriate aeronautical facility.

(b) Unless ATC authorizes an abbreviated flight plan—

(1) A flight plan for IFR flight must contain the information specified in [§ 91.169]; and

(2) A flight plan for VFR flight must contain the information specified in [§ 91.153 (a) (1) through (6)].

(3) If airport of departure is within the Alaskan ADIZ and there is no facility for filing a flight plan then:

(i) Immediately after takeoff or when within range of an appropriate aeronautical facility, comply with provisions of paragraph (b)(1) or (b)(2) as appropriate.

(ii) Proceed according to the instructions issued by the appropriate aeronautical facility.

(c) The pilot shall designate a flight plan for VFR flight as a DVFR flight plan.

§ 99.12 Transponder-on requirements.

[(a) *Aircraft transponder-on operation.* Each person operating an aircraft into or out of the United States into, within, or across an ADIZ designated in Subpart B of this part, if that aircraft is equipped with an operable radar beacon transponder, shall operate the transponder, including altitude encoding equipment if installed, and shall reply on the appropriate code or as assigned by ATC.

[(b) *ATC transponder equipment and use.* Effective September 7, 1990, unless otherwise authorized by ATC, no person may operate a civil aircraft into or out of the United States into, within, or across the contiguous U.S. ADIZ designated in Subpart B of this part unless that aircraft is equipped with a coded radar beacon transponder.

[(c) *ATC transponder and altitude reporting equipment and use.* Effective December 30, 1990, unless otherwise authorized by ATC, no person may operate a civil aircraft into or out of the United States into, within, or across the contiguous U.S. ADIZ unless that aircraft is equipped with a coded radar beacon transponder and automatic pressure altitude reporting equipment having altitude reporting capability that automatically replies to interrogations by transmitting pressure altitude information in 100-foot increments.

[(d) Paragraphs (b) and (c) of this section do not apply to the operation of an aircraft which was not originally certificated with an engine-driven electrical system and which has not subsequently been certified with such a system installed, a ballon, or a glider.]

§ 99.15 Arrival or completion notice.

The pilot in command of an aircraft for which a flight plan has been filed shall file an arrival or completion notice with an appropriate aeronautical facility, unless the flight plan states that no notice will be filed.

§ 99.17 Position reports; aircraft operating in or penetrating an ADIZ; IFR.

The pilot of an aircraft operating in or penetrating an ADIZ under IFR—

(a) In controlled airspace, shall make the position reports required in [§ 91.183]; and

(b) In uncontrolled airspace, shall make the position reports required in § 99.19.

§ 99.19 Position reports; aircraft operating in or penetrating an ADIZ; DVFR.

No pilot may penetrate an ADIZ under DVFR unless—

(a) That pilot reports to an appropriate aeronautical facility before penetration: The time, position, and altitude at which the aircraft passed the last reporting point before penetration and the estimated time of arrival over the next appropriate reporting point along the flight route;

(b) If there is no appropriate reporting point along the flight route, that pilot reports at least 15 minutes before penetration: The estimated time, position, and altitude at which he will penetrate; or

(c) If the airport departure is within an ADIZ or so close to the ADIZ boundary that it prevents his complying with paragraphs (a) or (b) of this section, that pilot has reported immediately after taking off: The time of departure, altitude, and estimated time of arrival over the first reporting point along the flight route.

§ 99.21 Position reports; aircraft entering the United States through an ADIZ; United States aircraft.

The pilot of an aircraft entering the United States through an ADIZ shall make the reports required in § 99.17 or § 99.19 to an appropriate aeronautical facility.

§ 99.23 Position reports; aircraft entering the United States through an ADIZ; foreign aircraft.

In addition to such other reports as ATC may require, no pilot in command of a foreign civil aircraft may enter the United States through an ADIZ unless that pilot makes the reports required in § 99.17 or § 99.19 or reports the position of the aircraft when it is not less than one hour and not more than two hours average direct cruising distance from the United States.

§ 99.27 Deviation from flight plans and ATC clearances and instructions.

(a) No pilot may deviate from the provisions of an ATC clearance or ATC instruction except in accordance with [§ 91.123] of this chapter.

(b) No pilot may deviate from the filed IFR flight plan when operating an aircraft in uncontrolled airspace unless that pilot notifies an appropriate aeronautical facility before deviating.

(c) No pilot may deviate from the filed DVFR flight plan unless that pilot notifies an appropriate aeronautical facility before deviating.

§ 99.29 Radio failure; DVFR.

If the pilot operating an aircraft under DVFR in an ADIZ cannot maintain two-way radio communications, the pilot may proceed in accordance with his original DVFR flight plan or land as soon as practicable. The pilot shall report the radio failure to an appropriate aeronautical facility as soon as possible.

§ 99.31 Radio failure; IFR.

If a pilot operating an aircraft under IFR in an ADIZ cannot maintain two-way radio communications, the pilot shall proceed in accordance with [§ 91.185] of this chapter.

Subpart B—Designated air defense identification zones

§ 99.41 General

The airspace above the areas described in this subpart is established as an ADIZ Defense Area. The lines between points described in this subpart are great circles except that the lines joining adjacent points on the same parallel of latitude are rhumb lines.

§ 99.42 Contiguous U.S. ADIZ.

(a) The area bounded by a line from 26°00'N, 96°35'W; 26°00'N, 95°00'W; 26°30'N, 95°00'W; then along 26°30'N to 26°30'N, 84°00'W; 24°00'N, 83°00'W; 24°00'N, 80°00'W; 24°00'N, 79°25'W; 25°40'N, 79°25'W; 27°30'N, 78°50'W; 30°45'N, 74°00'W; 39°30'N, 63°45'W; 43°00'N, 65°48'W; 41°15'N, 69°30'W; 40°32'N, 72°15'W; 39°55'N, 73°00'W; 39°38'N, 73°00'W; 39°36'30"N, 73°40'30"W; 39°30'N, 73°45'W; 37°00'N, 75°30'W; 36°10'N, 75°10'W; 35°10'N, 75°10'W; 32°01'N, 80°32'W; 30°50'N, 80°54'W; 30°05'N, 81°07'W; 27°59'N, 79°23'W; 24°49'N, 80°00'W; 24°49'N, 80°55'W; 25°10'N, 81°12'W; then along a line 3 nautical miles from the shoreline to 25°45'N, 81°27'W; 25°45'N, 82°07'W; 28°55'N, 83°30'W; 29°20'N, 85°00'W; 30°00'N, 86°00'W; 30°00'N, 88°30'W; 29°00'N, 89°00'W; 28°45'N, 90°00'W; 29°26'N, 94°00'W; 28°42'N, 95°17'W; 28°05'N, 96°30'W; 26°25'N, 96°30'W; 26°00'N, 96°35'W; 25°58'N, to 97°07'W;

(b) The area bounded by a line from 32°32'03"N, 117°07'25"W; 32°30'N, 117°20'W; 32°00'N, 118°24'W; 30°45'N, 120°50'W; 29°00'N, 124°00'W; 37°42'N, 130°40'W; 48°20'N, 132°00'W; 48°20'N, 128°00'W; 48°30'N, 125°00'W; 48°29'38"N, 124°43'35"W; 48°00'N, 125°15'W; 46°15'N, 124°30'W; 43°00'N, 124°40'W; 40°00'N, 124°35'W; 38°50'N, 124°00'W; 34°50'N, 121°10'W; 34°00'N, 120°30'W; 32°00'N, 118°24'W; 32°30'N, 117°20'W; 32°32'03"N, to 117°07'25"W; and

(c) A line extending from 32°32'03"N, 117°07'25"W; eastward along the United States-Mexico Border to 25°58'00"N, 97°07'00"W.

§ 99.43 Alaska ADIZ.

The area bounded by a line 54°00'N, 136°00'W; 56°57'N, 144°00'W; 57°00'N, 145°00'W; 53°00'N, 158°00'W; 50°00'N, 169°00'W; 50°00'N, 180°00'; 50°00'N, 170°00'E; 53°00'N, 170°00'E; 60°00'N, 180°00'; 65°00'N, 169°00'W; then along 169°00'W to 75°00'N, 169°00'W; then along the 75°00'N parallel to 75°00'N; 141°00'W to 69°50'N, 141°00'W; 71°18'N, 156°44'W; 69°52'N, 163°00'W; then south along 163°00'W to 54°00'N, 163°00'W; 56°30'N, 154°00'W; 59°20'N, 146°00'W; 59°30'N, 140°00'W; 57°00'N, 136°00'W; 54°35'N, 133°00'W; to point of origin.

§ 99.45 Guam ADIZ.

(a) *Inner boundary.* From a point 13°52'07"N, 143°59'16"E, counterclockwise along the 50-nautical-mile radius arc of the NIMITZ VORTAC located at 13°27'11"N, 144°43'51"E); to a point 13°02'08"N, 145°28'17"E; then to a point 14°49'07"N, 146°13'58"E; counterclockwise along the 35-nautical-mile radius arc of the SAIPAN NDB (located at 15°06'46"N, 145°42'42"E); to a point 15°24'21"N, 145°11'21"E; then to point of origin.

(b) *Outer boundary.* The area bounded by a circle with a

radius of 250 NM centered at latitude 13°32'41"N, longitude 144°50'30"E.

§ 99.47 Hawaii ADIZ.

(a) *Outer boundary.* The area included in the irregular octagonal figure formed by a line connecting 26°30'N, 156°00'W; 26°30'N, 161°00'W; 24°00'N, 164°00'W; 20°00'N, 164°00'W; 17°00'N, 160°00'W; 17°00'N, 156°00'W; 20°00'N, 153°00'W; 22°00'N, 153°00'W; to point of orgin.

(b) *Inner boundary.* The inner boundary to follow a line connecting 22°30'N, 157°00'W; 22°30'N, 160°00'W; 22°00'N, 161°00'W; 21°00'N, 161°00'W; 20°00'N, 160°00'W; 20°00'N, 156°30'W; 21°00'N, 155°30'W; to point of origin.

§ 99.49 Defense area.

All airspace of the United States is designated as Defense Area except that airspace already designated as Air Defense Identification Zone.

Part 103 – Ultralight vehicles

Subpart A—General

103.1 Applicability
103.3 Inspection requirements
103.5 Waivers
103.7 Certification and registration

Subpart B—Operating rules

103.9 Hazardous operations
103.11 Daylight operations
103.13 Operation near aircraft; right-of-way rules
103.15 Operations over congested areas
103.17 Operations in certain airspace
103.19 Operations in prohibited or restricted areas
103.20 Flight restrictions in the proximity of certain areas designated by notice to airmen
103.21 Visual reference with the surface
103.23 Flight visibility and cloud clearance requirements

Subpart A—General

Source: Docket No. 21631 (47 FR 38776, 9/2/82) effective 10/4/82, for each subpart, unless otherwise noted.

§ 103.1 Applicability.

This part prescribes rules governing the operation of ultralight vehicles in the United States. For the purposes of this part, an ultralight vehicle is a vehicle that:

(a) Is used or intended to be used for manned operation in the air by a single occupant;

(b) Is used or intended to be used for recreation or sport purposes only;

(c) Does not have any U.S. or foreign airworthiness certificate; and

(d) If unpowered, weighs less than 155 pounds; or

(e) If powered:

(1) Weighs less than 254 pounds empty weight, excluding floats and safety devices which are intended for deployment in a potentially catastrophic situation;

(2) Has a fuel capacity not exceeding 5 U.S. gallons;

(3) Is not capable of more than 55 knots calibrated airspeed at full power in level flight; and

(4) Has a power-off stall speed which does not exceed 24 knots calibrated airspeed.

§ 103.3 Inspection requirements.

(a) Any person operating an ultralight vehicle under this part shall, upon request, allow the Administrator, or his designee, to inspect the vehicle to determine the applicability of this part.

(b) The pilot or operator of an ultralight vehicle must, upon request of the Administrator, furnish satisfactory evidence that the vehicle is subject only to the provisions of this part.

§ 103.5 Waivers.

No person may conduct operations that require a deviation from this part except under a written waiver issued by the Administrator.

§ 103.7 Certification and registration.

(a) Notwithstanding any other section pertaining to certification of aircraft or their parts or equipment, ultralight vehicles and their component parts and equipment are not required to meet the airworthiness certification standards specified for aircraft or to have certificates of airworthiness.

(b) Notwithstanding any other section pertaining to airman certification, operators of ultralight vehicles are not required to meet any aeronautical knowledge, age, or experience requirements to operate those vehicles or to have airman or medical certificates.

(c) Notwithstanding any other section pertaining to registration and marking of aircraft, ultralight vehicles are not required to be registered or to bear markings of any type.

Subpart B—Operating rules

§ 103.9 Hazardous operations.

(a) No person may operate any ultralight vehicle in a manner that creates a hazard to other persons or property.

(b) No person may allow an object to be dropped from an ultralight vehicle if such action creates a hazard to other persons or property.

§ 103.11 Daylight operations.

(a) No person may operate an ultralight vehicle except between the hours of sunrise and sunset.

(b) Notwithstanding paragraph (a) of this section, ultralight vehicles may be operated during the twilight periods 30 minutes before official sunrise and 30 minutes after official sunset or, in Alaska, during the period of civil twilight as defined in the Air Almanac, if:

(1) The vehicle is equipped with an operating anticollision light visible for at least 3 statute miles; and

(2) All operations are conducted in uncontrolled airspace.

§ 103.13 Operation near aircraft; right-of-way rules.

(a) Each person operating an ultralight vehicle shall maintain vigilance so as to see and avoid aircraft and shall yield the right-of-way to all aircraft.

(b) No person may operate an ultralight vehicle in a manner that creates a collision hazard with respect to any aircraft.

(c) Powered ultralights shall yield the right-of-way to unpowered ultralights.

§ 103.15 Operations over congested areas.

No person may operate an ultralight vehicle over any congested area of a city, town, or settlement, or over any open air assembly of persons.

§ 103.17 Operations in certain airspace.

No person may operate an ultralight vehicle within Class A, Class B, Class C, or Class D airspace or within the lateral boundaries of the surface area of Class E airspace designated for an airport unless that person has prior authorization from the ATC facility having jurisdiction over that airspace.]

[(Amdt. 103-4, Eff. 9/16/93)]

§ 103.19 Operations in prohibited or restricted areas.

No person may operate an ultralight vehicle in prohibited or restricted areas unless that person has permission from the using or controlling agency, as appropriate.

§ 103.20 Flight restrictions in the proximity of certain areas designated by notice to airmen.

No person may operate an ultralight vehicle in areas designated in a Notice to Airmen under § 91.143 or § 91.141 of this chapter, unless authorized by ATC.

Docket No. 24454 (50 FR 4969) 2/5/85; (Amdt. 103-1, Eff. 2/5/85); (Amdt. 103-3, Eff. 8/18/90)

§ 103.21 Visual reference with the surface.

No person may operate an ultralight vehicle except by visual reference with the surface.

§ 103.23 Flight visibility and cloud clearance requirements.

[No person may operate an ultralight vehicle when the flight visibility or distance from clouds is less than that in the table found on page 510. All operations in Class A, Class B, Class C, and Class D airspace or Class E airspace designated for an airport must receive prior ATC authorization as required in § 103.17 of this part.

See basic VFR weather minimums on page 510.

BASIC VFR WEATHER MINIMUMS

Airspace	Flight visibility	Distance from clouds
Class A	Not Applicable	Not Applicable.
Class B	3 statute miles	Clear of Clouds.
Class C	3 statute miles	500 feet below. 1,000 feet above. 2,000 feet horizontal.
Class D	3 statute miles	500 feet below. 1,000 feet above. 2,000 feet horizontal.
Class E		
Less than 10,000 feet MSL	3 statute miles	500 feet below. 1,000 feet above. 2,000 feet horizontal.
At or above 10,000 feet MSL	5 statute miles	1,000 feet below. 1,000 feet above. 1 statute mile horizontal.
Class G		
1,200 feet or less above the surface (regardless of MSL altitude)		
Day, except as provided in §91.155(b)	1 statute mile	Clear of clouds.
Night, except as provided in §91.155(b)	3 statute miles	500 feet below. 1,000 feet above. 2,000 feet horizontal.
More than 1,200 feet above the surface but less than 10,000 feet MSL		
Day	1 statute mile	500 feet below. 1,000 feet above. 2,000 feet horizontal.
Night	3 statute miles	500 feet below. 1,000 feet above. 2,000 feet horizontal.
More than 1,200 feet above the surface and at or above 10,000 feet MSL	5 statute miles	1,000 feet below. 1,000 feet above. 1 statute mile horizontal.

Part 105 – Parachute jumping

Subpart A—General

105.1 Applicability

Subpart B—Operating rules

105.11 Applicability
105.13 General
105.14 Radio equipment and use requirements
105.15 Jumps over or into congested areas or open air assembly of persons
105.17 Jumps over or onto airports
[105.19 Jumps in or into Class A, Class B, Class C, and Class D airspace]
[105.20 Removed and Reserved]
[105.21 Removed and Reserved]
105.23 Jumps in or into other airspace
105.25 Information required, and notice of cancellation or postponement of jump
105.27 Jumps over or within restricted or prohibited areas
105.29 Flight visibility and clearance from clouds requirements
105.33 Parachute jumps between sunset and sunrise
105.35 Liquor and drugs
105.37 Inspections

Subpart C—Parachute equipment

105.41 Applicability
105.43 Parachute equipment and packing requirements

Subpart A—General

Source: Docket No. 1491 (27 FR 11636) 11/27/62, unless otherwise noted.

§ 105.1 Applicability.

(a) This part prescribes rules governing parachute jumps made in the United States except parachute jumps necessary because of an inflight emergency.

(b) For the purposes of this part, a *parachute jump* means the descent of a person, to the surface from an aircraft in flight, when he intends to use, or uses, a parachute during all or part of that descent.

Subpart B—Operating rules

§ 105.11 Applicability.

(a) Except as provided in paragraphs (b) and (c) of this section, this subpart prescribes operating rules governing parachute jumps to which this part applies.

(b) This subpart does not apply to a parachute jump necessary to meet an emergency on the surface, when it is made at the direction, or with the approval, of an agency of the United States, or of a State, Puerto Rico, the District of Columbia, or a possession of the United States, or of a political subdivision of any of them.

(c) Sections 105.13 through 105.17 and §§ 105.27 through 105.37 of this subpart do not apply to a parachute jump made by a member of an Armed Force:

(1) Over or within a restricted area when that area is under the control of an Armed Force; or

(2) In military operations in uncontrolled airspace.

(d) Section 105.23 does not apply to a parachute jump made by a member of an Armed Force within a restricted area that extends upward from the surface when that area is under the control of an Armed Force.

(Amdt. 105-4, Eff. 9/21/68)

§ 105.13 General.

No person may make a parachute jump, and no pilot in command of an aircraft may allow a parachute jump to be made from that aircraft, if that jump creates a hazard to air traffic or to persons or property on the surface.

§ 105.14 Radio equipment and use requirements.

(a) Except when otherwise authorized by ATC—

(1) No person may make a parachute jump, and no pilot in command of an aircraft may allow a parachute jump to be made from that aircraft, in or into controlled airspace unless, during that flight—

(i) The aircraft is equipped with a functioning two-way radio communications system appropriate to the ATC facilities to be used;

(ii) Radio communications have been established between the aircraft and the nearest FAA air traffic control facility or FAA flight service station at least 5 minutes before the jumping activity is to begin, for the purpose of receiving information in the aircraft about known air traffic in the vicinity of the jumping activity; and

(iii) The information described in paragraph (a)(1)(ii) of this section has been received by the pilot in command and the jumpers in that flight; and

(2) The pilot in command of an aircraft used for any jumping activity in or into controlled airspace shall, during each flight—

(i) Maintain or have maintained a continuous watch on the appropriate frequency of the aircraft's radio communications system from the time radio communications are first established between the aircraft and ATC, until he advises ATC that the jumping activity is ended from that flight; and

(ii) Advise ATC that the jumping activity is ended for that flight when the last parachute jumper from the aircraft reaches the ground.

(b) If, during any flight, the required radio communications system is or becomes inoperative, any jumping activity from the aircraft in or into controlled airspace shall be abandoned. However, if the communications system becomes inoperative in flight after receipt of a required ATC authorization, the jumping activity from that flight may be continued.

(Amdt. 105-2, Eff. 3/24/67)

§ 105.15 Jumps over or into congested areas or open air assembly of persons.

(a) No person may make a parachute jump, and no pilot in command of an aircraft may allow a parachute jump to be made from that aircraft, over or into a congested area of a city, town, or settlement, or an open air assembly of persons unless a certificate of authorization for that jump has been issued under this section. However, a parachutist may drift over that congested area or open air assembly with a fully deployed and properly functioning parachute if he is at a sufficient altitude to avoid creating a hazard to persons and property on the ground.

(b) An application for a certificate of authorization issued under this section is made in a form and in a manner prescribed by the Administrator and must be submitted to the FAA Flight Standards District Office having jurisdiction over the area in which the parachute jump is to be made, at least 4 days before the day of that jump.

(c) Each holder of a certificate of authorization issued under this section shall present that certificate for inspection upon the request of the Administrator, or any Federal, State, or local official.

(Amdt. 105-1, Eff. 12/4/64); (Amdt. 105-7, Eff. 6/26/78)

§ 105.17 Jumps over or onto airports.

Unless prior approval has been given by the airport management, no person may make a parachute jump, and no pilot in command of an aircraft may allow a parachute jump to be made from that aircraft—

(a) Over an airport that does not have a functioning control tower operated by the United States; or

(b) Onto any airport.

However, a parachutist may drift over that airport with a fully deployed and properly functioning parachute if he is at least 2,000 feet above that airport's traffic pattern, and avoids creating a hazard to air traffic or to persons and property on the ground.

Docket No. 4057 (29 FR 14920) Eff. 11/4/64; (Amdt. 105-1, Eff. 12/4/64)

[§ 105.19 Jumps in or into Class A, Class B, Class C, and Class D airspace.

[(a) No person may make a parachute jump, and no pilot in command may allow a parachute jump to be made from that aircraft, in or into Class A, Class B, Class C, and Class D airspace without, or in violation of, the terms of an ATC authorization issued under this section.

[(b) Each request for an authorization under this section must be submitted to the nearest FAA air traffic control facility or FAA flight service station and must include the information prescribed by § 105.25(a).]

[(Amdt. 105-10, Eff. 9/16/93)]

[§ 105.20 Removed and Reserved]

[Docket No. 23708 (50 FR 9259) Eff. 3/6/85; (Amdt. 105-8, Eff. 3/14/85); [(Amdt. 105-10, Eff. 9/16/93)]

[§ 105.21 Removed and Reserved]

(Amdt. 105-2, Eff. 3/24/67); (Amdt. 105-9, Eff. 7/19/86); [*(Amdt. 105-10, Eff. 9/16/93)*]

§ 105.23 Jumps in or into other airspace.

(a) No person may make a parachute jump, and no pilot in command of an aircraft may allow a parachute jump to be made from that aircraft, in or into airspace unless the nearest FAA air traffic control facility or FAA flight service station was notified of that jump at least 1 hour before the jump is to be made, but not more than 24 hours before the jumping is to be completed, and the notice contained the information prescribed in § 105.25(a).

(b) Notwithstanding paragraph (a) of this section, ATC may accept from a parachute jumping organization a written notification of a scheduled series of jumps to be made over a stated period of time not longer than 12 calendar months. The notification must contain the information prescribed by § 105.25(a), identify the responsible persons associated with that jumping activity, and be submitted at least 15 days, but not more than 30 days, before the jumping is to begin. ATC may revoke the acceptance of the notification for any failure of the jumping organization to comply with its terms.

(c) This section does not apply to parachute jumps in or into any airspace or place described in § 105.15, § 105.19, or § 105.21.

(Amdt. 105-2, Eff. 3/24/67)

§ 105.25 Information required, and notice of cancellation or postponement of jump.

(a) Each person requesting an authorization under § 105.19 or § 105.21, and each person submitting a notice under § 105.23, must include the following information (on an individual or group basis) in that request or notice:

(1) The date and time jumping will begin.

(2) The size of the jump zone expressed in nautical mile radius around the target.

(3) The location of the center of the jump zone in relation to—

(i) The nearest VOR facility in terms of the VOR radial on which it is located, and its distance in nautical miles from the VOR facility when that facility is 30 nautical miles or less from the drop zone target; or

(ii) The nearest airport, town, or city depicted on the appropriate Coast and Geodetic Survey WAC or Sectional Aeronautical chart, when the nearest VOR facility is more than 30 nautical miles from the drop zone target.

(4) The altitudes above mean sea level at which jumping will take place.

(5) The duration of the intended jump.

(6) The name, address, and telephone number of the person requesting the authorization or giving notice.

(7) The identification of the aircraft to be used.

(8) The radio frequencies, if any, available in the aircraft.

(b) Each person requesting an authorization under § 105.19 or § 105.21, and each person submitting a notice under § 105.23, must promptly notify the FAA air traffic control facility or FAA flight service station from which it requested authorization or which it notified, if the proposed or scheduled jumping activity is canceled or postponed.

(Amdt. 105-2, Eff. 3/24/67); (Amdt. 105-6, Eff. 11/26/76); (Amdt. 105-9, Eff. 7/19/86)

§ 105.27 Jumps over or within restricted or prohibited areas.

No person may make a parachute jump, and no pilot in command may allow a parachute jump to be made from that aircraft, over or within a restricted area or prohibited area unless the controlling agency of the area concerned has authorized that jump.

§ 105.29 Flight visibility and clearance from clouds requirements.

No person may make a parachute jump, and no pilot in command of an aircraft may allow a parachute jump to be made from that aircraft—

(a) Into or through a cloud; or

(b) When the flight visibility is less, or at a distance from clouds that is less, than that prescribed in the following table:

Altitude	Flight visibility (statute miles)	Distance from clouds
(1) 1,200 feet or less above the surface regardless of the MSL altitude.	3	500 feet below. 1,000 feet above. 2,000 feet horizontal.
(2) More than 1,200 feet above the surface but less than 10,000 feet MSL.	3	500 feet below. 1,000 feet above. 2,000 feet horizontal.
(3) More than 1,200 feet above the surface and at or above 10,000 feet MSL.	5	1,000 feet below. 1,000 feet above. 1 mile horizontal.

(Amdt. 105-1, Eff. 12/4/64); (Amdt. 105-5, Eff. 6/12/71)

§ 105.33 Parachute jumps between sunset and sunrise.

(a) No person may make a parachute jump, and no pilot in command of an aircraft may allow any person to make a parachute jump from that aircraft, between sunset and sunrise, unless that person is equipped with a means of producing a light visible for at least 3 statute miles.

(b) Each person making a parachute jump between sunset and sunrise shall display the light required by paragraph (a) of this section from the time that person exits the aircraft until that person reaches the surface.

(Amdt. 105-7, Eff. 6/12/78)

§ 105.35 Liquor and drugs.

No person may make a parachute jump while, and no pilot in command of an aircraft may allow a person to make a parachute jump from that aircraft if that person appears to be:

(a) Under the influence of intoxicating liquor; or

(b) Using any drug that affects his faculties in any way contrary to safety.

§ 105.37 Inspections.

The Administrator may inspect (including inspections at the jump site), any parachute jump operation to which this part applies, to determine compliance with the regulations of this part.

Subpart C—Parachute equipment

§ 105.41 Applicability.

(a) Except as provided in paragraph (b) of this section, this subpart prescribes rules governing parachute equipment used in parachute jumps to which this part applies.

(b) This subpart does not apply to a parachute jump made by a member of an Armed Force using parachute equipment of an Armed Force.

§ 105.43 Parachute equipment and packing requirements.

(a) No person may make a parachute jump, and no pilot in command of an aircraft may allow any person to make a parachute jump from that aircraft, unless that person is wearing a single harness dual parachute pack, having at least one main parachute and one approved auxiliary parachute that are packed as follows:

(1) The main parachute must have been packed by a certificated parachute rigger, or by the person making the jump, within 120 days before the date of its use.

(2) The auxiliary must have been packed by a certificated and appropriately rated parachute rigger:

(i) Within 120 days before the date of use, if its canopy, shroud, and harness are composed exclusively of nylon, rayon, or other similar synthetic fiber or material that is substantially resistant to damage from mold, mildew, or other fungi and other rotting agents propagated in a moist environment; or

(ii) Within 60 days before the date of use, if it is composed in any amount of silk, pongee, or other natural fiber, or material not specified in paragraph (a)(2)(i) of this section.

(b) No person may make a parachute jump using a static line attached to the aircraft and the main parachute unless an assist device, described and attached as follows, is used to aid the pilot chute in performing its function, or, if no pilot chute is used, to aid in the direct deployment of the main parachute canopy.

(1) The assist device must be long enough to allow the container to open before a load is placed on the device.

(2) The assist device must have a static load strength of—

(i) At least 28 pounds but not more than 160 pounds, if it is used to aid the pilot chute in performing its function; or

(ii) At least 56 pounds but not more than 320 pounds, if it is used to aid in the direct deployment of the main parachute canopy.

(3) The assist device must be attached—

(i) At one end, to the static line above the static line pins, or, if static pins are not used, above the static line ties to the parachute cone; and

(ii) At the other end, to the pilot chute apex, bridle cord or bridle loop, or, if no pilot chute is used, to the main parachute canopy.

(c) No person may attach an assist device required by paragraph (b) of this section to any main parachute unless he has a current parachute rigger certificate issued under Part 65 of this chapter or is the person who makes the jump with that parachute.

(d) For the purpose of this section, an *approved parachute* is:

(1) A parachute manufactured under a type certificate or a technical standard order (C-23 series); or

(2) A personnel-carrying military parachute (other than a high altitude, high-speed, or ejection kind) identified by an NAF, AAF, or AN drawing number, an AAF order number, or any other military designation or specification number.

(Amdt. 105-3, Eff. 8/7/68); (Amdt, 105-7, Eff. 6/26/78)

Part 121 – Appendix I—Drug testing program

This appendix contains the standards and components that must be included in an antidrug program required by this chapter.

I. *DOT Procedures.* Each employer shall ensure that drug testing programs conducted pursuant to 14 CFR parts 65, 121, and 135 complies with the requirements of this appendix and the "Procedures for Transportation Workplace Drug Testing Programs" published by the Department of Transportation (DOT) (49 CFR part 40). An employer may not use or contract with any drug testing laboratory that is not certified by the Department of Health and Human Services (DHHS) pursuant to the DHHS "Mandatory Guidelines for Federal Workplace Drug Testing Programs" (53 FR 11970; April 11, 1988 as amended by 59 FR 29908; June 9, 1994).

II. *Definitions.* For the purpose of this appendix, the following definitions apply:

Accident means an occurrence associated with the operation of an aircraft which takes place between the time any person boards the aircraft with the intention of flight and all such persons have disembarked, and in which any person suffers death or serious injury, or in which the aircraft receives substantial damage.

Annualized rate for the purposes of unannounced testing of employees based on random selection means the percentage of specimen collection and testing of employees performing a safety-sensitive function during a calendar year. The employer shall determine the annualized rate by referring to the total number of employees performing a safety-sensitive function for the employer at the beginning of the calendar year.

Employee is a person who performs, either directly or by contract, a safety-sensitive function for an employer, as defined below. Provided, however, that an employee who works for an employer who holds a part 135 certificate and who holds a part 121 certificate is considered to be an employee of the part 121 certificate holder for the purposes of this appendix.

Employer is a part 121 certificate holder, a part 135 certificate holder, an operator as defined in § 135.1 (c) of this chapter, or an air traffic control facility not operated by the FAA or by or under contract to the U.S. military. Provided, however, that an employer may use a person who is not included under that employer's drug program to perform a safety-sensitive function, if that person is subject to the requirements of another employer's FAA-approved antidrug program.

Performing (a safety-sensitive function): an employee is considered to be performing a safety-sensitive function during any period in which he or she is actually performing, ready to perform, or immediately available to perform such function.

Prohibited drug means marijuana, cocaine, opiates, phencyclidine (PCP), amphetamines, or a substance specified in Schedule I or Schedule II of the Controlled Substances Act, 21 U.S.C. 811, 812, unless the drug is being used as authorized by a legal prescription or other exemption under Federal, state, or local law.

Refusal to submit means that an individual failed to provide a urine sample as required in 49 CFR part 40, without a valid medical explanation, after he or she has received notice of the requirement to be tested in accordance with this appendix or engaged in conduct that clearly obstructed the testing process.

Safety-sensitive function means a function listed in section III of this appendix.

Substance abuse professional means a licensed physician (Medical Doctor or Doctor of Osteopathy), or a licensed or certified psychologist, social worker, employee assistance professional, or addiction counselor (certified by the National Association of Alcoholism and Drug Abuse Counselors Certification Commission), with knowledge of and clinical experience in the diagnosis and treatment of disorders related to drug use and abuse.

Verified negative drug test result means that the test result of a urine sample collected and tested under this appendix has been verified by a Medical Review Officer as negative in accordance with 49 CFR part 40.

Verified positive drug test result means that the test result of a urine sample collected and tested under this appendix has been verified by a Medical Review Officer as positive in accordance with 49 CFR part 40.

III. *Employees Who Must Be Tested.* Each person who performs a safety-sensitive function directly or by contract for an employer must be tested pursuant to an FAA-approved antidrug program conducted in accordance with this appendix:

A. Flight crewmember duties.

B. Flight attendant duties.

C. Flight instruction duties.

D. Aircraft dispatcher duties.

E. Aircraft maintenance or preventive maintenance duties.

F. Ground security coordinator duties.

G. Aviation screening duties.

H. Air traffic control duties.

IV. *Substances for Which Testing Must Be Conducted.* Each employer shall test each employee who performs a safety-sensitive function for evidence of marijuana, cocaine, opiates, phencyclidine (PCP), and amphetamines during each test required by section V of this appendix. As part of a reasonable cause drug testing program established pursuant to this part, employers may test for drugs in addition to those specified in this pad only with approval granted by the FAA under 49 CFR part 40 and for substances for which the Department of Health and Human Services has established an approved testing protocol and positive threshold.

V. *Types of Drug Testing Required.* Each employer shall conduct the following types of testing in accordance with the procedures set forth in this appendix and the DOT "Procedures for Transportation Workplace Drug Testing Programs" (49 CFR part 40):

A. *Pre-employment Testing.*

1. Prior to the first time an individual performs a safety-sensitive function for an employer, the employer shall require the individual to undergo testing for prohibited drug use.

2. An employer is permitted to require pre-employment testing of an individual if the following criteria are met.

(a) The individual previously performed a covered function for the employer;

(b) The employer removed the individual from the employer's random testing program conducted under this appendix for reasons other than a verified positive test result on an FAA-mandated drug test or a refusal to submit to such testing; and

(c) The individual will be returning to the performance of a safety-sensitive function.

3. No employer shall allow an individual required to undergo pre-employment testing under section V, paragraphs A.1 or A.2 of this appendix to perform a safety-sensitive function unless the employer has received a verified negative drug test result for the individual.

4. The employer shall advise each individual applying to perform a safety-sensitive function at the time of application that the individual will be required to undergo pre-employment testing to determine the presence of marijuana, cocaine, opiates, phencyclidine (PCP), and amphetamines, or a metabolite of those drugs in the individual's system. The employer shall provide this same notification to each individual required by the employer to undergo pre-employment testing under section V, paragraph A.(2) of this appendix.

B. *Periodic Testing.* Each employee who performs a safety-sensitive function for an employer and who is required to undergo a medical examination under part 67 of this chapter shall submit to a periodic drug test. The employee shall be tested for the presence of marijuana, cocaine, opiates, phencyclidine (PCP), and amphetamines, or a metabolite of those drugs during the first calendar year of implementation of the employer's antidrug program. The tests shall be conducted in conjunction with the first medical evaluation of the employee or in accordance with an alternative method for collecting periodic test specimens detailed in an employer's approved antidrug program. An employer may discontinue periodic testing of its employees after the first calendar year of implementation of the employer's antidrug program when the employer has implemented an unannounced testing program based on random selection of employees.

C. *Random Testing.* Each employer shall randomly select employees who perform a safety-sensitive function for the employer for unannounced drug testing. The employer shall randomly select employees for unannounced testing for the presence of marijuana, cocaine, opiates, phencyclidine (PCP), and amphetamines, or a metabolite of those drugs in an employee's system using a random number table or a computer-based, number generator that is matched with an employee's social security number, payroll identification number, or any other alternative method approved by the FAA.

(1) During the first 12 months following implementation of unannounced testing based on random selection pursuant to this appendix, an employer shall meet the following conditions:

(a) The unannounced testing based on random selection of employees shall be spread reasonably throughout the 12-month period.

(b) The last collection of specimens for random testing during the year shall be conducted at an annualized rate equal to not less than 50 percent of employees performing a safety-sensitive function.

(c) The total number of unannounced tests based on random selection during the 12 months shall be equal to not less than 25 percent of the employees performing a safety-sensitive function.

(2) Following the first 12 months, an employer shall achieve and maintain an annualized rate equal to not less than 50 percent of employees performing a safety-sensitive function.

D. *Post-accident Testing.* Each employer shall test each employee who performs a safety-sensitive function for the presence of marijuana, cocaine, opiates, phencyclidine (PCP), and amphetamines, or a metabolite of those drugs in the employee's system if that employee's performance either contributed to an accident or can not be completely discounted as a contributing factor to the accident. The employee shall be tested as soon as possible but not later than 32 hours after the accident. The decision not to administer a test under this section must be based on a determination, using the best information available at the time of the determination, that the employee's performance could not have contributed to the accident. The employee shall submit to post-accident testing under this section.

E. *Testing Based on Reasonable Cause.* Each employer shall test each employee who performs a safety-sensitive function and who is reasonably suspected of using a prohibited drug. Each employer shall test an employee's specimen for the presence of marijuana, cocaine, opiates, phencyclidine (PCP), and amphetamines, or a metabolite of those drugs. An employer may test an employee's specimen for the presence of other prohibited drugs or drug metabolites only in accordance with this appendix and the DOT "Procedures for Transportation Workplace Drug Testing Programs" (49 CFR part 40). At least two of the employee's supervisors, one of whom is trained in detection of the symptoms of possible drug use, shall substantiate and concur in the decision to test an employee who is reasonably suspected of drug use; provided, however, that in the case of an employer other than a part 121 certificate holder who employs 50 or fewer employees who perform safety-sensitive functions, one supervisor who is trained in detection of symptoms of possible drug use shall substantiate the decision to test an employee who is reasonably suspected of drug use. The decision to test must be based on a reasonable and articulable belief that the employee is using a prohibited drug on the basis of specific contemporaneous physical, behavioral, or performance indicators of probable drug use.

F. *Return to Duty Testing.* Each employer shall ensure that before an individual is returned to duty to perform a safety-sensitive function after refusing to submit to a drug test required by this appendix or receiving a verified positive drug test result on a test conducted under this appendix the individual shall undergo a drug test. No employer shall allow an individual required to undergo return to duty testing to perform a safety-sensitive function unless the employer has received a verified negative drug test result for the individual.

G. *Follow-up Testing.* Each employer shall implement a reasonable program of unannounced testing of each individual who has been hired to perform or who has been returned to the performance of a safety-sensitive function after refusing to submit to a drug test required by this appendix or receiving a verified positive drug test result on a test conducted under this appendix.

2. The number and frequency of such testing shall be determined by the employer's Medical Review Officer. In the case of any individual evaluated under this appendix and determined to be in need of assistance in resolving problems associated with illegal use of drugs, follow-up testing shall consist of at least six tests in the first 12 months following the employee's return to duty.

3. The employer may direct the employee to undergo testing for alcohol, in addition to drugs, if the Medical Review Officer determines that alcohol testing is necessary for the particular employee. Any such alcohol testing shall be conducted in accordance with the provisions of 49 CFR part 40.

4. Follow-up testing shall not exceed 60 months after the date the individual begins to perform or returns to the performance of a safety-sensitive function. The Medical Review Officer may terminate the requirement for follow-up testing at any time after the first six tests have been conducted, if the Medical Review Officer determines that such testing is no longer necessary.

VI. Administrative and other matters

A. *Collection, Testing, and Rehabilitation Records.* Each employer shall maintain all records related to the collection process, including all logbooks and certification statements, for two years. Each employer shall maintain records of employee confirmed positive drug test results, SAP evaluations, and employee rehabilitation for five years. The employer shall maintain records of negative test results for 12 months. The employer shall permit the Administrator or the Administrator's representative to examine these records.

B. *Laboratory Inspections.* The employer shall contract only with a laboratory that permits pre-award inspections by the employer before the laboratory is awarded a testing contract and unannounced inspections, including examination of any and all records at any time by the employer, the Administrator, or the Administrator's representative.

C. *Employee Request for Test of a Split Specimen.* Not later than 72 hours after receipt of notice of a verified positive test result, an employee may request that the MRO arrange for testing of the second, "split" specimen obtained during the collection of the primary specimen that resulted in the confirmed positive test result.

2. The split specimen shall be tested in accordance with the procedures in 49 CFR part 40.

3. The MRO shall not delay verification of the primary test result following a request for a split specimen test unless such delay is based on reasons other than the pendency of the split specimen test result. If the primary test result is verified as positive, actions required under this rule (e.g., notification to the Federal Air Surgeon, removal from safety-sensitive position) are not stayed during the 72-hour request period or pending receipt of the split specimen test result.

D. *Release of Drug Testing Information.* An employer shall release information regarding an employee's drug testing results, evaluation, or rehabilitation to a third party in accordance with the specific, written consent of the employee authorizing release of the information to an identified person, to the National Transportation Safety Board as part of an accident investigation upon written request or order, to the FAA upon request, or as required by this appendix. Except as required by law or this appendix, no employer shall release employee information.

E. *Refusal to Submit to Testing.* Each employer shall notify the FAA within 5 working days of any employee who holds a certificate issued under part 61, part 63, or part 65 of this chapter who has refused to submit to a drug test required under this appendix. Notification should be sent to Federal Aviation Administration, Aviation Standards National Field Office, Airmen Certification Branch, AVN-460, P.O. Box 25082, Oklahoma City, OK 73125.

2. Employers are not required to notify the above office of refusals to submit to pre-employment or return to duty testing.

F. *Permanent Disqualification from Service.* An employee who has verified positive drug test results on two drug tests required by Appendix I to part 121 of this chapter and conducted after September 19, 1994 is permanently precluded from performing for an employer the safety-sensitive duties the employee performed prior to the second drug test.

2. An employee who has engaged in prohibited drug use during the performance of a safety-sensitive function after September 19, 1994 is permanently precluded from performing that safety-sensitive function for an employer.

VII. Medical Review Officer/Substance Abuse Professional

The employer shall designate or appoint a Medical Review Officer (MRO) who shall be qualified in accordance with 49 CFR part 40 and shall perform the functions set forth in 49 CFR part 40 and this appendix. If the employer does not have a qualified individual on staff to serve as MRO, the employer may contract for the provision of MRO services as part of its drug testing program.

A. *MRO and Substance Abuse Professional Duties.* In addition to the functions delineated in 49 CFR part 40, the MRO shall perform the duties listed hereunder.

1. During the MRO's interview with an employee or applicant who has had a confirmed positive drug test result, the MRO shall inquire, and the individual must disclose, whether the individual holds an airman medical certificate issued under part 67 of this chapter or, if an applicant, would be required to hold such certificate in order to perform the duties of the position for which the applicant is applying.

2. The MRO must process employee requests for testing of split specimens in accordance with section VI, paragraph C, of this appendix.

3. The MRO shall advise each employee who receives a verified positive drug test result on or refuses to submit to a drug test required under this appendix of the resources available to the employee in evaluating and resolving problems associated with illegal use of drugs, including the names, addresses, and telephone numbers of substance abuse professionals (SAP) and counseling and treatment programs.

4. The MRO shall ensure that each employee who receives a verified positive drug test result on or refuses to submit to a drug test required under this appendix is evaluated by a SAP to determine if the employee is in need of assistance in resolving problems associated with illegal use of drugs. The MRO may perform this evaluation if the MRO is qualified as a SAP.

5. Prior to recommending that an employee be returned to the performance of a safety-sensitive function after the employee has received a verified positive drug test result on or refused to submit to a drug test required by this appendix, the MRO shall—

a. Ensure that an employee returning to the performance of a safety-sensitive function has received a return to duty verified negative drug test result on a test conducted under section V., paragraph F of this appendix;

b. Ensure that each employee has been evaluated in accordance, with section VII, paragraph A.4 of this appendix; and

c. Ensure that the employee demonstrates compliance with any rehabilitation program recommended following the evaluation required under section VII, paragraph A.4 of this appendix.

6. Prior to recommending that an individual be hired to perform a safety-sensitive function after such individual has received a verified positive drug test result on a pre-employment test or has refused to submit to a pre-employment drug test required by this appendix, the MRO shall—

a. Ensure that an individual has received a verified negative drug test result on a subsequent pre-employment test conducted under section V, paragraph A, of this appendix;

b. Evaluate the individual (if the MRO is qualified to be a SAP), or have the individual evaluated by a SAP, for drug use or abuse; and

c. Ensure that the individual has complied with the requirements of any rehabilitation program in which the individual participated following the verified positive pre-employment drug test result or the refusal to submit to a pre-employment test.

7. The MRO shall not recommend that a person who fails to satisfy the requirements in section VII, paragraph A.5 or A.6 of this appendix be hired to perform or returned to duty to perform a safety-sensitive function.

B. *MRO Determinations.* In the case of an employee or applicant who holds an airman medical certificate issued under part 67 of this chapter, or who is or would be required to hold such certificate in order to perform a safety-sensitive function for an employer, the MRO shall take the following actions after verifying a positive drug test result.

1. In addition to the evaluation required in section VII, paragraph A.4 of this appendix, the MRO shall make a determination of probable drug dependence or nondependence as specified in part 67 of this chapter within 10 working days of verifying the test result. If the MRO is unable to make such a determination, he or she should so state in the individual's records.

2. If the MRO determines that an individual is nondependent, the MRO may recommend that the individual be returned to duty or hired to perform safety-sensitive functions subject to the requirements of section VII, paragraph A.5 of this appendix. If the MRO makes a determination of probable drug dependence or cannot make a dependency determination, the MRO shall not recommend that the individual be returned to duty unless and until such individual has been found nondependent by or has received a

special issuance medical certificate from the Federal Air Surgeon.

3. After making the determinations in section VII, paragraphs B.1 and B.2 of this appendix, the MRO must forward the names of such individuals with identifying information, the determinations concerning dependence, SAP evaluation (if available), return to duty recommendations, and any supporting information to the Federal Air Surgeon within 12 working days after verifying the positive drug test result of such individuals.

4. All reports required under this section shall be forwarded to the Federal Air Surgeon, Federal Aviation Administration, Attn: Drug Abatement Division (AAM-800), 400 7th Street, SW., Washington, DC 20590.

C. *MRO Records.* Each MRO shall maintain records concerning drug tests performed under this rule in accordance with the following provisions:

1. All records shall be maintained in confidence and shall be released only in accordance with the provisions of this rule and 49 CFR part 40.

2. Records concerning drug tests confirmed positive by the laboratory shall be maintained for 5 years. Such records include the MRO copies of the custody and control form, medical interviews, documentation of the basis for verifying as negative test results confirmed as positive by the laboratory, any other documentation concerning the MRO's verification process, and copies of dependency determinations where applicable.

3. Records of confirmed negative test results shall be maintained for 12 months.

4. All records maintained pursuant to this rule by each MRO are subject to examination by the Administrator or the Administrator's representative at any time.

5. Should the employer change MROs for any reason, the employer shall ensure that the former MRO forwards all records maintained pursuant to this rule to the new MRO within 10 working days of receiving notice from the employer of the new MRO's name and address.

6. Any employer obtaining MRO services by contract, including a contract through a consortium, shall ensure that the contract includes a recordkeeping provision that is consistent with this paragraph, including requirements for transferring records to a new MRO.

D. *Evaluations and Referrals.* Each employer shall ensure that a substance abuse professional, including an MRO if he/she is qualified as a substance abuse professional, who determines that a covered employee requires assistance in resolving problems associated with illegal use of drugs does not refer the employee to the substance abuse professional's private practice or to a person or organization from which the substance abuse professional receives remuneration or in which the substance abuse professional has a financial interest. This paragraph does not prohibit a substance abuse professional from referring an employee for assistance provided through—

1. A public agency, such as a State, county, or municipality;

2. The employer or a person under contract to provide treatment for drug problems on behalf of the employer;

3. The sole source of therapeutically appropriate treatment under the employee's health insurance program; or

4. The sole source of therapeutically appropriate treatment reasonably accessible to the employee.

VIII. Employee Assistance Program (EAP)

The employer shall provide an EAP for employees. The employer may establish the EAP as a part of its internal personnel services or the employer may contract with an entity that will provide EAP services to an employee. Each EAP must include education and training on drug use for employees and training for supervisors making determinations for testing of employees based on reasonable cause.

A. *EAP Education Program.* Each EAP education program must include at least the following elements: display and distribution of informational material; display and distribution of a community service hot-line telephone number for employee assistance; and display and distribution of the employer's policy regarding drug use in the workplace. The employer's policy shall include information regarding the consequences under the rule of using drugs while performing safety-sensitive functions, receiving a verified positive drug test result, or refusing to submit to a drug test required under the rule.

B. *EAP Training Program.* Each employer shall implement a reasonable program of initial training for employees. The employee training program must include at least the following elements: The effects and consequences of drug use on personal health, safety, and work environment; the manifestations and behavioral cues that may indicate drug use and abuse; and documentation of training given to employees and employer's supervisory personnel. The employer's supervisory personnel who will determine when an employee is subject to testing based on reasonable cause shall receive specific training on specific, contemporaneous physical, behavioral, and performance indicators of probable drug use in addition to the training specified above. The employer shall ensure that supervisors who will make reasonable cause determinations receive at least 60 minutes of initial training. The employer shall implement a reasonable recurrent training program for supervisory personnel making reasonable cause determinations during subsequent years. The employer shall identify the employee and supervisor EAP training in the employer's drug testing plan submitted to the FAA for approval.

IX. Employer's antidrug program plan

A. *Schedule for Submission of Plans and Implementation.* Each employer shall submit an antidrug program plan to the Federal Aviation Administration, Office of Aviation Medicine, Drug Abatement Division (AAM-800), 400 7th Street, SW, Washington, DC 20590.

2. (a) Any person who applies for a certificate under the

provisions of part 121 or part 135 of this chapter after September 19, 1994 shall submit an antidrug program plan to the FAA for approval and must obtain such approval prior to beginning operations under the certificate. The program shall be implemented not later than the date of inception of operations. Contractor employees to a new certificate holder must be subject to an FAA-approved antidrug program within 60 days of the implementation of the employer's program.

(b) Any person who intends to begin sightseeing operations as an operator under 14 CFR 135.1(c) after September 19, 1994 shall, not later than 60 days prior to the proposed initiation of such operations, submit an antidrug program plan to the FAA for approval. No operator may begin conducting sightseeing flights prior to receipt of approval; the program shall be implemented concurrently with the inception of operations. Contractor employees to a new operator must be subject to an FAA-approved program within 60 days of the implementation of the employer's program.

(c) Any person who intends to begin air traffic control operations as an employer as defined in 14 CFR 65.46(a)(2) (air traffic control facilities not operated by the FAA or by or under contract to the U.S. military) after September 19, 1994 shall, not later than 60 days prior to the proposed initiation of such operations, submit an antidrug program plan to the FAA for approval. No air traffic control facility may begin conducting air traffic control operations prior to receipt of approval; the program shall be implemented concurrently with the inception of operations. Contractor employees to a new air traffic control facility must be subject to an FAA-approved program within 60 days of the implementation of the facility's program.

3. In accordance with this appendix, an entity or individual that holds a repair station certificate issued by the FAA pursuant to part 145 of this chapter and employs individuals who perform a safety-sensitive function pursuant to a primary or direct contract with an employer or an operator may submit an antidrug program plan (specifying the procedures for complying with this appendix) to the FAA for approval. Each certificated repair station shall implement its approved antidrug program in accordance with its terms.

4. Any entity or individual whose employees perform safety-sensitive functions pursuant to a contract with an employer (as defined in section II of this appendix), and any consortium may submit an antidrug program plan to the FAA for approval on a form and in a manner prescribed by the Administrator.

(a) The plan shall specify the procedures that will be used to comply with the requirements of this appendix.

(b) Each consortium program must provide for reporting changes in consortium membership to the FAA within 10 working days of such changes.

(c) Each contractor or consortium shall implement its antidrug program in accordance with the terms of its approved plan.

5. Each air traffic control facility operating under contract to the FAA shall submit an antidrug program plan to the FAA (specifying the procedures for all testing required by this appendix) not later than November 17, 1994. Each facility shall implement its antidrug program not later than 60 days after approval of the program by the FAA. Employees performing air traffic control duties by contract for the air traffic control facility (i.e., not directly employed by the facility) must be subject to an FAA-approved antidrug program within 60 days of implementation of the air traffic control facility's program.

6. Each employer, or contractor company that has submitted an antidrug plan directly to the FAA, shall ensure that it is continuously covered by an FAA-approved antidrug program, and shall obtain appropriate approval from the FAA prior to changing problems (e.g., joining another carrier's program, joining a consortium, or transferring to another consortium).

B. An employer's antidrug plan must specify the methods by which the employer will comply with the testing requirements of this appendix. The plan must provide the name and address of the laboratory which has been selected by the employer for analysis of the specimens collected during the employer's antidrug testing program.

C. An employer's antidrug plan must specify the procedures and personnel the employer will use to ensure that a determination is made as to the veracity of test results and possible legitimate explanations for an employee receiving a verified positive drug test result.

D. The employer shall consider its antidrug program to be approved by the Administrator, unless notified to the contrary by the FAA, within 60 days after submission of the plan to the FAA.

X. Reporting of antidrug program results

A. Annual reports of antidrug program results shall be submitted to the FAA in the form and manner prescribed by the Administrator by March 15 of the succeeding calendar year for the prior calendar year (January 1 through December 31) in accordance with the provisions below.

1. Each part 121 certificate holder shall submit an annual report each year.

2. Each entity conducting an antidrug program under an FAA-approved antidrug plan, other than a part 121 certificate holder, that has 50 or more employees performing a safety-sensitive function on January 1 of any calendar year shall submit an annual report to the FAA for that calendar year.

3. The Administrator reserves the right to require that aviation employers not otherwise required to submit annual reports prepare and submit such reports to the FAA. Employers that will be required to submit annual reports under this provision will be notified in writing by the FAA.

B. Each report shall be submitted, in the form and manner prescribed by the Administrator. No other form, including another DOT Operating Administration's form, is acceptable for submission to the FAA.

C. Each report shall be signed by the employer's antidrug program manager or other designated representative.

D. Each report with verified positive drug test results shall include all of the following informational elements:

1. Number of covered employees by employee Category.

2. Number of covered employees affected by the antidrug rule of another operating administration identified and reported by number and employee category.

3. Number of specimens collected by type of test and employee category.

4. Number of positive drug test results verified by a Medical Review Officer (MRO) by type of test, type of drug, and employee category.

5. Number of negative drug test results reported by an MRO by type of test and employee category.

6. Number of persons denied a safety-sensitive position based on a verified positive pre-employment drug test result reported by an MRO.

7. Action taken following a verified positive drug test result(s), by type of action.

8. Number of employees returned to duty during the reporting period after having received a verified positive drug test result on or refused to submit to a drug test required under the FAA rule.

9. Number of employees by employee category with tests verified positive for multiple drugs by an MRO.

10. Number of employees who refused to submit to a drug test and the action taken in response to the refusal(s).

11. Number of covered employees who have received required initial training.

12. Number of supervisory personnel who have received required initial training.

13. Number of supervisors who have received required recurrent training.

E. Each report with only negative drug test results shall include all of the following informational elements. (This report may *only* be submitted by employers with no verified positive drug test results during the reporting year.)

1. Number of covered employees by employee category.

2. Number of covered employees affected by the antidrug rule of another operating administration identified and reported by number and employee category.

3. Number of specimens collected by type of test and employee category.

4. Number of negative tests reported by an MRO by type of test and employee category.

5. Number of employees who refused to submit to a drug test and the action taken in response to the refusal(s).

6. Number of employees returned to duty during the reporting period after having received a verified positive drug test result on or refused to submit to a drug test required under the FAA rule.

7. Number of covered employees who have received required initial training.

8. Number of supervisory personnel who have received required initial training.

9. Number of supervisors who have received required recurrent training.

F. An FAA-approved consortium may prepare reports on behalf of individual aviation employers for purposes of compliance with this reporting requirement. However, the aviation employer shall sign and submit such a report and shall remain responsible for ensuring the accuracy and timeliness of each report prepared on its behalf by a consortium.

XI. Preemption

A. The issuance of 14 CFR parts 65, 121, and 135 by the FAA preempts any state or local law, rule, regulation, order, or standard covering the subject matter of 14 CFR parts 65, 121, and 135, including but not limited to, drug testing of aviation personnel performing safety-sensitive functions.

B. The issuance of 14 CFR parts 65, 121, and 135 does not preempt provisions of state criminal law that impose sanctions for reckless conduct of an individual that leads to actual loss of life, injury, or damage to property whether such provisions apply specifically to aviation employees or generally to the public.

XII. Employees located outside the territory of the United States

A. No individual shall undergo a drug test required under the provisions of this appendix while located outside the territory of the United States.

1. Each employee who is assigned to perform safety-sensitive functions solely outside the territory of the United States shall be removed from the random testing pool upon the inception of such assignment.

2. Each covered employee who is removed from the random testing pool under this paragraph A shall be returned to the random testing pool when the employee resumes the performance of safety-sensitive functions wholly or partially within the territory of the United States.

B. The provisions of this appendix shall not apply to any person who performs a function listed in section III of this appendix by contract for an employer outside the territory of the United States.

Appendix J—Alcohol misuse prevention program

This appendix contains the standards and components that must be included in an alcohol misuse prevention program required by this chapter.

I. General

A. *Purpose*. The purpose of this appendix is to establish programs designed to help prevent accidents and injuries resulting from the misuse of alcohol by employees who perform safety-sensitive functions in aviation.

B. *Alcohol testing procedures*. Each employer shall ensure that all alcohol testing conducted pursuant to this appendix complies with the procedures set forth in 49 CFR part 40. The provisions of 49 CFR part 40 that address alcohol testing are made applicable to employers by this appendix.

C. *Definitions*.

As used In this appendix—

Accident means an occurrence associated with the operation of an aircraft which takes place between the time any person boards the aircraft with the intention of flight and the time all such persons have disembarked, and in which any person suffers death or serious injury or in which the aircraft receives substantial damage.

Administrator means the Administrator of the Federal Aviation Administration or his or her designated representative.

Alcohol means the intoxicating agent in beverage alcohol, ethyl alcohol, or other low molecular weight alcohols, including methyl or isopropyl alcohol.

Alcohol concentration (or content) means the alcohol in a volume of breath expressed in terms of grams of alcohol per 210 liters of breath as indicated by an evidential breath test under this appendix.

Alcohol use means the consumption of any beverage, mixture, or preparation, including any medication, containing alcohol.

Confirmation test means a second test, following a screening test with a result 0.02 or greater, that provides quantitative data of alcohol concentration.

Consortium means an entity, including a group or association of employers or contractors, that provides alcohol testing as required by this appendix and that acts on behalf of such employers or contractors, provided that it has submitted an alcohol misuse prevention program certification statement to the FAA in accordance with this appendix.

Contractor company means a company that has employees who perform safety-sensitive functions by contract for an employer.

Covered employee means a person who performs, either directly or by contract, a safety-sensitive function listed in section II of this appendix for an employer as (defined below). For purposes of pre-employment only, the term "covered employee" includes a person applying to perform a safety-sensitive function.

DOT agency means an agency (or "operating administration") of the United States Department of Transportation administering regulations requiring alcohol testing (14 CFR parts 65, 121, and 135; 49 CFR parts 199, 219, and 382) in accordance with 49 CFR part 40.

Employer means a part 121 certificate holder; a part 135 certificate holder; an air traffic control facility not operated by the FAA or by or under contract to the U.S. military; and an operator as defined in 14 CFR 135.1(c).

Performing (a safety-sensitive function): an employee is considered to be performing a safety-sensitive function during any period in which he or she is actually performing, ready to perform, or immediately available to perform such functions.

Refuse to submit (to an alcohol test) means that a covered employee fails to provide adequate breath for testing without a valid medical explanation after he or she has received notice of the requirement to be tested in accordance with this appendix, or engages in conduct that clearly obstructs the testing progress.

Safety-sensitive function means a function listed in section II of this appendix.

Screening test means an analytical procedure to determine whether a covered employee may have a prohibited concentration of alcohol in his or her system.

Substance abuse professional means a licensed physician (Medical Doctor or Doctor of Osteopathy), or a licensed or certified psychologist, social worker, employee assistance professional, or an addiction counselor (certified by the National Association of Alcoholism and Drug Abuse Counselors Certification Commission) with knowledge of and clinical experience in the diagnosis and treatment of alcohol-related disorders.

Violation rate means the number of covered employees (as reported under section IV of this appendix) found during random tests given under this appendix to have an alcohol concentration of 0.04 or greater plus the number of employees who refused a random test required by this appendix, divided by the total reported number of employees in the industry given random alcohol tests under this appendix plus the total reported number of employees in the industry who refuse a random test required by this appendix.

D. *Preemption of State and local laws*.

1. Except as provided in subparagraph 2 of this paragraph, these regulations preempt any State or local law, rule, regulation, or order to the extent that:

(a) Compliance with both the State or local requirement and this appendix is not possible; or

(b) Compliance with the State or local requirement is an obstacle to the accomplishment and execution of any requirement in this appendix.

2. The alcohol misuse requirements of this title shall not be construed to preempt provisions of State criminal law that impose sanctions for reckless conduct leading to actual loss of life, injury, or damage to property, whether the provisions apply specifically to transportation employees or employers or to the general public.

E. *Other requirements imposed by employers.*

Except as expressly provided in these alcohol misuse requirements, nothing in these requirements shall be construed to affect the authority of employers, or the rights of employees, with respect to the use or possession of alcohol, including any authority and rights with respect to alcohol testing and rehabilitation.

F. *Requirement for notice.*

Before performing an alcohol test under this appendix, each employer shall notify a covered employee that the alcohol test is required by this appendix. No employer shall falsely represent that a test is administered under this appendix.

II. Covered employees

Each employee who performs a function listed in this section directly or by contract for an employer as defined in this appendix must be subject to alcohol testing under an FAA-approved alcohol misuse prevention program implemented in accordance with this appendix. The covered safety-sensitive functions are:

1. Flight crewmember duties.
2. Flight attendant duties.
3. Flight instruction duties.
4. Aircraft dispatcher duties.
5. Aircraft maintenance or preventive maintenance duties.
6. Ground security coordinator duties.
7. Aviation screening duties.
8. Air traffic control duties.

III. Tests required

A. *Pre-employment*

1. Prior to the first time a covered employee performs safety-sensitive functions for an employer, the employee shall undergo testing for alcohol. No employer shall allow a covered employee to perform safety-sensitive functions unless the employee has been administered an alcohol test with a result indicating an alcohol concentration less than 0.04. If a pre-employment test result under this paragraph indicates an alcohol concentration of 0.02 or greater but less than 0.04, the provisions of paragraph F of section V of this appendix apply.

2. An employer is not required to administer an alcohol test as required by this paragraph if:

(a) The employee has undergone an alcohol test required by this appendix or the alcohol misuse rule of another DOT agency under 49 CFR part 40 within the previous 6 months, with a result indicating an alcohol concentration less than 0.04; and

(b) The employer ensures that no prior employer of the covered employee of whom the employer has knowledge has records of a violation of § 65-46a, 121.458, or 135.253 of this chapter or the alcohol misuse rule of another DOT agency within the previous 6 months.

B. *Post-accident*

1. As soon as practicable following an accident, each employer shall test each surviving covered employee for alcohol if that employee's performance of a safety-sensitive function either contributed to the accident or cannot be completely discounted as a contributing factor to the accident. The decision not to administer a test under this section shall be based on the employer's determination, using the best available information at the time of the determination, that the covered employee's performance could not have contributed to the accident.

2. If a test required by this section is not administered within 2 hours following the accident, the employer shall prepare and maintain on file a record stating the reasons the test was not promptly administered. If a test required by this section is not administered within 8 hours following the accident, the employer shall cease attempts to administer an alcohol test and shall prepare and maintain the same record. Records shall be submitted to the FAA upon request of the Administrator or his or her designee.

3. A covered employee who is subject to post-accident testing shall remain readily available for such testing or may be deemed by the employer to have refused to submit to testing. Nothing in this section shall be construed to require the delay of necessary medical attention for injured people following an accident or to prohibit a covered employee from leaving the scene of an accident for the period necessary to obtain assistance in responding to the accident or to obtain necessary emergency medical care.

C. *Random testing*

1. Except as provided in paragraphs 2–4 of this section, the minimum annual percentage rate for random alcohol testing will be 25 percent of the covered employees.

2. The Administrator's decision to increase or decrease the minimum annual percentage rate for random alcohol testing is based on the violation rate for the entire industry. All information used for this determination is drawn from alcohol MIS reports required by this appendix. In order to ensure reliability of the data, the Administrator considers the quality and completeness of the reported data, may obtain additional information or reports from employers, and may make appropriate modifications in calculating the industry violation rate. Each year, the Administrator will publish in the **Federal Register** the minimum annual percentage rate for random alcohol testing of covered employees. The now minimum annual percentage rate for random alcohol testing will be applicable starting January 1 of the calendar year following publication.

3. (a) When the minimum annual percentage rate for random alcohol testing is 25 percent or more, the Administrator may lower this rate to 10 percent of all covered employees

if the Administrator determines that the data received under the reporting requirements of this appendix for two consecutive calendar years indicate that the violation rate is less than 0.5 percent.

(b) When the minimum annual percentage rate for random alcohol testing is 50 percent, the Administrator may lower this rate to 25 percent of all covered employees if the Administrator determines that the data received under the reporting requirements of this appendix for two consecutive calendar years indicate that the violation rate is less than 1.0 percent but equal to or greater than 0.5 percent.

4. (a) When the minimum annual percentage rate for random alcohol testing is 10 percent, and the data received under the reporting requirements of this appendix for that calendar year indicate that the violation rate is equal to or greater than 0.5 percent but less than 1.0 percent, the Administrator will increase the minimum annual percentage rate for random alcohol testing to 25 percent of all covered employees.

(b) When the minimum annual percentage rate for random alcohol testing is 25 percent or less, and the data received under the reporting requirements of this appendix for that calendar year indicate that the violation rate is equal to or greater than 1.0 percent, the Administrator will increase the minimum annual percentage rate for random alcohol testing to 50 percent of all covered employees.

5. The selection of employees for random alcohol testing shall be made by a scientifically valid method, such as a random-number table or a computer-based random number generator that is matched with employees' Social Security numbers, payroll identification numbers, or other comparable identifying numbers. Under the selection process used, each covered employee shall have an equal chance of being tested each time selections are made.

6. The employer shall randomly select a sufficient number of covered employees for testing during each calendar year to equal an annual rate not less than the minimum annual percentage rate for random alcohol testing determined by the Administrator. If the employer conducts random testing through a consortium, the number of employees to be tested may be calculated for each individual employer or may be based on the total number of covered employees who are subject to random alcohol testing at the same minimum annual percentage rate under this appendix or any DOT alcohol testing rule.

7. Each employer shall ensure that random alcohol tests conducted under this appendix are unannounced and that the dates for administering random tests are spread reasonably throughout the calendar year.

8. Each employer shall require that each covered employee who is notified of selection for random testing proceeds to the testing site immediately; provided, however, that if the employee is performing a safety-sensitive function at the time of the notification, the employer shall instead ensure that the employee ceases to perform the safety-sensitive function and proceeds to the testing site as soon as possible.

9. A covered employee shall only be randomly tested while the employee is performing safety-sensitive functions; just before the employee is to perform safety-sensitive functions; or just after the employee has ceased performing such functions.

10. If a given covered employee is subject to random alcohol testing under the alcohol testing rules of more than one DOT agency, the employee shall be subject to random alcohol testing at the percentage rate established for the calendar year by the DOT agency regulating more than 50 percent of the employee's functions.

11. If an employer is required to conduct random alcohol testing under the alcohol testing rules of more than one DOT agency, the employer may—

(a) Establish separate pools for random selection, with each pool containing the covered employees who are subject to testing at the same required rate; or

(b) Randomly select such employees for testing at the highest percentage rate established for the calendar year by any DOT agency to which the employer is subject.

D. *Reasonable Suspicion Testing*

1. An employer shall require a covered employee to submit to an alcohol test when the employer has reasonable suspicion to believe that the employee has violated the alcohol misuse prohibitions in § 65.46a, 121.458, or 135.253 of this chapter.

2. The employer's determination that reasonable suspicion exists to require the covered employee to undergo an alcohol test shall be based on specific, contemporaneous, articulable observations concerning the appearance, behavior, speech or body odors of the employee. The required observations shall be made by a supervisor who is trained in detecting the symptoms of alcohol misuse. The supervisor who makes the determination that reasonable suspicion exists shall not conduct the breath alcohol test on that employee.

3. Alcohol testing is authorized by this section only if the observations required by paragraph 2 are made during, just preceding, or just after the period of the work day that the covered employee is required to be in compliance with this rule. An employee may be directed by the employer to undergo reasonable suspicion testing for alcohol only while the employee is performing safety-sensitive functions; just before the employee is to perform safety-sensitive functions; or just after the employee has ceased performing such functions.

4. (a) If a test required by this section is not administered within 2 hours following the determination made under paragraph 2 of this section, the employer shall prepare and maintain on file a record stating the reasons the test was not promptly administered. If a test required by this section is not administered within 8 hours following the determination made under paragraph 2 of this section, the

employer shall cease attempts to administer an alcohol test and shall state in the record the reasons for not administering the test.

(b) Notwithstanding the absence of a reasonable suspicion alcohol test under this section, no covered employee shall report for duty or remain on duty requiring the performance of safety-sensitive functions while the employee is under the influence of or impaired by alcohol, as shown by the behavioral, speech, or performance indicators of alcohol misuse, nor shall an employer permit the covered employee to perform or continue to perform safety-sensitive functions until:

(1) An alcohol test is administered and the employee's alcohol concentration measures less than 0.02; or

(2) The start of the employee's next regularly scheduled duty period, but not less than 8 hours following the determination made under paragraph 2 of this section that there is reasonable suspicion that the employee has violated the alcohol misuse provisions in § 65.46a, 121.458, or 135.253 of this chapter.

(c) Except as provided in paragraph 4(b), no employer shall take any action under this appendix against a covered employee based solely on the employee's behavior and appearance in the absence of an alcohol test. This does not prohibit an employer with authority independent of this appendix from taking any action otherwise consistent with law.

E. *Return to Duty Testing*

Each employer shall ensure that before a covered employee returns to duty requiring the performance of a safety-sensitive function after engaging in conduct prohibited in 65.46a, 121.458, or 135.253 of this chapter, the employee shall undergo a return to duty alcohol test with a result indicating an alcohol concentration of less than 0.02.

F. *Follow-up Testing*

Following a determination under section VI, paragraph C.2 of this appendix that a covered employee is in need of assistance in resolving problems associated with alcohol misuse, each employer shall ensure that the employee is subject to unannounced follow-up alcohol testing as directed by a substance abuse professional in accordance with the provisions of section VI, paragraph C.3(b)(2) of this appendix. A covered employee shall be tested under this paragraph only while the employee is performing safety-sensitive functions; just before the employee is to perform safety-sensitive functions; or just after the employee has ceased performing such functions.

G. *Retesting of Covered Employees with an Alcohol Concentration of 0.02 or Greater but Less Than 0.04*

Each employer shall retest a covered employee to ensure compliance with the provisions of section V, paragraph F of this appendix. If the employer chooses to permit the employee to perform a safety-sensitive function within 8 hours following the administration of an alcohol test indicating an alcohol concentration of 0.02 or greater but less than 0.04.

IV. Handling of test results, record retention, and confidentiality

A. *Retention of Records*

1. *General Requirement.* Each employer shall maintain records of its alcohol misuse prevention program as provided in this section. The records shall be maintained in a secure location with controlled access.

2. *Period of Retention.* Each employer shall maintain the records in accordance with the following schedule:

(a) *Five years.* Records of employee alcohol test results with results indicating an alcohol concentration of 0.02 or greater, documentation of refusals to take required alcohol tests, calibration documentation, employee evaluations and referrals, and copies of any annual reports submitted to the FAA under this appendix shall be maintained for a minimum of 5 years.

(b) *Two years.* Records related to the collection process (except calibration or evidential breath testing devices) and training shall be maintained for a minimum of 2 years.

(c) *One year.* Records of all test results below 0.02 shall be maintained for a minimum of 1 year.

3. *Types of records.* The following specific records shall be maintained.

(a) Records related to the collection process:

(1) Collection logbooks, if used.

(2) Documents relating to the random selection process.

(3) Calibration documentation for evidential breath testing devices.

(4) Documentation of breath alcohol technician training.

(5) Documents generated in connection with decisions to administer reasonable suspicion alcohol tests.

(6) Documents generated in connection with decisions on post-accident tests.

(7) Documents verifying existence of a medical explanation of the inability of a covered employee to provide adequate breath for testing.

(b) Records related to test results:

(1) The employer's copy of the alcohol test form, including the results of the test;

(2) Documents related to the refusal of any covered employee to submit to an alcohol test required by this appendix.

(3) Documents presented by a covered employee to dispute the result of an alcohol test administered under this appendix.

(c) Records related to other violations of § 65.46a, 121.248, or 135.253 of this chapter.

(d) Records related to evaluations:

(1) Records pertaining to a determination by a substance abuse professional concerning a covered employee's need for assistance.

(2) Records concerning a covered employee's compliance with the recommendations of the substance abuse professional.

(3) Records of notifications to the Federal Air Surgeon

of violations of the alcohol misuse prohibitions in this chapter by covered employees who hold medical certificates issued under part 67 of this chapter.

(e) Records related to education and training:

(1) Materials on alcohol misuse awareness, including a copy of the employer's policy on alcohol misuse.

(2) Documentation of compliance with the requirements of section VI, paragraph A of this appendix.

(3) Documentation of training provided to supervisors for the purpose of qualifying the supervisors to make a determination concerning the need for alcohol testing based on reasonable suspicion.

(4) Certification that any training conducted under this appendix complies with the requirements for such training.

B. *Reporting of Results in a Management Information System*

1. Annual reports summarizing the results of alcohol misuse prevention programs shall be submitted to the FAA in the form and manner prescribed by the Administrator by March 15 of each year covering the previous calendar year (January 1 through December 31) in accordance with the provisions below.

(a) Each part 121 certificate holder shall submit an annual report each year.

(b) Each entity conducting an alcohol misuse prevention program under the provisions of this appendix, other than a part 121 certificate holder, that has 50 or more covered employees on January 1 of any calendar year shall submit an annual report to the FAA for that calendar year.

(c) The Administrator reserves the right to require employers not otherwise required to submit annual reports to prepare and submit such reports to the FAA. Employers that will be required to submit annual reports under this provision will be notified in writing by the FAA.

2. Each employer that is subject to more than one DOT agency alcohol rule shall identify each employee covered by the regulations of more than one DOT agency. The identification will be by the total number and category of covered function. Prior to conducting any alcohol test on a covered employee subject to the rules of more than one DOT agency, the employer shall determine which DOT agency rule or rules authorizes or requires the test. The test result information shall be directed to the appropriate DOT agency or agencies.

3. Each employer shall ensure the accuracy and timeliness of each report submitted.

4. Each report shall be submitted in the form and manner prescribed by the Administrator.

5. Each report shall be signed by the employer's alcohol misuse prevention program manager or other designated representative.

6. Each report that contains information on an alcohol screening test result of 0.02 or greater or a violation of the alcohol misuse provisions of § 65.46a, 121.458, or 135.253 of this chapter shall include the following informational elements:

(a) Number of covered employees by employee category.

(b) Number of covered employees in each category subject to alcohol testing under the alcohol misuse rule of another DOT agency, identified by each agency.

(c)(1) Number of screening tests by type of test and employee category.

(2) Number of confirmation tests, by type of test and employee category.

(d) Number of confirmation alcohol tests indicating an alcohol concentration of 0.02 or greater but less than 0.04 by type of test and employee category.

(e) Number of confirmation alcohol tests indicating an alcohol concentration of 0.04 or greater, by type of test and employee category.

(f) Number of persons denied a position as a covered employee following a pre-employment alcohol test indicating an alcohol concentration of 0.04 or greater.

(g) Number of covered employees with a confirmation alcohol test indicating an alcohol concentration of 0.04 or greater who were returned to duty in covered positions (having complied with the recommendations of a substance abuse professional as described in section V, paragraph E, and section VI, paragraph C of this appendix).

(h) Number of covered employees who were administered alcohol and drug tests in the same time, with both a positive drug test result and an alcohol test result indicating an alcohol concentration of 0.04 or greater.

(i) Number of covered employees who were found to have violated other alcohol misuse provisions of §§ 65.46a, 121.458, or 135.253 of this chapter, and the action taken in response to the violation.

(j) Number of covered employees who refused to submit to an alcohol test required under this appendix, the number of such refusals that were not random tests, and the action taken in response to each refusal.

(k) Number of supervisors who have received required training during the reporting period in determining the existence of reasonable suspicion of alcohol misuse.

7. Each report with no screening test results of 0.02 or greater or violations of the alcohol misuse provisions of §§ 65.46a, 121.458, or 135.253 of this chapter shall include the following informational elements. (This report may only be submitted if the program results meet these criteria.)

(a) Number of covered employees by employee category.

(b) Number of covered employees in each category subject to alcohol testing under the alcohol misuse rule of another DOT agency, identified by each agency.

(c) Number of screening tests by type of test and employee category.

(d) Number of covered employees who engaged in alcohol misuse who were returned to duty in covered positions (having complied with the recommendations of a substance abuse professional as described in section V, paragraph E, and section VI, paragraph C of this appendix).

(e) Number of covered employees who refused to submit to an alcohol test required under this appendix, and the action taken in response to each refusal.

(f) Number of supervisors who have received required training during the reporting period in determining the existence of reasonable suspicion of alcohol misuse.

8. An FAA-approved consortium may prepare reports on behalf of individual aviation employers for purposes of compliance with this reporting requirement. However, the aviation employer shall sign and submit such a report and shall remain responsible for ensuring the accuracy and timeliness of each report prepared on its behalf by a consortium.

C. *Access to Records and Facilities*

1. Except as required by law or expressly authorized or required in this appendix, no employer shall release covered employee information that is contained in records required to be maintained under this appendix.

2. A covered employee is entitled, upon written request, to obtain copies of any records pertaining to the employee's use of alcohol, including any records pertaining to his or her alcohol tests. The employer shall promptly provide the records requested by the employee. Access to an employee's records shall not be contingent upon payment for records other than those specifically requested.

3. Each employer shall make available copies of all results of alcohol testing conducted under this appendix and any other information pertaining to the employer's alcohol misuse prevention program, when requested by the Secretary of Transportation or any DOT agency with regulatory authority over the employer or covered employee.

4. When requested by the National Transportation Safety Board as part of an accident investigation, each employer shall disclose information related to the employer's administration of a post-accident alcohol test administered following the accident under investigation.

5. Records shall be made available to a subsequent employer upon receipt of written request from the covered employee. Disclosure by the subsequent employer is permitted only as expressly authorized by the terms of the employee's request.

6. An employer may disclose information required to be maintained under this appendix pertaining to a covered employee to the employee or to the decisionmaker in a lawsuit, grievance, or other proceeding initiated by or on behalf of the individual and arising from the results of an alcohol test administered under this appendix or from the employer's determination that the employee engaged in conduct prohibited under §§ 65.46a, 121.458, or 135.253 of this chapter (including, but not limited to, a worker's compensation, unemployment compensation, or other proceeding relating to a benefit sought by the employee).

7. An employer shall release information regarding a covered employee's records as directed by the specific, written consent of the employee authorizing release of the information to an identified person. Release of such information by the person receiving the information is permitted only in accordance with the terms of the employee's consent.

8. Each employer shall permit access to all facilities utilized in complying with the requirements of this appendix to the Secretary of Transportation or any DOT agency with regulatory authority over the employer or any of its covered employees.

V. Consequences for employees engaging in alcohol-related conduct

A. *Removal from Safety-sensitive Function*

1. Except as provided in section VI of this appendix, no covered employee shall perform safety-sensitive functions if the employee has engaged in conduct prohibited by §§ 65.46a, 121.458, or 135.253 of this chapter or an alcohol misuse rule of another DOT agency.

2. No employer shall permit any covered employee to perform safety-sensitive functions if the employer has determined that the employee has violated this paragraph.

B. *Permanent Disqualification from Service*

An employee who violates §§ 65.46a(c), 121.458(c), or 135.253(c) or who violates other alcohol misuse provisions of § 65.46a, 121.458, or 135.253 of this chapter and had previously engaged in conduct that violated the provisions of §§ 65.46a, 121.458, or 135.253 of this chapter after March 18, 1994 is permanently precluded from performing for an employer the safety-sensitive duties the employee performed before such violation.

C. *Notice to the Federal Air Surgeon*

1. An employer who determines that a covered employee who holds an airman medical certificate issued under part 67 of this chapter has violated the provisions of §§ 65.46a, 121.458, or 135.253 of this chapter shall notify the Federal Air Surgeon within 2 working days.

2. Each such employer shall forward to the Federal Air Surgeon a copy of the report of any evaluation performed under the provisions of section VI of this appendix within 2 working days of the employer's receipt of the report.

3. All documents shall be sent to the Federal Air Surgeon, Office of Aviation Medicine, Drug Abatement Division (AAM-800), 400 7th Street SW., Washington, DC 20590.

4. No covered employee who holds a part 67 airman medical certificate shall perform safety-sensitive duties for an employer following a violation until and unless the Federal Air Surgeon has recommended that the employee be permitted to perform such duties.

D. *Notice of Refusals*

1. Except as provided in subparagraph 2 of this paragraph, each employer shall notify the FAA of any covered employee who holds a certificate issued under part 61, part 63, or part 65 who has refused to submit to an alcohol test required under this appendix. Notifications should be sent

to: Federal Aviation Administration, Aviation Standards National Field Office, Airmen Certification Branch, AVN-460, P.O. Box 25082, Oklahoma City, OK 73125.

2. An employer is not required to notify the FAA of refusals to submit to pre-employment alcohol tests or refusals to submit to return to duty tests.

E. *Required Evaluation and Testing*

No covered employee who has engaged in conduct prohibited by §§ 65.46a, 121.458, or 135.253 of this chapter shall perform safety-sensitive functions unless the employee has met the requirements of section VI, paragraph C of this appendix. No employer shall permit a covered employee who has engaged in such conduct to perform safety-sensitive functions unless the employee has met the requirements of section VI, paragraph C of this appendix.

F. *Other Alcohol-Related Conduct*

1. No covered employee tested under the provisions of section III of this appendix who is found to have an alcohol concentration of 0.02 or greater but less than 0.04 shall perform or continue to perform safety-sensitive functions for an employer, nor shall an employer permit the employee to perform or continue to perform safety-sensitive functions, until:

(a) The employee's alcohol concentration measures less than 0.02; or

(b) The start of the employee's next regularly scheduled duty period, but not less than 8 hours following administration of the test.

2. Except as provided in subparagraph 1 of this paragraph, no employer shall take any action under this rule against an employee based solely on test results showing an alcohol concentration less than 0.04. This does not prohibit an employer with authority independent of this rule from taking any action otherwise consistent with law.

VI. Alcohol misuse information, training, and referral

A. *Employer Obligation to Promulgate a Policy on the Misuse of Alcohol*

1. *General requirements.* Each employer shall provide educational materials that explain these alcohol misuse requirements and the employer's policies and procedures with respect to meeting those requirements.

(a) The employer shall ensure that a copy of these materials is distributed to each covered employee prior to the start of alcohol testing under the employer's FAA-mandated alcohol misuse prevention program and to each person subsequently hired for or transferred to a covered position.

(b) Each employer shall provide written notice to representatives of employee organizations of the availability of this information.

2. *Required content.* The materials to be made available to employees shall include detailed discussion of at least the following:

(a) The identity of the person designated by the employer to answer employee questions about the materials.

(b) The categories of employees who are subject to the provisions of these alcohol misuse requirements.

(c) Sufficient information about the safety-sensitive functions performed by those employees to make clear what period of the work day the covered employee is required to be in compliance with these alcohol misuse requirements.

(d) Specific information concerning employee conduct that is prohibited by this chapter.

(e) The circumstances under which a covered employee will be tested for alcohol under this appendix.

(f) The procedures that will be used to test for the presence of alcohol, protect the employee and the integrity of the breath testing process, safeguard the validity of the test results, and ensure that those results are attributed to the correct employee.

(g) The requirement that a covered employee submit to alcohol tests administered in accordance with this appendix.

(h) An explanation of what constitutes a refusal to submit to an alcohol test and the attendant consequences.

(i) The consequences for covered employees found to have violated the prohibitions in this chapter, including the requirement that the employee be removed immediately from performing safety-sensitive functions, and the procedures under section VI of this appendix.

(j) The consequences for covered employees found to have an alcohol concentration of 0.02 or greater but less than 0.04.

(k) Information concerning the effects of alcohol misuse on an individual's health, work, and personal life: signs and symptoms of an alcohol problem; and available methods of evaluating and resolving problems associated with the misuse of alcohol; and intervening when an alcohol problem is suspected, including confrontation, referral to any available employee assistance program, and/or referral to management.

(l) *Optional provisions.* The materials supplied to covered employees may also include information on additional employer policies with respect to the use or possession of alcohol, including any consequences for an employee found to have a specified alcohol level, that are based on the employer's authority independent of this appendix. Any such additional policies or consequences must be clearly and obviously described as being based on independent authority.

B. *Training for Supervisors*

Each employer shall ensure that persons designated to determine whether reasonable suspicion exists to require a covered employee to undergo alcohol testing under section II of this appendix receive at least 60 minutes of training on the physical, behavioral, speech, and performance indicators of probable alcohol misuse.

C. *Referral, Evaluation, and Treatment*

1. Each covered employee who has engaged in conduct prohibited by §§ 65.46a, 121.458, or 135.253 of

this chapter shall be advised by the employer of the resources available to the employee in evaluating and resolving problems associated with the misuse of alcohol, including the names, addresses, and telephone numbers of substance abuse professionals and counseling and treatment programs.

2. Each covered employee who engages in conduct prohibited under §§ 65.46a, 121.458, or 135.253 of this chapter shall be evaluated by a substance abuse professional who must determine what assistance, if any, the employee needs in resolving problems associated with alcohol misuse.

3. (a) Before a covered employee returns to duty requiring the performance of a safety-sensitive function after engaging in conduct prohibited by §§ 65.46a, 121.458, or 135.253 of this chapter, the employee shall undergo a return-to-duty alcohol test with a result indicating an alcohol concentration of less than 0.02.

(b) In addition, each covered employee identified as needing assistance in resolving problems associated with alcohol misuse—

(i) Shall be evaluated by a substance abuse professional to determine whether the employee has properly followed any rehabilitation program prescribed under subparagraph 2 of this paragraph, and,

(ii) Shall be subject to unannounced follow-up alcohol tests administered by the employer following the employee's return to duty. The number and frequency of such follow-up testing shall be determined by a substance abuse professional, but shall consist of at least six tests in the first 12 months following the employee's return to duty. The employer may direct the employee to undergo testing for drugs (both return to duty and follow-up), in addition to alcohol testing, if the substance abuse professional determines that drug testing is necessary for the particular employee. Any such drug testing shall be conducted in accordance with the requirements of 49 CFR part 40. Follow-up testing shall not exceed 60 months from the date of the employee's return to duty. The substance abuse professional may terminate the requirement for follow-up testing at any time after the first six tests have been administered, if the substance abuse professional determines that such testing is no longer necessary.

4. Evaluation and rehabilitation may be provided by the employer, by a substance abuse professional under contract with the employer, or by a substance abuse professional not affiliated with the employer. The choice of substance abuse professional and assignment of costs shall be made in accordance with employer/employee agreements and employer policies.

5. Each employer shall ensure that a substance abuse professional who determines that a covered employee requires assistance in resolving problems with alcohol misuse does not refer the employee to the substance abuse professional's private practice or to a person or organization from which the substance abuse professional receives remuneration or in which the substance abuse professional has a financial interest. This paragraph does not prohibit a substance abuse professional from referring an employee for assistance provided through—

(a) A public agency, such as a State, county, or municipality;

(b) The employer or a person under contract to provide treatment for alcohol problems on behalf of the employer;

(c) The sole source of therapeutically appropriate treatment under the employee's health insurance program; or

(d) The sole source of therapeutically appropriate treatment reasonably accessible to the employee.

6. The requirements of this paragraph with respect to referral, evaluation, and rehabilitation do not apply to applicants who refuse to submit to pre-employment testing or have a pre-employment test with a result indicating an alcohol concentration of 0.04 or greater.

VII. Employer's alcohol misuse prevention program

A. *Schedule for Submission of Certification of Statements and Program Implementation*

1. Each employer shall submit an alcohol misuse prevention program (AMPP) certification statement as prescribed in paragraph B of section VII of this appendix, in duplicate, to the FAA, Office of Aviation Medicine, Drug Abatement Division (AAM-800), 400 7th Street SW., Washington, DC 20590, in accordance with the schedule below.

(a) Each employer that holds a part 121 certificate, each employer that holds a part 135 certificate and directly employs more than 50 covered employees, and each air traffic control facility affected by this rule shall submit a certification statement to the FAA by July 1, 1994. Each employer must implement an AMPP meeting the requirements of this appendix on January 1, 1995. Contractor employees to these employers must be subject to an AMPP meeting the requirements of this appendix by July 1, 1995.

(b) Each employer that holds a part 135 certificate and directly employs from 11 to 50 covered employees shall submit a certification statement to the FAA by January 1, 1995. Each employer must implement an AMPP meeting the requirements of this appendix on July 1, 1995. Contractor employees to these employers must be subject to an AMPP meeting the requirements of this appendix by January 1, 1996.

(c) Each employer that holds a part 135 certificate and directly employs ten or fewer covered employees, and each operator as defined in 14 CFR 135.1(c) shall submit a certification statement to the FAA by July 1, 1995. Each employer must implement an AMPP meeting the requirements of this appendix on January 1, 1996. Contractor employees to these employers must be subject to an AMPP meeting the requirements of this appendix by July 1, 1996.

2. A company providing covered employees by contract to employers may be authorized by the FAA to establish an

AMPP under the auspices of this appendix by submitting a certification statement meeting the requirements of paragraph B of section VII of this appendix directly to the FAA. Each contractor company that establishes an AMPP shall implement its AMPP in accordance with the provisions of this appendix.

(a) The FAA may revoke its authorization in the case of any contractor company that fails to properly implement its AMPP.

(b) No employer shall use a contractor company's employee who is not subject to the employer's AMPP unless the employer has first determined that the employee is subject to another FAA-mandated AMPP.

3. A consortium may be authorized to establish a consortium AMPP under the auspices of this appendix by submitting a certification statement meeting the requirements of paragraph B of section VII of this appendix directly to the FAA. Each consortium that so certifies shall implement the AMPP on behalf of the consortium members in accordance with the provisions of this appendix.

(a) The FAA may revoke its authorization in the case of any consortium that fails to properly implement the AMPP.

(b) Each employer that participates in an FAA-approved consortium remains individually responsible for ensuring compliance with the provisions of these alcohol misuse requirements and must maintain all records required under section IV of this appendix.

(c) Each consortium shall notify the FAA of any membership termination within 10 days of such termination.

4. Any person who applies for a certificate under the provisions of parts 121 or 135 of this chapter after the effective date of the final rule shall submit an alcohol misuse prevention program (AMPP) certification statement to the FAA prior to beginning operations pursuant to the certificate. The AMPP shall be implemented concurrently with beginning such operations or on the date specified in paragraph A.1 of this section, whichever is later. Contractor employees to a new certificate holder must be subject to an FAA-mandated AMPP within 180 days of the implementation of the employer's AMPP.

5. Any person who intends to begin air traffic control operations as an employer as defined in 14 CFR 65.46(a)(2) (air traffic control facilities not operated by the FAA or by or under contract to the U.S. military) after March 18, 1994 shall, not later than 60 days prior to the proposed initiation of such operations, submit an alcohol misuse prevention program certification statement to the FAA. The AMPP shall be implemented concurrently with the inception of operations or on the date specified in paragraph A.1 of this section, whichever is later. Contractor employees to a new air traffic control facility must be subject to an FAA-approved program within 180 days of the implementation of the facility's program.

6. Any person who intends to begin sightseeing operations as an operator under 14 CFR 135.1(c) after March 18, 1994 shall, not later than 60 days prior to the proposed initiation of such operations, submit an alcohol misuse prevention program (AMPP) certification statement to the FAA. The AMPP shall be implemented concurrently with the inception of operations or on the date specified in paragraph A.1 of this section, whichever is later. Contractor employees to a new operator must be subject to an FAA-mandated AMPP within 180 days of the implementation of the employer's AMPP.

7. The duplicate certification statement shall be annotated indicating receipt by the FAA and returned to the employer, contractor company, or consortium.

8. Each consortium that submits an AMPP certification statement to the FAA must receive actual notice of the FAA's receipt of the statement prior to performing services as an FAA-approved consortium under this appendix on behalf of employers or contractor companies.

9. Each employer, and each contractor company that submits a certification statement directly to the FAA, shall notify the FAA of any proposed change in status (*e.g.*, join a consortium or another carrier's program, change consortium, etc.) prior to the effective date of such change. The employer or contractor company must ensure that it is continuously covered by an FAA-mandated alcohol misuse prevention program.

B. *Required Content of AMPP Certification Statements*

1. Each AMPP certification statement submitted by an employer or a contractor company shall provide the following information:

(a) The name, address, and telephone number of the employer/contractor company and for the employer/contractor company AMPP manager;

(b) FAA operating certificate number (if applicable);

(c) The date on which the employer or contractor company will implement its AMPP;

(d) If the submitter is a consortium member, the identity of the consortium; and

(e) A statement signed by an authorized representative of the employer or contractor company certifying an understanding of and agreement to comply with the provisions of the FAA's alcohol misuse prevention regulations.

2. Each consortium certification statement shall provide the following information.

(a) The name, address, and telephone number of the consortium's AMPP manager;

(b) A list of the specific services the consortium will be providing in implementation of FAA-mandated AMPPs (e.g., random testing, SAP).

(c) A statement signed by an authorized representative of the consortium certifying an understanding of and agreement to comply with the provisions of the FAA's alcohol misuse prevention regulations.

VIII. Employees located outside the U.S.

A. No covered employee shall be tested for alcohol misuse while located outside the territory of the United States.

1. Each covered employee who is assigned to perform safety-sensitive functions solely outside the territory of the United States shall be removed from the random testing pool upon the inception of such assignment.

2. Each covered employee who is removed from the random testing pool under this paragraph shall be returned to the random testing pool when the employee resumes the performance of safety-sensitive functions wholly or partially within the territory of the United States.

B. The provisions of this appendix shall not apply to any person who performs a safety-sensitive function by contract for an employer outside the territory of the United States.

Part 135 – Air taxi operators and commercial operators

Subpart A—General

135.1 Applicability
135.2 Air taxi operations with large aircraft
135.3 Rules applicable to operations subject to this part
135.5 Certificate and operations specifications required
135.7 Applicability of rules to unauthorized operators
135.9 Duration of certificate
135.10 Compliance dates for certain rules
135.11 Application and issue of certificate and operations specifications
135.13 Eligibility for certificate and operations specifications
135.15 Amendment of certificate
135.17 Amendment of operations specifications
135.19 Emergency operations
135.21 Manual requirements
135.23 Manual contents
135.25 Aircraft requirements
135.27 Business office and operations base
135.29 Use of business names
135.31 Advertising
135.33 Area limitations on operations
135.35 Termination of operations
135.37 Management personnel required
135.39 Management personnel qualifications
135.41 Carriage of narcotic drugs, marijuana, and depressant or stimulant drugs or substances
135.43 Crewmember certificate: International operations: application and issue

Subpart B—Flight operations

135.61 General
135.63 Recordkeeping requirements
135.65 Reporting mechanical irregularities
135.67 Reporting potentially hazardous meteorological conditions and irregularities of communications or navigational facilities
135.69 Restriction or suspension of operations: Continuation of flight in an emergency
135.71 Airworthiness check
135.73 Inspections and tests
135.75 Inspectors credentials: Admission to pilots' compartment: forward observer's seat
135.77 Responsibility for operational control
135.79 Flight locating requirements
135.81 Informing personnel of operational information and appropriate changes
135.83 Operating information required
135.85 Carriage of persons without compliance with the passenger-carrying provisions of this part
135.87 Carriage of cargo including carry-on baggage
135.89 Pilot requirements: Use of oxygen
135.91 Oxygen for medical use by passengers
135.93 Autopilot: minimum altitudes for use
135.95 Airmen: limitations on use of services
135.97 Aircraft and facilities for recent flight experience
135.99 Composition of flight crew
135.100 Flight crewmember duties
135.101 Second in command required in IFR conditions
135.103 Exception to second in command requirement: IFR operations
135.105 Exception to second in command requirement: Approval for use of autopilot system
135.107 Flight attendant crewmember requirement
135.109 Pilot in command or second in command: Designation required
135.111 Second in command required in Category II operations
135.113 Passenger occupancy of pilot seat
135.115 Manipulation of controls
135.117 Briefing of passengers before flight
135.119 Prohibition against carriage of weapons
135.121 Alcoholic beverages
[135.122 Stowage of food, beverage, and passenger service equipment during aircraft movement on the surface, takeoff, and landing]
135.123 Emergency and emergency evacuation duties
135.125 Airplane security
135.127 Passenger information
[135.128 Use of safety belts and child restraint systems]
135.129 [Exit Seating]

Subpart C—Aircraft and equipment

135.141 Applicability
135.143 General requirements
135.145 Aircraft proving tests
135.147 Dual controls required
135.149 Equipment requirements: General
135.150 Public address and crewmember interphone systems
135.151 Cockpit voice recorders
135.152 Flight recorders
135.153 Ground proximity warning system
135.155 Fire extinguishers: Passenger-carrying aircraft
135.157 Oxygen equipment requirements
135.158 Pitot heat indication systems

135.159 Equipment requirements: Carrying passengers under VFR at night or under VFR over-the-top conditions
135.161 Radio and navigational equipment: Carrying passengers under VFR at night or under VFR over-the-top
135.163 Equipment requirements: Aircraft carrying passengers under IFR
135.165 Radio and navigational equipment: Extended overwater or IFR operations
135.167 Emergency equipment: Extended overwater operations
135.169 Additional airworthiness requirements
135.170 Materials for compartment interiors
135.171 Shoulder harness installation at flight crewmember stations
135.173 Airborne thunderstorm detection equipment requirements
135.175 Airborne weather radar equipment requirements
135.177 Emergency equipment requirements for aircraft having a passenger seating configuration of more than 19 passengers
[135.178 Additional emergency equipment]
135.179 Inoperable instruments and equipment
135.180 Traffic Alert and Collision Avoidance System
135.181 Performance requirements: Aircraft operated over-the-top or in IFR conditions
135.183 Performance requirements: Land aircraft operated over water
135.185 Empty weight and center of gravity: Currency requirement

Subpart D—VFR/IFR operating limitations and weather requirements

135.201 Applicability
135.203 VFR: Minimum altitudes
135.205 VFR: Visibility requirements
135.207 VFR: Helicopter surface reference requirements
135.209 VFR: Fuel supply
135.211 VFR: Over-the-top carrying passengers: Operating limitations
135.213 Weather reports and forecasts
135.215 IFR: Operating limitations
135.217 IFR: Takeoff limitations
135.219 IFR: Destination airport weather minimums
135.221 IFR: Alternate airport weather minimums
135.223 IFR: Alternate airport requirements
135.225 IFR: Takeoff, approach and landing minimums
135.227 Icing conditions: Operating limitations
135.229 Airport requirements

Subpart E—Flight crewmember requirements

135.241 Applicability
135.243 Pilot in command qualifications
135.244 Operating experience
135.245 Second in command qualifications
135.247 Pilot qualifications: Recent experience
135.249 Use of prohibited drugs
135.251 Testing for prohibited drugs

Subpart F—Flight crewmember flight time limitations and rest requirements

135.261 Applicability
135.263 Flight time limitations and rest requirements: All certificate holders
135.265 Flight time limitations and rest requirements: Scheduled operations
135.267 Flight time limitations and rest requirements: Unscheduled one- and two-pilot crews
135.269 Flight time limitations and rest requirements: Unscheduled three- and four-pilot crews
135.271 Helicopter hospital emergency medical evacuation service (HEMES)

Subpart G—Crewmember testing requirements

135.291 Applicability
135.293 Initial and recurrent pilot testing requirements
135.295 Initial and recurrent flight attendant crewmember testing requirements
135.297 Pilot in command: Instrument proficiency check requirements
135.299 Pilot in command: Line checks: Routes and airports
135.301 Crewmember: Tests and checks, grace provisions, training to accepted standards
135.303 [Removed]

Subpart H—Training

135.321 Applicability and terms used
135.323 Training program: General
135.325 Training program and revision: Initial and final approval
135.327 Training program: Curriculum
135.329 Crewmember training requirements
135.331 Crewmember emergency training
135.333 Training requirements: Handling and carriage of hazardous materials
135.335 Approval of aircraft simulators and other training devices
135.337 Training program: Check airmen and instructor qualifications
135.339 Check airmen and flight instructors: Initial and transition training
135.341 Pilot and flight attendant crewmember training programs
135.343 Crewmember initial and recurrent training requirements
135.345 Pilots: Initial, transition, and upgrade ground training

135.347 Pilots: Initial, transition, upgrade, and differences flight training
135.349 Flight attendants: Initial and transition ground training
135.351 Recurrent training
135.353 Prohibited drugs

Subpart I—Airplane performance operating limitations

135.361 Applicability
135.363 General
135.365 Large transport category airplanes: Reciprocating engine powered: Weight limitations
135.367 Large transport category airplanes: Reciprocating engine powered: Takeoff limitations
135.369 Large transport category airplanes: Reciprocating engine powered: En route limitations: All engines operating
135.371 Large transport category airplanes: Reciprocating engine powered: En route limitations: One engine inoperative
135.373 Part 25 transport category airplanes with four or more engines: Reciprocating engine powered: En route limitations: Two engines inoperative
135.375 Large transport category airplanes: Reciprocating engine powered: Landing limitations: Destination airports
135.377 Large transport category airplanes: Reciprocating engine powered: Landing limitations: Alternate airports
135.379 Large transport category airplanes: Turbine engine powered: Takeoff limitations
135.381 Large transport category airplanes: Turbine engine powered: En route limitations: One engine inoperative
135.383 Large transport category airplanes: Turbine engine powered: En route limitations: Two engines inoperative
135.385 Large transport category airplanes: Turbine engine powered: Landing limitations: Destination airports
135.387 Large transport category airplanes: Turbine engine powered: Landing limitations: Alternate airports
135.389 Large nontransport category airplanes: Takeoff limitations
135.391 Large nontransport category airplanes: En route limitations: One engine inoperative
135.393 Large nontransport category airplanes: Landing limitations: Destination airports
135.395 Large nontransport category airplanes: Landing limitations: Alternate airports
135.397 Small transport category airplane performance operating limitations
135.398 Commuter category airplane performance operating limitations
135.399 Small nontransport category airplane performance operating limitations

Subpart J—Maintenance, preventive maintenance, and alterations

135.411 Applicability
135.413 Responsibility for airworthiness
135.415 Mechanical reliability reports
135.417 Mechanical interruption summary report
135.419 Approved aircraft inspection program
135.421 Additional maintenance requirements
135.423 Maintenance, preventive maintenance, and alteration organization
135.425 Maintenance, preventive maintenance, and alteration programs
135.427 Manual requirements
135.429 Required inspection personnel
135.431 Continuing analysis and surveillance
135.433 Maintenance and preventive maintenance training program
135.435 Certificate requirements
135.437 Authority to perform and approve maintenance, preventive maintenance, and alterations
135.439 Maintenance recording requirements
135.441 Transfer of maintenance records
135.443 Airworthiness release or aircraft maintenance log entry

Appendix A—Additional airworthiness standards for 10 or more passenger airplanes

Appendix B—Airplane flight recorder specifications

Appendix C—Helicopter flight recorder specifications

Appendix D—Airplane flight recorder specifications

Appendix E—Helicopter flight recorder specifications

Special Federal Aviation Regulation 36 (*Rule*)
Special Federal Aviation Regulation 38-2—Certification and Operating Requirements Index.
Special Federal Aviation Regulation 38-2 (*Rule*)
Special Federal Aviation Regulation 50-2 (*Rule*)
Special Federal Aviation Regulation 52 (*Rule*)
Special Federal Aviation Regulation 58 (*Rule*)

Subpart A—General

§ 135.1 Applicability.

(a) Except as provided in paragraph (b) of this section, this part prescribes rules governing—

(1) Air taxi operations conducted under the exemption authority of Part 298 of this title;

(2) The transportation of mail by aircraft conducted under a postal service contract awarded under section 5402c of Title 39, United States Code;

(3) The carriage in air commerce of persons or property for compensation or hire as a commercial operator (not an air carrier) in aircraft having a maximum seating capacity of less than 20 passengers or a maximum payload capacity of less than 6,000 pounds, or the carriage in air commerce of persons or property in common carriage operations solely between points entirely within any state of the United States in aircraft having a maximum seating capacity of 30 seats or less or a maximum payload capacity of 7,500 pounds or less; and

(4) Each person who applies for provisional approval of an Advanced Qualification Program curriculum, curriculum segment, or portion of a curriculum segment under SFAR No. 58 and each person employed or used by an air carrier or commercial operator under this part to perform training, qualification, or evaluation functions under an Advanced Qualification Program under SFAR No. 58; and

(5) Each person who is on board an aircraft being operated under this part.

(b) Except as provided in paragraph (c) of this section, this part does not apply to—

(1) Student instruction;

(2) Nonstop sightseeing flights that begin and end at the same airport, and are conducted within a 25 statute mile radius of that airport;

(3) Ferry or training flights;

(4) Aerial work operations, including—

(i) Crop dusting, seeding, spraying, and bird chasing;

(ii) Banner towing;

(iii) Aerial photography or survey;

(iv) Fire fighting;

(v) Helicopter operations in construction or repair work (but not including transportation to and from the site of operations); and

(vi) Powerline or pipeline patrol, or similar types of patrol approved by the Administrator;

(5) Sightseeing flights conducted in hot air balloons;

(6) Nonstop flights conducted within a 25 statute mile radius of the airport of takeoff carrying persons for the purpose of intentional parachute jumps;

(7) Helicopter flights conducted within a 25 statute mile radius of the airport of takeoff, if—

(i) Not more than two passengers are carried in the helicopter in addition to the required flight crew;

(ii) Each flight is made under VFR during the day;

(iii) The helicopter used is certificated in the standard category and complies with the 100-hour inspection requirements of Part 91 of this chapter;

(iv) The operator notifies the FAA Flight Standards District Office responsible for the geographic area concerned at least 72 hours before each flight and furnishes any essential information that the office requests;

(v) The number of flights does not exceed a total of six in any calendar year;

(vi) Each flight has been approved by the Administrator; and

(vii) Cargo is not carried in or on the helicopter;

(8) Operations conducted under Part 133 or 375 of this title;

(9) Emergency mail service conducted under section 405(h) of the Federal Aviation Act of 1958; or

(10) This part does not apply to operations conducted under the provisions of § 91.321.

(c) For the purpose of §§ 135.249, 135.251, 135.253, 135.255, and 135.353, *operator* means any person or entity conducting non-stop sight-seeing flights for compensation or hire in an airplane or rotorcraft that begin and end at the same airport and are conducted within a 25 statute mile radius of that airport.

(d) [Notwithstanding the provisions of this part and Appendices I and J to Part 121 of this chapter, an operator who does not hold a Part 121 or Part 135 certificate is permitted to use a person who is otherwise authorized to perform aircraft maintenance or preventive maintenance duties and who is not subject to FAA-approved anti-drug and alcohol misuse prevention programs to perform—

(1) [Aircraft maintenance or preventive maintenance on the operator's aircraft if the operator would otherwise be required to transport the aircraft more than 50 nautical miles further than the repair point closest to operator's principal place of operation to obtain these services; or

(2) [Emergency repairs on the operator's aircraft if the aircraft cannot be safely operated to a location where an employee subject to FAA-approved programs can perform the repairs.]

(Amdt. 135-5, Eff. 7/1/80); (Amdt. 135-7, Eff. 2/1/81); (Amdt. 135-20, Eff. 1/6/87); (Amdt. 135-28, Eff. 12/21/88); (Amdt. 135-32, Eff. 8/18/90); (Amdt. 135-37, Eff. 10/1/90); (Amdt. 135-40, Eff. 10/5/91); [(Amdt. 135-48, Eff. 3/17/94)]

§ 135.2 Air taxi operations with large aircraft.

(a) Except as provided in paragraph (d) of this section, no person may conduct air taxi operations in large aircraft under an individual exemption and authorization issued by the Civil Aeronautics Board or under the exemption authority of Part 298 of this title, unless that person—

(1) Complies with the certification requirements for supplemental air carriers in Part 121 of this chapter, except that the person need not obtain, and that person is not eligible for, a certificate under that part; and

(2) Conducts those operations under the rules of Part 121 of this chapter that apply to supplemental air carriers.

However, the Administrator may issue operations specifications which require an operator to comply with the rules of Part 121 of this chapter that apply to domestic or flag air carriers, as appropriate, in place of the rules required by paragraph (a)(2) of this section, if the Administrator determines compliance with those rules is necessary to provide an appropriate level of safety for the operation.

(b) The holder of an operating certificate issued under this part who is required to comply with Subpart L of Part 121 of this chapter, under paragraph (a) of this section, may perform and approve maintenance, preventive maintenance, and alterations on aircraft having a maximum passenger seating configuration, excluding any pilot seat, of 30 seats or less and a maximum payload capacity of 7,500 pounds or less as provided in that subpart. The aircraft so maintained shall be identified by registration number in the operations specifications of the certificate holder using the aircraft.

(c) Operations that are subject to paragraph (a) of this section are not subject to §§ 135.21 through 135.43 of Subpart A and Subparts B through J of this part. Seaplanes used in operations that are subject to paragraph (a) of this section are not subject to § 121.291(a) of this chapter.

(d) Operations conducted with aircraft having a maximum passenger seating configuration, excluding any pilot seat, of 30 seats or less, and a maximum payload capacity of 7,500 pounds or less shall be conducted under the rules of this part. However, a certificate holder who is conducting operations on December 1, 1978, in aircraft described in this paragraph may continue to operate under paragraph (a) of this section.

(e) For the purposes of this part—

(1) "Maximum payload capacity" means:

(i) For an aircraft for which a maximum zero fuel weight is prescribed in FAA technical specifications, the maximum zero fuel weight, less empty weight, less all justifiable aircraft equipment, and less the operating load (consisting of minimum flight crew, foods and beverages and supplies and equipment related to foods and beverages, but not including disposable fuel or oil);

(ii) For all other aircraft, the maximum certificated takeoff weight of an aircraft, less the empty weight, less all justifiable aircraft equipment, and less the operating load (consisting of minimum fuel load, oil, and flight crew). The allowance for the weight of the crew, oil, and fuel is as follows:

(A) Crew—200 pounds for each crewmember required under this chapter.

(B) Oil—350 pounds.

(C) Fuel—the minimum weight of fuel required under this chapter for a flight between domestic points 174 nautical miles apart under VFR weather conditions that does not involve extended overwater operations.

(2) "Empty weight" means the weight of the airframe, engines, propellers, rotors, and fixed equipment. Empty weight excludes the weight of the crew and payload, but includes the weight of all fixed ballast, unusable fuel supply, undrainable oil, total quantity of engine coolant, and total quantity of hydraulic fluid.

(3) "Maximum zero fuel weight" means the maximum permissible weight of an aircraft with no disposable fuel or oil. The zero fuel weight figure may be found in either the aircraft type certificate data sheet or the approved Aircraft Flight Manual, or both.

(4) For the purposes of this paragraph, "justifiable aircraft equipment" means any equipment necessary for the operation of the aircraft. It does not include equipment or ballast specifically installed, permanently or otherwise, for the purpose of altering the empty weight of an aircraft to meet the maximum payload capacity specified in paragraph (d) of this section.

§ 135.3 Rules applicable to operations subject to this part.

Each person operating an aircraft in operations under this part shall—

(a) While operating inside the United States, comply with the applicable rules of this chapter; and

(b) While operating outside the United States, comply with Annex 2, Rules of the Air, to the Convention on International Civil Aviation or the regulations of any foreign country, whichever applies, and with any rules of Parts 61 and 91 of this chapter and this part that are more restrictive than that Annex or those regulations and that can be complied with without violating that Annex or those regulations. Annex 2 is incorporated by reference in § 91.703(b) of this chapter.

(Amdt. 135-32, Eff. 8/18/90)

§ 135.5 Certificate and operations specifications required.

No person may operate an aircraft under this part without, or in violation of, an air taxi/commercial operator (ATCO) operating certificate and appropriate operations specifications issued under this part, or for operations with large aircraft having a maximum passenger seating configuration, excluding any pilot seat, of more than 30 seats, or a maximum payload capacity of more than 7,500 pounds, without, or in violation of, appropriate operations specifications issued under Part 121 of this chapter.

§ 135.7 Applicability of rules to unauthorized operators.

The rules in this part which apply to a person certificated under § 135.5 also apply to a person who engages in any operation governed by this part without an appropriate certificate and operations specifications required by § 135.5.

§ 135.9 Duration of certificate.

(a) An ATCO operating certificate is effective until surrendered, suspended or revoked. The holder of an ATCO operative certificate that is suspended or revoked shall return it to the Administrator.

(b) Except as provided in paragraphs (c) and (d) of this section, an ATCO operating certificate in effect on December 1, 1978, expires on February 1, 1979. The certificate holder must continue to conduct operations under Part 135 and the operations specifications in effect on November 30, 1978, until the certificate expires.

(c) If the certificate holder applies before February 1, 1979, for new operations specifications under this part, the operating certificate held continues in effect and the certificate holder must continue operations under Part 135 and operations specifications in effect on November 30, 1978, until the earliest of the following—

(1) The date on which new operations specifications are issued; or

(2) The date on which the Administrator notifies the certificate holder that the application is denied; or

(3) August 1, 1979.

If new operations specifications are issued under paragraph (c)(1) of this paragraph, the ATCO operating certificate continues in effect until surrendered, suspended or revoked under paragraph (a) of this section.

(d) A certificate holder may obtain an extension of the expiration date in paragraph (c) of this section, but not beyond December 1, 1979, from the Director, Flight Standards Service, if before July 1, 1979, the certificate holder—

(1) Shows that due to the circumstances beyond its control it cannot comply by the expiration date; and

(2) Submits a schedule for compliance, acceptable to the Director, indicating that compliance will be achieved at the earliest practicable date.

(e) The holder of an ATCO operating certificate that expires, under paragraphs (b), (c), or (d) of this section, shall return it to the Administrator.

§ 135.10 Compliance dates for certain rules.

After January 2, 1991, no certificate holder may use a person as a flight crewmember unless that person has completed the windshear ground training required by §§ 135.345(b)(6) and 135.351(b)(2) of this part.

Docket No. 19110 (53 FR 37697) Eff. 9/27/88; (Amdt 135-1, Eff. 5/7/79) (Amdt. 135-6, Eff. 9/10/80); (Amdt. 135-9, Eff. 12/1/80) (Amdt. 135-13, Eff. 5/19/81); (Amdt. 135-27, Eff. 1/2/89)

§ 135.11 Application and issue of certificate and operations specifications.

(a) An application for an ATCO operating certificate and appropriate operations specifications is made on a form and in a manner prescribed by the Administrator and filed with the FAA Flight Standards District Office that has jurisdiction over the area in which the applicant's principal business office is located.

(b) An applicant who meets the requirements of this part is entitled to—

(1) An ATCO operating certificate containing all business names under which the certificate holder may conduct operations and the address of each business office used by the certificate holder; and

(2) Separate operations specifications, issued to the certificate holder, containing:

(i) The type and area of operations authorized.

(ii) The category and class of aircraft that may be used in those operations.

(iii) Registration numbers and types of aircraft that are subject to an airworthiness maintenance program required by § 135.411 (a)(2), including time limitations or standards for determining time limitations, for overhauls, inspections, and checks for airframes, aircraft engines, propellers, rotors, appliances, and emergency equipment.

(iv) Registration numbers of aircraft that are to be inspected under an approved aircraft inspection program under § 135.419.

(v) Additional maintenance items required by the Administrator under § 135.421.

(vi) Any authorized deviation from this part.

(vii) Any other items the Administrator may require or allow to meet any particular situation.

(c) No person holding operations specifications issued under this part may list on its operations specifications or on the current list of aircraft required by § 135.63(a)(3) any airplane listed on operations specifications issued under Part 125.

(Amdt. 135-24, Eff. 8/25/87)

§ 135.13 Eligibility for certificate and operations specifications.

(a) To be eligible for an ATCO operating certificate and appropriate operations specifications, a person must—

(1) Be a citizen of the United States, a partnership of which each member is a citizen of the United States, or a corporation or association created or organized under the laws of the United States or any state, territory, or possession of the United States, of which the president and two-thirds or more of the board of directors and other managing officers are citizens of the United States and in which at

least 75 percent of the voting interest is owned or controlled by citizens of the United States or one of its possessions; and

(2) Show, to the satisfaction of the Administrator, that the person is able to conduct each kind of operation for which the person seeks authorization in compliance with applicable regulations; and

(3) Hold any economic authority that may be required by the Civil Aeronautics Board.

However, no person holding a commercial operator operating certificate issued under Part 121 of this chapter is eligible for an ATCO operating certificate unless the person shows to the satisfaction of the Administrator that the person's contract carriage business in large aircraft, having a maximum passenger seating configuration, excluding any pilot seat, of more than 30 seats or a maximum payload capacity of more than 7,500 pounds, will not result directly or indirectly from the person's air taxi business.

(b) The Administrator may deny any applicant a certificate under this part if the Administrator finds—

(1) That an air carrier or commercial operator operating certificate under Part 121 or an ATCO operating certificate previously issued to the applicant was revoked; or

(2) That a person who was employed in a position similar to general manager, director of operations, director of maintenance, chief pilot, or chief inspector, or who has exercised control with respect to any ATCO operating certificate holder, air carrier, or commercial operator, whose operating certificate has been revoked, will be employed in any of those positions or a similar position, or will be in control of or have a substantial ownership interest in the applicant, and that the person's employment or control contributed materially to the reasons for revoking that certificate.

§ 135.15 Amendment of certificate.

(a) The Administrator may amend an ATCO operating certificate—

(1) On the Administrator's own initiative, under section 609 of the Federal Aviation Act of 1958 (49 U.S.C. 1429) and Part 13 of this chapter; or

(2) Upon application by the holder of that certificate.

(b) The certificate holder must file an application to amend an ATCO operating certificate at least 15 days before the date proposed by the applicant for the amendment to become effective, unless a shorter filing period is approved. The application must be on a form and in a manner prescribed by the Administrator and must be submitted to the FAA Flight Standards District Office charged with the overall inspection of the certificate holder.

(c) The FAA Flight Standards District Office charged with the overall inspection of the certificate holder grants an amendment to the ATCO operating certificate if it is determined that safety in air commerce and the public interest allow that amendment.

(d) Within 30 days after receiving a refusal to amend the operating certificate, the certificate holder may petition the Director, Flight Standards Service, to reconsider the request.

§ 135.17 Amendment of operations specifications.

(a) The FAA Flight Standards District Office charged with the overall inspection of the certificate holder may amend any operations specifications issued under this part if—

(1) It determines that safety in air commerce requires that amendment; or

(2) Upon application by the holder, that Flight Standards District Office determines that safety in air commerce allows that amendment.

(b) The certificate holder must file an application to amend operations specifications at least 15 days before the date proposed by the applicant for the amendment to become effective, unless a shorter filing period is approved. The application must be on a form and in a manner prescribed by the Administrator and be submitted to the FAA Flight Standards District Office charged with the overall inspection of the certificate holder.

(c) Within 30 days after a notice of refusal to approve a holder's application for amendment is received, the holder may petition the Director, Flight Standards Service to reconsider the refusal to amend.

(d) When the FAA Flight Standards District Office charged with the overall inspection of the certificate holder amends operations specifications, that Flight Standards District Office gives notice in writing to the holder of a proposed amendment to the operations specifications, fixing a period of not less than 7 days within which the holder may submit written information, views, and arguments concerning the proposed amendment. After consideration of all relevant matter presented, that Flight Standards District Office notifies the holder of any amendment adopted, or a rescission of the notice. The amendment becomes effective not less than 30 days after the holder receives notice of the adoption of the amendment, unless the holder petitions the Director, Flight Standards Service, for reconsideration of the amendment. In that case, the effective date of the amendment is stayed pending a decision by the Director. If the Director finds there is an emergency requiring immediate action as to safety in air commerce that makes the provisions of this paragraph impracticable or contrary to the public interest, the Director notifies the certificate holder that the amendment is effective on the date of receipt, without previous notice.

(Amdt. 135-6, Eff. 9/10/80); (Amdt. 135-33, Eff. 10/25/89)

§ 135.19 Emergency operations.

(a) In an emergency involving the safety of persons or property, the certificate holder may deviate from the rules of this part relating to aircraft and equipment and weather minimums to the extent required to meet that emergency.

(b) In an emergency involving the safety of persons or property, the pilot in command may deviate from the rules of this part to the extent required to meet that emergency.

(c) Each person who, under the authority of this section, deviates from a rule of this part shall, within 10 days, excluding Saturdays, Sundays, and Federal holidays, after the deviation, send to the FAA Flight Standards District Office charged with the overall inspection of the certificate holder a complete report of the aircraft operation involved, including a description of the deviation and reasons for it.

§ 135.21 Manual requirements.

(a) Each certificate holder, other than one who uses only one pilot in the certificate holder's operations, shall prepare and keep current a manual setting forth the certificate holder's procedures and policies acceptable to the Administrator. This manual must be used by the certificate holder's flight, ground, and maintenance personnel in conducting its operations. However, the Administrator may authorize a deviation from this paragraph if the Administrator finds that, because of the limited size of the operation, all or part of the manual is not necessary for guidance of flight, ground, or maintenance personnel.

(b) Each certificate holder shall maintain at least one copy of the manual at its principal operations base.

(c) The manual must not be contrary to any applicable Federal regulations, foreign regulation applicable to the certificate holder's operations in foreign countries, or the certificate holder's operating certificate or operations specifications.

(d) A copy of the manual, or appropriate portions of the manual (and changes and additions) shall be made available to maintenance and ground operations personnel by the certificate holder and furnished to—

(1) Its flight crewmembers; and

(2) Representatives of the Administrator assigned to the certificate holder.

(e) Each employee of the certificate holder to whom a manual or appropriate portions of it are furnished under paragraph (d)(1) of this section shall keep it up to date with the changes and additions furnished to them.

(f) Except as provided in paragraph (g) of this section, each certificate holder shall carry appropriate parts of the manual on each aircraft when away from the principal operations base. The appropriate parts must be available for use by ground or flight personnel.

(g) If a certificate holder conducts aircraft inspections or maintenance at specified stations where it keeps the approved inspection program manual, it is not required to carry the manual aboard the aircraft en route to those stations.

(Amdt. 135-18, Eff. 8/2/82)

§ 135.23 Manual contents.

Each manual shall have the date of the last revision on each revised page. The manual must include—

(a) The name of each management person required under § 135.37(a) who is authorized to act for the certificate holder, the person's assigned area of responsibility, the person's duties, responsibilities, and authority, and the name and title of each person authorized to exercise operational control under § 135.77;

(b) Procedures for ensuring compliance with aircraft weight and balance limitations and, for multiengine aircraft, for determining compliance with § 135.185;

(c) Copies of the certificate holder's operations specifications or appropriate extracted information, including area of operations authorized, category and class of aircraft authorized, crew complements, and types of operations authorized;

(d) Procedures for complying with accident notification requirements;

(e) Procedures for ensuring that the pilot in command knows that required airworthiness inspections have been made and that the aircraft has been approved for return to service in compliance with applicable maintenance requirements;

(f) Procedures for reporting and recording mechanical irregularities that come to the attention of the pilot in command before, during, and after completion of a flight;

(g) Procedures to be followed by the pilot in command for determining that mechanical irregularities or defects reported for previous flights have been corrected or that correction has been deferred;

(h) Procedures to be followed by the pilot in command to obtain maintenance, preventive maintenance, and servicing of the aircraft at a place where previous arrangements have not been made by the operator, when the pilot is authorized to so act for the operator;

(i) Procedures under § 135.179 for the release for, or continuation of, flight if any item of equipment required for the particular type of operation becomes inoperative or unserviceable en route;

(j) Procedures for refueling aircraft, eliminating fuel contamination, protecting from fire (including electrostatic protection), and supervising and protecting passengers during refueling;

(k) Procedures to be followed by the pilot in command in the briefing under § 135.117;

(l) Flight locating procedures, when applicable;

(m) Procedures for ensuring compliance with emergency procedures, including a list of the functions assigned each category of required crewmembers in connection with an emergency and emergency evacuation duties under § 135.123;

(n) En route qualification procedures for pilots, when applicable;

(o) The approved aircraft inspection program, when applicable;

(p) Procedures and instructions to enable personnel to recognize hazardous materials, as defined in Title 49 CFR, and if these materials are to be carried, stored, or handled, procedures and instructions for—

(1) Accepting shipment of hazardous material required by Title 49 CFR, to assure proper packaging, marking, labeling, shipping documents, compatibility of articles, and instructions on their loading, storage, and handling;

(2) Notification and reporting hazardous material incidents as required by Title 49 CFR; and

(3) Notification of the pilot in command when there are hazardous materials aboard, as required by Title 49 CFR;

(q) Procedures for the evacuation of persons who may need the assistance of another person to move expeditiously to an exit if an emergency occurs; and

(r) Other procedures and policy instructions regarding the certificate holder's operations, that are issued by the certificate holder.

(Amdt. 135-20, Eff. 1/6/87)

§ 135.25 Aircraft requirements.

(a) Except as provided in paragraph (d) of this section, no certificate holder may operate an aircraft under this part unless that aircraft—

(1) Is registered as a civil aircraft of the United States and carries an appropriate and current airworthiness certificate issued under this chapter; and

(2) Is in an airworthy condition and meets the applicable airworthiness requirements of this chapter, including those relating to identification and equipment.

(b) Each certificate holder must have the exclusive use of at least one aircraft that meets the requirements for at least one kind of operation authorized in the certificate holder's operations specifications. In addition, for each kind of operation for which the certificate holder does not have the exclusive use of an aircraft, the certificate holder must have available for use under a written agreement (including arrangements for performing required maintenance) at least one aircraft that meets the requirements for that kind of operation. However, this paragraph does not prohibit the operator from using or authorizing the use of the aircraft for other than air taxi or commercial operations and does not require the certificate holder to have exclusive use of all aircraft that the certificate holder uses.

(c) For the purposes of paragraph (b) of this section, a person has exclusive use of an aircraft if that person has the sole possession, control, and use of it for flight, as owner, or has a written agreement (including arrangements for performing required maintenance), in effect when the aircraft is operated, giving the person that possession, control, and use for at least 6 consecutive months.

(d) A certificate holder may operate in common carriage, and for the carriage of mail, a civil aircraft which is leased or chartered to it without crew and is registered in a country which is a party to the Convention on International Civil Aviation if—

(1) The aircraft carries an appropriate airworthiness certificate issued by the country of registration and meets the registration and identification requirements of that country;

(2) The aircraft is of a type design which is approved under a U.S. type certificate and complies with all of the requirements of this chapter (14 CFR Chapter 1) that would be applicable to that aircraft were it registered in the United States, including the requirements which must be met for issuance of a U.S. standard airworthiness certificate (including type design conformity, condition for safe operation, and the noise, fuel venting, and engine emission requirements of this chapter), except that a U.S. registration certificate and a U.S. standard airworthiness certificate will not be issued for the aircraft;

(3) The aircraft is operated by U.S.-certificated airmen employed by the certificate holder; and

(4) The certificate holder files a copy of the aircraft lease or charter agreement with the FAA Aircraft Registry, Department of Transportation, 6400 South MacArthur Boulevard, Oklahoma City, Oklahoma (Mailing address: P.O. Box 25504, Oklahoma City, Oklahoma 73125).

(Amdt. 135-8, Eff. 10/16/80)

§ 135.27 Business office and operations base.

(a) Each certificate holder shall maintain a principal business office.

(b) Each certificate holder shall, before establishing or changing the location of any business office or operations base, except a temporary operations base, notify in writing the FAA Flight Standards District Office charged with the overall inspection of the certificate holder.

(c) No certificate holder who establishes or changes the location of any business office or operations base, except a temporary operations base, may operate an aircraft under this part unless the certificate holder complies with paragraph (b) of this section.

§ 135.29 Use of business names.

No certificate holder may obtain an aircraft under this part under a business name that is not on the certificate holder's operating certificate.

§135.31 Advertising.

No certificate holder may advertise or otherwise offer to perform operations subject to this part that are not authorized by the certificate holder's operating certificate and operations specifications.

§ 135.33 Area limitations on operations.

(a) No person may operate an aircraft in a geographical area that is not specifically authorized by appropriate operations specifications issued under this part.

(b) No person may operate an aircraft in a foreign country unless that person is authorized to do so by that country.

§ 135.35 Termination of operations.

Within 30 days after a certificate holder terminates operations under this part, the operating certificate and operations specifications must be surrendered by the certificate holder to the FAA Flight Standards District Office charged with the overall inspection of the certificate holder.

§ 135.37 Management personnel required.

(a) Each certificate holder, other than one who uses only one pilot in a certificate holder's operations, must have enough qualified management personnel in the following or equivalent positions to ensure safety in its operations:

(1) Director of operations.

(2) Chief pilot.

(3) Director of maintenance.

(b) Upon application by the certificate holder, the Administrator may approve different positions or numbers of positions than those listed in paragraph (a) of this section for a particular operation if the certificate holder shows that it can perform its operations safely under the direction of fewer or different categories of management personnel.

(c) Each certificate holder shall—

(1) Set forth the duties, responsibilities, and authority of the personnel required by this section in the manual required by § 135.21;

(2) List in the manual required by § 135.21 the name of the person or persons assigned to those positions; and

(3) Within 10 working days, notify the FAA Flight Standards District Office charged with the overall inspection of the certificate holder of any change made in the assignment of persons to the listed positions.

(Amdt 135-18, Eff. 8/2/82)

§ 135.39 Management personnel qualifications.

(a) *Director of operations.* No person may serve as director of operations under § 135.37(a) unless that person knows the contents of the manual required by § 135.21, the operations specifications, the provisions of this part and other applicable regulations necessary for the proper performance of the person's duties and responsibilities and:

(1) The director of operations for a certificate holder conducting any operations for which the pilot in command is required to hold an airline transport pilot certificate must—

(i) Hold or have held an airline transport pilot certificate; and

(ii) Have at least 3 years of experience as pilot in command of an aircraft operated under this part, Part 121 or Part 127 of this chapter; or

(iii) Have at least 3 years of experience as director of operations with a certificate holder operating under this part, Part 121 or Part 127 of this chapter.

(2) The director of operations for a certificate holder who is not conducting any operation for which the pilot in command is required to hold an airline transport pilot certificate must—

(i) Hold or have held a commercial pilot certificate; and

(ii) Have at least 3 years of experience as a pilot in command of an aircraft operated under this part, Part 121 or Part 127 of this chapter; or

(iii) Have at least 3 years of experience as director of operations with a certificate holder operating under this part, Part 121 or Part 127 of this chapter.

(b) *Chief pilot.* No person may serve as chief pilot under § 135.37(a) unless that person knows the contents of the manual required by § 135.21, the operations specifications, the provisions of this part and other applicable regulations necessary for the proper performance of the person's duties, and:

(1) The chief pilot of a certificate holder conducting any operation for which the pilot in command is required to hold an airline transport pilot certificate must—

(i) Hold a current airline transport pilot certificate with appropriate ratings for at least one of the types of aircraft used; and

(ii) Have at least 3 years of experience as a pilot in command of an aircraft under this part, Part 121 or Part 127 of this chapter.

(2) The chief pilot of a certificate holder who is not conducting any operation for which the pilot in command is required to hold an airline transport pilot certificate must—

(i) Hold a current, commercial pilot certificate with an instrument rating. If an instrument rating is not required for the pilot in command under this part, the chief pilot must hold a current, commercial pilot certificate; and

(ii) Have at least 3 years of experience as a pilot in command of an aircraft under this part, Part 121 or Part 127 of this chapter.

(c) *Director of maintenance.* No person may serve as a director of maintenance under § 135.37(a) unless that person knows the maintenance sections of the certificate holder's manual, the operations specifications, the provisions of this part and other applicable regulations necessary for the proper performance of the person's duties, and—

(1) Holds a mechanic certificate with both airframe and powerplant ratings; and

(2) Has at least 3 years of maintenance experience as a certificated mechanic on aircraft, including, at the time of appointment as director of maintenance, the recent experience requirements of § 65.83 of this chapter in the same category and class of aircraft as used by the certificate holder, or at least 3 years of experience with a certificated airframe repair station including 1 year in the capacity of approving aircraft for return to service.

(d) Deviations from this section may be authorized if the person has had equivalent aeronautical experience.

The Flight Standards Division Manager in the region of the certificate holding district office may authorize a deviation for the director of operations, chief pilot, and the director of maintenance.

(Amdt 135-6, Eff. 9/10/80); (Amdt. 135-18, Eff. 8/2/82); (Amdt 135-20, Eff. 1/6/87); (Amdt. 135-33, Eff. 10/25/89)

§ 135.41 Carriage of narcotic drugs, marijuana, and depressant or stimulant drugs or substances.

If the holder of a certificate issued under this part allows any aircraft owned or leased by that holder to be engaged in any operation that the certificate holder knows to be in violation of § 91.19(a) of this chapter, that operation is a basis for suspending or revoking the certificate.

(Amdt 135-32, Eff. 8/18/90)

§ 135.43 Crewmember certificate: International operations: application and issue.

(a) This section provides for the issuance of a crewmember certificate to United States citizens who are employed by certificate holders as crewmembers on United States registered aircraft engaged in international air commerce. The purpose of the certificate is to facilitate the entry and clearance of those crewmembers into ICAO contracting states. They are issued under Annex 9, as amended, to the Convention on International Civil Aviation.

(b) An application for a crewmember certificate is made on FAA Form 8060-6 "Application for Crewmember Certificate," to the FAA Flight Standards District Office charged with the overall inspection of the certificate holder by whom the applicant is employed. The certificate is issued on FAA Form 8060-42, "Crewmember Certificate."

(c) The holder of a certificate issued under this section, or the certificate holder by whom the holder is employed, shall surrender the certificate for cancellation at the nearest FAA Flight Standards District Office or submit it for cancellation to the Airmen Certification Branch, AAC-260, P.O. Box 25082, Oklahoma City, Oklahoma 73125, at the termination of the holder's employment with that certificate holder.

Subpart B—Flight operations

§ 135.61 General.

This subpart prescribes rules, in addition to those in Part 91 of this chapter, that apply to operations under this part.

§ 135.63 Recordkeeping requirements.

(a) Each certificate holder shall keep at its principal business office or at other places approved by the Administrator, and shall make available for inspection by the Administrator the following—

(1) The certificate holder's operating certificate;

(2) The certificate holder's operations specifications;

(3) A current list of the aircraft used or available for use in operations under this part and the operations for which each is equipped; and

(4) An individual record of each pilot used in operations under this part, including the following information:

(i) The full name of the pilot.

(ii) The pilot certificate (by type and number) and ratings that the pilot holds.

(iii) The pilot's aeronautical experience in sufficient detail to determine the pilot's qualifications to pilot aircraft in operation under this part.

(iv) The pilot's current duties and the date of the pilot's assignment to those duties.

(v) The effective date and class of the medical certificate that the pilot holds.

(vi) The date and result of each of the initial and recurrent competency tests and proficiency and route checks required by this part and the type of aircraft flown during that test or check.

(vii) The pilot's flight time in sufficient detail to determine compliance with the flight time limitations of this part.

(viii) The pilot's check pilot authorization, if any.

(ix) Any reaction taken concerning the pilot's release from employment for physical or professional disqualification.

(x) The date of the completion of the initial phase and each recurrent phase of the training required by this part.

(b) Each certificate holder shall keep each record required by paragraph (a)(3) of this section for at least 6 months, and each record required by paragraph (a)(4) of this section for at least 12 months, after it is made.

(c) For multiengine aircraft, each certificate holder is responsible for the preparation and accuracy of a load manifest in duplicate containing information concerning the loading of the aircraft. The manifest must be prepared before each takeoff and must include—

(1) The number of passengers;

(2) The total weight of the loaded aircraft;

(3) The maximum allowable takeoff weight for that flight;

(4) The center of gravity limits;

(5) The center of gravity of the loaded aircraft, except that the actual center of gravity need not be computed if the aircraft is loaded according to a loading schedule or other approved method that ensures that the center of gravity of the loaded aircraft is within approved limits. In those

cases, an entry shall be made on the manifest indicating that the center of gravity is within limits according to a loading schedule or other approved method;

(6) The registration number of the aircraft or flight number;

(7) The origin and destination; and

(8) Identification of crewmembers and their crew position assignments.

(d) The pilot in command of the aircraft for which a load manifest must be prepared shall carry a copy of the completed load manifest in the aircraft to its destination. The certificate holder shall keep copies of completed load manifest for at least 30 days at its principal operations base, or at another location used by it and approved by the Administrator.

§ 135.65 Reporting mechanical irregularities.

(a) Each certificate holder shall provide an aircraft maintenance log to be carried on board each aircraft for recording or deferring mechanical irregularities and their correction.

(b) The pilot in command shall enter or have entered in the aircraft maintenance log each mechanical irregularity that comes to the pilot's attention during flight time. Before each flight, the pilot in command shall, if the pilot does not already know, determine the status of each irregularity entered in the maintenance log at the end of the preceding flight.

(c) Each person who takes corrective action or defers action concerning a reported or observed failure or malfunction of an airframe, powerplant, propeller, rotor, or appliance, shall record the action taken in the aircraft maintenance log under the applicable maintenance requirements of this chapter.

(d) Each certificate holder shall establish a procedure for keeping copies of the aircraft maintenance log required by this section in the aircraft for access by appropriate personnel and shall include that procedure in the manual required by § 135.21.

§ 135.67 Reporting potentially hazardous meteorological conditions and irregularities of communications or navigational facilities.

Whenever a pilot encounters a potentially hazardous meteorological condition or an irregularity in a ground communications or navigational facility in flight, the knowledge of which the pilot considers essential to the safety of other flights, the pilot shall notify an appropriate ground radio station as soon as practicable.

(Amdt. 135-1, Eff. 5/7/79)

§ 135.69 Restriction or suspension of operations: Continuation of flight in an emergency.

(a) During operations under this part, if a certificate holder or pilot in command knows of conditions, including airport and runway conditions, that are a hazard to safe operations, the certificate holder or pilot in command, as the case may be, shall restrict or suspend operations as necessary until those conditions are corrected.

(b) No pilot in command may allow a flight to continue toward any airport of intended landing under the conditions set forth in paragraph (a) of this section, unless in the opinion of the pilot in command, the conditions that are a hazard to safe operations may reasonably be expected to be corrected by the estimated time of arrival or, unless there is no safer procedure. In the latter event, the continuation toward that airport is an emergency situation under § 135.19.

§ 135.71 Airworthiness check.

The pilot in command may not begin a flight unless the pilot determines that the airworthiness inspections required by § 91.409 of this chapter, or § 135.419, whichever is applicable, have been made.

(Amdt. 135-32, Eff. 8/18/90)

§ 135.73 Inspections and tests.

Each certificate holder and each person employed by the certificate holder shall allow the Administrator, at any time or place, to make inspections or tests (including en route inspections) to determine the holder's compliance with the Federal Aviation Act of 1958, applicable regulations, and the certificate holder's operating certificate, and operations specifications.

§ 135.75 Inspectors credentials: Admission to pilots' compartment: Forward observer's seat.

(a) Whenever, in performing the duties of conducting an inspection, an FAA inspector presents an Aviation Safety Inspector credential, FAA Form 110A, to the pilot in command of an aircraft operated by the certificate holder, the inspector must be given free and uninterrupted access to the pilot compartment of that aircraft. However, this paragraph does not limit the emergency authority of the pilot in command to exclude any person from the pilot compartment in the interest of safety.

(b) A forward observer's seat on the flight deck, or forward passenger seat with headset or speaker must be provided for use by the Administrator while conducting en route inspections. The suitability of the location of the seat and the headset or speaker for use in conducting en route inspections is determined by the Administrator.

§ 135.77 Responsibility for operational control.

Each certificate holder is responsible for operational control and shall list, in the manual required by § 135.21, the name and title of each person authorized by it to exercise operational control.

§ 135.79 Flight locating requirements.

(a) Each certificate holder must have procedures established for locating each flight, for which an FAA flight plan is not filed, that—

(1) Provide the certificate holder with at least the information required to be included in a VFR flight plan;

(2) Provide for timely notification of an FAA facility or search and rescue facility, if an aircraft is overdue or missing; and

(3) Provide the certificate holder with the location, date, and estimated time for reestablishing radio or telephone communications, if the flight will operate in an area where communications cannot be maintained.

(b) Flight locating information shall be retained at the certificate holder's principal place of business, or at other places designated by the certificate holder in the flight locating procedures, until the completion of the flight.

(c) Each certificate holder shall furnish the representative of the Administrator assigned to it with a copy of its flight locating procedures and any changes or additions, unless those procedures are included in a manual required under this part.

§ 135.81 Informing personnel of operational information and appropriate changes.

Each certificate holder shall inform each person in its employment of the operations specifications that apply to that person's duties and responsibilities and shall make available to each pilot in the certificate holder's employ the following materials in current form:

(a) Airman's Information Manual (Alaska Supplement in Alaska and Pacific Chart Supplement in Pacific-Asia Regions) or a commercial publication that contains the same information.

(b) This part and Part 91 of this chapter.

(c) Aircraft Equipment Manuals, and Aircraft Flight Manual or equivalent.

(d) For foreign operations, the International Flight Information Manual or a commercial publication that contains the same information concerning the pertinent operational and entry requirements of the foreign country or countries involved.

§ 135.83 Operating information required.

(a) The operator of an aircraft must provide the following materials, in current and appropriate form, accessible to the pilot at the pilot station, and the pilot shall use them:

(1) A cockpit checklist.

(2) For multiengine aircraft or for aircraft with retractable landing gear, an emergency cockpit checklist containing the procedures required by paragraph (c) of this section, as appropriate.

(3) Pertinent aeronautical charts.

(4) For IFR operations, each pertinent navigational en route, terminal area, and approach and letdown chart.

(5) For multiengine aircraft, one-engine-inoperative climb performance data and if the aircraft is approved for use in IFR or over-the-top operations, that data must be sufficient to enable the pilot to determine compliance with § 135.181(a)(2).

(b) Each cockpit checklist required by paragraph (a)(1) of this section must contain the following procedures: (1) Before starting engines; (2) Before takeoff; (3) Cruise; (4) Before landing; (5) After landing; (6) Stopping engines.

(c) Each emergency cockpit checklist required by paragraph (a)(2) of this section must contain the following procedures as appropriate:

(1) Emergency operation of fuel, hydraulic, electrical, and mechanical systems.

(2) Emergency operation of instruments and controls.

(3) Engine inoperative procedures.

(4) Any other emergency procedures necessary for safety.

§ 135.85 Carriage of persons without compliance with the passenger-carrying provisions of this part.

The following persons may be carried aboard an aircraft without complying with the passenger-carrying requirements of this part:

(a) A crewmember or other employee of the certificate holder.

(b) A person necessary for the safe handling of animals on the aircraft.

(c) A person necessary for the safe handling of hazardous materials (as defined in Subchapter C of Title 49 CFR).

(d) A person performing duty as a security or honor guard accompanying a shipment made by or under the authority of the U.S. Government.

(e) A military courier or a military route supervisor carried by a military cargo contract, air carrier or commercial operator in operations under a military cargo contract, if that carriage is specifically authorized by the appropriate military service.

(f) An authorized representative of the Administrator conducting an en route inspection.

(g) A person, authorized by the Administrator, who is performing a duty connected with a cargo operation of the certificate holder.

§ 135.87 Carriage of cargo including carry-on baggage.

No person may carry cargo, including carry-on baggage, in or on any aircraft unless—

(a) It is carried in an approved cargo rack, bin, or compartment installed in or on the aircraft;

(b) It is secured by an approved means; or

(c) It is carried in accordance with each of the following:

(1) For cargo, it is properly secured by a safety belt or other tie-down having enough strength to eliminate the possibility of shifting under all normally anticipated flight and ground conditions, or for carry-on baggage, it is restrained so as to prevent its movement during air turbulence.

(2) It is packaged or covered to avoid possible injury to occupants.

(3) It does not impose any load on seats or on the floor structure that exceeds the load limitation for those components.

(4) It is not located in a position that obstructs the access to, or use of, any required emergency or regular exit, or the use of the aisle between the crew and the passenger compartment, or located in a position that obscures any passenger's view of the "seat belt" sign, "no smoking" sign, or any required exit sign, unless an auxiliary sign or other approved means for proper notification of the passengers is provided.

(5) It is not carried directly above seated occupants.

(6) It is stowed in compliance with this section for takeoff and landing.

(7) For cargo only operations, paragraph (c)(4) of this section does not apply if the cargo is loaded so that at least one emergency or regular exit is available to provide all occupants of the aircraft a means of unobstructed exit from the aircraft if an emergency occurs.

(d) Each passenger seat under which baggage is stowed shall be fitted with a means to prevent articles of baggage stowed under it from sliding under crash impacts severe enough to induce the ultimate inertia forces specified in the emergency landing condition regulations under which the aircraft was type certificated.

(e) When cargo is carried in cargo compartments that are designed to require the physical entry of a crewmember to extinguish any fire that may occur during flight, the cargo must be loaded so as to allow a crewmember to effectively reach all parts of the compartment with the contents of a hand fire extinguisher.

§ 135.89 Pilot requirements: Use of oxygen.

(a) *Unpressurized aircraft.* Each pilot of an unpressurized aircraft shall use oxygen continuously when flying

(1) At altitudes above 10,000 feet through 12,000 feet MSL for that part of the flight at those altitudes that is of more than 30 minutes duration; and

(2) Above 12,000 feet MSL.

(b) *Pressurized aircraft.*

(1) Whenever a pressurized aircraft is operated with the cabin pressure altitude more than 10,000 feet MSL, each pilot shall comply with paragraph (a) of this section.

(2) Whenever a pressurized aircraft is operated at altitudes above 25,000 feet through 35,000 feet MSL unless each pilot has an approved quick-donning type oxygen mask—

(i) At least one pilot at the controls shall wear, secured and sealed, an oxygen mask that either supplies oxygen at all times or automatically supplies oxygen whenever the cabin pressure altitude exceeds 12,000 feet MSL; and

(ii) During that flight, each other pilot on flight deck duty shall have an oxygen mask, connected to an oxygen supply, located so as to allow immediate placing of the mask on the pilot's face sealed and secured for use.

(3) Whenever a pressurized aircraft is operated at altitudes above 35,000 feet MSL, at least one pilot at the controls shall wear, secured and sealed, an oxygen mask required by paragraph (2)(i) of this paragraph.

(4) If one pilot leaves a pilot duty station of an aircraft when operating at altitudes above 25,000 feet MSL, the remaining pilot at the controls shall put on and use an approved oxygen mask until the other pilot returns to the pilot duty station of the aircraft.

§ 135.91 Oxygen for medical use by passengers.

(a) Except as provided in paragraphs (d) and (e) of this section, no certificate holder may allow the carriage or operation of equipment for the storage, generation or dispensing of medical oxygen unless the unit to be carried is constructed so that all valves, fittings, and gauges are protected from damage during that carriage or operation and unless the following conditions are met—

(1) The equipment must be—

(i) Of an approved type or in conformity with the manufacturing, packaging, marking, labeling and maintenance requirements of Title 49 CFR Parts 171, 172, and 173, except § 173.24(a)(1);

(ii) When owned by the certificate holder, maintained under the certificate holder's approved maintenance program;

(iii) Free of flammable contaminants on all exterior surfaces; and

(iv) Appropriately secured.

(2) When the oxygen is stored in the form of a liquid, the equipment must have been under the certificate holder's approved maintenance program since its purchase new or since the storage container was last purged.

(3) When the oxygen is stored in the form of a compressed gas as defined in Title 49 CFR § 173.300(a)—

(i) When owned by the certificate holder, it must be maintained under its approved maintenance program; and

(ii) The pressure in any oxygen cylinder must not exceed the rated cylinder pressure.

(4) The pilot in command must be advised when the equipment is on board, and when it is intended to be used.

(5) The equipment must be stowed, and each person using the equipment must be seated, so as not to restrict access to or use of any required emergency or regular exit, or of the aisle in the passenger compartment.

(b) No person may smoke and no certificate holder may allow any person to smoke within 10 feet of oxygen storage and dispensing equipment carried under paragraph (a) of this section.

(c) No certificate holder may allow any person other than a person trained in the use of medical oxygen equipment to connect or disconnect oxygen bottles or any other ancillary component while any passenger is aboard the aircraft.

(d) Paragraph (a)(1)(i) of this section does not apply when that equipment is furnished by a professional or medical emergency service for use on board an aircraft in a medical emergency when no other practical means of transportation (including any other properly equipped certificate holder) is reasonably available and the person carried under the medical emergency is accompanied by a person trained in the use of medical oxygen.

(e) Each certificate holder who, under the authority of paragraph (d) of this section, deviates from paragraph (a)(1)(i) of this section under a medical emergency shall, within 10 days, excluding Saturdays, Sundays, and Federal holidays, after the deviation, send to the FAA Flight Standards District Office charged with the overall inspection of the certificate holder a complete report of the operation involved, including a description of the deviation and the reasons for it.

§ 135.93 Autopilot: Minimum altitudes for use.

(a) Except as provided in paragraphs (b), (c), and (d) of this section, no person may use an autopilot at an altitude above the terrain which is less than 500 feet or less than twice the maximum altitude loss specified in the approved Aircraft Flight Manual or equivalent for a malfunction of the autopilot, whichever is higher.

(b) When using an instrument approach facility other than ILS, no person may use an autopilot at an altitude above the terrain that is less than 50 feet below the approved minimum descent altitude for that procedure, or less than twice the maximum loss specified in the approved Airplane Flight Manual or equivalent for a malfunction of the autopilot under approach conditions, whichever is higher.

(c) For ILS approaches, when reported weather conditions are less than the basic weather conditions in § 91.155 of this chapter, no person may use an autopilot with an approach coupler at an altitude above the terrain that is less than 50 feet above the terrain, or the maximum altitude loss specified in the approved Airplane Flight Manual or equivalent for the malfunction of the autopilot with approach coupler, whichever is higher.

(d) Without regard to paragraph (a), (b), or (c) of this section, the Administrator may issue operations specifications to allow the use, to touchdown, of an approved flight control guidance system with automatic capability, if—

(1) The system does not contain any altitude loss (above zero) specified in the approved Aircraft Flight Manual or equivalent for malfunction of the autopilot with approach coupler; and

(2) The Administrator finds that the use of the system to touchdown will not otherwise adversely affect the safety standards of this section.

(e) This section does not apply to the operations conducted in rotorcraft.

(Amdt. 135-32, Eff. 8/18/90)

§ 135.95 Airmen: Limitations on use of services.

No certificate holder may use the services of any person as an airman unless the person performing those services—

(a) Holds an appropriate and current airman certificate; and

(b) Is qualified, under this chapter, for the operation for which the person is to be used.

§ 135.97 Aircraft and facilities for recent flight experience.

Each certificate holder shall provide aircraft and facilities to enable each of its pilots to maintain and demonstrate the pilot's ability to conduct all operations for which the pilot is authorized.

§ 135.99 Composition of flight crew.

(a) No certificate holder may operate an aircraft with less than the minimum flight crew specified in the aircraft operating limitations or the Aircraft Flight Manual for that aircraft and required by this part for the kind of operation being conducted.

(b) No certificate holder may operate an aircraft without a second in command if that aircraft has a passenger seating configuration, excluding any pilot seat, of ten seats or more.

§ 135.100 Flight crewmember duties.

(a) No certificate holder shall require, nor may any flight crewmember perform, any duties during a critical phase of flight except those duties required for the safe operation of the aircraft. Duties such as company required calls made for such nonsafety related purposes as ordering galley supplies and confirming passenger connections, announcements made to passengers promoting the air carrier or pointing out sights of interest, and filling out company payroll and related records are not required for the safe operation of the aircraft.

(b) No flight crewmember may engage in, nor may any pilot in command permit, any activity during a critical phase of flight which could distract any flight crewmember from the performance of his or her duties or which could interfere in any way with the proper conduct of those duties. Activities such as eating meals, engaging in nonessential conversations within the cockpit and nonessential communications between the cabin and cockpit crews, and reading publications not related to the proper conduct of the flight are not required for the safe operation of the aircraft.

(c) For the purposes of this section, critical phases of

flight includes all ground operations involving taxi, takeoff and landing, and all other flight operations conducted below 10,000 feet, except cruise flight.

NOTE: Taxi is defined as "movement of an airplane under its own power on the surface of an airport."

(Amdt. 135-11, Eff. 5/18/81); (Amdt. 135-14, Eff. 6/18/81); (Amdt. 135-15, Eff. 6/11/81)

§ 135.101 Second in command required in IFR conditions.

Except as provided in §§ 135.103 and 135.105, no person may operate an aircraft carrying passengers in IFR conditions, unless there is a second in command in the aircraft.

§ 135.103 Exception to second in command requirement: IFR operations.

The pilot in command of an aircraft carrying passengers may conduct IFR operations without a second in command under the following conditions:

(a) A takeoff may be conducted under IFR conditions if the weather reports or forecasts, or any combination of them, indicate that the weather along the planned route of flight allows flight under VFR within 15 minutes flying time, at normal cruise speed, from the takeoff airport.

(b) En route IFR may be conducted if unforecast weather conditions below the VFR minimums of this chapter are encountered on a flight that was planned to be conducted under VFR.

(c) An IFR approach may be conducted if, upon arrival at the destination airport, unforecast weather conditions do not allow an approach to be completed under VFR.

(d) When IFR operations are conducted under this section:

(1) The aircraft must be properly equipped for IFR operations under this part.

(2) The pilot must be authorized to conduct IFR operations under this part.

(3) The flight must be conducted in accordance with an ATC IFR clearance.

IFR operations without a second in command may not be conducted under this section in an aircraft requiring a second in command under § 135.99.

§ 135.105 Exception to second in command requirement: Approval for use of autopilot system.

(a) Except as provided in §§ 135.99 and 135.111, unless two pilots are required by this chapter for operations under VFR, a person may operate an aircraft without a second in command, if it is equipped with an operative approved autopilot system and the use of that system is authorized by appropriate operations specifications. No certificate holder may use any person, nor may any person serve, as a pilot in command under this section of an aircraft operated by a Commuter Air Carrier (as defined in § 298.2 of this title) in passenger-carrying operations unless that person has at least 100 hours pilot in command flight time in the make and model of aircraft to be flown and has met all other applicable requirements of this part.

(b) The certificate holder may apply for an amendment of its operations specifications to authorize the use of an autopilot system in place of a second in command.

(c) The Administrator issues an amendment to the operations specifications authorizing the use of an autopilot system, in place of a second in command, if—

(1) The autopilot is capable of operating the aircraft controls to maintain flight and maneuver it about the three axes; and

(2) The certificate holder shows, to the satisfaction of the Administrator, that operations using the autopilot system can be conducted safely and in compliance with this part.

The amendment contains any conditions or limitations on the use of the autopilot system that the Administrator determines are needed in the interest of safety.

(Amdt. 135-3, Eff. 3/1/80)

§ 135.107 Flight attendant crewmember requirement.

No certificate holder may operate an aircraft that has a passenger seating configuration, excluding any pilot seat, of more than 19 unless there is a flight attendant crewmember on board the aircraft.

§ 135.109 Pilot in command or second in command: Designation required.

(a) Each certificate holder shall designate a—

(1) Pilot in command for each flight; and

(2) Second in command for each flight requiring two pilots.

(b) The pilot in command, as designated by the certificate holder, shall remain the pilot in command at all times during the flight.

§ 135.111 Second in command required in Category II operations.

No person may operate an aircraft in a Category II operation unless there is a second in command of the aircraft.

§ 135.113 Passenger occupancy of pilot seat.

No certificate holder may operate an aircraft type certificate after October 15, 1971, that has a passenger seating configuration, excluding any pilot seat, of more than eight seats if any person other than the pilot in command, a second in command, a company check airman, or an authorized representative of the Administrator, the National Transportation Safety Board, or the United States Postal Service occupies a pilot seat.

§ 135.115 Manipulation of controls.

No pilot in command may allow any person to manipulate the flight controls of an aircraft during flight conducted under this part, nor may any person manipulate the controls during such flight unless that person is—

(a) A pilot employed by the certificate holder and qualified in the aircraft; or,

(b) An authorized safety representative of the Administrator who has the permission of the pilot in command, is qualified in the aircraft, and is checking flight operations.

§ 135.117 Briefing of passengers before flight.

(a) Before each takeoff each pilot in command of an aircraft carrying passengers shall ensure that all passengers have been orally briefed on—

(1) Smoking. [Each passenger shall be briefed on when, where, and under what conditions smoking is prohibited (including, but not limited to, any applicable requirements of Part 252 of this title). This briefing shall include a statement that the Federal Aviation Regulations require passenger compliance with the lighted passenger information signs (if such signs are required), posted placards, areas designated for safety purposes as no smoking areas, and crewmember instructions with regard to these items. The briefing shall also include a statement (if the aircraft is equipped with a lavatory) that Federal law prohibits: tampering with, disabling, or destroying any smoke detector installed in an aircraft lavatory; smoking in lavatories; and, when applicable, smoking in passenger compartments.

(2) [The use of safety belts, including instructions on how to fasten and unfasten the safety belts. Each passenger shall be briefed on when, where, and under what conditions the safety belt must be fastened about that passenger. This briefing shall include a statement that the Federal Aviation Regulations require passenger compliance with lighted passenger information signs and crewmember instructions concerning the use of safety belts.]

(3) The placement of seat backs in an upright position before takeoff and landing;

(4) Location and means for opening the passenger entry door and emergency exits;

(5) Location of survival equipment;

(6) If the flight involves extended overwater operation, ditching procedures and the use of required flotation equipment;

(7) If the flight involves operations above 12,000 feet MSL, the normal and emergency use of oxygen; and

(8) Location and operation of fire extinguishers.

(b) Before each takeoff the pilot in command shall ensure that each person who may need the assistance of another person to move expeditiously to an exit if an emergency occurs and that person's attendant, if any, has received a briefing as to the procedures to be followed if an evacuation occurs. This paragraph does not apply to a person who has been given a briefing before a previous leg of a flight in the same aircraft.

(c) The oral briefing required by paragraph (a) of this section shall be given by the pilot in command or a crewmember.

(d) Notwithstanding the provisions of paragraph (c) of this section, for aircraft certificated to carry 19 passengers or less, the oral briefing required by paragraph (a) of this section shall be given by the pilot in command, a crewmember, or other qualified person designated by the certificate holder and approved by the Administrator.

(e) The oral briefing required by paragraph (a) shall be supplemented by printed cards which must be carried in the aircraft in locations convenient for the use of each passenger.

The cards must—

(1) Be appropriate for the aircraft on which they are to be used;

(2) Contain a diagram of, and method of operating, the emergency exits; and

(3) Contain other instructions necessary for the use of emergency equipment on board the aircraft.

(f) The briefing required by paragraph (a) may be delivered by means of an approved recording playback device that is audible to each passenger under normal noise levels.

(Amdt. 135-20, Eff. 1/6/87); (Amdt. 135-25, Eff. 4/23/88); [(Amdt. 135-44, Eff. 10/15/92)]

§ 135.119 Prohibition against carriage of weapons.

No person may, while on board an aircraft being operated by a certificate holder, carry on or about that person a deadly or dangerous weapon, either concealed or unconcealed. This section does not apply to—

(a) Officials or employees of a municipality or a State, or of the United States, who are authorized to carry arms; or

(b) Crewmembers and other persons authorized by the certificate holder to carry arms.

§ 135.121 Alcoholic beverages.

(a) No person may drink any alcoholic beverage aboard an aircraft unless the certificate holder operating the aircraft has served that beverage.

(b) No certificate holder may serve any alcoholic beverage to any person aboard its aircraft if that person appears to be intoxicated.

(c) No certificate holder may allow any person to board any of its aircraft if that person appears to be intoxicated.

[§ 135.122 Stowage of food, beverage, and passenger service equipment during aircraft movement on the surface, takeoff, and landing.

[(a) No certificate holder may move an aircraft on the surface, take off, or land when any food, beverage, or tableware furnished by the certificate holder is located at any passenger seat.

[(b) No certificate holder may move an aircraft on the surface, take off, or land unless each food and beverage tray and seat back tray table is secured in its stowed position.

[(c) No certificate holder may permit an aircraft to move on the surface, take off, or land unless each passenger serving cart is secured in its stowed position.

[(d) Each passenger shall comply with instructions given by a crewmember with regard to compliance with this system.]

[(Amdt. 135-44, Eff. 10/15/92)]

§ 135.123 Emergency and emergency evacuation duties.

(a) Each certificate holder shall assign to each required crewmember for each type of aircraft as appropriate, the necessary functions to be performed in an emergency or in a situation requiring emergency evacuation. The certificate holder shall ensure that those functions can be practicably accomplished, and will meet any reasonably anticipated emergency including incapacitation of individual crewmembers or their inability to reach the passenger cabin because of shifting cargo in combination cargo passenger aircraft.

(b) The certificate holder shall describe in the manual required under § 135.21 the functions of each category of required crewmembers assigned under paragraph (a) of this section.

§ 135.125 Airplane security.

Certificate holders conducting operations under this part shall comply with the applicable security requirements in Part 108 of this chapter.

(Amdt. 135-9, Eff. 12/1/80); (Amdt. 135-10, Eff. 4/1/81)

§ 135.127 Passenger information.

(a) No person may conduct a scheduled flight segment on which smoking is prohibited unless the "No Smoking" passenger information signs are lighted during the entire flight segment, or one or more "No Smoking" placards meeting the requirements of § 25.1541 are posted during the entire flight segment. If both the lighted signs and the placards are used, the signs must remain lighted during the entire flight segment.

Smoking is prohibited on scheduled flight segments—

(1) Between any two points within Puerto Rico, the United States Virgin Islands, the District of Columbia, or any State of the United States (other than Alaska or Hawaii) or between any two points in any one of the above-mentioned jurisdictions (other than Alaska or Hawaii);

(2) Within the State of Alaska or within the State of Hawaii; or

(3) Scheduled in the current Worldwide or North American Edition of the *Official Airline Guide* or 6 hours or less in duration and between any point listed in paragraph (a)(1) of this section and any point in Alaska or Hawaii, or between any point in Alaska and any point in Hawaii.

(b) [No person may smoke while a "No Smoking" sign is lighted or while "No Smoking" placards are posted, except that the pilot in command may authorize smoking on the flight deck (if it is physically separated from the passenger compartment) except during any movement of an aircraft on the surface, takeoff, and landing.]

(c) No person may smoke in any aircraft lavatory.

(d) After December 31, 1988, no person may operate an aircraft with a lavatory equipped with a smoke detector unless there is in that lavatory a sign or placard which reads: "Federal law provides for a penalty of up to $2,000 for tampering with the smoke detector installed in this lavatory."

(e) No person may tamper with, disable, or destroy any smoke detector installed in any aircraft lavatory.

(f) [On flight segments other than those described in paragraph (a) of this section, the "No Smoking" sign required by § 135.177(a)(3) of this part must be turned on during any movement of the aircraft on the surface, for each takeoff or landing, and at any other time considered necessary by the pilot in command.

[(g) The passenger information requirements prescribed in § 91.517(b) and (d) of this chapter are in addition to the requirements prescribed in this section.

[(h) Each passenger shall comply with instructions given him or her by crewmembers regarding compliance with paragraphs (b), (c), and (e) of this section.]

(Amdt. 135-25, Eff. 4/23/88); (Amdt. 135-35, Eff. 2/25/90); [(Amdt. 135-44, Eff. 10/15/92)]

[§ 135.128 Use of safety belts and child restraint systems.

[(a) Except as provided in this paragraph, each person on board an aircraft operated under this part shall occupy an approved seat or berth with a separate safety belt properly secured about him or her during movement on the surface, takeoff, and landing. For seaplane and float equipped rotorcraft operations during movement on the surface, the person pushing off the seaplane or rotorcraft from the dock and the person mooring the seaplane or rotorcraft at the dock are excepted from the preceding seating and safety belt requirements. A safety belt provided for the occupant of a seat may not be used by more than one person who has reached his or her second birthday. Notwithstanding the preceding requirements, a child may:

[(1) Be held by an adult who is occupying an approved seat or berth if that child has not reached his or her second birthday; or

[(2) Notwithstanding any other requirement of this chapter, occupy an approved child restraint system furnished by the certificate holder or one of the persons described in paragraph (a)(2)(i) of this section, provided:

[(i) The child is accompanied by a parent, guardian, or

attendant designated by the child's parent or guardian to attend to the safety of the child during the flight;

[(ii) The approved child restraint system bears one or more labels as follows:

[(A) Seats manufactured to U.S. standards between January 1, 1981, and February 25, 1985, must bear the label: "This child restraint system conforms to all applicable Federal motor vehicle safety standards." Vest- and harness-type child restraint systems manufactured before February 26, 1985, bearing such a label are not approved for the purposes of this section;

[(B) Seats manufactured to U.S. standards on or after February 26, 1985, must bear two labels:

[(*1*) "This child restraint system conforms to all applicable Federal motor vehicle safety standards"; and

[(*2*) "THIS RESTRAINT IS CERTIFIED FOR USE IN MOTOR VEHICLES AND AIRCRAFT" in red lettering;

[(C) Seats that do not qualify under paragraphs (a)(2)(ii)(A) and (a)(2)(ii)(B) of this section must bear either a label showing approval of a foreign government or a label showing that the seat was manufactured under the standards of the United Nations; and

[(iii) The certificate holder complies with the following requirements:

[(A) The restraint system must be properly secured to an approved forward-facing seat or berth;

[(B) The child must be properly secured in the restraint system and must not exceed the specified weight limit for the restraint system; and

[(C) The restraint system must bear the appropriate label(s).

[(b) No certificate holder may prohibit a child, if requested by the child's parent, guardian, or designated attendant from occupying a child restraint system furnished by the child's parent, guardian, or designated attendant, provided the child holds a ticket for an approved seat or berth, or such seat or berth is otherwise made available by the certificate holder for the child's use, and the requirements contained in paragraphs (a)(2)(i) through (a)(2)(iii) of this section are met. This section does not prohibit the certificate holder from providing child restraint systems or, consistent with safe operating practices, determining the most appropriate passenger seat location for the child restraint system.]

[(Amdt. 135-44, Eff. 10/15/92)]

§ 135.129 [Exit seating.]

(a)(1) [*Applicability*. This section applies to all certificate holders operating under this part, except for on-demand operations with aircraft having 19 or fewer passenger seats and commuter operations with aircraft having 9 or fewer passenger seats.

[(2) *Duty to make determination of suitability*. Each certificate holder shall determine, to the extent necessary to perform the applicable functions of paragraph (d) of this section, the suitability of each person it permits to occupy an exit seat. For the purpose of this section—

(i) *Exit seat* means—

(A) Each seat having direct access to an exit; and,

(B) Each seat in a row of seats through which passengers would have to pass to gain access to an exit, from the first seat inboard of the exit to the first aisle inboard of the exit.

(ii) A passenger seat having *direct access* means a seat from which a passenger can proceed directly to the exit without entering an aisle or passing around an obstruction.]

([3]) [*Persons designated to make determination.*] Each certificate holder shall make the passenger exit seating determinations required by this paragraph in a non-discriminatory manner consistent with the requirements of this section, by persons designated in the certificate holder's required operations manual.

([4]) [*Submission of designation for approval.*] Each certificate holder shall designate the exit seats for each passenger seating configuration in its fleet in accordance with the definitions in this paragraph and submit those designations for approval as part of the procedures required to be submitted for approval under paragraphs (n) and (p) of this section.

(b) No certificate holder may seat a person in a seat affected by this section if the certificate holder determines that it is likely that the person would be unable to perform one or more of the applicable functions listed in paragraph (d) of this section because—

(1) The person lacks sufficient mobility, strength, or dexterity in both arms and hands, and both legs:

(i) To reach upward, sideways, and downward to the location of emergency exit and exit-slide operating mechanisms;

(ii) To grasp and push, pull, turn, or otherwise manipulate those mechanisms;

(iii) To push, shove, pull, or otherwise open emergency exits;

(iv) To lift out, hold, deposit on nearby seats, or maneuver over the seatbacks to the next row objects the size and weight of over-wing window exit doors;

(v) To remove obstructions of size and weight similar over-wing exit doors;

(vi) To reach the emergency exit expeditiously;

(vii) To maintain balance while removing obstructions;

(viii) To exit expeditiously;

(ix) To stabilize an escape slide after deployment; or

(x) To assist others in getting off an escape slide;

(2) The person is less than 15 years of age or lacks the capacity to perform one or more of the applicable functions listed in paragraph (d) of this section without the assistance of an adult companion, parent, or other relative;

(3) The person lacks the ability to read and understand instructions required by this section and related to emer-

gency evacuation provided by the certificate holder in printed or graphic form or the ability to understand oral crew commands.

(4) The person lacks sufficient visual capacity to perform one or more of the applicable functions in paragraph (d) of this section without the assistance of visual aids beyond contact lenses or eyeglasses;

(5) The person lacks sufficient aural capacity to hear and understand instructions shouted by flight attendants. without assistance beyond a hearing aid;

(6) The person lacks the ability adequately to impart information orally to other passengers; or,

(7) The person has:

(i) A condition or responsibilities, such as caring for small children, that might prevent the person from performing one or more of the applicable functions listed in paragraph (d) of this section; or

(ii) A condition that might cause the person harm if he or she performs one or more of the applicable functions listed in paragraph (d) of this section.

(c) Each passenger shall comply with instructions given by a crewmember or other authorized employee of the certificate holder implementing exit seating restrictions established in accordance with this section.

(d) Each certificate holder shall include on passenger information cards, presented in the language in which briefings and oral commands are given by the crew, at each exit seat affected by this section, information that, in the event of an emergency in which a crewmember is not available to assist, a passenger occupying an exit seat may use if called upon to perform the following functions:

(1) Locate the emergency exit;

(2) Recognize the emergency exit opening mechanism;

(3) Comprehend the instructions for operating the emergency exit;

(4) Operate the emergency exit;

(5) Assess whether opening the emergency exit will increase the hazards to which passengers may be exposed;

(6) Follow oral directions and hand signals given by a crewmember;

(7) Stow or secure the emergency exit door so that it will not impede use of the exit;

(8) Assess the condition of an escape slide, activate the slide, and stabilize the slide after deployment to assist others in getting off the slide;

(9) Pass expeditiously through the emergency exit; and,

(10) Assess, select, and follow a safe path away from the emergency exit.

(e) Each certificate holder shall include on passenger information cards, at each exit seat—

(1) In the primary language in which emergency commands are given by the crew, the selection criteria set forth in paragraph (b) of this section, and a request that a passenger identify himself or herself to allow reseating if he or she—

(i) Cannot meet the selection criteria set forth in paragraph (b) of this section;

(ii) Has a nondiscernible condition that will prevent him or her from performing the applicable functions listed in paragraph (d) of this section;

(iii) May suffer bodily harm as the result of performing one or more of those functions; or

(iv) Does not wish to perform those functions; and,

(2) In each language used by the certificate holder for passenger information cards, a request that a passenger identify himself or herself to allow reseating if he or she lacks the ability to read, speak, or understand the language or the graphic form in which instructions required by this section and related to emergency evacuation are provided by the certificate holder, or the ability to understand the specified language in which crew commands will be given in an emergency;

(3) May suffer bodily harm as the result of performing one or more of those functions; or,

(4) Does not wish to perform those functions.

A certificate holder shall not require the passenger to disclose his or her reason for needing reseating.

(f) Each certificate holder shall make available for inspection by the public at all passenger loading gates and ticket counters at each airport where it conducts passenger operations, written procedures established for making determinations in regard to exit row seating.

(g) No certificate holder may allow taxi or pushback unless at least one required crewmember has verified that no exit seat is occupied by a person the crewmember determines is likely to be unable to perform the applicable functions listed in paragraph (d) of this section.

(h) Each certificate holder shall include in its passenger briefings a reference to the passenger information cards, required by paragraphs (d) and (e), the selection criteria set forth in paragraph (b), and the functions to be performed, set forth in paragraph (d) of this section.

(i) Each certificate holder shall include in its passenger briefings a request that a passenger identify himself or herself to allow reseating if he or she—

(1) Cannot meet the selection criteria set forth in paragraph (b) of this section.

(2) Has a nondiscernible condition that will prevent him or her from performing the applicable functions listed in paragraph (d) of this section;

(3) May suffer bodily harm as the result of performing one or more of those functions; or,

(4) Does not wish to perform those functions.

A certificate holder shall not require the passenger to disclose his or her reason for needing reseating.

(j) Removed and [Reserved]

(k) In the event a certificate holder determines in accordance with this section that it is likely that a passenger assigned to an exit seat would be unable to perform the functions listed in paragraph (d) of this section or a passenger requests a non-exit seat, the certificate holder shall expeditiously relocate the passenger to a non-exit seat.

(l) In the event of full booking in the non-exit seats and if necessary to accommodate a passenger being relocated from an exit seat, the certificate holder shall move a passenger who is willing and able to assume the evacuation functions that may be required, to an exit seat.

(m) A certificate holder may deny transportation to any passenger under this section only because—

(1) The passenger refuses to comply with instructions given by a crewmember or other authorized employee of the certificate holder implementing exit seating restrictions established in accordance with this section, or

(2) The only seat that will physically accommodate the person's handicap is an exit seat.

(n) In order to comply with this section certificate holders shall—

(1) Establish procedures that address:

(i) The criteria listed in paragraph (b) of this section;

(ii) The functions listed in paragraph (d) of this section;

(iii) The requirements for airport information, passenger information cards, crewmember verification of appropriate seating in exit seats, passenger briefings, seat assignments, and denial of transportation as set forth in this section;

(iv) How to resolve disputes arising from implementation of this section, including identification of the certificate holder employee on the airport to whom complaints should be addressed for resolution; and,

(2) Submit their procedures for preliminary review and approval to the principal operations inspectors assigned to them at the FAA Flight Standards District Offices that are charged with the overall inspection of their operations.

(o) Certificate holders shall assign seats prior to boarding consistent with the criteria listed in paragraph (b) and the functions listed in paragraph (d) of this section, to the maximum extent feasible.

(p) The procedures required by paragraph (n) of this section will not become effective until final approval is granted by the Director, Flight Standards Service, Washington, DC. Approval will be based solely upon the safety aspects of the certificate holder's procedures.

(Amdt. 135-36, Eff. 4/5/90); (Amdt. 135-45, Eff. 10/27/92); [(Amdt. 135-50, Eff. 7/29/94)]

Subpart C—Aircraft and equipment

§ 135.141 Applicability.

This subpart prescribes aircraft and equipment requirements for operations under this part. The requirements of this subpart are in addition to the aircraft and equipment requirements of Part 91 of this chapter. However, this part does not require the duplication of any equipment required by this chapter.

§ 135.143 General requirements.

(a) No person may operate an aircraft under this part unless that aircraft and its equipment meet the applicable regulations of this chapter.

(b) Except as provided in § 135.179, no person may operate an aircraft under this part unless the required instruments and equipment in it have been approved and are in an operable condition.

(c) ATC transponder equipment installed within the time periods indicated below must meet the performance and environmental requirements of the following TSO's.

(1) *Through January 1, 1992:*

(i) Any class of TSO-C74b or any class of TSO-C74c as appropriate, provided that the equipment was manufactured before January 1, 1990; or

(ii) The appropriate class of TSO-C112 (Mode S).

(2) *After January 1, 1992:* The appropriate class of TSO-C112 (Mode S). For purposes of paragraph (c)(2) of this section, "installation" does not include—

(i) Temporary installation of TSO-C74b or TSO-C74c substitute equipment, as appropriate, during maintenance of the permanent equipment;

(ii) Reinstallation of equipment after temporary removal for maintenance; or

(iii) For fleet operations, installation of equipment in a fleet aircraft after removal of the equipment for maintenance from another aircraft in the same operator's fleet.

(Amdt. 135-22, Eff. 5/26/87)

§ 135.145 Aircraft proving tests.

(a) No certificate holder may operate a turbojet airplane, or an aircraft for which two pilots are required by this chapter for operations under VFR, if it has not previously proved that aircraft or an aircraft of the same make and similar design in any operation under this part unless, in addition to the aircraft certification tests, at least 25 hours of proving tests acceptable to the Administrator have been flown by that certificate holder including—

(1) Five hours of night time, if night flights are to be authorized;

(2) Five instrument approach procedures under simulated or actual instrument weather conditions, if IFR flights are to be authorized; and

(3) Entry into a representative number of en route airports as determined by the Administrator.

(b) No certificate holder may carry passengers in an aircraft during proving tests, except those needed to make the tests and those designated by the Administrator to observe the tests. However, pilot flight training may be conducted during the proving tests.

(c) For the purposes of paragraph (a) of this section, an aircraft is not considered to be of similar design if an alteration includes—

(1) The installation of powerplants other than those of a type similar to those with which it is certificated; or

(2) Alterations to the aircraft or its components that materially affect flight characteristics.

(d) The Administrator may authorize deviations from this section if the Administrator finds that special circumstances make full compliance with this section necessary.

§ 135.147 Dual controls required.

No person may operate an aircraft in operations requiring two pilots unless it is equipped with functioning dual controls. However, if the aircraft type certification operating limitations do not require two pilots, a throwover control wheel may be used in place of two control wheels.

§ 135.14 Equipment requirements: General.

No person may operate an aircraft unless it is equipped with—

(a) A sensitive altimeter that is adjustable for barometric pressure;

(b) Heating or deicing equipment for each carburetor or, for a pressure carburetor, an alternate air source;

(c) For turbojet airplanes, in addition to two gyroscopic bank-and-pitch indicators (artificial horizons) for use at the pilot stations, a third indicator that is installed in accordance with the instrument requirements prescribed in § 121.305(j) of this chapter.

(d) [Reserved]

(e) For turbine powered aircraft, any other equipment as the Administrator may require.

(Amdt. 135-1, Eff. 5/7/79); (Amdt. 135-34, Eff. 11/27/89); (Amdt. 135-38, Eff. 11/26/90)

§ 135.150 Public address and crewmember interphone systems.

No person may operate an aircraft having a passenger seating configuration, excluding any pilot seat, of more than 19 unless it is equipped with—

(a) A public address system which—

(1) Is capable of operation independent of the crewmember interphone system required by paragraph (b) of this section, except for handsets, headsets, microphones, selector switches, and signaling devices;

(2) Is approved in accordance with § 21.305 of this chapter;

(3) Is accessible for immediate use from each of two flight crewmember stations in the pilot compartment;

(4) For each required floor-level passenger emergency exit which has an adjacent flight attendant seat, has a microphone which is readily accessible to the seated flight attendant, except that one microphone may serve more than one exit, provided the proximity of the exits allows unassisted verbal communication between seated flight attendants;

(5) Is capable of operation within 10 seconds by a flight attendant at each of those stations in the passenger compartment from which its use is accessible;

(6) Is audible at all passenger seats, lavatories, and flight attendant seats and work stations; and

(7) For transport category airplanes manufactured on or after November 27, 1990, meets the requirements of § 25.1423 of this chapter.

(b) A crewmember interphone system which—

(1) Is capable of operation independent of the public address system required by paragraph (a) of this section, except for handsets, headsets, microphones, selector switches, and signaling devices;

(2) Is approved in accordance with § 21.305 of this chapter;

(3) Provides a means of two-way communication between the pilot compartment and—

(i) Each passenger compartment; and

(ii) Each galley located on other than the main passenger deck level;

(4) Is accessible for immediate use from each of two flight crewmember stations in the pilot compartment;

(5) Is accessible for use from at least one normal flight attendant station in each passenger compartment;

(6) Is capable of operation within 10 seconds by a flight attendant at each of those stations in each passenger compartment from which its use is accessible; and

(7) For large turbojet-powered airplanes—

(i) Is accessible for use at enough flight attendant stations so that all floor-level emergency exits (or entryways to those exits in the case of exits located within galleys) in each passenger compartment are observable from one or more of those stations so equipped;

(ii) Has an alerting system incorporating aural or visual signals for use by flight crewmembers to alert flight attendants and for use by flight attendants to alert flight crewmembers;

(iii) For the alerting system required by paragraph (b)(7)(ii) of this section, has a means for the recipient of a call to determine whether it is a normal call or an emergency call; and

(iv) When the airplane is on the ground, provides a means of two-way communication between ground personnel and either of at least two flight crewmembers in the pilot compartment. The interphone system station for use by ground personnel must be so located that personnel using the system may avoid visible detection from within the airplane.

Docket No. 24995 (54 FR 43926) Eff. 10/27/89 (Amdt. 135-34, Eff. 11/27/89)

§ 135.151 Cockpit voice recorders.

(a) After October 11, 1991, no person may operate a multiengine, turbine-powered airplane or rotorcraft having a passenger seating configuration of six or more and for which two pilots are required by certification or operating rules unless it is equipped with an approved cockpit voice recorder that:

(1) Is installed in compliance with Part 23.1457 (a)(1) and (2), (b), (c), (d), (e), (f), and (g); § 25.1457(a)(1) and (2), (b), (c), (d), (e), (f), and (g); § 27.1457 (a)(1) and (2), (b), (c), (d), (e), (f), and (g); § 29.1457(a)(1) and (2), (b), (c), (d), (e), (f), and (g); of this chapter, as applicable; and

(2) Is operated continuously from the use of the check list before the flight to completion of the final check list at the end of the flight.

(b) After October 11, 1991, no person may operate a multiengine, turbine-powered airplane or rotorcraft having a passenger seating configuration of 20 or more seats unless it is equipped with an approved cockpit voice recorder that—

(1) Is installed in compliance with § 23.1457, § 25.1457, § 27.1457 or § 29.1457 of this chapter, as applicable; and

(2) Is operated continuously from the use of the check list before the flight to completion of the final check list at the end of the flight.

(c) In the event of an accident, or occurrence requiring immediate notification of the National Transportation Safety Board which results in termination of the flight, the certificate holder shall keep the recorded information for at least 60 days or, if requested by the Administrator or the Board, for a longer period. Information obtained from the record may be used to assist in determining the cause of accidents or occurrences in connection with investigations. The Administrator does not use the record in any civil penalty or certificate action.

(d) For those aircraft equipped to record the uninterrupted audio signals received by a boom or a mask microphone the flight crewmembers are required to use the boom microphone below 18,000 feet mean sea level. No person may operate a large turbine engine powered airplane manufactured after October 11, 1991, or on which a cockpit voice recorder has been installed after October 11, 1991, unless it is equipped to record the uninterrupted audio signal received by a boom or mask microphone in accordance with § 25.1457(c)(5) of this chapter.

(e) In complying with this section, an approved cockpit voice recorder having an erasure feature may be used, so that during the operation of the recorder, information:

(1) Recorded in accordance with paragraph (a) of this section and recorded more than 15 minutes earlier; or

(2) Recorded in accordance with paragraph (b) of this section and recorded more than 30 minutes earlier; may be erased or otherwise obliterated.

(Amdt. 135-23, Eff. 5/26/87); (Amdt. 135-26, Eff. 10/11/88)

§ 135.152 Flight recorders.

(a) No person may operate a multiengine, turbine-powered airplane or rotorcraft having a passenger seating configuration, excluding any pilot seat, of 10 to 19 seats, that is brought onto the U.S. register after October 11, 1991, unless it is equipped with one or more approved flight recorders that utilize a digital method of recording and storing data, and a method of readily retrieving that data from the storage medium. The parameters specified in Appendix B or C, as applicable, of this part must be recorded within the range, accuracy, resolution, and recording intervals as specified. The recorder shall retain no less than 8 hours of aircraft operation.

(b) After October 11, 1991, no person may operate a multiengine, turbine-powered airplane having a passenger seating configuration of 20 to 30 seats or a multiengine, turbine-powered rotorcraft having a passenger seating configuration of 20 or more seats unless it is equipped with one or more approved flight recorders that utilize a digital method of recording and storing data, and a method of readily retrieving that data from the storage medium. The parameters in Appendix D or E of this part, as applicable, that are set forth below, must be recorded within the ranges, accuracies, resolutions, and sampling intervals as specified:

(1) Except as provided in paragraph (b)(3) of this section for aircraft type certificated before October 1, 1969, the following parameters must be recorded:

(i) Time;

(ii) Altitude;

(iii) Airspeed;

(iv) Vertical acceleration;

(v) Heading;

(vi) Time of each radio transmission to or from air traffic control;

(vii) Pitch attitude;

(viii) Roll attitude;

(ix) Longitudinal acceleration;

(x) Control column or pitch control surface position; and

(xi) Thrust of each engine.

(2) Except as provided in paragraph (b)(3) of this section for aircraft type certificated after September 30, 1969, the following parameters must be recorded:

(i) Time;

(ii) Altitude;

(iii) Airspeed;

(iv) Vertical acceleration;

(v) Heading;

(vi) Time of each radio transmission either to or from air traffic control;

(vii) Pitch attitude;

(viii) Roll attitude;

(ix) Longitudinal acceleration;

(x) Pitch trim position;

(xi) Control column or pitch control surface position;

(xii) Control wheel or lateral control surface position;

(xiii) Rudder pedal or yaw control surface position;

(xiv) Thrust of each engine;

(xv) Position of each thrust reverser;

(xvi) Trailing edge flap or cockpit flap control position; and

(xvii) Leading edge flap or cockpit flap control position.

(3) For aircraft manufactured after October 11, 1991, all of the parameters listed in Appendix D or E of this part, as applicable, must be recorded.

(c) Whenever a flight recorder required by this section is installed, it must be operated continuously from the instant the airplane begins the takeoff roll or the rotorcraft begins the lift-off until the airplane has completed the landing roll or the rotorcraft has landed at its destination.

(d) Except as provided in paragraph (c) of this section, and except for recorded data erased as authorized in this paragraph, each certificate holder shall keep the recorded data prescribed in paragraph (a) of this section until the aircraft has been operating for at least 8 hours of the operating time specified in paragraph (c) of this section. In addition, each certificate holder shall keep the recorded data prescribed in paragraph (b) of this section for an airplane until the airplane has been operating for at least 25 hours, and for a rotorcraft until the rotorcraft has been operating for at least 10 hours, of the operating time specified in paragraph (c) of this section. A total of 1 hour of recorded data may be erased for the purpose of testing the flight recorder or the flight recorder system. Any erasure made in accordance with this paragraph must be of the oldest recorded data accumulated at the time of testing. Except as provided in paragraph (c) of this section, no record need be kept more than 60 days.

(e) In the event of an accident or occurrence that requires that immediate notification of the National Transportation Safety Board under 49 CFR Part 830 of its regulations and that results in termination of the flight, the certificate holder shall remove the recording media from the aircraft and keep the recorded data required by paragraphs (a) and (b) of this section for at least 60 days or for a longer period upon request of the Board or the Administrator.

(f) Each flight recorder required by this section must be installed in accordance with the requirements of §§ 23.1459, 25.1459, 27.1459, or 29.1459, as appropriate, of this chapter. The correlation required by paragraph (c) of §§ 23.1459, 25.1459, 27.1459, or 29.1459, as appropriate, of this chapter need be established only on one aircraft of a group of aircraft:

(1) That are of the same type;

(2) On which the flight recorder models and their installations are the same; and

(3) On which there are no differences in the type design with respect to the installation of the first pilot's instruments associated with the flight recorder. The most recent instrument calibration, including the recording medium from which this calibration is derived, and the recorder correlation must be retained by the certificate holder.

(g) Each flight recorder required by this section that records the data specified in paragraphs (a) and (b) of this section must have an approved device to assist in locating that recorder under water.

Docket No. 25530 (53 FR 26151) Eff. 7/11/88; (Amdt. 135-26, Eff. 10/11/88)

§ 135.153 Ground proximity warning system.

[(a) Except as provided in paragraph (b) of this section, after April 20, 1994, no person may operate a turbine-powered airplane having a passenger seating configuration, excluding any pilot seat, of 10 seats or more, unless it is equipped with an approved ground proximity warning system.

[(b) Any airplane equipped before April 20, 1992, with an alternative system that conveys warnings of excessive closure rates with the terrain and any deviations below glide slope by visual and audible means may continue to be operated with that system until April 20, 1996, provided that—

[(1) The system must have been approved by the Administrator;

[(2) The system must have a means of alerting the pilot when a malfunction occurs in the system; and

[(3) Procedures must have been established by the certification holder to ensure that the performance of the system can be appropriately monitored.

[(c) For a system required by this section, the Airplane Flight Manual shall contain—

[(1) Appropriate procedures for—

[(i) The use of the equipment;

[(ii) Proper flight crew action with respect to the equipment; and

[(iii) Deactivation for planned abnormal and emergency conditions; and

[(2) An outline of all input sources that must be operating.

[(d) No person may deactivate a system required by this section except under procedures in the Airplane Flight Manual.

[(e) Whenever a system required by this section is deactivated, an entry shall be made in the airplane maintenance record that includes the date and time of deactivation.]

(Amdt. 135-6, Eff. 9/10/80); (Amdt. 135-33, Eff. 10/25/89); [(Amdt. 135-42, Eff. 4/20/92)]

§ 135.155 Fire extinguishers: Passenger-carrying aircraft.

No person may operate an aircraft carrying passengers unless it is equipped with hand fire extinguishers of an approved type for use in crew and passenger compartments as follows—

(a) The type and quantity of extinguishing agent must be suitable for all the kinds of fires likely to occur;

(b) At least one hand fire extinguisher must be provided and conveniently located on the flight deck for use by the flight crew; and

(c) At least one hand fire extinguisher must be conveniently located in the passenger compartment of each aircraft having a passenger seating configuration, excluding any pilot seat, of at least 10 seats but less than 31 seats.

§ 135.157 Oxygen equipment requirements.

(a) *Unpressurized aircraft.* No person may operate an unpressurized aircraft at altitudes prescribed in this section unless it is equipped with enough oxygen dispensers and oxygen to supply the pilots under § 135.89(a) and to supply, when flying—

(1) At altitudes above 10,000 feet through 15,000 feet MSL, oxygen to at least 10 percent of the occupants of the aircraft, other than the pilots, for that part of the flight at those altitudes that is of more than 30 minutes duration; and

(2) Above 15,000 feet MSL oxygen to each occupant of the aircraft other than the pilots.

(b) *Pressurized aircraft.* No person may operate pressurized aircraft

(1) At altitudes above 25,000 feet MSL, unless at least a 10-minute supply of supplemental oxygen is available for each occupant of the aircraft, other than the pilots, for use when a descent is necessary due to loss of cabin pressurization; and

(2) Unless it is equipped with enough oxygen dispensers and oxygen to comply with paragraph (a) of this section whenever the cabin pressure altitude exceeds 10,000 feet MSL and, if the cabin pressurization fails, to comply with § 135.89(a) or to provide a 2-hour supply for each pilot, whichever is greater, and to supply when flying—

(i) At altitudes above 10,000 feet through 15,000 feet MSL, oxygen to at least 10 percent of the occupants of the aircraft, other than the pilots, for that part of the flight at those altitudes that is of more than 30 minutes duration; and

(ii) Above 15,000 feet MSL, oxygen to each occupant of the aircraft, other than the pilots, for one hour unless, at all times during flight above that altitude, the aircraft can safely descend to 15,000 feet MSL within four minutes, in which case only a 30-minute supply is required.

(c) The equipment required by this section must have a means—

(1) To enable the pilots to readily determine, in flight, the amount of oxygen available in each source of supply and whether the oxygen is being delivered to the dispensing units; or

(2) In the case of individual dispensing units, to enable each user to make those determinations with respect to that person's oxygen supply and delivery; and

(3) To allow the pilots to use undiluted oxygen at their discretion at altitudes above 25,000 feet MSL.

§ 135.158 Pitot heat indication systems.

(a) Except as provided in paragraph (b) of this section, after April 12, 1981, no person may operate a transport category airplane equipped with a flight instrument pitot heating system unless the airplane is also equipped with an operable pitot heat indication system that complies with § 25.1326 of this chapter in effect on April 12, 1978.

(b) A certificate holder may obtain an extension of the April 12, 1981, compliance date specified in paragraph (a) of this section, but not beyond April 12, 1983, from the Director, Flight Standards Service if the certificate holder—

(1) Shows that due to circumstances beyond its control it cannot comply by the specified compliance date; and

(2) Submits by the specified compliance date a schedule for compliance, acceptable to the Director, indicating that compliance will be achieved at the earliest practicable date.

(Amdt. 135-17, Eff. 9/30/81); (Amdt. 135-33, Eff. 10/25/89)

§ 135.159 Equipment requirements: Carrying passengers under VFR at night or under VFR over-the-top conditions.

No person may operate an aircraft carrying passengers under VFR at night or under VFR over-the-top unless it is equipped with—

(a) A gyroscopic rate-of-turn indicator except on the following aircraft:

(1) Airplanes with a third attitude instrument system usable through flight attitudes of 360 degrees of pitch-and-roll and installed in accordance with the instrument requirements prescribed in § 121.3056) of this chapter.

(2) Helicopters with a third attitude instrument system usable through flight attitudes of ±80 degrees of pitch and ±120 degrees of roll and installed in accordance with § 29.1303(g) of this chapter.

(3) Helicopters with a maximum certificated takeoff weight of 6,000 pounds or less.

(b) A slip skid indicator.

(c) A gyroscopic bank-and-pitch indicator.

(d) A gyroscopic direction indicator.

(e) A generator or generators able to supply all probable combinations of continuous in-flight electrical loads for required equipment and for recharging the battery.

(f) For night flights—

(1) An anticollision light system;

(2) Instrument lights to make all instruments, switches, and gauges easily readable, the direct rays of which are shielded from the pilot's eyes; and

(3) A flashlight having at least two size "D" cells or equivalent.

(g) For the purpose of paragraph (e) of this section, a continuous in-flight electrical load includes one that draws current continuously during flight, such as radio equipment, electrically driven instruments and lights, but does not include occasional intermittent loads.

(h) Notwithstanding provisions of paragraphs (b), (c), and (d), helicopters having a maximum certificated takeoff weight of 6,000 pounds or less may be operated until January 6, 1988, under visual flight rules at night without a slip skid indicator, a gyroscopic bank-and-pitch indicator, or a gyroscopic direction indicator.

Docket No. 24550 (51 FR 40709) Eff. 11/7/86; (Amdt. 135-20, Eff. 1/6/87); (Amdt. 135-38, Eff. 11/26/90)

§ 135.161 Radio and navigational equipment: Carrying passengers under VFR at night or under VFR over-the-top.

(a) No person may operate an aircraft carrying passengers under VFR at night, or under VFR over-the-top, unless it has two-way communications equipment able, at least in flight, to transmit to, and receive from, ground facilities 25 miles away.

(b) No person may operate an aircraft carrying passengers under VFR over-the-top unless it has radio navigational equipment able to receive radio signals from the ground facilities to be used.

(c) No person may operate an airplane carrying passengers under VFR at night unless it has radio navigational equipment able to receive radio signals from the ground facilities to be used.

§ 135.163 Equipment requirements: Aircraft carrying passengers under IFR.

No person may operate an aircraft under IFR, carrying passengers, unless it has—

(a) A vertical speed indicator;

(b) A free-air temperature indicator;

(c) A heated pitot tube for each airspeed indicator;

(d) A power failure warning device or vacuum indicator to show the power available for gyroscopic instruments from each power source;

(e) An alternate source of static pressure for the altimeter and the airspeed and vertical speed indicators;

(f) For a single-engine aircraft, a generator or generators able to supply all probable combinations of continuous inflight electrical loads for required equipment and for re-charging the battery;

(g) For multiengine aircraft, at least two generators each of which is on a separate engine, of which any combination of one-half of the total number are rated sufficiently to supply the electrical loads of all required instruments and equipment necessary for safe emergency operation of the aircraft except that for multiengine helicopters, the two required generators may be mounted on the main rotor drive train; and

(h) Two independent sources of energy (with means of selecting either), of which at least one is an engine-driven pump or generator, each of which is able to drive all gyroscopic instruments and installed so that failure of one instrument or source does not interfere with the energy supply to the remaining instruments or the other energy source, unless, for single-engine aircraft, the rate-of-turn indicator has a source of energy separate from the bank and pitch and direction indicators. For the purpose of this paragraph, for multiengine aircraft, each engine-driven source of energy must be on a different engine.

(i) For the purpose of paragraph (f) of this section, a continuous inflight electrical load includes one that draws current continuously during flight, such as radio equipment, electrically driven instruments, and lights, but does not include occasional intermittent loads.

§ 135.165 Radio and navigational equipment: Extended overwater or IFR operations.

(a) No person may operate a turbojet airplane having a passenger seating configuration, excluding any pilot seat, of 10 seats or more, or a multiengine airplane carrying passengers as a "Commuter Air Carrier" as defined in Part 298 of this title, under IFR or in extended overwater operations unless it has at least the following radio communication and navigational equipment appropriate to the facilities to be used which are capable of transmitting to and receiving from, at any place on the route to be flown, at least one ground facility:

(1) Two transmitters, (2) two microphones, (3) two headsets or one headset and one speaker, (4) a marker beacon receiver, (5) two independent receivers for navigation, and (6) two independent receivers for communications.

(b) No person may operate an aircraft other than that specified in paragraph (a) of this section, under IFR or in extended overwater operations unless it has at least the following radio communication and navigational equipment appropriate to the facilities to be used and which are capable of transmitting to, and receiving from, at any place on the route, at least one ground facility:

(1) A transmitter, (2) two microphones, (3) two headsets or one headset and one speaker, (4) a marker beacon receiver, (5) two independent receivers for navigation, (6) two independent receivers for communications, and (7) for extended overwater operations only, an additional transmitter.

(c) For the purpose of paragraphs (a)(5), (a)(6), (b)(5), and (b)(6) of this section, a receiver is independent if the function of any part of it does not depend on the functioning of any part of another receiver. However, a receiver that can receive both communications and navigational signals may be used in place of a separate communications receiver and a separate navigational signal receiver.

§ 135.167 Emergency equipment: Extended overwater operations.

(a) No person may operate an aircraft in extended overwater operations unless it carries, installed in conspicuously marked locations easily accessible to the occupants if a ditching occurs, the following equipment:

(1) An approved life preserver equipped with an approved survivor locator light for each occupant of the aircraft. The life preserver must be easily accessible to each seated occupant.

(2) Enough approved life rafts of a rated capacity and buoyancy to accommodate the occupants of the aircraft.

(b) Each life raft required by paragraph (a) of this section must be equipped with or contain at least the following:

(1) One approved survivor locator light.

(2) One approved pyrotechnic signaling device.

(3) Either—

(i) One survival kit, appropriately equipped for the route to be flown; or

(ii) One canopy (for sail, sunshade, or rain catcher);

(iii) One radar reflector;

(iv) One life raft repair kit;

(v) One bailing bucket;

(vi) One signaling mirror;

(vii) One police whistle;

(viii) One raft knife;

(ix) One CO± bottle for emergency inflation;

(x) One inflation pump;

(xi) Two oars;

(xii) One 75-foot retaining line;

(xiii) One magnetic compass;

(xiv) One dye marker;

(xv) One flashlight having at least two size "D" cells or equivalent;

(xvi) A two-day supply of emergency food rations supplying at least 1,000 calories a day for each person;

(xvii) For each two persons the raft is rated to carry, two pints of water or one sea water desalting kit;

(xviii) One fishing kit; and

(xix) One book on survival appropriate for the area in which the aircraft is operated.

(c) No person may operate an airplane in extended overwater operations unless there is attached to one of the life rafts required by paragraph (a) of this section, an approved survival type emergency locator transmitter. Batteries used in this transmitter must be replaced (or recharged, if the batteries are rechargeable) when the transmitter has been in use for more than 1 cumulative hour, or, when 50 percent of their useful life (or for rechargeable batteries, 50 percent of their useful life of charge) has expired, as established by the transmitter manufacturer under its approval. The new expiration date for replacing (or recharging) the battery must be legibly marked on the outside of the transmitter. The battery useful life (or useful life of charge) requirements of this paragraph do not apply to batteries (such as water-activated batteries) that are essentially unaffected during probable storage intervals.

(Amdt. 135-4, Eff. 9/9/80); (Amdt. 135-20, Eff. 1/6/87); [(Amdt. 135-49, Eff. 6/21/94)]

§ 135.169 Additional airworthiness requirements.

(a) Except for commuter category airplanes, no person may operate a large airplane unless it meets the additional airworthiness requirements of §§ 121.213 through 121.283, 121.307, and 121.312 of this chapter.

(b) No person may operate a reciprocating-engine or turbopropeller-powered small airplane that has a passenger seating configuration, excluding pilot seats, of 10 seats or more unless it is type certificated—

(1) In the transport category;

(2) Before July 1, 1970, in the normal category and meets special conditions issued by the Administrator for airplanes intended for use in operations under this part;

(3) Before July 19, 1970, in the normal category and meets the additional airworthiness standards in Special Federal Aviation Regulation No. 23;

(4) In the normal category and meets the additional airworthiness standards in Appendix A;

(5) In the normal category and complies with section 1.(a) of Special Federal Aviation Regulation No. 41;

(6) In the normal category and complies with section 1.(b) of Special Federal Aviation Regulation No. 41; or

(7) In the commuter category.

(c) No person may operate a small airplane with a passenger seating configuration, excluding any pilot seat, of 10 seats or more, with a seating configuration greater than the maximum seating configuration used in that type airplane in operations under this part before August 19, 1977. This paragraph does not apply to—

(1) An airplane that is type certificated in the transport category; or

(2) An airplane that complies with—

(i) Appendix A of this part provided that its passenger seating configuration, excluding pilot seats, does not exceed 19 seats; or

(ii) Special Federal Aviation Regulation No. 41.

(d) Cargo or baggage compartments:

(1) After March 20, 1991, each Class C or D compartment, as defined in § 25.857 of Part 25 of this chapter, greater than 200 cubic feet in volume in a transport category airplane type certificated after January 1, 1958, must have ceiling and sidewall panels which are constructed of:

(i) Glass fiber reinforced resin;

(ii) Materials which meet the test requirements of Part 25, Appendix F, Part III of this chapter; or

(iii) In the case of liner installations approved prior to March 20, 1989, aluminum.

(2) For compliance with this paragraph, the term "liner" includes any design feature, such as a joint or fastener, which would affect the capability of the liner to safely contain a fire.

(Amdt. 135-2, Eff. 10/17/79); (Amdt. 135-21, Eff. 2/17/87); (Amdt. 135-31, Eff. 3/20/89)

§ 135.170 Materials for compartment interiors.

No person may operate an airplane that conforms to an amended or supplemental type certificate issued in accordance with SFAR No. 41 for a maximum certificated takeoff weight in excess of 12,500 pounds, unless within one year after issuance of the initial airworthiness certificate under that SFAR, the airplane meets the compartment interior requirements set forth in § 25.853(a), (b), (b-1), (b-2), and (b-3) of this chapter in effect on September 26, 1978.

(Amdt. 135-2, Eff. 10/17/79)

§ 135.171 Shoulder harness installation at flight crewmember stations.

(a) No person may operate a turbojet aircraft or an aircraft having a passenger seating configuration, excluding any pilot seat, of 10 seats or more unless it is equipped with an approved shoulder harness installed for each flight crewmember station.

(b) Each flight crewmember occupying a station equipped with a shoulder harness must fasten the shoulder harness during takeoff and landing, except that the shoulder harness may be unfastened if the crewmember cannot perform the required duties with the shoulder harness fastened.

§ 135.173 Airborne thunderstorm detection equipment requirements.

(a) No person may operate an aircraft that has a passenger seating configuration, excluding any pilot seat, of 10 seats or more in passenger-carrying operations, except a helicopter operating under day VFR conditions, unless the aircraft is equipped with either approved thunderstorm detection equipment or approved airborne weather radar equipment.

(b) After January 6, 1988, no person may operate a helicopter that has a passenger seating configuration, excluding any pilot seat, of 10 seats or more in passenger-carry operations, under night VFR when current weather reports indicate that thunderstorms or other potentially hazardous weather conditions that can be detected with airborne thunderstorm detection equipment may reasonably be expected along the route to be flown, unless the helicopter is equipped with either approved thunderstorm detection equipment or approved airborne weather radar equipment.

(c) No person may begin a flight under IFR or night VFR conditions when current weather reports indicate that thunderstorms or other potentially hazardous weather conditions that can be detected with airborne thunderstorm detection equipment, required by paragraph (a) or (b) of this section, may reasonably be expected along the route to be flown, unless the airborne thunderstorm detection equipment is in satisfactory operating condition.

(d) If the airborne thunderstorm detection equipment becomes inoperative en route, the aircraft must be operated under the instructions and procedures specified for that event in the manual required by § 135.21.

(e) This section does not apply to aircraft used solely within the State of Hawaii, within the State of Alaska, within that part of Canada west of longitude 130 degrees W, between latitude 70 degrees N, and latitude 53 degrees N, or during any training, test, or ferry flight.

(f) Without regard to any other provision of this part, an alternate electrical power supply is not required for airborne thunderstorm detection equipment.

(Amdt. 135-20, Eff. 1/6/87)

§ 135.175 Airborne weather radar equipment requirements.

(a) No person may operate a large, transport category aircraft in passenger-carrying operations unless approved airborne weather radar equipment is installed in the aircraft.

(b) No person may begin a flight under IFR or night VFR conditions when current weather reports indicate that thunderstorms, or other potentially hazardous weather conditions that can be detected with airborne weather radar equipment, may reasonably be expected along the route to be flown, unless the airborne weather radar equipment required by paragraph (a) of this section is in satisfactory operating condition.

(c) If the airborne weather radar equipment becomes inoperative en route, the aircraft must be operated under the instructions and procedures specified for that event in the manual required by § 135.21.

(d) This section does not apply to aircraft used solely within the State of Hawaii, within the State of Alaska, within that part of Canada west of longitude 130 degrees W, between latitude 70 degrees N, and latitude 53 degrees N, or during any training, test, or ferry flight.

(e) Without regard to any other provision of this part, an alternate electrical power supply is not required for airborne weather radar equipment.

§ 135.177 Emergency equipment requirements for aircraft having a passenger seating configuration of more than 19 passengers.

(a) No person may operate an aircraft having a passenger seating configuration, excluding any pilot seat, of more than 19 seats unless it is equipped with the following emergency equipment:

(1) One approved first-aid kit for treatment of injuries likely to occur in flight or in a minor accident, which meets the following specifications and requirements:

(i) Each first-aid kit must be dust and moisture proof, and contain only materials that either meet Federal Specifications GGK-319a, as revised, or as approved by the Administrator.

(ii) Required first-aid kits must be readily accessible to the cabin flight attendants.

(iii) At time of takeoff, each first-aid kit must contain at least the following or other contents approved by the Administrator:

Contents **(Quantity)**
Adhesive bandage compressors, 1 in **(16)**
Antiseptic swabs **(20)**
Ammonia inhalents **(10)**
Bandage compressors, 4 in **(8)**
Triangular bandage compressors, 40 in **(5)**
Arm splint, noninflatable **(1)**
Leg splint, noninflatable **(1)**
Roller bandage, 4 in **(4)**
Adhesive tape, 1-in standard roll **(2)**
Bandage scissors **(1)**

(2) A crash axe carried so as to be accessible to the crew but inaccessible to passengers during normal operations.

(3) Signs that are visible to all occupants to notify them when smoking is prohibited and when safety belts must be fastened. The signs must be constructed so that they can be turned on during any movement of the aircraft on the surface, for each takeoff or landing, and at other times considered necessary by the pilot in command. "No smoking" signs shall be turned on when required by § 135.127.

(4) (Reserved)

(b) Each item of equipment must be inspected regularly under inspection periods established in the operations specifications to ensure its condition for continued serviceability and immediate readiness to perform its intended emergency purposes.

(Amdt. 135-25, Eff. 4/23/88); *[(Amdt. 135-43, Eff. 6/30/92)]; [(Amdt. 135-44, Eff. 10/15/92)]; [(Amdt. 135-47, Eff. 1/12/94)]

[§ 135.178 Additional emergency equipment.

[No person may operate an airplane having a passenger seating configuration of more than 19 seats, unless it has the additional emergency equipment specified in paragraphs (a) through (1) of this section.

[(a) *Means for emergency evacuation.* Each passenger-carrying landplane emergency exit (other than over-the-wing) that is more than 6 feet from the ground, with the airplane on the ground and the landing gear extended, must have an approved means to assist the occupants in descending to the ground. The assisting means for a floor-level emergency exit must meet the requirements of § 25.809(f)(1) of this chapter in effect on April 30, 1972, except that, for any airplane for which the application for the type certificate was filed after that date, it must meet the requirements under which the airplane was type certificated. An assisting means that deploys automatically must be armed during taxiing, takeoffs, and landings; however, the Administrator may grant a deviation from the requirement of automatic deployment if he finds that the design of the exit makes compliance impractical, if the assisting means automatically erects upon deployment and, with respect to required emergency exits, if an emergency evacuation demonstration is conducted in accordance with § 121.291(a) of this chapter. This paragraph does not apply to the rear window emergency exit of Douglas DC-3 airplanes operated with fewer than 36 occupants, including crewmembers, and fewer than five exits authorized for passenger use.

[(b) *Interior emergency exit marking.* The following must be complied with for each passenger-carrying airplane:

[(1) Each passenger emergency exit, its means of access, and its means of opening must be conspicuously marked. The identity and location of each passenger emergency exit must be recognizable from a distance equal to the width of the cabin. The location of each passenger emergency exit must be indicated by a sign visible to occupants approaching along the main passenger aisle. There must be a locating sign—

[(i) Above the aisle near each over-the-wing passenger emergency exit, or at another ceiling location if it is more practical because of low headroom;

[(ii) Next to each floor level passenger emergency exit, except that one sign may serve two such exits if they both can be seen readily from that sign; and

[(iii) On each bulkhead or divider that prevents fore and aft vision along the passenger cabin, to indicate emergency exits beyond and obscured by it, except that if this is not possible, the sign may be placed at another appropriate location.

[(2) Each passenger emergency exit marking and each locating sign must meet the following:

[(i) For an airplane for which the application for the type certificate was filed prior to May 1, 1972, each passenger emergency exit marking and each locating sign must be manufactured to meet the requirements of § 25.812(b) of this chapter in effect on April 30, 1972. On these airplanes, no sign may continue to be used if its luminescence (brightness) decreases to below 100 microlamberts. The colors may be reversed if it increases the emergency illumination of the passenger compartment. However, the Administrator may authorize deviation from the 2-inch background requirements if he finds that special circumstances exist that make compliance impractical and that the proposed deviation provides an equivalent level of safety.

[(ii) For an airplane for which the application for the type certificate was filed on or after May 1, 1972, each passenger emergency exit marking and each locating sign must be manufactured to meet the interior emergency exit marking requirements under which the airplane was type certificated. On these airplanes, no sign may continue to be used if its luminescence (brightness) decreases to below 250 microlamberts.

[(c) *Lighting for interior emergency exit markings.* Each passenger-carrying airplane must have an emergency lighting system, independent of the main lighting system; however, sources of general cabin illumination may be common to both the emergency and the main lighting systems if the power supply to the emergency lighting system is independent of the power supply to the main lighting system. The emergency lighting system must—

[(1) Illuminate each passenger exit marking and locating sign;

[(2) Provide enough general lighting in the passenger cabin so that the average illumination when measured at 40-inch intervals at seat armrest height, on the centerline of the main passenger aisle, is at least 0.05 foot-candle; and

[(3) For airplanes type certificated after January 1, 1958, include floor proximity emergency escape path marking which meets the requirements of § 25.812(e) of this chapter in effect on November 26, 1984.

[(d) *Emergency light operation.* Except for lights forming part of emergency lighting subsystems provided in compliance with § 25.812(h) of this chapter (as prescribed in paragraph (h) of this section) that serve no more than one assist means, are independent of the airplane's main emergency lighting systems, and are automatically activated when the assist means is deployed, each light required by paragraphs (c) and (h) of this section must:

[(1) Be operable manually both from the flightcrew station and from a point in the passenger compartment that is readily accessible to a normal flight attendant seat;

[(2) Have a means to prevent inadvertent operation of the manual controls;

[(3) When armed or turned on at either station, remain lighted or become lighted upon interruption of the airplane's normal electric power;

[(4) Be armed or turned on during taxiing, takeoff, and landing. In showing compliance with this paragraph, a transverse vertical separation of the fuselage need not be considered;

[(5) Provide the required level of illumination for at least 10 minutes at the critical ambient conditions after emergency landing; and

[(6) Have a cockpit control device that has an "on," "off," and "armed" position.

[(e) *Emergency exit operating handles.*

[(1) For a passenger-carrying airplane for which the application for the type certificate was filed prior to May 1, 1972, the location of each passenger emergency exit operating handle, and instructions for opening the exit, must be shown by a marking on or near the exit that is readable from a distance of 30 inches. In addition, for each Type I and Type II emergency exit with a locking mechanism released by rotary motion of the handle, the instructions for opening must be shown by—

[(i) A red arrow with a shaft at least three-fourths inch wide and a head twice the width of the shaft, extending along at least 70° of arc at a radius approximately equal to three-fourths of the handle length; and

[(ii) The word "open" in red letters 1 inch high placed horizontally near the head of the arrow.

[(2) For a passenger-carrying airplane for which the application for the type certificate was filed on or after May 1, 1972, the location of each passenger emergency exit operating handle and instructions for opening the exit must be shown in accordance with the requirements under which the airplane was type certificated. On these airplanes, no operating handle or operating handle cover may continue to be used if its luminescence (brightness) decreases to below 100 microlamberts.

[(f) *Emergency exit access.* Access to emergency exits must be provided as follows for each passenger-carrying airplane:

[(1) Each passageway between individual passenger areas, or leading to a Type I or Type II emergency exit, must be unobstructed and at least 20 inches wide.

[(2) There must be enough space next to each Type I or Type II emergency exit to allow a crewmember to assist in the evacuation of passengers without reducing the unobstructed width of the passageway below that required in paragraph (f)(1) of this section; however, the Administrator may authorize deviation from this requirement for an airplane certificated under the provisions of Part 4b of the Civil Air Regulations in effect before December 20, 1951, if he finds that special circumstances exist that provide an equivalent level of safety.

[(3) There must be access from the main aisle to each Type III and Type IV exit. The access from the aisle to these exits must not be obstructed by seats, berths, or other protrusions in a manner that would reduce the effectiveness of the exit. In addition, for a transport category airplane type certificated after January 1, 1958, there must be placards installed in accordance with 25.813(c)(3) of this chapter for each Type III exit after December 3, 1992.

[(4) If it is necessary to pass through a passageway between passenger compartments to reach any required emergency exit from any seat in the passenger cabin, the passageway must not be obstructed. Curtains may, however, be used if they allow free entry through the passageway.

[(5) No door may be installed in any partition between passenger compartments.

[(6) If it is necessary to pass through a doorway separating the passenger cabin from other areas to reach a required emergency exit from any passenger seat, the door must have a means to latch it in the open position, and the door must be latched open during each takeoff and landing. The latching means must be able to withstand the loads imposed upon it when the door is subjected to the ultimate inertia forces, relative to the surrounding structure, listed in § 25.561(b) of this chapter.

[(g) *Exterior exit markings.* Each passenger emergency exit and the means of opening that exit from the outside must be marked on the outside of the airplane. There must be a 2-inch colored band outlining each passenger emergency exit on the side of the fuselage. Each outside marking, including the band, must be readily distinguishable from the surrounding fuselage area by contrast in color. The markings must comply with the following:

[(1) If the reflectance of the darker color is 15 percent or less, the reflectance of the lighter color must be at least 45 percent.

[(2) If the reflectance of the darker color is greater than 15 percent, at least a 30 percent difference between its reflectance and the reflectance of the lighter color must be provided.

[(3) Exits that are not in the side of the fuselage must have the external means of opening and applicable instructions marked conspicuously in red or, if red is inconspicuous against the background color, in bright chrome yellow and, when the opening means for such an exit is located on only one side of the fuselage, a conspicuous marking to that effect must be provided on the other side. "Reflectance" is the ratio of the luminous flux reflected by a body to the luminous flux it receives.

[(h) *Exterior emergency lighting and escape route.*

[(1) Each passenger-carrying airplane must be equipped with exterior lighting that meets the following requirements:

[(i) For an airplane for which the application for the type certificate was filed prior to May 1, 1972, the requirements of § 25.812 (f) and (g) of this chapter in effect on April 30, 1972.

[(ii) For an airplane for which the application for the type certificate was filed on or after May 1, 1972, the exterior emergency lighting requirements under which the airplane was type certificated.

[(2) Each passenger-carrying airplane must be equipped with a slip-resistant escape route that meets the following requirements:

[(i) For an airplane for which the application for the type certificate was filed prior to May 1, 1972, the requirements of § 25.803(e) of this chapter in effect on April 30, 1972.

[(ii) For an airplane for which the application for the type certificate was filed on or after May 1, 1972, the slip-resistant escape route requirements under which the airplane was type certificated.

[(i) *Floor level exits.* Each floor level door or exit in the side of the fuselage (other than those leading into a cargo or baggage compartment that is not accessible from the passenger cabin) that is 44 or more inches high and 20 or more inches wide, but not wider than 46 inches, each passenger ventral exit (except the ventral exits on Martin 404 and Convair 240 airplanes), and each tail cone exit, must meet the requirements of this section for floor level emergency exits. However, the Administrator may grant a deviation from this paragraph if he finds that circumstances make full compliance impractical and that an acceptable level of safety has been achieved.

[(j) *Additional emergency exits.* Approved emergency exits in the passenger compartments that are in excess of the minimum number of required emergency exits must meet all of the applicable provisions of this section, except paragraphs (f)(1), (2), and (3) of this section, and must be readily accessible.

[(k) On each large passenger-carrying turbojet-powered airplane, each ventral exit and tailcone exit must be—

[(1) Designed and constructed so that it cannot be opened during flight; and

[(2) Marked with a placard readable from a distance of 30 inches and installed at a conspicuous location near the means of opening the exit, stating that the exit has been designed and constructed so that it cannot be opened during flight.

[(l) *Portable lights.* No person may operate a passenger-carrying airplane unless it is equipped with flashlight stowage provisions accessible from each flight attendant seat.]

[(Amdt. 135-43, Eff. 6/3/92)]

§ 135.179 Inoperable instruments and equipment.

(a) No person may take off an aircraft with inoperable instruments or equipment installed unless the following conditions are met:

(1) An approved Minimum Equipment List exists for that aircraft.

(2) The Flight Standards District Office having certification responsibility has issued the certificate holder operations specifications authorizing operations in accordance with an approved Minimum Equipment List. The flight crew shall have direct access at all times prior to flight to all of the information contained in the approved Minimum Equipment List through printed or other means approved by the Administrator in the certificate holders operations specifications. An approved Minimum Equipment List, as authorized by the operations specifications, constitutes an approved change to the type design without requiring recertification.

(3) The approved Minimum Equipment List must:

(i) Be prepared in accordance with the limitations specified in paragraph (b) of this section.

(ii) Provide for the operation of the aircraft with certain instruments and equipment in an inoperable condition.

(4) Records identifying the inoperable instruments and

equipment and the information required by (a)(3)(ii) of this section must be available to the pilot.

(5) The aircraft is operated under all applicable conditions and limitations contained in the Minimum Equipment List and the operations specifications authorizing use of the Minimum Equipment List.

(b) The following instruments and equipment may not be included in the Minimum Equipment List:

(1) Instruments and equipment that are either specifically or otherwise required by the airworthiness requirements under which the airplane is type certificated and which are essential for safe operations under all operating conditions.

(2) Instruments and equipment required by an airworthiness directive to be in operable condition unless the airworthiness directive provides otherwise.

(3) Instruments and equipment required for specific operations by this part.

(c) Notwithstanding paragraphs (b)(1) and (b)(3) of this section, an aircraft with inoperable instruments or equipment may be operated under a special flight permit under §§ 21.197 and 21.199 of this chapter.

(Amdt. 135-39, Eff. 6/20/91)

§ 135.180 Traffic alert and collision avoidance system.

(a) After February 9, 1995, no person may operate a turbine powered airplane that has a passenger seating configuration, excluding any pilot seat, of 10 to 30 seats unless it is equipped with an approved traffic alert and collision avoidance system.

(b) The airplane flight manual required by § 135.21 of this part shall contain the following information on the TCAS I system required by this section:

(1) Appropriate procedures for—

(i) The use of the equipment; and

(ii) Proper flightcrew action with respect to the equipment operation.

(2) An outline of all input sources that must be operating for the TCAS to function properly.

Docket No. 25355 (54 FR 951) Eff. 1/10/89;
(Amdt. 135-30, Eff. 2/9/89)

§ 135.181 Performance requirements: Aircraft operated over-the-top or in IFR conditions.

(a) Except as provided in paragraphs (b) and (c) of this section, no person may—

(1) Operate a single-engine aircraft carrying passengers over-the-top or in IFR conditions; or

(2) Operate a multiengine aircraft carrying passengers over-the-top or in IFR conditions at a weight that will not allow it to climb, with the critical engine inoperative, at least 50 feet a minute when operating at the MEAs of the route to be flown or 5,000 feet MSL, whichever is higher.

(b) Notwithstanding the restrictions in paragraph (a)(2) of this section, multiengine helicopters carrying passengers offshore may conduct such operations in over-the-top or in IFR conditions at a weight that will allow the helicopter to climb at least 50 feet per minute with the critical engine inoperative when operating at the MEA of the route to be flown or 1,500 feet MSL, whichever is higher.

(c) Without regard to paragraph (a) of this section—

(1) If the latest weather reports or forecasts, or any combination of them, indicate that the weather along the planned route (including takeoff and landing) allows flight under VFR under the ceiling (if a ceiling exists) and that the weather is forecast to remain so until at least 1 hour after the estimated time of arrival at the destination, a person may operate an aircraft over-the-top; or

(2) If the latest weather reports or forecasts, or any combination of them, indicate that the weather along the planned route allows flight under VFR under the ceiling (if a ceiling exists) beginning at a point no more than 15 minutes flying time at normal cruise speed from the departure airport, a person may—

(i) Take off from the departure airport in IFR conditions and fly in IFR conditions to a point no more than 15 minutes flying time at normal cruise speed from that airport;

(ii) Operate an aircraft in IFR conditions if unforecast weather conditions are encountered while en route on a flight planned to be conducted under VFR; and

(iii) Make an IFR approach at the destination airport if unforecast weather conditions are encountered at the airport that do not allow an approach to be completed under VFR.

(d) Without regard to paragraph (a) of this section, a person may operate an aircraft over-the-top under conditions allowing—

(1) For multiengine aircraft, descent or continuance of the flight under VFR if its critical engine fails; or

(2) For single-engine aircraft, descent under VFR if its engine fails.

(Amdt. 135-20, Eff. 1/6/87)

§ 135.183 Performance requirements: Land aircraft operated over water.

No person may operate a land aircraft carrying passengers over water unless—

(a) It is operated at an altitude that allows it to reach land in the case of engine failure;

(b) It is necessary for takeoff or landing;

(c) It is a multiengine aircraft operated at a weight that will allow it to climb, with the critical engine inoperative, at least 50 feet a minute, at an altitude of 1,000 feet above the surface; or

(d) It is a helicopter equipped with helicopter flotation devices.

§ 135.185 Empty weight and center of gravity: Currency requirement.

(a) No person may operate a multiengine aircraft unless the current empty weight and center of gravity are calculated from values established by actual weighing of the aircraft within the preceding 36 calendar months.

(b) Paragraph (a) of this section does not apply to—

(1) Aircraft issued an original airworthiness certificate within the preceding 36 calendar months; and

(2) Aircraft operated under a weight and balance system approved in the operations specifications of the certificate holder.

Subpart D—VFR/IFR operating limitations and weather requirements

§ 135.201 Applicability.

This subpart prescribes the operating limitations for VFR/IFR flight operations and associated weather requirements for operations under this part.

§ 135.203 VFR: Minimum altitudes.

Except when necessary for takeoff and landing, no person may operate under VFR—

(a) An airplane—

(1) During the day, below 500 feet above the surface or less than 500 horizontally from any obstacle; or

(2) At night, at an altitude less than 1,000 feet above the highest obstacle within a horizontal distance of 5 miles from the course intended to be flown or, in designated mountainous terrain, less than 2,000 feet above the highest obstacle within a horizontal distance of 5 miles from the course intended to be flown; or

(b) A helicopter over a congested area at an altitude less than 300 feet above the surface.

§ 135.205 VFR: Visibility requirements.

(a) No person may operate an airplane under VFR in uncontrolled airspace when the ceiling is less than 1,000 feet unless flight visibility is at least 2 miles.

(b) [No person may operate a helicopter under VFR in Class G airspace at an altitude of 1,200 feet or less above the surface or within the lateral boundaries of the surface areas of Class B, Class C, Class D, or Class E airspace designated for an airport unless the visibility is at least—]

(1) During the day—½ mile; or

(2) At night—1 mile.

[(Amdt. 135-41, Eff. 9/16/93)]

§ 135.207 VFR: Helicopter surface reference requirements.

No person may operate a helicopter under VFR unless that person has visual surface reference or, at night, visual surface light reference, sufficient to safely control the helicopter.

§ 135.209 VFR: Fuel supply.

(a) No person may begin a flight operation in an airplane under VFR unless, considering wind and forecast weather conditions, it has enough fuel to fly to the first point of intended landing and, assuming normal cruising fuel consumption—

(1) During the day, to fly after that for at least 30 minutes; or

(2) At night, to fly after that for at least 45 minutes.

(b) No person may begin a flight operation in a helicopter under VFR unless, considering wind and forecast weather conditions, it has enough fuel to fly to the first point of intended landing and, assuming normal cruising fuel consumption, to fly after that for at least 20 minutes.

§ 135.211 VFR: Over-the-top carrying passengers: Operating limitations.

Subject to any additional limitations in § 135.181, no person may operate an aircraft under VFR over-the-top carrying passengers, unless—

(a) Weather reports or forecasts, or any combination of them, indicate that the weather at the intended point of termination of over-the-top flight—

(1) Allows descent to beneath the ceiling under VFR and is forecast to remain so until at least 1 hour after the estimated time of arrival at that point; or

(2) Allows an IFR approach and landing with flight clear of the clouds until reaching the prescribed initial approach altitude over the final approach facility, unless the approach is made with the use of radar under § 91.175(f) of this chapter; or

(b) It is operated under conditions allowing—

(1) For multiengine aircraft, descent or continuation of the flight under VFR if its critical engine fails; or

(2) For single-engine aircraft, descent under VFR if its engine fails.

(Amdt. 135-32, Eff. 8/18/90)

§ 135.213 Weather reports and forecasts.

(a) Whenever a person operating an aircraft under this part is required to use a weather report or forecast, that person shall use that of the U.S. National Weather Service, a source approved by the U.S. National Weather Service, or a source approved by the Administrator. However, for operations under VFR, the pilot in command may, if such a report is not available, use weather information based on that pilot's own observations or on those of other persons competent to supply appropriate observations.

(b) For the purposes of paragraph (a) of this section, weather observations made and furnished to pilots to conduct IFR operations at an airport must be taken at the airport where those IFR operations are conducted, unless the Administrator issues operations specifications allowing the use of weather observations taken at a location not at the airport where the IFR operations are conducted. The Administrator issues such operations specifications when, after investigation by the U.S. National Weather Service and the FAA Flight Standards District Office charged with the overall inspection of the certificate holder, it is found that the standards of safety for that operation would allow the deviation from this paragraph for a particular operation for which an ATCO operating certificate has been issued.

§ 135.215 IFR: Operating limitations.

(a) Except as provided in paragraphs (b), (c) and (d) of this section, no person may operate an aircraft under IFR outside of controlled airspace or at any airport that does not have an approved standard instrument approach procedure.

(b) The Administrator may issue operations specifications to the certificate holder to allow it to operate under IFR over routes outside controlled airspace if—

(1) The certificate holder shows the Administrator that the flight crew is able to navigate, without visual reference to the ground, over an intended track without deviating more than 5 degrees or 5 miles, whichever is less, from that track; and

(2) The Administrator determines that the proposed operations can be conducted safely.

(c) A person may operate an aircraft under IFR outside of controlled airspace if the certificate holder has been approved for the operations and that operation is necessary to—

(1) Conduct an instrument approach to an airport for which there is in use a current approved standard or special instrument approach procedure; or

(2) Climb into controlled airspace during an approved missed approach procedure; or

(3) Make an IFR departure from an airport having an approved instrument approach procedure.

(d) The Administrator may issue operations specifications to the certificate holder to allow it to depart at an airport that does not have an approved standard instrument approach procedure when the Administrator determines that it is necessary to make an IFR departure from that airport and that the proposed operations can be conducted safely. The approval to operate at that airport does not include an approval to make an IFR approach to that airport.

§ 135.217 IFR: Takeoff limitations.

No person may take off an aircraft under IFR from an airport where weather conditions are at or above takeoff minimums but are below authorized IFR landing minimums unless there is an alternate airport within 1 hour's flying time (at normal cruising speed, in still air) of the airport of departure.

§ 135.219 IFR: Destination airport weather minimums.

No person may take off an aircraft under IFR or being an IFR or over-the-top operation unless the latest weather reports or forecasts, or any combination of them, indicate that weather conditions at the estimated time of arrival at the next airport of intended landing will be at or above authorized IFR landing minimums.

§ 135.221 IFR: Alternate airport weather minimums.

No person may designate an alternate airport unless the weather reports or forecasts, or any combination of them, indicate that the weather conditions will be at or above authorized alternate airport landing minimums for that airport at the estimated time of arrival.

§ 135.223 IFR: Alternate airport requirements.

(a) Except as provided in paragraph (b) of this section, no person may operate an aircraft in IFR conditions unless it carries enough fuel (considering weather reports or forecasts or any combination of them) to—

(1) Complete the flight to the first airport of intended landing;

(2) Fly from that airport to the alternate airport; and

(3) Fly after that for 45 minutes at normal cruising speed, or helicopters, fly after that for 30 minutes at normal cruising speed.

(b) Paragraph (a)(2) of this section does not apply if Part 97 of this chapter prescribes a standard instrument approach procedure for the first airport of intended landing and, for at least one hour before and after the estimated time of arrival, the appropriate weather reports or forecasts, or any combination of them, indicate that—

(1) The ceiling will be at least 1,500 feet above the lowest circling approach MDA; or

(2) If a circling instrument approach is not authorized for the airport, the ceiling will be at least 1,500 feet above the lowest published minimum or 2,000 feet above the airport elevation, whichever is higher; and

(3) Visibility for that airport is forecast to be at least three miles, or two miles more than the lowest applicable visibility minimums, whichever is the greater, for the instrument approach procedure to be used at the destination airport.

(Amdt. 135-20, Eff. 1/6/87)

§ 135.225 IFR: Takeoff, approach and landing minimums.

(a) No pilot may begin instrument approach procedure to an airport unless—

(1) That airport has a weather reporting facility operated by the U.S. National Weather Service, a source approved by U.S. National Weather Service, or a source approved by the Administrator; and

(2) The latest weather report issued by that weather reporting facility indicates that weather conditions are at or above the authorized IFR landing minimums for that airport.

(b) No pilot may begin the final approach segment of an instrument approach procedure to an airport unless the latest weather reported by the facility described in paragraph (a)(1) of this section indicates that weather conditions are at or above the authorized IFR landing minimums for that procedure.

(c) If a pilot has begun the final approach segment of an instrument approach to an airport under paragraph (b) of this section and a later weather report indicating below minimum conditions is received after the aircraft is—

(1) On an ILS final approach and has passed the final approach fix; or

(2) On an ASR or PAR final approach and has been turned over to the final approach controller; or

(3) On a final approach using a VOR, NDB, or comparable approach procedure; and the aircraft—

(i) Has passed the appropriate facility or final approach fix; or

(ii) Where a final approach fix is not specified, has completed the procedure turn and is established inbound toward the airport on the final approach course within the distance prescribed in the procedure; the approach may be continued and a landing made if the pilot finds, upon reaching the authorized MDA or DH, that actual weather conditions are at least equal to the minimums prescribed for the procedure.

(d) The MDA or DH and visibility landing minimums prescribed in Part 97 of this chapter or in the operator's operations specifications are increased by 100 feet and ½ mile respectively, but not to exceed the ceiling and visibility minimums for that airport when used as an alternate airport, for each pilot in command of a turbine-powered airplane who has not served at least 100 hours as pilot in command in that type of airplane.

(e) Each pilot making an IFR takeoff or approach and landing at a military or foreign airport shall comply with applicable instrument approach procedures and weather minimums prescribed by the authority having jurisdiction over the airport. In addition, no pilot may, at that airport—

(1) Take off under IFR when the visibility is less than 1 mile; or

(2) Make an instrument approach when the visibility is less than ½ mile.

(f) If takeoff minimums are specified in Part 97 of this chapter for the takeoff airport, no pilot may take off an aircraft under IFR when the weather conditions reported by the facility described in paragraph (a)(1) of this section are less than the takeoff minimums specified for the takeoff airport in Part 97 or in the certificate holder's operations specifications.

(g) Except as provided in paragraph (h) of this section, if takeoff minimums are not prescribed in Part 97 of this chapter for the takeoff airport, no pilot may take off an aircraft under IFR when the weather conditions reported by the facility described in paragraph (a)(1) of this section are less than that prescribed in Part 91 of this chapter or in the certificate holder's operations specifications.

(h) At airports where straight-in instrument approach procedures are authorized, a pilot may take off an aircraft under IFR when the weather conditions reported by the facility described in paragraph (a)(1) of this section are equal to or better than the lowest straight-in landing minimums, unless otherwise restricted, if—

(1) The wind direction and velocity at the time of takeoff are such that a straight-in instrument approach can be made to the runway served by the instrument approach;

(2) The associated ground facilities upon which the landing minimums are predicated and the related airborne equipment are in the normal operation; and

(3) The certificate holder has been approved for such operations.

§ 135.227 Icing conditions: Operating limitations.

(a) [No pilot may take off an aircraft that has frost, ice, or snow adhering to any rotor blade, propeller, windshield, wing, stabilizing or control surface, to a powerplant installation, or to an airspeed, altimeter, rate of climb, or flight attitude instrument system, except under the following conditions:

[(l) Takeoffs may be made with frost adhering to the wings, or stabilizing or control surfaces, if the frost has been polished to make it smooth.

[(2) Takeoffs may be made with frost under the wing in the area of the fuel tanks if authorized by the Administrator.]

[(b) No certificate holder may authorize an airplane to take off and no pilot may take off an airplane any time conditions are such that frost, ice, or snow may reasonably be expected to adhere to the airplane unless the pilot has completed all applicable training as required by § 135.341 and unless one of the following requirements is met:

[(1) A pretakeoff contamination check, that has been established by the certificate holder and approved by the Administrator for the specific airplane type, has been completed within 5 minutes prior to beginning takeoff. A pretakeoff contamination check is a check to make sure the wings and control surfaces are free of frost, ice, or snow.

[(2) The certificate holder has an approved alternative

procedure and under that procedure the airplane is determined to be free of frost, ice, or snow.

[(3) The certificate holder has an approved deicing/anti-icing program that complies with § 121.629(c) of this chapter and the takeoff complies with that program.]

([c]) Except for an airplane that has ice protection provisions that meet § 34 of Appendix A, or those for transport category airplane type certificate, no pilot may fly—

(1) Under IFR into known or forecast light or moderate icing conditions; or

(2) Under VFR into known light or moderate icing conditions; unless the aircraft has functioning deicing or anti-icing equipment protecting each rotor blade, propeller, windshield, wing, stabilizing or control surface, and each airspeed, altimeter, rate of climb, or flight attitude instrument system.

([d]) No pilot may fly a helicopter under IFR into known or forecast icing conditions or under VFR into known icing conditions unless it has been type certificated and appropriately equipped for operations in icing conditions.

([e]) Except for an airplane that has ice protection provisions that meet § 34 of Appendix A, or those for transport category airplane type certification, no pilot may fly an aircraft into known or forecast severe icing conditions.

([f]) If current weather reports and briefing information relied upon by the pilot in command indicate that the forecast icing condition that would otherwise prohibit the flight will not be encountered during the flight because of changed weather conditions since the forecast, the restrictions in paragraphs (b), (c), and (d) of this section based on forecast conditions do not apply.

(Amdt. 135-20, Eff. 1/6/87); [(Amdt. 135-46, Eff. 1/31/94)]

§ 135.229 Airport requirements.

(a) No certificate holder may use any airport unless it is adequate for the proposed operation, considering such items as size, surface, obstructions, and lighting.

(b) No pilot of an aircraft carrying passengers at night may take off from, or land on, an airport unless—

(1) That pilot has determined the wind direction from an illuminated wind direction indicator or local ground communications or, in the case of takeoff, that pilot's personal observations; and

(2) The limits of the area to be used for landing or takeoff are clearly shown—

(i) For airplanes, by boundary or runway marker lights;

(ii) For helicopters, by boundary or runway marker lights or reflective material.

(c) For the purpose of paragraph (b) of this section, if the area to be used for takeoff or landing is marked by flare pots or lanterns, their use must be approved by the Administrator.

Subpart E—Flight crewmember requirements

§ 135.241 Applicability.

This subpart prescribes the flight crewmember requirements for operations under this part.

§ 135.243 Pilot in command qualifications.

(a) No certificate holder may use a person, nor may any person serve, as pilot in command in passenger-carrying operations of a turbojet airplane, of an airplane having a passenger seating configuration, excluding any pilot seat, of 10 seats or more, or a multiengine airplane being operated by the "Commuter Air Carrier" (as defined in Part 298 of this title), unless that person holds an airline transport pilot certificate with appropriate category and class ratings and, if required, an appropriate type rating for that airplane.

(b) Except as provided in paragraph (a) of this section, no certificate holder may use a person, nor may any person serve, as pilot in command of an aircraft under VFR unless that person—

(1) Holds at least a commercial pilot certificate with appropriate category and class ratings and, if required, an appropriate type rating for that aircraft; and

(2) Has had at least 500 hours of flight time as a pilot, including at least 100 hours of cross-country flight time, at least 25 hours of which were at night; and

(3) For an airplane, holds an instrument rating or an airline transport pilot certificate with an airplane category rating; or

(4) For helicopter operations conducted VFR over-the-top, holds a helicopter instrument rating, or an airline transport pilot certificate with a category and class rating for that aircraft, not limited to VFR.

(c) Except as provided in paragraph (a) of this section, no certificate holder may use a person, nor may any person serve, as pilot in command of an aircraft under IFR unless that person—

(1) Holds at least a commercial pilot certificate with appropriate category and class ratings and, if required, an appropriate type rating for that aircraft; and

(2) Has had at least 1,200 hours of flight time as a pilot, including 500 hours of cross-country flight time, 100 hours of night flight time, and 75 hours of actual or simulated instrument time at least 50 hours of which were in actual flight; and

(3) For an airplane, holds an instrument rating or an airline transport pilot certificate with an airplane category rating; or

(4) For a helicopter, holds a helicopter instrument rat-

ing, or an airline transport pilot certificate with a category and class rating for the aircraft, not limited to VFR.

(d) Paragraph (b)(3) of this section does not apply when—

(1) The aircraft used is a single reciprocating-engine-powered airplane;

(2) The certificate holder does not conduct any operation pursuant to a published flight schedule which specifies five or more round trips a week between two or more points and places between which the round trips are performed, and does not transport mail by air under a contract or contracts with the United States Postal Service having total amount estimated at the beginning of any semiannual reporting period (January 1–June 30; July 1–December 31) to be in excess of $20,000 over the 12 months commencing with the beginning of the reporting period;

(3) The area, as specified in the certificate holder's operations specifications, is an isolated area, as determined by the Flight Standards district office, if it is shown that—

(i) The primary means of navigation in the area is by pilotage, since radio navigational aids are largely ineffective; and

(ii) The primary means of transportation in the area is by air;

(4) Each flight is conducted under day VFR with a ceiling of not less than 1,000 feet and visibility not less than 3 statute miles;

(5) Weather reports or forecasts, or any combination of them, indicate that for the period commencing with the planned departure and ending 30 minutes after the planned arrival at the destination the flight may be conducted under VFR with a ceiling of not less than 1,000 feet and visibility of not less than 3 statute miles, except that if weather reports and forecasts are not available, the pilot in command may use that pilot's observations or those of other persons competent to supply weather observations if those observations indicate the flight may be conducted under VFR with the ceiling and visibility required in this paragraph;

(6) The distance of each flight from the certificate holder's base of operation to destination does not exceed 250 nautical miles for a pilot who holds a commercial pilot certificate with an airplane rating without an instrument rating, provided the pilot's certificate does not contain any limitation to the contrary; and

(7) The areas to be flown are approved by the certificate-holding FAA Flight Standards district office and are listed in the certificate holder's operations specifications.

(Amdt. 135-15, Eff. 6/11/81)

§ 135.244 Operating experience.

(a) No certificate holder may use any person, nor may any person serve, as a pilot in command of an aircraft operated by a Commuter Air Carrier (as defined in § 298.2 of this title) in passenger-carrying operations, unless that person has completed, prior to designation as pilot in command, on that make and basic model aircraft and in that crewmember position, the following operating experience in each make and basic model of aircraft to be flown:

(1) Aircraft, single engine—10 hours.

(2) Aircraft multiengine, reciprocating engine-powered—15 hours.

(3) Aircraft multiengine, turbine engine-powered—20 hours.

(4) Airplane, turbojet-powered—25 hours.

(b) In acquiring the operating experience, each person must comply with the following:

(1) The operating experience must be acquired after satisfactory completion of the appropriate ground and flight training for the aircraft and crewmember position. Approved provisions for the operating experience must be included in the certificate holder's training program.

(2) The experience must be acquired in flight during commuter passenger-carrying operations under this part. However, in the case of an aircraft not previously used by the certificate holder in operations under this part, operating experience acquired in the aircraft during proving flights or ferry flights may be used to meet this requirement.

(3) Each person must acquire the operating experience while performing the duties of a pilot in command under the supervision of a qualified check pilot.

(4) The hours of operating experience may be reduced to not less than 50 percent of the hours required by this section by the substitution of one additional takeoff and landing for each hour of flight.

Docket No. 20011 (45 FR 7541) Eff. 2/4/80;

(Amdt. 135-3, Eff. 3/1/80); (Amdt. 135-9, Eff. 12/1/80)

§ 135.245 Second in command qualifications.

(a) Except as provided in paragraph (b), no certificate holder may use any person, nor may any person serve, as second in command of an aircraft unless that person holds at least a commercial pilot certificate with appropriate category and class ratings and an instrument rating. For flight under IFR, that person must meet the recent instrument experience requirements of Part 61 of this chapter.

(b) A second in command of a helicopter operated under VFR, other than over-the-top, must have at least a commercial pilot certificate with an appropriate aircraft category and class rating.

(Amdt. 135-1, Eff. 5/7/79)

§ 135.247 Pilot qualifications: Recent experience.

(a) No certificate holder may use any person, nor may any person serve, as pilot in command of an aircraft carrying passengers unless, within the preceding 90 days, that person has—

(1) Made three takeoffs and three landings as the sole manipulator of the flight controls in an aircraft of the same

category and class and, if a type rating is required, of the same type in which that person is to serve; or

(2) For operation during the period beginning 1 hour after sunset and ending 1 hour before sunrise (as published in the Air Almanac), made three takeoffs and three landings during that period as the sole manipulator of the flight controls in an aircraft of the same category and class and, if a type rating is required, of the same type in which that person is to serve.

A person who complies with paragraph (a)(2) of this section need not comply with paragraph (a)(1) of this section.

(b) For the purpose of paragraph (a) of this section, if the aircraft is a tailwheel airplane, each takeoff must be made in a tailwheel airplane and each landing must be made to a full stop in a tailwheel airplane.

§ 135.249 Use of prohibited drugs.

(a) This section applies to persons who perform a function listed in Appendix I to Part 121 of this chapter for a certificate holder or an operator. For the purpose of this section, a person who performs such a function pursuant to a contract with the certificate holder or the operator is considered to be performing that function for the certificate holder or the operator.

(b) No certificate holder or operator may knowingly use any person to perform, nor may any person perform for a certificate holder or an operator, either directly or by contract, any function listed in Appendix I to Part 121 of this chapter while that person has a prohibited drug, as defined in that appendix, in his or her system.

(c) No certificate holder or operator shall knowingly use any person to perform, nor shall any person perform for a certificate holder or operator, either directly or by contract, any safety-sensitive function if the person has a verified positive drug test result on or has refused to submit to a drug test required by Appendix I of Part 121 of this chapter and the person has not met the requirements of Appendix I to Part 121 of this chapter for returning to the performance of safety-sensitive duties.

Docket No. 25148 (54 FR 47061) Eff. 11/21/88; (Amdt. 135-28, Eff. 12/21/88)

§ 135.251 Testing for prohibited drugs.

(a) Each certificate holder or operator shall test each of its employees who performs a function listed in Appendix I to Part 121 of this chapter in accordance with that appendix.

(b) No certificate holder or operator may use any contractor to perform a function listed in Appendix I to Part 121 of this chapter unless that contractor tests each employee performing such a function for the certificate holder or operator in accordance with that appendix.

Docket No. 25148 (54 FR 47061) Eff. 11/21/88; (Amdt. 135-28, Eff. 1/21/88)

§ 135.253 Misuse of alcohol.

(a) This section applies to employees who perform a function listed in Appendix J to Part 121 of this chapter for a certificate holder or operator (*covered employees*). For the purpose of this section, a person who meets the definition of covered employee in Appendix J is considered to be performing the function for the certificate holder or operator.

(b) *Alcohol concentration.* No covered employee shall report for duty or remain on duty requiring the performance of safety-sensitive functions while having an alcohol concentration of 0.04 or greater. No certificate holder or operator having actual knowledge that an employee has an alcohol concentration of 0.04 or greater shall permit the employee to perform or continue to perform safety-sensitive functions.

(c) *On-duty use.* No covered employee shall use alcohol while performing safety-sensitive functions. No certificate holder or operator having actual knowledge that a covered employee is using alcohol while performing safety-sensitive functions shall permit the employee to perform or continue to perform safety-sensitive functions.

(d) *Pre-duty use.*

(1) No covered employee shall perform flight crewmember or flight attendant duties within 8 hours after using alcohol. No certificate holder or operator having actual knowledge that such an employee has used alcohol within 8 hours shall permit the employee to perform or continue to perform the specified duties.

(2) No covered employee shall perform safety-sensitive duties other than those specified in paragraph (d)(1) of this section within 4 hours after using alcohol. No certificate holder or operator having actual knowledge that such an employee has used alcohol within 4 hours shall permit the employee to perform or continue to perform safety-sensitive functions.

(e) *Use following an accident.* No covered employee who has actual knowledge of an accident involving an aircraft for which he or she performed a safety-sensitive function at or near the time of the accident shall use alcohol for 8 hours following the accident, unless he or she has been given a post-accident test under Appendix J of Part 121 of this chapter, or the employer has determined that the employee's performance could not have contributed to the accident.

(f) *Refusal to submit to a required alcohol test.* No covered employee shall refuse to submit to a post-accident, random, reasonable suspicion, or follow-up alcohol test required under Appendix J to Part 121 of this chapter. No operator or certificate holder shall permit a covered employee who refuses to submit to such a test to perform or continue to perform safety-sensitive functions.

[(Amdt. 135-48, Eff. 3/17/94)]

§ 135.255 Testing for alcohol.

(a) Each certificate holder and operator must establish an alcohol misuse prevention program in accordance with the provisions of Appendix J to Part 121 of this chapter.

(b) No certificate holder or operator shall use any person who meets the definition of "covered employee" in Appendix J to Part 121 to perform a safety-sensitive function listed in that appendix unless such person is subject to testing for alcohol misuse in accordance with the provisions of Appendix J.

[(Amdt. 135-48, Eff. 3/17/94)]

Subpart F—Flight crewmember flight time limitations and rest requirements

Source: Docket No. 23634 (50 FR 29320) Eff. 7/18/85

§ 135.261 Applicability.

Sections 135.263 through 135.271 prescribe flight time limitations and rest requirements for operations conducted under this part as follows:

(a) Section 135.263 applies to all operations under this subpart.

(b) Section 135.265 applies to:

(1) Scheduled passenger-carrying operations except those conducted solely within the state of Alaska. "Scheduled passenger-carrying operations" means passenger-carrying operations that are conducted in accordance with a published schedule which covers at least five round trips per week on at least one route between two or more points, includes dates or times (or both), and is openly advertised or otherwise made readily available to the general public, and

(2) Any other operation under this part, if the operator elects to comply with § 135.265 and obtains an appropriate operations specification amendment.

(c) Sections 135.267 and 135.269 apply to any operation that is not a scheduled passenger-carrying operation and to any operation conducted solely within the State of Alaska, unless the operator elects to comply with § 135.265 as authorized under paragraph (b)(2) of this section.

(d) Section 135.271 contains special daily flight time limits for operations conducted under the helicopter emergency medical evacuation service (HEMES).

§ 135.263 Flight time limitations and rest requirements: All certificate holders.

(a) A certificate holder may assign a flight crewmember and a flight crewmember may accept an assignment for flight time only when the applicable requirements of §§ 135.263 through 135.271 are met.

(b) No certificate holder may assign any flight crewmember to any duty with the certificate holder during any required rest period.

(c) Time spent in transportation, not local in character, that a certificate holder requires of a flight crewmember and provides to transport the crewmember to an airport at which he is to serve on a flight as a crewmember, or from an airport at which he was relieved from duty to return to his home station, is not considered part of a rest period.

(d) A flight crewmember is not considered to be assigned flight time in excess of flight time limitations if the flights to which he is assigned normally terminate within the limitations, but due to circumstances beyond the control of the certificate holder or flight crewmember (such as adverse weather conditions), are not at the time of departure expected to reach their destination within the planned flight time.

§ 135.265 Flight time limitations and rest requirements: Scheduled operations.

(a) No certificate holder may schedule any flight crewmember, and no flight crewmember may accept an assignment, for flight time in scheduled operations or in other commercial flying if that crewmember's total flight time in all commercial flying will exceed—

(1) 1,200 hours in any calendar year.

(2) 120 hours in any calendar month.

(3) 34 hours in any 7 consecutive days.

(4) 8 hours during any 24 consecutive hours for a flight crew consisting of one pilot.

(5) 8 hours between required rest periods for a flight crew consisting of two pilots qualified under this part for the operation being conducted.

(b) Except as provided in paragraph (c) of this section, no certificate holder may schedule a flight crewmember, and no flight crewmember may accept an assignment, for flight time during the 24 consecutive hours preceding the scheduled completion of any flight segment without a scheduled rest period during that 24 hours of at least the following:

(1) 9 consecutive hours of rest for less than 8 hours of scheduled flight time.

(2) 10 consecutive hours of rest for 8 or more but less than 9 hours of scheduled flight time.

(3) 11 consecutive hours of rest for 9 or more hours of scheduled flight time.

(c) A certificate holder may schedule a flight crewmember for less than the rest required in paragraph (b) of this section or may reduce a scheduled rest under the following conditions:

(1) A rest required under paragraph (b)(1) of this section may be scheduled for or reduced to a minimum of 8 hours if the flight crewmember is given a rest period of at least 10 hours that must begin no later than 24 hours after the commencement of the reduced rest period.

(2) A rest required under paragraph (b)(2) of this section may be scheduled for or reduced to a minimum of 8 hours if the flight crewmember is given a rest period of at least 11 hours that must begin no later than 24 hours after the commencement of the reduced rest period.

(3) A rest required under paragraph (b)(3) of this section may be scheduled for or reduced to a minimum of 9 hours if the flight crewmember is given a rest period of at least 12 hours that must begin no later than 24 hours after the commencement of the reduced rest period.

(d) Each certificate holder shall relieve each flight crewmember engaged in scheduled air transportation from all further duty for at least 24 consecutive hours during any 7 consecutive days.

§ 135.267 Flight time limitations and rest requirements: Unscheduled one- and two-pilot crews.

(a) No certificate holder may assign any flight crewmember, and no flight crewmember may accept an assignment, for flight time as a member of a one- or two-pilot crew if that crewmember's total flight time in all commercial flying will exceed—

(1) 500 hours in any calendar quarter.

(2) 800 hours in any two consecutive calendar quarters.

(3) 1,400 hours in any calendar year.

(b) Except as provided in paragraph (c) of this section, during any 24 consecutive hours the total flight time of the assigned flight when added to any other commercial flying by that flight crewmember may not exceed—

(1) 8 hours for a flight crew consisting of one pilot; or

(2) 10 hours for a flight crew consisting of two pilots qualified under this part for the operation being conducted.

(c) A flight crewmember's flight time may exceed the flight time limits of paragraph (b) of this section if the assigned flight time occurs during a regularly assigned duty period of no more than 14 hours and—

(1) If this duty period is immediately preceded by and followed by a required rest period of at least 10 consecutive hours of rest;

(2) If flight time is assigned during this period, that total flight time when added to any other commercial flying by the flight crewmember may not exceed—

(i) 8 hours for a flight crew consisting of one pilot; or

(ii) 10 hours for a flight crew consisting of two pilots; and

(3) If the combined duty and rest periods equal 24 hours.

(d) Each assignment under paragraph (b) of this section must provide for at least 10 consecutive hours of rest during the 24-hour period that precedes the planned completion time of the assignment.

(e) When a flight crewmember has exceeded the daily flight time limitations in this section, because of circumstances beyond the control of the certificate holder or flight crewmember (such as adverse weather conditions), that flight crewmember must have a rest period before being assigned or accepting an assignment for flight time of at least—

(1) 11 consecutive hours of rest if the flight time limitation is exceeded by not more than 30 minutes;

(2) 12 consecutive hours of rest if the flight time limitation is exceeded by more than 30 minutes, but not more than 60 minutes; and

(3) 16 consecutive hours of rest if the flight time limitation is exceeded by more than 60 minutes.

(f) The certificate holder must provide each flight crewmember at least 13 rest periods of at least 24 consecutive hours each in each calendar quarter.

(g) The Director, Flight Standards Service may issue operations specifications authorizing a deviation from any specific requirement of this section if he finds that the deviation is justified to allow a certificate holder additional time, but in no case beyond October 1, 1987, to bring its operations into full compliance with the requirements of this section. Each application for a deviation must be submitted to the Director, Flight Standards Service before October 1, 1986. Each applicant for a deviation may continue to operate under the requirements of Subpart F of this part as in effect on September 30, 1985, until the Director, Flight Standards Service has responded to the deviation request.

(Amdt. 135-33, Eff. 10/25/89)

§ 135.269 Flight time limitations and rest requirements: Unscheduled three- and four-pilot crews.

(a) No certificate holder may assign any flight crewmember, and no flight crewmember may accept an assignment, for flight time as a member of a three- or four-pilot crew if that crewmember's total flight time in all commercial flying will exceed—

(1) 500 hours in any calendar quarter.

(2) 800 hours in any two consecutive calendar quarters.

(3) 1,400 hours in any calendar year.

(b) No certificate holder may assign any pilot to a crew of three or four pilots, unless that assignment provides—

(1) At least 10 consecutive hours of rest immediately preceding the assignment;

(2) No more than 8 hours of flight deck duty in any 24 consecutive hours;

(3) No more than 18 duty hours for a three-pilot crew or 20 duty hours for a four-pilot crew in any 24 consecutive hours;

(4) No more than 12 hours aloft for a three-pilot crew or 16 hours aloft for a four-pilot crew during the maximum duty hours specified in paragraph (b)(3) of this section;

(5) Adequate sleeping facilities on the aircraft for the relief pilot;

(6) Upon completion of the assignment, a rest period of at least 12 hours;

(7) For a three-pilot crew, a crew which consists of at least the following:

(i) A pilot in command (PC) who meets the applicable flight crewmember requirements of Subpart E of Part 135;

(ii) A PC who meets the applicable flight crewmember requirements of Subpart E of Part 135, except those prescribed in §§ 135.244 and 135.247; and

(iii) A second in command (SIC) who meets the SIC qualifications of § 135.245.

(8) For a four-pilot crew, at least three pilots who meet the conditions of paragraph (b)(7) of this section, plus a fourth pilot who meets the SIC qualifications of § 135.245.

(c) When a flight crewmember has exceeded the daily flight deck duty limitation in this section by more than 60 minutes, because of circumstances beyond the control of the certificate holder or flight crewmember, that flight crewmember must have a rest period before the next duty period of at least 16 consecutive hours.

(d) A certificate holder must provide each flight crewmember at least 13 rest periods of at least 24 consecutive hours each in each calendar quarter.

§ 135.271 Helicopter hospital emergency medical evacuation service (HEMES).

(a) No certificate holder may assign any flight crewmember, and no flight crewmember may accept an assignment for flight time if that crewmember's total flight time in all commercial flying will exceed—

(1) 500 hours in any calendar quarter.

(2) 800 hours in any two consecutive calendar quarters.

(3) 1,400 hours in any calendar year.

(b) No certificate holder may assign a helicopter flight crewmember, and no flight crewmember may accept an assignment, for hospital emergency medical evacuation service helicopter operations unless that assignment provides for at least 10 consecutive hours of rest immediately preceding reporting to the hospital for availability for flight time.

(c) No flight crewmember may accrue more than 8 hours of flight time during any 24 consecutive hour period of a HEMES assignment, unless an emergency medical evacuation operation is prolonged. Each flight crewmember who exceeds the daily 8 hour flight time limitation in this paragraph must be relieved of the HEMES assignment immediately upon the completion of that emergency medical operation and must be given a rest period in compliance with paragraph (h) of this section.

(d) Each flight crewmember must receive at least 8 consecutive hours of rest during any 24 consecutive hour period of a HEMES assignment. A flight crewmember must be relieved of the HEMES assignment if he or she has not or cannot receive at least 8 consecutive hours of rest during any 24 consecutive hour period of a HEMES assignment.

(e) A HEMES assignment may not exceed 72 consecutive hours at the hospital.

(f) An adequate place of rest must be provided at, or in close proximity to, the hospital at which the HEMES assignment is being performed.

(g) No certificate holder may assign any other duties to a flight crewmember during a HEMES assignment.

(h) Each pilot must be given a rest period upon completion of the HEMES assignment and prior to being assigned any further duty with the certificate holder of—

(1) At least 12 consecutive hours for an assignment of less than 48 hours.

(2) At least 16 consecutive hours for an assignment of more than 48 hours.

(i) The certificate holder must provide each flight crewmember at least 13 rest periods of at least 24 consecutive hours each in each calendar quarter.

Subpart G—Crewmember testing requirements

§ 135.291 Applicability.

This subpart prescribes the tests and checks required for pilot and flight attendant crewmembers and for the approval of check pilots in operations under this part.

§ 135.293 Initial and recurrent pilot testing requirements.

(a) No certificate holder may use a pilot, nor may any person serve as a pilot, unless, since the beginning of the 12th calendar month before that service, that pilot has passed a written or oral test, given by the Administrator or an authorized check pilot, on that pilot's knowledge in the following areas—

(1) The appropriate provisions of Parts 61, 91, and 135 of this chapter and the operations specifications and the manual of the certificate holder;

(2) For each type of aircraft to be flown by the pilot, the aircraft powerplant, major components and systems, major appliances, performance and operating limitations, standard and emergency operating procedures, and the contents of the approved Aircraft Flight Manual or equivalent, as applicable;

(3) For each type of aircraft to be flown by the pilot, the

method of determining compliance with weight and balance limitations for takeoff, landing and en route operations;

(4) Navigation and use of air navigation aids appropriate to the operation or pilot authorization, including, when applicable, instrument approach facilities and procedures;

(5) Air traffic control procedures, including IFR procedures when applicable;

(6) Meteorology in general, including the principles of frontal systems, icing, fog, thunderstorms, and windshear, and, if appropriate for the operation of the certificate holder, high altitude weather;

(7) Procedures for—

(i) Recognizing and avoiding severe weather situations;

(ii) Escaping from severe weather situations, in case of inadvertent encounters, including low-altitude windshear (except that rotorcraft pilots are not required to be tested on escaping from low-altitude windshear); and

(iii) Operating in or near thunderstorms (including best penetrating altitudes), turbulent air (including clear air turbulence), icing, hail, and other potentially hazardous meteorological conditions; and

(8) New equipment, procedures, or techniques, as appropriate.

(b) No certificate holder may use a pilot, nor may any person serve as a pilot, in any aircraft unless, since the beginning of the 12th calendar month before that service, that pilot has passed a competency check given by the Administrator or an authorized check pilot in that class of aircraft, if single-engine airplane other than turbojet, or that type of aircraft, if helicopter, multiengine airplane, or turbojet airplane, to determine the pilot's competence in practical skills and techniques in that aircraft or class of aircraft. The extent of the competency check shall be determined by the Administrator or authorized check point conducting the competency check. The competency check may include any of the maneuvers and procedures currently required for the original issuance of the particular pilot certificate required for the operations authorized and appropriate to the category, class and type of aircraft involved. For the purposes of this paragraph, type, as to an airplane, means any one of a group of airplanes determined by the Administrator to have a similar means of propulsion, the same manufacturer, and no significantly different handling or flight characteristics. For the purposes of this paragraph, type, as to a helicopter, means a basic make and model.

(c) The instrument proficiency check required by § 135.297 may be substituted for the competency check required by this section for the type of aircraft used in the check.

(d) For the purpose of this part, competent performance of a procedure or maneuver by a person to be used as a pilot requires that the pilot be the obvious master of the aircraft, with the successful outcome of the maneuver never in doubt.

(e) The Administrator or authorized check pilot certifies the competency of each pilot who passes the knowledge or flight check in the certificate holder's pilot records.

(f) Portions of a required competency check may be given in an aircraft simulator for other appropriate training device, if approved by the Administrator.

(Amdt. 135-27, Eff. 1/2/89)

§ 135.295 Initial and recurrent flight attendant crewmember testing requirements.

No certificate holder may use a flight attendant crewmember, nor may any person serve as a flight attendant crewmember unless, since the beginning of the 12th calendar month before that service, the certificate holder has determined by appropriate initial and recurrent testing that the person is knowledgeable and competent in the following areas as appropriate to assigned duties and responsibilities—

(a) Authority of the pilot in command;

(b) Passenger handling, including procedures to be followed in handling deranged persons or other persons whose conduct might jeopardize safety;

(c) Crewmember assignments, functions, and responsibilities during ditching and evacuation of persons who may need the assistance of another person to move expeditiously to an exit in an emergency;

(d) Briefing of passengers;

(e) Location and operation of portable fire extinguishers and other items of emergency equipment;

(f) Proper use of cabin equipment and controls;

(g) Location and operation of passenger oxygen equipment;

(h) Location and operation of all normal and emergency exits, including evacuation chutes and escape ropes; and

(i) Seating of persons who may need assistance of another person to move rapidly to an exit in an emergency as prescribed by the certificate holder's operations manual.

§ 135.297 Pilot in command: Instrument proficiency check requirements.

(a) No certificate holder may use a pilot, nor may any person serve, as a pilot in command of an aircraft under IFR unless, since the beginning of the 6th calendar month before that service, that pilot has passed an instrument proficiency check under this section administered by the Administrator or an authorized check pilot.

(b) No pilot may use any type of precision instrument approach procedure under IFR unless, since the beginning of the 6th calendar month before that use, the pilot satisfactorily demonstrated that type of approach procedure. No pilot may use any type of nonprecision approach procedure under IFR unless, since the beginning of the 6th calendar month before that use, the pilot has satisfactorily demonstrated either that type of approach procedure or any other two different types of nonprecision approach procedures. The instrument approach procedure or procedures must include at least one straight-in approach, one circling approach, and one missed approach. Each type of approach procedure demonstrated must be conducted to published minimums for that procedure.

(c) The instrument proficiency check required by paragraph (a) of this section consists of an oral or written equipment test and a flight check under simulated or actual IFR conditions. The equipment test includes questions on emergency procedures, engine operation, fuel and lubrication systems, power settings, stall speeds, best engine-out speed, propeller and supercharger operations, and hydraulic, mechanical, and electrical systems, as appropriate. The flight check includes navigation by instruments, recovery from simulated emergencies, and standard instrument approaches involving navigational facilities which that pilot is to be authorized to use. Each pilot taking the instrument proficiency check must show that standard of competence required by § 135.293(d).

(1) The instrument proficiency check must—

(i) For a pilot in command of an airplane under § 135.243(a), include the procedures and maneuvers for an airline transport pilot certificate in the particular type of airplane, if appropriate; and

(ii) For a pilot in command of an airplane or helicopter under § 135.243(c), include the procedures and maneuvers for a commercial pilot certificate with an instrument rating and, if required, for the appropriate type rating.

(2) The instrument proficiency check must be given by an authorized check airman or by the Administrator.

(d) If the pilot in command is assigned to pilot only one type of aircraft, that pilot must take the instrument proficiency check required by paragraph (a) of this section in that type of aircraft.

(e) If the pilot in command is assigned to pilot more than one type of aircraft, that pilot must take the instrument proficiency check required by paragraph (a) of this section in each type of aircraft to which that pilot is assigned, in rotation, but not more than one flight check during each period described in paragraph (a) of this section.

(f) If the pilot in command is assigned to pilot both single-engine and multiengine aircraft, that pilot must initially take the instrument proficiency check required by paragraph (a) of this section in a multiengine aircraft, and each succeeding check alternately in single-engine and multiengine aircraft, but not more than one flight check during each period described in paragraph (a) of this section. Portions of a required flight check may be given in an aircraft simulator or other appropriate training device, if approved by the Administrator.

(g) If the pilot in command is authorized to use an autopilot system in place of a second in command, that pilot must show, during the required instrument proficiency check, that the pilot is able (without a second in command) both with and without using the autopilot to—

(1) Conduct instrument operations competently; and

(2) Properly conduct air-ground communications and comply with complex air traffic control instructions.

(3) Each pilot taking the autopilot check must show that, while using the autopilot, the airplane can be operated as proficiently as it would be if a second in command were present to handle air-ground communications and air traffic control instructions. The autopilot check need only be demonstrated once every 12 calendar months during the instrument proficiency check required under paragraph (a) of this section.

(h) [Deleted.]

(Amdt. 135-15, Eff. 6/11/81)

§ 135.299 Pilot in command: Line checks: Routes and airports.

(a) No certificate holder may use a pilot, nor may any person serve as a pilot in command of a flight unless, since the beginning of the 12th calendar month before that service, that pilot has passed a flight check in one of the types of aircraft which that pilot is to fly. The flight check shall—

(1) Be given by an approved check pilot or by the Administrator;

(2) Consist of at least one flight over one route segment; and

(3) Include takeoffs and landings at one or more representative airports. In addition to the requirements of this paragraph, for a pilot authorized to conduct IFR operations, at least one flight shall be flown over a civil airway, an approved off-airway route, or a portion of either of them.

(b) The pilot who conducts the check shall determine whether the pilot being checked satisfactorily performs the duties and responsibilities of a pilot in command in operations under this part, and shall so certify in the pilot training record.

(c) Each certificate holder shall establish in the manual required by § 135.21 a procedure which will ensure that each pilot who has not flown over a route and into an airport within the preceding 90 days will, before beginning the flight, become familiar with all available information required for the safe operation of that flight.

§ 135.301 Crewmember: Tests and checks, grace provisions, training to accepted standards.

(a) If a crewmember who is required to take a test or a flight check under this part, completes the test or flight check in the calendar month before or after the calendar month in which it is required, that crewmember is considered to have completed the test or check in the calendar month in which it is required.

(b) If a pilot being checked under this subpart fails any of the required maneuvers, the person giving the check may give additional training to the pilot during the course of the check. In addition to repeating the maneuvers failed, the person giving the check may require the pilot being checked to repeat any other maneuvers that are necessary to determine the pilot's proficiency. If the pilot being checked is unable to demonstrate satisfactory performance to the person conducting the check, the certificate holder may not use the pilot, nor may the pilot serve, as a flight crewmember in operations under this part until the pilot has satisfactorily completed the check.

§ 135.303 [(Removed)]

[(Amdt. 135-44, Eff. 10/15/92)]

Subpart H—Training

§ 135.321 Applicability and terms used.

(a) This subpart prescribes requirements for establishing and maintaining an approved training program for crewmembers, check airmen and instructors, and other operations personnel, and for the other training devices in the conduct of that program.

(b) For the purposes of this subpart, the following terms and definitions apply:

(1) *Initial training.* The training required for crewmembers who have not qualified and served in the same capacity on an aircraft.

(2) *Transition training.* The training required for crewmembers who have qualified and served in the same capacity on another aircraft.

(3) *Upgrade training.* The training required for crewmembers who have qualified and served as second in command on a particular aircraft type, before they serve as pilot in command on that aircraft.

(4) *Differences training.* The training required for crewmembers who have qualified and served on a particular type aircraft, when the Administrator finds differences training is necessary before a crewmember serves in the same capacity on a particular variation of that aircraft.

(5) *Recurrent training.* The training required for crewmembers to remain adequately trained and currently proficient for each aircraft, crewmember position, and type of operation in which the crewmember serves.

(6) *In flight.* The maneuvers, procedures, or functions that must be conducted in the aircraft.

§ 135.323 Training program: General.

(a) Each certificate holder required to have a training program under § 135.341 shall:

(1) Establish, obtain the appropriate initial and final approval of, and provide a training program that meets this subpart and that ensures that each crewmember, flight instructor, check airman, and each person assigned duties for the carriage and handling of hazardous materials (as defined in 49 CFR 171.8) is adequately trained to perform their assigned duties.

(2) Provide adequate ground and flight training facilities and properly qualified ground instructors for the training required by this subpart.

(3) Provide and keep current for each aircraft type used and, if applicable, the particular variations within the aircraft type, appropriate training material, examinations, forms, instructions, and procedures for use in conducting the training and checks required by this subpart.

(4) Provide enough flight instructors, check airmen, and simulator instructors to conduct required flight training and flight checks, and simulator training courses allowed under this subpart.

(b) Whenever a crewmember who is required to take recurrent training under this subpart completes the training in the calendar month before, or the calendar month after, the month in which that training is required, the crewmember is considered to have completed it in the calendar month in which it was required.

(c) Each instructor, supervisor, or check airman who is responsible for a particular ground training subject, segment of flight training, course of training, flight check, or competence check under this part shall certify as to the proficiency and knowledge of the crewmember, flight instructor, or check airman concerned upon completion of that training or check. That certification shall be made a part of the crewmember's record. When the certification required by this paragraph is made by an entry in a computerized recordkeeping system, the certifying instructor, supervisor, or check airman, must be identified with that entry. However, the signature of the certifying instructor, supervisor, or check airman, is not required for computerized entries.

(d) Training subjects that apply to more than one aircraft or crewmember position and that have been satisfactorily completed during previous training while employed by the certificate holder for another aircraft or another crewmember position, need not be repeated during subsequent training other than recurrent training.

(e) Aircraft simulators and other training devices may be used in the certificate holder's training program if approved by the Administrator.

§ 135.325 Training program and revision: Initial and final approval.

(a) To obtain initial and final approval of a training program, or a revision to an approved training program, each certificate holder must submit to the Administrator—

(1) An outline of the proposed or revised curriculum, that provides enough information for a preliminary evaluation of the proposed training program or revision; and

(2) Additional relevant information that may be requested by the Administrator.

(b) If the proposed training program or revision complies with this subpart, the Administrator grants initial approval in writing after which the certificate holder may conduct the training under that program. The Administrator then evaluates the effectiveness of the training program and advises the certificate holder of deficiencies, if any, that must be corrected.

(c) The Administrator grants final approval of the proposed training program or revision if the certificate holder shows that the training conducted under the initial approval in paragraph (b) of this section ensures that each person who successfully completes the training is adequately trained to perform that person's assigned duties.

(d) Whenever the Administrator finds that revisions are necessary for the continued adequacy of a training program that has been granted final approval, the certificate holder shall, after notification by the Administrator, make any changes in the program that are found necessary by the Administrator. Within 30 days after the certificate holder receives the notice, it may file a petition to reconsider the notice with the Administrator. The filing of a petition to reconsider stays the notice pending a decision by the Administrator. However, if the Administrator finds that there is an emergency that requires immediate action in the interest of safety, the Administrator may, upon a statement of the reasons, require a change effective without stay.

§ 135.327 Training program: Curriculum.

(a) Each certificate holder must prepare and keep current a written training program curriculum for each type of aircraft for each crewmember required for that type aircraft. The curriculum must include ground and flight training required by this subpart.

(b) Each training program curriculum must include the following:

(1) A list of principal ground training subjects, including emergency training subjects, that are provided.

(2) A list of all the training devices, mockups, systems trainers, procedures trainers, or other training aids that the certificate holder will use.

(3) Detailed descriptions or pictorial displays of the approved normal, abnormal, and emergency maneuvers, procedures and functions that will be performed during each flight training phase or flight check, indicating those maneuvers, procedures and functions that are to be performed during the inflight portions of flight training and flight checks.

§ 135.329 Crewmember training requirements.

(a) Each certificate holder must include in its training program the following initial and transition ground training as appropriate to the particular assignment of the crewmember:

(1) Basic indoctrination ground training for newly hired crewmembers including instruction in at least the—

(i) Duties and responsibilities of crewmembers as applicable;

(ii) Appropriate provisions of this chapter;

(iii) Contents of the certificate holder's operating certificate and operations specifications (not required for flight attendants); and

(iv) Appropriate portions of the certificate holder's operating manual.

(2) The initial and transition ground training in §§ 135.345 and 135.349, as applicable.

(3) Emergency training in § 135.331.

(b) Each training program must provide the initial and transition flight training in § 135.347, as applicable.

(c) Each training program must provide recurrent ground and flight training in § 135.351.

(d) Upgrade training in §§ 135.345 and 135.347 for a particular type aircraft may be included in the training program for crewmembers who have qualified and served as second in command on that aircraft.

(e) In addition to initial, transition, upgrade and recurrent training, each training program must provide ground and flight training, instruction, and practice necessary to ensure that each crewmember—

(1) Remains adequately trained and currently proficient for each aircraft, crewmember position, and type of operation in which the crewmember serves; and

(2) Qualifies in new equipment, facilities, procedures, and techniques, including modifications to aircraft.

§ 135.331 Crewmember emergency training.

(a) Each training program must provide emergency training under this section for each aircraft type, model, and configuration, each crewmember, and each kind of operation conducted, as appropriate for each crewmember and the certificate holder.

(b) Emergency training must provide the following:

(1) Instruction in emergency assignments and procedures, including coordination among crewmembers.

(2) Individual instruction in the location, function, and operation of emergency equipment including—

(i) Equipment used in ditching and evacuation;

(ii) First aid equipment and its proper use; and

(iii) Portable fire extinguishers, with emphasis on the type of extinguisher to be used on different classes of fires.

(3) Instruction in the handling of emergency situations including—

(i) Rapid decompression;

(ii) Fire in flight or on the surface and smoke control procedures with emphasis on electrical equipment and related circuit breakers found in cabin areas;

(iii) Ditching and evacuation;

(iv) Illness, injury, or other abnormal situations involving passengers or crewmembers; and

(v) Hijacking and other unusual situations.

(4) Review of the certificate holder's previous aircraft accidents and incidents involving actual emergency situations.

(c) Each crewmember must perform at least the following emergency drills, using the proper emergency equipment and procedures, unless the Administrator finds that, for a particular drill, the crewmember can be adequately trained by demonstration:

(1) Ditching, if applicable.

(2) Emergency evacuation.

(3) Fire extinguishing and smoke control.

(4) Operation and use of emergency exits, including deployment and use of evacuation chutes, if applicable.

(5) Use of crew and passenger oxygen.

(6) Removal of life rafts from the aircraft, inflation of the life rafts, use of life lines, and boarding of passengers and crew, if applicable.

(7) Donning and inflation of life vests and the use of other individual flotation devices, if applicable.

(d) Crewmembers who serve in operations above 25,000 feet must receive instruction in the following:

(1) Respiration.

(2) Hypoxia.

(3) Duration of consciousness without supplemental oxygen at altitude.

(4) Gas expansion.

(5) Gas bubble formation.

(6) Physical phenomena and incidents of decompression.

§ 135.333 Training requirements: Handling and carriage of hazardous materials.

(a) Except as provided in paragraph (d) of this section, no certificate holder may use any person to perform, and no person may perform, any assigned duties and responsibilities for the handling or carriage of hazardous materials (as defined in 49 CFR 171.8), unless within the preceding 12 calendar months that person has satisfactorily completed initial or recurrent training in an appropriate training program established by the certificate holder, which includes instruction regarding—

(1) The proper shipper certification, packaging, marking, labeling, and documentation for hazardous materials; and

(2) The compatibility, loading, storage, and handling characteristics of hazardous materials.

(b) Each certificate holder shall maintain a record of the satisfactory completion of the initial and recurrent training given to crewmembers and ground personnel who perform assigned duties and responsibilities for the handling and carriage of hazardous materials.

(c) Each certificate holder that elects not to accept hazardous materials shall ensure that each crewmember is adequately trained to recognize those items classified as hazardous materials.

(d) If a certificate holder operates into or out of airports at which trained employees or contract personnel are not available, it may use persons not meeting the requirements of paragraphs (a) and (b) of this section to load, offload, or otherwise handle hazardous materials if these persons are supervised by a crewmember who is qualified under paragraphs (a) and (b) of this section.

§ 135.335 Approval of aircraft simulators and other training devices.

(a) Training courses using aircraft simulators and other training devices may be included in the certificate holder's training program if approved by the Administrator.

(b) Each aircraft simulator and other training device that is used in a training course or in checks required under this subpart must meet the following requirements:

(1) It must be specifically approved for—

(i) The certificate holder; and

(ii) The particular maneuver, procedure, or crewmember function involved.

(2) It must maintain the performance, functional, and other characteristics that are required for approval.

(3) Additionally, for aircraft simulators, it must be—

(i) Approved for the type aircraft and, if applicable, the particular variation within type for which the training or check is being conducted; and

(ii) Modified to conform with any modification to the aircraft being simulated that changes the performance, functional, or other characteristics required for approval.

(c) A particular aircraft simulator or other training device may be used by more than one certificate holder.

[(d) In granting initial and final approval of training programs or revisions to them, the Administrator considers the training devices, methods, and procedures listed in the certificate holder's curriculum under § 135.327.

(Amdt. 135-1, Eff. 5/7/79)

§ 135.337 Training program: Check airmen and instructor qualifications.

(a) No certificate holder may use a person, nor may any person serve, as a flight instructor or check airman in a training program established under this subpart unless, for the particular aircraft type involved, that person—

(1) Holds the airman certificate and ratings that must be held to serve as a pilot in command in operations under this part;

(2) Has satisfactorily completed the appropriate training phases for the aircraft, including recurrent training, required to serve as a pilot in command in operations under this part;

(3) Has satisfactorily completed the appropriate proficiency or competency checks required to serve as a pilot in command in operations under this part;

(4) Has satisfactorily completed the applicable training requirements of § 135.339;

(5) Holds a Class I or Class II medical certificate required to serve as a pilot in command in operations under this part;

(6) In the case of a check airman, has been approved by the Administrator for the airman duties involved; and

(7) In the case of a check airman used in an aircraft simulator only, holds a Class III medical certificate.

(b) No certificate holder may use a person, nor may any person serve, as a simulator instructor for a course of training given in an aircraft simulator under this subpart unless that person—

(1) Holds at least a commercial pilot certificate; and

(2) Has satisfactorily completed the following as evidenced by the approval of a check airman—

(i) Appropriate initial pilot and flight instructor ground training under this subpart; and

(ii) A simulator flight training course in the type simulator in which that person instructs under this subpart.

§ 135.339 Check airmen and flight instructors: Initial and transition training.

(a) The initial and transition ground training for pilot check airmen must include the following:

(1) Pilot check airman duties, functions, and responsibilities.

(2) The applicable provisions of this chapter and certificate holder's policies and procedures.

(3) The appropriate methods, procedures, and techniques for conducting the required checks.

(4) Proper evaluation of pilot performance including the detection of—

(i) Improper and insufficient training; and

(ii) Personal characteristics that could adversely affect safety.

(5) The appropriate corrective action for unsatisfactory checks.

(6) The approved methods, procedures, and limitations for performing the required normal, abnormal, and emergency procedures in the aircraft.

(b) The initial and transition ground training for pilot flight instructors, except for the holder of a valid flight instructor certificate, must include the following:

(1) The fundamental principles of the teaching-learning process.

(2) Teaching methods and procedures.

(3) The instructor-student relationship.

(c) The initial and transition flight training for pilot check airmen and pilot flight instructors must include the following:

(1) Enough inflight training and practice in conducting flight checks from the left and right pilot seats in the required normal, abnormal, and emergency maneuvers to ensure that person's competence to conduct the pilot flight checks and flight training under this subpart.

(2) The appropriate safety measures to be taken from either pilot seat for emergency situations that are likely to develop in training.

(3) The potential results of improper or untimely safety measures during training.

The requirements of paragraphs (2) and (3) of this paragraph may be accomplished in flight or in an approved simulator.

§ 135.341 Pilot and flight attendant crewmember training programs.

(a) Each certificate holder, other than one who uses only one pilot in the certificate holder's operations, shall establish and maintain an approved pilot training program, and each certificate holder who uses a flight attendant crewmember shall establish and maintain an approved flight attendant training program, that is appropriate to the operations to which each pilot and flight attendant is to be assigned, and will ensure that they are adequately trained to meet the applicable knowledge and practical testing requirements of §§ 135.293 through 135.301. However, the Administrator may authorize a deviation from this section if the Administrator finds that, because of the limited size and scope of the operation, safety will allow a deviation from these requirements.

(b) Each certificate holder required to have a training program by paragraph (a) of this section shall include in that program ground and flight training curriculums for—

(1) Initial training;

(2) Transition training;

(3) Upgrade training;

(4) Differences training; and

(5) Recurrent training.

(c) Each certificate holder required to have a training program by paragraph (a) of this section shall provide current and appropriate study materials for use by each required pilot and flight attendant.

(d) The certificate holder shall furnish copies of the pilot and flight attendant crewmember training program, and all changes and additions, to the assigned representative of the Administrator. If the certificate holder uses training facilities of other persons, a copy of those training programs or appropriate portions used for those facilities shall also be furnished. Curricula that follow FAA published curricula may be cited by reference in the copy of the training program furnished to the representative of the Administrator and need not be furnished with the program.

(Amdt. 135-18, Eff. 8/2/82)

§ 135.343 Crewmember initial and recurrent training requirements.

No certificate holder may use a person, nor may any person serve, as a crewmember in operations under this part unless that crewmember has completed the appropriate initial or recurrent training phase of the training program appropriate to the type of operation in which the crewmember is to serve since the beginning of the 12th calendar month before that service. This section does not apply to a certificate holder that uses only one pilot in the certificate holder's operations.

(Amdt. 135-18, Eff. 8/2/82)

§ 135.345 Pilots: Initial, transition, and upgrade ground training.

Initial, transition, and upgrade ground training for pilots must include instruction in at least the following, as applicable to their duties:

(a) General subjects—

(1) The certificate holder's flight locating procedures;

(2) Principles and methods for determining weight and balance, and runway limitations for takeoff and landing;

(3) Enough meteorology to ensure a practical knowledge of weather phenomena, including the principles of frontal systems, icing, fog, thunderstorms, windshear and, if appropriate, high altitude weather situations;

(4) Air traffic control systems, procedures, and phraseology;

(5) Navigation and the use of navigational aids, including instrument approach procedures;

(6) Normal and emergency communication procedures;

(7) Visual cues before and during descent below DH or MDA; and

(8) Other instructions necessary to ensure the pilot's competence.

(b) For each aircraft type—

(1) A general description;

(2) Performance characteristics;

(3) Engines and propellers;

(4) Major components;

(5) Major aircraft systems (i.e., flight controls, electrical, and hydraulic), other systems, as appropriate, principles of normal, abnormal, and emergency operations, appropriate procedures and limitations;

(6) [Knowledge and] procedures for—

(i) Recognizing and avoiding severe weather situations;

(ii) Escaping from severe weather situations, in case of inadvertent encounters, including low-altitude windshear (except that rotorcraft pilots are not required to be trained in escaping from low-altitude windshear);

(iii) Operating in or near thunderstorms (including best penetrating altitudes), turbulent air (including clear air turbulence), icing, hail, and other potentially hazardous meteorological conditions; and

[(iv) Operating airplanes during ground icing conditions (i.e., any time conditions are such that frost, ice, or snow may reasonably be expected to adhere to the airplane), if the certificate holder expects to authorize takeoffs in ground icing conditions, including:

[(A) The use of holdover times when using deicing/anti-icing fluids;

[(B) Airplane deicing/anti-icing procedures, including inspection and check procedures and responsibilities;

[(C) Communications;

[(D) Airplane surface contamination (i.e., adherence of frost, ice, or snow) and critical area identification, and knowledge of how contamination adversely affects airplane performance and flight characteristics;

[(E) Types and characteristics of deicing/anti-icing fluids, if used by the certificate holder;

[(F) Cold weather preflight inspection procedures;

[(G) Techniques for recognizing contamination on the airplane;]

(7) Operating limitations;

(8) Fuel consumption and cruise control;

(9) Flight planning;

(10) Each normal and emergency procedure; and

(11) The approved Aircraft Flight Manual, or equivalent.

(Amdt. 135-27, Eff. 1/2/89); [(Amdt. 135-46, Eff. 1/31/94)]

§ 135.347 Pilots: Initial, transition, upgrade, and differences flight training.

(a) Initial, transition, upgrade, and differences training for pilots must include flight and practice in each of the maneuvers and procedures in the approved training program curriculum.

(b) The maneuvers and procedures required by paragraph (a) of this section must be performed in flight, except to the extent that certain maneuvers and procedures may be performed in an aircraft simulator, or an appropriate training device, as allowed by this subpart.

(c) If the certificate holder's approved training program includes a course of training using an aircraft simulator or other training device, each pilot must successfully complete—

(1) Training and practice in the simulator or training device in at least the maneuvers and procedures in this subpart that are capable of being performed in the aircraft simulator or training device; and

(2) A flight check in the aircraft or a check in the simulator or training device to the level of proficiency of a pilot in command or second in command, as applicable, in at least the maneuvers and procedures that are capable of being performed in an aircraft simulator or training device.

§ 135.349 Flight attendants: Initial and transition ground training.

Initial and transition ground training for flight attendants must include instruction in at least the following—

(a) General subjects—

(1) The authority of the pilot in command; and

(2) Passenger handling, including procedures to be followed in handling deranged persons or other persons whose conduct might jeopardize safety.

(b) For each aircraft type—

(1) A general description of the aircraft emphasizing physical characteristics that may have a bearing on ditching, evacuation, and inflight emergency procedures and on other related duties;

(2) The use of both the public address system and the means of communicating with other flight crewmembers, including emergency means in the case of attempted hijacking or other unusual situations; and

(3) Proper use of electrical galley equipment and the controls for cabin heat and ventilation.

§ 135.351 Recurrent training.

(a) Each certificate holder must ensure that each crewmember receives recurrent training and is adequately trained and currently proficient for the type aircraft and crewmember position involved.

(b) Recurrent ground training for crewmembers must include at least the following:

(1) A quiz or other review to determine the crewmember's knowledge of the aircraft and crewmember position involved.

(2) [Instruction as necessary in the subjects required for initial ground training by this subpart, as appropriate, including low-altitude windshear training and training on operating during ground icing conditions, as prescribed in §135.341 and described in §135.345, and emergency training.]

(c) Recurrent flight training for pilots must include, at least, flight training in the maneuvers or procedures in this subpart, except that satisfactory completion of the check required by § 135.293 within the preceding 12 calendar months may be substituted for recurrent flight training.

(Amdt. 135-27, Eff. 1/2/89); [(Amdt. 135-46, Eff. 1/31/94)]

§ 135.353 Prohibited drugs.

(a) Each certificate holder or operator shall provide each employee performing a function listed in Appendix I to Part 121 of this chapter and his or her supervisor with the training specified in that appendix.

(b) No certificate holder or operator may use any contractor to perform a function specified in Appendix I to Part 121 of this chapter unless that contractor provides each of its employees performing that function for the certificate holder or the operator and his or her supervisor with the training specified in that appendix.

Docket No. 25148 (53 FR 47061) Eff. 11/21/88; (Amdt. 135-28, Eff. 12/21/88)

Subpart I—Airplane performance operating limitations

§ 135.361 Applicability.

(a) This subpart prescribes airplane performance operating limitations applicable to the operation of the categories of airplanes listed in § 135.363 when operated under this part.

(b) For the purpose of this subpart, "effective length of the runway," for landing means the distance from the point at which the obstruction clearance plane associated with the approach end of the runway intersects the centerline of the runway to the far end of the runway.

(c) For the purpose of this subpart, "obstruction clearance plane" means a plane sloping upward from the runway at a slope of 1:20 to the horizontal, and tangent to or clearing all obstructions within a specified area surrounding the runway as shown in a profile view of that area. In the plan view, the centerline of the specified area coincides with the centerline of the runway, beginning at the point where the obstruction clearance plane intersects the centerline of the runway and proceeding to a point at least 1,500 feet from the beginning point. After that the centerline coincides with the takeoff path over the ground for the runway (in the case of takeoffs) or with the instrument approach counterpart (for landings), or, where the applicable one of these paths has not been established, it proceeds consistent with turns of at least 4,000-foot radius until a point is reached beyond which the obstruction clearance plane clears all obstructions. This area extends laterally 200 feet on each side of the centerline at the point where the obstruction clearance plane intersects the runway and continues at this width to the end of the runway; then it increases uniformly to 500 feet on each side of the centerline at a point 1,500 feet from the intersection of the obstruction clearance plane with the runway; after that it extends laterally 500 feet on each side of the centerline.

§ 135.363 General.

(a) Each certificate holder operating a reciprocating engine powered large transport category airplane shall comply with §§ 135.365 through 135.377.

(b) Each certificate holder operating a turbine engine powered large transport category airplane shall comply with §§ 135.379 through 135.387, except that when it operates a turbo-propeller-powered large transport category airplane certificated after August 29, 1959, but previously type certificated with the same number of reciprocating engines, it may comply with §§ 135.365 through 135.377.

(c) Each certificate holder operating a large nontransport category airplane shall comply with §§ 135.389 through 135.395 and any determination of compliance must be based only on approved performance data. For the purpose of this subpart, a large nontransport category airplane is an airplane that was type certificated before July 1, 1942.

(d) Each certificate holder operating a small transport category airplane shall comply with § 135.397.

(e) Each certificate holder operating a small nontransport category airplane shall comply with § 135.399.

(f) The performance data in the Airplane Flight Manual applies in determining compliance with §§ 135.365 through 135.387. Where conditions are different from those on which the performance data is based, compliance is determined by interpolation or by computing the effects of change in the specific variables, if the results of the interpolation or computations are substantially as accurate as the results of direct tests.

(g) No person may take off a reciprocating engine powered large transport category airplane at a weight that is more than the allowable weight for the runway being used (determined under the runway takeoff limitations of the transport category operating rules of this subpart) after taking into account the temperature operating correction factors in § 4a.749a-T or § 4b.117 of the Civil Air Regulations in effect on January 31, 1965, and in the applicable Airplane Flight Manual.

(h) The Administrator may authorize in the operations specifications deviations from this subpart if special circumstances make a literal observance of a requirement unnecessary for safety.

(i) The 10-mile width specified in §§ 135.369 through 135.373 may be reduced to 5 miles, for not more than 20 miles, when operating under VFR or where navigation facilities furnish reliable and accurate identification of high ground and obstructions located outside of 5 miles, but within 10 miles, on each side of the intended track.

(j) Each certificate holder operating a commuter category airplane shall comply with § 135.398.

(Amdt. 135-21, Eff. 2/17/87)

§ 135.365 Large transport category airplanes: Reciprocating engine powered: Weight limitations.

(a) No person may take off a reciprocating engine powered large transport category airplane from an airport located at an elevation outside of the range for which maximum takeoff weights have been determined for that airplane.

(b) No person may take off a reciprocating engine powered large transport category airplane for an airport of intended destination that is located at an elevation outside of the range for which maximum landing weights have been determined for that airplane.

(c) No person may specify, or have specified, an alternate airport that is located at an elevation outside of the range for which maximum landing weights have been determined for the reciprocating engine powered large transport category airplane concerned.

(d) No person may take off a reciprocating engine powered large transport category airplane at a weight more than the maximum authorized takeoff weight for the elevation of the airport.

(e) No person may take off a reciprocating engine powered large transport category airplane if its weight on arrival at the airport of destination will be more than the maximum authorized landing weight for the elevation of that airport, allowing for normal consumption of fuel and oil en route.

§ 135.367 Large transport category airplanes: Reciprocating engine powered: Takeoff limitations.

(a) No person operating a reciprocating engine powered large transport category airplane may take off that airplane unless it is possible—

(1) To stop the airplane safely on the runway, as shown by the accelerate-stop distance data, at any time during takeoff until reaching critical engine failure speed;

(2) If the critical engine fails at any time after the airplane reaches critical-engine failure speed V_1, to continue the takeoff and reach a height of 50 feet, as indicated by the takeoff path data, before passing over the end of the runway; and

(3) To clear all obstacles either by at least 50 feet vertically (as shown by the takeoff path data) or 200 feet horizontally within the airport boundaries and 300 feet horizontally beyond the boundaries, without banking before reaching a height of 50 feet (as shown by the takeoff path data) and after that without banking more than 15 degrees.

(b) In applying this section, corrections must be made for any runway gradient. To allow for wind effect, takeoff data based on still air may be corrected by taking into account not more than 50 percent of any reported headwind component and not less than 150 percent of any reported tailwind component.

§ 135.369 Large transport category airplanes: Reciprocating engine powered: En route limitations: All engines operating.

(a) No person operating a reciprocating engine powered large transport category airplane may take off that airplane at a weight, allowing for normal consumption of fuel and oil, that does not allow a rate of climb (in feet per minute), with all engines operating, of at least 6.90 Vs_0 (that is, the number of feet per minute obtained by multiplying the number of knots by 6.90) at an altitude of at least 1,000 feet above the highest ground or obstruction within ten miles of each side of the intended track.

(b) This section does not apply to large transport category airplanes certificated under Part 4a of the Civil Air Regulations.

§ 135.371 Large transport category airplanes: Reciprocating engine powered: En route limitations: One engine inoperative.

(a) Except as provided in paragraph (b) of this section, no person operating a reciprocating engine powered large transport category airplane may take off that airplane at a weight, allowing for normal consumption of fuel and oil, that does not allow a rate of climb (in feet per minute), with one engine inoperative, of at least (0.079–0.106/N) Vs_0^2 (where N is the number of engines installed and Vs_0 is expressed in knots) at an altitude of least 1,000 feet above the highest ground or obstruction within 10 miles of each side of the intended track. However, for the purposes of this paragraph the rate of climb for transport category airplanes certificated under Part 4a of the Civil Air Regulations is 0.026 Vs_0^2.

(b) In place of the requirements of paragraph (a) of this section, a person may, under an approved procedure, operate a reciprocating engine powered large transport category airplane at an all-engine-operating altitude that allows the airplane to continue, after an engine failure, to an alternate airport where a landing can be made under § 135.377, allowing for normal consumption of fuel and oil. After the assumed failure, the flight path must clear the ground and any obstruction within five miles on each side of the intended track by at least 2,000 feet.

(c) If an approved procedure under paragraph (b) of this section is used, the certificate holder shall comply with the following:

(1) The rate of climb (as prescribed in the Airplane Flight Manual for the appropriate weight and altitude) used in calculating the airplane's flight path shall be diminished by an amount in feet per minute equal to (0.079–0.106/N) Vs_0^2 (when N is the number of engines installed and Vs_0 is expressed in knots) for airplanes certificated under Part 25 of this chapter and by 0.026 Vs_0^2 for airplanes certificated under Part 4a of the Civil Air Regulations.

(2) The all-engines-operating altitude shall be sufficient so that in the event the critical engine becomes inoperative at any point along the route, the flight will be able to proceed to a predetermined alternate airport by use of this procedure. In determining the takeoff weight, the airplane is assumed to pass over the critical obstruction following engine failure at a point no closer to the critical obstruction than the nearest approved radio navigational fix, unless the Administrator approves a procedure established on a different basis upon finding that adequate operational safeguards exist.

(3) The airplane must meet the provisions of paragraph (a) of this section at 1,000 feet above the airport used as an alternate in this procedure.

(4) The procedure must include an approved method of accounting for winds and temperatures that would otherwise adversely affect the flight path.

(5) In complying with this procedure, fuel jettisoning is allowed if the certificate holder shows that it has an adequate training program, that proper instructions are given to the flight crew, and all other precautions are taken to ensure a safe procedure.

(6) The certificate holder and the pilot in command shall jointly elect an alternate airport for which the appropriate weather reports or forecasts, or any combination of them, indicate that weather conditions will be at or above the alternate weather minimum specified in the certificate holder's operations specifications for that airport when the flight arrives.

§ 135.373 Part 25 transport category airplanes with four or more engines: Reciprocating engine powered: En route limitations: Two engines inoperative.

(a) No person may operate an airplane certificated under Part 25 and having four or more engines unless—

(1) There is no place along the intended track that is more than 90 minutes (with all engines operating at cruising power) from an airport that meets § 135.377; or

(2) It is operated at a weight allowing the airplane, with the two critical engines inoperative, to climb at 0.013 Vs_0^2 feet per minute (that is, the number of feet per minute obtained by multiplying the number of knots squared by 0.013) at an altitude of 1,000 feet above the highest ground or obstruction within 10 miles on each side of the intended track, or at an altitude of 5,000 feet, whichever is higher.

(b) For the purposes of paragraph (a)(2) of this section, it is assumed that—

(1) The two engines fail at the point that is most critical with respect to the takeoff weight;

(2) Consumption of fuel and oil is normal with all engines operating up to the point where the two engines fail with two engines operating beyond that point;

(3) Where the engines are assumed to fail at an altitude above the prescribed minimum altitude, compliance with the prescribed rate of climb at the prescribed minimum altitude need not be shown during the descent from the cruising altitude to the prescribed minimum altitude, if those requirements can be met once the prescribed minimum altitude is reached, and assuming descent to be along a net flight path and the rate of descent to be 0.013 Vs_0^2

greater than the rate in the approved performance data; and

(4) If fuel jettisoning is provided, the airplane's weight at the point where the two engines fail is considered to be not less than that which would include enough fuel to proceed to an airport meeting § 135.377 and to arrive at an altitude of at least 1,000 feet directly over that airport.

§ 135.375 Large transport category airplanes: Reciprocating engine powered: Landing limitations: Destination airports.

(a) Except as provided in paragraph (b) of this section, no person operating a reciprocating engine powered large transport category airplane may take off that airplane, unless its weight on arrival, allowing for normal consumption of fuel and oil in flight, would allow a full stop landing at the intended destination within 50 percent of the effective length of each runway described below from a point 50 feet directly above the intersection of the obstruction clearance plane and the runway. For the purposes of determining the allowable landing weight at the destination airport the following is assumed:

(1) The airplane is landed on the most favorable runway and in the most favorable direction in still air.

(2) The airplane is landed on the most suitable runway considering the probable wind velocity and direction (forecast for the expected time of arrival), the ground handling characteristics of the type of airplane, and other conditions such as landing aids and terrain, and allowing for the effect of the landing path and roll of not more than 50 percent of the headwind component or not less than 150 percent of the tailwind component.

(b) An airplane that would be prohibited from being taken off because it could not meet paragraph (a)(2) of this section may be taken off if an alternate airport is selected that meets all of this section except that the airplane can accomplish a full stop landing within 70 percent of the effective length of the runway.

§ 135.377 Large transport category airplanes: Reciprocating engine powered: Landing limitations: Alternate airports.

No person may list an airport as an alternate airport in a flight plan unless the airplane (at the weight anticipated at the time of arrival at the airport), based on the assumptions in § 135.375(a)(1) and (2), can be brought to a full stop landing within 70 percent of the effective length of the runway.

§ 135.379 Large transport category airplanes: Turbine engine powered: Takeoff limitations.

(a) No person operating a turbine engine powered large transport category airplane may take off that airplane at a weight greater than that listed in the Airplane Flight Manual for the elevation of the airport and for the ambient temperature existing at takeoff.

(b) No person operating a turbine engine powered large transport category airplane certificated after August 26, 1957, but before August 30, 1959 (SR422, 422A), may take off that airplane at a weight greater than that listed in the Airplane Flight Manual for the minimum distance required for takeoff. In the case of an airplane certificated after September 30, 1958 (SR422A, 422B), the takeoff distance may include a clearway distance but the clearway distance included may not be greater than one-half of the takeoff run.

(c) No person operating a turbine engine powered large transport category airplane certificated after August 29, 1959 (SR422B), may take off that airplane at a weight greater than that listed in the Airplane Flight Manual at which compliance with the following may be shown:

(1) The accelerate-stop distance, as defined in § 25.109 of this chapter, must not exceed the length of the runway plus the length of any stopway.

(2) The takeoff distance must not exceed the length of the runway plus the length of any clearway except that the length of any clearway included must not be greater than one-half the length of the runway.

(3) The takeoff run must not be greater than the length of the runway.

(d) No person operating a turbine engine powered large transport category airplane may take off that airplane at a weight greater than that listed in the Airplane Flight Manual—

(1) For an airplane certificated after August 26, 1957, but before October 1, 1958 (SR422), that allows a takeoff path that clears all obstacles either by at least (35 + 0.01 D) feet vertically (D is the distance along the intended flight path from the end of the runway in feet), or by at least 200 feet horizontally within the airport boundaries and by at least 300 feet horizontally after passing the boundaries; or

(2) For an airplane certificated after September 30, 1958 (SR422A, 422B), that allows a net takeoff flight path that clears all obstacles either by a height of at least 35 feet vertically, or by at least 200 feet horizontally within the airport boundaries and by at least 300 feet horizontally after passing the boundaries.

(e) In determining maximum weights, minimum distances and flight paths under paragraphs (a) through (d) of this section, correction must be made for the runway to be used, the elevation of the airport, the effective runway gradient, and the ambient temperature and wind component at the time of takeoff.

(f) For the purposes of this section, it is assumed that the airplane is not banked before reaching a height of 50 feet, as shown by the takeoff path or net takeoff flight path data (as appropriate) in the Airplane Flight Manual, and after that the maximum is not more than 15 degrees.

(g) For the purposes of this section, the terms, "takeoff distance," "takeoff run," "net takeoff flight path," have the same meanings as set forth in the rules under which the airplane was certificated.

§ 135.381 Large transport category airplanes: Turbine engine powered: En route limitations: One engine inoperative.

(a) No person operating a turbine engine powered large transport category airplane may take off that airplane at a weight, allowing for normal consumption of fuel and oil, that is greater than that which (under the approved, one engine inoperative, en route net flight path data in the Airplane Flight Manual for that airplane) will allow compliance with subparagraph (1) or (2) of this paragraph, based on the ambient temperatures expected en route.

(1) There is a positive slope at an altitude of at least 1,000 feet above all terrain and obstructions within five statute miles on each side of the intended track, and, in addition, if that airplane was certificated after August 29, 1958 (SR422B), there is a positive slope at 1,500 feet above the airport where the airplane is assumed to land after an engine fails.

(2) The net flight path allows the airplane to continue flight from the cruising altitude to an airport where a landing can be made under § 135.387 clearing all terrain and obstructions within five statute miles of the intended track by at least 2,000 feet vertically and with a positive slope at 1,000 feet above the airport where the airplane lands after an engine fails, or, if that airplane was certificated after September 30, 1958 (SR422A, 422B), with a positive slope at 1,500 feet above the airport where the airplane lands after an engine fails.

(b) For the purpose of paragraph (a)(2) of this section, it is assumed that—

(1) The engine fails at the most critical point en route;

(2) The airplane passes over the critical obstruction, after engine failure at a point that is no closer to the obstruction than the approved radio navigation fix, unless the Administrator authorizes a different procedure based on adequate operational safeguards;

(3) An approved method is used to allow for adverse winds;

(4) Fuel jettisoning will be allowed if the certificate holder shows that the crew is properly instructed, that the training program is adequate, and that all other precautions are taken to ensure a safe procedure;

(5) The alternate airport is selected and meets the prescribed weather minimums; and

(6) The consumption of fuel and oil after engine failure is the same as the consumption that is allowed for in the approved net flight path data in the Airplane Flight Manual.

§ 135.383 Large transport category airplanes: Turbine engine powered: En route limitations: Two engines inoperative.

(a) Airplanes certificated after August 26, 1957, but before October 1, 1958 (SR422). No person may operate a turbine engine powered large transport category airplane along an intended route unless that person complies with either of the following:

(1) There is no place along the intended track that is more than 90 minutes (with all engines operating at cruising power) from an airport that meets § 135.387.

(2) Its weight, according to the two-engine-inoperative, en route, net flight path data in the Airplane Flight Manual, allows the airplane to fly from the point where the two engines are assumed to fail simultaneously to an airport that meets § 135.387, with a net flight path (considering the ambient temperature anticipated along the track) having a positive slope at an altitude of at least 1,000 feet above all terrain and obstructions within five statute miles on each side of the intended track, or at an altitude of 5,000 feet, whichever is higher.

For the purposes of paragraph (2) of this paragraph, it is assumed that the two engines fail at the most critical point en route, that if fuel jettisoning is provided, the airplane's weight at the point where the engines fail includes enough fuel to continue to the airport and to arrive at an altitude of at least 1,000 feet directly over the airport, and that the fuel and oil consumption after engine failure is the same as the consumption allowed for in the net flight path data in the Airplane Flight Manual.

(b) Airplanes certificated after September 30, 1958, but before August 30, 1959 (SR422A). No person may operate a turbine engine powered large transport category airplane along an intended route unless that person complies with either of the following:

(1) There is no place along the intended track that is more than 90 minutes (with all engines operating at cruising power) from an airport that meets § 135.387.

(2) Its weight, according to the two-engine-inoperative, en route, net flight path data in the Airplane Flight Manual allows the airplane to fly from the point where the two engines are assumed to fail simultaneously to an airport that meets § 135.387 with a net flight path (considering the ambient temperatures anticipated along the track) having a positive slope at an altitude of at least 1,000 feet above all terrain and obstructions within five statute miles on each side of the intended track, or at an altitude of 2,000 feet, whichever is higher.

For the purpose of paragraph (2) of this paragraph, it is assumed that the two engines fail at the most critical point en route, that the airplane's weight at the point where the engines fail includes enough fuel to continue to the airport, to arrive at an altitude of at least 1,500 feet directly over the airport, and after that to fly for 15 minutes at cruise power or thrust, or both, and that the consumption of fuel and oil after engine failure is the same as the consumption allowed for in the net flight path data in the Airplane Flight Manual.

(c) Aircraft certificated after August 29, 1959 (SR422B). No person may operate a turbine engine powered large transport category airplane along an intended route unless that person complies with either of the following:

(1) There is no place along the intended track that is more than 90 minutes (with all engines operating at cruising power) from an airport that meets § 135.387.

(2) Its weight, according to the two-engine-inoperative, en route, net flight path data in the Airplane Flight Manual, allows the airplane to fly from the point where the two engines are assumed to fail simultaneously to an airport that meets § 135.387, with the net flight path (considering the ambient temperatures anticipated along the track) clearing vertically by at least 2,000 feet all terrain and obstructions within five statute miles on each side of the intended track. For the purposes of this paragraph, it is assumed that—

(i) The two engines fail at the most critical point en route;

(ii) The net flight path has a positive slope at 1,500 feet above the airport where the landing is assumed to be made after the engines fail;

(iii) Fuel jettisoning will be approved if the certificate holder shows that the crew is properly instructed, that the training program is adequate, and that all other precautions are taken to ensure a safe procedure;

(iv) The airplane's weight at the point where the two engines are assumed to fail provides enough fuel to continue to the airport, to arrive at an altitude of at least 1,500 feet directly over the airport, and after that to fly for 15 minutes at cruise power or thrust, or both; and

(v) The consumption of fuel and oil after the engines fail is the same as the consumption that is allowed for in the net flight path data in the Airplane Flight Manual.

§ 135.385 Large transport category airplanes: Turbine engine powered: Landing limitations: Destination airports.

(a) No person operating a turbine engine powered large transport category airplane may take off that airplane at a weight that (allowing for normal consumption of fuel and oil in flight to the destination or alternate airport) the weight of the airplane on arrival would exceed the landing weight in the Airplane Flight Manual for the elevation of the destination or alternate airport and the ambient temperature anticipated at the time of landing.

(b) Except as provided in paragraph (c), (d), or (e) of this section, no person operating a turbine engine powered large transport category airplane may take off that airplane unless its weight on arrival, allowing for normal consumption of fuel and oil in flight (in accordance with the landing distance in the Airplane Flight Manual for the elevation of the destination airport and the wind conditions anticipated there at the time of landing), would allow a full stop landing at the intended destination airport within 60 percent of the effective length of each runway described below from a point 50 feet above the intersection of the obstruction clearance plane and the runway. For the purpose of determining the allowable landing weight at the destination airport the following is assumed:

(1) The airplane is landed on the most favorable runway and in the most favorable direction, in still air.

(2) The airplane is landed on the most suitable runway considering the probable wind velocity and direction and the ground handling characteristics of the airplane, and considering other conditions such as landing aids and terrain.

(c) A turbopropeller powered airplane that would be prohibited from being taken off because it could not meet paragraph (b)(2) of this section, may be taken off if an alternate airport is selected that meets all of this section except that the airplane can accomplish a full stop landing within 70 percent of the effective length of the runway.

(d) Unless, based on a showing of actual operating landing techniques on wet runways, a shorter landing distance (but never less than that required by paragraph (b) of this section) has been approved for a specific type and model airplane and included in the Airplane Flight Manual, no person may take off a turbojet airplane when the appropriate weather reports or forecasts, or any combination of them indicate that the runways at the destination airport may be wet or slippery at the estimated time of arrival unless the effective runway length at the destination airport is at least 115 percent of the runway length required under paragraph (b) of this section.

(e) A turbojet airplane that would be prohibited from being taken off because it could not meet paragraph (b)(2) of this section may be taken off if an alternate airport is selected that meets all of paragraph (b) of this section.

§ 135.387 Large transport category airplanes: Turbine engine powered: Landing limitations: Alternate airports.

No person may select an airport as an alternate airport for a turbine engine powered large transport category airplane unless (based on the assumption in § 135.385(b)) that airplane, at the weight anticipated at the time of arrival, can be brought to a full stop landing within 70 percent of the effective length of the runway for turbopropeller-powered airplanes and 60 percent of the effective length of the runway for turbojet airplanes, from a point 50 feet above the intersection of the obstruction clearance plane and the runway.

§ 135.389 Large nontransport category airplanes: Takeoff limitations.

(a) No person operating a large nontransport category airplane may take off that airplane at a weight greater than the weight that would allow the airplane to be brought to a safe stop within the effective length of the runway, from any point during the takeoff before reaching 105 percent of minimum control speed (the minimum speed at which an airplane can be safely controlled in flight after an engine becomes inoperative) or 115 percent of the power off stalling speed in the takeoff configuration, whichever is greater.

(b) For the purposes of this section—

(1) It may be assumed that takeoff power used on all engines during the acceleration;

(2) Not more than 50 percent of the reported headwind component, or not less than 150 percent of the reported tailwind component, may be taken into account;

(3) The average runway gradient (the difference between the elevations of the endpoints of the runway divided by the total length) must be considered if it is more than one-half of one percent;

(4) It is assumed that the airplane is operating in standard atmosphere; and

(5) For takeoff, "effective length of the runway" means the distance from the end of the runway at which the takeoff is started to a point at which the obstruction clearance plane associated with the other end of the runway intersects the runway centerline.

§ 135.391 Large nontransport category airplanes: En route limitations: One engine inoperative.

(a) Except as provided in paragraph (b) of this section, no person operating a large nontransport category airplane may take off that airplane at a weight that does not allow a rate of climb of at least 50 feet a minute, with the critical engine inoperative, at an altitude of at least 1,000 feet above the highest obstruction within five miles on each side of the intended track, or 5,000 feet, whichever is higher.

(b) Without regard to paragraph (a) of this section, if the Administrator finds that safe operations are not impaired, a person may operate the airplane at an altitude that allows the airplane, in case of engine failure, to clear all obstructions within five miles on each side of the intended track by 1,000 feet. If this procedure is used, the rate of descent for the appropriate weight and altitude is assumed to be 50 feet a minute greater than the rate in the approved performance data. Before approving such a procedure, the Administrator considers the following for the route, route segment, or area concerned:

(1) The reliability of wind and weather forecasting.

(2) The location and kinds of navigation aids.

(3) The prevailing weather conditions, particularly the frequency and amount of turbulence normally encountered.

(4) Terrain features.

(5) Air traffic problems.

(6) Any other operational factors that affect the operations.

(c) For the purposes of this section, it is assumed that—

(1) The critical engine is inoperative;

(2) The propeller of the inoperative engine is in the minimum drag position;

(3) The wing flaps and landing gear are in the most favorable position;

(4) The operating engines are operating at the maximum continuous power available;

(5) The airplane is operating in standard atmosphere; and

(6) The weight of the airplane is progressively reduced by the anticipated consumption of fuel and oil.

§ 135.393 Large nontransport category airplanes: Landing limitations: Destination airports.

(a) No person operating a large nontransport category airplane may take off that airplane at a weight that—

(1) Allowing for anticipated consumption of fuel and oil, is greater than the weight that would allow a full stop landing within 60 percent of the effective length of the most suitable runway at the destination airport; and

(2) Is greater than the weight allowable if the landing is to be made on the runway—

(i) With the greatest effective length in still air; and

(ii) Required by the probable wind, taking into account not more than 50 percent of the headwind component or not less than 150 percent of the tailwind component.

(b) For the purpose of this section, it is assumed that—

(1) The airplane passes directly over the intersection of the obstruction clearance plane and the runway at a height of 50 feet in a steady gliding approach at a true indicated airspeed of at least 1.3 Vs_0;

(2) The landing does not require exceptional pilot skill; and

(3) The airplane is operating in standard atmosphere.

§ 135.395 Large nontransport category airplanes: Landing limitations: Alternate airports.

No person may select an airport as an alternate airport for a large nontransport category airplane unless that airplane (at the weight anticipated at the time of arrival), based on the assumptions in § 135.393(b), can be brought to a full stop landing within 70 percent of the effective length of the runway.

§ 135.397 Small transport category airplane performance operating limitations.

(a) No person may operate a reciprocating engine powered small transport category airplane unless that person complies with the weight limitations in § 135.365, the takeoff limitations in § 135.367 (except paragraph (a)(3)), and the landing limitations in §§ 135.375 and 135.377.

(b) No person may operate a turbine engine powered small transport category airplane unless that person complies with the takeoff limitations in § 135.379 (except paragraphs (d) and (f)) and the landing limitations in §§ 135.385 and 135.387.

§ 135.398 Commuter category airplane performance operating limitations.

(a) No person may operate a commuter category airplane unless that person complies with the takeoff weight limitations in the approved Airplane Flight Manual.

(b) No person may take off an airplane type certificated in the commuter category at a weight greater than that listed in the Airplane Flight Manual that allows a net takeoff flight path that clears all obstacles either by a height of at least 35 feet vertically, or at least 200 feet horizontally within the airport boundaries and by at least 300 feet horizontally after passing the boundaries.

(c) No person may operate a commuter category airplane unless that person complies with the landing limitations prescribed in §§ 135.385 and 135.387 of this part. For purposes of this paragraph, §§ 135.385 and 135.387 are applicable to all commuter category airplanes notwithstanding their stated applicability to turbine-engine-powered large transport category airplanes.

(d) In determining maximum weights, minimum distances and flight paths under paragraphs (a) through (c) of this section, correction must be made for the runway gradient, and ambient temperature, and wind component at the time of takeoff.

(e) For the purpose of this section, the assumption is that the airplane is not banked before reaching a height of 50 feet as shown by the new takeoff flight path data in the Airplane Flight Manual and thereafter the maximum bank is not more than 15 degrees.

Docket No. 23516 (52 FR 1836) Eff. 1/15/87; (Amdt. 135-21, Eff. 2/17/87)

§ 135.399 Small nontransport category airplane performance operating limitations.

(a) No person may operate a reciprocating engine or turbopropeller-powered small airplane that is certificated under § 135.169(b)(2); (3), (4), (5), or (6) unless that person complies with the takeoff weight limitations in the approved Airplane Flight Manual or equivalent for operations under this part, and, if the airplane is certificated under § 135.169(b)(4) or (5) with the landing weight limitations in the Approved Airplane Flight Manual or equivalent for operations under this part.

(b) No person may operate an airplane that is certificated under § 135.169(b)(6) unless that person complies with the landing limitations prescribed in §§ 135.385 and 135.387 of this part. For purposes of this paragraph, §§ 135.385 and 135.387 are applicable to reciprocating and turbopropeller-powered small airplanes notwithstanding their stated applicability to turbine engine powered large transport category airplanes.

(Amdt. 135-2, Eff. 10/17/79)

Subpart J—Maintenance, preventive maintenance, and alterations

§ 135.411 Applicability.

(a) This subpart prescribes rules in addition to those in other parts of this chapter for the maintenance, preventive maintenance, and alterations for each certificate holder as follows:

(1) Aircraft that are type certificated for a passenger seating configuration, excluding any pilot seat, of nine seats or less, shall be maintained under Parts 91 and 43 of this chapter and §§ 135.415, 135.417, and 135.421. An approved aircraft inspection program may be used under § 135.419.

(2) Aircraft that are type certificated for a passenger seating configuration, excluding any pilot seat, of ten seats or more, shall be maintained under a maintenance program in §§ 135.415, 135.417, and 135.423 through 135.443.

(b) A certificate holder who is not otherwise required, may elect to maintain its aircraft under paragraph (a)(2) of this section.

§ 135.413 Responsibility for airworthiness.

(a) Each certificate holder is primarily responsible for the airworthiness of its aircraft, including airframes, aircraft engines, propellers, rotors, appliances, and parts, and shall have its aircraft maintained under this chapter, and shall have defects repaired between required maintenance under Part 43 of this chapter.

(b) Each certificate holder who maintains its aircraft under § 135.411(a)(2) shall—

(1) Perform the maintenance, preventive maintenance, and alteration of its aircraft, including airframe, aircraft engines, propellers, rotors, appliances, emergency equipment and parts, under its manual and this chapter; or

(2) Make arrangements with another person for the performance of maintenance, preventive maintenance or alteration. However, the certificate holder shall ensure that any maintenance, preventive maintenance, or alteration that is performed by another person is performed under the certificate holder's manual and this chapter.

§ 135.415 Mechanical reliability reports.

(a) Each certificate holder shall report the occurrence or detection of each failure, malfunction, or defect in an aircraft concerning—

(1) Fires during flight and whether the related fire-warning system functioned properly;

(2) Fires during flight not protected by related fire-warning system;

(3) False fire-warning during flight;

(4) An exhaust system that causes damage during flight to the engine, adjacent structure, equipment, or components;

(5) An aircraft component that causes accumulation or circulation of smoke, vapor, or toxic or noxious fumes in the crew compartment or passenger cabin during flight;

(6) Engine shutdown during flight because of flameout;

(7) Engine shutdown during flight when external damage to the engine or aircraft structure occurs;

(8) Engine shutdown during flight due to foreign object ingestion or icing;

(9) Shutdown of more than one engine during flight;

(10) A propeller feathering system or ability of the system to control overspeed during flight;

(11) A fuel or fuel-dumping system that affects fuel flow or causes hazardous leakage during flight;

(12) An unwanted landing gear extension or retraction or opening or closing of landing gear doors during flight;

(13) Brake system components that result in loss of brake actuating force when the aircraft is in motion on the ground;

(14) Aircraft structure that requires major repair;

(15) Cracks, permanent deformation, or corrosion of aircraft structures, if more than the maximum acceptable to the manufacturer or the FAA; and

(16) Aircraft components or systems that result in taking emergency actions during flight (except action to shutdown an engine).

(b) For the purpose of this section, "during flight" means the period from the moment the aircraft leaves the surface of the earth on takeoff until it touches down on landing.

(c) In addition to the reports required by paragraph (a) of this section, each certificate holder shall report any other failure, malfunction, or defect in an aircraft that occurs or is detected at any time if, in its opinion, the failure, malfunction, or defect has endangered or may endanger the safe operation of the aircraft.

(d) Each certificate holder shall send each report required by this section, in writing, covering each 24-hour period beginning at 0900 hours local time of each day and ending at 0900 hours local time on the next day to the FAA Flight Standards District Office charged with the overall inspection of the certificate holder. Each report of occurrences during a 24-hour period must be mailed or delivered to that office within the next 72 hours. However, a report that is due on Saturday or Sunday may be mailed or delivered on the following Monday and one that is due on a holiday may be mailed or delivered on the next work day. For aircraft operated in areas where mail is not collected, reports may be mailed or delivered within 72 hours after the aircraft returns to a point where the mail is collected.

(e) The certificate holder shall transmit the reports required by this section on a form and in a manner prescribed by the Administrator, and shall include as much of the following as is available:

(1) The type and identification number of the aircraft.

(2) The name of the operator.

(3) The date.

(4) The nature of the failure, malfunction, or defect.

(5) Identification of the part and system involved, including available information pertaining to type designation of the major component and time since last overhaul, if known.

(6) Apparent cause of the failure, malfunction or defect (e.g., wear, crack, design deficiency, or personnel error).

(7) Other pertinent information necessary for more complete identification, determination of seriousness, or corrective action.

(f) A certificate holder that is also the holder of a type certificate (including a supplemental type certificate), a Parts Manufacturer Approval, or a Technical Standard Order Authorization, or that is the licensee of a type certificate need not report a failure, malfunction, or defect under this section if the failure, malfunction, or defect has been reported by it under § 21.3 or § 37.17 of this chapter or under the accident reporting provisions of Part 830 of the regulations of the National Transportation Safety Board.

(g) No person may withhold a report required by this section even though all information required by this section is not available.

(h) When the certificate holder gets additional information, including information from the manufacturer or other agency, concerning a report required by this section, it shall expeditiously submit it as a supplement to the first report and reference the date and place of submission of the first report.

§ 135.417 Mechanical interruption summary report.

Each certificate holder shall mail or deliver, before the end of the 10th day of the following month, a summary report of the following occurrences in multiengine aircraft for the preceding month to the FAA Flight Standards District Office charged with the overall inspection of the certificate holder:

(a) Each interruption to a flight, unscheduled change of aircraft en route, or unscheduled stop or diversion from a route, caused by known or suspected mechanical difficulties or malfunctions that are not required to be reported under § 135.415.

(b) The number of propeller featherings in flight, listed by type of propeller and engine and aircraft on which it was installed. Propeller featherings for training, demonstration, or flight check purposes need not be reported.

§ 135.419 Approved aircraft inspection program.

(a) Whenever the Administrator finds that the aircraft inspections required or allowed under Part 91 of this chapter are not adequate to meet this part, or upon application by a certificate holder, the Administrator may amend the certificate holder's operations specifications under § 135.17, to require or allow an approved aircraft inspection program for any make and model aircraft of which the certificate holder has the exclusive use of at least one aircraft (as defined in § 135.25(b)).

(b) A certificate holder who applies for an amendment of its operations specifications to allow an approved aircraft inspection program must submit that program with its application for approval by the Administrator.

(c) Each certificate holder who is required by its operations specifications to have an approved aircraft inspection program shall submit a program for approval by the Administrator within 30 days of the amendment of its operations specifications or within any other period that the Administrator may prescribe in the operations specifications.

(d) The aircraft inspection program submitted for approval by the Administrator must contain the following:

(1) Instructions and procedures for the conduct of aircraft inspections (which must include necessary tests and checks), setting forth in detail the parts and areas of the airframe, engines, propellers, rotors and appliances, including emergency equipment, that must be inspected.

(2) A schedule for the performance of the aircraft inspections under paragraph (1) of this paragraph expressed in terms of the time in service, calendar time, number of system operations, or any combination of these.

(3) Instructions and procedures for recording discrepancies found during inspections and correction or deferral of discrepancies including form and disposition of records.

(e) After approval, the certificate holder shall include the approved aircraft inspection program in the manual required by § 135.21.

(f) Whenever the Administrator finds that revisions to an approved aircraft inspection program are necessary for the continued adequacy of the program, the certificate holder shall, after notification by the Administrator, make any changes in the program found by the Administrator to be necessary. The certificate holder may petition the Administrator to reconsider the notice to make any changes in a program. The petition must be filed with the representatives of the Administrator assigned to it within 30 days after the certificate holder receives the notice. Except in the case of an emergency requiring immediate action in the interest of safety, the filing of the petition stays the notice pending a decision by the Administrator.

(g) Each certificate holder who has an approved aircraft inspection program shall have each aircraft that is subject to the program inspected in accordance with the program.

(h) The registration number of each aircraft that is subject to an approved aircraft inspection program must be included in the operations specifications of the certificate holder.

§ 135.421 Additional maintenance requirements.

(a) Each certificate holder who operates an aircraft type certificated for a passenger seating configuration, excluding any pilot seat, of nine seats or less, must comply with the manufacturer's recommended maintenance programs, or a program approved by the Administrator, for each aircraft engine, propeller, rotor, and each item of emergency equipment required by this chapter.

(b) For the purpose of this section, a manufacturer's maintenance program is one which is contained in the maintenance manual or maintenance instructions set forth by the manufacturer as required by this chapter for the aircraft, aircraft engine, propeller, rotor or item of emergency equipment.

§ 135.423 Maintenance, preventive maintenance, and alteration organization.

(a) Each certificate holder that performs any of its maintenance (other than required inspections), preventive maintenance, or alterations, and each person with whom it arranges for the performance of that work, must have an organization adequate to perform the work.

(b) Each certificate holder that performs any inspections required by its manual under § 135.427(b)(2) or (3) (in this subpart referred to as "required inspections"), and each person with whom it arranges for the performance of that work, must have an organization adequate to perform that work.

(c) Each person performing required inspections in addition to other maintenance, preventive maintenance, or alterations, shall organize the performance of those functions so as to separate the required inspection functions from the other maintenance, preventive maintenance, and alteration functions. The separation shall be below the level of administrative control at which overall responsibility for the required inspection functions and other maintenance, preventive maintenance, and alteration functions is exercised.

§ 135.425 Maintenance, preventive maintenance, and alteration programs.

Each certificate holder shall have an inspection program and a program covering other maintenance, preventive maintenance, and alterations, that ensures that—

(a) Maintenance, preventive maintenance, and alterations performed by it, or by other persons, are performed under the certificate holder's manual;

(b) Competent personnel and adequate facilities and equipment are provided for the proper performance of maintenance, preventive maintenance and alterations; and

(c) Each aircraft released to service is airworthy and has been properly maintained for operation under this part.

§ 135.427 Manual requirements.

(a) Each certificate holder shall put in its manual the chart or description of the certificate holder's organization required by § 135.423 and a list of persons with whom it has arranged for the performance of any of its required inspections, other maintenance, preventive maintenance, or alterations, including a general description of that work.

(b) Each certificate holder shall put in its manual the programs required by § 135.425 that must be followed in performing maintenance, preventive maintenance, and alterations of that certificate holder's aircraft, including airframes, aircraft engines, propellers, rotors, appliances, emergency equipment, and parts, and must include at least the following:

(1) The method of performing routine and nonroutine maintenance (other than required inspections), preventive maintenance, and alterations.

(2) A designation of the items of maintenance and alteration that must be inspected (required inspections) including at least those that could result in a failure, malfunction, or defect endangering the safe operation of the aircraft, if not performed properly or if improper parts or materials are used.

(3) The method of performing required inspections and a designation by occupational title of personnel authorized to perform each required inspection.

(4) Procedures for the reinspection of work performed under previous required inspection findings ("buy-back procedures").

(5) Procedures, standards, and limits necessary for required inspections and acceptance or rejection of the items required to be inspected and for periodic inspection and calibration of precision tools, measuring devices, and test equipment.

(6) Procedures to ensure that all required inspections are performed.

(7) Instructions to prevent any person who performs any item of work from performing any required inspection of that work.

(8) Instructions and procedures to prevent any decision of an inspector regarding any required inspection from being countermanded by persons other than supervisory personnel of the inspection unit, or a person at the level of administrative control that has overall responsibility for the management of both the required inspection functions and the other maintenance, preventive maintenance, and alterations functions.

(9) Procedures to ensure that required inspections, other maintenance, preventive maintenance, and alterations that are not completed as a result of work interruptions are properly completed before the aircraft is released to service.

(c) Each certificate holder shall put in its manual a suitable system (which may include a coded system) that provides for the retention of the following information—

(1) A description (or reference to data acceptable to the Administrator) of the work performed;

(2) The name of the person performing the work if the work is performed by a person outside the organization of the certificate holder; and

(3) The name or other positive identification of the individual approving the work.

§ 135.429 Required inspection personnel.

(a) No person may use any person to perform required inspections unless the person performing the inspection is appropriately certificated, properly trained, qualified, and authorized to do so.

(b) No person may allow any person to perform a required inspection unless, at the time, the person performing that inspection is under the supervision and control of an inspection unit.

(c) No person may perform a required inspection if that person performed the item of work to be inspected.

(d) In the case of rotorcraft that operate in remote areas or sites, the Administrator may approve procedures for the performance of required inspection items by a pilot when no other qualified person is available, provided—

(1) The pilot is employed by the certificate holder;

(2) It can be shown to the satisfaction of the Administrator that each pilot authorized to perform required inspections is properly trained and qualified;

(3) The required inspection is a result of a mechanical interruption and is not a part of a certificate holder's continuous airworthiness maintenance program;

(4) Each item is inspected after each flight until the item has been inspected by an appropriately certificated mechanic other than the one who originally performed the item of work; and

(5) Each item of work that is a required inspection item that is part of the flight control system shall be flight tested and reinspected before the aircraft is approved for return to service.

(e) Each certificate holder shall maintain, or shall determine that each person with whom it arranges to perform its required inspections maintains, a current listing of persons who have been trained, qualified, and authorized to conduct required inspections. The persons must be identified by name, occupational title and the inspections that they are authorized to perform. The certificate holder (or person with whom it arranges to perform its required inspections) shall give written information to each person so authorized, describing the extent of that person's responsibilities, authorities, and inspectional limitations. The list shall be made available for inspection by the Administrator upon request.

(Amdt. 135-20, Eff. 1/6/87)

§ 135.431 Continuing analysis and surveillance.

(a) Each certificate holder shall establish and maintain a system for the continuing analysis and surveillance of the performance and effectiveness of its inspection program and the program covering other maintenance, preventive maintenance, and alterations and for the correction of any deficiency in those programs, regardless of whether those programs are carried out by the certificate holder or by another person.

(b) Whenever the Administrator finds that either or both of the programs described in paragraph (a) of this section does not contain adequate procedures and standards to meet this part, the certificate holder shall, after notification by the Administrator, make changes in those programs requested by the Administrator.

(c) A certificate holder may petition the Administrator to reconsider the notice to make a change in a program. The petition must be filed with the FAA Flight Standards District Office charged with the overall inspection of the certificate holder within 30 days after the certificate holder receives the notice. Except in the case of an emergency requiring immediate action in the interest of safety, the filing of the petition stays the notice pending a decision by the Administrator.

§ 135.433 Maintenance and preventive maintenance training program.

Each certificate holder or a person performing maintenance or preventive maintenance functions for it shall have a training program to ensure that each person (including inspection personnel) who determines the adequacy of work done is fully informed about procedures and techniques and new equipment in use and is competent to perform that person's duties.

§ 135.435 Certificate requirements.

(a) Except for maintenance, preventive maintenance, alterations, and required inspections performed by repair stations certificated under the provisions of Subpart C of Part 145 of this chapter, each person who is directly in charge of maintenance, preventive maintenance, or alterations, and each person performing required inspections must hold an appropriate airman certificate.

(b) For the purpose of this section, a person "directly in charge" is each person assigned to a position in which that person is responsible for the work of a shop or station that performs maintenance, preventive maintenance, alterations, or other functions affecting airworthiness. A person who is "directly in charge" need not physically observe and direct each worker constantly but must be available for consultation and decision on matters requiring instruction or decision from higher authority than that of the person performing the work.

§ 135.437 Authority to perform and approve maintenance, preventive maintenance, and alterations.

(a) A certificate holder may perform, or make arrangements with other persons to perform, maintenance, preventive maintenance, and alterations as provided in its maintenance manual. In addition, a certificate holder may perform these functions for another certificate holder as provided in the maintenance manual of the other certificate holder.

(b) A certificate holder may approve any airframe, aircraft engine, propeller, rotor, or appliance for return to service after maintenance, preventive maintenance, or alterations that are performed under paragraph (a) of this section. However, in the case of a major repair or alteration, the work must have been done in accordance with technical data approved by the Administrator.

§ 135.439 Maintenance recording requirements.

(a) Each certificate holder shall keep (using the system specified in the manual required in § 135.427) the following records for the periods specified in paragraph (b) of this section:

(1) All the records necessary to show that all requirements for the issuance of an airworthiness release under § 135.443 have been met.

(2) Records contain the following information:

(i) The total time in service of the airframe, engine, propeller, and rotor.

(ii) The current status of life-limited parts of each airframe, engine, propeller, rotor, and appliance.

(iii) The time since last overhaul of each item installed on the aircraft which are required to be overhauled on a specified time basis.

(iv) The identification of the current inspection status of the aircraft, including the time since the last inspections required by the inspection program under which the aircraft and its appliances are maintained.

(v) The current status of applicable airworthiness directives, including the date and methods of compliance, and, if the airworthiness directive involves recurring action, the time and date when the next action is required.

(vi) A list of current major alterations and repairs to each airframe, engine, propeller, rotor, and appliance.

(b) Each certificate holder shall retain the records required to be kept by this section for the following periods:

(1) Except for the records of the last complete overhaul of each airframe, engine, propeller, rotor, and appliance the records specified in paragraph (a)(1) of this section shall be retained until the work is repeated or superseded by other work or for one year after the work is performed.

(2) The records of the last complete overhaul of each airframe, engine, propeller, rotor, and appliance shall be retained until the work is superseded by work of equivalent scope and detail.

(3) The records specified in paragraph (a)(2) of this section shall be retained and transferred with the aircraft at the time the aircraft is sold.

(c) The certificate holder shall make all maintenance records required to be kept by this section available for inspection by the Administrator or any representative of the National Transportation Safety Board.

§ 135.441 Transfer of maintenance records.

Each certificate holder who sells a United States registered aircraft shall transfer to the purchaser, at the time of the sale, the following records of that aircraft, in plain language form or in coded form which provides for the preservation and retrieval of information in a manner acceptable to the Administrator.

(a) The records specified in § 135.439(a)(2).

(b) The records specified in § 135.439(a)(1) which are not included in the records covered by paragraph (a) of this section, except that the purchaser may allow the seller to keep physical custody of such records. However, custody of records by the seller does not relieve the purchaser of its responsibility under § 135.439(c) to make the records available for inspection by the Administrator or any representative of the National Transportation Safety Board.

§ 135.443 Airworthiness release or aircraft maintenance log entry.

(a) No certificate holder may operate an aircraft after maintenance, preventive maintenance, or alterations are performed on the aircraft unless the certificate holder prepares, or causes the person with whom the certificate holder arranges for the performance of the maintenance, preventive maintenance, or alterations, to prepare—

(1) An airworthiness release; or

(2) An appropriate entry in the aircraft maintenance log.

(b) The airworthiness release or log entry required by paragraph (a) of this section must—

(1) Be prepared in accordance with the procedure in the certificate holder's manual;

(2) Include a certificate that—

(i) The work was performed in accordance with the requirements of the certificate holder's manual;

(ii) All items required to be inspected were inspected by an authorized person who determined that the work was satisfactorily completed;

(iii) No known condition exists that would make the aircraft unairworthy;

(iv) So far as the work performed is concerned, the aircraft is in condition for safe operation; and

(3) Be signed by an authorized certificated mechanic or repairman, except that a certificated repairman may sign the release or entry only for the work for which that person is employed and for which that person is certificated.

Notwithstanding paragraph (b)(3) of this section, after maintenance, preventive maintenance, or alterations performed by a repair station certified under the provisions of Subpart C of Part 145, the airworthiness release or log entry required by paragraph (a) of this section may be signed by a person authorized by that repair station.

(c) Instead of restating each of the conditions of the certification required by paragraph (b) of this section, the certificate holder may state in its manual that the signature of an authorized certificated mechanic or repairman constitutes that certification.

(Amdt. 135-29, Eff. 12/22/88)

Appendix A—
Additional airworthiness standards for 10 or more passenger airplanes

APPLICABILITY

1. *Applicability.* This appendix prescribes the additional airworthiness standard required by § 135.169.

2. *References.* Unless otherwise provided references in this appendix to specific sections of Part 23 of the Federal Aviation Regulations (FAR Part 23) are to those sections of Part 23 in effect on March 30, 1967.

FLIGHT REQUIREMENTS

3. *General.* Compliance must be shown with the applicable requirements of Subpart B of FAR Part 23, as supplemented or modified in §§ 4 through 10.

PERFORMANCE

4. *General.* (a) Unless otherwise prescribed in this appendix, compliance with each applicable performance requirement in §§ 4 through 7 must be shown for ambient atmospheric conditions and still air.

(b) The performance must correspond to the propulsive thrust available under the particular ambient atmospheric conditions and the particular flight condition. The available propulsive thrust must correspond to engine power or thrust, not exceeding the approved power or thrust less—

(1) Installation losses; and

(2) The power or equivalent thrust absorbed by the accessories and services appropriate to the particular ambient atmospheric conditions and the particular flight condition.

(c) Unless otherwise prescribed in this appendix, the applicant must select the takeoff, en route, and landing configurations for the airplane.

(3) The airplane configuration may vary with weight, altitude, and temperature, to the extent they are compatible

with the operating procedures required by paragraph (e) of this section.

(e) Unless otherwise prescribed in this appendix, in determining the critical engine inoperative takeoff performance, the accelerate-stop distance, takeoff distance, changes in the airplane's configuration, speed, power, and thrust must be made under procedures established by the applicant for operation in service.

(f) Procedures for the execution of balked landings must be established by the applicant and included in the Airplane Flight Manual.

(g) The procedures established under paragraphs (e) and (f) of this section must—

(1) Be able to be consistently executed in service by a crew of average skill;

(2) Use methods or devices that are safe and reliable; and

(3) Include allowance for any time delays, in the execution of the procedures, that may reasonably be expected in service.

5. *Takeoff*—(a) *General.* Takeoff speeds, the accelerate-stop distance, the takeoff distance, and the one-engine-inoperative takeoff flight path data (described in paragraphs (b), (c), (d), and (f) of this section), must be determined for—

(1) Each weight, altitude, and ambient temperature within the operational limits selected by the applicant;

(2) The selected configuration for takeoff;

(3) The center of gravity in the most unfavorable position;

(4) The operating engine within approved operating limitations; and

(5) Takeoff data based on smooth, dry, hard-surface runway.

(b) *Takeoff speeds.* (1) The decision speed V_1 is the calibrated airspeed on the ground at which, as a result of engine failure or other reasons, the pilot is assumed to have made a decision to continue or discontinue the takeoff. The speed V_1 must be selected by the applicant but may not be less than—

(i) 1.10 V_{S1};

(ii) 1.10 V_{MC};

(iii) A speed that allows acceleration to V_1 and stop under paragraph (c) of this section; or

(iv) A speed at which the airplane can be rotated for takeoff and shown to be adequate to safely continue the takeoff, using normal piloting skill, when the critical engine is suddenly made inoperative.

(2) The initial climb out speed V_2, in terms of calibrated airspeed, must be selected by the applicant so as to allow the gradient of climb required in § 6(b)(2), but it must not be less than V_1 or less than 1.2 V_{S1}.

(3) Other essential takeoff speeds necessary for safe operation of the airplane.

(c) *Accelerate-stop distance.* (1) The accelerate-stop distance is the sum of the distances necessary to—

(i) Accelerate the airplane from a standing start to V_1; and

(ii) Come to a full stop from the point at which V_1 is reached assuming that in the case of engine failure, failure of the critical engine is recognized by the pilot at the speed V_1.

(2) Means other than wheel brakes may be used to determine the accelerate-stop distance if that means is available with the critical engine inoperative and—

(i) Is safe and reliable;

(ii) Is used so that consistent results can be expected under normal operating conditions; and

(iii) Is such that exceptional skill is not required to control the airplane.

(d) *All engines operating takeoff distance.* The all engine operating takeoff distance is the horizontal distance required to take off and climb to a height of 50 feet above the takeoff surface under the procedures in FAR 23.51(a).

(e) *One-engine-inoperative takeoff.* Determine the weight for each altitude and temperature within the operational limits established for the airplane, at which the airplane has the capability, after failure of the critical engine at V_1 determined under paragraph (b) of this section, to take off and climb at not less than V_2, to a height 1,000 feet above the takeoff surface and attain the speed and configuration at which compliance is shown with the en route one-engine-inoperative gradient of climb specified in § 6(c).

(f) *One-engine-inoperative takeoff flight path data.* The one-engine-inoperative takeoff flight path data consist of takeoff flight paths extending from a standing start to a point in the takeoff at which the airplane reaches a height 1,000 feet above the takeoff surface under paragraph (e) of this section.

6. *Climb*—(a) *Landing climb: All-engines-operating.* The maximum weight must be determined with the airplane in the landing configuration, for each altitude, and ambient temperature within the operational limits established for the airplane, with the most unfavorable center of gravity, and out-of-ground effect in free air, at which the steady gradient of climb will not be less than 3.3 percent, with:

(1) The engines at the power that is available 8 seconds after initiation of movement of the power or thrust controls from the minimum flight idle to the takeoff position.

(2) A climb speed not greater than the approach speed established under § 7 and not less than the greater of 1.05 V_{MC} or 1.10 V_{S1}.

(b) *Takeoff climb: one-engine-inoperative.* The maximum weight at which the airplane meets the minimum climb performance specified in subparagraphs (1) and (2) of this paragraph must be determined for each altitude and ambient temperature within the operational limits established for the airplane, out of ground effect in free air, with the airplane in the takeoff configuration, with the most unfavorable center of gravity, the critical engine inoperative, the remaining engines at the maximum takeoff power or thrust, and the propeller of the inoperative engine windmilling with the propeller controls in the normal position except that, if an approved automatic feathering system is installed, the propellers may be in the feathered position:

(1) *Takeoff landing gear extended.* The minimum steady gradient of climb must be measurably positive at the speed V_1.

(2) *Takeoff landing gear retracted.* The minimum steady gradient of climb may not be less than 2 percent at speed V_2. For airplanes with fixed landing gear this requirement must be met with the landing gear extended.

(c) *En route climb: one-engine-inoperative.* The maximum weight must be determined for each altitude and ambient temperature within the operational limits established for the airplane, at which the steady gradient of climb is not less than 1.2 percent at an altitude 1,000 feet above the takeoff surface, with the airplane in the en route configuration, the critical engine inoperative, the remaining engine at the maximum continuous power or thrust, and the most unfavorable center of gravity.

7. *Landing.* (a) The landing field length described in paragraph (b) of this section must be determined for standard atmosphere at each weight and altitude within the operational limits established by the applicant.

(b) The landing field length is equal to the landing distance determined under FAR 23.75 (a) divided by a factor of 0.6 for the destination airport and 0.7 for the alternate airport. Instead of the gliding approach specified in FAR 23.75(a)(1), the landing may be preceded by a steady approach down to the 50-foot height at a gradient of descent not greater than 5.2 percent (3°) at a calibrated airspeed not less than 1.3 V_{S1}.

TRIM

8. *Trim*—(a) *Lateral and directional trim.* The airplane must maintain lateral and directional trim in level flight at a speed of V_H or V_{MO}/M_{MO}, whichever is lower, with landing gear and wing flaps retracted.

(b) *Longitudinal trim.* The airplane must maintain longitudinal trim during the following conditions, except that it need not maintain trim at a speed greater than V_{MO}/M_{MO}:

(1) In the approach conditions specified in FAR 23.161(c)(3) through (5), except that instead of the speeds specified in those paragraphs, trim must be maintained with a stick force of not more than 10 pounds down to a speed used in showing compliance with § 7 or 1.4 V_{S1} whichever is lower.

(2) In level flight at any speed from V_H or V_{MO}/M_{MO}, whichever is lower, to either V_X or 1.4 V_{S1}, with the landing gear and wing flaps retracted.

STABILITY

9. *Static longitudinal stability.* (a) In showing compliance with FAR 23.175(b) and with paragraph (b) of this section, the airspeed must return to within 7½ percent of the trim speed.

(b) *Cruise stability.* The stick force curve must have a stable slope for a speed range of ±50 knots from the trim speed except that the speeds need not exceed V_{FC}/M_{FC} or be less than 1.4 V_{S1}. This speed range will be considered to begin at the outer extremes of the friction band and the stick force may not exceed 50 pounds with—

(1) Landing gear retracted;

(2) Wing flaps retracted;

(3) The maximum cruising power as selected by the applicant as an operating limitation for turbine engines or 75 percent of maximum continuous power for reciprocating engines except that the power need not exceed that required at V_{MO}/M_{MO};

(4) Maximum takeoff weight; and

(5) The airplane trimmed for level flight with the power specified in subparagraph (3) of this paragraph.

V_{FC}/M_{FC} may not be less than a speed midway between V_{MO}/M_{MO} and V_{DF}/M_{DF}, except that, for altitudes where Mach number is the limiting factor, M_{FC} need not exceed the Mach number at which effective speed warning occurs.

(c) *Climb stability (turbopropeller powered airplanes only).* In showing compliance with FAR 23.175(a), an applicant must, instead of the power specified in FAR 23.175(a)(4), use the maximum power or thrust selected by the applicant as an operating limitation for use during climb at the best rate of climb speed, except that the speed need not be less than 1.4 V_{S1}.

STALLS

10. *Stall warning.* If artificial stall warning is required to comply with FAR 23.207, the warning device must give clearly distinguishable indications under expected conditions of flight. The use of a visual warning device that requires the attention of the crew within the cockpit is not acceptable by itself.

CONTROL SYSTEMS

11. *Electric trim tabs.* The airplane must meet FAR 23.677 and in addition it must be shown that the airplane is safely controllable and that a pilot can perform all the maneuvers and operations necessary to effect a safe landing following any probable electric trim tab runaway which might be reasonably expected in service allowing for appropriate time delay after pilot recognition of the runaway. This demonstration must be conducted at the critical airplane weights and center of gravity positions.

INSTRUMENTS: INSTALLATION

12. *Arrangement and visibility.* Each instrument must meet FAR 23.1321 and in addition:

(a) Each flight, navigation, and powerplant instrument for use by any pilot must be plainly visible to the pilot from the pilot's station with the minimum practicable deviation from the pilot's normal position and line of vision when the pilot is looking forward along the flight path.

(b) The flight instruments required by FAR 23.1303 and by the applicable operating rules must be grouped on the instrument panel and centered as nearly as practicable about the vertical plane of each pilot's forward vision. In addition—

(1) The instrument that most effectively indicates the attitude must be in the panel in the top center position;

(2) The instrument that most effectively indicates the airspeed must be on the panel directly to the left of the instrument in the top center position;

(3) The instrument that most effectively indicates altitude must be adjacent to and directly to the right of the instrument in the top center position; and

(4) The instrument that most effectively indicates direction of flight must be adjacent to and directly below the instrument in the top center position.

13. *Airspeed indicating system.* Each airspeed indicating system must meet FAR 23.1323 and in addition:

(a) Airspeed indicating instruments must be of an approved type and must be calibrated to indicate true airspeed at sea level in the standard atmosphere with a minimum practicable instrument calibration error when the corresponding pitot and static pressures are supplied to the instruments.

(b) The airspeed indicating system must be calibrated to determine the system error, i.e., the relation between IAS and CAS, in flight and during the accelerate-takeoff ground run. The ground run calibration must be obtained between 0.8 of the minimum value of V_1 and 1.2 times the maximum value of V_1, considering the approved ranges of altitude and weight. The ground run calibration is determined assuming an engine failure at the minimum value of V_1.

(c) The airspeed error of the installation excluding the instrument calibration error, must not exceed 3 percent or 5 knots whichever is greater, throughout the speed range from V_{MO} to 1.3 V_{S1}, with flaps retracted and from 1.3 V_{SO} to V_{FE} with flaps in the landing position.

(d) Information showing the relationship between IAS and CAS must be shown in the Airplane Flight Manual.

14. *Static air vent system.* The static air vent system must meet FAR 23.1325. The altimeter system calibration must be determined and shown in the Airplane Flight Manual.

OPERATING LIMITATIONS AND INFORMATION

15. *Maximum operating limit speed V_{MO}/M_{MO}.* Instead of establishing operating limitations based on V_{NE}/V_{NO}, the applicant must establish a maximum operating limit speed V_{MO}/M_{MO} as follows:

(a) The maximum operating limit speed must not exceed the design cruising speed V_C and must be sufficiently below V_D/M_D or V_{DF}/M_{DF} to make it highly improbable that the latter speeds will be inadvertently exceeded in flight.

(b) The speed V_{MO} must not exceed 0.8 V_D/M_D or 0.8 V_{DF}/M_{DF} unless flight demonstrations involving upsets as specified by the Administrator indicates a lower speed margin will not result in speeds exceeding V_D/M_D or V_{DF}. Atmospheric variations, horizontal gusts, system and equipment errors, and airframe production variations are taken into account.

16. *Minimum flight crew.* In addition to meeting FAR 23.1523, the applicant must establish the minimum number and type of qualified flight crew personnel sufficient for safe operation of the airplane considering—

(a) Each kind of operation for which the applicant desires approval;

(b) The workload on each crewmember considering the following:

(1) Flight path control.

(2) Collision avoidance.

(3) Navigation.

(4) Communications.

(5) Operation and monitoring of all essential aircraft systems.

(6) Command decisions; and

(c) The accessibility and ease of operation of necessary controls by the appropriate crewmember during all normal and emergency operations when at the crewmember flight station.

17. *Airspeed indicator.* The airspeed indicator must meet FAR 23.1545 except that, the airspeed notations and markings in terms of V_{NO} and V_{NH} must be replaced by the V_{MO}/M_{MO} notations. The airspeed indicator markings must be easily read and understood by the pilot. A placard adjacent to the airspeed indicator is an acceptable means of showing compliance with FAR 23.1545(c).

AIRPLANE FLIGHT MANUAL

18. *General.* The Airplane Flight Manual must be prepared under FARs 23.1583 and 23.1587, and in addition the operating limitations and performance information in §§ 19 and 20 must be included.

19. *Operating limitations.* The Airplane Flight Manual must include the following limitations—

(a) *Airspeed limitations.* (1) The maximum operating limit speed V_{MO}/M_{MO} and a statement that this speed limit may not be deliberately exceeded in any regime of flight (climb, cruise, or descent) unless a higher speed is authorized for flight test or pilot training;

(2) If an airspeed limitation is based upon compressibility effects, a statement to this effect and information as to any symptoms, the probable behavior of the airplane, and the recommended recovery procedures; and

(3) The airspeed limits, shown in terms of V_{MO}/M_{MO} instead of V_{NO} and V_{NE}.

(b) *Takeoff weight limitations.* The maximum takeoff weight for each airport elevation, ambient temperature, and available takeoff runway length within the range selected by the applicant may not exceed the weight at which—

(1) The all-engine-operating takeoff distance determined under § 5(b) or the accelerate-stop distance determined under § 5(c), whichever is greater, is equal to the available runway length;

(2) The airplane complies with the engine-inoperative takeoff requirements specified in § 5(e); and

(3) The airplane complies with the one-engine-inoperative takeoff and en route climb requirements specified in §§ 6(b) and (c).

(c) *Landing weight limitations.* The maximum landing weight for each airport elevation (standard temperature) and available landing runway length, within the range selected by the applicant. This weight may not exceed the weight at which the landing field length determined under § 7(b) is equal to the available runway length. In showing compliance with this operating limitation, it is acceptable to assume that the landing weight at the destination will be equal to the takeoff weight reduced by the normal consumption of fuel and oil en route.

20. *Performance information.* The Airplane Flight Manual must contain the performance information determined under the performance requirements of this appendix. The information must include the following:

(a) Sufficient information so that the takeoff weight limits specified in § 19(b) can be determined for all temperatures and altitudes within the operation limitations selected by the applicant.

(b) The conditions under which the performance information was obtained, including the airspeed at the 50-foot height used to determine landing distances.

(c) The performance information (determined by extrapolation and computed for the range of weights between the maximum landing and takeoff weights) for—

(1) Climb in the landing configuration; and

(2) Landing distance.

(d) Procedure established under § 4 related to the limitations and information required by this section in the form of guidance material including any relevant limitations or information.

(e) An explanation of significant or unusual flight or ground handling characteristics of the airplane.

(f) Airspeeds, as indicated airspeeds, corresponding to those determined for takeoff under § 5(b).

21. *Maximum operating altitudes.* The maximum operating altitude to which operation is allowed, as limited by flight, structural, powerplant, functional, or equipment characteristics, must be specified in the Airplane Flight Manual.

22. *Stowage provision for airplane flight manual.* Provision must be made for stowing the Airplane Flight Manual in a suitable fixed container which is readily accessible to the pilot.

23. *Operating procedures.* Procedures for restarting turbine engines in flight (including the effects of altitude) must be set forth in the Airplane Flight Manual.

AIRFRAME REQUIREMENTS FLIGHT LOADS

24. *Engine Torque.* (a) Each turbopropeller engine mount and its supporting structure must be designed for the torque effects of:

(1) The conditions in FAR 23.361(a).

(2) The limit engine torque corresponding to takeoff power and propeller speed multiplied by a factor accounting for propeller control system malfunction, including quick feathering action, simultaneously with 1*g* level flight loads. In the absence of a rational analysis, a factor of 1.6 must be used.

(b) The limit torque is obtained by multiplying the mean torque by a factor of 1.25.

25. *Turbine engine gyroscopic loads.* Each turbopropeller engine mount and its supporting structure must be designed for the gyroscopic loads that result, with the engines at maximum continuous r.p.m. under either—

(a) The conditions in FARs 23.351 and 23.423; or

(b) All possible combinations of the following:

(1) A yaw velocity of 2.5 radians per second.

(2) A pitch velocity of 1.0 radians per second.

(3) A normal load factor of 2.5.

(4) Maximum continuous thrust.

26. *Unsymmetrical loads due to engine failure.*

(a) Turbopropeller powered airplanes must be designed for the unsymmetrical loads resulting from the failure of the critical engine including the following conditions in combination with a single malfunction of the propeller drag limiting system, considering the probable pilot corrective action on the flight controls:

(1) At speeds between V_{MO} and V_D, the loads resulting from power failure because of fuel flow interruption are considered to be limit loads.

(2) At speeds between V_{MO} and V_C, the loads resulting from the disconnection of the engine compressor from the turbine or from loss of the turbine blades are considered to be ultimate loads.

(3) The time history of the thrust decay and drag buildup occurring as a result of the prescribed engine failures must be substantiated by test or other data applicable to the particular engine-propeller combination.

(4) The timing and magnitude of the probable pilot corrective action must be conservatively estimated, considering the characteristics of the particular engine-propeller-airplane combination.

(b) Pilot corrective action may be assumed to be initiated at the time maximum yawing velocity is reached, but not earlier than 2 seconds after the engine failure. The magnitude of the corrective action may be based on the control forces in FAR 23.397 except that lower forces may be assumed where it is shown by analysis or test that these forces can control the yaw and roll resulting from the prescribed engine failure conditions.

GROUND LOADS

27. *Dual wheel landing gear units.* Each dual wheel landing gear unit and its supporting structure must be shown to comply with the following:

(a) *Pivoting.* The airplane must be assumed to pivot about one side of the main gear with the brakes on that side locked. The limit vertical load factor must be 1.0 and the coefficient of friction 0.8. This condition need apply only to the main gear and its supporting structure.

(b) *Unequal tire inflation.* A 60-40 percent distribution of the loads established under FAR 23.471 through FAR 23.483 must be applied to the dual wheels.

(c) *Flat tire.* (1) Sixty percent of the loads in FAR 23.471 through FAR 23.483 must be applied to either wheel in a unit.

(2) Sixty percent of the limit drag and side loads and 100 percent of the limit vertical load established under FARs 23.493 and 23.485 must be applied to either wheel in a unit except that the vertical load need not exceed the maximum vertical load in paragraph (c)(1) of this section.

FATIGUE EVALUATION

28. *Fatigue evaluation of wing and associated structure.* Unless it is shown that the structure, operating stress levels, materials and expected use are comparable from a fatigue standpoint to a similar design which has had substantial satisfactory service experience, the strength, detail design, and the fabrication of those parts of the wing, wing carry-through, and attaching structure whose failure would be catastrophic must be evaluated under either—

(a) A fatigue strength investigation in which the structure is shown by analysis, tests, or both to be able to withstand the repeated loads of variable magnitude expected in service; or

(b) A fail-safe strength investigation in which it is shown by analysis, tests, or both that catastrophic failure of the structure is not probable after fatigue, or obvious partial failure, of a principal structural element, and that the remaining structure is able to withstand a static ultimate load factor of 75 percent of the critical limit load factor at V_C. These loads must be multiplied by a factor of 1.15 unless the dynamic effects of failure under static load are otherwise considered.

DESIGN AND CONSTRUCTION

29. *Flutter.* For multiengine turbopropeller powered airplanes, a dynamic evaluation must be made and must include—

(a) The significant elastic, inertia, and aerodynamic forces associated with the rotations and displacements of the plane of the propeller; and

(b) Engine-propeller-nacelle stiffness and damping variations appropriate to the particular configuration.

LANDING GEAR

30. *Flap operated landing gear warning device.* Airplanes having retractable landing gear and wing flaps must be equipped with a warning device that functions continuously when the wing flaps are extended to a flap position that activates the warning device to give adequate warning before landing, using normal landing procedures, if the landing gear is not fully extended and locked. There may not be a manual shut off for this warning device. The flap position sensing unit may be installed at any suitable location. The system for this device may use any part of the system (including the aural warning devices).

PERSONNEL AND CARGO ACCOMMODATIONS

31. *Cargo and baggage compartments.* Cargo and baggage compartments must be designed to meet FAR 23.787 (a) and (b), and in addition means must be provided to protect passengers from injury by the contents of any cargo or baggage compartment when the ultimate forward inertia force is 9 *g*.

32. *Doors and exits.* The airplane must meet FAR 23.783 and FAR 23.807 (a)(3), (b), and (c), and in addition:

(a) There must be a means to lock and safeguard each external door and exit against opening in flight either inadvertently by persons, or as a result of mechanical failure. Each external door must be operable from both the inside and the outside.

(b) There must be means for direct visual inspection of the locking mechanism by crewmembers to determine whether external doors and exits, for which the initial opening movement is outward, are fully locked. In addition, there must be a visual means to signal to crewmembers when normally used external doors are closed and fully locked.

(c) The passenger entrance door must qualify as a floor level emergency exit. Each additional required emergency exit except floor level exits must be located over the wing or must be provided with acceptable means to assist the occupants in descending to the ground. In addition to the passenger entrance door:

(1) For a total seating capacity of 15 or less, an emergency exit as defined in FAR 23.807(b) is required on each side of the cabin.

(2) For a total seating capacity of 16 through 23, three emergency exits as defined in FAR 23.807(b) are required with one on the same side as the door and two on the side opposite the door.

(d) An evacuation demonstration must be conducted utilizing the maximum number of occupants for which certification is desired. It must be conducted under simulated night conditions utilizing only the emergency exits on the most critical side of the aircraft. The participants must be representative of average airline passengers with no previous practice or rehearsal for the demonstration. Evacuation must be completed within 90 seconds.

(e) Each emergency exit must be marked with the word "Exit" by a sign which has white letters 1 inch high on a red background 2 inches high, be self-illuminated or independently internally electrically illuminated, and have a minimum luminescence (brightness) of at least 160 microlamberts. The colors may be reserved if the passenger compartment illumination is essentially the same.

(f) Access to window type emergency exits must not be obstructed by seats or seat backs.

(g) The width of the main passenger aisle at any point between seats must equal or exceed the values in the following table:

	Total seating capacity	
Minimum main passenger aisle width	**Less than 25 inches from floor**	**25 inches and more from floor**
10 through 23	9 inches	15 inches

MISCELLANEOUS

33. *Lightning strike protection.* Parts that are electrically insulated from the basic airframe must be connected to it through lightning arrestors unless a lightning strike on the insulated part—

(a) Is improbable because of shielding by other parts; or

(b) Is not hazardous.

34. *Ice protection.* If certification with ice protection provisions is desired, compliance with the following must be shown:

(a) The recommended procedures for the use of the ice protection equipment must be set forth in the Airplane Flight Manual.

(b) An analysis must be performed to establish, on the basis of the airplane's operational needs, the adequacy of the ice protection system for the various components of the airplane. In addition, tests of the ice protection system must be conducted to demonstrate that the airplane is capable of operating safely in continuous maximum and intermittent maximum icing conditions as described in Appendix C of Part 25 of this chapter.

(c) Compliance with all or portions of this section may be accomplished by reference, where applicable because of similarity of the designs, to analysis and tests performed by the applicant for a type certificated model.

35. *Maintenance information.* The applicant must make available to the owner at the time of delivery of the airplane the information the applicant considers essential for the proper maintenance of the airplane. That information must include the following:

(a) Description of systems, including electrical, hydraulic, and fuel controls.

(b) Lubrication instructions setting forth the frequency and the lubricants and fluids which are to be used in the various systems.

(c) Pressures and electrical loads applicable to the various systems.

(d) Tolerances and adjustments necessary for proper functioning.

(e) Methods of leveling, raising, and towing.

(f) Methods of balancing control surfaces.

(g) Identification of primary and secondary structures.

(h) Frequency and extent of inspections necessary to the proper operation of the airplane.

(i) Special repair methods applicable to the airplane.

(j) Special inspection techniques, such as X-ray, ultrasonic, and magnetic particle inspection.

(k) List of special tools.

PROPULSION

GENERAL

36. *Vibration characteristics.* For turbopropeller powered airplanes, the engine installation must not result in vibration characteristics of the engine exceeding those established during the type certification of the engine.

37. *In-flight restarting of engine.* If the engine on turbopropeller powered airplanes cannot be restarted at the maximum cruise altitude, a determination must be made of the altitude below which restarts can be consistently accomplished. Restart information must be provided in the Airplane Flight Manual.

38. *Engines.* (a) *For turbopropeller powered airplanes.* The engine installation must comply with the following:

(1) *Engine isolation.* The powerplants must be arranged and isolated from each other to all operation, in at least one configuration, so that the failure or malfunction of any engine, or of any system that can affect the engine, will not—

(i) Prevent the continued safe operation of the remaining engines; or

(ii) Require immediate action by any crewmember for continued safe operation.

(2) *Control of engine rotation.* There must be a means to individually stop and restart the rotation of any engine in flight except that engine rotation need not be stopped if continued rotation could not jeopardize the safety of the airplane. Each component of the stopping and restarting system on the engine side of the firewall, and that might be exposed to fire, must be at least fire resistant. If hydraulic propeller feathering systems are used for this purpose, the feathering lines must be at least fire resistant under the op-

erating conditions that may be expected to exist during feathering.

(3) *Engine speed and gas temperature control devices.* The powerplant systems associated with engine control devices, systems, and instrumentation must provide reasonable assurance that those engine operating limitations that adversely affect turbine rotor structural integrity will not be exceeded in service.

(b) *For reciprocating engine powered airplanes.* To provide engine isolation, the powerplants must be arranged and isolated from each other to allow operation, in at least one configuration, so that the failure or malfunction of any engine, or of any system that can affect that engine, will not—

(1) Prevent the continued safe operation of the remaining engines; or

(2) Require immediate action by any crewmember for continued safe operation.

39. *Turbopropeller reversing systems.* (a) Turbopropeller reversing systems intended for ground operation must be designed so that no single failure or malfunction of the system will result in unwanted reverse thrust under any expected operating condition. Failure of structural elements need not be considered if the probability of this kind of failure is extremely remote.

(b) Turbopropeller reversing systems intended for in flight use must be designed so that no unsafe condition will result during normal operation of the system, or from any failure (or reasonably likely combination of failures) of the reversing system, under any anticipated condition of operation of the airplane. Failure of structural elements need not be considered if the probability of this kind of failure is extremely remote.

(c) Compliance with this section may be shown by failure analysis, testing, or both for propeller systems that allow propeller blades to move from the flight low-pitch position to a position that is substantially less than that at the normal flight low-pitch stop position. The analysis may include or be supported by the analysis made to show compliance with the type certification of the propeller and associated installation components. Credit will be given for pertinent analysis and testing completed by the engine and propeller manufacturers.

40. *Turbopropeller drag-limiting systems.* Turbopropeller drag-limiting systems must be designed so that no single failure or malfunction of any of the systems during normal or emergency operation results in propeller drag in excess of that for which the airplane was designed. Failure of structure elements of the drag-limiting systems need not be considered if the probability of this kind of failure is extremely remote.

41. *Turbine engine powerplant operating characteristics.* For turbopropeller powered airplanes, the turbine engine powerplant operating characteristics must be investigated in flight to determine that no adverse characteristics (such as stall, surge, or flameout) are present to a hazardous degree, during normal and emergency operation within the range of operating limitations of the airplane and of the engine.

42. *Fuel flow.* (a) For turbopropeller powered airplanes—

(1) The fuel system must provide for continuous supply of fuel to the engines for normal operation without interruption due to depletion of fuel in any tank other than the main tank; and

(2) The fuel flow rate for turbopropeller engine fuel pump systems must not be less than 125 percent of the fuel flow required to develop the standard sea level atmospheric conditions takeoff power selected and included as an operating limitation in the Airplane Flight Manual.

(b) For reciprocating engine powered airplanes, it is acceptable for the fuel flow rate for each pump system (main and reverse supply) to be 125 percent of the takeoff fuel consumption of the engine.

FUEL SYSTEM COMPONENTS

43. *Fuel pumps.* For turbopropeller powered airplanes, a reliable and independent power source must be provided for each pump used with turbine engines which do not have provisions for mechanically driving the main pumps. It must be demonstrated that the pump installations provide a reliability and durability equivalent to that in FAR 23.991(a).

44. *Fuel strainer or filter.* For turbopropeller powered airplanes, the following apply:

(a) There must be a fuel strainer or filter between the tank outlet and the fuel metering device of the engine. In addition, the fuel strainer or filter must be—

(1) Between the tank outlet and the engine driven positive displacement pump inlet, if there is an engine-driven positive displacement pump;

(2) Accessible for drainage and cleaning and, for the strainer screen, easily removable; and

(3) Mounted so that its weight is not supported by the connecting lines or by the inlet or outlet connections of the strainer or filter itself.

(b) Unless there are means in the fuel system to prevent the accumulation of ice on the filter, there must be means to automatically maintain the fuel-flow if ice-clogging of the filter occurs; and

(c) The fuel strainer or filter must be of adequate capacity (for operating limitations established to ensure proper service) and of appropriate mesh to ensure proper engine operation, with the fuel contaminated to a degree (for particle size and density) that can be reasonably expected in service. The degree of fuel filtering may not be less than that established for the engine type certification.

45. *Lightning strike protection.* Protection must be provided against the ignition of flammable vapors in the fuel vent system due to lightning strikes.

COOLING

46. *Cooling test procedures for turbopropeller powered airplanes.* (a) Turbopropeller powered airplanes must be shown to comply with FAR 23.1041 during takeoff, climb, en route, and landing stages of flight that correspond to the applicable performance requirements. The cooling tests must be conducted with the airplane in the configuration, and operating under the conditions that are critical relative to cooling during each stage of flight. For the cooling tests a temperature is "stabilized" when its rate of change is less than 2°F per minute.

(b) Temperatures must be stabilized under the conditions from which entry is made into each stage of flights being investigated unless the entry condition is not one during which component and engine fluid temperatures would stabilize, in which case, operation through the full entry condition must be conducted before entry into the stage of flight being investigated to allow temperatures to reach their natural levels at the time of entry. The takeoff cooling test must be preceded by a period during which the powerplant component and engine fluid temperatures are stabilized with the engines at ground idle.

(c) Cooling tests for each stage of flight must be continued until—

(1) The component and engine fluid temperatures stabilize;

(2) The stage of flight is completed; or

(3) An operating limitation is reached.

INDUCTION SYSTEM

47. *Air induction.* For turbopropeller powered airplanes—

(a) There must be means to prevent hazardous quantities of fuel leakage or overflow from drains, vents, or other components of flammable fluid systems from entering the engine intake systems; and

(b) The air inlet ducts must be located or protected so as to minimize the ingestion of foreign matter during takeoff, landing, and taxiing.

48. *Induction system icing protection.* For turbopropeller powered airplanes, each turbine engine must be able to operate throughout its flight power range without adverse effect on engine operation or serious loss of power or thrust, under the icing conditions specified in Appendix C of Part 25 of this chapter. In addition, there must be means to indicate to appropriate flight crewmembers the functioning of the powerplant ice protection system.

49. *Turbine engine bleed air systems.* Turbine engine bleed air systems of turbopropeller powered airplanes must be investigated to determine—

(a) That no hazard to the airplane will result if a duct rupture occurs. This condition must consider that a failure of the duct can occur anywhere between the engine port and the airplane bleed service; and

(b) That, if the bleed air system is used for direct cabin pressurization, it is not possible for hazardous contamination of the cabin air system to occur in event of lubrication system failure.

EXHAUST SYSTEM

50. *Exhaust system drains.* Turbopropeller engine exhaust systems having low spots or pockets must incorporate drains at those locations. These drains must discharge clear of the airplane in normal and ground attitudes to prevent the accumulation of fuel after the failure of an attempted engine start.

POWERPLANT CONTROLS AND ACCESSORIES

51. *Engine controls.* If throttles or power levers for turbopropeller powered airplanes are such that any position of these controls will reduce the fuel flow to the engine(s) below that necessary for satisfactory and safe idle operation of the engine while the airplane is in flight, a means must be provided to prevent inadvertent movement of the control into this position. The means provided must incorporate a positive lock or stop at this idle position and must require a separate and distinct operation by the crew to displace the control from the normal engine operating range.

52. *Reverse thrust controls.* For turbopropeller powered airplanes, the propeller reverse thrust controls must have a means to prevent their inadvertent operation. The means must have a positive lock or stop at the idle position and must require a separate and distinct operation by the crew to displace the control from the flight regime.

53. *Engine ignition systems.* Each turbopropeller airplane ignition system must be considered an essential electrical load.

54. *Powerplant accessories.* The powerplant accessories must meet FAR 23.1163, and if the continued rotation of any accessory remotely driven by the engine is hazardous when malfunctioning occurs, there must be means to prevent rotation without interfering with the continued operation of the engine.

POWERPLANT FIRE PROTECTION

55. *Fire detector system.* For turbopropeller powered airplanes, the following apply:

(a) There must be a means that ensures prompt detection of fire in the engine compartment. An overtemperature switch in each engine cooling air exit is an acceptable method of meeting this requirement.

(b) Each fire detector must be constructed and installed to withstand the vibration, inertia, and other loads to which it may be subjected in operation.

(c) No fire detector may be affected by any oil, water, other fluids, or fumes that might be present.

(d) There must be means to allow the flight crew to check, in flight, the functioning of each fire detector electric circuit.

(e) Wiring and other components of each fire detector system in a fire zone must be at least fire resistant.

56. *Fire protection, cowling and nacelle skin.* For reciprocating engine powered airplanes, the engine cowling must be designed and constructed so that no fire originating in the engine compartment can enter either through openings or by burn through, any other region where it would create additional hazards.

57. *Flammable fluid fire protection.* If flammable fluids or vapors might be liberated by the leakage of fluid systems in areas other than engine compartments, there must be means to—

(a) Prevent the ignition of those fluids or vapors by any other equipment; or

(b) Control any fire resulting from that ignition.

EQUIPMENT

58. *Powerplant instruments.* (a) The following are required for turbopropeller airplanes:

(1) The instruments required by FAR 23.1305(a) (1) through (4), (b)(2) and (4).

(2) A gas temperature indicator for each engine.

(3) Free air temperature indicator.

(4) A fuel flowmeter indicator for each engine.

(5) Oil pressure warning means for each engine.

(6) A torque indicator or adequate means for indicating power output for each engine.

(7) Fire warning indicator for each engine.

(8) A means to indicate when the propeller blade angle is below the low-pitch position corresponding to idle operation in flight.

(9) A means to indicate the functioning of the ice protection system for each engine.

(b) For turbopropeller powered airplanes, the turbopropeller blade position indicator must begin indicating when the blade has moved below the flight low-pitch position.

(c) The following instruments are required for reciprocating engine powered airplanes:

(1) The instruments required by FAR 23.1305.

(2) A cylinder head temperature indicator for each engine.

(3) A manifold pressure indicator for each engine.

SYSTEMS AND EQUIPMENTS

GENERAL

59. *Function and installation.* The systems and equipment of the airplane must meet FAR 23.1301, and the following:

(a) Each item of additional installed equipment must—

(1) Be of a kind and design appropriate to its intended function;

(2) Be labeled as to its identification, function, or operating limitations, or any applicable combination of these factors, unless misuse or inadvertent actuation cannot create a hazard;

(3) Be installed according to limitations specified for that equipment; and

(4) Function properly when installed.

(b) Systems and installations must be designed to safeguard against hazards to the aircraft in the event of their malfunction or failure.

(c) Where an installation, the functioning of which is necessary in showing compliance with the applicable requirements, requires a power supply, that installation must be considered an essential load on the power supply, and the power sources and the distribution system must be capable of supplying the following power loads in probable operation combinations and for probable durations:

(1) All essential loads after failure of any prime mover, power converter, or energy storage device.

(2) All essential loads after failure of any one engine on two-engine airplanes.

(3) In determining the probable operating combinations and durations of essential loads for the power failure conditions described in subparagraphs (1) and (2) of this paragraph, it is permissible to assume that the power loads are reduced in accordance with a monitoring procedure which is consistent with safety in the types of operations authorized.

60. *Ventilation.* The ventilation system of the airplane must meet FAR 23.831, and in addition, for pressurized aircraft, the ventilating air in flight crew and passenger compartments must be free of harmful or hazardous concentrations of gases and vapors in normal operation and in the event of reasonably probable failures or malfunctioning of the ventilating, heating, pressurization, or other systems, and equipment. If accumulation of hazardous quantities of smoke in the cockpit area is reasonably probable, smoke evacuation must be readily accomplished.

ELECTRICAL SYSTEMS AND EQUIPMENT

61. *General.* The electrical systems and equipment of the airplane must meet FAR 23.1351, and the following:

(a) *Electrical system capacity.* The required generating capacity, and number and kinds of power sources must—

(1) Be determined by an electrical load analysis; and

(2) Meet FAR 23.1301.

(b) *Generating system.* The generating system includes electrical power sources, main power busses, transmission cables, and associated control, regulation and protective devices. It must be designed so that—

(1) The system voltage and frequency (as applicable) at the terminals of all essential load equipment can be maintained within the limits for which the equipment is designed, during any probable operating conditions;

(2) System transients due to switching, fault clearing, or other causes do not make essential loads inoperative, and do not cause a smoke or fire hazard;

(3) There are means, accessible in flight to appropriate crewmembers, for the individual and collective disconnection of the electrical power sources from the system; and

(4) There are means to indicate to appropriate crewmembers the generating system quantities essential for the safe operation of the system, including the voltage and current supplied by each generator.

62. *Electrical equipment and installation.* Electrical equipment, controls, and wiring must be installed so that operation of any one unit or system of units will not adversely affect the simultaneous operation of any other electrical unit or system essential to the safe operation.

63. *Distribution system.* (a) For the purpose of complying with this section, the distribution system includes the distribution busses, their associated feeders, and each control and protective device.

(b) Each system must be designed so that essential load circuits can be supplied in the event of reasonably probable faults or open circuits, including faults in heavy current carrying cables.

(c) If two independent sources of electrical power for particular equipment or systems are required under this Appendix, their electrical energy supply must be ensured by means such as duplicate electrical equipment, throw-over switching, or multi-channel or loop circuits separately routed.

64. *Circuit protective devices.* The circuit protective devices for the electrical circuits of the airplane must meet FAR 23.1357, and in addition circuits for loads which are essential to safe operation must have individual and exclusive circuit protection.

Docket No. 25530 (53 FR 26152) Eff. 7/11/88; (Amdt. 135-26, Eff. 10/11/88)

Appendix B—Airplane flight recorder specifications

Parameters	Range	Installed system[1] minimum accuracy (to recovered data)	Sampling interval (per second)	Resolution[4] read out
Relative time (from recorded on prior to takeoff).	8 hr minimum	±0.125% per hour	1	1 sec.
Indicated airspeed.	V_{SO} to V_D (KIAS)	±5% or ±10 kts., whichever is greater. Resolution 2 kts. below 175 KIAS.	1	1%[3]
Altitude.	–1,000 ft. to max cert. alt. of A/C.	±100 to ±700 ft. (see Table 1, TSO C51-a).	1	25 to 150
Magnetic heading.	360°	±5°	1	1°
Vertical acceleration.	–3g to +6g	±0.2g in addition to ±0.3g maximum datum.	4 (or 1 per second where peaks, ref. to 1g are recorded).	0.03g
Longitudinal acceleration.	±1.0g	±1.5% max. range excluding datum error of ±5%.	2	0.01g
Pitch attitude.	100% if usable	±2°	1	0.8°
Roll attitude.	±60° or 100% of usable range, whichever is greater.	±2°	1	0.8°
Stabilizer trim position,	Full range	±3% unless higher uniquely required.	1	1%[3]
Or				
Pitch control position.	Full range	±3% unless higher uniquely required.	1	1%[3]

Parameters	Range	Installed system[1] minimum accuracy (to recovered data)	Sampling interval (per second)	Resolution[4] read out
Engine Power, Each Engine				
Fan or N_1 speed or EPR or cockpit indications used for aircraft certification.	Maximum range	±5%	1	1%[3]
Or				
Prop. speed and torque (sample once/sec as close together as practicable).			1 (prop speed), 1 (torque).	
Altitude rate[2] (need depends on altitude resolution).	±8,000 fpm	±10%. Resolution 250 fpm below 12,000 ft. indicated.	1	250 fpm below 12,000
Angle of attack[2] (need depends on altitude resolution).	–20° to 40° or of usable range.	±2°	1	0.8%[3]
Radio transmitter keying (discrete).	On/Off		1	
TE flaps (discrete or analog).	Each discrete position (U, D, T/O, AAP).		1	
	Or			
	Analog 0–100% range	±3°	1	1%[3]
LE flaps (discrete or analog).	Each discrete position (U, D, T/O, AAP).		1	
	Or			
	Analog 0–100% range	±3°	1	1%[3]
Thrust reverser, each engine (discrete).	Stowed or full reverse		1	
Spoiler/speedbrake (discrete).	Stowed or out		1	
Autopilot engaged (discrete).	Engaged or disengaged		1	

[1] When data sources are aircraft instruments (except altimeters) of acceptable quality to fly the aircraft the recording system excluding these sensors (but including all other characteristics of the recording system) shall contribute no more than half of the values in this column.
[2] If data from the altitude encoding altimeter (100 ft. resolution) is used, then either one of these parameters should also be recorded. If however, altitude is recorded at a minimum resolution of 25 feet, then these two parameters can be omitted.
[3] Percent of full range.
[4] This column applies to aircraft manufactured after October 11, 1991.

Docket No. 25530 (53 FR 26152) Eff. 7/11/88;
(Amdt. 135–26, Eff. 10/11/88)

Appendix C—Helicopter flight recorder specifications

Parameters	Range	Installed system[1] minimum accuracy (to recovered data)	Sampling interval (per second)	Resolution[3] read out
Relative time (from recorded on prior to takeoff).	8 hr minimum	±0.125% per hour	1	1 sec.
Indicated airspeed.	V_m in to V_D (KIAS) (minimum airspeed signal attainable with installed pilot-static system).	±5% or ±10 kts., whichever is greater.	1	1 kt.
Altitude.	–1,000 ft. to 20,000 ft. pressure altitude.	±100 to ±700 ft. (see Table 1, TSO C51-a).	1	25 to 150 ft.
Magnetic heading.	360°	±5°	1	1°
Vertical acceleration.	–3g to +6g	±0.2g in addition to ±0.3g maximum datum.	4 (or 1 per second where peaks, ref to 1g are recorded).	0.05g
Longitudinal acceleration.	±1.0g	±1.5% max. range excluding datum error of ±5%.	2	0.03g
Pitch attitude.	100% of usable range	±2°	1	0.8°
Roll attitude.	±60% or 100% of usable range, whichever is greater.	±2°	1	0.8°
Altitude rate.	±8,000 fpm	±10% resolution 250 fpm below 12,000 ft. indicated.	1	250 fpm below 12,000
Engine Power, Each Engine				
Main rotor speed.	Maximum range	±5%	1	1%[2]
Free or power turbine.	Maximum range	+5%	1	1%[2]
Engine torque.	Maximum range	±5%	1	1%[2]
Flight Control—Hydraulic Pressure				
Primary (discrete).	High/low		1	
Secondary—if applicable (discrete).	High/low		1	
Radio transmitter keying (discrete).	On/off		1	
Autopilot engaged (discrete).	Engaged or disengaged		1	
SAS status—engaged (discrete).	Engaged/disengaged		1	

Parameters	Range	Installed system[1] minimum accuracy (to recovered data)	Sampling icnterval (per second)	Resolution[3] read out
SAS fault status (discrete).	Fault/OK		1	
Flight Controls				
Collective.	Full range	±3%	2	1%[2]
Pedal position.	Full range	±3%	2	1%[2]
Lat. cyclic.	Full range	±3%	2	1%[2]
Long. cyclic.	Full range	±3%	2	1%[2]
Controllable stabilator position.	Full range	±3%	2	1%[2]

[1] When data sources are aircraft instruments (except altimeters) of acceptable quality to fly the aircraft the recording system excluding these sensors (but including all other characteristics of the recording system) shall contribute no more than half of the values in this column.
[2] Percent of full range.
[3] This column applies to aircraft manufactured after October 11, 1991.

Docket No. 25530 (53 FR 26152) Eff. 7/11/88;
(Amdt. 135–26, Eff. 10/11/88)

Appendix D—Airplane flight recorder specifications

Parameters	Range	Accuracy sensor input to DFDR readout	Sampling interval (per second)	Resolution[4] read out
Time (GMT or Frame Counter) (range 0 to 4095, sampled 1 per frame).	24 hrs	±0.125% per hour	0.25 (1 per 4 seconds).	1 sec.
Altitude.	–1,000 ft to max certificated altitude of aircraft.	±100 to ±700 ft (See Table 1, TSO)–C51a).	1	5' to 35'[1]
Airspeed.	50 KIAS to V_{SO1} to V_{SO} to 1.2 V_D.	±5%, ±3%	1	1 kt.
Heading.	360°	±2°	1	0.5°
Normal Acceleration (Vertical).	–3g to +6g	±1% of max range excluding datum error or ±5%.	8	0.01g
Pitch Attitude.	±75°	±2°	1	0.5°
Roll Attitude.	±180°	±2°	1	0.5°
Radio Transmitter Keying.	On-Off (Discrete)		1	
Thrust/Power on Each Engine.	Full range forward	±2%	1 (per engine)	0.2%[2]
Trailing Edge Flap or Cockpit Control Selection.	Full range or each discrete position.	±3° or as pilot's indicator	0.5	0.5%[2]
Leading Edge Flap on or Cockpit Control Selection.	Full range or each discrete position.	±3° or as pilot's indicator	0.5	0.5%[2]

Parameters	Range	Accuracy sensor input to DFDR readout	Sampling interval (per second)	Resolution[4] read out
Thrust Reverser Position.	Stowed, in transit, and reverse (discrete).		1 (per 4 seconds per engine).	
Ground Spoiler Position/ Speed Brake Selection.	Full range or each discrete position.	±2% unless higher accuracy uniquely required.	1	0.22[2]
Marker Beacon Passage.	Discrete		1	
Autopilot Engagement.	Discrete		1	
Longitudinal Acceleration.	±1g	±1.5% max range excluding datum error of ±5%.	4	0.01g
Pilot Input and/or Surface Position-Primary Controls (Pitch, Roll, Yaw)[3].	Full range	±2° unless higher accuracy uniquely required.	1	0.2%[2]
Lateral Acceleration.	±1g	±1.5% max range excluding datum error of ±5%.	4	0.01g
Pitch Trim Position.	Full range	±3% unless higher accuracy uniquely required.	1	0.3%[2]
Glideslope Deviation.	±400 Microamps	±3%	1	0.3%[2]
Localizer Deviation.	±400 Microamps	±3%	1	0.3%[2]
AFCS Mode and Engagement Status.	Discrete		1	
Radio Altitude.	–20 ft to 2,500 ft	±2 ft or ±3% whichever is greater below 500 ft and ±5% above 500 ft.	1	1 ft + 5%[2] above 500'
Master Warning.	Discrete		1	
Main Gear Squat Switch Status.	Discrete		1	
Angle of Attack (if recorded directly).	As installed	As installed	2	0.3%[2]
Outside Air Temperature or Total Air Temperature.	–50°C to +90°C	±2°C	0.5	0.3°C
Hydraulics, Each System Low Pressure.	Discrete		0.5	Or 0.5%[2]
Groundspeed.	As installed	Most accurate systems installed (IMS equipped aircraft only).	1	0.2%[2]
Drift angle.	When available. As installed.	As installed	4	
Wind Speed and Direction.	When available. As installed.	As installed	4	
Latitude and longitude.	When available. As installed.	As installed	4	
Brake pressure/Brake pedal position.	As installed	As installed	1	

Parameters	Range	Accuracy sensor input to DFDR readout	Sampling interval (per second)	Resolution[4] read out
EPR	As installed	As installed	1 (per engine)	
N1	As installed	As installed	1 (per engine)	
N2	As installed	As installed	1 (per engine)	
EGT	As installed	As installed	1 (per engine)	
Throttle Lever Position	As installed	As installed	1 (per engine)	
Fuel Flow	As installed	As installed	1 (per engine)	
TCAS:				
TA	As installed	As installed	1	
RA	As installed	As installed	1	
Sensitivity level (as selected by crew).	As installed	As installed	2	
GPWS (ground proximity warning system).	Discrete		1	
Landing gear or gear selector position.	Discrete		0.25 (1 per 4 seconds).	
DME 1 and 2 Distance.	0–200 NM;	As installed	0.25	1 mi.
Nav 1 and 2 Frequency Selection.	Full range	As installed	0.25	

If additional recording capacity is available, recording of the following parameters is recommended. The parameters are listed in order of significance:
[1] When altitude rate is recorded. Altitude rate must have sufficient resolution and sampling to permit the derivation of altitude to 5 feet.
[2] Percent of full range.
[3] For airplanes that can demonstrate the capability of deriving either the control input on control movement (one from the other) for all modes of operation and flight regimes, the "or" applies. For airplanes with non-mechanical control systems (fly-by-wire) the "and" applies. In airplanes with split surfaces, suitable combination of inputs is acceptable in lieu of recording each surface separately.
[4] This column applies to aircraft manufactured after October 11, 1991.

Docket No. 25530 (53 FR 26152) Eff. 7/11/88;
(Amdt. 135–26, Eff. 10/11/88)

Appendix E—Helicopter flight recorder specifications

Parameters	Range	Accuracy sensor input to DFDR readout	Sampling interval (per second)	Resolution[2] read out
Time (GMT).	24 hrs	±0.125% per hour	0.25 (1 per 4 seconds).	1 sec.
Altitude.	–1,000 ft to max certificated altitude of aircraft.	±100 to ±700 ft (See Table 1, TSO–C51a).	1	5' to 30'
Airspeed.	As the installed measuring system.	±3%	1	1 kt.
Heading.	360°	±2°	1	0.5°
Normal Acceleration (Vertical).	–3g to +6g	±1% of max range excluding datum error of ±5%.	8	0.01g

Parameters	Range	Accuracy sensor input to DFDR readout	Sampling interval (per second)	Resolution[2] read out
Pitch Attitude.	±75°	±2°	2	0.5°
Roll Attitude.	±180°	±2°	2	0.5°
Radio Transmitter Keying.	On-Off (Discrete)		1	0.25 sec.
Power in Each Engine: Free Power Turbine Speed *and* Engine Torque.	0–130% (power Turbine Speed) Full range (Torque).	±2%	1 speed 1 torque (per engine).	0.2%[1] to 0.4%[1]
Main Rotor Speed.	0–130%	±2%	2	0.3%[1]
Altitude Rate.	±6,000 ft/min	As installed	2	0.2%[1]
Pilot Input—Primary Controls (Collective, Longitudinal Cyclic, Lateral Cyclic, Pedal).	Full range	±3%	2	0.5%[1]
Flight Control Hydraulic Pressure Low.	Discrete, each circuit		1	
Flight Control Hydraulic Pressure Selector Switch Position, 1st and 2nd stage.	Discrete		1	
AFCS Mode and Engagement Status.	Discrete (5 bits necessary).		1	
Stability Augmentation System Engage.	Discrete		1	
SAS Fault Status.	Discrete		0.25	
Main Gearbox Temperature Low.	As installed	As installed	0.25	0.5%[1]
Main Gearbox Temperature High.	As installed	As installed	0.5	0.5%[1]
Controllable Stabilator Position.	Full Range	±3%	2	0.4%[1]
Longitudinal Acceleration.	±1g	±1.5% max range excluding datum error of ±5%.	4	0.01g
Lateral Acceleration.	±1g	±1.5% max range excluding datum of ±5%.	4	0.01g
Master Warning.	Discrete		1	
Nav 1 and 2 Frequency Selection.	Full Range	As installed	0.25	
Outside Air Temperature.	–50° C+90° C	±2° C	0.5	0.3 °C

[1] Percent of full range.
[2] This column applies to aircraft manufactured after October 11, 1991.

Docket No. 25530 (53 FR 26152) Eff. 7/11/88;
(Amdt. 135–26, Eff. 10/11/88)

Special Federal Aviation Regulation 36. Operations review program amendment no. 2A; development of major repair data

[1. *Definitions.* For purposes of this Special Federal Aviation Regulation—

(a) A product is an aircraft, airframe, aircraft engine, propeller, or appliance;

(b) An article is an airframe, powerplant, propeller, instrument, radio, or accessory; and

(c) A component is a part of a product or article.

2. *General.*

(a) Contrary to provisions of § 121.379(b) of the Federal Aviation Regulations notwithstanding, the holder of an air carrier operating or commercial operating certificate, or the holder of an air taxi operating certificate that operates large aircraft, that has been issued operations specifications for operations required to be conducted in accordance with 14 CFR Part 121, may perform a major repair on a product, as described in § 121.379(a), using technical data that have not been approved by the Administrator, and approve that product for return to service, if authorized in accordance with this Special Federal Aviation Regulation.

(b) Contrary to provisions of § 127.40(b) of the Federal Aviation Regulations notwithstanding, the holder of an air carrier operating certificate that has been issued operations specifications for operations required to be conducted in accordance with 14 CFR Part 127 may perform a major repair on a product as described in § 127.140(a), using technical data that have not been approved by the Administrator, and approve that product for return to service, if authorized in accordance with this Special Federal Aviation Regulation.

(c) Contrary to provisions of § 145.51 of the Federal Aviation Regulations notwithstanding, the holder of a domestic repair station certificate under 14 CFR Part 145 may perform a major repair on an article for which it is rated, using technical data not approved by the Administrator, and approve that article for return to service, if authorized in accordance with this Special Federal Aviation Regulation. If the certificate holder holds a rating limited to a component of a product or article, the holder may not, by virtue of this Special Federal Aviation Regulation, approve that product or article for return to service.

3. *Major repair data and return to service.*

(a) As referenced in section 2 of this Special Federal Aviation Regulation, a certificate holder may perform a major repair on a product or article using technical data that have not been approved by the Administrator, and approve that product or article for return to service, if the certificate holder—

(1) Has been issued an authorization under, and a procedures manual that complies with, Special Federal Aviation Regulation No. 36, effective on January 23, 1994;

(2) Has developed the technical data in accordance with the procedures manual;

(3) Has developed the technical data specifically for the product or article being repaired; and

(4) Has accomplished the repair in accordance with the procedures manual and the procedures approved by the Administrator for the certificate.

(b) For purposes of this section, an authorization holder may develop technical data to perform a major repair on a product or article and use that data to repair a subsequent product or article of the same type as long as the holder—

(1) Evaluates each subsequent repair and the technical data to determine that performing the subsequent repair with the same data will return the product or article to its original or properly altered condition, and that the repaired product or article conforms with applicable airworthiness requirements; and

(2) Records each evaluation in the records referenced in paragraph (a) of section 13 of this Special Federal Aviation Regulation.]

[4. *Application.* The applicant for an authorization under this Special Federal Aviation Regulation must submit an application, in writing and signed by an officer of the applicant, to the FAA Flight Standards District Office charged with the overall inspection of the applicant's operations under its certificate. The application must contain—

(a) If the applicant is

(1) The holder of an air carrier operating or commercial operating certificate, or the holder of an air taxi operating certificate that operates large aircraft, the—

(i) The applicant's certificate number; and

(ii) The specific product(s) the applicant is authorized to maintain under its certificate, operations specifications, and maintenance manual; or

(2) The holder of a domestic repair station certificate—

(i) The applicant's certificate number;

(ii) A copy of the applicant's operations specifications; and

(iii) The specific article(s) for which the applicant is rated;

(b) The name, signature, and title of each person for whom authorization to approve, on behalf of the authorization holder, the use of technical data for major repairs is requested; and

(c) The qualifications of the applicant's staff that show compliance with section 5 of this Special Federal Aviation Regulation.

5. *Eligibility.*

(a) To be eligible for an authorization under this Special Federal Aviation Regulation, the applicant, in addition to having the authority to repair products or articles must—

(1) Hold an air carrier, commercial, or air taxi operating certificate, and have been issued operations specifications for operations required to be conducted in accordance with 14 CFR Part 121 or 127, or § 135.2, or hold a domestic repair station certificate under 14 CFR Part 145;

(2) Have an adequate number of sufficiently trained personnel in the United States to develop data and repair the products that the applicant is authorized to maintain under its operating certificate or the articles for which it is rated under its domestic repair station certificate;

(3) Employ, or have available, a staff of engineering personnel that can determine compliance with the applicable airworthiness requirements of the Federal Aviation Regulations.

(b) At least one member of the staff required by paragraph (a)(3) of this section must—

(1) Have a thorough working knowledge of the applicable requirements of the Federal Aviation Regulations;

(2) Occupy a position on the applicant's staff that has the authority to establish a repair program that ensures that each repaired product or article meets the applicable requirements of the Federal Aviation Regulations;

(3) Have at least one year of satisfactory experience in processing engineering work, in direct contact with the FAA, for type certification or major repair projects; and

(4) Have at least eight years of aeronautical engineering experience (which may include the one year of experience in processing engineering work for type certification or major repair projects).

(c) The holder of an authorization issued under this Special Federal Aviation Regulation shall notify the Administrator within 48 hours of any change (including a change of personnel) that could affect the ability of the holder to meet the requirements of this Special Federal Aviation Regulation.]

[6. *Procedures Manual.*

(a) A certificate holder may not approve a product or article for return to service under section 2 of this Special Federal Aviation Regulation unless the holder—

(1) Has a procedures manual that has been approved by the Administrator as complying with paragraph (b) of this section; and

(2) Complies with the procedures contained in the procedures manual.

(b) The approved procedures manual must contain—

(1) The procedures for developing and determining the adequacy of technical data for major repairs;

(2) The identification (names, signatures, and responsibilities) of officials and of each staff member described in section 5 of this Special Federal Aviation Regulation who—

(i) Has the authority to make changes in procedures that require a revision to the procedures manual; and

(ii) Prepares or determines the adequacy of technical data, plans or conducts tests, and approves, on behalf of the authorization holder, test results; and(3) A "log of revisions" page that identifies each revised item, page, and date of revision, and contains the signature of the person approving the change for the Administrator.

(c) The holder of an authorization issued under this Special Federal Aviation Regulation may not approve a product or article for return to service after a change in staff necessary to meet the requirements of section 5 of this regulation or a change in procedures from those approved under paragraph (a) of this section, unless that change has been approved by the FAA and entered in the procedures manual.

7. *Duration of Authorization.* Each authorization issued under this Special Federal Aviation Regulation is effective from the date of issuance until January 23, 1999, unless it is earlier surrendered, suspended, revoked, or otherwise terminated. Upon termination of such authorization, the terminated authorization holder must:

(a) Surrender to the FAA all data developed pursuant to Special Federal Aviation Regulation No. 36; or

(b) Maintain indefinitely all data developed pursuant to Special Federal Aviation Regulation No. 36, and make that data available to the FAA for inspection upon request.

8. *Transferability.* An authorization issued under this special Federal Aviation Regulation is not transferable.

9. *Inspections.* Each holder of an authorization issued under this Special Federal Aviation Regulation and each applicant for an authorization must allow the Administrator to inspect its personnel, facilities, products and articles, and records upon request.

10. *Limits of Applicability.* An authorization issued under this Special Federal Aviation Regulation applies only to—

(a) A product that the air carrier, commercial, or air taxi operating certificate holder is authorized to maintain pursuant to its continuous airworthiness maintenance program or maintenance manual; or

(b) An article for which the domestic repair station certificate holder is rated. If the certificate holder is rated for a component of an article, the holder may not, in accordance with this Special Federal Aviation Regulation, approve that article for return to service.

11. *Additional Authorization Limitations.* Each holder of an authorization issued under this Special Federal Aviation Regulation must comply with any additional limitations prescribed by the Administrator and made a part of the authorization.]

[12. *Data Review and Service Experience.* If the Administrator finds that a product or article has been approved for return to service after a major repair has been performed under this Special Federal Aviation Regulation, that the product or article may not conform to the applicable airworthiness requirements or that an unsafe feature or characteristic of the product or article may exist, and that the nonconformance or unsafe feature or characteristic may be attributed to the repair performed, the holder of the authorization, upon notification by the Administrator, shall—

(a) Investigate the matter;

(b) Report to the Administrator the results of the investigation and any action proposed or taken; and

(c) If notified that an unsafe condition exists, provide within the time period stated by the Administrator, the information necessary for the FAA to issue an airworthiness directive under Part 39 of the Federal Aviation Regulations.

13. *Current Records.* Each holder of an authorization issued under this Special Federal Aviation Regulation shall maintain, at its facility, current records containing—

(a) For each product or article for which it has developed and used major repair data, a technical data file that includes all data and amendments thereto (including drawings, photographs, specifications, instructions, and reports) necessary to accomplish the major repair;

(b) A list of products or articles by make, model, manufacturer's serial number (including specific part numbers and serial numbers of components) and, if applicable, FAA Technical Standard Order (TSO) or Parts Manufacturer Approval (PMA) identification, that have been repaired under the authorization; and

(c) A file of information from all available sources on difficulties experienced with products and articles repaired under the authorization.

This Special Federal Aviation Regulation terminates January 23, 1999.]

[(SFAR 36-6, Eff. 1/23/94)]

Special Federal Aviation Regulation 38-2.

Index

1. Applicability.
(a)(1) Certificates.
(a)(2) Certification requirements.
(a)(3) Operating requirements.
(b) Operations conducted under more than one paragraph.
(c) Prohibition against operating without certificate or in violation of operations specifications.
2. Certificates and foreign air carrier operations specifications.
(a) Air Carrier Operating Certificate.
(b) Operating Certificate.
(c) Foreign air carrier operations specifications.
3. Operations specifications.
4. Air carriers and those commercial operators engaged in scheduled intrastate common carriage.
(a)(1) Airplanes, more than 30 seats/7,500 pounds payload, scheduled within 48 States,
(a)(2) Airplanes, more than 30 seats/7,500 pounds payload, scheduled outside 48 States.
(a)(3) Airplanes, more than 30 seats/7,500 pounds payload, not scheduled and all cargo.
(b) Airplanes, 30 seats or less/7,500 or less pounds payload.
(c) Rotorcraft, 30 seats or less/7,500 pounds or less payload.
(d) Rotorcraft, more than 30 seats/more than 7,500 pounds payload.
5. Operations conducted by a person who is not engaged in air carrier operations, but is engaged in passenger operations, cargo operations, or both, as a commercial operator.
(a) Airplanes, 20 or more seats/6,000 or more pounds payload.
(b) Airplanes, less than 20 seats/less than 6,000 pounds payload.
(c) Rotorcraft, 30 seats or less/7,500 pounds or less payload.
(d) Rotorcraft, more than 30 seats/more than 7,500 pounds payload.
6. Definitions.
(a) Terms in FAR.
(1) Domestic/flag/supplemental/commuter.
(2) ATCO.
(b) FAR references to:
(1) Domestic air carriers.
(2) Flag air carriers.
(3) Supplemental air carriers.
(4) Commuter air carriers.
(c) SFAR terms.
(1) Air carrier.
(2) Commercial operator.
(3) Foreign air carrier.
(4) Scheduled operations.
(5) Size of aircraft.
(6) Maximum payload capacity.
(7) Empty weight.
(8) Maximum zero fuel weight.
(9) Justifiable aircraft equipment.

Certification and operating requirements

Contrary provisions of Parts 121, 125, 127, 129, and 135 of the Federal Aviation Regulations notwithstanding—

1. *Applicability.*

(a) This Special Federal Aviation Regulation applies to persons conducting commercial passenger operations, cargo operations, or both, and prescribes—

(1) The types of operating certificates issued by the Federal Aviation Administration;

(2) The certification requirements an operator must meet in order to obtain and hold operations specifications for each type of operation conducted and each class and size of aircraft operated; and

(3) The operating requirements an operator must meet in conducting each type of operation and in operating each class and size of aircraft authorized in its operations specifications.

A person shall be issued only one certificate and all operations shall be conducted under that certificate, regardless of the type of operation or the class or size of aircraft operated. A person holding an air carrier operating certificate may not conduct any operations under the rules of Part 125.

(b) Persons conducting operations under more than one paragraph of this SFAR shall meet the certification requirements specified in each paragraph and shall conduct operations in compliance with the requirements of the Federal Aviation Regulations specified in each paragraph for the operation conducted under that paragraph.

(c) Except as provided under this SFAR, no person may operate as an air carrier or as a commercial operator without, or in violation of, a certificate and operations specifications issued under this SFAR.

2. *Certificates and foreign air carrier operations specifications.*

(a) Persons authorized to conduct operations as an air carrier will be issued an Air Carrier Operating Certificate.

(b) Persons who are not authorized to conduct air carrier operations, but who are authorized to conduct passenger, cargo, or both; operations as a commercial operator will be issued an Operating Certificate.

(c) FAA certificates are not issued to foreign air carriers. Persons authorized to conduct operations in the United States as a foreign air carrier who hold a permit issued under Section 402 of the Federal Aviation Act of 1958, as amended (49 U.S.C. 1372), or other appropriate economic or exemption authority issued by the appropriate agency of the United States of America will be issued operations specifications in accordance with the requirements of Part 129 and shall conduct their operations within the United States in accordance with those requirements.

3. *Operations specifications.*

The operations specifications associated with a certificate issued under paragraph 2(a) or (b) and the operations specifications issued under paragraph 2(c) of this SFAR will prescribe the authorizations, limitations and certain procedures under which each type of operation shall be conducted and each class and size of aircraft shall be operated.

4. *Air carriers, and those commercial operators engaged in scheduled intrastate common carriage.*

Each person who conducts operations as an air carrier or as a commercial operator engaged in scheduled intrastate common carriage of persons or property for compensation or hire in air commerce with—

(a) Airplanes having a passenger seating configuration of more than 30 seats, excluding any required crewmember seat, or a payload capacity of more than 7,500 pounds, shall comply with the certification requirements in Part 121, and conduct its—

(1) Scheduled operations within the 48 contiguous states of the United States and the District of Columbia, including routes that extend outside the United States that are specifically authorized by the Administrator, with those airplanes in accordance with the requirements of Part 121 applicable to domestic air carriers, and shall be issued operations specifications for those operations in accordance with those requirements.

(2) Scheduled operations to points outside the 48 contiguous states of the United States and the District of Columbia with those airplanes in accordance with the requirements of Part 121 applicable to flag air carriers, and shall be issued operations specifications for those operations in accordance with those requirements.

(3) All-cargo operations and operations that are not scheduled with those airplanes in accordance with the requirements of Part 121 applicable to supplemental air carriers, and shall be issued operations specifications for those operations in accordance with those requirements; except the Administrator may authorize those operations to be conducted under paragraph (4)(a)(1) or (2) of this paragraph.

(b) Airplanes having a maximum passenger seating configuration of 30 seats or less, excluding any required crewmember seat, and a maximum payload capacity of 7,500 pounds or less, shall comply with the certification requirements in Part 135, and conduct its operations with those airplanes in accordance with the requirements of Part 135, and shall be issued operations specifications for those operations in accordance with those requirements; except that the Administrator may authorize a person conducting operations in transport category airplanes to conduct those operations in accordance with the requirements of paragraph 4(a) of this paragraph.

(c) Rotorcraft having a maximum passenger seating configuration of 30 seats or less and a maximum payload capacity of 7,500 pounds or less shall comply with the certification requirements in Part 135, and conduct its operations with those aircraft in accordance with the requirements of Part 135, and shall be issued operations specifications for those operations in accordance with those requirements.

(d) Rotorcraft having a passenger seating configuration of more than 30 seats or a payload capacity of more than 7,500 pounds shall comply with the certification requirements in Part 135, and conduct its operations with those aircraft in accordance with the requirements of Part 135, and shall be issued special operations specifications for those operations in accordance with those requirements and this SFAR.

5. *Operations conducted by a person who is not engaged in air carrier operations, but is engaged in passenger operations, cargo operations, or both, as a commercial operator.*

Each person, other than a person conducting operations under paragraph 2(c) or 4 of this SFAR, who conducts operations with—

(a) Airplanes having a passenger seating configuration of 20 or more, excluding any required crewmember seat, or a maximum payload capacity of 6,000 pounds or more, shall comply with the certification requirements in Part 125, and conduct its operations with those airplanes in accordance with the requirements of Part 125, and shall be issued operations specifications in accordance with those requirements, or shall comply with an appropriate deviation authority.

(b) Airplanes having a maximum passenger seating configuration of less than 20 seats, excluding any required crewmember seat, and a maximum payload capacity of less than 6,000 pounds shall comply with the certification requirements in Part 135, and conduct its operations in those airplanes in accordance with the requirements of Part 135, and shall be issued operations specifications in accordance with those requirements.

(c) Rotorcraft having a maximum passenger seating configuration of 30 seats or less and a maximum payload capacity of 7,500 pounds or less shall comply with the certification requirements in Part 135, and conduct its operations in those aircraft in accordance with the requirements of Part 135, and shall be issued operations specifications for those operations in accordance with those requirements.

(d) Rotorcraft having a passenger seating configuration of more than 30 seats or a payload capacity of more than 7,500 pounds shall comply with the certification requirements in Part 135, and conduct its operations with those aircraft in accordance with the requirements of Part 135, and shall be issued special operations specifications for those operations in accordance with those requirements and this SFAR.

6. *Definitions.*

(a) Wherever in the Federal Aviation Regulations the terms—

(1) "Domestic air carrier operating certificate," "flag air carrier operating certificate," "supplemental air carrier operating certificate," or "commuter air carrier" (in the context of Air Carrier Operating Certificate) appears, it shall be deemed to mean an "Air Carrier Operating Certificate" issued and maintained under this SFAR.

(2) "ATCO operating certificate" appears, it shall be deemed to mean either an "Air Carrier Operating Certificate" or "Operating Certificate," as is appropriate to the context of the regulation. All other references to an operating certificate shall be deemed to mean an "Operating Certificate" issued under this SFAR unless the context indicates the reference is to an Air Carrier Operating Certificate.

(b) Wherever in the Federal Aviation Regulations a regulation applies to—

(1) "Domestic air carriers," it will be deemed to mean a regulation that applies to scheduled operations solely within the 48 contiguous states of the United States and the District of Columbia conducted by persons described in paragraph 4(a) (1) of this SFAR.

(2) "Flag air carriers," it will be deemed to mean a regulation that applies to scheduled operations to any point outside the 48 contiguous states of the United States and the District of Columbia conducted by persons described in paragraph 4(a)(2) of this SFAR.

(3) "Supplemental air carriers," it will be deemed to mean a regulation that applies to charter and all-cargo operations conducted by persons described in paragraph 4(a) (3) of this SFAR.

(4) "Commuter air carriers," it will be deemed to mean a regulation that applies to scheduled passenger carrying operations, with a frequency of operations of at least five round trips per week on at least one route between two or more points according to the published flight schedules, conducted by persons described in paragraph 4(b) or (c) of this SFAR. This definition does not apply to Part 93 of this chapter.

(c) For the purpose of this SFAR, the term—

(1) "Air carrier" means a person who meets the definition of an air carrier as defined in the Federal Aviation Act of 1958, as amended.

(2) "Commercial operator" means a person, other than an air carrier, who conducts operations in air commerce carrying persons or property for compensation or hire.

(3) "Foreign air carrier" means any person other than a citizen of the United States, who undertakes, whether directly or indirectly or by lease or any other arrangement, to engage in foreign air transportation.

(4) "Scheduled operations" means operations that are conducted in accordance with a published schedule for passenger operations which includes dates or times (or both) that is openly advertised or otherwise made readily available to the general public.

(5) "Size of aircraft" means an aircraft's size as determined by its seating configuration or payload capacity, or both.

(6) "Maximum payload capacity" means:

(i) For an aircraft for which a maximum zero fuel weight is prescribed in FAA technical specifications, the maximum zero fuel weight, less empty weight, less all justifiable aircraft equipment, and less the operating load (consisting of minimum flight crew, foods and beverages, and supplies and equipment related to foods and beverages, but not including disposable fuel or oil).

(ii) For all other aircraft, the maximum certificated takeoff weight of an aircraft, less the empty weight, less all justifiable aircraft equipment, and less the operating load (consisting of minimum fuel load, oil, and flightcrew). The allowance for the weight of the crew, oil, and fuel is as follows:

(A) Crew—200 pounds for each crewmember required by the Federal Aviation Regulations.

(B) Oil—350 pounds.

(C) Fuel—the minimum weight of fuel required by the applicable Federal Aviation Regulations for a flight between domestic points 174 nautical miles apart under VFR weather conditions that does not involve extended overwater operations.

(7) "Empty weight" means the weight of the airframe, engines, propellers, rotors, and fixed equipment. Empty weight excludes the weight of the crew and payload, but includes the weight of all fixed ballast, unusable fuel supply, undrainable oil, total quantity of engine coolant, and total quantity of hydraulic fluid.

(8) "Maximum zero fuel weight" means the maximum permissible weight of an aircraft with no disposable fuel or oil. The zero fuel weight figure may be found in either the aircraft type certificate data sheet, or the approved Aircraft Flight Manual, or both.

(9) "Justifiable aircraft equipment" means any equipment necessary for the operation of the aircraft. It does not include equipment or ballast specifically installed, permanently or otherwise, for the purpose of altering the empty weight of an aircraft to meet the maximum payload capacity.

This Special Federal Aviation Regulation No. 38-2 terminates June 1, [1993], or the effective date of the codification of SFAR 38-2 into the Federal Aviation Regulations, whichever occurs first.

Special Federal Aviation Regulation 50-2. Special flight rules in the vicinity of the Grand Canyon National Park, AZ

Section 1. *Applicability.* This rule prescribes special operating rules for all persons operating aircraft in the following airspace, designated as the Grand Canyon National Park Special Flight Rules Area:

That airspace extending upward from the surface up to but not including 14,500 feet MSL within an area bounded by a line beginning at lat. 36°09'30" N., long. 114°03'00" W.; northeast to lat. 36°14'00" N., long. 113°09'50" W.; thence northeast along the boundary of the Grand Canyon National Park to lat. 36°24'47" N., long. 112°52'00" W.; to lat. 36°30'30" N., long. 112°36'15" W.; to lat. 36°21'30" N., long. 112°00'00" W.; to lat. 36°35'30" N., long. 111°53'10" W.; to lat 36°53'00" N., long. 111°36'45" W.; to lat 36°53'00" N., long. 111°33'00" W.; lat. 36°19'00" N., long. 111°50'50" W.; to lat. 36°17'00" N., long. 111°42'00" W.; to lat. 35°59'30" N., long. 111°42'00" W; to lat. 35°57'30" N., long. 112°03'55" W.; thence counterclockwise via the 5-statute mile radius of the Grand Canyon Airport reference point(lat. 35°57'09" N., long. 112°08'47" W.) to lat. 35°57'30" N., long. 112°14'00" W.; to lat. 35°57'30" N., long. 113°11'00" W.; to lat. 35°42'30" N., long. 113°11'00" W.; to lat. 35°38'30" N., long. 113°27'30" W.; thence counterclockwise via the 5-statute mile radius of the Peach Springs VORTAC to lat. 35°41'20" N., long. 113°36'00" W; to lat. 35°55'25" N., long. 113°49'10" W.; to lat. 35°57'45" N., long. 113°45'20" W.; thence northwest along the park boundary to lat. 36°02'20" N., long. 113°50'15" W.; to lat. 36°00'10" N., long. 113°53'45" W.; thence to the point of beginning.

Section 2. *Definitions.* For the purposes of this special regulation:

"Flight Standards District Office" means the FAA Flight Standards District Office with jurisdiction for the geographical area containing the Grand Canyon.

"Park" means the Grand Canyon National Park.

"Special Flight Rules Area" means the Grand Canyon National Park Special Flight Rules Area.

Section 3. *Aircraft operations: general.* Except in an emergency, no person may operate an aircraft in the Special Flight Rules Area under VFR on or after September 22, 1988, or under IFR on or after April 6, 1989, unless the operation—

(a) Is conducted in accordance with the following procedures:

NOTE.—The following procedures do not relieve the pilot from see-and-avoid responsibility or compliance with FAR 91.119.

(1) Unless necessary to maintain a safe distance from other aircraft or terrain—

(i) remain clear of the areas described in Section 4; and

(ii) remain at or above the following altitudes in each sector of the canyon:

Eastern section from Lees Ferry to North Canyon and North Canyon to Boundary Ridge: as prescribed in Section 5.

Boundary Ridge to Supai Point (Yumtheska Point): 10,000 feet MSL.

Supai Point to Diamond Creek: 9,000 feet MSL.

Western section from Diamond Creek to the Grand Wash Cliffs: 8,000 feet MSL.

(2) Proceed through the four flight corridors described in Section 4 at the following altitudes unless otherwise authorized in writing by the Flight Standards District Office:

Northbound: 11,500 or 13,500 feet MSL

Southbound: 10,500 or 12,500 feet MSL

(b) Is authorized in writing by the Flight Standard District Office and is conducted in compliance with the conditions contained in that authorization. Normally authorization will be granted for operation in the area described in Section 4 or below the altitudes listed in Section 5 only for operations of aircraft necessary for law enforcement, firefighting, emergency medical treatment/evacuation of persons in the vicinity of the Park; for support of Park maintenance or activities; or for aerial access to and maintenance of other property located within the Special Flight Rules Area. Authorization may be issued on a continuing basis.

(c)(1) Prior to November 1, 1988, is conducted in accordance with a specific authorization to operate in that airspace incorporated in the operator's Part 135 operations specifications in accordance with the provisions of SFAR 50-1, notwithstanding the provisions of Sections 4 and 5; and

(2) On or after November 1, 1988, is conducted in accordance with a specific authorization to operate in that airspace incorporated in the operator's Part 135 operations specifications and approved by the Flight Standards District Office in accordance with the provisions of SFAR 50-2.

(d) Is a search and rescue mission directed by the U.S. Air Force Rescue Coordination Center.

(e) Is conducted within 3 nautical miles of Whitmore Airstrip, Pearce Ferry Airstrip, North Rim Airstrip, Cliff Dwellers Airstrip, or Marble Canyon Airstrip at an altitude less than 3,000 feet above airport elevation, for the purpose of landing at or taking off from that facility. *Or*

(f) Is conducted under an IFR clearance and the pilot is acting in accordance with ATC instructions. An IFR flight plan may not be filed on a route or at an altitude that would require operation in an area described in Section 4.

Section 4. *Flight-free zones.* Except in an emergency or if otherwise necessary for safety of flight, or unless otherwise authorized by the Flight Standards District Office for a purpose listed in Section 3(b), no person may operate an aircraft in the Special Flight Rules Area within the following areas:

(a) *Desert View Flight-Free Zone.* Within an area bounded by a line beginning at Lat. 35°59'30" N., Long. 111°46'20" W.; to 35°59'30" N., Long. 111°52'45" W.; to Lat. 36°04'50" N., Long. 111°52'00" W.; to Lat. 36°06'00" N., Long. 111°46'20" W.; to the point of origin; but not including the airspace at and above 10,500 feet MSL within 1 mile of the western boundary of the zone. The area between the Desert View and Bright Angel Flight-Free Zones is designated the "Zuni Point Corridor."

(b) *Bright Angel Flight-Free zone.* Within an area bounded by a line beginning at Lat. 35°59'30" N, Long. 111°55'30" W.; to Lat. 35°59'30" N., Long. 112°04'00" W.; thence counterclockwise via the 5-statute-mile radius of the Grand Canyon Airport point (Lat. 35°57'09" N., Long. 112°08'47" W.) to Lat. 36°01'30" N., Long. 112°11'00" W.; to Lat. 36°06'15" N., Long. 112°12'50" W.; to Lat. 36°14'40" N., Long. 112°08'50" W.; to Lat. 36°14'40" N., Long. 111°57'30" W.; to Lat. 36°12'30" N., Long. 111°53'50" W.; to the point of origin; but not including the airspace at and above 10,500 feet MSL within 1 mile of the eastern boundary between the southern boundary and Lat. 36°04'50" N. or the airspace at and above 10,500 feet MSL within 2 miles of the northwest boundary. The area bounded by the Bright Angel and Shinumo Flight-Free Zones is designated the "Dragon Corridor."

(c) *Shinumo Flight-Free Zone.* Within an area bounded by a line beginning at Lat. 36°4'00" N., Long. 112'16'40" W.; northwest along the park boundary to a point at Lat. 36°12'47" N., Long. 112°30'53" W.; to Lat. 36°21'15" N., Long. 112°20'20" W.; east along the park boundary to Lat. 36°21'15" N., Long. 112°13'55" W.; to Lat. 36°14'40" N., Long. 112°11'25" W.; to the point of origin. The area between the Thunder River/Toroweap and Shinumo Flight-Free Zones is designated the "Fossil Canyon Corridor."

(d) *Toroweap/Thunder River Flight-Free Zone.* Within an area bounded by a line beginning at Lat. 36°22'45" N., Long. 112°20'35" W.; thence northwest along the boundary of the Grand Canyon National Park to Lat. 36°17'48" N., Long. 113°09'15" W.; to Lat. 36°15'00" N., Long. 113°07'10" W.; to Lat. 36°10'00" N., Long. 113°07'10" W.; thence east along the Colorado River to the confluence of Havasu Canyon (Lat. 36°18'40" N., Long. 112°45'45" W.); including that area within a 1.5-nautical-mile radius of Toroweap Overlook (Lat. 36°12'45" N., Long. 113°03'30" W.) to the point of origin; but not including the following airspace designated as the "Tuckup Corridor": at or above 10,500 feet MSL within 2 nautical miles either side of a line extending between Lat. 36°24'47" N., Long. 112°48'50" W.; and Lat. 36°17'10" N., Long. 112°48'50" W.; to the point of origin.

Section 5. *Minimum flight altitudes.* Except in an emergency or if otherwise necessary for safety of flight, or unless otherwise authorized by the Flight Standards District Office for a purpose listed in Section 3(b), no person may operate an aircraft in the Special Flight Rules Area at an altitude lower than the following:

(a) Eastern section from Lees Ferry to North Canyon: 5,000 feet MSL.

(b) Eastern section from North Canyon to Boundary Ridge: 6,000 feet MSL.

(c) Boundary Ridge to Supai (Yumtheska) Point: 7,500 feet MSL.

(d) Supai Point to Diamond Creek: 6,500 feet MSL.

(e) Western section from Diamond Creek to the Grand Wash Cliffs: 5,000 feet MSL.

Section 6. *Commercial sightseeing flights.*

(a) Notwithstanding the provisions of Federal Aviation Regulations § 135. 1 (b)(2), nonstop sightseeing flights that begin and end at the same airport, are conducted within a 25-statute-mile radius of that airport, and operate in or through the Special Flight Rules Area during any portion of the flight are governed by the provisions of Part 135.

(b) No person holding or required to hold an operating certificate under Part 135 may operate an aircraft in the Special Flight Rules Area except as authorized by operations specifications issued under that part.

Section 7. *Minimum terrain clearance.* Except in an emergency, when necessary for takeoff or landing, or unless authorized by the Flight Standards District Office for a purpose listed in Section 3(b), no person may operate an aircraft within 500 feet of any terrain or structure located between the north and south rims of the Grand Canyon.

Section 8. *Communications.* Except when in contact with the Grand Canyon National Park Airport Traffic Control Tower during arrival or departure or on a search and rescue mission directed by the U.S. Air Force Rescue Coordination Center, no person may operate an aircraft in the Special Flight Rules Area unless he monitors the appropriate frequency continuously while in that airspace.

Section 9. *Termination date.* This Special Federal Aviation Regulation expires on June 15, [1995].

Authority: 49 U.S.C. 1303, 1348, 1354(a), 1421, and 1422; 16 U.S.C. 228g; Pub. L. 100-91, August 18, 1987 49 U.S.C. 106(g) (Revised Pub. L. 97-449, January 12, 1983).

[(SFAR 50-2, Eff. 6/15/92)]

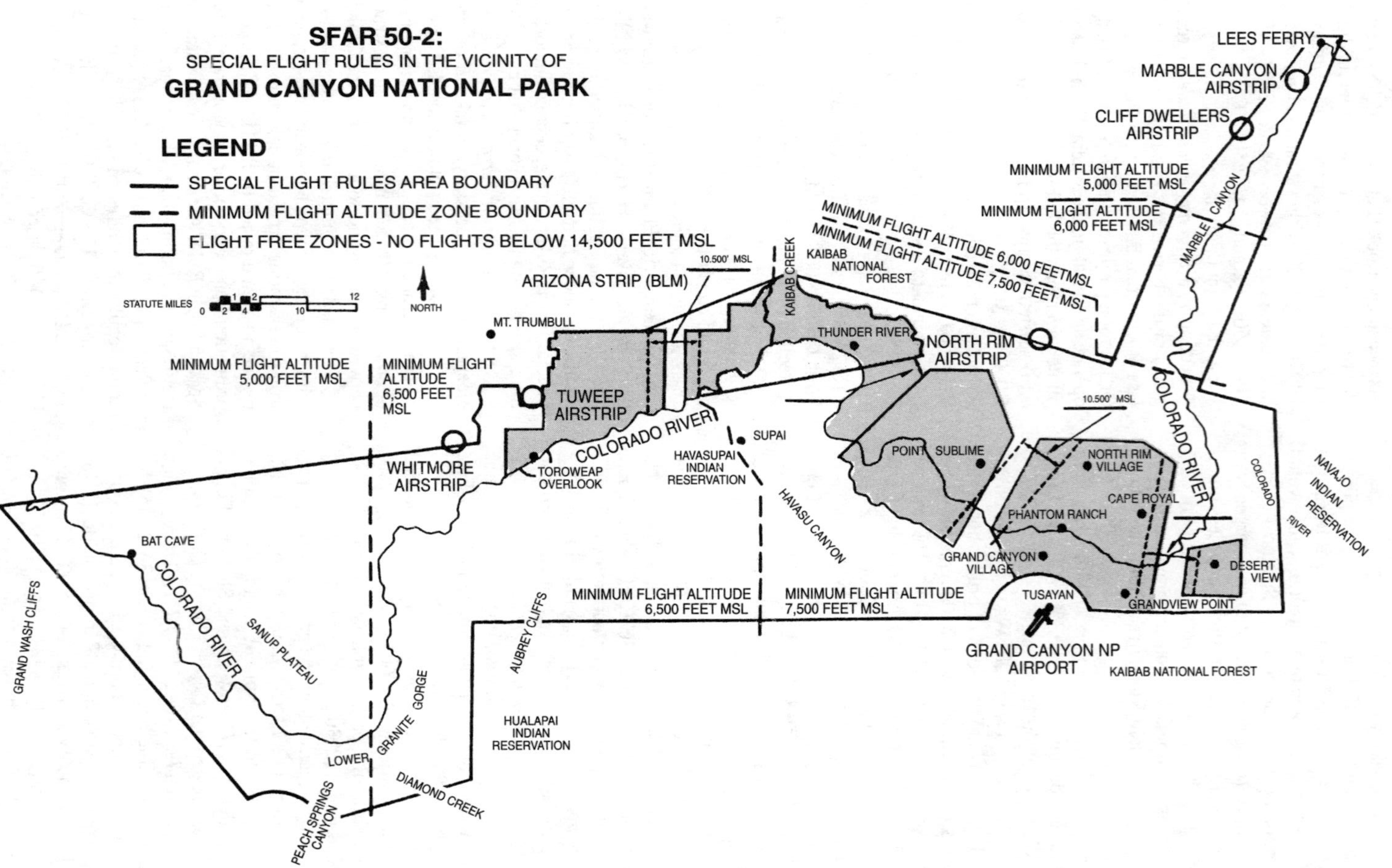
SFAR 50-2:
SPECIAL FLIGHT RULES IN THE VICINITY OF
GRAND CANYON NATIONAL PARK
LEGEND
SPECIAL FLIGHT RULES AREA BOUNDARY
MINIMUM FLIGHT ALTITUDE ZONE BOUNDARY
FLIGHT FREE ZONES - NO FLIGHTS BELOW 14,500 FEET MSL
STATUTE MILES 0 1 2 2 4 10 12
NORTH
LEES FERRY
MARBLE CANYON AIRSTRIP
CLIFF DWELLERS AIRSTRIP
MINIMUM FLIGHT ALTITUDE 5,000 FEET MSL
MINIMUM FLIGHT ALTITUDE 6,000 FEET MSL
MARBLE CANYON
MINIMUM FLIGHT ALTITUDE 6,000 FEETMSL
MINIMUM FLIGHT ALTITUDE 7,500 FEET MSL
KAIBAB NATIONAL FOREST
KAIBAB CREEK
10,500' MSL
ARIZONA STRIP (BLM)
MT. TRUMBULL
THUNDER RIVER
NORTH RIM AIRSTRIP
MINIMUM FLIGHT ALTITUDE 5,000 FEET MSL
MINIMUM FLIGHT ALTITUDE 6,500 FEET MSL
TUWEEP AIRSTRIP
COLORADO RIVER
10,500' MSL
COLORADO RIVER
COLORADO RIVER
NAVAJO INDIAN RESERVATION
SUPAI
POINT SUBLIME
NORTH RIM VILLAGE
CAPE ROYAL
WHITMORE AIRSTRIP
TOROWEAP OVERLOOK
HAVASUPAI INDIAN RESERVATION
HAVASU CANYON
PHANTOM RANCH
GRAND CANYON VILLAGE
DESERT VIEW
BAT CAVE
COLORADO RIVER
GRAND WASH CLIFFS
SANUP PLATEAU
MINIMUM FLIGHT ALTITUDE 6,500 FEET MSL
MINIMUM FLIGHT ALTITUDE 7,500 FEET MSL
TUSAYAN
GRANDVIEW POINT
AUBREY CLIFFS
GRAND CANYON NP AIRPORT
KAIBAB NATIONAL FOREST
LOWER GRANITE GORGE
HUALAPAI INDIAN RESERVATION
DIAMOND CREEK
PEACH SPRINGS CANYON

Special Federal Aviation Regulation 58. Advanced qualification program

1. *Purpose and Eligibility.*

(a) This Special Federal Aviation Regulation provides for approval of an alternate method (known as "Advanced Qualification Program" or "AQP") for qualifying, training, certifying, and otherwise ensuring competency of crewmembers, aircraft dispatchers, other operations personnel, instructors, and evaluators who are required to be trained or qualified under Parts 121 and 135 of the FAR or under this SFAR.

(b) A certificate holder is eligible under this Special Federal Aviation Regulation if the certificate holder is required to have an approved training program under § 121.401 or § 135.341 of the FAR, or elects to have an approved training program under § 135.341.

(c) A certificate holder obtains approval of each proposed curriculum under this AQP as specified in section 10 of this SFAR.

(d) A curriculum approved under the AQP may include elements of present Part 121 and Part 135 training programs. Each curriculum must specify the make, model, and series aircraft (or variant) and each crewmember position or other positions to be covered by that curriculum. Positions to be covered by the AQP must include all flight crewmember positions, instructors, and evaluators and may include other positions, such as flight attendants, aircraft dispatchers, and other operations personnel.

(e) Each certificate holder that obtains approval of an AQP under this SFAR shall comply with all of the requirements of that program.

2. *Definitions.*

As used in this SFAR:

"Curriculum" means a portion of an Advanced Qualification Program that covers one of three program areas: (1) indoctrination, (2) qualification, or (3) continuing qualification. A qualification or continuing qualification curriculum addresses the required training and qualification activities for a specific make, model, and series aircraft (or variant) and for a specific duty position.

"Evaluator" means a person who has satisfactorily completed training and evaluation that qualifies that person to evaluate the performance of crewmembers, instructors, other evaluators, aircraft dispatchers, and other operations personnel.

"Facility" means the physical environment required for training and qualification (e.g., buildings, classrooms).

"Training center" means an independent organization that provides training under contract or other arrangement to certificate holders. A training center may be a certificate holder that provides training to another certificate holder, an aircraft manufacturer that provides training to certificate holders, or any non-certificate holder that provides training to a certificate holder.

"Variant" means a specifically configured aircraft for which the FAA has identified training and qualification requirements that are significantly different from those applicable to other aircraft of the same make, model, and series.

3. *Required Curriculums.*

Each AQP must have separate curriculums for indoctrination, qualification, and continuing qualification as specified in §§ 4, 5, and 6 of this SFAR.

4. *Indoctrination Curriculums.*

Each indoctrination curriculum must include the following:

(a) For newly hired persons being trained under an AQP: Company policies and operating practices and general operational knowledge.

(b) For newly hired flight crewmembers and aircraft dispatchers: General aeronautical knowledge.

(c) For instructors: The fundamental principles of the teaching and learning process; methods and theories of instruction; and the knowledge necessary to use aircraft, flight training devices, flight simulators, and other training equipment in advanced qualification curriculums.

(d) For evaluators: Evaluation requirements specified in each approved curriculum; methods of evaluating crewmembers and aircraft dispatchers and other operations personnel; and policies and practices used to conduct the kinds of evaluations particular to an advanced qualification curriculum (e.g., proficiency and on-line).

5. *Qualification Curriculums.*

Each qualification curriculum must include the following:

(a) The certificate holder's planned hours of training, evaluation, and supervised operating experience.

(b) A list of and text describing the training, qualification, and certification activities, as applicable for specific positions subject to the AQP, as follows:

(1) *Crewmembers, aircraft dispatchers, and other operations personnel.* Training, evaluation, and certification activities which are aircraft- and equipment-specific to qualify a person for a particular duty position on, or duties related to the operation of a specific make, model, and series aircraft (or variant); a list of and text describing the knowledge requirements, subject materials, job skills, and each maneuver and procedure to be trained and evaluated; the practical test requirements in addition to or in place of the requirements of Parts 61, 63, and 65; and a list of and text describing supervised operating experience.

(2) *Instructors.* Training and evaluation to qualify a person to impart instruction on how to operate, or on how to ensure the safe operation of a particular make, model, and series aircraft (or variant).

(3) *Evaluators.* Training, evaluation, and certification activities that are aircraft and equipment specific to qualify a person to evaluate the performance of persons who operate or who ensure the safe operation of, a particular make, model, and series aircraft (or variant).

6. *Continuing Qualification Curriculums.*

Continuing qualification curriculums must comply with the following requirements:

(a) *General.* A continuing qualification curriculum must be based on—

(1) A continuing qualification cycle that ensures that during each cycle each person qualified under an AQP, including instructors and evaluators, will receive a balanced mix of training and evaluation on all events and subjects necessary to ensure that each person maintains the minimum proficiency level of knowledge, skills, and attitudes required for original qualification; and

(2) If applicable, flight crewmember or aircraft dispatcher recency of experience requirements.

(b) *Continuing Qualification Cycle Content.* Each continuing qualification cycle must include at least the following:

(1) *Evaluation period.* An evaluation period during which each person qualified under an AQP must receive at least one training session and a proficiency evaluation at a training facility. The number and frequency of training sessions must be approved by the Administrator. A training session, including any proficiency evaluation completed at that session, that occurs any time during the two calendar months before the last date for completion of an evaluation period can be considered by the certificate holder to be completed in the last calendar month.

(2) *Training.* Continuing qualification must include training in all events and major subjects required for original qualification, as follows:

(i) For pilots in command, seconds in command, flight engineers, and instructors and evaluators: Ground training including a general review of knowledge and skills covered in qualification training, updated information on newly developed procedures, and safety information.

(ii) For crewmembers, aircraft dispatchers, instructors, evaluators, and other operation personnel who conduct their duties in flight: Proficiency training in an aircraft, flight training device, or flight simulator on normal, abnormal, and emergency flight procedures and maneuvers.

(iii) For instructors and evaluators who are limited to conducting their duties in flight simulators and flight training devices: Proficiency training in a flight training device and/or flight simulator regarding operation of this training equipment and in operational flight procedures and maneuvers (normal, abnormal, and emergency).

(3) *Evaluations.* Continuing qualification must include evaluation in all events and major subjects required for original qualification, and online evaluations for pilots in command and other eligible flight crewmembers. Each person qualified under an AQP must successfully complete a proficiency evaluation and, if applicable, an online evaluation during each evaluation period. An individual's proficiency evaluation may be accomplished over several training sessions if a certificate holder provides more than one training session in an evaluation period. The following evaluation requirements apply:

(i) Proficiency evaluations as follows:

(A) For pilots in command, seconds in command, and flight engineers: A proficiency evaluation, portions of which may be conducted in an aircraft, flight simulator, or flight training device as approved in the certificate holder's curriculum which must be completed during each evaluation period.

(B) For any other persons covered by an AQP a means to evaluate their proficiency in the performance of their duties in their assigned tasks in an operational setting.

(ii) On-line evaluations as follows:

(A) For pilots in command: An on line evaluation conducted in an aircraft during actual flight operations under Part 121 or Part 135 or during operationally (line) oriented flights, such as ferry flights or proving flights. An on-line evaluation in an aircraft must be completed in the calendar month that includes the midpoint of the evaluation period. An on-line evaluation that is satisfactorily completed in the calendar month before or the calendar month after the calendar month in which it becomes due is considered to have been completed during the calendar month it became due. However, in no case is an on-line evaluation under this paragraph required more often than once during an evaluation period.

(B) During the on-line evaluations required under paragraph (5)(3)(ii)(A) of this section, each person performing duties as a pilot in command, second in command, or flight engineer for that flight, must be individually evaluated to determine whether he or she—(1) Remains adequately trained and currently proficient with respect to the particular aircraft, crew position, and type of operation in which he or she serves; and (2) Has sufficient knowledge and skills to operate effectively as part of a crew.

(4) *Recency of experience.* For pilots in command and seconds in command, and, if the certificate holder elects, flight engineers and aircraft dispatchers, approved recency of experience requirements.

(c) *Duration periods.* Initially the continuing qualification cycle approved for an AQP may not exceed 26 calendar months and the evaluation period may not exceed 13 calendar months. Thereafter, upon demonstration by a certificate holder that an extension is warranted, the Administrator may approve extensions of the continuing qualification cycle and the evaluation period in increments not exceeding 3 calendar months. However, a continuing qualification cycle may not exceed 39 calendar months and an evaluation period may not exceed 26 calendar months.

(d) *Requalification.* Each continuing qualification curriculum must include a curriculum segment that covers the requirements for requalifying a crewmember, aircraft dispatcher, or other operations personnel who has not maintained continuing qualification.

7. *Other Requirements.* In addition to the requirements

of sections 4, 5, and 6, each AQP qualification and continuing qualification curriculum must include the following requirements:

(a) Approved Cockpit Resource Management (CRM) Training applicable to each position for which training is provided under an AQP.

(b) Approved training on and evaluation of skills and proficiency of each person being trained under an AQP to use their cockpit resource management skills and their technical (piloting or other) skills in an actual or simulated operations scenario. For flight crewmembers this training and evaluation must be conducted in an approved flight training device or flight simulator.

(c) Data collection procedures that will ensure that the certificate holder provides information from its crewmembers, instructors, and evaluators that will enable the FAA to determine whether the training and evaluations are working to accomplish the overall objectives of the curriculum.

8. *Certification.* A person enrolled in an AQP is eligible to receive a commercial or airline transport pilot, flight engineer, or aircraft dispatcher certificate or appropriate rating based on the successful completion of training and evaluation events accomplished under that program if the following requirements are met:

(a) Training and evaluation of required knowledge and skills under the AQP must meet minimum certification and rating criteria established by the Administrator in Parts 61, 63, or 65. The Administrator may accept substitutes for the practical test requirements of Parts 61, 63, or 65, as applicable.

(b) The applicant satisfactorily completes the appropriate qualification curriculum.

(c) The applicant shows competence in required technical knowledge and skills (e.g., piloting) and cockpit resource management knowledge and skills in scenarios that test both types of knowledge and skills together.

(d) The applicant is otherwise eligible under the applicable requirements of Part 61, 63, or 65.

9. *Training Devices and Simulators.*

(a) *Qualification and approval of flight training devices and flight simulators.*

(1) Any training device or simulator that will be used in an AQP for one of the following purposes must be evaluated by the Administrator for assignment of a flight training device or flight simulator qualification level:

(i) Required evaluation of individual or crew proficiency.

(ii) Training activities that determine if an individual or crew is ready for a proficiency evaluation.

(iii) Activities used to meet recency of experience requirements.

(iv) Line Operational Simulations (LOS).

(2) To be eligible to request evaluation for a qualification level of a flight training device or flight simulator an applicant must—

(i) Hold an operating certificate; or

(ii) Be a training center that has applied for authorization to the Administrator or has been authorized by the Administrator to conduct training or qualification under an AQP.

(3) Each flight training device or flight simulator to be used by a certificate holder or training center for any of the purposes set forth in paragraph (a)(1) of this section must—

(i) Be, or have been, evaluated against a set of criteria established by the Administrator for a particular qualification level of simulation;

(ii) Be approved for its intended use in a specified AQP; and

(iii) Be part of a flight simulator or flight training device continuing qualification program approved by the Administrator.

(b) *Approval of other Training Equipment.*

(1) Any training device that is intended to be used in an AQP for purposes other than those set forth in paragraph (a)(1) of this section must be approved by the Administrator for its intended use.

(2) An applicant for approval of a training device under this paragraph must identify the device by its nomenclature and describe its intended use.

(3) Each training device approved for use in an AQP must be part of a continuing program to provide for its serviceability and fitness to perform its intended function as approved by the Administrator.

10. *Approval Of Advanced Qualification Program.*

(a) *Approval Process.* Each applicant for approval of an AQP curriculum under this SFAR shall apply for approval of that curriculum. Application for approval is made to the certificate holder's FAA Flight Standards District Office.

(b) *Approval Criteria.* An application for approval of an AQP curriculum will be approved if the program meets the following requirements:

(1) It must be submitted in a form and manner acceptable to the Administrator.

(2) It must meet all of the requirements of this SFAR.

(3) It must indicate specifically the requirements of Parts 61, 63, 65, 121 or 135, as applicable, that would be replaced by an AQP curriculum. If a requirement of Parts 61, 63, 65, 121, or 135 is replaced by an AQP curriculum, the certificate holder must show how the AQP curriculum provides an equivalent level of safety for each requirement that is replaced. Each applicable requirement of Parts 61, 63, 65, 121 or 135 that is not specifically addressed in an AQP curriculum continues to apply to the certificate holder.

(c) *Application and Transition.* Each certificate holder that applies for one or more advanced qualification curriculums or for a revision to a previously approved curriculum must comply with § 121.405 or § 135.325, as applicable, and must include as part of its application a proposed transition plan (containing a calendar of events) for moving from its present approved training to the advanced qualification training.

(d) *Advanced Qualification Program Revisions or Rescissions of Approval.* If after a certificate holder begins operations under an AQP, the Administrator finds that the certificate holder is not meeting the provisions of its approved AQP, the Administrator may require the certificate holder to make revisions in accordance with § 121.405 or § 135.325, as applicable, or to submit and obtain approval for a Plan (containing a schedule of events) that the certificate holder must comply with and use to transition to an approved Part 121 or Part 135 training program, as appropriate.

11. *Approval of Training, Qualification, or Evaluation by a Person who Provides Training by Arrangement.*

(a) A certificate holder under Part 121 or Part 135 may arrange to have AQP required training, qualification, or evaluation functions performed by another person (a "training center") if the following requirements are met:

(1) The training center's training and qualification curriculums, curriculum segments, or portions of curriculum segments must be provisionally approved by the Administrator. A training center may apply for provisional approval independently or in conjunction with, a certificate application for AQP approval. Application for provisional approval must be made to the FAA's Flight Standards District Office that has responsibility for the training center.

(2) The specific use of provisionally approved curriculums, curriculum segments, or portions of curriculum segments in a certificate holder's AQP must be approved by the Administrator as set forth in Section 10 of this SFAR.

(b) An applicant for provisional approval of a curriculum, curriculum segment, or portion of a curriculum segment under this paragraph must show that the following requirements are met:

(1) The applicant must have a curriculum for the qualification and continuing qualification of each instructor or evaluator employed by the applicant.

(2) The applicant's facilities must be found by the Administrator to be adequate for any planned training, qualification, or evaluation for a Part 121 or Part 135 certificate holder.

(3) Except for indoctrination curriculums, the curriculum, curriculum segment, or portion of a curriculum segment must identify the specific make, model, and series aircraft (or variant) and crewmember or other positions for which it is designed.

(c) A certificate holder who wants approval to use a training center's provisionally approved curriculum, curriculum segment, or portion of a curriculum segment in its AQP, must show that the following requirements are met:

(1) Each instructor or evaluator used by the training center must meet all of the qualification and continuing qualification requirements that apply to employees of the certificate holder that has arranged for the training, including knowledge of the certificate holder's operations.

(2) Each provisionally approved curriculum, curriculum segment, or portion of a curriculum segment must be approved by the Administrator for use in the certificate holder's AQP. The Administrator will either provide approval or require modifications to ensure that each curriculum, curriculum segment, or portion of a curriculum segment is applicable to the certificate holder's AQP.

12. *Recordkeeping Requirements.*

Each certificate holder and each training center holding AQP provisional approval shall show that it will establish and maintain records in sufficient detail to establish the training, qualification, and certification of each person qualified under an AQP in accordance with the training, qualification, and certification requirements of this SFAR.

13. *Expiration.* This Special Federal Aviation Regulation terminates on October 2, 1995, unless sooner terminated.

Part 141 – Pilot schools

Subpart A—General

141.1 Applicability
141.3 Certificate required
141.5 Pilot school certificate
141.7 Provisional pilot school certificate
141.9 Examining authority
141.11 Pilot school ratings
141.13 Application for issuance, amendment, or renewal
141.15 Location of facilities
141.17 Duration of certificates
141.18 Carriage of narcotic drugs, marihuana, and depressant or stimulant drugs or substances
141.19 Display of certificate
141.21 Inspections
141.23 Advertising limitations
141.25 Business office and operations base
141.27 Renewal of certificates and ratings
141.29 [Reserved]

Subpart B—Personnel aircraft and facilities requirements

141.31 Applicability
141.33 Personnel
141.35 Chief instructor qualifications
[141.36 Assistant chief instructor qualifications]
141.37 Airports
141.39 Aircraft
141.41 Ground trainers and training aids
141.43 Pilot briefing areas
141.45 Ground training facilities

Subpart C—Training course outline and curriculum

141.51 Applicability
141.53 Training course outline: General
141.55 Training course outline: Contents
141.57 Special curricula

Subpart D—Examining authority

141.61 Applicability
141.63 Application and qualification
141.65 Privileges
141.67 Limitations and reports

Subpart E—Operating rules

141.71 Applicability
141.73 Privileges
141.75 Aircraft requirements
141.77 Limitations
141.79 Flight instruction
141.81 Ground training
141.83 Quality of instruction
141.85 Chief instructor responsibilities
141.87 Change of chief instructor
141.89 Maintenance of personnel, facilities, and equipment
141.91 Satellite bases
141.93 Enrollment
141.95 Graduation certificate

Subpart F—Records

141.101 Training records

Appendix A. Private pilot certification course (Airplanes)
Appendix B. Private test course (airplanes)
Appendix C. Instrument rating course (airplanes)
Appendix D. Commercial pilot certification course (airplanes)
Appendix E. Commercial test course (airplanes)
Appendix F. Rotorcraft, gliders, lighter-than-air-aircraft and aircraft rating courses
Appendix G. Pilot ground school course
Appendix H. Test preparation courses

Subpart A—General

§ 141.1 Applicability.

This Part prescribes the requirements for issuing pilot school certificates, provisional pilot school certificates, and associated ratings and the general operating rules for the holders of those certificates and ratings.

§ 141.3 Certificate required.

No person may operate as a certificated pilot school without, or in violation of, a pilot school certificate or provisional pilot school certificate issued under this part.

§ 141.5 Pilot school certificate.

An applicant is issued a pilot school certificate with associated ratings for that certificate if—

(a) It meets the pertinent requirements of Subparts A through C of this part; and

(b) Within the 24 months before the date of application, it has trained and recommended for pilot certification and rating tests, at least 10 applicants for pilot certificates and ratings and at least 8 of the 10 most recent graduates tested by an FAA inspector or designated pilot examiner, passed that test the first time.

§ 141.7 Provisional pilot school certificate.

An applicant is issued a provisional pilot school certificate with associated ratings if it meets the pertinent requirements of Subparts A through C of this part, but does not meet the recent training activity requirement specified in § 141.5 (b).

§ 141.9 Examining authority.

An applicant is issued an examining authority for its pilot school certificate if it meets the requirements of Subpart D of this part.

§ 141.11 Pilot school ratings.

Associated ratings are issued with a pilot school certificate or a provisional pilot school certificate, specifying each of the following courses that the school is authorized to conduct:

(a) *Certification courses.*

(1) Private pilot.

(2) Private test course.

(3) Instrument rating.

(4) Commercial pilot.

(5) Commercial test course.

(6) Additional aircraft rating.

(b) *Pilot ground school course.*

(1) Pilot ground school.

(c) *Test preparation courses.*

(1) Flight instructor certification.

(2) Additional flight instructor rating.

(3) Additional instrument rating.

(4) Airline transport pilot certification.

(5) Pilot refresher course.

(6) Agricultural aircraft operations course.

(7) Rotorcraft external load operations course.

§ 141.13 Application for issuance, amendment, or renewal.

(a) Application for an original certificate and rating, for an additional rating, or for the renewal of a certificate under this part is made on a form and in a manner prescribed by the Administrator.

(b) An application for the issuance or amendment of a certificate or rating must be accompanied by three copies of the proposed training course outline for each course for which approval is sought.

§ 141.15 Location of facilities.

Neither a pilot school certificate nor a provisional pilot school certificate is issued for a school having a base or other facilities located outside the United States unless the Administrator finds that the location of the base or facilities at that place is needed for the training of students who are citizens of the United States.

§ 141.17 Duration of certificates.

(a) Unless sooner surrendered, suspended, or revoked, a pilot school certificate or a provisional pilot school certificate expires—

(1) At the end of the twenty-fourth month after the month in which it was issued or renewed; or

(2) Except as provided in paragraph (b) of this section, on the date that any change in ownership of the school or the facilities upon which its certification is based occurs; or

(3) Upon notice by the Administrator that the school has failed for more than 60 days to maintain the facilities, aircraft, and personnel required for at least one of its approved courses.

(b) A change in the ownership of a certificated pilot school or provisional pilot school does not terminate that certificate if within 30 days after the date that any change in ownership of the school occurs, application is made for an appropriate amendment to the certificate and no change in the facilities, instructor, personnel or training course is involved.

(c) An examining authority issued to the holder of a pilot school certificate expires on the date that the pilot school certificate expires, or is surrendered, suspended, or revoked.

§ 141.18 Carriage of narcotic drugs, marahuana, and depressant or stimulant drugs or substances.

If the holder of a certificate issued under this part permits any aircraft owner or leased by that holder to be engaged in any operation that the certificate holder knows to be in violation of [§ 91.19(a)] of this chapter, that operation is a basis for suspending or revoking the certificate.

§ 141.19 Display of certificate.

(a) Each holder of a pilot school certificate or a provisional pilot school certificate shall display that certificate at a place in the school that is normally accessible to the public and is not obscured.

(b) A certificate shall be made available for inspection upon request by the Administrator, or an authorized representative of the National Transportation Safety Board, or of any Federal, State, or local law enforcement office.

§ 141.21 Inspections.

Each holder of a certificate issued under this part shall allow the Administrator to inspect its personnel, facilities, equipment, and records to determine its compliance with the Federal Aviation Act of 1958, and the Federal Aviation Regulations, and its eligibility to hold its certificate.

§ 141.23 Advertising limitations.

(a) The holder of a pilot school certificate or a provisional pilot school certificate may not make any statement relating to its certification and ratings which is false or designed to mislead any person contemplating enrollment in that school.

(b) The holder of a pilot school certificate or a provisional pilot school certificate may not advertise that the school is certificated unless it clearly differentiates between courses that have been approved and those that have not.

(c) The holder of a pilot school certificate or a provisional pilot school certificate—

(1) That has relocated its school shall promptly remove from the premises it has vacated all signs indicating that the school was certificated by the Administrator; or

(2) Whose certificate has expired, or has been surrendered, suspended, or revoked shall promptly remove all indications (including signs), wherever located, that the school is certificated by the Administrator.

§ 141.25 Business office and operations base.

(a) Each holder of a pilot school or a provisional pilot school certificate shall maintain a principal business office with a mailing address in the name shown on its certificate. The business office shall have facilities and equipment that are adequate to maintain the required school files and records and to operate the business of the school. The office may not be shared with, or used by, another pilot school.

(b) Each certificate holder shall, before changing the location of its business office or base of operations, notify the FAA Flight Standards District Office having jurisdiction over the area of the new location. The notice shall be submitted in writing at least 30 days before the change. For a change in the holder's base of operations, the notice shall be accompanied by any amendments needed for the holder's approved training course outline.

(c) No certificate holder may conduct training at an operations base other than the one specified in its certificate, until—

(1) The base has been inspected and approved by the FAA Flight Standards District Office having jurisdiction over the school for use by the certificate holder; and

(2) The course of training and any needed amendments thereto have been approved for training at that base.

§ 141.27 Renewal of certificates and ratings.

(a) *Pilot school certificates.* The holder of a pilot school certificate may apply for a renewal of the certificate not less than 30 days before the certificate expires. If the school meets the requirements of this part for the issuance of the certificate, its certificate is renewed for 24 months.

(b) *Pilot school ratings.* Each pilot school rating on a pilot school certificate may be renewed with that certificate for another 24 months if the Administrator finds that the school meets the requirements prescribed in this part for the issuance of the rating.

(c) *Provisional pilot school certificates.*

(1) A provisional pilot school certificate and any ratings on that certificate may not be renewed. However, the holder of that certificate may apply for a pilot school certificate with appropriate ratings not less than 30 days before the provisional certificate expires. The school is issued a pilot school certificate with appropriate ratings, if it meets the appropriate requirements of this part.

(2) The holder of a provisional pilot school certificate may not reapply for a provisional pilot school certificate for at least 180 days after the date of its expiration.

§ 141.29 [Reserved]

Subpart B—Personnel, aircraft and facilities requirements

§ 141.31 Applicability.

This subpart prescribes the personnel and aircraft requirements for a pilot school or a provisional pilot school certificate. It also prescribes the facilities an applicant must have available to him on a continuous use basis to hold a pilot school or provisional pilot school certificate. As used in this subpart, a person has the continuous use of a facility, including an airport, if it has the use of the facility when needed as the owner, or under a written agreement giving it that use for at least 6 calendar months from the date of the application for the initial certificate or a renewal of that certificate.

§ 141.33 Personnel.

(a) An applicant for a pilot school or provisional pilot school certificate must show that—

(1) It has adequate personnel and authorized instructors, including a chief instructor for each course of training, who are qualified and competent to perform the duties to which they are assigned;

(2) Each dispatcher, aircraft handler, line crewman, and serviceman to be used has been instructed in the procedures and responsibilities of his employment. (Qualified operations personnel may serve in more than one capacity with a pilot school or provisional pilot school); and

(3) Each instructor to be used for ground or flight instruction holds a flight or ground instructor certificate, as appropriate, with ratings for the course of instruction and any aircraft used in that course.

(b) An applicant for a pilot school or a provisional pilot school certificate shall designate a chief instructor for each course of training who meets the requirements of a § 141.35 of this part. Where necessary, the applicant shall also designate at least one instructor to assist the chief instructor and serve for the chief instructor in his absence. A chief instructor or his assistant may be designated to serve in that capacity for more than one approved course but not for more than one school.

§ 141.35 Chief instructor qualifications.

(a) [To be eligible for a designation as a chief flight instructor for a course of training, a person must meet the following requirements:

[(1) Possess a commercial pilot or airline transport pilot certificate and a valid flight instructor certificate,

[(2) Meet the pilot-in-command recent flight experience requirements of § 61.57 of this chapter,

[(3) Pass an oral test on teaching methods, applicable provisions of the Airman's Information Manual, Parts 61, 91, and 141 of this chapter, and the objectives and approved course completion standards of the course for which the person seeks to obtain designation,

[(4) Pass a flight test demonstrating satisfactory performance of and the ability to instruct on the flight procedures and maneuvers appropriate to that course, and

[(5) Meet the applicable requirements of paragraph (b), (c), and (d) of this section. However, a chief flight instructor for a course of training for gliders, free balloons, or airships is only required to have 40 percent of the hours required in paragraphs (b) and (c) of this section.

(b) [For a course of training leading to the issuance of a private pilot certificate or rating, a chief flight instructor must have—

(1) [At least a commercial pilot or airline transport pilot certificate and a valid flight instructor certificate, each with a rating for the category and class of aircraft used in the course;]

(2) At least 1,000 hours as pilot in command;

(3) Primary flight instruction experience, acquired as either a certificated flight instructor or an instructor in a military pilot primary flight training program, or a combination thereof, consisting of at least—

(i) Two years and a total of 500 flight hours; or

(ii) [1,000 flight hours.]

(c) [For a course of training leading to the issuance of an instrument rating or a rating with instrument privileges, a chief flight instructor must have—

(1) [At least a commercial pilot or airline transport pilot certificate and a valid flight instructor certificate, each with an appropriate instrument rating;]

(2) At least 100 hours of flight time under actual or simulated instrument conditions;

(3) At least 1,000 hours as pilot in command;

(4) Instrument flight instructor experience, acquired as either a certificated instrument flight instructor or an instructor in a military pilot basic or instrument flight training program, or a combination thereof, consisting of at least—

(i) Two years and a total of 250 flight hours; or

(ii) [4000 flight hours.

(d) [For a course of training other than those that lead to the issuance of a private pilot certificate or rating, or an instrument rating or a rating with instrument privileges, a chief flight instructor must have—]

(1) [At least a commercial pilot or airline transport pilot certificate and a valid flight instructor certificate, each with a rating for the category and class of aircraft used in the course of training and, for a course of training using airplanes or airships, an instrument rating on the instructor's commercial pilot certificate;]

(2) At least 2,000 hours as pilot in command;

(3) Flight instruction experience, acquired as either a certificated flight instructor or an instructor in a military pilot primary or basic flight training program or a combination thereof, consisting of at least—

(i) Three years and a total of 1,000 flight hours; or

(ii) [1,500 flight hours.

(e) [To be eligible for a designation as a chief instructor for a ground school course, a person must have one year of experience as a ground school instructor in a certificated pilot school.]

[§ 141.36 Assistant chief instructor qualifications.

[(a) To be eligible for a designation as an assistant chief flight instructor for a course of training, a person must meet the following requirements:

[(1) Possess a commercial pilot or airline transport pilot certificate and a valid flight instructor certificate,

[(2) Meet the pilot-in-command recent flight experience requirements of § 61.57 of this chapter,

[(3) Pass an oral test on teaching methods, applicable provisions of the Airman's Information Manual, Parts 61, 91, and 141 of this chapter, and the objectives and approved course completion standards of the course for which the person seeks to obtain designations,

[(4) Pass a flight test on the flight procedures and maneuvers appropriate to that course, and

[(5) Meet the applicable requirements of paragraphs (b), (c), and (d) of this section. However, an assistant chief flight instructor for a course of training for gliders, free balloons, or airships is only required to have 40 percent of the hours required in paragraphs (b) and (c) of this section.

[(b) For a course of training leading to the issuance of a private pilot certificate or rating, an assistant chief flight instructor must have—

[(1) At least a commercial pilot or airline transport pilot certificate and a valid flight instructor certificate, each with a rating for the category and class of aircraft used in the course;

[(2) At least 500 hours as pilot in command;

[(3) Primary flight instruction experience, acquired as either a certificated flight instructor or an instructor in a military pilot primary flight training program, or a combination thereof, consisting of at least—

[(i) One year and a total of 250 flight hours; or

[(ii) 500 flight hours.

[(c) For a course of training leading to the issuance of an instrument rating or a rating with instrument privileges, and assistant chief flight instructor must have—

[(1) At least a commercial pilot or airline transport pilot certificate and a valid flight instructor certificate, each with an appropriate instrument rating;

[(2) At least 50 hours of flight time under actual or simulated instrument conditions;

[(3) At least 500 hours as pilot in command;

[(4) Instrument flight instructor experience, acquired as either a certificated instrument flight instructor or an instructor in a military pilot basic or instrument flight training program, or a combination thereof, consisting of at least—

[(i) One year and a total of 125 flight hours; or

[(ii) 200 flight hours.

[(d) For a course of training other than those that lead to the issuance of a private pilot certificate or rating, or an instrument rating or a rating with instrument privileges, an assistant chief flight instructor must have—

[(1) At least a commercial pilot or airline transport pilot certificate and a valid flight instructor certificate, each with a rating for the category and class of aircraft used in the course of training and, for a course of training using airplanes or airships, an instrument rating on the instructor's commercial pilot certificate;

[(2) At least 1,000 hours as pilot in command;

[(3) Flight instruction experience, acquired as either a certificated flight instructor or an instructor in a military pilot primary or basic flight training program or a combination thereof, consisting of at least—

[(i) One and one half years and a total of 500 flight hours; or

[(ii) 750 flight hours.

[(e) To be eligible for a designation as an assistant chief instructor for a ground school course, a person must have one year of experience as a ground school instructor in a certificated pilot school.]

§ 141.37 Airports.

(a) An applicant for a pilot school certificate or a provisional pilot school certificate must show that it has continuous use of each airport at which training flights originate.

(b) Each airport used for airplanes and gliders must have at least one runway or takeoff area that allows training aircraft to make a normal takeoff or landing at full gross weight—

(1) Under calm wind (not more than five miles per hour) conditions and temperatures equal to the mean high temperature for the hottest month of the year in the operating area;

(2) Clearing all obstacles in the takeoff flight path by at least 50 feet;

(3) With the powerplant operation and landing gear and flap operation, if applicable, recommended by the manufacturer; and

(4) With smooth transition from liftoff to the best rate of climb speed without exceptional piloting skills or techniques.

(c) Each airport must have a wind direction indicator that is visible from the ends of each runway at ground level.

(d) Each airport must have a traffic direction indicator when the airport has no operating control tower and UNICOM advisories are not available.

(e) Each airport used for night training flights must have permanent runway lights.

§ 141.39 Aircraft.

An applicant for a pilot school or provisional pilot school certificate must show that each aircraft used by that school for flight instruction and solo flights meets the following requirements:

(a) It must be registered as a civil aircraft of the United States.

(b) Except for aircraft used for flight instruction and solo flights in a course of training for agricultural aircraft operations, external load operations and similar aerial work operations, it must be certificated in the standard airworthiness category.

(c) It must be maintained and inspected in accordance with the requirements of Part 91 of this chapter that apply to aircraft used to give flight instruction for hire.

(d) For use in flight instruction, it must be at least a two place aircraft having engine power controls and flight controls that are easily reached and that operate in a normal manner from both pilot stations.

(e) For use in IFR en route operations and instrument approaches, it must be equipped and maintained for IFR operations. However, for instruction in the control and precision maneuvering of an aircraft by reference to instruments, the aircraft may be equipped as provided in the approved course of training.

§ 141.41 Ground trainers and training aids.

An applicant for a pilot school or a provisional pilot school certificate must show that its ground trainers, and training aids and equipment meet the following requirements:

(a) *Pilot ground trainers.*

(1) Each pilot ground trainer used to obtain the maximum flight training credit allowed for ground trainers in an approved pilot training course curriculum must have—

(i) An enclosed pilot's station or cockpit which accommodates one or more flight crewmembers;

(ii) Controls to simulate the rotation of the trainer about three axes;

(iii) The minimum instrumentation and equipment required for powered aircraft in § 91.205 of this chapter, for the type of flight operations simulated;

(iv) For VFR instruction, a means for simulating visual flight conditions, including motion of the trainer, or projections, or models operated by the flight controls; and

(v) For IFR instruction, a means for recording the flight path simulated by the trainer.

(2) Pilot ground trainers other than those covered under paragraph (a) (1) of this section must have—

(i) An enclosed pilot's station or cockpit, which accommodates one or more flight crewmembers;

(ii) Controls to simulate the rotation of the trainer about three axes; and

(iii) The minimum instrumentation and equipment required for powered aircraft in § 91.205 of this chapter, for the type of flight operations simulated.

(b) *Training aids and equipment.*

Each training aid, including any audio-visuals, mockup, chart, or aircraft component listed in the approved training course outline must be accurate and appropriate to the course for which it is used.

§ 141.43 Pilot briefing areas.

(a) An applicant for a pilot school or provisional pilot school certificate must show that it has the continuous use of a briefing area located at each airport at which training flights originate, that is—

(1) Adequate to shelter students waiting to engage in their training flights;

(2) Arranged and equipped for the conduct of pilot briefings; and

(3) For a school with an instrument or commercial pilot course rating, equipped with private landline or telephone communication to the nearest FAA Flight Service Station, except that this communication equipment is not required if the briefing area and the flight service station are located on the same airport and are readily accessible to each other.

(b) A briefing area required by paragraph (a) of this section may not be used by the applicant if it is available for use by any other pilot school during the period it is required for use by the applicant.

§ 141.45 Ground training facilities.

An applicant for a pilot school or provisional pilot school certificate must show that each room, training booth, or other space used for instructional purposes is heated, lighted, and ventilated to conform to local building, sanitation, and health codes. In addition, the training facility must be so located that the students in that facility are not distracted by the instruction conducted in other rooms, or by flight and maintenance operations on the airport.

Subpart C—Training course outline and curriculum

§ 141.51 Applicability.

This subpart prescribes the curriculum and course outline requirements for the issuance of a pilot school or provisional pilot school certificate and ratings.

§ 141.53 Training course outline: General

(a) *General.* An applicant for a pilot school or provisional pilot school certificate must obtain the Administrator's approval of the outline of each training course for which certification and rating is sought.

(b) *Application.* An application for the approval of an initial or amended training course outline is made in triplicate to the FAA Flight Standards District Office having jurisdiction over the area in which the operations base of the applicant is located. It must be made at least 30 days before any training under that course, or any amendment thereto, is scheduled to begin. An application for an amendment to an approved training course must be accompanied by three copies of the pages in the course outline for which an amendment is requested.

§ 141.55 Training course outline: Contents.

(a) *General.* The outline for each course of training for which approval is requested must meet the minimum curriculum for that course prescribed in the appropriate appendix of this part, and contain the following information:

(1) A description of each room used for ground training, including its size and the maximum number of students that may be instructed in the room at one time.

(2) A description of each type of audiovisual aid, projector, tape recorder, mockup, aircraft component and other special training aid used for ground training.

(3) A description of each pilot ground trainer used for instruction.

(4) A listing of the airports at which training flights originate and a description of the facilities, including pilot briefing areas that are available for use by the students and operating personnel at each of those airports.

(5) A description of the type of aircraft including any special equipment, used for each phase of instruction.

(6) The minimum qualifications and ratings for each instructor used for ground or flight training.

(b) *Training syllabus.* In addition to the items specified in paragraph (a) of this section, the course outline must include a training syllabus for each course of training that includes at least the following information:

(1) The pilot certificate and ratings, if any; the medical certificate, if necessary; and the training, pilot experience and knowledge, required for enrollment in the course.

(2) A description of each lesson, including its objectives and standards and the measurable unit of student accomplishment or learning to be derived from the lesson or course.

(3) The stage of training (including the standards therefor) normally accomplished within each training period of not more than 90 days.

(4) A description of the tests and checks used to measure a student's accomplishment for each stage of training.

§ 141.57 Special curricula.

An applicant for a pilot school or provisional pilot school certificate may apply for approval to conduct a special course of pilot training for which a curriculum is not prescribed in the appendixes to this part, if it shows that the special course of pilot training contains features which can be expected to achieve a level of pilot competency equivalent to that achieved by the curriculum prescribed in the appendixes to this part or the requirements of Part 61 of this chapter.

Subpart D—Examining authority

§ 141.61 Applicability.

This subpart prescribes the requirements for the issuance of an examining authority to the holder of a pilot school certificate and the privileges and limitations of that authority.

§ 141.63 Application and qualification.

(a) Application for an examining authority is made on a form and in a manner prescribed by the Administrator.

(b) To be eligible for an examining authority an applicant must hold a pilot school certificate. In addition, the applicant must show that—

(1) It has actively conducted a certificated pilot school for at least 24 months before the date of application; and

(2) Within the 24 months before the date of application for the examining authority, at least 10 students were graduated from the course for which the authority is requested, and at least 9 of the most recent 10 graduates of that course, who were given an interim or final test by an FAA inspector or a designated pilot examiner, passed that test the first time.

§ 141.65 Privileges.

The holder of an examining authority may recommend graduates of the school's approved certification courses for pilot certificates and ratings except flight instructor certificates, airline transport pilot certificates and ratings, and turbojet type ratings, without taking the FAA flight or written test, or both, in accordance with the provisions of this subpart.

§ 141.67 Limitations and reports.

(a) The holder of an examining authority may not recommend any person for the issuance of a pilot certificate or rating without taking the FAA written or flight test unless that person has—

(1) Been enrolled by the holder of the examining authority in its approved course of training for the particular pilot certificate or rating recommended; and

(2) Satisfactorily completed all of that course of training at its school.

(b) Each final written or flight test given by the holder of an examining authority to a person who has completed the approved course of training must be at least equal in scope, depth, and difficulty to the comparable written or flight test prescribed by the Administrator under Part 61 of this chapter.

(c) A final ground school written test may not be given by the holder of an examining authority to a student enrolled in its approved course of training unless the test has been approved by the FAA Flight Standards District Office having jurisdiction over the area in which the holder of the examining authority is located. In addition, an approved test may not be given by the holder of an examining authority when—

(1) It knows or has reason to believe that the test has been compromised; or

(2) It has been notified that the Flight Standards District Office knows or has reason to believe that the test has been compromised.

(d) The holder of an examining authority shall submit to the FAA Flight Standards District Office a copy of the appropriate training record for each person recommended by it for a pilot certificate or rating.

Subpart E—Operating rules

§ 141.71 Applicability.

This subpart prescribes the operating rules that are applicable to a pilot school or provisional pilot school certificated under the provisions of this part.

§ 141.73 Privileges.

(a) The holder of a pilot school or a provisional pilot school certificate may advertise and conduct approved pilot training courses in accordance with the certificate and ratings that it holds.

(b) A certificated pilot school holding an examining authority for a certification course may recommend each graduate of that course for the issuance of a pilot certificate and rating appropriate to that course without the necessity of taking an FAA written or flight test from an FAA inspector or designated pilot examiner.

§ 141.75 Aircraft requirements.

(a) A pretakeoff and prelanding checklist, and the operator's handbook for the aircraft (if one is furnished by the manufacturer) or copies of the handbook if furnished to each student using the aircraft, must be carried on each aircraft used for flight instruction and solo flights.

(b) Each aircraft used for flight instruction and solo flight must have a standard airworthiness certificate, except that an aircraft certificated in the restricted category may be used for flight training and solo flights conducted under special courses for agricultural aircraft operation, external load operations, and similar aerial work operations if its use for training is not prohibited by the operating limitations for the aircraft.

§ 141.77 Limitations.

(a) The holder of a pilot school or a provisional pilot school certificate may not issue a graduation certificate to a student, nor may a certificated pilot school recommend a student for a pilot certificate or rating, unless the student has completed the training therefor specified in the school's course of training and passed the required final tests.

(b) The holder of a pilot school or a provisional pilot school certificate may not graduate a student from a course of training unless he has completed all of the curriculum requirements of that course. A student may be credited, but not for more than one-half of the curriculum requirements, with previous pilot experience and knowledge, based upon an appropriate flight check or test by the school. Course credits may be transferred from one certificated school to another. The receiving school shall determine the amount to be transferred, based on a flight check or written test, or both, of the student. Credit for training and instruction received in another school may not be given unless—

(1) The other school holds a certificate issued under this Part and certifies to the kind and amount of training and to the result of each stage and final test given to that student;

(2) The training and instruction was conducted by the other school in accordance with that school's approved training course; and

(3) The student was enrolled in the other school's approved training course before he received the instruction and training.

§ 141.79 Flight instruction.

(a) No person other than a flight instructor who has the ratings and the minimum qualifications specified in the approved training course outline may give a student flight instruction under an approved course of training.

(b) No student pilot may be authorized to start a solo practice flight from an airport until the flight has been approved by an authorized flight instructor who is present at that airport.

(c) Each chief flight instructor must complete at least once each 12 months, an approved flight instructor refresher course consisting of ground or flight instruction, or both.

(d) Each flight instructor for an approved course of training must satisfactorily accomplish a flight check given to him by the designated chief flight instructor for the school by whom he is employed. He must also satisfactorily accomplish this flight check each 12 months from the month in which the initial check is given. In addition, he must satisfactorily accomplish a flight check in each type of aircraft in which he gives instruction.

(e) An instructor may not be used in an approved course of training until he has been briefed in regard to the objectives and standards of the course by the designated chief instructor or his assistant.

§ 141.81 Ground training.

(a) Except as provided in paragraph (b) of this section, each instructor used for ground training in an approved course of training must hold a flight or ground instructor certificate with an appropriate rating for the course of training.

(b) A person who does not meet the requirements of paragraph (a) of this section may be used for ground training in an approved course of training if—

(1) The chief instructor for that course of training finds him qualified to give that instruction; and

(2) The instruction is given under the direct supervision of the chief instructor or the assistant chief instructor who is present at the base when the instruction is given.

(c) An instructor may not be used in an approved course of training until he has been briefed in regard to the objectives and standards of that course by the designated chief instructor or his assistant.

§ 141.83 Quality of instruction.

(a) Each holder of a pilot school or provisional pilot school certificate must comply with the approved course of training and must provide training and instruction of such quality that at least 8 out of the 10 students or graduates of that school most recently tested by an FAA inspector or designated pilot examiner, passed on their first attempt either of the following tests:

(1) A test for a pilot certificate or rating, or for an operating privilege appropriate to the course from which the student graduated; or

(2) A test given to a student to determine his competence and knowledge of a completed stage of the training course in which he is enrolled.

(b) The failure of a certificated pilot school or provisional pilot school to maintain the quality of instruction specified in paragraph (a) of this section is considered to be the basis for the suspension or revocation of the certificate held by that school.

(c) The holder of a pilot school or provisional pilot school certificate shall allow the Administrator to make any test, flight check, or examination of its students to determine compliance with its approved course of training and the quality of its instruction and training. A flight check conducted under the provisions of this paragraph is based upon the standards prescribed in the school's approved course of training. However, if the student has completed a course of training for a pilot certificate or rating, the flight test is based upon the standards prescribed in Part 61 of this chapter.

§ 141.85 Chief instructor responsibilities.

(a) Each person designated as a chief instructor for a certificated pilot school or provisional pilot school shall be responsible for—

(1) Certifying training records, graduation certificates, stage and final test reports, and student recommendations;

(2) Conducting an initial proficiency check of each instructor before he is used in an approved course of instruction and, thereafter, at least once each 12 months from the month in which the initial check was conducted;

(3) Conducting each stage or final test given to a student enrolled in an approved course of instruction; and

(4) Maintaining training techniques, procedures, and standards for the school that are acceptable to the Administrator.

(b) [The chief instructor or his designated assistant chief instructor shall be available at the pilot school or, if away from the premises, by telephone, radio, or other electronic means during the time that instruction is given for an approved course of training.]

§ 141.87 Change of chief instructor.

(a) The holder of a pilot school or provisional pilot school certificate shall immediately notify in writing the FAA Flight Standards District Office having jurisdiction over the area in which the school is located, of any change in its designation of a chief instructor of an approved training course.

(b) The holder of a pilot school or provisional pilot school certificate may, after providing the notification required in paragraph (a) of this section and pending the designation and approval of another chief instructor, conduct training or instruction without a chief instructor for that course of training for a period of not more than 60 days. However, during that time each stage or final test of a student enrolled in that approved course of training must be given by an FAA inspector, or a designated pilot examiner.

§141.89 Maintenance of personnel, facilities, and equipment.

The holder of a pilot school or provisional pilot school certificate may not give instruction or training to a student who is enrolled in an approved course of training unless—

(a) Each airport, aircraft, and facility necessary for that instruction or training meets the standards specified in the holder's approved training course outline and the appropriate requirements of this part; and

(b) Except as provided in § 141.87, each instructor or chief instructor meets the qualifications specified in the holder's approved course of training and the appropriate requirements of this part.

§ 141.91 Satellite bases.

The holder of a pilot school or provisional pilot school certificate may conduct ground or flight training and instruction in an approved course of training at a base other than its main operations base if—

(a) [An assistant chief instructor is designated for each satellite base, and that assistant chief instructor shall be available at the satellite pilot school or, if away from the premises, by telephone, radio, or other electronic means during the time that instruction is given for an approved course of training;]

(b) The airport, facilities, and personnel used at the satellite base meets the appropriate requirements of Subpart B of this part and its approved training course outline;

(c) [The instructors are under the direct supervision of the chief instructor or assistant chief flight instructor for the appropriate training course, who is readily available for consultation in accordance with § 141.85 (b); and]

(d) The FAA Flight Standards District Office having jurisdiction over the area in which the school is located is notified in writing if training or instruction is conducted there for more than seven consecutive days.

§ 141.93 Enrollment.

(a) The holder of a pilot school or a provisional pilot school certificate shall furnish each student, at the time he is enrolled in each approved training course, with the following:

(1) A certificate of enrollment containing—

(i) The name of the course in which he is enrolled; and

(ii) The date of that enrollment.

(2) A copy of the training syllabus required under § 141.55 (b).

(3) A copy of the safety procedures and practices developed by the school covering the use of its facilities and the operation of its aircraft, including instructions on the following:

(i) The weather minimums required by the school for dual and solo flights.

(ii) The procedures for starting and taxiing aircraft on the ramp.

(iii) Fire precautions and procedures.

(iv) Redispatch procedures after unprogrammed landings, on and off airports.

(v) Aircraft discrepancies and write offs.

(vi) Securing of aircraft when not in use.

(vii) Fuel reserves necessary for local and cross-country flights.

(viii) Avoidance of other aircraft in flight and on the ground.

(ix) Minimum altitude limitations and simulated emergency landing instructions.

(x) Description and use of assigned practice areas.

(b) The holder of a pilot school or provisional pilot certificate shall, within 5 days after the date of enrollment, forward a copy of each certificate of enrollment required by paragraph (a) (1) of this section to the FAA Flight Standards District Office having jurisdiction over the area in which the school is located.

§ 141.95 Graduation certificate.

(a) The holder of a pilot school or provisional pilot school certificate shall issue a graduation certificate to each student who completes its approved course of training.

(b) The certificate shall be issued to the student upon his completion of the course of training and contain at least the following information:

(1) The name of the school and the number of the school certificate.

(2) The name of the graduate to whom it was issued.

(3) The course of training for which it was issued.

(4) The date of graduation.

(5) A statement that the student has satisfactorily completed each required stage of the approved course of training including the tests for those stages.

(6) A certification of the information contained in the certificate by the chief instructor for that course of training.

(7) A statement showing the cross-country training the student received in the course of training.

Subpart F—Records

§ 141.101 Training records.

(a) Each holder of a pilot school or provisional pilot school certificate shall establish and maintain a current and accurate record of the participation and accomplishment of each student enrolled in an approved course of training conducted by the school (the student's logbook is not acceptable for this record). The record shall include—

(1) The date the student was enrolled;

(2) A chronological log of the student's attendance, subjects, and flight operations covered in his training and instruction, and the names and grades of any tests taken by the student; and

(3) The date the student graduated, terminated his training, or transferred to another school.

(b) Whenever a student graduates, terminates his training, or transfers to another school, his record shall be certified to that effect by the chief instructor.

(c) The holder of a certificate for a pilot school or a provisional pilot school shall retain each student record required by this section for at least 1 year from the date that the student graduates from the course to which the record pertains, terminates his enrollment in that course, or transfers to another school.

(d) The holder of a certificate for a pilot school or a provisional pilot school shall, upon request of a student, make a copy of his record available to him.

Appendix A to Part 141—Private pilot certification course (airplanes)

1. *Applicability.* This appendix prescribes the minimum curriculum for a private pilot certification course (airplanes) required by § 141.55.

2. *Ground training.* The course must consist of at least 35 hours of ground training in the following subjects:

(a) The Federal Aviation Regulations applicable to private pilot privileges, limitations, and flight operations; the rules of the National Transportation Safety Board pertaining to accident reporting; the use of the Airman's Information Manual; and the FAA Advisory Circular System.

(b) VFR navigation using pilotage, dead reckoning, and radio aids.

(c) The recognition of critical weather situations from the ground and in flight and the procurement and use of aeronautical weather reports and forecasts.

(d) The safe and efficient operation of airplanes, including high density airport operations, collision avoidance precautions, and radio communication procedures.

[(e) Stall awareness, spin entry, spins, and spin recovery techniques.]

3. *Flight training.*

(a) The course must consist of at least 35 hours of the flight training listed in this section and section 4 of this appendix. Instruction in a pilot ground trainer that meets the requirements of § 141.41(a) (1) may be credited for not more than 5 of the required 35 hours of flight time. Instruction in a pilot ground trainer that meets the requirement of § 141.41(a) (2) may be credited for not more than 2.5 hours of the required 35 hours of flight time.

(b) Each training flight must include a preflight briefing and a postflight critique of the student by the instructor assigned to that flight.

(c) Flight training must consist of at least 20 hours of instruction in the following subjects:

(1) Preflight operations, including weight and balance determination, line inspection, starting and runups, and airplane servicing.

(2) Airport and traffic pattern operations, including operations at controlled airports, radio communications, and collision avoidance precautions.

(3) Flight maneuvering by reference to ground objects.

(4) [Flight at slow airspeeds with realistic distractions, recognition of and recovery from stalls entered from straight flight and from turns.]

(5) Normal and crosswind takeoffs and landings.

(6) Control and maneuvering an airplane solely by reference to instruments, including emergency descents and climbs using radio aids or radar directives.

(7) Cross-country flying using pilotage, dead reckoning, and radio aids, including a two-hour dual flight at least part of which must be on Federal airways.

(8) Maximum performance takeoffs and landings.

(9) Night flying, including 5 takeoffs and landings as sole manipulator of the controls, and VFR navigation.

(10) Emergency operations, including simulated aircraft and equipment malfunctions, lost procedures, and emergency go-arounds.

4. *Solo flights.* The course must provide at least 15 hours of solo flights, including:

(a) *Solo practice.*

[Directed solo practice on all VFR flight operations for which flight instruction is required (except simulated emergencies) to develop proficiency, resourcefulness, and self-reliance.]

(b) *Cross-country flights.*

[(1) Ten hours of cross-country flights, each flight with a landing at a point more than 50 nautical miles from the original departure point. One flight must be of at least 300 nautical miles with landings at a minimum of three points, one of which is at least 100 nautical miles from the original departure point.]

(2) If a pilot school or a provisional pilot school shows that it is located on an island from which cross-country flights cannot be accomplished without flying over water more than 10 nautical miles from the nearest shoreline, it need not include cross-country flights under subparagraph (1) of this paragraph. However, if other airports that permit civil operations are available to which a flight may be made without flying over water more than 10 nautical miles from the nearest shoreline, the school must include in its course, two round trip solo flights between those airports that are farthest apart, including a landing at each airport on both flights.

5. *Stage and final tests.*

(a) Each student enrolled in a private pilot certification course must satisfactorily accomplish the stage and final test prescribed in this section. The written tests may not be credited for more than 3 hours of the 35 hours of required ground training, and the flight tests may not be credited for more than 4 hours of the 35 hours of required flight training.

(b) Each student must satisfactorily accomplish a written examination at the completion of each stage of training specified in the approved training syllabus for the private certification course and a final test at the conclusion of that course.

(c) Each student must satisfactorily accomplish a flight test at the completion of the first solo flight and at the completion of the first solo cross-country flight and at the conclusion of that course.

Appendix B to Part 141—Private test course (airplanes)

1. *Applicability.* This appendix prescribes the minimum curriculum for a private test course (airplanes) required by § 141.55.

2. *Experience.* For enrollment as a student in a private test course (airplanes) an applicant must—

(a) Have logged at least 30 hours of flight time as a pilot; and

(b) Have such experience and flight training that upon completion of his approved private test course (airplanes) he will meet the aeronautical experience requirements prescribed in Part 61 of this chapter for a private pilot certificate.

3. *Ground training.* The course must consist of at least 35 hours of ground training in the subjects listed in § 2 of Appendix A of this part.

4. *Flight training.*

(a) The course must consist of a total of at least 10 hours of flight instruction in the subjects listed in § 3 (c) of Appendix A of this part.

(b) Each training flight must include a preflight briefing and a postflight critique of the student by the instructor assigned to that flight.

5. *Stage and final tests.* Each student enrolled in the course must satisfactorily accomplish the final tests prescribed in § 5 of Appendix A of this part. Written tests may not be credited for more than 3 hours of the required 35 hours of ground training, and the flight tests may not be credited for more than 2 hours of the required 10 hours of flight training.

Appendix C to Part 141—Instrument rating course (airplanes)

1. *Applicability.* This appendix prescribes the minimum curriculum for a training course for an Instrument Rating Course (airplanes) required by § 141.55.

2. *Ground training.* The course must consist of at least 30 hours of ground training instruction in the following subjects:

(a) The Federal Aviation Regulations that apply to flight under IFR conditions, the IFR air traffic system and procedures, and the provisions of the Airman's Information Manual pertinent to IFR flights.

(b) Dead reckoning appropriate to IFR navigation, IFR navigation by radio aids using the VOR, ADF, and ILS systems, and the use of IFR charts and instrument approach procedure charts.

(c) The procurement and use of aviation weather reports and forecasts, and the elements of forecasting weather trends on the basis of that information and personal observation of weather conditions.

(d) The function, use, and limitations of flight instruments required for IFR flight, including transponders, radar and radio aids to navigation.

3. *Flight training.* The course must consist of at least 35 hours of instrument flight instruction given by an appropriately rated flight instructor, covering the operations listed in paragraphs (a) through (d) of this section. Instruction given by an authorized instructor in a pilot ground trainer which meets the requirements of § 141.41 (a) (1) may be credited for not more than 15 hours of the required flight instruction. Instruction in a pilot ground trainer that meets the requirements of § 141.41 (a) (2) may be credited for not more than 7.5 of the required 35 hours of flight time.

(a) Control and accurate maneuvering of an airplane solely by reference to flight instruments.

(b) IFR navigation by the use of VOR and ADF systems, including time, speed and distance computations and compliance with air traffic control instructions and procedures.

(c) Instrument approaches to published minimums using the VOR, ADF, and ILS systems (instruction in the use of the ILS glide slope may be given in an instrument ground trainer or with an airborne ILS simulator).

(d) Cross-country flying in simulated or actual IFR conditions, on Federal airways or as routed by ATC, including one such trip of at least 250 nautical miles including VOR, ADF, and ILS approaches at different airports.

(e) Emergency procedures appropriate to the maneuvering of an airplane solely by reference to flight instruments.

4. *Stage and final tests.*

(a) Each student must satisfactorily accomplish a written test at the completion of each stage of training specified in the approved training syllabus for the instrument rating course. In addition, he must satisfactorily accomplish a final written test at the conclusion of that course. The written tests may not be credited for more than 5 hours of the 30 hours of required ground training.

(b) Each student must satisfactorily accomplish a flight stage test at the completion of each operation listed in paragraphs (a), (b), and (c) of § 3 of this Appendix. In addition, he must satisfactorily accomplish a final flight test at the completion of the course. The stage and final tests may not be credited for more than 5 hours, of the required 35 hours of flight training.

Appendix D to Part 141—Commercial pilot certification course (airplanes)

1. *Applicability.* This appendix prescribes the minimum curriculum for a commercial pilot certification course (airplanes) required by § 141.55.

2. *Ground training.* The course must consist of at least 100 hours of ground training instruction in the following subjects:

(a) The ground training subjects prescribed in § 2 of Appendix A of this part for a private pilot certification course, except the private pilot privileges and limitations of paragraph (a) of that section.

(b) The ground training subjects prescribed in § 2 of Appendix C of this part for an Instrument Rating Course.

(c) The Federal Aviation Regulations covering the privileges, limitations, and operations of a commercial pilot, and the operations for which an air taxi/commercial operator, agricultural aircraft operator, and external load operator certificate, waiver, or exemption is required.

(d) Basic aerodynamics, and the principles of flight which apply to airplanes.

(e) The safe and efficient operation of airplanes, including inspection and certification requirements, operating limitations, high altitude operations and physiological considerations, loading computations, the significance of the use of airplane performance speeds, the computations involved in runway and obstacle clearance and crosswind component considerations, and cruise control.

3. *Flight training.*

(a) *General.* The course, must consist of at least 190 hours of the flight training and instruction prescribed in this section. Instruction in a pilot ground trainer that meets the requirements of § 141.41 (a) (1) may be credited for not more than 40 hours of the required 190 hours of flight time. Instruction in a pilot ground trainer that meets the requirements of § 141.41 (a) (2) may be credited for not more than 20 hours of the required 190 hours of flight time.

(b) *Flight instruction.* The course must consist of at least 75 hours of instruction in the operations listed in subparagraphs (1) through (6) of this paragraph. Instruction in a pilot ground trainer that meets the requirements of § 141.41 (a) (1) may be credited for not more than 20 hours of the required 75 hours. Instruction in a pilot ground trainer that meets the requirements of § 141.41 (a) (2) may be credited for not more than 10 hours of the required 75 hours.

(1) The pilot operations for the Private Pilot Course prescribed in § 3 of Appendix A of this part.

(2) The IFR operations for the Instrument Rating Course prescribed in § 3 of Appendix C of this part.

(3) Ten hours of flight instruction in an airplane with retractable gear, flaps, a controllable propeller, and powered by at least 180 hp. engine.

(4) Night flying, including a cross-country night flight with a landing at a point more than 100 miles from the point of departure.

(5) Normal and maximum performance takeoffs and landings using precision approaches and prescribed airplane performance speeds, including operation at maximum authorized takeoff weight.

(6) Emergency procedures appropriate to VFR and IFR flight and to the operation of complex airplane systems.

(c) *Solo practice.* The course must consist of at least 100 hours of the flights listed in subparagraphs (1) through (4) of this paragraph. Flight time as pilot in command of an airplane carrying only those persons who are pilots assigned by the school to specific flight crew duties on the flight may be credited for not more than 50 hours of that requirement.

(1) Directed solo practice on each VFR operation for which flight instruction is required (except simulated emergencies).

[(2) At least 40 hours of solo cross-country flights, each flight with a landing at a point more than 50 nautical miles from the original departure point. One flight must have landings at a minimum of three points, one of which is at least 150 nautical miles from the original departure point if the flight is conducted in Hawaii, or at least 250 nautical miles from the original departure point if it is conducted elsewhere.]

(3) At least 5 hours of pilot in command time in an airplane described in paragraph (b) (3) of this section, including not less than 10 takeoffs and 10 landings to a full stop.

(4) At least 5 hours of night flight, including at least 10 takeoffs and 10 landings to a full stop.

4. *Stage and final tests.*

(a) *Written examinations.* Each student enrolled in the course must satisfactorily accomplish a written test upon the completion of each stage of training specified in the approved training syllabus for the commercial pilot certification course. In addition, he must satisfactorily accomplish a final stage test at the completion of all of that course. The stage and final tests may be credited for not more than 6 hours of the required 100 hours of ground training.

(b) *Flight tests.* Each student enrolled in a commercial pilot certification course (airplanes) must satisfactorily accomplish a stage flight test at the completion of each of the stages listed in subparagraphs (1), (2), (3), (4), and (5), of this paragraph. In addition, he must satisfactorily accomplish a final test at the completion of all of those stages. The stage and final tests may not be credited for more than 10 hours of the required 190 hours of flight training.

(1) Solo.

(2) Cross-country.

(3) High performance airplane operations.

(4) IFR operations.

(5) Commercial Pilot Course test, VFR and IFR.

Appendix E to Part 141—Commercial test course (airplanes)

1. *Applicability.* This appendix prescribes the minimum curriculum for a commercial test course (airplanes) required by § 141.55.

2. *Experience.* For enrollment as a student in a commercial test course (airplanes) an applicant must—

(a) Hold a valid private pilot certificate;

(b) Hold a valid instrument rating, or be enrolled in an approved instrument rating course; and

(c) Have such experience and flight training that upon completion of his approved commercial test course he will meet the aeronautical experience requirements prescribed in Part 61 of this chapter for a commercial pilot certificate.

3. *Ground training.* The course must consist of at least 50 hours of ground training instruction in the following subjects:

(a) A review of the ground training subjects prescribed in § 2 of Appendix A of this part for a private pilot certification.

(b) A review of the ground training subjects prescribed in § 2 of Appendix C of this part for an instrument rating course.

(c) The Federal Aviation Regulations covering the privileges, limitations, and operations of a commercial pilot, and the operations for which an air taxi/commercial operator, agricultural aircraft operator, and external load operator certificate, waiver or exemption is required.

(d) Basic aerodynamics, and the principles of flight that apply to airplanes.

(e) The safe and efficient operation of airplanes, including inspection and certification requirements, operating limitations, high altitude operations and physiological considerations, loading computations, the significance and use of airplane performance speeds, and computations involved in runway and obstacle clearance and crosswind component considerations.

4. *Flight training.*

(a) *General.* The course must consist of at least 25 hours of flight training prescribed in this section. Instruction in a pilot ground trainer that meets the requirements of § 141.41 (a) (1) may be credited for not more than 20 percent of the total number of hours of flight time. Instruction in a pilot ground trainer that meets the requirements of § 141.41 (a) (2) may be credited for not more than 10 percent of the total number of hours of flight time.

(b) *Flight instruction.* The course must consist of at least 20 hours of flight instruction in the subjects listed in subparagraphs (1) through (3) of this paragraph. Instruction in a ground trainer that meets the requirements of § 141.41 (a) (1) may be credited for not more than 4 hours of the required 20 hours. Instruction in a ground trainer that meets the requirements of § 141.41 (a) (2) may be credited for not more than 2 hours of the required 20 hours.

(1) A review of the VFR operations prescribed in § 3 of Appendix A of this part for a private course.

(2) A review of the IFR operations prescribed in § 3 of Appendix C of this part for an instrument rating course.

(3) A review of the VFR operations prescribed in § 3 (b) (3) through (6) of Appendix D of this part for a commercial pilot certification course.

(c) *Directed solo practice.* If the course includes directed solo practice necessary to develop the flight proficiency of each student, the practice may not exceed a ratio of 3 hours of directed solo practice for each hour of the flight instruction required by the school's approved course outline.

5. *Stage and final tests.*

(a) *Written tests.* Each student enrolled in the course must satisfactorily accomplish a stage test upon the completion of each stage of training specified in the approved training syllabus for the commercial test course. In addition, he must satisfactorily accomplish a final test at the conclusion of that course. The stage and final tests may not be credited for more than 4 hours of the required 50 hours of ground training.

(b) *Flight tests.* Each student enrolled in the course must satisfactorily accomplish a final test at the completion of the course. However, if the approved course of training exceeds 35 hours he must be given a test at an appropriate stage prior to completion of 35 hours of flight training. The flight tests may not be credited for more than 3 of the required hours of flight training.

(c) *Total flight experience.* The approved training course outline must specify the minimum number of hours of flight instruction and directed solo practice (if any) that is provided for each student under the requirements of paragraphs (b) and (c) of § 4 of this appendix. The total number of hours of all flight training given to a student under this section and the minimum experience required for enrollment under § 2 of this appendix must meet the minimum aeronautical experience requirements of § 61.129 of this chapter for the issuance of a commercial pilot certificate.

Appendix F to Part 141—Rotorcraft, gliders, lighter-than-air aircraft and aircraft rating courses

A. *Applicability.* This appendix prescribes the minimum curriculum for a pilot certification course for a rotorcraft, glider, lighter-than-air aircraft, or aircraft rating, required by § 141.55.

B. *General Requirements.* The course must be comparable in scope, depth, and detail with the curriculum prescribed in Appendixes A through D of this part for a pilot certification course (airplanes) with the same rating. Each course must provide ground and flight training covering the aeronautical knowledge and skill items required by Part 61 of this chapter for the certificate or rating concerned. In addition, each course must meet the appropriate requirements of this appendix.

C. *Rotocraft.*

I. *Kinds of rotocraft pilot certification courses.* An approved rotorcraft pilot certification course includes—

(a) A helicopter or gyroplane course—private pilots;

(b) A helicopter or gyroplane course—commercial pilots; and

(c) An instrument rating—helicopter.

II. *Helicopter or gyroplane course: private pilots.*

(a) A private pilot certification course for helicopters or gyroplanes must consist of at least the following:

(1) Ground training—35 hours.

(2) Flight training—35 hours, including the following:

(i) Flight instruction—20 hours.

(ii) Solo practice—10 hours, including a flight with landings at three points, each of which is more than 25 nautical miles from the other two points..

(b) Stage and final tests may be credited for not more than 3 hours of the 35 hours of ground training, and for not more than 4 hours of the 35 hours of flight training required by paragraphs (a) (1) and (a) (2) of this section.

III. *Helicopter or gyroplane course—commercial pilots.*

(a) A commercial pilot certification course of training for helicopters or gyroplanes must consist of at least the following:

(1) Ground training—65 hours.

(2) Flight training—150 hours, of flight training at least 50 hours of which must be in helicopters or gyroplanes. The flight training must include the following:

(i) Flight instruction—50 hours.

(ii) Directed solo—100 hours (including a cross-country flight with landings at three points, each of which is more than 50 nautical miles from the other two points).

(b) Stage and final tests may be credited for not more than 5 hours of the required 65 hours of ground training, and for not more than 7 hours of the required 150 hours of flight training prescribed in paragraphs (a) (1) and (a) (2) of this section.

IV. *Instrument rating—helicopter course.*

(a) An instrument rating — helicopter course of training must consist of at least the following:

(1) Ground training—35 hours.

(2) Instrument flight training — 35 hours. Instrument instruction in a pilot ground trainer that meets the requirements of § 141.41 (a) (1) may be credited for not more than 10 hours of the required 35 hours of flight training. Instruction in a ground trainer that meets the requirements of § 141.41 (a) (2) may be credited for not more than 5 hours of the required 35 hours. The instrument flight instruction must include a 100 mile simulated or actual IFR cross-country flight, and 25 hours of flight instruction.

(3) Stage and final tests may be credited for not more than 5 hours of the 35 hours of required ground training, and not more than 5 hours of the 35 hours of instrument training.

D. *Gliders.*

I. *Kinds of glider pilot certification courses.*

An approved glider certification course includes—

(a) A glider course—private pilots; and

(b) A glider course—commercial pilots.

II. *Glider course: private pilot.*

A private pilot certification course for gliders must consist of at least the following:

(a) Ground training—15 hours.

(b) Flight training—8 hours (including 35 flights if ground tows are used, or 20 flights if aero tows are used). The flight training must include the following:

(1) Flight instruction—2 hours (including 20 flights if ground tows are used or 15 flights if aero tows are used).

(2) Directed solo—5 hours (including at least 15 flights if ground tows are used or 5 flights if aero tows are used).

(c) Stage and flight tests may be credited for not more than one hour of the 15 hours of ground training, and for not more than one-half hour of the 2 hours of flight instruction required by paragraphs (a) (1) and (a) (2) of this section.

III. *Glider course: commercial pilot.*

(a) An approved commercial pilot certification course for gliders must consist of at least the following:

(1) Ground training—25 hours.

(2) Flight training—20 hours of flight time in gliders (consisting of at least 50 flights), including the following;

(i) Flight instruction—8 hours.

(ii) Directed solo—10 hours.

(b) Stage and final tests may be credited for not more than 2 hours of the 25 hours of ground training, and for not more than 2 hours of the 20 hours of flight training required by paragraphs (a) (1) and (a) (2) of this section.

E. *Lighter-than-air aircraft.*

I. *Kinds of lighter-than-air pilot certification courses.*

An approved lighter-than-air pilot certification course includes—

(a) An airship course—private pilot;

(b) A free balloon course—private pilot;

(c) An airship course—commercial pilots; and

(d) A free balloon course—commercial pilot.

II. *Airship—private pilot.*

(a) A private pilot certification course for an airship must consist of at least the following:

(1) Ground training—35 hours.

(2) Flight training—50 hours (45 hours must be in airships), including the following:

(i) Flight instruction—20 hours in airships.

(ii) Directed solo, or performing the functions of a pilot in command of an airship for which more than one pilot is required—10 hours.

(b) Stage and final tests may be credited for not more than 5 hours of the 35 hours of ground training, and not more than 5 hours of the 50 hours of flight training required by paragraphs (a) (1) and (a) (2) of this section.

III. *Free balloon course; private pilot.*

(a) A private pilot course for a free balloon must consist of at least the following:

(1) *Ground training*—10 hours.

(2) Flight training—6 free flights, including—

(i) Two flights of one hour duration each if a gas balloon is used, or of 30 minutes duration if a hot air balloon is used;

(ii) At least one solo flight; and

(iii) One ascent under control to 5,000 feet above the point of takeoff if a gas balloon is used, or 3,000 feet above the point of takeoff if a hot air balloon is used.

(b) The written and stage checks may be credited for not more than one hour of the ground training, and not more than one of the 6 flights required by paragraph (a) (1) and (a) (2) of this section.

IV. *Airship course—commercial pilot.*

(a) A commercial pilot course for an airship must consist of at least the following:

(1) Ground training—100 hours.

(2) Flight training—190 hours in airships as follows:

(i) Flight instruction—80 hours, including 30 hours instrument time.

(ii) 100 hours of solo time, or flight time performing the functions or a pilot in command in an airship that requires more than one pilot, including 10 hours of cross-country flying and 10 hours of night flying.

(b) Stage and final tests may be credited for not more than 6 hours of the 100 hours of ground training, and not more than 10 hours of the 190 hours of flight training required by paragraphs (a) (1) and (a) (2), respectively, of this section.

V *Free balloon course; commercial pilot.*

(a) A commercial pilot certification course for free balloons must consist of the following:

(1) Ground training—20 hours.

(2) Flight training—8 free flights, including—

(i) 2 flights of more than 2 hours duration if a gas balloon is used, or 2 flights of more than 1 hour duration if a hot air balloon is used;

(ii) 1 ascent under control to more than 10,000 feet above the takeoff point if a gas balloon is used, or to more than 5,000 feet above the takeoff point if a hot air balloon is used; and

(iii) 2 solo flights.

(b) Stage and final tests may be credited for not more than 2 hours of the 20 hours of ground training, and not more than one of the flights required by paragraphs (a) (1) and (a) (2), respectively, of this section.

F. *Aircraft rating course.*

I. *Kinds of aircraft rating courses.* An approved aircraft rating course includes—

(a) An aircraft category rating;

(b) An aircraft class rating; and

(c) An aircraft type rating.

II. *Aircraft category rating.* An aircraft rating course must include at least the ground training and flight instruction required by Part 61 of this chapter for the issuance of a pilot certificate with a category rating appropriate to the course. However, the Administrator may approve a lesser number of hours of ground training, or flight instruction, or both, if the course provides for the use of special training aids, such as ground procedures, trainers, systems mockups, and audio-visual training materials, or requires appropriate aeronautical experience of the students as a prerequisite for enrollment in the course.

III. *Aircraft class rating.* An aircraft class rating course must include at least the flight instruction required by Part 61 of this chapter for the issuance of a pilot certificate with a class rating appropriate to the course.

IV. *Aircraft type rating.*

[(a) An aircraft type rating course must include at least 10 hours of ground training on the aircraft systems, performance, operation, and loading. In addition, it must include at least 10 hours of flight instruction. Instruction in a pilot ground trainer that meets the requirements of § 141.41 (a) (1) may be credited for not more than 5 of the 10 hours of required flight instruction. Instruction in a pilot ground trainer that meets the requirements of § 141.41 (a) (2) may be credited for not more than 2.5 of the 10 hours of required flight instruction.

[(b) For airplanes that require type ratings, the aircraft type rating course must include ground flight training on the maneuvers and procedures of Part 61, Appendix A that is appropriate to the airplane for which a type rating is sought.]

Appendix G to Part 141—Pilot ground school course

1. *Applicability.* This appendix prescribes the minimum curriculum for a pilot ground school course required by § 141.55.

2. *General requirements.* An approved course of training for a pilot ground school course must contain the instruction necessary to provide each student with adequate knowledge of those subjects needed to safely exercise the privileges of the pilot certificate sought.

3. *Ground training instruction.* A pilot ground school course must include at least the subjects and the numbers of hours of ground training specified in the ground training section of the curriculum prescribed in the appendixes to this part for the certification or test preparation course to which the ground school course is directed.

4. *Stage and final tests.* Each student must pass a written test at the completion of each stage of training specified in the approved training syllabus for each ground training course in which he is enrolled. In addition, he must pass a final written test at the completion of the course. The stage and final tests may be credited towards the total ground training time required for each certification and test preparation course as provided in the curriculum prescribed in the appendixes to this part for that course.

Appendix H to Part 141—Test preparation courses

1. *Applicability.* This appendix prescribes the minimum curriculum required under § 141.55 for this part for each test preparation course listed in § 141.11.

2. *General requirements.*

(a) A test preparation course is eligible for approval if the Administrator determines that it is adequate for a student enrolled in that course, upon graduation, to safely exercise the privileges of the certificate, rating, or authority for which the course is conducted.

(b) Each course for a test preparation must be equivalent in scope, depth, and detail with the curriculum for the corresponding test course prescribed in Appendixes A, B, C, and D of this part. However, the number of hours of ground training and flight training included in the course must meet the curriculum prescribed in this appendix. (The minimums prescribed in this appendix for each test preparation course are based upon the amount of training that is required for students who meet the total flight experience requirements prescribed in Part 61 of this chapter at the time of enrollment.)

(c) Minimum experience, knowledge, or skill, requirements necessary as a prerequisite for enrollment are prescribed in the appropriate test preparation courses contained in this appendix.

3. *Flight instructor certification course.*

(a) An approved course of training for a flight instructor certification course must contain at least the following;

(1) *Ground training*—40 hours.

(2) *Instructor training*—25 hours, including—

(i) [10 hours of flight instruction in the analysis and performance of flight training maneuvers, which for students enrolled in a flight instructor airplane certification course and a flight instructor glider certification course includes the satisfactory demonstration of stall awareness, spin entry, spins, and spin recovery techniques in an aircraft of the appropriate category that is certificated for spins;]

(ii) 5 hours of practice ground instruction; and

(iii) 10 hours of practice flight instruction (with the instructor in the aircraft).

(b) *Credit for previous training of experience:* A student may be credited with the following training and experience acquired before his enrollment in the course.

(1) Satisfactory completion of two years of study on the principles of education in a college or university may be credited for 20 hours of the required 40 hours of ground training prescribed in paragraph (a) (1) of this section.

(2) One year of experience as a fulltime instructor in an institution of secondary or advanced education may be credited for 5 hours of the required practice ground instruction prescribed in paragraph (a) (2) of this section.

(c) *Prerequisite for enrollment.* To be eligible for enrollment each student must hold—

(1) A commercial pilot certificate;

(2) A rating for the aircraft used in the course; and

(3) An instrument rating for enrollment in an airplane instructor rating course.

4. *Additional flight instructor rating courses.*

(a) An approved course of training for an additional flight instructor rating course must consist of at least the following:

(1) *Ground training*—20 hours.

(2) *Instructor training (with an instructor in the aircraft).* 20 hours, including—

(i) [10 hours, or 10 flights in a glider in the case of a glider instructor rating course, performing analysis of flight training maneuvers, which in the case of an airplane instructor rating course and a glider instructor rating course includes the satisfactory demonstration of stall awareness, spin entry, spins, and spin recovery techniques in an aircraft of the appropriate category that is certificated for spins; and]

(ii) 10 hours of practice flight instruction, or, in the case of glider instructor rating course, 10 flights in a glider.

5. *Additional instrument rating course (airplane or helicopter).*

(a) An approved training course for an additional instrument rating course must include at least the following:

(1) *Ground training*—15 hours.

(2) *Flight instruction*—15 hours.

(b) *Prerequisites for enrollment.* To be eligible for enrollment each student must hold a valid pilot certificate with an instrument rating, and an aircraft rating for the aircraft used in the course.

6. *Airline transport pilot test course.*

(a) An approved training course for an airline transport pilot test course must include at least the following:

(1) *Ground training*—40 hours.

(2) *Flight instruction*—25 hours, including at least 15 hours of instrument flight instruction.

[(3) In airplanes that require type ratings, the course must include ground and flight training on the maneuvers and procedures of Part 61, Appendix A that are appropriate to the airplane for which a type rating is sought.]

(b) *Prerequisites for enrollment.* To be eligible for enrollment each student must—

(1) Hold a commercial pilot certificate with an instrument rating and a rating for the aircraft used in the course; and

(2) Meet the experience requirements of Part 61 of this chapter for the issuance of an airline transport pilot certificate.

7. *Pilot certificate, aircraft or instrument rating refresher course.*

(a) An approved refresher training course for a pilot certificate, aircraft rating or an instrument rating must contain at least the following:

(1) *Ground training*—4 hours.

(2) *Flight instruction*—6 hours, which may include not more than 2 hours of directed solo or pilot in command practice.

(b) *Prerequisites for enrollment.* To be eligible for enrollment each student must hold a valid pilot certificate with ratings appropriate to the refresher course.

8. *Agricultural aircraft operations course.*

(a) An approved training course for pilots of agricultural aircraft must include at least the following:

(1) *Ground training*—25 hours, including at least 15 hours on the handling of agricultural and industrial chemicals.

(2) *Flight instruction*—15 hours, which may include not more than 5 hours of directed solo practice.

(b) *Prerequisite for enrollment.* To be eligible for enrollment each student must hold a valid commercial pilot certificate with a rating for the aircraft used in the course.

9. *Rotorcraft external-load operation course.*

(a) An approved training course for pilots of a rotorcraft with an external-load must contain at least the following:

(1) *Ground training*—10 hours.

(2) *Flight instruction*—15 hours.

(b) *Prerequisite for enrollment.* To be eligible for enrollment each student must hold a valid commercial pilot certificate with a rating for the rotorcraft used in the course.

Part 143 – Ground instructors

143.1 Applicability
143.3 Application and issue
143.5 Temporary certificate
143.7 Duration of certificate
143.8 Change of name; replacement of lost or destroyed certificate
143.9 Eligibility requirements: general
143.11 Knowledge requirements
143.15 Tests: general procedures
143.17 Retesting after failure
143.18 Written tests: cheating or other unauthorized conduct
143.19 Recent experience
143.20 Applications, certificates, logbooks, reports, and records: falsification, reproduction, or alteration
143.21 Display of certificate
143.23 Change of address

§ 143.1 Applicability.

This part prescribes the requirements for issuing ground instructor certificates and associated ratings and the general operating rules for the holders of those certificates and ratings.

§ 143.3 Application and issue.

(a) An application for a certificate and rating, or for an additional rating, under this part, is made on a form and in a manner prescribed by the Administrator. However, a person whose ground instructor certificate has been revoked may not apply for a new certificate for a period of one year after the effective date of the revocation unless the order of revocation provides otherwise.

(b) An applicant who meets the requirements of this part is entitled to an appropriate certificate with ratings naming the ground school subjects that he is authorized to teach.

(c) Unless authorized by the Administrator, a person whose ground instructor certificate is suspended may not apply for any rating to be added to that certificate during the period of suspension.

(d) Unless the order of revocation provides otherwise, a person whose ground instructor certificate is revoked may not apply for any ground instructor certificate for one year after the date of revocation.

§ 143.5 Temporary certificate.

A certificate or rating effective for a period of not more than 90 days may be issued to a qualified applicant, pending the issue of the certificate or rating for which he applied.

§ 143.7 Duration of certificate.

(a) A certificate or rating issued under this part, is effective until it is surrendered, suspended, or revoked.

(b) The holder of any certificate issued under this part that is suspended or revoked shall, upon the Administrator's request, return it to the Administrator.

§ 143.8 Change of name; replacement of lost or destroyed certificate.

(a) An application for a change of name on a certificate issued under this part must be accompanied by the applicant's current certificate and the marriage license, court order, or other document verifying the change. The documents are returned to the applicant after inspection.

(b) An application for a replacement of a lost or destroyed certificate is made by letter to the Department of Transportation, Federal Aviation Administration, Airman Certification Branch, P.O. Box 25082, Oklahoma City, Okla. 73125. The letter must—

(1) Contain the name in which the certificate was issued, the permanent mailing address (including zip code), social security number (if any), and date and place of birth of the certificate holder, and any available information regarding the grade, number, and date of issue of the certificate, and the ratings on it; and

(2) Be accompanied by a check or money order for $2.00, payable to the Federal Aviation Administration.

(c) A person whose certificate issued under this part has been lost may obtain a telegram from the FAA confirming that it was issued. The telegram may be carried as a certificate for a period not to exceed 60 days pending his receiving a duplicate certificate under paragraph (b) of this section, unless he has been notified that the certificate has been suspended or revoked. The request for such a telegram may be made by prepaid telegram, stating the date upon which a duplicate certificate was requested, or including the request for a duplicate and a money order for the necessary amount. The request for a telegraphic certificate should be sent to the office prescribed in paragraph (b) of this section.

§ 143.9 Eligibility requirements: general.

To be eligible for a certificate under this Part, a person must be at least 18 years of age, be of good moral character, and comply with § 143.11.

§ 143.11 Knowledge requirements.

Each applicant for a ground instructor certificate must show his practical and theoretical knowledge of the subject for which he seeks a rating by passing a written test on that subject.

§ 143.15 Tests: general procedures.

(a) Tests prescribed by or under this part are given at times and places, and by persons, designated by the Administrator.

(b) The minimum passing grade for each test is 70 percent.

§ 143.17 Retesting after failure.

An applicant for a ground instructor rating who fails a test under this part may apply for retesting—

(a) After 30 days after the date he failed that test; or

(b) Upon presenting a statement from a certificated ground instructor, rated for the subject of the test failed, certifying that he has given the applicant at least five hours additional instruction in that subject and now considers that he can pass the test.

§ 143.18 Written tests: cheating or other unauthorized conduct.

(a) Except as authorized by the Administrator, no person may—

(1) Copy, or intentionally remove, a written test under this part;

(2) Give to another, or receive from another, any part or copy of that test;

(3) Give help on that test to, or receive help on that test from, any person during the period that test is being given;

(4) Take any part of that test in behalf of another person;

(5) Use any material or aid during the period that test is being given; or

(6) Intentionally cause, assist, or participate in any act prohibited by this paragraph.

(b) No person who commits an act prohibited by paragraph (a) of this section is eligible for any airman or ground instructor certificate or rating under this chapter for a period of one year after the date of that act. In addition, the commission of that act is a basis for suspending or revoking any airman or ground instructor certificate or rating held by that person.

§ 143.19 Recent experience.

The holder of a ground instructor certificate may not perform the duties of a ground instructor unless, within the 12 months before he intends to perform them—

(a) He has served for at least three months as a ground instructor; or

(b) The Administrator has determined that he meets the standards prescribed in this part for the certificate and rating.

§ 143.20 Applications, certificates, logbooks, reports, and records: falsification, reproduction, or alteration.

(a) No person may make or cause to be made—

(1) Any fraudulent or intentionally false statement on any application for a certificate or rating under this part;

(2) Any fraudulent or intentionally false entry in any logbook, record, or report that is required to be kept, made, or used, to show compliance with any requirement for any certificate or rating under this part;

(3) Any reproduction, for fraudulent purpose, of any certificate or rating under this part; or

(4) Any alteration of any certificate or rating under this part.

(b) The commission by any person of an act prohibited under paragraph (a) of this section is a basis for suspending or revoking any airman or ground instructor certificate or rating held by that person.

§ 143.21 Display of certificate.

Each person who holds a ground instructor certificate shall keep it readily available to him while instructing and shall present it for inspection upon the request of the Administrator or an authorized representative of the National Transportation Safety Board, or of any Federal, State, or local law enforcement officer.

§ 143.23 Change of address.

Within 30 days after any change in his permanent mailing address, the holder of a ground instructor certificate shall notify the Department of Transportation, Federal Aviation Administration, Airman Certification Branch, P.O. Box 25082, Oklahoma City, Okla. 73125 in writing, of his new address.

NTSB Part 830 – Notification and reporting of aircraft accidents or incidents and overdue aircraft, and preservation of aircraft wreckage, mail, cargo, and records

Subpart A—General

830.1 Applicability
830.2 Definitions

Subpart B—Initial notification of aircraft accidents, incidents, and overdue aircraft

830.5 Immediate notification
830.6 Information to be given in notification

Subpart C—Preservation of aircraft wreckage, mail, cargo, and records

830.10 Preservation of aircraft wreckage, mail, cargo, and records

Subpart D—Reporting of aircraft accidents, incidents and overdue aircraft

830.15 Reports and statement to be filed

Subpart E—Reporting of public aircraft accidents and incidents

830.20 Reports to be filed

Authority: 49 U.S.C. 1441 and 1901 et seq.
Amended: June 21, 1989

Subpart A—General

§ 830.1 Applicability.

This part contains rules pertaining to:

(a) Notification and reporting aircraft accidents and incidents and certain other occurrences in the operation of aircraft when they involve civil aircraft of the United States wherever they occur, or foreign civil aircraft when such events occur in the United States, its territories or possessions.

(b) Reporting aircraft accidents and listed incidents in the operation of aircraft when they involve certain public aircraft.

(c) Preservation of aircraft wreckage, mail, cargo, and records involving all civil aircraft in the United States, its territories or possessions.

§ 830.2 Definitions.

As used in this part the following words or phrases are defined as follows:

"Aircraft accident" means an occurrence associated with the operation of an aircraft which takes place between the time any person boards the aircraft with the intention of flight and all such persons have disembarked, and in which any person suffers death or serious injury, or in which the aircraft receives substantial damage.

"Civil aircraft" means any aircraft other than a public aircraft.

"Fatal injury" means any injury which results in death within 30 days of the accident.

"Incident" means an occurrence other than an accident associated with the operation of an aircraft, which affects or could affect the safety of operations.

"Operator" means any person who causes or authorizes the operation of an aircraft, such as the owner, lessee, or bailee of an aircraft.

"Public aircraft" means an aircraft used exclusively in the service of any government or of any political subdivision thereof, including the government of any State, Territory, or possession of the United States, or the District of Columbia, but not including any government-owned aircraft engaged in carrying persons or property for commercial purposes. For purposes of this section "used exclusively in the service of" means, for other than the Federal Government, an aircraft which is owned and operated by a governmental entity for other than commercial purposes or which is exclusively leased by such governmental entity for not less than 90 continuous days.

"Serious injury" means any injury which: (1) Requires hospitalization for more than 48 hours, commencing within 7 days from the date of the injury was received; (2) results in a fracture of any bone (except simple fractures of fingers, toes, or nose); (3) causes severe hemorrhages, nerve, muscle, or tendon damage; (4) involves any internal organ; or (5) involves second- or third-degree burns, or any burns affecting more than 5 percent of the body surface.

"Substantial damage" means damage or failure which adversely affects the structural strength, performance, or flight characteristics of the aircraft, and which would normally require major repair or replacement of the affected component. Engine failure or damage limited to an engine if only one engine fails or is damaged, bent fairings or cowling, dented skin, small punctured holes in the skin or fabric, ground damage to rotor or propeller blades, and damage to landing gear, wheels, tires, flaps, engine accessories, brakes, or wingtips are not considered "substantial damage" for the purpose of this part.

Subpart B—Initial notification of aircraft accidents, incidents and overdue aircraft

§ 830.5 Immediate notification.

The operator of an aircraft shall immediately, and by the most expeditious means available, notify the nearest National Transportation Safety Board (Board), field office[1] when:

(a) An aircraft accident or any of the following listed incidents occur:

(1) Flight control system malfunction or failure;

(2) Inability of any required flight crewmember to perform normal flight duties as a result of injury or illness;

(3) Failure of structural components of a turbine engine excluding compressor and turbine blades and vanes;

(4) In-flight fire; or

(5) Aircraft collide in flight.

(6) Damage to property, other than the aircraft, estimated to exceed $25,000 for repair (including materials and labor) or fair market value in the event of total loss, whichever is less.

(7) For large multiengine aircraft (more than 12,500 pounds maximum certificated takeoff weight):

(i) In-flight failure of electrical systems which requires the sustained use of an emergency bus powered by a back-up source such as a battery, auxiliary power unit, or air-driven generator to retain flight control or essential instruments;

(ii) In-flight failure of hydraulic systems that results in sustained reliance on the sole remaining hydraulic or mechanical system for movement of flight control surfaces;

(iii) Sustained loss of the power or thrust produced by two or more engines; and

(iv) An evacuation of aircraft in which an emergency egress system is utilized.

(b) An aircraft is overdue and is believed to have been involved in an accident.

§ 830.6 Information to be given in notification.

The notification required in § 830.5 shall contain the following information, if available;

(a) Type, nationality, and registration marks of the aircraft;

(b) Name of owner, and operator of the aircraft;

(c) Name of the pilot-in-command;

(d) Date and time of the accident;

(e) Last point of departure and point of intended landing of the aircraft;

(f) Position of the aircraft with reference to some easily defined geographical point;

(g) Number of persons aboard, number killed, and number seriously injured;

(h) Nature of the accident, the weather and the extent of damage to the aircraft, so far as is known; and

(i) A description of any explosives, radioactive materials, or other dangerous articles carried.

Subpart C—Preservation of aircraft wreckage, mail, cargo, and records

§ 830.10 Preservation of aircraft wreckage, mail, cargo, and records.

(a) The operator of an aircraft involved in an accident or incident for which notification must be given is responsible for preserving to the extent possible any aircraft wreckage, cargo, and mail aboard the aircraft, and all records, including all recording mediums of flight, maintenance, and voice recorders, pertaining to the operation and maintenance of the aircraft and to the airmen until the Board takes custody thereof or a release is granted pursuant to § 831.12(b) of this chapter.

(b) Prior to the time the Board or its authorized representative takes custody of aircraft wreckage, mail, or cargo, such wreckage, mail, or cargo may not be disturbed or moved except to the extent necessary:

(1) To remove persons injured or trapped;

(2) To protect the wreckage from further damage; or

(3) To protect the public from injury.

(c) Where it is necessary to move aircraft wreckage, mail or cargo, sketches, descriptive notes, and photographs shall be made, if possible, of the original positions and condition of the wreckage and any significant impact marks.

(d) The operator of an aircraft involved in an accident or incident shall retain all records, reports, internal documents, and memoranda dealing with the accident or incident, until authorized by the Board to the contrary.

Subpart D—Reporting of aircraft accidents, incidents, and overdue aircraft

§ 830.15 Reports and statements to be filed.

(a) *Reports.* The operator of an aircraft shall file a report on Board Form 6120.1 (OMB No. 3147-005) or Board Form 7120.2 (OMB No. 3147-0001)[2] within 10 days after an accident, or after 7 days if an overdue aircraft is still missing. A report on an incident for which notification is required by § 830.5(a) shall be filed only as requested by an authorized representative of the Board.

(b) *Crewmember statement.* Each crewmember, if physically able at the time the report is submitted, shall attach a statement setting forth the facts, conditions, and circumstances relating to the accident or incident as they appear to him. If the crewmember is incapacitated, he shall submit the statement as soon as he is physically able.

(c) *Where to file the reports.* The operator of an aircraft shall file any report with the field office of the Board nearest the accident or incident.

Subpart E—Reporting of public aircraft accidents and incidents

§ 830.20 Reports to be filed.

The operator of a public aircraft other than an aircraft of the Armed Forces or Intelligence Agencies shall file a report on NTSB Form 6120.1 (OMB No. 3147-001)[3] within 10 days after an accident or incident listed in § 830.5(a). The operator shall file the report with the filed office of the Board nearest the accident or incident.[4]

Signed at Washington, DC, on this 16th day of September 1988.

James L. Kolstad,
Acting Chairman
[FR Doc. 88-21705 Filed 9-22-88; 8:45 am]

---

[1] The National Transportation Safety Board field offices are listed under U.S. Government in the telephone directories in the following cities: Anchorage, Alaska; Atlanta Ga.; Chicago, Ill.; Denver, Colo.; Fort Worth, Tex.; Kansas City, Mo.; Los Angeles, Calif.; Miami, Fla.; New York, N.Y.; Seattle, Wash.

[2] Forms are available from the Board field offices, (see footnote 1), the National Transportation Safety Board, Washington, DC 20594, and the Federal Aviation Administration, Flight Standards District Office

[3] To obtain this form, see footnote 2.

[4] The locations or the Board's field offices are set forth in footnote 1.

Flight forum

• Phonetic facts

The new classes of airspace, a, bee, cee, dee, e, and gee, all sound pretty much the same; especially, during congested times in the cockpit, on the edge of ATIS reception, or busy ATC periods. Why can't we use the phonetic alphabet to avoid confusion?

We can. FAA Order 7110.65, *Air Traffic Control*, paragraph 2-85, instructs controllers to use the ICAO pronunciation of numbers, and as necessary to clarify individual letters; paragraph 2-90, Airspace Classes states, "Classes A, B, C, D, E, and G airspace are pronounced in individual letter form. Only use ICAO phonetic pronunciation of letters to clarify the class of airspace. Examples: "Cessna 1-2-3 Mike Romeo cleared to enter Class B airspace." "Sikorsky 1-2-3 Tango Sierra cleared to enter New York Class Bravo airspace." Similar guidance can be found in paragraph 4-37 of the *Airman's Information Manual.* The criticality of pilot/controller communications demands complete understanding by all parties. Regardless of what method the controller uses, you are authorized to use the geographic and phonetic designators during your request and again during your acknowledgement. For example, you call San Antonio approach for clearance through the Class C, "San Antonio approach, Cessna two-three-four-six-Victor fifteen miles east, request clearance through the Class Charlie westbound at two-thousand five-hundred." San Antonio responds with: "Cleared through the Class C." You may reply, "Roger, Cessna four-six-Victor is cleared through the San Antonio Class Charlie." Thanks for flying safe.

• IFR Currency Requirements

There are a few questions that keep coming up about instrument currency. If an instrument rated pilot does not have the six hours and six approaches in six months required by FAR § 61.57(e)(1) but passes an instrument competency check as required by FAR § 61.57(e)(2), can he then fly IFR or does he still need the six hours and six approaches? If he passes an instrument rating check ride, type rating check ride, Airline Transport Pilot check ride or FAR 135 instrument check ride, does he still need the six hours and six approaches?

If a pilot is past the one year requirement of FAR § 61.57(e)(2) and passes a check ride for an ATP, type rating or FAR Part 135 IFR, is he required to have a separate sign off to meet the requirements of FAR § 61.57(e)(2) or is the paperwork for these other check rides sufficient?

A pilot who passes an instrument competency check required by FAR § 61.57(e)(2) does not need six IFR hours and six approaches in addition to passing the instrument competency check. The competency check suffices.

Any check ride in which the pilot demonstrates instrument competency may serve as an instrument competency check if the check ride meets the requirements of FAR § 61.57(e)(2). Passing an ATP check ride reestablishes instrument currency.

Regarding record keeping, although the FAR require only a reliable record to show compliance with the FAR, a pilot should ask for an appropriate entry in his or her log book. The reason is some check flight records such as a company FAR Part 135 checkride may not be entered in the pilot's logbook.

• High Performance

If a student pilot takes his private check ride in a high performance airplane (such as a Cessna 182), does he still need a high performance airplane sign off in accordance with FAR § 61.31(e) or is the check ride itself sufficient? Note that FAR § 61.31(e) specifically states that it applies to private or commercial pilots (how about ATP?) and not student pilots, so therefore the instructor could not sign him off before the checkride. If a separate sign off is required, can it be done by the examiner or must it be done by an instructor after the check ride?

How about the case of a private pilot without a high performance sign off getting an additional rating such as multiengine land or single-engine sea where no solo time is required? How would this sign off be accomplished?

What about a student pilot who has been signed off for solo flight in a tailwheel airplane? Does he need an additional sign off to meet the requirements of FAR § 61.31 (g) either before or after his private check ride? What about a commercial pilot taking a FAR Part 135 checkride in a tailwheel airplane who does not have this sign off? Keep in mind that not all 135 instructors and check airmen are also CFIs.

Believe it or not, these have come up recently. To be safe, we signed him off after his checkride, but are still not sure if this is required.

Every applicant must come to a checkride with all required instruction and endorsements. This includes appropriate aircraft and instructor endorsements. The easiest way to handle this situation is to make an appropriate logbook endorsement when the appropriate instruction is given.

• Landing Minima

The Landing Minima page of the IAP booklet refers to meteorological visibility and to RVR. Flight visibility is not mentioned. FAR §§ 91.175 (c)(2) and (d) state that operations below DH/MDA and landing are not permitted when flight visibility is less than the visibility minimum for the particular procedure. The FAR does not mention either RVR or meteorological visibility. Can a Pilot legally land an aircraft on a runway for which reported RVR is below the prescribed minimum, if the pilot estimates that the flight visibility is greater than the minimum RVR prescribed for the particular runway?

Yes, for FAR Part 91 operations.

• Uncontrolled IFR

Can instrument meteorological conditions legally be entered in uncontrolled airspace (e.g., below 1,200 feet AGL) without an ATC clearance, if pilot and airplane are legal for instrument flight?

Yes. A qualified pilot can fly IFR in uncontrolled airspace without an ATC clearance as long as he or she complies with the FAR. For example, IFR altitude rules based upon direction of flight and altitude apply as well as minimum IFR altitude rules based upon terrain and flight path. Certain IFR rules apply regardless of type of airspace. Although every flight or situation can be different, this question is an example of what may be legal may not be wise. In the eastern part of the U.S. it is hard to find enough uncontrolled (Class G) airspace for such a flight without entering controlled airspace. And all airspace is controlled in the Continental U.S. above 14,500 feet.

A cursory review of some IFR en route charts for the Nevada, Utah, and Montana areas show that much of the uncontrolled airspace left in the west is crossed or cut by controlled airspace. Entry into controlled airspace in IMC without an ATC IFR clearance is grounds for FAA enforcement action as well as being unsafe. Unauthorized IFR entry into controlled airspace in IMC also violates the trust that all pilots have with each other to operate according to a given set of rules (the FAR) for the safety of all.

Finally, ATC provides many important safety services to IFR pilots on an IFR flight plan in controlled airspace that an IFR pilot in uncontrolled airspace may not receive.

• Importing Aircraft

I am a pilot and I am also an aircraft broker. My question deals with importing an aircraft into the U.S. from England. If I was buying a Citation 1SP in England and wanted to import it into the United States, what inspections would be needed to be performed? I have heard two different stories. Some dealers say you need to do a Phase 1-5 inspection, and others say you only need to do a Phase 1-4 inspection. Also, if the Phase 1-4 or 1-5 inspection was done within the last 90 days do you need to do it again for export?

Rather than trying to answer your question regarding a specific Citation maintenance procedure, we want to refer you to the following advisory circulars (AC). AC 21-23, *Airworthiness Certification of Aircraft, Engines, Propellers, and Related Products Imported to the United States*, dated 7/7/87, and AC-21-2G, *Export Airworthiness Approval Procedures*, dated 7/9/92, should provide you the information you need. Both ACs provide additional references. You need to pay particular attention to the various Bilateral Airworthiness Agreements, if any, between the United States and the country you want to import from or export to. The agreements outline what standards you must meet when importing or exporting aircraft. You can contact your local Flight Standards district office for more information. AC-21-2G is available for sale from the Government Printing Office. AC-21-23 is available from FAA. For complete information on ordering ACs you can request AC No. 00-2.6, *Advisory Circular Checklist (and Status of Other FAA Publications)* from the U.S. Department of Transportation, General Services Section, M-443.2, Washington, DC 20590.

• Controller Responsibility

What responsibility, if any, does an ATC "tower" controller have to inform an FSDO/ACDO of an aircraft landing (IFR) when the reported visibility is below that prescribed for the approach?

Air traffic controllers' responsibility is to report incidents that are in violation of the Code of Federal Regulations (CFR). If an aircraft landing below the prescribed visibility for an approach violated a CFR and the controller was aware of the violation, then he/she is required to report the incident.

INSTRUMENT CORNER

• Avionics questions

Because the following avionics questions are commonly asked, we decided to expand our explanation to help others understand the relationship between FAA and FCC regulations concerning aircraft avionics equipment.—FAA Aviation News *Editor*

I have several questions about avionics equipment and licensing based upon the following situations. I have seen many advertisements for avionics equipment that offer the equipment at low prices, but I am not sure who can install it or which type equipment I might need. I have also passed up an opportunity to buy an aircraft that had radios installed for which there were no FAA Form 337s because I was uncomfortable with the undocumented installation. And I would like to use a handheld radio in the aircraft I rent, what must I do?

Please answer the following questions based upon the above information:

Who can install which avionics, and what FAA and FCC licenses or certificates do they need?

If I rent an airplane with no radios that does not have an FCC station license what can I do to operate legally a handheld transceiver?

What avionics need be TSO'd and under what circumstances?

Are avionics considered instruments for maintenance purposes?

Your questions involve both the Federal Communications Commission (FCC) and the FAA. Basically, the FCC requires all transmitters used in an aircraft to be licensed as well as anyone making any internal adjustments to those transmitters. It may or may not require operators of that equipment to be licensed. Examples of aircraft transmitters include transponders, ELTs, radios, and airborne telephones. Normally, all of the transmitters aboard an aircraft are listed on the aircraft's radio station license. Although some aviation related frequencies are available for all airborne stations, the FCC has select frequencies for specific classes of aircraft operators or operations such as air carrier; private aircraft; flight test activities; aviation support which includes flying schools, soaring and free ballooning operations, ramp operations, and safety related activities; and public service applications. In the case of air carrier and private aircraft, you can only apply for one category of aircraft service on the same application.

If you use your handheld transceiver only as a backup on an aircraft that has a radio station license, you do not need a radio station license for the handheld. If the aircraft does not have a radio station license, then you will need to apply to the FCC for a "portable" radio station license before you can transmit using your handheld transceiver. But before you mail in your radio station application, you should check your aircraft. If the aircraft has an ELT installed for example, it should have a radio station license. If the aircraft is not required to have an ELT aboard, FAR § 91.207 applies, and you could not find any other transmitter that requires a radio station license, then you should apply for a "portable" radio station license for your handheld transmitter using FCC Aircraft Radio Station Application (FCC 404) form.

Because you would be asking for a portable license, you will be asked for a statement about why you need a portable station license. Once issued, though, a radio station license is valid for five years. The current application fee is $115 per aircraft per application (subject to change). The FCC 404 application form's directions explain about the required comment and how to compute the required fee. FCC 404 is available from the FCC through its regional offices listed in your telephone book or by writing to the FCC Private Radio Bureau, Licensing Division, Consumer Assistance Branch, 1270 Fairfield Rd., Gettysburg, PA 17325-7245. Telephone (717) 337-1212.

The FCC publishes fact sheets that list requirements for its various licenses. For example, Private Radio Bureau, PR-5000, Fact Sheet Number 4, March 1993, *Aircraft Radio Stations*, states, "No Restricted Radiotelephone Operators Permit is required to operate VHF radio equipment on board an aircraft when that aircraft is flown domestically. However, when operating other than VHF radio equipment on domestic flights and when operating any radio equipment on international flights, the operator must hold at least a Restricted Radiotelephone Operators Permit (FCC Form 753)." A $35 fee is required for the permit.

Regarding your question about TSO'd equipment, TSO'd equipment is only required to be installed if specified in an FAA regulation. For example, FAR § 91.215 lists TSO requirements for general aviation transponders. If not required by FAR, you can install only those items which meet the aircraft's certification basis (FAR Parts 21, 23, or 25).

Regarding the term "instrument," if you are asking is a radio considered an instrument, the answer is no. FAR Part 1 states the word *instrument* "means a device using an internal mechanism to show visually or aurally the attitude, altitude, or operation of an aircraft or aircraft part. It includes electronic devices for automatically controlling an aircraft in flight." The FAR Part 1 definition for the term *appliance* applies. The broadly defined term *appliance* includes communication equipment. As defined, an aircraft appliance can be just about anything installed in or attached to an aircraft that is not part of an aircraft's airframe, engine, or propeller.

Regarding who can install a piece of avionics gear, an airframe rated mechanic with a valid Inspection Authorization (IA) can install avionics equipment and return the aircraft to service. A better installation choice is an FAA-certificated Repair Station with the appropriate radio rating. An appropriately rated Repair Station can also perform maintenance on the installed equipment.

Generally, the FAA certificates qualified persons who can install a piece of avionics equipment in an aircraft. The FCC sets the standards for transmitting equipment and determines who can make any internal adjustments to that item. The FAA prescribes maintenance and installation standards for avionics components and systems.

Finally, regarding your question about an "undocumented" installation, only major alterations and repairs require FAA Form 337. Either an appropriately rated Repair Station or an A&P mechanic with an IA can review the installation, and if an installation meets appropriate airworthiness standards, the installation can be signed off with an appropriate maintenance record entry. A new aircraft weight and balance must be computed and the equipment list revised when installing or removing avionics components. Each installation of new equipment must also be documented in the aircraft's equipment list and weight and balance records.

Are you ready for METAR?
The upcoming meteorological code change

by Dan Gudgel in FAA's *Aviation News*

The advent of flight in the United States at the turn of the century brought about a coinciding and incipient development of aviation meteorological services in the 1930s. The aeronautical meteorological codes in use today by the National Weather Service (NWS) for the domestic United States evolved because of limited transmission speed on teletype systems. However, different codes evolved for the remainder of the world, resulting in two sets of aeronautical meteorology codes.

The Aviation (Surface) Weather Observation (SAO) and Terminal Forecast (FT) products use the North American aviation meteorological codes while the remainder of the world uses the Aviation Routine Weather Report (METAR) and Terminal Aerodrome Forecast (TAF) formats, respectively. With the deregulation of the air transport industry, rapid expansion in the 1980s, and subsequent operational changes, air commerce dictated the standardization of aeronautical meteorological codes.

The North American codes used in the SAO and FT formats have long been a favorite airman test item for the Federal Aviation Administration (FAA) as the information they contain contributes to the assurance that airmen will be able to conduct a safe flight (FAR § 91.103, "Preflight Action").

Although the North American aeronautical meteorological codes have experienced only minor changes through the past 60 years, new codes will be implemented in the United States on January 1, 1996—resulting in major changes.

These code changes are not occurring overnight, and they are not limited to North America. On July 1, 1993, several changes were implemented:

1. Two hundred fifty landing rights airports in the United States had their Aviation Weather Observations converted by computer software to the new METAR code for international dissemination. The change is transparent within North America, but international air transport interests are able to receive the product using the new METAR code.
2. The world aviation community switched from the "old" METAR and TAF codes to the "new" METAR and TAF code.
3. The approximately 80 internationally-required TAFs issued by the NWS began to use the new TAF code format.
4. Other international aviation forecasts for part of the North Atlantic, Gulf of Mexico/Caribbean area, and Pacific routes were prepared in the new METAR and TAF code form, i.e., weather abbreviations and contractions.

So what has this all to do with aviation within the United States? The changes in aeronautical meteorological codes which only impact those in the international community currently will come to fruition for the "domestic" community on January 1, 1996! The weather information gathered by meteorologists, pilots, and other related interests will see the following occur at that time:

1. The United States will convert completely to the new METAR and TAF codes. The unique North American observation and forecast codes will no longer be used.
2. Other U.S. domestic aviation forecasts will convert to the new METAR and TAF code form, i.e., weather abbreviations and contractions found in TWEBs, Area Forecasts, etc.

The accompanying figures provide a comparison between the TAF/METAR codes as opposed to the North American FT/SAO codes for aviation forecast products and aviation weather observations. Providing you with the current codes format, Fig. 7-27[1] on page 221 depicts updated information commonly seen on the venerable "key card" describing NWS Aviation Weather Forecasts and the Manual Aviation Weather Observation code. (Note: This card now specifies "Manual" Aviation Weather Observations differentiating from the automated surface observations coming on-line.) The tables on pages 648 and 649 reflect the new TAF/METAR code format and associated decoding.

For example, the Manual Aviation Weather Observation for Pittsburgh, PA, is first listed in the North American SAO code, and then in METAR. For familiarity, practice reading the SAO and then check with the decode card. Take a brief look at the following codification in the METAR format for the same observation.

North American

PIT SA 1955 M10 OVC 3/4TRW
103/65/61/2215G25/992/R22LVR27

METAR

METAR KPIT 1955Z 22015G25KT
3/4SM R22L/2700FT TSRA
OVC010CB 18/16 A2992

While not covering every nuance of the new METAR code, I make specific references to some of the more pronounced differences between it and the North American code:

• While the Station Identifier and Time groups are still listed first, subsequent weather observation parameters in the METAR have been altered in their presentation.

• Within the METAR the Wind Group utilizes five digits, with additional digits/alphanumeric characters for gusts. Rather than listing the wind direction as only two digits

Continued on page 650.

KEY to NEW INTERNATIONAL AERODROME FORECAST (TAF) and NEW AVIATION ROUTINE WEATHER REPORT (METAR)

TAF
KPIT 091720Z 1818 22020KT 3SM -SHRA BKN020 FM20 30015G25KT 3SM SHRA OVC015 PROB40 2022 1/2SM TSRA OVC008CB FM23 27008KT 5SM -SHRA BKN020 OVC040 TEMPO 0407 00000KT 1SM -RA FG FM10 22010KT 5SM -SHRA OVC020 BECMG 1315 20010KT P6SM NSW SKC
METAR
KPIT 1955Z 22015G25KT 3/4SM R22L/2700FT TSRA OVC010CB 18/16 A2992

Forecast	Explanation	Report
TAF	Message type: TAF-routine and TAF AMD-amended forecast, METAR-hourly and SPECI-special report	**METAR**
KPIT	ICAO location indicator	**KPIT**
091720Z	Issuance time: ALL times in UTC "Z", 2-digit date and 4-digit time for TAF, 4-digit time for METAR	**1955Z**
1818	Valid period: first 2 digits begins and last 2 ends forecast	
22020KT	Wind: first 3 digits mean true-north direction, nearest 10 degrees, (or VaRiaBle;) next 2 digits mean speed and unit, KT, (KMH or MPS); as needed, Gust and 2-digit maximum speed; 00000KT for calm; (for reports only, if direction varies 60 degrees or more, Variability appended, e.g., 180V260)	**22015G25KT**
3SM	Prevailing visibility: In U.S., Statute Miles & fractions; above 6 miles in TAF Plus6SM. (Or, 4-digit minimum visibility in meters and as required, lowest value with direction)	**3/4SM**
	Runway Visual Range: R; 2-digit runway designator Left, Center, or Right as needed; "/"; Minus or Plus in U.S., 4-digit value, FeeT in U.S., (usually meters elsewhere); Variability (and tendency Down, Up or No change)	**R22L/2700FT**
-SHRA	Significant present, forecast and recent weather: see table	**TSRA**
BKN020	Cloud amount, height and type: SKy Clear, SCaTtered, BroKeN, OVerCast; 3-digit height in hundreds of feet; and either Towering CUmulus or CumulonimBus. Or Vertical Visibility for obscured sky and height "VV004", or unknown height "VV///". More than one layer may be forecast or reported. CLeaR for "clear below 12 thousand feet" at automated observation sites.	**OVC010CB**
	Temperature: in degress Celsius; first 2 digits, temperature "/" last 2 digits, dew-point temperature; below zero reported with Minus, e.g., M06	**18/16**
	Altimeter setting: indicator and 4 digits; in U.S., A-inches and hundredths; (Q-hectoPascals, e.g., Q1013)	**A2992**

KEY to NEW INTERNATIONAL AERODROME FORECAST (TAF) and NEW AVIATION ROUTINE WEATHER REPORT (METAR)

	Supplementary information for report: (Wind Shear in lower layers, LanDinG or TaKeOFf, RunWaY, 2-digit designator; REcent weather of operational significance.) ReMarK: Automated Observation or AOAugmented; wind shear; recent weather and time; and TORNADO, FUNNEL CLOUD, WATERSPOUT.	
FM20	FroM and 2-digit hour: indicates significant change	
PROB40 2022	PROBability and 2-digit percent: probable condition during 2-digit beginning and 2-digit ending time period	
TEMPO 0407	TEMPOrary: changes expected for less than 1 hour and in total, less than half of 2-digit beginning and 2-digit ending time period	
BECMG 1315	BECoMinG: change expected during 2-digit beginning and 2-digit ending time period	

Table of Significant Present, Forecast and Recent Weather - Grouped in categories and used in the order listed below; or as needed in TAF, No Significant Weather.

QUALIFIER

Intensity or Proximity

- Light "no sign" Moderate + Heavy

VC Vicinity: but not at aerodrome; in U.S., 5-10SM from center of runway complex (elsewhere within 8000m)

Descriptor

MI Shallow	BC Patches	DR Drifting	TS Thunderstorm
BL Blowing	SH Showers	FZ Supercooled/freezing	

WEATHER PHENOMENA

Precipitation

DZ Drizzle	RA Rain	SN Snow	SG Snow grains
IC Diamond dust	PE Ice pellets	GR Hail	GS Small hail/snow pellets
UP Unknown precipitation in automated observations			

Obscuration

BR Mist	FG Fog	FU Smoke	VA Volcanic ash
SA Sand	HZ Haze	PY Spray	DU Widespread dust

Other

SQ Squall	SS Sandstorm	DS Duststorm	PO Well developed dust/sand whirls
FC Funnel cloud/tornado/waterspout			

Explanations in parentheses "( )" indicate different worldwide practices. Ceiling defined as the lowest broken or overcast layer, or the vertical visibility. Automated Observations may be Augmented manually for certain weather phenomena. TAFs exclude temperature, turbulence and icing forecasts and METARs exclude trend forecasts. Although not used in U.S., Ceiling And Visibility OK replaces visibility, weather and clouds if: visibility is 10 kilometers or more; no cloud below 1500 meters (5000 feet) or below the highest minimum sector altitude, whichever is greater and no cumulonimbus; and no precipitation, thunderstorm, duststorm, sandstorm, shallow fog, or low drifting dust, sand or snow.

March 1993 **UNITED STATES DEPARTMENT OF COMMERCE**
NOAA/PA 93054 **National Oceanic and Atmospheric Administration**—National Weather Service

(rounded to the nearest 10 degrees), the METAR gives the full three-digit wind direction. The next two digits represent the wind speed with the units listed as "KT" for knots. For example, a wind from 220 degrees at 15 knots with gusts to 25 knots is encoded as "22015G25KT."

• The units of the observed weather parameter are listed in most cases in the METAR whereas none are given in the North American SAO, e.g., "3/4" denotes three-quarters statute mile visibility in the SAO while METAR lists it as "3/4SM." The reason for this is that the U.S. and Canada use English units of measure, whereas the rest of the world uses metric. Therefore, the unit of measure used is noted to avoid possible confusion.

• METAR Runway Visual Range (RVR) is listed next to the prevailing visibility rather than in Remarks.

• No longer will the Current Weather be limited to two characters. Four characters describe the type of precipitation and/or current weather along with a symbol for intensity, e.g., "TSRA" means a thunderstorm with moderate rain is occurring at the station.

• The Cloud Amount and Height is listed together within a six-character group, e.g., "OVC010CB" denotes that the sky condition is overcast, the base of the clouds is at 1,000 feet above the station, and the cloud type is cumulonimbus.

• The Temperature and Dew Point group is listed as 18/16, respectively, and it is in degrees **Celsius** rather than degrees Fahrenheit. Below zero temperatures will be preceded by the letter "M."

• Rather than encoded as only a three-digit group, the altimeter setting lists all four digits and is preceded by the letter "A," e.g., an altimeter setting of 29-92 inches of mercury is encoded as "A2992."

• Sea Level Pressure is not reported in the METAR.

• Any subsequent Remarks are designated by "RMK" preceding the information.

In addition to the changes listed above in METAR, the TAF incorporates new "Change Groups" such as From (FM), Probability (PROB), Temporary (TEMPO), and Becoming (BECMG) rather than Occasional (OCNL), Chance (CHC), or Slight Chance (SLGT CHC) as currently used in the FT.

Keep in mind that change has already come to the world meteorological community regarding the new codes but the United States and the remainder of North America are being given another two-plus years to prepare. I encourage attendance at any FAA Accident Prevention Program where the TAF/METAR codes will be presented as well as studying the figures presented in this article.

Complete information regarding the TAF/METAR codes is published by the World Meteorological Organization in WMO Handbook No. 782, *Aerodrome Reports and Forecasts, A User's Handbook to the Codes*, dated 1992. It can be ordered for $16, plus $5 for shipping and handling from the American Meteorological Society, 45 Beacon Street, Boston, MA 02108; Attention: Publications Order.

Figures 3 and 4 are available as a free, informational key card. This card can be ordered through the Government Printing Office, Washington, DC 20402. Refer to NOAA/PA 93054, "Key to New International Aerodrome Forecast (TAF) and New Aviation Routine Weather Report (METAR)," dated March 1993. The card is also available from any NWS Office. In addition to the NWS, the FAA Accident Prevention Program will be printing these "TAF/METAR" cards for distribution. Airmen publications which will contain TAF/METAR codes in the near future are the *Airman's Information Manual* and the upcoming revision of *Aviation Weather Services*, AC-00-45D.

While agreeing that this change will cause transition difficulties because of the re-education it requires, the U.S. aviation community is already closely aligned with the international aviation community in several domestic aviation programs. The end result of the implementation of this "standard" meteorological code will be to continue to closely tie the United States aviation community to the world through this vehicle of common interest, the weather.

Mr. Gudgel works for the National Weather Service in the Aviation Services Branch, NWS Headquarters, Silver Spring, MD. He has been an active participant in the FAA's Accident Prevention Program since 1979, where he began working with the program in the Fresno, CA FSDO. Most recently he works with the Baltimore, MD FSDO. A professional pilot and flight instructor with ASEL, AMEL, Instrument—Airplane, Rotorcraft—Helicopter, he is also a Designated Pilot Examiner (Gliders) and holds an agricultural pilot certificate from the state of California.

Index

Federal Aviation Regulation page references are in boldface.
Glossary terminology used by pilots and controllers is in quotation marks.

A

abbreviations, FAR, **341-342**
accidents
  aircraft and incident reporting, 241-242
  cause factors, 236
  definitions, **641-642**
  initial notification, **642**
  judgment aspects of collision avoidance, 251-252
  near midair collision, 242, 251-252, **450, 563**
  potential flight hazards, 236-240
  preservation of wreckage, **642**
  reporting, **641-643**
  reporting overdue aircraft, 183, **642-643**
advanced qualification program (AQP), **617-620**
advisory circular (AC), vii
aerobatic flight, 251, **451**
aeronautical charts, 253-263
  aerospace planning (ASC), 261
  air distance/geography, 262
  airport obstruction (OC), 261
  Alaska enroute, 255
  auxiliary, 263
  charted VFR flyway planning, 255
  continental entry (CEC), 261
  enroute high altitude, 255
  enroute low altitude, 254
  global navigation, 263
  Gulf of Mexico and Caribbean planning, 255
  helicopter route, 259
  instrument approach procedure (IAP), 257
  jet navigation, 261-263
  LORAN C navigation, 262
  LORAN navigation & consol LORAN navigation, 261
  North Atlantic route, 255
  North Pacific Oceanic route, 255
  obtaining civil aeronautical, 253
  operational navigation, 262
  planning, 255
  related publications, 259-263
  standard instrument departure (SID), 257
  standard terminal arrival (STAR), 257
  terminal procedures publication (TPP), 257
  types available, 253
  VFR terminal area, 254
  weather plotting, 262
  world, 254, 262
aeronautical light beacons, 44
aircraft
  agricultural, **473**
  airborne inspection, ix, 240
  air taxi (*see* air taxi operations)
  airworthiness (*see* airworthiness)
  arresting devices, 64-65
  call signs, 108-109
  civil (*see* civil aircraft)
  ditching procedures, 190-193
  downed, 188
  dropping objects from, **428**
  emergencies (*see* emergencies)
  equipment (*see* equipment)
  fire fighting, **473**
  flight authorizations for noise restrictions, **498-499**
  increased weight for planes in Alaska, **454**
  interception signals, 176-178
  interchange, 109
  large nontransport limitations, **585-586**
  leased, 109
  lights, 125
  maintaining/overhauling/rebuilding, **343-355**
  noise (*see* noise)
  operating near other, **431**
  performance operating limitations, **580-587**
  performance requirements, **563-564**
  primary category, **454**
  proving tests, **552-553**
  ratings (*see* ratings)
  reckless/careless operation of, **427**
  seaplanes, ix, 238-240
  search and rescue, 181-184
  small nontransport limitations, **587**
  speed (*see* speed)
  transport (*see* transport aircraft)
  turbine-powered multiengine (*see* turbine-powered multiengine aircraft)
  waivers, **478**
  weight limitations, **464**
aircraft conflict alert, 171-172
air defense identification zones, **506-507**
airframe (*see also* airworthiness; maintenance)
  major alterations, **348**
  major repairs, **349**
  requirements, **596**
airline transport pilots, **391-394**
  airplane rating/aeronautical experience, **392**
  airplane rating/aeronautical knowledge, **391**
  airplane rating/aeronautical skill, **392-393**
  eligibility requirements, **391**
  instruction in air transportation service, **394**
  limitations, **394**
  privileges, **394**
  rotorcraft rating/aeronautical knowledge, **393**
  rotorcraft rating/aeronautical skill, **393**
  rotorcraft with helicopter class rating, **393-394**
  test requirements, **394, 397-410**
AIRMET, 199-202
air navigation lighting, 44
air navigation radio aids, 1-33
  distance measuring equipment (DME), 3, 7
  Doppler radar, 32
  flight management system (FMS), 32
  global positioning system (GPS), 32-33
  inertial navigation system (INS), 32
  instrument landing system (ILS), 6-9
  interim standard microwave landing system (ISMLS), 14
  LORAN, viii, 15-31
  marker beacon, 5-7

air navigation radio aids *continued*
  microwave landing system (MLS), 10-14
  NAVAID identifier removal during maintenance, 14
  NAVAID service volumes, 3, 5
  NAVAID with voice, viii, 15
  nondirectional radio beacon (NDB), 1
  OMEGA and OMEGA/VLF systems, 31-32
  tactical navigation (TACAN), 2
  user reports on NAVAID performance, 15
  VHF direction finder (VHF/DF), 32
  VHF omni-directional range (VOR), 1
  VHF omni-directional range/tactical air navigation (VORTAC), 2-3
  VOR receiver check, 1-2
air piracy, 193-194
Airport/Facility Directory, vii, 259
airport reservations office (ARO), 104-105
airports
  IFR requirements, **567**
  lighting aids, 38-43
    approach light systems (ALS), 38
    controlling, 42-43
    in-runway lighting, 41-42
    rotating beacons, 43
    runway edge light systems, 41
    runway end identifier lights (REIL), 40-41
    visual glideslope indicators, 38-40
  Los Angeles International, **490**
  marking aids, 44-65
    holding position markers, 52, 54, 57
    miscellaneous, 57-60
    pavement markings, 45
    runway markings, 45-52
    taxiway markings, 47, 52
  Mode C veil, 83-90
  operating vicinity of airport in Class E airspace, **433-434**
  operating vicinity of airport in Class G airspace, **433**
  reservation operations/procedures, viii, 104-106
  signs, 60-65
    destination, 64
    direction, 63-64
    information, 64
    location, 61-63
    mandatory instruction, 60-61
  special operating restrictions, **483-484**
  tower controlled, viii, 113-114
  traffic advisory practices without control towers, viii, 92-94
  transponder suspension from, **491-498**
  visual aids, 38-65
  visual indicators at uncontrolled, 114-115
airport surveillance radar (ASR), 34, 163-164
air route surveillance radar (ARSR), 34
air route traffic control center (ARTCC), 91
  en route communications, 150-157
airship rating, for commercial pilots, **390**
airspace, 66-90
  airport advisory area, 76
  alert area, 75
  Class A, 68, **417-418, 436**
  Class B, 68-69, 80, **380, 418, 435**
  Class C, viii, 70-71, 72-73, **418, 435**
  Class D, viii, 71, **418, 434-435**
  Class E, 71, **419-420, 433-434**
  Class G, 74, **433, 510**
  controlled, 67-74
  controlled firing area, 75
  flight limitations/prohibitions, 78
  flying VFR in congested, 236
  military operations area (MOA), 75
  military training route (MTR), 76
  North Atlantic (NAT), **468, 482**
  operating in vicinity of designated Mode C veil airports, 83-90
  parachute jump aircraft operations, 78
  prohibited areas, 74-75, **422-423, 436**
  restricted areas, 75, **421-422, 436**
  segment dimensions, 66-67
  special use, **421-423**
  suspension of transponder/automatic altitude reporting equipment requirements, 83-90
  temporary flight restrictions, 76-78
  terminal radar service area (TRSA), 80, 83
  VFR routes, 66-67, 78-80
  warning area, 75
airspeed (*see* speed)
air taxi operations, 121, **532-620**
  aircraft requirements, **540**
  airworthiness check, **543**
  autopilot, **546**
  business considerations
    advertising, **540**
    area limitations on operations, **540**
    management personnel, **541-542**
    office and operations base, **540**
    personnel, **544**
    recordkeeping, **542-543**
    terminating operations, **541**
    use of business name, **540**
  certification, **534-538**
  crewmembers (*see* crewmembers)
  emergencies, **538-539, 549**
  equipment on aircraft, **552-564**
  flight attendants (*see* crewmembers)
  flight locating requirements, **544**
  food service, **548-549**
  inspections and tests, **543**
  maintenance, **587-592**
  manual contents/requirements, **538-539**
  mechanical concerns, **543**
  oxygen, **545-546**
  passengers
    cargo, **544-555**
    carry-on baggage, **544**
    carrying weapons, **548**
    exit seating, **550-552**
    flight information/rules, **549-550**
    manipulation of controls by, **548**
    occupancy of pilot seat, **547**
    oxygen for, **545-546**
    pre-flight briefing, **548**
    provisions for, **544**
    seat belts, **550**
  second-in-command (*see* crewmembers)
  security, **549**
  weather conditions, **543**
  with large aircraft, **535-536**
air traffic control (ATC), 91-136
  airport operations, 113-129

approach control service, 91-92
clearances, 130-134
ground vehicle operations, 94
IFR approaches, 94
in-flight weather avoidance assistance, 212-213
light signals, 119, **433**
operation raincheck, 91
operation takeoff, 91
pilot services, 91-106
procedures, vii, 137-179
arrival, 157-169
departure, 147-150
en route, 150-157
national security and interception, 174-175
pilot/controller roles and responsibilities, 169-174
preflight, 137-146, 197-199
radio communications (*see* radio communications)
recommended communication procedures, 92
recording and monitoring, 91
roles and responsibilities, 169-174
safety alerts, 97-98
separations, 134-136
services (*see* specific service names of)
towers, 91
UNICOM/MULTICOM frequencies, 95
visiting, 91
air traffic control radar beacon system (ATCRBS), 34, 102
interrogator, 34
radarscope, 34
transponder, 34
airways, 152-153
reporting changes in, 153-154
airworthiness, **592-602**
airframe, **596**
airplane flight manual, **595-596**
cargo compartments, **597**
civil aircraft, **427**
control systems, **594**
cooling system, **600**
design and construction, **597**
electrical system, **601-602**
exhaust system, **600**
fatigue evaluation, **597**
flight requirements, **592**
fuel system components, **599**
ground loads, **597**
induction system, **600**
instruments, **594-595**
landing gear, **597**
limitations, **347**
miscellaneous items, **598**
operating limitations and information, **595**
passenger seating/accommodations, **597-598**
performance, **592-294**
powerplant controls/accessories, **600**
powerplant fire protection, **600-601**
propulsion, **598-599**
requirements, **558-559**
responsibility for, **587**
stability, **594**
stalls, **594**
systems and equipment, **601**
trim, **594**
Alaska Supplement, vii, 260
alcohol, 246, **412, 413, 414, 428, 542, 548, 569**
offenses involving, **361**
parachute jumping and, **513**
refusal to submit to test for, **360-361**
refusing to submit test results, **361**
alcohol misuse prevention program, **522-531**
(*see also* drug testing program; health and fitness)
administration/recordkeeping, **525-527**
consequences for employees engaging in alcohol, **527**
covered employees, **523**
definitions, **522**
employer requirements, **523**
employer's program, **529-530**
information for training/referral, **528-529**
post-accident testing, **523**
pre-employment testing, **523**
preemption of state/local laws, **522-523**
purpose, **522**
random testing, **523-524**
reasonable suspicion testing, **524-525**
return to duty testing, **525**
alphabet, phonetic, 110
altimeter, 228-230
errors, 230
high barometric pressure, 230, **437**
low barometric pressure, 230
setting procedures, 229-230
settings, **432**
system test and inspection, **353-354**
tests and inspections, **456-457**
altitude
alerting systems, **450**
decompression sickness, 248
density, 238
ear block, 248
effects of, 247-248
hypoxia, 247-248
instrument flight rules (IFR), 74
cruising, **443**
minimums, **442-443**
minimum for autopilot, **546**
minimum safe, 159, **432**
minimum vectoring, 159-160
reporting equipment, **449-450**
rules for turbine-powered multiengine aircraft, **461**
sinus block, 248
VFR cruising, 67, **439-440**
VFR minimums, **564**
appliances
major alterations, **348-349**
major repairs, **349**
approach clearance, viii, 160
approach control, 158
radar, 158
approaches
(*see also* instrument approach procedure; arrival procedures; instrument landing system; landing; microwave landing system)
"cleared for the option" procedure, 125
contact, 170
IFR minimums, **566**
instrument, 123-125, 158-159, 160-161, 170-171
instrument procedure charts, 159-160
low, 119

approaches *continued*
 minimums, 166-167
 missed, 167, 171, **441**
 no-gyro, 164
 overhead approach maneuver, 169
 parallel ILS/MLS, 165
 radar, 163-164
 radar monitoring of instrument, 164
 side-step maneuver, 165-166
 simultaneous converging instrument, 165
 simultaneous ILS/MLS, 164-165
 timed from holding fix, 162
 visual, viii, 167-168, 172-173
approach light systems (ALS), 38
approved maintenance organization (AMO), **347**
arresting devices, 64-65
arrival procedures 157-169 (*see also* approaches; landing)
 approach control, 158
 charted visual flight procedures (CVFP), 168
 contact approach, 168-169
 landing priority, 169
 local flow traffic management program, 157-158
 overhead approach maneuver, 169
 procedure turn, 161-162
 standard terminal arrival (STAR), 157
ATCRBS (*see* air traffic control radar beacon system)
aural speed warning device, **464**
automated radar terminal system (ARTS), 34
automated surface observation system (ASOS), 206-211
automated weather observing system (AWOS), 204-206
automatic meteorological observing station (AMOS), 203
automatic observing station (AUTOB), 204
automatic pressure altitude reporting equipment, **450**
automatic terminal information service (ATIS), 95-96
autopilot, minimum altitudes for, **546**
aviation safety reporting program, **429**
AWW (*see* severe weather forecast alerts)

B
baggage (*see also* equipment)
 carry-on, **462, 544**
 compartment, **597**
balloons, unmanned free, 237
barometric pressure, 212, 230
 flight restrictions during high, **437**
beacons
 aeronautical light, 44
 code, 44
 marker, 5-7
 nondirectional radio (NDB), 1, 5
 rotating airport, 43
birds (*see* wildlife)
braking action reports/advisories, viii, 117
briefings
 passenger, 137-146, **461-462, 548**
 preflight weather, 197-199

C
cabin pressure, **448**
call signs
 aircraft, 108-109
 ground station, 110
carbon monoxide poisoning, 249
cargo, **462, 544-555**
Category II
 approving instruments/equipment, **479-481**
 certificate of authorization for operation, **445**
 duration of pilot authorization, **362**
 maintenance program, **481**
 manual, **444-445, 479**
 operating rules, **444, 479-481**
 pilot authorization requirements, **372-373**
 required instruments/equipment, **479**
Category III, operating rules, **444**
ceiling (*see* clouds; meteorology; visibility)
center radar automated radar terminal system processing (CENRAP), 34
center weather advisory (CWA), 199, 200
center weather service unit (CWSU), 197
certificates and certification, **356-410** (*see also* pilot schools)
 address changes, **370**
 advanced qualification program (AQP), **617-620**
 aircraft ratings, **370-376**
 airline transport pilots, **391-394**
 air taxi operator, **534-538**
 and operating requirements, **611-614**
 application, **360**
 civil aircraft, **445**
 commercial pilots, **387-391, 633**
 exchanging, **362**
 exchanging obsolete for current, **359-360**
 experimental, **453**
 expired, **360**
 falsifying/reproducing/altering documents, **370**
 flight instruction from noncertified FAA instructors, **365**
 flight instructors, **358, 362, 394-397**
 flight review, **368-369**
 foreign flight instructors, **357**
 foreign pilots, **357**
 ground instructors, **639-640**
 instrument rating, **632**
 issued under Part 61, **358-359**
 limitations, **363-364**
 maintenance, **591**
 medical, **362**
  delegation of authority, **415-416**
  denial, **416**
  examiner requirements, **415, 416**
   falsifying/reproducing/altering records, **415**
  first class, **411-413**
  foreign airmen, **411**
  procedures, **415-416**
  records, **416**
  second class, **413-414**
  special issue of, **415**
  standards for, **411-416**
  third class, **414-415**
 name changes, **362**
 obsolete, **359**
 operations during medical deficiency, **368**
 pilot logbooks, **367-368**
 pilot-in-command, **369-370**
 pilots
  certificate requirements, **357-359**
  duration, **362**
  duration of Category II, **362**

foreign, **375**
recreational, **380-383**
student, **377-380**
private, **383-387**
qualifications, **360**
reissuing expired, **360**
replacing lost/destroyed certificates, **363**
second-in-command qualifications, **368**
special, **370-376**
surrendering, **362**
temporary, **361**
tests,
airline transport pilot, **397-404**
cheating or unauthorized conduct, **365**
general procedures for flight, **365**
general procedures for written, **364**
passing grades for written, **364**
prerequisites for flight, **365**
prerequisites for written, **364**
rating, **397**
required aircraft and equipment, **366**
retesting after failure, **367**
rotorcraft airline transport pilot with helicopter class, **405**
status of FAA inspectors and flight examiners, **366**
changeover point (COP), 154
charted visual flight procedures (CVFP), 168
charts, aeronautical, 253-263
civil aeronautical charts, obtaining, 253
civil aircraft
airworthiness, **427**
altitude reporting equipment, **449-450**
ATC transponders, **449-450**
automatic pressure altitude reporting equipment, **450**
certifications required, **445**
emergency locator transmitters, **447**
foreign, **468-469**
flight authorizations, **469**
inoperative instruments/equipment, **448-449**
instrument/equipment requirements, **445-446**
lights, **447-448**
limited category, **452**
operating outside of U.S., **468-469**
operation of Cuban registry, **469**
provisionally certificated, **452-453**
restricted category, **452**
supplemental oxygen, **448**
traffic alert/collision-avoidance system, **450**
turbojet-powered, **450**
Class A airspace, 68, **417-418, 436**
bearings/radials/mileages, **418**
overlapping designations, **418**
reporting points, **417-418**
Class B airspace, 68-69, 80, **418, 435**
student pilots and, **380**
Class C airspace, viii, 70-71, 72-73, **418, 435**
Class D airspace, viii, 71, **418, 434-435**
Class E airspace, 71, 73-74, **419-420**
classification of Federal airways, **419**
designation of VOR Federal airways, **420**
extent of Federal airways, **419**
operating on or in vicinity of airport in, **433-434**
Class G airspace, 74
operating on or in vicinity of airport in, **433**

ultralight flying requirements, **510**
clear air turbulence (CAT), 217
clearances, 130-134
abbreviated IFR, 147-148
adherence to, 133-134
air traffic, 170
altitude data, 130
amended, 131
approach, viii, 160
ATC and IFR flight plan requirements, **441**
"clearance," 276
"clearance void if not off by (time)," 277
"cleared (type of) approach," 277
"cleared approach," 277
"cleared as filed," 277
"cleared for takeoff," 277
"cleared for the option," 277
"cleared through," 277
"cleared to land," 277
complying with ATC, **433**
definition, 130
departure procedures, 130
deviating from, **506**
"expect further clearance (time)," 284
"expect further clearance via (airways/routes/fix)," 284
holding, 148
holding instructions, 131
IFR with VFR-on-top, 132
items contained in, 130-131
landing priority, 169
obstacle, 166
pilot responsibility, 132
pre-taxi procedures, 147
route of flight, 130
special VFR, viii, 131-132
taxi, 147
VFR/IFR flights, 133
visual procedures, 135-136
clouds
ceiling and sky cover, 206
parachute jumping requirements, **513**
reporting height of, 214
TAF codes, 228
ultralight flying requirements, **509**
cockpit voice recorders, **465-466, 554**
collision avoidance system, **450** (*see also* accidents; near midair collision)
commercial aircraft (*see* air taxi operations; transport aircraft; turbine-powered mutliengine aircraft)
commercial pilots, **387-391**
aeronautical knowledge, **387-388**
airplane rating/aeronautical experience, **389-390**
airship rating/aeronautical experience, **390**
applicability, **387**
eligibility requirements, **387**
flight proficiency, **388-389**
free balloon rating/aeronautical experience, **390-391**
glider rating/aeronautical experience, **390**
privileges and limitations, **391**
rotorcraft rating/aeronautical experience, **390**
common traffic advisory frequency (CTAF), 93-94
communications (*see* radio communications)
commuter planes, performance operating limitations, **587**
compass locator, 7

computerized voice reservation system (CVRS), 104, 106
contact approach, 168-169
 pilot/controller responsibilities, 170
contracts, truth-in-leasing clause, **428-429**
control towers (*see* air traffic control)
controlled airspace, 67-74
 Class A, 68, **417-418, 436**
 Class B, 68-69, 80, **380, 418, 435**
 Class C, viii, 70-71, 72-73, **418, 435**
 Class D, viii, 71, **418, 434-435**
 Class E, 71, 73-74, **419-420, 433-434**
 definition, 67
convective SIGMET, 200, 202
cooling system, **600**
course lights, 44
crewmembers
 alcohol use, **569-570**
 drug use, **569**
 flight attendants, **463, 547, 573, 580**
 flight engineers, **463**
 flight time limitations and rest requirements, **570-572**
 interfering with, **427**
 international operations certificate, **542**
 operating experience, **568**
 pilots, **568-569, 572-573**
 pilot-in-command, **369, 427, 567-568, 574-575**
 requirements, **567-570**
 responsibilities at stations, **430**
 second-in-command qualifications, **368, 463, 547, 568**
 training, **575-580**
cross-country flight
 private pilots and, **385**
 requirements for student pilots, **379-380**

D

data communications, microwave landing system (MLS), 13
Defense Mapping Agency Aerospace Center (DMAAC), 261
defense VFR (DVFR), 140
 radio communications failure, **506**
definitions, FARs, **335-341, 641-642**
density altitude, 238
department of transportation (DOT), drug testing program procedures, **515**
departure (*see also* takeoff)
 abbreviated IFR clearance, 147-148
 clearance void times, viii, 148
 hold for release, viii, 148
 instrument, 149-150, 173-174
 pre-taxi clearance, 147
 procedures, 147-150
 release times, viii, 148
 taxi clearance, 147
departure control, 148-149
dew point, 210, 212
Digital Aeronautical Chart Supplement (DACS), 260
direct user access terminal system (DUATS), 197, 222
direction finder (DF)
 emergency services, 181
 VHF, 32
distance measuring equipment (DME), 3
 holding, 156
 instrument landing system (ILS) and, 7
 standard service volumes, 3
Doppler radar, 32
drizzle, reporting intensity of, 215
drugs, **412, 413, 414, 428, 542, 569, 580, 622**
 offenses involving, **361**
 parachute jumping and, **513**
 smuggling, **428, 542**
 test for, **360-361**
drug testing program, **515-531** (*see also* alcohol misuse prevention program)
 administration/recordkeeping, **517-518**
 department of transportation procedures, **515**
 employee assistance program (EAP), **519**
 employees located outside territory of U.S., **521**
 employer's antidrug program, **519-521**
 follow-up testing, **517**
 medical review of results, **518-519**
 periodic testing, **516**
 post-accident testing, **516**
 pre-employment testing, **516**
 preemption, **521**
 random testing, **516**
 reasonable cause testing, **516-517**
 return to duty testing, **517**
 types of testing, **516**
 who must be tested, **515-516**

E

electrical system, **601-602**
electronic devices, portable, **428**
emergencies, 180-195 (*see also* safety alert)
 airborne inspection, ix, 240
 aircraft ditching procedures, 190-193
 aircraft exits, **464-465**
 air piracy/terrorism/highjacking, 193-194
 air taxi, **538-539**
 air traffic rules, **437**
 ATC operation, **490-491**
 direction finding instrument approach procedure, 181
 distress and urgency procedures, 189-194
 equipment needed for, **461**
 equipment requirements for 19+ passenger planes, **559-562**
 evacuation duties, **549**
 explosive detection, 182
 fire extinguishers, **556**
 fuel dumping, 194
 ground proximity warning system, **555**
 intercept and escort, 181-184
 observance of downed aircraft, 188
 obtaining assistance, 189-190
 overwater equipment, **558**
 pilot responsibility and authority, 180
 radar service for VFR aircraft in difficulty, 180
 radio communications failure, 194-195, **443-444**
 radio equipment for overwater operations, **460**
 requesting immediate assistance, 180
 search and rescue, 181-184
 security, **504-505**
 services available to pilots, 180-188
 signals for, 184-188
 survival equipment for overwater operations, **460**
 training for crewmembers, **576**
 transponder operation, 180-181

emergency arresting gear, 64-65
emergency locator transmitter (ELT), 181-182, **447**
  false alarms, 181
  in-flight monitoring and reporting, 182
  testing, 181
employee assistance program (EAP), **519**
en route flight advisory service (EFAS), 199
en route procedures, 150-157
  additional reports, 152
  airways and route systems, 152-153
  ARTCC communications, 150-151
  changeover points, 154
  holding, viii, 154-157
  position reporting, 151-152
engines (*see* maintenance)
equipment, **552-564**
  airborne weather radar, **559**
  child restraints, **430-431**
  cockpit voice recorders, **465-466, 554**
  dual controls, **553**
  emergency, **559-562**
  fire extinguishers, **556**
  flight recorders, **465-466, 484-487, 554-555, 602-608**
  ground proximity warning system, **555**
  inoperable, **562-563**
  materials for compartment interiors, **467, 559**
  overwater operations, **460, 558**
  oxygen, **448, 545-546, 556**
  parachutes, **451, 513-514**
  pitot heat indication system, **556**
  public address system, **553-554**
  seat belts/shoulder harnesses, **430-431, 550, 559**
  thunderstorm detection, **559**
exhaust system, **600**
explosives, detecting, 182

F

fan marker (FM), 5
fatigue, 246-247
Federal Aviation Administration (FAA), vi
  weather services, 197 (*see also* meteorology)
ferry flight, **466-467**
fire extinguishers, **556**
flight attendants, **463, 547, 573, 580**
flight engineers, requirements, **463**
Flight Forum, **643-646**
flight information publications (FLIP), 262
flight inspection, 125
flight instructors
  aeronautical knowledge, **395**
  authorizations, **395-396**
  certificate
    expiration, **397**
    renewal, **396**
  certification, **394-397**
    requirements, **358**
  eligibility requirements, **394-395**
  flight proficiency, **395**
  limitations, **396**
  ratings, **395, 397**
  recordkeeping, **395**
flight management system (FMS), 32
flight manual, compliance with, **427**
flight path
  base leg, 114
  downwind leg, 114
  final approach, 114
  upwind leg, 114
flight plan
  cancelling IFR, 146
  change in, 146
  change in proposed departure time, 146
  closing VFR/DVFR, 146
  composite (VFR/IFR flights), 140-141
  defense VFR (DVFR), 140
  deviating from, **506**
  flights outside United States and U.S. territories, 145-146
  IFR flights, 141-145, **440-445**
  IFR operations to high altitude destinations, 145
  mountain flying, 237-239
  VFR, viii, 139-140, **438**
flight recorders, **465-466, 554-555**
  specifications for aircraft, **484-485, 602-603, 605-607**
  specifications for helicopter, **485-487, 604-605, 607-608**
flight rules, **429-464** (*see also* operating rules)
  aircraft speed, **432**
  altimeter settings, **432**
  ATC emergency operation, **490-491**
  ATC light signals, **433**
  compliance with ATC clearances/instructions, **433**
  crewmembers, **430**
  emergency air traffic rules, **437**
  flight authorization for noise restricted aircraft, **498-499**
  general, **429-437**
  Grand Canyon National Park, **487-490, 614-616**
  instrument (*see* instrument flight rules)
  limitations in proximity of space flight operations, **437**
  Los Angeles International Airport, **490**
  minimum safe altitude, **432**
  operating in Class A airspace, **436**
  operating in Class B airspace, **435**
  operating in Class C airspace, **435**
  operating in Class D airspace, **434-435**
  operating near other aircraft, **431**
  operating on or in vicinity of airport in Class E airspace, **433-434**
  operating on or in vicinity of airport in Class G airspace, **433**
  restricted and prohibited areas, **436**
  restrictions, **436-437**
  restrictions during high barometric pressure conditions, **437**
  restrictions in Hawaii national disaster areas, **437**
  restrictions in proximity of President and others, **437**
  right-of-way, **431-432**
  seat belts/shoulder harnesses/child restraints, **430-431**
  simulated instruction, **431**
  transponder suspension requirement for certain airports, **491-498**
  visual (*see* visual flight rule)
flights
  operations outside of U.S., **468**
  operations within North Atlantic minimum navigation performance specifications (MNPS), **468, 482**
  U.S. to/from Afghanistan, **499-500**
  U.S. to/from Canada, **468**

flights *continued*
U.S. to/from Cuba, **468**
U.S. to/from Haiti, **500**
U.S. to/from Mexico, **468**
U.S. to/from Yemen, **500**
U.S. to/from Yugoslavia, **499**
flight safety (*see* safety)
flight service station (FSS), 91, 197
flight simulation, **431, 577**
food service, **463-464, 548-549**
free balloon rating
for commercial pilots, **390-391**
for private pilots, **387**
frequency
common traffic advisory (CTAF), 93-94
MULTICOM, 95
UNICOM, 95
very low frequency (VLF), 30
fuel
dumping, 194
minimum fuel advisories, ix, 174
system, **599**
VFR requirements, **438**
VFR supply, **564**

G
gliders
certification course, **635-636**
rating
for commercial pilots, **390**
for private pilots, **386**
towing, **373, 451-452**
glide path, 6-7
glideslope, 6-7
critical area, 8
glideslope indicators
alignment of elements systems, 40
precision approach path indicator (PAPI), 39
pulsating systems, 39-40
tri-color systems, 39
visual, 38-40
visual approach slope indicator (VASI), 38
global positioning system (GPS), 32-33
Grand Canyon National Park, **487-490, 614-616**
ground instructors, **639-640**
application, **639**
change of address, **640**
change of name, **639**
cheating or unauthorized conduct while testing, **640**
displaying certificate, **640**
duration of certificate, **639**
eligibility, **639**
falsifying/reproducing/altering records, **640**
knowledge requirements, **640**
replacement of lost/destroyed certificate, **639**
retesting after failure, **640**
temporary certificate, **639**
test procedures, **640**
ground proximity warning system, **555**

H
hand signals, 126-129
hazardous area reporting service, 102-104
health and fitness, 246-252
aerobatic flight, 251
alcohol and drugs, 246, **360-361, 412, 413, 414, 428, 513, 542, 548, 569-570, 580, 622**
alcohol misuse prevention program, **522-531**
carbon monoxide poisoning, 249
cardiovascular, **412, 413, 414**
certification, **358, 411-416**
decompression sickness, 248
drug testing program, **515-531**
ear/nose/throat, **411-412, 413, 414**
effects of altitude, 247-248
emotions, 247
fatigue, 246-247
flying after scuba diving, 248
hyperventilation in flight, 248-249
hypoxia, 247-248
illusions
in flight, 249-250
leading to landing errors, 249-250
leading to spatial disorientation, 249
judgment aspects of collision avoidance, 251-252
medical certificate, 246, **362**
medical standards, **411-416**
neurological, **412, 413, 414**
operations during medical deficiency, **368**
personal checklist, 247
sinus block, 248
stress, 247
vision, **411, 413, 414**
vision in flight, 250-251
helicopter hospital emergency medical evacuation service (HEMES), **572**
helicopters
flight recorders, **485-487, 604-605, 607-608**
landing areas, 60
landing procedures, 121-122
route charts, 259
takeoff procedures, 121-122
taxiing, 121
VFR operations at controlled airports, 121-122
VFR surface reference requirements, **564**
wake turbulence, 233-234
high density traffic airport (HDTA), reservations, 104-105
high intensity runway lights (HIRL), 41
highjacking, 193-194
holding, viii, 154-157
DME, 156
entry procedures, 156
examples of, 155
maximum airspeed, 155
nonstandard pattern, 157
pilot action during, 156-157
timed approaches from, 162
timing, 156
hot-air balloons, 68
hover taxi, 121
hyperventilation, 248-249
hypoxia, 247-248

I
icing, 200, **462-463, 566-567**
airframe, 216
TAF codes, 228

illness (*see* health and fitness)
illusions, 249-250
ILS course distortion, 8
induction system, **600**
inertial navigation system (INS), 32
initial approach fix (IAF), 160
inspections, **455-456, 588-589**
  air taxi, **543**
  altimeter, **353-354, 456**
  annual and 100-hour, **347**
  changes to aircraft programs, **457**
  emergency airborne of other aircraft, ix, 240
  flight checks, 125
  items included in annual/100-hour, **351-352**
  parachute jumping, **513**
  performance rules, **346-347**
  personnel, **590**
  progressive, **347**
  rotorcraft, **346**
  transponder, **354-355, 457**
  ultralights, **508**
instrument approach procedure (IAP), 159-161, **501-503**
  chart, 257
  direction finding, 181
  military field, 167
  minimum safe altitude (MSA), 159
  minimum vectoring altitude (MVA), 159-160
  pilot/controller responsibilities, 170-171
  procedures, **502-503**
  radar monitoring of, 164
  simultaneous converging, 165
  symbols and terms, **501-502**
  terminal procedures, **503**
instrument departures, pilot/controller responsibilities, 173-174
instrument flight rules (IFR), 201, **358, 440-445**
  airport requirements, **567**
  alternate airport requirements, **565**
  alternate airport weather minimums, **565**
  altitudes and flight levels, 74
  approach minimums, **566**
  ATC clearance, **441**
  Category II operations, **444-445**
  Category III operations, **444**
  course to be flown, **443**
  cruising altitude, **443**
  destination airport weather minimums, **565**
  flight level, 74, **443**
  flight simulation, **431**
  icing, **566-567**
  information required on flight plan, **440**
  landing minimums, **566**
  landing under, **441-442**
  minimum altitudes, **442-443**
  operating limitations, **565**
  operation in controlled airspace, **444**
  radio communications, **443**
    two-way failure, **443-444, 506**
  requirements, **371-372**
  reservations, 105
  takeoff limitations, **565**
  takeoff minimums, **566**
  takeoff under, **441-442**
  VOR equipment check, **440-441**
  weather requirements, **565-567**
instrument landing system (ILS), 6-9
  compass locator, 7
  components, **442**
  course distortion, 8
  critical area boundary sign, 62-63
  critical area holding position sign, 61
  distance measuring equipment (DME), 7
  frequency, 7
  glide path, 6-7
  glideslope, 6-7
  glideslope critical area, 8
  holding position markings, 57
  inoperative components, 8
  localizer, 6
  localizer critical area, 8
  localizer-type directional aid (LDA), 6
  marker beacons, 7
  minimums, 7-8
  parallel approach with MLS, 165
  simultaneous approach with MLS, 164-165
interim standard microwave landing system (ISMLS), viii, 14
International Civil Aviation Organization (ICAO), 189, 222-228

L

landing (*see also* approaches; arrival procedures; instrument landing system; microwave landing system)
  designated helicopter areas, 60
  exiting runway after, 123
  flight paths, 114
  helicopters, 121-122
  IFR minimums, **566**
  illusions leading to errors in, 249-250
  instrument approaches, 123-125
  interim standard MLS, 14
  large nontransport aircraft limitations, **586**
  low approaches and, 119
  microwave landing system, 10-14
  minimums, 166-167
  priority of, 169
  transport aircraft limitations, **583**
  turbine-powered multiengine aircraft limitations, **585**
  under IFR, **441**
landing direction indicator, 115
landing gear, **597**
landing strip indicator, 115
large aircraft (*see* airships; air taxi operations; transport aircraft; turbine-powered multiengine aircraft)
law enforcement, civil and military organizations, 176
lighter-than-air rating
  certification course, **635-636**
  for private pilots, **386**
lighting, 38-43
  aircraft, 125
  air navigation, 44
  alignment of elements systems, 40
  approach light systems (ALS), 38
  ATC signals, 119, **433**
  aviation red, 44
  civil aircraft, **447-448**
  controlled by control tower, 42
  controlled by pilots, 42-43

lighting *continued*
course, 44
dual, 44
high intensity runway lights (HIRL), 41
high intensity white, 44
low intensity runway lights (LIRL), 41
medium intensity runway lights (MIRL), 41
obstruction, 44
precision approach path indicator (PAPI), 39
pulsating systems, 39-40
radio control system, 43
rotating airport beacons, 43
runway centerline lighting (RCLS), 41
runway end identifier lights (REIL), 40-41
taxiway, 43
centerline, 43
edge, 43
turnoff lights, 42
touchdown zone lighting (TDZL), 41
tri-color systems, 39
visual approach slope indicator (VASI), 38
line-of-position (LOP), 18
local airport advisory (LAA), 93
local flow traffic management program, 157-158
localizer, 6
critical area, 8
localizer-type directional aid (LDA), 6
logbooks, pilot, **367-368**
LOP, 18
LORAN, viii, 15-31
chains, 15, 18-30
navigation, 18, 20
receiver, 18
Los Angeles International Airport, **490**
low IFR (LIFR), 201
low intensity runway light (LIRL), 41
low level wind shear alert system (LLWAS), 117
low powered fan marker (LFM), 5
luggage (*see* baggage)

M

Mach 1, authorizations to exceed, **481-482**
maintenance, **343-355, 454-458, 587-592** (*see also* airworthiness)
alteration organization and, **589**
altimeter system test and inspection, **353-354**
approval for return to service after, **344**
authority to perform and approve, **591**
Category II operations, **481**
certificate requirements, **591**
continuing analysis and surveillance, **590**
development of major repair data, **609-611**
falsifying/reproducing/altering records, **346**
inspection performance rules, **346-347**
inspections (*see* inspections)
major alterations, **348-350**
major repairs, **348-350**
manual requirements, **589-590**
mechanical interruption summary reports, **588**
mechanical reliability reports, **587-588**
operation after, **455**
performed by Canadian persons on U.S. products, **347-348**
persons authorized to approve aircraft, **345**
persons authorized to perform work, **344**
preventive, **349-350**
recordkeeping, **343-344, 345-346, 457-458**
airworthiness release, **592**
major repairs/alterations, **350-351**
rebuilt engine records, **458, 591**
transferring records, **458, 592**
required, **454-455**
training program, **591**
transponder test and inspection, **354-355**
marginal VFR (MVFR), 201
marker beacon, 5-6
instrument landing system (ILS) and, 7
markings, airport (*see* airports, marking aids; runways, pavement markings; taxiways, pavement markings)
MAYDAY, 189, 295
medical (*see* health and fitness)
medium intensity runway lights (MIRL), 41
meteorology, 196-228
AIRMET, 199-202
altimeter setting procedures, 228-230
ATC in-flight weather avoidance assistance, 212-213
automated surface observation system (ASOS), 206-211
automated weather observing system (AWOS), 204-206
automatic meteorological observing station (AMOS), 203
automatic observing station (AUTOB), 204
barometric pressure, 212, 230, **437**
categorical outlooks, 201
ceiling and sky cover, 206
center weather advisory (CWA), 199, 200
clear air turbulence (CAT), 217
clouds, 214, 228
convective SIGMET, 199, 200, 202
dew point, 210, 212
drizzle intensity reporting, 215
en route flight advisory service (EFAS), 199
FAA weather services, 197
icing, 200, 216, 228, **462-463, 566-567**
IFR weather minimums, **565**
IFR weather requirements, **564-567**
in-flight weather advisories, 199-201
in-flight weather broadcasts, 202-203
international civil aviation organization (ICAO), 222-228
key to aviation weather observations and forecasts, 221-222
low level wind shear alert system (LLWAS), 117
microbursts, 218-219
national aviation weather advisory unit (NAWAU), 196
national weather service (NWS), 196
pilot weather report (PIREP), 215-218
pilots automatic telephone weather answering service (PATWAS), 196, 201
precipitation, 212
precipitation intensity reporting, 214
preflight briefing, 197-199
prevailing visibility, 214
runway visual range (RVR), 213-214, **442**
severe weather bulletin (WW), 199
severe weather forecast alert (AWW), 199, 202
SIGMET, 199-200, 202
snow intensity estimates, 215
terminal forecast (TAF), 222-228
thunderstorms, 199, 219-221, **559**

tornadoes, 199
transcribed weather broadcast (TWEB), 196, 201-202
turbulence, 200, 216, 228
VFR weather minimums, **438-439**
VFR weather requirements, **564-567**
visibility, 206, 208, 212
volcanic activity report, 244
volcanic ash, 208, 209
wake turbulence, 231-234
weather message switching center (WMSC), 196
weather observing programs, 203-211
weather radar services, 211-212
wind, 200, 210, 212
wind shear, 117, 216-217
microbursts, 218-219
microwave landing system (MLS), 10-14
approach azimuth guidance, 10
back azimuth guidance, 10
data communications, 13
elevation guidance, 13
interim standard (ISMLS), viii, 14
operational flexibility, 13-14
parallel approach with ILS, 165
range guidance, 13
simultaneous approach with ILS, 164-165
military operations area (MOA), 75
military training route (MTR), 76
minimum safe altitude (MSA), 159
minimum vectoring altitude (MVA), 159-160
minimums
altitude, **432**
altitude for autopilot, **546**
circling, 166
instrument flight rules (IFR)
altitudes, **442-443**
approaches, **566**
landing, **566**
takeoff, **566**
instrument landing system (ILS), 7-8
landing, 166
published approach, 166
side-step maneuver, 166
straight-in, 166
visual flight rules (VFR)
altitude, **564**
weather, 67, **438-439**
missed approach, 167, 171, **441**
pilot/controller responsibilities, 171
Mode C transponder requirements, 83-90, 102
mountain flying, 237-239
MULTICOM frequency, 95

N

National Airspace System (NAS), 91
National Aviation Weather Advisory Unit (NAWAU), 196
National Driver Register, **411**
National Ocean Service, 253, 261
National Transportation Safety Board (NTSB), **641-643**
National Weather Service (NWS)
aviation products, ix, 196
radar network, 211-212
NAVAID (*see also* air navigation radio aids)
identifier removal during maintenance, 14
service volumes, 3, 5
user reports on performance, 15
voice equipped, viii, 15
navigation, 1-37
air navigation radio aids 1-33
charts (*see* aeronautical charts)
radar services and procedures, 33-37
radio aids for air, 1-33
tactical, 2
near midair collision (NMCA), **450, 563**
judgment aspects of, 251-252
reporting, 242
NOAA Aeronautical Chart User's Guide, 260
NOAA distribution branch, 253, 261
noise, 117, **469-478**
acoustical technology/strategy codes, **473**
agricultural and fire fighting planes, **473**
airplanes imported to points outside U.S., **475**
annual progress reports, **477-478**
base level, **475**
carry-forward compliance, **476**
civil aircraft, sonic boom, **473**
definitions, **474**
designation of applicable regulations, **470**
entry and nonaddition rule, **474**
flight authorizations for noise restricted aircraft, **498-499**
modification to meet Stage 3 levels, **475**
phased compliance for new entrants, **476**
phased compliance for operators with base level, **475-476**
replacement airplanes, **471**
service to small communities exemption, **471**
sonic boom, **473**
subsonic airplanes
final compliance, **470, 474**
phased compliance, **470**
U.S. operations of, **471-473**
supersonic airplanes, **473-474**
transfers of Stage 2 with base level, **475**
waivers from final compliance, **477**
waivers from interim compliance requirements, **476-477**
nondirectional radio beacon (NDB), 1
standard service volumes, 5
North Atlantic (NAT) minimum navigation performance specifications (MNPS), **468, 482**
notice to airmen (NOTAM), vii, 138-139

O

obstruction alert, 171
obstruction lighting, 44
obstruction lights
aviation red, 44
dual, 44
high intensity white, 44
obstructions, 236-237
charts, 261
mountain flying, 237-239
unmanned free balloons, 237
utility lines, 237
volcanic ash, ix, 240
wildlife, 235-236
oceans (*see* water)
OMEGA navigation system, 31-32
OMEGA/VLF navigation system, 31-32

operating rules (*see also* flight rules)
air taxi, **532-620**
Category II, **444, 479-481**
Category III, **444**
foreign aircraft, **468-469**
noise, **469-478**
parachute jumping, **511-513**
right-of-way for ultralights, **509**
special flight, **451-454**
turbine-powered multiengine aircraft, **458-464**
U.S. civil aircraft outside of U.S., **468-469**
ultralights, **508-509**
operation raincheck, 91
operation takeoff, 91
overhauling (*see* maintenance)
overhead approach maneuver, 169
oxygen, **545-546**
equipment requirements, **556**
supplemental, **448**

P

Pacific Chart Supplement, vii, 260
PAN-PAN, 189
parachute jumping, 68, 78, **451, 511-514**
airports, **512**
alcohol/drugs and, **513**
between sunset and sunrise, **513**
congested areas, **512**
information required before, **513**
inspections, **513**
operating rules, **511-513**
other airspace, **512**
prohibited areas, **513**
radio equipment and use requirements, **511-512**
restricted areas, **513**
visibility/cloud clearance, **513**
parachutes, **451**
equipment, **513-514**
packing requirements, **514**
passengers
cargo, **544-555**
carry-on baggage, **462, 544**
carrying weapons, **548**
emergency exits, **464-465**
exit seating, **550-552**
flight briefing, 137-146, **461-462, 548**
flight information/rules, **549-550**
information, **461**
manipulation of controls by, **548**
occupancy of pilot seat, **547**
oxygen for, **545-546**
provisions for, **544**
seat belts/shoulder harnesses, **462, 550**
phonetic alphabet, 110
pilot/controller communications, 264
"abeam," 264
"abort," 264
"acknowledge," 264
"advise intentions," 265
"affirmative," 266
"AIRMET," 267
"altitude readout," 270
"altitude restrictions are canceled," 271
"appropriate obstacle clearance minimum altitude," 271
"appropriate terrain clearance minimum altitude," 271
"blind transmission," 275
"braking action (good/fair/poor/nil)," 275
"chase," 276
"circle to runway (runway number)," 276
"clearance," 276
"clearance void if not off by (time)," 277
"cleared (type of) approach," 277
"cleared approach," 277
"cleared as filed," 277
"cleared for takeoff," 277
"cleared for the option," 277
"cleared through," 277
"cleared to land," 277
"climb to VFR," 277
"contact approach," 278
"convective SIGMET," 279
"cross (fix) at (altitude)," 280
"cross (fix) at or above (altitude)," 280
"cross (fix) at or below (altitude)," 280
"cruise," 280
"direct," 282
"emergency," 283
"execute missed approach," 284
"expect (altitude) at (time) or (fix)," 284
"expect further clearance (time)," 284
"expect further clearance via (airways/routes/fix)," 284
"expedite," 284
"final," 285
"flight level," 286
"FRC," 287
"go ahead," 288
"have numbers," 289
"heavy," 289
"homing," 290
"how do you hear me?," 290
"I say again," 293
"if no transmission received for (time)," 290
"immediately," 291
"increase speed to (speed)," 291
"low altitude alert, check your altitude immediately," 295
"maintain," 295
"make short approach," 295
"MAYDAY," 295
"missed approach," 298
"negative," 299
"negative contact," 299
"no gyro approach," 299
"no gyro vector," 299
"NORDO," 300
"numerous targets vicinity (location)," 301
"out," 302
"over," 303
"pilot's discretion," 303
"radar contact," 306
"radar contact lost," 306
"radar service terminated," 307
"read back," 308
"reduce speed to (speed)," 308
"report," 309
"request full route clearance," 309
"resume own navigation," 309
"roger," 309

"runway heading," 310
"say again," 311
"say altitude," 311
"say heading," 311
"SIGMET," 312
"significant meteorological information," 312
"speak slower," 313
"squawk (mode/code/function)," 313
"stand by," 314
"stop altitude squawk," 314
"stop burst," 314
"stop buzzer," 314
"stop squawk (mode or code)," 314
"stop stream," 314
"taxi into position and hold," 316
"that is correct," 317
"traffic alert, advise you turn left/right (specific heading if appropriate), AND/OR, CLIMB/DESCEND (specific altitude if appropriate) immediately," 318
"traffic in sight," 318
"traffic no factor," 318
"traffic no longer observed," 318
"transmitting in the blind," 319
"unable," 320
"verify," 320
"verify specific direction or takeoff (or turns)," 320
"VFR conditions," 320
"VFR not recommended," 321
"VFR-on-top," 321
"when able," 323
"wilco," 323
"words twice," 323

pilots
airline transport certification, **391-394**
ATC services for, 91-106
Category II, **362, 372-373**
certification (*see* certification, pilots)
commercial certification, **387-391**
military, **374-375**
pilot-in-command qualifications, **427, 567-568, 574-575**
private certification, **383-387**
qualifications, **568-569**
recreational certification, **380-383**
requirements, **572-573**
roles and responsibilities, 169-174
schools for (*see* pilot schools; training)
student certification, **377-380**

pilots automatic telephone weather answering service (PATWAS), 196, 201

pilot schools, **621-638**
advertising limitations, **623**
aircraft, **625, 628**
airports, **625**
application, **622**
assistant chief instructor qualifications, **624-625**
briefing areas, **626**
business office and operations base, **623**
certificate, **622**
duration, **622**
renewal, **623**
chief instructor
changing, **629**
qualifications, **624**
responsibilities, **629**
commercial pilot certification course, **633**
commercial test course, **634**
course outline, **626-627**
curriculum, **626-627**
drug use, **622**
enrollment, **630**
examining authority, **622, 627**
facility maintenance, **629**
flight instruction, **628**
glider course, **635-636**
graduation certificate, **630**
ground school course, **637**
ground training, **625-626, 628-629**
inspections, **623**
instrument rating course, **632**
lighter-than-air aircraft course, **635-636**
limitations, **628**
location, **622**
operating rules, **628-630**
personnel, **623-624**
private pilot certification course, **631**
private test course, **632**
privileges, **628**
quality of instruction, **629**
ratings, **622, 623**
records, **630**
rotorcraft course, **635-636**
satellite bases, **629**
test preparation courses, **637-638**
training aids, **625-626**

pilot weather report (PIREP), 215-218
related to airframe icing, 216
related to clear air turbulence (CAT), 217
related to turbulence, 216-217
related to windshear, 216-217

pitot heat indication system, **556**

portable electronic devices, **428**

powerplant
controls/accessories, **600**
fire protection, **600-601**
major alterations, **348**
major repairs, **349**

precipitation, 212 (*see also* meteorology)
estimating intensity of, 214

precision approach path indicator (PAPI), 39

precision approach radar (PAR), 36, 163

preflight procedures, 137-146
briefings, 137
changing flight plans, 146
composite (VFR/IFR flights), 140-141
defense VFR (DVFR), 140
flight rules, **429-430**
flights outside United States and U.S. territories, 145-146
following IFR when operating VFR, 138
IFR flights, 141-145
IFR operations to high altitude destinations, 145
NOTAM system, 138-139
VFR flight plan, viii, 139-140
weather briefings, 197-199

private pilots, **383-387**
aeronautical knowledge, **383-384**
airplane rating/aeronautical experience, **385**
applicability, **383**
certification course, **631**

private pilots *continued*
cross-country flights, **385**
eligibility requirements, **383**
flight proficiency, **384-385**
free balloon rating/limitations, **387**
glider rating/aeronautical experience, **386**
lighter-than-air rating/aeronautical experience, **386**
privileges and limitations, **386-387**
rotorcraft rating/aeronautical experience, **385-386**
procedure turn, 161-162
propellers
major alterations, **348**
major repairs, **349**
propulsion, **598-599**
public address system, **553-554**
publications, 259-263
Airport/Facility Directory, vii, 259
Alaska Supplement, vii, 260
Digital Aeronautical Chart Supplement (DACS), 260
Flight Information Publications (FLIP), 262
NOAA Aeronautical Chart User's Guide, 260
Pacific Chart Supplement, vii, 260

R

radar
airborne weather equipment, **559**
airport surveillance radar (ASR), 34, 163-164
air route surveillance radar (ARSR), 34
air traffic control radar beacon system (ATCRBS), 34, 102
ATC assistance to VFR aircraft, 98-100
automated radar terminal system (ARTS), 34
capabilities, 32
center radar automated radar terminal system processing (CENRAP), 34
Doppler, 32
emergency service for VFR aircraft in difficulty, 180
limitations, 32
precision approach radar (PAR), 36, 163
services and procedures, 33-37
surveillance, 34, 36
terminal radar service area (TRSA), 80, 83
weather services, 211-212
radar approach control, 158
radar traffic information service, 96-97
radar vectors, pilot/controller responsibilities, 171
radio communications
altitudes and flight levels, 111
before takeoff, 119-120
between pilot/controller (*see* pilot/controller communications)
call signs
aircraft, 108-109
ground station, 110
common traffic advisory frequency (CTAF), 93-94
contact procedures, 107-108
description of interchange/leased aircraft, 109
directions, 111
distress and urgency, 189
en route, 150-157
equipment needed for overwater operations, **460**
failure, 194-195, **443-444**
DVFR, **506**
IFR, **506**
figures, 111
gate holding due to departure delays, 120
IFR, **443**
MAYDAY, 189, 295
MULTICOM frequency, 95
obtaining assistance in emergencies, 189-190
PAN-PAN, 189
parachute jumping, **511-512**
phonetic alphabet, 110
position reporting, 151-152
recommended ATC, 92
reestablishing after failure, 195
reporting to ATC or FSS, 152-153
speeds, 112
techniques, 107
time, 112
transponder operation during two-way failure, 195
VFR flights, 113
with tower when transmitter/receiver inoperative, 112-113
UNICOM frequency, 95
radio control, lighting systems, 42-43
ratings
airplane
for airline transport pilots, **391-392**
for commercial pilots, **389-390**
for private pilots, **385**
for recreational pilots, **381**
airship for commercial pilots, **390**
applicability, **370**
Category II, **362, 372-373**
flight instructor, **395, 397**
foreign pilot license, **375**
free balloon
for commercial pilots, **390-391**
for private pilots, **387**
glider
for commercial pilots, **390**
for private pilots, **386**
graduates of pilot schools, **373-374**
instrument requirements, **371-372**
lighter-than-air for private pilots, **386**
military pilots, **374-375**
operation of U.S. plane leased by non-U.S. citizen, **376**
rotorcraft
for airline transport pilots, **393**
for commercial pilots, **390**
for private pilots, **385-386**
for recreational pilots, **382**
with helicopter class, **393-394**
rebuilding (*see* maintenance)
recreational pilots, **380-383**
aeronautical knowledge, **380-381**
airplane rating/aeronautical experience, **381**
eligibility requirements, **380**
flight proficiency, **381**
limitations, **382-383**
privileges, **382-383**
rotorcraft rating/aeronautical experience, **382**
reporting points, **420**
reports (*see also* accidents; maintenance)
aircraft and incident, 241-242
aircraft safety reporting program, 241
aviation safety, **429**
mechanical interruption summary, **588**

mechanical reliability, **587-588**
near midair collision, 242
volcanic activity, 244
wildlife, 243-244
reservations
airport, viii, 104-106
computer modem users, 106
HDTA, 104-105
IFR, 105
telephone users, 106
using computerized voice system, 104, 106
VFR, 105-106
rotorcraft
certification course, **635-636**
inspecting, **346**
rating
for airline transport pilots, **393**
for commercial pilots, **390**
for private pilots, **385-386**
for recreational pilots, **382**
route systems, 152-153
reporting course changes, 153-154
rules (*see* flight rules; instrument flight rules; operating rules; visual flight rules)
runway centerline lighting (RCLS), viii, 41
runway end identifier lights (REIL), 40-41
runway separation, 135
runway visual range (RVR), 213-214, **442**
runways
emergency arresting gear, 64-65
exiting after landing, 123
friction reports/advisories, 118
lighting
end identifier, 40-41
high intensity, 41
low intensity, 41
medium intensity, 41
taxiways, 42-43
touchdown zone, 41
pavement markings, viii, 45-52
aiming point, 45
centerline, 45
chevrons, 47
demarcation bar, 47
designators, 45
holding position, 52, 54, 57
holding position for ILS, 57
permanently closed, 59
side strip, 45
temporarily closed, 59-60
threshold, 45, 47
threshold bar, 47
touchdown zone, 45
signs
approach area holding position, 61
boundary, 62
distance remaining, 64
holding position, 60
ILS critical area boundary, 62-63
ILS critical area holding position, 61
location, 62
no entry, 61
simultaneous operations on intersecting, 118-119
use of with declared distances, 117

S

safety
aviation reporting program, 241
bird hazards and, 235-236
meteorology, 196-228
mountain flying, 237-239
potential flight hazards, 236-240
seaplane, ix, 238-240
safety alert, 97-98 (*see also* emergencies)
aircraft conflict, 171-172
minimum fuel advisories, ix, 174
obstruction, 171
pilot/controller responsibilities, 171-172
see and avoid, 172
terrain, 171
safety belts, shoulder harnesses and, **462, 550**
schools (*see* pilot schools)
SDF (*see* simplified directional facility)
seaplane flying, ix, 238-240
search and rescue, 181-184
Air Force centers, 183
Alaskan Air Command centers, 183
Coast Guard centers, 183
joint coordination centers, 183
national plan, 182
overdue aircraft, 183, **642-643**
survival equipment, 184
VFR protection, 183
security, 174-179, **504-507, 549**
ADIZ flight plan requirements, **505**
air defense identification zones, **506-507**
arrival/completion notice, **505**
deviation from flight plans/ATC clearness, **506**
dog/handler team, 182
emergency situations, **504-505**
explosive detection, 182
interception procedures, 175-176
interception signals, 176-178
law enforcement operations, 176
national, 174-175
position reports
aircraft entering through ADIZ; foreign aircraft, **506**
aircraft entering through ADIZ; U.S. aircraft, **505**
aircraft operating ADIZ; DVFR, **505**
aircraft operating ADIZ; IFR, **505**
radio requirements, **505**
transponder-on requirements, **505**
separations
air traffic wake turbulence, 234
IFR standards, 134
runway, 135
speed adjustments, 134-135
visual, 135, 173
severe weather bulletin (WW), 199
severe weather forecast alert (AWW), 199, 202
side-step maneuver, 165-166
minimums, 166
SIGMET, 199-200, 202
signals
aircraft interception, 176-178
ATC light, **433**
emergency, 184-188
hand, 126-129
signs, airport, 60-65

simplified directional facility (SDF), 10
simulated flight, **431**
smoking, **549-550**
smuggling, drugs, **428, 542**
snow, estimating intensity of, 215 (*see also* meteorology)
spatial disorientation, 249-250
special traffic management program (STMP), 106
speed
  adjusting, viii-ix, 134-135, 172
  aircraft, **432**
  authorizations to exceed Mach 1, **481-482**
  maximum airspeed while holding, 155
  warning device, **464**
stalls, **594**
standard instrument departure (SID), 149-150
  chart, 257
standard service volume (SSV), 3, 5
  distance measuring equipment (DME), 3
  nondirectional radio beacon (NDB), 5
  tactical navigation (TACAN), 3
  VHF omni-directional range (VOR), 3
standard terminal arrival (STAR), 157
  chart, 257
steward/stewardess (*see* crewmembers; flight attendants)
stress, 247
student pilots, **377-383**
  application, **377**
  cross-country flight requirements, **379-380**
  eligibility requirements, **377**
  limitations, **378-379**
  operations in Class B airspace, **380**
  solo flight requirements, **377-378**
supplemental weather service location (SWSL), 91, 197
survival equipment, 184
symbols
  FARs, **342**
  instrument approach procedure, **501**

T

TACAN (*see* tactical navigation)
tactical navigation (TACAN), 2
  standard service volumes, 3
takeoff
  helicopters, 121-122
  instrument flight rules (IFR), **441**
    limitations, **565**
    minimums, **566**
  intersection, 118
  large nontransport aircraft limitations, **585-586**
  transport aircraft limitations, **581**
  turbine-powered multiengine aircraft limitations, **583**
  visual clearing procedures before, 135-136
taxiing
  clearance for, 147
  during low visibility, 123
  exiting runway after landing, 123
  helicopters, 121
  hover taxi, 121
  pre-taxi clearance procedures, 147
  radio communications and, 122-123
taxiways
  lighting
    centerline, 43
    edge, 43
    turnoff, 42
  pavement markings
    centerline, 47
    edge, 47
    geographic position markings, 52
    holding position, 52, 54, 57
    permanently closed, 59
    shoulder, 47, 52
    surface painted direction signs, 52
    surface painted location signs, 52
    temporarily closed, 59-60
  signs, direction, 64
terminal forecast (TAF), 222-228 (*see also* meteorology)
  cloud codes, 228
  icing codes, 228
  turbulence codes, 228
  visibility conversion TAF code to miles, 223
  weather codes, 224-226
terminal instrument approach procedure (TERP), **503**
terminal procedures publication (TPP), 257
terminal radar service area (TRSA), 80, 83, 98-100
terrain alert, 171
terrorism, 193-194
tests and testing
  airline transport pilot, **397-404**
  altimeter, **353-354, 456-457**
  certification, **364-367**
  drugs and alcohol, **360-361, 515-531**
  emergency locator transmitter (ELT), 181
  rotorcraft airline transport pilot with helicopter class rating, **405-410**
  transponders, **354-355, 457**
threshold crossing height (TCH), 7
thunderstorms, 199, 219-220
  detection equipment, **559**
  flying in, 220-221
time difference (TD), 18
tornadoes, 199
touchdown zone lighting (TDZL), 41
tower en route control (TEC), 100
towing, gliders, **373, 451-452**
traffic advisories
  pilot/controller responsibilities, 172
  practices without control towers, viii, 92-94
traffic alert, **563**
traffic alert and collision avoidance system (TCAS), 136
traffic pattern indicator, 115
traffic patterns, viii, 114
  tower controlled airports, 113-114
  uncontrolled airports, 114-115
  unexpected maneuvers in, 116-117
training, **575-580**
  advanced qualification program (AQP), **617-620**
  airmen qualifications, **577-578**
  crewmembers, **576, 578-579**
  curriculum, **576**
  differences, **575**
  emergency, **576-577**
  flight attendants, **580**
  flight simulators, **577**
  handling/carrying hazardous material, **577**
  in flight, **575**
  initial, **575**

instructor qualifications, **577-578**
operations, 136
pilot programs, **578-579**
pilot schools, **621-638**
recurrent, **575, 580**
transition, **575**
upgrade, **575**
transcribed weather broadcast (TWEB), 196, 197, 201
transmitters, emergency locator, 181-182, **447**
transponders
ATC, **449-450**
emergency operation, 180-181
Mode C requirements, 83-90, 102
operating during communications failure, 195
operating under VFR, 102
operation, 101-102
security requirements, **505**
suspension of requirement for certain airports, **491-498**
tests and inspections, **354-355, 457**
transport aircraft (*see also* air taxi operations; turbine-powered multiengine aircraft)
aural speed warning device, **464**
cockpit voice recorders, **465-466**
commuter performance operating limitations, **587**
emergency exits, **464-465**
en route limitations, **581-583**
ferry flight with one engine inoperative, **466-467**
flight recorders, **465-466**
landing limitations, **583**
limitations for small, **586**
materials for compartment interiors, **467**
passengers, emergency exits, **464-465**
performance operating limitations, **580-587**
takeoff limitations, **581**
weight limitations, **464, 581**
transporting, Federal election candidates, **453**
truth-in-leasing clause, **428-429**
turbine-powered multiengine aircraft, **458-464**
carrying cargo, **462**
emergency equipment, **460, 461**
en route limitations, **584-585**
equipment required, **459-460**
flight altitude rules, **461**
flight attendant requirements, **463**
flight engineer requirements, **463**
food service, **463-464**
landing limitations, **585**
operating in icing conditions, **462-463**
operating limitations, **460**
passengers,
briefings, **461-462**
carry-on baggage, **462**
information, **461**
shoulder harness, **462**
radio equipment for overwater operations, **460**
second-in-command requirements, **463**
survival equipment for overwater operations, **460**
takeoff limitations, **583**
turbulence, 200, 216-217 (*see also* meteorology)
clear air (CAT), 217
microbursts, 218-219
TAF codes, 228
wake, 231-234
wind shear, 200, 216-217

U

U.S. Department of Transportation, vii
ultralights, 68, **508-510**
airspace limitations, **509**
certifications/registration, **508**
congested airspace, **509**
daylight operations, **508-509**
flight visibility/cloud clearance, **509**
hazardous operations, **508**
inspection requirements, **508**
operations near aircraft, **509**
prohibited areas, **509**
restricted areas, **509**
right-of-way rules, **509**
visual reference with ground surface, **509**
waivers, **508**
UNICOM frequency, 95

V

very low frequency (VLF), 30
VFR-on-top
IFR clearances with, 132
pilot/controller responsibilities, 173
VHF direction finder (VHF/DF), 32
VHF omni-directional range (VOR), 1
equipment check for IFR operations, **440-441**
standard service volumes, 3
VHF omni-directional range/tactical air navigation (VORTAC), 2-3
visibility, 206, 208, 212
parachute jumping requirements, **513**
prevailing, 214
reporting, 214
runway visual range (RVR), 213-214, **442**
surface, 214
ultralight flying requirements, **509**
VFR requirements, **564**
vision, 250-251
visual aids
airport, 38-65
lighting, 38-43
visual approach, viii, 167-168
pilot/controller responsibilities, 172-173
visual approach slope indicator (VASI), 38
visual flight rules (VFR), 66-67, 201, **358, 438-440**
approach control service, 91-92
ATC radar assistance, 98-100
Class B transition routes, 80
civil aircraft, **445-446**
congested areas and flying, 236
corridors, 80
cruising altitudes, 67, **439-440**
flights in terminal areas, 120-121
flight levels, 67, **439-440**
flight plan information, **438**
flyways, 78, 80
fuel, **438, 564**
helicopters at controlled airports, 121-122
helicopter surface reference requirements, **564**
minimum altitudes, **564**
operating limitations, **564-565**

visual flight rules (VFR) *continued*
over-the-top carry passengers limitations, **564**
radio communications, 113
reservations, 105-106
special clearances for, viii, 131-132
terminal area routes, 80
transponder operation under, 102
two-way radio failure, **443**
visibility requirements, **564**
weather requirements, **564-565**
weather minimums, 67, **438-439**
visual separation, 135
pilot/controller responsibilities, 173
voice recorders, **465-466, 554**
volcanic activity report (VAR), 244
volcanic ash, flight operations in, ix, 240
VOR (*see* VHF omni-directional range)
VOR receiver check, 1-2, 58-59
VOR test facility (VOT), 1-2
vortex (*see* wake turbulence)

W
waivers, **476-478**
ultralights, **508**
wake turbulence, 231-234
air traffic separations, 234
helicopters, 233-234
pilot responsibility, 234
problem areas, 232-233
vortex
avoidance procedures, 233
behavior, 232
generation, 231
strength, 231
water
landing plane in, 190-193
radio equipment for overwater operations, **460**
seaplane flying, ix, 238-240
survival equipment for overwater operations, **460**
weapons, carrying on planes, **548**
weather (*see* meteorology)
weather message switching center (WMSC), 196
weather service office (WSO), 197
wildlife
bird strike reports, 243-244
flights over charted U.S. refuges/parks/forest service areas, 235-236
migratory bird activity, 235
pilot advisories, 235
reducing bird strike risks, 235
reporting activities of, 235
reporting bird strikes, ix, 235
wind, 200, 210, 212 (*see also* turbulence)
character, 210
speed, 210
wind direction indicator, 115
wind shear, 117, 216-217
WW (*see* severe weather bulletin)

Z
Z marker, 5-6